COMPTON GAMMA-RAY OBSERVATORY

AIP CONFERENCE PROCEEDINGS 280

COMPTON GAMMA-RAY OBSERVATORY

ST. LOUIS, MO 1992

EDITORS:
MICHAEL FRIEDLANDER
WASHINGTON UNIVERSITY
ST. LOUIS, MO

NEIL GEHRELS
DARYL J. MACOMB
NASA/GODDARD SPACE
FLIGHT CENTER

American Institute of Physics   New York

Authorization to photocopy items for internal or personal use, beyond the free copying permitted under the 1978 U.S. Copyright Law (see statement below), is granted by the American Institute of Physics for users registered with the Copyright Clearance Center (CCC) Transactional Reporting Service, provided that the base fee of $2.00 per copy is paid directly to CCC, 27 Congress St., Salem, MA 01970. For those organizations that have been granted a photocopy license by CCC, a separate system of payment has been arranged. The fee code for users of the Transactional Reporting Service is: 0094-243X/87 $2.00.

© 1993 American Institute of Physics.

Individual readers of this volume and nonprofit libraries, acting for them, are permitted to make fair use of the material in it, such as copying an article for use in teaching or research. Permission is granted to quote from this volume in scientific work with the customary acknowledgment of the source. To reprint a figure, table, or other excerpt requires the consent of one of the original authors and notification to AIP. Republication or systematic or multiple reproduction of any material in this volume is permitted only under license from AIP. Address inquiries to Series Editor, AIP Conference Proceedings, AIP, 335 East 45th Street, New York, NY 10017-3483.

L.C. Catalog Card No. 93-71830
ISBN 1-56396-104-0
DOE CONF-9210245

Printed in the United States of America.

Contents

Preface xxi
    M. Friedlander

Arthur Holly Compton xxii

OPENING REMARKS

The Compton Observatory in Perspective 3
    N. Gehrels, E. Chipman, and D. A. Kniffen

DIFFUSE EMISSION

Initial Results from COMPTEL—An Overview 21
    V. Schönfelder, W. Collmar, R. Diehl, G. G. Lichti, H. Steinle, A. Strong,
    M. Varendorff, H. Bloemen, H. de Boer, R. van Dijk, J. W. den Herder,
    W. Hermsen, L. Kuiper, B. Swanenburg, C. de Vries, A. Connors, D. Forrest,
    M. McConnell, D. Morris, J. Ryan, J. G. Stacy, K. Bennet, B. G. Taylor,
    and C. Winkler

COMPTEL Observations of the Inner Galaxy 30
    H. Bloemen, H. de Boer, R. van Dijk, J. W. den Herder, W. Hermsen,
    L. Kuiper, B. N. Swanenburg, C. P. de Vries, W. Collmar, R. Diehl,
    G. G. Lichti, V. Schönfelder, H. Steinle, A. W. Strong, M. McConnell,
    A. Connors, D. Morris, G. Stacy, and K. Bennett

Diffuse Galactic Continuum Emission Measured by COMPTEL 35
    A. W. Strong, R. Diehl, V. Schönfelder, M. Varendorff, G. Youssefi,
    H. Bloemen, H. De Boer, W. Hermsen, B. Swanenburg, C. de Vries,
    D. Morris, J. G. Stacy, and K. Bennett

COMPTEL Measurements of 1.809 MeV Gamma-Ray Line Emission
from the Galactic Plane 40
    R. Diehl, W. Collmar, G. Lichti, V. Schönfelder, A. Strong, H. Bloemen,
    C. Dupraz, C. de Vries, W. Hermsen, B. N. Swanenburg, D. Morris,
    M. Varendorff, and C. Winkler

Nucleosynthesis of ^{26}Al in AGB Stars and Wolf-Rayet Stars 47
    G. Bazan, L. E. Brown, D. D. Clayton, M. F. El Eid, D. H. Hartmann,
    and J. W. Truran

Radioactive ^{26}Al from Massive Stars: Production and Distribution
in the Galaxy 52
    N. Prantzos

Galactic Chemical Evolution and ^{26}Al Production by Supernovae 64
    F. X. Timmes, S. E. Woosley, and T. A. Weaver

Diffuse Galactic Low Energy Gamma-Ray Continuum Emission 70
    J. Skibo and R. Ramaty

Galactic Annihilation Radiation and the Galactic Nucleosynthesis Rate 75
  K. -W. Chan, R. E. Lingenfelter, and R. Ramaty
Gamma-Ray Line Emission from AGB Star Ejecta 80
  G. Bazan
Galactic Gamma-Ray Emission from Pulsars.. 85
  N. G. Schnepf, L. E. Brown, D. H. Hartmann, D. D. Clayton, J. Cordes, and A. Harding
The Search for the ^{7}Be Line at 478 keV and its Implications 90
  M. Signore, G. Vedrenne, F. Melchiorri, P. de Bernardis, and P. Encrenaz

GALACTIC CENTER

Positron Annihilation Radiation from the Galactic Center 97
  J. Tueller
OSSE Observations of Galactic 511 keV Annihilation Radiation....................... 107
  W. R. Purcell, D. A. Grabelsky, M. P. Ulmer, W. N. Johnson, R. L. Kinzer, J. D. Kurfess, M. S. Strickman, and G. V. Jung
GRIS Detections of the Galactic Center 511 keV Line in 1992........................ 119
  M. Leventhal, S. D. Barthelmy, N. Gehrels, B. J. Teegarden, J. Tueller, and L. M. Bartlett
The Number of Positron Sources in the Galactic Center Region....................... 124
  Ph. Durouchoux and P. Wallyn
The Annihilation of Positrons in a Molecular Cloud.. 129
  P. Wallyn and Ph. Durouchoux

SUPERNOVAE AND NOVAE

Compton Observatory OSSE Studies of Supernovae and Novae....................... 137
  M. D. Leising, D. D. Clayton, L.-S. The, W. N. Johnson, J. D. Kurfess, R. L. Kinzer, R. A. Kroeger, M. S. Strickman, J. E. Grove, D. A. Grabelsky, W. R. Purcell, M. P. Ulmer, R. A. Cameron, and G. V. Jung
Search for Gamma-Ray Line Emission from Supernova SN 1991T with COMPTEL 147
  G. G. Lichti, W. Collmar, R. Diehl, V. Schönfelder, A. Strong, D. Morris, J. Ryan, H. Bloemen, H. de Boer, R. van Dijk, W. Hermsen, K. Bennett, M. Busetta, and C. Winkler
X-ray and Gamma-Ray Transport in Clumpy Supernova Ejecta 149
  L.-S. The, D. H. Hartmann, D. D. Clayton, and M. D. Leising
A Search for Radiative Neutrino Decay from Supernovae 154
  R. S. Miller and R. C. Svoboda
Spectrum of Gamma-Ray Lines from SN1987A.. 159
  A. Stupp
Gamma Rays and X Rays from Classical Novae... 163
  J. W. Truran and S. Starrfield
The Production of ^{22}Na and ^{26}Al in Nova Explosions: The Observational Picture 168
  S. Starrfield, S. N. Shore, G. Sonneborn, R. Gonzalez-Riestra, W. M. Sparks, J. W. Truran, M. A. Dopita, and R. E. Williams

PULSARS

EGRET Observations of Pulsars 177
D. A. Kniffen, D. L. Bertsch, C. E. Fichtel, R. C. Hartman, S. D. Hunter,
J. R. Mattox, D. J. Thompson, K. T. S. Brazier, G. Kanbach,
H. A. Mayer-Hasselwander, C. Von Montigny, K. Pinkau, J. Chiang,
J. M. Fierro, Y. C. Lin, P. F. Michelson, P. L. Nolan, E. Schneid, H. I. Nel,
P. Sreekumar, Z. Arzoumanian, V. M. Kaspi, D. Nice, J. Taylor, M. Bailes,
S. Johnston, R. N. Manchester, A. G. Lyne, N. D'Amico, and L. Nicastro

Pulsar Searches with COMPTEL 189
K. Bennett, M. Busetta, A. Carramiñana, W. Collmar, R. Diehl, G. Lichti,
V. Schönfelder, A. Strong, W. Hermsen, L. Kuiper, B. Swanenburg,
A. Connors, J. Ryan, and R. Buccheri

GRIS Observations of the Crab 194
L. M. Bartlett, S. D. Barthelmy, N. Gehrels, M. Leventhal, B. J. Teegarden,
J. Tueller, and C. J. MacCallum

Time Variability Analysis of the COMPTEL Crab Light Curve 199
A. Carramiñana, K. Bennett, M. Busetta, V. Schönfelder, A. Strong,
W. Hermsen, L. Kuiper, A. Connors, J. Ryan, and R. Buccheri

COMPTEL Detections of the Vela Pulsar: First Energy Spectra 204
W. Hermsen, H. Bloemen, L. Kuiper, B. N. Swanenburg, R. Diehl,
G. G. Lichti, V. Schönfelder, A. W. Strong, A. Connors, D. Morris,
J. M. Ryan, M. Varendorff, K. Bennett, M. Busetta, A. Carramiñana,
C. Winkler, and R. Buccheri

OSSE Observations of the Vela Pulsar 209
M. S. Strickman, J. E. Grove, W. N. Johnson, R. L. Kinzer, R. A. Kroeger,
J. D. Kurfess, D. A. Graabelsky, S. M. Matz, W. R. Purcell, and M. P. Ulmer

Ginga Observations of Isolated Pulsars: PSR 1509-58 and the Crab Pulsar 213
N. Kawai, R. Okayasu, and Y. Sekimoto

The Contribution of the Northern Cross Observations to the Gamma-Ray Observatory Pulsar Timing Support 218
N. D'Amico, L. Nicastro, C. Bortolotti, A. Cattani, G. Grueff,
A. Maccaferri, S. Montebugnoli, and F. Fauci

Search for TeV Gamma Rays from Geminga 223
D. J. Fegan, C. W. Akerlof, A. C. Breslin, M. F. Cawley, M. Chantell,
S. Fennell, J. A. Gaidos, J. Hagan, A. M. Hillas, A. D. Kerrick, R. C. Lamb,
M. A. Lawrence, D. A. Lewis, D. I. Mayer, G. Mohanty, K. S. O'Flaherty,
M. Punch, P. T. Reynolds, A. Rovero, M. Schubnell, G. Sembroski,
T. C. Weekes, M. West, T. Whitaker, and C. Wilson

Possible Detection of the Geminga Pulsar in the Energy Range 0.15–0.48 MeV with the FIGARO II Experiment 228
E. Massaro, B. Agrinier, E. Barouch, R. Comte, E. Costa, G. C. Cusumano,
G. Gerardi, D. Lemoine, P. Mandrou, J. L. Masnou, G. Matt, T. Mineo,
M. Niel, J. F. Olive, B. Parlier, B. Sacco, M. Salvati, and L. Scarsi

Understanding GEMINGA: Past and Future Observations 233
G. F. Bignami, P. A. Caraveo, and S. Mereghetti

Search for Fast Galactic Gamma-Ray Pulsars 238
P. Hertz, J. E. Grove, J. D. Kurfess, W. N. Johnson, M. S. Strickman,
S. Matz, and M. P. Ulmer

Search for Hard X-Ray Emission from Millisecond Pulsars and Compact
Binaries in the Globular Cluster 47 Tuc.. 243
    J. E. Grindlay, R. P. Manandhar, K. S. K. Lum, M. Krockenberger,
    S. Eikenberry, and C. E. Covault

An OSSE Search for the Binary Radio Pulsar 1259-63 249
    P. S. Ray, J. E. Grove, J. D. Kurfess, T. A. Prince, and M. P. Ulmer

TeV Observations of Gamma-Ray Pulsars: A Test for Pulsar Models 254
    H. I. Nel, O. C. De Jager, L. J. Haasbroek, B. C. Raubenheimer, C. Brink,
    P. J. Meintjes, A. R. North, G. Van Urk, and B. Visser

Gamma-Ray Pulsars and Geminga 259
    M. Ruderman, K. Chen, K. S. Cheng, and J. P. Halpern

Unveiling Galactic Gamma-Ray Sources: Isolated vs Binary Pulsars................. 272
    M. Tavani

Polarization and Emission Geometry of the Crab Pulsar 278
    K. Chen and C. Ho

Gamma Rays from Pulsar Outer Gaps... 283
    J. Chiang, R. W. Romani, and C. Ho

Magnetic Compton Scattering in the Magnetospheres
of Radio Pulsars 284
    S. J. Sturner, C. D. Dermer, and F. C. Michel

X-RAY BINARIES

Observations of Isolated Pulsars and Disk-Fed X-Ray Binaries......................... 291
    R. B. Wilson, G. J. Fishman, M. H. Finger, G. N. Pendleton, T. A. Prince,
    and D. Chakrabarty

OSSE Observations of Galactic Sources During Phase 1.................................... 303
    J. D. Kurfess, W. N. Johnson, R. L. Kinzer, R. A. Kroeger, M. S. Strickman,
    J. E. Grove, D. A. Grabelsky, S. M. Matz, W. R. Purcell, M. P. Ulmer,
    M. D. Leising, R. A. Cameron, and G. V. Jung

Earth Occultation Measurements of Galactic Hard X-Ray/Gamma-Ray
Sources: A Survey of BATSE Results... 314
    B. A. Harmon, C. A. Wilson, M. N. Brock, R. B. Wilson, G. J. Fishman,
    C. A. Meegan, W. S. Paciesas, G. N. Pendleton, B. C. Rubin,
    and M. H. Finger

Detection of Quasi-Periodic Oscillations (QPO) from Cyg X-1
and GRO J0422+32 319
    C. Kouveliotou, M. H. Finger, G. J. Fishman, C. A. Meegan, R. B. Wilson,
    W. S. Paciesas, T. Minamitani, and J. van Paradijs

Ultraviolet, Optical, and Radio Observations of the Transient X-Ray
Source GRO J0422+32 324
    C. R. Shrader, R. M. Wagner, S. G. Starrfield, R. M. Hjellming,
    and X. H. Han

Disk Oscillations and their Application to GRO J0422+32................................. 330
    C. Luo and E. Liang

COMPTEL Observations of Cygnus X-1 335
    M. McConnell, A. Connors, D. Forrest, J. Ryan, W. Collmar, R. Diehl,
    V. Schönfelder, H. Steinle, A. Strong, H. Bloemen, R. van Dijk, W. Hermsen,
    L. Kuiper, B. Swanenburg, and C. Winkler

Long-Term Temporal and Spectral Variation of Cygnus X-1
Observed by BATSE 340
    J. C. Ling, N. F. Ling, R. T. Skelton, W. A. Wheaton, B. A. Harmon,
    G. J. Fishman, C. A. Meegan, R. B. Wilson, W. S. Paciesas, G. N. Pendleton,
    and B. C. Rubin

OSSE Spectral Observations of GX 339-4 and Cyg X-1 345
    D. A. Grabelsky, S. M. Matz, W. R. Purcell, M. P. Ulmer, W. N. Johnson,
    R. L. Kinzer, R. A. Kroeger, J. D. Kurfess, M. S. Strickman, J. E. Grove,
    R. A. Cameron, G. V. Jung, and M. D. Leising

Observation of a Hard State Outburst in the GX339-4 System 350
    B. A. Harmon, C. A. Wilson, R. B. Wilson, G. J. Fishman, C. A. Meegan,
    W. S. Paciesas, G. N. Pendleton, B. C. Rubin, M. H. Finger, W. A. Wheaton,
    J. C. Ling, and R. T. Skelton

Hard X-Ray Observation of GX 339-4 with Welcome-1 355
    N. Y. Yamasaki, S. Gunji, M. Hirayama, T. Kamae, S. Miyazaki,
    Y. Sekimoto, T. Takahashi, T. Tamura, M. Tanaka, T. Yamagami,
    M. Nomachi, H. Murakami, J. Braga, and J. A. Neri

IR and Optical Observations of Be/X-Ray Binaries in Conjunction
with BATSE 360
    M. J. Coe, C. Everall, A. J. Norton, P. Roche, S. J. Unger, J. Fabregat,
    V. Reglero, J. M. Grunsfeld, and T. A. Prince

EGRET Observation of the Constellation Cygnus 365
    J. R. Mattox, D. L. Bertsch, B. L. Dingus, C. E. Fichtel, R. C. Hartman,
    S. D. Hunter, P. Sreekumar, D. J. Thompson, P. W. Kwok, K. Brazier,
    T. Halaczek, G. Kanbach, H. A. Mayer-Hasselwander, C. von Montigny,
    K. Pinkau, H. D. Radecke, H. Rothermel, M. Sommer, J. Fierro, Y. C. Lin,
    P. F. Michelson, P. L. Nolan, J. Chiang, D. A. Kniffen, and E. Schneid

Sigma Observations of the Burst Source GX 354-00 366
    A. Goldwurm, A. Claret, B. Cordier, J. Paul, J. P. Roques, L. Bouchet,
    M. C. Schmitz-Fraysse, P. Mandrou, R. Sunyaev, E. Churazov, M. Gilfanov,
    N. Khavenson, A. Dyachkov, B. Novikov, R. Kremnev, and V. Kovtunenko

BATSE Observations of EXO 2030+375 371
    M. T. Stollberg, G. N. Pendleton, W. S. Paciesas, M. H. Finger,
    G. J. Fishman, R. B. Wilson, C. A. Meegan, B. A. Harmon, and C. A. Wilson

BATSE Observations of GS 0834-430 376
    C. A. Wilson, B. A. Harmon, R. B. Wilson, G. J. Fishman, C. A. Meegan,
    D. Chakrabarty, J. M. Grunsfeld, T. A. Prince, M. H. Finger, W. S. Paciesas,
    and G. N. Pendleton

BATSE Observations of the Massive X-Ray Binary
4U1700-37/HD153919 381
    B. C. Rubin, B. A. Harmon, G. J. Fishman, C. A. Meegan, R. B. Wilson,
    M. S. Briggs, W. S. Paciesas, and M. H. Finger

BATSE Observations of Cen X-3 386
    M. H. Finger, R. B. Wilson, B. A. Harmon, G. J. Fishman, C. A. Meegan,
    and W. S. Paciesas

Correlated Optical Observations of Sco X-1 Using the Batse
Spectroscopy Detectors 391
    B. McNamara, G. Fitzgibbons, G. J. Fishman, C. A. Meegan, R. B. Wilson,
    B. A. Harmon, W. S. Paciesas, B. C. Rubin, and M. H. Finger

Thermal Hard X-Ray and Gamma-Ray Emission of Disk-Accreting
Black Holes 396
    E. P. Liang
Compton Backscattered Annihilation Line from the Nova Muscae 408
    X.-M. Hua and R. E. Lingenfelter
Time Variable Multiple Backscattered 511 keV Photons from Black
Holes 413
    Ph. Durouchoux and P. Wallyn
Spectral Modeling of Gamma Rays from Black Hole Candidates 418
    E. P. Liang
The Black Hole Mass in X-Ray Nova Muscae: Constraint from SIGMA
Annihilation Line Data 423
    W. Chen and N. Gehrels
Magnetic Field Reconnection and High-Energy Emission from Accreting
Neutron Stars 428
    M. Tavani and E. Liang
Gamma-Ray Production by Neutron Stars Accreting from a Disk 433
    M. C. Miller, F. K. Lamb, and R. J. Hamilton
Acceleration of Particles in the Magnetospheres of Accreting Neutron
Stars 438
    R. J. Hamilton, F. K. Lamb, and M. C. Miller
Electrodynamics of Neutron Stars Accreting from a Disk 443
    F. K. Lamb, R. J. Hamilton, and M. C. Miller
Compton Scattering Polarization in Accretion Discs 448
    J. Poutanen and O. Vilhu
Compton Scattering Matrix for Relativistic Maxwellian Electrons 453
    J. Poutanen

ACTIVE GALAXIES

High-Energy Gamma-Ray Emission from Active Galactic Nuclei
Observed by the Energetic Gamma-Ray Experiment Telescope (EGRET) 461
    G. E. Fichtel, D. L. Bertsch, B. L. Dingus, R. C. Hartman, S. D. Hunter,
    P. W. Kwok, J. R. Mattox, P. Sreekumar, D. J. Thompson, D. A. Kniffen,
    Y. C. Lin, P. L. Nolan, P. F. Michelson, G. Kanbach,
    H. A. Mayer-Hasselwander, C. von Montigny, K. Pinkau, H. Rothermel,
    M. Sommer, and E. J. Schneid
Monitoring the Long-Term Behavior of Active Galactic Nuclei Using
BATSE 473
    W. S. Paciesas, B. A. Harmon, C. A. Wilson, G. J. Fishman, C. A. Meegan,
    R. B. Wilson, G. N. Pendleton, and B. C. Rubin
OSSE Observations of Active Galaxies and Quasars 478
    R. A. Cameron, J. E. Grove, W. N. Johnson, J. D. Kurfess, R. L. Kinzer,
    R. A. Kroeger, M. S. Strickman, M. Maisack, C. H. Starr, G. V. Jung,
    D. A. Grabelsky, W. R. Purcell, and M. P. Ulmer

Search for Gamma-Ray Emission from AGN with COMPTEL 483
  W. Collmar, R. Diehl, G. G. Lichti, V. Schönfelder, H. Steinle, A. W. Strong,
  H. Bloemen, J. W. den Herder, W. Hermsen, B. N. Swanenburg, C. de Vries,
  M. McConnell, J. Ryan, G. Stacy, K. Bennett, O. R. Williams,
  and C. Winkler

Detection of TeV Photons from Markarian 421 488
  M. Punch, C. W. Akerlof, M. F. Cawley, M. Chantell, D. J. Fegan, S. Fennell,
  J. A. Gaidos, J. Hagan, A. M. Hillas, Y. Jiang, A. D. Kerrick, R. C. Lamb,
  M. A. Lawrence, D. A. Lewis, D. I. Meyer, G. Mohanty, K. S. O'Flaherty,
  P. T. Reynolds, A. C. Rovero, M. S. Schubnell, G. Sembroski, T. C. Weekes,
  T. Whitaker, and C. Wilson

OSSE Observations of NGC 4151 493
  M. Maisack, W. N. Johnson, R. L. Kinzer, M. S. Strickman, J. D. Kurfess,
  G. V. Jung, D. A. Grabelsky, W. R. Purcell, and M. P. Ulmer

**Observation of the Starburst Galaxy NGC 253 with the OSSE
Instrument** 498
  D. Bhattacharya, N. Gehrels, J. D. Kurfess, W. N. Johnson, R. L. Kinzer,
  M. S. Strickman, L.-S. The, G. V. Jung, D. A. Grabelsky, W. R. Purcell,
  and M. P. Ulmer

OSSE Observations of Starburst Galaxy M82 503
  L.-S. The, D. D. Clayton, M. D. Leising, J. D. Kurfess, W. N. Johnson,
  R. L. Kinzer, M. S. Strickman, G. V. Jung, D. A. Grabelsky, W. R. Purcell,
  and M. P. Ulmer

**Search for TeV Gamma-Ray Emission from AGN's Using the Whipple
Imaging Telescope** 508
  S. Fennell, C. W. Akerlof, M. F. Cawley, M. Chantell, D. J. Fegan,
  J. A. Gaidos, J. Hagan, A. M. Hillas, A. D. Kerrick, R. C. Lamb,
  M. A. Lawrence, D. A. Lewis, D. I. Meyer, G. Mohanty, K. S. O'Flaherty,
  P. T. Reynolds, A. C. Rovero, M. S. Schubnell, G. Sembroski, T. C. Weekes,
  T. Whitaker, and C. Wilson

The Optical-UV Spectrum of 3C 279 During Outburst 513
  C. R. Shrader, J. R. Webb, T. J. Balonek, M. S. Brotherton, B. J. Wills,
  D. Wills, S. D. Godlin, and B. McCollum

EGRET Upper Limits for the Blazars 3C 345, BL Lac and 3C 371 518
  C. von Montigny, G. Kanbach, H. A. Mayer-Hasselwander, K. Pinkau,
  H. Rothermel, M. Sommer, D. L. Bertsch, C. E. Fichtel, R. C. Hartman,
  S. D. Hunter, P. W. Kwok, J. R. Mattox, P. Sreekumar, D. J. Thompson,
  Y. C. Lin, P. F. Michelson, P. L. Nolan, D. A. Kniffen, and E. Schneid

Hard X-Ray Observation of Cen A and a Break in the Energy Spectrum ... 523
  T. Takahashi, S. Gunji, M. Hirayama, T. Kamae, S. Miyazaki, Y. Sekimoto,
  M. Tanaka, T. Tamura, N. Y. Yamasaki, H. Inoue, T. Kanou, T. Yamagami,
  M. Nomachi, H. Murakami, J. Braga, and J. A. Neri

AGN Emission Above 100 keV: The Hard X-Ray Detection Problem 528
  M. Maisack, K. S. Wood, and D. E. Gruber

Gamma-Ray Jets from Active Galactic Nuclei 533
  R. D. Blandford

Gamma Rays from Active Galactic Nuclei 541
  C. D. Dermer

Gamma Rays from Hot Accretion Disks in AGN 554
  M. Kafatos and P. A. Becker

High-Energy Emission from Ultrarelativistic Jets 559
    P. S. Coppi, J. F. Kartje, and A. Königl
X-Ray and Gamma-Ray Emission from Active Galactic Nuclei 564
    K. Y. Ding, K. S. Cheng, and K. N. Yu
Gamma Rays from the Strong Magnetic Field AGN Central Engine 569
    H. D. Greyber
Gamma-Ray Emission from Star-Accretion Disk Collisions in
Active Galactic Nuclei 574
    O. Heinrich, G. Shaviv, and R. Wehrse
Examining the Synchrotron Self-Compton Model for Blazars 578
    S. D. Bloom and A. P. Marscher
A Model of the Cosmic X- and Gamma-Ray Background 583
    M. Matsuoka, N. Terasawa, and M. Hattori
A Test of the Unified Model for Active Galactic Nuclei 588
    A. Owens, S. Sembay, P. Nandra, and I. George
Starburst and Reflection-Dominated AGN Contributions to the
Diffuse X-Ray Background 593
    P. M. Ricker and P. Mészáros
Comptonization of External Radiation in Blazars 598
    M. Sikora, M. C. Begelman, and M. J. Rees
Gamma Rays and Neutrinos from Point Sources 603
    Y. Tomozawa
Comptonization of a Soft Photon Disk Spectrum by Moving Blobs as
a Source of Gamma Rays in AGN 608
    M. Zbyszewska
Why So Many (So Few) EGRET AGN's, or, Which Ones Next? 613
    G. F. Bignami, P. A. Caraveo, and P. Ciliegi

SOLAR FLARES

OSSE Observations of Solar Flares 619
    R. J. Murphy, G. H. Share, J. E. Grove, W. N. Johnson, R. L. Kinzer,
    R. A. Kroeger, J. D. Kurfess, M. S. Strickman, S. M. Matz, D. A. Grabelsky,
    W. R. Purcell, M. P. Ulmer, R. A. Cameron, G. V. Jung, C. M. Jensen,
    W. T. Vestrand, and D. J. Forrest
COMPTEL Gamma-Ray and Neutron Measurements of Solar Flares 631
    J. Ryan, D. Forrest, J. Lockwood, M. Loomis, M. McConnell, D. Morris,
    W. Webber, G. Rank, V. Schönfelder, B. N. Swanenburg, K. Bennett,
    L. Hanlon, C. Winkler, and H. Debrunner
High-Energy Processes in Solar Flares 643
    R. Ramaty and N. Mandzhavidze
Solar Flare Neutron Spectra and Accelerated Ion Pitch-Angle
Scattering 656
    R. E. Lingenfelter, X.-M. Hua, B. Kozlovsky, and R. Ramaty
Observations of the 1991 June 11 Solar Flare with COMPTEL 661
    G. Rank, R. Diehl, G. G. Lichti, V. Schönfelder, M. Varendorff,
    B. N. Swanenburg, D. Forrest, J. Macri, M. McConnell, J. Ryan, K. Bennett,
    L. Hanlon, and C. Winkler

GAMMA-RAY BURSTS: DISTRIBUTIONS

BATSE Observations of Gamma-Ray Bursts 669
    G. J. Fishman

The Spatial Distribution of Gamma-Ray Bursts Observed by BATSE 681
    C. Meegan, G. Fishman, R. Wilson, M. Brock, J. Horack, W. Paciesas,
    G. Pendleton, and C. Kouveliotou

**Limitations on Detecting Dipole and Quadrupole Anisotropies
in BATSE's Gamma-Ray Burst Locations** 686
    M. S. Briggs, W. S. Paciesas, M. N. Brock, G. J. Fishman, C. A. Meegan,
    and R. B. Wilson

**Coordinate-System-Independent Tests of Isotropy Applied to BATSE's
Gamma-Ray Burst Locations** 691
    M. S. Briggs, W. S. Paciesas, M. N. Brock, G. J. Fishman, C. A. Meegan,
    and R. B. Wilson

**BATSE Observations of Gamma-Ray Bursts in Sun-Referenced
Coordinate Systems** 694
    J. M. Horack, S. D. Storey, G. J. Fishman, C. A. Meegan, R. B. Wilson,
    T. M. Koshut, R. S. Mallozzi, and W. S. Paciesas

**Preliminary Angular Correlation Analyses of Gamma-Ray Bursts
Detected by BATSE** 699
    J. M. Horack, G. J. Fishman, C. A. Meegan, R. B. Wilson, M. N. Brock,
    J. Hakkila, W. S. Paciesas, G. N. Pendleton, and M. S. Briggs

**Constraints on Galactic Gamma-Ray Burst Models from BATSE
Angular and Intensity Distributions** 704
    J. Hakkila, C. A. Meegan, G. J. Fishman, R. B. Wilson, M. N. Brock,
    J. M. Horack, G. N. Pendleton, and W. S. Paciesas

**Improvements in Measuring the Direction to Gamma-Ray Bursts with
BATSE** 709
    M. N. Brock, G. J. Fishman, C. A. Meegan, R. B. Wilson, G. N. Pendleton,
    M. T. Stollberg, and W. S. Paciesas

**Effects of Location Uncertainties on the Observed Distribution
of Gamma-Ray Bursts Detected by BATSE** 714
    J. M. Horack, C. A. Meegan, G. J. Fishman, R. B. Wilson, M. N. Brock,
    W. S. Paciesas, A. G. Emslie, and G. N. Pendleton

A Search for Untriggered Gamma-Ray Bursts in the BATSE Data 719
    B. C. Rubin, J. M. Horack, M. N. Brock, C. A. Meegan, G. J. Fishman,
    R. B. Wilson, W. S. Paciesas, and J. van Paradijs

Some Statistical Properties of 153 BATSE Gamma-Ray Bursts 724
    W. Kluźniak

**Gamma-Ray Burst Redshifts: Zero or One? In Search of a Bend in
Log N–Log S** 729
    R. A. M. J. Wijers and L. M. Lubin

The Use of V/V_{max} in the Study of Gamma-Ray Bursts 734
    D. Band

**The Consistency of Standard Cosmology and the BATSE Number vs
Brightness Relation** 739
    W. A. D. T. Wickramasinghe, R. J. Nemiroff, C. Kouveliotou, J. P. Norris,
    G. J. Fishman, C. A. Meegan, R. B. Wilson, and W. S. Paciesas

The Normalization of the PVO Log N–Log P Distribution 744
    E. E. Fenimore, R. W. Klebesadel, J. Laros, C. Lacey, C. Madras, M. Meier, and G. Schwarz

Inferring the Spatial and Energy Distribution of Burst Sources from Peak Count Rate Data 749
    T. Loredo and I. Wasserman

Distribution of Peak Counts of Gamma-Ray Bursts 754
    V. Petrosian, W. J. Azzam, and B. Efron

Statistics of Cosmic Gamma-Ray Bursts: Evolution of Different Populations of Galactic Sources 761
    I. G. Mitrofanov, A. A. Kozlenkov, A. M. Chernenko, V. Sh. Dolidze, A. M. Pozanenko, D. A. Ushakov, J.-L. Atteia, C. Barat, M. Niel, and G. Vedrenne

GAMMA-RAY BURSTS: COUNTERPARTS

Gamma-Ray Burst Locations While-You-Wait: First Results of the Third Interplanetary Network Burst Watch 769
    K. Hurley, M. Sommer, T. Cline, M. Boër, M. Niel, E. Fenimore, R. Klebesadel, J. Laros, G. Fishman, C. Kouveliotou, C. Meegan, and R. Wilson

Gamma-Ray Burst Observations with the Compton/Ulysses/Pioneer–Venus Network 774
    T. L. Cline, K. C. Hurley, M. Sommer, M. Boer, M. Niel, G. J. Fishman, C. Kouveliotou, C. A. Meegan, W. S. Paciesas, R. B. Wilson, E. E. Fenimore, J. G. Laros, and R. W. Klebesadel

A Search for Long-Lived Emission from Well-Localized Gamma-Ray Bursts using the BATSE Occultation Technique 778
    J. M. Horack, B. A. Harmon, G. J. Fishman, C. A. Meegan, R. B. Wilson, W. S. Paciesas, G. N. Pendleton, and C. Kouveliotou

The E_{max} Distribution of Gamma-Ray Bursts Observed by the PHEBUS Experiment 783
    J.-P. Dezalay, C. Barat, R. Talon, R. Sunyaev, O. Terekhov, and A. Kuznetsov

Gamma-Ray Burst Results from DMSP Satellites 788
    J. Terrell, P. Lee, R. W. Klebesadel, and J. W. Griffee

Search for Correlations of BATSE GRB's with Known Objects 793
    S. Howard, G. J. Fishman, C. A. Meegan, R. B. Wilson, and W. S. Paciesas

ROSAT Wide Field Camera Search for XUV Counterparts of GRB and Optical Transients 798
    A. Owens, S. Sembay, M. Sims, A. Wells, and B. E. Schaefer

Gamma-Ray Burster Distances from Soft X-Ray Observations 803
    B. E. Schaefer

An X-Ray Counterpart to the 5 March 1979 Gamma-Ray Burst? 808
    R. E. Rothschild, R. E. Lingenfelter, F. D. Seward, and O. Vancura

ROSAT X-Ray Observations of GBS J0815-3245 = GRB 920501 813
    M. Boer, K. Hurley, J. Greiner, M. Sommer, M. Niel, G. Fishman, C. Kouveliotou, C. Meegan, W. Paciesas, R. Wilson, E. Fenimore, R. Klebesadel, J. Laros, and T. Cline

Gamma-Ray Burst Astrometry IV: The Atteia *et al.* Catalog 818
    L. G. Taff, J. H. Scott, and S. T. Holfeltz
**The Search for Gamma-Ray Burst Counterparts: A Coordinated
COMPTEL/BATSE Rapid Response Approach** 823
    R. M. Kippen, J. Macri, J. Ryan, B. McNamara, and C. Meegan
Simultaneous Optical/Gamma-Ray Observations of GRB's 828
    J. Greiner, W. Wenzel, R. Hudec, P. Pravec, T. Rezek, M. Nesvara,
    E. I. Moskalenko, A. Karnashov, V. Metlov, N. S. Chernych, V. S. Getman,
    C. Kouveliotou, G. J. Fishman, C. A. Meegan, W. S. Paciesas,
    and R. B. Wilson
**A Method for Searching the Whipple Observatory Gamma-Ray
Database for Evidence of GRB's** 833
    M. Chantell, C. W. Akerlof, M. F. Cawley, V. Connaughton, D. J. Fegan,
    S. Fennell, J. A. Gaidos, J. Hagan, A. M. Hillas, A. D. Kerrick, R. C. Lamb,
    M. A. Lawrence, D. A. Lewis, D. I. Meyer, G. Mohanty, K. S. O'Flaherty,
    P. T. Reynolds, A. C. Rovero, M. S. Schubnell, G. Sembroski, T. C. Weekes,
    T. Whitaker, and C. Wilson
**A Search for Neutrino and Gamma-Ray Burst Temporal Correlations
with the IMB Detector** 838
    R. Becker-Szendy, C. B. Bratton, J. Breault, D. Casper, S. T. Dye,
    W. Gajewski, M. Goldhaber, T. J. Haines, P. G. Halverson, D. Kielczewska,
    W. R. Kropp, J. G. Learned, S. Matsuno, J. Matthews,
    G. McGrath, C. McGrew, R. S. Miller, L. Price, F. Reines, J. Schultz,
    D. Sinclair, H. W. Sobel, J. L. Stone, L. R. Sulak, and R. Svoboda

GAMMA-RAY BURSTS: SPECTROSCOPY

**Gamma-Ray Burst Studies by COMPTEL During its First Year of
Operation** 845
    C. Winkler, K. Bennett, L. Hanlon, O. R. Williams, W. Collmar, R. Diehl,
    V. Schönfelder, H. Steinle, M. Varendorff, J. W. den Herder, W. Hermsen,
    L. Kuiper, B. N. Swanenburg, C. de Vries, A. Connors, D. Forrest,
    M. Kippen, M. McConnell, and J. Ryan
EGRET Observations of Gamma-Ray Bursts 850
    E. J. Schneid, D. L. Bertsch, B. L. Dingus, C. E. Fichtel, R. C. Hartman,
    S. D. Hunter, G. Kanbach, D. A. Kniffen, P. W. Kwok, Y. C. Lin,
    J. R. Mattox, H. A. Mayer-Hasselwander, P. F. Michelson, C. von Montigny,
    P. L. Nolan, K. Pinkau, H. Rothermel, M. Sommer, P. Sreekumar,
    and D. J. Thompson
**EGRET Observations of Gamma-Ray Bursts on June 1, 1991 and
August 14, 1991** 855
    P. W. Kwok, D. L. Bertsch, B. L. Dingus, C. E. Fichtel, R. C. Hartman,
    S. D. Hunter, J. R. Mattox, P. Sreekumar, D. J. Thompson, E. J. Schneid,
    G. Kanbach, H. A. Mayer-Hasselwander, C. von Montigny, K. Pinkau,
    H. Rothermel, M. Sommer, Y. C. Lin, P. F. Michelson, P. L. Nolan,
    and D. A. Kniffen

**Search for Gamma-Ray Burst Spectral Features in the Compton
GRO BATSE Data**.. 860
    B. J. Teegarden, T. L. Cline, D. Palmer, B. E. Schaefer, G. J. Fishman,
    C. Meegan, R. B. Wilson, M. Briggs, W. S. Paciesas, G. Pendleton,
    P. Lestrade, D. L. Band, L. Ford, and J. L. Matteson

**Searching for GRB Spectral Features in the BATSE Large Area
Detector Data**... 867
    R. D. Preece, G. J. Fishman, C. A. Meegan, R. B. Wilson, M. N. Brock,
    W. S. Paciesas, G. N. Pendleton, M. S. Briggs, and B. Teegarden

BATSE Observations of Gamma-Ray Burst Spectral Diversity 872
    D. Band, J. Matteson, L. Ford, B. Schaefer, D. Palmer, B. Teegarden,
    T. Cline, G. Fishman, C. Meegan, R. Wilson, W. Paciesas, G. Pendleton,
    and P. Lestrade

A Search for Untriggered Low-Energy Events in the BATSE Database 877
    J. van Paradijs, B. C. Rubin, C. Kouveliotou, J. M. Horack, G. J. Fishman,
    C. A. Meegan, R. B. Wilson, W. S. Paciesas, W. H. G. Lewin,
    and M. van der Klis

BATSE Observations of a Soft Gamma Repeater (SGR) 882
    C. Kouveliotou, G. J. Fishman, C. A. Meegan, R. B. Wilson, R. D. Preece,
    J. M. Horack, W. S. Paciesas, M. S. Briggs, T. M. Koshut, and J. van Paradijs

**Observability of Line Features in BATSE Spectroscopy Detector
Observations of Gamma-Ray Bursts**... 887
    L. A. Ford, D. L. Band, J. L. Matteson, D. M. Palmer, B. E. Schaefer,
    B. J. Teegarden, T. L. Cline, R. D. Preece, G. J. Fishman, C. A. Meegan,
    R. B. Wilson, M. S. Briggs, W. S. Paciesas, and G. N. Pendleton

**The Sensitivity of the BATSE Spectroscopy Detectors to Cyclotron Lines
in Gamma-Ray Bursts**... 892
    D. M. Palmer, B. Teegarden, B. Schaefer, T. Cline, G. Fishman, C. Meegan,
    R. Wilson, W. Paciesas, G. Pendleton, M. Briggs, D. Band, L. Ford,
    J. Matteson, and J. P. Lestrade

**Establishing the Existence of Harmonically Spaced Lines in
Gamma-Ray Burst GB870303, Using Bayesian Inference**...................... 897
    C. Graziani, D. Q. Lamb, T. J. Loredo, E. E. Fenimore, T. Murakami,
    and A. Yoshida

Spectral Variability in Gamma-Ray Bursts on Millisecond Time Scales............ 902
    T. M. Koshut, W. S. Paciesas, G. N. Pendleton, C. Kouveliotou,
    G. J. Fishman, C. A. Meegan, and R. B. Wilson

Rapid Spectral Evolution Analysis of BATSE GRB's 907
    V. E. Kargatis, E. P. Liang, G. Fishman, C. Meegan, R. Wilson, W. Paciesas,
    B. Schaefer, B. Teegarden, J. Matteson, and D. Band

**Spectral Evolution Studies of a Sub-Class of Gamma-Ray Bursts
Observed by BATSE**... 912
    P. N. Bhat, G. J. Fishman, C. A. Meegan, R. B. Wilson, C. Kouveliotou,
    W. S. Paciesas, G. N. Pendleton, and B. E. Schaefer

The Ginga Rate of Observing Cyclotron Lines.. 917
    E. E. Fenimore, G. Schwarz, D. Q. Lamb, P. Freeman, and T. Murakami

**Sensitivity of the BATSE Spectroscopy Detector to Gamma-Ray
Burst Spectral Lines Like Those Seen in Ginga** 922
    P. E. Freeman, D. Q. Lamb, and E. E. Fenimore

Compton Scattering of Gamma-Ray Burst Spectra 927
    X.-M. Hua and R. E. Lingenfelter
Gamma-Ray Bursts: Galactic or Cosmological in Origin? 932
    D. Q. Lamb

GAMMA-RAY BURSTS: PULSE PROFILES

Possible Detection of Signature Consistent with Time Dilation
in Gamma-Ray Bursts 947
    J. P. Norris, R. J. Nemiroff, C. Kouveliotou, G. J. Fishman, C. A. Meegan,
    R. B. Wilson, and W. S. Paciesas
Morphological Study of Short Gamma-Ray Bursts Observed by BATSE 953
    P. N. Bhat, G. J. Fishman, C. A. Meegan, R. B. Wilson, and W. S. Paciesas
Fractal Analysis of the GRB Light Curves 958
    N. I. Shakura, E. I. Moskalenko, K. A. Postnov, M. E. Prokhorov,
    and N. N. Shakura
Deconvolution of Pulse Shapes in Bright Gamma-Ray Bursts 959
    J. P. Norris, S. P. Davis, C. Kouveliotou, G. J. Fishman, C. A. Meegan,
    R. B. Wilson, and W. S. Paciesas
Calibration of an Algorithm for Deconvolution of Overlapping Pulses in
Gamma-Ray Bursts 964
    S. P. Davis and J. P. Norris
An Analysis of the Structure of Gamma-Ray Burst Time Histories 969
    J. P. Lestrade, G. J. Fishman, C. A. Meegan, R. B. Wilson, W. S. Paciesas,
    G. N. Pendleton, P. Moore, and H. E. Cody
Search for Gravitational Lens Echoes in Gamma-Ray Bursts 974
    R. J. Nemiroff, J. M. Horack, J. P. Norris, W. A. D. T. Wickramasinghe,
    C. Kouveliotou, G. J. Fishman, C. A. Meegan, R. B. Wilson,
    and W. S. Paciesas

GAMMA-RAY BURSTS: MODELS

Gamma-Ray Bursts 981
    B. Paczyński
The Impact of Relativistic Fireballs on an External Medium: A New
Model for "Cosmological" Gamma-Ray Burst Emission 987
    M. J. Rees and P. Mészáros
Contribution to Panel Discussion on Gamma-Ray Bursts 992
    R. D. Blandford
Stellar Collapse and Gamma-Ray Bursts 995
    S. E. Woosley
On the Extended Halo Origin of Gamma-Ray Bursts 1003
    D. H. Hartmann, E. V. Linder, and L.-S. The
A Possible Contribution of Local ($<$ kpc) Neutron Stars to the
Gamma-Ray Bursts 1010
    J. C. Higdon
High Lorentz-Factor $e^{\pm}$ Jets in Gamma-Ray Burst Sources 1015
    P. Mészáros and M. J. Rees

Disk Plus Halo Models of Gamma-Ray Burst Sources 1020
    I. A. Smith and D. Q. Lamb
Gamma-Ray Bursts as a Probe of Large-Scale Structure in the Universe 1025
    D. Q. Lamb and J. M. Quashnock
Gamma-Ray Bursts from Sheared Alfvén Waves in the Magnetosphere of Extragalactic Radio Pulsars 1030
    M. Fatuzzo and F. Melia
Galactic Halo Model for Gamma-Ray Bursts from High-Velocity Neutron Stars 1035
    H. Li, C. D. Dermer, and E. P. Liang
A Study of Gamma-Ray Burst Continuum Properties Presenting Evidence for Two Spectral States in Bursts 1040
    G. N. Pendleton, W. S. Paciesas, R. S. Mallozzi, T. M. Koshut,
    G. J. Fishman, C. A. Meegan, R. B. Wilson, B. A. Harmon, and J. P. Lestrade
A Semi-Analytic Model for Cyclotron Line Formation 1045
    J. C. L. Wang, I. Wasserman, and D. Q. Lamb
Gamma-Ray Bursts: Magnetosphere Around Neutron Stars with Comets, Planets, and Black Holes 1049
    H. Hanami
On the Halo Neutron Star Origin of the Gamma-Ray Bursts: Origin of the Halo Neutron Stars and Metal Enrichment of the Intracluster Medium 1054
    M. Hattori and N. Terasawa
Gamma-Ray Bursts from Collisions of Primordial Small-Mass Black Holes with Comets 1059
    K. F. Bickert and J. Greiner
Models of Gravitationally Accelerated Matter in Relativistic Motion as Gamma-Ray Burst Sources 1064
    J. S. Graber
Re-Ignition of Dead Pulsars: A Possible Source of Gamma-Ray Bursts 1069
    K. Y. Ding and K. S. Cheng
Halo Beaming Models for Gamma-Ray Bursts 1074
    R. C. Duncan, H. Li, and C. Thompson
Energy Storage in Old Neutron-Star Crusts 1080
    P. C. Mock and P. C. Joss
X-Ray Emission From Neutron Stars with Supercritical Magnetic Fields: A Model for the Soft Gamma Repeaters 1085
    C. Thompson and R. C. Duncan
A Burst of Speculation 1090
    J. I. Katz
Two Population—Disk & Halo—Gamma-Ray Burst Models 1095
    J. C. Higdon and R. E. Lingenfelter
Models for the Spatial Distribution of Gamma-Ray Bursts 1099
    H. Li and E. Liang

INSTRUMENTATION

Calculation of the Induced Radioactivity Background in OSSE 1107
S. J. R. Battersby, J. J. Quenby, C. S. Dyer, P. R. Truscott,
N. D. A. Hammond, C. Comber, J. D. Kurfess, W. N. Johnson, R. L. Kinzer,
M. S. Strickman, G. V. Jung, W. R. Purcell, D. A. Grabelsky,
and M. P. Ulmer

COMPTEL High-Energy Neutron Response 1112
T. J. O'Neill, F. Ait-Ouamer, T. A. Roth, O. T. Tumer, R. S. White,
and A. D. Zych

BATSE Burst Triggers from Cygnus X-1 Fluctuations 1117
C. Meegan, G. Fishman, R. Wilson, and W. Paciesas

Identification of Events Observed by BATSE 1122
R. S. Mallozzi, W. S. Paciesas, C. A. Meegan, G. J. Fishman,
and R. B. Wilson

Modeling the Gamma-Ray Background on BATSE 1127
B. C. Rubin, B. A. Harmon, M. N. Brock, G. J. Fishman, C. A. Meegan,
R. B. Wilson, W. S. Paciesas, M. H. Finger, J. C. Ling, R. T. Skelton,
W. A. Wheaton, and D. E. Gruber

Selection Biases in Gamma-Ray Burst Detectors 1132
G. Pizzichini

The Rapidly Moving Telescope (RMT): A First-Light Report 1137
S. D. Barthelmy, T. L. Cline, B. J. Teegarden, and T. T. von Rosenvinge

The Design of a Gamma-Ray Burst Polarimeter 1142
M. McConnell, D. Forrest, K. Levenson, and W. T. Vestrand

An Instrument Called Prometheus 1147
R. C. Haymes, M. J. Moss, and P. W. Walker

**An Updating About FIP: A Photometer Devoted to the Search for
Optical Flashes from Gamma-Ray Bursters** 1152
A. Piccioni, C. Bartolini, C. Cosentino, A. Guarnieri, S. Ricca Rosellini,
A. Di Cianno, A. Di Paolantonio, C. Giuliani, E. Micolucci,
and G. Pizzichini

**The Transient Gamma-Ray Spectrometer: A New High Resolution
Detector for Gamma-Ray Burst Spectroscopy** 1156
H. Seifert, R. Baker, T. L. Cline, N. Gehrels, J. Jermakian, T. Nolan,
R. Ramaty, D. A. Sheppard, G. Smith, D. E. Stilwell, B. J. Teegarden,
J. Trombka, A. Owens, C. P. Cork, D. A. Landis, P. N. Luke, N. W. Madden,
D. Malone, R. H. Pehl, H. Yaver, K. Hurley, S. Mathias, and A. H. Post, Jr.

**Figaro IV: 16 Square Meter Balloon Borne Telescope to Study Rapid
Variabilities and Transient Phenomena at Energies above 50 MeV** 1161
B. Sacco, B. Agrinier, G. Agnetta, B. Biondo, O. Catalano, M. N. Cinti,
E. Costa, G. Cusumano, N. D'Amico, G. D'Alí, R. Di Raffaele, G. Gerardi,
M. Gros, J. M. Lavigne, M. C. Maccarone, A. Mangano, B. Martino,
J. L. Masnou, E. Massaro, G. Matt, G. Medici, T. Mineo, E. Morelli,
G. Natali, L. Nicastro, F. Pedichini, A. Rubini, L. Scarsi, and M. Tripiciano

GRIS Background Reduction Results Using Isotopically Enriched Ge 1166
S. D. Barthelmy, L. M. Bartlett, N. Gehrels, M. Leventhal, B. J. Teegarden,
J. Tueller, S. Belyaev, V. Lebedev, and H. V. Klapdor-Kleingrothaus

GRANITE—A Stereoscopic Imaging Cherenkov Telescope System 1171
    M. Schubnell, C. W. Akerlof, M. F. Cawley, M. Chantell, D. J. Fegan,
    S. Fennell, K. S. O'Flaherty, S. Freeman, D. Frishman, J. A. Gaidos,
    J. Hagan, K. Harris, A. M. Hillas, A. D. Kerrick, R. C. Lamb, T. Lappin,
    M. A. Lawrence, H. Levy, D. A. Lewis, D. I. Meyer, G. Mohanty, M. Punch,
    P. T. Reynolds, A. C. Rovero, G. Sembroski, C. Weaverdyck, T. C. Weekes,
    T. Whitaker, and C. Wilson

Calibration of an Atmospheric Cherenkov Telescope Using Muon
Ring Images 1176
    A. C. Rovero, K. Harris, Y. Jiang, M. A. Lawrence, D. A. Lewis, M. Urban,
    and T. C. Weekes

Results from the CYGNUS Extensive Air Shower Array 1181
    D. A. Williams

SOFTWARE AND ARCHIVES

Status of the BATSE Enhanced Earth Occultation Analysis Package
for Studying Point Sources 1189
    R. T. Skelton, J. C. Ling, N. F. Ling, R. Radocinski, and W. A. Wheaton

Preliminary EGRET Source Catalog 1194
    C. E. Fichtel, D. L. Bertsch, B. L. Dingus, R. C. Hartman, S. D. Hunter,
    P. W. Kwok, J. R. Mattox, P. Sreekumar, D. J. Thompson, D. A. Kniffen,
    Y. C. Lin, P. L. Nolan, P. F. Michelson, G. Kanbach,
    H. A. Mayer-Hasselwander, C. von Montigny, K. Pinkau, H. Rothermel,
    M. Sommer, and E. J. Schneid

Accessing the BATSE Catalog of Bursts 1202
    S. Howard, C. A. Meegan, G. J. Fishman, R. B. Wilson, and W. Paciesas

The Compton Observatory Archive 1205
    T. McGlynn, E. Chipman, J. Jordan, N. Ruggiero, D. Jennings,
    and T. Serlemitsos

Appendix A: Conference Program 1213
Appendix B: List of Participants 1217
Author Index 1223

Preface

In his 1958 paper *Il Nuovo Climento*, Philip Morrison drew attention to the astrophysical information that was probably being carried by gamma rays. He suggested that "the extension of observations to such a new domain will in the end repay considerable effort," and commented that his paper was "intended mainly to attract to this problem the attention of those experimenters skilled in the required arts."

Some 35 years later, the vigorous state of gamma-ray astronomy and the breadth of its reach were evident at the Compton Symposium held at Washington University in St. Louis, October 15–17, 1992. The scheduling of the Symposium was timely—around a year and a half after the launch of NASA's Gamma-Ray Observatory. Soon after its launch, the observatory was named after Arthur Holly Compton whose centennial was being celebrated. The papers and posters at the Symposium were not confined to the results from the Compton Observatory, but included complementary observations from the GRANAT and Ulysses vehicles as well as from balloon-borne systems.

In our planning, we had anticipated an attendance of approximately 150 scientists, but with the torrent of new gamma-ray observations and the parallel flow of theoretical interpretation and speculation, we ended up with an attendance of over 300. Because the duration of the Symposium had been firmly set and the equally firm decision to try to avoid parallel sessions had been made, the program schedule was inevitably tight. We must again express our appreciation for both the good-natured acceptance of this schedule by the contributors and the effectiveness of our colleagues who chaired the sessions.

The range of topics covered can be gauged from the Table of Contents. As expected, the sessions on gamma-ray bursts generated the greatest interest and the least agreement. This can be seen in the equivocal results of the opinion poll on the probable location of the bursts: solar system, galactic halo, or cosmological. As with all polls, we note the margin of error as about $\pm 5\%$, disregarding systematic effects.

For the successful planning and operation of the Symposium, we are grateful to Dr. Trudi Spigel, Ms. Nancy Galofre, and Mrs. Barbara Wilcox (at Washington University), Ms. Kim Pollock, Ms. Sandy Barnes, and Ms. Meta Hutchinson-Frost (at NASA), and to Dr. Don Kniffen, the Project Scientist for GRO when the Symposium was first discussed. We are grateful for the support received from Washington University and its McDonnell Center for the Space Sciences, and especially to the James S. McDonnell Foundation for a generous grant. The contributions of the Scientific Organizing Committee: Neil Gehrels (GSFC, Chairman), M. W. Friedlander (WU, Local Chairman), E. G. Chipman (GSFC), Ph. Durouchoux (Saclay), C. E. Fichtel (GSFC), G. J. Fishman (MSFC), J. E. Grindlay (CfA), D. A. Kniffen (Hampden-Sydney), J. D. Kurfess (NRL), D. Q. Lamb (Chicago), M. Leventhal (GSFC), E. P. Liang (Rice), T. A. Prince (CIT), J. M. Ryan (New Hampshire), V. Schönfelder (MPE), and R. Sunyaev (IKI) are gratefully acknowledged. Finally, special thanks to Dr. Chris Shrader whose work on the organization of the program and co-editing of these proceedings have been invaluable.

M. Friedlander

ARTHUR HOLLY COMPTON (1892–1962)

Arthur Holly Compton was born on September 12, 1892 in Wooster, Ohio. His undergraduate studies were at the College of Wooster where his father was a Professor of Philosophy; his Ph.D. degree was obtained from Princeton University in 1916. After teaching at the University of Minnesota for a year, he spent two years working for the Westinghouse Lamp Company followed by a year in Cambridge University as one of the first National Research Council Fellows. On his return to the U.S. in 1920 he was appointed Wayman Crow Professor and head of the Department of Physics at Washington University, a position he held until 1923 when he went to the University of Chicago.

During the Second World War, Compton played a major role in the atomic bomb project, directing the Metallurgical Laboratory at the University of Chicago where controlled fission was achieved for the first time in 1942. At the end of the war, Compton returned to Washington University as Chancellor and served in that capacity until his retirement in 1953.

During the early years in St. Louis he carried out his major x-ray experiments. The results were presented at the December 1922 meeting of the American Physical Society and appeared in print in May 1923 in the *Physical Review*. For his x-ray research he was awarded the Nobel Prize in Physics in 1927. While at the University of Chicago, Compton continued his x-ray research but then switched to the study of cosmic rays. In the early 1930s his world-wide survey clearly demonstrated the latitude effect, which showed that the primary cosmic rays were electrically charged particles and not photons, as Millikan had suggested.

Compton scattering—the scattering of photons off of electrons—is an important process in many astrophysical circumstances. It is an interesting coincidence that the importance of the inverse process—the scattering of electrons from photons—was first pointed out by two of Compton's younger faculty colleagues at Washington University, Eugene Feenburg and Henry Primakoff, in a 1948 paper.

Arthur Holly Compton, with one of the ionization chambers used in his cosmic-ray survey.

OPENING REMARKS

THE COMPTON OBSERVATORY IN PERSPECTIVE

N. Gehrels, E. Chipman*, D. A. Kniffen**
Laboratory for High Energy Astrophysics,
NASA/Goddard Space Flight Center, Greenbelt, Maryland 20771

ABSTRACT

The Arthur Holly Compton Gamma Ray Observatory (*Compton*) was launched by the Space Shuttle Atlantis on 5 April 1991. The spacecraft and instruments are in good health and returning exciting results. The mission provides nearly six orders of magnitude in spectral coverage, from 15 keV to 30 GeV, with sensitivity over the entire range an order of magnitude better than that of previous observations. The 16,000 kilogram observatory contains four instruments on a stabilized platform. The mission began normal operations on 16 May 1991 and has now almost completed a full-sky survey. The mission duration is expected to be from six to ten years. A Science Support Center has been established at Goddard Space Flight Center for the purpose of supporting a vigorous Guest Investigator Program. New scientific results to date include: (1) the establishment of the isotropy, combined with spatial inhomogeneity, of gamma-ray bursts on the sky; (2) the discovery of intense high energy (100 MeV) gamma-ray emission from 3C 279 and other AGNs and BL Lac objects, making these the most distant and luminous gamma-ray sources ever detected; (3) the observation of intense nuclear and positron-annihilation gamma-ray lines and neutrons from several large solar flares; (4) the discovery and multi-wavelength observation of a very bright hard X-ray transient in August 1992, and detection of several other transients; and (5) the detection of three new gamma-ray pulsars, including the discovery that Geminga is a pulsar emitting most of its energy as gamma rays.

INTRODUCTION

The Compton Gamma Ray Observatory (*Compton*) is the second, following the Hubble Space Telescope, in NASA's series of Great Observatories. The mission goal is to perform broad-band gamma-ray observations with better angular resolution and an order of magnitude better sensitivity than previous missions, and to perform the first full sky gamma-ray survey. The scientific theme of *Compton* is the study of physical processes taking place in the most dynamic sites in the universe, including

---

* Also, Science Programs, Computer Sciences Corp., Greenbelt, Maryland 20771
** Also, Hampden-Sydney College, Virginia 23943

supernovae, novae, pulsars, black holes, active galaxies, gamma-ray bursters, and solar flares.

Compton was announced by NASA in 1977 as an opportunity for gamma-ray experiments on a free-flying observatory to be launched by the space shuttle. The nominal mission was chosen to be two years, limited by the propellant gas for orbit maintenance. A later decision to manifest *Compton* as a dedicated payload on the high-performance Atlantis Shuttle allowed direct launch to 450 kilometers and extended the mission duration to six to ten years. The extension of the lifetime tremendously enhances the expected scientific return from *Compton*, and makes it possible to expand the scientific involvement beyond the original Principal Investigators and their Co-Investigators. This not only increases the access to the data and observing opportunities, but it also allows for the infusion of ideas from a broader community. The Compton Science Support Center has been established at the Goddard Space Flight Center to assist Guest Investigators in accessing the data from *Compton*.

A more detailed presentation of the instruments and science objectives is given in the proceedings of the First GRO Science Workshop[9]. Some of the initial results from the *Compton* mission have been presented at the 22nd International Cosmic Rays Conference held in Dublin in August 1991, the Compton Observatory Science Workshop, held in Annapolis in September 1991[20], and the Huntsville Gamma-Ray Burst Workshop, held in October 1991[18]. A collection of papers on the ground calibration of each instrument is also planned, the first of which are in press.

THE SPACECRAFT

Compton is a 16,000 kg spacecraft (Figure 1) containing a complement of four instruments to make observations of the gamma-ray sky over the wide energy range from about 15 keV to 30 GeV. This very wide dynamic range requires the use of different instruments with a number of detection techniques. The Z-axis of the spacecraft may be pointed to any region of the sky at any time to an accuracy of 0.5°. However, once chosen, the X-axis must be selected to satisfy sun angle constraints which affect the power from the solar panels and the thermal environment of various spacecraft components. The attitude knowledge is maintained to an accuracy of 2 arc minutes. Absolute timing is accurate to better than 0.1 millisecond. The spacecraft is in general performing well and is meeting or exceeding all design goals. Two anomalies have occurred that should be mentioned. The spacecraft tape recorders are both unusable, which has necessitated the use of real-time downlinks only for telemetry retrieval. The present average level of data coverage in this mode is approximately 65%, with plans underway to increase to about 82% early in Phase 3. Also, two of the battery cells in one of the redundant Modular Power Supplies are showing signs of premature aging, but this is not affecting operations at this time.

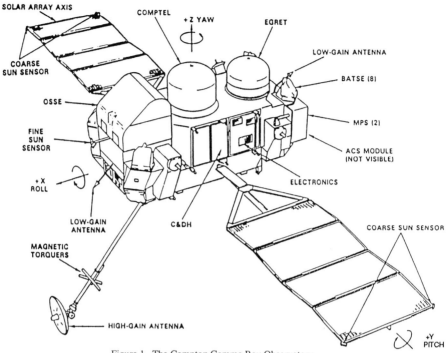

Figure 1. The Compton Gamma Ray Observatory

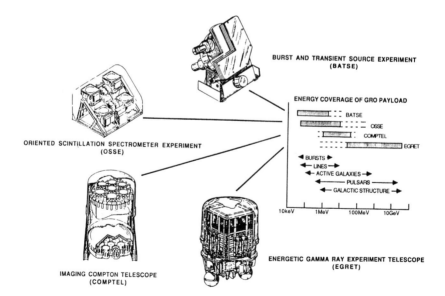

Figure 2. Spectral Coverage of the Compton Gamma Ray Observatory Instruments

THE INSTRUMENTS

Figure 2 shows the spectral coverage of the four instruments. They are the Burst and Transient Source Experiment (BATSE), the Imaging Compton Telescope (COMPTEL), the Energetic Gamma Ray Experiment Telescope (EGRET) and the Oriented Scintillation Spectrometer Experiment (OSSE). The BATSE experiment is designed to monitor the entire sky continuously for bursts and transient gamma-ray events using eight uncollimated, wide-field detectors placed at all corners of the spacecraft, top and bottom. The OSSE experiment, with a relatively narrow field of view, is primarily sensitive to discrete sources, the dominant feature of the low energy gamma-ray sky. The COMPTEL and EGRET experiments are wide-field instruments designed to provide a high sensitivity survey of the medium and high energy gamma-ray sky. A summary of the characteristics of each instrument is given in Table 1.

BATSE

BATSE is optimized to measure brightness variations on timescales down to milliseconds. To accomplish this, eight detector modules of identical configuration are used. The main detector is a large area sodium iodide (NaI) scintillation crystal (50 cm in diameter by 1.3 cm thick). It detects photons in the energy range from 30 keV to 1.9 MeV. Light produced as the incident gamma rays interact in this scintillator is sensed by three large photomultiplier tubes placed behind it, while in front is a plastic scintillator used to reduce the background due to charged particles. The plastic scintillator is virtually transparent to gamma rays. Upon detection of a fast, intense transient event, such as a gamma-ray burst, BATSE processes large amounts of gamma-ray data for later transmission to the ground. It also sends a signal to the three other telescopes on *Compton* that such an event has occurred so that it can be studied over a wider energy band. A smaller spectroscopy detector, optimized for broad energy coverage (15 keV to 110 MeV) and fine energy resolution is contained within each of the 8 detectors.

Dr. Gerald R. Fishman of the NASA Marshall Space Flight Center (MSFC) is the Principal Investigator of the BATSE experiment with Co-Investigators at MSFC, the University of Alabama, Huntsville, Goddard Space Flight Center (GSFC) and the University of California, San Diego.

OSSE

OSSE is designed to undertake comprehensive observations of astrophysical sources in the 0.1 to 10 MeV range, and also includes limited capability above 10 MeV, primarily for solar gamma-ray and neutron observations. The instrument consists of four identical detector systems, each of which is articulated to provide a

	OSSE	COMPTEL	EGRET	BATSE LARGE AREA	BATSE SPECTROSCOPY
ENERGY RANGE (MeV)	0.06 to 10.0	1.0 to 30.0	20 to 3×10^4	0.03 to 1.9	0.015 to 110
ENERGY RESOLUTION (FWHM)	12.5% at 0.2 MeV 6.8% at 1.0 MeV 4.0% at 5.0 MeV	8.8% at 1.27 MeV 6.5% at 2.75 MeV 6.3% at 4.43 MeV	~20% 100 to 2000 MeV	32% at 0.06 MeV 27% at 0.09 MeV 20% at 0.66 MeV	8.2% at 0.09 MeV 7.2% at 0.66 MeV 5.8% at 1.17 MeV
EFFECTIVE AREA (cm²)	2013 at 0.2 MeV 1480 at 1.0 MeV 569 at 5.0 MeV	25.8 at 1.27 MeV 29.3 at 2.75 MeV 29.4 at 4.43 MeV	1200 at 100 MeV 1600 at 500 MeV 1400 at 3000 MeV	1000 ea. at 0.03 MeV 1800 ea. at 0.1 MeV 550 ea. at 0.66 MeV	100 ea. at 0.3 MeV 127 ea. at 0.2 MeV 52 ea. at 3 MeV
POSITION LOCALIZATION	10 arc min square error box (special mode; 0.1 × Crab spectrum)	8.5 arc min (90% confidence at 2.75 MeV – 20σ source)	5 to 10 arc min (1σ radius; 0.2 × Crab spectrum)	3° (strong burst)	—
FIELD OF VIEW	3.8° × 11.4°	~ 64° (~ 1 sr)	~ 0.6 sr	4 π sr	4 π sr
MAXIMUM EFFECTIVE GEOMETRIC FACTOR (cm² sr)	13	30	1050 (~ 500 MeV)	15000	5000
ESTIMATED SOURCE SENSITIVITY (5×10^5 sec; on source, off Galactic Plane) LINE	$(3-8) \times 10^{-5}$ cm^{-2} s^{-1}	1.5×10^{-5} to 6×10^{-5} cm^{-2} s^{-1}			0.4% equivalent width (5 sec integration)
CONTINUUM	3×10^{-7} cm^{-2} s^{-1} keV^{-1} (@1 MeV)	1.6×10^{-4} cm^{-2} s^{-1} (3σ detection, 1 – 30 MeV)	7×10^{-8} cm^{-2} s^{-1} (> 100 MeV) 2×10^{-8} cm^{-2} s^{-1} (> 1000 MeV)	6×10^{-8} erg cm^{-2} (10 sec–burst)	—

Table 1. Summary of Compton Gamma Ray Observatory Detector Characteristics

192° rotation capability about the Y-axis of the spacecraft. Normally, two detectors observe the source, and two a nearby off-source region. The combination is reversed at regular intervals, and the difference signal represents the net source flux. Each detector system includes a 33-cm diameter sodium iodide–cesium iodide (NaI-CsI) phoswich assembly which provides the basic gamma-ray detection capability. Active shielding is provided by an annular NaI shield and the CsI section of the phoswich. The acceptance aperture of 3.8° x 11.4° is defined by a passive tungsten alloy collimator. The offset pointing feature also allows the observation of targets away from the prime viewing axis of *Compton*. This permits viewing secondary sources when the main spacecraft axis is occulted by the Earth, and allows for quick response observations of transient phenomena such as solar flares and novae without impacting the Observatory viewing program in a major way.

Dr. James D. Kurfess of the U.S. Naval Research Laboratory (NRL) is the Principal Investigator of the OSSE experiment with Co-Investigators at NRL, Northwestern University, Clemson University and the Royal Aircraft Establishment.

COMPTEL

In COMPTEL, gamma rays are detected by the occurrence of two successive interactions: the first one is a compton scattering collision in a detector of low-Z material (liquid scintillator) followed by a second interaction in a detector with high-Z material (NaI), in which, ideally, the scattered gamma ray is totally absorbed. The sequence of collisions is established by a time-of-flight measurement between the two detectors. Many neutron-induced background events, which may simulate double scatter gamma-ray events, are identified and rejected on the basis of the shape of the light pulse emitted in the first scatter detector. This method, in combination with charged particle shield detectors, is used to suppress the instrumental background.

COMPTEL is designed to perform a sensitive survey in the energy range from 1 to 30 MeV. The spatial resolution in the two detectors, together with the well-defined kinematics of the interaction, permits the reconstruction of discrete and extended sky images over a wide field of view with a resolution on the order of 1°. The minimum source detectability is a few percent of the total emission from the Crab nebula. The first or upper detector of COMPTEL contains 7 cylindrical modules of liquid scintillator. Each module is approximately 28 cm in diameter and 8.5 cm thick and viewed by eight photomultipliers. The second detector consists of 14 cylindrical NaI blocks of 7.5 cm thickness and 28 cm diameter. Each block is viewed by seven photomultipliers from below. The distance between the first and second detectors is 1.5 meters. Each detector is surrounded by a thin anticoincidence shield of plastic scintillator.

The Principal Investigator for the COMPTEL instrument is Dr. Volker Schönfelder of the Max Planck Institute for Extraterrestrial Physics (MPE) in Garching, Germany. Co-Investigators are located at MPE, at the University of New

Hampshire, at the Laboratory for Space Research, Leiden, and at the Space Sciences Department of the European Space Agency.

EGRET

EGRET, the highest energy instrument on *Compton*, covers the broadest energy range from about 20 MeV to about 30 GeV. Like COMPTEL, it is an instrument with a wide field of view, good angular resolution and very low background. Because it is designed for high energy studies, the detector is optimized to detect gamma rays when they interact by the dominant high energy pair-production process which forms an electron and a positron within the upper EGRET spark chamber. This unique signature allows a positive identification of the parent gamma ray. The basic imaging portion of this instrument consists of two sections of spark chambers separated by a layer of segmented plastic scintillators. The upper section consists of 28 spark chamber modules interleaved with 27 thin foils of high-Z target material (tantalum) for the conversion of the gamma rays to electron-positron pairs. After the electrons are formed, their trajectory is followed through the remainder of the upper and lower spark chamber sections and into the large NaI crystal where the rest of their energy is deposited. The NaI counter provides good precision in estimating the energy of the gamma ray over the large sensitive range of this instrument. A central scintillator matrix, in coincidence with a similar layer below the bottom spark chamber section, identifies the presence of the electrons, applies the high voltage to the spark chambers and initiates the read-out of the digitized "picture" of the particle trajectories. A large plastic scintillator dome over the instrument identifies and rejects events containing particles which are already charged on entry into the detector.

Co-Principal Investigators of the EGRET instrument are Dr. Carl E. Fichtel of GSFC, and Dr. Klaus Pinkau of the Max Planck Institute in Germany. Co-Investigators are located at each of these institutions, at Stanford University, and at the Grumman Aerospace Corporation.

THE COMPTON GAMMA RAY OBSERVATORY MISSION

Compton was launched by the Space Shuttle Atlantis (STS-37) on 5 April 1991 from the Kennedy Space Center in Florida and deployed on 7 April 1991. After a month-long activation and checkout period, routine science operations began on 16 May 1991. The orbital altitude will be maintained between 350 and 450 km (the launch altitude) by an on-board propulsion system. At the end of the mission the spacecraft will either be retrieved by the Space Shuttle, or will reenter in a controlled manner.

During early operations, data were recorded on magnetic tapes on the spacecraft, and telemetered via the Tracking and Data Relay Satellite (TDRS) once every other orbit. Currently all data is transmitted in real time at 32 kilobits per second. Uplink commanding to the experiments and the spacecraft is also sent during the real-time TDRS contacts. The Control Center is at the Goddard Space Flight Center. Data are received in packetized form with all ancillary information required to analyze the data already inserted on board the spacecraft. Time ordering, overlap elimination and quality checks are done on the ground before the data are forwarded, within 48 hours of receipt, to the experiment data analysis facility sites.

The viewing program for the first 15 months of operation (Phase 1) is currently in period 42 of a total of 44 observing periods which vary in length from 1 to 3 weeks. This phase will provide a nearly uniform full-sky survey of two-week exposures for the two wide-field instruments (EGRET and COMPTEL) (Figure 3). The narrow aperture instrument (OSSE) has selected approximately 55 primary and secondary targets for discrete source studies during this 15-month period and the burst instrument (BATSE) covers the entire unocculted sky at all times. The viewing program for Phase 1 is given in Table 2. The planned viewing program for Phase 2, which will run from November 1992 to August 1993, is given in Table 3, with the z-axis pointings shown in Figure 4. In both Phase 1 and Phase 2, a considerable effort has been placed on coordinating with both space- and Earth-based observations at other wavelengths.

GUEST INVESTIGATOR PROGRAM AND THE SCIENCE SUPPORT CENTER

The *Compton* Guest Investigator Program was instituted by NASA in the U.S. and by the German Ministry for Research and Technology (BMFT) in Germany to enhance the scientific return by broadening the scientific participation in the analysis of data, expanding the scope of observations, and conducting correlative and theoretical research that is closely tied to the *Compton* observations[15,16]. Although the instrument development teams retain proprietary rights to the sky survey data obtained in Phase 1 of the mission for a period of one year, significant Guest Investigator participation has nonetheless been supported. Fifty-one proposals were selected for support during Phase 1 through a peer review process, and a total of 121 proposals were selected during Phase 2. During phase 2, 30% of the observing time has been allocated to Guest Investigators. A NASA Research Announcement soliciting Guest Investigations for Phase 3 of the mission will be issued in December 1992, with proposals due in March 1993. The Guest Investigator share of the observing time for the first phases of the mission is shown in Figure 5.

The central point of contact for Guest Investigator support is the *Compton* Science Support Center (*Compton* SSC) at the Goddard Space Flight Center. In addition to its function of aiding Guest Investigators in accessing *Compton* data, the *Compton* SSC is a source of information for potential users at all stages of their involvement, from the pre-proposal stage through the analysis phase. The Center provides information on the availability and timing of opportunities, and gives

Table 2. Compton Gamma Ray Observatory Phase 1 viewing plan as of 9 September 1992.

VIEW PER.	START	+Z-AXIS* TARGET	GRO Z-AXIS* RA	GRO Z-AXIS* DEC	Z-AXIS LONG	Z-AXIS LAT	GALACTIC*	GRO X-AXIS* RA	GRO X-AXIS* DEC	OSSE PRIMARY TARGET	OSSE PRIMARY TARGET RA	OSSE PRIMARY TARGET DEC	OSSE SECONDARY TARGET	OSSE SECONDARY TARGET RA	OSSE SECONDARY TARGET DEC
1	5/16/91	CRAB PULSAR	88.07	17.14	190.92	-4.74		339.12	46.48	CRAB PULSAR	83.52	22.02	PSR 1957+20	299.84	20.81
2.0	5/30/91	CYG X-1	301.39	36.58	73.28	2.56		60.31	33.09	CYG X-1	299.59	35.21	SUN	-	-
2.5	6/8/91	SUN	87.83	12.47	194.86	-7.29		338.68	56.03	SUN	-	-	CYG X-1	299.59	35.21
3	6/15/91	SN 1991T	191.54	2.62	299.76	65.46		101.49	1.04	SN 1991T	188.54	2.66	QSO 0736+016	114.82	1.62
4	6/28/91	NGC 4151	179.84	41.52	156.19	72.08		57.38	31.23	NGC 4151	182.63	39.41	3C 111	64.59	38.02
5	7/12/91	G CENTER	270.39	-30.96	0.00	-4.00		196.56	24.91	G CENTER	266.40	-28.94	MRK 421	166.11	38.21
6	7/26/91	SN 1987A	91.28	-67.97	278.00	-29.32		153.97	10.52	SN 1987A	83.86	-69.27	MRK 421	166.11	38.21
7.0	8/8/91	CYG X-3	310.05	28.06	70.44	-8.31		143.74	61.25	CYG X-3	308.10	40.96	M 82	148.96	69.67
7.5	8/15/91	GAL 025-14	291.98	-13.27	25.00	-14.00		208.11	24.38	G PLANE 25	279.22	-7.05	NGC 5548	214.50	25.13
8	8/22/91	VELA PULSAR	124.96	-46.35	262.94	-5.67		198.99	14.71	VELA PULSAR	128.92	-45.18	SN 1991T	188.54	2.66
9.0	9/5/91	GAL 339-84	8.34	-32.31	338.94	-83.50		244.20	-41.58	GX 339-4	255.71	-48.79	NGC 253	11.89	-25.29
9.5	9/12/91	HER X-1	251.28	36.89	59.67	40.29		142.05	23.69	NOVA MUSCAE	171.61	35.34	MRK 421	166.11	38.21
10	9/19/91	FAIRALL 9	30.91	-60.66	287.85	-54.31		190.28	-27.75	3C 273	187.28	-68.68	3C 279	194.05	-5.79
11	10/3/91	3C 273	189.02	1.06	294.25	63.67		278.42	-29.71	CEN A	201.38	2.05	G CENTER	266.40	-28.94
12	10/17/91	CEN A	202.29	-40.09	310.71	22.21		221.48	48.29	CEN A	187.28	-43.01	3C 390.3	280.56	79.76
13.0	10/31/91	GAL 025-14	291.98	-13.27	25.00	-14.00		208.11	24.38	G PLANE 25	279.22	-7.05	NGC 5548	214.50	25.13
13.5	11/7/91	GAL 339-84	8.34	-32.31	338.94	-83.50		244.20	-41.58	GX 339-4	255.71	-48.79	ESO 141-55	290.31	-58.67
14	11/14/91	ETA CAR	156.83	-58.51	285.04	-0.74		274.30	-15.78	G CENTER	266.40	-28.94	ETA CAR	161.95	-59.98
15	11/28/91	NGC 1275	52.00	40.24	152.63	-13.44		293.36	29.53	NGC 1275	49.96	41.51	CYG X-1	299.59	35.21
16	12/12/91	SCO X-1	248.36	-17.20	0.00	20.29		351.62	-36.55	G CENTER	266.40	-28.94	SCO X-1	244.97	-15.63
17	12/27/91	SN 1987A	83.48	-72.27	283.21	-31.62		266.22	-17.72	SN 1987A	83.86	-69.27	G CENTER	266.40	-28.94
18	1/10/92	M 82	154.61	72.04	137.47	40.49		292.51	13.52	M 82	148.96	69.67	PSR 1929+10	293.05	10.99
19	1/23/92	GAL 058-43	331.41	-1.93	58.15	-43.00		242.82	36.31	G PLANE 58.1	294.98	22.28	HER X-1	254.46	35.34
20	2/6/92	SS 433	285.28	6.37	39.70	0.76		12.01	-27.04	SS 433	287.95	4.99	NGC 253	11.89	-25.29
21	2/20/92	G CENTER	39.09	-1.24	171.52	-53.90		308.14	-37.66	G CENTER	266.40	-28.94	NGC 1068	40.67	-0.01
22	3/5/92	MRK 279	216.00	70.74	112.47	44.47		329.43	7.91	NOVA CYG 92	307.63	52.63	MRK 279	208.24	69.31
23	3/19/92	CIR X-1	227.43	-54.62	322.14	3.01		11.72	-29.97	CIR X-1	230.23	-57.18	PSR 1509-58	228.48	59.14
24.0	4/2/92	GAL 010+57	223.34	11.03	9.53	57.15		302.52	-43.94	G CENTER	266.40	-28.94	NGC 7582	349.60	-42.37
24.5	4/9/92	GAL 010+57	223.34	11.03	9.53	57.15		304.56	-38.07	G PLANE 5	269.34	-24.53			
26	4/16/92	GAL 007+48	229.85	4.47	6.84	48.09		315.25	-45.69	G CENTER	266.40	-28.94	NGC 7582	349.60	-42.37
26	4/23/92	MRK 335	1.59	20.20	108.77	-41.43		73.89	-39.58	MRK 335	1.59	20.21	VELA PULSAR	128.92	-45.18
27	4/28/92	4U 1543-47	241.11	-49.06	332.23	2.52		348.99	-14.92	4U 1543-47	237.00	-47.80	NGC 7314	338.94	-26.05
28	5/7/92	MRK 335	1.59	20.20	108.77	-41.43		73.89	-39.58	MRK 335	1.59	20.21	VELA PULSAR	128.92	-45.18
29	5/14/92	GAL 224-40	68.97	-25.09	224.00	-40.00		53.03	64.04	3C 390.3	280.56	79.76	SN 1987A	83.86	-69.27
30	6/4/92	NGC 2992	149.50	14.73	252.41	30.65		61.41	7.22	NGC 2992	146.43	-14.33	3C 120	68.29	5.35
31	6/11/92	MCG +8-11-11	88.87	49.44	163.09	11.92		86.22	-40.53	MCG +8-11-11	88.73	46.44	SN 1987A	83.86	-69.27
32	6/25/92	NGC 3783	171.17	-36.81	284.20	22.89		92.97	15.28	NGC 3783	174.76	-37.74	CRAB PULSAR	83.52	22.02
33	7/2/92	NGC 2992	149.50	14.73	252.41	30.65		61.41	7.22	NGC 2992	146.43	-14.33	3C 120	68.29	5.35
34	7/16/92	CAS A	345.77	57.49	108.75	-2.37		100.24	14.79	CAS A	350.87	58.81	GEMINGA	98.48	17.77
35	8/6/92	ESO 141-55	287.12	-61.21	335.10	-25.56		144.32	-23.64	ESO 141-55	290.31	-58.67	MCG -5-23-16	146.92	-30.96
36.0	8/11/92	GRO J0422+32	68.99	30.42	169.84	-11.35		141.11	-27.60	GRO J0422+32	65.43	32.91	MCG -5-23-16	146.92	-30.96
36.5	8/12/92	GRO J0422+32	69.39	32.90	168.17	-9.46		168.57	13.86	GRO J0422+32	65.43	32.91	3C 273	187.28	2.05
37	8/20/92	MRK 335	358.75	18.82	104.83	-42.06		98.42	26.22	GRO J0422+32	65.43	32.91	MRK 335	1.58	20.20
38	8/27/92	ESO 141-55	287.12	-61.21	335.10	-25.56		144.32	-23.64	ESO 141-55	290.31	-58.67	MCG -5-23-16	146.92	-30.96
39	9/1/92	GRO J0422+32	68.87	33.82	167.18	-9.18		168.45	13.95	GRO J0422+32	65.43	32.91	3C 273	187.28	2.05

Table 2. (Continued)

		+Z-AXIS*	GRO Z-AXIS*		GRO Z-AXIS* GALACTIC		GRO X-AXIS*		OSSE PRIMARY TARGET			OSSE SECONDARY TARGET		
		TARGET	RA	DEC	LONG	LAT	RA	DEC	TARGET	RA	DEC	TARGET	RA	DEC
40	9/17/92	MCG +5-23-16	140.88	30.40	195.90	44.71	224.38	-10.90	NGC 4388	186.44	12.66	G CENTER	266.40	-28.94
41	10/8/92	GAL 228+03	112.43	-12.05	228.02	2.84	210.43	-33.09	MCG -6-30-15	203.98	-34.29	1A 0620-00	95.69	-0.35
42	10/15/92	PKS 2155-304	319.72	-41.67	359.98	-44.59	189.66	-35.87	PKS 2155-304	329.72	-30.22	PSR 1509-58	228.49	-59.14
43	10/29/92	MRK 509	307.83	-13.95	31.13	-28.33	204.08	-43.75	MRK 509	311.04	-10.72	CEN A	201.37	-43.02
44	11/3/92	GAL 228+03	112.43	-12.05	228.02	2.84	210.43	-33.09	MCG -6-30-15	203.98	-34.29	1A 0620-00	95.69	-0.35
	11/17/92	end Phase 1												

* Z-axis is pointing direction for COMPTEL and EGRET. OSSE points in X-Z plane. RA and DEC are J2000.

Table 3. Compton Gamma Ray Observatory Phase 2 viewing plan as of 15 September 1992.

VIEW PER.	START	+Z-AXIS* TARGET	GRO Z-AXIS* RA	DEC	GALACTIC LONG	LAT	GRO X-AXIS* RA	DEC	OSSE PRIMARY TARGET	RA	DEC	OSSE SECONDARY TARGET	RA	DEC
201	11/17/92	HER X-1	253.15	42.26	66.79	39.28	266.02	-47.01	HER X-1	254.46	35.34	G PLANE 355	263.30	-33.17
202	11/24/92	HER X-1	251.55	45.40	70.85	40.50	270.04	-43.08	HER X-1	254.46	35.34	G CENTER	266.40	-28.94
203	12/01/92	CYG X-3	309.72	42.56	81.85	0.70	261.28	-35.85	CYG X-3	308.11	40.96	G CENTER	266.40	-28.94
204	12/22/92	3C 273	188.99	-0.74	294.71	61.88	279.50	-34.03	3C 273	187.28	2.05	G PLANE 355	263.30	-33.17
205	12/29/92	3C 273	188.84	-1.03	294.46	61.58	279.32	-24.79	3C 273	187.28	2.05	G PLANE 5	269.27	-24.64
206	01/05/93	3C 273	188.99	-0.74	294.71	61.88	279.50	-34.30	3C 273	187.28	2.05	G PLANE 355	263.30	-33.17
207	01/12/93	IC 4329A	203.86	-30.41	314.06	31.51	293.35	0.86	IC 4329A	207.33	-30.31	PSR 1822-09	276.25	-9.00
208	02/02/93	NGC 4507	198.47	-41.93	299.64	20.75	296.86	-9.24	NGC 4507	188.90	-39.91	NGC 6814	295.67	-10.31
209	02/09/93	2CG 010-31	305.69	-40.81	0.24	-34.01	297.95	48.93	2CG 010-31	304.75	-31.84	3C 390.3	280.56	79.76
210	02/22/93	GAL CENTER	257.65	-29.10	355.62	6.28	351.03	-6.03	1E 1740-2942	265.98	-29.72	MCG -2-58-22	346.18	-8.69
211	02/25/93	GAL 123-05	18.38	58.05	125.86	-4.70	8.08	-31.53	4U 0115+634	19.86	62.49	47 TUC	5.83	-72.07
212	03/09/93	WR 140	298.65	50.54	84.33	11.43	341.61	-31.07	WR 140	305.12	43.85	NGC 7314	338.94	-26.05
213	03/23/93	CRAB PULSAR	80.30	22.29	182.62	-8.22	348.47	4.44	CRAB PULSAR	83.52	22.02	MRK 509	311.04	-10.72
214	03/29/93	GAL CENTER	257.65	-29.10	355.62	6.28	351.03	-6.03	1E 1740-2942	265.98	-29.72	MCG -2-58-22	346.18	-8.69
215	04/01/93	CEN A	203.26	-39.28	311.66	22.89	53.56	-46.55	CEN A	201.37	-43.02	SN 1987A	83.86	-69.27
216	04/20/93	NGC 4151	178.02	42.45	157.78	70.63	60.65	26.85	NGC 4151	182.64	39.41	MCG +8-11-11	88.73	46.44
217	05/04/93	GAL CENTER	255.39	-28.84	354.64	8.03	349.27	-7.01	1E 1740-2942	265.98	-29.72	MCG -2-58-22	346.18	-8.69
218	05/07/93	SMC	24.11	-72.07	298.09	-44.63	69.97	12.70	SMC X-1	19.18	-73.44	3C 120	68.30	5.35
219	05/25/93	NGC 4151	178.02	42.25	157.78	70.63	60.65	26.85	NGC 4151	182.64	39.41	MCG +8-11-11	88.73	46.44
220	05/31/93	CAL CENTER	265.98	-29.72	359.14	-0.09	316.33	48.19	1E 1740-2942	265.98	-29.72	CYG X-3	308.11	40.96
221	06/03/93	CRAB PULSAR	85.21	19.46	187.52	-5.88	326.68	53.49	CRAB PULSAR	83.52	22.02	PSR 1951+32	298.24	32.89
222	06/15/93	GAL 347-01	258.74	-39.78	347.48	-0.70	14.94	-27.93	GPLANE 347.5	258.02	-39.35	EMPTY SPACE	13.15	-28.73
223	06/22/93	GAL 347-00	258.02	-39.35	347.50	-0.00	194.38	28.44	GPLANE 347.5	258.02	-39.35	COMA CLUSTER	194.90	27.96
224	06/29/93	GAL CENTER	266.40	-28.94	359.99	-0.00	192.86	27.13	G CENTER	266.40	-28.94	COMA CLUSTER	194.90	27.96
225	07/08/93	GAL CENTER	236.79	-7.75	359.99	35.00	153.50	40.63	G CENTER	266.40	-28.94	COMA CLUSTER	194.90	27.96
226	07/13/93	VELA PULSAR	129.38	-48.16	266.17	-4.28	121.03	41.53	VELA PULSAR	128.92	-45.18	MRK 78	115.67	65.18
227	07/27/93	NGC 6814	289.94	-15.33	22.22	-13.08	186.93	-39.38	NGC 6814	295.67	-10.31	NGC 4507	188.90	-39.91
228	08/03/93	GAL 139+66	182.84	49.50	139.29	66.34	120.91	-21.90	QSO 1028+313	157.75	31.05	PSR 0740-28	115.71	-28.38
229	08/20/93	GAL CENTER	269.66	-29.98	0.54	-2.97	181.62	3.40	1E 1740-2942	265.98	-29.72	3C 279	194.05	-5.79

* Z-axis is pointing direction for COMPTEL and EGRET. OSSE points in X-Z plane. RA and DEC are J2000.

N. Gehrels et al. 13

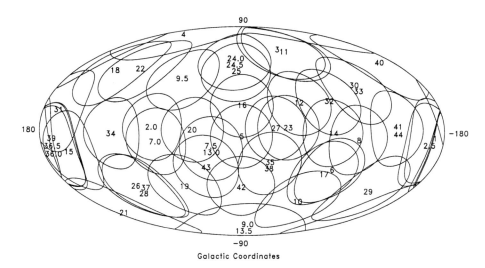

Figure 3. Sky coverage for COMPTEL and EGRET during Phase 1 (May 1991 to November 1992). The circles are 25 degrees in radius to represent the fields of view of COMPTEL and EGRET. The portion of the survey completed thus far corresponds to those viewing areas with numbers up to 41.

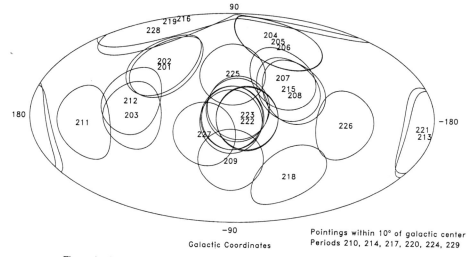

Figure 4. Sky coverage for COMPTEL and EGRET for the viewing plan of Phase 2 (November 1992 to August 1993). The circles are 25 degrees in radius to represent the fields of view of COMPTEL and EGRET.

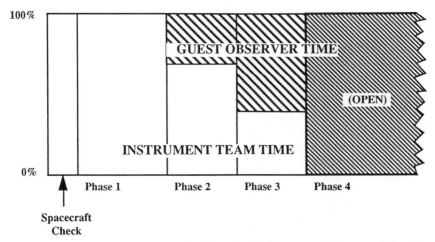

Figure 5. Share of the observing time for the Instrument Teams and for Guest Investigators during the first phases of the *Compton* mission.

technical support in the implementation of chosen guest investigations. It is the source for technical information on the *Compton* instruments, for scientific and technical information on the spacecraft and mission, for catalogs of data for *Compton* and other astrophysics and astronomy observations, for the status of *Compton* observations, and for availability of useful analysis software.

EARLY *COMPTON* RESULTS

A number of important results have been announced as a result of observations during the first 17 months of the *Compton* mission thus far. The instruments are performing well and a rich set of data has been obtained from the sources and pointing directions listed in Table 2. We summarize some of the early results here, and refer the reader to the published papers and to the proceedings of previous conferences and of this Conference for more details.

Gamma-ray bursts are being observed by BATSE at an order of magnitude better sensitivity than ever before. Meegan *et al.*[12] and more recent updates of the BATSE burst statistics[13] have shown that these data are consistent with an isotropic distribution of gamma-ray bursts, but that the number versus intensity distribution does not follow the -3/2 power law expected for a spatially homogeneous distribution of sources. This finding has led to a significant re-evaluation of possible sources of gamma-ray bursts, with special consideration being paid to sources which could potentially satisfy the apparent constraints imposed by the BATSE data, such as cosmological or galactic halo models. Many papers being presented at this workshop deal with this question. Some individual bursts detected by the Compton instruments are also remarkable, including the detection by BATSE on 18 July 1991 of the shortest duration (~10 msec) gamma-ray burst ever seen and the occurrence on 3 May

1991 of a bright gamma-ray burst in the fields of view of COMPTEL and EGRET whose spectrum was measured at energies up to 200 MeV.[19,24]

Compton has found other surprises in the gamma-ray sky during its survey. In a major new finding, EGRET detected intense high energy (~100 MeV) gamma radiation from the quasar 3C 279.[7] At a redshift z = 0.54 (distance ~1500 Mpc), at the time of its detection this was much more distant than any gamma-ray source observed before *Compton*. If the emission is isotropic, the inferred luminosity is ~10^{48} ergs s^{-1}, or about 10,000 times the total luminosity of our galaxy. Since then EGRET has detected 15 additional AGN's and BL Lacs,[4,5,8,14] some at even greater distances than 3C 279. Many of these objects, including 3C 279, show evidence for strong luminosity variations on time scales of days to months.

We are fortunate to have had an active gamma-ray sky during *Compton's* early time in orbit. Several different transient events have occurred that motivated changes in the original viewing plan. In early June 1991, active region 6659 on the Sun spawned six large solar flares over a period of two weeks. The Sun was declared a *Compton* Target of Opportunity on June 8 and the sky survey was interrupted for one week of solar viewing. Several large flares were observed by all instruments on *Compton*, giving a unique data set. Spectral lines from nuclear interactions and positron annihilation (including a particularly intense line at 2.22 MeV from deuterium formation) were detected, along with high energy gamma rays and solar neutrons.

More recently, BATSE detected a very bright X-ray nova on August 5, 1992. This source, named GRO J0422+32 by its discoverers,[17] was declared a Compton Target of Opportunity and was observed for several weeks by all three pointed instruments.[2] Initial results of these observations are reported at this Symposium. Other transient sources have been found by BATSE, using the occultation technique to detect sources as they disappear behind the Earth's limb and a Fourier transform technique to detect pulsed sources. An outburst of the X-ray source GX 339-4 was reported in August 1991.[6] The X-ray pulsar EXO 2030+375 was seen in outburst in February 1992.[23]

Steady (non-transient) pulsed sources have also been observed by BATSE. Wilson *et al*.[22] announced the detection of the 150-ms radio pulsar PSR 1509-58. This is only the second radio pulsar to be seen in this energy range of 20 keV to 2 MeV, after the Crab. EGRET has found two new gamma-ray pulsars, one being the newly discovered radio pulsar PSR 1706-44,[21] and the other being the well-known source Geminga,[1] whose pulse period was determined by ROSAT, and which was subsequently shown to be a radio-quiet pulsar emitting most of its energy as gamma rays.

The COMPTEL Team has begun mapping the diffuse distribution of ^{26}Al in the galaxy, via observations of the 1.8 Mev line[3]; ^{26}Al is a component of the interstellar medium whose source is not yet known. Other diffuse emissions are being observed by the OSSE Team, which has a major objective of mapping the diffuse and point-source emission of the 511-keV annihilation line in the galactic plane and near the galactic center. Their initial measurements and detection of the 511-keV line in the galactic center region were reported by Kurfess et al.[10] Further studies of this region by OSSE will continue in Phase 2 of the mission. OSSE has also detected emission in the 122-keV line of ^{57}Co from the supernova 1987A[11] at a level which has important theoretical implications for models of the supernova outburst.

SUMMARY

The Compton Gamma Ray Observatory represents a dramatic increase in capability over previous gamma-ray missions. The Observatory was launched on 5 April 1991 and is in good health. Important discoveries have already been made in the first 17 months of the mission. With a mission duration of six to ten years, *Compton* promises to revolutionize gamma-ray astronomy and significantly advance our understanding of the high energy universe.

REFERENCES

1. Bertsch, D.L., *et al.*, 1992, Nature, **357**, 306.
2. Cameron, R.A., *et al.*, 1992, IAU Circular No. 5587.
3. Diehl, R., 1992, Astr. Astrophys. Supp., in press.
4. EGRET Team on Compton Observatory, 1992, IAU Circular No. 5431.
5. Fichtel, C.E., *et al.*, 1992, IAU Circular No. 5460.
6. Fishman, G., *et al.*, 1991, IAU Circular No. 5327.
7. Hartman, R.C., *et al.*, 1992, Ap. J. Lett., **385**, L1.
8. Hunter, S.D., *et al.*, 1992, IAU Circular No. 5594.
9. Johnson, W.N., ed., 1989, Proceedings of the GRO Science Workshop, April 10–12, 1989, Goddard Space Flight Center, Greenbelt, Maryland.
10. Kurfess, J.D., *et al.*, 1991, IAU Circular No. 5323.
11. Kurfess, J.D., *et al.*, 1992, Ap. J. Lett., in press.
12. Meegan, C.A., *et al.*, 1992, Nature, **355**, 143.
13. Meegan, C.A., *et al.*, 1992, IAU Circular No. 5641.
14. Michelson, P.F., *et al.*, 1992, IAU Circular No. 5470.
15. NASA NRA 90-OSSA-4, January 30, 1990, NASA Research Announcement for the GRO Phase 1 Guest Investigator Program.
16. NASA NRA 91-OSSA-22, September 16, 1991, NASA Research Announcement for the GRO Phase 2 Guest Investigator Program.
17. Paciesas, W.S., *et al.*, 1992, IAU Circular No. 5580.
18. Paciesas, W.S. and Fishman, G.J., 1992, eds, Gamma-Ray Bursts, Huntsville, AL, 1991; AIP Conference Proceedings 265, Am. Inst. of Physics, New York.
19. Schneid, E.J., *et al.*, 1992, Astr. Astrophys., **255**, L13.

20. Shrader, C.R., Gehrels, N. and Dennis, B., 1992, eds, The Compton Observatory Science Workshop, NASA CP-3137.
21. Thompson, D.J., *et al.*, 1992, Nature, **359**, 615.
22. Wilson, R.B., *et al.*, 1992, IAU Circular No. 5429.
23. Wilson, R.B., *et al.*, 1992, IAU Circular No. 5454.
24. Winkler, C., *et al.*, 1992, Astr. Astrophys. 255, L9.

DIFFUSE EMISSION

INITIAL RESULTS FROM COMPTEL - AN OVERVIEW

V. Schönfelder, W. Collmar, R. Diehl, G.G. Lichti, H. Steinle
A. Strong, M. Varendorff
Max-Planck-Institut für extraterrestrische Physik 8046 Garching bei München, FRG

H. Bloemen, H. de Boer, R. v. Dijk[*], J. W. den Herder, W. Hermsen, L. Kuiper
B. Swanenburg, C. de Vries
SRON-Leiden, P.O. Box 9504, 2300 RA Leiden, The Netherlands

A. Connors, D. Forrest, M. McConnell, D. Morris, J. Ryan, G. Stacy[**]
University of New Hampshire, Institute for the Study of Earth Oceans and Space
Durham NH 03824, USA

K. Bennett, B.G. Taylor, C. Winkler
Astrophysics Division, ESTEC, 2200 AG Noordwijk, The Netherlands.

ABSTRACT

COMPTEL is presently completing the first full sky survey in MeV gamma-ray astronomy (0.7 to 30 MeV). An overview of initial results from the survey is given: among these are the observations of the Crab and Vela pulsars with unprecedented accuracy, the observation of the black hole candidates Cyg X-1 and Nova Persei 1992, an analysis of the diffuse Galactic continuum emission from the Galactic centre region, the broad scale distribution of the 1.8 MeV line from radioactive ^{26}Al, upper limits on gamma-ray line emission from SN 1991T, observations of the three quasars 3C273, 3C279 and PKS 0528+134 and the radio galaxy Cen A, measurements of energy spectra, time histories and locations of a number of cosmic gamma-ray bursts, and gamma-ray and neutron emission from solar flares.

INTRODUCTION

COMPTEL covers the middle energy range of the four GRO-instruments, namely 0.7 to 30 MeV. This is one of the most difficult spectral ranges to explore in astronomy. Prior to the launch of GRO only very few celestial objects were detected in this part of the electromagnetic spectrum. With COMPTEL the field of MeV gamma-ray astronomy can now be explored.
COMPTEL is the first imaging MeV gamma-ray telescope ever flown on a satellite. It has a large field-of-view of about 1 steradian. Different sources within this field can be resolved if they are separated by more than ≈ 3° to 5°. With its energy resolution of 5 % to 10 % FWHM, COMPTEL is well suited to study continuum and line emission. COMPTEL has an unprecedented sensitivity: at 1 MeV sources about 10-times weaker than the Crab can be detected in a 2-week observation period. In addition to gamma rays, solar neutrons above 15 MeV can also be measured. A comprehensive description of the capabilities and characteristics of COMPTEL is given in /1/ and /2/.

---

[*] Anstronomical Institute, University of Amsterdam, The Netherlands
[**] Compton Observatory Science Support Center, GSFC, MD, USA

Together with EGRET, COMPTEL is at present performing a complete sky survey - the first in gamma-ray astronomy. Most of the pointings have lasted two weeks each. The analysis of the data from these observations is an arduous and difficult process. This is due to the fact that the arrival direction of each photon detected by COMPTEL is not defined unambiguously, but is only known to lie on a circle on the sky (see Fig. 1). Most of the scientific analysis is still preliminary. An overview of the most important results obtained from this analysis is given here.

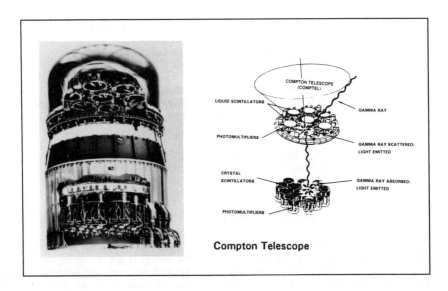

Fig. 1. Schematic view of COMPTEL. A gamma-ray is detected by a Compton collision in an upper detector consisting of 7 modules of liquid scintillator NE 213 and a subsequent interaction in a lower detector consisting of 14 modules of Na (Tl). The center of each event circle is defined by the direction of the scattered gamma ray, the radius of the circle by the energy losses in both interactions.

RESULTS

The preliminary results from COMPTEL can be grouped under the following headings:

1. Composite sky map of the inner part of the Galaxy
2. Observation of the anticentre of the Galaxy with the Crab nebula and its pulsar
3. Observation of the Vela pulsar
4. Search for gamma-ray emission from other pulsars
5. Observation of the black hole candidates Cyg X-1 and Nova Persei 1992
6. Study of the diffuse Galactic continuum emission
7. Study of the 1.8 MeV gamma-ray line from radioactive ^{26}Al
8. Search for other gamma-ray lines
9. Observations of the quasars 3C273, 3C279 and PKS 0528+134, and the radio galaxy Cen A
10. Localization of cosmic gamma-ray bursts and measurements of burst spectra and time profiles
11. Observation of gamma-ray and neutron emission from solar flares.

Each of these topics is briefly discussed.

2.1 Map of the Galactic Plane in the Central Region

A COMPTEL map of the entire Galactic plane in the light of continuum gamma radiation does not yet exist; first, the full sky survey is not yet completed and second, only a fraction of the observations along the Galactic plane has been analyzed so far. Nonetheless, preliminary maps of the central part of the plane do exist already. They were derived by combining data from different GRO-observations. Examples are shown in /3/. The maps clearly show the emission to be concentrated towards the Galactic plane. There seem to be localized sources as well as diffuse Galactic emission. The identification of the sources needs further study.

2.2 The Crab

The Crab is by far the strongest steady source in the sky so far seen by COMPTEL. The pulsar analysis of 4 weeks of data yields a light curve with strong emission between the two peaks, resembling very much that seen at hard x-ray energies /4/ and /5/. The pulsed fraction of the total Crab emission is about 25 % to 35 %. No significant differences in the shape of the light curves in the 4 observations of the Crab in 1991 have been found /6/. The photon energy spectra of the total, the pulsed and unpulsed emission, can all be fitted by single power-law spectra over the entire COMPTEL energy range /7/.

2.3 The Vela Pulsar

The Vela pulsar has been detected by COMPTEL between 3 and 30 MeV. The light curve in the 10 - 30 MeV range from 4 combined observations (0, 6, 8, and 14) clearly shows the two main peaks separated by 0.4 in phase as seen at higher energies. There is no statistically significant interpulse emission between the 2 peaks /5/. The energy spectrum of the pulsar shows a bending of the high-energy power law spectrum at MeV-energies /8/.

2.4 Other Pulsars

A search for pulsed emission from other radio pulsars has so far turned-up negative results. In particular, PSR 1706-44 and PSR 1509-58 are not seen by COMPTEL at this stage of analysis /9/. Furthermore, no signal (steady or pulsed) was observed from the Geminga pulsar. An upper limit to the total Geminga emission in the COMPTEL energy range is given in /7/.

2.5 Black-Hole Candidates

One characteristic signature of black hole candidates in x-ray binary systems may be the existence of temporary strong soft gamma-ray emission possibly extending into the MeV-range.

With COMPTEL we have been able to observe MeV-emission from two such black hole candidates. One is Cyg X-1, which has clearly been seen between 0.75 and 3 MeV in observation 7 (August 1992), and also in observation 2 (June 1992), see /10/. The MeV-flux measured by COMPTEL is more than a factor of 10 lower than the MeV-bump, measured by HEAO-C /11/. This is interesting in that Cyg X-1 was in a very low hard x-ray state during the time of observation 7. Therefore, a low hard x-ray flux seems not to be significant condition for MeV-emission to exist, as has been suggested /11/.

The other COMPTEL observation of a black-hole candidate is that of Nova Persei 1992 (GRO J0422+32). This low mass x-ray binary transient was discovered by BATSE on August 5, 1992. Its energy spectrum has been measured by OSSE up to about 600 keV with high accuracy (see these proceedings). The composite COMPTEL maximum likelihood map of Fig. 2 contains that transient in the 1 to 2 MeV range. During observation 36 (August 1992) the nova was seen up to 2 MeV, in observation 39 (September 1992) up to 1 MeV.

2.6 Diffuse Galactic Continuum Emission

The continuum gamma-ray emission from interstellar space in the 1 - 30 MeV range is produced by interactions of cosmic ray electrons, mainly via the bremsstrahlung process, and to a smaller amount by inverse Compton scattering. The contribution of the π°-decay component can be neglected in this spectral range. First attempts have been made to derive the gamma-ray emissivity (number of gamma rays produced per H-atom ster sec MeV) and the differential gamma-ray flux per radian from COMPTEL observations towards the Galactic centre region (see /12/). The gamma-ray fluxes derived by COMPTEL in the inner Galaxy are consistent with previous measurements but more in agreement with the lowest end of the range of published values. In this context it has to be noted that the COMPTEL emissivities should strictly be regarded as upper limits, because point sources may contribute significantly to the observed emission. The COMPTEL results are in reasonable agreement with extrapolations of the COS-B measurements /13/ and with calculations of the bremsstrahlung emissivity, using a "leaky box" propagation model for the cosmic ray electrons /12/, /14/.

2.7 Galactic 1.8 MeV ^{26}Al Gamma-Ray Line

The 1.8 MeV gamma-ray line from radioactive ^{26}Al was discovered more than 10 years ago by HEAO-C /15/. Little information was available from those measurements regarding the location of the line emission except that it originates from the general direction of the Galactic centre region.

^{26}Al is an isotope with a radioactive decay time of $1.04 \cdot 10^6$ years. Therefore, one can expect to see the line from the accumulation of all ^{26}Al formation sites over the last million years. It has been suggested that supernovae, novae and peculiar massive stars (like Wolf-Rayet stars) might be the sites in which ^{26}Al is produced and then ejected into interstellar space.

Obviously, a map of the entire Galactic plane in the light of the line is of utmost importance to constrain the various models.

COMPTEL has detected the line /16/ and first images of the central part of the Galaxy ($-40° < l < +30°$) have been derived /17/. The Galactic plane is clearly visible in this part of the sky. The emission covers the entire longitude range investigated so far, although the emission is not uniform: instead, there are regions with indications of enhanced emission. During this conference (/18/) evidence for line-emission even outside this longitude range is provided. Especially the direction of Vela is a source of the ^{26}Al-line, as well. Bootstrap analyses are at present being applied to assess the statistical significance of the excesses.

2.8 Search for Other Gamma-Ray Lines

A search for gamma-ray lines from SN 1991T has yielded a negative result. The type Ia supernova, which occurred on or shortly before April 10, 1991 in the spiral galaxy NGC 4527 at a distance of about 13.5 Mpc, was observed by COMPTEL in June and October, 1991. Preliminary 2σ upper limits to the ^{56}Co lines at 846 keV and 1.238 MeV derived by Lichti et al. /19/, are close to predicted line fluxes, however, they do not yet constrain different theoretical models /20/.

Efforts have now been taken to improve the COMPTEL limits by making use of the full knowledge of the response function of the instrument. The new limits are roughly two-times lower than the previous limits. Now, a few of the proposed theoretical models can be ruled out /21/.

2.9 Active Galactic Nuclei

Prior to the launch of GRO, quasars and other nuclei of active galaxies were thought to be promising objects in the COMPTEL energy range. Many have hard x-ray spectra, from which one might conclude that at least some of them have their peak luminosity at MeV-energies. The two quasars 3C273 and 3C279 were the first AGN's detected by COMPTEL /22/. Though both objects were rather weak during the COMPTEL observation in June 1991 (about 10 % of the Crab flux, between 3 and 10 MeV), their detection was statistically significant (7σ and 4σ, respectively).

During a second observation in October 1991 3C273 was roughly two-times weaker than in June /23/. In the COMPTEL energy range (0.7 to 30 MeV) 3C273 has a significantly softer spectrum than 3C279. The spectral difference becomes more evident, when combining COMPTEL, EGRET and OSSE-results. The energy spectra of both quasars can be described by two power-law components which show a break or steepening between 1 and 3 MeV (3C273) and near 10 MeV (3C279). Whereas the peak luminosity of 3C273 lies between 1 and 3 MeV, that of 3C279 ranges from 10 MeV to 5 GeV /22/.

Since its discovery last year, the origin of the gamma-ray emission from 3C279 has been widely discussed in the literature. In most cases (e.g. /24/, /25/ and /26/) it is proposed that the gamma-ray emission is not produced in the central nucleus, but in the jets. The beamed jet-emission (due to the relativistic Doppler-Lorentz factor) would require in a 10^4-times smaller luminosity than isotropic core-emission.

A search for COMPTEL-detections of other EGRET-AGN's resulted in the discovery of gamma-ray emission from PKS 0528+134. This object has one of the steepest energy spectra of all EGRET-AGN's (differential energy spectrum $\approx E^{-2.4}$). In so far, it was indeed one of the most promising candidates to be seen by COMPTEL. The fact that the source is seen by COMPTEL only between 3 - 30 MeV, but not below 3 MeV, suggest that also this AGN-object has a gamma-ray spectrum, which bends at MeV-energies like those of 3C273 and 3C279 /23/.

The only other AGN-object, seen by COMPTEL so far is the radio galaxy Centaurus A. The flux derived from COMPTEL at $\approx$ 1 MeV is consistent with a power law extrapolation of the OSSE-observations (see /23/ and /27/).

2.10   Gamma-Ray Bursts

During the all-sky survey, a number of gamma-ray bursts and solar flares could be detected within the COMPTEL field-of-view. To this date, the positions of nine of the measured gamma-ray bursts have been derived from their maximum-likelihood maps /28/.

Energy spectra and time profiles of some of these bursts are described in /28/ and /29/. The combination of the COMPTEL error boxes with the one-dimensional localisation by triangulation using Ulysses and GRO burst arrival times, leads to elongated error boxes of the burst positions, which are a few degrees wide in one dimension and a few arcminutes wide in the other dimension. Based on such an error box, a counterpart search for GRB 910503 has been performed by using ROSAT data /30/. No positive identification is reported from this search. The counterpart search continues for the other bursts contained in the COMPTEL field-of-view.

2.10   Solar Flares

On June 9, 11, and 15, 1991, COMPTEL observed three x-class solar flares within its field-of-view. Preliminary results from these flares are described in /31/. In all three flares the gamma-ray spectra show continuum and line components. Lines are seen at 1.6 MeV (^{20}Ne), 2.2 MeV (neutron capture line), and weakly at 4.4 MeV (^{12}C). The June 15 flare still showed observable MeV-emission 90 minutes after the onset of the flare, suggesting a correspondingly long-lasting particle acceleration time. The detection of the 2.2 MeV neutron-capture line in all three flares indicates that neutrons were produced in these flares. In the cases of the June 9 and June 15 flare, these neutrons have already been detected by COMPTEL in the energy range 15 - 80 MeV. The simultaneous measurements of the 2.2 MeV line and the neutron flux provide a powerful diagnostic tool to study the emission processes and geometries.

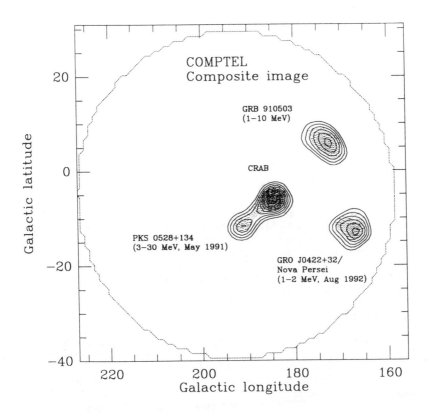

Fig. 2. Composite COMPTEL image of the anticentre region of the galaxy.

CONCLUSION

COMPTEL is the first imaging MeV gamma-ray telescope ever flown on a satellite. Its imaging capabilities are of great use in exploring the nearly unknown field of MeV-astronomy. The quality of images obtained by COMPTEL is illustrated in Fig. 2, which shows a composition of images obtained from different observations of the anticentre region of the galaxy. The Crab, the AGN PKS 0528+134, the transient GRO J 0422+32, and the gamma-ray burst GRB 910503 are all within a sphere of roughly 25 degrees in diameter.

The first results from COMPTEL have demonstrated that a multitude of phenomena can be studied at MeV-energies.

LITERATURE

1. V. Schönfelder, R. Diehl, G.G. Lichti, H. Steinle, B.N. Swanenburg, A.J.M. Deerenberg, H. Aarts, J. Lockwood, W. Webber, J. Macri, J. Ryan, G. Simpson, B.G. Taylor, K. Bennett, and M. Snelling, IEEE-Trans. on Nucl. Sci., Vol NS-31, No. 1, p. 766 (1984)

2. V. Schönfelder, H. Aarts, K. Bennett, H. de Boer, J. Clear, W. Collmar, A. Connors, A. Deerenberg, R. Diehl, A. v. Dordrecht, J.W. den Herder, W. Hermsen, M. Kippen, L. Kuiper, G. Lichti, J. Lockwood, J. Macri, M. McConnell, D. Morris, R. Much, J. Ryan, G. Simpson, M. Snelling, G. Stacy, H. Steinle, A. Strong, B.N. Swanenburg, B. Taylor, C. de Vries, C. Winkler, Ap.J. Suppl. (1992), in press

3. H. Bloemen, K. Bennett, H. de Boer, W. Collmar, A. Connors, R. Diehl, R. v. Dijk, J.W. den Herder, W. Hermsen, L. Kuiper, G. Lichti, M. McConnell, D. Morris, V. Schönfelder, B.N. Swanenburg, G. Stacy, H. Steinle, A.W. Strong, C.P. de Vries 1992, this issue

4. K. Bennett, H. Aarts, H. Bloemen, R. Buccheri, M. Busetta, W. Collmar, A. Connors, A. Carramiñana, R. Diehl, H. de Boer, J.W. den Herder[2], W. Hermsen, L. Kuiper, G. Lichti, J. Lockwood, J. Macri, M. McConnell, D. Morris, R. Much, J. Ryan, V. Schönfelder, G. Simpson, G. Stacy, H. Steinle, A. Strong, B. Swanenburg, B. Taylor, M. Varendorff, C. de Vries, W. Webber and C. Winkler, A & A, Suppl. (1992), in press

5. R. Buccheri, K. Bennett, M. Busetta, A. Carramiñana, W. Collmar, A. Connors, W. Hermsen, L. Kuiper, G.G. Lichti, V. Schönfelder, J.G. Stacy, A.W. Strong, C. Winkler, COSPAR World Space Congress, Washington (1992), in press

6. A. Carramiñana, K. Bennett, R. Buccheri, M. Busetta, A. Connors, W. Hermsen, L. Kuiper, J. Ryan, V. Schönfelder, A. Strong, 1992, this issue

7. A. Strong, K. Bennett, H. Bloemen, H. de Boer, R. Buccheri, M. Busetta, W. Collmar, A. Connors, R. Diehl, J.W. den Herder, W. Hermsen, L. Kuiper, J. Lockwood, G.G. Lichti, J. Macri, M. McConnell, D. Morris, R. Much, J. Ryan, V. Schönfelder, G. Simpson, J.G. Stacy, H. Steinle, B. Swanenburg, M. Varendorff, C. Winkler and C. de Vries, A & A Suppl. (1992), in press

8. W. Hermsen, K. Bennett, R. Buccheri, M. Busetta, A. Carramiñana, A. Connors, R. Diehl, L. Kuiper, G. Lichti, D. Morris, J. Ryan, V. Schönfelder, A. Strong, B.N. Swanenburg, M. Varendorff, C. Winkler, 1992, this issue

9. K. Bennett, R. Buccheri, M. Busetta, A. Carramiñana, W. Collmar, A. Connors, R. Diehl, W. Hermsen, L. Kuiper, G. Lichti, J. Ryan, V. Schönfelder, A. Strong, B.N. Swanenburg, 1992, this issue

10. M. McConnell, H. Bloemen, A. Connors, W. Collmar, R. Diehl, R. v. Dijk, D. Forrest, W. Hermsen, L. Kuiper, J. Ryan,, V. Schönfelder, H. Steinle, A. Strong, B.N. Swanenburg, C. Winkler, 1992, this issue

11. J. Ling et al., AP.J. (Letters) 321, L117 (1987)

12. A. Strong, K. Bennett, H. Bloemen, H. de Boer R. Diehl, W. Hermsen, D. Morris, V. Schönfelder, G. Stacy, B.N. Swanenburg, M. Varendorff, C. de Vries, G. Youssefi, 1992, this issue

13. A. Strong, J.B.G.M. Bloemen, T.M. Dame, I.A. Grenier, W. Hermsen, F. Lebrun, L.-A. Nyman, A.M.T. Pollock, P. Thaddeus, A & A 207, 1-15 (1988)

14. J. Skibo and R. Ramaty, A & A Suppl. (1992), in press

15. W.A. Mahoney, J.C. Ling, WM.A. Wheaton, A.S. Jacobson, Ap.J. 286, 578-585 (1984)

16. R. Diehl, K. Bennett, H. Bloemen, H. de Boer, M. Busetta, W. Collmar, A. Connors, J. W. den Herder, C. de Vries, W. Hermsen, J. Knödlseder, L. Kuiper, G.G. Lichti, J.A. Lockwood, J. Macri, M. McConnell, D. Morris, R. Much, J. Ryan, G. Stacy, H. Steinle, A. Strong, B.N. Swanenburg, M. Varendorff, P. von Ballmoos, W. Webber, C. Winkler, A & A Suppl. (1992), in press

17. R. Diehl, K. Bennett, H. Bloemen, W. Collmar, W. Hermsen, G.G. Lichti, M. McConnell, D. Morris, J. Ryan, V. Schönfelder, H. Steinle, A.W. Strong, B.N. Swanenburg, M. Varendorff, C. Winkler, COSPAR World Space Congress, Washinton (1992), in press

18. R. Diehl, H. Bloemen, W. Collmar, W. Hermsen, G. Lichti, D. Morris, V. Schönfelder, B.N. Swanenburg, M. Varendorff, C. de Vries, C. Winkler, 1992, this issue

19. G.G. Lichti, K. Bennett, H. Boloemen, H. deBoer, M. Busetta, W. Collmar, A. Connors, R. Diehl, R. van Dijk, J.W. den Herder, W. Hermsen, L. Kuiper, J. Lockwood, J. Macri, M. McConnell, D. Morris, R. Much, J. Ryan, V. Schönfelder, G. Simpson, J.G. Stacy, H. Steinle, A.W. Strong, B.N. Swanenburg, M. Varendorff, C. de Vries, C. Winkler, A & A Suppl. (1992), in press

20. P. Ruiz-Lapunte, R. Lehoucq, R. Canal & M. Cassé, Gamma-Ray Spectra from Fast Deflagration Models of SNIa, Proc. of 'Origin and Evolution of the Elements in Honor of H. Reeves 60th Birthday', ed: N. Prantzos and E. Vanglioni-Flarn, Cambr. Univ. Press (1992)

21. G. Lichti, K. Bennett, H. Bloemen, H. de Boer, M. Busetta, W. Collmar, R. Diehl, R. v. Dijk, W. Hermsen, D. Morris, J. Ryan, V. Schönfelder, A. Strong, C. Winkler, 1992, this issue

22. W. Hermsen, H.J.M. Aarts, K. Bennett, H. Bloemen, H. de Boer, W. Collmar, A. Connors, R. Diehl, R. van Dijk, J.W. den Herder, L. Kuiper, G.G. Lichti, J.A. Lockwood, J. Macri, M. McConnell, D. Morris, J.M. Ryan, V. Schönfelder, G. Simpson, H. Steinle, A.W. Strong, B.N. Swanenburg, C. de Vries, W.R. Webber, O.R. Williams, C. Winkler, A & A Suppl. (1992a), in press

23. W. Collmar, K. Bennett, H. Bloemen, R. Diehl, J.W. den Herder, W. Hermsen, G. Lichti, M. McConnell, J. Ryan, V. Schönfelder, G. Stacy, H. Steinle, A. Strong, B.N. Swanenburg, C. de Vries, O. Williams, C. Winkler, 1992, this issue

24. C.O. Dermer, R. Schlickeiser, and A. Mastichiadis A & A, 256, L27 (1992)

25. M. Camenzind and O. Dreisigacker, A & A (1992), in press

26. K. Mannheim and P.L. Biermann, A & A 253, L21 (1992)

27. H. Steinle, H. Bloemen, W. Collmar, R. Diehl, W. Hermsen, G. Lichti, M. McConnell, J. Ryan, V. Schönfelder, G. Stacy, A. Strong, B.N. Swanenburg, M. Varendorff, O.R. Williams, COSPAR World Space Congress, Washington (1992), in press

28. C. Winkler, K. Bennett, W. Collmar, A. Connors, R. Diehl, D. Forrest, L. Hanlon, J.W. den Herder, W. Hermsen, M. Kippen, L. Kuiper, M. McConnell, J. Ryan, V. Schönfelder, H. Steinle, B.N. Swanenburg, M. Varendorff, C. de Vries, O.R. Williams, 1992, this issue

29. W. Collmar, K. Bennett, H. Bloemen, H. de Boer, M. Busetta, A. Connors, R. Diehl, J. Greiner, L. Hanlon, J.W. den Herder, W. Hermsen, L. Kuiper, G.G. Lichti, J. Lockwood, J. Macri, M. McConnell, D. Morris, R. Much, J. Ryan, V. Schönfelder, G. Stacy, H. Steinle, A. Strong, B. Swanenurg, B.G. Taylor, M. Varendorff, C. de Vries, W. Webber, O.R. Willams and C. Winkler, A & A, Suppl. (1992), in press

30. M. Boer, J. Greiner, P. Kahabka, C. Motch, W. Voges, A & A Suppl. (1992), in press

31. J. Ryan, K. Bennett, H. Debrunner, D. Forrest, L. Hanlon, J. Lockwood, M. Loomis, M. McConnell, D. Morris, V. Schönfelder, B.N. Swanenburg, W. Webber, and C. Winkler, 1992, this issue

COMPTEL OBSERVATIONS OF THE INNER GALAXY

H. Bloemen[1], H. de Boer, R. van Dijk[2], J.W. den Herder, W. Hermsen,
L. Kuiper, B.N. Swanenburg, C.P. de Vries
SRON-Leiden, P.O. Box 9504, 2300 RA Leiden, The Netherlands

W. Collmar, R. Diehl, G.G. Lichti, V. Schönfelder, H. Steinle, A.W. Strong
Max-Planck Institut für Extraterrestrische Physik, D-8046 Garching, F.R.G.

M. McConnell, A. Connors, D. Morris, G. Stacy[3]
Space Science Center, Univ. of New Hampshire, Durham, NH 03824, U.S.A.

K. Bennett
Astrophysics Division, ESTEC/ESA, 2200 AG Noordwijk, The Netherlands

[1] Leiden Observatory, The Netherlands
[2] Astronomical Institute, University of Amsterdam, The Netherlands
[3] Compton Observatory Science Support Center, NASA GSFC, U.S.A.

ABSTRACT

This paper presents a first global study of COMPTEL observations of the inner Galaxy in the energy range 0.75–10 MeV. Preliminary findings demonstrate COMPTEL's capabilities for mapping the observed gamma radiation and disentangling the contributions from point sources and diffuse emission.

INTRODUCTION

The COMPTEL telescope aboard the Compton Observatory provides for the first time extensive imaging possibilities at MeV energies with a source-location accuracy of typically 1°. This imaging capability is of particular value for studies of the Galactic disk, where a variety of point sources as well as diffuse emission can be expected to contribute. Strong et al. (this volume) found the total flux measured by COMPTEL from the central radian of the Galactic disk to be near the lower bound of a variety of previous flux estimates. This paper describes our initial attempts to map the observed emission and to disentangle the contributions from point sources and diffuse emission, using a combination of ten observations of the first and fourth Galactic quadrants obtained during the first half of the COMPTEL all-sky survey (Table 1). All analyses presented here were applied to three energy intervals, $0.75-1$ MeV, $1-3$ MeV, and $3-10$ MeV, but the findings for the latter two have been combined.

MAPPING OF SOURCES AND DIFFUSE EMISSION

Each incident photon detected by COMPTEL is first Compton scattered in the upper layer of detectors and then absorbed in the lower layer[1]. The energy deposits in the two layers define the Compton scatter angle (as well as the energy of the incident photon) and the interaction locations define the scatter direction. For a selected energy range, the photons can thus be binned in a 3-D dataspace

Table 1. COMPTEL observations used in present analysis.

Observation	Pointing ℓ	b	Start yy-mm-dd	End yy-mm-dd
2.0	73.3°	+2.6°	91-05-30	91-06-08
5	0.0°	−4.0°	91-07-12	91-07-26
7.0	70.4°	−8.3°	91-08-08	91-08-15
7.5	25.0°	−14.0°	91-08-15	91-08-22
12	310.7°	+22.2°	91-10-17	91-10-31
13.0	25.0°	−14.0°	91-10-31	91-11-07
14*	285.0°	−0.7°	91-11-14	91-11-28
16	0.0°	+20.2°	91-12-12	91-12-27
20*	39.7°	+0.8°	92-02-06	92-02-20
23*	322.1°	+3.0°	92-03-19	92-04-02

* Only part of the available data included here.

consisting of the scatter direction and the scatter angle. The search for sources, model fitting, etc., are all carried out in this 3-D dataspace.

Two main imaging methods are applied: a maximum-entropy method for general imaging purposes[2] and a likelihood method optimized to search for point sources[3] (including determination of statistical significance, flux, error region, etc.). This paper presents results from both methods.

The likelihood method in particular requires a careful estimate of the instrumental background, which is the dominant contributor to the total number of observed photons. The background models used in the present work are determined by applying a filter technique to the 3-D dataspace, which smooths the photon distribution and eliminates source signatures in first-order approximation. The likelihood algorithms for source mapping and model fitting use a modified instrument response to take this filtering into account. Formally, this technique is limited to point sources and line sources. It will be discussed in detail in a forthcoming paper.

MODEL FITTING AND SIMULATIONS OF DIFFUSE EMISSION

In order to illustrate the imaging potential of COMPTEL and to compare the observations with the results expected from diffuse emission, we also applied our mapping tools to Monte Carlo simulations of a simple model of the diffuse (bremsstrahlung) emission (see Strong et al. in this volume). This model is based on the assumption that the cosmic-ray (CR) electron density is constant throughout the Galaxy, so the γ-ray intensity distribution is approximately given by $I_\gamma = \varepsilon_\gamma N_\mathrm{H}$, where N_H is the column-density distribution of the interstellar gas (derived from various HI surveys and the CO survey by Dame et al.[4], using a CO-to-H_2 conversion factor of 2.3×10^{20} molecules cm^{-2} (K km s^{-1})$^{-1}$ — see Fig. 1). First, the value of the γ-ray emissivity, ε_γ, was determined for each energy interval by maximum-likelihood fitting of the model to the binned photons in the 3-D data space. The values are on average $\sim$20% lower than found by Strong et al., but considered to be in comfortable agreement in view of the different sets of observations used and the differing treatments of the background. Using these emissivities, simulated 3-D dataspaces of the diffuse emission were made, to which background simulations were added.

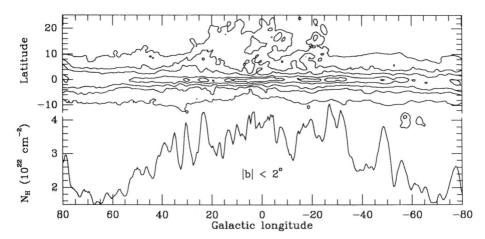

Figure 1: Part of the gas-column-density map (and longitude profile) used for model fitting and simulations of the diffuse γ-ray emission. Contour levels: 0.2, 0.5, 1, 3, and 6 $\times 10^{22}$ cm^{-2}. The CO-to-H$_2$ conversion factor for the strong CO emission from the Galactic centre was assumed to be ten times lower than for the Galactic disk[5]. The data were slightly smoothed for pictorial purposes.

RESULTS

Figures 2 and 3 show our mapping results in the energy ranges 1–10 MeV and 0.75–1 MeV, both for the actual observations (top) and for the simulations of the diffuse emission (bottom). Only likelihood maps are shown for the 0.75–1 MeV range — the maximum-entropy maps are very similar.

The likelihood maps show the quantity $-2\ln\lambda$, where λ is the maximum likelihood ratio L(background) / L(source+background). Formally, in a search for sources, $-2\ln\lambda$ has a chi-square distribution with 3 degrees of freedom (for instance, a 3σ detection corresponds to $-2\ln\lambda = 13.8$). Simulations have shown, however, that our method of background modelling requires a somewhat more stringent criterion for source detection (discussed in a future paper).

Figures 2 and 3 give quite a different impression of the global appearance of the gamma-ray sky in the two energy bands. Fig. 2 (1-10 MeV) shows a clear ridge of emission. The likelihood maps and maximum-entropy maps are very similar, the latter showing more detailed structure because of the deconvolution inherent in the maximum-entropy method. The global appearance of the observations is similar to that of the simulated observations of the diffuse emission, but detailed differences seem to be present.

In Fig. 3 (0.75-1 MeV), only a few excesses are seen in addition to those at the positions of the black-hole candidate Cyg X-1 and the radio galaxy Cen A (both at flux levels at least an order of magnitude below some previous detections — see McConnell et al. and Collmar et al. in this volume). Although one of the other excesses coincides with the Galactic-centre source 1E1740.7-2942 (with a flux of $\sim 1 \times 10^{-4}$ cm^{-2} s^{-1}, at least an order of magnitude above extrapolations of all published spectra of this source), it cannot be excluded that this excess is actually due to a peak in the diffuse emission, visible in the lower map of Fig. 3. Only a few features due to diffuse emission can apparently be expected to be seen by COMPTEL in this narrow energy band. The observed excesses

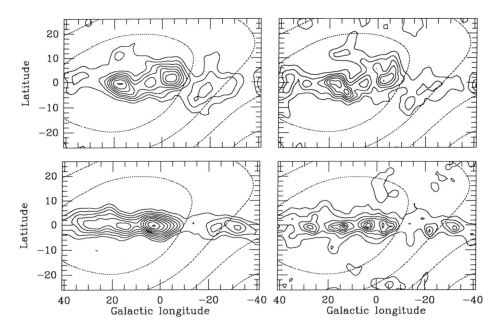

Figure 2: *Top*: Likelihood map (left) and maximum-entropy map (right) of combined COMPTEL observations (Table 1) of the inner Galaxy (1–10 MeV). *Bottom*: Same for simulated observations of the diffuse emission, with emissivity from model fitting. Contour levels: (left) start 15, step 12; (right) start 4×10^{-3}, step 3×10^{-3} cm^{-2} s^{-1} sr^{-1}. Contours of effective exposure, at 20% intervals from the peak value, are superimposed.

only partly coincide with the expected diffuse excesses. This may either hint at the presence of sources (or local CR-density enhancements), implying that we overestimated the contribution from diffuse emission in the model fitting, or it may be attributable to the limitations of our simple diffuse model.

The obvious question is whether the apparent differences between observations and simulations are indeed *significant*. This was studied by including the diffuse model in the likelihood search for sources. Fig. 4 shows the resulting map of $-2\ln\lambda$, where λ is now the maximum likelihood ratio L(diffuse+background) / L(source+diffuse+background). Two main excesses can be seen. The one near $\ell = 15° - 20°$ coincides with a similar extended feature seen in both the COS-B data[6] and the EGRET data (Mayer-Hasselwander et al., this volume). There is no obvious counterpart for the second excess ($\ell \simeq 345°$, $b \simeq -6°$, 95% error radius of $\sim 1.5°$); its position is consistent with that of the LMXRB 1735-44. The location of these 'sources' *outside* the Galactic disk may indicate that an ensemble of sources along the plane is hidden because we have overestimated the contribution from diffuse emission. This issue will be addressed in future work by fitting a number of sources simultaneously with diffuse emission.

It can be concluded that COMPTEL's potential to disentangle the MeV emission from the inner Galaxy is promising. We emphasize, however, that this paper should mainly be regarded as a summary of our analysis strategy; the results are preliminary.

COMPTEL Observations of the Inner Galaxy

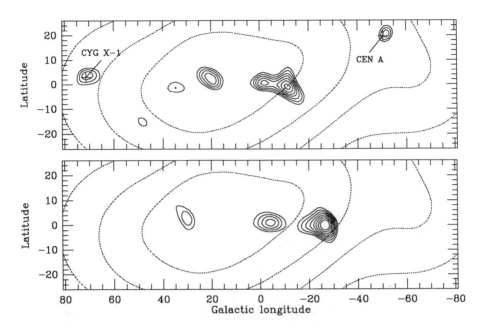

Figure 3: *Top*: Likelihood map (0.75–1 MeV) of combined COMPTEL observations of the inner Galaxy (Table 1). *Bottom*: Same for simulated observations of the diffuse emission, with emissivity from model fitting. Contour levels: start at $-2\ln\lambda = 15$, step 5. Exposure contours as in Fig. 2.

Figure 4:
Likelihood map (0.75–10 MeV) showing enhanced emission not explained by our model of the diffuse emission (see text). This map is a sum of the maps for the three individual energy intervals. Contours: start 20, step 8.

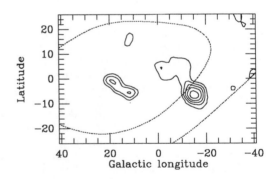

REFERENCES

1. V. Schönfelder et al., Ap. J. Suppl., in press (1993)
2. A.W. Strong et al., in Data Analysis in Astronomy IV, eds. V. Di Gesù et al., Plenum Press, New York (1992)
3. H. de Boer et al., in Data Analysis in Astronomy IV (see ref. 2), p. 241
4. T.M. Dame et al., Ap. J. **322**, 706 (1987)
5. L. Blitz et al., A&A **143**, 267 (1985)
6. H. Bloemen, in The Evolution of the Interstellar Medium, ed. L. Blitz, ASP Conference Series, p. 79 (1990)

DIFFUSE GALACTIC CONTINUUM EMISSION MEASURED BY COMPTEL

A.W. Strong, R. Diehl, V. Schönfelder, M. Varendorff, G. Youssefi
Max-Planck Institut für extraterrestrische Physik,
D-8046 Garching, FRG

H. Bloemen, H. de Boer, W. Hermsen, B. Swanenburg, C. de Vries
SRON-Leiden, P.B. 9504,
NL-2300 RA Leiden, The Netherlands

D. Morris, J.G. Stacy
University of New Hampshire, Institute for the Study
of Earth, Oceans and Space, Durham NH 03824, USA

K. Bennett
Astrophysics Division, Space Science Department of ESA,
ESTEC, 2200 AG Noordwijk, The Netherlands

ABSTRACT

Diffuse galactic continuum γ-ray emission in the 0.75-10 MeV range from the inner Galaxy has been studied using data from COMPTEL on the Compton Observatory. Several observations of the inner Galaxy from the first year of operation have been used. The imaging properties of COMPTEL enable spatial analysis of the γ-ray distribution using model fitting. A model based on atomic and molecular gas distributions in the Galaxy has been used to derive the emissivity spectrum of the γ-ray emission and this spectrum is compared with theoretical estimates of bremsstrahlung emission from cosmic-ray electrons.

INTRODUCTION

The diffuse continuum γ-ray emission from the Galactic plane at MeV energies is believed to originate mainly in bremsstrahlung interactions of cosmic-ray electrons with interstellar gas. Measurements in this range are difficult and until now have mainly been made with non-imaging detectors which only allow a wide-angle average of the emission to be measured. The COMPTEL instrument on the Compton Observatory has sufficient imaging capability to allow a sensitive measurement of the diffuse radiation. Although a detailed mapping has yet to be carried out (first results are presented by Bloemen et al., this volume), a correlation analysis with Galactic gas distributions shows that the Galaxy can be detected and the intensity of the emission determined by model fitting.

METHOD

The principles of COMPTEL measurements are given in detail by Schönfelder et al[1]. For each incident photon detected by the instrument, the energy deposits and interaction locations in the upper and lower detectors are measured. The Compton scatter angle $\bar{\varphi}$ is computed from the measured energy deposits, and the photons are binned in a 3-D dataspace consisting of the direction of the scattered γ-ray and $\bar{\varphi}$. The response of the instrument in this dataspace to a given incident γ-ray intensity distribution is computed on the basis of the knowledge of the instrument configuration as a function of time, and responses from the pre-launch calibration and simulations. Hence given a parameterized model of the sky, the values of the parameters can be determined by model fitting in the dataspace described above.

We assume the continuum radiation originates in cosmic-ray interactions with interstellar gas, and therefore use the relation

$$I_\gamma = \frac{q}{4\pi} N_H + I_B \qquad (1)$$

where I_γ is the γ-ray intensity (cm^{-2}sr^{-1} s^{-1}), $q/4\pi$ the γ-ray emissivity (atom^{-1} sr^{-1} s^{-1}), N_H the total hydrogen gas column density and I_B a celestial background term. N_H is taken from HI and CO surveys, assuming that the molecular hydrogen column density is related to the integrated CO temperature by $N_{H_2}/W_{CO} = 2.3\ 10^{20}$ molecules cm^{-2} (K km s^{-1})$^{-1}$ as determined from COS-B data [2]. The free parameters of the model are q, I_B and the instrumental background. Since the instrumental background is rather high and not known *a priori* its treatment is critical in obtaining a reliable result. We use high-latitude observations as the basis for our estimates, assuming that the shape of this background is the same in the instrument system for all observations. To obtain a good estimate of the average background shape several high-latitude observations were summed and the result slightly smoothed. In the fits about 90% of the total observed γ-rays are assigned to this component, and about 10% to celestial origin.

RESULTS

Table I summarizes the five observations used in this work. Observations 7.5 and 13.0 have exactly the same pointing so were analysed as a single observation.

Table I. Summary of COMPTEL observations used

Observation	Target	l	b	Start yymmdd	End yymmdd
2.0	Cyg X-1	73.3°	+02.6°	910530	910608
5	G. Centre	00.0°	−04.0°	910712	910726
7.0	Cyg X-3	70.4°	−08.3°	910808	910815
7.5	Gal 025-14	25.0°	−14.0°	910815	910822
13.0	Gal 025-14	25.0°	−14.0°	911031	911107

The fit to equation (1) gives a positive value for q for all the observations analysed; we conclude from this that the Galactic continuum radiation was indeed detected and the emissivities can be reasonably interpreted in terms of emission from the interstellar medium. The values obtained were averaged over the observations; Table II summarizes the results.

Table II. Inner Galaxy emissivities for integral energy ranges

Energy range (MeV)	0.75-1.0	1.0-3.0	3.0-10.0
$q/4\pi$ (10^{-25}sr^{-1}s^{-1})	2.6	4.8	1.6

Fig 1 shows the emissivity spectrum of the inner Galaxy based on the present results and also values obtained from the application of a similar method to COS-B data[2]. The COMPTEL points are consistent with a spectral index of -2.0; this slope is used to convert the integral points to a differential spectrum in this plot. Errors on the COMPTEL points are estimated to be about 30%, including statistical and systematic errors. The spectrum is in agreement with a smooth connection to the COS-B spectrum, although our values represent strictly an upper limit since there may be some contribution from point sources in the Galactic plane (see Bloemen et al., this volume).

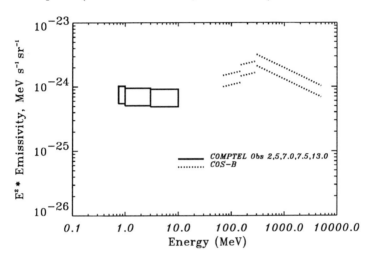

Fig 1: Emissivity spectrum of the inner Galaxy

In order to compare with other experimental results for the Galactic centre region, where the results are normally given as a flux per radian integrated over latitude, we compute, for the COMPTEL q-values, the quantity

$$J_\gamma = \frac{q}{4\pi} \int_{-20°}^{+20°} N_H db \qquad (2)$$

averaged over $-30° < l < +30°$. Fig 2 shows the longitude dependence of $\int_{-20°}^{+20°} N_H db$ for reference. Fig 3 shows a compilation of results from several instruments for the flux in the inner Galaxy; note that such a comparison is at best approximate because of the very different fields of view of the various instruments and the fact that the observations refer to different regions of the Galactic plane. Nevertheless the COMPTEL result is consistent with the overall spectrum; it suggests however that the true spectrum is up to a factor 2 lower than the 'best guess' of Gehrels and Tueller[3]. The difficulty of comparison underlines the need for a uniform analysis method for different instruments; note that the approach used here could be applied easily to other data and would allow direct comparison of results (as is the case for the COMPTEL and COS-B analysis shown in Fig 1).

38 Diffuse Galactic Continuum Emission

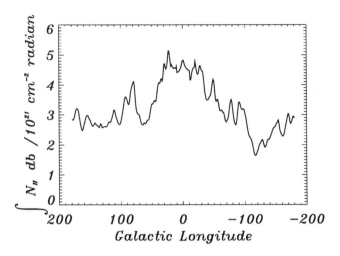

Fig 2: Total gas column density integrated over -20°<b<+20°

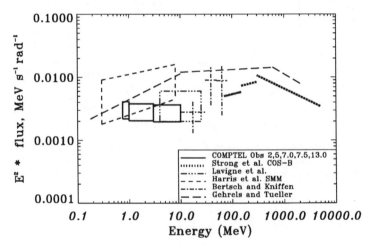

Fig 3: Compilation of flux measurements for inner Galaxy

GAMMA-RAY SPECTRUM FROM BREMSSTRAHLUNG

The interstellar electron spectrum is not directly measurable due to solar modulation, so we have to rely on model calculations to obtain predictions of the γ-ray emissivity from bremsstrahlung. Spectra for 'leaky-box' models of cosmic-ray propagation have been calculated [4,5] for particular choices of model parameters. We compute the emission for one set of parameters which fits the cosmic-ray nucleon composition data [6] : mean path length $(\lambda) = 10.8\beta$ g cm^{-2} for E_e <4GeV, $= 24.9E^{-0.6}$ g cm^{-2} for E_e >4GeV, and escape time $\tau = 2.5 \times 10^7$ years at 4 GeV. The shape of the spectrum over the 1-1000 MeV range is effectively determined by λ. We have therefore computed models for various λ to illustrate to what extent the observations constrain this parameter (Fig 4). The electron injection spectrum is assumed to have a constant power law index of -2.4; the interstellar spectrum is normalized to the value at 1 GeV derived from an analysis of COS-B data [7]: $2\times10^{-5}\rm{cm}^{-2}\rm{sr}^{-1}\rm{s}^{-1}\rm{MeV}^{-1}$; the π^{o} emissivity is taken from Dermer [8]. The value of λ(<4 GeV) which best fits the data is in the range 4-6g cm^{-2} which is somewhat lower than the value derived from cosmic-ray secondary data [6]. The difference could be accomodated in our model by using a slightly steeper electron injection spectrum, otherwise an additional component (perhaps to to point sources) might be required.

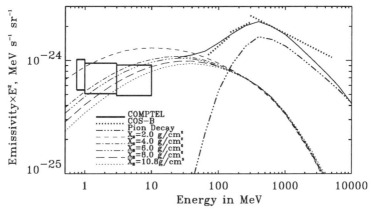

Fig 4: Predicted γ-ray emissivity from bremsstrahlung. Models for various λ compared with COMPTEL and COS-B results. The solid line is the sum of bremsstrahlung (λ=6 g cm^{-2}) and π^{o} emission.

CONCLUSIONS

The present diffuse γ-ray emissivity values were derived using only a small part of the COMPTEL sky-survey, and already significant constraints can be placed on the interstellar electron spectrum. This analysis will be extended to the entire Galactic plane; on the basis of the present analysis a further improvement in our knowledge of the low-energy γ-ray continuum can be expected. Combination with complementary OSSE and EGRET results will then be an important step in unravelling the components of the diffuse γ-ray emission from the Galaxy.

REFERENCES

1. V. Schönfelder et al., Ap. J. Supp. (in press) (1992)
2. A.W. Strong et al., Astron. Astrophys. 207, 1, (1988)
3. N. Gehrels, J. Tueller, preprint, (1992)
4. W.-H. Ip, W.I. Axford, Astron. Astrophys. 149, 7, (1985)
5. J.G. Skibo, R. Ramaty, Astron. Astrophys. Supp. (in press) (1992)
6. W.R. Webber, Proc XXI Int. Cosmic Ray Conference, 3, 393, (1990)
7. A.W. Strong, Proc XIX Int. Cosmic Ray Conference, 1, 333, (1985)
8. C.D. Dermer, Astron. Astrophys. 157, 223, (1986)
9. M.J. Harris et al., Ap. J. 362, 135, (1990)
10. J.M. Lavigne et al., Ap. J. 308, 370, (1986)
11. D.L Bertsch, D.A. Kniffen, Ap. J. 270, 305, (1983)

COMPTEL MEASUREMENTS OF 1.809 MeV GAMMA-RAY LINE EMISSION FROM THE GALACTIC PLANE

R. Diehl, W. Collmar, G. Lichti, V. Schönfelder, A.Strong
Max Planck Institut für extraterrestrische Physik, 8046 Garching, FRG

H. Bloemen, C. Dupraz*, C. deVries, W. Hermsen, B.N. Swanenburg
SRON Leiden, P.B. 9504 2300 RA Leiden, The Netherlands
(* also: Ecole Normale Supérieure, 75231 Paris, France)

D. Morris, M. Varendorff
University of New Hampshire, Institute for EOS Studies, Durham, NH 03824, USA

C. Winkler
Astrophysics Division, ESTEC, 2200 AG Noordwijk, The Netherlands

ABSTRACT

The COMPTEL experiment on the Compton Gamma-Ray Observatory (CGRO) is designed to image celestial gamma radiation in the energy range from 0.75 to 30 MeV within a field of view of 1 steradian. It can locate strong point sources with an accuracy better than 1° and is capable of mapping diffuse emission at the same time. The Galactic plane was observed by CGRO for several periods in 1991 and 1992. Emission in the 1.8 MeV line (attributed to radioactive ^{26}Al) along the Galactic disc is observed towards the Galactic centre region. In addition, the observations away from the central region of the Galaxy reveal evidence for 1.8 MeV emission. These preliminary results provide new data for determination of the ^{26}Al source origin: detailed mapping in the Galactic-centre region may be supplemented by study of other less distant nucleosynthesis regions, such as the Vela region and Carina arm.

INTRODUCTION

The 1.809 MeV gamma-ray line originating from the decay of radioactive ^{26}Al ($\tau = 1.04 \cdot 10^6$ years) was predicted by Ramaty and Lingenfelter in 1977 and first detected by the HEAO-C instrument[10] in the direction of the Galactic centre. Many other measurements have been made since then, concentrating on the Galactic centre region, and the existence of the line was firmly established by the SMM measurements[16] in 1985. The decay of ^{26}Al is believed to take place in the interstellar medium (as inferred from the narrow line width measured by HEAO-C). The formation of ^{26}Al occurs in nucleosynthesis sites such as novae, supernovae, and the interior of massive stars[1,13]. The Galactic distribution of these progenitors can be derived with some uncertainty from other measurements (radio, optical, infrared). Imaging of the 1.8 MeV emission thus provides the basis to distinguish among the potential ^{26}Al sources in the Galaxy.

Attempts to image the 1.809 MeV line emission have been reported by different groups: MPE's balloone-borne Compton telescope[23], SMM exploiting earth occultation[14], and the GRIS telescope using an on-off-source pointing technique[20]. Limitations in sky exposure or instrumental capabilities have resulted in inconclusive spatial determination. The data do indicate however that the emission region is extended along the plane rather than 'pointlike'. The predicted differences in spatial signatures of the potential sources of ^{26}Al in the Galactic centre region are small[12] compared to the spatial resolution of these measurements. The COMPTEL imaging telescope aboard the Compton Gamma-Ray Observatory[17] has adequate sensitivity and spatial resolution to provide new insight into the origin of ^{26}Al; in particular observations along the Galactic plane, beyond the Galactic centre region, provide valuable additional information on 1.8 MeV emissivity (spiral arms or other candidate nucleosynthesis regions). First COMPTEL results on the Galactic centre region from spectral and imaging analyses have been reported[3,4]; this paper adds first results from observations along the Galactic plane.

ANALYSIS AND RESULTS

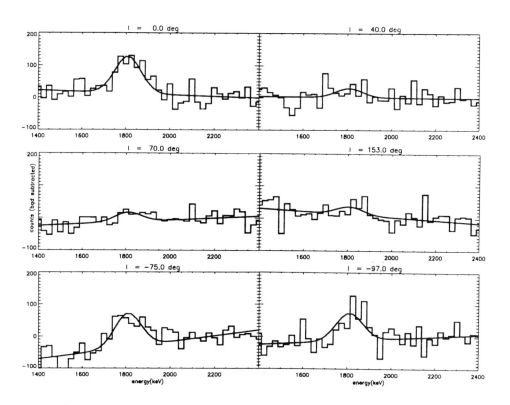

Figure 1. Background corrected energy spectra for regions near l=0° (a), l=40° (b), l=70° (c), l=153° (d), l=285° (e) and l=263° (f).

The Galactic plane was observed during the CGRO all-sky survey as listed in table I. With an instrumental energy resolution of 8.5% FWHM at 1.8 MeV and a high photopeak fraction at MeV energies, the 1.8 MeV ^{26}Al line can easily be seen in the raw data[3]. Using the imaging information of the measured telescope signal, such spectra have been generated for selected sky areas along the Galactic plane. Similarly, an average background spectrum was derived from observations at high Galactic latitudes. Figure 1 presents background subtracted energy spectra for six observations along the plane of the Galaxy in the regime of the 1.8 MeV line. One region of about 10° effective width was selected for each observation, centered on the Galactic plane (b=0) at the longitudes indicated in the Figure. A Gaussian fit with fixed energy and the instrumental line width (plus a linear background) is shown for comparison. The derived 1.8 MeV intensities are listed in Table I for our samples along the Galactic plane (normalized to the Galactic-centre region observation).

Most interestingly, the samples in the direction of the Vela region (l≃263°) and the Carina arm (l≃285°) show evidence for a 1.8 MeV signal. The residual broadband features in these background-subtracted spectra indicate the variation of the overall background continuum with selected region in the field of view and with different observation periods. Further improvement of the accuracy of the background assessment is in progress in order to firmly establish the line intensities in those regions. More samples are required to provide a full mapping of the Galactic plane.

Table I. COMPTEL Galactic Plane 1.8 MeV Observations.				
Obs. No.	Pointing Longitude(°)	Pointing Latitude(°)	Target Name	1.8 MeV Intensity (relative)
2.0	73.3	2.6	Cyg X-1	0.27 ± 0.24
5	0.0	-4.0	Galactic Center	1.00 ± 0.27
7.0	70.4	-8.3	Cyg X-3	0.56 ± 0.28
7.5/13.0	25.0	-14.0	Gal 025-14	0.21 ± 0.23
8	262.9	-5.7	Vela Pulsar	0.78 ± 0.26
14	285.0	-0.7	ETA CAR	1.14 ± 0.26
15	152.6	-13.4	NGC 1275	0.28 ± 0.23
20	39.7	0.8	SS 433	0.32 ± 0.23
27	332.2	2.5	4U 1543-47	0.89 ± 0.34
31	163.1	11.9	MCG+8-11-11	0.12 ± 0.23

In order to fully utilize the full imaging capability of the telescope, events in the energy interval of interest (1.7-1.9 MeV) were binned in a 3-D dataspace of measured scatter direction and scatter angle. The background in this dataspace was derived by a similar binning of events from the adjacent energy bands

1.5-1.7 and 1.9-2.1 MeV of the same observations, thus subtracting instrumental background and celestial continuum emission. Using appropriate response and exposure matrices, a sky image can be constructed through maximum-entropy or maximum-likelihood deconvolution. Figure 2 shows the deconvolved images based on the four Galactic-centre region observations for both independent methods. (Maximum entropy deconvolved imaging varies the sky intensity in all pixels simultaneously until the best fit with maximized entropy is found; maximum likelihood deconvolution displays the fit quality with a single point source as a function of hypothesized source position, on top of instrumental background, as a sky map.) An extension of the emission along the Galactic plane is clearly seen, with some localized enhancements along the plane, the brightest feature being somewhat offset from the the Galactic centre itself. We emphasize that only the bright global structure in this map is significant, while individual structural details in this map reflect the level of statistical noise and residual background uncertainty; bootstrap analyses are being applied to assess statistical uncertainties, and maximum likelihood techniques are used for diffuse model fitting; both these will be discussed in a forthcoming paper.

A similar detailed analysis of the Vela region is in progress. Assuming that the excess observed in spectral analysis (see Figure 1) can be assigned to the ^{26}Al line, the derived first images indicate an excess structure extended about 10° and including the nearby Vela region objects (Vela supernova remnant, and Wolf Rayet star binary), somewhat south of the Galactic plane.

DISCUSSION AND CONCLUSIONS

The COMPTEL data from ten CGRO observations along the plane of the Galaxy show evidence for 1.8 MeV line emission (attributed to radioactive ^{26}Al), with several bright regions, but also wide regions where the 1.8 MeV emission is below the COMPTEL sensitivity limit. The Galactic-centre region data show extended emission over a longitude range of about 40° and a latitude extent of more than 5°, with some evidence for localized enhancements in the intensity distribution, and a particularly prominent peak somewhat offset from the Galactic centre. Beyond the central region of the Galaxy, indications of 1.8 MeV emission from the Vela and Carina region are most interesting. Although our findings so far should be regarded as preliminary, COMPTEL data will clearly provide stringent constraints on the origin of ^{26}Al.

The nucleosynthesis sites of ^{26}Al (such as novae, type II supernovae, and massive stars) must be sufficiently hot and sufficiently enriched in ^{26}Al seed nuclei for the ^{26}Al production to be effective. In addition, the destruction of ^{26}Al due to photodisintegration in such hot environments should be sufficiently small so that freshly produced ^{26}Al survives. Two classes of nucleosynthesis processes fulfil those conditions. One of them is non-equilibrium nuclear burning: explosive nova nucleosynthesis in the convective zone on the surface of metal enriched O-Ne-Mg white dwarfs, or core collapse supernovae, where rapid propagation of a nuclear burning zone through the seed matter ensures that the ^{26}Al generated in the burning zone survives and is diluted in the interstellar medium. An alternative process is core nuclear burning in the convective atmospheres of massive stars, where the freshly synthesized ^{26}Al is rapidly convected away from the hot inner burning region.

Precise yield calculations are very difficult and require accurate knowledge of nuclear cross sections in the relevant temperature regimes and adequate treat-

44 1.809 MeV Gamma-Ray Line Emission

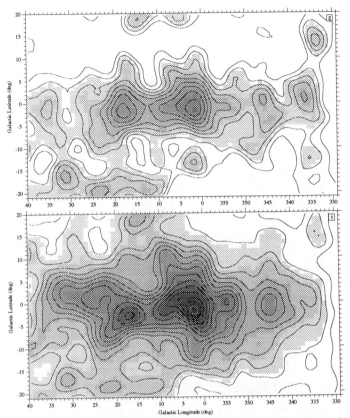

Figure 2. Images of the Galactic-centre region in the 1.8 MeV gamma-ray line, derived from a combination of four COMPTEL observations, using maximum entropy deconvolution (above), and maximum likelihood deconvolution (below)

ment of the hydrodynamics of the nucleosynthesis region in 3 dimensions. In addition, it is known from theoretical calculations that the ^{26}Al yield varies by orders of magnitude with composition of the burning material, i.e. metallicity of the progenitors plays an important role. The predicted yields from all candidate processes are in the range compatible with the observed ^{26}Al mass of ~ 1.5 M$_\odot$ as inferred from the Galactic-centre region data, as shown[12,19,24] in Table II.

Constraints on the origin of ^{26}Al can be obtained from correlation studies of the 1.8 MeV intensity distribution and (often indirectly inferred) Galactic distributions of potential sources. Notice that a few individual nucleosynthesis events, of more local origin, could contribute a substantial fractions of the total ^{26}Al emission, in addition to the expected smooth steady-state contribution from supernovae, novae, and massive stars throughout the Galaxy. Also, if the Galactic nucleosynthesis history includes regions of locally enhanced activity, irregular and clumpy 1.8 MeV intensity distribution along the Galactic plane can be expected.

The present results already suggest some preliminary conclusions:
- If novae are the main source of Galactic ^{26}Al, a distribution following the spheroidal component in the Galaxy (population I objects) seems ruled out by our result of an extended ridge of 1.8 MeV emission along the Galactic plane. (A detailed analysis of the latitude extent of the emission is in progress and will provide constraints on the ratio of disk to spheroid contribution for alternate nova model distributions.)
- The 1.8 MeV emission peak observed near the Galactic centre coincides with the peak of the Galactic gas distribution as found[2] in CO, also slightly offset from the Galactic centre. The interpretation of the CO peak in the vicinity of the Galactic centre is unclear; if it testifies to enhanced column density of gas and hence star forming activity, a correlation with a supernova origin of ^{26}Al might be suggested by COMPTEL data.
- The additional excess in the direction of Vela suggests that the COMPTEL observations do not exclude a (partly) local ^{26}Al origin. This could imply that the total ^{26}Al mass derived from the 1.8 MeV line intensity may be well less than the canonical value of 1-3 M$_\odot$.

Further analysis of these data, and additional observations in other regimes of the Galaxy are underway and can be expected to set stringent constraints on the nature of the ^{26}Al sources.

Table II. Characteristics of Possible ^{26}Al Sources				
Source	Process	Rate	max. yield per event (M$_\odot$)	Galactic Contribution (M$_\odot$)
Novae	explosive hydrogen burning on O-Ne-Mg WD	10-30 yr^{-1}	$\sim$ few $\cdot$ 10^{-7}	0.4-5
Supernovae	core collapse with C-Ne burning in SN shock wave	2 (100y)$^{-1}$	10^{-4}	0.5-1.5
Wolf Rayet Stars	convective hydrogen burning	($\sim$ 13% of all stars)	10^{-4}	$\sim$ 0.5
Asymptotic Giant Branch Stars	H shell burning of core ^{25}Mg	(n.a.)	10^{-8}	0.1-30

REFERENCES

1. Clayton D.D., Ap.J. **280**, 144 (1984).
2. Dame T.M., Ungerechts H., Cocn R.S., deGeus E., Grenier I., May J., Murphy D.C., Nyman L.A., Thadeus P., Ap.J. **322**, 706 (1987).
3. Diehl R., et al., Astr.& Astr. , (in press) (1992).
4. Diehl R., et al., Adv. in Sp.Res. , (in press) (1992).
5. Forrestini M., Paulus G., Arnould M., Astr.& Astr. , (1991).
6. Green D.A., Astr.& Sp.Sci. **148**, 3 (1988).
7. Higdon J., Fowler W., Ap.J. **339**, 956 (1989).
8. Leising M., Clayton D.D., Ap.J. **294**, 591 (1985).
9. MacCallum C.J., Huters A.F., Stang P.D., Leventhal M., Ap.J. **317**, 877 (1987).
10. Mahoney W.A., Ling J.C., Wheaton W.A., Jacobson A.S., Ap.J. **286**, 578 (1984).
11. Malet I., Montmerle T., von Ballmoos P., AIP Conf. Proceedings 232 (eds.P. Durouchoux and N. Prantzos, 1991), p. 123.
12. Prantzos N., AIP Conf. Proceedings 232 (eds.P. Durouchoux and N. Prantzos, 1991), p. 129.
13. Prantzos N., Astr.& Astr. , (in press) (1992).
14. Purcell W.R., Ulmer M.P., Share G.H., Kinzer R.L., GRO Science Workshop Proceedings (NASA Publ., 1989), p. 4-327.
15. Ramaty R., Lingenfelter R.E., Nature **278**, 127 (1979).
16. Share G., Kinzer R.L., Chupp E.L., Forrest D.J., Rieger E., Ap.J. Lett. **292**, L61 (1985).
17. Schönfelder V. et al., 2nd GRO Science Workshop (NASA Rep. No. CP-3137, 1991), p. 76.
18. Schönfelder V., et al., IEEE Trans. on Nucl.Sci. $NS-31, 1$, 76 (1984).
19. Starrfield S., private communication , (1992).
20. Teegarden B.J., Barthelmy S.D., Gehrels N., Tueller J, Leventhal M., MacCallum C., AIP Conf. Proceedings 232 (eds.P. Durouchoux and N. Prantzos, 1991), p. 116.
21. Varendorff M. and Schönfelder V., Ap.J. **395**, 158-165 (1992).
22. van der Hucht K.A., Hidayat B., Admiranto A.G., Supelli K.R., Doom C., Astr.& Astrophys. **199**, 217 (1988).
23. von Ballmoos P., Diehl R., Schönfelder V., Ap.J. **312**, 657 (1987).
24. Woosley S.E., 'Nucleosynthesis and Chemical Evolution' (Eds B. Hauck, A. Maeder, and G. Meynet, 1986), p. 78.

Nucleosynthesis of ^{26}Al in AGB Stars and Wolf-Rayet Stars

Grant Bazan[*], Lawrence E. Brown[†], Donald D. Clayton[†],
Mounib F. El Eid[†,‡], Dieter H. Hartmann[†], and
James W. Truran[§]

ABSTRACT

We report on our recent studies of ^{26}Al production from stars initially on the upper end of the main sequence: the intermediate mass AGB stars and the presupernova stars. Hot hydrogen burning is the source of the ^{26}Al in both objects, although it is known that explosive burning creates modest amounts also in the shock heated C and Ne shells. We examine the amounts created by hydrogen burning that may escape in stellar winds that play an important role in both AGB and Wolf-Rayet star evolution. We will also discuss simple models of the Galactic distribution of the associated diffuse γ-ray line emission.

AGB MODELS

When a star ascends the asymptotic giant branch (AGB) it develops a deep convective envelope. For stars of mass $M \gtrsim 4 M_\odot$ (depending on metallicity), the lower regions of this envelope can reach temperatures high enough to initiate significant CNO processing and the creation of ^{26}Al by the reaction ^{25}Mg $(p,\gamma)^{26}$Al *in the envelope*. This ^{26}Al is continuously mixed to the surface of the star throughout its whole AGB lifetime. If ^{25}Mg enriched material from the convective helium burning shell is dredged up into the convective envelope, ^{26}Al production can even exceed the initial abundance of ^{25}Mg

The main features of our models are: Masses are in the range 2-8$M_\odot$. Hot Bottom Burning in the convective envelope is assumed to be the sole production mechanism. We ignore any dredgeup of ^{26}Al from the H burning shell[1] because it is highly dependent on the treatment of semiconvection. Additionally, it may only be important in low mass stars (main production occurs at higher masses) if it occurs at all. Core evolution, luminosity, mass loss, dredgeup etc. are treated in parametrized form, but we compute the envelope structure from hydrostatic equilibrium. Envelope nucleosynthesis uses full convective diffusion.[2]

AGB RESULTS

Our basic conclusions are: ^{26}Al production is strongly dependent on mass loss rate, mixing length, and metallicity. Opacities below 7,000 K are extremely important owing to their influence on envelope structure. Increasing the low temperature opacity

---

[*]McDonald Observatory, University of Texas at Austin, Austin, TX 78712
[†]Department of Physics and Astronomy, Clemson University, Clemson, SC
[‡]Universitäts-Sternwarte, Geismarlandstrasse 11, D-3400 Göttingen, FRG
[§]Department of Astronomy and Astrophysics, University of Chicago, Chicago, IL

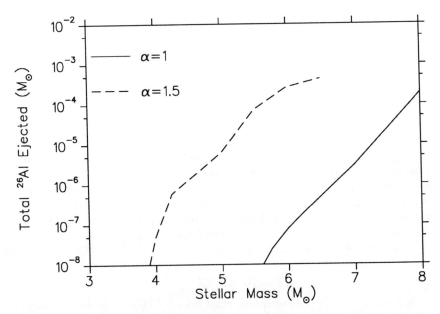

Figure 1: *The production of ^{26}Al in AGB star envelopes as a function of initial stellar mass. We plot the total amount ejected over the whole AGB lifetime. For the short (AGB) lifetimes of these intermediate mass stars, this is close to the total observable ^{26}Al mass. The parameter α is the ratio of the mixing length to the pressure scale height.*

results in cooler envelopes. Using the large opacities derived by Hoeflich[3] completely quenches ^{26}Al production in our models. For other reasons, particularly the high surface abundance of ^{7}Li observed in some AGB stars, we suspect that these opacities are too high, since the same effect prevents ^{7}Li production. We use Cox-Stewart opacities for the following results. We find efficient production for stellar masses $M \gtrsim 5 M_\odot$ (Figure 1). The maximum mass of ^{26}Al ejected from a single star is $\sim 5.0 \times 10^{-5} M_\odot$. We can make a rough estimate of the total mass of ^{26}Al in the Galaxy by integrating our yields (Figure 1) over the initial mass function (IMF) of Scalo.[4] (We assume a constant star formation rate over a 15×10^9 yr lifetime of the Galaxy for deriving the IMF from the present day mass function.) An approximate average mass fraction of ^{26}Al in the interstellar medium is given by:

$$X_{26} \approx \frac{b}{M_{ISM} \lambda_{26}} \frac{\int M_{26ej}(m) \phi(m) dm}{\int m \phi(m) dm} \quad (1)$$

where X_{26} is the average mass fraction of ^{26}Al in the Galaxy, b is the current star formation rate in $M_\odot$ per year, M_{ISM} is the mass of the Galactic interstellar medium, λ_{26} is the decay constant of ^{26}Al, $M_{26ej}(m)$ is the mass of ^{26}Al ejected by a star of mass M, $\phi(m)$ is the IMF in $M_\odot^{-1} pc^{-2}$, and the integrals are taken over all possible stellar masses. If we assume $10^{10} M_\odot$ as the mass of the ISM and a birthrate of $3.5 M_\odot$ per year, we derive $4.4 M_\odot$ masses of ^{26}Al in the Galaxy. (To represent the effect of changing parameter values, for $\alpha = 1.0$ we produce $0.6 M_\odot$, and for the highest current

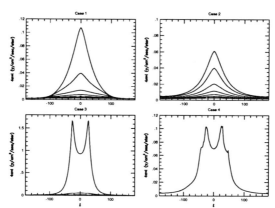

Figure 2: *Four possible simple models of the diffuse Galactic glow in* ^{26}Al *line emission. We have made models of the Galactic emissivity of the 1809 keV line by assuming that AGB stars are correlated with some observable or calculated quantity. Case 1: AGB star emission is correlated with the IR flux distribution of the Galaxy. Case 2: AGB star emission is predicted using the OH/IR star kinematical models of te Lintel Hekkert.[5] Case 3: AGB star emission is predicted using direct counts of the OH/IR star distribution of Herman and Habing.[6] Case 4: AGB star emission is tied to spiral density wave induced star formation. See next figure.*

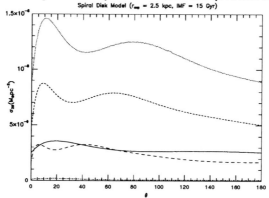

Figure 3: *The* ^{26}Al *distribution in the presence of spiral density waves. The features of the model are: Gas mass surface density falls off exponentially in azimuth with a scale distance of 2.5 kpc. Star formation goes as surface density but efficiency goes as the relative number of observed HII regions as a function of galactic radius Spiral pattern is logarithmic with pitch angle of 7 degrees Constant angular rotation speed of 250 km/sec. We show the resulting surface density of* ^{26}Al *as a function of azimuthal angle. The lines correspond to radii 3 kpc (solid), 5 kpc (dotted), 6 kpc (dashed), 8 kpc (long dashed), and 12 kpc (dot-dashed).*

Table 1: Massive Star Presupernova Yields

Mass ($M_\odot$)	Mass ^{26}Al ($M_\odot$)	X_{26}
30	3.21e-5	2.03e-6
40	6.37e-5	3.01e-6

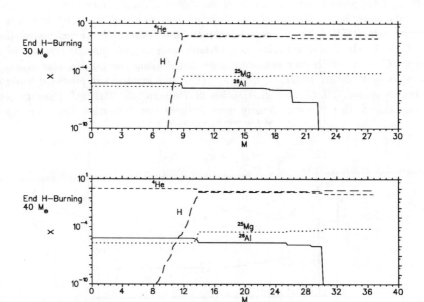

Figure 4: *The composition (mass fraction) of 30 and 40 $M_\odot$ stars at the end of core hydrogen burning. Material outside the hydrogen exhausted core should survive and be ejected either in the subsequent Wolf-Rayet stage or in a supernova.*

estimates of low temperature opacities we produce none). We present some preliminary models of the diffuse 1.8MeV γ-ray line emission in Figures 2 and 3.

MASSIVE STAR MODELS

Another proposed site for ^{26}Al production is massive stars. For masses larger than 20 $M_\odot$, ^{26}Al is produced in the convective H burning core and is left behind later in the outer stellar material when the core shrinks. Later it is ejected into the ISM in the Wolf-Rayet wind or in a supernova. Most of the ^{26}Al inside the H exhausted region of the star will be destroyed by He burning. Additional production in the C and Ne shells of supernovae also occurs[7] but we do not model this here. Enhanced ^{26}Al yields are expected when neutrino processing is included.[8]

Stellar models were calculated with the 1D hydrodynamic evolution code developed at Göttingen Observatory[9,10,11] to trace the production of ^{26}Al during core hydrogen burning. We use the most recent OPAL opacities.[12] Preliminary results for $30M_\odot$ and $40M_\odot$ models (see Table 1 and Figure 4) imply that the presupernova production of ^{26}Al results in a mass fraction of $\sim 2 \times 10^{-6}$. If we assume that all stars more massive than $15M_\odot$ create a mass fraction of $X_{26} \sim 2 \times 10^{-6}$ in their ejecta, we find $\sim 0.59 M_\odot$ in the ISM. This may not be the sole source of galactic ^{26}Al, but it is close. Our results confirm those of Weaver and Woosley.[13]

Part of this work was supported by NASA grant NAG 5-1578 (DH).

REFERENCES

1. Forestini, M., Paulus, G., and Arnould, M. (1991) A&A, 252, 597

2. Sackmann, I.J., Smith, R.L., & Despain, K.H. (1974) ApJ, 187, 555

3. Hoeflich, P.A. (1992) private communication

4. Scalo, J. (1986) in *Fund. of Cosmic Physics* (Gordan and Breach) pp. 1-278

5. te Lintel Hekkert, P. (1992) preprint

6. Herman, J. & Habing, H.J. (1985) PhysRep 124,255

7. Woosley, S.E. (1991) in *Gamma-Ray Line Astrophysics*, ed P. Durouchoux and N. Prantzos (AIP: New York) pp. 270-290

8. Woosley, S.E., Hartmann, D., Hoffman, R., Haxton, W. (1990) ApJ, 356, 272

9. Ober, W., El Eid, M.F., & Fricke, K.J. (1983) A&A, 119, 61

10. El Eid, M.F. & Langer, N. (1986) A&A, 167, 274

11. Baraffe, I. & El Eid, M.F. (1991) A&A, 245, 548

12. Rogers, F.J., & Iglesias, C.A. (1992) ApJS, 79, 507

13. Weaver, T. and Woosley, S.E. (1992) PhysRep, in press

RADIOACTIVE ^{26}Al FROM MASSIVE STARS: PRODUCTION AND DISTRIBUTION IN THE GALAXY

N. Prantzos

Institut d' Astrophysique de Paris
and
Service d' Astrophysique, CE Saclay

ABSTRACT

The production of radioactive ^{26}Al by massive stars in our Galaxy is reviewed in the light of recent theoretical results. Stars exploding as Type II supernovae (SNII) seem to be the most promising candidates to explain the galactic emission at 1.8 MeV, due to the decay of ^{26}Al; however, considerable uncertainties in current nucleosynthesis models do not allow definite conclusions yet. On the other hand, recent observations by the Compton Observatory can be interpreted as evidence for a young stellar population at the origin of that emission. We present numerical simulations of the expected flux profile as a function of galactic longitude, adopting a realistic spiral pattern for our Galaxy. We discuss the recent observations in the light of our results.

INTRODUCTION

^{26}Al is unstable to positron emission with a mean lifetime of $\tau_{26} \sim 10^6$ y. Its decay feeds the first excited state of ^{26}Mg at 1.809 MeV, the deexcitation of which gives rise to a gamma-ray photon. The detection of the 1.8 MeV γ-ray line from the galactic center (GC) direction by the HEAO-3 satellite (Mahoney et al. 1984), is a discovery of paramount importance (Clayton 1984). Indeed, ^{26}Al is the first radioactive nucleus ever seen in extrasolar gamma-ray astronomy and its detection clearly demonstrates that nucleosynthesis is currently active in the Galaxy (since its lifetime is very short compared to the time-scale of galactic chemical evolution); it offers a rare opportunity to confront nucleosynthesis theories with observational data.

The original detection has been confirmed by several balloon and satellite experiments up to now (see Schoenfelder and Varendorff 1991, for a summary of the observational situation just prior to the Compton launching). The detected 1.8 MeV flux ($\sim 4\ 10^{-4}$ photons cm^{-2} s^{-1}) implies that $M_{26} \sim$ 2-3 $M_\odot$ of ^{26}Al (depending somewhat on the unknown source distribution) are ejected in the interstellar medium (ISM) of our Galaxy every $\sim 10^6$ years. Although several astrophysical sites (novae, Wolf-Rayet stars, Type II supernovae, asymptotic giant branch stars) can produce considerable amounts of ^{26}Al, it seems difficult to explain the observations with current nucleosynthesis models (see e.g. Clayton and Leising 1987 and Prantzos 1991 for reviews). Uncertainties in the modelling of all the sources, as well as in their frequency of occurrence in the Galaxy, do not allow any definite conclusion yet.

In view of that difficulty encountered by theory, it has been expected that observations with good angular resolution could help, mapping the distribution of the 1.8 MeV emission in the Galaxy and thus revealing the distribution and the nature of the underlying sources (making the implicit, and quite plausible, assumption that during its $\sim 10^6$ y lifetime ^{26}Al does not move very far away from its source). Such a map is now being obtained with the COMPTEL instrument aboard the Compton Gamma-Ray Observatory. Preliminary results (R. Diehl, this meeting) show a relatively flat but asymmetric longitude profile with an important emission from the directions of both the galactic center and the Carina arm. Those results can tentatively be interpreted as evidence for an underlying young population (i.e. massive stars), in agreement with theoretical expectations (Prantzos 1991, 1993; Ramaty and Prantzos 1991).

In this work we review the production of ^{26}Al from massive stars and evaluate the total amount ejected in the Galaxy in the past $\sim 10^6$ years in the light of recent theoretical models. We present then numerical simulations of the expected flux profile as a function of galactic longitude, taking into account a realistic pattern for the spiral structure of our Galaxy. Finally we discuss some implications of the recent COMPTEL observations in the light of our results.

PRODUCTION OF ^{26}Al IN MASSIVE STARS

The main production mechanism of ^{26}Al in astrophysical environments is the reaction ^{25}Mg(p,γ)^{26}Al, operating in hydrostatic or explosive hydrogen, carbon, or neon burning. For the gamma-ray photons of its decay to be observable, ^{26}Al has to be ejected in the ISM before destruction. This can be easily achieved in the case of an explosive site, like e.g. a nova or a supernova, as well as in the case of an object suffering extensive mass loss, like a Wolf-Rayet or an AGB (Asymptotic Giant Branch) star. In the following we shall consider only massive stars, in view of the motivations mentioned in the Introduction. Indeed, the production of ^{26}Al in models of low mass progenitors, like novae and AGB stars, seems up to now insufficient to explain the observed emission (see Prantzos 1991 and references therein); notice, however, that the corresponding uncertainties (due mostly to the treatment of convection) are much larger than in the case of massive stars. Starrfield et al.(1993) find large amounts of ^{26}Al in recent hydrodynamic models of O-Ne rich novae; but the "patchy" and relatively flat COMPTEL profile of the 1.8 MeV emission (if confirmed) does not favour an old source population, expected to give a relatively smooth profile, peaked towards the galactic center (Cassé and Prantzos 1993).

Hydrostatic core H burning can lead to the production of significant amounts of ^{26}Al if the central temperature of the star is $T_c > 35\text{-}40\ 10^6$ K, i.e. for stars more massive than ~ 30 M$_\odot$. For temperatures $T < 45\ 10^6$ K, ^{26}Al is predominantly destroyed by ^{26}Al(β^+)^{26}Mg, while for higher temperatures ^{26}Al(p,γ) is the main destruction mechanism in that environment. As the convective stellar core gradually retreats (while the star still burns H on the main sequence) it leaves behind ^{26}Al that has been previously produced and mixed in the core.

Thus, some ^{26}Al is found in the stellar envelope, where it β^+ decays to ^{26}Mg. Whether or not it will appear at the stellar surface depends on the mass loss. For stars with M>30-40 M$_\odot$ (depending on metallicity) radiatively driven winds induce such high mass loss rates that the initial envelope of the star is expelled and regions processed by H burning finally emerge at the surface: this is the standard senario for the formation of single *Wolf-Rayet* (WR) stars. Obviously, the higher the mass loss rate and/or the extent of the convective core, the more ^{26}Al will be ejected. During the subsequent phase of central He burning, neutrons released through ^{13}C(α,n) efficiently destroy the remaining ^{26}Al in the stellar core, through (n,α) and (n,p) reactions. However, ^{26}Al continues to be ejected from the stellar envelope, and disappears only when the He burning products appear, in their turn, at the surface. Despite the use of quite different stellar evolution codes and different prescriptions for convection and mass loss, all recent calculations of ^{26}Al production by WR stars lead to rather similar results (to better than 30%): a few 10^{-5} - 10^{-4} M$_\odot$ of ^{26}Al are ejected in the ISM during the $\sim$2-5 10^5 y of the WR stage (see Fig. 2 and Prantzos 1991 for references); the corresponding 1.8 MeV luminosity (averaged over the duration of the WR stage) is L$_\gamma$ $\sim$$10^{37}$-$10^{38}$ ph s^{-1} (Prantzos and Cassé 1986).

After hydrogen is exhausted in the stellar core, a thin *H-burning shell* is established, moving outwards as the star evolves. ^{26}Al can also be produced in that site and eventually ejected by the final supernova explosion. However, in the most massive stars (i.e. those that evolve in the WR stage and explode probably as SNIb), mass loss prevails the development of an important H-shell. In stars with 10< M/M$_\odot$ <35, which do develop an important H-shell, the amount of ^{26}Al found in the only consistent calculation up to now (Woosley 1991) is much smaller than the corresponding amount in the core of WR stars or in the C and Ne zones (see Fig. 1 in Woosley 1991).

During *hydrostatic shell carbon and neon burning* in massive stars, at temperatures T$\sim$1.5 10^9 K, densities ρ $\sim$$10^5$ g cm^{-3} and timescales $\sim$a few months or years, ^{26}Al is produced by ^{25}Mg(p,γ). Background protons, coming mainly from reactions on ^{12}C or ^{20}Ne nuclei, are captured on Mg nuclei to produce ^{26}Al; background neutrons, liberated essentially through ^{13}C(α,n) and ^{22}Ne(α,n) reactions, are now the main destruction agent of ^{26}Al, instead of (p,γ) reactions or β decays. The amount of ^{26}Al produced hydrostatically in that site depends crucially on the treatment of convection and semiconvection, as can be seen in Fig. 1 where the recent results of Weaver and Woosley (1992) are plotted for various stellar masses: adopting a "nominal" or "restricted" semiconvection may lead to large variations (by a factor of $\sim$10) in the resulting amount of ^{26}Al.

After the collapse of the Fe core of the star, the passage of the shock wave heats the stellar regions that have experienced shell C-Ne burning (or were convectively coupled to those shells) to peak temperatures T$_p$ $\sim$2-3 10^9 K, depending on the stellar mass (and the adopted model). *Explosive neon burning* takes place at those temperatures, at peak densities ρ_p $\sim$ a few 10^5

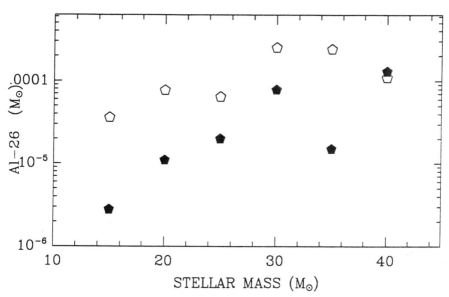

FIGURE 1: Quantities of ^{26}Al present in massive stars *prior* to their explosion (i.e. produced during hydrostatic nucleosynthesis) as a function of stellar mass (from Weaver and Woosley 1992). The results depend sensitively on the treatment of semi-convection. *Open symbols:* normal semi-convection; *filled symbols:* restricted semi-convection.

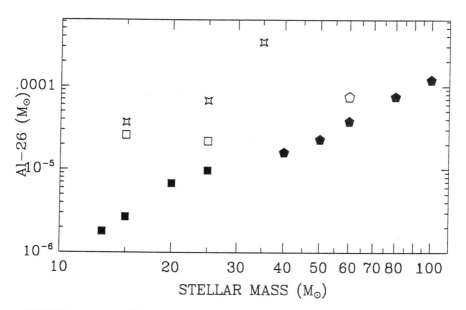

FIGURE 2: Yields of ^{26}Al from massive stars as a function of stellar mass. *Pentagons:* Wolf-Rayet stars *(filled:* Prantzos 1991; *open* : Meynet and Arnould 1993). *All other symbols:* Type II supernovae [*filled squares:* explosive nucleosynthesis only, from Thielemann et al. 1993; *open squares:* explosive nucleosynthesis only, from Weaver and Woosley 1992; *asteriscs:* total (i.e. pre-explosive + explosive) from Weaver and Woosley 1992].

g cm^{-3} and timescales $\tau \sim 1$ s. In the hotter regions (T> 2.5 10^9 K) ^{20}Ne is completely exhausted, and ^{26}Al may be destroyed by photodisintegration, while in the colder ones neon burning is incomplete, and ^{26}Al is produced. Slightly before that moment, the neutrinos liberated from the collapsed Fe core may react with the nuclei of the stellar onion, producing also some ^{26}Al in the neon shell (Woosley et al. 1990). The extent of this *neutrino induced nucleosynthesis* depends crucially on the adopted neutrino spectrum, being favoured by the most energetic neutrinos (i.e. ~ 8 MeV) and is quite uncertain at the present.

The total amount of ^{26}Al produced by massive stars is difficult to estimate, since it depends on several parameters, like the stellar mass, initial composition and mass loss, the treatment of convection and semiconvection, the rate of the ^{12}C(α,γ) reaction (which determines the extent and position of the various stellar zones in SNII) etc. In Fig. 2 we present the results of the most recent calculations of the the production of ^{26}Al in massive stars as a function of the stellar mass. Obviously, the yield of ^{26}Al increases rapidly with stellar mass, due to the fact that the more massive the star, the larger the zone experiencing appropriate burning conditions for the production of ^{26}Al. For M>40 M$_\odot$ the results (Prantzos 1991, and Meynet and Arnould 1993) correspond to the pre-explosive phase, i.e. the ^{26}Al produced in the H-core and ejected by the WR winds. Eventually, ^{26}Al will also be produced in the C and Ne shells of those stars, but their fate is still not quite clear: they may either explode as SNIb (in which case the numbers given in Fig. 2 should considerably increase) or collapse into black holes (Maeder 1992), without giving any more ^{26}Al. For less massive stars, exploding as SNII, the situation is even less clear. Besides the uncertainties of convection during hydrostatic C and Ne shell burning (see Fig. 1), there are also considerable uncertainties as to the amount of ^{26}Al produced by explosive nucleosynthesis: there is a factor of ~ 3 difference between the results of Thielemann et al. (1993) and Weaver and Woosley (1992).

Notice that the results in Fig. 2 are obtained in stars with solar metallicity. In order to evaluate the production of ^{26}Al at the galactic level, one should take into account the effect of the decrease of metallicity (z) with galactocentric radius (r). The *galactic metallicity gradient* is: $d(logz)/dr = -0.07 dex/kpc$ for oxygen (Shaver et al. 1983). Indeed, metallicity affects in various ways the production of ^{26}Al in massive stars. In WR stars the production of ^{26}Al is doubly favoured by an increased metallicity, because of larger available ^{25}Mg and of an earlier ejection of the more metal rich and opaque stellar envelope. As a result, the yield of ^{26}Al in that site scales with z^k with $1 < k < 2$ (values of k close to 2 are favoured by recent calculations of Meynet, private communication). On the other hand, the dependence on metallicity of the ^{26}Al yields of SNII is less clear. The production of ^{26}Al in the carbon and neon shell should not depend on metallicity, but its destruction (mainly by (n,p) and (n,α) reactions) could depend on the amount on the neutron source and thereof on z. Actually it has been proposed (Woosley and Weaver 1980) that the *net production* of ^{26}Al in SNII might even be *inversely proportional to metallicity*. One then should adopt $-1 < k < 0$ for that site, although larger values of k could not be excluded.

MASS AND DISTRIBUTION OF ^{26}Al IN THE GALAXY

The total quantity of ^{26}Al ejected by massive stars in the Galaxy in the last $\tau_{26} \sim 10^6$ y is formally given by:

$$M_{26} = \int_0^R 2\pi r dr \; \sigma(r) \int_{M_1}^{M_2} \Phi(M) Y(M,r) dM \qquad (1)$$

where r is the galactocentric radius, $R=15$ kpc the radius of the Galaxy, $\sigma(r)$ the adopted radial surface density of the sources (see below), $Y(M)$ at $r = r_\odot = 8.5$ kpc are the yields of Fig. 2 (calculated with solar metallicity $z_\odot$), $Y(M,r) = Y(M) * (z(r)/z_\odot)^k$, where $z(r)$ is the metallicity [such as $d(\log z)/dr = -0.07 \; dex/kpc, z(r_\odot) = z_\odot, z(r < 3 \; kpc) = 3z_\odot$], and $k=1$ or 2 for WR stars and $k=0$ or -1 for SN. $\Phi(M)$ is the adopted initial mass function for massive stars, between $M_1 \sim 10 \; M_\odot$ and $M_2 \sim 100 \; M_\odot$. It is of the Salpeter type ($\Phi(M) \propto M^\alpha$ with α=-2.5 to -2.7), and is normalized to : $\int_0^R 2\pi r dr \; \sigma(r) \int_{M_1}^{M_2} \Phi(M) dM = f_{SN} \tau_{26}$, i.e. to the total number of massive stars that exploded in the Galaxy in the last τ_{26} years (Signore and Dupraz 1990). The adopted frequency of massive star explosions in our Milky Way is $f_{SN} = f_{SNII} + f_{SNIb} \sim 3$ per century (see Tutukov et al. 1992).

The radial surface distribution of massive stars in our Galaxy is not well known. A distribution following the one of giant molecular clouds (Fig. 3), presumed site of massive star formation, seems plausible on theoretical grounds. A recent study of the radial distribution of supernovae in external galaxies (Bartunov et al. 1992) indicates, however, a "swallower" density gradient, comparable to the one of visual light and similar to the one obtained by Li et al. (1992) for the supernovae remnants of our Galaxy. Such studies are, however, limited for the moment by small samples and selection effects. In fact, the most important uncertainty concerns the star formation in the inner regions ($r < 2$ kpc) of the Galaxy. It is very difficult, indeed, to estimate the H_2 density there (the central H_2 concentration in Fig. 3 is certainly overestimated) and it is even more difficult to evaluate the corresponding star formation rate: the existence of important turbulence and/or magnetic fields, suggested by observations, could prevent star formation even in the presence of large amounts of molecular gas (Gusten 1989). Notice that these considerations affect little (less than a factor of two) the total amount of galactic ^{26}Al in Eq. (1), which depends essentially on the adopted supernova frequency; they are however important for the derived flux profiles as a function of galactic longitude, to be considered below. Notice also that the COMPTEL results seem to indicate a relatively *flat* radial distribution of ^{26}Al sources, and we have included it in our analysis as an extreme case.

Taking into account the above remarks, i.e. the effects of metallicity, stellar initial mass function and source radial distribution, one obtains through Eq. (1) values of $M_{26} \sim 0.2$-$1.4 \; M_\odot$ from SNII (the larger yields are obtained when the

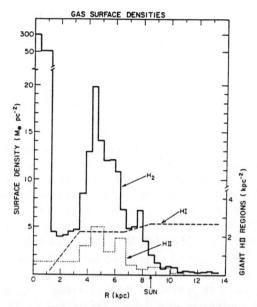

FIGURE 3: Surface density of giant molecular clouds (H_2) as a function of galactocentric radius, according to Scoville and Sanders (1987). The amount of H_2 in the inner Galaxy is certainly overestimated and in this work we adopted a value three times smaller than in the figure.

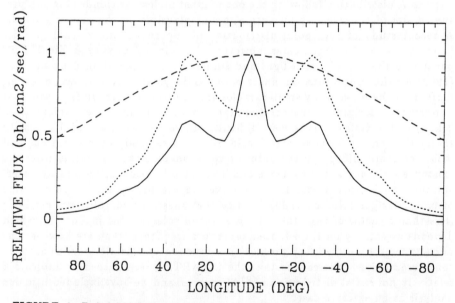

FIGURE 4: Emissivity of the galactic plane as a function of galactic longitude for three axisymmetric distributions of the sources of emission. *Solid line:* H_2 distribution of Fig. 3, with a source density in the inner Galaxy reduced by a factor of three; *Dotted line:* Same as before, but with no contribution from the inner Galalaxy; *Dashed line:* Uniform source density in the whole Galaxy. Results are normalised to maximum emissivity =1.

results of Weaver and Woosley are extrapolated up to the larger stellar masses, i.e. when SNIb are included) and 0.2-0.5 $M_\odot$ from the winds of WR stars alone. It is clear that, with current models, WR stars can produce only up to $\sim$20 % of galactic ^{26}Al, while supernovae can produce more than half of it; however, the uncertainties of the ^{26}Al yields are more important in the case of supernovae than in the case of WR.

Obviously, the radial source distributions discussed in the previous paragraph lead to different flux profiles on Earth as a function of galactic longitude (see Fig. 4). A relatively flat longitude profile is obtained from a radially flat source distribution; a steeper longitude profile with a pronounced peak in the direction of the galactic center is obtained if an important star formation is adopted for the inner Galaxy; and a longitude profile hollow in the direction of the galactic center with prominent features at $l \sim \pm 35°$ (corresponding to the molecular "ring at $r \sim$3.5 kpc) is obtained if the star formation is assumed to be negligible in the inner Galaxy (see Leising and Clayton 1985 and Prantzos and Cassé 1986, for early modelizations).

None of those profiles, obtained with an axisymmetric radial source distribution, corresponds to the observations of the galactic 1.8 MeV emission by COMPTEL (Diehl, this meeting) Indeed, these preliminary results seem to indicate some asymmetry between the first and fourth galactic quadrants, as well as an important "feature" at longitude $l \sim$-75°, i.e. in the direction of the Carina arm. Prantzos (1991, 1993) and Ramaty and Prantzos (1991) have suggested that this kind of profile would indeed be expected if the corresponding sources were confined inside the spiral arms of our Galaxy; obviously this concerns essentially the progeny of massive stars, i.e. SNII or SNIb and WR stars (and the most massive AGB stars also). Unfortunately, the poor knowledge of the spiral structure of our Galaxy (even the number of spiral arms is under debate) does not allow to make very accurate predictions of the resulting longitude profile and to compare to observations. Preliminary attempts (Prantzos 1991, 1993) adopted a 2-arm spiral pattern with logarithmic spirals and should therefore be considered only as indicative; they showed clearly, however, that an asymmetric longitude profile with several superimposed "spikes" (corresponding to the position of spiral arms) is the *generic feature* of an underlying source distribution with a spiral pattern.

In this work we adopt a more "realistic", 4-arm spiral pattern for our Galaxy (Fig. 5, from Taylor and Cordes 1992); it is based on observations of HII regions by Georgelin and Georgelin (1976) and takes into account more recent work by Downes (1980) and Caswell and Haynes (1987). We treat as a free parameter the interarm/arm emissivity ratio f, i.e. we allow for some ^{26}Al to be distributed outside the spiral arms. This allows to treat the inner Galaxy in the same way as the ISM outside the arms (of course, one may adopt a different treatment for the inner Galaxy). Obviously, with f=0., i.e. no sources in the inner Galaxy or between the arms, a longitude profile hollow around $l \sim$0° is obtained, with clearly detached spikes at the position of the

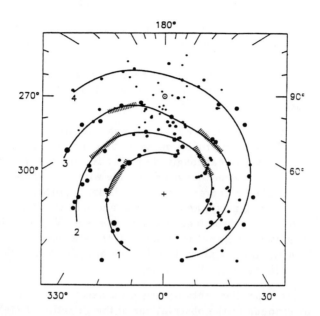

FIGURE 5: Spiral structure of our Galaxy according to Taylor and Cordes (1992). Enhanced emission is expected when the line of sight is tangent to the spiral arms.

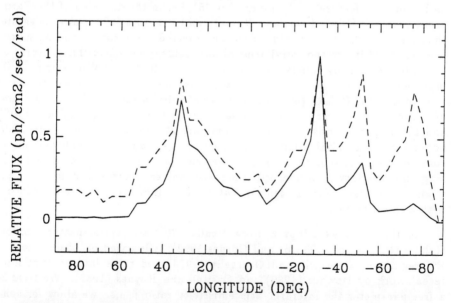

FIGURE 6: Same as Fig. 4, but this time the 4-arm spiral structure of Fig. 5 is superimposed on the radial distributions of Fig. 3. The interarm/arm intensity ratio is $f = 0$. i.e. no sources are assumed outside the spiral arms, neither in the inner Galaxy (the two H_2 distributions of Fig. 4 coincide in that case). Notice the feature at $l=-75°$ (Carina arm), prominent in the case of a flat radial distribution, but very weak in the case of the H_2 distribution.

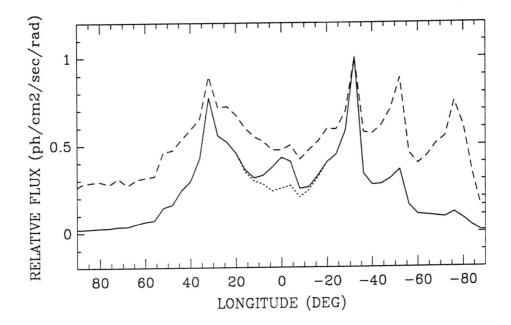

FIGURE 7: Same as Fig. 6, with $f = 0.1$.

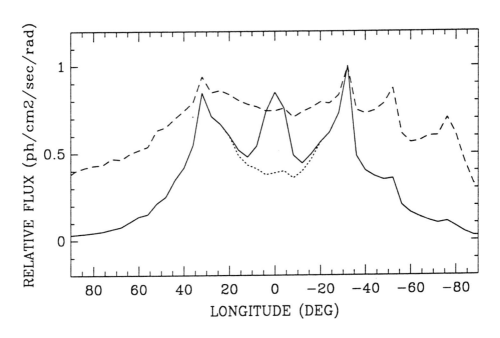

FIGURE 8: Same as Fig. 6, with $f = 0.3$.

spiral arms and especially at $l \sim -75°$, the direction of the Carina arm. Notice that the intensity of this latter feature is important in the case of a flat radial distribution, while it decreases considerably in the case of the H_2 distribution of Fig. 3. Notice also that in both cases the molecular ring at $l \sim \pm 35°$ is prominent. Increasing f to 0.1 and 0.3 leads to the longitude profiles of Fig. 7 and 8 respectively. Now, some star formation in the inner Galaxy is allowed and the "hole" at $l \sim 0°$ can be "filled"; this corresponds better to the COMPTEL results showing an important emission from the direction of the galactic center. The remarks concerning the feature at $l \sim -75°$ still hold, i.e. a radially flat distribution leads to an important emission from that direction (in agreement with COMPTEL) while a radially decreasing distribution gives a very weak emission from the Carina arm.

CONCLUSION

The detection of a galactic emission at 1.8 MeV boosted theoretical and observational work on the origin of the 2-3 $M_\odot$ of radioactive ^{26}Al currently present in the ISM.

After more than ten years of intense theoretical effort, considerable uncertainties still exist on the yields of the various candidate sources (novae, AGB stars, SNII and WR stars). Massive stars seem to be the most promising candidates on theoretical grounds, with SNII (+SNIb in case they don't collapse to a black hole) giving ~ 0.2 - 1.4 $M_\odot$ and WR stars giving ~ 0.2 - 0.5 $M_\odot$ of ^{26}Al per 10^6 years. This idea is supported by the recent COMPTEL results, indicating some asymmetry between the first and fourth galactic quadrants and an enhanced emission from the Carina arm. Indeed, a distribution of older and more numerous sources (like novae and AGB stars), not following the spiral arm pattern expected for massive stars, should give a relatively smooth and symmetric longitude profile; moreover, in that case a prominent contribution from the galactic bulge population of those sources (i.e. from the direction of the galactic centre) is also expected.

On the other hand, the relatively flat radial source profile indicated by the results of COMPTEL is somewhat puzzling; indeed, it is not compatible with a steeply decreasing radial source distribution, like that of H_2 clouds (presumed site of massive star formation). If this important result is confirmed, and if massive stars are really the main source of galactic ^{26}Al, some new idea would be required. One possibility is to assume again that massive stars are born inside giant molecular clouds, and that the ^{26}Al yield of SNII is indeed inversely proportional to metallicity; in that case the emissivity of the outer Galaxy is enhanced and a relatively flat emission profile results without any further assumption (Prantzos 1993, in preparation).

In any case, an accurate map of the galactic ^{26}Al emission, when finally obtained and properly unfolded, will give us one of the most direct estimates of the current sites and rate of formation of massive stars in the Galaxy.

ACKNOWLEDGMENTS: I warmly thank Michel Cassé and Stan Woosley for enlightening discussions; also Stan Woosley for allowing me to present some results of his recent work with T. Weaver.

REFERENCES

Bartunov O., Makarova J., Tsvetkov D. 1992, *A. A.*, **264**, 428
Cassé M., Prantzos N. 1993, *Origin and Evolution of the Elements*, Eds. N. Prantzos, E. Vangioni-Flam, M. Cassé (Cambridge University Press), in press
Caswell J., Haynes R. 1987, *A. A.*, **171**, 261
Clayton D. D. 1984, *Ap. J.*, **280**, 144
Clayton D. D., Leising M. D. 1987 *Phys. Rep.*, **144**, 1
Downes D., Wilson T., Bieging J., Wink J. 1980, *A. A. Sup.*, **40**, 379
Georgelin Y. M., Georgelin Y. P. 1976, *A. A.*, **49**, 57
Gusten R. 1989, *The Center of the Galaxy*, Ed. M. Morris (IAU Symp.), p. 89
Leising M. D., Clayton D. D. 1985, *Ap. J.*, **294**, 591
Li Z., Wheeler J. C., Bash F., Jefferys W. 1991, *Ap. J.*, **378**, 93
Maeder A. 1992, *A. A.*, **264**, 105
Mahoney W., Ling J., Wheaton W., Jacobson A. 1984, *Ap. J.*, **286**, 578
Meynet G., Arnould M. 1993, *Origin and Evolution of the Elements*, Eds. N. Prantzos, E. Vangioni-Flam, M. Cassé (Cambridge Univ. Press), in press
Prantzos N. 1991, *Gamma-Ray Line Astrophysics*, Eds. Ph. Durouchoux, N. Prantzos (AIP), p. 129
Prantzos N. 1993, *Advances in High Energy Astrophysics*, Ed. P. Mandrou, *A. A. Sup.*, in press
Prantzos N., Cassé M. 1986 *Ap. J.*, **307**, 324
Ramaty R., Prantzos N. 1991, *Comments on Astrophysics*, **XV**, 301
Schoenfelder V., Varendorf J. 1991, *Gamma-Ray Line Astrophysics*, Eds. Ph. Durouchoux, N. Prantzos (AIP), p. 101
Scoville N. Z., Sanders D. B. 1987, in *Interstellar Processes*, eds. H. Thronson and D. Hollenbach (Reidel), p. 21
Shaver P. A. et al. 1983, *M. N. R. A. S.*, **204**, 53
Signore M., Dupraz C. 1990, *A. A.*, **234**, L15
Starrfield S. et al. 1993, *Origin and Evolution of the Elements*, Eds. N. Prantzos, E. Vangioni-Flam, M. Cassé (Cambridge Univ. Press), in press
Taylor J. H., Cordes J. M. 1992, *Ap. J.*, submitted
Thieleman F.-K., Nomoto K., Hashimoto M. 1993, *Les Houches Lectures: Supernovae*, Ed. R. Mochkovitch, in press
Tutukov A., Yungelson L., Iben I. Jr. 1992, *Ap. J.*, **386**, 197
Weaver T. A., Woosley S. E. 1992, *Phys. Rep.*, in press
Woosley S. E. 1991, *Gamma-Ray Line Astrophysics*, Eds. Ph. Durouchoux, N. Prantzos (AIP), p. 270
Woosley S. E., Weaver T. A. 1980, *Ap. J.*, **238**, 1017
Woosley S. E., Hartmann D. H., Hoffman R. D., Haxton W. C. 1990, *Ap. J.*, **356**, 272

GALACTIC CHEMICAL EVOLUTION AND ^{26}AL PRODUCTION BY SUPERNOVAE

F. X. Timmes[1], S. E. Woosley[1,2], and Thomas A. Weaver[2]

[1]Board of Studies in Astronomy and Astrophysics, UCO/Lick Observatory, University of California at Santa Cruz, Santa Cruz, CA 95064

[2]General Studies Group, Physics Department, Lawrence Livermore National Laboratory, Livermore, CA 94550

Abstract. Nucleosynthesis in the intermediate mass range (carbon through nickel) has been calculated for a grid of stellar masses between 10 and 40 $M_\odot$ for solar metallicity, and 12 and 75 $M_\odot$ for zero metallicity, with a total of 26 stars evolved to the presupernova state. Explosions have been simulated in 13 of these and the final nucleosynthetic yields, including the ν-process, determined. Except for the products of the neutrino process (fluorine and boron), the presupernova abundances of isotopes lighter than about A = 40 (including ^{26}Al), closely resemble the final yields. These results, when incorporated into a multi-zone model for Galactic chemical evolution that includes contributions from lower mass stars and Type Ia supernovae, give present day abundances that are in good agreement with those observed in the solar system when a particular choice is made for the $^{12}C(\alpha, \gamma)^{16}O$ reaction rate (S(300 keV) = 0.17 MeV barns). We find the production rate of ^{26}Al in the present epoch to be 1.78 - 2.14 $M_\odot$ Myr^{-1}, depending on the choice and range of the initial mass function. This result, which is consistent with the *HEAO-3*, *SMM*, and so far the Comptel observations, suggests that the dominant source of ^{26}Al in the Galaxy is from the supernova of Types II and Ib.

I. The Input Stellar Nucleosynthesis

Massive stars of 12, 13, 15, 18, 20, 22, 25, 30, 35, 40, 50, and 75 $M_\odot$ were considered for the zero metallicity series. Initially they are assumed to have a Big Bang composition by mass : 76% H, 0.0091% ^{2}H, 0.0034% ^{3}He, 24% ^{4}He, and 8 × 10^{-8} % ^{7}Li. In addition, stars of 11, 12, 13, 15, 18, 19, 20, 21, 22, 25, 27, 30, 35, and 40 $M_\odot$ were considered

(Weaver & Woosley 1992) for the solar metallicity series, which were constructed to have an Anders & Grevesse (1989) composition. A Ledoux based convection criteria was used, with semiconvection incorporated in a simple parametric way. The stellar models used in the Galactic chemical evolution studies employ a nominal amount of semiconvective mixing. The adopted $^{12}C(\alpha,\gamma)^{16}O$ rate is 1.7 times the Caughlan & Fowler (1988) value. That is, 1.7 times 100 keV barns for the S-factor at 300 keV for all components of the cross section. Mass loss was not included in these models. However, mass loss should not significantly change our results. Explosion was simulated by a piston at a selected mass cut in 13 stars and the final nucleosynthetic yields, including the ν-process, determined. Stellar yields for values of a stellar mass and metallicity not in the grid are interpolated bilinearly. The Renzini & Voli (1981) case A tables have been used for the intermediate and low mass stellar yields, while the popular W7 model of Nomoto et al. (1984) is taken as representative of Type Ia nucleosynthesis.

II. Galactic Chemical Evolution

A simple dynamical model is used for the mass of the galaxy in both time (infall of primordial gas) and space (nuclear bulge + exponential disk). The dynamical model is similiar to the model suggested by Chiosi (1980). The instantaneous recycling approximation is not invoked in the chemical evolution equations, so that the full set of integro-differential equations that describe zone models of chemical evolution are solved. Details of the dynamical and isotopic evolution models may be found in Timmes, Woosley & Weaver (1993) and Woosley, Timmes & Weaver (1992).

Figure 1 shows the resulting nucleosynthesis from the presupernova models and the intermediate mass models in the solar vicinity during formation of the solar nebula, where the solar mass fractions are taken from Anders and Grevesse (1989). Explosive processing in massive stars and Type 1a supernova have been included in Figure 1. Isotopes of the same element are connected by lines and the most abundant isotope indicated by an asterisk. The horizontal dashed band is a range of a factor of two about the solar value and points found in the band are regarded as successes. ^{15}N is expected to originate from hot hydrogen burning, with environments typified by classical novae, while ^{10}B is probably made by spallation reactions in the interstellar medium, and both classical novae and spallation reactions are not included in our chemical evolution model.

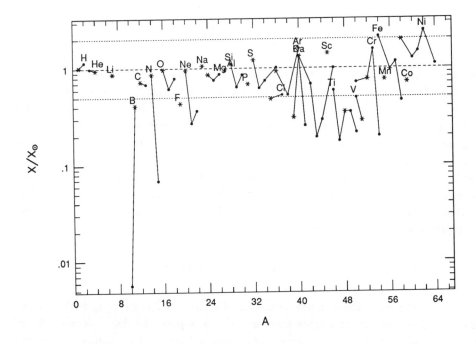

Fig. 1.— The computed isotopic composition of the Solar zone at the time when the Sun was born compared to the Anders and Grevesse (1988) mass fractions.

It is difficult to measure the supernova rates in our Galaxy because of the lack of quality historical observations and the difficulty of estimating the total Galactic blue luminosity. Estimates of the supernova rates for a subset of the spiral galaxies in the Shapley-Ames catalog were derived by van den Bergh, McClure & Evans (1987), van den Bergh (1988), and recently summarized by van den Bergh & Tammann (1991). Assuming a blue luminosity of 2.3×10^{10} $L_\odot$, and a Hubble constant of 75 km s^{-1} Mpc^{-1}, the resulting supernova rates from Table 11 of van den Bergh & Tammann (1991) are about 4 per century for Type II + Ib supernova varietals, and about 0.6 per century for Type Ia supernova. The temporal evolution of the supernova rates per century as given by our computation is given in Figure 2, and the present epoch supernova rates are in excellent accord with the above supernova rate estimates.

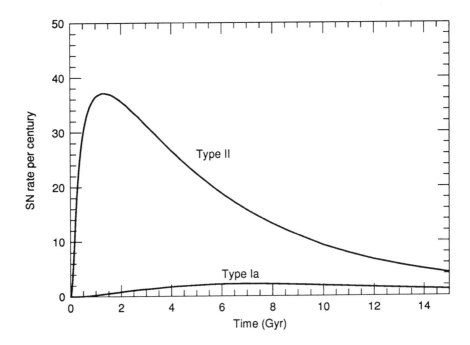

Fig. 2.— The Type II and Type Ia supernova rates per century as a function of time.

Figure 3 shows the surface mass density of ^{26}Al times $2\pi r$, in the present epoch, as a function of Galactic radius. This surface mass density curve is a reflection of our assumption about the spatial distribution of the gas in the Galaxy, and should not be interpreted as an a priori determination of the ^{26}Al distribution. Integration of the curve in Figure 3 then yields the total amount of ^{26}Al. Depending on the total mass of the Galaxy, and the upper limit of the initial mass function, the integrated values range from 1.7 to 2.2 $M_\odot$ Myr^{-1}. These values for the present injection rate of ^{26}Al are consistent with the Comptel observations reported at this meeting. Notice that ^{27}Al is not overproduced, as evidenced by Figure 1. Thus, it would appear quite plausible that the majority, if not all of the ^{26}Al in the Galaxy is attributable to massive stars.

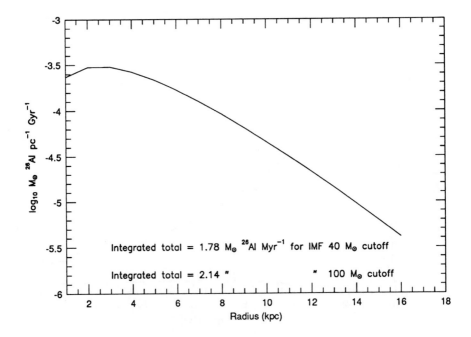

Fig. 3.— The radial distribution of Alumininum 26 in the present epoch.

The stellar models and chemical evolution models presented are in excellent agreement with almost all observations (both relative and absolute) of the Milky Way galaxy. These include the Anders & Grevesse mass fractions in the solar vicinity; the age-metallicity relationship; the time history of the CNO, alpha chain, and odd Z isotopes; the radial gradients present in the Galaxy; the present epoch Type II and Type Ia supernova rates in the Galaxy, and most of the inferred amount of aluminum 26 in the Galaxy.

This work has been supported by the Clemson NASA grant SC00680-74-328 and by the NASA grant NAGW 2525. It is a pleasure to thank Rob Hoffman for help in keeping our nuclear reaction network current, and valuable discussions of nucleosynthetic processes. We would also like to thank Ruth Peterson for valuable assistance and discussion concerning observational abundance determinations.

REFERENCES

Anders, E. & Grevesse, N. 1989, Geochim. Cosmochin. Acta, 53, 197
Chiosi, C. 1980, AA, 83, 206
Nomoto, K. Thielemann, F. K., & Yokoi, Y. 1984, ApJ, 286, 644
Renzini, A. & Voli, M. 1981, AA, 94, 175
Timmes, F. X., Woosley, S. E., & T. A. Weaver 1993, in Proceedings of the VI Advanced School of Astrophysics in São Paulo, Brazil, ed. Barbuy, B., Freitas Pacheco, J. A. & Janot-Pacheo, E. (São Paulo: IAGUSP), 000
van den Bergh, S., McClure, R. D. & Evans, R. 1987, ApJ, 323, 44
van den Bergh, S. 1988, Comments on Astrophysics, 7, 13
van den Bergh, S., & Tammann, G. A. 1991, in ARA&A Vol. 29, ed. G. Burbidge (Palo Alto: Annual Reviews), 363
Weaver, T. A., & Woosley, S. E. 1992, ApJ, in press
Woosley, S. E., Timmes, F. X., & T. A. Weaver 1992, in Nuclei in the Cosmos 1992, ed. Käppeler, F. (Berlin: Springer-Verlag), 000
Woosley, S. E., & Weaver, T. A. 1992, in Les Houches, Session LIV 1990: Supernovae, ed. J. Audouze, S. Bludman, R. Mochkovitch, J. Zinn-Justin (Elsevier Science Publishers: BV), 000

DIFFUSE GALACTIC LOW ENERGY GAMMA RAY CONTINUUM EMISSION

Jeff Skibo and Reuven Ramaty
Laboratory for High Energy Astrophysics
NASA/GSFC, Greenbelt, MD 20771

ABSTRACT

We investigate the origin of the diffuse low energy Galactic gamma ray continuum. We calculate gamma ray emission via bremsstrahlung and inverse Compton scattering by propagating an unbroken electron power law injection spectrum and employing a Galactic emissivity model for the distributions of gas and radiation. To maintain the low energy electron population capable of producing the observed continuum via bremsstrahlung, a total power input of $\sim 10^{41}$ erg s^{-1} is required. This exceeds earlier estimates of the total power supplied to the nuclear cosmic rays by about an order of magnitude. These electrons could be the agent responsible for energetically maintaining the warm ($\sim 10^4$ K) ionized component of the interstellar medium.

INTRODUCTION

Gamma ray continuum observations from the direction of the Galactic Center span a broad range of energies from about 0.1 MeV to over a GeV. Above about 10 MeV a large fraction of the observed emission results from cosmic ray interactions with interstellar gas and radiation (pion production, bremsstrahlung, inverse Compton scattering). At lower energies the origin of the gamma ray continuum is less certain. An analysis of the observations made with the balloon borne Ge detector GRIS from the direction of the Galactic center and a direction in the plane away from the Galactic center (l=335°) suggests that the emission below 1 MeV is of diffuse origin and has a relatively broad longitude distribution over the central radian of the Galaxy[1]. Furthermore, hard X-ray observations from the direction of the Galactic Center with detectors of moderate fields of view (15° to 30°) show[2] that, even though the observed fluxes vary in time, they have a lower envelope which coincides with the flux observed with GRIS from $l = 335°$. However, between about 0.5 and 3 MeV, observations[3] with HEAO-3 have indicated that the observed flux may be time variable, suggesting contributions from unidentified discrete sources. In addition, recent measurements made with COMPTEL[4,5] show that the diffuse emission around 1 MeV is somewhat lower than that previously estimated[6] from an ensemble of satellite and balloon measurements.

The most likely mechanisms for producing the diffuse low energy gamma ray emission are electron bremsstrahlung and inverse Compton scattering. As in our previous paper[7] we derive a cosmic ray electron spectrum by propagating in the Galaxy an unbroken power law injection spectrum and show that the resultant equilibrium spectrum is consistent with the high energy electron observations (which are free from modulation effects) and radio observations from the direction of the Galactic pole. We then expand upon our previous analysis by modeling the Galactic distributions of the atomic, molecular and ionized gas, of the infrared and optical radiation, and of the cosmic ray electrons using recent data.

ANALYSIS

We assume a primary component of electrons whose source function consists of a single power law in kinetic energy over the entire range from 10^{-1} to 10^6 MeV. This primary component constitutes the bulk of the relativistic electrons in the interstellar medium. In addition we include secondary electrons and positrons[8] produced, via pion decay and Coulomb upscatter, in cosmic ray interactions with interstellar gas.

We take the cosmic ray escape path to be constant at 6 g cm^{-2} below 3 GeV and decreasing as $E^{-0.49}$ at higher energies[8]. This escape path and the cosmic ray lifetime[9] of 8.8×10^6 yr below a GeV imply a mean hydrogen density along a typical electron trajectory of 0.3 cm^{-3}. Assuming a scale height of 500 pc for the electrons and normalizing to the local > 30 GeV electron measurements[10,11], we find that a single injection spectral index of 2.42 leads to an equilibrium spectrum consistent with both the electron observations above 3 GeV and the polar 1 - 100 MHz synchrotron radio spectrum[12]. This requires a Galactic magnetic field of 9 μG, which is also the value that we use in the electron transport. The resultant electron spectrum is shown by the solid curve in Figure 1 along with other calculations[13,14].

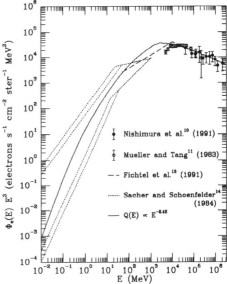

Fig. 1. The local Galactic electron spectrum.

We use a three component model for the spatial distribution of interstellar gas. We adopt a model[15] for the distribution of atomic gas derived from 21 cm observations. For the molecular gas we use a model[16,17] derived from the CO data[18]. In addition to the atomic and molecular components we include, as in our previous study[19], the distribution of warm partially ionized gas ($\sim 10^4$ K) derived from pulsar dispersion measures and optical emission lines[20,21,22,23]. This gas has a much larger scale height than the HI or H$_2$ components and, therefore, could contribute significantly to electron bremsstrahlung at high Galactic latitudes. We divide the ambient interstellar photons into five components: the 2.7°K cosmic microwave background, the infrared emission from Galactic dust and three diluted blackbody optical components having temperatures 3000°K, 4000°K and 7500°K. We derived the infrared energy density distribution from the 60 μm and 100 μm IRAS data[24] and the optical energy density distributions from Galactic emissivity and dust opacity distributions[25].

To model the cosmic ray distributions in the Galaxy, we assume galactocentric radial gradients of the form $\exp[\alpha(r - R_\odot)]$, with $\alpha = 0.026$ kpc^{-1} for the nuclear component and $\alpha = 0.2078$ kpc^{-1} for the electrons[26].

We calculate the bremsstrahlung in both the neutral and ionized components using the available cross sections.[27,28,29] For the inverse Compton calculation we use the full Klein-Nishina cross section. We also include the contribution of π° decay using the previously calculated emissivity per H atom[30]. Using the models for the target media and cosmic rays given above, we calculate the gamma ray fluxes by computing line of site integrals through the Galaxy. The results, integrated over all latitudes and the central radian in longitude, are shown in Figure 2. Here, in addition to the total bremsstrahlung, π° decay and inverse Compton contributions, we also show the inverse Compton components from the various target radiations. The contributions of the atomic H, molecular H and warm ionized gas are 31%, 51% and 18% for the bremsstrahlung, and 32%, 43% and 25% for the π° decay.

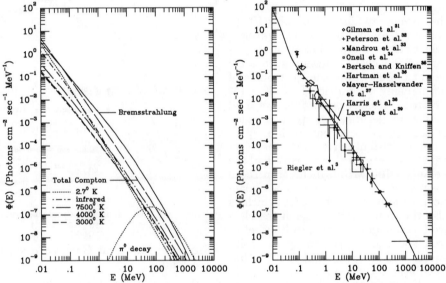

Fig. 2. The various components of the calculated flux from the central radian of the Galaxy.

Fig. 3. The calculated total central radian flux compared with the data.

In Figure 3 we compare these calculations with data. The calculated curve represents the total bremsstrahlung, inverse Compton and π^0 components, to which we have added the contribution of thermal bremsstrahlung from the Galactic ridge hot plasma and the orthopositronium continuum resulting from positron annihilation in the interstellar medium. For the hot plasma we extrapolate into the hard X-ray range a thermal bremsstrahlung spectrum with $kT = 10$ keV and derive a spatial model for this emission[19] from the Galactic plane data[40] normalized to the Galactic center observations[41]. We obtain the orthopositronium continuum directly from measurements with SMM[38]. It is evident that our calculated curve fits the data reasonably well over many decades of energy. However, at low energies there are some discrepancies,[31,33] perhaps due to the presence of hard X-ray sources in the Galactic center region. Also, our calculated continuum runs lower than the continuum measured with SMM by about 1 σ. However, preliminary reports[4,5] indicate that this is in accordance with the continuum measured with COMPTEL.

CONCLUSION

A very large power is required to maintain the population of low energy electrons if the low energy gamma ray continuum is of diffuse origin. This situation is similar to that in solar flares, where the observed hard X-ray emission is known to be nonthermal electron bremsstrahlung and where the energy contained in the 0.01 to 0.1 MeV electrons is comparable to the total flare energy[42]. For our assumed injection spectrum, we obtain a local power input of 1.0×10^{-26} erg s^{-1} cm^{-3} due to electrons of energies greater than 0.05 MeV. The corresponding local ionization rate is $\zeta = 6.0 \times 10^{-16}$ s^{-1} for a density of 0.3 cm^{-3}. This value is not inconsistent with recent estimates[43]. That is, the electrons responsible for the bremsstrahlung will not over ionize the interstellar medium.

The electrons responsible for the production of the low energy diffuse gamma ray continuum could provide the ionization and heating required to maintain the warm ($T = 10^4$) ionized component of the interstellar medium[44]. The power required to maintain this ionized gas is 7×10^{-27} to 7×10^{-25} erg s^{-1} cm^{-3}, depending on the temperature of the gas[45]. Our calculated local power input of 1.0×10^{-26} erg s^{-1} cm^{-3} falls within this range. We calculate the total power input into the Galaxy. The result, 1.6×10^{41} erg s^{-1}, exceeds earlier estimates of the total power supplied to the nuclear cosmic rays by about an order of magnitude.

From a dissertation to be submitted to the graduate school, University of Maryland, by J. G. Skibo in partial fulfillment of the requirements for the Ph.D. degree in physics.

REFERENCES

1. N. Gehrels, S. D. Barthelmy, B. J. Teegarden, J. Tueller, M. Leventhal and C. J. MacCallum, Ap. J., 375, L13 (1991).
2. N. Gehrels and J. Tueller, in The Compton Observatory Science Workshop, ed. C. R. Shrader, N. Gehrels and B. Dennis (NASA Conf. Publ. 3137, 1992), p. 446.
3. G. R. Riegler, J. C. Ling, W. A. Mahoney, W. A. Wheaton and A. S. Jacobson, Ap. J.(Letters) 294, L13 (1985).
4. H. Bloemen et al., these proceedings (1993)
5. A. W. Strong et al., these proceedings (1993)
6. N. Gehrels and J. Tueller, Ap. J., in press (1992)
7. J. G. Skibo and R. Ramaty, Astr. and Ap., in press (1992)
8. R. Ramaty and J. Westergaard, Ap. and Sp. Sci., 45, 143 (1976).
9. M. M. Shapiro, J. R. Letaw, R. Silberberg and C. H. Tsao, 22nd Internat. Cosmic Ray Conf. Papers. 2, 304 (1991).
10. L. Nishimura, M. Fujii, T. Kobayashi, H. Aizu, Y. Komori, M. Kazuno and T. Taira, 21st Internat. Cosmic Ray Conf. Papers. 3, 213 (1991).

11. D. Müller and K. Tang, Ap. J., 312, 183 (1987).
12. W. R. Webber, G. A. Simpson and H. V. Cane, Ap. J., 236, 448 (1980).
13. C. E. Fichtel, M. E. Özel, R. G. Stone and P. Sreekumar, Ap. J., 374, 134 (1991).
14. W. Sacher and V. Schönfelder, Ap. J., 279, 817 (1984).
15. M. A. Gordon and W. B. Burton, Ap. J., 208, 346 (1976).
16. L. Bronfman et al., Ap. J., 324, 248 (1988).
17. A. W. Strong et al., Astr. and Ap., 207, 1 (1988).
18. T. M. Dame et al., Ap. J., 322, 706 (1987).
19. J. G. Skibo, R. Ramaty and M. Leventhal, Ap. J., 397, 135 (1992).
20. A. G. Lyne, R. N. Manchester and J. H. Taylor, MNRAS, 213, 613 (1985).
21. R. J. Reynolds, The Interstellar Disk-Halo Connection in Galaxies, ed. H. Bloemen (IAU Symp. 144, 1991), p. 67.
22. D. A. Frail, J. M. Cordes, T. H. Hankins and J. M. Weisberg, Ap. J., 382, 168 (1991).
23. J. M. Cordes, J. M. Weisberg, D. A. Frail, S. R. Spangler and M. Ryan, Nature, 354, 121 (1991).
24. J. C. Good and N. Z. Scoville, private communication (1989)
25. J. S. Mathis, P. G. Metzger and N. Panagia, Astr. and Ap., 128, 212 (1983).
26. J. B. G. M. Bloemen et al., Astr. and Ap., 154, 25 (1986).
27. H. W. Koch and J.W. Motz, Rev. Mod. Phys., 31, 920 (1959).
28. G. R. Blumenthal and R. J. Gould, Rev. Mod. Phys., 42, 237 (1970).
29. E. Haug, Zs. Naturforsch., 30a, 1099 (1975).
30. C. D. Dermer, Astr. and Ap., 157, 223 (1986).
31. D. Gilman, A. E. Metzger, R. H. Parker and J. I. Trombka, NASA Tech. Memo No. 76619, ed. T. L. Cline, R. Ramaty (1978), p. 90.
32. L. E. Peterson, D. E. Gruber, G. V. Jung and J. L. Matteson, 21st Internat. Cosmic Ray Conf. Papers. 1, 44 (1990).
33. P. Mandrou, A. Bui-Van, G. Vedrenne and M. Niel, Ap. J., 237, 431 (1980).
34. T. O'Neill, B. Dayton, J. Long, E. Zanrosso, A. Zych and R. S. White, 18th Internat. Cosmic Ray Conf. Papers. 9, 45 (1983).
35. D. L. Bertsch and D. A. Kniffen, Ap. J., 270, 305 (1983).
36. R. C. Hartman, D. A. Kniffen, D. J. Thomson, C. E. Fichtel, H. B. Ogelman, T. Turner and M. E. Özel, Ap. J., 230, 597 (1979).
37. H. A. Mayer-Hasselwander et al., Astr. and Ap., 105, 164 (1982).
38. M. J. Harris, G. H. Share, M. D. Leising, R. L. Kinzer and D. C. Messina, Ap. J., 362, 135 (1990).
39. J. M. Lavigne, P. Mandrou, M. Niel, B. Agrinier, E. Bonfand and B. Parlier, Ap. J., 308, 370 (1986).
40. K. Koyama, H. Awaki, H. Kunieda, S. Takano, Y. Tawara, S. Yamauchi, I. Hatsukade, F. Nagase, Nature, 339, 603 (1989).
41. S. Yamauchi, M. Kawada, K. Koyama, H. Kunieda, Y. Tawara and I. Hatsukade, Ap. J., 365, 532 (1990).
42. B. R. Dennis, Solar Physics118, 49 (1988).
43. R. J. Reynolds, private communication (1992)
44. X. Chi and A. W. Wolfendale, The Interstellar Disk-Halo Connection in Galaxies, ed. H. Bloemen (IAU Symp. 144, 1991), p. 197.
45. R. J. Reynolds, Ap. J.(Letters) 349, L17 (1990).

GALACTIC ANNIHILATION RADIATION AND THE GALACTIC NUCLEOSYNTHESIS RATE

Kai-Wing Chan
Laboratory for High Energy Astrophysics,
Goddard Space Flight Center, Greenbelt, MD 20771

Richard E. Lingenfelter
Center for Astrophysics and Space Sciences, C-0111,
University of California, San Diego, La Jolla, CA 92093

Reuven Ramaty
Laboratory for High Energy Astrophysics,
Goddard Space Flight Center, Greenbelt, MD 20771

ABSTRACT

We discuss here recent calculations of the survival of positrons in supernova ejecta, expected from current supernova models. These calculations show that positrons from the β^+-decay of nucleosynthetic ^{56}Co$\to^{56}$Fe, ^{44}Sc$\to^{44}$Ca and ^{26}Al$\to^{26}$Mg, can easily account for the observed diffuse galactic annihilation radiation, and they suggest a present galactic rate of ^{56}Fe nucleosynthesis of $\sim (0.8 \pm 0.6) M_\odot$ per 100 years.

INTRODUCTION

Observations of positron annihilation radiation from the Galactic Center region indicate that there are two components of the radiations: a steady, diffuse galactic disk component and a variable component from discrete sources. The existence of a diffuse galactic positron annihilation radiation was first inferred from the correlation between the line fluxes and the fields of view of different detectors; and has more recently been demonstrated by the GRIS[1] and OSSE[2] observations which measured the annihilation radiation in the galactic plane at different longitudes.

Potential sources for these galactic positrons have long been suggested[3-7] to be the β^+-decay of ^{56}Co, ^{44}Sc and ^{26}Al produced by various processes of galactic nucleosynthesis, primarily in supernovae. Here we discuss recent calculations that show[8,9] that a fraction of these positrons can survive the supernova expansion into the interstellar medium. There the slowing down and annihilation lifetime of positrons is $\sim 10^5$ years which is much longer than the average supernovae occurrence time of $\sim 10^2$ years. Thus, positrons from thousands of supernovae throughout the Galaxy could combine to produce the observed steady diffuse emission. The observed 511 keV line flux in turn place constraints on the present rate of galactic ^{56}Fe nucleosynthesis.

POSITRONS FROM GALACTIC NUCLEOSYNTHESIS

Comparison with various expected spatial distributions of galactic positron annihilation radiation shows[2,10] that the observed distribution is most consistent with that of galactic novae[11,12], which are thought to be the progenitors of Type Ia supernovae. Although a significant contribution from variable discrete sources cannot be ruled out, if all of the line flux observed from the Galactic Center direction is diffuse, the observed annihilation line fluxes would require[10] in steady state a total galactic positron production rate of $(1.6 \pm 0.5) \times 10^{43}$ e^+ s^{-1}.

Part or all of these galactic positrons should come from the β^+-decay of nucleosynthetic ^{56}Ni$\to^{56}$Co$\to^{56}$Fe, ^{44}Ti$\to^{44}$Sc$\to^{44}$Ca and ^{26}Al$\to^{26}$Mg. Existence of the diffuse galactic gamma ray line at 1.809 MeV from ^{26}Al→^{26}Mg decay shows that accompanying positrons are also produced in the same radioactivity. Measurements[13,14] of the diffuse 1.809 MeV line flux, however, limit[10] the contribution of the positrons from ^{26}Al decay to about $13 \pm 5\%$ of the total annihilating galactic disk positrons, depending on the assumed galactic distribution.

For the decay of ^{56}Co and ^{44}Sc, the expected galactic positron production rate by these nucleosynthesis sources can simply be expressed in terms of the present galactic rate of ^{56}Fe nucleosynthesis by supernovae and the fraction of positrons that escape. Positrons are produced in 95% of the decays of ^{44}Ti→^{44}Sc→^{44}Ca, which has a mean life of 78 years. Essentially all of these positrons should escape from the supernovae since the ejecta should become transparent to the positrons on a time scale much less than the decay mean life. It is generally assumed that the galactic ^{44}Ca all come from ^{44}Ti→^{44}Sc→^{44}Ca decay and that the average galactic ^{44}Ca to ^{56}Fe mass ratio, X_{44}/X_{56}, is 1.23×10^{-3}, equal to that in the solar neighborhood[15]. Thus, the galactic production rate from ^{44}Ti decay can be written as $Q_{44} = 1.0 f_{44} \dot{M}_{56}$, where $\dot{M}_{56}$ is the present rate of galactic ^{56}Fe nucleosynthesis in $M_\odot/100$ years; and f_{44}, generally equal to unity, is the mean escape fraction of positrons from ^{44}Sc decay, and Q_{44} is in units of 10^{43} e$^+$ s^{-1}.

Because of the much shorter ^{56}Co decay mean life of 111.4 days, most of the positrons, produced in 19% of the ^{56}Co decays, lose their energy and annihilate inside the supernova ejecta before it becomes thin enough for them to escape or rarefied enough for them to survive. The contribution of ^{56}Co to the galactic positrons therefore depends strongly on the survival fraction f_{56}. The positron production rate from ^{56}Co decay is $Q_{56} = 130 f_{56} \dot{M}_{56}$, where the important factor f_{56} has to be determined.

SURVIVAL OF ^{56}Co AND ^{44}Sc POSITRONS IN SUPERNOVAE

The probability that a positron survives in a supernova ejecta depends both on the local density where it is produced and on its initial energy. A positron emitted at a time and place deep in the ejecta where the local density is high can lose all of its kinetic energy and thermalize in the ejecta, subsequently depending on the ejecta temperature and density, it may annihilate in the ejecta, or it may survive annihilation, if the continuously decreasing density of the ejecta has already become low enough. On the other hand, if a positron is emitted at shallower depths or at later times, such that it cannot completely slow down before the local medium becomes too rarefied, or if it propagates farther out to such a region, then it may escape from the ejecta. The probability that a positron escapes from a supernova also depends on how it is transported in the ejecta. In a non-magnetized medium, the trajectory is essentially a straight path. However, even a small magnetic field in the ejecta will have a significant effect on the transport of the charged positrons.

To better explore these possibilities, detailed calculations have been made[8,9] of the escape and survival fractions of positrons from ^{56}Co and ^{56}Sc decay, in current supernova models, considering the effects of magnetic fields, turbulence and possible mixing in the ejecta. In particular, for Type Ia supernovae, calculations were made for both the carbon deflagration model W7 of Nomoto et al.[16] for accreting carbon-oxygen white dwarfs in binary systems and the delayed detonation model of Woosley and Weaver[17] for such stars, both of which seem to be able to produce the observed supernova abundances and spectra. For Type Ib supernovae, calculations were made for the core-bounce models 4B and 6C of Ensman and Woosley[18] for Wolf-Rayet stars;

for Type Ip the detonation model 4 of Woosley et al.[19] for helium dwarfs; and for Type II the core-bounce model 15A of Weaver and Woosley[20] for massive stars.

We also consider possible mixing of ^{56}Ni into the outer parts of the ejecta, which would enhance the probability of positron survival. To obtain a rough upper bound on the effects of such mixing, we have calculated the positron survival fractions for each of these models after artificially mixing the ^{56}Ni uniformly throughout the ejecta, which is clearly an extreme example. Two limits of positron propagation within the supernova ejecta are also considered. For the lower limit on the survival fraction, we assume that the magnetic field within the ejecta is so tangled that the positrons are trapped locally with a small diffusion mean free path $(l = 0)$; while for the upper limit, we assume that the magnetic field in the supernova ejecta is 'combed out' radially by the differential expansion of the ejecta, whose kinetic energy density much exceeds that of the field, as first suggested by Colgate et al.[21,22], thus giving the positrons a very long diffusion mean free path ($l = \infty$). The survival fractions calculated[9] are shown in Table 1, and the positron survival probabilities as a function of interior mass for the Type Ia models are shown in Figure 1.

Table 1. Survival Fractions of Supernova e^+

Supernova Type	^{56}Co e^+ Survival f_{56} (%)			
	$l = 0$		$l = \infty$	
	unmix.	mix.	unmix.	mix.
Ia (DDT) †	0.48	4.7	8.2	15
Ia (DFN) †	0.08	2.5	5.2	13
Ip	8.6		27	
Ib/c	0.0	1.6	0.0	5.8
II	0.0	0.0	0.0	0.7

† Positron survival fractions from SN Ia are calculated for the delayed detonation (DDT) and deflagration (DFN) models.

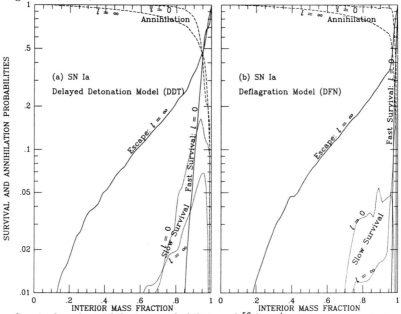

Fig. 1. Survival and annihilation probabilities of ^{56}Co e^+ in supernova ejecta as a function of interior mass fraction for two Type Ia models[16,17]. Probabilities of thermal and nonthermal survival, and annihilation for the two limiting case of e^+ propagation are shown at the coordinate where e^+ are produced.

PRESENT RATE OF GALACTIC IRON NUCLEOSYNTHESIS

With the survival fractions of positrons from ^{56}Co calculated above, for various types of supernovae, a weighted mean value of the survival fraction appropriate to the overall galactic rate of ^{56}Fe nucleosynthesis can be calculated, $f\dot{M}_{56} = (f\eta M)_{Ia} + (f\eta M)_{Ib} + (f\eta M)_{Ip} + (f\eta M)_{II}$, where η is the galactic occurrence rate and M the mean ^{56}Ni mass per supernova of each type of supernova. Similarly the overall galactic rate of ^{56}Fe nucleosynthesis can be estimated as $\dot{M}_{56} = (\eta M)_{Ia} + (\eta M)_{Ib} + (\eta M)_{Ip} + (\eta M)_{II}$.

The galactic rates, η, of Type Ia, Ib and II supernovae have been estimated by van den Bergh and Tammann[23] at $1.1h^2$, $1.2h^2$ and $6.1h^2$ per 100 years, where h is the Hubble constant in 100 km s^{-1} Mpc^{-1}. The relative rate of occurrence of the peculiar Type Ip supernova has been estimated[24] to be about $13 \pm 7\%$ of all Type I supernovae, or a rate of $0.3h^2$. However, there is considerable uncertainty here. In fact, if we adopt the helium dwarf detonation model for such supernovae, as proposed by Woosley and Pinto[5], such a rate would slightly overproduce ^{44}Ca relative to ^{56}Fe. From this consideration, an upper limit of $0.26h^2$ can be set on Type Ip supernovae that result from He dwarf detonations, and we also adopt an effective 2σ lower limit[9] at $0.08h^2$.

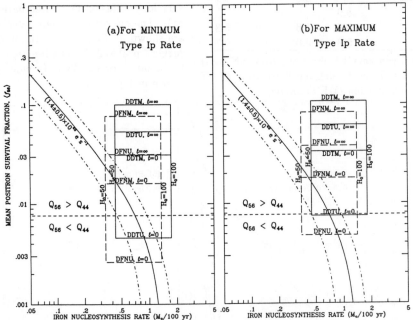

Fig. 2. Constraints on the present rate of galactic ^{56}Fe nucleosynthesis. Constraints from the observation of galactic diffuse 511 keV emission are shown by the curves. Calculations of the present rate of galactic ^{56}Fe nucleosynthesis and the mean survival fraction of ^{56}Co e$^+$ are shown by the boxes, which cover a range of ^{56}Ni mixing, positron propagation, Hubble constant ($0.5 \leq h \leq 1.0$) and Type Ip occurrence rates, for different models of Type Ia supernovae.

The total galactic positron production rate Q is an upper bound on the nucleosynthetic positron production rate Q_{nuc}, so that $Q \geq Q_{nuc} = Q_{26} + Q_{44} + Q_{56}$. With the rate Q_{26} determined directly from observation and an effective f_{44} of unity (see reference 9, however), the inequality above becomes $Q - Q_{26} = 1.4 \pm 0.5 \geq (1.0 + 130 f_{56}) \dot{M}_{56}$, where f_{56} and $\dot{M}_{56}$ are the weighted mean values computed as discussed above. The constraints that this relationship sets on the positron survival fraction and the iron nucleosynthesis rate are shown in Figure 2.

We see that within the range of current supernova models and supernova rates, positrons from the decay of nucleosynthetic ^{56}Co, ^{44}Ti and ^{26}Al can fully account for the observed diffuse galactic annihilation radiation. We also see that the extreme upper limits of uniform mixing and positron streaming are effectively ruled out by the observations. On the other hand, even the minimum positron survival fractions, based on the most conservative assumptions of negligible ejecta mixing and local trapping of the positrons can very easily account for the observed diffuse galactic annihilation radiation, for either the deflagration or delayed detonation models of Type Ia supernovae and for any Type Ip rate between the limits set. For these minimum conditions the present galactic rate of ^{56}Fe nucleosynthesis is $(0.8 \pm 0.5) M_\odot$ per 100 years.

Acknowledgments. The work of R.E.L was supported under NASA Grant NAGW-1970 and K.W.C. is an NRC Associate.

REFERENCES

1. N. Gehrels, et al. Ap. J., 375, L13 (1991).
2. W.A. Purcell, et al. in Compton Observatory Science Workshop, eds. C. Shrader, N. Gehrels & B. Dennis, (Washington: NASA) p.431 (1992).
3. D.D. Clayton, Nature Phys. Sci., 224, 137 (1973).
4. R. Ramaty and R.E. Lingenfelter, Nature, 278, 127 (1979).
5. S.E. Woosley and P.A. Pinto, in Nuclear Spectroscopy of Astrophysical Sources, eds. N. Gehrels and G.H. Share (New York: AIP) p.98 (1988).
6. M. Signore and G. Vedrenne, Astron. Ap., 201, 379 (1988).
7. R.E. Lingenfelter and R. Ramaty, Nucl. Phys. B (Supp.), 10B, 67 (1989).
8. K.W. Chan and R.E. Lingenfelter, 21st Int. Cosmic Ray Conf., 3, 253 (1991).
9. K.W. Chan and R.E. Lingenfelter, Ap. J., 405, in press (1993).
10. J.G. Skibo, R. Ramaty and M. Leventhal, Ap. J., 397, 135 (1992).
11. J.C. Higdon and W.A. Fowler, Ap. J., 317, 710 (1987).
12. M. Leising and D.D. Clayton, Ap. J., 294, 591 (1985).
13. W.A. Mahoney, et al., Ap. J., 286, 578 (1984).
14. B.J. Teegarden et al., Ap. J., 375, L9 (1991).
15. A.G.W. Cameron, in Essays in Nuclear Astrophysics, ed. C. Barnes, D. Clayton, and D. Schramm (Cambridge: CUP), p.23 (1982).
16. K. Nomoto, et al., Ap. J., 286, 644 (1984).
17. S.E. Woosley and T.A. Weaver, in Supernovae, eds. J. Audouze, et al. (New York: Elsevier), (1992).
18. L.M. Ensman and S.E. Woosley, Ap. J., 333, 754 (1988).
19. S.E. Woosley, et al., Ap. J., 301, 601 (1986).
20. T.A. Weaver and S.E. Woosley, in Supernova Spectra, eds. R. Meyerott and G.H. Gillespie (New York: AIP) p.15 (1980).
21. S.A. Colgate, Ap. Sp. Sci., 8, 457 (1970).
22. S.A. Colgate, et al., Ap. J., 237, L81 (1980).
23. S. van den Bergh and G. Tammann, Ann. Rev. Astron. Ap. 29, 363 (1991).
24. D. Branch, Ap. J., 300, L51 (1986).

γ-Ray Line Emission From AGB Star Ejecta

Grant Bazan
McDonald Observatory, Austin, TX 78712

ABSTRACT

We examine the production of radionuclides via the s-process in asymptotic giant branch (AGB) stars. Schematic models of AGB star evolution are combined with four neutron source scenarios to determine radionuclide abundances in the atmospheres and winds of $1 - 4$ $M_\odot$ AGB stars of solar metallicity. We show that the maximum γ-ray line fluxes from AGB stellar winds in any neutron source scenario are below the current observational sensitivity of the Compton-GRO satellite. In addition, we calculate the total mass yields of long-lived radionuclides for use in one-zone galactic chemical evolution models. Depending on the neutron source scenario, we determine that AGB stars should be considerable sources of ^{40}K and ^{60}Fe in the galaxy, and, in two separate neutron source scenarios, the resulting γ-ray line fluxes may be visible by COMPTEL.

INTRODUCTION

AGB stars are known both observationally and theoretically to be the site of the s-process. From successive captures of neutrons and β-decay or e⁻ capture by light elements, mostly Fe-group nuclei, heavier nuclei are produced in the interiors of AGB stars during thermal pulses. Among these nuclei are several γ-ray-emitting radionuclei decay lifetimes greater than 1 year (see Table 1). With the observations corroborating most of the theoretical results, there seems to be an understanding of the transition from M giants to S and C stars.[1]

Despite the controversy over which neutron source model is the appropriate representation of reality, it must be seen whether or not significant quantities of γ-ray-emitting radionuclides can be made in any of the models. Table 1 shows a list of nuclei likely to be enhanced in our models with their halflives and three strongest γ-ray lines with relative intensities. We look at four neutron sources in schematic AGB evolution models and calculate the total yields for $1 - 4$ $M_\odot$ AGB stars of solar metallicity. We then incorporate one-zone chemical evolution models to estimate the galactic content of long-lived radionuclides and their observability.

CALCULATIONS

Schematic models of AGB evolution are based on earlier work[2] where the results of several independent studies of AGB evolution are cast into analytical formulae. A stellar envelope code is used in conjunction to estimate surface temperature, mass loss, and pulsational period properties. 632 nuclei from ^{4}He to ^{210}Po are evolved via a chosen neutron source in successive thermal pulses until mass loss depletes the envelope to 0.2 M$_\odot$, the average mass of planetary nebulae. Along the way, dredge-up mixes this material with the surface and mass loss establishes an enriched stellar wind.

We examine four separate neutron source models. The first one involves the ^{22}Ne rate at its lower limit based on experimentally observed resonances. The second scenario is where the ^{22}Ne reaction is set at its upper limit based on low-energy resonances not seen in direct capture experiments, but inferred through neutron capture experiments on ^{25}Mg.[3] The third model is a derivative of the ^{13}C model found in a model of a low mass, low metallicity AGB star.[4] The final model is that of the 'classical' model[5] with parameters set to achieve the observed relative distribution of s-only isotopes.[6] Instead of evolving the nuclei according to the stellar conditions in this final scenario, we calculate the abundances beforehand using a separate nucleosynthesis code and dredge-up and lose mass according to the schematic evolution model involving the lower limit ^{22}Ne rate.

With our mass loss prescription, we examine whether the abundances of the radionuclei are sufficient to establish the ejecta of these stars as point sources of line emission. The intensities are determined by the mass of a particular radionuclide, its decay rate, and the distance to the circumstellar ejecta, which is set to 100 pc. At the end of the stars' evolution, the wind abundances are studied for concentrations of γ-emitting nuclei that could impact observations on a galactic scale. Once we have these yields, we employ them in a galactic chemical evolution model.[7]

Once a local mass density and emissivity are determined, we must adopt a galactic distribution of the nuclei. Because of the short decay timescales (compared to galactic timescales), these nuclei will be distributed as their stellar sources. We employ the IR flux distribution of the galaxy as the observed AGB star distribution in the galaxy.[8] Integrating this function with the emissivity normalized to our position at 8.5 kpc results in $dI/d\Omega$ at selected values of galactic longitude and latitude. Intensity measurements by Compton-GRO are estimated by integrating $dI/d\Omega$ over the instrument seeing function, which we assume to be Gaussian.

RESULTS

Figure 1 shows the maximum intensities for short-lived radionuclides in AGB stellar winds of initial masses 1.5, 2, 3, and 4 M$_\odot$. Only results for the 'classical' model have been examined. The nuclei are, specifically, ^{60}Co (solid line), ^{85}Kr (dotted line), ^{134}Cs (dashed line), ^{137}Cs (long dashed line), ^{154}Eu (dot-dashed line), and ^{155}Eu (dot-long dashed line). The dominant γ-ray sources, ^{60}Co and ^{85}Kr, are seen to be two orders of magnitude below OSSE sensitivity ($\sim 10^5$ cm^{-2} s^{-1}) and one order of magnitude below COMPTEL sensitivity ($\sim 3 \times 10^6$ cm^{-2} s^{-1}); thus we feel AGB star winds or circumstellar shells are not likely places for point source emission from nuclei made in the s-process.

Figure 2 shows the total yields for abundant long-lived nuclei in product with an initial mass function.[9] All four neutron source models are considered. In our calculations, we find the most abundant, long-lived, γ-ray-emitting nuclei to be ^{40}K, ^{60}Fe, ^{129}I, and ^{182}Hf. We see that the dominant contributions for these nuclei are usually near an initial mass of 2 M$_\odot$. However, the peak contributing stellar mass to a particular nucleus is a strong function of the adopted neutron source model. This is mostly due to a combination of neutron density characteristics in a thermal pulse, dredge-up, and mass loss. In comparison with an estimate of ^{60}Fe yields in massive stars,[10] AGB stars are comparable sources. The integral of yield $\times$ IMF has a maximum value of 3.1×10^{-5} M$_\odot$pc^{-2} for the enhanced ^{22}Ne model. Adopting He core sizes and the 'average' ^{60}Fe mass fraction from the earlier work gives a maximum yield $\times$ IMF integral of 3.4×10^{-5} M$_\odot$pc^{-2}, about the same as our maximum AGB value. Note that the massive star value is probably lower in reality due to ^{60}Fe decay in evolution to core collapse.

The total amount of radiation expected from the top four abundant radionuclides is summarized in Table 2. Different values of expected dI/dl and I are shown for our assumed galactic distribution and the field of views for OSSE and COMPTEL in the direction of the galactic center. Of all the models considered, none can give measurable fluxes by OSSE. Only ^{40}K in the 'classical' model and ^{60}Fe in the enhanced ^{22}Ne model could be observable by COMPTEL.

CONCLUSIONS

We have examined the ability of AGB stars to produce γ-ray-emitting radionuclides via the s-process. As point sources, AGB stellar winds and circumstellar shells are unlikely to be seen by Compton-GRO at any appreciable distance. However, depending on the neutron source model, AGB stars can supply enough ^{40}K and ^{60}Fe that their emissions may be visible by COMPTEL.

REFERENCES

1. V. V. Smith and D. L. Lambert, Astrophys. J. Supp. Ser., 72, 387 (1990).
2. G. Bazan, Ph. D. Thesis, University of Illinois (1991).
3. K. Wolke, V. Harms, H. W. Becker, J. W. Hammer, K.-L. Kratz, C. Rolfs, U. Schroder, H. P. Trautvetter, M. Wiescher, and A. Wohr, Zeis. fur Phys. A, 334, 491 (1989).
4. D. E. Hollowell, Ph. D. Thesis, University of Illinois (1988).

5. P. A. Seeger, W. A. Fowler, and D. D. Clayton, Astrophys. J. Supp. Ser., 11, 121 (1965).
6. F. Kappeler, R. Gallino, M. Busso, G. Picchio, and C. M. Raiteri, Astrophys. J., 354, 630 (1990).
7. G. J. Mathews, G. Bazan, and J. J. Cowan, Astrophys. J., 391, 317 (1992).
8. H. J. Habing, Astron. Astrophys., 200, 40 (1988).
9. G. E. Miller and J. M. Scalo, Astrophys. J. Supp. Ser., 41, 513 (1979).
10. N. Prantzos, Proceedings of the 5th Workshop on Nuclear Astrophysics, (MPI, Garching, 1989), p. 76.

Nucleus	$t_{1/2}$(yr)	E_γ(MeV)(%)	E_γ(MeV)(%)	E_γ(MeV)(%)
^{40}K	1.204×10^9	1.4608(100)		
^{60}Fe	3.000×10^5	0.0586(99.8)		
^{60}Co	5.272	1.1732(100)	1.3325(100)	
^{85}Kr	10.70	0.5140(100)		
^{129}I	1.574×10^7	0.0396(100)		
^{134}Cs	2.063	0.6047(98)	0.7958(88)	0.5693(14)
^{137}Cs	30.17	0.6616(85)		
^{151}Sm	87.90	0.0216(1.7)		
^{154}Eu	8.550	0.1229(40)	1.2745(37)	0.7233(21)
^{155}Eu	4.960	0.0866(72.4)	0.1054(48.2)	
^{182}Hf	9.200×10^6	0.2704(80.5)		

Table 1: Radionuclei Produced In the s-Process

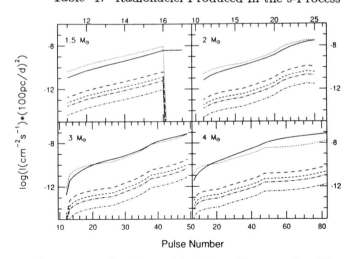

Figure 1: γ-Ray Intensities From Circumstellar Ejecta

Gamma-Ray Line Emission from AGB Star Ejecta

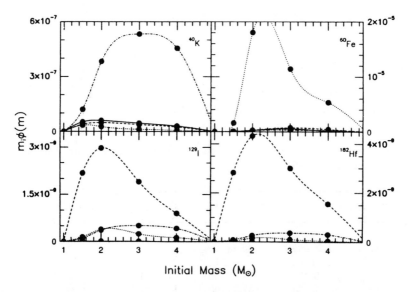

Figure 2: Yields × IMF for Dominant Radionuclei

Nucleus	model	$\sigma_\odot$ $M_\odot pc^{-2}$	$\frac{dI}{dl}$(OSSE) $cm^{-2}s^{-1}rad^{-1}$	I(OSSE) $cm^{-2}s^{-1}$	$\frac{dI}{dl}$(COMPTEL) $cm^{-2}s^{-1}rad^{-1}$	I(COMPTEL) $cm^{-2}s^{-1}$
^{40}K	^{22}Ne	9.947×10^{-9}	1.121×10^{-7}	2.228×10^{-8}	2.185×10^{-7}	1.950×10^{-7}
	^{22}Ne*	6.995×10^{-9}	7.887×10^{-8}	1.567×10^{-8}	1.537×10^{-7}	1.371×10^{-7}
	^{13}C	9.023×10^{-9}	1.017×10^{-7}	2.021×10^{-8}	1.981×10^{-7}	1.769×10^{-7}
	classical	4.999×10^{-8}	5.952×10^{-7}	1.119×10^{-7}	1.098×10^{-6}	9.799×10^{-7}
^{60}Fe	^{22}Ne	7.945×10^{-12}	2.395×10^{-7}	4.759×10^{-8}	4.668×10^{-7}	4.167×10^{-7}
	^{22}Ne*	2.840×10^{-10}	8.559×10^{-6}	1.705×10^{-6}	1.669×10^{-5}	1.489×10^{-5}
	^{13}C	1.310×10^{-11}	7.460×10^{-7}	1.482×10^{-7}	7.698×10^{-7}	6.870×10^{-7}
	classical	8.000×10^{-12}	2.412×10^{-7}	4.793×10^{-8}	4.701×10^{-7}	4.196×10^{-7}
^{129}I	^{22}Ne	1.135×10^{-14}	3.034×10^{-12}	6.029×10^{-13}	5.913×10^{-12}	5.278×10^{-12}
	^{22}Ne*	3.116×10^{-13}	8.331×10^{-11}	1.655×10^{-11}	1.623×10^{-10}	1.449×10^{-10}
	^{13}C	3.505×10^{-12}	9.371×10^{-10}	1.862×10^{-10}	1.826×10^{-9}	1.630×10^{-9}
	classical	6.369×10^{-13}	1.703×10^{-10}	2.601×10^{-11}	3.318×10^{-10}	2.961×10^{-10}
^{182}Hf	^{22}Ne	1.988×10^{-15}	6.446×10^{-12}	1.281×10^{-13}	1.256×10^{-12}	1.126×10^{-12}
	^{22}Ne*	7.724×10^{-14}	2.503×10^{-11}	4.975×10^{-12}	4.878×10^{-11}	4.376×10^{-11}
	^{13}C	3.008×10^{-12}	2.821×10^{-10}	1.937×10^{-10}	1.900×10^{-9}	1.696×10^{-9}
	classical	2.669×10^{-13}	8.650×10^{-11}	1.719×10^{-11}	1.686×10^{-10}	1.505×10^{-10}

Table 2: γ-Ray Emission From ISM Due To AGB Stars

Galactic Gamma-ray Emission from Pulsars

Neil G. Schnepf*, Lawrence E. Brown*, Dieter H. Hartmann*
Donald D. Clayton*, J. Cordes[†], and Alice Harding[‡]

ABSTRACT

A significant fraction of the diffuse galactic γ-ray emission could be due to γ-ray emission from individually undetected pulsars. Using the polar cap model of pulsar γ-ray emission, the contribution of aging radio pulsars to the diffuse galactic glow is estimated. We calculate the γ-ray flux from an evolving pulsar population using stellar orbits in a galactic potential. Using COS-B data, we show that the full skymap provides more stringent constraints on the diffuse glow from pulsars than latitude-integrated flux profiles. Detailed observations of individual γ-ray pulsars are most desirable, but here we emphasize the value of analyzing the collective contribution of pulsars that are individually too faint to be detected by the Compton Observatory.

INTRODUCTION

Observations of COS-B and SAS 2 have long identified the plane of the Galaxy as the predominant source of high energy γ-rays. The origins of these emissions have been attributed to diffuse emission processes which involve interactions between high energy cosmic rays and interstellar matter[7]. However, a significant fraction of the diffuse γ-ray glow from the galactic plane could be due to unresolved point sources. Pulsars, known to be γ-ray sources, could provide such an unresolved population[3]. The galactic distribution of pulsars is fairly well known, so that we may accurately estimate their contribution to the diffuse flux if their emission processes were understood. Overall, the trend in recent years has been towards a down-scaling of their contribution to the total detected γ-ray emission. However, because no generally accepted model for γ-ray emission from pulsars exists at present we investigate whether existing data can be used to provide constraints. In particular, we constrain the overall Pulsar Birth Frequency (PBF) by requiring that nowhere on the sky the pulsar contribution exceeds that observed by the COS-B experiment.

---

*Department of Physics and Astronomy, Clemson University, Clemson, SC
[†]Cornell University, Ithaca NY, Astronomy Department
[‡]Goddard Space Flight Center, Lab for High Energy Astrophysics

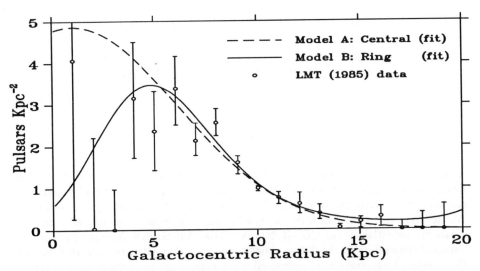

Figure 1: *Pulsar surface density (Pulsars kpc $^{-2}$) vs. Galactocentric Radius (kpc). Fits were made to the data provided in LMT (1985) in the form $e^{f(x)}$.*
Model A: $f(x) = ax^{-2}+bx+c$; $a = -.0186283$ $b = .0374768$ $c = 1.5625$
Model B: $f(x) = ax^{-3}+bx^{-2}+cx+d$; $a = .0033870$ $b = -.10978$ $c = .82617$ $d = -.56767$

MODEL

A number of contributions in the last decade have established the framework for any investigations into the nature of pulsar distributions and evolution. In particular, the work of Emmering & Chevalier[2], Lyne, Manchester & Taylor[4], and Narayan[6] have provided comprehensive studies of the overall characteristics of pulsar populations. Drawing from these and the polar cap model of γ-ray emission[3], we extend a study undertaken by Bailes & Kniffen[1] (BK) to estimate the pulsar contribution. As in BK, we adopted a Monte Carlo simulation as an effective means for sampling a distribution of random point sources. Initial birth velocities were taken to be gaussian with 1-d dispersions of 70 km s^{-1}. Originating positions were selected from a density profile that is exponential in z, with a scale height of 100 pc, and from either one of two fits to the observed radial distribution[5] (Figure 1). We traced the evolution of 5x10^5 pulsars with lifetimes randomly selected to a maximum age of 2.5x10^7 years (to recover a PBF of 1 every 50 years). We consider exact orbits in a Miyamoto-Nagai Potential. Initial magnetic field strengths and periods were assumed to be gaussian in the log (log q_0 = 11.8 ; log P_0 = -.4)[2] with an exponential field decay timescale of τ = 5x10^6 years. Evolution of the initial parameters followed, for the period,

$$P = P_0^2 + B_0^2\tau\frac{8\pi^2 R^6}{3Ic^3}(1 - e^{-\frac{2t}{\tau}}) \qquad (1)$$

for the rate of change of the pulsar period,

$$\dot{P} = \frac{Q_0^2 e^{-\frac{2t}{\tau}}}{P\tau} \qquad (2)$$

the magnetic field,

$$B = \frac{3Ic^3}{8\pi R^6} P\dot{P} \qquad (3)$$

and the total γ-ray luminosity,

$$L_\gamma(> 100 MeV) = 1.2x10^{35} B_{12}^{0.95} P^{-1.7} \text{ photons s}^{-1}. \qquad (4)$$

We assumed that the γ-ray beaming fraction is unity. It is certainly more realistic to assume that the radiation is beamed into solid angles that are significantly smaller than 4π. We approximately compensate for this by neglecting the diffuse γ-ray emission caused by cosmic-ray interactions. Cosmic ray models[7] (CR) fit the observed γ-ray glow very well, leaving little space for an unresolved point source contribution. So ignoring the CR contribution is similar to assuming a small beaming fraction. We employed the efficiency of γ-ray output from the polar cap model of Harding[3]

$$n_\gamma = \frac{20}{4\pi^2} B_{12}^{-1.05} P^{2.3} \qquad (5)$$

to generate Monte Carlo realizations. According to this relation, more than 100% of the pulsar spin-down energy can be extracted in the form of γ-rays. Obviously, this equation must break down eventually. We used the equation for efficiencies less than one and 'forced' the γ-ray production to zero thereafter. Since the total luminosity of the pulsar is low when $n_\gamma = 1$, ignoring possible γ-ray emission in this latter regime did not effect our results.

CONCLUSION

The analysis of our simulations naturally divides between 1-d and 2-d interpretations. Comparing latitude-integrated fluxes (figure 2) illustrates that, indeed, pulsars could provide a significant fraction of the overall γ-ray emission from the Galactic plane. A nominal PBF of 1/50 years appears to be consistent with the COS-B data since no flux excess is apparent anywhere along the plane. The models over-produce γ-rays predominantly towards the galactic center, as evidenced by the pronounced 'bulge' shown in the skymaps of figure 3. As the PBF increases, over-production near the Galactic center sets the limits. For the model under consideration, no more than 1 pulsar per 30 years can be tolerated. Comparing the models in two dimensions yields a refined constraint since we are now using the full angular information that reflects the kinematic evolution of the pulsar population. Figure 4 shows the fraction of all the bins between -10 and 10 degrees in latitude that have flux in excess of the corresponding COS-B value as a function of pulsar birth rate. The 1-d limit of 1/30 years violates more than a third of all the pixels. If one were to adopt a tolerance level of 10% the limiting PBF is 1/70 years, clearly more stringent than in the 1-d case. These results suggest

88 Galactic Gamma-Ray Emission from Pulsars

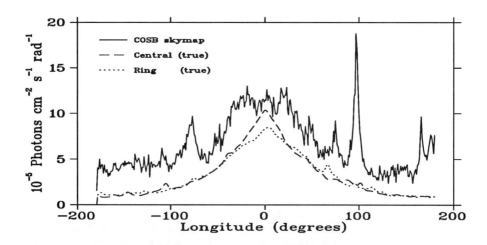

Figure 2: *Latitude integrated flux from indicated models presented against the COS-B skymap. Each of the models were filtered through the COS-B PSF, with summation ranging over -10 to 10 degrees in 1 degree bins.*

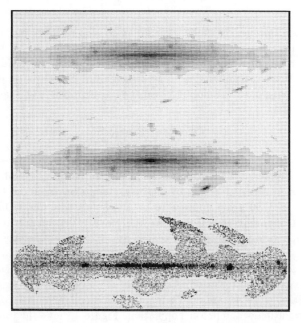

Figure 3: *PSF smoothed skymaps for the two models and the COS-B map.*

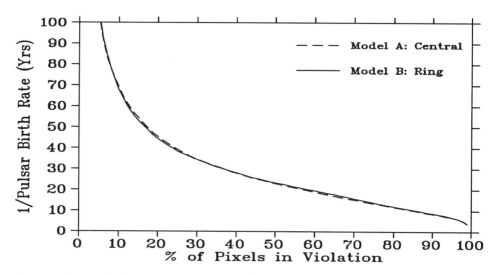

Figure 4: *Each of the curves represents a 2-d comparison of our orbit-integrated models to the COS-B skymap. Percent violation measures the fraction that, when enhanced by some scaling factor, exceed the associated bin of the COS-B data.*

that it is important to incorporate the 2-d aspects of this kind of study. The EGRET detector aboard Compton Observatory will provide such a map with unprecedented sensitivity. Combining that information with the knowledge we will gain from Compton data on individual pulsars holds great potential for constraining radio pulsar evolution, γ-ray pulsar emission mechanisms, or the birth rate of galactic neutron stars. This study has shown that one must include detailed pulsar orbits to obtain accurate skymaps.

REFERENCES

1. Bailes, M. & Kniffen, D.A., ApJ, 391, 659 (1992).
2. Emmering, R.T. & Chevalier, R.T., ApJ, 304, 140 (1986).
3. Harding, A.K., ApJ, 247, 639 (1981).
4. Lyne, A.G., Machester, R.N. & Taylor, J.H., MNRAS, 213, 613 (1985).
5. Manchester, R.N. & Taylor, J.H., Pulsars, 1977.
6. Narayan, R., ApJ, 319, 162 (1987).
7. Strong, A.W., et al., A&A 207, 1 (1988).

The search for the ^{7}Be line at 478 keV and its implications

M.Signore
Ecole Normale Supérieure, Paris

G.Vedrenne
CESR, Toulouse

F.Melchiorri, P.de Bernardis,
La Sapienza, Roma

P.Encrenaz
Observatoire de Meudon

ABSTRACT

The search for the ^{7}Be line at 478 keV with the OSSE telescope on board the COMPTON Gamma Ray Observatory (GRO) is described as well as i/ the search for the redshifted LiH line at 1.3mm, ii/ their implications in cosmology, in chemical history of the Galaxy, in explosive nucleosynthesis models.

I.- INTRODUCTION

Lithium is a key element in cosmology.

In the model [1] of "Standard Big Bang Nucleosynthesis" (SBBN), the predicted abundances depend on only one free parameter $\eta = n_b/n_\gamma$, the ratio of baryons to photons or Ω_b, the fraction of the cosmic critical density in baryons. In particular, the knowledge of the primordial ^{7}Li abundance leads to constraints on η, Ω_b, the number of neutrino families. The predicted ^{7}Li abundance has an upper bound $(^7Li/H)_{prim} \leq 1\text{-}2 \ 10^{-10}$ for $\eta \leq 1.2\text{-}5 \ 10^{-10}$ and $\Omega_b \leq (0.004 - 0.018) \ h^{-2} \ (T_o/2.74)^3$, where $H_o = 100h$ km s^{-1} Mpc^{-1} and T_o is the current temperature of the microwave background. In SBBN, primordial nucleosynthesis begins when $t > 100$ s and abruptly ends at $t \sim 1000$ s - when nuclear reactions are quenched by Coulomb barriers.

In models [2] of "Inhomogeneous Big Bang Nucleosynthesis" (IBBN), nucleosynthesis depends on a number of new free parameters related to the scale and geometry of the inhomogeneities. It seems - although not definitive because of incompleteness in reaction networks - that ^{7}Li is much more copious in IBBN than in SBBN : $(^7Li/H)_{prim} \sim 10^{-9}$ while Ω_b is pushed near 1. Let us only note that the knowledge of abundances in IBBN can provide information back from the first few microseconds - instead of back from the first three minutes in SBBN.

^{7}Li abundances have been observed in various astrophysical sites. Let us only recall that for the ratio of the number of ^{7}Li atoms to the number of H atoms [3] :

i/ values observed in low metallicity Pop II stars are in the range : $(^7Li/H)_{obs\ Pop\ II\ *} \sim 1 \ 10^{-10}$

ii/ values observed in younger Pop I stars are typically such that : $(^7Li/<H>)_{obs\ Pop\ I\ *} \sim 1\text{-}2.10^{-9}$

The "lithium problem" arises in relating the observed abundances to the "cosmic abundances". It is the subject of an ongoing debate :

a/ in "SBBN" : the primordial ^{7}Li abundance is at the level of that observed in old Pop II stars. This would require a galactic enrichment of ^{7}Li- by a factor of 10 - from galactic sources.

b/ in "IBBN": the primordial ^{7}Li abundance is at the level of that observed in young Pop I stars. This would require a uniform and large (10%) depletion in old and warm Pop II stars.

In 1988, Vauclair [4] sugggested this solution. But Deliyannis et al [5] calculated that the necessary lithium destruction in Pop II stars cannot occur unless meridional circulation - or some form of mixing - is invoked to take surface lithium down to where it will be burned through to helium. Until very recently this latest conclusion was the firmest objection to IBBN and - a contrario - the firmest confirmation of SBBN. However, preliminary results of a recent [6] "study of rotation induced mixing and lithium depletion in stars" show that a lithium depletion was possible in Pop II stars and therefore that these stars were born with a ^{7}Li abundance such that :

$$2 \cdot 10^{-10} < (^7Li/H)_{\text{Pop II}^*\text{ at format}} < 5 \cdot 10^{-10}$$

Because previous discussions (a,b) supposed that ^{7}Li abundances remained unaltered between the BBN epoch and the formation epoch of Pop II stars, we have undertaken a millimetric search for LiH lines at high redshifts - see section II .

Moreover, we will search for possible recent galactic lithium sources with the OSSE telescope on board the COMPTON GRO since a ^{7}Li enrichment is necessary to account for observed Pop I star abundances - see section III :

- of at most a factor 2 with Vauclair group's conclusion [6] :

$$(Li/H)_{\text{gal at format}} \sim (Li/H)_{\text{PopII}^* \text{ at format}} \leq 5 \cdot 10^{-10}$$

- of at most a factor (5-10) within SBBN framework (a)

$$(Li/H)_{\text{prim}} \sim (Li/H)_{\text{gal at format}} \sim (Li/H)_{\text{Pop II}^* \text{ at format}} \sim (Li/H)_{\text{PopII now}} \leq 1\text{-}2 \cdot 10^{-10}$$

- of even more within IBBN framework (b) plus the heretic assumption of Kurucz[7] who notes that there is no necessary relation between the primordial ^{7}Li abundance and the observed ^{7}Li abundance in Pop II stars and suggests that : i/ most of the primordial lithium is destroyed, ii/ the lithium we see now has been made by secondary methods.

II.- THE SEARCH FOR LITHIUM ABUNDANCES AT HIGH REDSHIFTS

1.- The region of the universe between $z \sim 5$ and $z \sim 1000$

In the framework of the gravitational instability theory each proto-structure started as a tiny local over-density i.e. as a large primordial cloud, a protocloud. There are many studies of dynamics and evolution of these protoclouds. In all scenarios these protoclouds (from 10^6 to 10^{11} M$_\odot$) cool and collapse fast enough - within a Hubble time - to survive. But one can say very little about the fate of a collapsing primordial cloud.

For example : does the lack of heavy elements (SBBN) help (via opacity) or hinder (via cooling) fragmentation for a realistic, non uniform, non spherical collapse ? Among the different possible scenarios of evolution, the protoclouds have various possible fates : direct collapse to one SMO (supermassive object), into VMO's (very massive objects), into compact stellar remmants - neutron stars - plus some low mass stars, fragmentation into brown dwarfs etc.

Moreover, as the universe cools, at some point, conditions can occur for the formation of molecules; this happens when the kinetic energy of the matter becomes significantly lower than the binding energy of molecules and photodissociation processes become ineffective. Due to initial cosmic abundances, favorite species are[8] : H_2, HD, HD^+, HeH^+, LiH (at z<200), H_2D^+ (at z<20), H_3^+.

During the primordial cloud evolution, molecular radiative transitions can play a major role - absorbing or dissipating heat. In particular, they can facilitate Population III star formation as cooling agents in collapsing protoclouds - at least in simple models of protocloud collapse .

Let us also note that if dark matter is provided by neutrinos [9] their radiative decay should dissociate efficiently molecules - such as LiH- at high redshifts.

We search for lines expected to be emitted in collapsing protoclouds - through a method similar to that proposed for observations of the redshifted 21 cm line by Hogan and Rees in 1979 [10].

2.- The search for LiH primordial lines at $z \leq 180$

The rotational and vibrational lines of the above molecules will be observable provided that [8] :

a/ the product of molecule abundances times the line strength is high enough
b/ the gas temperature is significantly different from the radiation temperature
c/ the gas is significantly clumped, so that differential methods can be used to observe molecular emission from a gas cloud against the background
d/ the optical depth through the cloud is high enough

From the observational point of view, the LiH molecule is well suited for detection since:
- LiH has a high dipole moment (5.88D)
- abundances of LiH can be more than 10% of lithium abundances
- LiH can form at z < 200, where there is a maximum temperature difference between gas and radiation.
- at z ~ 200, the roto-vibrational lines at about 6.5 µm are redshifted at about 1.3 mm where there is an atmospheric window

- . a D = 2kpc - cloud at z ~ 180 is observable with an angular size of $\vartheta = DH_0\Omega z/2c$ ~(14h)"

. the linewidth is such that : $\Delta v/v = \Delta z /(1 + z) \sim 2 \vartheta z^{1/2}/\Omega^{1/2}$;

for $\vartheta \sim 10"$, $z \sim 180$, $\Omega \sim 1$, one has $\Delta v/v \sim 1.3 \cdot 10^{-3}$;

at $v \sim 230$ GHz, the spectral resolution is 0.3 GHz.

. It results that the detection of a single cloud is possible with an instrument having both an angular resolution of 10" and a spectral resolution of 0.3 GHz

. The IRAM 30 m telescope featuring a 12" FWHM beam at 1.3 mm is ideally suited for this observation.

A preliminary search for LiH lines has been carried out at the IRAM 30m telescope on April 1992. These observations set a 95% upper limit of the order of 20 mK for the intensity of the LiH lines. Corresponding upper limits on LiH abundance and on protocloud density are discussed in de Bernardis et al[8] (1992).

The purpose of this search is to provide information on :
- the ^7Li abundance at z ~ 180 - part of the "lithium problem"

- the protocloud evolution at $z \sim 180$ and its fate,
- dark matter theories with decaying neutrinos.

III.- THE SEARCH FOR GALACTIC SOURCES OF ^{7}Li

^{7}Li is often regarded as a product of BBN (at the level of Population II stars abundances) plus a product of spallation by galactic cosmic rays on the interstellar medium. But no spallation model can explain the Population I star abundance of ^{7}Li.

Many authors - see Harris et al [12] - have considered the possible production of ^{7}Li during nova outbursts by destruction of accreted ^{3}He through : ^{3}He$(\alpha\gamma)^7$Be. The ^{7}Li synthesis by the same reaction appears to be unlikely during explosive hydrogen burning in SNII[13]. But recently the "ν - process " - the effect of inelastic neutrino scattering on explosive nucleosynthesis - has been introduced as a new mode of nucleosynthesis in SNII [13]. ^{7}Li and its progenitor ^{7}Be can be made from μ and τ neutrinos interacting with helium : either in the helium shell by
^{4}He $(\nu \nu'n)$ ^{3}He $(\alpha\gamma)^7$Be or with that helium present during cooling phase of ejecta during high temperature nuclear statistical equilibrium. ^{7}Be and ^{7}Li yields are given by Weaver and Woosley [14] :

M (M_O)	15	25	35
Y^7_{Li} ($10^{-7} M_O$)	2.6	4.4	2
Y^7_{Be} ($10^{-7} M_O$)	5	4.1	8

a/ The flux at earth in the 478 keV line, from SN II located at R, is given by :

$$F_{15-25} \sim \frac{1.1 \ 10^{-2}}{R^2(kpc)} \text{ ph cm}^{-2}\text{s}^{-1} \ ; \ F_{35} \sim \frac{2 \ 10^{-2}}{R^2 (kpc)} \text{ ph cm}^{-2} \text{ s}^{-1}$$

OSSE - with a sensitivity of 3 10^{-5} ph cm^{-2} s^{-1} for a line at 0.5 MeV - can detect the 478 keV line if the SN II is within a radius $R \sim 19$ kpc for a 15 - 25 M_O star or within a radius $R \sim 26$ kpc for a 35 M_O star.

b/ First, let us note that [Fe/H] (or better [O/H] for massive stars) - the logarithm of the ratio of the number of Fe atoms (or O atoms) to the number of H atoms in the star divided by the same ratio in the sun - is treated for the overall metal abundance Z in the star. As Y^7_{Li} and Y^7_{Be} do not change with the initial metallicity Z of the star -

because ^{7}Li and ^{7}Be are produced by the ν - process and not by spallation - it follows that :

i/ in the framework of the standard model of the galactic history, the ^{7}Li enrichment by Type II-SN would be early and at the same rate as now - since the yields Y^7_{Be} and Y^7_{Li} are identical for $Z=Z_o$ and $Z \sim o$ and since the star formation rate

(SFR) s is constant for 12 Gyr in this standard model.

ii/ the galactic distribution of these galactic ^{7}Li sources is effectively a CO-distribution [15] with an effective surface $S_{eff} \sim 1.27 \ 10^9$ pc^2 - since the yields Y^7_{Be}

Moreover from this S_{eff} value and from the galactic rate of 4 core collapse supernovae per century - expected on grounds of extragalactic supernovae counts - we estimate a local rate of core collapse supernovae of [15]: $t_{II} \sim 3.2\ 10^{-11}\ pc^{-2}\ yr^{-1}$. Then, we compare the lithium production rate from core collapse supernovae with the observed abundance by mass fraction in the sun [16]: $X_{70} = 1.13\ 10^{-8}$. Here, we follow Tinsley [17] and make a crude estimate of the abundance rate of ^7Li ejected from core-collapse supernovae by : $X_7 = <Y_7>\ t_{II}/s$ where $<Y_7>$ is the total production of ^7Li - i.e. the sum of ^7Li itself plus ^7Be : $<Y_7> = <Y^7_{Be}> + <Y^7_{Li}>$ - ejected by each supernova. For $s = 2.7\ 10^{-9}\ M_O\ pc^{-2}\ yr^{-1}$ (Scalo private communication) and for typical core collapse supernova yields :

$<Y^7_{Li}> \sim 4\ 10^{-7}\ Mo$; $<Y^7_{Be}> \sim 6\ 10^{-7}\ Mo$, one finds $X_7 \sim 1.15\ 10^{-8}$

We conclude that - within uncertainties reflecting those of SNII rates, nucleosynthesis models of SNII, the SFR value s - core-collapse supernovae can account for the observed ^7Li abundance of the solar neighborhood.

The main galactic ^7Li sources are :

1/ Novae - for peculiar conditions of temperature, time scales and ^3He enrichment of their envelope[12];

2/ Core-collapse supernovae - if the ν-process is confirmed [13,14].

Their detection, in the 478 keV line, are difficult but they are accepted as targets of opportunity for the OSSE telescope within the framework of the COMPTON GRO GUEST INVESTIGATOR PROGRAM.

The purpose of this search is to provide information on :
- the recent galactic ^7Li sources - part of the "lithium problem"
- the chemical history of the Galaxy
- the nucleosynthesis models of novae and SNII with a signature for the ν-process

REFERENCES

1. - L.M.Krauss and P.Romanelli ApJ 358, 47 (1990) and references therein
2. - C.J.Hogan in "Primordial Nucleosynthesis" (Eds., W.J.Thompson, B.W. Carney, H.J..Karwowski; World Scientific, 1990) p.15
3. - R.T. Rood id p.36 (and references therein)
4. - S.Vauclair, ApJ. 335, 971 (1988)
5. - C.P. Deliyannis , P..Demarque, S.Kawaler, Ap.J. Sup.73,21 (1990)
6. - C.Charbonnel, S..Vauclair and J.P.Zahn, A&A 255, 191 (1992)
7. - R.L.Kurucz, Com on Astroph. 16, 1, (1992)
8. - P.de Bernardis, V. Dubrovitch, P.Encrenaz, R. Maoli, S. Masi, G.Mastrantonio, B.Melchiorri, F.Melchiorri, M. Signore,P.E.Tanzilli and references therein , A&A , in press.
9. - D.Sciama in "IVth Rome Meeting on Astrophysics Cosmology and Particle Physics (Eds.,L. Maiani and F. Melchiorri ; Ed. Frontières, 1993) in press
10. - C.J. Hogan and M.J. Rees, MNRAS, 188, 791 (1979)
11. - G.Steigman and T.P. Walker ApJ 385, L13 (1992) and references therein.
12. - M.J.Harris, M.D.Leising and G.H. Share ApJ 375, 216 (1991) and references therein.
13. - S.W. Woosley, D.H. Hartmann, R.D.Hoffman, W.C. Haxton Ap.J. 356, 279(1990)
14. - T.A. Weaver and S.E. Woosley Phys. Rep. in press
15. - M.Signore and C.Dupraz, A&A Suppl. in press
16. - A.G.W.Cameron in "Essays in Nuclear Astrophysics"(Eds C.A. Barnes, D.D.Clayton,D.N. Schramm; Cambridge University Press, 1982) p.23
17. - B.M. Tinsley in "Type I - supernovae" (Ed.J.C. Wheeler; Austin: Iniv.of Texas, 1980) p.196.

GALACTIC CENTER

POSITRON ANNIHILATION RADIATION FROM THE GALACTIC CENTER

Jack Tueller
NASA/Goddard Space Flight Center Code 661, Greenbelt, MD 20771

ABSTRACT

Since the 1970's, it has been known that the region near the center of our galaxy contained a strong source of positron annihilation radiation. The gamma-ray line at 511 keV was discovered by balloon instruments (Rice Univ. & Bell/Sandia) and confirmed by satellite instruments (HEAO-3 and SMM). Since its discovery, this source has been one of the primary objectives of every gamma-ray astrophysics mission, yet the exact location and nature of the source or sources of the 511 keV line still eludes our best efforts. At the beginning of the 1980's, dramatic evidence of variability from the HEAO-3 germanium spectrometer suggested a point-source origin of the line. As the decade progressed, results from SMM confirmed the presence of the line, but the high flux observed by SMM contradicted the low flux observed in the second HEAO-3 scan and confirmed by balloon instruments in 1981 and 1984. These results stimulated the creation of a new class of models with a compact source near the Galactic Center and a diffuse source, which could be combined to explain all the observations. Recently, results from a new generation of balloon (GRIS & HEXAGONE) and satellite (GRANAT/SIGMA, ROSAT & GRO) instruments have provided exciting new insights about the Galactic Center. In particular, results from GRANAT/SIGMA strongly suggest that the Einstein source, 1E1740.7-2942, is a variable source of positrons. Even with these new results, all the problems are far from solved and continued observations combined with new approaches will be required to reach a full understanding of this complex phenomena. We expect the Galactic Center positron story to continue to surprise and fascinate us in the future.

INTRODUCTION

The history of the Galactic Center 511 keV line is almost the entire history of gamma-ray line astronomy. Until recently, the only other confirmed sources of gamma-ray lines were our Sun during the brightest solar flares and the galactic diffuse 1809 keV line from the decay of ^{26}Al. This feature has been the focus of intense scrutiny since its discovery in the 1970's as each new gamma-ray instrument has turned its eye to the center of our galaxy to see what contribution it could make to this puzzle. Nearly all of these instruments have confirmed an excess in 511 keV line emission from the center region (see figure 2). Nevertheless, we are still uncertain about the source or sources of the positrons that produce the 511 keV line. Some of this uncertainty results from the limitations of gamma-ray instrumentation. During the time since the discovery of the line, the best efforts of gamma-ray astronomers have produced only about a factor of 10 improvement in sensitivity of gamma-ray spectrometers. The results of the first truly imaging gamma-ray instruments have only recently been available and their sensitivities are only about the same as the wide field-of-view (FOV) spectrometers that discovered the line. The puzzle is also complicated by the unique physics of the positron annihilation process, which are briefly reviewed in the next section.

POSITRON ANNIHILATION

The positron is the anti-particle of the electron. Although the positron is a stable particle, it will eventually encounter an electron and annihilate to produce 2 photons with the rest mass of an electron/positron (511.0 keV). It was first discovered in showers of particles produced by the interactions of high-energy cosmic rays in the atmosphere. The annihilation process requires that positrons be newly synthesized and provides an indicator of their existence (511 keV gamma rays). Positrons can be produced by high-energy interactions of charged particles or gamma-gamma interactions in astrophysical sites like jets or pair plasmas. They can also be produced by +ß-decay of freshly synthesized radioactive elements. The most astrophysically significant isotopes for the production of positrons are ^{56}Co, ^{26}Al, and ^{44}Sc. Each of these mechanisms will produce other signatures that can be used to identify the source. Pair plasmas should produce significant gamma-ray continuum with a characteristic cutoff at 1 MeV due to self absorption by the production of pairs. Jets are identifiable by radio imaging of the well-known linear structures and symmetric radio lobes. Radioactive ß-decays produce gamma-ray lines other than 511 keV, such as the 847 and 1238 keV lines from ^{56}Co that were observed from SN1987A or the diffuse galactic 1809 keV line from the decay of ^{26}Al.

All of these production processes produce positrons with energies $\gtrsim$ 1 MeV. Before the positron can annihilate it must lose most of that energy. The time required to lose this energy is strongly dependent on the properties of the stopping medium and can vary from nearly instantaneous in a thick accretion disk to >10^5 years in the hot, low-density phase of the interstellar medium. The longer times imply that positrons produced in a point source or a distribution of point sources can escape into the interstellar medium before annihilating. This yields a diffuse source.

Not all positrons decay by direct annihilation emitting two 511 keV photons in opposite directions. The positron can combine with an electron to form a hydrogen-like atom of positronium. Three quarters of the time this atom will form in the ortho-positronium state with the spin of the electron and positron aligned. This state decays by emitting 3-gamma rays in a plane with a summed energy of 1022 keV. This 3 photon decay produces a low-energy tail on the peak with a characteristic shape and a maximum flux equal to 4.5 times the flux in the peak[1]. The fraction of positrons that decay by positronium formation is very sensitive to the stopping medium. At high temperatures or densities this fraction can be very small, but in most astrophysical environments it will be significant.

Because the positron is a light particle, it is likely to come to rest before annihilating and its high thermal velocities at rest make the shape of the annihilation line a sensitive probe of conditions in the stopping medium. Parameters like temperature, ionization state, and dust content can have a dramatic influence in the line profile. The different paths to annihilation can result in complex line profiles which cannot be reduced to an equivalent Gaussian shape[2]. This sensitivity to the annihilation medium makes high-resolution spectroscopy of the positron annihilation line especially interesting.

HISTORY

The earliest measurements of the gamma-ray spectrum of the Galactic Center that indicated a 511 keV excess were made by the Rice University balloon instrument with NaI detectors[3,4,5]. The relatively poor energy resolution and uncertainties in the energy calibration of this instrument made the identification of the excess as a positron annihilation line uncertain. The first successful high-resolution measurements by the Bell/Sandia group with cooled germanium detectors in the fall of 1977 immediately clarified these uncertainties (see figure 1)[6]. They observed a narrow line (< 3 keV FWHM) with a centroid that was consistent with the laboratory rest energy of 511.0

keV. An excess below the line in the form of a positronium "like" tail and a possible backscatter feature at 170 keV were also observed for the first time. This measurement set the context for future observations and their interpretation, but it left unanswered the question of the source. The power emitted in this line was 7×10^{36} ergs s^{-1}, which was much stronger than any contemporaneous predictions for possible 511 keV line sources. This immediately suggested the possibility of a supermassive black hole at the Galactic Center, which had been previously suggested by infrared studies, as a prime candidate. The lack of a gravitational redshift and the narrow line widths (which imply a relatively cold annihilation medium) required this model to transport the positrons away from the source. The key to confirmation of this model was localization of the line source to the exact center of our galaxy (Sag A*) using gamma-ray imaging techniques. If the 511 keV line came from a compact point source, the proof would be an observation of variability in the line flux or profile. These two measurements became the primary goals for Galactic Center gamma-ray spectroscopy.

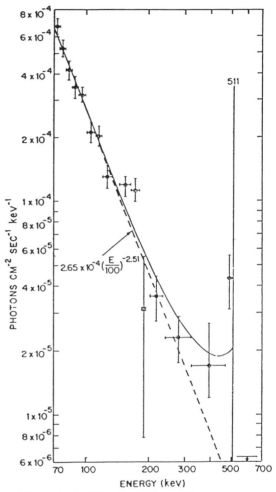

Figure 1 Bell/Sandia Balloon Measurement

The next important contributions to our understanding of (or confusion about) the Galactic Center 511 keV source came from a pair of satellite experiments. The germanium spectrometer built by JPL for the HEAO-3 satellite made two scans of the Galactic plane, first in the fall of 1979 and again in the Spring of 1980[7]. The first of these two measurements confirmed the balloon observations. The flux observed was consistent with the balloon measurements, though on the high side, and the source was localized to within a few degrees of the Galactic Center. The continuum spectrum observed during this first scan showed a high level of emission in the hard x-ray range and an excess above the line that extended to 1 MeV with a very flat spectrum[8]. This MeV hump was predicted by Liang and Dermer[21] from gamma-gamma interactions in a compact source. The 35° FOV of the HEAO-3 spectrometer prevented a clear distinction between the exact center, any of the numerous x-ray sources that cluster near the center, or a compact diffuse distribution. The second Galactic plane scan produced dramatic confirmation of a point-source origin. The HEAO-3 results gave strong evidence (3.5 σ) of variability of the line[7]. The observed variability was confirmed by balloon flights in 1981 and 1984 by Bell/Sandia and GSFC[9,10,11]. At this point, it was widely accepted that the Galactic Center 511 keV line was probably due entirely or at least mostly to a single, strong point source.

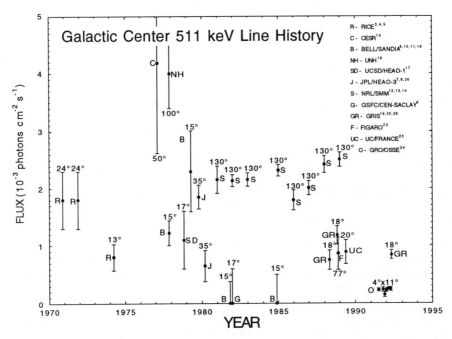

Figure 2 History of Galactic Center 511 keV line measurements.

In 1980, the Solar Maximum Mission (SMM) was launched and included a NaI gamma-ray spectrometer designed for the study of solar flares. This instrument had a wide FOV (~130°) and remained continuously pointed at the Sun throughout the 1980's. Although this instrument was not designed for astrophysical work (in particular, it lacked a simple technique for accurate determination of the background) a remarkable effort led by the group at NRL was able to extract significant new results on a number of astrophysical questions. Excess 511 keV line emission from a broad region towards the center of our galaxy was clearly seen by SMM but the flux observed by SMM was much higher than observed by the balloon instruments[12,13,14]. This was especially true for the low levels observed simultaneously with SMM by balloon instruments in 1981 and 1984. Figure 2 summarizes all the Galactic Center 511 keV line flux measurements. The SMM data did not show clear evidence of variability due to the large systematic errors in the background determination, but if only the statistical errors are used, the data were not inconsistent with significant variability. The high flux observed by SMM and some other broad FOV instruments strongly suggested that another broadly distributed component was contributing to their results, but the large FOV of SMM severely limited its ability to determine the shape of any diffuse emission distribution or to distinguish truly diffuse emission from a population of point sources. At the same time a reanalysis of the HEAO-3 results by the JPL group, which included the effects of possible diffuse emission from the Galactic plane, significantly increased the systematic uncertainties of the observed variability[26]. The Galactic Center 511 keV source had eluded our attempts to clearly pin down its origin and new, more complex models with multiple sources had been introduced.

NEW OBSERVATIONS

In the spring of 1988, the first of a new generation of sensitive high-resolution balloon spectrometers, the Gamma-Ray Imaging Spectrometer (GRIS), was rushed to completion to observe SN1987A in the Large Magellanic Cloud. This source was fortuitously located in the Southern Hemisphere almost exactly opposite the Galactic Center in Right Ascension. This allowed the Galactic Center and the supernova to be observed in a single, long flight from Alice Springs, Australia. GRIS used an array of seven very large n-type germanium detectors (total of ~1800 cm^3) and a very thick (15 cm) NaI active shield to achieve about a factor of five improvement in sensitivity over the previous generation of instruments. Observations in the spring and fall of 1988 made the highest significance measurements of the Galactic Center 511 keV line and indicated a level similar to that observed in the 1970's[19,20]. These measurements provided the first indication of the intrinsic width of the line with a 2 σ lower limit of

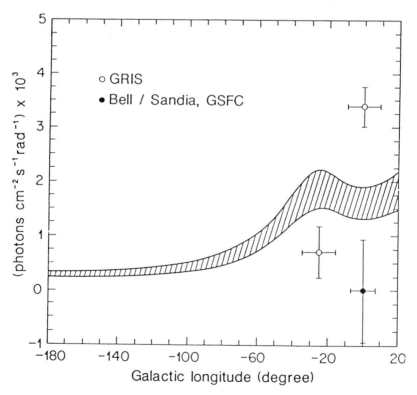

Figure 3 GRIS observations of the 511 keV line from the Galactic Center and Galactic plane.

1.7 keV. Due to an extraordinarily long flight in the fall of 1988 GRIS was able to make a second observation on the Galactic plane at l = 335°. The 511 keV line observed from the plane was much lower than from the center (2 σ upper limit 4.0 x 10^{-4} ph cm^{-2} s^{-1}) showing clearly that the distribution of the 511 keV line must be much more peaked towards the center than the models that had been used by NRL to interpret the SMM data (see figure 3)[12]. In the spring of 1990 an instrument similar to GRIS was flown by the UC/France collaboration (HEXAGONE). It observed the Galactic Center in a distinctly different and lower flux hard x-ray state[23]. The HEXAGONE hard x-ray spectrum shows an unusual edge feature near 170 keV, which has been attributed to backscatter of the 511 keV line[27]. At the same time, the narrow 511 keV line was in essentially the same state as in the GRIS observations.

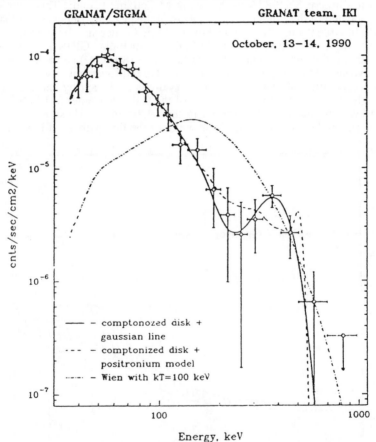

Figure 4 SIGMA spectrum of 1E1740.7-2942 transient.

The Galactic Center had been imaged in the hard x-ray and gamma-ray energy ranges by balloon and satellite instruments[28,29,30]. It had been determined that the strongest hard x-ray source in the region was usually 1E1740.7-2942, but all the Galactic Center sources are variable in this energy range. These instruments did not have the sensitivity to image the narrow 511 keV line, particularly if it was diffuse or divided into several point sources. With the launch of the French-built SIGMA gamma-ray imaging instrument on the Russian GRANAT satellite, an exciting new clue to the origin of the Galactic Center 511 keV line was discovered. The coded aperture system in SIGMA could image a 4° x 4° region around the Galactic Center with a position resolution of 15 arcmin. The hard x-ray spectrum of the 1E1740.7-2942 was well fit by a Comptonized spectrum of the Sunyaev-Titarchuk model[31], which was similar to the spectrum of known black-hole candidates like Cyg X-1. During one particular 12 hour observation a strong (~1 x 10^{-2} ph cm^{-2} s^{-1}) transient feature was observed near 511 keV[32,33]. This transient line feature was clearly imaged by the SIGMA experiment and was found to be broader than the energy resolution of the SIGMA NaI detector. The SIGMA teams have fit the feature with a 180-240 keV FWHM line centered at 410-480 keV (see figure 4). The redshift, width, and transient nature of this feature fit well with expectations for annihilation of positrons in the inner edge of the accretion disk. There was no significant change in the hard x-ray continuum during the transient, which suggests that the emission of positrons is an on going process and the transient was due to the positrons being briefly directed onto the disk. Since the 1E1740.7-2942 transient, SIGMA has also observed a 511 keV transient from the x-ray nova, Nova Muscae, which is also a black-hole candidate[34,35]. The SIGMA results are the first clear evidence for the possible identification of point sources of the positrons and suggest that accreting black holes may make a significant contribution to the galactic 511 keV line.

The SIGMA results sparked new interest in 1E1740.7-2942 and it has been intensively observed in the radio, infrared, and x-ray wavelengths. Radio observations show that the x-ray position corresponds to a large molecular cloud which could provide the annihilation medium for the positrons[36]. New x-ray observations with the ROSAT satellite have provided an exact position for 1E1740.7-2942, which allowed high spatial resolution observations that show radio jets extending from the point source[37]. Combined with the SIGMA results, this represents an exciting confirmation of models that produce positrons from pair jets.

In 1991 a new era for gamma-ray astronomy began with the launch of the Compton Gamma-Ray Observatory (GRO). This is a NASA Great Observatory-class mission with a suite of four instruments capable of covering the entire range from 10 keV to 1 GeV. The instrument most relevant to the Galactic Center 511 keV line question is the Oriented Scintillation Spectrometer Experiment (OSSE), which is a large area NaI spectrometer with a 4° x 11° FWHM FOV. This instrument is capable of making sensitive measurements of the 511 keV line from relatively small areas of the plane. It has just begun to map the 511 keV line in the Galactic Center region. Confirming the trend we have observed in previous measurements, OSSE sees less flux with a smaller FOV[24]. OSSE has yet to see significant evidence for variability at the Galactic Center or at the position of 1E1740.7-2942. Attempts to map the emission near the center are consistent with a point source with intensity about 2 x 10^{-5} ph cm^{-2} s^{-1}. The position of this source is still consistent with the Galactic Center, but the best fit is closer to the position of 1E1740.7-2942, about 0.9° away. The OSSE maps can also be fit by a concentrated diffuse clump near the center but too compact to be resolved by OSSE (galactic bulge model). A significant uncertainty is introduced in the OSSE measurements by the necessity to use pointings near the center (typically around 5° to 10° away) to determine background. For diffuse distributions, OSSE could be subtracting a significant part of the source. Nearly simultaneous measurements by the GRIS balloon instrument in the spring of 1992 show a flux of 9 x 10^{-4} ph cm^{-2} s^{-1} for a 19° FWHM region around the Galactic Center[25]. The low level of the OSSE source leaves a lot of signal seen by the balloon instruments and SMM to be explained. OSSE has also observed 511 keV line emission from the Galactic plane at the few x 10^{-5} ph cm^{-2} s^{-1} level.

DISCUSSION

The OSSE team prefers the galactic bulge model of a compact clump of diffuse emission because it increases the total flux observed and helps to alleviate the discrepancies with the fluxes measured by wider FOV instruments, but the GRIS and OSSE measurements on the Galactic plane indicate there must be another source which is broadly distributed along the plane (disk component). The bulge model still requires two components of diffuse emission with independent distributions to fit all the data. These distributions are not clearly identified with a proposed source of the positrons and would probably require multiple types of positron sources to make significant contributions to the total galactic emission (at least one for each component). Obviously, we need to also consider models that involve populations of point sources or point sources combined with diffuse emission as possible explanations of the galactic 511 keV line results. For all such multi-component models, it is possible that even a high spatial resolution map of the distribution of 511 keV emission from the galaxy will not clearly identify the sources of positrons. Even in this case, there are still many reasons to make such a map, especially if the map can be made with high-resolution spectroscopy, which has the potential to map the physical parameters of the annihilation medium.

Some have argued that we should ignore the indications of variability in the narrow-line emission and assume that the Galactic Center 511 keV line is truly diffuse and constant over human timescales until more conclusive evidence of variability is detected, but significant evidence for variability exists from both balloon and satellite instruments. The instruments providing the most significant evidence for variability in the narrow line used high-resolution detectors, which yield higher signal-to-background ratios and consequently better systematics. The exact significance of this evidence can be calculated in a variety of ways, but different workers have produced estimates of the significance near 4 σ (probability of chance variation ~1 x 10^{-4})[19,38]. The possibility of variability in the narrow 511 keV line emission from the Galactic Center must still be taken very seriously and the search for this variability with OSSE, GRIS, and new gamma-ray spectroscopy instruments should be continued. The timescale for variability in the narrow line is poorly known at best and the importance of this test for identifying the sources of positrons requires a continuing search on all possible time-scales. The identification of transients like the fast, strong SIGMA transients is of particular importance. These transients seem to directly identify at least some of the point sources of positrons in our galaxy that are required to produce the narrow 511 keV line.

CONCLUSIONS

What is known about the Galactic Center positron annihilation line?

1)   Nearly all observers agree that there is a 511 keV gamma-ray line excess from the Galactic Center region. This is arguably the best established gamma-ray line from outside our solar system with many different instrumental techniques and observers confirming the result.

2) The line is very narrow. Since the first high-resolution germanium spectrometer measurements by Bell/Sandia, we have known that the line is < 3 keV FWHM. Although the line can only be marginally resolved in the best germanium measurements, it is narrow enough to provide real constraints on the annihilation region. The energy of the line is exactly 511 keV within the capabilities of any instrument to measure it. (The shift is < 0.3 keV for GRIS fall '88). Both of these constraints do not apply to the SIGMA transient and, therefore, this kind of transient cannot directly explain the 511 keV line seen by other observers, although it may indicate the source of positrons.

3) There is a low energy tail on the line. This tail is frequently attributed to the 3 photon decay of the positronium atom. The formation of positronium is dependent on the annihilation medium but at least some of the positrons will decay by positronium in most astrophysical settings. Further, the positronium fraction for the Galactic Center gamma-ray spectrum has never been observed to be greater than 100%, which would violate the theoretical limit or with a shape that is inconsistent with the 3 photon spectrum. It must be noted that a similarly shaped spectrum could be produced in point sources by Compton scattering in a thick target like an accretion disk or by a spectrum of different redshifts near a massive object (a black hole or a neutron star).

4) No single component model can adequately describe all the measurements. The wide range of flux values observed by instruments with different FOVs is not consistent with a single point source. Recent, nearly simultaneous observations by GRO/OSSE and the GRIS balloon instrument confirm the earlier results at high significance. Statistically significant evidence for variability in the line parameters exists, which would suggest that at least part of the line flux comes from compact objects. Models with a variable point source and broadly distributed diffuse source can readily fit all the data, but so can models with a distribution of point sources. While it is still possible to invent time-independent, purely diffuse models that are only marginally inconsistent with all the results (at the 3 to 4 σ level), these models must incorporate at least 2 components (a sharp central bulge to fit the OSSE results and a broader diffuse component to provide the rest of the flux observed by wide-FOV instruments). It should be noted that all theories produce the positrons in point sources. Diffuse models allow the positrons a sufficient lifetime to escape from the sources and spread out into the interstellar medium before annihilating. We can conclude that it is unlikely that any simple model with one source or type of source will provide a satisfactory explanation of all the results.

Controversy about the origin of the narrow 511 keV line from the Galactic Center will probably continue until unequivocal evidence of variability is detected by one of the new generation of more sensitive spectrometers now in operation or until a new, much more sensitive, coded aperture instrument images the 511 keV line with sufficient spatial resolution to distinguish point sources from diffuse clumps. A joint ESA/NASA program (INTEGRAL) has been proposed to do simultaneous high-resolution spectroscopy and high-resolution imaging of astrophysical sources. Even a very detailed map of the 511 keV line emission from the Galactic plane might not provide conclusive evidence for a particular model of the positrons' origin. The key to understanding the origin of the positrons that produce the annihilation line is identifying the point sources. The GRANAT/SIGMA transient features near 511 keV from 1E1470.7-2942, Nova Muscae, and perhaps the Crab pulsar are the most promising new clues in this search. These results suggest that a modest sized instrument, designed to monitor the Galactic plane continuously for bright transient features, might be able to map the actual positron sources and solve the puzzle that has eluded gamma-ray astronomers for more than 25 years. The Germanium Galactic Plane Patrol (GGAPP) is an instrument of this kind that will be proposed as a Small Explorer mission[39].

REFERENCES

1. M. Leventhal, Ap. J., **183**, L147 (1973).
2. R. Ramaty & R. E. Lingenfelter, in Gamma-Ray Line Astrophysics, ed Ph. Durouchoux & Nikos Prantzos, (New York: AIP), 67 (1991).
3. W. N. Johnson, F. R. Harnden, & R. C. Haymes, Ap. J., **172**, L1 (1972).
4. W. N. Johnson, & R. C. Haymes, Ap. J., **184**, 103 (1973).
5. R. C. Haymes et al., Ap. J., **201**, 593 (1975).
6. M. Leventhal, C. J. MacCallum, & P. D. Stang, Ap. J., **225**, L11 (1978).
7. G. R. Reigler et al., Ap. J., **248**, L13 (1981).
8. G. R. Reigler et al., Ap. J., **294**, L13 (1985).
9. W. S. Paciesas et al., Ap. J., **260**, L7 (1982).
10. M. Leventhal et al., Ap. J., **260**, L1 (1982).
11. M. Leventhal et al., Ap. J., **302**, 459 (1986).
12. G. H. Share et al., Ap. J., **326**, 717 (1988).
13. G. H. Share et al., Ap. J., **358**, L45 (1990).
14. M. J. Harris et al., Ap. J., **362**, L35 (1990).
15. F. Albernhe et al., A. & A., **94**, 214 (1981).
16. B. M. Gardner et al., in The Galactic Center, ed G. R. Riegler & R. D. Blanford (New York: AIP) **144** (1982).
17. M. S. Briggs, Observations of Galactic Positron Annihilation Radiation, PhD thesis, (San Diego:UCSD), (1991).
18. M. Leventhal et al., Ap. J., **240**, 338 (1980).
19. M. Leventhal et al., Nature, **339**, 36 (1989).
20. N. Gehrels et al., Ap. J., **375**, L13 (1991).
21. E. P. Liang & C. D. Dermer, Ap. J., **325**, L39 (1988).
22. M. Niel et al., Ap. J., **356**, L21 (1990).
23. P. Wallyn et al., submitted to Ap. J. (1991).
24. W. Purcell (this volume).
25. M. Leventhal et al. (this volume).
26. W. A. Mahoney et al., in Nuclear Spectroscopy of Astrophysical Sources, eds. N. Gehrels & G. H. Share, (New York: AIP), 149 (1988).
27. R. E. Lingenfelter & X.-M. Hua, Ap. J., **381**, 426 (1991).
28. G. K. Skinner et al., Nature, **330**, 544 (1987).
29. W. R. Cook et al., Ap. J., **372**, L75 (1991).
30. G. K. Skinner et al., A. & A., **252**, 172 (1991).
31. R. Sunyaev & L. Titarchuk, A. & A., **86**, 121 (1980).
32. L. Bouchet et al., Ap. J., **383**, L45 (1991).
33. R. Sunyaev et al., Ap. J., **383**, L49 (1991).
34. R. Sunyaev et al., Ap. J., **389**, L75 (1992).
35. R. Sunyaev, private communication.
36. J. Bally & M. Leventhal, Nature, **353**, 284 (1991).
37. I. F. Mirabel et al., Nature, **358**, 215 (1992).
38. J. G. Skibo, R. Ramaty, & M. Leventhal, Ap. J., **397**, 135 (1992).
39. J. Tueller, N. Gehrels, & M. Leventhal, in Gamma-Ray Bursts, eds. W. S. Paciesas & G. J. Fishman, (New York: AIP), 390 (1988).

OSSE OBSERVATIONS OF GALACTIC 511 keV ANNIHILATION RADIATION

W. R. PURCELL, D. A. GRABELSKY, AND M. P. ULMER
Northwestern University, Evanston, IL

W. N. JOHNSON, R. L. KINZER, J. D. KURFESS, AND M. S. STRICKMAN
Naval Research Laboratory, Washington DC

AND

G. V. JUNG
Universities Space Research Association, Washington DC

ABSTRACT

The Oriented Scintillation Spectrometer Experiment (OSSE) on the *Compton Gamma-Ray Observatory* has performed numerous observations of the galactic plane and galactic center region to measure the distribution of galactic 511 keV positron annihilation radiation and to search for time variability of the emission. These observations show conclusive evidence for a narrow 511 keV line and positronium continuum. For the first OSSE galactic center observation, the fitted 511 keV line flux was $(2.3 \pm 0.3) \times 10^{-4}$ γ cm^{-2} s^{-1} and the positronium continuum flux was $(8.8 \pm 0.7) \times 10^{-4}$ γ cm^{-2} s^{-1}, corresponding to a positronium fraction of (0.96 ± 0.04). The quoted uncertainty in the positronium flux does not include the effect of the underlying continuum model on the fitted positronium flux. No significant time variability of the line flux has been observed; the 3σ upper limit to daily variations from the mean is 3×10^{-4} γ cm^{-2} s^{-1}. The galactic distribution of the 511 keV line emission is found to be most consistent with a 2-component diffuse distribution consisting of a galactic disk and a nuclear bulge component. This model is also consistent with most of the observations performed by other instruments. The fitted fluxes in the disk and bulge components are (0.5 ± 0.2) and $(1.8 \pm 0.2) \times 10^{-3}$ γ cm^{-2} s^{-1} within the inner radian of the Galaxy, respectively.

INTRODUCTION

Positron annihilation radiation was first detected during observations of the galactic center region in the mid-1970's and has since been observed by several balloon and satellite-borne experiments (see Lingenfelter and Ramaty 1989 for a recent review). The observations performed to date, however, have not yet been able to determine the location or distribution of the emission. The situation has been further complicated by the possibility that the 511 keV annihilation line may vary in intensity by as much as $\sim 10^{-3}$ γ cm^{-2} sec^{-1}, leading to the suggestion[1] that the observed emission is composed of two separate sources: 1) a steady state diffuse galactic component, and 2) a time-variable point source near the galactic center. The source of the diffuse galactic component is thought to be the β^+-decay products from radioactive nuclei[2] produced by supernovae, novae or Wolf-Rayet stars, while γ-γ interactions in the vicinity of an accreting black hole has been suggested[3] as a possible source of time variable annihilation emission. At this time, however, few details about the source or sources of the positrons is known.

One of the major scientific goals of the OSSE team is to map the distribution of positron annihilation radiation and to search for time variability of the emission. The investigation of the distribution of the annihilation radiation was begun during the GRO Phase I observations and will be continued throughout the subsequent phases of the mission.

OBSERVATIONS

The OSSE instrument is one of four experiments on NASA's *Compton Gamma-Ray Observatory* (GRO) satellite. The GRO spacecraft was launched on-board the space shuttle Atlantis on 5 April 1991 and was deployed on 7 April 1991 into a nearly circular orbit with an altitude of 457 km and an inclination of 28.5°. After an activation period of approximately four weeks, the GRO Phase I science observations were begun on 16 May 1991 and continued through 17 November, 1992. This interval was divided into viewing periods representing different orientations of the GRO spacecraft. Prior to the tape recorder failure in March, 1992, the length of the viewing periods was nominally two weeks. Because of the reduction in available telemetry following the tape recorder failure, the length of the nominal viewing period was increased to three weeks beginning with viewing period 24 on 2 April, 1992.

The OSSE instrument[4] consists of four separate, nearly identical detectors. The primary detecting element of each detector is a large area NaI(Tl)-CsI(Na) phoswich crystal, providing gamma-ray spectral information over the energy range 0.05 – 10 MeV. The nominal energy resolution at 0.5 MeV is $\sim$ 9%, with a photopeak effective area of $\sim$ 500 cm^2 for each detector. The phoswich is actively shielded and passively collimated. Tungsten collimators provide a field-of-view which is 3.8° $\times$ 11.4° full-width at half-maximum (FWHM), with the long axis of the collimators oriented parallel to the spacecraft Y-axis. Each detector has a separate elevation control system which provides independent positioning of the detectors through an angle of 192° about an axis parallel to the spacecraft Y-axis. The total positional uncertainty in the detector pointing direction, including the aspect uncertainties in the spacecraft orientation, is $\sim 0.1°$.

Table 1: OSSE Phase I Galactic Center / Galactic Plane Observations

Viewing Period	Target Position	Position Angle	Background Offset	Observation Interval	On-Source Time (sec)
5[a]	(0°, 0°)	−90°	±10°	12 − 26 July, 1991	1.8×10^5
7.5	(25°, 0°)	−90°	±10°	15 − 22 Aug, 1991	8×10^4
9.0[b]	(339°, 0°)	−90°	+4.3°/ − 8.5°	5 − 12 Sept, 1991	5×10^4
10[c]	(300°, 0°)	−84°	+3.8°/ − 12°	19 Sept − 3 Oct, 1991	1.2×10^5
11[d]	(0°, 0°)	66°	±12°	3 − 17 Oct, 1991	1.2×10^5
13.0	(25°, 0°)	−90°	±10°	31 Oct − 7 Nov, 1991	1.1×10^5
13.5[b]	(339°, 0°)	−90°	+4.3°/ − 8.5°	7 − 14 Nov, 1991	7×10^4
14[d]	(0°, 0°)	0°	±12°	14 − 27 Nov, 1991	1.6×10^5
16[a]	(0°, 0°)	90°	±10°	12 − 27 Dec, 1991	2.3×10^5
17[e,f]	(0°, 0°)	−32°	±6° − ±12°	27 Dec, 1991 − 10 Jan, 1992	2.0×10^5
19[f]	(58°, 0°)	−90°	±10°	23 Jan − 6 Feb, 1992	1.6×10^5
20[g,f]	(40°, 0°)	90°	±10°	6 − 20 Feb, 1992	1.1×10^5
21[h]	(0°, 0°)	−96°	±10°	20 Feb − 5 Mar, 1992	1.8×10^4
24.0[i,f]	(0°, 0°)	96°	±10°	2 − 9 Apr, 1992	6×10^4
24.5[j,f]	(5°, 0°)	93°	±10°	9 − 16 Apr, 1992	4×10^4
25[i,f]	(0°, 0°)	96°	±10°	16 − 23 Apr, 1992	6×10^4

[a] With scan angles of 0°, ±1.5° and ±3.0°.
[b] Included observations of GX 339−4.
[c] Included observations of Nova Muscae.
[d] With scan angles of 0°, ±1°, ±2°, ±3°, ±4° and ±5°.
[e] With scan angles of 0°, ±2.0° and ±4.0°.
[f] Detector gain at twice nominal value.
[g] With scan angles of 0°, ±2.3°.
[h] Only observed by detectors 3 and 4.
[i] With scan angles of 0°, ±4.8°.
[j] With scan angles of 0°, ±4.5°.

During source observations, periodic background measurements are performed by offset-pointing the detectors from the target. Source and background observations are typically 131 seconds in length, with the observations alternating between source and background measurements. Background observations are generally performed by alternately offset-pointing on each side of the source position to minimize potential systematic effects. In addition, to maximize the OSSE science data, a second independent target may be observed during the periods in which the first target is occulted by the earth.

During the GRO Phase I operations, OSSE performed 16 separate observations of the galactic center and galactic plane. A description of these observations is given in Table 1. The orientation of the OSSE collimator on the sky is characterized by the position angle, which represents the angle between the long axis of the collimator field-of-view and Galactic North. The galactic plane observations were typically performed with a position angle of 90° (e.g., with the long axis of the collimator oriented parallel to the galactic plane) to maximize the instrument response to a diffuse galactic distribution. The background offset angles represent the scan angles from the target position at which the background observations were performed. For many of the galactic center observations, additional source pointings were performed at various scan angles (relative to the target along the OSSE scan plane) to provide information about the extent of the emission and to search for possible point sources of radiation. The on-source time represents the total accumulation time for all of the source

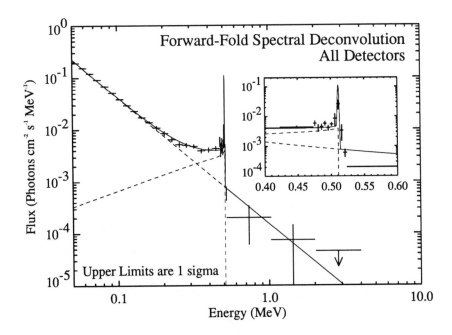

Figure 1: The OSSE galactic center spectrum from viewing period 5. The spectrum represents the sum of all four detectors and was accumulated over the period 13–24 July, 1991. The fitted function consists of a single power law, a photopeak line fixed in energy at 511 keV and in width at 2 keV, and a positronium continuum component.

scan positions. Several of the observations were performed with a detector gain of twice the nominal value, providing a spectral energy range of 0.05 – 5.0 MeV.

ANALYSIS AND RESULTS

The first step in the spectral analysis process consists of screening the data. The initial data screening flags data having low live-times, telemetry errors or in which the detector zenith angle was greater than 108°. Additional data screening was also performed to identify spectra containing cosmic-ray or high-voltage events. Cosmic-ray events, produced when a high-energy cosmic-ray deposits a large amount ($\gtrsim 100$ GeV) of energy in the phoswich crystal, are typically $\lesssim 0.1$ seconds in duration and are characterized by a very soft, low energy ($\lesssim 100$ keV) continuum excess. High-voltage events, produced by the occasional excursion of a PMT high-voltage for one of the detector's, are also short in duration ($\lesssim 4$ seconds) and result in a temporary degradation in the detector's energy resolution. Spectral data containing any of these events were identified and excluded from the subsequent analysis.

Background estimation and subtraction was performed for each detector separately. In order for a source observation to have been included in the analysis at least three background measurements were required, with the source observation bracketed by the background observations. Four background measurements were used, if they were available. For these source observations, the background spectrum was estimated by fitting the background spectra, channel-by-channel, using a quadratic function in time and evaluating the fitted function at the time of the source observation. Additional data screening was performed on the individual background-subtracted spectra to further eliminate low-level cosmic-ray and high-voltage events. The background-subtracted spectra passing the final screening were then summed for each detector separately.

The resulting summed spectra were then fitted using forward-folding with non-linear least-squares fitting techniques. Forward-folding consists of folding an input photon spectrum through the instrument response and comparing the resultant count spectrum with the observed spectrum. The photon model parameters are then modified to minimize the χ^2 for the fit. Since OSSE consists of four detectors with slightly different responses, the spectral fitting was performed for all four detectors simultaneously using the same input photon spectrum.

The spectrum for the viewing period 5 observation of the galactic center is shown in Figure 1. This spectrum shows strong evidence for a 511 keV line and positronium continuum. The position of the line is consistent with an energy of 511 keV and the line width is consistent with the instrumental resolution. The spectral fit shown in Figure 1 was performed over the energy range 0.05 – 5.0 MeV and consisted of a single power-law, a photopeak line fixed in energy at 511 keV and in width at 2 keV, and a positronium continuum component. The spectral fit resulted in a 511 keV line flux of $(2.3 \pm 0.3) \times 10^{-4}$ γ cm^{-2} s^{-1} and a total positronium flux of $(8.8 \pm 0.7) \times 10^{-4}$ γ cm^{-2} s^{-1}. Using the equation for the positronium fraction[5]:

$$f = \frac{2.0}{(2.25 \times (f_{511}/f_{\rm pos}) + 1.5)} \quad (1)$$

where f_{511} is the 511 keV line flux and $f_{\rm pos}$ is the positronium continuum flux, the corresponding positronium fraction is (0.96 ± 0.04). It should be noted, however, that the fitted positronium flux is dependent on the continuum model used in the fit. Since the low energy continuum spectrum is significantly different for each of the galactic center observations, a detailed analysis of the positronium fraction will require a more rigorous treatment of the underlying continuum. The shape and intensity of the continuum, however, does not significantly effect the fitted intensity of the 511 keV line.

Time variability of the 511 keV line flux was investigated by summing the background-subtracted spectra for each of the galactic center observation intervals separately. These observation-averaged spectra were then fitted using

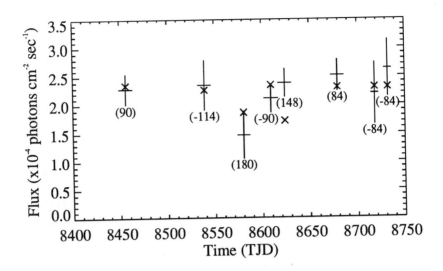

Figure 2: The viewing period averaged 511 keV line flux for the OSSE galactic center observations. The numbers in parenthesis represent the OSSE collimator position angle for the associated observation interval. The "X" symbols represent the expected 2-component diffuse model flux.

the model described above. The resulting 511 keV line fluxes are shown in Figure 2. While there is considerable variation in the low energy continuum flux, partly due to the different observation configurations (e.g., collimator position angle), no significant variability of the 511 keV line flux is observed. When corrections are made for the different observation configurations, the 3σ upper limit to variations from the mean is 1.5×10^{-4} γ cm^{-2} s^{-1}. A similar study was performed to search for daily variability; the 3σ upper limit to daily variations from the mean is 3×10^{-4} γ cm^{-2} s^{-1}.

The spectral fitting model described above was also used to fit the separate, observation-averaged spectra for each galactic plane and galactic center observation. The results of these fits are shown in Figures 3 and 4. As can be seen in Figure 3, the data are sharply peaked in longitude near the galactic center, consistent with the reported GRIS observations[6]. The galactic center scan observations shown in Figure 4 also indicate that the emission is concentrated in the vicinity of the galactic center. It should be pointed out that the observed OSSE fluxes are always measured relative to the flux present in the background pointed positions. For example, since the 180° position angle observation of the galactic center had background observations at $(-12°, 0°)$ and $(12°, 0°)$, the OSSE flux for this observation is measured relative to any emission present at these positions. The model simulations include the effect of the offset-pointed

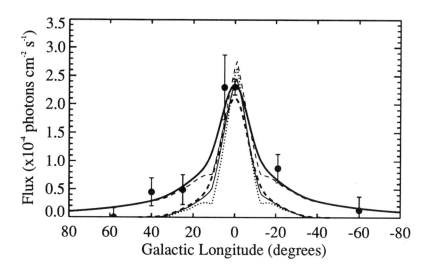

Figure 3: The fitted 511 keV line flux for the galactic plane observations. The curves represent the expected OSSE responses for several galactic distribution models, fitted simultaneously to all of the galactic center and galactic plane data. The models displayed are the point source model (dotted curve), bulge model (thick dashed curve), bulge + point source model (thin solid curve), disk + point source model (thin dashed curve), and the disk + bulge model (thick solid curve). The results of the model fits are given in Table 2.

background observations, as shown in Figures 2 and 4.

Figures 3 and 4 also show the expected OSSE responses for several galactic distribution models. The diffuse models investigated in this study include a model for the CO distribution in the Galaxy[7], the observed galactic CO distribution[8], a model for the distribution of galactic novae based on observations of novae in M31[7], and a model for the visual luminosity distribution in the Galaxy[9]. The visual luminosity model is made up of two components – a disk component which describes the distribution of the galactic disk and a bulge component which describes the distribution of the galactic bulge. The ratio of the intensities for the disk and bulge components was fixed for the visual luminosity model. The disk and bulge distributions from the visual luminosity model were also used separately to fit the OSSE data. A point source of emission located at the position of the X-ray source 1E 1740.7−2942 (359.14°, −0.09°) was also investigated. Each model was fitted separately to all of the data shown. In addition, the point source component was added to each of the diffuse distributions. Finally, a 2-component diffuse distribution consisting of the disk and bulge components was also fitted to the data, with the intensities

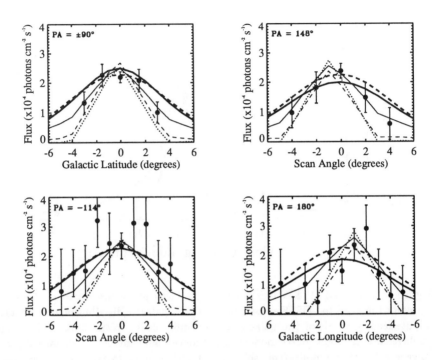

Figure 4: The fitted 511 keV line flux for scans through the galactic center with various position angles. The collimator position angle is indicated in each panel. The curves shown in these figures are the same as described in Figure 3.

of the disk and bulge components each free parameters in the fit. The results of these fits are given in Table 2.

DISCUSSION

The OSSE data suggest that the distribution of the 511 keV line emission is composed of two components: 1) a component which is concentrated near the galactic center and is approximately symmetric in galactic longitude and latitude with a FWHM $\lesssim 10°$ (as seen in the similarity of the scan profiles having different position angle shown in Figure 4), and 2) a galactic disk component producing significant emission at longitudes up to $\sim \pm 40°$ (as shown in Figure 3).

If all of the observed emission is assumed to originate from a diffuse galactic source, the OSSE data are inconsistent at the $> 3\sigma$ level with the CO model, CO distribution, visual luminosity and disk models. Further, the fitted OSSE intensity for the M31 novae model is inconsistent with the reported SMM flux[10] of $(2.3 ^{+0.5}_{-0.8}) \times 10^{-3} \gamma$ cm^{-2} s^{-1}. The uncertainty in the SMM value represents

Table 2: OSSE Fitted Distribution Model Results

Model	Single Model Fits			Double Model Fits			
	χ^2	$P(\chi^2, 52)$	Flux[a]	χ^2	$P(\chi^2, 51)$	Diffuse Flux[a]	1E Flux[b]	TOTAL Flux[a]
CO Model	336.78	$< 10^{-15}$	1.20 ± 0.07	59.49	0.19	0.30 ± 0.05	0.25 ± 0.03	0.55 ± 0.06
CO Distribution	114.75	$< 10^{-6}$	0.96 ± 0.04	57.02	0.26	0.34 ± 0.05	0.19 ± 0.03	0.53 ± 0.06
M31 Novae Model	68.51	0.06	0.99 ± 0.04	50.98	0.47	0.53 ± 0.06	0.14 ± 0.04	0.67 ± 0.07
Visual Luminosity	180.34	$< 10^{-15}$	1.86 ± 0.09	56.18	0.29	0.56 ± 0.08	0.22 ± 0.02	0.78 ± 0.08
Disk	197.33	$< 10^{-15}$	1.80 ± 0.08	57.28	0.25	0.51 ± 0.07	0.22 ± 0.02	0.73 ± 0.07
Bulge	64.60	0.11	2.26 ± 0.09	45.07	0.71	1.25 ± 0.11	0.14 ± 0.05	1.39 ± 0.12
1E 1740.7−2942[b]	70.96	0.04	0.28 ± 0.01	–	–	–	–	–
Disk + Bulge	–	–	–	49.67	0.53	Disk 0.52 ± 0.19	Bulge 1.81 ± 0.17	2.33 ± 0.25

[a] Units are $(\times 10^{-3} \, \gamma \, cm^{-2} \, sec^{-1})$ for the inner radian of the Galaxy.
[b] Units are $(\times 10^{-3} \, \gamma \, cm^{-2} \, sec^{-1})$.

the total allowed range (statistical + systematic) in the 511 keV line flux so, if all of the observed emission originates from a diffuse galactic source, the combined OSSE and SMM data are inconsistent with the M31 novae model at the $> 3\sigma$ level. The bulge model provides an acceptable fit to the OSSE data and is consistent with the SMM flux; however, as shown in Figure 3, it systematically under-estimates the galactic plane contribution, suggesting the existence of a more diffuse galactic longitude component of the distribution. Finally, while the 1E 1740.7–2942 point source model provides an acceptable fit to the OSSE data, it would require variability of the point source flux by $\gtrsim 1.5 \times 10^{-3}$ γ cm^{-2} s^{-1} to be consistent with the earlier observations of the galactic center region. This model also systematically under-estimates the galactic plane contribution, again suggesting the existence of a more diffuse galactic longitude component of the distribution.

Adding a point source component to any of the diffuse models provides an acceptable fit to the OSSE data, as shown in Table 2. For these models, the fitted point source flux is in the range $1-3 \times 10^{-4}$ γ cm^{-2} s^{-1}, depending on the diffuse model, and the total flux within the inner radian of the Galaxy is in the range $0.5-1.4 \times 10^{-3}$ γ cm^{-2} s^{-1}. These models would require variability of the point source flux by as much as $\sim 1.5 \times 10^{-3}$ γ cm^{-2} s^{-1} to be consistent with the earlier observations of the galactic center region. A nearly simultaneous observation of the galactic center by the OSSE and GRIS instruments occurred in late April, 1992. The GRIS instrument observed[11] an average 511 keV line flux of $(8.3 \pm 0.8) \times 10^{-4}$ γ cm^{-2} s^{-1}; this flux, which is consistent with the 1988 GRIS observations[6], would be inconsistent at the $> 3\sigma$ level with all of these models with the exception of the bulge + point source model. For this model, the expected GRIS flux is $(6.5 \pm 0.7) \times 10^{-4}$ γ cm^{-2} s^{-1}. As shown in Figure 3, this model also systematically under-estimates the galactic plane contribution.

The final distribution model investigated was a 2-component diffuse model consisting of disk and bulge components. This model was found to fit the OSSE data as well as the diffuse + point source models. The expected GRIS flux for this model is $(8.4 \pm 0.8) \times 10^{-4}$ γ cm^{-2} s^{-1}, in good agreement with the observed flux. This 2-component diffuse model is also consistent with the average flux observed by SMM, as well as many of the other previous measurements. This is shown in Figure 5 where the diffuse component (based on the OSSE best-fit disk + bulge model given in Table 2) was subtracted from the flux reported for each observation. Note that the uncertainties shown in Figure 5 represent only the statistical uncertainties in the measurements. Assuming the variations in the residuals for the pre-OSSE observations are not due to systematic effects in the measurements, they suggest a limit to the time-variable component of the 511 keV line flux of $\sim 5 \times 10^{-4}$ γ cm^{-2} s^{-1}. As shown in Figure 5, the OSSE data are well fitted by the 2-component diffuse model and do not require a time-variable component.

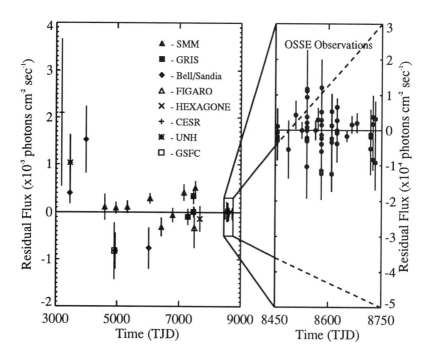

Figure 5: The history of the 511 keV line flux from the galactic center region as observed by various balloon and satellite instruments, including the recent OSSE results. Note that the uncertainties represent the statistical uncertainties only. The diffuse contribution (based on the OSSE best-fit disk + bulge model) has been subtracted from the reported flux for each observation. See text for details. References: SMM – (12); GRIS – (6); Bell/Sandia – (13, 14, 15, 16); FIGARO – (17); HEXAGONE – (18); CESR – (19); UNH – (20); GSFC – (21).

CONCLUSIONS

The OSSE observations of the galactic center and galactic plane show conclusive evidence for a narrow 511 keV line and positronium continuum. The distribution of the line emission is found to be concentrated near the galactic center. There is no evidence for time variability of the 511 keV line flux during the galactic center observations; the 3σ upper limit to variations from the mean is 1.5×10^{-4} γ cm^{-2} s^{-1} for weekly variability and 3×10^{-4} γ cm^{-2} s^{-1} for daily variability.

The galactic distribution of the 511 keV line emission has been investigated by fitting various distribution models to the OSSE data. A 2-component diffuse distribution consisting of disk and bulge components is found to fit the OSSE data well and is also consistent with most of the observations by other instruments, including the nearly simultaneous observation of the galactic center region by the GRIS instrument. The fitted fluxes in the disk and bulge components are (0.5 ± 0.2) and $(1.8 \pm 0.2) \times 10^{-3}$ γ cm^{-2} s^{-1} within the inner radian of the Galaxy, respectively. When applied to previous observations by other instruments, this model suggests a limit to the time-variable component of the 511 keV line flux of $\sim 5 \times 10^{-4}$ γ cm^{-2} s^{-1}.

This research was supported under NASA grant DPR S-10987C.

REFERENCES

1. Lingenfelter, R. E., and Ramaty, R., *Ap. J.*, **343**, 686 (1989).
2. Clayton, D. D., *Nature Phys. Sci.*, **244**, 137, (1973).
3. Lingenfelter, R. E., and Ramaty, R., in *The Galactic Center*, ed. G. R. Riegler and R. D. Blandford (New York:AIP), p. 148 (1982).
4. Johnson, W. N., et al., *Ap. J. Suppl.*, in press, (1993).
5. Brown, B. L., and Leventhal, M., *Ap. J.*, **319**, 637 (1987).
6. Gehrels, N., et al., *Ap. J. (Letters)*, **375**, L13 (1991).
7. Leising, M. D., and Clayton, D. D., *Ap. J.*, **294**, 591, (1985).
8. Dame, T. M., et al., *Ap. J.*, **322**, 706 (1987).
9. Bahcall, J. N., and Soneira, R. M., *Ap. J. Suppl.*, **44**, 73 (1980).
10. Harris, M. J., et al., *Ap. J.*, **362**, 135 (1990).
11. Leventhal, M., et al., this volume, (1993).
12. Share, G. H., et al., *Ap. J. (Letters)*, **358**, L45 (1990).
13. Leventhal, M., et al., *Ap. J. (Letters)*, **225**, L11 (1978).
14. Leventhal, M., et al., *Ap. J.*, **240**, 338 (1980).
15. Leventhal, M., et al., *Ap. J. (Letters)*, **260**, L1 (1982).
16. Leventhal, M., et al., *Ap. J.*, **302**, 459 (1986).
17. Neil, M., et al., *Ap. J. (Letters)*, **356**, L21 (1990).
18. Chapuis, C. G. L., et al., in *Int. Symp. on Gamma Ray Line Astrophysics*, in press, (1991).
19. Albernhe, F., et al., *Astr. Ap.*, **94**, 214 (1981).
20. Gardner, B. M., et al., in *The Galactic Center*, ed. G. R. Riegler and R. D. Blandford (New York:AIP), p. 144 (1982).
21. Paciesas, W. S., et al., *Ap. J. (Letters)*, **260**, L7 (1982).

GRIS DETECTIONS OF THE GALACTIC CENTER 511 keV LINE IN 1992

M. Leventhal, S. D. Barthelmy, N. Gehrels, B. J. Teegarden and J. Tueller
NASA/Goddard Space Flight Center, Greenbelt, MD 20771

L. M. Bartlett
University of Maryland, College Park, MD 20742

ABSTRACT

The Gamma Ray Imaging Spectrometer (GRIS) was flown on balloons over Alice Springs, Australia on 26 April and 7 May 1992. A full Galactic Center transit (~12 hours) was achieved on both flights with the instrument working nominally. The electron/positron annihilation line was detected on both flights at the 7-8 sigma level. A preliminary analysis of these results is presented and compared to earlier GRIS results and the results of GRO/OSSE observations of the Galactic Center. The GRIS 1992 results represent the first time that successive high resolution balloon measurements have been achieved on a time scale of days.

INTRODUCTION

For the past 20 years various balloon and satellite experiments have reported 511 keV electron/positron annihilation line radiation from the general direction of our own Galactic Center (GC). The history of these observations and its implications are reviewed in another talk at this workshop [1]. The GC 511 keV source appears to consist of at least two components, a steady diffuse source extended over many degrees and a point source, probably a stellar mass black hole, located near the GC. The argument for an extended source rests largely on the fact that instruments with increasing fields-of-view (FOV) report increasing fluxes, whereas the argument for a point source rests largely on the fact that narrow FOV (~15°) Ge experiments have reported variable

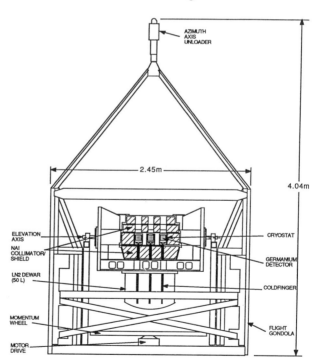

Figure 1. Schematic diagram of the GRIS instrument.

© 1993 American Institute of Physics

signals. This is a preliminary report on the latest chapter in this ongoing saga, namely the joint balloon and satellite GC observations in the spring of 1992. In particular, the 1992 GRIS results will be presented and compared to earlier GRIS results and to nearly contemporaneous OSSE observations and some tentative conclusions drawn. The results of additional balloon (HEXAGONE) and satellite (GRANAT/SIGMA) observations that were part of this joint campaign are not yet available.

Figure 1 shows the GRIS balloon instrument [2]. The instrument consists of 7 of the largest available high-purity Ge detectors cryogenically cooled by a liquid nitrogen dewar. The detectors are surrounded by a thick (15 cm) NaI shield in active anticoincidence which also serves to collimate the FOV of each detector to ~18° at 511 keV. It contains ~2000 cm^3 of Ge having an effective area ~100 cm^2 at 511 keV. The energy resolution of the system is ~2 keV at 511 keV. It is an azimuth over elevation system and points with an accuracy of a few tenths of a degree. GRIS was flown twice in 1988 and twice in 1992 over Alice Springs, Australia to observe the GC which transits very nearly through the zenith over Alice Springs. In each of these flights about 12 hour GC observations were achieved. In addition the Oct 1988 flight was long enough to allow a 6 hour observation of a point in the Galactic Plane 25° west of the GC.

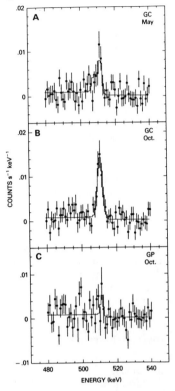

Figure 2. Spectra and model fits of the 511 keV line for the 1988 GRIS observations.

RESULTS

Figure 2 shows the results of the 1988 flights [3]. The May 1988 flight was the maiden flight of GRIS and an electronics problem degraded the energy resolution. The problem was fixed for the second flight in October. Nevertheless the GC annihilation line was easily detected on both flights. In the October observation the 511 keV line was resolved in energy for the first time. A striking decrease in the signal occurred when the telescope was pointed away from the GC confirming the fact that most of the emission is well localized near the GC. The residual signal seen in the off-center pointing is believed to be of diffuse origin. The 511 keV line flux derived for each GC flight is listed in Table I. There is a modest suggestion here, at the 1.9 sigma level, of an intensity variation between May and October 1988. The tabulated uncertainties are of statistical origin only which are believed to dominate over systematic uncertainties.

The Ge array flown in 1992 differed from that in 1988: on average the detectors were larger, and one of the detectors was isotopically enriched in Ge^{70} (a report on the performance of this detector is given in another talk at this workshop [4]). Figure 3 shows spectra and model fits for our April 26 and May 7, 1992 flights. The data have been accumulated in alternating ~20 minute target background pairs with the background taken at the same elevation angle but the instrument rotated in azimuth so as to maximize the distance from the Galactic Plane. The instrument worked well aside from a single detector failure. Complete GC transits were observed. Spectra from each detector were gain-corrected, compressed into 1 keV bins, and summed together

TABLE I

GRIS Galactic Center 511 keV Line Results

Flight	Flux (photons $cm^{-2} s^{-1}$)	FWHM (keV)	Centroid (keV)
May 1, 1988	(7.5 +/- 1.7) x 10^{-4}	≤ 3.6	511.46 +/- 0.38
October 29, 1988	(11.8 +/- 1.6) x 10^{-4}	2.9 +/- 0.6	510.97 +/- 0.26
April 26, 1992	(7.7 +/- 1.2) x 10^{-4}	1.3 +/- 0.7	511.34 +/- 0.18
May 7, 1992	(8.9 +/- 1.1) x 10^{-4}	3.6 +/- 1.0	511.37 +/- 0.30
Weighted Mean	(8.8 +/- 0.7) x 10^{-4}	2.5 +/- 0.4	
Chi-Squared	1.65 (3 D. O. F.)	2.29 (2 D. O. F.)	

into a single composite spectrum for each ~20 minute interval. Background intervals were subtracted from target intervals after scaling for live-time, and the difference was corrected to the mean atmospheric slant-range depth of the observation. These individual flux estimates were then averaged over the entire observation to obtain the final spectra shown in Figure 3. Preliminary least-squares curve fitting has been done to obtain estimates of the line fluxes and widths. The solid lines represent model fits to the sum of a Gaussian line plus a constant continuum with an adjustable "step" at 511 keV, to allow for a positronium-like [5] or Compton [6] continuum below the line. Considerable evidence exists in the data from both GRIS 1992 flights for a low energy "tail" on the 511 keV line. This is under study and will be addressed in a future publication. In the past it has usually been interpreted as a three-photon positronium continuum and its intensity relative to the 511 keV line used to infer a large positronium fraction [7]. However, a similar tail could be generated by Compton scattering from an attenuating medium or by gravitational redshifts near a compact object. The shape and intensity of these tails should allow us to constrain possible models.

The model fits shown in Figure 3 were derived by multiplying the model photon spectrum by the instrument response matrix and adjusting parameters in the model to minimize chi-squared as described by Tueller et al. [2]. The response matrix has the effective area times the atmospheric attenuation convolved with the instrument resolution on the diagonal and terms for Compton scattering in the instrument and atmosphere off the diagonal. A Gaussian was used for the instrument resolution function.

There is an apparent bump at ~ 518 keV in the May 7 spectra. In an attempt to test for the reality of this feature these data were also fit to the sum of two Gaussians plus constant but unequal continua above and below 511 keV. The F test [8] was applied to see if the observed improvement in chi-squared was more than that expected statistically. In fact the probability of getting the observed improvement statistically was only 0.009. However, at the present time, we are not inclined to believe in an astrophysical origin for the feature. Figure 4 shows the sum of our target data and background data plotted separately for the May 7 flight. It is apparent that the 518 keV bump is due to a dip in the background spectra rather than an excess in the target spectra. No such dip is seen in the background taken with other targets on this same flight.

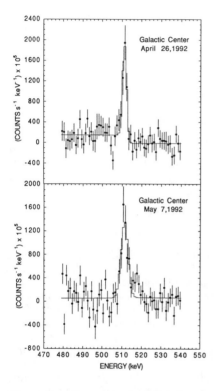 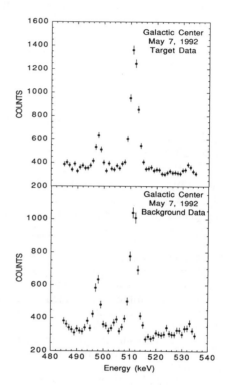

Figure 3. Spectra and model fits of the 511 keV line for the 1992 GRIS observations.

Figure 4. Target and background data for the May 7, 1992 GRIS GC observation. Instrumental lines at 497 keV and 511 keV are evident.

CONCLUSIONS

Table I lists our preliminary results for the 1992 GC 511 keV line. The uncertainty indicates 1 sigma statistical error bars calculated by finding the deviation of the parameter of interest required to increase the minimum value of chi-squared by 1, with all other parameters free to vary [9]. A fundamental question to ask about these results is whether or not line flux or line width changes were detected between these flights and/or the earlier GRIS flights. A quantitative answer to this question may be obtained by forming weighted means of the two quantities and assuming that GRIS is just detecting statistical fluctuations about these means. One can then calculate chi-squared values for this model and decide if the scatter is reasonable. The weighted means and reduced chi-squared values are listed in Table I. The probability of getting chi-squared values this large are 0.18 and 0.10 for the line fluxes and line widths respectively. *Hence the evidence for source variability (and the inferred presence of a point source) within the GRIS data set alone is suggestive but not compelling.*

The OSSE GC 511 keV line results are reviewed in another talk at this workshop [10]. As of April 1992, OSSE had viewed the GC for 7 ~two week observing periods. The last period ended on April 23, 1992, three days before the first GRIS 1992 balloon flight. The OSSE data suggest that they are looking at a steady source within a few degrees of the GC be it a point source, a diffuse source or some combination with flux ~2-3 x 10^{-4} ph cm^{-2} s^{-1}. *The obvious conclusion to reach from this result is that the GRIS data can not be understood in terms of emission from a single point source such as 1E1740.7-2942 within a few degrees of the GC.* Additional signal is required either from an extended diffuse source seen more effectively with the larger GRIS FOV and/or other point sources out of the OSSE FOV but within the GRIS FOV [11]. The variability reported by other narrow FOV experiments in the 1970's and 1980's implies that the known sources such as 1E1740.7-2942 exhibited greater variability in the past and/or the presence of additional variable point sources near the GC. Indeed one such additional source may already be known (V1223 Sgr) [12] and a second (EXS1737.9-2952) [13] suggested based upon observational data.

We wish to acknowledge the outstanding efforts of the other members of the GRIS team: M. Colbert, S. Derdeyn, C. Miller, K. Patel, and S. Snodgrass. Launch and recovery support was provided by an NSBF crew led by D. Gage. One of us, M. L., would like to acknowledge useful conversations with C. J. MacCallum concerning the observing program.

REFERENCES

1. J. Tueller, these proceedings (1992).
2. J. Tueller, S. D. Barthelmy, L. M. Bartlett, N. Gehrels, M. Leventhal, D. M. Palmer and B. J. Teegarden, Ap. J. Sup., submitted (1992).
3. N. Gehrels, S. D. Barthelmy, B. J. Teegarden, J. Tueller, M. Leventhal and C. J. MacCallum, Ap. J. 375, L13 (1991).
4. S. D. Barthelmy, L. M. Bartlett, N. Gehrels, M. Leventhal, J. Tueller and B. J. Teegarden, these proceedings (1992).
5. B. L. Brown and M. Leventhal, Ap. J. 319, 637 (1987).
6. R. E. Lingenfelter and X. M. Hua, Ap. J. 381, 426 (1991).
7. M. J. Harris, G. H. Share, M. D. Leising, R. L. Kinzer and D. C. Messina, Ap. J. 362, 135 (1990).
8. P. R. Bevington, Data Reduction and Error Analysis for the Physical Sciences (McGraw-Hill, N. Y. 1969), p. 158.
9. M. Lampton, B. Margon and S. Bower, Ap. J. 208, 117 (1976).
10. W. R., Purcell, these proceedings (1992).
11. J. G. Skibo, R. Ramaty and M. Leventhal, Ap. J. 397, 135 (1992).
12. M. S. Briggs, Phd thesis, University of California San Diego (1991).
13. J. E. Grindlay, C. E. Covault and R. P. Manandhar, Astron. & Ap., submitted (1992).

THE NUMBER OF POSITRON SOURCES

IN THE GALACTIC CENTER REGION

Ph. Durouchoux and P. Wallyn
Service d'Astrophysique, Centre d'Etudes Nucléaires de Saclay
91191 Gif sur Yvette Cedex, FRANCE.

ABSTRACT

On May 22 1989, HEXAGONE observed a broad region (19° FWHM) in the direction of the center of the Galaxy, and detected a narrow 511 keV line[1] and also a broad emission[2] around 170 keV. It is supposed in this paper that the observed positron annihilation features take place at least in three different regions: (1) The interstellar medium where positrons annihilate to form the diffuse[3] 511 keV component, (2) The molecular cloud G-0.86-0.08[4,5] where the positrons emitted by the hard-X-ray source 1E1740.7-2942[6] annihilate and contribute to the time-variable 511 keV line component, and (3) a new transient source EXS1737.9-2952[7] which showed a bump around 102 keV explained in terms of double-backscattered annihilation photons. The observations support the view that the scattering of annihilation photons is varying versus time, with a period of a few hours to a few weeks.

We propose a simple semi-quantitative model which can mimic the bumps as well as its time variations and emphasize the strong similarities between EXS1737.9-2952 and Nova Muscae: EXS1737.9-2952 is also supposed in this paper to be a low-mass X-ray binary containing a black hole with an important accretion rate fueled by a companion star, with instabilities which periodically heat the inner parts of the disk ($T>10^9$ K) where pair plasma is formed. We suppose here that outflows of matter during such supercritical states[8] are the sites where both the annihilation of the positrons and the subsequent scatterings of the photons take place. The features at 102 and 170 keV are then explained, assuming that they are emitted by the same source: the transient EXS1737.9-2952.

INTRODUCTION

The Galactic center has been observed, in hard X-rays and gamma-rays, many times in more than two decades and we still do not have a comprehensive picture of this region. Nevertheless, due to the refinement of the experiments, mainly in the accuracy of the positioning (GRANAT[9] observatory, POKER, GRIP and EXITE ballon-borne experiments) and the use of high-resolution detectors (GRIS and HEXAGONE germanium detectors balloon-borne experiments), our understanding of the phenomena which take place in the region of the Galactic center is increasing.

It is fairly well established that the 511 keV annihilation line has two origins[3]: i) a steady diffuse emission, which follows more or less the type I supernova distribution along the Galactic plane, and ii) a variable point source, assuming the positrons are emitted by 1E1740.7-2942 (1E hereinafter) and annihilating in a nearby cold molecular cloud.[4,5]

Here we question the validity of these assumptions: analyzing carefully the balloon data successfully gathered in spring 1989 (GRIP, EXITE, POKER and HEXAGONE, with performances listed in table 1), we investigate the possibility that the bumps (around 170 keV, 102 keV and possibly 72 keV) which appear in some of the spectra (and not in others) can be explained if we assume that 1E is not the positron source which produces the backscattered photons.

NAME	DATE	Performances	73 KeV	102 KeV	170 KeV
GRIP	1989 April 12th	FOV 14° RES 1.1° 0.03-10MeV	NO BUMP IN 1E SOURCE		
EXITE	May 9th	FOV 3.4° RES 22' 20-250 KeV	NO (*)	BUMP 83-111 KeV	NO
POKER	May 17th	FOV 1.9° 15-150 KeV	NO (**)	NO	OUT OF RANGE
HEXAGONE	May 22th	FOV 19° 0.02-10 MeV	POSSIBLE 2σ	NO	YES
SIGMA (Nova Muscae)	Jan 91 1st 7 hours	15' images 35-1300 keV	NO	YES	NO
	2nd 7 hours		NO	NO	YES
	last 6 hours		NO	NO	YES

Table 1: EXS1739.9-2952 (April/May 1989) and Nova Muscae (1991 Jan 20-21) observations.

(*) 4σ detection in low X-ray band
(**) an absorption feature at 47 keV or emission feature at 58 keV interpreted as cyclotron line
FOV : Field of view; RES: Angular resolution

Among the four balloon experiments mentioned above, the first three have positioning capabilities, whereas the fourth has high-energy resolution. GRIP detected on April 12 the 1E source in its normal state, EXITE (May 9) also, but found a "new" source: EXS1737.9-2959 (40' west of 1E, EXS hereinafter), which exhibits a feature in the range 83-111 keV, as well as a bump in the soft X-ray band (20-30 keV).

Futhermore, a bump around 170 keV[1] in the spectrum corresponding to a large region of the center of the Galaxy was detected on May 23 and interpreted as backscattered 511 keV photons.[10] For clarification, we summarized (Table 1) these April/May 1989 observations having 1E and EXS sources in their fields of view. Furthermore, due to similarities in their spectra, we added in Table 1 the results concerning Nova Muscae obtained with the SIGMA telescope[9] although it is not a source sitting in the GC region. Novae Muscae has been monitored by the GRANAT observatory in the 3-1300 keV band and during the January 20/21 observation features around 102, 170, and 480 keV have also been observed and their successive appearances reported.[11] This observation has been separated into three sub-sequences of about the same duration; In the first sub-sequence a prominent 102 keV bump is observed, while during the second sub-sequence this bump has disappeared and a broad 170 keV feature emerged as well as an other around the energy of 480 keV. The third sub-sequence shows a narrowing of the 170 keV feature and a decrease in the 480 keV line flux. We have here the simultaneous presence of both features: a possible redshifted annihilation line and the 170 keV feature. Unfortunately the next sequence took place one week later on February 1 and neither feature nor annihilation line are seen. The typical time scale of the rising time of the annihilation line is clearly established to be of the order of a few hours[11] reducing the probability to frequently observe this kind of annihilation outburst.

It must also be pointed out that the 102 keV feature appears before the 170 keV bump in Nova Muscae as well as in EXITE and HEXAGONE GC observations. This behavior was not expected since a single backscattered feature has a travelling time shorter than a double backscattered feature.

MODEL AND MONTE CARLO SIMULATION

We are now trying to explain this timing sequence with a model taking into account the evolution of the features versus time: Shakura & Sunyaev[8] have investigated the evolution of an accretion disk around a black hole related to the variation of the accretion rate. During supercritical inflows of matter, a strong anisotropic mass outflow due to the radiation pressure exceeding the gravitational attraction can be observed in a small cone near the axis of the rotational disk. The matter flows out in a spiral with a speed decreasing when this bubble moves away from the compact source (Figure 1). We thus propose that Nova Muscae and the 1989 observations can be tentatively explained in a roughly similar framework: a high activity state in the inner part of the accretion disk (following a variation in the accretion rate) creates the conditions for a hot plasma $T>10^9$ K) and the subsequent emission of pairs. This stream of electron-positron particles will create a hot spot in the outflowing matter when it penetrates this region. The location of this hot spot compared with the observing direction (θ as shown on fig. 1) varies versus time, since the bubble is spiraling.

We performed a Monte Carlo simulation based on this simple geometry, varying the values of θ, τ (the Compton scattering optical depth at 511 keV, equivalent to $3.5 \; 10^{24}$ electrons.cm^{-2}) and the photon arrival time at the detector level.

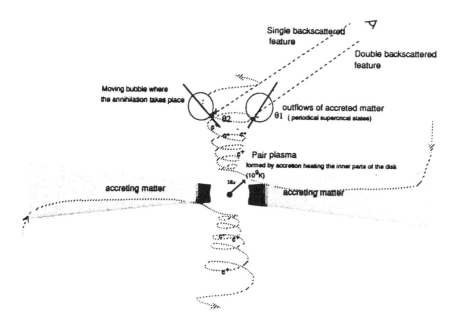

Fig 1: Geometry of a spherical cloud of matter coming from a supercritical state, with a hot spot annihilation source on its surface. The spectrum evolution is studied by varying both the observing angle and the arrival time at the detector. For $\theta = \theta_1$, the cloud is located behind both the observer and the annihilation region: in the scattered spectrum a 170 keV feature and no forward scattered photons must be present. For $\theta = \theta_2$, a part of the cloud is located between the annihilation region and the observer, and therefore no 170 keV feature must be seen.

The evolution of the spectra, versus time, for different observing angles is presented in fig. 2. One can notice that first (i.e., for high θ value: e.g. 150°), a double backscattering feature (102 keV) clearly appears, and then (for lower values of θ: e.g. 30°) a single backscattered line (170 keV) is visible and no more 102 keV bump is present in this later spectrum. This timing sequence reproduces satisfactorily well both the Nova Muscae and EXS source behaviors.

CONCLUSION

The model we have developed here is able to take into account the behavior of both the Nova Muscae and the EXS source if we suppose we observed it with HEXAGONE. Multiple positron point sources might be present in the galactic disk, belonging or not to a binary system with a black hole, with different geometries, accretion rates, etc. Many of these sources could have much lower luminosities than Nova Muscae or EXS and not be detected as a point source but be a part of the 511 keV diffuse component.

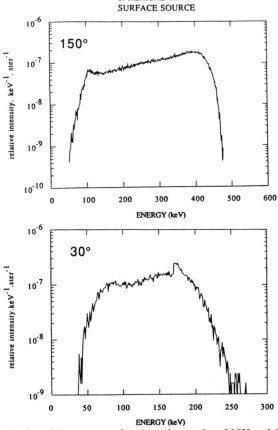

Fig 2: Time evolution of the spectrum for observing angles of 150° and 30° (one can clearly see the 102 keV and then the 170 keV appearances)

REFERENCES

1. Ph. Durouchoux et al. ,1992, Astr. Ap. Sup. Ser., in press
2. D. Smith, et al., 1989, submitted to Ap.J.
3. R. Lingenfelter & R. Ramaty , 1989, Ap.J., 343, 686
4. J. Bally & M. Leventhal, 1991, Nature, 353, 234
5. F. Mirabel et al., 1991, Astr.Ap.(Letters), 251, L43
6. L. Bouchet et al., Ap.J. (Letters), 383, L45
7. J. Grindlay et al., 1992, Astr. Ap. Sup. Ser., in press
8. N. Shakura & R. Sunyaev., Astr. Ap., 1973, 24, 337
9. J. Paul et al., 1991, Adv. in Space Research, 11.8, 289
10. R. Lingenfelter & X. Hua., 1991, Ap. J., 381, 426
11. A. Goldwurm et al., 1992, Astr. Ap. Sup. Ser., in press

THE ANNIHILATION OF POSITRONS IN

A MOLECULAR CLOUD

P. Wallyn and Ph. Durouchoux
Service d'Astrophysique, Centre d'Etudes Nucléaires de Saclay
91191 Gif sur Yvette Cedex FRANCE

ABSTRACT

We reevaluated positron cross-sections with atomic and molecular hydrogen and performed a Monte Carlo simulation for the e^+ charge exchange in a molecular cloud. A slowing down time is derived for different initial e^+ energies and cloud densities.

For positron energies ranging from 300 keV to 1 MeV, we calculated slowing down times on approximately of 1 month to 1 year for H_2 medium densities of 10^5 cm^{-3}. The e^+ annihilation time for the same energy range varies from 1 to 4 days.

Taking into account the 6 month variation in the annihilation line intensity observed by the spectroscopic experiment onboard the HEAO 3 satellite, we derived a maximum injection energy for the positrons of about 0.5 MeV for a density of approximately of 10^5 cm^{-3}. If we consider the molecular cloud G-0.86-0.08 as the slowing down and the annihilation site of the e^+ emitted by the source 1E1740-2942, and assume a mean density in this cloud of approximately of 10^3 cm^{-3}, one derives an initial positron energy of the order 130 keV with a duration of annihilation of about 1 month.

INTRODUCTION

The behavior of positrons emitted by various astrophysical objects and the understanding of annihilation radiation physics require a good knowledge of the interactions of positrons with ambient interstellar medium (ISM) and especially the measurements of the various cross-sections over a large energy band.

The slowing down and annihilation of positrons and the formation of the bound state of positronium was first investigated for the physical conditions occurring in solar flares[1], then in the ISM [Bussard, Ramaty & Drachman[2] (BRD)], the dusty ISM[3], superstrong magnetic fields[4] and hot electron-positron plasmas[5]. These studies are particularly important since a cold and dense molecular cloud in the vicinity of the positron source 1E1740-2942 has been suggested as a possible annihilation site[6,7].

Recently, calculations of the evolution with time of the 511 keV line width and positronium fraction from the time-variable component in the

Galactic center region for various annihilation media (cold and ionized) has been performed [8].

Following the pioneering works, we performed more accurate calculations and studied the behavior of positrons in a cold phase, reevaluating the e^+-H and e^+-H_2 cross-sections.

POSITRON-H_2 INTERACTION-CROSS SECTIONS

We used the P_s formation experimental cross-section results, which are the sum of the available ground and excited state values. Below 26 eV, we took into account accurate experimental results[9] and, very close to the threshold of Ps-formation (E_{ps} = 8.63 eV), we fit by a steep straight line[10]. Between 20 and 35 eV, where the cross-section is maximum, theoretical results of BRD overestimate these experimental results even if we take slightly larger values[11] which are also in fair agreement with the values of Diana et al.[12] (Figure 1). Between 26 and 250 eV we average the Bielefeld group[9] and the Arlington group[11,12] experimental data which give consistent results.

Beyond 250 eV no experimental data have yet been published and we connect the low energy values with those of a recent theoretical calculation of Biswas et al.[13], who redid the calculation of Sural et al.[14] (up to 1000 eV) that was used by BRD to fit the data at energies greater than 50 eV. Only the ground state cross section is calculated but the relative intensity of this cross-section at high energies is small compared to the ionization and does not change the results drastically (Figure 1). The method we used for the positron-impact ionization cross-section is described in Wallyn & Durouchoux[15]. Then we substracted the charge exchange, ionization, and elastic cross-sections from the experimental data for the total cross-section and assume the remainder represents the excitation cross-section for H_2.

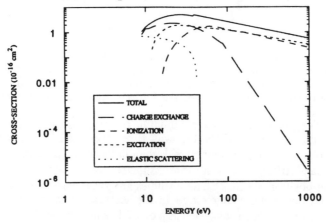

Fig 1: e^+ - H_2 cross-sections

POSITRON - H CROSS-SECTIONS

The theoretical values used by BRD for the Ps cross-section[16] did not take into account any excited states, but calculation of the charge exchange cross-section is receiving increasing attention from theoretical investigators and several results are now available. Unfortunately there are some important discrepancies[17, 18, 19, 20]. We applied the same method (Wallyn & Durouchoux[15]) to calculate the excitation cross-section (Figure 2). Our values are consistently greater than the results with electron interactions and in accordance with the behavior for H_2 excitation cross-sections.

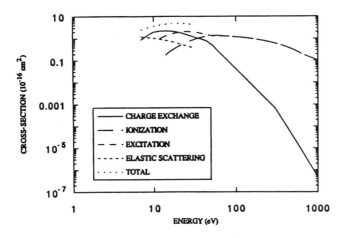

Fig 2: e^+ - H cross-sections

MONTE CARLO SIMULATION AND DISCUSSION

A positron emitted in a neutral medium loses its energy by excitation and ionization of hydrogen. In our simulation, we neglected the energy loss during each elastic scattering and, based on the above cross-sections, we modeled the lifetime before annihilation by charge exchange in a molecular hydrogen cloud with a density of 10^5 cm^{-3}, of positrons with various initial energies. As shown in Figure 3, the range of the slowing down time depends on the initial energy of the positron. This time is approximatly 2 days for 600-keV positrons. One can deduce from Figure 3 an average time of annihilation versus initial energy.

Taking into account the 6-month variation (which implies that the slowing down and annihilation of positrons by charge exchange takes a shorter time), we deduce a maximum energy of the positrons of 0.5 MeV.
Nevertheless, molecular clouds are not likely to be of constant density and the observations of G-0.86-0.08 in the millimeter wavelengths cannot set a stringent constraint. Low densities (of up to 10^3 cm^{-3}), are expected (Mirabel, private communication 1992). Under these conditions the maximum initial energy for positrons is around 130 keV, with a duration of annihilation of 1 month.

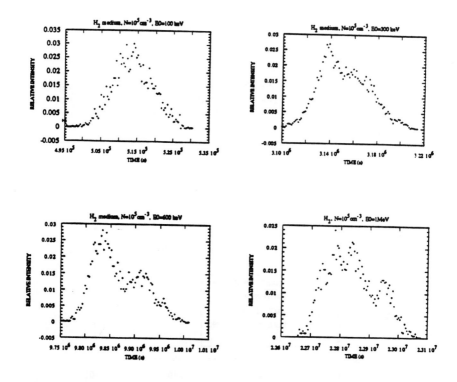

Fig 3 : Slowing down time before annihilation by charge exchange and duration of the annihilation time in a H_2 media containing 10^5 particles cm^{-3} versus the initial energy of the positrons.

CONCLUSION

This framework does not imply any correlation [like the similarities between the high states flux of 1E1740-2942 in the radio and gamma energy bands (Mirabel et al. 1992b)] between the observation of a burst from 1E1740-2942 and a possible detection of a high state of the narrow 511 keV line due to the travel and slowing down of the positrons before annihilation. A delay of typically 4 years (1pc travel plus the slowing down in the hot spots) is expected before the contribution to the narrow 511 keV line of the positrons emitted from 1E1740-2942 burst can be detected.

REFERENCES

1. C. J. Crannel et al. 1976, Ap.J., 210, 582.
2. R. W. Bussard et al. 1979, Ap.J., 228, 928.
3. W. H. Zurek., 1985, Ap. J., 289, 603.
4. J. K. Daugherty & R. W. Bussard., 1980, Ap. J., 238, 296.
5. R. Ramaty & P. Mészaros., 1981, Ap. J., 250, 384.
6. J. Bally & M. Leventhal., 1991, Nature., 353, 234.
7. I.F. Mirabel et al., 1991, Astr. Ap. (Letters), 251, L43.
8. R. Ramaty et al., 1992, Ap. J.(Letters), 392, L63.
9. D. Fromme et al., in Atomic Physics with positrons, eds. J. W. Humberston, and E. A. G. Armour, Plenum Press. 1986.p 407.
10. W. Raith, in Atomic Physics with positrons eds. J. W. Humberston, and E. A. G. Armour, Plenum Press. 1986.p 1.
11. L. S. Fornary, L. M. Diana, P. G. Coleman, 1983, Phys. Rev. Letters, 51, 2276.
12. L. M. Diana, P. G. Coleman, D. L. Brooks, P. K. Pendleton, and D. M. Norman., 1986, Phys. Rev. A, 34, 2731.
13. P. K. Biswas, T. Mukherjee, and A. S. Ghosh., 1991, J. Phys. B: At. Mol. Opt. Phys., 24, 2601.
14. D. P. Sural, S. C. Mukherjee., 1970, Physics, 49, 249.
15. P. Wallyn & Ph. Durouchoux, submitted to Ap. J., 1992.
16. R. J. Drachman, 1983, in Positron-Electron Pairs in Astrophysics, eds. M. L. Burns, A. k. Harding, and R. Ramaty (New York: AIP 101), p. 242.
17. M. A. Abdel-Raouf et al., 1984, Phys. Rev. Letters, A.100, 353.
18. P. Kahn & A. S. Ghosh., 1983, Phys. Rev.. A. 27, 1904.
19. J. W. Humberston., 1986, Adv. At. Mol. Phys. 22, 1.
20. E. Ficocelli Varracchio, & M. D. Girardeau., 1983, J. Phys. B: At. Mol. Phys. 16, 1097.

SUPERNOVAE AND NOVAE

COMPTON OBSERVATORY OSSE STUDIES OF SUPERNOVAE AND NOVAE

M. D. Leising, D. D. Clayton, and L.-S. The
Clemson University, Clemson SC 29634-1911

W. N. Johnson, J. D. Kurfess, R. L. Kinzer, R. A. Kroeger,
M. S. Strickman, and J. E. Grove
Naval Research Laboratory, Washington DC 20375

D. A. Grabelsky, W. R. Purcell, and M. P. Ulmer
Northwestern University, Evanston IL 60201

R. A. Cameron and G. V. Jung
Universities Space Research Association, Washington DC 20024

ABSTRACT

A primary objective of the Compton Observatory is the direct study of explosive nucleosynthesis in supernovae and classical novae. We have been fortunate in that three rare events have coincided, relatively speaking, with the Compton Observatory launch. Supernova 1987A, roughly a once per century event, was only 4 years old at launch and so the γ-ray flux from ^{57}Co decay was not much past its peak value. Supernova 1991T, a SN Ia which exploded within a few days of launch, is a once in a decade event. It offers as good a chance as we could reasonably expect to detect the ^{56}Ni and ^{56}Co decays which are supposed to be responsible for the impressive SN Ia display. Nova Cygni 1992, also a once in a decade event, might be our best chance to detect γ-rays from ^{22}Na, a unique nucleosynthesis byproduct of the explosive hydrogen burning thought to power classical novae.

The OSSE has detected 122 keV line and Compton scattered continuum photons from ^{57}Co decay in SN 1987A. The total flux of 9×10^{-5} cm^{-2} s^{-1} corresponds to a ^{57}Ni/^{56}Ni production of 1.5 ± 0.5 relative to solar Fe, which is in conflict with previous interpretations of optical and infrared data. OSSE and COMPTEL upper limits on ^{56}Co γ-ray lines from SN 1991T at 3-4 $\times 10^{-5}$ cm^{-2} s^{-1} are not in conflict with published models of SNIa at distances $\geq$10 Mpc. However if those models are correct, the distance must be $\leq$10 Mpc to give the observed optical luminosity, and are then in conflict with the γ-ray limits. The OSSE upper limit from N Cyg 1992 near 10^{-4} cm^{-2} s^{-1} at 1.275 MeV corresponds to an ejected ^{22}Na mass of 8×10^{-8} M$_\odot$ at a distance of 1.5 kpc.

I. INTRODUCTION

The prospect of detection of short-lived radioactivity just following explosive events has been a cornerstone of astrophysical γ-ray spectroscopy since its beginning. The best hope, ^{56}Co in supernovae,[1] was realized first, somewhat surprisingly, in the Type II supernova SN 1987A.[2] We looked forward to the Compton Observatory, still awaiting the first detections of radioactivity in a Type Ia supernova, which should be intrinsically much brighter, and in a classical nova,[3] which are much more frequent and therefore likely to be very nearby. The Compton Observatory launch was just soon enough that we still hoped to be able to detect ^{57}Co in SN 1987A, and to measure the mass of that isotope ejected. This would for the first time allow us to determine the ratio of abundances of two isotopes produced in the deepest, highest temperature layers ejected.

II. OSSE OBSERVATIONS OF SN 1987A

Details of two OSSE observations[4] of SN 1987A and their interpretation[5] have been recently reported. Here we summarize the important features. Two 2-week averages of OSSE difference spectra are shown in Figure 1. The first observation began on July 25 1991 (1613 days after the explosion), the second on Jan 10 1992 (1768 days). Each of these spectra was fit with a model, consisting of an exponential continuum plus a ^{57}Co template, convolved with the OSSE instrument response. The template is derived from a supernova model evolved to this time and includes the 122 and 136 keV lines plus a Compton-scattered continuum. In the fit shown the model is based on the model 10HMM of Pinto and Woosley[6] which has roughly equal numbers of photons in the 122 keV line and in its Compton-scattered continuum at these times.

Both observations show an excess consistent with this template, but atop very different continua. The flux in the template, lines plus continuum, is $(9.0 \pm 2.2) \times 10^{-5}$ for the first observation, and essentially the same, although with a larger uncertainty, for the second observation. For a variety of continuum models and supernova templates, the best-fit value of the flux varies by less than the 1σ statistical uncertainty. Fitting the 122 keV line only, without its scattered continuum, plus an exponential continuum yields a line flux consistent with one-half that measured for the template. We have interpreted this as a detection of ^{57}Co in the SN 1987A ejecta.

In the model of Pinto and Woosley, as in all previously published models of the ejecta of SN 1987A, essentially all ($\simeq$98%) of the ^{57}Co decay photons escape at these late times, either directly in the lines or after very few scatters. Assuming this is correct, we can convert the flux into the ejected mass of ^{57}Co, which is 2.7×10^{-3} M$_\odot$ extrapolated back to t=0, assuming a distance of 50 kpc. Compared to the ^{56}Co mass of 0.075 M$_\odot$ inferred from the early UVOIR light curve, the production ratio of the parents, X(^{57}Ni)/X(^{56}Ni), is 1.4$\pm$0.35 times the

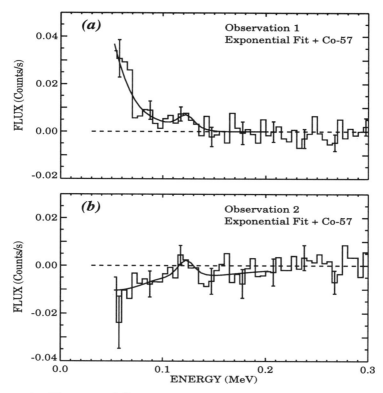

Figure 1: *The mean difference spectra for SN 1987A for the two-week viewing periods (a) No. 6 and (b) No. 17. The solid lines are the best-fit exponential continua plus ^{57}Co line and continuum template from model 10HMM.*

solar ratio of the daughters, $(^{57}Fe)/(^{56}Fe)_\odot$. This measurement is inconsistent with earlier inferences[7,8] that this production ratio was five times the solar value to explain the late optical/infrared luminosity in the context of the same supernova model, 10HMM.

The OSSE flux and the ground-based measurements can be reconciled with a 57/56 production ratio roughly twice the solar value *and* a supernova model which is thicker to ^{57}Co γ-rays than published models. The thicker supernova, probably a result of lower expansion velocity in the inner ejecta, would yield a lower escaping hard flux per ^{57}Co nucleus because of increased photoelectric absorption, and would have a larger fraction of the ^{57}Co power thermalized (roughly 40% at the OSSE epoch rather that 20% in, e.g.,10HMM).

However, the late-time optical/infrared luminosity is not necessarily derived entirely from ^{57}Co. Other power sources, each of which, except for the last, should be present at some level include

- Delayed release of a small fraction of the earlier large power from ^{56}Co.

- Conversion of a fraction ($\simeq 10^{-13}$ s^{-1}) of the mechanical energy of the ejecta into infrared emission.

- Other radioactivity, e.g., ^{44}Ti.

- Accretion onto or rotational energy extracted from a compact remnant.

The first effect should exist because the processes between γ-ray emission and optical escape – Compton scattering, then electron energy loss (via ionization), and subsequent recombination – have inherent delays. The most important is probably recombination, because the rapidly declining density in the ejecta causes the recombination time to grow progressively longer, delaying the release of the power.[5]

The second source, mechanical energy is available only if the ejecta interact with external material. Even the blue supergiant wind, if it blew at 2×10^{-6} M$_\odot$/yr at 500 km/s prior to the explosion could cause significant dissipation. The ejecta, at 10^4 km/s, is sweeping up this wind at 4×10^{-5} M$_\odot$/yr, and conservation of momentum demands that $\simeq 10^{39}$ erg/s is being dissipated. If 10% of this is emitted as light, all the luminosity in excess of ^{56}Co power can be accounted for. The rest of this dissipation luminosity might be detectable in another band (e.g., radio, x-ray), or might go into PdV work.

As for the last two effects in the list, we doubt, based on straightforward nucleosynthesis arguments,[9] that ^{44}Ti is a dominant contributor to the luminosity, and we find no conclusive evidence for a contribution from a compact object. However, we can not rule out that the soft continuum we detect in the first observation comes from within SN 1987A.

III. OSSE OBSERVATIONS OF SN 1991T

Prior to SN 1987A, it was generally considered that a Type Ia supernovae would yield the first detection of these γ-rays, because it is thought that the energy of these events is entirely thermonuclear in origin.[10] The Type Ia light curve is supposed to be powered by ^{56}Ni decay initially and later ^{56}Co. This is the result of the thermonuclear disruption of a white dwarf near the Chandrasekhar mass, a large fraction of which is converted to ^{56}Ni. Calculations indicate that 0.5–1.0 M$_\odot$ of ^{56}Ni is ejected, depending on the nature of the burning, which remains uncertain.[11-15] In addition to the large mass of ^{56}Ni, very large expansion velocities are expected and observed in Type Ia supernova, so they become thin to γ-rays in a matter of months, before significant decay.

Supernova 1991T was discovered on April 13 1991[16] in NGC 4527, a spiral galaxy on the edge of the Virgo cluster. Its premaximum spectrum, which lacked lines of intermediate mass elements, was unusual for a Type Ia supernova, but its

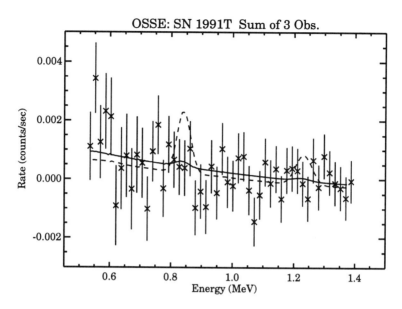

Figure 2: *The mean difference spectra for SN 1991T for all three viewing periods, all with equal weights. The solid line shows the best-fit power-law continuum plus 847 and 1238 keV lines with relative amplitudes equal to the laboratory branching ratio. The dashed line shows the same model but is fit with the 847 keV flux fixed at 5×10^{-5} cm^{-2} s^{-1}.*

postmaximum spectrum and its light curves were quite similar to normal Type Ia's throughout.[17,18] SN 1991T peaked at about V=11.5 on about April 30 1991,[18] making it the brightest SNIa since SN 1972E.

The Compton Observatory, launched only a week before the SN 1991T discovery, was thus provided with an opportunity to test the idea of SNIa's as thermonuclear explosions. By chance the OSSE team happened to contact supernova observers just hours after the discovery. With much cooperation from the Compton Observatory flight operations team and among the 4 experiments, plans were quickly made to observe SN 1991T as the third target of the observatory. The spacecraft axis was pointed toward the supernova during the periods June 15-29 1991 and October 3-17, so both the OSSE and the COMPTEL experiment could observe it. The OSSE was also able to observe SN 1991T as a "secondary" target during August 22 - September 5 1991.

The OSSE has good sensitivity for the two strongest lines of ^{56}Co decay at 847 keV and 1238 keV. Standard OSSE analyses techniques[19] were applied to the data from all three viewing periods, and a mean difference spectrum was obtained for each. Each spectrum shows evidence for excess low-energy continuum photons from the source location. We presume these come from the quasar 3C 273 which is only 1.4 degrees from SN 1991T. We find no evidence for either line in any of the three spectra. Figure 2 shows a simple live-time weighted mean of the three difference spectra. The solid curve is the best-fit of a model consisting of a power-law continuum plus the two lines where the ratio of the line amplitudes is fixed at the laboratory value. Also shown for illustration, as the dashed curve, is the fit when the flux in the 847 keV line is fixed at 5.0×10^{-5} cm^{-2} s^{-1}. That the flux is this high is statistically very implausible.

To draw the firmest conclusions from these combined observations, we really need to appeal to supernova models, because the fluxes are expected to vary with time, as is the ratio of the fluxes of the two lines. However, for no reasonable model is the upper limit on the 847 keV flux in the first observation very different from this simple determination, and in all cases we have tested it is $\leq 5 \times 10^{-5}$ cm^{-2} s^{-1}. Formally, 99.5% confidence upper limits, quoted for the 847 keV line flux in the first observation, vary from 3.0×10^{-5} cm^{-2} s^{-1} to 4.0×10^{-5} cm^{-2} s^{-1}. Similar limits are obtained from the COMPTEL observations during each of two observing periods (Lichti et al., this volume).

The probability of detecting the γ-ray lines from ^{56}Co decay depends mainly on the mass of ^{56}Ni produced and on the distance to the supernova. The distance to NGC 4527 is uncertain, even relative to the uncertain Virgo cluster center distance.[20] SN 1991T was somewhat brighter at maximum than the typical Virgo Cluster Type Ia supernovae (V=12.0, corrected for extinction and distance from the cluster center;[21,22]). This could be because it had a peak luminosity typical of other SNIa but is somewhat closer,[18] or because it was anomalously luminous and at nearly the same distance as the cluster center.[17] Significant extinction would require even smaller distance or higher luminosity. Regardless of these uncertainties, we expect the γ-ray line flux also to be larger than for typical Virgo cluster SNIa. However, we do not know those fluxes are, because we do not have any entirely successful model and we do not know the distance to Virgo. Still, any theoretical model is safe from the contradiction by this upper limit as long as the distance can be large enough. However, as pointed out by Arnett[23] both optical and γ-ray fluxes scale the same way with distance and approximately the same way with ^{56}Ni mass, so the ratio of γ-ray to optical flux at their respective peaks should be roughly constant for Type Ia supernovae.

A pertinent question is then: are any models consistent with a Type Ia supernova being this bright optically but with γ-line fluxes this low? Figure 3 illustrates this relationship. It shows the peak 847 keV flux versus the extinction-corrected blue magnitude at maximum. Although we do not know exactly when the peak γ-ray flux should occur, the flux at the times of our first two measurements, at least, should be very near the peak. Shown in Figure 3 are two lines of varying distance for fixed combinations of γ-ray flux/blue magnitude. One is based on an analytic description,[23,24] another on model N21 of Khokhlov.[14,25,26] The latter model is shown as a point at distance 13.5 Mpc, which has been listed for NGC 4527.[27] Also shown at this distance is the model W7,[11] which, although it has a much lower ^{56}Ni mass and is fainter, because it lies so close to the line for model

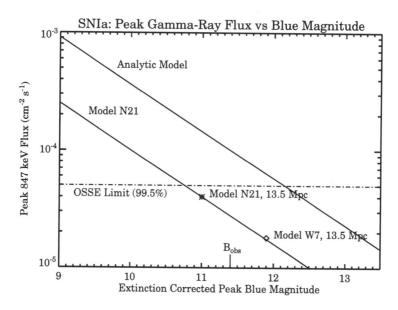

Figure 3: *The peak 847 keV flux versus peak intrinsic blue magnitude for various supernova models. Solid lines: an analytic description[24], and model N21 [14] at varying distances. Points are for N21 and model W7 [11] at distance 13.5 Mpc.*

N21, demonstrates the validity of Arnett's idea if not his numerical result.

We can use the optical observations to determine the distance and therefore predict the peak γ-ray flux for a given model. For example, the model N21, if at a distance such that at maximum B=11.4, predicts a peak 847 keV flux of 2.8×10^{-5} cm^{-2} s^{-1}. This is consistent with the OSSE data at the 7% level of confidence. Model W7 predicts very nearly the same flux for B=11.4, but it would be at a smaller distance. The analytic model shown clearly overestimates the γ-ray line flux, and several of its assumptions are brought into question.

Significant reddening and extinction would imply that a given model should be placed at a still smaller distance, with a higher γ-ray flux, to give the observed peak B magnitude. Filipenko et al.[17] estimate extinction as high as 0.7 mag, but settle on 0.4 mag as a more probable value. Then B_{max}=11.0 mag intrinsically , and we expect the 847 keV line flux to be 4.0×10^{-5} cm^{-2} s^{-1} for either model N21 or W7. This is ruled out at the 97% level of confidence. Ruiz-Lapuente et al.[28] find $\simeq$ 1.0 mag of extinction, which implies B_{max} =10.4 and predicts $F_{max}(847) = 7 \times 10^{-5}$ cm^{-2} s^{-1} - well above our current limit.

It is unlikely that current Type Ia models can satisfy both the optical and γ-ray measurements of SN 1991T. However, the conflict is not great enough that we can rule out the basic thermonuclear explosion paradigm. Models which more efficiently convert the decay power into optical luminosity at maximum, possibly by storing for a short time the large premaximum ^{56}Ni decay power, can probably be developed.

IV. Nova Cygni 1992

N Cyg 1992 was discovered[29] on February 19, and reached visual magnitude V≃4 on 29 February. As this was the brightest classical nova since 1975 and likely to be relatively close, we decided to observe it with the OSSE, and did so from 5–20 March 1992. The most likely detectable radioactive byproducts of the thermonuclear explosion at that time were ^{7}Be ($\tau_{1/2}$=53 days; E_γ=478 keV in 10% of decays) and ^{22}Na ($\tau_{1/2}$=2.6 yr; E_γ=1275 keV). The prospects for detecting ^{22}Na were improved by observations that the ejecta of N Cyg 1992 were highly enriched in neon,[30,31] the seed of ^{22}Na production.

Standard analysis techniques were applied to the OSSE data, however very restrictive screening was performed. This was needed because the Compton Observatory tape recorder errors were at a very high rate at that time. Useable data were greatly reduced (less than one-half the data are included here), especially during the last week of these observations.

Figure 4 shows the mean difference spectrum for 5–20 March. We find no evidence for any line emission, including the two mentioned above and the 511 keV line which should accompany the 1275 keV line because positrons are emitted in 92% of ^{22}Na decays. Our 99.5% confidence upper limit on 1275 keV emission, assuming also that the 511 keV line flux is 0.5 of that at 1275 keV (i.e., that the positronium fraction is ≥90%) is 1.3×10^{-4} cm^{-2} s^{-1}. The 99.5% confidence upper limit on 478 keV line emission is 1.2×10^{-4} cm^{-2} s^{-1}. The limits might ultimately be somewhat lower if all the data can be corrected for tape recorder errors.

Assuming that the ejecta are thin by late March, and assuming a distance to the nova, we can convert these flux limits to limits on the ejected masses of the radioactive nuclei. The first assumption is valid if ejection velocities are ≥100 km/s, even if as much as 10^{-4} M$_\odot$ of matter are ejected. Based on the decline of the light curve we estimate that the distance is in the range 1 to 3 kpc. For illustration, we assume a distance of 1.5 kpc in the following. Our limit corresponds to an ejected mass of ^{22}Na of $\leq 8\times10^{-8}$ M$_\odot$. This is less than given by some models, which have ^{22}Na mass-fractions of 10^{-3} (see Starrfield et al., this volume), if the ejected mass is 10^{-4} M$_\odot$. However, those models do not eject so much total mass and might not be appropriate to slow or moderately-slow classical novae (such as N Cyg 1992). If significant ^{22}Na is produced but not ejected explosively, it might be subsequently ejected in wind-driven mass-loss, in which case it might have been thick to γ-rays during our observation. This possibility and the poor data quality argue for another observation of N Cyg 1992 in the next year or two. Our 478 keV flux limit corresponds to 1.9×10^{-8} M$_\odot$ of ^{7}Be ejected at t=0. This is about the most one expects from a nova,[32] but the uncertainties in production make the upper limit much less interesting than a

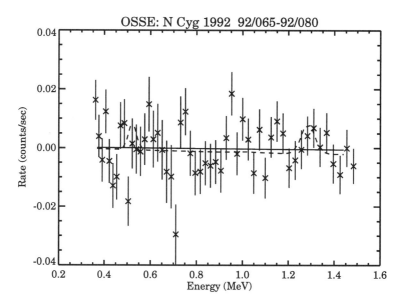

Figure 4: *The mean difference spectrum for N Cyg 1992 for the period 5–20 March 1992. The dashed line shows how a line flux of 10^{-4} cm^{-2} s^{-1} at 1275 keV and 1/2 that at 511 keV would appear.*

detection would have been.

We have been very fortunate indeed to have such an interesting observation of a Type II supernova, SN 1987A; SN 1991T was tantalizingly close to the distance where we must be able to detect current Type Ia models, but is not inconsistent with them; and N Cyg 1992, a reasonably good chance to confirm nova ^{22}Na production, might yet be detected.

References

[1] D. D. Clayton, S. A. Colgate, and G.J. Fishman. *Astrophys. J.*, 155:75, 1969.

[2] M. D. Leising and G. H. Share. *Astrophys. J.*, 357:638, 1990.

[3] D. D. Clayton and F. Hoyle. *Astrophys. J. Letters*, 187:L101, 1974.

[4] J. D. Kurfess et al. *Astrophys. J. Letters*, 399:L137, 1992.

[5] D. D. Clayton, M. D. Leising, L. S. The, W. N. Johnson, and J. D. Kurfess. *Astrophys. J. Letters*, 399:L141, 1992.

[6] P. A. Pinto and S. E. Woosley. *Astrophys. J.*, 329:820, 1988.

[7] N. B. Suntzeff, M. M. Phillips, J. H. Elias, D. L. Depoy, and A. R. Walker. *Astrophys. J.*, 384:L33, 1992.

[8] E. Dwek, S. H. Moseley, W. Glaccum, J. R. Graham, R. F. Loewenstein, R. F. Silverberg, and R. K. Smith. *Astrophys. J.*, 389:L21, 1992.

[9] S. E. Woosley and R. D. Hoffman. *Astrophys. J.*, 368:L31, 1991.

[10] S. E. Woosley and T. A. Weaver. *Ann. Rev. Astron. and Astrophys.*, 24:205, 1986.

[11] K. Nomoto, F. K. Thielemann, and K. Yokoi. *Astrophys. J.*, 286:644, 1984.

[12] S. E. Woosley, R. E. Taam, and T. A. Weaver. *Astrophys. J.*, 301:601, 1986.

[13] S. E. Woosley and P. A. Pinto. In J. Audouze, R. Mochkovitch, and J. Zinn-Justin, editors, *Supernovae*, number Session LVI in Les Houches Lectures, page in press, Amsterdam, 1991. Elsevier.

[14] A. M. Khokhlov. *Astron. Astrophys.*, 245:L25, 1991.

[15] H. Yamaoka, K. Nomoto, T. Shigeyama, and F. K. Thielemann. *Astrophys. J.*, 393:L55, 1992.

[16] E. Waagen and S. Knight. *IAU Circ.*, 5239:1, 1991.

[17] A. V. Filippenko et al. *Astrophys. J.*, 384:L15, 1991.

[18] M. M. Phillips, L. A. Wells, N. B. Suntzeff, M. Hamuy, B. Leibundgut, R. P. Kirshner, and C. B. Foltz. *Astron. J.*, 103:1632, 1992.

[19] W. N. Johnson et al. *Astrophys. J. Suppl.*, 1993 in press.

[20] R. F. Peletier and S. P. Willner. *Astrophys. J.*, 382:382, 1991.

[21] B. Leibundgut and G. A. Tammann. *Astron. and Astrophys.*, 230:81, 1990.

[22] B. Leibundgut. In S. E. Woosley, editor, *Supernovae*, pages 751–759, New York, 1991. Springer-Verlag.

[23] W. D. Arnett. *Astrophys. J.*, 230:L37, 1979.

[24] N. Gehrels, M. Leventhal, and C. J. MacCallum. *Astrophys. J.*, 322:215, 1987.

[25] E. Muller, P. Hoflich, and A. Khokhlov. *Astron. Astrophys.*, 249:L1, 1991.

[26] P. Hoflich, E. Muller, and A. Khokhlov. *Astron. Astrophys.*, page submitted, 1992.

[27] R. B. Tully. *Catalog of Nearby Galaxies*. Cambridge University Press, Cambridge, 1988.

[28] P. Ruiz-lapuente, E. Cappellaro, M. Tutatto, C. Gouiffes, I. J. Danziger, M. Della valle, and L. B. Lucy. *Astrophys. J.*, 387:L33, 1992.

[29] P. Collins and B. A. Skiff. *IAU Circ.*, 5454:1, 1992.

[30] S. J. Austin, S. G. Starrfield, M. Wagner, R. Bertram, B. M. Peterson, M. Houdashelt, and S. N. Shore. *IAU Circ.*, 5522:1, 1992.

[31] S. N. Shore, S. G. Starrfield, and S. J. Austin. *IAU Circ.*, 5523:1, 1992.

[32] M. D. Leising. In P. Durouchoux and N. Prantzos, editors, *Gamma-Ray Line Astrophysics*, number 232 in AIP Proceedings, pages 173–182, New York, 1991. American Institute of Physics.

SEARCH FOR GAMMA-RAY LINE EMISSION FROM SUPERNOVA SN 1991T WITH COMPTEL

G. G. Lichti, W. Collmar, R. Diehl, V. Schönfelder, A. Strong
Max-Planck-Institut für extraterrestrische Physik,
8046 Garching, Germany

D. Morris, J. Ryan
University of New Hampshire, Institute for Studies of Earth,
Oceans and Space, Durham, NH 03824, USA

H. Bloemen, H. de Boer, R. van Dijk, W. Hermsen
SRON-Leiden, 2300 RA Leiden, The Netherlands

K. Bennett, M. Busetta, C. Winkler
Astrophysics Division, Space-Science Department of ESA, ESTEC,
2200 AG Noordwijk, Netherlands

ABSTRACT

The imaging Compton telescope COMPTEL on the Compton Observatory CGRO measures γ-rays in the energy range 0.7 - 30 MeV with an energy resolution of 9.7% FWHM at 1 MeV. From June 15 to 28 and again from October 3 to 17, 1991 the region containing the supernova SN 1991T was observed. A search for γ-ray line emission from the supernova was performed. No line emission was detected. Upper limits for the two predicted lines at 847 keV and at 1.238 MeV could be derived. These limits were compared with the predictions of several theoretical models. Two of the models predict flux levels just above our upper limits.

MEASUREMENTS AND RESULTS

On April 10, 1991, the explosion of one of the closest and therefore brightest type Ia supernovae in recent years[1] was observed in the Sb galaxy NGC 4527. Since this supernova also showed a very peculiar chemical evolution, measurements of γ-ray line emission at 846.8 keV and at 1.2383 MeV from ^{56}Co decay are of great interest. A first observation of SN 1991T took place from June 15 to June 28, 1991, and a second one occurred four months later (from October 3 to 17, 1991).

Since the bulk of the emission from the induced radioactivity is in the low-energy γ-ray range, COMPTEL is very well suited to measure this radiation[2]. Based on counting-rate statistics, preliminary upper limits on the line emission from SN 1991T have been published earlier[3]. The results presented here are based on a maximum-likelihood method[4] which takes into account the full energy response[5] (a more detailed paper on the method and the results is in preparation[6]).

Maximum-likelihood skymaps were produced in ± 2σ energy bands around each line energy (0.779 - 0.914 MeV for the 847-keV line and 1.152 - 1.324 MeV for the 1.2-MeV line). The maps in these two energy bands do not show any evidence for a source at the position of the supernova. From the error of the source counts measured from the sky pixel containing SN 1991T, upper limits for the two lines could be derived. In Table I the 2σ upper limits obtained by this method for the two line energies and the two observations are presented.

line energy [MeV]	first observation	second observation
0.847	3.2	3.3
1.238	2.5	2.6

Table I: 2σ upper limits (in units of 10^{-5} cm^{-2} s^{-1}) from the two observations for the two line energies.

A comparison with theoretical line intensities calculated by various authors for a distance to the supernova of 13.5 Mpc shows that two models predict intensities just above our upper limits. These are the C/O dwarf detonation model of Chan and Lingenfelter[7] and the delayed detonation model N21 of Höflich, Khokhlov and Müller[8]. Their predicted line intensities for the first observation are $4.4 \cdot 10^{-5}$ cm^{-2} s^{-1} and $3.6 \cdot 10^{-5}$ cm^{-2} s^{-1} for the 847-keV line and $4.1 \cdot 10^{-5}$ cm^{-2} s^{-1} and $2.6 \cdot 10^{-5}$ cm^{-2} s^{-1} for the 1.2-MeV line, respectively.

REFERENCES

1. Burrows, A., A. Shankar and K. A. van Riper, Ap. J. 379, L7 (1991)
2. Schönfelder, V., et al., in Data Analysis in Astronomy IV, Vol. 59, 185 (1992)
3. Lichti, G. G., et al., A&A Suppl., in press
4. de Boer, H, et al., in Data Analysis in Astronomy IV, Vol. 59, 241 (1992)
5. Diehl, R., et al., in Data Analysis in Astronomy IV, Vol. 59, 201 (1992)
6. Lichti, G. G., et al., in preparation (to be submitted to A&A)
7. Chan, K. W., and R. E. Lingenfelter, Ap. J. 368, 515 (1991)
8. Höflich, P., A. Khokhlov and E. Müller, A&A 259, 243 (1992)

X-Ray and Gamma-Ray Transport in Clumpy Supernova Ejecta

Lih-Sin The, Dieter H. Hartmann, Donald D. Clayton, & Mark D. Leising

Department of Physics & Astronomy, Clemson University, Clemson, SC 29634-1911

ABSTRACT

We investigate γ-ray transport in spherically expanding supernova ejecta in the presence of density inhomogeneities. We compare the emergent photon spectra for the clumpy model with those predicted by the uniform density model and by an effective model with reduced opacity. We find that the γ-ray line fluxes in a model with uniformly distributed clumps can be calculated analytically by the reduced opacity method of Bowyer and Field. However, the Comptonized x-ray continuum significantly differs from the predictions of the reduced opacity model if the optical depth is large. The differences gradually disappear as the optical depth of the ejecta decreases.

INTRODUCTION

The UVOIR, hard x-ray, and γ-ray light curves of SN1987A suggest that matter in the supernova ejecta was mixed from the slower moving inner regions into the faster moving outer layers.[1-6] Hydrodynamic calculations have shown that mixing of heavy elements could involve significant inhomogeneities in the density (blobs, clumps, bubbles, fingers) and chemical composition of the medium.[2,4,6,7] Two-dimensional hydrodynamic simulations for SN1987A, carried out by three independent groups, show that Rayleigh-Taylor instabilites grow into mushroom-like fingers within the first few hours after the explosion (as anticipated[8]) if substantial nonspherical seed pertubations are present. Detailed x-ray and γ-ray Monte Carlo transport calculations for multidimensional hydrodynamic models should be performed,[9] but insight can be gained from the simple model of clumpy media considered here.

Kumagai et al.,[2] in an attempt to fit the 16-28 keV flux of SN1987A observed by Ginga through day 520 after the explosion,[10,11] treat the effect of clumps in the supernova by reducing the photoelectric opacity in the core. This effective opacity method was adopted from a similar technique developed by Bowyer and Field[12] in the context of interstellar absorption of the soft x-ray background in our galaxy. The same method was also used by Nagase et al.[13] in a study of circumstellar x-ray absorption from x-ray pulsars. Within the framework of a simple clumpy model, we test predictions based on effective models in which a homogeneous medium with reduced opacity is studied according to the prescription of Bowyer and Field.

THE MODEL

We consider a very simple inhomogeneous model in which clumps are superimposed on a sphere of constant density. The element composition of the ejecta has been simplified to include only hydrogen, oxygen, and iron. We let the hydrogen mass fraction represent the combined mass fractions of hydrogen and helium in the W10hmm model of SN1987A, the fraction of oxygen similarly represents all intermediate elements in W10hmm, and the fraction of iron corresponds directly to the iron abundance in W10hmm. The ejecta have a total mass of 15 $M_\odot$ and were ejected with an explosion energy of 10^{51} ergs, which leads to the surface expansion velocity[14] of $\sim$ 3300 km s^{-1}. The total mass fraction locked up in clumps, f_{cl}, the density of each clump, ρ_{cl}, the radius of each clump, r_{cl}, and the total number of clumps, n_{cl}, are related to the mean density of the supernova ejecta, ρ_o, and its size, R, by

$$n_{cl} \frac{4\pi}{3} \rho_{cl} r_{cl}^3 = f_{cl} \frac{4\pi}{3} \rho_o R^3. \qquad (1)$$

Two-dimensional hydrodynamical models[15] show that the column density along various lines of sight varies by factors from 2 to 5. The density contrast in the core material of the simulations by Hachisu et al.[4] is about 5-6. We use this density contrast in the supernova debris to guide us in setting our model density ratio between clumps and background to values between 5 and 10. From Figure 2 of Reference 2 we estimate the approximate size of clumps relative to the size of the expanding supernova envelope to be of order 0.1 to 0.02. We vary the total mass fraction in clumps between 0.2 and 0.9. The number of clumps in the ejecta follows from these parameters. We assume that the clumps expand analogously to the entire ejecta, maintaining constant density contrast and radius ratio during the evolution of the model. For this exploratory calculation we take the clumps to have the same composition as the background medium.

To further simplify the study, we consider a central source of monoenergetic photons. We employ a variance-reduction γ-ray transport scheme.[16,17] The reduced opacity model developed by Bowyer and Field[12] for uniformly distributed clumps around a central point source implies that photon transport in this clumpy medium can be treated like transport in a homogeneous medium in which the opacity is reduced by the factor

$$r_\tau = (1 - f_{cl}) + f_{cl} \left(\frac{1 - exp(-\tau_{cl})}{\tau_{cl}} \right), \qquad (2)$$

which coincides with equation (9) of Nagase et al.[13] in the limit $f_{cl} \to 1$.

RESULTS

In Figure 1 we show the emergent x-ray spectrum of a clumpy model with $f_{cl} = 0.9$, density ratio 10, and radius ratio 0.05. These x-rays are entirely due to Compton scattering of the centrally emitted 1 MeV γ-ray line. We also show the corresponding results for a homogeneous model and a reduced opacity model. We find that at early times the emergent x-ray spectrum of the clumpy model is always larger than that of the homogeneous model. The figure also shows that the calculated x-ray spectrum of the effective model does not agree with that of the clumpy model. These differences are emphasized in Figure 2. We do confirm that the γ-ray line fluxes of the effective model trace the γ-ray line fluxes of the clumpy model very well for models with uniformly distributed clumps. Energy deposition due to scattering and photoelectric absorption

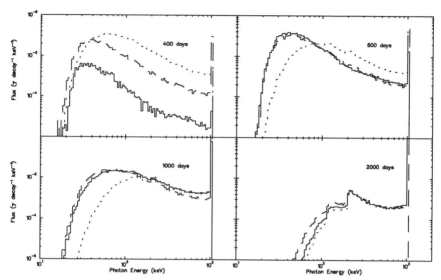

Figure 1: *Emergent photon continuum flux per keV due to a central 1 MeV γ-ray line source. Solid lines, dashed lines, and dotted lines represent fluxes from homogeneous model, clumpy model with $f_{cl}=0.9$, $\frac{R}{r_{cl}}=20$, $\frac{\rho_{cl}}{\rho_o}=10$, and reduced-opacity effective model, respectively.*

in the ejecta is inversely related to the emerging photon flux. However, the percentage differences of deposition energy between the homogeneous, effective, and clumpy models are not as noticeable as those in the emergent photon spectra.

Increasing the f_{cl}, while keeping the radius and density ratios fixed, enhances the differences between the models. Increasing the density of each clump, but keeping f_{cl} and the radius ratio fixed, increases the emergent x- and γ-ray fluxes. However, increasing the size of clumps, while keeping f_{cl} and the density ratio fixed, only slightly increases the emergent photon fluxes. We also performed simulations in which the clump concentration was enhanced toward the core. In that case, the γ-ray line fluxes of the clumpy model cannot be treated with the effective model. We find that the γ-ray line fluxes of the clumpy model can be smaller than the fluxes predicted from the homogeneous model, depending on the degree of central condensation of the clumps.

DISCUSSION

If some fraction of supernova ejecta is locked into uniformly distributed clumps, the emergent x-ray and γ-ray fluxes are both increased, thus reducing the fraction of deposited radiactive power that drives the UVOIR light curve. For uniformly distributed clumps, γ-ray line fluxes can be estimated using a reduced opacity. However, the x-ray continuum cannot be treated in the same way, because the reduced opacity method only applies to the central point source case, while the emergent x-ray continuum is due to multiply scattered photons that originate from locations throughout the ejecta. The

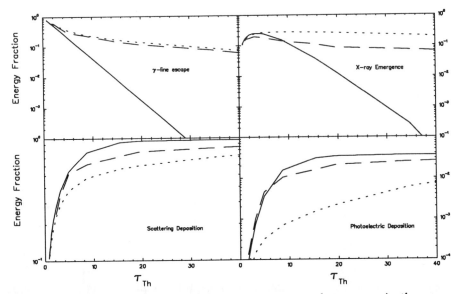

Figure 2: *Top panel: fraction of the photon source energy that emerge in the x-ray continuum and 1 MeV γ-ray line. Bottom panel: fraction of the source energy deposited by scattering and by photoelectric absorption as a function of Thomson optical depth. The calculated γ-ray line flux from the effective model (dotted line) traces the γ-ray line flux from the clumpy model (dashed line) very well, but the x-ray flux from the effective model deviates from the clumpy model at early times when optical depth is large.*

effective opacity approximation was used by Kumagai et al.[2] to explain the elevated 16-28 keV Ginga fluxes of SN1987A. Clayton and The[18] show that reducing the photoelectric opacity in the core of model SN11E1 for SN1987A increases the x-ray fluxes between 10 to 60 keV only, while reducing the photoelectric opacity in the envelope does not increase the x-ray flux above 20 keV.

The effect of density inhomogeneities on the emergent photon flux and the deposited energy also depends on the assumed distribution of clumps in the ejecta. Our preliminary results show that if the clumps are more concentrated to the core, photon emergence could even be reduced with respect to that of the homogeneous model. The detection of ^{57}Co in SN1987A by the OSSE detector[19,20] indicates that the inner core of SN1987A may be denser than predicted in the standard model of SN1987A, so that the observed γ-ray flux due to decaying ^{57}Co could be consistent with the UVOIR light curve around day 1600.[21,22] These observations could also be consistent with the interpretation that clumps in SN1987A are more concentrated to the core, so that the γ-ray emergence is smaller than in a model without clumps while the energy deposition efficiency is larger.

Although the clumps considered in the current simple model may not evolve like the instabilities encountered in hydrodynamic simulations, our results should provide some insight for the basic effects clumping will have on the transport of high energy photons in supernova ejecta. In future work we will consider radioactive sources that are distributed throughout the ejecta. We will also calculate γ-ray line shapes for radioactive sources embedded in clumps to investigate whether the γ-ray line shapes observed in SN1987A with the GRIS experiment[23-25] can be explained with our simple clumpy model.

REFERENCES

1. Arnett, W.D., Fryxell, B., & Müller, E. 1989, ApJ, 341, L63
2. Kumagai, S., Shigeyama, T., Nomoto, K., Itoh, M., Nishimura, J., & Tsuruta, S. 1989, ApJ, 345, 412
3. Woosley, S.E., Pinto, P.A., & Hartmann, D. 1989. ApJ 346, 395
4. Hachisu, I., Matsuda, T., Nomoto, K., & Shigeyama, T. 1990, ApJ, 358, L57
5. Sunyaev, R. A., et al. 1991, in *High Energy Astrophysics*, ed. W. H. G. Lewin, G. W. Clark & R. A. Sunyaev, National Acad. Press, Washington D.C., p. 368
6. Herant, M. & Benz, W. 1992, ApJ, 387, 294
7. Herant, M., Benz, W., & Colgate, S.A. 1992, ApJ, 395, 642
8. Clayton, D.D. 1974, ApJ, 188, 155
9. Burrows, A. 1991, in Gamma-Ray Line Astrophysics. AIP Conf. Proc. 232, ed. N. Prantzos, & P. Durouchoux (New York, AIP), p.297
10. Tanaka, Y. 1988a, in IAU Colloquium 108, Atmospheric Diagnostics of Stellar Evolution, ed. K. Nomoto, Springer-Verlag, p.399
11. Tanaka, Y. 1988b, in Physics of Neutron Stars and Black Holes, ed. Y. Tanaka, Universal Acad. Press: Tokyo, p.431
12. Bowyer, C.S., & Field, G.B. 1969, Nature, 223, 573
13. Nagase, F., Hayakawa, S., Sato, N., Masai, K., & Inoue, H. 1986, Publ. Astron. Soc. Japan, 38, 747
14. Bussard, R.W., Burrows, A., & The, L.-S. ApJ, 1989, 341, 401
15. Fryxell, B., Müller, E., & Arnett, W.D. 1991, ApJ, 367, 619
16. Pozdnyakov, L.A., Sobol, I.M., & Sunyaev, R.A. 1983, Ap. Space Phys. Rev., 2, 189
17. The, L.-S., Burrows, A., & Bussard, R.W. 1990, ApJ 352, 731
18. Clayton, D.D. & The, L.-S. 1991, ApJ, 375, 221
19. Kurfess, J.D. et al. 1992, ApJ in press
20. Clayton, D.D. et al. 1992, ApJ in press
21. Bouchet, P., Phillips, M.M., Suntzeff, N.B., Gouiffes, C., Hanuschik, R.W., & Wooden, D.H. 1991, A&A, 245, 490
22. Suntzeff, N.B., Phillips, M.M., Elias, J.H., DePoy, D.L., & Walker, A.R. 1992, ApJ, 384, L33
23. Tueller, J. et al. 1990, ApJ, 351, L41
24. Teegarden, B. J., et al. 1989, Nature, 339, 122
25. Teegarden, B. J., 1991, Adv. Space Res., in press

A SEARCH FOR RADIATIVE NEUTRINO DECAY FROM SUPERNOVAE

R.S. Miller and R.C. Svoboda
Department of Physics and Astronomy
Louisiana State University
Baton Rouge, LA 70803-4001

Abstract

If a massive neutrino species exists then it is possible that it has a radiative decay mode. Cosmological arguments require that a neutrino of mass $> 100eV$ be unstable. Motivated by these considerations, a 3-D model of radiative neutrino decay has been developed and used to simulate the decay evolution of supernova neutrinos. A sensitive search for the decay gamma rays using extragalactic supernovae and SN1987a as neutrino sources is proceeding using the COMPTEL instrument aboard the Compton Gamma Ray Observatory.

I. The Search for Massive Neutrinos

The search for the existence of a massive neutrino species is an area of intense theoretical and experimental research. Until recently, laboratory studies indicated that the observed properties of the neutrino were consistent with zero rest mass. The notable exceptions have been the experiments studying solar neutrinos and nuclear beta decay [1].

Very massive neutrinos ($> 100eV$) are expected to be unstable. Astrophysical limits based on the requirement that the mass density of the universe not exceed $\Omega = 1$ require $\Sigma m_\nu < 100eV$ [2]. Thus a massive ν_τ should decay. A very reasonable decay mode to expect would then be:

$$\nu_\tau \to \nu_{e,\mu} + \gamma \qquad (1)$$

This mode is the simplest two-body decay that does not violate any known conservation laws (except lepton flavor). No new particles are required and angular momentum sum rules are satisfied.

II. A Search for Radiative Neutrino Decay

We are currently performing a sensitive search for radiative neutrino decay in conjuction with the COMPTEL instrument team. The recently begun search uses type-II supernovae as the source of neutrinos. The high angular resolution and sensitivity of the COMPTEL gamma ray telescope [3] provides a unique oportunity to study any supernova gamma ray emissions. By observing SN1987A and extragalactic supernovae, any observed emissions can be analyzed and the relevant neutrino mass and lifetime computed (assuming a radiative decay mode) as described in the following sections. The non-observation of gamma rays from these sources will place significant limits on neutrino mass and lifetime.

III. Supernova Neutrinos

The detection of neutrinos from SN1987A [4] confirmed the central theoretical prediction of neutrino production in type-II supernovae: approximately 3×10^{53} ergs of binding energy released primarily as neutrinos on a time scale of a few seconds. The neutrinos detected by the IMB and Kamiokande collaborations were very likely $\bar{\nu}_e$, based on the relevant neutrino-water cross sections. It is widely believed, however, that since the $\bar{\nu}_e$'s were probably produced by e^+e^- scattering within the proto-neutron star, a roughly equal number of all six neutrino types should have been generated. If the tau neutrino, for example, is composed primarily of a massive component then this "neutrino laboratory", in the form of a type-II supernova, provides a copious number of more than 6×10^{57} ν_τ to study (with an equal number of $\bar{\nu}_\tau$) [5].

IV. Radiative Neutrino Decay Model

The computer model developed is a full 3-D simulation including parent neutrino spectra, relativistic kinematics, and angular dependencies. The model allows the neutrinos to stream from the surface of the proto-neutron star and decay in flight. An expanding spherically-symmetric shell is assumed and relativistic kinematics appropriate to two-body decays in flight are used. Relative time since the SN, resultant gamma ray energies, and angle between the gamma-ray arrival direction and the SN-Earth axis are recorded. In this way, arrival-time distributions can be made on selected gamma-ray energy slices for assumed values of m_ν and τ_ν. In addition, energy spectra can be produced on given time windows along with angular distribution histograms.

V. Search Method

Due to the parent neutrino energy spectrum and the assumed finite neutrino lifetime, the energy spectra of the decay gamma rays will evolve over time in a complex (but predictable) manner, examples of which are shown in Figure 1 for two mass values. Because of this spectral evolution, the gamma ray fluence detected - and therefore the mass/lifetime limits obtained - depends on when, in the history of the supernova, an observation occurs. The most sensitive limits can be achieved by observing the source during a period when the gamma ray fluence is

Radiative Neutrino Decay from Supernovae

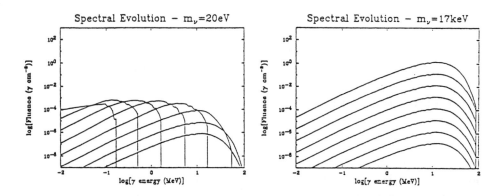

Figure 1: Spectral Evolution - $m_\nu = 20eV$ and $17keV$, D=20 Mpc, $\tau_\nu = 3.5 \times 10^{14}$ (17keV), $\tau_\nu = 5.6 \times 10^{16}$ (20eV), each curve represents the spectrum for a different time interval. The bottom curve represents the spectrum for arrival times 0-10 seconds. The second curve for 10-100 seconds, etc.

Table 1: Estimates of Gamma-Ray Flux

m_ν	D	0 s delay	10^5 s delay	10^7 s delay
20 eV	55 Kpc	5.6	0.0	0.0
17 keV	55 Kpc	2.4×10^4	2.6×10^4	3.5×10^4
20 eV	20 Mpc	1.6×10^{-2}	1.3×10^{-3}	0.0
17 KeV	20 Mpc	0.17	0.18	0.92

at maximum. The gamma ray arrival times can be distributed over a few seconds or many years, depending on the mass and lifetime of the decaying neutrinos. The arrival time distribution for a sample source is shown in Figure 2.

It is clear that there exists an optimal observational delay which will provide the most stringent limit. In the current search however, the ability to choose the optimal delay does not exist. Except for SN1987a observations which are an integral part of the Compton observatory's viewing program, we rely on past supernovae - or if we're lucky a new supernova - appearing in the COMPTEL instrument's field of view (approximately 64°). Thus, the observational delay since a supernova occurred will dictate the m_ν/τ_ν parameter space being sampled in any given observation. Table 1 shows estimates of the gamma ray fluence (ϕ_γ (cm^{-2})) expected in COMPTEL for different observational delays (assuming an exposure of 5×10^5 s).

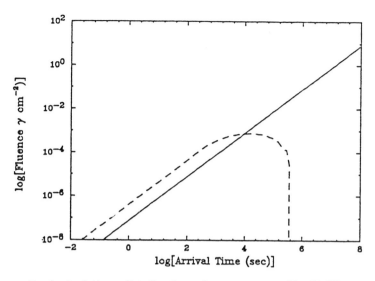

Figure 2: Arrival time distribution of gamma-rays, D=20 Mpc, $\tau_\nu = 3.5 \times 10^{14}$ (17keV,solid), $\tau_\nu = 5.6 \times 10^{16}$ (20eV,dash)

VI. Past Searches

SN1987A provides a unique opportunity to study neutrino properties. At the time of the supernova there were a number of spacecraft capable of performing gamma ray observations. The most sensitive instrument was the Gamma Ray Spectrometer (GRS) aboard the Solar Maximum Mission (SMM) satellite [6]. No gamma ray pulse was seen within 10 seconds of the IMB neutrino burst time. Thus, limits were set on the neutrino lifetime.

The limits published from the SMM observations are based on a simple 1-D model of radiative neutrino decay using monoenergetic supernova neutrinos. For a light neutrino species the limit from Bludman [7] is $\tau_\nu = 2.8 \times 10^{15} m_\nu$ sec while for a heavy neutrino the limit is $\tau_\nu = 6.0 \times 10^{18} m_\nu^{-1}$ sec, both assuming a branching ratio to a radiative decay mode of 1.

Detailed calculations show that COMPTEL, because of improved sensitivity and source exposure, should be able to improve these limits from 1-6 orders of magnitude for very massive neutrinos, while for light neutrinos the limits achievable may be slightly worse (unless a Galactic supernova is detected). Recent analysis from the COBE experiment puts very significant limits on the lifetime and branching ratio of radiative neutrino decay [8]. However these results, in addition to other cosmological arguments, require assumptions about the early universe and thus are not a direct test of the stability of massive neutrinos.

References

1. H.A.Bethe, *Phys. Rev. Lett.* **63** (1989) 837; M.Cherry, *Nature* **347** (1990) 708; J.Bahcall, Neutrino Astrophysics, Cambridge Press, Cambridge (1990);

   A.Hime and N.A.Jelly, Oxford preprint (1991); B.Sur, *et al.*, LBL Report (1991); J.J.Simpson, *Phys. Rev. Lett.* **54** (1985) 1891; J.J.Simpson and A.Hime, *Phys. Rev.* **D39** (1989) 1825; J.J.Simpson and A.Hime, *Phys. Rev.* **D39** (1989) 1837.

2. E.Kolb and M.Turner, The Early Universe, Addison-Wesley, New York (1990) 139.

3. W.Neil Johnson *ed.*, Proceedings of the Gamma Ray Observatory Science Workshop (1989).

4. R.Bionta, *et al.*, *Phys. Rev. Lett.* **58** (1987) 1494; K.Hirata *et al.*, *Phys. Rev. Lett.* **58** (1987) 1490.

5. A.Burrows, Univ. of Arizona preprint no. 90-09; D.Kielczewska, *Phys. Rev.* **D41** (1990) 2967.

6. E.Chupp, W.T.Vestrand, and C.Reppin, *Phys. Rev. Lett.* **62** (1989) 505.

7. S.A.Bludman, to be published *Phys. Rev. D* (1992).

8. B.B.Nath, *Phys. Rev. D* **46** (1992) 2341; B.Wang *Phys. Rev. D* **46** (1992) 2362.

SPECTRUM OF GAMMA-RAY LINES FROM SN1987A

Amnon Stupp
School of Physics and Astronomy, Faculty of Exact Sciences, Tel Aviv University,
Tel Aviv, Israel

ABSTRACT

847 Kev and 1238 Kev gamma lines from SN1987A were observed to be approximately ±2500 km/sec wide with small central doppler shifts.[1] I present a non-symmetric model of the SNR which reproduces these features by assuming the density changes as $1/R^3(\theta)$, $R(\theta)$ being the dimension of the SNR in direction θ.

INTRODUCTION

847 Kev and 1238 Kev γ lines from the decay of ^{56}Co were observed with unexpected features: early emergence, early peak times, large width, and the center of the line is at small doppler shifts. In addition, most observational results also give red shifts for the center of the line. All these observations were contrary to most accepted predictions, and the red shift especially is difficult to understand.[1]

The red shift is especially difficult because it seems obvious that photons from the far side of the SNR have to travel through larger optical depth than photons from the near side, and therefore the red side of the line should be attenuated, giving a blue shift (relative to the velocity of the center of the SNR).

Most models before SN1987A assumed a spherically symmetric homogenous SNR.[2] This is obviously an idealization and the natural thing to do was to try and introduce either a non-symmetric model, or a fragmented model, and attempt to explain the γ–ray features through these complications.

My work has concentrated on the non-symmetric model described below.

THE MODEL

My models are characterized by 3 parameters:
1) Geometric structure, wnich can be described in polar coordinates as Vmax(θ, φ) - Vmax being the maximum velocity in a certain direction.
2) Density distribution of Co as a function of (v,θ, φ) for every velocity up to Vmax.
3) Electron density distribution.

An additional parameter of importance is the angle α at which the SNR is viewed from earth. This is important for asymmetric SNR because from different directions the SNR looks different.

For the non-symmetric model I call the Elipsoid model I assumed a dependence of Vmax on theta alone according to the formula

$$\mathbf{Vmax = A - Bcos(\theta)} \tag{1}$$

In general B may be <0 and then the longer axis will be in the z direction.
The important feature of this model is the assumption of density dependence on Vmax, and through it on theta.

I took both the electron and the Co density to depend on Vmax as

$$\rho \propto 1/V_{max}^3 \qquad (2)$$

This dependence makes the total amount of matter within a volume $V_{max}^3 d\Omega$ constant.

The angle α at which we view the SNR is taken relative to the z axis, and because of the symmetry in the ϕ direction it suffices to determine the way the SNR will look.

The simplest model is one with a constant density. For the Elipsoid this means a density distribution

$$\rho = \text{Const.}/V_{max}^3 \qquad (3)$$

For the Co I assumed a cut-off at a fixed fraction of Vmax, for example 1/4, creating a Co elipsoid within the larger electron elipsoid.

In the calculation of the γ line spectrum I assumed the following:
1) Homologous expansion after a short time, so that for every point r, $v = r/t$.
2) Line width is the result of doppler shift alone. The doppler shift is the result of the homologous bulk velocity, and all other velocities (thermal, turbulent) are negligible.
3) Optical depth results from compton scattering from electrons, and all other effects are negligible. The electrons are assumed to be free.
4) A single compton scattering is sufficient to remove a photon from the line, and the contribution of photons scattered into the line is negligible.

In the calculations themselves I neglected the effects of the expansion of the SNR on the scattering cross section and the source function, in effect freezing the SNR during the time of flight of a photon through it.

Sample calculations for some cases in which those effects were taken into account gave results which differed by less then 1% from calculations done without them. I take this to mean that neglecting these effects does not change the results in general.

RESULTS

The models are tested by their predictions for the time of first γ – ray emergence, by the time of maximum of flux, by the width of the lines, and by the position of the center of the line.

In the table below is an example for a model with the following parameters:
1) Elipsoid with $A=10^9$ cm/sec, $B=5 \cdot 10^8$ cm/sec.
2) Co reaching to 1/4 Vmax for every θ.
3) Electron density distribution such that the total mass, assuming 1 proton for every electron, is 12 Sun masses.
4) The SNR is seen at an angle of 30 degrees.

The results are for the 847 Kev line, but are essentially the same for the 1238 Kev line.

Results of Elipsoid Model viewed at 30 degrees

Days from Explosion	Flux in units of photons/cm^2/sec times 10^4	Calculated center of line	Doppler cell with the maximum no. of photons	Width of line in Kev
200	3.630	848.28	849.82	5.6
300	25.795	847.74	849.26	6.5
400	33.427	847.47	849.26	6.9
500	24.366	847.30	848.69	7.0
600	13.879	847.20	848.69	7.1

The calculation was made on a grid of 120 velocity divisions, 360 theta divisions (every half degree), and 12 phi divisions (every 30 degrees). Doubling the resolution in all dimensions of the grid resulted in a change of less then 5% in significant cells, and no change of the shape of the graphs.

As can be seen from the table and from figure 1 the center of the lines is close to zero, a shift of only 0.5 Kev at day 400 against a shift of 3 Kev calculated by Bussard[2] for the spherically symmetric models of Arnett and Woosley.

However, it is still a blue shift, and not a red shift.

CONCLUSION

Of the features of SN1987A the early emergence, early peak times and the large width can be explained as the result of mixing of Co to high velocity, in my model to 1/4 of the way out of the SNR. As is well known now inclusion of mixing in spherical models reproduces these features equally well.[2]

The contribution of my model is that the line center can be shifted close to zero doppler. This brings us closer to the observations.

I have also found that it is possible to produce a red shift by assuming that the Co is limited in theta, for example between 85 degrees and 95 degrees. This limitation seems to me artificial, and a better way to model this would be by assuming a fragmented model with perhaps excess Co in certain directions.

REFERENCES

1) Tueller, J., in Proceedings of AIP conference on Gamma-Ray Line Astrophysics, AIP Proceedings 232, p.199.
2) Bussard, R.W., Burrows, A., and The, L. S., *Ap. J.* 341, p. 401.

Spectrum of Gamma-Ray Lines

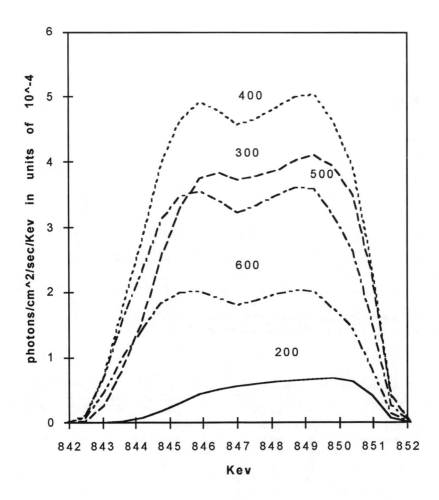

Fig 1: Elipsoid at 30 degree angle.

GAMMA RAYS AND X-RAYS FROM CLASSICAL NOVAE

James W. Truran
Department of Astronomy and Astrophysics
Enrico Fermi Institute
University of Chicago

and

Sumner Starrfield
Department of Physics and Astronomy
Arizona State University

ABSTRACT

The thermonuclear runaways occurring in accreted hydrogen shells on white dwarfs in close binary systems, that define the outbursts of the classical novae, provide an environment in which explosive hydrogen burning can produce significant concentrations of the interesting longer lived radioactive isotopes ^{7}Be ($\tau_{1/2}$ = 53.28 days), ^{22}Na ($\tau_{1/2}$ = 2.605 years), and ^{26}Al ($\tau_{1/2}$ = 7.2x10^5 years), as well as of the short-lived positron-unstable isotopes ^{13}N, ^{14}O, ^{15}O, ^{17}F, and ^{18}F. It follows that nearby novae may produce flux levels of gamma rays, during several phases of their evolution, that are detectable with the Compton Gamma Ray Observatory. We briefly review the theoretical and observational picture with regard to the identified potential sources of gamma rays from novae. In particular, we reexamine the current situation regarding the production of ^{7}Be, ^{22}Na, and ^{26}Al in nova outbursts, and provide updated estimates of the expected fluxes of ^{7}Be and ^{22}Na gamma rays from individual nova events. Expectations for X-rays from novae are also noted.

INTRODUCTION

The outbursts of classical novae are now understood to result from the occurrences of thermonuclear runaways in accreted hydrogen-rich envelopes on the white dwarf components of close binary system.[1,2] The peak temperatures achieved in the runaways typically reach values ~200-300 °K, consistent with the operation of the hot hydrogen burning sequences and the concomitant formation of a number of unstable short-lived nuclear species which may, upon decay, yield detectable fluxes of gamma rays. During the early stages of the runaway, prior to the achievement of visual maximum, substantial abundances of the isotopes ^{13}N, ^{14}O, ^{15}O, ^{17}F, and ^{18}F are transported rapidly to the surface regions by convection; their subsequent decays may give rise to a detectable level of gamma ray emission. During the later stages of nova outbursts occurring on ONeMg white dwarfs, detectable flux levels of 1.275 MeV gamma rays from ^{22}Na decay may be produced. Comptonization of these gamma rays may, under some circumstances, give rise to the production of hard X-rays. Classical novae associated with ONeMg white dwarfs may also have contributed significantly to the production of ^{26}Al in the Galaxy. Brief discussions of each of these potential gamma ray sources are presented in the next section. A brief review of possible X-ray emission from novae is presented in a subsequent section.

NUCLEAR DECAY GAMMA RAYS FROM NOVAE

In the context of the thermonuclear runaway model for classical nova outbursts, the formation of significant concentrations of a number of relatively short-lived radioactive nuclear species can occur, as a consequence of the hot hydrogen burning sequences that proceed in this environment.

Early (Pre-Maximum) Gamma Ray Emission. Radioactive decays in the outermost regions of the nova envelope can possibly produce a strong gamma-ray flux from a nova prior to the realization of visual maximum (see, e.g. Leising and Clayton [3]). Explosive hydrogen burning during the earliest stages of the thermonuclear runaway produces large concentrations of the short lived positron-unstable isotopes ^{13}N ($\tau_{1/2}$ = 9.97 m), ^{14}O ($\tau_{1/2}$ = 70.5 s), ^{15}O ($\tau_{1/2}$ = 122 ms), ^{17}F ($\tau_{1/2}$ = 64.5 s), and ^{18}F ($\tau_{1/2}$ = 109.8 m). The convective burning region, at this early stage, ultimately extends outward to encompass the entire envelope. The convective timescale of the nova envelope is sufficiently short to allow the transport of the longest lived of these radioactive isotopes to the surface regions, prior to decay. Since the surface convective region subsequently retreats rapidly due, at least in part, to the energy deposition arising from these decays, it will thus be possible to observe these gamma rays only very early, if at all. Fishman et al.[4] have performed calculations which indicate that this gamma ray event should be of duration ~ 10^4-10^5 seconds and be characterized by a relatively hard spectrum (~ 20-100 keV). They also estimate that such an event from a typical nova may be detectible with BATSE to a distance of order ~1 kpc. To the best of our knowledge, no such early burst of gamma ray emission was detected for the recent nova, Nova Cygni 1992, for which current estimates indicate a distance ~1.5 kpc.

^{7}Be Production and Decay. A question of considerable interest to nucleosynthesis and cosmology is whether nova explosions are an important site for ^{7}Li production. In this environment, as in red giant envelopes, nuclei of mass A=7 are formed predominantly via the ^{3}He$(\alpha,\gamma)^7$Be reaction as ^{7}Be, which is then transported rapidly out of the burning region to the surface by convection. Here again, the convective history and timescale of the nova envelope are critical issues. In an early numerical study of this problem, in the context of full hydrodynamic models of nova explosions, Starrfield et al.[5] found that ^{7}Li(^{7}Be) abundances of ~100-1000 times solar were possible (see the discussion by Boffin, Paulus, and Arnould [6]). For the specific (and favorable) choice of a factor 1000 enrichment, we obtain the following expression for the expected flux of gamma rays from ^{7}Be decay:

$$F_{0.478} = 2 \times 10^{-5} \, cm^{-2} \, sec^{-1} \, \left(\frac{M_{ej}}{10^{-5} M_\odot}\right) \left(\frac{X_7}{10^{-5}}\right) \left(\frac{e^{-t/0.21}}{(D/1kpc)^2}\right)$$

In view of the importance of questions concerning ^{7}Li production and destruction mechanisms, with regard to the implied constraints on the cosmological production of ^{7}Li, a new examination of the problem of ^{7}Li synthesis in novae seems appropriate.

^{22}Na Production and Decay. The extent to which ^{22}Na and ^{26}Al production may be expected to occur in classical nova explosions has been examined in several recent papers. Weiss and Truran [7] and Nofar, Shaviv, and Starrfield [8] have calculated nucleosynthesis accompanying nova explosions for representative temperature histories extracted from hydrodynamic models. Politano et al.[9] have performed detailed studies of nova nucleosynthesis with an extended nuclear reaction network coupled directly to the hydrodynamics. All of these studies confirm that significant ^{22}Na and ^{26}Al production can occur in envelope matter characterized by large initial concentrations of elements in the range from neon to aluminum.

The significance of this latter finding arises from the fact that the ejecta of a large fraction of well studied novae have been found to be enriched in nuclei in the Ne-Al region (see, e.g. the review by Truran [10]). Since the source of these large abundance enrichments is assumed to be matter dredged up from the underlying white dwarf, this is taken to reflect the fact that such systems involve massive ONeMg white dwarfs, rather than CO white dwarfs. The high frequency of occurrence of ONeMg white dwarfs in nova systems observed in outburst is understood to arise from selection effects [11,12] associated with the fact that more massive white dwarfs require less accreted matter to achieve runaway conditions.

The calculations by Politano et al.[9] reveal that ^{22}Na mass fractions of the order of $\sim 10^{-3}$ may typically be achieved, with values as high as $\sim 5 \times 10^{-3}$ being possible. The expected flux of 1.275 MeV gamma rays from an individual nova event may then be written:

$$F_{1.275} = 4 \times 10^{-5} \text{ cm}^{-2} \text{ sec}^{-1} \left(\frac{M_{ej}}{10^{-5} M_\odot}\right) \left(\frac{X_{22}}{10^{-3}}\right) \left(\frac{e^{-t/3.75}}{(D/1kpc)^2}\right)$$

Here M_{ej} is the mass of the nova ejecta and X_{22} is the ^{22}Na mass fraction in the ejecta. By comparison, we note that the quoted OSSE upper limit[13] on 1.275 MeV line emission from Nova Cygni 1992 of $\sim 10^{-4}$ cm^{-2} s^{-1}, assuming a distance of 1.5 kpc, corresponds to an ejected ^{22}Na mass of $\sim 6 \times 10^{-8}$ $M_\odot$. We believe that a second observation of Nova Cygni 1992, a nearby neon-rich nova, is justified.

^{26}Al Production and Decay. The principal question of interest regarding ^{26}Al is whether novae can have contributed significantly to the production of the ~ 3 $M_\odot$ of ^{26}Al in the Galaxy that is required to explain the 1.809 MeV gamma ray line emissions reported by Mahoney et al.[14,15] The mass of ^{26}Al in the Galaxy that is attributable to nova explosions can be estimated from the following expression (see, e.g. Weiss and Truran [7]):

$$M_{26} = 0.2 M_\odot \left(\frac{M_{ej}}{10^{-5} M_\odot}\right) \left(\frac{X_{26}}{10^{-3}}\right) \left(\frac{R_{novae}}{40 yr^{-1}}\right) \left(\frac{f_{26}}{0.50}\right)$$

Here, we have assumed a rate of occurrence of nova events of 40 yr^{-1}, a fraction 0.50 of all nova events that produce ^{26}Al, a ^{26}Al mass fraction in the ejecta of 10^{-3}, and an ejected mass of 10^{-5} $M_\odot$ per event, for these more massive systems. This result appears to fall short of the approximately 3 $M_\odot$ required to account for the gamma ray observations. However, if we choose rather a value $X_{26} = 10^{-2}$, consistent with the largest mass fractions obtained in the recent numerical hydrodynamic studies of Politano et al.[9], we obtain a value M_{26} = 2 $M_\odot$ Myr^{-1}, comparable with the range for supernovae of 1.78-2.14 $M_\odot$ Myr^{-1} quoted by Timmes, Woosley, and Weaver[14]. Further information concerning specifically the distribution of ^{26}Al in the Galaxy may ultimately be able to distinguish between these possible sources.

X-RAY EMISSION FROM NOVAE IN OUTBURST

Classical nova events may also be expected to produce detectable levels of X-rays. MacDonald, Fujimoto, and Truran [17] have argued that an extended period of EUV and soft X-ray emission should occur subsequent to visual maximum, when all novae are predicted to experience a phase of evolution at constant bolometric luminosity, associated with hydrogen shell burning at near Eddington luminosities. A gradual hardening of the radiation field to first EUV and then soft X-ray wavelengths occurs, as the photospheric radius decreases, while nuclear burning continues at approximately constant bolometric luminosity. The timescales for both the onset and the duration of this phase are functions of the timescale for the depletion of the residual hydrogen envelope by some combination of nuclear burning, wind-driven mass loss, and common-envelope-driven mass loss.[17] It is this stage of evolution and its associated soft X-ray emission that, presumably, was detected by Ögelman, Krautter, and Beuermann [18] with EXOSAT, for several novae (including Nova GQ Muscae 1983). More recently, Ögelman et al.[19] have reported the detection of Nova GQ Muscae 1983 at soft X-ray wavelengths with the ROSAT satellite, now nine years into the outburst. The observed spectrum is very soft, and is found to be generally consistent with black body emission from a ~$M_\odot$ white dwarf, burning at a near Eddington luminosity and an effective temperature ~ 3.4×10^5 °K. Such a soft X-ray signature is entirely consistent with expectations for the long term evolution of classical novae in outburst, as discussed by MacDonald et al.[17] The question remains as to why GQ Muscae is the only classical nova known to exhibit this behavior.

We also note that the radioactive decay of ^{22}Na may play a role in the production of hard X-rays. Livio et al.[20] have calculated the X-ray flux that may be expected to result from Compton degradation of gamma rays produced in ^{22}Na decay. They conclude that interesting flux levels of hard X-rays may, under some circumstances, be produced in this manner.

CONCLUDING REMARKS

Significant gamma-ray and X-ray emission is predicted to accompany the outbursts of classical novae. In particular, an early burst of gamma-ray emission, characterized by a relatively hard spectrum (~20-100 keV) may be detectable with BATSE for nearby novae to a distance 1 kpc.[4] Fluxes of 1.275 MeV gamma rays from ^{22}Na decay are also predicted to be consistent with detection by the Compton Observatory, for nova outbursts involving ONeMg white dwarfs at distances up to ~ 1 kpc; in this regard, Nova Cygni 1992 remains a potentially interesting target.[13] Novae can also contribute to the observed level of ^{26}Al in the Galaxy. Finally, we note that an extended phase of soft X-ray emission is an expected feature of the late stages of nova outbursts.[17,18,19]

REFERENCES

1. J.W. Truran, in Essays in Nuclear Astrophysics, ed. C.A. Barnes, D.D. Clayton, and D.N. Schramm (Cambridge: Cambridge University Press, 1982), p. 467.
2. S. Starrfield, in Classical Novae, ed. M.F. Bode and A. Evans (New York: Wiley, 1989), p. 39.
3. M.D. Leising & D.D. Clayton, Ap. J. 323, 159 (1987).
4. G.J. Fishman, et al., in Gamma Ray Line Astrophysics, ed. P. Durouchoux and N. Prantzos (New York: AIP Conference Proceedings No. 232, 1990), p. 190.
5. S. Starrfield, J.W. Truran, W.M. Sparks, & M. Arnould, Ap. J. 222, 600 (1978).
6. H.M.J. Boffin, G. Paulus, & M. Arnould, preprint (1992).
7. A. Weiss & J.W. Truran, A & A 238, 178 (1990).
8. I. Nofar, G. Shaviv, & S. Starrfield, Ap. J. 369, 440 (1991).
9. M. Politano, S. Starrfield, J.W. Truran, W.M. Sparks, & A. Weiss, preprint (1992).
10. J.W. Truran, in The Physics of Classical Novae, ed. A. Cassatella and R. Viotti (Heidelberg: Springer-Verlag, 1990), p. 373.
11. J.W. Truran & M. Livio, Ap. J. 308, 721 (1986).
12. H. Ritter, M. Politano, M. Livio, & R.F. Webbink, Ap. J. 376, 177 (1991).
13. M.D. Leising, these proceedings (1992).
14. W.A. Mahoney, J.C. Ling, A.S. Jacobson, & R.E. Liggenfelter, Ap. J. 262, 742 (1982).
15. W.A. Mahoney, J.C. Ling, W.A. Wheaton, & A.S. Jacobson, Ap. J. 286, 578 (1984).
16. F. Timmes, S.E. Woosley, & T.A. Weaver, these proceedings (1992).
17. J. MacDonald, M.Y. Fujimoto, & J.W. Truran, Ap. J. 294, 263 (1985).
18. H. Ögelman, J. Krautter, & K. Beuermann, Astr. Ap. 177, 110 (1987).
19. H. Ögelman, M. Orio, J. Krautter, & S. Starrfield, preprint (1992).
20. M. Livio, A. Mastichiadis, H. Ögelman, & J.W. Truran, Ap. J. 394, 217 (1992).

THE PRODUCTION OF ^{22}Na AND ^{26}Al IN NOVA EXPLOSIONS: THE OBSERVATIONAL PICTURE

S. Starrfield[1,9], S. N. Shore[2,9], G. Sonneborn[3,9], R. Gonzalez-Riestra[4], W. M. Sparks[5,6], J. W. Truran[6], M. A. Dopita[7], and R. E. Williams[8]

[1]Dept. of Physics and Astronomy, ASU, Tempe, AZ 85287-1504
[2]GHRS Science Team-CSC, Code 681, GSFC
[3]Laboratory for Astronomy and Solar Physics, Code 681, GSFC
[4]IUE Observatory, ESA, Madrid
[5]Applied Theoretical Physics Division, Los Alamos National Laboratory
[6]Dept. of Astronomy and Astrophysics and Enrico Fermi Inst., U. of Chicago
[7]Mt. Stromlo and Siding Spring Observatory, ANU
[8]Cerro Tololo Interamerican Observatory
[9]Guest Observer, International Ultraviolet Explorer Observatory

ABSTRACT

International Ultraviolet Explorer (**IUE**) studies of Nova V693 CrA showed that its ejecta were enriched in the intermediate mass nuclei from nitrogen to silicon and, thereby defined the ONeMg outburst as a compositional class of novae. The observational studies found that both sodium and aluminum were $\sim$100 times overabundant with respect to solar material while the theoretical studies predicted that a significant fraction of the sodium and aluminum should be ^{22}Na and ^{26}Al. Therefore, nearby ONeMg novae should be observed by **COMPTON** for possible detection of γ-rays from the decay of ^{22}Na and novae are probably responsible for the ^{26}Al observed in the ISM and solar system. Both Nova Her 1991 and Nova Cyg 1992 (the nearest ONeMg nova) are close enough so that they should be observed by **COMPTON**.

INTRODUCTION

Observational studies of ONeMg novae have reported large enhancements of the intermediate mass nuclei (V1500 Cyg: Ferland and Shields 1978; V693 CrA: Williams et al. 1985; V1370 Aql: Snijders et al 1987; QU Vul: Saizar et al. 1992; LMC 1990 #1: Dopita et al. 1992). Unfortunately, these novae are either too old or too distant for **COMPTON** to detect the γ-ray's emitted by the ^{22}Na in their ejecta. However, in the past two years we have observed two ONeMg novae which should be close enough to be detected by **COMPTON**. These are Nova Her 1991 (Starrfield et al. 1992) and Nova Cyg 1992.

Nova Her 1991 was one of the optically fastest novae ever discovered. It was first seen on 24 March 1991 at V $\sim$5 and then began a rapid decline. We began observations with the **IUE** satellite almost immediately and found that it's ejecta were still optically thick in the continuum. However, the expansion velocities were $\sim$7000 km s^{-1} and the expanding material quickly became optically thin. A very striking result of the optical studies was that the spectra were very similar to early spectra obtained of V1500 Cygni 1975. Dopita, Ryder, and Vassiliadis (1991) confirmed that Nova Her 1991 was an ONeMg nova when they reported that the [Ne III] 3868Å line was present at about twice the strength of Hβ. They also found lines from [NeV] in their spectra and reported that their spectra were very similar to those of other ONeMg novae (Dopita et al. 1992).

Nova Cygni 1992 was discovered in outburst at V~6.5 on 20 February 1992. Over the next few days it rose to V~4 which makes it the brightest nova since V1500 Cygni reached V~1.8 in September 1975. The first set of **IUE** spectra showed that it had been caught in the expanding "fireball" stage of evolution before UV maximum. It quickly evolved into the optically thick, "iron curtain" phase where the UV spectrum is produced by overlapping iron lines. **IUE** spectra showed that material had been ejected with velocities of ~3000 km s^{-1}. By April 1992 [Ne II] 12.8μm emission was detected which suggested that this was an ONeMg nova (Gehrz et al. 1992). This classification was confirmed both by **IUE** and optical spectra. The first ROSAT observation, on 22 April, showed a relatively hard spectrum with the nova emitting at ~10^{32} erg s^{-1} (Krautter, Ögelman, and Starrfield 1992). The X-ray flux has since both softened and increased to ~10^{34} erg s^{-1} (Krautter, Ögelman, Starrfield, and Orio 1993, in preparation). There was a two week pointing by **COMPTON** at about the same time. A 3-σ upper limit to the ^{22}Na emission was ~2×10^{-4} cm^{-2} s^{-1} (Leising 1993, these proceedings).

Distance estimates place it between ~1.5kpc and 2kpc, with a small reddening, which makes this nova the closest confirmed ONeMg nova. This distance also implies that at maximum it was emitting at a very super-Eddington luminosity. Estimates of the ejected mass indicate that more than $10^{-4} M_\odot$ were ejected which allows us to predict the ^{22}Na and ^{26}Al production by this nova (Shore et al. 1993, in preparation).

It is important to emphasize that, while the abundance analysis of Nova Cyg 1992 has not yet been completed, the presence of large enhancements of the intermediate mass nuclei in novae ejecta is firmly based on observational studies of other ONeMg novae in which nebular emission lines have been used to determine abundances. Nova Cygni 1992 shows all the lines that are present in other ONeMg novae and the spectra of ONeMg novae appear very different from those of CO novae. In addition, the light curve of Nova Cygni 1992 closely follows the light curves of other ONeMg novae. Thus, there is no doubt either in the existence of two compositional classes of novae (Starrfield et al. 1992) or that Nova Cygni 1992 is an ONeMg nova.

However, one problem is that the UV line at ~2069Å, attributed to a forbidden sodium line (Williams et al. 1985), actually comes from permitted Ne III (Persson et al 1991). Therefore, it is not now possible to determine the abundance of sodium in the ejected material.

ACCRETION ONTO ONeMg NOVAE

The existence of ONeMg novae and the successes of the nucleosynthesis calculations (Weiss and Truran 1990; Nofar, Shaviv, and Starrfield 1991), required that we examine the consequences of accretion of hydrogen rich material onto ONeMg white dwarfs. This has now been done, using a hydro-nucleosynthesis code which follows the changes in abundance of 78 nuclei as it evolves the white dwarf and accreted matter through the explosion (Kutter and Sparks 1972; Weiss and Truran 1990; Nofar, Shaviv, and Starrfield 1991; Starrfield et al. 1992). The results of one set of calculations as applied to Nova Her 1991 can be found in Starrfield et al. (1992). Here we apply the theoretical results to the outburst of Nova Cyg 1992.

In work to be reported in Politano et al. (1993, in preparation), we evolved TNR s in accreted hydrogen rich layers of white dwarfs with masses of $1.0 M_\odot$, $1.25 M_\odot$, and $1.35 M_\odot$. For all sequences, we assumed that the rate of accretion onto the white dwarf was 10^{17} gm s^{-1} (1.6×10^{-9} $M_\odot$yr^{-1}) and that the initial abundance of ONeMg nuclei was equal to 50% of the envelope material (by mass). The remaining 50% consisted of a solar mixture of the elements. *It was assumed that this composition resulted from the mixing of the accreted layers with core material.*

The abundance results are given in Table 2 of Starrfield et al. (1992). We note, first, that the abundance of ^{26}Al declines while the abundance of ^{22}Na increases as the mass of the white dwarf increases. This implies that those novae which produce the largest amount of ^{26}Al will not necessarily be the same novae that produce enhanced ^{22}Na. In addition, Starrfield et al. (1992) shows that as the mass of the white dwarf increases, the abundances of ^{31}P, ^{32}S, and ^{36}Ar increase to very large values. All of these nuclei are produced from "slow" (relative to the rates which dictate energy generation) proton captures on ^{24}Mg over the few minute lifetime of the explosion.

γ-RAY EMISSION FROM THE DECAY OF ^{22}NA

In this section we predict the ^{22}Na γ-ray flux from Nova Her 1991 and Nova Cyg 1992 for the Summer of 1993. Our development follows Starrfield et al. (1992 and references therein). ^{22}Na has a half-life of $\tau_{1/2} = 2.6$yr and produces a γ-ray of energy $E_\gamma = 1.275$ MeV in its decay to ^{22}Ne. The γ-ray emission of a nova is

$$n_\gamma = 4.0 \times 10^{-5} \times \left(\frac{M_{ej}}{10^{-5} M_\odot}\right) \times \left(\frac{X_{22_{Na}}}{10^{-3}}\right) \times \left(\frac{D}{kpc}\right)^{-2} \times e^{-\frac{t}{3.75yr}} \; cm^{-2} \; s^{-1}$$

where X(^{22}Na) is the mass fraction of ^{22}Na in the ejecta and M_{ej} is the ejected envelope mass ($M_\odot$). For Nova Her 1991, M_{ej} was $\sim 7 \times 10^{-5} M_\odot$ (Woodward et al. 1992) and it is at a distance of $\sim$3kpc (Woodward et al. 1992; Starrfield et al. 1992). The studies of its outburst indicate that it occurred on a very massive white dwarf ($M_{wd} > 1.35 M_\odot$), so we estimate (Table 2 in Starrfield et al. 1992) that the ejected mass fraction of ^{22}Na is $\sim 6 \times 10^{-3}$. Summer 1993 will be about 2 years after the outburst so it should be emitting at $\sim 9 \times 10^{-5}$ cm^{-2} s^{-1}.

For Nova Cyg 1992, M_{ej} (estimated) was $\sim 10^{-4} M_\odot$ to $\sim 5 \times 10^{-4} M_\odot$ (Shore et al. 1993, in preparation) and it is at a distance of $\sim$1.5kpc. Since the time scale of the outburst is much longer than for Nova Her 1991, since it was optically thick in the UV for more than 6 weeks, and since its expansion velocities were smaller than seen in Nova Her 1991, we predict that the outburst took place on a lower mass white dwarf and, therefore, assume that the ejected mass fraction of ^{22}Na was only $\sim 10^{-3}$. Summer 1993 will be about 1.5 yr after the outburst and so the γ-ray emission from this nova should be $\sim 1.2 \times 10^{-4}$ cm^{-2} s^{-1}.

Our results predict that a 1.35M$_\odot$ ONeMg nova can emit as much as 10^{34} erg s^{-1} in γ-rays if it ejects a mass of 10^{-5}M$_\odot$. Here we note that the presence of large amounts of ^{22}Na nuclei, in an optically thick (to γ-rays) expanding envelope, would Compton scatter a small fraction of them down in energy to hard X-ray wavelengths (Hollowell et al. 1993, in preparation). The hard X-rays in Nova Cyg 1992 were observed at a time when the expanding envelope was optically thick in the UV and we suggest that some of the X-rays were Compton scattered ^{22}Na γ-rays (Starrfield et al. 1992). We note that the X-rays were observed when Nova Cygni 1992 was also being observed by **COMPTON** and only an upper limit to the γ-ray emission was obtained (Leising 1992; these proceedings).

THE ABUNDANCE OF ^{26}Al IN THE ISM

^{26}Al has a half-life of $\tau_{1/2} = 7.3 \times 10^5$ yr and it is expected that this nucleus will be mixed through the ISM. The observations of the ^{26}Al distribution by **COMPTON** are consistent with production by novae (Diehl et al. 1992; these proceedings). Calculations of the abundance of this nucleus in the ISM, required to reproduce the observed flux, can be found in Mahoney et al. (1982, 1984), Weiss and Truran (1990), and Nofar, Shaviv, and Starrfield (1991). There are significant differences, however, in the predictions of the latter two papers: Weiss and Truran predict that a single nova cannot produce more than .2% of ^{26}Al (mass fraction) while Nofar, Shaviv, and Starrfield found that the abundance of ^{26}Al could reach 5%. The higher value is supported by the hydro-nucleosynthesis calculations (Starrfield et al. 1992). If the temperature in the nuclear burning region does not become too high, then virtually every ^{24}Mg nucleus will be converted to ^{26}Al. We estimate the amount of ^{26}Al produced by novae from

$$M_{26}(Novae) = 0.4 M_\odot \times \left(\frac{R_{nova}}{40 yr^{-1}}\right) \times \left(\frac{F_{ONeMg}}{0.25}\right) \times \left(\frac{M_{ej}}{2 \times 10^{-5} M_\odot}\right) \times \left(\frac{X_{26}}{2 \times 10^{-3}}\right)$$

where the variables are, R_{nova}, the nova rate in the galaxy, F_{ONeMg}, the fraction of novae that are ONeMg novae, M_{ej}, the mass ejected per event, and, X_{26}, the ejected mass fraction of ^{26}Al. F_{ONeMg} has been discussed in a number of papers (Truran and Livio 1986; Ritter et al. 1991) and the estimates range from 0.25 to 0.6. Our calculations predict that a low mass white dwarf will both accrete and eject more mass than a high mass dwarf. The range is from $\sim 10^{-4}$M$_\odot$ (1.0M$_\odot$ white dwarf) to $\sim 10^{-7}$M$_\odot$ (1.38M$_\odot$ white dwarf). The calculations predict that the amount of ^{26}Al produced in the outburst also depends on the white dwarf mass (see Starrfield et al. 1992, Table 2) with lower mass white dwarfs producing the most ^{26}Al. This is an extremely important result, if novae are responsible for ^{26}Al, since it means that low mass novae, which eject the most mass, are the same novae which produce the largest amount of ^{26}Al.

Observations suggest that Nova Cygni 1992 ejected $\sim 10^{-4} M_\odot$ indicating that the explosion occurred on a lower mass white dwarf. Nova Cyg 1992 appears to be a slow ONeMg nova and it exhibited strong Al III lines early in the outburst. The calculations reported in Starrfield et al. (1992) predict that $\sim 1\%$ of the ejecta is ^{26}Al. Therefore, the contribution of ONeMg novae, such as Nova Cyg 1992, can be expected to reach $\sim 10 M_\odot$ of ^{26}Al which is more than is observed (Mahoney et al. 1982). Since the fastest ONeMg novae (V693 CrA and LMC 1990 #1) eject only $\sim 10^{-6} M_\odot$ or less (Williams et al. 1985; Dopita et al. 1992), the amount of ^{26}Al produced by these novae will be far smaller than the amount ejected by slow ONeMg novae. *Therefore, the entire amount of ^{26}Al in the galaxy could come from slow ONeMg novae such as Nova Cygni 1992.*

SUMMARY AND DISCUSSION

In this paper we have used the observations of Nova Her 1991 and Nova Cyg 1992, in combination with the hydro-nucleosynthesis studies of accretion onto ONeMg white dwarfs, to predict both the ^{22}Na emission from these novae and, in addition, to examine the contribution of ONeMg novae to the ^{26}Al in the ISM. Specifically: 1) Hot hydrogen burning on ONeMg white dwarfs can produce as much as 2% of the ejected material in ^{26}Al and 3% in ^{22}Na, but not on the same white dwarf. 2) The largest amount of ^{26}Al is produced in the lowest mass white dwarfs, which according to our evolutionary calculations, should eject the largest amount of material. The observations of Nova Cyg 1992, a slow ONeMg nova, indicate that it has ejected $> 10^{-4} M_\odot$. 3) Nova Cyg 1992 should have produced sufficient ^{22}Na during its explosion so that **COMPTON** will detect the γ-rays from ^{22}Na decay in Summer 1993 when the ejected material will be completely optically thin to the γ-rays. 4) There should not be large numbers of QU Vul or Nova Cygni type novae in the galaxy since, if there were, the observed γ-ray emission from ^{26}Al would be far higher. However, in the past decade, we have observed 3 such novae: QU Vul 1984, Nova Pup 1991, and Nova Cyg 1992.

This work was supported in part by NSF and NASA grants to the University of Illinois and Arizona State University and by the DOE.

REFERENCES

Dopita, M., Ryder, S., and Vassiliadis E. 1991, IAU Circ., 5262.
Dopita, M. A., Meatheringham, S. J., Sutherland, R., Williams, R. E., Starrfield, S., Sonneborn, G. and Shore, S. N 1992, in New Aspects of Magellanic Cloud Research, ed. G. Klare, in press.
Ferland, G.J. and Shields, G.A. 1978, ApJ, **226**, 172.
Gehrz, R. D., Lawrence, G., and Jones, T. J. 1992, IAU Circ., 5463.
Krautter, J., Ögelman, H., and Starrfield, S. 1992, IAU Circ., 5550.
Kutter, G. S., and Sparks, W. M. 1972, *Ap. J.*, **175**, 407.
Mahoney W.A., Ling J.C., Jacobson A.S., and Lingenfelter, R.E. 1982, *Ap. J.*, **262**,

742.
Mahoney W.A., Ling J.C., Wheaton, W. M, and Jacobson, A.S. 1984, *Ap. J.*, **286**, 578.
Nofar, I., Shaviv, G., and Starrfield, S. 1991, *ApJ.*, **369**, 440.
Ögelman, H., Krautter, J., and Beuermann, K. 1987, *A. & A.*, **177**, 110.
Persson, W., Wahlstrom, C.G., Jonsson, L., and Di Rocco, H.D. 1991, Phys. Rev. A, **43**, 4791.
Politano, M., Starrfield, S., Truran, J. W., Sparks, W. M., and Weiss, A. 1993, in preparation.
Ritter, H., Politano, M., Livio, M., and Webbink, R. 1991, *Ap. J.*, **376**, 177.
Saizar, P. Starrfield, S., Ferland, G. J., Wagner, R. M., Truran, J. W., Kenyon, S. J., Sparks, W. M., Williams, R. E., and Stryker, L. L. 1992, *Ap. J.*, in press.
Snijders, M.A.J., Batt, T.J., Roche, P.F., Seaton, M.J., Morton, D.C., Spoelstra, T.A.T., and Blades, J.C. 1987, MNRAS, **228**, 329.
Starrfield, S., Shore, S. N., Sparks, Sonneborn, G., W. M., Politano. M., and Truran, J. W. 1992, *Ap. J. Letters*, **391**, L71.
Truran, J. W. and Livio, M. 1986, *Ap. J.*, **308**, 721.
Weiss, A. and Truran J. W. 1990, *A&A*, **238**, 178.
Williams, R. E., Sparks, W. M., Ney, E. P., Starrfield, S., Truran, J. W. 1985, MNRAS, **212**, 753.
Woodward, C.E., Gehrz, R.D., Jones, T.J., Lawrence, G.F. 1992, ApJ, 384, L41.

PULSARS

EGRET OBSERVATIONS OF PULSARS

D.A. Kniffen[a]
Hampden-Sydney College, Hampden-Sydney, VA 23943

D.L. Bertsch, C.E. Fichtel, R.C. Hartman, S.D. Hunter,
J.R. Mattox[b], and D.J. Thompson
NASA Goddard Space Flight Center, Greenbelt, MD 20771

K.T.S. Brazier[c], G. Kanbach, H.A. Mayer-Hasselwander,
C. v.Montigny, and K. Pinkau
Max-Planck-Institut für Extraterrestrische Physik, D-8046, Garching, Germany

J. Chiang, J.M. Fierro, Y.C. Lin, P.F. Michelson, and P.L. Nolan
W.W. Hansen Experimental Physics Laboratory and Dept. of Physics, Stanford University, Stanford, CA 94305

E. Schneid
Grumman Aerospace Corporation, Bethpage, NY 11714

H.I. Nel[d], P. Sreekumar
Univ. Space Research Association Research Associate

Z. Arzoumanian, V.M. Kaspi, D. Nice, and J. Taylor
Department of Physics, Princeton University, Princeton, NJ 08544

M. Bailes, S. Johnston, and R.N. Manchester
Australia Telescope National Facility, CSIRO
PO Box 76, Epping, NSW 2121, Australia

A.G. Lyne
Department of Physics, University of Manchester, Jodrell Bank Macclesfield, SK11, 9DL, UK

N. D'Amico[e], L. Nicastro
Instituto di Radioastronomie del CNR, I-40126 Bologna, Italy

a   Also Universities Space Research Association
b   Also *Compton* Observatory Science Support Center, operated by Computer Sciences Corporation
c   Royal Society Fellowship
d   On leave from Department of Physics, Potchefstroom University, Potchefstroom 2520, South Africa
e   Instituto di Fisca dell'Universita' di Palermo, Italy

EGRET Observations of Pulsars

ABSTRACT

Ever since the detection of the Crab and Vela pulsars as high-energy gamma-ray sources, there has been speculation that many pulsars might be gamma-ray emitters, some of which might be detectable with more sensitive instruments. Based on theoretical considerations, most interest has focused on young pulsars where a relatively high fraction of the spin down energy might be expected to be carried away in gamma rays. Millisecond pulsars have also been suggested as possible detectable gamma-ray sources. Consequently, the Compton Gamma Ray Observatory was designed with the ability to time tag gamma rays to an accuracy of a tenth of a millisecond in absolute time. This is needed in view of the long integration times required because of the low expected fluxes. In addition, a cooperative effort has been organized with radio astronomers who are making contemporary measurements of the ephemerides of a relatively large fraction of known pulsars which might be expected to be candidate gamma-ray "pulsars" based on theoretical considerations. In this way contemporary determinations of the period and period derivative necessary for the gamma-ray analysis are available. In addition to confirming the gamma-ray emission from the Crab and Vela pulsars, the Energetic Gamma Ray Experiment Telescope (EGRET) has the sensitivity to do pulse phase spectroscopy on strong sources such as these. The most significant discovery of a periodic source with EGRET has been the identification of Geminga with the recently discovered x-ray pulsar EGRET has also detected pulsed gamma-ray emission from the recently discovered radio pulsar PSR 1706-44 (2CG 342-02). When the systematic search is completed, EGRET will provide important new tools with which to evaluate the processes giving rise to the subclass of pulsars which are also emitting gamma-rays at the radio periods.

INTRODUCTION

The first gamma-ray source detected and verified by many different observers[1-8] was the Crab nebula pulsar now known as PSR 0531+21. Shortly thereafter the Vela pulsar (PSR 0833-45) was added to this class of objects[9-12]. The latter was a particular surprise at the time because it was not observable in the x-ray or low energy bands[13], but in both cases the energy implications were enormous. Beyond this beginning the high-energy gamma-ray satellite experiments SAS-2 and COS-B added no new pulsars beyond these two, although the 2CG catalog[14] contained many unidentified sources. A likely possibility for at least some of these was that they were undetected pulsars. Over the seven years of the COS-B mission, it was possible to search for time variability in the phase histograms of these sources and variability in various features of the emission were reported for both the Crab and Vela pulsars[8,10].

Following the end of the COS-B mission, there were no opportunities to make additional searches for gamma-ray-emitting pulsars until the launch of the Gamma-1 experiment in July 1990[15] and the *Compton* Observatory in April 1991. Because PSR 0531+21 was known to be observable over the six decades of energy covered with the instruments on the *Compton* Observatory, the region containing this source was chosen as the pointing direction for the verification phase of the mission just prior to Viewing Period 1 of the Phase I Plan on May 16, 1991. Because of its wide field of view, EGRET was also able to observe the high-energy gamma-ray source Geminga simultaneously with the Crab pulsar. Since the region was also the target for the first Viewing Period, EGRET has been able to obtain an exposure with excellent sensitivity to these two sources.

In one of the major discoveries of *Compton*, the long-standing mystery of Geminga was solved when this source was observed by EGRET[16] to be emitting gamma-rays at the 237 ms x-ray period discovered recently by Halpern and Holt[17] using ROSAT observations. Though these observations have shown that Geminga has the properties generally associated with an isolated pulsar, the lack of an observed radio signal gives significant new information on the different properties of the emission at different wavelengths.

Finally, a fourth periodic source, PSR 1706-44 has been reported by EGRET[18]. The source, discovered in a recent high frequency southern hemisphere galactic survey[19], is positionally consistent with the COS-B high-energy gamma-ray source 2CG 342-02[14]. Similar to the Vela pulsar in many ways, the gamma-ray light curve is quite different, again providing important information on the characteristics of the emitting region.

In the following sections the capabilities of EGRET for periodic source searches will be discussed briefly, and the details of each of the observed periodic sources discussed. Finally, some implications of the combined set of observations will be presented.

THE OBSERVATIONS

EGRET[20,21] has the conventional components of a high-energy gamma-ray telescope, covering the range from 20 MeV to 30 GeV. Using the unique signature of the pair production interaction for the positive identification of the ambient gamma-ray, EGRET has the following components for detecting and identifying the event: an anti-coincidence dome surrounding the upper portion of the instrument to reduce the number of charged particle induced triggers; an array of spark chambers interspersed with tantalum foils providing for the interaction of the incident gamma-rays and the imaging of the tracks from the electron-positron pair; two arrays of scintillation counters which form the coincidence trigger to apply the spark chamber high voltage and initiate the core readout; and a NaI (*Tl*) energy calorimeter to provide measurements over the broad EGRET energy range.

An important feature of EGRET and *Compton* for the study of pulsars is the good timing. Flight experience has shown that the design goal of 100 millisecond precision in the determination of the arrival time of each gamma-ray is being met with margin. This refers to absolute time, and includes all timing errors as well as error in the precision of the orbital determination. This is vital because of the long exposure times dictated by the low fluxes for all but the strongest pulsars. The correction to the barycenter adds only trivial additional error beyond the orbital uncertainty.

Another important factor in the study of pulsars with EGRET is the timing ephemeris for the sources. There simply are too few source photons at these energies to make an independent search for a period and period derivative for the fast young pulsars most likely to be emitting gamma-rays. For this reason the EGRET team joined the other instrument teams on the *Compton* Observatory to develop a cooperation with an international network of pulsar radio astronomers who are making observations of over 200 pulsars during *Compton* operations to ensure that contemporary epherimedes are available for the analysis. Radio telescopes at the National Radio Astronomy Observatory, Greenbank, WV; at the Nuffield Radio Astronomy Laboratories, Jodrell Bank; The Australia Telescope National Facility, Parkes and at the Northern Cross Radiotelescope near Bologna, operated by the Institute of Radioastronomy, CNR. Because of their major contributions, the principal observers are co-authors of this paper.

The technique used for determining the periodic flux of gamma-rays involves the selection of photons using an energy-dependent cone

$$\theta \leq 5.85° \times (E/100)^{-0.534}$$

about the known position of the source. This selection is based on the instrument point-spread function and is chosen so that 67 percent of the events originating from the source are accepted. The gamma rays at each energy are then binned according to the phase obtained from the calculated barycenter arrival time of each gamma-ray relative to the peak of the dispersion corrected radio pulse.

The energy spectra of the pulsars are obtained by fitting a trial power law spectrum to the data, allowing for the instrument energy uncertainty. The normalization and spectral index are parameters to be fit by iterating until a normal chi-squared function is minimized.

In the following sections the essential observed features of each of the four identified EGRET pulsars will be discussed in detail.

THE CRAB PULSAR

The Crab nebula pulsar (PSR 0531+21) was observed by EGRET during three different time intervals in the early phases of the Compton Observatory mission: 23 April-7 May 1991, 16-30 May 1991 and 8-15 June 1991. The total exposure to the source at 200 MeV exceeds 10^9 cm^2 s. The details of the techniques and results of the analysis are given by Nolan et al.[22] The results will be summarized here. The excellent exposure allows pulse phase spectroscopy, detailed studies for possible time variations in the phase histograms and very accurate determinations of the overall spectrum.

The pulse phase histogram for the >100 MeV pulsed emission from PSR 0531+21 is shown in Figure 1 together with that from the other three EGRET high-energy gamma-ray pulsars. The characteristic two peaks seen throughout the electromagnetic spectrum are evident, with a phase separation of 0.40 ± 0.02, in agreement with the COS-B

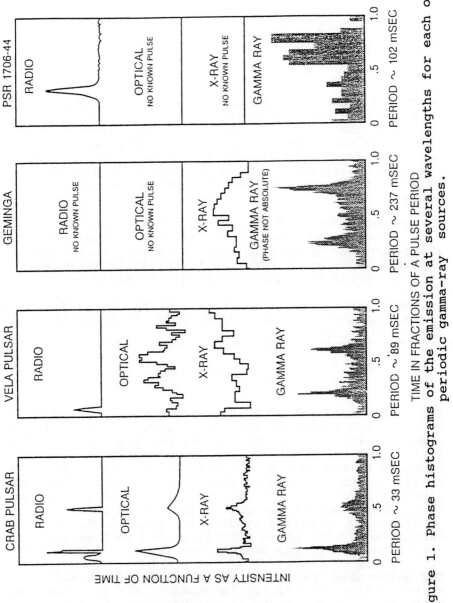

Figure 1. Phase histograms of the emission at several wavelengths for each of the periodic gamma-ray sources.

value[8]. There is no phase separation, within the errors, between the main pulse and the dispersion corrected radio main pulse. The second peak is clearly much smaller than the first. The width of this pulse is greater than that reported by COS-B, but it is not clear if this is a real effect, or a result of the better statistics and lower background of EGRET.

The time-averaged pulsed differential flux is well fitted by a power law of the form

$$(2.92 \pm 0.15) \times 10^{-9} \; (E/288)^{-2.23 \pm 0.05} \; \text{photons/cm}^{-2} \; s^{-1} \; \text{Mev}^{-1}$$

over the range from 50 MeV to 10 GeV. The instantaneous pulsed flux is about a factor of 1.6 larger. Nolan, et al.[22] have shown that the interpulse spectrum is much harder than the spectrum in the peaks. The pulse phases remain well aligned over the entire energy range.

The mechanisms proposed to explain the high energy gamma-ray emission include the polar cap and outer gap models[23,24]. Neither are able to explain all of the observed features.

A significant unpulsed emission is also observed from the Crab region.

THE VELA PULSAR

The gamma-ray emission from Vela is dominated by the pulsed component and only weak pulsed emission is seen in the x-ray range. Monitored extensively for seven years with COS-B until 1982, in addition to the EGRET observations, the source was until recently monitored almost continuously above 100 MeV with the Russian-French experiment[15], Gamma-1. The overall Vela pulse profile has remained remarkably stable over this period, but there are suggestions of short-term variability in the flux.

The EGRET observations were obtained from two dedicated pointings on 10-16 May 1991 and 22 August-5 September 1991, and four other viewings between July 1991 and June 1992 where it was well off the instrument pointing axis. The pulse phase histogram can be seen by reference to Figure 1. The same double peak structure is seen as for the Crab pulsar, separated in this case by 0.424 ± 0.002 in phase, in agreement with the Crab pulsar value. Kanbach et al.[25] have found that the phase is stable over the full energy range of EGRET, with no significant trends since it was last monitored. The observations encompass a large glitch which occurred 20 June 1991. Also shown on this figure is the recently reported observation of a weak flux of pulsed x-rays from Vela, using ROSAT observations[26].

GEMINGA

The gamma-ray source known as Geminga was discovered to be a gamma-ray "pulsar" during an observation of the galactic anticenter region. Discovered by SAS-2[27,28] and studied extensively by COS-B[29,30], the source remained a mystery until 1983 when Bignami, Caraveo and Lamb[31] discovered a candidate x-ray source in the COS-B error box using the Einstein Observatory. The source, named 1E 0630+178, was later identified with a blue star[32,33] in the COS-B error box.

The EGRET observations of Geminga include the same as those for the Crab plus low sensitivity observations in June, October and September 1992 when the source was well away from the instrument axis. Following the detection of periodic x-ray emission from this source at a period of 237 ms with ROSAT data[17], the EGRET data were successfully examined for periodicity[16] of this apparently isolated spinning neutron star. Using the timing solutions obtained with EGRET data, the COS-B data have been successfully examined for the periodic signal[34,35]. The periodicity has also been found in the SAS-2 data[36]. The long time base allows an excellent determination of the ephemeris for Geminga, one of only two "pulsars" for which other than radio data are used to determine the timing of the rotating neutron star.

The phase histogram is seen in Figure 1, together with that observed at other wavelengths. Again the double peak structure is seen, but with a separation of 0.50 ± 0.03. The two peaks are comparable and there is significant interpulse emission. As pointed out by Bertsch, et al.[16], assuming the energy source for the gamma-ray emission is the spin-down energy of the neutron star, a limit can be put on the distance of the source. For reasonable assumptions, it appears the object lies at a distance between about 40 and 400 parsecs.

The spectrum of the total pulsed emission from Geminga is quite hard with a flux above 100 MeV of about 4×10^{-6}/cm^{2}sec and a differential slope of 1.60±0.06.

PSR 1706-44

PSR 1706-44 is a short period radio pulsar discovered since the end of the COS-B mission, but its position is consistent within the uncertainties with that of the COS-B source 2CG 342-02. Considering its distance of about 1.5 kpc and assuming the same proportion of spin-down energy goes to high-energy gamma-ray production as the Vela pulsar, this source would be expected to be observable.

EGRET observed this source on several occasions. The most extensive exposure resulted from an exposure to the galactic center during the period 12-26 July 1991. More limited exposures occurred during 12-17 December 1991, 10 March - 2 April 1992, and 28 April - 7 May 1992. The source is detectable in all of these exposures. Figure 1 shows the phase histogram obtained for PSR 1706-44. This pulsar does not show an obvious double peak structure. The leading edge of the broad gamma-ray pulse follows the radio pulse by 0.22 ± 0.02 phase.

The time-averaged pulsed differential energy spectrum is given by

$(3.12 \pm 0.27) \times 10^{-10} (E/622.4)^{-1.72 \pm 0.08}$ photons/cm^{2}s MeV.

The instantaneous pulsed flux is a factor of 2.86 times greater.

This spectrum is harder than that given for the Crab pulsar above, and is more similar to the hard spectrum of Geminga and, at least at times, from Vela. The flux above 100 MeV is $(1.0 \pm 0.3) \times 10^{-6}$ ph/cm^{2}sec. The energy amounts to somewhat less than 2 percent of the spin-down energy. So, in many ways this pulsar is similar to PSR 0833-45. The exception is the phase histogram which is presumably an indicator of the region in which the emission is taking place. Although a double pulse structure is not ruled out, a phase separation of .4 to .5 phase separation probably is.

DISCUSSION

The number of known high-energy gamma-ray emitting pulsars is still only four, but the additional information provided by the EGRET observations has provided important new constraints of emission models. A fifth pulsar, PSR 1509-58, discovered to be a low-energy gamma-ray emitter, with energies up to 1.8 MeV,[37] adds additional insight.

There are two fundamentally different pieces of information which are derived from the observations of the pulsed emission. The pulse phase histogram tells of the geometry of the emitting region and the energy spectrum of the pulsed flux provides information on the emission process. The gamma-ray observations, based on the Crab and Vela pulsars, have been interpreted in terms of emission by some combination of curvature radiation, Compton scattering and synchrotron radiation in the magnetosphere of a highly magnetized rotating neutron star. The process takes place either near the surface at the polar caps[23], or in the vacuum gaps in the outer magnetosphere[24] While it is generally though not universally believed that the radio emission is produced in the polar cap region, there is even less consensus or evidence to make such a determination for the gamma-ray emission.

Neither the polar cap nor outer gap model is sufficiently developed to explain all of the observed features of the four pulsars observed by the EGRET instrument. The double peaked structure of the Crab, Vela and Geminga rotators is not clearly present in PSR 1706-44. It will be crucially important to determine with more extensive observations if the single broad peak is resolved into two more closely spaced peaks. This will address the question of whether the two peaks represent emission from the two poles of the pulsar or from different positions in the same polar region.

Additional information is obtained from the energy spectra of the pulsed emission. Overall, the spectra of each of the sources is well represented by a power law over most of the observed energy range. It will be important to see if additional observations and analysis show deviations from a power law. The two models for pulsar emission predict some curvature for the spectral distribution, but the spectra are not sufficiently precise to distinguish between them. Of particular importance are the phase-resolved spectra. For the first time, EGRET has been able to obtain the spectrum with good precision for the main peaks and the interpulse[22,25,38]. This will address the question of the differing processes that are thought to be important for different parts of the phase[39,40]. Also the extended energy range will allow an examination of the predictions of the differing mechanisms that might dominate in different portions of the spectrum[41,42].

ACKNOWLEDGMENTS

The authors gratefully acknowledge support from the following: Bundesministerium für Forschung and Technologie, Grant 50 QV 9095 (MPE); NASA Grant NAG5-1742 (HSC); NASA Grant NAG5-1605 (SU); and NASA Contract NAS 5-31210 (GAC).

REFERENCES

1. R. Browning, D. Ramsden, P.J. Wright, Nature Phys. Sci. 232, 99 (1971).
2. P. Albatts et al., Nature 240, 221 (1972).
3. R.L. Kinzer, G.H. Share, N. Seeman, Ap. J. 180, 547 (1973).
4. B. McBreen et al., Ap. J. 184, 571 (1973).
5. B. Parlier et al., Nature Phys. Sci. 242, 117 (1973).
6. D.A. Kniffen et al., Nature 251, 397 (1974).
7. K. Bennett et al., A & A 61, 279 (1977).
8. J. Clear et al., A & A 174, 85 (1987).
9. D.J. Thompson et al., Ap. J. Lett. 200, L79 (1975).
10. K. Bennett et al., A & A 61, 157 (1977).
11. G. Kanbach et al., A & A 90, 163 (1980).
12. I.A. Grenier, W. Hermsen, J. Clear, A & A 204, 117 (1988).
13. M.P. Ulmer et al., Ap. J. 369, 485 (1991)
14. B.N. Swanenburg et al., Ap. J. 243, L69 (1981).
15. V.V. Akimov et al., Proc. Int. Cos. Ray Conf. 1, 153 (1991).
16. D.L. Bertsch et al., Nature 357, 306 (1992).
17. J.P. Halpern and S.S. Holt, Nature 357, 222 (1992).
18. D.J. Thompson et al., Nature 359, 615 (1992).
19. S. Johnston et al., MNRAS 255, 401 (1992).
20. E. B. Hughes et al., IEEE Trans. Nucl. Sci. NS-27, 364 (1980).
21. D.J. Thompson et al., Ap. J. Suppl., in press (1992).
22. P.L. Nolan et al., in preparation (1992).
23. J.K. Daugherty, A.K. Harding, Ap. J. 252, 337 (1982).
24. K.S. Cheng, C. Ho, M. Ruderman, Ap. J. 300, 500 (1986).
25. K. Kanbach et al. (1993, in preparation).
26. H. Ogelman et al. (in preparation) (1992).
27. D. Kniffen et al., Proc. 14th Int. Cos. Ray Conf. 1, 100 (1975).
28. D.J. Thompson et al. Ap. J. 213, 213 (1977).

29. K. Bennett et al., A & A 56, 469 (1977).
30. J.L. Masnou et al., Proc. 17th Int. Cos. Ray Conf. 1, 177 (1981).
31. G.F. Bignami, P.A. Caraveo, and R.C. Lamb, 272, L9 (1083).
32. G.F. Bignami et al., Ap. J. 319, 358 (1987).
33. J.P. Halpern and D. Tytler, Ap. J. 330, 201 (1988).
34. G.F. Bignami and P.A. Caraveo, Nature 357, 287 (1992).
35. W. Hermsen et al., IAU circular 5541 (1992).
36. J.R. Mattox, Ap. J., in press (1993).
37. R.B. Wilson et al., Isolated Pulsars, ed. K.A. Van Riper and C. Ho (1992, in press).
38. H.A. Mayer-Hasselwander et al. (1993, in preparation).
39. J. Chiang, R.W. Romani, Proc. Compton Obs. Science Workshop, NASA Conference Pub. 3137, ed. C.R. Shrader, N. Gehrels and B. Dennis 237 (1992).
40. A.K. Harding and J. Daugherty, Isolated Pulsars, ed. K.A. Van Riper and C. Ho (1992, in press).
41. K.S. Cheng, C. Ho and M. Ruderman, Ap. J. 300, 522 (1986).
42. C. Ho, Ap. J. 342, 396 (1989).

PULSAR SEARCHES WITH COMPTEL

K. Bennett, M. Busetta, A. Carramiñana
Astrophysics Division, ESTEC, Noordwijk, The Netherlands

W. Collmar, R. Diehl, G. Lichti, V. Schönfelder, A. Strong
Max-Planck Institut für extraterrestrische Physik, Garching, FRG

W. Hermsen, L. Kuiper, B. Swanenburg
SRON-Leiden, Leiden, The Netherlands

A. Connors, J. Ryan
ISEOS, University of New Hampshire, Durham, USA

R. Buccheri
IFCAI/CNR, Piazza G. Verdi 6, 90139 Palermo, Italy

INTRODUCTION

During the first year of observations by the Compton Observatory, the number of known γ-ray pulsars has more than doubled. The five known pulsars have different energy spectra and light-curve shapes and, surprisingly, a given pulsar may not be seen by all of the GRO instruments. It is entirely possible that it will be observable by just one of the GRO instruments. With this in mind COMPTEL data has been scanned to look for pulsation from the "New" γ-ray pulsars and all likely entries in the radio pulsar catalogues. The result of this search is presented.

COMPTEL performs γ-ray astronomy in the 0.7-30 MeV energy range. Its ability to image γ-ray sources has been clearly demonstrated by images of the Crab pulsar, AGNs, γ-ray bursts and solar-flares[1]. The time resolution for each γ-ray event is 0.125 ms which has permitted detection of the previously established γ-ray pulsars PSR0531+21 (Crab) and PSR0833-45 (Vela). If a source pulsates at a known frequency then the selection of events arriving within the pulse interval can improve imaging sensitivity by reducing the background to only that in the pulsed interval, as already demonstrated for Vela[2].

Pulsar phases are calculated for events from a particular direction after translating the on-board arrival times to arrival times at the Solar System Barycentre.[3] The event phases which are determined using the contemporary radio ephemerides supplied in the public pulsar data-base maintained by Princeton[4] are then folded.

THE CLASSICAL PULSARS

So far, the Crab pulsar has been observed in the wide field of view of COMPTEL seven times during the GRO all-sky survey, four times by intent and fortuitously during three target-of-opportunity observations. Figure 1 shows the light curve obtained by combining the first 5 of these periods where the phase alignment has been made on the basis of the EGRET light curve for the first four periods and by eye for the latest period. The light curve has a shape quite distinct from that observed by COS-B and EGRET. In this energy interval the two pulses are equally strong and the bridging radiation is of equal strength. At high energies the first peak dominates and the bridging radiation is very weak. At lower energies e.g. in the 50 keV régime of OSSE and BATSE, the second pulse is again stronger than the primary and the relative contribution of the bridging radiation is less than for COMPTEL. Phase-resolved spectroscopy of all four GRO instruments is underway and has the potential to constrain pulsar production models. Despite the impressive light curve in the 1 − 10 MeV energy range, COMPTEL cannot measure statistically significant pulsed emission from the Crab above 10 MeV in a single observation period. This is consistent[6] with its having a spectral index of roughly $E^{-2.2}$, so this pulsar can be thought of as *soft*. We have optimised the event selection parameters in the 4 standard COMPTEL energy ranges so as to give the best signal-to-noise for the observed pulsed emission from the Crab as measured by the Z_3^2 parameter[5]. These criteria include *inter alia* Compton scatter angle and angular distance from the pulsar direction. We apply these selection criteria when searching for pulsations from other pulsars thought to be *soft*.

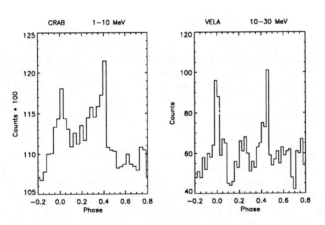

Figure 1 COMPTEL light curve for Crab (1-10 MeV) and Vela (10-30 MeV)

The Vela pulsar has also been viewed several times during 1991. In contrast to the Crab, the most significant light curve, shown in figure 1, is obtained in the highest standard analysis energy range, namely 10-30 MeV. This is consistent with the observed spectral index of $E^{-1.8}$ at high energies[7], although pulsations are also observed with lower significance throughout the energy range as shown by Hermsen et al.[8] We again optimised this pulsed signal and established a set of selection criteria appropriate for *hard* pulsars.

THE NEW PULSARS

PSR1509-58 is a fast juvenile pulsar observed by BATSE[9] and OSSE[10] at energies above 50 keV and possibly as high as few MeV. Being a *soft* object, we applied the *soft* set of selection criteria in looking at this source. During observation 23 in which PSR1509-58 was within 4.5° of the COMPTEL pointing axis, a 3σ deviation from a flat light curve (based on the Z_3^2 statistic) was observed in the 0.75-1 MeV energy interval. This preliminary result seen in figure 2 is reinforced by the correct phase of the broad "pulse", but given the trials made one merely concludes that this is an object which needs deeper study by GRO, especially as observation 23 was severely compromised by data transmission problems.

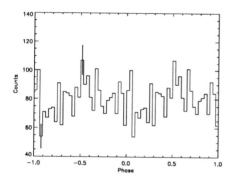

Figure 2 COMPTEL 0.7-1 MeV light curve for PSR1509-58

Geminga was established as a pulsar by ROSAT[11], and EGRET[12] observations combined with COS-B data[13] have allowed a highly refined ephemeris to be derived (recall that this is **not** a radio object). It is clearly a *hard* pulsar and the Vela *hard* selections were applied for our analysis. An independent search of the COMPTEL data has been made in each of the three observations. No significant pulsed emission is observed in any of these data. A first estimate of an upper limit (2σ) to the total DC emission from Geminga was provided by Strong et al.[6] : $0.14 \cdot 10^{-4} cm^{-2} sec^{-1}$ in the 10-30 MeV band.

Combining all three observations while paying due regard to the duty cycle of 15% observed by EGRET[14] we shall, so long as no pulsation appears be able to reduce the upper limits, thereby constraining the energy of the spectral break needed to fit the EGRET data and demanded by the COS-B data[15].

PSR1706-44 is a *very hard* γ-ray pulsar discovered by EGRET[16] and thereby confirming the existence of the COS-B source 2CG342-2: it has a flat spectrum $E^{-1.72}$. The single broad pulse distinguishes this from the other *hard* pulsars and reduces the sensitivity for detection. Using the criteria for a *hard* source we have searched in the COMPTEL data from two observations but no evidence for pulsed signal was so far found. Having such a flat spectrum, and being about a tenth as bright as Vela, we have not yet put effort into deriving an upper limit, which awaits the combination of the multiple observation periods.

THE UNDISCOVERED PULSARS

The Classical γ-ray pulsars, Crab and Vela, are radio pulsars, neutron stars whose emission is probably powered by the loss of rotational energy. An upper limit to the release of energy in a particular region of the spectrum can be inferred if the pulse period P and its time derivative $\dot{P}$ are known. As the total rotational energy is $E_{rot} = I\Omega^2/2$ its rate of dissipation is given by:

$$\dot{E}_{rot} = -I\Omega\dot{\Omega} \qquad (1)$$

where $\Omega = 2\pi/P$ is the angular velocity. The moment of inertia I is taken as:

$$I = 1.15 \times 10^{45}\, g \cdot cm^2\, \xi\, (M_*/1.44 M_\odot)\, (R_*/10 km)^2 \qquad (2)$$

The dimensionless factor ξ accounts for the radial dependence of the matter density ρ within the neutron star. For ρ decreasing with radius $\xi < 1$; for example $\rho \propto r^{-2}$ implies $\xi = 5/9$. We use here, not the calculated value of the moment of inertia, but a constant I of $I = 0.57 \times 10^{45}\, g \cdot cm^2$ (i.e. assuming $\xi = 0.5$) for all neutron stars in order to compare them. The total energy flux one may expect at the Earth is given by:

$$F_{lim} = \dot{E}_{rot}/4\pi D^2 \qquad (3)$$

Models for γ-ray emission are biased towards high values of $\dot{P}$ as they are most probably related to intense magnetic fields. However, no model is considered here and the only implicit assumption is that the fraction of energy released in γ-rays is the same in all pulsars. This favours the inclusion of millisecond pulsars in the list although the main argument in favour of not using a model is that the list is an "upper bound". If a radio pulsar is powered by rotational losses then equation (3) gives a limit on its observable flux of energy and on the possibility of detecting it as a γ-ray source.

Table I. COMPTEL Pulsar-Search Priority List

Pulsar	l	b	P ms	$\dot{P} * 10^{15}$ s/s	D kpc	$\dot{E}_{rot}$ (log)	Flux (log)
0531+21	184.6	-5.8	33.342	421.2	2.00	38.41	-6.27
0833-45	263.6	-2.8	89.286	124.3	0.50	36.60	-6.88
Geminga	195.1	4.3	237.097	11.40	(0.04)	34.29	-7.00
1706-44	343.1	-2.7	102.442	93.04	1.46	36.29	-8.12
1259-63	304.2	-1.0	47.762	18.50	2.34	36.58	-8.23
1509-58	320.3	-1.2	150.652	1537.	4.40	37.01	-8.36
1929+10	47.4	-3.9	226.517	1.157	0.08	33.35	-8.48
1951+32	68.8	2.8	39.530	5.849	2.50	36.33	-8.54
1046-58*	287.4	0.6	123.646	95.93	2.60	36.06	-8.85
Geminga	195.1	4.3	237.097	11.40	(0.40)	34.29	-9.00
1823-13*	18.0	-0.7	101.441	75.22	3.98	36.21	-9.07
1800-21*	8.4	0.2	133.588	134.1	3.90	36.10	-9.16
1758-24	5.3	-0.9	124.874	127.8	4.35	36.17	-9.18
0740-28	243.8	-2.5	166.752	16.83	1.10	34.91	-9.25
1937+21[No]	57.5	-0.3	1.558	.0001	3.72	35.80	-9.42
1821-24	7.8	-5.6	3.054	.0016	5.50	36.11	-9.45
1727-33	354.1	0.1	139.447	85.06	4.13	35.85	-9.46

This priority list, is compiled from the Australian Catalogue[17] (AusCat) of containing 548 known radio pulsars. The short-list is limited to objects with more than $Crab/1000$ "predicted flux" as being the most serious candidates.

Reassuringly, the Classical and "New" pulsars are near the top of the list given in Table I (Geminga appears twice under two assumptions of distance) which encourages us to believe the near neighbours are good candidates. We have thus searched for the evidence of pulsed emission from each of the objects in the list (except where flagged "No") using the Princeton ephemerides or, where no entry exists, the AusCat values (flagged "*"). This shows the importance of the Radio Astronomers' vigilance in monitoring these objects. A lack of contemporary ephemeris limits the conclusions we can draw from the gamma-ray data given the properties of "timing-noise" and glitches.

Using the selections optimised for the Crab, we detected no statistically significant pulsed signal from any of these objects within a single observation data-set i.e. $Z_3^2 < 3\sigma$. For the Crab shape a Z_3^2 of 3σ is equivalent to $P = 0.01$. This allows us to put conservative limits on their pulsed emission (based on a single observation) of about 25% of the Crab **pulsed** component in the 1-10 MeV range and of about 30% of Vela between 10-30 MeV. Already this makes this the most sensitive and complete search for pulsars in the MeV energy régime.

SUMMARY

COMPTEL has observed two radio pulsars with unprecedented sensitivity in the energy range 0.7-30 MeV but has so far been unable to register pulsed emission from the "New" γ-ray pulsars or putative radio pulsars suggested by the parameters in the available radio catalogues. Determination of accurate upper limits in COMPTEL energy range remain essential in order to constrain the spectra of pulsars and determine the break energies. With the "New" γ-ray pulsars having markedly differing spectra, it is reasonable to expect undiscovered pulsars to be luminous in COMPTEL's energy régime and searches continue. Filling the gaps in the catalogues is encouraged for nearby, short-period objects such as PSR1905+07 and PSR1915+15 which, given a reasonable $\dot{P}$, could be prime candidates for pulsed γ-ray emission.

REFERENCES

1. V. Schönfelder et al., Advances in Space Research, in press (1993).
2. M. Busetta et al., Workshop on Isolated Neutron Stars, in press (1992).
3. K. Bennett et al., A&A Suppl., in press (1992).
4. S. Johnston, V. Kaspi, R.N. Manchester, N. D'Amico, J.M. Cordes, R.S. Foster, J.M. Weisberg, A.G. Lyne, S.L. Shemar, R.S. Pritchard, Z. Arzoumanian, D. Nice, J.H. Taylor, GRO/radio timing data base, Princeton University (1992).
5. R. Buccheri & B. Sacco, Data Analysis in Astronomy (Plenum Press, 1985).
6. A. Strong et al., A&A Suppl., in press (1992).
7. P. Nolan et al., Bull. AAS **23** : 4, 67.02 (1991).
8. W. Hermsen et al., These proceedings.
9. R. Wilson et al., Bull. AAS **23** : 4, 57.07 (1991).
10. M. Ulmer et al., These proceedings.
11. J. Halpern & S. Holt, Nature **357**, 222 (1992).
12. D. Bertsch et al., Nature **357**, 306 (1992).
13. W. Hermsen et al., IAU circ. 5541 (1992).
14. D. Kniffen et al., These proceedings.
15. I. Grenier, W. Hermsen & R. Henrikson, A&A, in press (1992).
16. D. Thompson et al., Nature **359**, 615 (1992).
17. R. Manchester, Private communication (1992).

GRIS OBSERVATIONS OF THE CRAB

L. M. Bartlett, S. D. Barthelmy,
N. Gehrels, M. Leventhal, B. J. Teegarden and J. Tueller
NASA/Goddard Space Flight Center, Greenbelt, MD 20771

C. J. MacCallum
Department of Physics and Astronomy, University of New Mexico,
Albuquerque, NM 87131

ABSTRACT

The status of work in progress of the GRIS observations of the Crab is presented. We present lightcurves produced from epoch folding as evidence of GRIS pulsar analysis capability. The agreement of the GRIS data with the Jodrell Bank Crab Pulsar Timing Results indicates that our timing procedure is working properly. The lightcurve agrees well with the 'canonical' Crab lightcurve, with peaks 0.4 rotation apart. The energy dependence of the lightcurve is displayed as a series of lightcurves folded in specific energy ranges. Finally, we present our total Crab spectrum, from which we have seen no spectral lines.

INTRODUCTION

The Gamma Ray Imaging Spectrometer (GRIS) observed the Crab from Ft. Sumner, New Mexico on 31 May 1990. This balloon-borne high resolution germanium spectrometer achieved a full pass on the Crab (~10.5 hrs) just a few weeks prior to the FIGARO II observation[1] of the target. Our work in progress is multifold. We have demonstrated GRIS timing analysis, thus allowing high spectral resolution work on pulsars. Investigation of reported lightcurve features and spectral lines from the Crab during this epoch is proceeding using phase-resolved spectroscopy.

INSTRUMENT

The GRIS instrument is a heavily shielded balloon-borne germanium spectrometer. Seven large high-purity n-type Ge detectors are surrounded by a 15 cm NaI shield acting in anticoincidence to detect gamma rays in the 20 keV to 8 MeV range. The shielding allows a 17° FWHM low-energy FOV to be seen by the liquid nitrogen cooled detectors as they are pointed by the azimuth-elevation gondola.[2]

The instrument carried 1937 cm^3 of Ge with an effective area of 208 cm^2 at 100 keV. This is the largest effective area high spectral resolution observation of the Crab to date. The FWHM of the instrumental background lines was ~1.5 keV at 100 keV. The average float altitude of the instrument was 4.33 g/cm^2.

The observation was performed using alternating 20 minute target and 20 minute background segments. The background segments are taken at the same elevation angle as the target segments but offset in azimuth. Pointing, verified by a star camera and sun sensor, was in error by less than 2 degrees in azimuth during the flight.

Onboard timing was maintained by an oven-controlled oscillator. Drift in this oscillator was monitored by comparing the arrival of a 1 pulse per second signal with a rubidium oscillator 1 pps on the ground. Timing resolution was 144 microsecond.

ANALYSIS

The data analysis consisted of producing a monotonically increasing time tag for each event. This monotime was calculated relative to the reference epoch 48042.0 MJD. The event time was then corrected to the solar system barycenter using the JPL DE200 ephemeris to determine the Earth's position relative to the barycenter. The balloon's altitude, latitude and longitude allowed the instrument position to be determined relative to the geocenter. The barycentric corrected data was then epoch folded over a series of trial periods using the following total phase calculation for each event.

$$\text{TOTAL PHASE} = (\text{event time/trial period}) - (\text{event time/trial period})^2 \times (\text{period derivative})$$

The period derivative of 4.210892×10^{-13} s/s was calculated from the Jodrell Bank Crab ephemeris[3] for the reference epoch. The fractional part of the total phase was then binned for each trial period and the χ^2 value of the distribution was calculated with respect to the distribution's mean value. By this process, the true lightcurve distribution is in excess of the average of a random distribution, maximizing χ^2. Hence the true lightcurve will be indicated by a maximum in χ^2.

RESULTS

The results of the epoch folding are displayed in the following figures. The lightcurve corresponding to the maximum χ^2 is shown in Figure 1. This lightcurve covers the energy range from 25 to 300 keV. The shape of the lightcurve agrees well with the lightcurves of other instruments, many of which have had much better statistics from larger area and longer exposure times. The energy dependence of the lightcurve is shown in the four panels of Figure 2. The interpulse flux is seen to increase relative to the peaks flux as energy increases for the two lowest energy panels. Above 250 keV, the lightcurve is not as apparent as at lower energies and is not discernible above 540 keV.

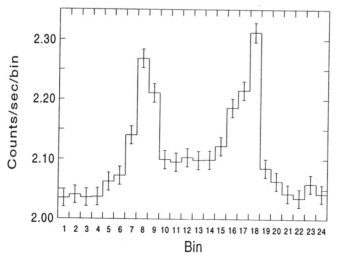

Figure 1. Lightcurve of the Crab observed by GRIS.

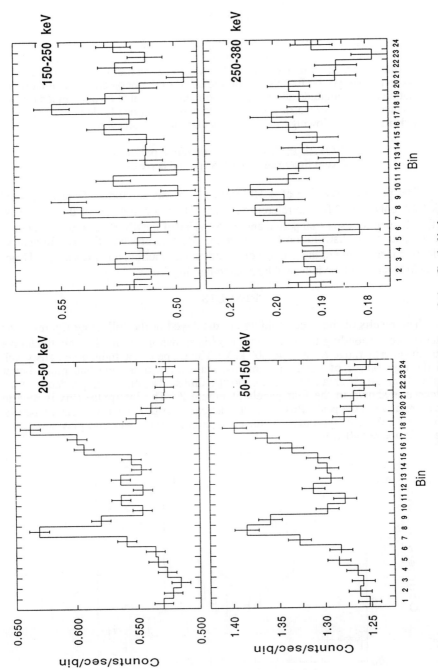

Figure 2. Energy dependence of the Crab lightcurve.

The χ^2 distribution indicates that the pulsar period was found and is in agreement with the ephemeris period. The distribution sampled periods between 0.033377650 and 0.033378050 sec at 2 nsec intervals.

Reference Epoch: 48042.0 MJD
    Ephemeris Period = 0.03337784693 sec
    Observed Period = 0.033377847 sec ± 1.5 nsec

The absolute phase of the first peak was interpolated to the reference epoch using the Crab ephemeris period and period derivative. Our observed phase of the first peak was determined to be 0.76 ± 0.02. The ephemeris phase was calculated to be 0.875 ± 0.003. The small discrepancy may be attributed to a correction of 0.12 rotation due to oscillator drift and is under investigation.

The energy spectrum shown in Figure 3 is our total Crab spectrum. It includes the nebula plus the pulsar. Strong instrumental background lines such as 24, 53, 66, 138, 198 keV have been excluded from this continuum spectrum. Two sigma upper limits are displayed for the high-energy data. The quoted errors are statistical and are smaller than the systematic uncertainty. The spectrum observed by GRIS is fit best by

$$\text{flux} = 4.16 \pm 0.04 \times 10^{-4} (E/100 \text{ keV})^{-2.18 \pm 0.01} \quad \text{photons/cm}^2\text{/s/keV}$$

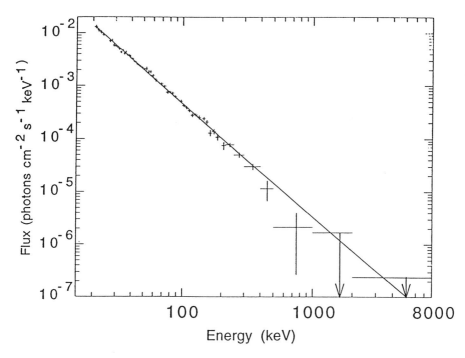

Figure 3. Total Crab spectrum observed by GRIS. This spectrum is comprised of the nebula plus pulsar.

A search for spectral lines was done on this data and we present two-sigma upper limits for previously reported line features.

Energy	Two sigma upper limit	Previously reported flux
73 keV	3.6×10^{-4} photons/cm^2/s	3.8×10^{-4} photons/cm^2/s [4]
440 keV	3.7×10^{-4} photons/cm^2/s	8.6×10^{-5} photons/cm^2/s [5]
545 keV	6.3×10^{-4} photons/cm^2/s	4.5×10^{-3} photons/cm^2/s [6]

CONCLUSION

We have demonstrated that GRIS can do timing analysis on pulsars. Our lightcurve is in agreement with those observed by other instruments. The observed pulsar period is in very good agreement with the Crab ephemeris period. The pulsar absolute phase is close to that predicted by the ephemeris and an investigation is underway to account for the small discrepancy.

Our two-sigma upper limits would have allowed us to see the 73 keV and the broad 545 keV feature if they had been present at their reported flux values during our observation. Our observation further confirms the transient nature of isolated gamma ray sources; even the Crab apparently displays this behavior. Phased-resolved spectroscopy will allow greater sensitivity to search for these and other line features at lower flux levels. The results of more detailed lightcurve work and the phase-resolved spectroscopy will be forthcoming.

REFERENCES

1. Agrinier, B., et al., Ap. J., **355**, 645 (1990).
2. Tueller, J., S. D. Barthelmy, L. M. Bartlett, N. Gehrels, M. Leventhal, D. M. Palmer, B. J. Teegarden, Ap. J. Supp., Submitted (1992).
3. Lyne, A. G. and Pritchard, R. S., Jodrell Bank Crab Pulsar Timing Results Monthly Ephemeris, private communications.
4. Ling, J.C., Mahoney, W.A., Willett, J. B., and Jacobson, A. S., Ap.J., **231**, 896 (1976).
5. Massaro, E., et al., Ap. J. Lett., **376**, L11 (1991).
6. Sunyaev, R. et al., IAU Circular No. 5481 (1992).

TIME VARIABILITY ANALYSIS OF THE COMPTEL CRAB LIGHT CURVE

A. Carramiñana, K. Bennett, M. Busetta
Astrophysics Division, ESTEC, Noordwijk, The Netherlands

V. Schönfelder, A. Strong
Max-Planck Institut für Extraterrestrische Physik, Garching, FRG

W. Hermsen, L. Kuiper
SRON-Leiden, Leiden, The Netherlands

A. Connors, J. Ryan
ISEOS, University of New Hampshire, Durham, USA

R. Buccheri
IFCAI/CNR, Piazza G. Verdi 6, 90139 Palermo, Italy

ABSTRACT

During the first year of observation by the Compton Gamma Ray Observatory, the Crab pulsar was in the field of view of COMPTEL four times. COMPTEL is sensitive to gamma radiation in the energy interval 0.7–30 MeV. The light curves that were obtained for each observation showed slight differences in appearance. A statistical analysis of the signal and the noise in the light curve has been performed to check whether the variation in shape is due to intrinsic source behaviour, which would obviously have theoretical implications, or simply due to statistical variations. The result of this analysis shows the latter to be the case.

INTRODUCTION

As far as we know, the Crab pulsar emits a constant flux of γ-rays, a fraction of them as a pulsed signal, providing a unique calibration tool for instruments and important clues to the behaviour of rotating neutron stars[1]. Fluxes of other objects are often referred to this "standard candle". The finding of time variability would result in a major re-thinking of these issues.

The Crab pulsar has rarely been studied in the MeV range[2] and the COMPTEL experiment on board GRO[3] gives the opportunity to make the most detailed analysis at these energies so far. Here we study the shape of its light curve in the energy range 1–10 MeV using the first four COMPTEL observations of the Crab pulsar. Specifically, we study the mutual statistical compatibility of these curves with each one and with the composite light curve. This question was recently addressed also by Buccheri et al.[4] using a different method.

THE CRAB LIGHT CURVE AS OBSERVED BY COMPTEL

The first three of these observations were presented elsewhere[5]. A summary of the exposures is given in Table I. Following standard practice, nearly 3×10^5 on-source arrival times, the majority of them background events, were folded into 100-bin light curves using contemporaneous radio ephemerides[6]. The 33-bin versions of these curves are shown in Figure 1. These observations were separated by a minimum and maximum of 18 and 214 days, respectively. As shown in Tables I and II, the different exposure times and angular distances to the pointing direction gave rise to different background levels and source sensitivities. Therefore the light curves cannot be directly compared with a χ^2 test. Any comparison analysis must take these differences into account.

Table I. Summary of COMPTEL observations of PSR0531+21

Observation Period	Target	TDJ (Start, End)	Duration 10^6 secs	Distance to axis	Total Counts
0	Crab pulsar	(8374,8383)	0.775	7.1°*	55970
1	Crab pulsar	(8392,8406)	1.199	6.4°	86847
2.5	Sun	(8415,8422)	0.667	10.5°	22075
15	NCG 1275	(8588,8602)	1.224	32.4°	134727
ALL		(8374,8602)	3.865	19.2°*	299269

* indicates more than one pointing direction and time average of the cosine of the distance to the axis

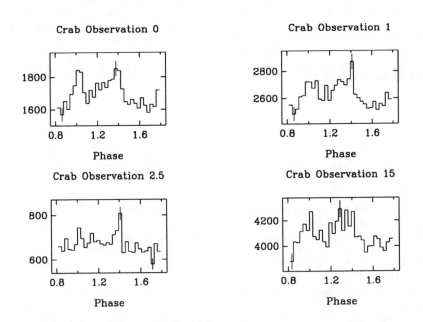

Figure 1. The 33-bin Crab light curves 1–10 MeV for each of the four observation periods

BACKGROUND LEVEL AND PULSED SIGNAL

The method chosen here was to compare observations in pairs. In order to do this, we deduce a shape consistent with the pulsed component of observation A, bootstrap events from the resulting distribution, and combine them with events drawn from a flat distribution in order to mimic the signal and noise levels of some other observation B. This new set contains the shape information of observation A with the normalization corresponding to observation B with which it can be directly compared using a χ^2 test.

A relatively simple filter was designed to estimate the background level and find a shape of the signal compatible with the observation. This filter is based on the comparison of the data with a collection of random numbers drawn from a gaussian distribution of mean 0 and standard deviation 1. It follows three steps:

(i) a first estimate of the background level is obtained by comparing the lowest values of both data and random sets (a maximun duty cycle is assumed); the randomly generated "datons" are renormalized using this estimate of the mean background level,

(ii) regions with statistically significant count excesses relative to the underlying background (renormalized random set) are "marked". In the remaining unpulsed regions "datons" are shifted to fit the background,

(iii) using this model of the shape of the pulsed signal, one revises the mean background level.

The process can be repeated several times (typically 10000 runs) with different random sets. Statistical estimates for the background level and the shape of the pulsed component are then obtained.

The filter was used with 100-bin and 33-bin light curves. We note that the first three observations are compatible with a 3% pulsed component while observation 15 has 1.6% pulsed component (see Table II). This is consistent with the Crab having been farther away from the pointing direction during this observation period and might be seen as a measure of the off-axis sensitivity of COMPTEL rather than an indication of variability in the γ-ray flux of the pulsar. Note that the pulsed strength is relative to the superposition of the background (dominant) and Crab DC component (secondary) and therefore does not reflect any feature intrinsic to the Crab. Figure 2 shows the pulse profiles extracted from the raw light curves.

Table II. background level and S/N level

Observation	Counts	Background	Excess	Pulsed/Background(%)
0	55970	54122 ± 495	1848	3.4 ± 0.8
1	86487	84282 ± 551	2207	2.6 ± 0.7
2.5	22075	21414 ± 225	661	3.1 ± 1.1
15	134727	132578 ± 750	2149	1.6 ± 0.6
ALL	299269	291139 ± 814	8130	2.8 ± 0.3

COMPARISON OF OBSERVATIONS

A bootstrap strategy was used to compare the shape of the pulse extracted from observation A with the light curve from observation B containing N_B counts with a pulsed ratio q: $q \cdot N_B$ events are drawn from the distribution defined by pulse shape A and the remaining $(1-q) \cdot N_B$ are taken from a flat distribution. This artificial dataset has the underlying distribution of shape A with the normalization of observation B. A direct comparison using a χ^2 test is then feasible. This procedure can be repeated many times in order to obtain a distribution of χ^2 values for each pair of observations compared. A statistical measure (probability) of observation B being compatible with pulse shape A can then be deduced. Individual observations are designated by their number 0, 1, 2.5 and 15, while ALL represents their combination. We will refer to the pulse extracted from observation A as "pulse A".

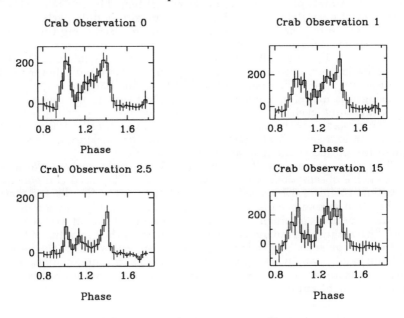

Figure 2. The pulse shapes extracted from the Crab light curves shown in Fig. 1.

To measure the degree of compatibility between observations A and B a 95% confidence level was used (Prob$(x \geq \chi^2) \leq 5\% \Rightarrow$ incompatibility) and the following criteria were used:
  (i) If A and B are compatible, then the data are consistent with a constant shape for the light curve,
  (ii) If A is not compatible with B nor with the combined light curve, then the data are inconsistent with a constant shape suggesting time variability.

The analysis was carried out for the entire 33-bin light curves and for four phase regions defined prior to the analysis. These regions are: the first pulse, interpulse, second pulse and background, according to the definitions by COS-B[1] and hereafter designated as FIRST, INTER, SECOND and BCKG. The combined light curve (WHOLE) was used as a fifth dataset for comparison. For each comparison 10000 bootstraps were generated. The phase regions labelled FIRST and SECOND cover 6 bins; INTER, 8 bins; BCKG, 13 bins; and WHOLE, 33 bins. It can be argued whether the number of degrees of freedom is N_{bins} or $N_{bins} - 1$ as both the observation and the bootstrap have the same normalization; however, while this is true for the WHOLE curve, it is unclear whether the same reasoning applies to individual regions as they were not normalized individually. We computed probabilities using $N_{dof} = N_{bins}$, which in any case would slightly underestimate incompatibilities.

RESULTS

As a total of 125 χ^2 distributions were computed (25 for each region of the light curve including the whole light curve itself) it is not possible to display the full results in detail. The following general results were noted:

- the χ^2 distributions found for the background region had associated probabilities between 34% and 83%, which indicate that all background regions are compatible, as expected:
- in the four pulsed regions (FIRST, INTER, SECOND and WHOLE) self-comparison ("pulse A" versus "observation A") gave systematically χ^2 values about 1 σ below expectation; this is consistent with the presence of an underlying structured shape:
- only 9 out of the 125 pairs showed incompatibility probabilities above the 95% level. These concerned mainly observation 2.5, which has the lowest signal-to-noise ratio and is therefore expected to produce the worst template. No hint of incompatibility was noted when using the template pulse ALL.

CONCLUSIONS AND FUTURE WORK

As there was no case of two mutually incompatible datasets we can conclude that the data are consistent with no time variability. This corroborates the complementary analysis by Buccheri et al.[4] on the same data. Future exposures by COMPTEL will be compared with the ones presented here allowing:
 (i) extension of this work to larger timescales,
 (ii) new observations which will contribute to the statistical significance of the combined light curve and a better template will then be obtained,
 (iii) groups of observations can be compared in order to be more sensitive to changes in the light curve. An upper limit on the degree of time variability that may exist is desirable.

Two planned applications of the method employed for this analysis are:
 (i) spectral phase analysis,
 (ii) empirical modelling of the combined light curve.

REFERENCES

1. J. Clear et al., A & A **174**, 85 (1987).
2. B. Agrinier et al., Ap. J. **355**, 645 (1990).
3. R. Diehl, Spa. Sci. Rev **49**, 85 (1988).
4. R. Buccheri et al., Advances in Space Research, in press (1993).
5. K. Bennett et al., A & A Suppl. Ser., in press (1992).
6. Z. Arzoumanian, D. Nice & J.H. Taylor, GRO/radio timing data base, Princeton University (1992).

COMPTEL DETECTIONS OF THE VELA PULSAR: FIRST ENERGY SPECTRA

W. Hermsen, H. Bloemen, L. Kuiper, B.N. Swanenburg
SRON-Leiden, Leiden, The Netherlands

R. Diehl, G.G. Lichti, V. Schönfelder, A.W. Strong
Max-Planck Institut für extraterrestrische Physik, Garching, F.R.G.

A. Connors, D. Morris, J.M. Ryan, M. Varendorff
ISEOS, University of New Hampshire, Durham, U.S.A.

K. Bennett, M. Busetta, A. Carramiñana, C. Winkler
Astrophysics Division, ESTEC, Noordwijk, The Netherlands

R. Buccheri
IFCAI/CNR, Piazza G. Verdi 6, 90139 Palermo, Italy

ABSTRACT

The COMPTEL telescope aboard the Compton Gamma-Ray Observatory, sensitive to low-energy γ-rays between about 0.7 and 30 MeV, has detected the pulsed signal of the Vela pulsar (PSR0833-45) in the 10–30 MeV range with a significance of 12 σ. The two main peaks in the light curve resemble the sharp peaks detected earlier in high-energy γ-rays. A combination of timing and spatial analysis, lowering the COMPTEL source detection threshold, enabled the detection of PSR0833-45 down to 0.75 MeV. The energy spectrum of the total pulsed emission connects smoothly to that measured simultaneously at higher energies by EGRET (and the time-averaged COS-B spectrum) and to the OSSE spectrum and ROSAT detection at lower energies. The combined spectrum cannot be represented by a single power law: bends/breaks are required at two or three energies to represent the data from the EGRET high-energy γ-rays down to the ROSAT soft X-rays. Spectra between 1 and 30 MeV are also shown for the sum of the emission from the two main peaks in the light curve as well as for the first peak alone.

INTRODUCTION

The COMPTEL telescope aboard the Compton Gamma Ray Observatory (CGRO) is sensitive to low-energy γ-rays between about 0.7 MeV and 30 MeV. It covers the middle energy range, with EGRET sensitive to high-energy γ-rays and OSSE and BATSE predominantly to hard X-rays. Prior to the launch of CGRO, only two radio pulsars were established γ-ray pulsars at energies above 1 MeV: the Crab pulsar (PSR0531+21) and the Vela pulsar (PSR0833-45), exhibiting very different γ-ray spectra in contrast to the similarity in the shape of their γ-ray light curves. The spectrum of PSR0531+21 could be represented by a single power law, whereas the spectrum of PSR0833-45 showed a break around about 300 MeV and was time variable, as seen in the COS-B data[1]. In fact, at MeV energies the source could hardly be detected: from two balloon flights one detection[2] and an upper limit[3] are reported. However, the latter at a lower flux level than the first.

Table 1. COMPTEL observations used in the present analysis.

Observation	Pointing ℓ	b	Start yy-mm-dd	End yy-mm-dd	View Angle
0.6	266.7°	.47°	91-05-10	91-05-16	4.5°
6	278.0°	−29.3°	91-07-25	91-08-08	30.2°
8	262.9°	−5.7°	91-08-22	91-09-05	3.0°
14	285.0°	−0.7°	91-11-14	91-11-28	21.7°

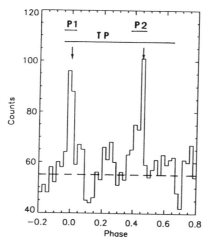

Figure 1: Vela light curve for energies 10–30 MeV; data from observations 0.6, 6, 8 and 14. The phase selections Total Pulsed (TP), Peak 1 (P1) and Peak 2 (P2) are indicated. The background level is shown by the dashed line.

Three more γ-ray pulsars have been discovered[4,5,6] since the launch of CGRO, more than doubling the examples with which the two main 'schools' for pulsar modelling ('outer-gap'[7] and 'polar-cap'[8] models) can confront their predictions. More importantly, each pulsar can now be studied in more detail, due to improved statistics and a wider coverage in energy. In particular, the evaluation of the spectra for different pulsar-phase intervals can provide important diagnostic tools revealing the underlying production mechanisms. A detailed analysis for the Vela pulsar at high energies indeed disclosed spectral variations and time variability with phase[1]. In this paper we present results from such a spectral analysis of the COMPTEL data.

OBSERVATIONAL DATA AND METHODOLOGY

The Vela pulsar was in the ∼1 sr wide field of view of COMPTEL during four observations in 1991 (see Table 1). The most sensitive observation is 8, in which the source was first detected: between 10 and 30 MeV a significant light curve was found, but at lower energies only hints for the presence of the main peaks could be seen[9,10]. Analysis of all four observations individually showed the source to be visible between 10 and 30 MeV in each of them. Figure 1

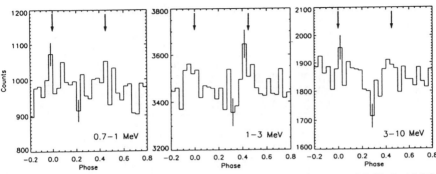

Figure 2: Vela light curves as in Figure 1 for 0.7–1 MeV, 1–3 MeV, 3–10 MeV.

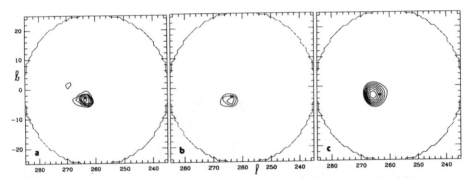

Figure 3: Maximum Likelihood images showing Vela (observation 8 only): a) Total pulsed emission for 10–30 MeV; b) P1 + P2 for 3–10 MeV; c) P1 + P2 for 0.75–1 MeV. The crosses mark the position of Vela. Contours at 15 step 3.

presents the sum of these light curves, representing a 12 σ detection. They have been summed by visually aligning the first peaks at phase 0. The sharp main peaks resemble, in widths and separation, the peaks seen at higher γ-ray energies. Figure 1 also indicates the definition of the phase windows for the two main peaks (P1 and P2) and the total pulsed emission (TP), as defined earlier in the analysis of COS-B data[1].

Figure 2 shows, for three energy intervals below 10 MeV, summed light curves which do not render significant detections ($\leq 2.2\ \sigma$ for each curve). However, selection of events arriving within the pulse interval improves the imaging sensitivity by reducing the background to only that in the pulse interval. This approach has been used to derive energy spectra: for a selected energy interval events were selected in a pulsed phase interval (e.g. TP, P1 or P2) and sorted in a 3D-data space defined by the scatter direction in the Compton telescope and the scatter angle (see Schönfelder et al.[11]). Then a search for sources in this 3D-data space was made using a maximum likelihood model fitting, which provides quantitative information regarding detection significance, source location and flux[12]. In principle, the detection significance can only be determined unambiguously if an independent estimate of the background structure (predominantly instrumental in origin) can be made in the 3D-data space. As this is not possible yet, the background structure was derived from the measured

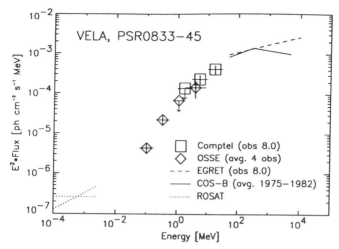

Figure 4: Differential Vela spectrum ($\times E^2$) for the Total Pulsed emission. For references see text. The ROSAT flux value is given for two assumed spectral shapes: power index -1.6 and -2.0.

data by an averaging technique efficiently suppressing point-source signals, yet preserving the general background structure.

Figure 3 presents examples of likelihood maps for observation 8. Shown is the quantity $-2ln\lambda$, where λ is the maximum likelihood ratio L(background) / L(source + background). Formally, a 3σ source detection corresponds to $-2ln\lambda = 13.8$, however, simulations have shown that our method of background modelling requires a somewhat higher threshold (discussed in future paper). The positions of the excesses are consistent with the position of Vela. Figure 3c proves the detection of PSR0833-45 down to the lowest COMPTEL energies.

ENERGY SPECTRA

In this paper we present preliminary spectra using only data from observation 8. The systematic uncertainties are dominated by uncertainties in the normalization and shape of the 3D-response function. For the set of simulated response functions used, we estimate the systematic uncertainties to be smaller than the statistical errors. Because the simulated response below 1 MeV is not yet available, no flux values are given below this energy.

Figure 4 shows the Vela spectrum ($\times E^2$) for the TP phase interval, together with the simultaneous EGRET[13] measurement and an OSSE[14] spectrum averaged over four observations. Also the average COS-B[1] spectrum and the ROSAT[15] detection are given. The COMPTEL and OSSE values are consistent with a power index of $\sim$-1.3, however, a break at MeV energies seems probable.

A selection of events of the main peaks (P1+P2) gives the COMPTEL spectrum shown in Figure 5a together with the time-averaged OSSE spectrum (same as in Figure 4, because the OSSE data did not show interpulse emission). The first peak (P1) appeared sufficiently intense in observation 8 to construct its spectrum (Figure 5b). The latter seems to be harder than the summed spectrum. In both figures, also the time-averaged COS-B spectrum is indicated for identical pulsar-phase selections.

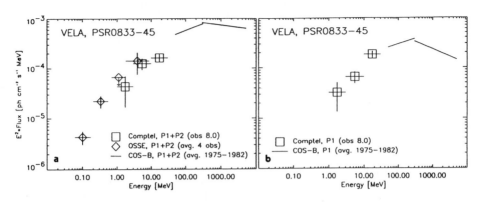

Figure 5: Vela energy spectra ($\times E^2$) for P1 + P2 *(a)*, and P1 only *(b)*.

DISCUSSION

The CGRO results appear to be self consistent and clearly indicate that the total spectrum of PSR0833-45 does not exhibit a power-law shape. Bends/breaks are required to link the EGRET and COMPTEL spectra (change in index $\sim$+0.4), probably also at MeV energies, and again in the extrapolation to the flux reported by ROSAT (index change possibly $\sim$-0.4) In the framework of 'polar cap'and 'outer gap'models detailed spectra have been calculated recently[16,17,18] and can now be tested over this wide energy interval. These calculations should also be confronted with the phase-resolved spectra. It is interesting to note that the latter COMPTEL spectra for the main peaks smoothly link to the same time averaged COS-B spectra. In further analysis a direct comparison with the phase-resolved EGRET and OSSE spectra will be of particular importance.

REFERENCES

1. I.A. Grenier, W. Hermsen, J. Clear, A&A **204**, 117 (1988).
2. O.T. Tümer, et al., Nature **310**, 214 (1984).
3. B. Sacco et al., ApJ **349**, L21 (1990).
4. R. Wilson et al., Bull. AAS **23**, 4, 57.07 (1991).
5. D.L. Bertsch et al., Nature **357**, 306 (1992).
6. D. Thompson et al., Nature **359**, 615 (1992).
7. K.S. Cheng, C. Ho, M. Ruderman, ApJ **300**, 522 (1986).
8. J.K. Daugherty, A.K. Harding, ApJ **252**, 337 (1982).
9. K. Bennett et al., A&A Suppl. (in press, 1992).
10. M. Busetta et al., Proc. Taos workshop "Physics of Isolated Pulsars', 1992, eds. Van Riper and Ho (Cambridge Univ. Press, in press).
11. V. Schönfelder et al., ApJ Suppl. (in press, 1993).
12. H. de Boer et al., Data Analysis in Astronomy IV, eds. V. di Gésu et al. (Plenum Press, New York, 1992), p. 241.
13. C. Fichtel, Presentation AAS Meeting (Atlanta, January, 1992).
14. M. Strickman et al., IAU Circ. (5557, 1992).
15. H. Ögelman et al., BAAS **23**, 1349 (1991).
16. J. Chiang, R.W. Romani, ApJ (in press, 1992).
17. A.K. Harding, J.K. Daugherthy, Proc. Taos workshop (see ref. 10, 1992).
18. C. Ho, Proc. Taos workshop (see ref. 10, 1992).

OSSE OBSERVATIONS OF THE VELA PULSAR

M.S. Strickman, J.E. Grove, W.N. Johnson, R.L. Kinzer, R.A. Kroeger, & J.D. Kurfess
Naval Research Laboratory, Washington, DC 20375-5320

D.A. Grabelsky, S.M. Matz, W.R. Purcell, & M.P. Ulmer
Northwestern University, Evanston, IL 60208

ABSTRACT

The OSSE detector on board the Compton Gamma Ray Observatory observed the Vela Pulsar (PSR 0833-45) during August-September 1991 and April-May 1992. Pulsed emission was detected at the 4-5σ level in the 0.06-0.57 MeV band in the sum of the two observing periods, as well as in each individual observation at lower significance. There is no significant variability observed. Light curves have a peak structure similar to that observed at higher energies. The spectrum is hard at lower energies and appears to require a break in the 0.5-2 MeV region.

OBSERVATIONS

The OSSE detectors on board the Compton Gamma Ray Observatory (CGRO) were used to observe the Vela Pulsar (PSR 0833-45) on three occasions during phase one of CGRO operations. The first observation, referred to as viewing period 8 (vp8), occurred during the interval 22 August – 5 September 1991. The second and third observations, which were analyzed together and are referred to as viewing period 26/28 (vp26/28), occurred during the intervals 23 April – 28 April 1992 and 7 May – 14 May 1992.

The vp8 observation consisted of three OSSE detectors staring in the direction of the pulsar. No background chopping was performed with these detectors. The total live time was $\sim 1 \times 10^6$ detector-seconds. The instrument was in a mode that supplied rates every 4 msec in four energy bands, all three detectors being summed together into the bands. The vp26/28 observations were similar to vp8 except that only two detectors stared at the pulsar and the live time was $\sim 5 \times 10^5$ detector-seconds. The energies associated with the four bands into which data were summed, averaged over detectors and observations, were 0.07 – 0.18 MeV, 0.21 – 0.57 MeV, 0.71 – 1.88 MeV, and 1.88 – 9.04 MeV.

ANALYSIS

The analysis procedure consisted of epoch-folding rate data to produce light curves for each viewing period and energy band. Arrival times at the solar system barycenter were computed for each 4 msec rate bin and the phase of that bin determined using a radio ephemeris supplied by Princeton University[1] as part of the CGRO pulsar monitoring program. The ephemeris used is constructed so as to be valid for both the vp8 and vp26/28 observations. Corrections were applied to account for a 2.042 second clock offset in the times supplied by the CGRO spacecraft. Given these corrections, the radio ephemeris adequately predicted both period and radio peak absolute phase.

We epoch-folded all data into 22 phase bins, yielding approximately one time resolution unit per phase bin. The bin phases were not varied in any way to optimize the observed signal, nor were multiple periods attempted, hence the number of trial light curves was kept to a minimum, preventing degradation of the significance of the result due to a large number of trials. The quality of the radio ephemeris was judged to be such that multiple trials were not necessary in any event.

The epoch-folding routine produced light curves such that phase 0.0 was the center of the radio peak for each viewing period. We then summed the resulting light curves assuming that the phase relation between radio and gamma-ray peaks did not vary over time. Light curves from the lowest two energy bands were summed to generate an optimum signal-to-noise light curve for determination of detection significance. For each light curve, the mean and χ^2 vs. the mean were generated. We then fit each light curve with a model consisting of two peaks and a constant unpulsed flux. Use of this model allowed the entire light curve to be fit simultaneously without having to make arbitrary decisions concerning the extent of "off-pulse" regions of the light curve.

We modeled the peaks as circular normal functions[2] given by :

$$Flux = \frac{A}{I_0(K)} \times e^{K \cos(2\pi(\phi - \phi_{center}))} \qquad (1)$$

where I_0 is a modified Bessel function of the first kind, order zero, which normalizes the function such that amplitude (A) is the integral of the peak. "Compactness" (K) is a peak width parameter that decreases with increasing peak width. This function has the property that it is periodic and is approximately gaussian in shape. The separation of the two peaks in the model was constrained to be 0.44 in phase, in agreement with results at higher energies[3].

To generate spectra, we divided the best fit peak amplitudes and uncertainties by the product of the exposure factor and the photopeak efficiency integrated over each energy band. Uncertainties in each spectral point were determined by χ^2-mapping of the amplitude parameters. The significance of each peak was determined from the value of the F statistic resulting from the addition of that peak to the best fit model without the peak (see RESULTS).

RESULTS

We detected the pulsar during both the vp8 and vp26/28 observations. We observed no evidence of significant variability between the two observations, so we report the sum of both viewing periods. For a light curve summed over the lowest two energy bands (figure 1), adding both peaks to no-peak model improves the fit at the $\sim 4.4\sigma$ level, as determined by the F-test. Note that χ^2/dof for the no-peak model is 1.56 for 21 degrees of freedom, allowing the rejection of that model at only the 2σ level. None of the peaks in the individual energy bands exceeds the 3.5σ level of significance and we detect no significant flux in the 0.71-1.88 MeV band.

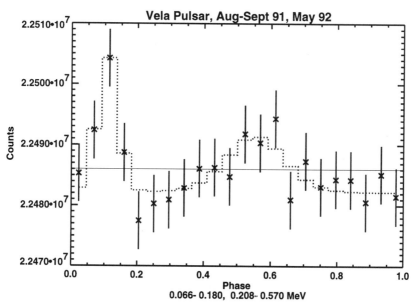

Figure 1: Vela Pulsar light curve, sum of viewing periods 8 and 26/28. The dotted line is the best fit model; the solid line is the mean.

The light curve looks qualitatively like the light curves measured at higher energies[3,4,5], in that it has two peaks at the similar absolute phase (first peak phase 0.11 after radio peak) and separation (~ 0.44 phase). The first peak is sharp but the second peak, which is detected at only the $\sim 3.5\sigma$ level as determined from the F-test, appears rather broad. We see no evidence for an interpulse between the peaks or peak leaders or trailers, unless the breadth of the second peak is interpreted as a narrow peak with a leader and trailer. The statistics available in the second peak do not allow us to make this distinction.

The spectrum of the first and second peaks together (figure 2), averaged over the entire light curve, indicates our detection of the pulsar in the two lower energy bands and a marginal detection in the highest energy band. The spectrum is very hard below 0.57 MeV, a power law drawn between the two points yielding an index $\sim -0.3 \pm 0.5$. The high energy[5] data shown in figure 1 is the total pulsed flux from COS-B averaged over all their observations of the pulsar. The dashed line is the extrapolation of their best fit power law below 300 MeV for the data shown. The dotted lines indicate the best fit power law spectra below 300 MeV for each individual observation. These extrapolated spectra are consistent with the OSSE data, at least within their range of variation. Our data, by themselves or taken together with the high energy data, require a break in the spectrum in the 0.5-2 MeV region.

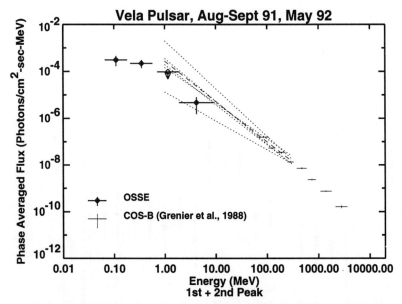

Figure 2: Vela Pulsar spectrum, sum of viewing periods 8 and 26/28. Flux is the sum of both peaks, averaged over the entire light curve. The high energy data and model spectra are taken from Grenier et al., 1988.

REFERENCES

1. J. Taylor, D. Nice, and Z. Azourmainian, GRO/Radio Timing Database, Maintained by Princeton University (1992).

2. P.M. Lee, Baysian Statistics: An Introduction (Oxford University Press, N.Y., 1989), p. 110.

3. D. Thompson, private communication (1992).

4. R. Buccheri, P. Caraveo, N. D'Amico, W. Hermsen, G. Kanbach, G.G. Lichti, L. Masnou, R.D. Wills, R.N. Manchester, and L.M. Newton, Astron. Astrophys. 69, 141 (1978).

5. I.A. Grenier, W. Hermsen, and J. Clear, Astron. Astrophys 204, 117 (1988).

GINGA OBSERVATIONS OF ISOLATED PULSARS: PSR1509–58 AND THE CRAB PULSAR

N. KAWAI & R. OKAYASU
The Institute of Physical and Chemical Research (RIKEN),
Wako, Saitama 351-01, Japan

Y. SEKIMOTO
Department of Physics, University of Tokyo,
7-3-1, Hongo, Bunkyo-ku, Tokyo 113, Japan

ABSTRACT

The energy spectra and the pulse light curve of PSR1509–58 in the X-ray range 2 – 60 keV are presented. The X-ray pulse spectrum can be fitted by a power-law model with a photon index of $\sim$ 1.3. The non-pulse component is softer, and has a photon index of 2.2 if modeled by a power-law. The Crab pulsar was observed by Ginga in March 1987 and September 1991. The variation of single pulse intensity was evaluated with a best statistics in the 2 – 11 keV X-ray, but no significant deviation from the Poisson fluctuation was found. The long-term variation of the relative intensity of the two peaks in the pulse profile was studied using the data taken 4.5 years apart. Unlike in the gamma-ray range, no significant variation of the pulse peaks were found. In this respect the soft X-ray pulse in the Crab pulsar shares more characteristics in common with the optical pulse rather than the gamma-ray.

ENERGY SPECTRA OF PSR1509–58

PSR1509–58 is the second brightest rotation-powered pulsar in the sky in X-ray range. If the spectral measurements from different spacecrafts could be combined, it will be the third example of the detailed measurements of the wide-band spectra after the Crab and Vela pulsars.

PSR1509–58 was observed by Ginga during 1990 March 13 – 17. We obtained 28 data segments of durations 4 – 20 min with a time resolution of 7.8 ms. These data have 12 energy channels in the energy range of 2 – 35 keV (March 13, 14) and 2 – 60 keV (March 15, 17). We also obtained $\sim$ 1000s of data with 48 energy channels in the energy range of 2 – 60 keV on March 17. Those data with 48 energy channels have a low time resolution, and are used to obtain the phase averaged spectrum. The field of view (1° × 2°) contains both the SNR nebula and the pulsar. The folded pulse profiles of PSR1509–58 are shown in Fig. 1 for various energy ranges. Even in the energy range above 30 keV, the 150 ms X-ray pulsation was detected significantly. Dependence of the pulse profile on the photon energy is not clear within our observation energy range.

214 PSR 1509-58 and the Crab Pulsar

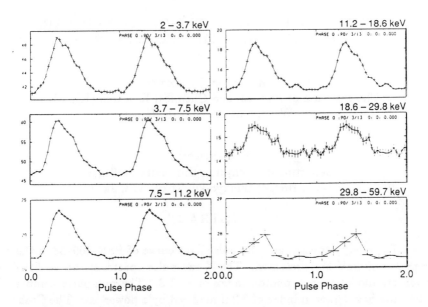

Figure 1: Folded light curve of PSR1509–58 in six energy bands.

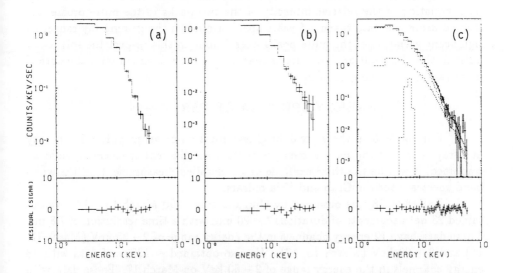

Figure 2: Energy Specra of PSR1509–58: (a) and (b) pulse component observed in different gain settings, (c) phase-averaged spectrum. Crosses denote the observation, and the histograms indicate the fitting solution.

	(a) Pulse	(b) Pulse	(b) Non-Pulse
Normalization	15.7 ± 2.1	13.7 ± 1.3	316. ± 16.
Photon Index	1.33 ± 0.06	1.30 ± 0.05	2.15 ± 0.02
log N_H	22.30 ± 0.13	22.30 (fixed)	21.88 ± 0.08

Table 1: Model Parameters of PSR1509–58 for the pulse and non-pulse components shown in Figure 2. The power-law normalization is the unabsorbed flux at 1 keV in $photons\,s^{-1}keV^{-1}$ for $4000\,cm^2$.

Three energy spectra of PSR1509–58 are shown in Fig. 2: (a) pulse component in normal-gain mode (2 – 35 keV) on March 13 and 14, (b) pulse component in low-gain mode (2 – 60 keV) on March 15 and 17, and (c) phase averaged spectrum in low-gain with fine energy binning (48 channels). The data are plotted with crosses, and the power-law model count spectra are plotted with histograms. The solution parameters of the model fitting are shown in Table 1. The two spectra taken with different gains yielded a consistent model parameters. The photon index of 1.3 is much smaller than that of the Crab pulsar[1] (~ 1.8). The normalization is the unabsorbed photon density at 1 keV for the phase-averaged pulse flux.

For the phase averaged spectrum, a simple power-law model with some N_H gives an acceptable fit to the data. Inclusion of a weak Fe line at 6.7 keV improves the fit significantly. The Fe line may be emitted from the thermal SNR shell found in the Einstein image, or it may be the galactic ridge emission. In Table 1, only the parameters for the non-pulse component are shown, where the pulse component with the parameters fixed to the value for spectrum (b) is subtracted from the phase-averaged spectrum. The non-pulse component is much softer than the pulse component, and has a different N_H when it is fitted to a simple power-law model.

As the normalization of the pulse component was determined, our results would be useful for constructing a wide-band spectrum by combining the results in hard X-ray ranges obtained by the balloon[2] and Compton[3] observations.

STABILITY OF PULSED X-RAY EMISSION

The intensity of individual pulses from radio pulsars are known to be very variable. The Crab pulsar exhibits an extreme case of this phenomenon, producing "giant pulses" in radio[4]. On the other hand, no such violent variation has been found in optical pulses, indicating a very different emission mechanism for the two wave bands. The study of individual pulses has been difficult in X-ray and gamma-ray energy range because the number of collected photons in a single pulse is usually too small. With the Ginga LAC, however, we can test the deviation from the Poisson statistics thanks to its large effective area and wide energy band.

The EGRET observations[5] of the Crab pulsar confirmed that the relative height of the two peaks in the Crab pulse in high energy gamma-ray has changed

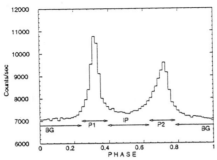

Figure 3: Pulse profile of the Crab pulsar in 1 – 6 keV. The definition of phase intervals is shown by arrows.

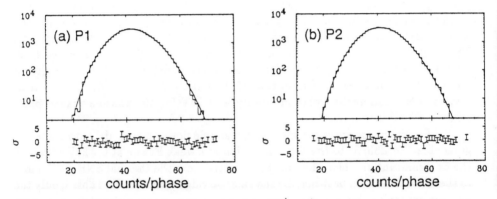

Figure 4: Distribution of pulses are plotted in histograms against the counts contained in the phase intervals P1 (a) and P2 (b). Solid lines show the expected Poisson distribution.

on a time scale of years since the early observations by SAS-2 and COS-B. It is also interesting to search for such pulse shape variation in X-ray range, but is difficult because of the pulse shape dependence of the Crab on the energy range. Extreme care has to be taken in comparing the data from two different instruments taken many years apart. During its 4.5 yr mission life, the performance of Ginga LAC was quite stable. Its data offer the best opportunity to study the long-term pulse shape variation with a single stable instrument.

The observations of the Crab pulsar were made on 1987 March 13 and 1991 September 19. The data in the range of 1 – 6 keV and 6 – 20 keV were accumulated with the time resolution of 0.98 ms and 1.95 ms respectively.

We analyzed a sample of ∼ 50000 pulses. These pulses are sorted by the counts contained in the phases shown in Fig. 3 (P1, P2, IP, and BG). The distribution of pulse samples against the number of counts contained in each phase are shown in Fig. 4. The model distribution predicted by the Poisson statistics and the residual from the model are also shown. We did not find any significant

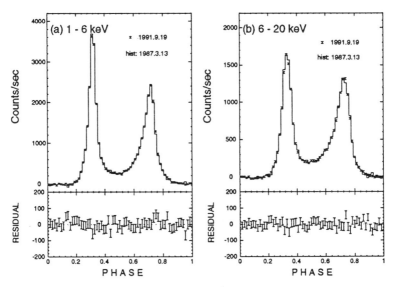

Figure 5: Pulse profiles of the Crab pulsar in (a) 1 – 6 keV and (b) 6 – 20 keV measured in March 1987 and September 1991. Ordinates for the two light curves are arbitrarily normalized. The difference in the pulse shape in each bins are plotted as residuals in the bottom panel.

deviation from the Poisson statistics in this test.

We compared the folded pulse profiles on 1987 March 13 and that on 1991 September 19. In Fig 4, the folded pulse profiles are fitted to each other with the scaling and the background adjusted as free parameters. As seen in the residual, we did not find any significant difference between the pulses shapes taken with 4.5 year interval. We also compared the ratio of the two peaks of the pulse. The change of ratio (P2+IP)/P1 was less than 2 % in 4.5 years. This long-term stability of the pulse profile found in X-ray is clearly different from the characteristics found in gamma-ray. The optical pulsation from the Crab pulsar is known to be very stable on any time scale. In such sense, the X-ray pulse has more characteristics in common with the optical pulse than the gamma-ray pulse.

REFERENCES

1. F. K. Knight 1982, *Astrophys.J.*,260, 538.
2. T. Takahashi et al. 1992, *this symposium*.
3. R. B. Wilson 1992, *Los Alamos Workshop on Isolated Pulsars* (Taos 1992).
4. S. C. Lundgren et al. 1991, *Compton Observatory Science Workshop* (Annapolis 1991), p. 260.
5. D. J. Thompson 1992, *Los Alamos Workshop on Isolated Pulsars* (Taos 1992).

The Contribution of the Northern Cross Observations to the Gamma Ray Observatory Pulsar Timing Support

N. D'Amico[1,2], L. Nicastro[1], C. Bortolotti[1], A. Cattani[1], G. Grueff[1,3], A.Maccaferri[1], S. Montebugnoli[1] and F. Fauci[2]

(1) Istituto di Radioastronomia del CNR, Bologna, Italy

(2) Istituto di Fisica dell'Universita' di Palermo, Italy

(3) Dipartimento di Astronomia dell'Universita' di Bologna, Italy

Abstract

The search for gamma-ray pulsations from known radio pulsars requires updated timing parameters in order to fold the gamma-ray data over a long time interval. A pulsar timing network set up by the Compton Observatory pulsar team provides continuous radio observations of a large sample of pulsars. The Northern Cross radiotelescope is part of this network, and about 50 pulsars are periodically observed using a new hardware system at the Cross. This paper describes the main characteristics of the system and the results of nearly 2 years of timing observations.

In the last few years a major technical effort has been devoted to exploit the large collecting area available at Northern Cross radiotelescope for pulsar observations (1,2). Now, a sensitive pulsar observing system is available at the Cross.

The hardware system (fig.1) is essentially designed to be used as a back-end for the E-W arm of the Northern Cross which operates at a the frequency of 408 MHz.

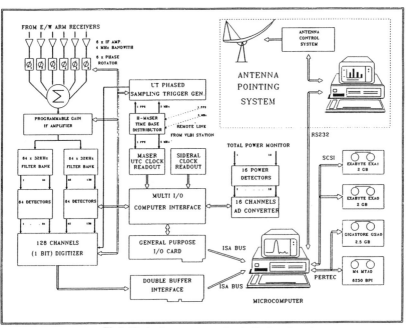

Fig.1 The hardware system used at the Northern Cross for pulsar observations.

As can be seen from the figure, the 6 IF channels derived from the corresponding antenna sections are summed with a computer controlled phase gradient to track the source within the E-W arm primary beam. The resulting total power IF signal is then split into a 128 32 KHz filterbank, and the individual signals after detection are filtered and digitized. The sampling triggers are supplied by a special UT-phased generator (3) which is driven by the H-Maser Medicina Master clock. The computer control system is based on a uP80386 microprocessor and MS-DOS operating system. Observations can be made completely automatically, and the data quality can be easily checked onsite using a quick-look facility. Offline data reduction of timing observations can be made using the standard TEMPO software package running under Unix on a Convex computer system at the main computer center of the Institute of Radioastronomy in Bologna.

The high performance of the Medicina pulsar system, the good sensitivity level available at the Cross, and the large range of declinations observable, are ideal for a long-term timing program. The Northern Cross also has participated in the GRO pulsar timing network. The timing observing program at the Cross started early in March 1991, before the expected GRO launch date, and has been in continuous operation since. Fig 2 shows a display of the Cross timing observing log.

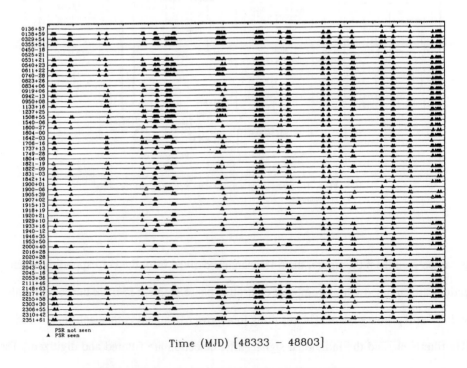

Fig.2 A display of the Pulsar Timing log file of the Northern Cross.

The sample of pulsars to the Italian group by GRO contains about 40 objects, selected on the basis of flux, low DM, and declination. As can be seen from the observing log display, some other pulsars are observed occasionally for calibration purposes, etc. For all the GRO pulsars, an accurate phase formula was obtained at the Cross after the first year of timing observations, and updated pulsar parameters (Table I) were transferred to the GRO public pulsar database at Princeton University.

Table I : Northern Cross Pulsar parameters

PSR	RA(J2000)	Dec(J2000)	data span	To (MJD)	F (Hz)	$\dot{F}$ (Hz/s)
0138+59	01 41 40.038	60 09 32.07	48334 48621	48334.546879771	0.81769592916	-2.88865D-16
0329+54	03 32 59.420	54 34 43.36	48334 48621	48334.625007856	1.39954089361	-4.04250D-15
0355+54	03 58 53.692	54 13 13.83	48334 48621	48334.640626627	6.39460066560	-1.79789D-13
0540+23	05 43 10.422	23 23 22.35	48300 48621	48300.000000723	4.06545906058	-2.55047D-13
0611+22	06 14 16.952	22 29 46.08	48300 48621	48300.000003806	2.98549439135	-5.31285D-13
0740-28	07 42 49.177	-28 22 40.63	48300 48621	48300.000000311	5.99661309943	-6.05352D-13
0834+06	08 37 5.777	06 10 19.47	48300 48621	48300.000005434	0.78507229432	-4.23376D-15
0919+06	09 22 14.074	06 38 25.04	48334 48621	48472.000001896	2.32224604743	-7.40649D-14
0942-13	09 44 29.060	-13 54 38.37	48334 48619	48334.882818766	1.75356180698	-1.64175D-16
0950+08	09 53 9.292	07 55 36.61	48334 48621	48334.886720469	3.95155076487	-3.57494D-15
1133+16	11 36 3.203	15 51 6.30	48334 48621	48334.960944967	0.84181200224	-2.66061D-15
1508+55	15 09 25.787	55 31 33.00	48335 48621	48335.101565865	1.35193369629	-9.10740D-15
1540-06	15 43 30.125	-06 20 46.99	48334 48621	48334.000004379	1.41030995600	-1.76827D-15
1600-27	16 03 7.716	-27 13 32.96	48330 48621	48330.000007297	1.28483050263	-5.22914D-15
1642-03	16 45 2.041	-03 17 58.30	48330 48631	48330.000003463	2.57938058603	-1.19146D-14
1706-16	17 09 26.441	-16 40 53.06	48330 48621	48330.000006366	1.53126530220	-1.48811D-14
1737+13	17 40 7.328	13 11 56.97	48330 48621	48330.000004847	1.24525205136	-2.27024D-15
1749-28	17 52 58.697	-28 06 35.60	48330 48621	48330.000004147	1.77759153839	-2.55798D-14
1821-19	18 24 0.450	-19 45 51.00	48330 48621	48330.000001526	5.28166344268	-1.46277D-13
1822-09	18 25 30.539	-09 35 22.79	48330 48619	48330.000007860	1.30042534733	-8.86196D-14
1831-03	18 33 41.976	-03 39 4.54	48330 48621	48330.000007027	1.45624098100	-8.80607D-14
1842+14	18 44 54.893	14 54 14.62	48330 48621	48330.000002119	2.66337677617	-1.32044D-14
1900-06	19 03 38.160	-06 32 26.41	48330 48621	48330.000000410	2.31542267731	-1.79203D-14
1900+01	19 03 29.976	01 35 38.20	48330 48632	48330.000001307	1.37117084981	-7.59828D-15
1907+02	19 09 38.370	02 54 49.22	48330 48620	48330.000000724	1.01027392447	-5.59939D-15
1915+13	19 17 39.787	13 53 56.77	48330 48632	48330.000000257	5.13795165014	-1.89999D-13
1920+21	19 22 53.551	21 10 42.28	48330 48620	48330.000008634	0.92270983595	-7.03778D-15
1929+10	19 32 13.898	10 59 30.68	48330 48632	48330.000000823	4.41466379226	-2.23905D-14
1933+16	19 35 47.835	16 16 40.80	48330 48621	48330.000002345	2.78753885580	-4.66499D-14
1940-12	19 43 25.498	-12 37 44.31	48330 48620	48330.000010254	1.02835292869	-1.73578D-15
2000+40	20 02 44.040	40 50 54.70	48330 48621	48330.000004382	1.10489108276	-2.17536D-15
2043-04	20 46 0.120	-04 21 25.30	48330 48588	48330.000001008	0.66643829174	-6.17968D-16
2053+36	20 55 31.370	36 30 21.18	48330 48619	48330.000000673	4.51450607088	-7.55529D-15
2148+63	21 49 58.638	63 29 44.27	48330 48620	48330.000003727	2.63060693896	-1.18422D-15
2217+47	22 19 48.123	47 54 53.92	48330 48620	48330.000002751	1.85711631061	-9.54004D-15
2255+58	22 57 57.795	59 09 14.68	48330 48621	48330.000001336	2.71556747725	-4.24671D-14
2303+30	23 05 58.356	31 00 3.24	48330 48621	48330.000015794	0.63456357058	-1.20863D-15
2306+55	23 08 13.829	55 47 35.24	48330 48618	48330.000003153	2.10496328513	-7.88183D-16
2310+42	23 13 8.604	42 53 12.67	48330 48619	48330.000000015	2.86177334499	-8.88668D-16
2351+61	23 54 4.700	61 55 48.20	48330 48632	48330.000001527	1.05844480837	-1.82817D-14

References

1. D'Amico N., Bortolotti C., Cattani A., Fauci F., Grueff G., Maccaferri A., Montebugnoli S., Roma M. & Tomassetti Technical Report IRA 137/90

2. D'Amico N., Il Nuovo Cimento, 1990, **13** C,505

3. D'Amico and Maccaferri, 1992, Experimental Astronomy, Submitted.

SEARCH FOR TeV GAMMA RAYS FROM GEMINGA

D.J.Fegan[1], C.W.Akerlof[2], A.C.Breslin[1], M.F.Cawley[3], M.Chantell[6],
S.Fennell[1,6], J.A.Gaidos[7], J.Hagan[7], A.M.Hillas[4], A.D.Kerrick[5],
R.C.Lamb[5], M.A.Lawrence[6], D.A.Lewis[5], D.I.Mayer[2], G.Mohanty[5],
K.S.O'Flaherty[1], M.Punch,[1,6] P.T.Reynolds[5], A.Rovero[6], M.Schubnell[2],
G.Sembroski[7], T.C.Weekes[6], M.West[4], T.Whitaker[6] and C.Wilson[7]

[1] Physics Department, University College, Dublin, Ireland.
[2] Physics Department, University of Michigan, Ann Arbor, Michigan, U.S.A.
[3] Physics Department, St.Patricks College, Maynooth, Co.Kildare, Ireland.
[4] Physics Department, University of Leeds, Leeds, England.
[5] Physics Department, Iowa State University, Ames, Iowa, U.S.A.
[6] Whipple Observatory, Harvard-Smithsonian CfA, Amado, Arizona, U.S.A.
[7] Physics Department, Purdue University, West Lafayette, Indiana, U.S.A.

ABSTRACT

Recently the Tata group have reported (1) the detection of TeV γ-rays from Geminga. Results of a search by the Whipple Observatory Collaboration are presented here, based on observations made during 1989-90 and 1990-91, using the 10m high resolution imaging Cerenkov camera.

1. SHOWER IMAGE PARAMETERIZATION

The high resolution Cerenkov imaging camera has been described in detail elsewhere (2). The energy threshold of the camera is 500 GeV. Following pre-processing stages described in (2), individual images are parameterized (3, 4) using simple moment fitting routines. To a first approximation the images are elliptical. The WIDTH is a measure of the r.m.s extent of the image along the minor axis and LENGTH is a similar measure along the major axis. MISS is the perpendicular distance between the image major axis and center of the field of view while DIST is the spacing between the image centroid and the field center. ALPHA is a derived parameter defined below. The image parameterization scheme of Hillas (4) forms the practical basis of TeV γ-ray selections. Since the hadronic shower images are on average considerably broader and longer than their γ-ray counterparts, discrimination becomes feasible both on the basis of shower shape and shower orientation, with preferential selection of γ-rays being possible based on predicted properties of γ-ray images from Monte Carlo simulations. The "supercuts" (5) γ-ray selection procedure has been established on the Crab Nebula database and forms the basis of the successful detections of γ-rays from both that object (5,6) and the giant elliptical galaxy

Markarian 421 (7). The basis of selection using supercuts is as follows,

Shape selection: $0.51^0 < \text{DIST} < 1.1^0$
                        $0.073^0 < \text{WIDTH} < 0.15^0$
                        $0.16^0 < \text{LENGTH} < 0.30^0$
Orientation selection: $0.51^0 < \text{DIST} < 1.1^0$
                        $\alpha = \sin^{-1}(\text{MISS}/\text{DIST}) < 15^0$.

In the analysis of the Geminga database, these image selections were applied both independently and in combination.

2. DATABASE, ANALYSIS AND RESULTS

Geminga has been extensively observed by the Whipple Observatory 10m reflector during the winters 1989-90 and 1990-91. A majority of the database consists of 28 minute ON/OFF scan pairs. In this mode the source is located in the center of the field of view of the camera and tracked for a 28 minute ON scan before slewing the reflector in right ascension to subsequently track a 28 minute comparative OFF region. Some scans were also taken in this mode where the durations were 38 minutes. An appreciable amount of data was also taken in OFFSET mode, whereby the true source location was deliberately offset by ± 0.5° from the center of the field of view, giving simultaneous monitoring of both ON and OFF regions in 18 minute scans. The observational database is summarized in Table (1).

TABLE (1)

DATASET	MODE	OBSERVATION	DURATION
1	ON/OFF	891218-910215	1685 min.
2	OFFSET	910306-910416	1120 min.

Analysis based on searching for a net excess of TeV γ-rays from Geminga is summarised in Table (2). Selection is based on the application of supercuts to 1685 minutes of ON/OFF data and yields a net excess of 2.35σ which is not significant. The ON/OFF excess in the unselected data is 0.63σ. This result indicated that there is no evidence for steady emission of TeV γ-rays from this source. Using a collection area of 3.5×10^8 cm^2 and the values quoted in the final column of Table (2) leads to a three standard deviation upper limit for steady emission from Geminga of 8.8×10^{-12} cm^{-2}s^{-1} (E >0.5 TeV) or approximately 125 milli-Crab at the same energy.

TABLE (2)

ANALYSIS	UNSELECTED	SELECTED ON SHAPE	ORIENTATION SELECTION	SELECTED ON BOTH
ON	327676	10942	25380	1749
OFF	327170	10587	25164	1613
DIFFERENCE	506	355	216	136
σ	0.63	2.42	0.96	2.35

Although no evidence was found for steady emission of TeV γ-rays, nevertheless the Geminga ON-scans from Dataset .1 and the OFFSET scans from Dataset 2 were subjected to periodic analysis with a view to searching for periodic emission from the pulsar. Taking the source position (J2000.0) as RA=$06^h33^m54.02^s$, DEC = $17°46'11.52"$, individual event times were barycentered and coverted to phases using elements of the EGRET ephemeris reported in (8),

T_0 = MJD 48400.0   : Epoch.
f_0 = 4.2176749957 s^{-1} : Frequency.
f_1 = -1.9508 x 10^{-13} s^{-2} : 1st derivative.
f_2 = 4.0 x 10^{-25} s^{-3} : 2nd derivative.

This search was based both on individual scans and on the two winters' databases, each treated as a composite phase linked sample. Phases of the 3050 supercut selected events were subjected to a Rayleigh test at the fundamental pulsar period. No statistically significant modulation was observable in the lightcurves for either years data, Fig. (1). Phase bin zero corresponds with alignment to the first peak in the EGRET lightcurve which occurs 0.21 cycles after 0hr. U.T. on May 24 1991 as seen at the geocentre. Since the X-ray and low energy γ-ray light curves are double peaked with 0.5 phase separation, similar tests were applied at the 2nd harmonic of the pulsar period, again no significant periodic emission was detected. Assuming a 20% signal modulation, the five standard deviation upper limit for pulsed emission from Geminga is 6.54 x 10^{-12} $cm^{-2}s^{-1}$ (E >0.5 TeV).

Using the Z^2_n test (which sums the first n harmonics, here n = 5) a search for a signal in a ± 150 nHz range about the frequency given by the EGRET ephemeris was performed based on a Rayleigh analysis. The errors on the ephemeris are on the order of ± 5 nHz for data in 1989. For the single year datasets, the 1989-90 observation span corresponds to 1.5 independent Fourier samples (IFS) while for 1990-91 the observation span corresponds to 2.6 IFS view. Figure (2) summarises results of the Z^2_5 test, for each season's data. No significant features are observable in raw, shape, orientation or supercuts.

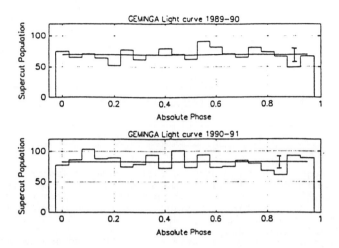

Fig.1. GEMINGA light curves at the fundamental pulsar period.

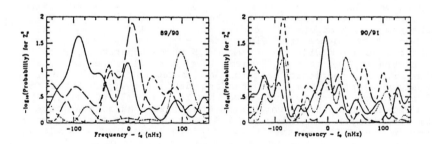

Fig.2. Z^2_5 distributions for selections : Raw(....); Shape(- - -); Orientation (-- --) and Supercut(———).

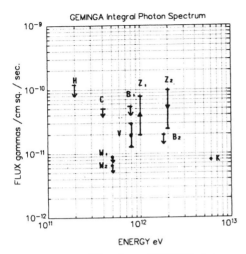

Fig.3. GEMINGA Integral TeV photon spectrum.

In conclusion the Whipple Observatory Collaboration have failed to find evidence of TeV γ-ray emission from Geminga based on observations made during 1989-90 and 1990-91. Upper limits calculated on the basis of both steady and pulsed emission are below upper limits and detections reported by other groups, Fig (3). The possibility of source variablity at TeV energies cannot be ruled out.

ACKNOWLEDGEMENTS

We thank K.Harris and T.Lappin for help in obtaining these observations. We acknowledge support from the U.S. Department of Energy, NASA, the Smithsonian Scholarly Studies Fund and EOLAS, the Scientific funding agency of Ireland.

REFERENCES

1. Vishwanath, P.R. et al., IAU Telegram 5612, Sept 14th (1992).
2. Cawley,M.F. et al., Experimental Astronomy, 1, 173 (1990).
3. MacKeown,P.K. et al., Proc.18th ICRC (Bangalore), 9, 175, (1983).
4. Hillas,A.M., Proc.19th ICRC (La Jolla), 3, 445 (1985).
5. Punch,M. et al., Proc.22nd ICRC (Dublin),1 , 464 (1991).
6. Vacanti,G. et al., Ap.J., 377, 467, (1991).
7. Punch,M. et al., Nature, 358, 447 (1992).
8. Mattox,J.R., et al., IAU Telegram 5583, Aug 11th (1992).
9. Weekes,T.C. and Helmken,H.F., Proc. 12th ESLAB Symp., 124, 39, (1977).
10. Cawley,M.F., et al., Proc. 19th ICRC (La Jolla), 1, 173 (1985).
11. Bhat,P.N., et al., Proc. 19th ICRC (La Jolla), 1, 181 (1985).
12. Zyskin,Y.L. and Mukanov,D.B., Prov. 19th ICRC (La Jolla), 1, 177 (1985).
13. Kaul,R.K., et al., Proc. 19th ICRC (La Jolla), 1, 165 (1985).

POSSIBLE DETECTION OF THE GEMINGA PULSAR IN THE ENERGY RANGE (0.15-0.48) MeV WITH THE FIGARO II EXPERIMENT

E.Massaro[1], B. Agrinier[2], E. Barouch[2], R. Comte[2], E. Costa[3],
G.C. Cusumano[4], G.Gerardi[5], D. Lemoine[2], P. Mandrou[6],
J.L. Masnou[7], G. Matt[1], T. Mineo[4], M. Niel[6], J.F. Olive[6],
B. Parlier[2], B. Sacco[4], M. Salvati[8] and L. Scarsi[4]

[1] Istituto Astronomico, Universitá "La Sapienza", Roma (Italy)

[2] Service d' Astrophysique, D.A.Ph.P.E., CEN, Saclay (France)

[3] Istituto di Astrofisica Spaziale, CNR, Frascati (Italy)

[4] Istituto di Fisica Cosmica e Appl. Inform., CNR, Palermo (Italy)

[5] Istituto di Fisica, Universitá di Palermo (Italy)

[6] CESR, Université P. Sabatier, Toulouse (France)

[7] UPR176 CNRS, DARC, Observatoire de Paris, Meudon (France)

[8] Osservatorio Astrofisico di Arcetri, Firenze (Italy)

ABSTRACT

The gamma-ray source Geminga (2CG195+04) has been observed in the energy range (0.15-4) MeV with the FIGARO II experiment on July 9th, 1990, during a transmediterranean balloon flight. The data were folded at the period value derived from the recent EGRET ephemeris. The resulting light curve shows a broad peak (phase width ~ 0.3) in the energy range (0.15-0.48) MeV. The chance occurrence probability turns out to be about 2.8 gaussian standard deviations.

INTRODUCTION

The FIGARO II (French Italian GAmma Ray Observatory) is a balloon borne experiment specifically designed to study sources with a well established time signature in the low energy γ rays. The main characteristics of the detector are the large area (3600 cm^2) and the wide field of view (about 77°); a more detailed description is given in Agnetta et al.[1].

In the course of the last transmediterranean balloon flight on July 9, 1990 the instrument was pointed at the direction of the Crab for more than seven hours. Because of the large field of view the γ-ray source Geminga, at an angular distance from Crab of about fifteen degrees, was also observed with only a small reduction of the exposed area. After the discovery that the X and γ-ray emission of Geminga is periodic at 237 ms [2,3], we searched the FIGARO data for a pulsed signal at the expected period for the epoch of our observation. In this contribution we report preliminary results of this analysis indicating the occurrence of a weak signal in the energy band 0.145-0.480 MeV.

DATA ANALYSIS AND RESULTS

The pulsed light curve of Geminga was obtained by folding the arrival time (converted to the solar system barycenter by means of the JPL Ephemeris DE200) of each accepted event with the instantaneous period. The pulsar parameters for the observation epoch (J.D.=2448081.5) were derived from the frequency and its first and second derivatives as given by Mattox et al.[4], which are expected to provide the phase of the pulsed signal with an uncertainty not larger than 0.1 in the time interval 1990.5-1992.9 (Mattox, private communication).

Two resulting phase histograms in five and fifteen bins are shown in Fig. 1a and 1b, respectively: a broad peak with a duty cycle of about 0.3 and roughly centred at phase 0.6 is apparent. The statistical significance of this feature is not high: a simple χ^2 test gives for the curve of Fig. 1a a significance of 2.6 gaussian standard deviations. Because of the broad structure of the signal, the χ^2 test is not the most efficient method to estimate the probability. We applied therefore to the light curve of Fig 1b the non-parametric "run" test: considering that the excess of the counting rate is concentrated in only six consequent bins over fifteen, the probability to have a chance effect is 4.9×10^{-3} We computed also the Fourier power associated with the fundamental frequency and obtained a value whose chance occurence probability is 4.6×10^{-3}. In conclusion, we have evidence for the detection of a pulsed signal at a level of about 2.8 gaussian standard deviations.

To evaluate the pulsed flux of the source we considered the source signal within the phase interval (0.4-0.7) and subtracted from it the mean value of the counts from the other bins, corresponding to the off-pulse region.

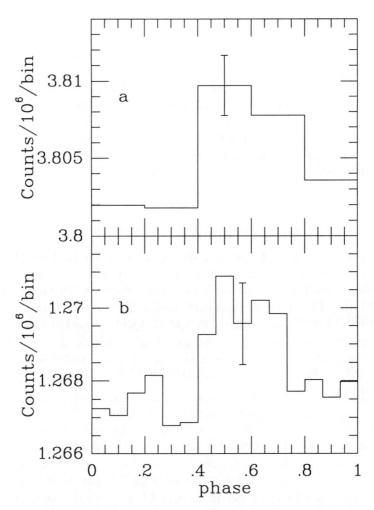

Fig. 1a,b - Phase histograms of Geminga in the energy band 0.145-0.480 MeV in five (panel a) and fifteen (panel b) bins.

The photon flux in the band (0.15-0.48) MeV, taking into account the average atmospheric thickness and the dead time corrections, and assuming an E^{-2} spectrum, is 2.7 ($\pm$0.9) 10^{-4} photons cm^{-2} s^{-1}. In Fig. 2 we have plotted the corresponding flux density, multiplied by E^2, in comparison with the integrated flux of ROSAT[2], the COS B data[6] and the upper limits of SIGMA[7] and COMPTEL[8].

DISCUSSION

The recent discovery that the puzzling Geminga is a γ-ray pulsar[2,3] and that its emission above 50 MeV shows a phase pattern quite similar to those of Crab and Vela is a major step towards explaining the nature of the galactic γ-ray sources. The physical mechanisms responsible for the pulsar emission in the various bands of the electromagnetic spectrum, however, are still poorly understood, particularly in the low energy γ rays where the instrumental limits are more severe.

The FIGARO data provide a first indication of a pulsed low energy γ-ray emission from Geminga; the statistical significance of this result is not compelling and a confirmation from future measurements is necessary. Preliminary results of the BATSE and OSSE experiments (Wilson et al., Ulmer et al., this symposium) give upper limits below the present estimate of the possible flux. An explanation of this disagreement could be that the effect is limited within a rather narrow energy band. A few comments are nonetheless in order.

As a first remark we see that the light curve of Fig. 1 does not show the double peak pattern as in the EGRET data[3], while it seems ideed more similar to the low energy X-ray curve of the ROSAT measurement[2]. These data are explained as thermal emission from a hot region of the neutron star surface: the spectral shape, in fact, can be fitted with a black body law with the exception of a tail in the high energy part of the band. One could try to explain the FIGARO data as an extension of the hard tail; the extension would have a very flat slope in order to pass through the FIGARO point, and would violate the COMPTEL upper limits unless a very sharp cutoff were invoked. Alternatively, and perhaps more likely, one could relate the light curve of Fig. 1 with those observed at higher energies, provided that in the FIGARO passband the ratio of the two peaks is very unbalanced. Such an unbalance is indeed seen in the Crab pulsar at energies around 1 MeV [5]; if the analogy is correct, more sensitive observations in the low energy γ rays should detect another peak at phase 0.1.

Acknowledgements. This research was financially supported by CNES and ASI. We are grateful to many persons from CESR, CNES, CEN, CNR and ASI who contributed to the payload and to the flight campaigns and particularly to A. Soubrier and O. Cosentino and to E. Myles-Standish Jr. for kindly providing DE200 Ephemeris.

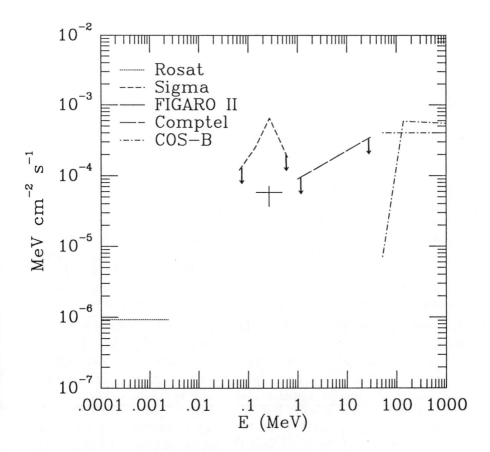

Fig. 2 - The energy spectrum of GEMINGA from X to γ rays

REFERENCES

1. Agnetta, G., et al., Nucl. Instr. Methods A 281, 197 (1989).
2. Halpern, J.P. and Holt, S.S., Nature 357, 222 (1992).
3. Bertsh, D.L., et al., Nature 357, 306 (1992).
4. Mattox, J.R., et al., I.A.U. Circ. 5583 (1992).
5. Agrinier, B. et al., Ap.J. 355, 645 (1990).
6. Grenier, I.A., Hermsen, W. and Hote, C., Adv. Space Res. 11, 107 (1991).
7. Lebrun, F. et al., Proc. 22nd ICRC, vol. 1, p. 165 (1991).
8. Strong, A.W. et al., COSPAR Symp. Proc. in press (1992).

Understanding GEMINGA: Past and Future Observations

Giovanni F. Bignami[2,1], Patrizia A. Caraveo[1] and Sandro Mereghetti[1]

1 Istituto di Fisica Cosmica del CNR
   Via Bassini, 15 - 20133 Milano (Italy)

2 Dipartimento di Ingegneria Industriale, Università di Cassino,
   Via Zamosh 43 - 03043 Cassino (Italy)

Introduction

Twenty years have gone by since the discovery of the γ-ray source 195+5, the first UGO (Unidentified Gamma Object) seen by the NASA SAS-2 satellite. These years have been characterized by an endless quest for an identification of this puzzling object first in the γ-ray domain, with the ESA COS-B satellite (1975-82), then in the X-ray domain, with the NASA Einstein Observatory (78-81) and ESA EXOSAT (83-86) missions, finally in the optical (1983-today) using all of the big telescopes of the world. Unfortunately, every step down in energy costed a factor of 1000 in photon flux (see table) and, adjusting the observing time, we ended up with ~1000 photons in each energy range. While wading through the intricacies of multiwavelength astronomy, the source acquired its multi-language name **GEMINGA**, officially for **G**Amma source in **GEMIN**i, in reality a pun from the milanese argot*, making Geminga the first celestial object named in a dialect.

The strong gamma-ray source GEMINGA was first associated with 1E 0630+178, a peculiar X-ray source (Bignami et al, 1983), and later (tentatively) with a very faint optical object G", first seen with the CFHT 3.5m and confirmed with the ESO 3.6m (Bignami et al. 1987,1988) and with the 5m Palomar (Halpern and Tytler,1988). The γ-ray-X-ray association rested on the strong similarity with the Vela pulsar, while the X-ray-optical one on the peculiar color of G". Geminga was supposed to be an isolated neutron star, since only this object could explain at once the strong γ emission and the faintness of the optical counterpart in spite of the X-ray emission. Back in 1983, we had also proposed that it ought to be very close,~100 pc (Bignami, Caraveo, Lamb, 1983). All this was very reasonable but rested purely on theoretical grounds rather than on hard facts. During 1992, however, new γ-X and optical observations provided the facts needed to secure the identification of Geminga.

---

* Gh'è minga means "is not there" to most northern Italians

The X-ray-γ-ray Link

First came the discovery of the **237 msec periodicity** in the ROSAT X-ray data (Halpern and Holt,1992). The same periodicity was immediately found in the contemporary GRO/EGRET γ-ray data (Bertsch et al, 1992) and in the old, archival COS-B data (Bignami and Caraveo, 1992) which, covering a long time span, provided the best estimate of the period derivative, $\dot{P}= 1.099\ 10^{-14}$ s/s. Apart from clinching the **identification of Geminga with 1E 0630+178**, this allows to compute the parameters of the rotating neutron star responsible for the X/γ-ray emission. The magnetic field turns out to be $1.5\ 10^{12}$ G, a rather typical value, while the standard formula $\dot{E} = I\Omega\dot{\Omega}$ gives $3.2\ 10^{34}$ erg/sec for the overall rotational energy loss of the pulsar.

This brings up the matter of distance: from the value of $\dot{E}$ one can immediately derive an absolute upper limit to the distance, in the assumption that all $\dot{E}$ goes into γ-rays. This upper limit is ~340 parsecs, and for a γ-ray production efficiency of, e.g. 10^{-2}, similar to that of PSR 0833-45 (see e.g. Bignami and Hermsen, 1982), this would imply that Geminga is < 40 pc from us.

Thus Geminga could be the neutron star nearest to us and, given the high velocities normal for pulsars, a measure of the **proper motion** ($\mu = 0.2\ v_{100}\ d_{100}^{-1}$ " yr^{-1}, with v in units of 100 km/sec and d in units of 100 pc) was the next sensible thing to do, as suggested by Bignami and Caraveo, 1992.

The Proper Motion of G"

The ESO Director General granted this project one night on the New Technology Telescope and, on November 4-5 1992, the Geminga field was observed, in service mode, with the SUper Seeing Imager (SUSI). Ten SUSI V frames, of 15 minutes each, were secured under very good seeing conditions (0.6-0.8").

The resulting image was compared with two others of the same field, obtained respectively at the CFHT 3.5m instrument in January 84 and at the ESO 3.6m in January 87 (Bignami et al. 1987, 1988). Fig.1 shows a composite of the three images, where the 84 and 87 ones have been re-binned and tilted to match the scale and orientation of the ESO 92 frame. The motion of G" to the NE is apparent, showing an 84-92 displacement of about 12 pixels, with the 87 data at the correct angle and position.

Using as reference star positions extracted from the original Hubble Space Telescope Guide Star Catalog, kindly supplied to us by D. Golombek, we have computed the coordinates of G" at the three epochs using the original (not rebinned) data. These are as follows:

1992, $\alpha_{(1950)}$= 6^h 30^m 59.15^s $\delta_{(1950)}$= 17° 48' 33.6"   +/- 0.16"
1987, $\alpha_{(1950)}$= 6^h 30^m 59.10^s $\delta_{(1950)}$= 17° 48' 33.0"   +/- 0.68"
1984, $\alpha_{(1950)}$= 6^h 30^m 59.06^s $\delta_{(1950)}$= 17° 48' 32.7"   +/- 0.46"

where the quoted uncertainties take into account the r.m.s.errors of the astrometry fits: 0.10", 0.12", and 0.19", respectively in the 1992, 1987 and 1984 data, as well as the error in the centering of G" (~1 pixel in each data set).
A linear fit to the derived coordinates gives the following components for the proper motion of G" : $\mu_\alpha = 0.14"/y$, $\mu_\delta = 0.10"/y$; for a total of $\mu = 0.17"/y$ +/- $0.05"/y$.

The reported evidence for a large proper motion of G" ($m_v = 25.5$) can only be interpreted in two ways: either the object is a solar system body, or it is a subluminous, truly faint star. The first possibility cannot be discarded lightly, in view of the low ecliptic latitude of Geminga. Arguing strongly against it, however, are the extremely slow motion (for a solar system body) at a large angle with the ecliptic plane and, of course, the low probability of finding such an event, in view of the very small solid angle considered.
Interpreting G" as a star, for a proper motion similar to that observed, one obtains an approximate distance figure of 100 pc for a velocity in the plane of the sky of 100 km/sec, not far from the mean for radio pulsars (see e.g. Lyne et al. 1982). At 100 pc, the object would have an $M_v = 20.5$, to be compared, e.g., with the Vela pulsar's 15, i.e. with an "underluminosity" only comprehensible for a neutron star. Anything more luminous would have to be correspondingly more distant and thus faster. For comparison, the Vela pulsar, at an accepted distance of 450 pc, has been measured to have a proper motion of about 0.05"/y (Bailes et al. 1989; Ögelman et al; 1989). Altogether, no known object other than a neutron star can explain the properties of G". It is thus inescapable to conclude that **the observed motion is proof of the optical identification of Geminga, the neutron star nature of which is by now firmly established from the γ-ray and X-ray data.**
Geminga then becomes the third neutron star identified in the optical, after the Crab and Vela pulsars. The LMC pulsar 0540-69, although definitely seen to pulsate at optical wavelengths (Middleditch et al. 1987), has so far only a probable identification through imaging (Caraveo et al. 1992). Geminga is **the first object discovered and identified through its gamma-ray emission and the first isolated neutron star studied without the help of radio astronomy,** and is surely the prototype of a class whose properties are now open for a better understanding
As to the physical nature of the optical emission, the data presented above do not add information, except for the possible constraint on Geminga's absolute magnitude. As discussed in the literature, thermal as well as non thermal mechanisms may contribute to the emission, which could be pulsed at 237 msec, with a still unknown duty cycle. The period-luminosity dependence originally proposed by Pacini (1971), and rediscussed more recently by Pacini and Salvati, 1987, can be applied, assuming that Geminga has an optical duty-cycle similar to that of the Vela pulsar, as has been seen to be the case at higher energies. This yields an $M_v \sim 28$ which would place Geminga at ~3 pc.
The observed motion of G" could also have a bearing in explaining some of the difficulties encountered recently with the timing parameters of the object (see IAU Circ 5649). In particular, the second derivative of the period, when computed over a long

time history to include both GRO and COS-B data (1991-1975), might be affected not only by period glitches, but also by a different position.

What next? The **parallax** measurement (e.g. 0.02" for 100 pc) is the next challenge. Director Discretionary time has been granted for the observation of Geminga with the Planetary Camera on the HST. The observation is planned for March 1993 with the purpose of pinpointing the position of G" to the best possible accuracy allowed by the current PSΓ. Repeating such measurements six months apart, something which cannot be easily done from the ground, might conceivably lead to a parallax measurement, thus also bringing to an end the distance problem.

References
- Bailes M., Manchester R.N., Kesteven,M.J., Norris R.P., and Reynolds J.E. Ap.J. 343,L53 (1989)
- Bertsch et al. Nature **357**, 306 (1992)
- Bignami, G.F. and Hermsen W., Ann. Rev. Astr. Ap. **21**, 67 (1983)
- Bignami, G.F.,Caraveo,P.A.,Lamb, R.C. Ap. J. Lett. **272**, L9 (1983)
- Bignami,G.F., Caraveo,P.A., Paul, J.A., Salotti,L. and Vigroux, L. Ap.J. **319**, 359 (1987)
- Bignami G.F., Caraveo, P.A. and Paul J.A. Astr. Ap. **202**, L1 (1988)
- Bignami G.F. and Caraveo P.A. Nature **357**, 287 (1992)
- Caraveo P.A., Bignami,G.F., Mereghetti S. Mombelli M. Ap.J. Lett **395**, L103 (1992a)
- Halpern, J.H. and Tytler D. Ap.J., **330** ,201 (1988)
- Halpern, J.P. and Holt S.S., Nature **357**, 222(1992)
- Lyne A.G., Anderson B., Salter M.J. M.N.R.A.S. **201**, 503 (1982)
- Middleditch, J., Pennypacker, C.R. and Burns, M.S. Ap.J. **315**,142 (1987)
- Ogelman H., Koch-Miramond, L., Auriere M. Ap.J. **342**,L83 (1989)
- Pacini, F., Ap.J. Letters **163**, L17 (1971)
- Pacini ,F., and Salvati, M.; Ap.J. **321**, 447 (1987)

Table 1

Name	ν	# photons	Obs. time	Flux erg/cm² sec
Geminga	γ-ray	~1,000	80 days	2 10⁻⁹
1972--->				
1E0630+178	X-ray	800+200	10,000 sec	2 10⁻¹²
1983--->				
G"	optical	1,600	few hours	3 10⁻¹⁶
1987--->				
--------	radio 21 cm	------	deep search	< 5 10⁻²⁰

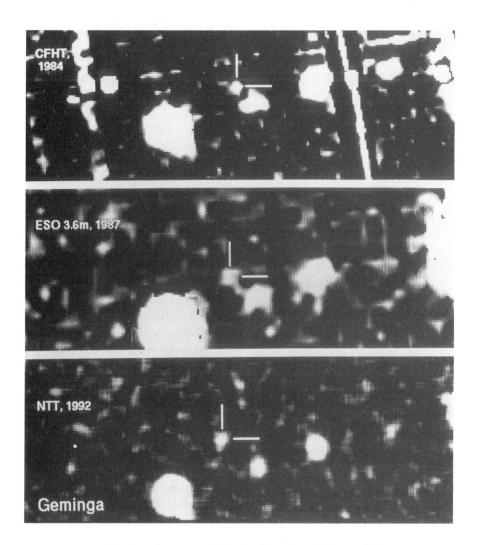

Figure 1 Images of the three data sets used, showing the motion of G" over ~8.8 years. The overall displacement between the first and the last observation is ~12 SUSI pixel (0.13" each), with the 87 data set showing G" at an intermediate position.
BOTTOM Nov.4, 1992 data. Stack of 10 V exposures of 15 min each, taken at the ESO NTT with SUSI. Seeing conditions were very good (0.6-0.8").
MIDDLE Jan. 28, 1987 data. Stack of 8 V exposures of 15 min each, taken at the ESO 3.6m equipped with EFOSC (pixel size 0.675"). Seeing conditions were mediocre ~1.6". The original data have been re-binned and tilted to match the SUSI field.
TOP Jan.7, 1984 data. Stack of 12 r exposures of 15 min each, taken under good seeing conditions (0.9") at the CFHT. The original data (with a pixel size of 0.412") have been re-binned and tilted to match the SUSI field. G" was seen for the first time in this observation (Bignami et al. 1987)

SEARCH FOR FAST GALACTIC GAMMA RAY PULSARS

P. Hertz[†], J. E. Grove, J. D. Kurfess, W. N. Johnson, & M. S. Strickman
Naval Research Laboratory, Washington, DC 20375-5320

S. Matz & M. P. Ulmer
Northwestern University, Evanston, IL 60208

ABSTRACT

We have undertaken a program to search for fast gamma ray pulsars ($P < 1$ s) in OSSE observations of the galactic center, galactic plane, LMC, and selected sources. We have used search strategies optimized for both isolated and binary pulsars. Applied to OSSE observations, these techniques are sensitive to isolated Crab pulsars at the galactic center and binary Crab pulsars in the local spiral arms. To date we have searched for pulsations from (i) known fast pulsars PSR1613-509 in RCW 103 and PSR0540-693 in the LMC, (ii) the gamma ray transient GRO J0422+32 and SN87A in the LMC, (iii) isolated pulsars in the galactic center, LMC, and galactic plane fields in Cygnus and Carina, and (iv) binary pulsars in these same fields. No pulsations have been detected at frequencies between 1 Hz and 4 kHz, with pulse fraction limits as low as 0.1% of total received count rate.

INTRODUCTION

NRL's Oriented Scintillation Spectrometer Experiment (OSSE), on board the Compton Gamma Ray Observatory, provides spectroscopic and photometric observations of gamma ray sources in the 0.05–10 MeV energy range. One of the goals of the OSSE mission is the search for gamma ray pulsars. Since gamma ray emission is an important aspect of theoretical models of pulsars, the detection or non-detection of these gamma rays will have a significant impact on our understanding of pulsar emission. Prior to the launch of the Compton GRO, the Crab pulsar was the only non-accretion powered pulsar positively detected in the OSSE low energy gamma ray band. Subsequently pulsations from the Vela pulsar[1] and PSR1509-58 in Circinus[2] have been detected with OSSE.

We have undertaken a program to search for fast gamma ray pulsars (pulse period $P < 1$ s) in OSSE observations of known or suspected fast pulsars, the galactic center, the galactic plane, and the LMC. No dedicated OSSE observations have been undertaken for this program; rather pulsar mode data from selected existing observations are analyzed for the presence of coherent signals. In order to increase sensitivity by using large data volumes, and to increase frequency coverage, this analysis has been carried out on a massively parallel computer.

---

† Compton Gamma Ray Observatory Guest Investigator

OBSERVATIONS

The four OSSE detectors have 3.8° × 11.4° (FWHM) fields-of-view; each detector has an independent offset pointing capability to permit background measurements and multi-target observations[3]. High time resolution gamma ray data is provided in the OSSE pulsar data modes. Up to eight energy bands can be included in the telemetry alloted to the pulsar mode data packets. Gamma ray events qualified as being in one of these eight energy bands are then processed in either (i) EBE (event-by-event) mode, where a maximum of ∼ 500 selected events per 2.048 s packet are time tagged with an accuracy of 125 μs, or (ii) rate mode, where selected events are binned into samples with a selectable sample size between 4 ms and 512 ms.

ANALYSIS

There are two classes of potential gamma ray pulsars, each requiring a different search algorithm. For the class of known or suspected isolated pulsars, the optimum strategy is to Fourier transform the longest coherent data stretches possible, and to incoherently sum as many of these as are available. Data stretches ranging between a single Compton Observatory orbit (∼ 2500 s) to over 6 days are used depending on the source.

For blind searches, the pulsar has an unknown period and may be located in a binary system with an unknown orbit. If the orbital period is short enough, Doppler shifts in the pulse period will destroy coherence over the term of the observation resulting in the pulse signal being spread over many channels in the power spectrum. We apply a coherence recovery technique[4,5], in which an optimal one dimensional search through the phase space of possible binary orbits is conducted, and corrections for trial orbits are applied to the data before the Fourier transform is calculated. Coherent data stretches are used that are short relative to trial orbital periods, and no incoherent summations to increase sensitivity are possible; see below for details. We choose data stretches corresponding to a single Compton Observatory orbit; this limits the search to binary systems with orbital periods $P_{\rm orb} > 7.1$ hr.

Standard data processing, including data selection and formatting, are performed on the OSSE VAX cluster. The data is then transferred to the NRL Connection Machine Facility. All subsequent analysis, including correction to solar system barycenter, binning of data, coherence recovery transforms, Fourier transforms, and selection of peaks in the power spectrum, are performed on NRL's 16,384-node parallel Connection Machine CM-200 computer with extended memory[5].

The software has been tested using OSSE observations of the Crab pulsar (isolated pulsar case) and Her X-1 (binary pulsar case). In both tests, the known pulse was recovered at the expected amplitude and significance.

COHERENCE RECOVERY TECHNIQUE

Pulsars in a binary system have their pulse frequency Doppler shifted due to the orbital motion of the pulsar. Without correction, this results in a broadening of the pulse signal into many channels of a standard Fourier power spectrum. In the cases under study here, very short pulse periods and very long observations, the signal can be spread over thousands of channels.

Correcting the signal for orbital motion requires accurate knowledge of the orbit. A general circular orbit can be specified by three parameters: projected semimajor axis $a_\perp$, orbital period P_{orb}, and orbital phase at the time origin ϕ_0. In the case where the orbit is unknown, searching all orbits would require searching a three dimensional phase space. This strategy is computationally impractical.

We have used a quadratic coherence recovery technique which involves searching a one dimensional phase space[4,5]. The time of arrival for data, t, is quadratically transformed to $t' = t + \alpha t^2$, where $\alpha = a_\perp \Omega_{orb}^2 \sin(\phi_0)$ and $\Omega_{orb} = 2\pi/P_{orb}$. This is mathematically equivalent to approximating the arc of the pulsar's orbit during the observation with a parabola. A one dimensional search over all possible values of α is then performed — for each value of α, the data is quadratically transformed and then Fourier transformed. The extreme values of α searched depend on the assumed mass function for the binary, but are less than 10^{-7} s^{-1} for systems comparable to close X-ray binaries with $P_{orb} > 7.1$ hr. The quadratic approximation is only good for relatively short data stretches; if the duration of the data stretch analyzed, T, exceeds $P_{orb}/4\pi$ then the quadratic transform will not completely recover the coherent signal into a single spectral channel.

RESULTS

The limits to the pulse fraction are determined using standard statistical techniques[6]. Each power spectrum is normalized to a value of 2 for Poisson noise (Leahy normalization[7]). If we adopt a $1 - \epsilon = 95\%$ confidence limit for detection of a coherent signal, then the detection threshold P_{thresh} for the highest peak detected in the power spectrum is given by $\epsilon/N_{trial} = Q(MP_{thresh}|2M)$, where $Q(\chi^2|\nu)$ is the probability of exceeding a value of χ^2 in a χ^2-distribution with ν degrees of freedom, M is the number of Fourier transforms incoherently summed, and N_{trial} is the number of independent power spectrum peaks searched. For an isolated pulsar, N_{trial} is the number of frequency channels searched N_{freq}, while for binary pulsar searches involving N_α quadratic transformations, $N_{trial} = N_{freq} N_\alpha / 4$ (ref. 5).

When no significant peak is detected, the upper limit to the pulse fraction A_{UL} is given by $A_{UL} = \sqrt{(P_{max} - P_{noise})/N_{ph}}$. Here P_{max} is the peak normalized power detected, P_{noise} is the $1 - \delta = 95\%$ confidence lower limit for the Poisson noise power given by $1 - \delta = Q(MP_{\text{noise}}|2M)$, and N_{ph} is the average number of counts analyzed in each of the M power spectra. Note that this is the pulse fraction of the total (source plus background) signal. The upper limit to the pulsed flux is $F_{\text{UL}} = A_{\text{UL}} N_{\text{ph}}/T A_{\text{eff}}$ for detectors with an effective area A_{eff}.

RESULTS

To date, OSSE observations of the galactic center, the Large Magellanic Cloud, several regions of the galactic plane, and several individual sources have been analyzed for the presence of fast gamma ray pulsars. No pulsars, either isolated or in binary systems, have been detected to date. We give upper limits to a selection of our searches in Table 1 and a typical power spectrum in Figure 1.

A typical OSSE observation during the sky survey portion of the Compton Observatory mission last two weeks. A complete search for isolated pulsars requires approximately 2 CPU hours on the CM-200 for each two week observation. A search for binary pulsars requires approximately 6.0 CPU hours for each Compton GRO orbit analyzed at 125 μs resolution, and CPU time scales approximately inversely with time resolution.

The techniques used, when applied to typical OSSE observations, are sensitive enough to detect the Crab pulsar at the distance of the galactic center. The sensitivity to binary pulsars is significantly less, due to the necessity of searching a large volume of phase space corresponding to possible binary orbits. Even so, the Crab pulsar could be detected in a binary within the nearby spiral arms ($D < 1$ kpc).

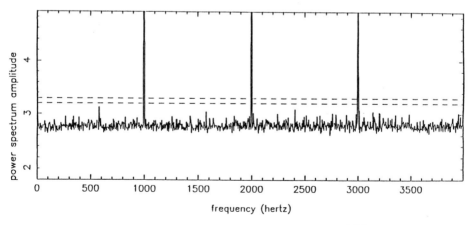

Figure 1 — Power spectrum for the LMC supernova SN87a. This power spectrum is made from incoherently summing 117 power spectra, each of which is the FFT of a 32M point time series with 125 μs resolution. Plotted are the 99% and 90% confidence limits for a single peak. The peaks at 1, 2, and 3 kHz are instrumental.

Table 1 — Flux Limits on Fast Gamma Ray Pulsars

	P (ms)	energy (keV)	A_{UL}	F_{UL} ‡
known pulsars:				
PSR0540-693	50.4	60 – 210	< 0.1 %	< 0.34
PSR1613-509	69.3	40 – 190	< 0.09 %	< 0.54
isolated pulsars:				
GRO J0422+32	> 0.25	40 – 170	< 0.2 %	< 0.64
SN87a	> 0.25	60 – 210	< 0.2 %	< 0.44
Galactic Center	> 8	40 – 150	< 0.3 %	< 0.74
Galactic Center	> 2	40 – 165	< 0.7 %	< 1.28
Gal Plane (Car)	> 8	50 – 165	< 0.3 %	< 0.41
Gal Plane (Cyg)	> 8	40 – 135	< 0.7 %	< 1.52
LMC	> 0.25	60 – 210	< 0.2 %	< 0.32
binary pulsars:				
GRO J0422+32	> 0.25	40 – 110	< 1.0 %	< 2.7
Galactic Center	> 8	40 – 150	< 1.0 %	< 2.9
Galactic Center	> 0.5	40 – 165	< 1.7 %	< 5.0
Gal Plane (Car)	> 8	50 – 165	< 0.7 %	< 1.3
Gal Plane (Cyg)	> 8	40 – 135	< 3.5 %	< 9.2

‡ 10^{-3} γ s^{-1} cm^{-2}

REFERENCES

1. M. S. Strickman et al., these proceedings (1993).
2. M. P. Ulmer et al., these proceedings (1993).
3. W. N. Johnson et al., Ap. J. Sup., in press (1993).
4. K. S. Wood et al., Ap. J., 379, 295 (1991).
5. P. Hertz et al., Ap. J., 354, 267 (1990).
6. M. van der Klis, in Timing Neutron Stars, ed. H. Ogelman & E. P. J. van den Heuvel (Kluwar: Dordrecht), p. 27 (1989).
7. D. A. Leahy et al., Ap. J., 266, 160 (1983).

SEARCH FOR HARD X-RAY EMISSION FROM MILLISECOND PULSARS AND COMPACT BINARIES IN THE GLOBULAR CLUSTER 47 TUC

J.E. Grindlay, R.P. Manandhar, K.S.K. Lum, M. Krockenberger, S. Eikenberry
Harvard-Smithsonian Center for Astrophysics, Cambridge MA 02138

C.E. Covault
Enrico Fermi Institute, University of Chicago, Chicago IL 60637

ABSTRACT

Millisecond pulsars are potential hard X-ray/soft gamma-ray sources, particularly for those cases where the pulsar is emitting a strong relativistic wind which ablates material from a close binary companion. Such systems appear to be concentrated in globular clusters. The massive globular cluster, 47 Tuc, for example, is known to contain at least a dozen millisecond pulsars. We report hard x-ray observations of the globular cluster 47 Tuc using the Energetic X-ray Imaging Telescope Experiment (EXITE), a coded-aperture telescope with sensitivity in the 20-300 keV energy band and 22 arcmin imaging over a 3.4 degree field-of-view. EXITE observed 47 Tuc for 3 hours during a flight from Alice Springs, Australia, on 9 May 1989. We present image and spectral analysis for potential sources located within the globular cluster. Our flux upper limits provide a constraint on the hard x-ray luminosity of the population of millisecond pulsars and low luminosity x-ray sources (e.g. CVs or quiescent LMXBs). More sensitive followup observations (e.g. CGRO/OSSE) should help to constrain models of the origin and evolution of compact objects within such clusters.

INTRODUCTION

The last few years have seen a series of major developments in the study of neutron stars, compact binaries and millisecond pulsars: these objects have all been discovered in unexpected profusion in globular clusters. Previously only low mass x-ray binaries (LMXBs) in about 10 globulars as well as low luminosity x-ray sources, possibly cataclysmic variables (CVs) in another 8 clusters observed with Einstein, were known (e.g. review by Grindlay 1988). Now there are millisecond pulsars (MSPs) detected in at least 12 globulars (Bhattacharya and Van den Heuvel 1991). Out of the some 38 MSPs currently known in the Galaxy, at least 28 are in globulars, with 12 in the globular 47 Tuc alone. The number of low luminosity x-ray sources in globulars is also now increasing rapidly with extensive ROSAT studies.

This new wealth of cluster MSPs is of great significance for understanding the physics of pulsars (the clusters provide a rich sample), the origin of neutron stars (particularly in such ancient systems as globulars), and the remarkable evolutionary history of compact binaries (the MSPs are almost certainly spun up in accreting binaries) both in and out of globulars. We have obtained the first hard x-ray imaging observation of the globular cluster 47 Tuc with the EXITE coded aperture imaging telescope. Preliminary results were reported by Grindlay et al. 1990. More sensitive results, incorporating improved calibrations and processing techniques as well as aspect corrections, are reported here.

HARD X-RAY PROBES OF CLUSTER BINARIES

Our primary objective is to establish whether MSPs, LMXBs and CVs are sources of hard x-rays and γ-rays and the degree to which any of these objects are particle accelerators. Whereas it is an observational fact that at least some MSPs can ablate their binary companions (e.g. PSR1957+20), it is not at all clear whether this is due to heating from a relativistic wind (e.g. Phinney et al. 1988) or other causes. Phinney et al. (1988) estimated that a pair wind shocked by the evaporating companion star wind would cool by inverse Compton and synchrotron emission to yield a $\sim$1 MeV flux of $\sim$1.4 x $10^{-5} f_\gamma$ photons/cm^2-s for a photon conversion efficiency factor of f_γ. In the case of LMXBs, Ruderman et al. (1989) have proposed that they may be MeV sources as well and may be significant contributors to the integrated MeV emission from the galactic plane. If the possible identification of the CV V1223 Sgr with the HEAO-A4 source of 511 keV emission (Briggs et al. 1991) is correct, then accreting white dwarfs may also be (under certain conditions) significant high energy sources. Finally, if hard x-ray spectra (including variable 511 keV sources–cf. Sunyaev et al. 1991) are indeed signatures of black holes, then the otherwise unknown BH content of globulars might be investigated through hard x-ray studies.

Because the radio emission from PSR1744-24A is sometimes quiescent, Tavani 1991 has speculated that this may represent a class of MSPs which are only sometimes detectable (as radio objects) and are otherwise engulfed in a cocoon. Such "hidden MSPs" could be revealed by their hard x-ray or γ-ray emission; they may also appear optically as variable and anomalously blue "blue stragglers" (BSs) possibly of the sort that Auriere et al. (1990) have discovered in the central cusp of NGC 6397, and which may be similar to the BSs found in the core of 47 Tuc with HST (Paresce et al. 1991). The Tavani (1991) picture would thus predict a hard x-ray/soft γ-ray flux from the large MSP and BS population in 47 Tuc.

Brink et al. (1990) have reported marginal evidence for emission from PSR1957+20 with COS B data and upper limits from the MPE balloon-borne prototype of COMPTEL (1-10 MeV). The latter upper limits are about 2 orders of magnitude above the estimated 1 MeV flux of Phinney et al. (1988). With our own balloon-borne hard x-ray imaging telescope EXITE, we have reported the first upper limits for the hard x-ray flux (30-150 keV) from PSR1957+20 which corresponds to a luminosity fraction of about 0.3 the spin-down luminosity (Grindlay et al. 1990).

Verbunt et al. (1984) have proposed that the faint x-ray sources in globulars discovered by Hertz and Grindlay (1983=HG) are generally LMXBs in quiescence.

Such LMXBs, with very low mass transfer rates, have been proposed to be luminous γ-ray emitters (Ruderman et al. 1989). At relatively low accretion rates, there might be significant dynamo action and particle acceleration by the interaction of the $\sim 10^{8-9}$ gauss magnetic field of the neutron star with the inner disk. In this case, a γ-ray luminosity for the LMXB comparable to the spin-down torque exerted by the disk on the star will be expected (Kluzniak et al. 1988). Thus the γ-ray luminosity of "quiescent" LMXBs, in which mass transfer has nearly ceased as the companion shrinks well within its Roche lobe may be greatly in excess of the residual x-ray (accretion) luminosity. Therefore, globular clusters where quiescent x-ray transients (LMXBs) are likely to be found (given the high rate of formation of LMXBs) are perhaps the ideal setting to test models of LMXB γ-ray production.

OBSERVATIONS

The EXITE instrument is a balloon-borne hard x-ray telescope sensitive in the range 20–300 keV. EXITE uses a coded-aperture image system with 22′ resolution within a 3.4° (FWHM) field-of-view. The detector is made of a thin (6.4 mm) NaI(Tl) scintillation crystal bonded to the entrance window of a large (34 cm diameter) position-sensitive two-stage image intensifier tube. The detector and gondola systems have been described by Grindlay et al. (1986) and Garcia et al. (1986). Flight performance and calibrations have been described by Braga et al. (1989). Image reconstruction and spectral analysis have been described in detail by Covault (1991). Results on several sources, such as the black hole candidate GX339-4 (Covault et al. 1992) and sources in the galactic center region (Grindlay et al. 1993) have been reported.

Observations of 47 Tuc were obtained during the $\sim$30 hour balloon flight from Alice Springs, Australia, on 9-10 May 1989. The telescope was pointed at 47 Tuc from UT 2244-0200, with a total of 181 minutes of data collected during the observation. Telescope aspect was derived from an on-board sunsensor, for azimuth, and a shaft angle-encoder and inclinometer for elevation. Pointing aspect corrections to within 10 arcmin were applied to individual data segments of 40 seconds each.

RESULTS

In **Figure 1** we show a broad-band image (26-152 keV) of the field containing 47 Tuc. The cross ('+') marks the expected location of the source. The size of the cross indicates the uncertainty on the absolute pointing ($\lesssim 10′$), as calibrated by an early pointing on the Crab x-ray source. No source is located at the position of 47 Tuc in the broad band image, nor in any image made for any individual energy band. The 3.5 σ fluctuation at the bottom of the image is consistent with noise fluctuations; the *a priori* probability of finding such a peak somewhere in the image is about 15 %.

We can constrain the spectrum by sampling the image at the expected location of 47 Tuc in several non-overlapping energy bands. **Figure 2** shows a plot of the band-wise flux from 47 Tuc relative to the Crab x-ray source (1000 mCrab $\equiv$ 1 Crab $\equiv 8.6 E^{-2.1}$ photons s^{-1} cm^{-2} keV^{-1}). Each energy band is consistent with zero flux, to within 2σ uncertainty. **Figure 3** indicates the band-wise flux upper limits at 90 % confidence, again in terms of the Crab flux. If we assume a form for the source spectrum, such as a Crab-like power law, then we can determine an upper limit flux envelope spectrum, as shown in **Figure 4**.

DISCUSSION

We obtain a composite flux limit of 80 mCrab (90% confidence) from 47 Tuc over the the 30-150 keV range. This limit provides (weak) constraints on the population of compact objects and binaries in the cluster. First, this energy flux limit ($\sim 1 \times 10^{-9}$ erg/cm^2-sec) corresponds to a luminosity limit (at the 4 kpc distance of the cluster) of about 1.4×10^{36} erg/sec. This is approximately 10 times the spin-down luminosity of the eclipsing MSP system, PSR1957+20. Thus $\lesssim 30$ of the $\gtrsim 100$ MSPs estimated to be in the cluster could be like PSR1957+20 if similar conversion efficiency limits (i.e. $\lesssim 30\%$ of spin-down into hard x-rays– cf. Grindlay et al 1990) apply. Second, the single "faint" cluster x-ray source (actually, one of the two brightest of this population) found (HG) in 47 Tuc with soft x-ray flux (0.5-4.5 keV) ~ 30 times fainter cannot have a correspondingly more luminous (i.e. $30 \times$) hard x-ray spectrum as might be expected if it were in fact a quiescent LMXB and these are efficient γ-ray sources. Third, the recent evidence that this soft source is consistent with being a magnetic CV (Paresce *et al.* 1992) suggests that its spectrum (and thus, not surprisingly, that of other DQ Her type CVs) does not have an excess in the 30-150 keV band. A hard Comptonized continuum might be expected if CVs produce 511 keV emission features, as suggested by Briggs et al 1991 for V1223 Sgr. Finally, our hard x-ray luminosity limit is ~ 10-$100 \times$ below the hard x-ray luminosity observed from BH transients in outburst. Since the hard x-ray spectrum appears to be much longer lived than the soft component after outbursts, BH transients should be looked for in hard x-ray surveys. Such systems (e.g. Nova Muscae 1991) may be expected in globular clusters if clusters contain a population of stellar-mass BHs. Since such BHs in globulars must arise from post SN-collapse of massive primordial stars in the cluster, whereas NSs could be created later by accretion-induced collapse of white dwarfs, still more sensitive hard x-ray studies of globulars could provide useful limits on the (primordial) initial mass function of globular clusters.

This work was supported by NASA grant NAGW-624 at Harvard. C.E. Covault gratefully acknowledges the support of the Grainger Fellowship at Chicago.

REFERENCES

1. Auriere, M. et al. 1990, *Nature*, **344**, 638.
2. Bhattacharya, D. and van den Heuvel, E. 1991, *Physics Reports*, **203**, Numbers 1 and 2.
3. Briggs, M. et al. 1991, *IAU Circ. No. 5229*.
4. Brink, C. et al. 1990, *Ap. J. Letters*, **364**, L37.
5. Covault, C.E. 1991, *Ph.D. Thesis*, Harvard University.
6. Covault, C.E., Grindlay, J.E. and Manandhar, R.P. 1992, *Ap. J. Letters*, **388**, L65.
7. Garcia, M.R., et al. 1986, *IEEE Trans. Nucl. Sci.*, **33**, 735.
8. Grindlay, J.E., et al. 1986, *IEEE Trans. Nucl. Sci.*, **33**, 750.
9. Grindlay, J. 1988a, *Proc. IAU 126*, (J. Grindlay and A. Phillip, eds.), p. 347.
10. Grindlay, J., Covault, C. and Manandhar, R. 1990, *BAAS*, **22**, No. 4, 1286.
11. Grindlay, J., Covault, C. and Manandhar, R. 1993, *Astr. Ap. Suppl. Ser.* in press.
12. Hertz, P. and Grindlay, J. 1983, *Ap. J.*, **275**, 105.
13. Lugger, P. et al. 1990, in *ASP Conf. Series*, **13**, 414.
14. Lyne, A.G. et al. 1990, *Nature*, **347**, 650.
15. Patterson, J. and Raymond, J. 1985, *Ap. J.*, **292**, 535.
16. Paresce, F. et al. 1991, *Nature*, **352**, 297.
17. Paresce, F. et al. 1992, *Nature*, in press.
18. Phinney, S. et al. 1988, *Nature*, **333**, 832.
19. Ruderman, M., Shaham, J. and Tavani, M. 1989, *Ap. J.*, **336**, 507.
20. Tavani, M. 1991, *Ap. J. Letters*, **379**, L69.
21. Verbunt, F. et al. 1984, *MNRAS*, **210**, 899.

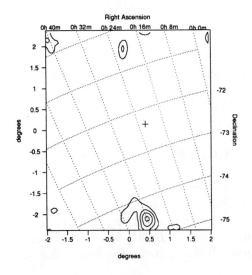

Figure 1: Broad-band image (26-153 keV) of the region containing the globular cluster 47 Tuc. The cross ('+') indicates the expected position of the source. Contours range from 2.0σ to 3.2σ by steps of 0.5σ.

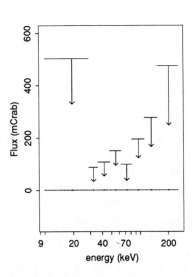

Figure 3: Band-wise flux upper limits at 90% confidence from 47 Tuc. Flux is given in units of mCrab.

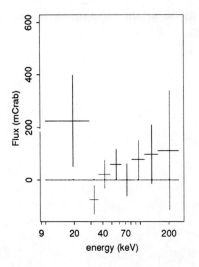

Figure 2: Flux relative to the Crab from the image position at 47 Tuc. Flux is given in units of mCrab (1000 mCrab $\equiv$ 1 Crab $\equiv 3.2 E^{-2.1}$ photons s^{-1} cm^{-2} keV^{-1}).

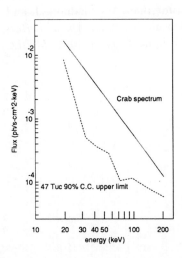

Figure 4: Spectral envelope corresponding to flux upper limits from 47 Tuc assuming a Crab-like power-law spectrum.

AN OSSE SEARCH FOR THE BINARY RADIO PULSAR 1259-63

P. S. RAY[1], J. E. GROVE[2], J. D. KURFESS[2], T. A. PRINCE[1], M. P. ULMER[3]

[1] Division of Physics, Mathematics, and Astronomy, Caltech, Pasadena, CA 91125

[2] E. O. Hulburt Center for Space Research, Naval Research Laboratory, Washington, DC 20375

[3] Northwestern University, Evanston, IL 60208

ABSTRACT

We have searched data from the Oriented Scintillation Spectrometer Experiment (OSSE) on the *Compton Gamma Ray Observatory* (GRO) for evidence of low-energy γ-ray emission from the binary radio pulsar PSR 1259-63. This 47 ms pulsar is in a long-period, highly eccentric orbit around a Be stellar companion and was observed by OSSE approximately 400 days after periastron. The period derivative allowed by the published radio ephemeris (Johnston et al. 1992) suggests that the pulsar might be relatively young, and therefore a γ-ray source. However, the ephemeris is not sufficiently accurate to allow the traditional epoch-folding technique over the full OSSE observation. Instead, the OSSE data were analyzed using Fourier transform spectral techniques after applying trial accelerations to correct for a range of possible orbital accelerations. We searched 48 accelerations; each FFT was 2^{29} points sampled at 2 ms, spanning $\sim 10^6$ seconds of observation time. There was no evidence of pulsed emission in the 64–150 keV band, with a 99.9% confidence upper limit of 6×10^{-3} photons cm^{-2} s^{-1} MeV^{-1} or ~ 40 mCrab pulsars, which suggests that the pulsar's intrinsic period derivative is small and its magnetic field weak. This work was performed on the Concurrent Supercomputing Consortium's Intel Touchstone Delta parallel supercomputer as part of a GRO Phase 1 Guest Investigation.

1. BACKGROUND ON PSR 1259-63

PSR 1259-63 is a recently discovered radio pulsar (Johnston et al. 1992) and is the only radio pulsar known to have a massive, non-degenerate companion. The stellar companion has been optically identified as the Be star SS 2883. The pulsar is in a highly eccentric orbit and has gone through only one periastron passage since its discovery. As a consequence, the published timing solution could not completely discriminate between orbital Doppler shift-induced period changes and the intrinsic period derivative ($\dot{P}_{int}$) of the pulsar. The best-fit solution gives $\dot{P}_{int} = 2 \times 10^{-14}$, but even $\dot{P}_{int} = 0$ cannot be excluded. An intrinsic $\dot{P}_{int} \sim 10^{-14}$ implies a young age for the pulsar. With pulse period and distance comparable to those of the Crab pulsar, such a large $\dot{P}_{int}$ would make it a likely candidate for pulsed γ-ray emission.

Table 1, reproduced from Johnston et al. (1992), lists the fitted pulsar and orbital parameters with two different assumed values for $\dot{P}_{int}$. The ambiguity in $\dot{P}_{int}$ will be removed by the time the pulsar reaches periastron again since the orbital period will be determined.

TABLE 1
Parameters of PSR 1259-63

Right acension (J2000)	$13^h02^m47^s.72 \pm 0^s.03$	
Declination (J2000)	$-63°50'08''.5 \pm 0''.2$	
Dispersion measure ($pc\,cm^{-3}$)	146.7 ± 0.2	
Distance (kpc)	2.3	
Pulse period (ms)	47.76164	47.76219
Period derivative	0.0	2×10^{-14}
Orbital period (d)	1133 ± 24	2150 ± 100
Epoch of ascending node (MJD)	48043 ± 2	48027 ± 3
Epoch of periastron (MJD)	48120 ± 2	48117 ± 3
$a \sin i$ (ls)	3480 ± 1900	3450 ± 1000
Longitude of periastron (deg)	164 ± 9	158 ± 6
Eccentricity	0.976 ± 0.025	0.967 ± 0.017
Mass function ($M_\odot$)	35	10

(from Johnston et al. 1992)

2. DETAILS OF THE OBSERVATION

The characteristics and performance of the OSSE instrument have been described by Johnson et al. (1993). The instrument consists of four identical large-area NaI(Tl)–CsI(Na) phoswich detector systems, each with a 3.8° × 11.4° (FWHM) field-of-view defined by a tungsten collimator.

The pulsar was observed serendipitously in OSSE's N Mus 91/Galactic plane pointing in September 1991, spanning MJD 48518–48532, about 400 days after periastron and immediately following the epoch studied by Johnston et al. (1992). The pulsar was about 8.2° off-axis, which reduced the detector response to $\sim 25\%$ of the on-axis value.

The OSSE observation strategy is implemented by alternately pointing each detector at source and background positions on a time scale (131 seconds) that is short with respect to typical Earth orbital background variations. The difference spectrum derived from these source and background pointings for the PSR 1259-63 field shows a significant soft-spectrum excess, presumably due to the sum of diffuse emission from the plane and several point sources. Temporal signatures can be used to resolve the ambiguity over the source of this excess.

In addition to the spectral data, time-tagged data (resolution = 0.125 ms) were collected in the 64–150 keV band, with a total livetime of $\sim 3.3 \times 10^5$ detector-seconds. These data are the subject of the search described here.

3. SEARCH ALGORITHM

In order to search for pulsed γ-ray emission from PSR 1259-63, we generated a binned time series from the OSSE data to which we applied our standard radio pulsar search algorithms (see Anderson et al. 1990). This was accomplished by first converting the spacecraft photon arrival times to Solar System barycentric arrival times using the known position of the pulsar. These arrival times were then used to generate a 2^{29} point time series consisting of 2 ms bins. This yields a Nyquist frequency of 250 Hz, so the power spectrum contains the first 11 harmonics of the 47 ms pulsar without aliasing.

This search utilized large 1-dimensional Fast Fourier Transforms (FFTs) to search for periodic signals in the data. Because the frequency bins in a 2^{29} point time series are very narrow, a signal must have a constant frequency to within a very narrow tolerance to keep from being smeared over multiple bins in the spectral domain with a concomitant loss in sensitivity. For a pulse period of ~ 50 ms, the largest period derivative that can be searched over a 10^6 s interval is $\dot{P} \sim 10^{-16}$. We correct for an assumed $\dot{P}$ by applying a quadratic stretch to the time series. A quadratic stretch is an adequate model of either linear spin-down of the pulsar or Doppler shifts over a short portion of a binary orbit. Since the published ephemeris was not accurate enough for us to know exactly what trial "acceleration" to use, we estimated an envelope of possible total period derivatives (intrinsic and orbital) from the trend of P with time from Figure 2 of Johnston et al. (1992). We then selected 48 trial accelerations spanning the range $1.8 \times 10^{-14} < \dot{P} < 1.5 \times 10^{-13}$. The accelerations were spaced such that the difference in corrected arrival times was < 10 ms anywhere in the time series. Thus the maximum residual pulse smearing was $< 1/4$ period, and up to 4 harmonics could have been significant.

The complete search was performed in about 1.5 hours on 512 nodes of the Concurrent Supercomputing Consortium's (CSCC) Intel Touchstone Delta Supercomputer at Caltech. The unstretched, barycentered time series was loaded onto the Delta's 90 GB Concurrent File System packed at 1 byte per 2 ms bin, using 1/2 GB of storage. Since the Delta has more than 4 GB of usable memory available after the OS and programs have been loaded, we could store both the unstretched time series and one trial stretched time series. So, the time series need only be read off disk once at the beginning of the run. The analysis then proceeds by making a new stretched copy of the time series in memory, performing a parallel FFT in place, generating a power spectrum and searching for significant peaks in the region of the spectrum where the pulsar should appear. Harmonic folds of 2, 4, and 8 harmonics were also used to search for the pulsar signal. This process is repeated for each acceleration trial.

The 48 trials consisted of about 4×10^{12} floating point operations performed at an average rate of about 1 GFLOP/s by the Delta. This analysis would have been considerably more cumbersome and much slower on a machine that had less than 2 GB of usable main memory, since the running time would be dominated by the tremendous amount of disk access needed to do out-of-core FFTs.

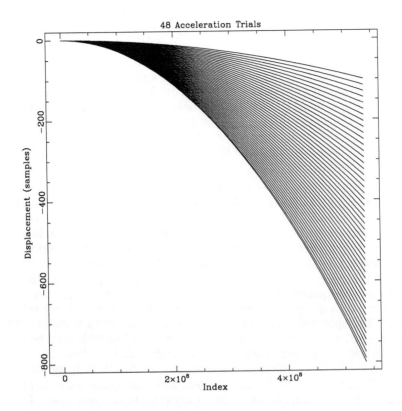

Figure 1: Plot of delay in samples vs. sample number for the 48 acceleration trials used in the search.

4. SENSITIVITY AND RESULT OF THE SEARCH

Sensitivity was assessed by Monte Carlo simulation. To the actual data stream, we added counts from a simulated pulsar with pulse period of 50 ms and pulse FWHM = 15% at various intensities. The pulsed flux corresponding to the 99.9% confidence upper limit was then simply interpolated from the observed powers.

No statistically significant powers were seen in any of the 48 acceleration trials. The resulting upper limit is $\sim$ 40 mCrab pulsar flux units (see Table 2). For comparison, the sensitivity of an epoch-folding search with known frequency and phase would be $\sim$ 20 mCrab pulsar flux units, for a pulse FWHM = 25%.

TABLE 2
Search Sensitivity

N_{freq}	N_{accel}	ΔE (keV)	A_{eff} (cm^2)	Flux limit (99.9% conf)
48 × 75000	48	64–150	400	$< 6 \times 10^{-3}$ ph cm^{-2} s^{-1} MeV^{-1}
				~ 40 mCrab pulsars

5. CONCLUSIONS

The known radio pulsars with significant low-energy γ-ray emission (Crab, Vela, PSR 1509-58) are young, with reasonably short periods and large intrinsic period derivatives. By virtue of the similarity of the period and distance of PSR 1259-63 to the Crab, and the possibility that its period derivative is also comparable to that of the Crab, PSR 1259-63 is a likely candidate for γ-ray emission.

We have placed a limit on the 64–150 keV flux from PSR 1259-63 of $F \lesssim 40$ mCrab pulsar flux units, which corresponds to a luminosity of ~ 5% of the Crab pulsar in the same band. This suggests that the pulsar's intrinsic period derivative, and therefore its magnetic field, are substantially smaller than the Crab's. Further radio observations will produce a definitive value of $\dot{P}_{int}$ by the time of the next periastron.

This observational limit lends support to the suggestion that this pulsar may not be a young pulsar within ~ 10^4 years of its birth, but an older pulsar whose short period is due to spin-up by accreting matter from its companion at each periastron passage. Since it is widely believed that neutron star magnetic fields decay during accretion, PSR 1259-63 would then be expected to have a weak magnetic field and no significant γ-ray emission.

If this scenario is the case, then the pulsar must be experiencing substantial accretion episodes, and at the next periastron passage one would expect the matter being accreted to eclipse the pulsar's radio emission, while making it an X-ray and γ-ray source. So, searches for transient hard X-ray emission near the time of the next periastron passage, including searches for pulsed emission, are warranted.

REFERENCES

Anderson, S.B., et al. 1990, Nature, **346**, 42.

Johnson, W.N., et al. 1993, accepted for publ. in Ap. J. Suppl.

Johnston, S., et al. 1992, Ap. J. Lett., **387**, L37.

TEV OBSERVATIONS OF GAMMA-RAY PULSARS: A TEST FOR PULSAR MODELS

H.I. Nel, O.C. De Jager, L.J. Haasbroek, B.C. Raubenheimer, C. Brink,
P.J. Meintjes, A.R. North, G. Van Urk, and B. Visser
Dept. of Physics, Potchefstroom University, Potchefstroom 2520, South Africa

ABSTRACT

We present the first results of TeV γ-ray observations of PSR 1706-44, indicating a sharp spectral break between $\sim$ 10 GeV and 1000 GeV. Our and other results confirm similar breaks from other γ-ray pulsars such as Vela, PSR 1509-58 and Geminga. Both the polar cap- and outer gap models can account for this break, although a much higher TeV flux from Geminga is expected if it is in the post-Vela phase, given that the gap is controlled by the inverse Compton process. Relatively flat (possibly time variable) spectra from these pulsars are also seen between X-rays and medium energy γ-rays, which can also be explained by the outer gap model.

INTRODUCTION

Previous TeV γ-ray observations of radio pulsars[25] were disappointing, leading to either marginal- or non-detections, but with the launch of the CGRO, our knowledge about γ-rays from isolated pulsars has improved. With the availability of contemporary radio parameters of the pulse period we are now in a position to conduct more sensitive searches for pulsed TeV γ-rays from pulsars which are known to be emitters of γ-rays in the MeV to GeV range, leading to flux limits (or detections) which are at levels lower than reached previously. In all pulsar models primary γ-rays with energies $E_\gamma > 10^{12}$ eV are necessary for the development of electron-photon cascades. Ground-based TeV observations lead to valuable constraints on the opacity for these γ-rays in the pulsar magnetosphere.

The TeV γ-ray flux upper limits derived from observations on 18 pulsars[25] were all lower than the predicted fluxes from the outer gap model[13]. With the help of CGRO and other satellites more complete spectra can now be constructed, so that pulsar models can be tested in greater detail.

Table 1: Pulsar observations at Nooitgedacht during 1992.

Pulsar	Time when observed (MJD)	Number of observations	Number of events	Duration (min)
Vela	8715-8746	12	32 134	1891
1509-58	8715-8806	20	70 289	3333
1706-44	8715-8858	45	192 041	10544

OBSERVATIONS

We have made extensive observations of Vela, PSR 1706-44 and PSR 1509-58 using the Nooitgedacht TeV γ-ray telescope[14] between April and August 1992 during dark moonless nights (see Table 1). Seasonal effects allowed best coverage (and hence sensitivity) for PSR 1706-44. Absolute UTC calibration was available on a daily basis to allow an in-phase analysis for all the data.

RESULTS

Contemporary radio parameters for the pulsations[32] were used to construct a phasogram for each pulsar at the expected period (Fig.1a-c.). No pulsed emission was seen from any of the three objects.

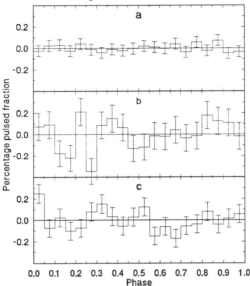

Fig. 1: Light curves of a) PSR 1706-44, b)Vela and c)PSR 1509-58. The zero level represents the average cosmic ray noise level. The same vertical scale was used for each pulsar to enable visual comparison of the sensitivity levels reached.

Assuming the same pulse profile as seen at lower γ-ray energies, 3σ upper limits on the pulsed strength and hence γ-ray flux around 10^{12} eV were calculated (Fig. 2a-c). Adding the TeV γ-ray upper limits to previously known spectra leads to the following results:

PSR 1706-44: This pulsar was identified by Egret[33] in the range 100 MeV to $\sim$ 5 GeV (Fig. 2a.). Our TeV γ-ray upper limit is lower than the extrapolated Egret spectrum[33] and indicates a significant spectral break between 5 GeV and 1000 GeV. The X-ray measurement[2] and Batse upper limits[37] indicate that the X-ray to low energy γ-ray spectrum is flatter than the extrapolated Egret spectrum.

Vela Pulsar: Vela (Fig. 2b) also shows a clear spectral flattening below a few MeV, and a spectral break above a few GeV, as inferred from various independent TeV measurements. If the 4σ detection[6] of Vela at 4 TeV were correct, our results from Nooitgedacht and other[9,17] TeV results would indicate a spectral flattening below 4 TeV. This flattened spectrum was marginally detected (3σ) at 300 GeV[17] (indicated as an upper limit on Fig. 2b).

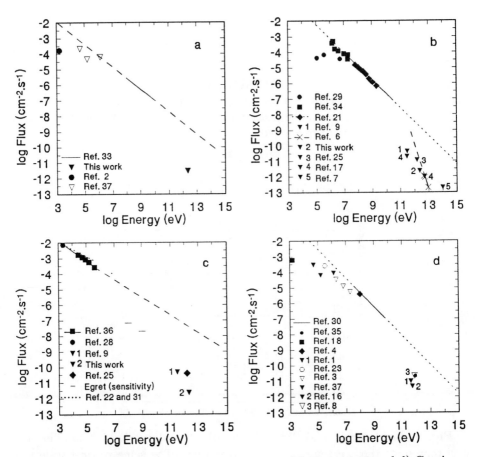

Fig. 2: Spectra for a) PSR 1706-44, b) Vela, c) PSR 1509-58 and d) Geminga. Inverted triangles represent $\sim 3\sigma$ upper limits.

PSR 1509-58: Due to the availability of contemporary radio parameters a better sensitivity could be reached for this pulsar than previously[15,24]. No transient emission was seen in 1992 as in the previous work. On Fig. 2c one can see that the X-ray[28] and low energy γ-ray[36] spectrum below 2 MeV can be fitted by a single power law. Another set of X-ray measurements[22,31], not coinciding in time with those in Ref. 28, shows a flatter spectrum, indicating possible time variability. The X-ray spectrum extrapolated to TeV energies indicates a significant spectral break between 2 MeV and 1 TeV. Results from Egret are awaited on the status of the 100 MeV emission from this pulsar.

Geminga: The COS-B[30] Geminga spectrum extrapolated to TeV energies as well as possible detections[35,8] and upper limits[1,16] in the TeV range also indicate a spectral break between 10 and 300 GeV for this pulsar (Fig. 2d). The soft X-ray, hard X-ray and low energy γ-ray[3,18,23,37] results indicate another spectral break between 3 keV and 10 MeV.

INTERPRETATION

Both the polar cap[20] and outer gap[10] models predict the spectral break above a few GeV. The lack of or lower than expected TeV γ-ray fluxes from these pulsars are consistent with the spectral break above a few GeV. The X-ray to medium energy γ-ray flux values are also consistent with a flattened spectrum relative to the extrapolated high energy γ-ray spectrum. The total photon spectra are qualitatively consistent with the outer gap model, but TeV observations constrain the model somewhat.

A newly born γ-ray pulsar is considered to be in the Crab-like phase. It reaches maturity in the Vela-like phase[10,13,26]. The earlier prediction[26] of an increase in the ratio of the γ-ray luminosity L_γ to spindown power $I\Omega\dot\Omega$ as the pulsar evolves from the Crab to post-Vela phase is confirmed by observations: For Crab we have $L_\gamma/I\Omega\dot\Omega \ll 1\%$, Vela ($\sim 1\%$), PSR 1706-44 ($\sim 5\%$) and Geminga ($\sim 50\%$ given the distance of 250 ± 150 pc[19]). From their P and $\dot P$ values one can show that Vela, PSR 1706-44 and PSR 1509-58 are Vela-like pulsars[13,26], and considering a variable geometry, one can also show that Geminga may be a post-Vela pulsar[27].

The flattened spectrum below medium energy γ-rays can be explained in terms of the residence time of secondary outer gap electrons which is less than their synchrotron lifetime, so that they cannot radiate down to the optical[11]. However, in the Crab the situation is reversed and we see strong optical and X-ray pulsed emission.

With the inverse Compton controlling mechanism for the Vela-like outer gap[10] it was shown that the post-Vela phase should be TeV bright since the mean free path of a primary TeV γ-ray is comparable to the size of the outer magnetosphere[13]. However, this is shown to be inconsistent with TeV observations[25]. This problem is avoided with the suggestion that the gap controlling mechanism for Geminga is probably curvature radiation instead of inverse Compton scattering[27].

If pulsars contribute to the diffuse galactic γ-ray background, a spectral depression should be seen above a few GeV. It is interesting to note that an excess above the predicted nuclear contribution is visible below ~ 5 GeV in a preliminary analysis of the Egret data[5] of this background. Better statistics may tell us in future whether there is a depression at energies ≥ 5 GeV, which would allow a direct measure of the contribution from galactic γ-ray pulsars to the diffuse background.

CONCLUSIONS

We have presented the first results of a search for TeV emission from PSR 1706-44, which shows clear evidence of a spectral break above 5 GeV. This same behaviour is verified for Vela, Geminga and PSR 1509-58. Both the outer gap- and polar cap models can qualitatively account for this spectral break.

The spectral flattening towards lower energies can be accounted for by the outer gap model in terms of the effect of the synchrotron lifetime which exceeds the electron residence time in the magnetosphere.

Although the outer gap model can explain the spectral break above 5 GeV, it still predicts larger fluxes in the TeV γ-ray range than is observed from both Vela-like and post Vela-like pulsars, indicating that the model for both types of pulsars has to be revised[25]. This is confirmed at an increased sensitivity by these new results.

REFERENCES

1. Akerlof, C.W. et al. 1992, *Proc. 22nd ICRC*, Dublin, **1**, 324.
2. Becker, W., Predehl, P., Trümper, J. 1992, IAU Circ. 5554.
3. Bennet, K. et al. 1992, these proceedings.
4. Bertsch, D.L. et al. 1992, *Nature*, **357**, 306.
5. Bertsch, D.L. et al. 1992, these proceedings.
6. Bhat, P.N. et al. 1987, *Astron. Astrophys.*, **178**, 242.
7. Bond, I.A. et al. 1990, *Proc. 21st ICRC, Adelaide*, **2**, 210.
8. Bowden, C.C.G. et al. 1992, private communication.
9. Brazier, K.T.S. et al. 1990, *Proc. 21st ICRC, Adelaide*, **2**, 304.
10. Cheng, K.S., Ho, C., Ruderman, M. 1986, *Ap.J.*, **300**, 500.
11. Cheng, K.S., 1990, private communication.
12. Cheng, K.S., Ho, C., 1990, private communication.
13. Cheng, K.S., De Jager, O.C. 1990, *Nuclear Phys. B* **14A**, 28.
14. De Jager, H.I. et al. 1986, *South African J. of Phys.* **9**, 107.
15. De Jager, O.C. et al. 1988, *Ap.J.*, **329**, 831.
16. Fegan, D.J. et al. 1992, these proceedings.
17. Grindlay, J.E. et al. 1975, *Ap.J.*, **201**, 82.
18. Halpern, J.P., Holt, S.S. 1992, *Nature*, **357**, 222.
19. Halpern, J.P., 1992, private communication.
20. Harding, A.K. 1981, *Ap.J.*, **245**, 267.
21. Kanbach, G. et al. 1980, *Astron. Astrophys.* **90**, 163.
22. Kawai, N. et al. 1992, these proceedings.
23. Massaro, E. et al. 1992, these proceedings.
24. Nel, H.I. et al. 1990, *Ap.J.*, **361**, 181.
25. Nel, H.I. et al. 1992, to appear in *Ap.J.*, **398**.
26. Ruderman, M., Cheng, K.S. 1988, *Ap.J.*, **335**, 306.
27. Ruderman, M., Chen, K., 1992, these proceedings.
28. Seward, F.D., Harnden, F.R. 1982, *Ap.J.*, **256**, L45.
29. Strickman, M. et al. 1992, IAU Circ. 5557.
30. Swanenburg, B.N. et al. 1981, *Ap.J.*, **243**, L69.
31. Takahashi, T. et al. 1992, these proceedings.
32. Taylor, J.H. et al. 1992, public database.
33. Thompson, D.J. et al. 1992, to appear in *Nature*.
34. Tümer, J. et al. 1984, *Nature*, **310**, 214.
35. Vishwanath, P.R. et al. 1992, IAU Circ. 5612.
36. Wilson, R.B. et al. 1992, to appear in *Isolated Pulsars*, eds. K.A. van Riper and C. Ho, Cambridge Univ. Press.
37. Wilson, R.B. et al. 1992, these proceedings.

GAMMA-RAY PULSARS AND GEMINGA

M. Ruderman
Department of Physics & Columbia Astrophysics Laboratory
Columbia University

K. Chen
Los Alamos National Laboratory

K.S. Cheng
Department of Physics, University of Hong Kong

and

J.P. Halpern
Department of Astronomy & Columbia Astrophysics Laboratory
Columbia University

ABSTRACT

Observed properties of γ-ray pulsars are related to those of the accelerators which power their radiation. It is argued that the relatively slowly spinning Geminga is a strong γ-ray source only because its magnetic dipole is more inclined than that of the more rapidly spinning Vela. This would also account for special Geminga properties including 180° subpulse separation, soft X-ray spectra and intensities, and suppression of radio emission.

1. INTRODUCTION

A number of common properties suggest that the family of strongly emitting γ-ray pulsars are all powered by similar accelerators. In most, and perhaps all of these pulsars, the "light curve" for emission of γ-rays consists of two subpulses. The EGRET γ-ray pulse profile for the Vela radiopulsar is shown in Fig. 1. The phase separation for it and others in this family ($\Delta\phi$ on a scale of 0 to 1) are shown in Table 1.

Although spectral breaks seem to occur at various energies below 10^2 MeV, above that energy the γ-ray flux per unit energy (E) varies as E^{-2} up to energies well above a GeV.

Some total γ-ray luminosities (for an assumed fan beam geometry) are indicated in Fig. 2. Although these absolute luminosities (L_γ) seem to decrease with increasing period (P), the ratio of γ-ray power to the total stellar spin-down power ($I\Omega\dot{\Omega}$) seems to increase. A probable Geminga distance, 250 ± 150 pc[1], somewhat less than that to the Vela pulsar, suggests that its L_γ is comparable to that of the Vela pulsar

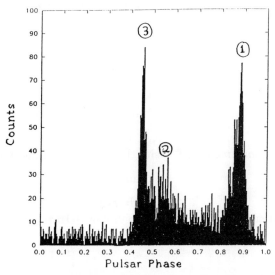

Fig. 1. The γ-ray light curve of the Vela radiopulsar as observed by EGRET[2] during Aug. – Sept. 1991. The photon energies > 50 MeV.

Table 1. Subpulse Separation in γ-ray Pulsars

Pulsar	Period (s)	$\Delta\phi$
Crab	0.033	0.4
PSR 0540 – 69[a]	0.050	0.2
Vela	0.089	0.4
PSR 1706 – 44	0.102	≤ 0.2 ?
Geminga	0.237	0.5

[a]PSR 0540 – 69 appears to be extremely similar to the Crab pulsar in almost every way but it is so distant that it has not yet been confirmed as a γ-ray source. The broad PSR 1706 – 44 γ-ray pulse has not been resolved.

despite a neutron star spin-down power which is two orders of magnitude less.

These γ-ray pulsar family similarities support the hypothesis that the γ-ray emission may be understood as a consequence of a common accelerator which differs among these pulsars because of their different ($\vec{\Omega}$), angle between $\vec{\Omega}$ and the stellar magnetic dipole moment $\vec{\mu}$, and

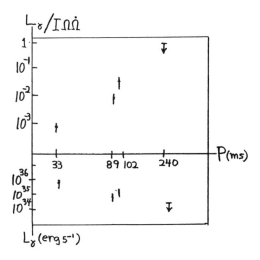

Fig. 2. The γ-ray luminosity (L_γ) and the ratio of L_γ to total spin-down power ($I\Omega\dot\Omega$) for the Crab, Vela, and Geminga pulsars and PSR 1706 − 44.

the direction to the observers. (The magnitude of $\vec{\mu}$ does not seem to vary much among the γ-ray pulsars). For the Crab pulsar, and then presumably for all, the γ-ray emission/accelerator region should be in the star's *outer*-magnetosphere. The Crab pulsar has strong optical pulses whose double subpulse structure coincides in shape and arrival time with those of its γ-rays. Therefore this emission is expected to come from the same local source as that for the γ-rays. The Crab's linearly polarized optical/IR luminosity is over 10^{-2} that of its γ-rays and over 10^{-5} that of its total spin-down power. Such an efficiency for optical radiation is not achieved within the magnetosphere by curvature acceleration, linear acceleration, or inverse Compton scattering on the pulsar radio beam[3]. This (as well as the polarization pattern[4] and the ratio of the optical to the X-ray intensity which has the same light curve) indicates that synchrotron radiation is the main optical source. The magnetic field which supports efficient sharply beamed synchrotron optical-IR emission must satisfy $eB\hbar/mc \ll 1$ eV. This inequality can be satisfied only at distances (r) from the Crab neutron star which satisfy the inequality $r \gg 3 \cdot 10^7$ cm $= 0.2 r_{\ell c}$ where $r_{\ell c} \equiv c\Omega^{-1}$ is the "light cylinder radius" where an exactly corotating magnetosphere would move at the speed of light.

A supporting argument, which also places the accelerator/emission region far from the star in the region near its light cylinder, follows from the Crab pulsar's total X-ray/γ-ray power. The maximum net current which can flow between the star and its light cylinder is $J \sim \mu\Omega^2 c^{-1}$ and the upper bound for the power into γ-ray emission is $J\Delta V$ where ΔV is the potential drop (along $\vec{B}$) through the accelerator. For the Crab pulsar (with an assumed fan beam emission geometry) $\Delta V > 10^{14} V$. Such a large ΔV along $\vec{B}$ could not be maintained near the star; it would be quenched by an avalanche of $e^{\pm}$ pairs from γ-rays (curvature radiated by those same pairs) converting in the large polar cap magnetic field.

2. OUTER-MAGNETOSPHERE ACCELERATORS

The outer-magnetosphere near the light cylinder is still a region for which there is not yet a consensus on a detailed description. There are idealizations which unrealistically assume alignment of μ and $\vec{\Omega}$, no current flow, and no $e^{\pm}$ production regardless of the magnitude of the local electric field ($\vec{E}$) along $\vec{B}$[4,5]. These give a corotating magnetosphere near the star with a charge density[6]

$$\rho \simeq \vec{\Omega} \cdot \vec{B}/2\pi c, \qquad (1)$$

and a completely empty region ("gap") much further out. In the former $\vec{E} \cdot \vec{B} = 0$ and (except for centripital acceleration) there is no acceleration of the charge separated magnetosphere plasma there. In the gap $\vec{E} \cdot \vec{B} \neq 0$ but there is no plasma to be accelerated. The gapless charge distribution around a non-aligned rotator when current flow is not permitted is shown in Fig. 3. When there are no magnetospheric sources of charge, but charge is allowed to flow from the magnetosphere, a different corotating stationary charge distribution is achieved. The expected new charge distribution is sketched in Fig. 4. When all of the unrealistic constraints are abandoned e^- or e^+ may be expected to fill most of the outer-magnetosphere gap region in those pulsars which can self sustain $e^{\pm}$ pair production there. Such pair production would involve a bootstrapped symbiosis between γ-ray or X-ray production by synchrotron, inverse Compton, and curvature radiation by e^-/e^+ from $e^{\pm}$ production by intersecting γ-ray and γ-ray, X-ray or IR beams.[7,8,9] Unquenched particle acceleration regions, where $\vec{E} \cdot \vec{B} \neq 0$ despite the copious $e^{\pm}$ production in the outer magnetosphere, can exist only in places which are not effectively reached by crossing beams of radiation. The resulting "outer-gap" accelerator geometry is sketched in Fig. 5.

The inner boundary of the accelerator lies near the intersection of the "null-surface" where $\rho = 0$ and the boundary of the "closed" field lines of the star on which magnetosphere current does not flow. The thickness of an outer-gap accelerator will grow until the $e^{\pm}$ production it sustains

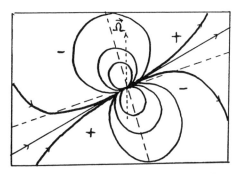

Fig. 3. Charge distribution around a rotating unaligned neutron star. The magnetosphere charge density vanishes on the "null-surface" indicated by the dashed lines on which $\vec{\Omega} \cdot \vec{B} = 0$.

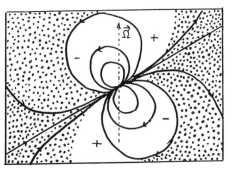

Fig. 4. The expected charge density around an inclined rotating neutron star with no magnetosphere currents. The dotted regions are empty of plasma.

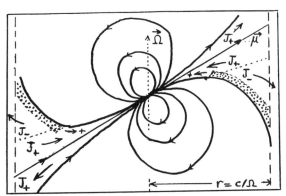

Fig. 5. Geometry and current flow of an outer-magnetosphere accelerator. One boundary of the gap is the last of the "closed" magnetic field lines which return to the stellar surface and do not reach the light cylinder. J is the current flow.

limits it while supplying the charged particles needed to complete the current flow in the magnetosphere through the null surface where $\vec{\Omega} \cdot \vec{B} = 0$.

Expected radiation beams from such an outer-gap accelerator are sketched in Fig. 6. Because of the $e^{\pm}$ production near the upper (lower) boundary of the RHS (LHS) outer-gap accelerator of Fig. 5, there is an approximate symmetry between radiation from the e^- and from the e^+ which flow oppositely within the accelerator and also from unseparated pairs which are formed outside of it. The accelerator and also radiation beams are emitted in both directions parallel to $\vec{B}$ from around the upper (lower) accelerator boundary. All beams with the same declination angle with respect to $\vec{\Omega}$ can be observable during a stellar rotation. Three potentially observable beams are indicated by 1, 2, and 3 in Figs. 6 and 7:

a) Beam 2 comes from curvature radiation by e^+ (or e^- if $\vec{\Omega} \cdot \vec{\mu}$ has the opposite sign) coming *toward* the stellar polar cap out of the near end of the RHS accelerator. It should have a spectral break near γ-ray energies of $\sim 10^2$ MeV or larger.

b) Beam 1 consists of some curvature radiation directly from *outflowing* e^- in the accelerator (or from e^- flowing out from its far end) together with synchrotron radiation (and some inverse Compton scattering) from pairs produced above the accelerator. The synchrotron radiation will have a much lower energy spectral break (because of the interval between pair production and emission of such radiation) than that of beam 2 curvature radiation.

c) Beam 3 is an approximate mirror image of beam 1 from the e^+ and $e^{\pm}$ pairs of the LHS accelerator which move along $\vec{B}$ *toward* the star.

Calculated[8] and observed[9] spectra suggest associating beams 1 and 3 of Fig. 6 with those of Fig. 1 and beam 2 of Fig. 6 with the "interpulse" 2 of Fig. 1.[10] Because of plasma inertia and gap induced $\vec{E} \cdot \vec{B}$, the velocity directions of γ-ray emitting e^- and e^+ are not known on the open field lines where r approaches the light cylinder radius. Consequently, accurate arrival phases for beams 1, 2, and 3 are not, at present, predictable. An oversimplified model[6] suggests that the large $\Delta\phi \sim 0.4$ of the Crab and Vela pulsars (and also their cusp-like sub-pulse shapes) comes from an emission radius $r \sim 2r_{\ell c}/3$. [Romani, Chiang, and Ho[11] assume a $\vec{B}$ near the light cylinder exactly equal to that from a rotating dipole with no magnetosphere at all. Then a single field line from the star to the light cylinder may pass through the same inclination angle (with respect to $\vec{\Omega}$) several times before leaving the magnetosphere (e.g. at the points 1' in Figs. 6 and 7). In their model the sources for emission

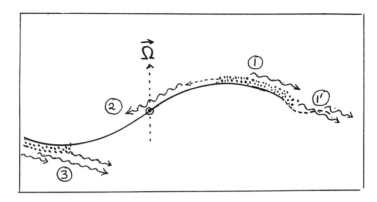

Fig. 6. Possible radiation beams from an outer-gap accelerator. All beams shown can have the same dip angle with respect to $\vec{\Omega}$.

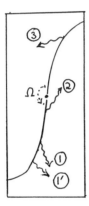

Fig. 7. A view of the emission beams of Fig. 6 from the spin axis direction which suggests the different aberration angles and effects of $\vec{B}$-line bending.

into beams 1, 2, and 3 of Fig. 1 are assumed to be from particle *outflow* from these special places on the same field line.]

3. DEATH VALLEY

As Ω decreases in a spinning-down pulsar the light cylinder radius expands and a self-sustaining outer-gap accelerator grows both in absolute thickness and in the fraction of the open field line bundle it occupies. A larger fraction of the stellar spin-down power must be expanded to maintain the $e^{\pm}$ production needed to sustain magnetospheric current flow between the star's polar cap and light cylinder. Almost all of this expended power is radiated away as γ-rays. Estimates[12,13] for the ratio $L_\gamma/I\Omega\dot{\Omega}$ of pulsars with aligned $\vec{\mu}$ and $\vec{\Omega}$ are shown in Fig. 8. As the

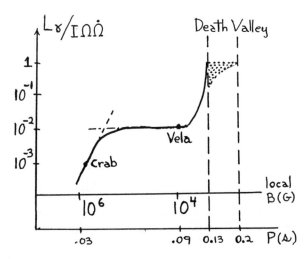

Fig. 8. A model calculation of the evolution of the ratio $L_\gamma/I\Omega\dot\Omega$ for a nearly aligned pulsar as it spins-down[12,13]. The two death lines correspond to inserting the accelerator's magnetic field B at $r = c/\Omega$ ($P = 0.13$ s) for which the Crab evolution is plotted and at $r = c/2\Omega$ ($P = 0.2$ s). The dotted region of the plot within Death Valley contains those values of P and L_γ which would give $L_\gamma \sim 4 \cdot 10^{35} - 5 \cdot 10^{36}$ erg s^{-1}, the γ-ray luminosities indicated for the remaining unidentified Cos B sources[17].

spin down period P slows to about 10^{-1} s L_γ approaches $I\Omega\dot\Omega$. The absolute value of L_γ is then around that estimated for most of the still unidentified Cos·B sources[14]. At still longer periods L_γ is expected to be largely quenched as the accelerator can no longer sustain itself. Just where the strong γ-ray emission should expire depends sensitively on the local B at the accelerator. Different death lines are found if B is evaluated as that at $r = r_{\ell c}$ or that at $r = r_{\ell c}/2$. Both are shown. When the angle between $\vec\mu$ and $\vec\Omega$ is large both underestimate the period at which an outer-gap accelerator is no longer a powerful γ-ray source.

4. GEMINGA

Several questions are raised by the identification of Geminga as a γ-ray pulsar.

a) Why is it such a strong γ-ray source, probably with an L_γ comparable to that of Vela, when its longer period and smaller $|\vec\mu|$ put it somewhat past the Death Valley of Fig. 8[13]?

b) Why is its subpulse separation $\Delta\phi$ so close to 0.5?

c) Why is it not seen as a radiopulsar?

Simple answers to all of these would follow from assuming that Geminga is, in most respects, a Vela-like pulsar except that its $\vec{\mu}$ is much more inclined to its $\vec{\Omega}$.

An outer-gap accelerator begins where the null surface ($\vec{\Omega} \cdot \vec{B} = 0$) intersects the boundary between a star's open and closed field lines. For an aligned $\vec{\mu}$, $r_{\rm ns}$, the minimum r where this occurs is $2r_{\ell c}/3$. When the angle between $\vec{\mu}$ and $\vec{\Omega}$ is large

$$r_{\rm ns} \sim \frac{4|\vec{\mu} \cdot \vec{\Omega}|^2}{9|\vec{\mu} \times \vec{\Omega}|^2} r_{\ell s} \qquad (2)$$

The magnitude of the local $\vec{B}$ around the beginning of an outer-magnetosphere accelerator gap can be about the same for Geminga and Vela if $\vec{\mu}$ is differently inclined with respect to $\vec{\Omega}$ in each. If the inclination angle is, say, 30° in Vela and about 65° in Geminga, B at $r = r_{\rm ns}$ is about the same for both. To sustain the needed outer-magnetosphere pair production and accelerator current (J_a) in Vela requires an accelerator there which spans about 10^{-1} of the open field line bundle between the star and the polar cap[8]. The electric field along $\vec{B}$ in the accelerator ($\vec{E}_a \cdot \hat{B}$) is also about 10^{-1} of that which would exist if the accelerator gap spanned the entire available open field line bundle. One consequence is that for the radiated power from Vela's outer-gap $L_\gamma \sim 10^{-1} \times 10^{-1} I \dot{\Omega} \Omega$. But if Vela sustains its needed outer-magnetosphere accelerator in the local B with those values of accelerator J_a and $\vec{E}_a \cdot \hat{B}$, the more inclined Geminga could do the same, as long as the fraction of available open field line flux needed to achieve them is less than one. Equivalently, Geminga can sustain a Vela-like outer-magnetosphere accelerator of the same power as Vela's if Geminga and Vela have similar $\mu r_{\rm ns}^{-3}$ and L_γ. For Geminga this would need $L_\gamma \sim I\Omega\dot{\Omega} \sim 5 \cdot 10^{34}$ erg s^{-1}.

The smaller $r/r_{\ell c}$ and Ω of Geminga would result in some significant differences in geometry and spectra relative to Vela's outer-gap emission.

a) Because Geminga's gap fills it accessible open field line bundle while Vela's fills only 10^{-1} of its, the relevant strengths of beams 1, 2, and 3 in Fig. 4 will change. The intensity of beam 2 would be expected to increase relative to that of the synchrotron part of beams 1 and 3. (Synchrotron radiation comes mainly from the boundary region *outside* the accelerator, and may no longer be the dominant contribution to L_γ.)

b) Because the radial distance to Geminga's emission region (r_e) is so much smaller relative to its $r_{\ell c}$ than is the case in Vela, the aberration and time of flight effects which determine the phase separation of beams 1 and 3 are greatly diminished. For Geminga beams 1 and 3 will merge with a separation

$$\delta \sim 2r_e/\pi r_{\ell c} \qquad (3)$$

and for beams 1 and 2

$$\Delta\phi = 0.5 + \delta. \qquad (4)$$

Geminga's observed $\Delta\phi$ and pulse widths suggest $r_e \lesssim 5 \cdot 10^{-2} r_{\ell c} \sim 5 \cdot 10^7$ cm which would need Geminga's inclination angle to be at least 65°.

c) If curvature radiation does become the dominant contributor to the observed beams, the low energy spectral breaks in Geminga's spectrum should be at or above 10^2 MeV, much higher than those in Vela's two (mainly synchrotron) subpulses. Geminga's spectrum should resemble that of Vela's "interpulse" (region 2 in Fig. 1). This may already have been observed by Cos B[10].

5. SURFACE X-RAY RADIATION

Beams of very energetic e^+ (or e^-) will flow toward the Geminga or Vela neutron star polar caps from the starward ends of their outer-magnetosphere accelerators. While moving to the star such leptons will lose most of their energy to curvature γ-rays (beam 2 of Fig. 6) but will still hit each of the two polar caps with a power $L_{pc} \sim 2 \cdot 10^{32}$ erg s^{-1}. The impacted part of each polar cap would then have a temperature $kT_{pc} \sim 0.5$ keV. The total soft X-ray luminosity from Vela[16] is, indeed, about $5 \cdot 10^{32}$ erg s^{-1} and that from Geminga[15] seems likely to be quite comparable. However, the effective area for most of this emission is nearly the whole surface area of the star instead of that of the very much smaller (by a factor $\sim 10^{-4}$) polar caps. For Geminga almost all of the soft X-ray emission is at $kT \sim 0.04$ keV. Only a small fraction, about $4 \cdot 10^{-2}$, is emitted with $kT \sim 0.3$ keV (Fig. 9)[15].

Just such a transfer of power from a hot polar cap to the cooler rest of the neutron star surface is expected when $\vec{\mu}$ is strongly inclined with respect to $\vec{\Omega}$. When this is the case, parts of beams 2 and 3 of Fig. 6 can pass within 10^7 cm of the neutron star. There, beam γ-rays of energy $\gtrsim 3 \cdot 10^2$ MeV will be converted into $e^\pm$ pairs by the stellar magnetic field. These pairs will, in turn, synchrotron radiate lower energy γ-rays (typically with energy $\gtrsim 3$ MeV) very many of whch will convert to produce a larger, second generation of pairs much closer to the star. The circumstellar pair density would not be sufficient to

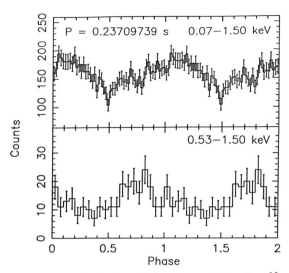

Fig. 9. Soft X-ray light curves of Geminga[15].

affect keV polar cap X-rays were it not for the strong magnetic field near the star. At $r \sim 3R$ where $B \sim 10^9$G the cross section for X-ray scattering at $\hbar\omega = 1$ keV is resonant with $\int \sigma d\omega = (2\pi)^2 e^2/mc$. This leads to estimates supporting the following account of the escape of keV polar cap X-rays from Geminga (and perhaps Vela).

1) Almost all keV polar cap X-rays are backscattered by resonant scattering on abundant $e^\pm$ pairs at $r \sim 3R$.

2) These X-rays, usually after multiple reflections, hit the stellar surface which reemits them at $kT \sim 10^{-1} kT_{\text{pc}} \sim 10^{-1}$keV. The much softer X-rays from the whole surface pass easily through the reflecting screen at $r \sim 3R$ and, perhaps after some reflection, through their own resonant $e^\pm$ scattering layer at $r \sim 6R$.

3) That small fraction of keV polar cap X-rays whose polarization is almost exactly parallel to the $\vec{B}$ they traverse at $r \sim 3R$ will have only the non-resonant Thomson cross section for scattering on e^-/e^+. These will, therefore, escape without degradation in energy. This may be the observed weak hotter ($kT \sim 0.3$ KeV) pulsed component from Geminga whose luminosity is less than 10^{-1} that of the softer ($kT \sim 4 \cdot 10^{-2}$ KeV) component. (Its single component light curve is consistent with a sunspot-like geometry for the polar caps. Such a geometry has also been suggested for other reasons[17].)

A pair-filled magnetosphere, which seems a necessary consequence of a powerful outer-gap accelerator in a sufficiently inclined γ-ray pulsar, may be the origin of other features in the observed emission from Geminga.

a) The densest near-magnetosphere pair density would be expected around (and within) the open field line bundle. The soft 0.1 KeV X-ray $e^{\pm}$ resonant cross section at $r \sim 6R$ should then result in a partial stellar eclipse during each rotation (Fig. 10) which reduces the observed X-ray luminosity by a fraction $\sim (\Omega R/c)(r/R)^3 \sim 0.2$. This may be the origin of the light curve minimum in Fig. 9 at phase $\phi = 0.5$. If so, that dip should be found to coincide in phase with one of Geminga's two γ-ray subpulses.

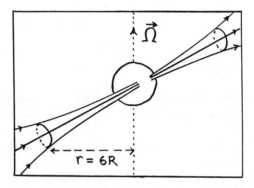

Fig. 10. Soft X-ray eclipse geometry of the Geminga neutron star by the open field line resonant reflection disk at $r \sim 6R \sim 6 \cdot 10^6$ cm.

b) If the near-magnetosphere $e^{\pm}$ density within the entire open field line bundle exceeds the Goldreich-Julian density there, then any polar cap accelerator which might otherwise form in that bundle would be quenched. The radio emission which is generally thought to be associated with such accelerators would then be absent. (This might also explain the relative absence of observed radiopulsars with $B \sim$ several $\cdot 10^{12}$ G and periods slightly less than Geminga's.)

ACKNOWLEDGEMENTS

This is contribution number 510 of the Columbia Astrophysics Laboratory. MR wishes to acknowledge support for this work under NASA Grant NAG5-2016 and NSF Grant AST89-01681.

REFERENCES

1. J. Halpern, in preparation (1992).
2. C. Fichtel *et al.*, these proceedings.
3. M. Ruderman in *Particle Acceleration Near Accreting Compact Objects*, J. van Paradijs, M van der Klis and A. Achtenberg, eds. (North Holland Amsterdam) (1991).
4. J. Krause-Polstorff and F.C. Michel, *M.N.R.A.S.*, **213**, 43 (1985).
5. K.S. Cheng and M. Ruderman, *Ap.J.*, **373**, 187 (1991).
6. P. Goldreich and W. Julian, *Ap.J.*, **157**, 869 (1969).
7. K.S. Cheng, Cheng Ho, and M. Ruderman, *Ap.J.*, **300**, 500 (1986).
8. K.S. Cheng, Cheng Ho, and M. Ruderman, *Ap.J.*, **300**, 502 (1986).
9. Cheng Ho, Los Alamos Workshop on Isolated Pulsars, Taos, NM, in press (1992).
10. I. Grenier, W. Hermsen, and C. Hote, *Adv. Space. Res.*, **11 (8)**, 107 (1991).
11. R. Romani, J. Chiang, Cheng Ho, these proceedings.
12. M. Ruderman and K.S. Cheng, *Ap.J.*, **335**, 306 (1988).
13. K. Chen and M. Ruderman, *Ap.J.*, in press (1992).
14. H.A. Mayer-Hasselwander and G. Simpson, *Adv. Space Res.*, **10(2)**, 89, (1990).
15. J. Halpern and S. Holt, *Nature*, **357**, 222 (1992).
16. H. Ögelman, J.P. Finley, and H.U. Zimmerman, preprint (1992).
17. K. Chen, K.S. Cheng, and M. Ruderman, in preparation.

UNVEILING GALACTIC GAMMA-RAY SOURCES: ISOLATED vs. BINARY PULSARS

Marco Tavani
Joseph Henry Laboratories and Department of Physics
Princeton University, Princeton, NJ 08544

ABSTRACT

We discuss the high-energy emission from isolated and binary pulsars. Models of magnetospheric gamma-ray emission from isolated pulsars will be constrained by CGRO observations of 'COS-B-like' sources. In addition to isolated pulsars, CGRO observations will constrain models for the interaction of pulsar winds with their nebular environments. We briefly describe the characteristics of unpulsed X-ray and gamma-ray emission expected in binary interacting pulsar systems where a radiative termination shock of the pulsar wind is produced. Two binary pulsar systems are candidate for high energy emission: young pulsars in massive binaries and old millisecond pulsars with low-mass companion stars in compact binaries.

INTRODUCTION

Unveiling the nature of Geminga[8,2] confirmed that pulsars are among the most important gamma-ray emitters in the Galaxy. Two formerly unidentified 'COS-B sources' are associated with pulsars (Geminga[2] and PSR 1706-44[23]) and more COS-B/CGRO sources may turn out to be isolated pulsars. Having unveiled the nature of Geminga and 2CG 342-02 is the prelude to a more comprehensive and possibly global understanding of galactic high energy sources. We consider here isolated and binary pulsars which are expected to produce high energy emission potentially observable by CGRO instruments.

We briefly discuss the impact on models of magnetospheric emission of detections vs. non-detections of known isolated pulsars. We also suggest that CGRO observations may valuably constrain models of pulsar winds interacting with their gaseous environments in binaries which act as calorimeters. Interacting binary pulsars include young pulsars interacting with mass outflows from massive companions and old millisecond pulsars interacting with mass outflows from low-mass companions possibly driven by pulsar irradiation. A class of galactic gamma-ray sources may turn out to be hidden pulsars in the radio band, and only optical, X-ray or γ-ray emission can be used to deduce their existence.

ISOLATED PULSARS

Current theoretical models of pulsar magnetospheres give efficiencies ϵ for γ-ray production in the range from 10^{-3} to 10^{-2} of the total pulsar spindown power luminosity L_p for Vela-like pulsars[9,4] and larger for Geminga-like pulsars[18]. The efficiency is a sensitive function of pulsar spin period P, pulsar age $\tau = P/2\dot{P}$ and emission geometry. Fig. 1 shows for the brightest known isolated pulsars the distribution of the expected γ-ray flux Φ_C (in units of $\gamma\,\text{cm}^{-2}\,\text{s}^{-1}$) for photon's energy $E > 100$ MeV) computed assuming the same spectrum and efficiency ϵ of conversion of spindown power into gamma-ray luminosity as the Crab pulsar. We notice that several pulsars with spin periods 100 ms $\lesssim P \lesssim$ 240 ms are potentially observable by EGRET if their $\epsilon \gtrsim 10$ and if their high energy spectrum is similar to the Crab.

Table 1

pulsar	P (s)	Φ_C	age (yr)	B_s (Gauss)	$\eta_{C,1}$	$\eta_{C,2}$	$\eta_{C,3}$
Vela	0.0893	0.90E-06	0.11E+05	0.33E+13	10.3	10.1	9.1
1706-44	0.1024	0.52E-07	0.17E+05	0.31E+13	15.0	23.3	13.9
Geminga	0.2371	0.11E-06	0.33E+06	0.16E+13	180.4	$> \epsilon_C^{-1}$	262.0
0740-28	0.1668	0.39E-08	0.16E+06	0.17E+13	80.3	906.7	125.0
1046-58	0.1236	0.97E-08	0.20E+05	0.34E+13	20.7	48.0	16.2
1509-58	0.1502	0.30E-07	0.15E+04	0.15E+14	7.8	-	1.2
1727-33	0.1394	0.24E-08	0.26E+05	0.34E+13	27.0	87.4	20.6
1758-24	0.1249	0.45E-08	0.15E+05	0.40E+13	18.3	37.4	12.3
1800-21	0.1336	0.48E-08	0.16E+05	0.42E+13	20.2	-	12.6
1823-13	0.1014	0.59E-08	0.21E+05	0.28E+13	16.3	27.7	17.0

The efficiency for high energy production normalized to the Crab can be parametrized in general as a function of the spin period P and pulsar age, $\eta_C \equiv \epsilon/\epsilon_C \sim P^\alpha \tau^\beta$. Theories of γ-ray production in pulsar magnetospheres predict different values for α and β. We consider the following phenomenological models: *model 1*, based on the 'polar cap' model of pulsar γ-ray production[9] characterized by the choice: $\alpha = 1.3$, $\beta = 0.47$; *model 2*, based on the 'outer gap model'[4] with $\alpha = 3.0$, $\beta = 1.0$ for pulsars with $P > 0.080$ msec and magnetic field in the range $1 \cdot 10^{12}\,G \leq B_s \leq 4 \cdot 10^{12}\,G$; *model 3*, based on a particular application of the outer gap model[3] characterized by the choice $\alpha = 0.0$ and $\beta = 1.0$. Table 1 gives the model-dependent normalized γ-ray efficiency η_C for the γ-ray pulsars already detected and for the brightest CGRO γ-ray pulsar candidates.

CGRO observations of this population of pulsars will valuably constrain the spectrum and high energy efficiencies of known isolated pulsars. CGRO observations of PSR 0740-28, PSR 1046-58, and PSR 1727-33 are particularly important since the models differ in their predicted efficiencies by a sizeable amount. A interesting example of a pulsar not seen by EGRET[13] whose pulsed emission should have been detected if its spectrum were similar to the Crab is PSR 1509-58.

INTERACTING PULSARS IN BINARIES

Binary interacting pulsars offer a unique possibility of studying the interaction of pulsar winds with circumbinary material produced by mass outflows from their companion stars. High energy emission is expected from the termination shock near the contact discontinuity where the pulsar wind interact with nebular material. The resulting non-thermal shock emission is similar to the 'nebular' unpulsed high energy emission (with photon energy $E \gtrsim 30$ MeV) detected from the Crab pulsar interacting with its nebula by COS-B[5] and EGRET[17]. Rapidly rotating pulsars produce a radially expanding relativistic wind of $e^{\pm}$ pairs and possibly ions[10] in addition to the comoving Poynting flux of electromagnetic energy in a MHD approximation[12,10]. The MHD pulsar wind is characterized by the Lorentz factor γ_1 and by the ratio σ of electromagnetic to kinetic energy of wind partcles. Detailed models for the Crab nebula require[12,10] $\sigma \sim 10^{-2} - 10^{-3}$ and $\gamma_1 \sim 10^6$. However, the composition of pulsar winds may vary as a function of pulsar parameters and it might substantially differ from that one of the Crab for different pulsars. At present, there are no high energy observations, except those of the Crab nebula, which constrain the properties of pulsar winds. CGRO detections of high energy shock emission in interacting binary pulsars combined with X-ray observations may be of great relevance in gaining information about the properties of pulsar winds.

The radiation spectrum emitted at the shock depends on synchrotron emission, inverse Compton scattering and thermal emission of the relativistic shock. Observations of the unpulsed high energy emission from the Crab nebula[5] and detailed theoretical calculations[10] give an efficiency of conversion of upstream pulsar wind energy into shock radiation near 20 %. The typical synchrotron energy is $E_{syn} \simeq (10\ keV)\gamma_{e\pm,6}^2 B_s \sqrt{\sigma/(1+\sigma)}$, where $\gamma_{e\pm} = 10^6 \gamma_{e\pm,6}$ and $B_s \simeq (440\ G) B_9/(P_{-3}^2 r_{s,11})$, with $P = 10^{-3} P_{-3}$ s, B_9 the surface magnetic field in units of 10^9 G, and $r_{s,11} = r_s/(10^{11}\ cm)$ with r_s the shock distance from the pulsar. For a variety of shock acceleration models and radiative conditions, the effective emissivity of the shock is a power law[10,1] $j_\epsilon^{(s)}/\epsilon \propto \epsilon^{-\tilde{\beta}-1}$ (in units of $\gamma\ \sec^{-1}\ cm^{-3}$), with $\tilde{\beta} \sim 1$ extending from $\epsilon_1^{(s)} \simeq 0.3 \cdot E_{syn}$ to $\epsilon_s^{(s)}$. If ions contribute to a substantial part of the pulsar wind kinetic energy the radiation spectrum is expected to be a power law of index s from a few keV to a few MeV and possibly GeV. Alternately, if ions do not appreciably contribute to the pulsar wind, most of the pulsar wind energy is expected[10,1] to be radiated thermally at lower energies peaked around E_{syn}. A non-negligible fraction of the total energy radiated at the shock may be in the ROSAT energy range. If ions contribute a substantial fraction of the particle wind kinetic energy, the soft X-ray luminosity can be 10^{-3} of the total shock luminosity. Alternatively, if a pure $e^{\pm}$ pulsar wind is produced, the fraction of the total energy radiated in the X-ray band may approach 50% of the total[1]. Therefore, by combining observations of interacting binary pulsars of different CGRO instruments as well as X-ray instruments such as ROSAT and ASTRO-D it will be possible to obtain spectral information on pulsar termination shocks that will valuably constrain theoretical models.

By combining the results of the shock acceleration theory and the properties of pulsar winds we can estimate the integrated photon spectrum per unit area and unit time for the case of non-thermal shock acceleration (ε is the photon energy) $dN_\gamma(E > \varepsilon)/dA\,dt \sim (3 \cdot 10^{-4}\ cm^{-2} s^{-1})(L_{p,36}/D_{kpc}^2)\,F\,f_s\,(\tilde{\varepsilon}/0.2)\,(1\ MeV/\varepsilon)^{\tilde{\beta}}$ where $\tilde{\varepsilon}$ is the shock efficiency of conversion of pulsar wind energy into radiation F the fraction of the solid angle from the pulsar with isotropic particle wind intersected by the nebular shock ($F = 1$ for totally enshrouded pulsars), f_s the fraction of the total shock luminosity radiated in a given spectral band, $\tilde{\beta}$ an exponent of order unity which depends whether the nebular shock emission is characterized by slow ($\tilde{\beta} = 1/2$) or fast $\tilde{\beta} = 1$) synchrotron cooling[1], $L_{p,36} = L_p/(10^{36} \text{ erg s}^{-1})$, and $D_{kpc} = D/(1 \text{ kpc})$. An isolated pulsar with the same $L_{p,36}/D_{kpc}^2$ typically produces a high energy flux smaller by a factor $(10^2 - 10^3) F$ than an interacting pulsar.

(a) YOUNG PULSARS IN MASSIVE BINARIES

Young pulsars with 30 ms $\lesssim P \lesssim$ 100 ms and $L_p \sim 10^{36} - 10^{37} \text{ erg s}^{-1}$ are expected to exist with massive early type companion such as OB or Be stars. The recent remarkable discovery[11] of the high eccentricity ($e = 0.97$) binary pulsar PSR 1259-63 with $P = 47.7$ ms orbiting around a Be star with $P_{orb} \simeq 1130$ days offers a unique possibility of testing the properties of shock emission. For a constant mass outflow from the Be star, the shape of the pulsar cavity and of the shocked region at the pulsar wind termination shock vary greatly as a function of orbital phase. For an estimated luminosity $L_p \sim 10^{36} \text{ erg s}^{-1}$ and a distance $d \sim 2.3$ kpc, PSR 1259-63 should be visible by CGRO instruments near periastron.

A class of pulsars enshrouded in mass outflows in massive binaries with $P_{orb} \sim$ 10 days can be even more difficult to detect in the radio, and their shock high energy emission can be the only way of unveiling them. An interacting pulsar candidate is the radio source LSI 61°303 possibly associated with the COS-B source 2CG 135+1.

(b) OLD PULSARS IN LOW-MASS BINARIES

There are at present four eclipsing pulsars with low-mass companions and Table 2 summarizes their pulsar and orbital properties as well as the length of the eclipsed fraction of the orbit $\Delta L/L$.

Table 2: Eclipsing Pulsars

PSR	P_{spin} (msec)	P_{orb} (hours)	m_c ($M_\odot$)	$\Delta L/L$	eclipse characteristics	Ref.
1957+20	1.60	9.2	~0.02	~1/10	stable	[6]
1744-24A	11.56	1.8	~0.10	1/3 - 1/2 - 1	highly variable	[14]
0021-72J	2.10	2.8	~0.02	1/4	stable	[16]
1726-19	1000.	6.1	~0.12	~1/4 − 2/3	freq. dependent	[15]

The enshrouding material from the companion star is probably driven by irradiation[19] and extensive theoretical work is in progress for the study of the formation of mass outflows in this systems and their hydrodynamic properties[21]. Given the selection effects against the detection of millisecond pulsars in compact binaries filled with enshrouding material provided by irradiated companion stars, we can expect that the number of partially or totally 'hidden' pulsars is not negligible and interesting for the interpretation of galactic high energy sources[20] with luminosities

$L \sim 10^{34} - 10^{35}\,\mathrm{erg\,s^{-1}}$. High energy emission produced in the inner shock of PSR 1957+20 is calculated[1] to be too low by a factor of ~ 10 to be observable by CGRO. However, recent ROSAT observations of PSR 1957+20 confirm the existence of a soft X-ray flux[7] with radiative efficiency near 10^{-3} as expected by the theory of radiative shocks[1]. In these systems, a possible modulation of the high energy flux with the orbital period together with soft X-ray emission and modulated optical- UV emission from the irradiated companion star can provide useful signatures for the interpretation of unidentified sources[20].

A particularly interesting search for γ-ray emission from hidden millisecond pulsars can be carried out in globular clusters where a relatively large population of hidden pulsars is predicted[22]. Collective emission from isolated cluster pulsars is not large enough to be detectable by CGRO and only sufficiently luminous enshrouded pulsars can produce detectable high energy emission from globular clusters.

We thank J. Arons, A. Lyne, R. Manchester, and J. Taylor for discussions and sharing of information. Research supported by NASA grant GRO/PFP-91-23.

REFERENCES

[1] Arons, J. and Tavani, M., 1993, *Ap.J.*, in press.
[2] Bertsch, D., *et al.*, 1992, *Nature*, **357**, 306.
[3] Buccheri, R., D'Amico, N., Massaro, E., Scarsi, L., 1978, *Nature*, **274**, 572.
[4] Cheng, K. S. and Ho, C. and Ruderman, M., 1986, *Ap.J.*, **300**, 500.
[5] Clear, J. *et al.*, 1987, *A.&A.*, **174**, 85.
[6] Fruchter, A.S., Stinebring, D.R., Taylor, J.H., 1988, *Nature*, **333**, 237.
[7] Fruchter, *et al.*, 1992, *Nature*, **359**, 303.
[8] Halpern, J. and Holt, S.S., 1992, *Nature*, **357**, 222.
[9] Harding, A., 1981, *Ap.J.*, **245**, 369.
[10] Hoshino, M., Arons, J., Gallant, Y.A., Langdon A.B., 1992, *Ap.J.*, **390**, 454.
[11] Johnston, S., *et al.*, 1992, *Ap.J.(Letters)*, **387**, L37.
[12] Kennel, C. F. & Coroniti, F. V., 1984, *Ap.J.*, **283**, 694.
[13] Kniffen, D., 1992, these Proceedings.
[14] Lyne, A.G. *et al.*, 1990, *Nature*, **347**, 650.
[15] Lyne, A.G., Biggs, J.D., Harrison, P.A. & Bailes, M., 1992, preprint.
[16] Robinson, C., Manchester, R.N., *et al.*, 1992, in preparation.
[17] Mattox, J., De Jager, O.C., Harding, A.K., 1992, preprint.
[18] Ruderman, M., 1992, these Proceedings.
[19] Ruderman, M., Shaham, J. & Tavani, M., 1989, *Ap.J.*, **336**, 507.
[20] Tavani, M., 1991, *Ap.J.(Letters)*, **379**, L69.
[21] Tavani, M. and Brookshaw, L., 1991, *Ap.J.(Letters)*, **381**, L21.
[22] Tavani, M., 1993, *Ap.J.*, in press.
[23] Thompson, D.J., *et al.*, 1992, *Nature*, **359**, 615.

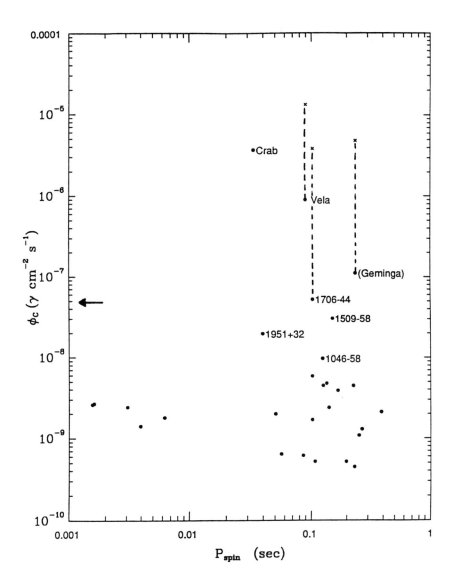

FIG. 1 - Distribution of estimated γ-ray emission above 100 MeV normalized to the Crab pulsar Φ_C as a function of pulsar spin period for the brightest pulsars. The distance of Geminga is assumed to be equal to 100 pc. Crosses give the actual COS-B measurements of pulsed γ-ray emission from Vela, Geminga, and PSR 1706-44. The lengths of the dashed lines indicate the enhancement of γ-ray emission compared to the Crab. The arrow indicates the EGRET nominal sensitivity limit. Magnetospheric emission from known millisecond pulsars is not expected to be detectable except for close ones with distance $d \lesssim 100$ pc and $\eta_C \sim 1$.

POLARIZATION AND EMISSION GEOMETRY OF THE CRAB PULSAR

Kaiyou Chen and Cheng Ho
Los Alamos National Laboratory, Los Alamos, NM 87545

ABSTRACT

Optical emission of the Crab pulsar can best be understood as synchrotron radiation of relativistic particles from the outer magnetosphere of the neutron star. The outer gap model was developed specifically to address energy balance and double-pulsed emission (from optical to high-energy gamma-ray) of young pulsars like the Crab. In this paper, we present the polarization properties of the optical pulses calculated from the outer gap model. We found that the theoretical light curves exhibit the same qualitative behavior as observations.

I. INTRODUCTION

With discoveries of more gamma-ray pulsars and their exhibited varieties in spectra and pulse profile, it is becoming evident that the intrinsic geometry (the inclination angle between the rotation axis and the magnetic axis) and the observational geometry (the viewing angle with respect to the rotation axis) are important to explain what we see. Unfortunately, there is no direct clue from observations on what these two angles are. Other than the radiation spectrum and pulse profile, the polarization of the emission appears to be a promising probe for the study of the pulsar geometry. The linear polarization of optical emission from the Crab pulsar exhibits significant variation across the pulse profile[1]. In principle, these observed polarization properties could be used as a diagnostic for probing the emission mechanism and geometry.

To explain the observed polarization properties of canonical radio pulsars, Radhakrishnan and Cooke (RC) proposed a simple geometry model where the polarization angle reflects the direction of the local magnetic field, thus providing clues to the underlying pulsar geometry. One could study the polarization properties of the Crab optical emission in the context of the RC model. However, it should be kept in mind that there are major differences between the Crab optical pulses and radio pulses from canonical pulsars. The RC model assumes coherent curvature radiation as the emission mechanism. In this case, radio waves are emitted along the direction of the magnetic field line. Furthermore, the radio waves have to propagate through regions loaded with dense plasma: the propagation could modulate the polarization[2]. In contrast, the optical emission from the Crab pulsar appears to be closely associated with the very broadband high-energy emission up to GeV, based on similarity in spectral characteristics and pulse profile. It has been argued that Crab's optical pulses are most likely incoherent synchrotron radiation in the outer magnetosphere[3,4].

In this paper, we study the polarization properties of the Crab optical emission based on the outer gap model.[5,6,7] In this model, a region devoid of charges is formed in the outer magnetosphere and serves as a powerful energy source for charge and photon production. This model yields a generic double pulse structure and broadband high-energy emission, qualitatively consistent with the observations of the Crab pulsar. We report here preliminary results of the polarization properties of high-energy pulses predicted by the outer gap model.

To study the pulse profile and polarization, we adopt the following simple geometrical scheme. The pulse phase is a reflection of the pulsar's rotational phase. As the pulsar rotates, we sample different regions of the magnetic fields lines which direct the emission towards us. It is also true that for a beamed emission with a finite opening angle from a fixed source location, the emission will be sampled differently at different rotational phases. Thus the contribution one receives at one particular phase will include emissions that are directed at a different angle but within the emission beam. The contribution can be decomposed into two components: 1) pencil beam, which is the beamed radiation emanating from the local relativistic charges, and 2) source distribution, which is the distribution of charges at field lines with different tangents. In principle, these two components are mixed and one needs a detailed geometry modeling. In this work, we assume that these two components can be decoupled. Furthermore, we assume that the source distribution can be separated into longitudinal (varying rotation phase) and latitudinal (along the same rotational phase) components. The pulse profile and polarization pattern are calculated by convolving the pencil beam with the source distribution. In the following, we concentrate on the optical emission from the Crab pulsar as predicted by the outer gap model.

II. CALCULATIONS

<u>Single Particle Emission</u> For a particle with energy γmc^2, we express the energy in terms of $\gamma = \gamma_\| \gamma_\perp$, where $\gamma_\perp = [(\gamma v_\perp)^2 + 1]^{1/2}$ and $\gamma_\| = (1 - v_\|^2)^{-1/2}$ with $(v_\|, v_\perp)$ being the parallel and perpendicular component of the velocity with respect to the local magnetic field line. A relativistic charge with energy $(\gamma_\|, \gamma_\perp)$ in a magnetic field B emits synchrotron radiation power in two polarizations:[8]

$$\frac{dP_\perp}{d\omega d\Omega} = k\,\gamma\gamma_\|^2 \left(\frac{1}{\gamma^2} + \theta^2\right)^2 K_{2/3}^2(\eta),$$

$$\frac{dP_\|}{d\omega d\Omega} = k\,\gamma\gamma_\|^2 \left(\frac{1}{\gamma^2} + \theta^2\right)\theta^2 K_{1/3}^2(\eta),$$

where

$$\eta = \frac{\omega}{3\omega_B}\gamma\gamma_\| \left(\frac{1}{\gamma^2} + \theta^2\right)^{3/2}, \quad \omega_B = \frac{eB}{mc}, \quad k = \frac{1}{6\pi^3}\frac{e^2}{\hbar c}\frac{\hbar\omega^2}{\omega_B},$$

and $K_{2/3}$ and $K_{1/3}$ are modified Bessel functions of 2/3 and 1/3 order. Here the polarization is with respect to the magnetic field line projected into the photon propagation direction. Emitted power of a single relativistic particle is mainly confined in a small cone with the pitching angle $\alpha \sim 1/\gamma_{\|}$.

<u>Pencil Beam</u> We adopt the particle distribution function in momentum space $F(\gamma_{\perp}, \gamma_{\|})$ calculated for the Crab pulsar from the outer gap model[7] to calculate pencil beam power:

$$\mathcal{P}_{\perp}(\theta, \eta) = \int\int d\gamma_{\|}\, d\gamma_{\perp} F(\gamma_{\perp}, \gamma_{\|}) P_{\perp},$$

$$\mathcal{P}_{\|}(\theta, \eta) = \int\int d\gamma_{\|}\, d\gamma_{\perp} F(\gamma_{\perp}, \gamma_{\|}) P_{\|}.$$

The total emitted power and degree of linear polarization of the pencil beam are $\mathcal{P} = \mathcal{P}_{\perp} + \mathcal{P}_{\|}$ and $\delta = (\mathcal{P}_{\perp} - \mathcal{P}_{\|})/(\mathcal{P}_{\perp} + \mathcal{P}_{\|})$, respectively. The pencil beam pattern $\mathcal{P}$ and δ for 1 eV photons are shown in Figure 1, with the magnetic field strength taken to be that at the light cylinder of the Crab pulsar ($\sim 10^6$ G).

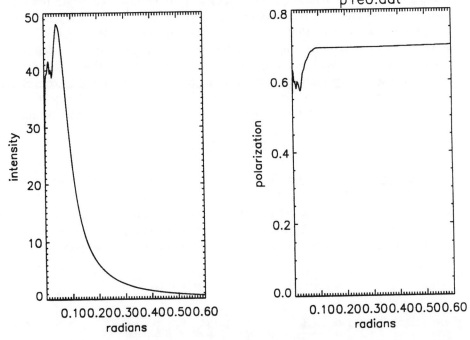

Fig 1: Emission of the "pencil beam" for 1 eV photons with the parameters of the Crab pulsar: (a) total emitted power; (b) the percentage of linear polarization.

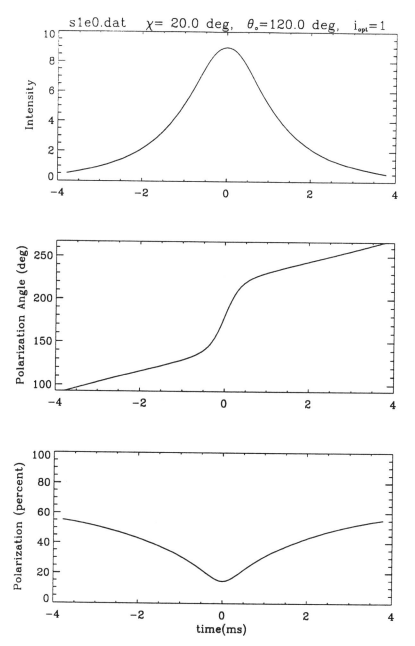

Fig 2: The theoretical light curve and its polarization properties of the outer gap model: (a) light curve; (b) position angle of polarization; (c) the degree of linear polarization.

<u>Source Distribution</u> We assume that all of the relevant field lines are loaded with particles with a momentum distribution following the same functional form. The particle distribution along a field line (latitudinal distribution) is assumed to be a constant. We take the longitudinal distribution to follow an exponential fall-off, yielding an overall particle distribution of

$$N(\phi_B, \gamma_\perp, \gamma_\parallel) \propto e^{-|\phi_B|/\sigma} F(\gamma_\perp, \gamma_\parallel),$$

where ϕ_B is the magnetic co-longitude. We have taken $\sigma = 0.1$ radians. Furthermore, we have integrated only the part of the pencil beam that points away from the sweep in the latitudinal direction, in order to mimic the effect of synchrotron self-absorption, as evidenced by the turn-over in the pulsed radiation spectrum.

III. RESULTS AND SUMMARY

By convolving the pencil beam with the source distribution function, we obtain the theoretical light curve and its polarization properties as a function of rotation phase angle ψ. The theoretical light curves shown in Figure 2 exhibit the same qualitative behavior as shown in observations.[1]

These preliminary results indicate that the outer gap model is capable of reproducing the qualitative behavior of the observed optical light curve and its polarization properties of the Crab pulsar. Furthermore, the qualitative polarization behavior shown in Fig 2 is generic of the outer gap model and is similar to a fairly large range of the pulsar parameters (tilt angle of the oblique rotator, and observer's viewing angle). Thus, based on the qualitative behavior alone, the pulsar geometry is not well constrained. More work is required before a quantitative fit can be performed to establish what the actual pulsar geometry is.

This work was supported in part by the GRO Guest Investigator Program and the NASA High-Energy Theory and Data Analysis Program. This work was done under the auspices of the US Department of Energy. We thank S. Barnes for her help on the manuscript.

V. REFERENCES

1. Manchester, R. N. & Taylor, J. H., 1977, *Pulsars* (Freeman, San Fransisco)
2. Barnard, J. J. & Arons, J., 1982, ApJ, 254, 713
3. Ruderman, M., 1990, in *Particle Acceleration near Accreting Compact Objects*
4. Chen, K. & Ruderman, M., 1993, ApJ, 402, 264
5. Cheng, K. S., Ho, C. & Ruderman, M., 1986, ApJ, 300, 500
6. Cheng, K. S., Ho, C. & Ruderman, M., 1986, ApJ, 300, 522
7. Ho, C., 1989, ApJ, 342, 396
8. Rybicki, G. B. & Lightman, A. P., 1979, *Radiative Processes in Astrophysics* (Wiley, New York)

GAMMA RAYS FROM PULSAR OUTER GAPS

James Chiang and Roger W. Romani
Department of Physics, Stanford University, Stanford, CA 94305-4060

and

Cheng Ho
Space Astronomy and Astrophysics Group, MS D436
Los Alamos National Laboratory, NM 87545

ABSTRACT

We describe a gamma ray pulsar code which computes the high energy photon emissivities from vacuum gaps in the outer magnetosphere, after the model outlined by Cheng, Ho and Ruderman (1986) and Ho (1989). Pair-production due to photon-photon interactions and radiation processes including curvature, synchrotron and inverse Compton processes are computed with an iterative scheme which converges to self-consistent photon and particle distributions for a sampling of locations in the outer magnetosphere. We follow the photons from these distributions as they propagate through the pulsar magnetosphere toward a distant observer. We include the effects of relativistic aberration, time-of-flight delays and reabsorption by photon-photon pair-production to determine an intensity map of the high energy pulsar emission on the sky. Using data from radio and optical observations to constrain the geometry of the magnetosphere as well as the possible observer viewing angles, we derive light curves and phase dependent spectra which can be directly compared to data from the *Compton Observatory*. Observations for Crab, Vela and the recently identified gamma ray pulsars Geminga, PSR1706-44 and PSR1509-58 will provide important tests of our model calculations, help us to improve our picture of the relevant physics at work in pulsar magnetospheres and allow us to comment on the implications for future pulsar discoveries.

A complete description of our work is in preparation and will be submitted to Astrophysical Journal.

MAGNETIC COMPTON SCATTERING IN THE MAGNETOSPHERES OF RADIO PULSARS

S. J. Sturner
Dept. of Space Physics and Astronomy, Rice University, Houston, TX 77251-1892

C. D. Dermer
E. O. Hulburt Center for Space Research, Code 7653,
Naval Research Laboratory, Washington, D.C. 20375-5000

and

F. C. Michel
Dept. of Space Physics and Astronomy, Rice University, Houston, TX 77251-1892

ABSTRACT

We investigate magnetic Compton scattering in the magnetospheres of neutron stars in an effort to explain the spectra of pulsed gamma-ray emission from radio pulsars such as the Crab and Vela and radio-quiet pulsars such as Geminga. Our model consists of a Monte Carlo simulation in which there is Compton upscattering of either a power-law or thermal soft photon distribution emitted from the surface of the neutron star. These upscattered photons then initiate a pair cascade via γ-B pair production. In our calculations the polarization averaged magnetic Compton cross section in the Thomson limit is used. We find that a satisfactory spectral fit is obtained for Vela and we discuss the applicability of our model to other pulsars.

INTRODUCTION

The gamma-ray spectra of radio pulsars such as the Crab and Vela have been studied many times with various detectors (e.g. OSO-8,[1] HEAO-1,[2] and COS-B[3,4]). Recent observations by ROSAT and GRO have added Geminga,[5,6] PSR 1509-58,[7] and PSR 1706-44[8] to the list of pulsars that emit observable gamma-rays.

The spectra of the Crab, Vela, and Geminga have one feature in common: a high-energy power-law form out to several GeV with a spectral index between 1.8 and 2.1. The major difference between the spectra of these objects is the location or even the existence of a break. The spectrum of Geminga shows a distinct flattening below ~100 MeV in the COS-B data, while Vela has a break at ~1 MeV. If a break exists in the Crab's spectrum, it occurs below ~10 keV.

Generally, models for gamma-ray production near radio pulsars have invoked the pair-cascade mechanism first discussed by Sturrock[9,10] in which high energy curvature radiation photons γ-B pair produce in the neutron-star magnetosphere. These initial high energy photons are often assumed to be produced in a charge-depleted gap near the neutron star pole.[11,12] These pairs then radiate synchrotron radiation which can initiate further pair production.

MODEL

Our model consists of a magnetic Compton-induced pair cascade near the polar caps of highly magnetized, rapidly rotating neutron stars. In large magnetic fields such as those found near neutron stars, the Compton cross section differs dramatically from the

Thomson cross section. The magnetic Compton cross section exhibits resonances at the local cyclotron energy and its harmonics. When the local magnetic field is « 4.414×10^{13} G, and the electron Lorentz factor γ is » the soft photon energy, the magnetic Compton cross section can be treated as having only one resonance at the local cyclotron energy.[13] The cross section approaches the Thomson value above the local cyclotron energy.

We treat a system where a continuum of soft photons produced near the neutron star surface is resonantly upscattered by highly relativistic electrons flowing outward from the polar cap along open magnetic field lines. Once scattered, these very energetic photons, many with energies » 1 GeV, are directed nearly parallel to the direction of the magnetic field at the location where they were produced. These scattered photons may then undergo γ-B attenuation as they propagate outward, thereby producing a pair cascade. From this point the calculation is very similar to a curvature-induced pair cascade[14] with the difference being the shape of the underlying curvature or Compton scattered spectrum.

We have used Monte Carlo techniques to model this problem using two possible sources of soft photons: a power-law distribution which is directed along the surface magnetic field and a thermal distribution emitted by a warm neutron star surface. In both cases the soft unscattered photons are assumed to originate at the neutron star surface. This model includes the synchrotron radiation from the first three generations of pairs with synchrotron radiation from the third generation undergoing attenuation.

The spectra displayed in Figures 1 through 4 are calculated assuming that the Lorentz factor of the scattering electrons is constant. This constraint will be relaxed in future work. At this point the model does not accommodate the enhancement of the density of the scattering electrons with those electrons and positrons produced in the cascade. This lack of self-consistency will be addressed in the future.

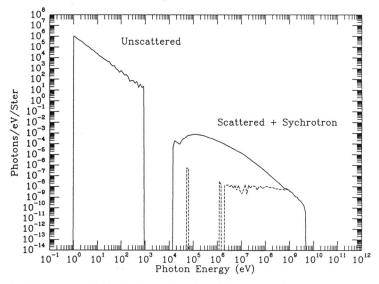

Figure 1: The scattered (dashed) and total (solid) Monte Carlo spectra are illustrated here for the case of a power-law distribution of soft photons ($\alpha = 5/3$) beamed along the local surface magnetic field direction. The parameter values are: $\gamma = 10^6$, the surface polar magnetic field = 4×10^{12} G, and an electron current of 10^{34} electrons/s.[15] Note the "flat top" nature of the scattered spectrum.

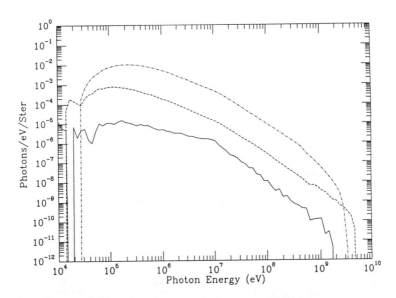

Figure 2: Comparison of the total output spectra for the case of a power-law soft photon spectrum with index $\alpha = 5/3$ that is beamed along the local surface magnetic field direction. The three curves represent the spectra where the Lorentz factor of the scattering electrons was 10^5 (solid), 10^6 (dashed), and 10^7 (dot-dashed). The surface polar magnetic field was taken to be 4×10^{12} G and the current from the polar cap was taken to be 10^{34} electrons/s.

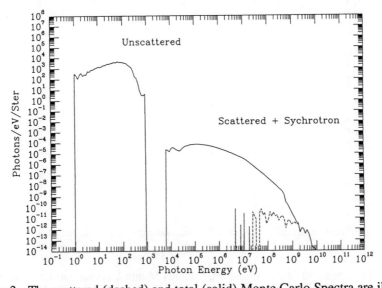

Figure 3: The scattered (dashed) and total (solid) Monte Carlo Spectra are illustrated here for the case of a thermal distribution of soft photons with kT = 51.1 eV. The parameter values are: $\gamma = 10^6$, the surface polar magnetic field = 4×10^{12} G, and an electron current of 10^{34} electrons/s.

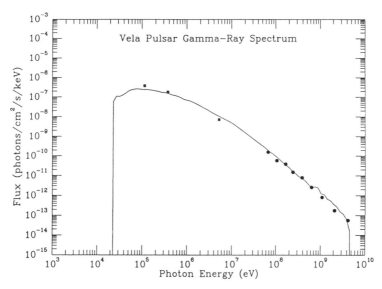

Figure 4: Here we have scaled the output of the Monte Carlo model to fit the observed spectrum of the Vela pulsar. Here the soft photon distribution was taken to be a power-law with index $\alpha = 5/3$. The parameters values are: $\gamma = 10^6$, surface polar magnetic field = 3.5×10^{12} G, and an electron current of 10^{33} electrons/s. The data plotted in here is from Grenier et al.[4] (circles) and Compton Observatory Science Report #73[16] (squares).

DISCUSSION

Note the shape of the underlying scattered spectrum in Figures 1 and 3. In the case of the power-law input spectrum (Fig. 1), the scattered spectrum is flat up to the highest energy to which a 1 eV photon (the arbitrary low energy cut-off used in the model) can be scattered. At that point the spectrum breaks to a power-law of approximately $\alpha+1/3$. The spectrum is attenuated at higher energies due to the γ-B process. The spectrum that results from scattering a thermal distribution (Fig. 3) is also flat but with a gradual turn over and an eventual pair attenuation break. The gaps in the spectra between 10^3 and 10^4 eV are due to the limited number of pair generations calculated, as well as assuming that all synchrotron photons are emitted at a mean energy that is a simple function of the particle energy and the local cyclotron energy.

The natural "flat top" shape of the scattered spectrum results in spectra that break at progressively lower photon energies as the electron Lorentz factors are increased. This is a consequence of more energetic synchrotron photons being produced, resulting in a total gamma-ray spectrum that becomes progressively less like a pure scattered spectrum. This is illustrated in Figure 2 where the scattered spectrum produced with electron Lorentz factors of 10^5 is only slightly changed by the addition of synchrotron photons below 10 MeV.

This trend can be checked against observations. In a simple model of particle acceleration near a neutron star polar cap, the accelerating electric field seen by the scattering electrons is approximated as being proportional to the electrostatic charge of the neutron star. This electrostatic charge has been estimated by Michel,[12,13] and is proportional to B/P, where B is the surface polar magnetic field strength and P is the period. Thus pulsars with smaller values of B/P should have lower energy electrons in their magnetospheres and, according to our model, have spectral breaks at higher photon energies. It should be noted that this is true only as long as a pair cascade takes place. Once the electron Lorentz factor falls below that required for a pair cascade, any further reduction of the Lorentz factor reduces the break energy by reducing the energy at which either the α+1/3 power-law begins, for the case of a power-law soft photon distribution, or the gradual turn over begins for the thermal distribution case. Michel and Dermer[17] have calculated the values of B/P for Vela and the Crab which are ≈ 38 and 114×10^{12} G s^{-1}, respectively. The corresponding value for Geminga is 7×10^{12} G s^{-1}. According to our model we would predict that Geminga has the highest break energy and the Crab the lowest, as is observed. To illustrate a fit of our model to observed pulsar spectra, we show a comparison of a model spectrum for electrons with an intermediate Lorentz factor, 10^6, and the observed spectrum of Vela in Figure 4.

In summary, we have shown that resonant Compton scattering can initiate pair cascades in pulsar magnetospheres with scattering electron Lorentz factors as low as a few times 10^5 as compared to Lorentz factors near 10^7 which are required for curvature radiation cascades. These cascades can successfully reproduce pulsar gamma-ray spectra and explain the differences observed between those spectra.

REFERENCES

1. Pravdo, S. H., and Serlemitsos, P. J., Ap. J. **246**, 484 (1981).
2. Knight, F. K., Ap. J., **260**, 538 (1982).
3. Clear, J., Bennett, K., Buccheri, R., Grenier, I. A., Hermsen, W., Mayer-Hasselwander, H. A., and Sacco, B., Astr. Astrophys., **174**, 85 (1987).
4. Grenier, I. A., Hermsen, W., and Clear, J., Astr. Astrophys., **204**, 117 (1988).
5. Bertsch, D. L. et al., Nature, **357**, 306 (1992).
6. Halpern, J. P., and Holt, S. S., Nature, **357**, 222 (1992).
7. Wilson, R. B. et al., IAU Circ. No. 5429 (1992).
8. Thompson, D. J. et al., Nature, 359, 615 (1992).
9. Sturrock, P. A., Ap. J., **164**, 529 (1971).
10. Sturrock, P. A., Nature, **227**, 465 (1970).
11. Harding, A. K., Ap. J., **245**, 267 (1981).
12. Daugherty, J. K. and Harding, A. K., Ap. J., **252**, 337 (1982).
13. Dermer, C. D., Ap. J., **360**, 197 (1990).
14. Michel, F. C., Ap. J., **383**, 808 (1991).
15. Michel, F. C., Theory of Neutron Star Magnetospheres (University of Chicago Press, Chicago, 1991), p. 112.
16. Compton Observatory Science Report #73 (1992).
17. Michel, F. C. and Dermer, C. D., Nature, **356**, 483 (1992).

X-RAY BINARIES

OBSERVATIONS OF ISOLATED PULSARS AND DISK-FED X-RAY BINARIES

R. B. Wilson and G. J. Fishman
NASA/Marshall Space Flight Center-ES62, AL 35812

M. H. Finger
Computer Sciences Corporation (NASA/MSFC-ES62)

G. N. Pendleton
Dept. of Physics, University of Alabama in Huntsville, AL 35899

T. A. Prince and D. Chakrabarty
Division of Physics, Mathematics, and Astronomy, Cal Tech, CA 91125

ABSTRACT

The BATSE experiment on the Compton Gamma Ray Observatory provides data with suitable time and energy resolution to monitor numerous pulsed systems. Using on-board folding of data, sources with periods as short as 4 ms can be observed. To date, 14 objects have been detected, encompassing gamma-ray emitting radio pulsars, disk-fed and wind-fed X-ray binary systems, and Be/transient systems. This paper summarizes results obtained to date for radio pulsars and disk-fed systems. Spectral measurements of the Crab pulsar and PSR1509-58 are presented, along with upper limits for other candidate objects. Observations by BATSE of the disk-fed X-ray binary systems Her X-1, GX 1+4, and 4U 1626-67 are presented.

INTRODUCTION

The full-sky, uncollimated field of view of the BATSE detectors enables nearly continuous viewing of objects which can be distinguished from the background by intrinsic source periodicity. For periods greater than a few seconds, data for this purpose is contained in all telemetered data packets. For shorter period objects, data can be collected for an optimal detector combination and folded on-board.

For sources requiring the folded-on-board data types, the counting rate summed over a selected subset of BATSE LADs in each of 16 energy channels is phase-resolved into 64 bins, typically integrated over an 8.132 s period, and telemetered. In subsequent analysis on the ground, individual accumulations are barycentered using the midpoint of the accumulation, and deposited in phase bins fixed in the barycenter, using the best-known ephemeris. A more detailed discussion of this analysis has previously appeared.[1,2]

For sources with longer periods, BATSE data with the same 16 energy channel resolution is available from each detector with 2.048s time resolution. This data is also barycentered using the midpoint time of each accumulation.

ISOLATED PULSARS

CRAB PULSAR

Data for the Crab pulsar has been collected on a daily basis since the beginning of the mission, principally as a "standard candle" to allow monitoring of detector gains and accuracy of the LAD detector response matrices.

Data has been combined over multiple observations for a time interval for which a single set of radio ephemeris parameters fit the contemporaneous radio data. The ephemeris parameters used are from the Princeton public database.[3] The profile obtained, shown in Figure 1, is consistent with previous observations in this energy range.[4,5] The spectral differences as a function of pulse phase are qualitatively apparent - softest at the main pulse, hardest in the "bridge" region between the pulses. The phase of the main and interpulse do not vary as a function of energy.

Phase resolved spectroscopy was performed for each GRO pointing interval, by folding a trial power-law spectrum through the detector response matrices, and comparing in counts space to the net source counting rate, obtained by subtracting the off-pulse background counting rate. The fits have values of χ^2 per degree of freedom of about 3. The systematic error still present at this stage of the analysis are likely due to uncertainties in the channel-to-energy calibration (especially at energies below 30 keV), atmospheric scattering into the uncollimated detectors (especially in the 50 - 150 keV interval), and response matrices that may not yet precisely fit the actual detector response for all angles and energies.

A weighted average of the best-fit values for the spectral index was then obtained. Figure 2 shows the behavior versus phase, and measurements from the HEAO A-4 experiment.[6] The variations versus phase obtained are very similar, but the observations differ in the spectral indices obtained. The BATSE values should be treated as preliminary results - there are known to be residual errors in the BATSE channel-to-energy calibration that are being evaluated using test observations, occultation observations of the Crab nebula, and solar flares.

PSR1509-58

This young radio pulsar, having the highest magnetic field of any known pulsar, and having been observed by *Ginga*[7] at above 30 keV, was perhaps the best candidate prior to CGRO launch for a new gamma-ray pulsar to be detectable by BATSE. It had not been observed in the BATSE energy range by HEAO 3.[8] BATSE first detected the source in Nov. 1991.[9] To date, a livetime of $2.2 \cdot 10^6$s has been obtained. The pulse profile as a function of energy is shown in Figure 3. The source is detectable at up to 750 keV, with weak evidence for emission above 1.8 MeV. The smooth, broad pulse is similar to that previously observed[7] at lower energies, but some structure is present in the 70-120 keV and 230-750 keV bands.

Spectral analysis was performed as described above for the Crab pulsar. The mean spectral index obtained for a power law fit is 1.6 ± 0.1. The indices obtained for the rising edge, peak, and falling edge are consistent with each other and this value.

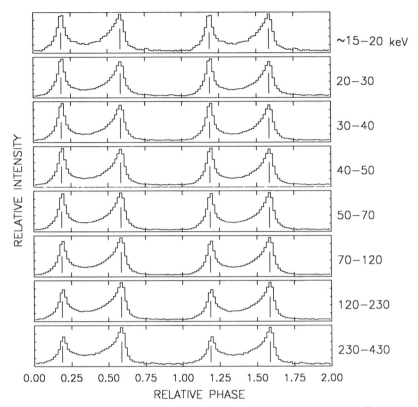

Figure 1. The pulse profile of the Crab pulsar obtained between Truncated Julian Day (TJD) 8505 and 8617, for 8 different energy intervals. Registration marks have been placed at the center of the main and interpulse peaks in each subfigure.

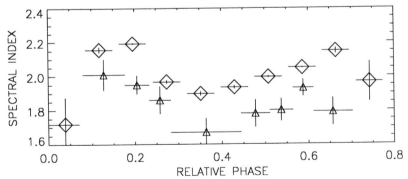

Figure 2. Spectral index of a power law fit to the Crab pulsar spectrum as a function of relative phase, as defined in Figure 1. The BATSE data are marked by diamonds; the data from Reference 6 are marked by triangles. Background was defined as the phase interval 0.78 to 0.99.

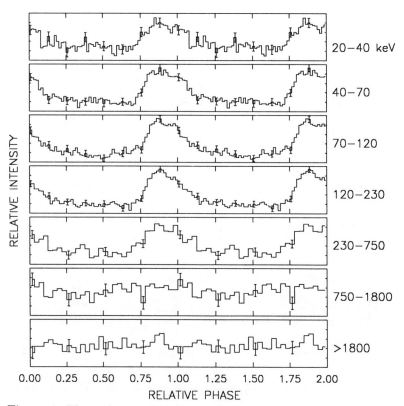

Figure 3. The pulse profile of PSR1509-58 obtained between TJD 8560 and 8866, for 7 different energy intervals.

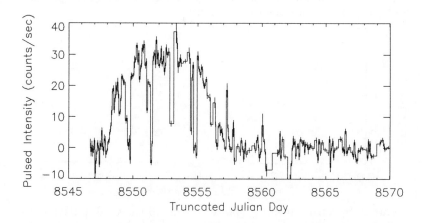

Figure 4. Pulsed intensity in counts per second, over the energy range 15 - 70 keV, during one 35d cycle of Her X-1.

UPPER LIMITS

Data has been collected and processed for a set of candidate objects. The 3σ upper limits given in Table 1 below have been obtained using sensitivities determined by a representative exposure of the Crab pulsar, and are thus for an assumed Crab-like pulse profile, and a presumed spectral index of 2. The limit for the Vela pulsar has been obtained using only 1/4 of the available data, since reprocessing of the data prior to correction of the GRO clock offset from UTC is not yet complete.

Table 1. BATSE Pulsed Source Upper Limits (3σ)				
Source	Processed Single-Epoch Exp. Time (10^4s)	Flux (photons cm^{-2} s^{-1} keV^{-1})		
		20-75 keV ($\times 10^{-5}$)	75 - 235 keV ($\times 10^{-6}$)	235 - 1120 keV ($\times 10^{-7}$)
PSR0656+14	50.0	0.4	0.3	0.7
Vela Pulsar	31.0	0.6	0.4	0.9
PSR1706-44	27.1	0.6	0.4	1.0
PSR2334+61	26.3	0.6	0.5	1.0
Geminga	25.8	0.6	0.5	1.0
PSR0611+22	8.0	1.1	0.8	1.8
PSR1737-30	6.0	1.2	0.9	2.1
PSR0959-54	5.7	1.3	1.0	2.1
PSR0540-693	4.2	1.5	1.1	2.5
PSR1951+32	3.1	1.7	1.3	2.9
PSR1822-09	2.6	1.9	1.4	3.2

DISK-FED X-RAY BINARIES

The period histories of low-mass x-ray binary systems (LMXRB) that are believed to be powered by material accreted to the neutron star via a disk fed by Roche-lobe overflow from the companion vary more smoothly with time than those of the wind fed systems.[10] The timing measurements made over the last 20 years of these systems have been by necessity sporadic and infrequent, since they were made by pointed instruments. Like the earlier Uhuru experiment, BATSE now is able to provide nearly continuous histories for a significant number of the brighter systems. In addition to the disk-fed sources discussed below, a report of BATSE observations of Cen X-3 also appears in these proceedings.[11]

HER X-1

Observations have been scheduled by BATSE throughout the CGRO mission of Her X-1, at all phases of the 35 day cycle. We detect the source only during the main-on portion of the cycle, for which coverage totals ~100 days, from part or all of 15 35d cycles. The livetime obtained from these intervals is about 10^6s. The mean value of the cycle duration observed (from a visual inspection) is 34.4±0.4 days, somewhat lower but consistent with the best previous determination.[12] The pulsed intensity, defined as the magnitude of the correlation coefficient of the mean pulse template and the measured pulse profile for a given $\sim 5600s$ elapsed time (one CGRO orbit), is shown in Figure 4 for one of the cycles with complete coverage. Each data point plotted is for all the data available from one GRO orbit, typically 1200s. The eclipses in the system are readily detected. Two pre-eclipse dips are observed, early in TJD 8551, and in the middle of TJD 8554, during which the pulse profiles show an additional interpulse component separated in phase by about 0.5. Emission at this pulse phase as been previously observed *after* eclipse egress[13], and during short-on states.[14]

The mean pulse profile obtained during this full main-on cycle is shown in Figure 5, with a shape similar to that previously observed[15], with the "shoulder" on the main peak most pronounced in the 30-40 keV interval. The phase-averaged spectrum obtained is shown in Figure 6. With the BATSE sensitivity and coarse energy binning, a power law is an adequate fit to the data, with an index of -4.5±0.2. In order to better measure sources of this type that are not detectable by BATSE above 100 keV, plans are being developed to reassign the energy boundaries used for the 16 energy channels available to a smaller range, approximately 15 - 200 keV, to maintain a binning smaller than the resolution of the detector throughout that interval.

HER X-1 ORBIT DETERMINATION

Using orbital parameters from Deeter, *et al.*[16], period and period epoch values chosen to eliminate observable phase drift during the TJD 8546-8560 observation, and the historical mean value of the source period derivative, data were epoch-folded for a GRO orbital time. These integrations typically consist of 600-1200s of exposure to the source.

A correlation analysis with a mean template (also formed from the TJD 8546-8560 interval) was used to determine a point of constant phase from each integration. A perturbation analysis was then performed, holding the orbital period and period derivative constant, allowing the epoch of 90° longitude in the orbit, $a_x sin(i)$, $e \cdot cos(\omega)$, and $e \cdot sin(\omega)$ to vary, minimizing χ^2. Each of the 10 intervals were analyzed separately, ranging in duration from 3 to 10 days. A weighted average of each of the varying quantities was then obtained. The errors quoted below in Table 2 have been increased above the statistical errors, to agree with the observed scatter in the parameters. The errors are 3.5-4 times larger than the computed statistical error.

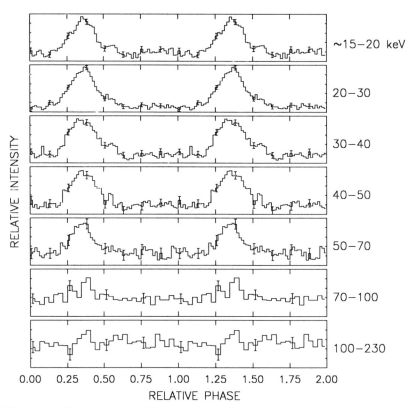

Figure 5. The pulse profile of Her X-1 obtained between TJD 8546 and 8560, for 7 different energy intervals.

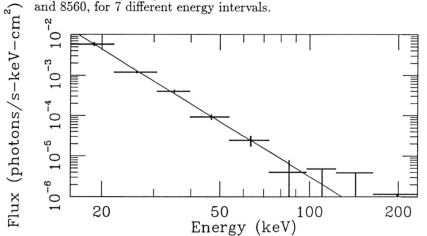

Figure 6. The phase-averaged pulsed spectrum of Her X-1 obtained between TJD 8546 and 8560. The spectral index from the power law fit is -4.5±0.2.

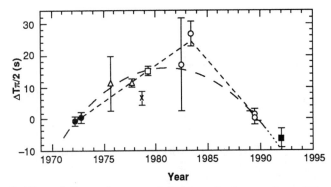

Figure 7. Plot of time of mean orbital longitude equal to 90° versus year (from Reference 16). The parabolic curve which displays the effect of the orbital period derivative has been extended beyond 1992 as the short dashes. The BATSE data point is the filled square at year 1991.966.

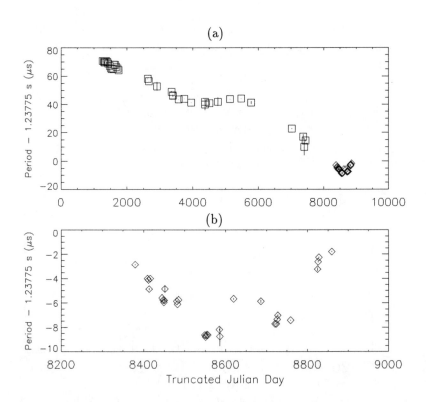

Figure 8. Barycentric period history of Her X-1. (a) Measurements since detection of the source. (b) Observations by BATSE over the duration of the CGRO mission.

This analysis has used a fixed pulse profile throughout the observation, disregarding the variations known to occur[15]; it is anticipated that a more careful analysis will reduce these errors.

The value obtained here of the binary epoch for mean orbital longitude (M.O.L) of 90° is compared with the measurement and previous data found in Reference 16 in Figure 7. Our measurement differs by only 1.2 σ, for this preliminary analysis.

The pulse period epoch and pulse period were obtained from a linear fit of the observed pulse phase versus time for all the main-on intervals, using 1.7 day intervals to reduce any systematic effects of binary orbital components not removed in the initial phase model. The entry in Table 2 is the measurement closest to the time of the orbital epoch.

Table 2. Her X-1 Parameters	
Epoch (M.O.L. = 90°)	JD 2448551.88789 (3) TDB
$a_x sin(i)$	13.1853 (15) s
Eccentricity	< 5.7 ×10^{-4} (3σ U. L.)
Epoch (Pulse Period)	JD 2448552.15617 TDB
Period	1.23774134 (4)

The period history obtained is shown in Figures 8(a) and 8(b). The BATSE measurements are very similar in quality and continuity of measurement to those obtained by Uhuru about 20 years ago, and will be available as long as BATSE remains operational and pulsed observations of Her X-1 are scheduled.

GX 1+4

The LMXRB system GX 1+4, whose long-term period history is shown in Figure 9(a), was found to have changed from a spin-up mode to spin-down by *Ginga*.[17,18] This trend was also observed by the MIR/HEXE instrument.[19] As shown in Figure 9(b), BATSE finds that this behavior still continues, but with a greater rate of spindown than previously observed. The source is usually bright enough to permit period determination on a daily basis.

An example of the pulse profile is shown in Figure 10, during a time when the source was moderately intense (similar to the HEXE observation in Oct. 1987[19]). The spectrum obtained is shown in Figure 11, with the best-fit OTTB spectrum having a temperature of 26.5±4.3 keV, cooler than at the time of the above HEXE observation.

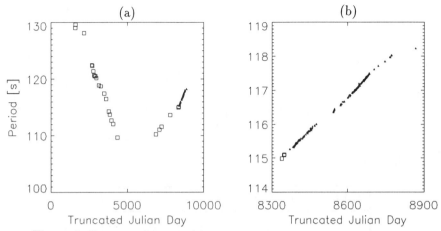

Figure 9. Barycentric period history of GX 1+4. (a) Measurements since detection of the source. (b) Observations by BATSE over the duration of the CGRO mission.

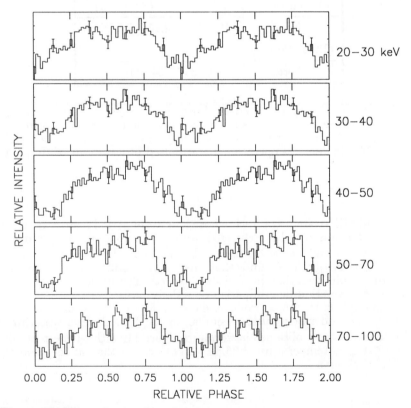

Figure 10. The pulse profile of the GX 1+4 obtained between TJD 8542 and 8546, for 5 different energy intervals.

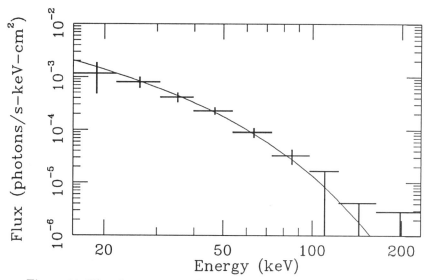

Figure 11. The phase-averaged pulsed spectrum of GX 1+4 obtained between TJD 8542 and 8546. The temperature obtained from the OTTB fit is 26.5±4.3 keV.

4U 1626-67

The 7.7s accreting pulsar 4U1626-67 has been studied by numerous spacecraft since its discovery by SAS-3 in 1977. In spite of these many observations, no orbital Doppler shifts have been detected in the pulsed X-ray signal, setting limits of 10 lt-ms on the semimajor axis of the neutron star orbit.[20] Observers in the optical range have reported[21] a modulation of 7.7s optical pulsations and have inferred an orbital period of 2485 s and a projected semimajor axis of the companion of 1 lt-s.

Observations prior to 1991 showed that 4U1626-27 was spinning up with a constant period derivative of approximately $4.5 \cdot 10^{-11}$ s per s. Our BATSE observations since May 1991, shown in Figure 12, indicate that 4U1626-67 has undergone a distinct transition, entering a spin-down phase with constant rate of approximately $5.5 \cdot 10^{-11}$ s per s. It thus appears that 4U1626-67 has two distinct accretion states, with equal and opposite period derivatives. These states seem to be stable on a time scale of at least 1 year.

ACKNOWLEDGEMENTS

We greatly appreciate the high quality work of the dedicated programmers who have produced the pulsar software analysis tools (L. Gibby, C. Wilson, R. Hunt), and helped to process and display a very large volume of data (K. Hagedon and S. Harris). We also are in debt to the entire BATSE operations team, specifically for data quality screening daily, to ensure the pulsar results are not contaminated by bursts, flares, or other transient episodes.

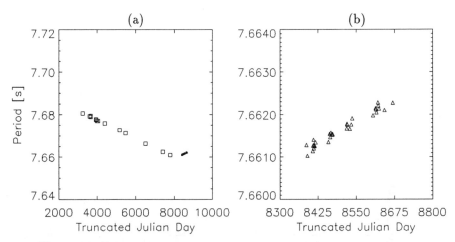

Figure 12. Barycentric period history of 4U 1626-67. (a) Measurements since detection of the source. (b) Observations by BATSE over the duration of the CGRO mission.

REFERENCES

1. R. B. Wilson et al., in *Proc. Isolated Pulsar Workshop*, (ed. K. A. van Riper, R. Epstein, and C. Ho., in press), Cambridge University Press, p. 257.
2. R. B. Wilson et al., in *Proc. Isolated Pulsar Workshop*, (ed. K. A. van Riper, R. Epstein, and C. Ho., in press), Cambridge University Press, p. 380.
3. Z. Arzoumanian et al., GRO/radio timing database, Princeton University (1992).
4. R. B. Wilson and G. J. Fishman, Ap. J. **269**, 273 (1983).
5. W. A. Mahoney et al., Ap. J. **278**, 784 (1984).
6. F. K. Knight, Ap. J. **260**, 538 (1984).
7. N. Kawai et al., Ap. J. Lett. **383**, L65 (1991).
8. M. P. Ulmer et al., Ap. J. **369**, 485 (1991).
9. R. B. Wilson et al., IAU Circ. 5429 (1992).
10. F. Nagase, PASJ **41**, 1 (1989).
11. M. H. Finger et al., these proceedings.
12. H. Ogelman, Astron. Astrophys. **172**, 79 (1987).
13. Y. T. Ushimaru, PASJ **41**, 441 (1989).
14. J. Trumper et al., Ap. J. Lett. **300**, L63 (1986).
15. Y. Soong et al., Ap. J. **348**, 634 (1990).
16. J. E. Deeter et al., Ap. J. **383**, 324 (1991).
17. D. Makishima et al., Nature **333**, 746 (1988).
18. T. Dotani et al., PASJ **41**, 427 (1989).
19. B. Mony et al., Astron. Astrophys. **247**, 405 (1991).
20. A. Levine et al., Ap. J. **327**, 732 (1988).
21. J. Middleditch et al., Ap. J. **244**, 1001 (1981).

OSSE OBSERVATIONS OF GALACTIC SOURCES DURING PHASE 1

J.D. Kurfess, W.N. Johnson, R.L. Kinzer, R.A. Kroeger
M.S. Strickman, J.E. Grove
E.O. Hulburt Center for Space Research, Naval Research Laboratory,
Washington DC 20375

D.A. Grabelsky, S.M. Matz, W.R. Purcell, M.P. Ulmer
Dept. Physics and Astronomy, Northwestern University, Evanston, IL 60201

M.D. Leising
Dept. Physics and Astronomy, Clemson University, Clemson, SC 29634

R.A. Cameron, and G.V. Jung
Universities Space Research Association, Washington, DC 20024

ABSTRACT

The Oriented Scintillation Spectrometer Experiment (OSSE) on the COMPTON Gamma Ray Observatory has undertaken comprehensive observations of astrophysical sources during the eighteen-month Phase 1 of the mission. These include investigations of many galactic sources, including binary X-ray sources, pulsars, several transient X-ray sources observed as Targets-of-Opportunity, and Nova Cygni 1992. Multiple observations of the galactic center region were undertaken to map the diffuse galactic emission and search for point sources. An overview of the galactic source observations and some preliminary results are presented.

INTRODUCTION

During Phase 1 of the COMPTON Observatory mission, the OSSE instrument has been used to undertake observations of a variety of galactic sources. The OSSE instrument covers the energy range from 50 keV to 10 MeV when operated in nominal gain (see Johnson et al.[1] for a detailed description of the OSSE instrument). For several of the galactic source observations the instrument was operated at twice the nominal gain and covered an energy range from 40 keV to 5.0 MeV. The field-of-view of the OSSE detectors is large: 3.8° x 11.4° FWHM. This is a compromise between a small field-of-view that is better suited for discrete source observations, and a large field-of-view that is preferred for study of the diffuse emission from the galactic plane.

Phase 1 observations consisted of a sequence of 2-week viewing periods (VP) during which the observatory was maintained in a fixed orientation in inertial space. This strategy was implemented to enable the large field-of-view EGRET and COMPTEL instruments to carry out a complete sky survey. The plan was followed from the start of the science program in May 1991 through March 1992, at which

time the effective failure of
the observatory tape
recorders forced a
conversion to real time
data acquisition though the
TDRS system. To make
up for the limited real-time
coverage through TDRS,
most viewing periods for
the remainder of Phase 1
were extended to three
weeks.

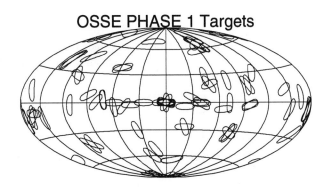

Table I provides a
list of the discrete galactic
sources observed during
Phase 1. Each entry in the
list provides source
location, the period of time that the source was observed, the GRO viewing period
number, and comments related to the observations. Only those sources which were
specific OSSE objectives are listed. While there were often additional sources, e.g.
soft X-ray sources, in the large OSSE field-of-view, such sources are unlikely to
contribute in the OSSE energy range above 50 keV. Likewise, while known radio
pulsars often populate these fields, Table I only lists those that are likely candidates
for detection based on loss of rotational energy and distance[2]. Figure 1 shows a map
of OSSE field-of-view locations during Phase 1. Background observations were
typically acquired at offset angles of ±4.5 degrees from the locations shown in Fig.
1. In some regions, most notably the galactic center, scans were undertaken over an
extended region. See Table 1 of Purcell et al.[3] for details.

Figure 1. Target fields viewed by OSSE during Phase 1.
Detector response is shown at the 90% level.

TABLE I: OSSE PHASE 1 GALACTIC SOURCE OBSERVATIONS

OBJECT	RA (J2000.0)	Dec	View Period	Observation Dates		Comments
CRAB PULSAR	83.52	22.02	1	16-MAY-1991	30-MAY-1991	
			32	25-JUN-1992	2-JUL-1992	
VELA PULSAR	128.92	45.18	8	22-AUG-1991	5-SEP-1991	first low-energy
			26	23-APR-1992	28-APR-1992	gamma-ray detection
			28	7-MAY-1992	14-MAY-1992	of Vela Pulsar
PSR 1509-58	228.48	59.14	23	19-MAR-1992	2-APR-1992	confirm BATSE obs.
			42	15-OCT-1992	29-OCT-1992	
GEMINGA	98.48	17.77	34	6-JUL-1992	6-AUG-1992	no detection
PSR 1957+20	299.84	20.81	1	16-MAY-1991	30-MAY-1991	no detection
PSR 1929+10	293.05	10.99	18	10-JAN-1992	23-JAN-1992	no detection

TABLE I (con't)

PSR 0950+08	86.03	-69.27	6	26-JUL-1991	8-AUG-1991	no detection
PSR 1706-44	257.25	-44.45	9	5-SEP-1991	12-SEP-1991	no detection
PSR 1951+32	297.21	32.89	15	28-NOV-1991	12-DEC-1991	no detection
PSR 1937+21	295.12	21.19	19	23-JAN-1992	6-FEB-1992	no detection
CYG X-1	299.59	35.21	2	30-MAY-1991	15-JUN-1991	low hard X-ray state
			7	8-AUG-1991	15-AUG-1991	low hard X-ray state
			15	28-NOV-1991	12-DEC-1991	low hard X-ray state
CYG X-3	308.10	40.96	2	30-MAY-1991	8-JUN-1991	before radio flare
			7	8-AUG-1991	8-AUG-1991	after radio flare
			15	28-NOV-1991	12-DEC-1991	
HER X-1	254.46	35.34	9	12-SEP-1991	19-SEP-1991	spans X-ray turn-on
			19	23-JAN-1992	6-FEB-1992	
SS 433	287.95	4.99	20	6-FEB-1992	20-FEB-1992	no detection
CIR X-1	230.23	-57.18	23	19-MAR-1992	2-APR-1992	no detection
GX 339-4	255.71	-48.79	9	5-SEP-1991	12-SEP-1991	high state
			13	7-NOV-1991	14-NOV-1991	low state
NOVA MUSCAE	171.61	-68.68	10	9-SEP-1991	3-OCT-1991	very weak
4U 1543-47	237.00	-47.80	27	28-APR-1992	7-MAY-1992	very weak
GRO J0422+32	65.43	32.91	36	11-AUG-1992	27-AUG-1992	intense transient
			38	1-SEP-1992	17-SEP-1992	
ETA CAR	161.95	-59.98	14	14-NOV-1991	28-NOV-1991	complex region
NOVA CYG 92	307.63	52.62	22	5-MAR-1992	19-MAR-1992	no detection
CAS A	350.87	58.81	34	16-JUL-1992	6-AUG-1992	no detection
G CENTER REGION	266.40	-28.94	5	12-JUL-1991	26-JUL-1991	sources obs. include:
			11	3-OCT-1991	17-OCT-1991	
			14	14-NOV-1991	28-NOV-1991	1E1740.7-2942
			16	12-DEC-1991	27-DEC-1991	GX1+4
			17	27-DEC-1991	10-JAN-1992	GX354+0
			21	20-FEB-1992	5-MAR-1992	GS1758-258
			24	2-APR-1992	9-APR--1992	
			25	16-APR-1992	23-APR-1992	coord. obs. with SIGMA
			40	17-SEP-1992	8-OCT-1992	Oct. '91 and Apr. '92
G PLANE 25	279.22	-7.05	7	15-AUG-1991	22-AUG-1991	
			12	31-OCT-1991	7-NOV-1991	
G PLANE 58.1	294.98	22.28	9	23-JAN-1992	6-FEB-1992	
G G PLANE 5+0	269.34	-24.53	24	9-APR-92	16-APR-1992	

PULSARS

OSSE has detected three radio pulsars in the low-energy gamma-ray band: the Crab Pulsar, the Vela Pulsar and PSR 1509-58. For pulsar observations, OSSE was operated in either an event-by-event mode or a rate mode to acquire high time resolution data. In the event-by-event mode the throughput is limited to about 200 counts/s and this restricts the energy range which can be investigated: the time resolution is either 1/8 ms or 1 ms. In the rate mode, broadband rates can be acquired in up to eight energy bands, with a best time resolution of 4 ms. For observations of the Crab pulsar and pulsars with comparable or shorter periods, the event-by-event mode is required. Preliminary results for the Crab pulsar were presented in Ulmer et al.[4] and spectral and temporal results are in preparation.[5]

OSSE observations of the Vela Pulsar have provided the first detection of low-energy gamma-ray emission from this object. Strickman et al.[6] discuss the results and implications of this detection.

PSR 1509-58, a radio pulsar with a period of 150 ms, was first detected by BATSE[7] in the low-energy gamma-ray region, and subsequently observed by OSSE during VP 23. The OSSE observations show a clear detection in the 60-570 keV region and are in general agreement with the BATSE results.

Geminga has recently been found to be both an X-ray pulsar[8] and a gamma-ray pulsar.[9] OSSE observed Geminga for a total of three weeks; however, we have not found clear evidence for a pulsed signal. The similarity between the X-ray and high-energy gamma-ray signatures of Geminga and the Vela pulsar indicates that additional observations of Geminga are warranted.

OSSE data have been used to search for gamma-ray emission from several millisecond pulsars with known periods, including PSR 1957+20 and PSR 1937+21. We find no evidence for gamma-ray emission from either object.

TRANSIENTS

OSSE was used to undertake extended observations of three transient sources as Targets-of-Opportunity following detections of outbursts by BATSE. These sources are GX339-4, 4U1543-47, and X-ray Nova Per = GRO J0422+32. GX339-4 is a well known transient X-ray source and black hole candidate which exhibits hard and soft spectral states similar to Cyg X-1, and QPO behavior. The OSSE observations of GX339-4 are discussed by Grabelsky et al.[10] OSSE obtained two observations of GX339-4. The first occurred near the end of the two-month "on" period when the source intensity was near the peak intensity observed during the outburst (300 mCrab). The second observation occurred after the intensity had rapidly decayed and was at a level of several mCrab. The spectrum observed during the first observation extends to several hundred keV and is similar to that of other black hole candidates. The GX339-4 spectrum is shown in Fig. 2 along with those of Cyg X-1 and GRO J0422+32.

The outburst of GRO J0422+32 was detected by BATSE on 1992 Aug 5[11] and a Target-of-Opportunity was implemented to enable OSSE to begin viewing the source within 24-hrs. of the BATSE report. At that time the source was already near its maximum intensity at about 3X Crab. OSSE undertook extensive observations of the source for 35 days of the next 40-day interval. The initial BATSE position, with a location uncertainty of about 1.5 degrees, was used to provide OSSE targeting for the first several days. During this period, OSSE performed observations at two scan angles to assist in obtaining a better location for the source. Using these data and additional BATSE occultation data, an improved location was obtained with an error radius of 0.2 degrees.[12] This assisted in the optical identification of the source with a low-mass X-ray binary system.[13,14]

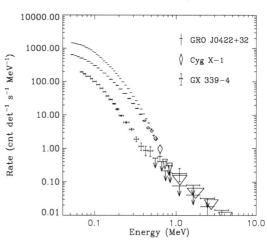

Figure 2. Energy spectra for three black hole candidates: Cyg X-1, GX 339-4, and X Nova Per.

The time history of the OSSE data is shown in Fig 3. The intensity of the source decreases rather monotonically over the period of the OSSE observations. Excellent spectral and timing information have been obtained by OSSE. The spectrum is hard and is well fit by a 100-keV thermal bremsstrahlung model. The spectrum for days 91/225-232 is shown in Fig 2. There may be a hint of an excess above this fit in the region above about 400 keV. However, this appears to be a hardening of the continuum rather than evidence for broad line emission such as has been reported for several other galactic sources. The spectrum hardened further as the intensity decreased toward the end of the OSSE observations. The limit on narrow line emission near 0.5

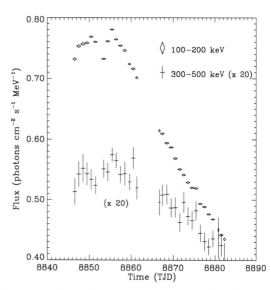

Figure 3. Time history for gamma-ray flux from X Nova Per.

MeV on a daily basis is < 0.1 times that reported by the SIGMA experiment for X-ray Nova Muscae on 1991 Jan 20.[15] Also, GRO J0422+32 reached an intensity of about six times that of Nova Muscae, so the limit for 0.5 MeV emission relative to the hard X-ray continuum flux is about 100 times lower in GRO J0422+32 than that observed for Nova Muscae on Jan 20, 1991.[15] OSSE has also detected QPO behavior in GRO J0422+32.

The third transient observed by OSSE is the source 4U1543-47. This source was reported by BATSE to be in outburst in April 1992[16], and re-orientation of GRO was undertaken to permit OSSE observations starting 1992 April 28. Unlike the other two transients discussed above, 4U1543-47 decayed rapidly and was nearly undetectable by BATSE at the time of the OSSE observation. Nevertheless, OSSE was able to observe the source for about one additional week and acquired temporal and spectral data during this period.

X-RAY BINARY SOURCES

In addition to the transient sources observed as Targets-of Opportunity, OSSE has also obtained observations of several other discrete galactic sources, most of which fall into the category of X-ray binary systems. These include the several XRB sources listed in Table I and several additional sources for which OSSE obtained significant exposure during the several viewing periods devoted to the galactic center region. Because of the large OSSE field-of-view, many other galactic sources were observed in the course of the Phase 1 observations for which the exposure was limited. These are not included in the table.

Cygnus X-1 had significant exposure for viewing periods 2 and 15 and a more limited exposure during VP 7. VP 7 was a Target-of-Opportunity following a reported radio flare in Cyg X-3, and Cyg X-1 was observed for a limited period of time with an effective exposure of only about 25%. All of the Cyg X-1 observations occurred when the source was near its lower intensity levels by historical standards. The energy spectrum observed during VP 2 is shown in Fig 2. The spectrum is observed up to energies of about 800 keV and is well represented by a two-temperature Sunyaev-Titarchuk spectrum.[10] Similar fits are also found for the other Cyg X-1 observations. We have seen no evidence for the gamma-1 state with excess MeV emission as reported by Ling et al.[17] If the low intensity of the hard X-ray emission is taken as evidence for the gamma-1 state with an accompanying MeV emission, the MeV component should have been present at levels ten times above the OSSE limits.

Cyg X-3 was also observed during the three viewing periods VP 2, VP 7 and VP 15. The VP 7 observation was undertaken as a Target-of-Opportunity following a major radio flare,[18] peaking at 17 Jy, which occurred in July 1991. This major flare was preceded by a smaller (5 Jy) flare in late June 1991. The VP 2 OSSE observation ended about one week prior to the precurser flare and the VP 7 observations occurred during the decay phase of the major flare when the radio emission had declined to below 2 Jy. The OSSE data provide the opportunity to investigate any potential correlation between the radio flaring and the high-energy gamma-ray emission. Analysis of these data is in progress. Figure 4 shows the time history of the hard X-ray flux during the period of time covering the three OSSE observations. It is seen that a dramatic decrease in the gamma-ray emission was observed during the VP 2 observation (prior to the precursor radio flare), dropping by a factor of three or more in a period of two days. The intensities during each of the latter observations, which occurred during the decay phase of the radio flare and during a radio quiet period resp., were higher and relatively constant. It is not clear whether there is any connection between the gamma-ray variability observed and the radio flaring. It is most plausible that radio flaring would be delayed from the higher energy activity, which is presumably indicative of processes much nearer to the central compact object.

Her X-1 was observed 1991 Sept 12-19, a period of time which included the X-ray turn-on during the 35-day period. OSSE detected the 1.24 sec. pulsar period at energies up to about 100 keV. 1.7-day eclipses are also seen in the data. The OSSE energy threshold (40-50 keV) is not low enough to see the previously reported cyclotron feature in the Her X-1 spectrum.[19] The OSSE data will be carefully analyzed for higher energy harmonics and phase-dependent spectral characteristics. However no information is available on these yet.

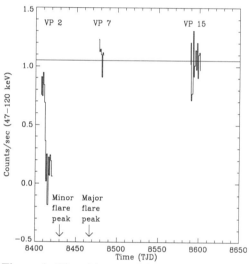

Figure 4. Time history of the gamma-ray emission from Cyg X-3. The OSSE observations during VP 7 were a Target-of-Opportunity following radio activity in June and July 1991.

GALACTIC CENTER SOURCES

The galactic center region has been observed for a total of 17 weeks during Phase 1 of the mission. The times of these observations are listed in Table 1. These observations have several objectives, including mapping the distribution of the 0.511 and 1.809 MeV emission, mapping the diffuse galactic continuum emission which is primarily due to contributions from electron bremsstrahlung and inverse compton scattering by cosmic ray electrons, searching for evidence of variable point source(s) of 0.51 MeV emission, and the detection and study of other discrete low-energy sources in the region. Purcell et al.[3] discuss the galactic center observations in some detail and give the OSSE results on the 0.511 MeV emission. In this article, we comment on the status of discrete sources in the region.

The galactic center observations were undertaken with several different position angles of the OSSE 3.8° x 11.4° collimators with respect to the galactic plane (see Purcell et al.[3] for details). At low energies, the challenge for the OSSE team is to extract the diffuse and discrete components in a region where there is considerable source confusion. The approach for doing this is to undertake scans of the region at several position angles. Analysis of the data is complicated by the transient nature of many of the hard X-ray sources in the region. To assist in this task, we have undertaken several correlative observations with the GRANAT/SIGMA experiment. SIGMA can obtain hard X-ray and low-energy gamma-ray images, but at a sensitivity which is about 5 times poorer that the OSSE limiting sensitivity. Nevertheless, undertaking joint analysis of these data is proving very beneficial.

OSSE has detected several sources in the galactic center region, including GX1+4, 1E1740.7-2942, and GX354+0. These have all been confirmed as SIGMA sources during several of the joint observations.[20] For the OSSE data, preliminary spectra of the individual sources can be obtained. The quality of these results will continue to improve as the mapping studies provide an improved distribution of the diffuse hard X-ray continuum, thereby permitting extraction of the discrete source contributions. Two examples illustrate the capabilities for

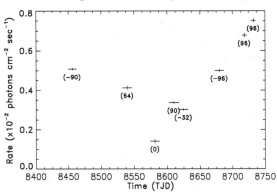

Figure 5. Time History of the 50-100 keV flux from the galactic center region. Observations were during VP 5, 11, 14, 16, 17, 21, 24 and 25. The position angle of the OSSE collimators is shown.

OSSE to detect and study discrete sources in this complex region. These observations were typically undertaken at several scan angle positions separated by 1 or 2 degrees. Background observations were usually acquired at scan positions of 10 to 12 degrees from the center of the scan region. See Purcell et al.[3] for details. Figure 5 shows the time history of the 80-150 keV emission during the first eight galactic center observations in Phase 1. Shown is the flux in the OSSE field-of-view for the scan position which was centered on the galactic center. The position angle for each of the viewing periods is shown. A position angle of ±90° corresponds to the long axis of the OSSE collimators being aligned parallel to the galactic plane. Note that the collimator was aligned along the galactic plane or close to it, with the exception of VP 14, when the long axis of the collimator was perpendicular to the plane, and VP 17 at a position angle of 32°. This results in a reduced exposure to a galactic plane component during VP 14. The lowest flux observed in these galactic center views was in VP 14, suggesting that the overall emission is dominated by a diffuse component or multiple discrete sources along the galactic plane, although it should be noted that the background observations for the VP 14 datum were also positioned on the galactic plane at galactic longitudes of ±12°. Note, also, that viewing periods 5, 16, 21, 24, and 25 all have collimator position angles of 80-90 degrees. The dramatic decrease between viewing periods 5 and 16 and subsequent increase in viewing periods 21 and 24 indicate that one or more sources are highly variable. Correlative SIGMA observations in VP 24 indicate that 1E1740.7-2942 was in a higher intensity state at that time.[20]

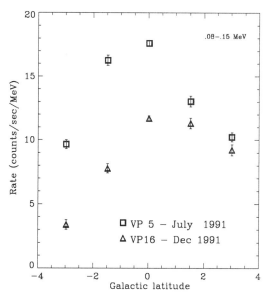

Figure 6. Comparison of the galactic latitude scan data at galactic longitude 0 degrees for VP 5 and VP 16. The observation strategies were identical for these viewing periods.

Figure 6 compares the flux vs. scan angle for viewing periods 5 and 16. These observations had identical exposures to the galactic center regions so any differences are due to variability of discrete sources in the region. The difference between the VP 16 data and the VP 5 data is shown in Figure 7. It is clear that there is a variable source near a galactic latitude of -1.5 to -2.0 degrees. Candidates for this object may be GS 1758-258[21] or 1H1743-32.[22]

OTHER GALACTIC SOURCES

Several other galactic sources that OSSE has observed in Phase 1 include Nova Cygni 1992, Cas A and the Eta Carina region. Leising et al.[23] present upper limits to the OSSE search for ^{22}Na emission from Nova Cygni. Additional observations of this source may be warranted due to the long half life of ^{22}Na and the possibility that Nova Cygni 92 was still optically thick to gamma rays at that time of the OSSE observation.[24] The Eta Carina region is interesting from several viewpoints. In addition to Eta Carina, it is a region of considerable current star formation and includes several Wolf-Rayet stars. Enhanced continuum emission is also likely from the galactic arm, thereby complicating the analysis for this confusing region. OSSE Phase 1 and Phase 2 data will be analyzed together to determine the emission, if any, from this region.

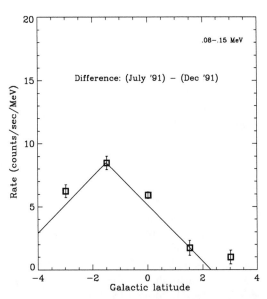

Figure 7. The difference between the VP 5 and VP 16 scans indicates a discrete source at galactic latitude -2 degrees was on during VP 5 but off, or at a lower intensity, during VP 16.

SUMMARY

OSSE has undertaken dedicated observations of about 20 galactic sources during Phase 1. Several of these have been Targets-of-Opportunity following initial discovery by BATSE, demonstrating the complementary nature of the GRO instruments. During Phase 1, multiple observations were devoted to the galactic center region in support of a number of scientific objectives and several sources in the galactic center region have been observed. Understanding the spectral and temporal characteristics of these sources will continue to improve as the spatial distribution of the diffuse component is better determined.

This work was supported under NASA DPR S-10987C.

REFERENCES

1. W.N. Johnson, et al., to be publ. Ap.J. Supp., 1993.
2. J.H. Taylor, Proc. Gamma Ray Observatory Science Workshop, 1989, **4**-143.

3. W.R. Purcell, et al., these proceedings.
4. M.P. Ulmer, et al., Proc. COMPTON Observatory Science Workshop, 1992, 253.
5. M.P. Ulmer, et al., in preparation.
6. M.S. Strickman, et al., these proceedings.
7. R.B. Wilson, et al., IAUC 5429.
8. J.P. Halpern and S.S. Holt, Nature, **357**, 222, (1992)
9. D. Bertsch, et al., Nature, **358**, 306, (1992)
10. D.A. Grabelsky, et al., these proceedings.
11. W.S. Paciesas, et al., IAUC 5580.
12. B.A. Harmon, et al., IAUC 5584.
13. A.J. Castro-Tirado, et al., IAUC 5588.
14. C.R. Shrader, et al., IAUC 5591.
15. R. Sunyaev, et al., Ap.J.,**389**,l75,(1992)
16. B.A. Harmon, et al., IAUC 5504.
17. J.C. Ling, et al., Ap.J., **321**, L117, (1987)
18. R. Fiedler, private communication.
19. J. Trumper, et al., Ann. N.Y. Acad. Sci., **302**, 538 (1977)
20. J.-P. Roques, private communication.
21. R. Sunyaev, et al., Astron. and Astrophys., **247**, L29, (1991)
22. A.M. Levine, et al., Ap.J. Supp. **54**, 581 (1984)
23. M.D. Leising, et al., these proceedings.
24. S. Starrfield, these proceedings.

EARTH OCCULTATION MEASUREMENTS OF GALACTIC HARD X-RAY/GAMMA-RAY SOURCES: A SURVEY OF BATSE RESULTS

B. A. Harmon, C. A. Wilson, M. N. Brock, R. B. Wilson, G. J. Fishman,
C. A. Meegan, W. S. Paciesas,[1] G. N. Pendleton,[1] B. C. Rubin,[2]
and M. H. Finger[3]

NASA – Marshall Space Flight Center, ES-64, Huntsville, AL 35812

ABSTRACT

The large area detectors on the Burst and Transient Source Experiment (BATSE) on the Compton Gamma Ray Observatory (CGRO) are being used to monitor about 40 sources in the energy range of 20–1800 keV and to search for transients on a daily basis. We present an overview of the sources which have been detected with the BATSE instrument, and discuss some of the properties of these sources which can be studied. Time histories of the flux and energy spectra of recently detected transient hard x-ray sources are presented.

INTRODUCTION

Table I is a list of the x-ray/gamma-ray sources which have been detected in the BATSE large area detectors using the Earth occultation technique. These sources are bright enough to be observed in one day of data with a 3σ sensitivity of 100 mCrab in the 30 to 100 keV energy band. Some indication of the temporal behavior of these objects is given along with the source name. Not all sources have been characterized as to their intrinsic variability, but some that have clear evidence of apparently random flaring behavior are: Cen X-3, 4U 1700−377, Vela X-1, Sco X-1, and OAO 1657−415. Other systems, like GX 301-2 and GS 0834−43, have regular outbursts that are correlated with the orbital period. Of the approximately 40 sources which have been monitored to date, about half of those added to the daily monitor list at various times through the mission do not show measureable flux. A sizeable fraction of these (10) are transient pulsars, x-ray novae or flare stars which have not been detected at the current sensitivity level since the start of the CGRO mission. Five transients have been observed so far; their outburst times are indicated in the table. Some notable non-detections are Cir X-1, A 0535+262 and Aql X-1. References to other presentations at this conference which incorporate occultation measurements using BATSE are indicated in Table I. In particular, detections of active galaxies are discussed by Paciesas et al.[1]

The Earth occultation technique consists of measuring the size of step-like features in the gamma-ray counting rate as a point source in the sky alternately rises above and sets below the Earth's limb. Measuring the size of the step gives a nearly continuous sampling of source intensity and spectrum as a function of time, typically 15–30 times a day. Recent improvements in the software include occultation step modeling with atmospheric extinction effects and fitting of individual steps. The improved code is now being used to reprocess large area

[1] University of Alabama in Huntsville
[2] Universities Space Research Association
[3] Computer Sciences Corporation

detector (LAD) data with an expanded source list from the beginning of the mission (Truncated Julian Date 8369), and has now reached TJD 8500. Work is continuing to understand systematic effects and to deal with source confusion in crowded sky regions.[2,3]

Sources Detected By Earth Occultation with BATSE

Crab				Transients		
Cyg X-1	(11)	++				
Cyg X-3				GX 339-4	(9)	(Jun-Oct 1991)
GX 301-2		+		GRO J0422+32		(Aug 1992- ?)
Her X-1		+		4U 1543-47		(Apr-Jul 1992)
Cen X-3	(12)		++	GRS 1915+105	(10)	(Jun 1992- ?)
4U 1700-377	(13)	+	++	Nova Muscae		(May-Sep 1991)
GS 0834-43	(14)	+				
OAO 1657-415		+	++	**Extragalactic Sources** (1)		
1E 1740-29						
GX 1+4				Cen A		
EXO 2030+375	(15)	+	++	NGC 4151		
Sco X-1	(16)		++	3C 273		
Vela X-1		+	++			

( ) Refer to other conference presentations (see references).
+ Flux variation correlated with binary orbit (regular flaring episodes or eclipses detectable).
++ Transient or aperiodic flaring episodes.

Table I. Sources detected by BATSE in the large area detectors (20–1800 keV) using Earth occultation at the current one-day sensitivity level of 0.1 Crab (3σ).

RESULTS

In Figure 1, a count rate history of the high mass x-ray binary GX 301-2 is presented. Periodic gaps in the coverage, due to orbital precession of the spacecraft, occur at times when the source is too high above the orbital plane for Earth occultation. GX 301-2 has a well-determined orbital period of 41.506 days[4] and is one of several wind-driven accretion systems observable by BATSE. X-ray flares occur at 1–2 days prior to periastron passage (time of periastron is shown by the arrows). Note, however, that the source is detected between outburst periods, illustrating the coverage which can be obtained with this technique. Comparisons of the pulsed flux from epoch-folding analysis and occultation measurements of the total flux is in progress.

BATSE TRANSIENTS

BATSE has detected several transients since launch in April 1991. These are listed in Table I. In Figure 2 we show the flux histories of two transient sources, 4U 1543-47, whose initial outburst was in April 1992, and the recently detected source GRO J0422+32. 4U 1543-47 had a risetime (10–90%) of about 2 days and e-folding decay time of 3 days in the 20–300 keV energy band. There may also be a low significance detection from about 8792 to 8860. The e-folding

316 A Survey of BATSE Results

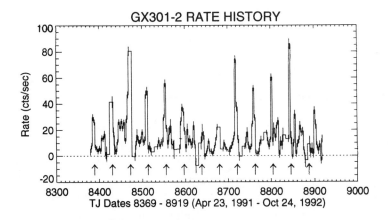

Figure 1. Count rate history (20–70 keV) for the high mass x-ray binary GX 301-2. Periastron passages (from ephemeris of Sato et al.[4]) are shown by arrows.

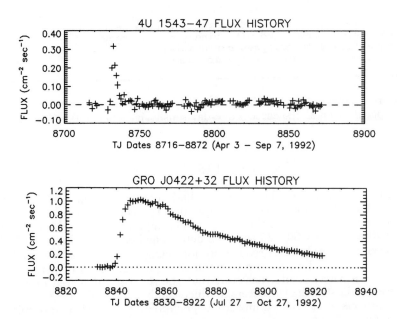

Figure 2. Flux histories (20–300 keV) for 4U 1543–47 and GRO J0422+32. The Crab nebula flux is approximately 0.3 photons/cm^2-s.

time in previous observations[5,6] of 4U 1543–47 in the 1–6 keV energy band was about 85 days. A recent measurement by ROSAT[7] showed a flux in the 0.1–1 keV band of 500 mCrab, which suggests that the decay time is highly energy-dependent. GRO J0422+32 (lower part of figure) is a newly discovered x-ray transient, or x-ray nova. It is also not found on Palomar sky plates prior to August 1992. GRO J0422+32 has a risetime of approximately 5 days, but a much longer decay time (20–300 keV band) at 41 days. No evidence of an up-turn in the flux intensity after maximum, as in the case of Nova Muscae,[8] and possibly 4U 1543–47, has been seen. We contrast the hard x-ray light curves of these sources with two other detected transients (GX 339–4[9] and GRS 1915+105[10]), which had slow rises of about 1–2 months, with no evidence of a fast-rising, exponentially-falling phase. It may be that this phase has not been detected previously for these transients, although in the case of the well-studied GX 339-4, this is unlikely.

In Figure 3 we show spectra for GRO J0422+32, GRS 1915+105, 4U 1543–47 and Nova Muscae, the latter being discovered in reprocessing of early mission BATSE data. All spectra were taken around the maximum of the outburst, except for Nova Muscae, which is from June 1991. Both Nova Muscae, 6 months after the initial outburst, and GRO J0422+32, have considerably harder spectra in the 20–100 keV band than do 4U 1543–47 and GRS 1915+105.

SUMMARY

About 22 sources have been detected on a daily basis in the BATSE data using the occultation technique. A variety of objects, both transient and persistent, have been observed, which include pulsars, black hole candidates, and active galaxies. This technique will be very valuable as a companion to pulsed flux measurements, correlated studies in other wavelength bands, and already has proven itself as an all-sky monitor.

REFERENCES

1. W. S. Paciesas et al., these proceedings.
2. B. A. Harmon et al., Compton Observatory Science Workshop, ed. C. R. Schrader, N. Gehrels, and B. Dennis (NASA CP 3137, 1992), p. 69.
3. W. S. Paciesas et al., Astr. Ap. Suppl., in press (1992).
4. N. Sato et al., Ap. J. **304**, 241 (1986).
5. H. van der Woerd et al., Ap. J. **344**, 320 (1989).
6. S. Kitamoto et al., Publ. Astron. Soc. Japan **36**, 799 (1984).
7. J. Greiner, private communication (1992).
8. R. A. Syunyaev et al., Sov. Astron. Lett. **17(6)**, 409 (1991).
9. B. A. Harmon et al., these proceedings.
10. C. Kouveliotou et al., these proceedings.
11. J. C. Ling et al., these proceedings.
12. M. H. Finger et al., these proceedings.
13. B. C. Rubin et al., these proceedings.
14. C. A. Wilson et al., these proceedings.
15. M. T. Stollberg et al., these proceedings.
16. B. McNamara et al., these proceedings.

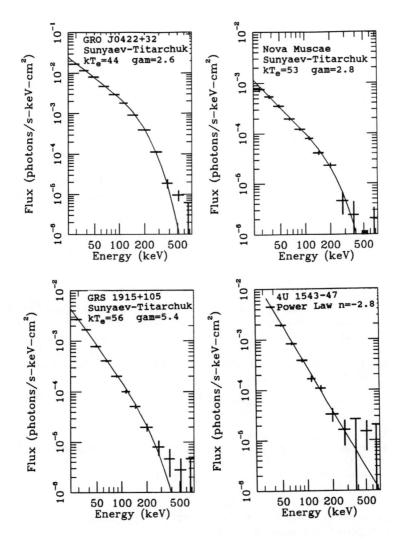

Figure 3. Energy spectra for four transients seen by BATSE. The best fit models are also shown. All sources were near maximum intensity, except Nova Muscae (six months after maximum).

DETECTION OF QUASI-PERIODIC OSCILLATIONS (QPO) FROM CYG X-1 AND GRO J0422+32

Chryssa Kouveliotou
Universities Space Research Association, Huntsville, AL 35812

Mark H. Finger[†], Gerald J. Fishman, Charles A. Meegan, Robert B. Wilson
NASA/MSFC, Huntsville, AL 35812

William S. Paciesas, Takahisha Minamitani
University of Alabama in Huntsville, AL 35899

Jan van Paradijs
Astronomical Institute "Anton Pannekoek" & CHEAF, Amsterdam

ABSTRACT

We have detected quasi-periodic oscillations (QPO) from the black-hole candidates Cyg X-1 and GRO J0422+32. The QPO are present in three separate energy bands covering 20 – 300 keV. Their average centroid frequencies are near 40 mHz for Cyg X-1, and near 30 mHz and 200 mHz for GRO J0422+32.

INTRODUCTION

We report here some preliminary results on the variability of the hard X-ray intensity of Cyg X-1 and GRO J0422+32 observed by BATSE. These results are based on Fourier analysis of the 240 s intervals of high time resolution discriminator rate data produced after each onboard trigger, and Fourier analysis of longer intervals of the continuously recorded 1.024 s time resolution discriminator rate data.

BATSE consists of eight Large Area Detectors (LADs) and eight smaller spectroscopic detectors. The combined surfaces of the LADs form a regular octahedron to provide all-sky monitoring in the 20 keV – 2 MeV energy range. Discriminator rates in four energy channels are collected with 1.024 s resolution continuously. In the burst aquisition mode, discriminator rates for four energy channels are collected with 64 ms resolution for an interval of 240 s. This burst mode is triggered when the counting rate in two or more detectors, as measured independently in time intervals of 64, 256, or 1024 ms, exceeds the average value observed in a 17.4 s window by more than 5.5 σ. Many other data types are also collected. A detailed description of the experiment has been given by Fishman et al.[1].

On average, BATSE triggers $\sim$ 4 times per day. A small fraction of the triggers is caused by emission from bright hard X-ray sources, either by the rise of the source above the Earth horizon or because of an intense fluctuation of its brightness.

A particularly active source is Cyg X-1. During the interval TJD 8450 - 8700 (TJD = JD - 2,440,000.5) it caused a total of 64 triggers, 31 of which were due to source rises and 33 due to its intensity fluctuations. The temporal

† Astronomy Programs, Computer Sciences Corporation

distribution of these triggers is not uniform; 60% of them occurred between TJD 8560 - 8600. During this interval the range of the daily average X-ray intensity of Cyg X-1 (in the 20 keV - 2 Mev energy range) was not different from that during the remainder of the interval TJD 8450 - 8700.

Another source which for some time gave rise to a large number of onboard triggers is the transient black-hole candidate GRO J0422+32 (Paciesas et al.[2]). In September 1992 the number of such triggers became so large that it was decided to raise the onboard trigger level for a few days from 5.5 σ to 10 σ.

CYG X-1

Visual inspection of the intensity curves for some of the triggered 240 s intervals showed that Cyg X-1 exhibits a lot of variability on a time-scale of the order of tens of seconds (Figure 1). This was subsequently confirmed by the results of Fourier analysis which for each of the intervals produced a Power Density Spectrum (PDS). Many of the power density spectra showed excess power, almost invariably in the frequency range between $\sim$ 30 and 50 mHz.

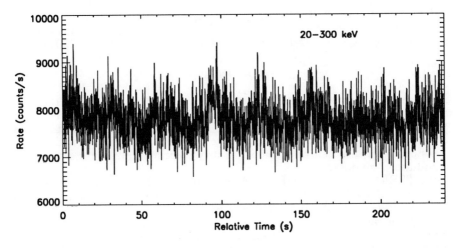

Figure 1. Light curve of the X-ray intensity of Cyg X-1 on during a 240 s interval triggered by an intense fluctuation (BATSE trigger 613). The counts are recorded with 64 ms resolution and integrated between 20 - 300 keV. Notice the 20 - 25 s fluctuations of the intensity profile.

Two examples of these power density spectra are shown in Figure 2; the first contains a strong QPO peak, with centroid frequency of 32 mHz, and FWHM = 5 mHz (as obtained from a Gaussian fit). In the second PDS no QPO peak is discernable near this frequency. The power spectra have been normalized so that the counting noise power level is 2.

From Gaussian fits to possible QPO peaks in the PDS (in the frequency range 10 - 110 mHz) we find that QPO are present in 31 of the 64 triggered intervals, with the majority of their centroid frequencies concentrated in the range 20 - 50 mHz. When present, the QPO are detected in the energy ranges

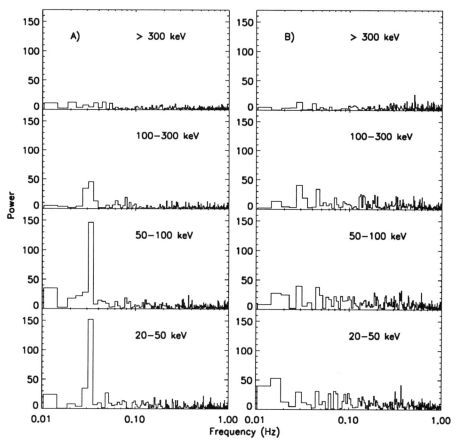

Figure 2. Power Density Spectra in four energy channels for two 240 s intervals triggered by Cyg X-1 intensity fluctuations (BATSE triggers 613 and 1094). A) The PDS exhibits a strong QPO peak in the first three energy channels centered at 32 mHz. B) No QPO is evident from the source in this data set.

20 - 50 keV, 50 - 100 keV, and 100 - 300 keV. These results show that QPO in the hard X-ray intensity, with centroid frequency near 40 mHz, are a fairly regular phenomenon in Cyg X-1 (Vikhlinin et al.[3], Kouveliotou et al.[4]). These QPO were first discovered by Frontera and Fuligni[5], but were by and large forgotten when QPO in X-ray binaries caught the attention of X-ray astronomers (i.e. after 1985, see Lewin et al.[6], and van der Klis[7] for reviews on QPOs).

GRO J0422+32

The transient X-ray source GRO J0422+32 which was discovered in August 1992 (Paciesas et al.[2]) is a low-mass X-ray binary as is apparent from the large increase in its optical brightness accompanying the X-ray outburst (Shrader

322 Detection of Quasi-Periodic Oscillations

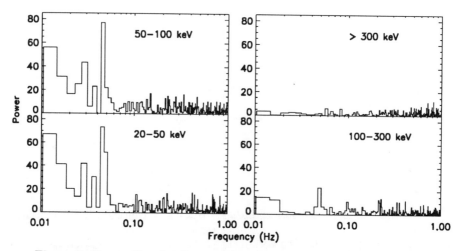

Figure 3. Power Density Spectrum in four energy channels of a 240 s interval triggered by an intensity fluctuation of the X-ray transient GRO J0422+32 (BATSE trigger 1801 on 12 August 1992). A strong QPO peak centered at 45 mHz is evident up to 300 keV in the spectrum.

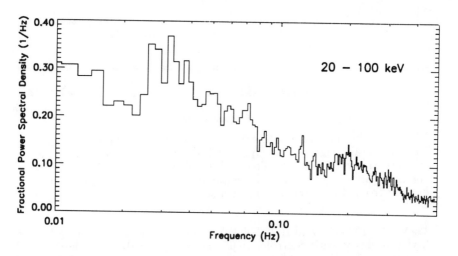

Figure 4. Average Power Density Spectrum of the transient GRO J0422+32 on 12 August 1992. This was derived by averaging the power spectra from 79 separate 262 s intervals of 1.024 s resolution discriminator rates, correcting for counting noise and binning, and normalized using the source flux determined by earth occultation. Notice the broad QPO peaks centered at 35 and 200 mHz.

et al.[8]). Its X-ray spectrum contains a hard power-law component extending to the MeV range (Grove et al.[9]), which makes the compact star in the binary a black-hole candidate (see e.g. Tanaka and Lewin[10]). During the period of maximum intensity following the onset of the outburst the source caused many onboard triggers on BATSE. To explore the fast variability characteristics of the source we have Fourier analysed a limited number of these triggered 240 s data sets. We found that in about half of their PDS a strong QPO peak near 30 mHz is present (an example is shown in Figure 3).

In addition, we Fourier analyzed one day of continuously recorded data (time resolution 1.024 s), selecting only the time intervals when the source was above the Earth horizon. A dynamic power spectrum shows that during this day the PDS also contained a variable broad peak with centroid frequency near 200 mHz. The average PDS for the whole day (12 August 1992) is shown in Figure 4.

DISCUSSION

Cyg X-1 and GRO J0422+32 have been added to the small list of black-hole candidates that have shown QPO in their X-ray intensity variations. Previously, QPO were found in LMC X-1[11] and GX339-4[12,13], and Nova Muscae[14] with centroid frequencies of $\sim$ 75 mHz, 50 mHz and 6 Hz, and 10 Hz respectively. Thus it appears that QPOs may be a common property of accreting black-holes.

ACKNOWLEDGEMENTS

JvP thanks the BATSE group at the Marshall Space Flight Center where most of this work was done, and the University of Alabama at Huntsville, for their hospitality. He acknowledges financial support from the Leids Kerkhoven Bosscha Fonds.

REFERENCES

1. G. J. Fishman et al., Proc. GRO Science Workshop (ed. W. N. Johnson, NASA/Goddard Space Flight Center, 1989), p. 39.
2. W. S. Paciesas et al., IAU Circular 5580 (1992).
3. A. Vikhlinin et al., IAU Circular 5576 (1992).
4. C. Kouveliotou et al., IAU Circular 5576 (1992).
5. F. Frontera and F. Fuligni, Ap. J. Letters **198**, 105 (1975).
6. W. G. H. Lewin et al., Sp. Sci. Rev. **46**, 273 (1988).
7. M. van der Klis, Annu. Rev. Astron. Astrophys. **27**, 517 (1989).
8. C. R. Shrader et al., IAU Circular 5591 (1992).
9. J. E. Grove et al., these proceedings (1993).
10. Y. Tanaka and W. H. G. Lewin, in "X-ray Binaries" (eds. W. H. G. Lewin, J. van Paradijs and E. van der Heuvel, Cambridge University Press, in press).
11. K. Ebisawa et al., PAS J. **41**, 519 (1989).
12. S. A. Grebenev et al., Sov. Astron. Lett. **17**, 413 (1991).
13. S. Miyamoto et al., Ap. J. **383**, 784 (1991).
14. S.A. Grebenev et al., Proc. Workshop on Nova Musacae 1991 (ed. S. Brandt, Danish Space Research Institute, 1991), p. 19.

ULTRAVIOLET, OPTICAL, AND RADIO OBSERVATIONS OF THE TRANSIENT X-RAY SOURCE GRO J0422+32

C.R. Shrader
Laboratory for High Energy Astrophysics, NASA Goddard Space Flight Center and Computer Sciences Corporation

R.M. Wagner
Department of Astronomy, Ohio State University

S.G. Starrfield
Department of Physics and Astronomy, Arizona State University

R.M. Hjellming and X.H. Han
National Radio Astronomy Observatory

ABSTRACT

We have monitored the early evolution of the transient X-ray source GRO J0422+32 from approximately 2 weeks post-discovery and into its early decline phase at ultraviolet, optical, and radio wavelengths. Optical and ultraviolet spectra, obtained with the Perkins 1.8-m telescope and with IUE respectively, exhibit numerous, but relatively weak, high-excitation emission lines such as those arising from He II, N III, N V, and C IV superposed on a very blue continuum. We find that the ultraviolet and optical characteristics of GRO J0422+32 as well as its radio evolution, are similar to other recent well-observed X-ray novae or soft X-ray transients such as Cen X-4, V616 Mon, and Nova Muscae 1991, which suggests that GRO J0422+32 is also a member of that subclass of low mass X-ray binaries (LMXBs). Further observations as GRO J0422+32 declines into quiescence will be required to determine whether the compact object is a neutron star or a black hole.

INTRODUCTION

The BATSE experiment on-board the Compton GRO reported the detection of a new hard X-ray transient in the direction of the constellation Perseus on 5 August 1992[1]. Within a week of the discovery, the source increased in strength and reached a peak intensity of 3 Crab in the 20-300 keV range. Subsequent observations using the OSSE collimated instruments provided a positional error box sufficiently small to facilitate searches for an optical counterpart. An optical identification was made on 15 August 1992[2]. We subsequently confirmed this identification spectroscopically in the optical and ultraviolet on 16 August[3,4].

In this contribution, which describes work performed as part of a Phase 2 *Compton* Guest Investigator Program, we describe some preliminary results of a multi-wavelength campaign to monitor the evolution of GRO J0422+32 from approximately 10 days post-outburst into its early decline phase. We describe our observations and data analysis, offering preliminary interpretation and drawing comparison between the early development of GRO J0422+32 and other recent X-ray novae.

OBSERVATIONS

Optical spectroscopy and photometry of GRO J0422+32 have been obtained with the

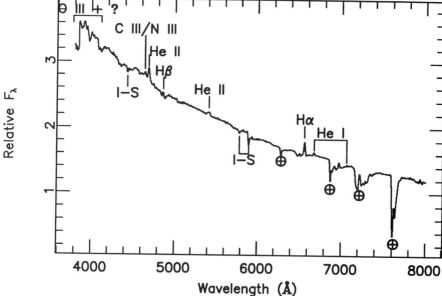

Figure 1. Composite optical spectrum.

Perkins 1.8-m telescope of the Ohio Wesleyan and Ohio State Universities and with the J.S. Hall 1.1-m telescope at the Lowell Observatory beginning on 16 August, when we confirmed the optical identification by Castro-Tirado et al.[2], and continuing through mid-October. Spectroscopy has been obtained on 15 nights using the Ohio State University CCD spectrograph on the Perkins telescope. Coverage includes the spectral region $\lambda\lambda 3300$-8500 at a resolution of approximately 10Å and the Hβ region, including He II $\lambda 4686$, at a resolution of approximately 2Å, the latter allowing us to examine the temporal behavior of the line profiles and to search for radial velocity variations. In addition, time-resolved differential photometry was first obtained in the V-band on 30 August using the Lowell Observatory RCA CCD camera at the Hall telescope simultaneously with high dispersion spectra at the Perkins telescope. Additional V-band differential photometry was obtained on 15, 16 and 17 September at the Perkins telescope.

Promptly, following the discovery of GRO J0422+32, we activated a target of opportunity guest observer program using the IUE. The long and short wavelength spectrographs were used in low resolution mode to obtain ultraviolet coverage between $\lambda\lambda 1150$-3250 at 5-8Å resolution. The first ultraviolet observations were obtained 1.6 days after the announcement of an optical counterpart and only 5 hours after the first optical spectra. We subsequently sampled the source at 5-10 day intervals over the next 2 months.

We observed GRO J0422+32 with the VLA on 13, 14, 17, 29 August; 1 September 1, 5 October and 25 November. Coverage included four frequency bands centered on 1.4, 4.9, 8.4 and 15 Ghz. For the epochs which preceded the optical identification, the beam center was aligned with the best available error-box center[5]. The source was thus within the half-power beam width for the lowest two frequencies.

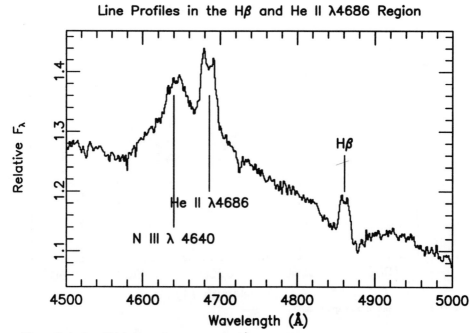

Figure 2. Averaged high dispersion spectrum (Δ≈2Å) obtained on 30 August 1992.

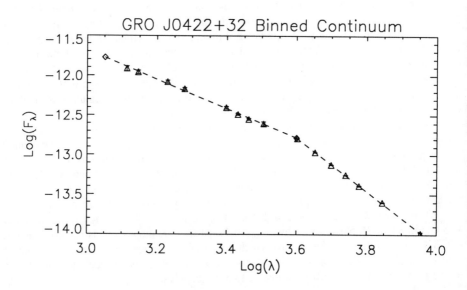

Figure 3. UV-optical binned continuum measurements. Two powerlaw fit with break at 4000Å overlaid.

RESULTS AND DISCUSSION

In Figure 1, we present a composite optical spectrum formed by averaging the individual spectra obtained on 6 nights between 16 August and 25 September. The outburst and early decline spectra of GRO J0422 are dominated by emission lines of the Balmer series of hydrogen; He I $\lambda 5876$, $\lambda 6678$, and $\lambda 7065$; He II $\lambda 4686$ and $\lambda 5411$; the C III/N III blend $\lambda\lambda 4640\text{-}4650$; N III $\lambda 4541$; and numerous lines due to O III in the $\lambda\lambda 3700\text{-}3900$ region arising from the Bowen mechanism. Weaker emission from C IV near $\lambda 5800$ may also be present. Numerous interstellar absorption lines are also present including the diffuse bands at $\lambda 4430$ and $\lambda 5780$ as well as NaD $\lambda\lambda 5890\text{-}5896$. The mean equivalent width of the $\lambda 5780$ interstellar feature implies that $E(B-V)=0.22\pm 0.11$.

The average of the high dispersion spectra in the Hβ region obtained on 30 August is shown in Figure 2. The Balmer lines appear as broad absorptions with double-peaked emission cores. The absorption wings extend approximately ± 2500 km s^{-1} from line center. The line profile of He II $\lambda 4686$ also shows a double-peaked profile and suggests that GRO J0422+32 might be viewed at a relatively high inclination. In our time resolved spectra, variations in the profile of He II $\lambda 4686$ on time scales of half an hour are quite evident.

Time resolved differential light curves of GRO J0422+32 in the V-band show stochastic flickering on a time scale of minutes or less with an amplitude of 5-10% as well as broader variations of similar amplitude on time scales of tens of minutes. Hourly trends in the photometry are also evident. These light curves are similar to those observed in cataclysmic variables and LMXBs.

The ultraviolet continuum has been used to obtain an independent estimate of the source reddening. We applied the galactic reddening law of Seaton[6] varying the color excess E(B-V) to remove the 2200Å feature. In this manner we derived $E(B-V)=0.40\pm 0.06$, which is discrepant with the value derived from measurement of interstellar lines. We are investigating this discrepancy. The de-reddened UV continuum exhibits a substantial blue excess which is approximated by a $\alpha \approx 2.0$ powerlaw (where $f_\lambda \propto \lambda^{-\alpha}$). There is a distinct break at approximately 4000Å with a two powerlaw fit required to satisfactorily fit the combined UV-optical continuum (Figure 3). The powerlaw index is somewhat sensitive to E(B-V), being steeper for larger color excesses; however, even use of our 1-σ upper limit leads to a continuum slope of $\alpha=2.2$; it also seems to steepen in the optical.

The shape of the UV continuum is similar to that seen in outburst spectra of the transient Cen X-4[7]. It is inconsistent with a simple steady-state accretion disk ($\alpha=2.3$), as has been noted in many dwarf novae outbursts and also in Nova Muscae 1991[8,9]. One must use caution however in interpretation; an extrapolation of the simple disk models, fitted to the UV data, to the soft X-ray band exceeds the measured flux by an order of magnitude in both Nova Muscae and Cen X-4. It is likely that the steady-state thermal accretion disk scenario must be supplemented to include effects such as X-ray illumination of the disk, and depending on the viewing geometry absorption of the UV continuum, perhaps by cool material in an outer toroidal extension of the disk. Some of the line profiles seen in the optical are suggestive of a highly inclined system, furthermore, the UV spectra exhibit a complex structure of absorption and/or a pseudo-continuum of blended emission line, particularly between about $\lambda\lambda 1250\text{-}1800$. Also, in an edge on viewing situation, the effects of gravitational focusing can affect the observed spectrum, particularly for Kerr geometries[10].

UV emission lines include N V $\lambda 1240$, C IV $\lambda 1550$, He II $\lambda 1640$, and possibly N IV $\lambda 1486$, each of which are typical of LMXB sources observed by IUE. The line-to-

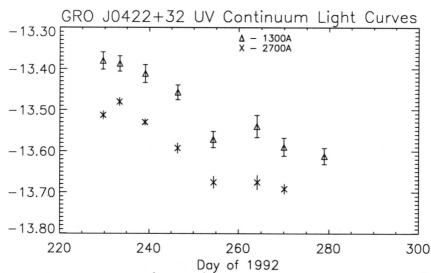

Figure 4. UV light curves for 1300Å and 2700Å continuum. Vertical axis is log(F), ergs/cm²/sec/Å.

continuum flux ratio however is lower than in Nova Muscae 1991 or Cen X-4. The C IV feature has a width of about 1400 km/sec, which is only marginally resolvable above instrumental broadening.

Pavlenko et al[11]. first suggested the presence of a weak star on the POSS red (E) print with a magnitude of about 16.5±0.7 and a clustering of grains on the blue (O) print near their measurement of the source location. Later, Mueller reported an outburst position, but noted that the object *did not* appear on POSS prints[12]. We examined the Lowell Observatory copy of the POSS O and E prints at the location of GRO J0422+32 and confirm the identification of an object near the location of the outburst position on the E print, but with a magnitude that is near the plate limit of about 20 and not 16.5 mag as first suggested. An object is also present on the O print at the same location and near the plate limit of about 21 mag, but its appearance is marred by a defect. We performed astrometry of the candidate on the E print and find that the position of the object is identical to the outburst position within the errors. We conclude that the outburst amplitude was about 7 magnitudes.

Distance estimates at present are highly speculative, but given an extinction of $A_V \approx 1.25$ based on the color excess derived from removal of the 2200Å feature, the intrinsic apparent magnitude of the progenitor is approximately 19. For a mid-K dwarf, this would imply a distance of about 2.4 kpc. The optical-UV luminosity near outburst peak is $2.5 \times 10^{36} (d_{2.4})^2$. By comparison the outburst UV-optical luminosity of Cen X-4 was $1.4 \times 10^{36} (d_{2.2})^2$ ergs/cm²/sec[7].

Light curves for several UV continuum bands are shown in Figure 4. The light curve is initially flat, followed by a decay with time-scale of about 40 days. A sharp upturn in the 1300Å light curve occurs approximately 50 days post-outburst with an associated flattening in the 2700Å band. The discontinuity has been seen in the soft X-ray decay light curves for various transients, and it was seen in the UV light curve of Nova Muscae 1991, where a similar hardening of the UV continuum occurred[8]. This type of behavior may be difficult to reconcile with the induced mass-loss model where the cool, outer disk region should respond first, while the time-scales involved may be hard to understand in the context of the accretion disk instability model.

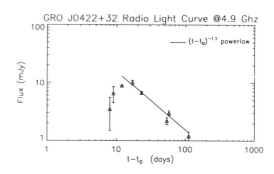

Figure 5. Radio light curve at 4.9 Ghz. Vertical axis is flux in mJy, horizontal is days post-outburst.

The radio spectrum is relatively flat, with fluxes in the 5-10 mJy range at about 10 days post-outburst. The decay light curve is shown in Figure 5. The flux rises slowly from mid-to-late August, and then decays in a manner approximated by a $(t-t_0)^{-1.0}$ powerlaw. The model is overlaid for comparison. This is similar to the decay light curve seen in V404 Cyg following the decay of its initial fast radio-transient[13], but our light curve is sparsely sampled.

REFERENCES

1. Paciesas, W.S. et al. 1992, *IAU Circ.* no. 5580.
2. Castro-Tirado, A.J., et al. 1992, *IAU Circ.* no. 5588.
3. Wagner, R.M., Bertram, R. Starrfield, S.G. and Shrader, C.R. 1992, *IAU Circ.* no. 5589.
4. Shrader, C.R., Wagner, R.M. and Starrfield, S.G. 1992, *IAU Circ.* no. 5591.
5. Harmon, B.A. et al. 1992, *IAU Circ.* no. 5584.
6. Seaton, M.J. 1979, *M.N.R.A.S.* **187**, 73p.
7. Blair, W.P., et al. 1984, *Ap. J.* **391**, 298.
8. Szkody, P. 1985, in *proc. Recent Results on Cataclysmic Variables*, Bamberg: ESA),p.39.
9. Shrader, C.R., et al., 1993 *Ast. and Astrop. Supp.*, (in press).
10. Czerny, B., Czerny, M., and Grindlay, J.E. 1986, *Ap. J.*, **311**, 241.
11. Pavlenko, E.P., Shlyapnikov, A.A., and Castro-Tirado, A.J., 1992, *IAU circ.* no. 5594.
12. Mueller, J. 1992, *IAU circ.* no. 5597.
13. Han, X. and Hjellming, R.M. 1990, in *Accretion-Powered Compact Binaries,* ed. C.W. Mauche, Cambridge Univ. Press, 25.

DISK OSCILLATIONS AND THEIR APPLICATION TO GRO J 0422+32

Chuan Luo and Edison Liang
Department of Space Physics and Astronomy, Rice University
Houston, TX 77251-1892

ABSTRACT

We have performed calculations on accretion disks using the formula bridging the optically thin case and optically thick case proposed by Liang and Wandel[1]. Using the α model, the steady structure solutions fall into two groups, one for optically thin, hot solutions and the other for optically thick, cold solutions. We notice that there are propagating modes corresponding to relatively short wavelength only in the cold solutions with a typical frequency ranging from several Hz to tens of Hz and that their frequencies change with accretion rate. This result may be applied to the QPOs observed in x-ray sources and thus sheds some light on the disk structure itself.

INTRODUCTION

As we know, a black hole can accrete its surrounding material through viscosity-induced angular momentum transport. We adopt the α disk model, which assumes that the viscous stress is proportional to the total pressure of the disk and that the disk is cooled through radiation transfer (though some other cooling mechanisms may be nonnegligible).
Our basic assumptions are:
    a. The disk is geometrically thin (half thickness H is much smaller than radius R).
    b. Hydrodynamic time scale is much shorter than thermal time scale and radial drift time scale; thermal time scale is much shorter than radial drift time scale[2].
    c. α disk assumption: viscous stress is proportional to total pressure p of the disk.
    d. The disk is cooled through radiation transfer.
    e. Electron scattering is the major source of opacity in the fully ionized proton electron plasma.
    f. Electrons and ions are thermal and coupled by Coulomb collisions.

BASIC EQUATIONS

Now we consider an accretion disk with half thickness H, surface density Σ around a black hole of mass M. Under the above assumptions, we have the energy equation[3]

$$\frac{\partial(\varepsilon H)}{\partial t} + p\frac{\partial H}{\partial t} = Q^+ - Q^- \qquad (1)$$

and the mass-transport equation

$$2\pi R\frac{\partial \Sigma}{\partial t} = \frac{\partial \dot{M}}{\partial R} \qquad (2)$$

where the accretion rate $\dot{M}=-2\pi R\Sigma V_r$.

Radial drift velocity V_r is determined by the angular momentum transport equation. For quasi-keplerian orbit, we have

$$V_r = -\frac{2}{\omega R^2 \Sigma}\frac{d}{dR}\left(W_{r\varphi}R^2\right)$$

and $W_{r\varphi}=2\alpha Hp$ is the shearing stress which is proportional to the total pressure p, while the pressure $p=\frac{1}{6}\Sigma\omega^2 H$ is determined by the balance between pressure gradient and tidal force in vertical direction. Internal energy density is

$$\varepsilon = \frac{3\rho k(T_i+T_e)}{2m_p} + 3p_r = \frac{3\rho k(T_i+T_e)}{2m_p} + \frac{3Q^-\tau_R}{c}$$

Total pressure

$$p = \frac{\rho k(T_i+T_e)}{m_p} + \frac{Q^-\tau_R}{c}$$

Heating is, however, limited by the ion-electron Coulomb coupling rate:

$$Q^+ = \frac{3}{2}C_{ie}H\rho^2\frac{kT_e^{\frac{3}{2}}}{m_p}(T_i-T_e)\left(1+\sqrt{\frac{kT_e}{m_e c^2}}\right) \qquad (3)$$

where $C_{ie}=4.8\times 10^{22}$ in cgs units.

The dissipation rate is

$$Q^+ = -\frac{1}{2}W_{r\varphi}R\frac{d\omega}{dR}$$

For the cooling rate, we use the interpolative formula proposed by Liang and Wandel (1991), which is valid in both optically thick and optically thin cases:

$$Q^- = \frac{4\sigma T_e^4}{3\tau_R\left[1+(e^{\tau_*}-1)^{-1}A^{-1}\tau_*^{-1}\frac{F_{cb}}{F_{cb}+F_{cs}}\right]} - F_s$$

where τ_R is the Rosseland mean optical depth while F_{cb} and F_{cs} are the Comptonized bremsstrahlung and soft photon flux, respectively.

$$\tau_* = \sqrt{\tau_{es}\tau_{ff}} = 1.72\times 10^{12} H\rho^{\frac{3}{2}} T_e^{-\frac{7}{4}}$$

is the Planck mean effective optical depth and A is the luminosity enhancement factor of Comptonized thermal bremsstrahlung.

ANALYSIS AND RESULTS

To get the steady structure solution for the disk, we set all the equations time-independent and solve them numerically.

All the steady structure solutions we get fall into two groups, one corresponding to the optically thick, cold solution and the other to the optically thin, hot solution. They merge together as the accretion rate $\dot{M}$ rises to a certain point, above which no steady structure exists. However, if the angular momentum transport is less effective and thus the dissipation rate decreases, this limit can exceed the Eddington limit. The thickness of both branches increases with accretion rate. The surface density of the disk, Σ, increases with $\dot{M}$ all the way to the critical accretion rate for the hot branch, while for the cold branch, it increases to a maximum with increasing accretion rate and then decreases to merge with the hot branch at this limit. See Fig. 1 and Fig. 2 for different cases.

By plotting the accretion rate $\dot{M}$ against the surface density Σ, we can study the stabilities of accretion disks. From the disk equations, we can prove that the secular stability condition is simply

$$\frac{d\dot{M}}{d\Sigma} > 0$$

and each passage through a maximum or minimum of $\dot{M}$ corresponds to a shift from secularly unstable solution to secularly stable solution, while each maximum or minimum Σ corresponds to a shift not only in secular stability, but also in thermal stability.

To get the numerical results of the disk oscillations, we now can perturb the disk by a small sinusoidal displacement on the disk thickness and surface density, i. e., $\sin(R/\Lambda)\exp(\Omega t)$ type perturbations, with the restriction $H<\Lambda<R$.

The perturbation is oscillatory (Ω is imaginary) only when the two unstable modes (real $\Omega>0$) merge into a single mode at short wavelengths (cf. Figs. 3 and 4). The calculated oscillation frequency is around several Hz to tens of Hz for typical keV black-body-like component, but can become very small around maximum accretion rate (see Figs. 3-5 for a gradual change in oscillation frequency), when the disk is optically thin and hot, which may be relevant to the very low QPO frequency at 100 keV of GRO J 0422+32.

Once the accretion rate passes the critical point for stability, we expect the disk oscillation to saturate at a certain nonlinear magnitude. This also predicts the millisecond time delay in the QPO high energy tail due to the Comptonization of soft photons from the outer part of the disk. Another interesting result is that the oscillation frequency decreases almost linearly as a function of α and this can at least be used to restrict the value of α to a pretty narrow range by observations. (Figs. 6 and 7 are plotted near the critical accretion rate.)

From the numerical results for the growth rate and the oscillation frequency which is right at the range of the observed QPO frenquencies and has a similar dependence on the luminosity (accretion rate) and the spectral hardness of the observed photons, we can hope to determine the viscous parameter α., which can be checked against the value obtained from steady disk models.

REFERENCES

1. Liang, E. P., & Wandel, A. 1991, ApJ, 376, 746
2. Lightman, A. P., & Eardley, D. M. 1974, ApJ, 187, L1
3. Shakura, N. I., & Sunyaev, R. A. 1976, MNRAS, 175, 613

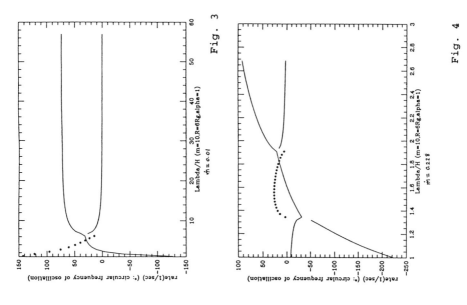

Fig. 1

Fig. 3

$\dot{m}$-Σ plot for different cases

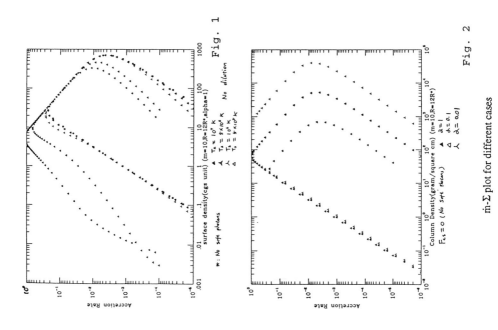

Fig. 2

Fig. 4

Growth rates and oscillation frequencies for different accretion rates

334 Disk Oscillations

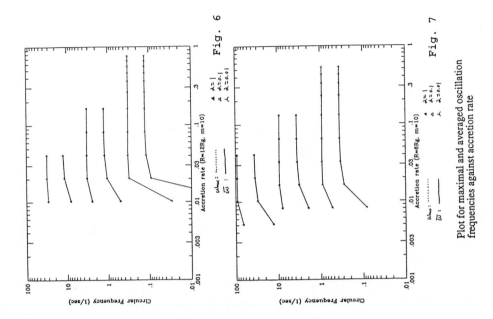

Fig. 6

Fig. 7

Plot for maximal and averaged oscillation frequencies against accretion rate

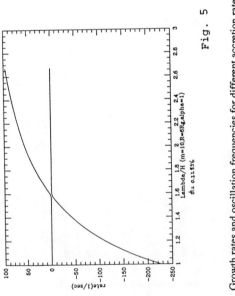

Fig. 5

Growth rates and oscillation frequencies for different accretion rates

COMPTEL OBSERVATIONS OF CYGNUS X-1

M. McConnell, A. Connors, D. Forrest, J. Ryan
Space Science Center, University of New Hampshire, Durham, NH 03824

W. Collmar, R. Diehl, V. Schönfelder, H. Steinle, A. Strong
Max Planck Institute for Extraterrestrial Physics, 8046 Garching,
Federal Republic of Germany

H. Bloemen, R. van Dijk[*], W. Hermsen, L. Kuiper, B. Swanenburg
SRON-Leiden, 2300 RA Leiden, The Netherlands

C. Winkler
Astrophysics Division, European Space Research and Technology Center,
2200 AG Noordwijk, The Netherlands

ABSTRACT

We report on the latest results from an analysis of data on Cygnus X-1 as collected by the COMPTEL experiment on the Compton Gamma-Ray Observatory (CGRO). COMPTEL detected the source around 1 MeV during two separate one-week observations in 1991. The measured flux levels indicate variability on a time scale of weeks. The results also indicate flux levels which are well below the levels reported by some investigators for an excess of emission in the energy range $\geq$ 1 MeV.

INTRODUCTION

The COMPTEL experiment on the Compton Gamma-Ray Observatory is designed to measure gamma-radiation in the energy range from 0.75-30 MeV. This is a particularly interesting energy range for the black-hole candidate Cygnus X-1. Various observers have reported detections of excess emission at energies around 1 MeV beyond that which would be expected based on an extrapolation of the hard x-ray spectrum[1,2,3]. The reported flux levels are more than two orders of magnitude above the COMPTEL detection threshold, suggesting that COMPTEL is capable of studying such emission in detail. During the first year of CGRO orbital operations, Cygnus X-1 was within the COMPTEL field-of-view on two separate occasions. The first observation took place in early June of 1991. The second observation took place some 1-1/2 months later, in August, 1991. Cygnus X-1 is clearly seen in data collected during both observations, with a weaker signal detected during the first observation. Here we shall present the latest results from the analysis of these data along with a comparison between the COMPTEL data and earlier measurements at these energies.

---

[*] Astronomical Institute "Anton Pannekoek", University of Amsterdam

OBSERVATIONS

During the first year of orbital operations, the COMPTEL experiment observed the Cygnus region on two separate occasions. The first observation, which was part of the planned 15-month sky survey, took place from May 30 to June 8, 1991 (Viewing Period 2.0). Although planned as a full two-week exposure, this observation was interrupted after only nine days due to the declaration of a solar target-of-opportunity for CGRO. A second opportunity to observe the Cygnus region came only 1-1/2 months later, when another target-of-opportunity was declared to observe Cygnus X-3 (which had recently exhibited an intense outburst of radio emission). This observation (Viewing Period 7.0) lasted from August 8 to August 15, 1991. The net result was a total exposure which exceeded the standard 14-day observation. More importantly, these data provided an opportunity to study time variability during the intervening 1-1/2 month time interval.

During the early phases of the GRO mission, the COMPTEL instrument was not operating at full capacity. This was due primarily to out-gassing effects which were noted in certain detector modules. For this reason, the affected modules remained turned off for most of the early part of the mission. During V.P. 2.0, the instrument configuration changed frequently. For the purposes of the present anlysis, only eight days of data were used. Three of the 14 D2 modules were not active during this period; this amounts to a reduction in overall COMPTEL efficiency by 21%. During V.P. 7.0, one D1 and one D2 module were inactive; this is equivalent to a reduction of about 20% in the overall COMPTEL efficiency.

ANALYSIS

The analysis of the COMPTEL imaging data for these two observation periods has been performed using the maximum likelihood method.[5] This method provides quantitative information regarding the source location and flux. For the present analysis, no independent estimate of the background (predominantly instrumental in origin) was available for the maximum likelihood analysis. Therefore, an estimate of the background was derived directly from the source data by an averaging technique which suppresses point-source signals, but preserves the general background structure.

For the present analysis, the response information comes from a COMPTEL simulation model. The simulation model is based on the CERN GEANT code and is used to simulate a large number of events which can then be used to define the PSF of the COMPTEL instrument. The present results are somewhat limited in terms of the number of simulated events which have been used to generate the PSFs. Therefore, it can be expected that the present results may be somewhat modified as the PSF statistics are improved with additional simulations.

It is also important to ensure that the PSF be defined for a spectrum which accurately represents the spectrum of the observed source. For the results described here, the PSFs were generated for an input Wien-type spectrum, which represents the high energy limit of the Sunyaev-Titarchuk inverse-Compton spectrum.[6] In addition to the normalization, this spectrum has only a single parameter - the electron temperature (kT) of the accreting plasma. Given the relatively high energy threshold of the COMPTEL data, the resulting observations are relatively insensitive to the Compton scattering optical depth.

RESULTS

In Figure 1 we show a maximum likelihood map of the Cygnus region as derived from the 0.75-1.0 MeV data collected during V.P. 7.0. Cygnus X-1 dominates the image. There is no apparent signature in the direction of Cygnus X-3; this is important to note, given that V.P. 7.0 was also the Cygnus X-3 target-of-opportunity. Upper limits for Cygnus X-3 have not yet been derived from these data.

Positive flux measurements are found for both observation periods in the range of 0.75-3.0 MeV. Only upper limits are available at higher energies. The flux levels found in both the 0.75-1.0 MeV band and the 1.0-3.0 MeV band are about a factor of two lower during V.P. 2.0 than those in V.P. 7.0.

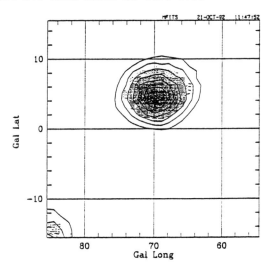

Fig. 1. Maximum likelihood image of the Cygnus region for the 0.75-1.0 MeV data collected during Viewing Period 7.0. The response from Cygnus X-1 (at $l=71.3°$, $b=3.1°$) stands out clearly whereas there appears to be no evidence for emission from Cygnus X-3 (at $l=79.9°$, $b=0.7°$).

For V.P. 7.0, we have crudely estimated the electron plasma temperature (kT) by comparing the measured flux ratio in the two lowest energy bands with that expected for the Wien spectrum used in the corresponding PSFs. It is found that the derived flux ratio agrees fairly well with that predicted for an electron temperature in

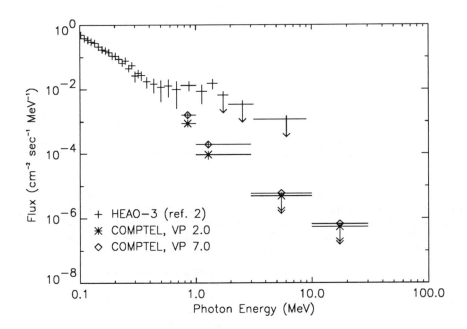

Fig. 2. The COMPTEL spectra compared with the γ_1 spectrum of HEAO-3. COMPTEL data below 3 MeV are plotted based on an 80 keV Wien spectrum.

the range of 80-100 keV. This is consistent with many previous measurements of the Cygnus X-1 electron temperature at hard x-ray energies.[7] Figure 2 shows the two COMPTEL measurements relative to the γ_1 spectrum as measured by Ling et al. (1987). The COMPTEL measurements are more than one order of magnitude below the HEAO-3 data.

Several relevant balloon observations are shown in Figure 3 along with the COMPTEL results for V.P. 7.0 (the higher of the two COMPTEL flux measurements). As with the HEAO-3 data, the COMPTEL data points fall far below some of these data. The UCR Compton telescope upper limits are, however, compatible with the COMPTEL results.

DISCUSSION AND SUMMARY

The present COMPTEL results are interesting in that they permit a high-sensitivity measurement of the Cygnus X-1 spectrum to energies

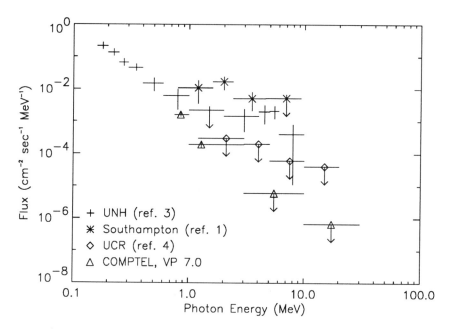

Fig. 3. The COMPTEL spectra compared with various balloon measurements. COMPTEL data below 3 MeV are plotted based on an 80 keV Wien spectrum.

above 1 MeV. These data show no evidence for any excess emission near 1 MeV as has often been reported in the past. It is interesting to note that during Viewing Period 2.0, the OSSE spectral data indicate that the hard x-ray flux was in a relatively low state.[8] More specifically, it was in a level comparable to that of the γ_1 state as defined by the HEAO-3 measurements[2]. The excess of emission near 1 MeV observed by HEAO-3 was found during a correspondingly low level of hard x-ray flux. The fact that the COMPTEL results show no evidence for any excess emission suggests that the low hard x-ray flux may be a necessary condition for MeV emission, but it is certainly not a sufficient condition.

REFERENCES

1. R. Baker et al., Nature **245**, 18 (1973).
2. J. Ling et al., Ap. J. (Letters) **321**, L117 (1987).
3. M. McConnell et al., Ap. J. **343**, 317 (1989).
4. R. White et al., Nature **284**, 608 (1980)
5. H. de Boer et al., in "Data Analysis in Astronomy IV", ed. V. di Gesù et al. (Plenum Press, 1992), p. 241.
6. R. Sunyaev and L. Titarchuk, Astron. Astrophys., in press (1992).
7. A. Owens and M. McConnell, Comments in Astrophys., in press (1992).
8. W. Johnson et al., Astron. Astrophys., in press (1992)

LONG-TERM TEMPORAL AND SPECTRAL VARIATION OF CYGNUS X-1 OBSERVED BY BATSE

J. C. Ling, N. F. Ling, R. T. Skelton,* and Wm. A. Wheaton
Jet Propulsion Laboratory
California Institute of Technology, Pasadena, CA 91109

B. A. Harmon, G. J. Fishman, C. A. Meegan, and R. B. Wilson
NASA/Marshall Space Flight Center ES-64, AL 35812

W. S. Paciesas and G. N. Pendleton
Dept. of Physics, University of Alabama, Huntsville, AL 35899

B. C. Rubin
Universities Space Research Association
NASA/Marshall Space Flight Center ES-64, AL 35812

ABSTRACT

Preliminary BATSE results for Cygnus X-1, analyzed by Earth occultation with the standard Mission Operations System (MOPS), show dramatic temporal variations of the broad-band (20—2000 keV) emission with time scale of the order of days. We report here preliminary results of a study of the temporal and spectral variability of the source using the newly developed "Enhanced Earth Occultation Analysis" developed at JPL.[1] From 25 October to 27 November 1991, variations in spectral form between a Comptonized and power-law shape were noted. The average photon flux in the 45—140 keV band is near the γ_2 level.

INTRODUCTION

Cygnus X-1, the premiere galactic black hole candidate, is well known to exhibit complex temporal and spectral variability.[2] The historical record is unfortunately fragmentary; however, the Burst and Transient Source Experiment (BATSE) on the Compton Gamma Ray Observatory (CGRO) is providing nearly uninterrupted long-term monitoring of Cygnus X-1 (as well as other sources) by the Earth occultation technique. The power of the Earth occultation technique is perhaps best illustrated by the number of significant results[3,4,5,6] obtained by the BATSE Mission Operations team with a fairly simple model. The JPL authors are developing a more physical multi-component background model[1] and other analysis improvements which improve the sensitivity by allowing more of the data to be utilized.

---

*National Academy of Sciences/National Research Council Senior Associate

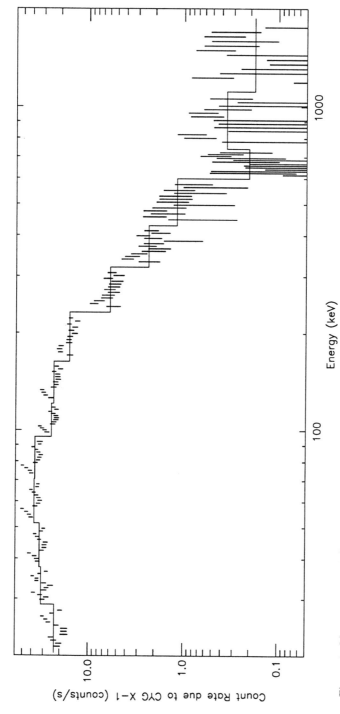

Figure 1. Measurements of Cygnus X-1 count rates for each of 13 consecutive days, in 14 energy bands. Each vertical bar represents the count rate in the energy bin for one day. These error bars include non-statistical effects. Temporal variations are visible in these data. See text.

342 Cygnus X-1 Observed by BATSE

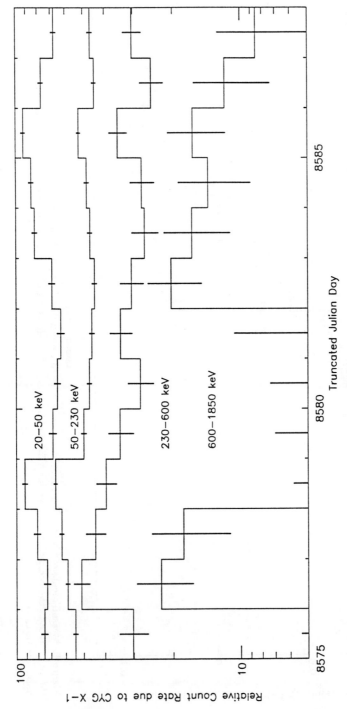

Figure 2. Count rates for Cygnus X-1 during a 13-day period in four broad energy bands. The outbursts in the 600–1850 keV band do not correspond to any parallel activity in the lower 2 energy bands, so they clearly represent a hardening of the spectrum. The high energy outburst on day 8576 is paralleled in the 230–600 keV band, but the one starting on day 8582 does not have any particular correlation with any of the other three bands.

METHOD

BATSE continuous* data are received in 16 different energy bands at 2-second time resolution from the Large Area Detectors. For purposes of our analysis, the data have been re-binned to 16-second resolution and the inner 14 energy bands are analyzed. The data are fit to a physical model[1] which estimates the count rate due to Cygnus X-1 for each energy band. At lower energies, statistically significant determinations can be made at each rise/set edge, twice per orbit; in the 1100–1850 keV band, significant answers require one day's data. For the present work, we have analyzed an entire day's data by a single fit for each energy band to get a single set of count rate estimates for the day. These count rates are to be deconvolved with the detector energy and angular response characteristics to yield source photon spectra. However, deconvolution is not yet fully operational at JPL. Hence we present our preliminary assessment of the spectral and temporal variability of the source emission based on count rates rather than source photon spectra. The variabilities are clearly visible in the count rate spectra.

RESULTS AND ANALYSIS

Preliminary analysis for Cygnus X-1 has been completed for data obtained during the period 25 October 1991 to 27 November 1991 (Truncated Julian Dates 8554 to 8587). In this period the Compton Observatory was in four different orientations, with the first three lasting for one week each and the fourth lasting for two weeks. Each dataset has been treated separately so that detector response characteristics enter identically in points to be compared. All datasets show evidence of temporal and spectral variation on the time scale of days, with the clearest evidence in the last and longest dataset. Figure 1 shows the count rate spectrum for each of 13 consecutive days, TJD 8575 to 8587. The histogram shows average count rates in each of the 14 energy bands. Each vertical bar represents the $\pm 1\sigma$ values for an individual day, and these are plotted sequentially within each energy band. The error bars include a non-statistical contribution, which is dominant at lower energies and has been estimated from reduced χ^2 values for the Crab.[1] A number of planned refinements to the analysis method are expected to reduce this non-statistical contribution significantly. Energy bands spanning 38 keV to 160 keV show a day-by-day increase for the first four days followed by an abrupt drop. This trend is not paralleled at high energy, where apparent flaring activity occurs at different times. In Figure 2, the 14 energy bands have been combined into 4 broader ones, and the count rate for each broader energy band is shown for the 13-day observation period. Error bars again include non-statistical scatter. For plotting purposes, count rates have been scaled arbitrarily so that low to high energies proceed from top to bottom of the plot. A steady increase over

---

* *I.e.*, not interrupted by burst trigger

the first four days is visible at low energies, but not in the 230—600 keV band. The 600–1850 keV band shows flaring on days 8576 to 8577, and again on days 8582 to 8586. The flaring is not paralleled at lower energy and clearly indicates a hardening of the spectrum on these days. Although the method of deriving the photon spectra is known to contain some inaccuracies, we believe some features will persist when a correct deconvolution is operational. For example, the photon spectra show a clear deviation from a Comptonized model on these days in the form of a high-energy tail. Since the Comptonized model rolls off rapidly above 600 keV, and since counts in the highest two BATSE energy bands indicate photons at or above those energies, this high energy tail should be real if the count rate estimated uncertainties are accurate. We therefore tentatively conclude that Cygnus X-1 exhibits high-energy flaring behavior. The inferred flux in the 45–140 keV band is around the γ_2 level; it is unlikely that the flux computed with a more accurate method will differ enough to alter this estimate.

SUMMARY AND CONCLUSIONS

Temporal and spectral variations in the hard X-ray and gamma-ray spectra from Cygnus X-1 on a scale of days are detectable using the Earth occultation technique. High energy outbursts on a similar time scale occur which reflect a hardening in the spectrum. When the correct instrument response function is installed, it will be possible to compare directly data from different detectors and different orientations of the spacecraft on a daily basis up to the MeV region. This will enable long-term, high sensitivity monitoring of the hard X-ray and gamma-ray emission characteristics of Cygnus X-1.

ACKNOWLEDGEMENTS

The research described in this paper was carried out by the Jet Propulsion Laboratory, California Institute of Technology, under contract to the National Aeronautics and Space Administration.

REFERENCES

1. R. T. Skelton et al., "Status of the BATSE Enhanced Earth Occultation Analysis Package for Studying Point Sources," these proceedings (1992).
2. J. C. Ling, in **Nuclear Spectroscopy of Astrophysical Sources**, AIP Conference Proceedings #170, ed. N. Gehrels and G. J. Share (New York: AIP), p. 315 (1988).
3. G. J. Fishman, R. B. Wilson, C. A. Meegan, B. A. Harmon, and M. Brock, IAUC Telegram #5327, *GX-339-4* (1991).
4. B. A. Harmon et al., "Observation of a Hard State Outburst in the GX339-4 System," these proceedings (1992).
5. B. A. Harmon et al., "Earth Occultation Measurements of Galactic Hard X-Ray/Gamma-Ray Sources: A Survey of the BATSE Results," these proceedings (1992).
6. B. C. Rubin et al., "BATSE Observations of the Massive X-Ray Binary 4U1700-377/HD 153919," these proceedings (1992).

OSSE SPECTRAL OBSERVATIONS OF GX 339-4 AND CYG X-1

D.A. Grabelsky, S.M. Matz, W.R. Purcell, and M.P. Ulmer
Northwestern University, Evanston, IL 60208

W.N. Johnson, R.L. Kinzer, R.A. Kroeger, J.D. Kurfess, M.S. Strickman,
J.E. Grove[1], R.A. Cameron[2], and G.V. Jung[2]
Naval Research Lab, Washington, DC 20375

M.D. Leising
Clemson University, Clemson SC 29634

ABSTRACT

The Oriented Scintillation Spectrometer Experiment on the *Compton Gamma Ray Observatory* has carried out spectral and timing observations of Galactic black hole candidates GX 339-4 and Cyg X-1. GX 339-4 was observed as a target of opportunity in 1991 September, in response to the outburst reported by BATSE and SIGMA. The source was detected from 50 to 400 keV, at a level relative to the Crab of ∼30%. During a follow-up observation made in 1991 November, the intensity of the source below 100 keV had dropped by nearly two orders of magnitude, and it was no longer detected above 100 keV. The observations of Cyg X-1 were made during three different observing periods between 1991 June and November. The OSSE time-averaged spectrum of Cyg X-1 is about 10 times brighter than that of GX 339-4, but remarkably similar in shape over the energy range in which both are detected. No significant emission is seen above about 1 MeV, and there is no evidence for any bumps or narrow lines near 0.511 MeV. The spectra of both sources are decribed reasonably well by a Comptonization model.

INTRODUCTION

The Galactic X-ray sources GX 339-4 and Cyg X-1 have long been considered prime black hole candidates (Nolan *et al.* 1982; Liang and Nolan 1984; Makishima *et al.* 1986; and references therein). One of the observational properties of suspected black holes is an extremely hard spectrum, out to several hundred keV. While a hard spectrum does not by itself confirm the black hole hyphothesis, high quality spectral measurements are nevertheless important to the characterization of such objects.

The Oriented Scintillation Spectrometer Experiment on the *Compton Gamma Ray Observatory* observed both of these interesting sources during several *CGRO* viewing

---

[1] NRC/NRL Resident Research Associate
[2] Universities Space Research Association, Washington, DC

periods in 1991. GX 339-4 was observed as a target of opportunity, in response to the 1991 August outburst reported by BATSE and SIGMA (*IAU Circulars* 5327, 5342, and 5352). Observations of Cyg X-1 were carried out as part of the OSSE viewing plan. This paper describes the observations, and discusses breifly some of the inital results.

OBSERVATIONS

OSSE observations of GX 339-4 were carried out during 1991 September 5 – 12 (viewing period 9), and 1991 November 7 – 14 (viewing period 13). In both viewing periods, the observations during the unocculted portion of each orbit were split evenly between GX 339-4 and a Galactic plane pointing. Each of the four detectors independently accumulated source and background spectra, chopping every two minutes. Background pointings were made at Galactic latitudes +4.5 and -8.5 degrees. The total on-source observing time (per detector) during the two viewing periods was 6.5×10^4 *sec* and 5.2×10^4 *sec*, respectively. In addition to the two-minute spectra, count-rate timing data with 8 *msec* resolution were collected, covering an energy range of 50 – 200 keV.

During the first observation of GX 339-4, made soon after the peak of the outburst reported by BATSE, the source was detected from 50 to 400 keV at a level of roughly 30% of the Crab. The count-flux at 100 keV was $\sim 1.2 \times 10^{-4}$ cm^{-2} s^{-1} keV^{-1}. No day-to-day variations in intensity or spectral shape were evident, although a ratio of the 80 – 200 keV to 200 – 400 keV energy bands shows a slight hint of softening over the course of the week. By the time of the second observation, GX 339-4 was no longer detectable above 100 keV, and its intensity below 100 keV had dropped by nearly two orders of magnitude.

Cyg X-1 was observed during three separate view priods: 1991 May 30 – June 15 (viewing period 2); 1991 August 9 – 15 (viewing period 7); and 1991 December 5 – 11 (viewing period 15). The instrument configuration during these observations included various combinations of detector coverage, as well as differing gain adjustments and offset pointing angles. The total on-source observation time ranged from 4.5×10^4 *sec* to 1.2×10^5 *sec*, per detector. During the first observation, count-rate timing data with 4 *msec* resolution were collected. During each viewing period, Cyg X-1 was observed at a count-flux of $\sim 7 \times 10^{-4}$ cm^{-2} s^{-1} keV^{-1} at 100 keV; the emission is detectable out to approximately 1 MeV. No spectral features are observed near 511 keV. The data do suggest variability on all time scales so far examined.

ANALYSIS AND RESULTS

Time-averaged count spectra for both sources were obtained by subtracting an estimated background spectrum from each two-minute source spectrum, then summing the resulting background-subtracted spectra, detector-by-detector. The background for each two-minute source spectrum was estimated from a quadratic fit to the three (or four) temporally nearest offset-pointed observations. Spectra averaged over intervals as short as one day, to as long as entire viewing periods, were produced in this way. Spectral fits were performed using a *forward-folding* technique, in which a model photon spectrum was convolved with the OSSE instrument response, resulting in a model count spectrum. The parameters of the model photon spectrum were adjusted so as to achieve a (least-squares) best fit of the model count spectrum to the measured count spectrum. Model photon spectra obtained by this fitting procedure are useful in evaluating specific source models, but cannot be used to fit other source models; the forward folding procedure must be applied separately to each prospective model.

Spectra from each viewing period of Cyg X-1 and from the bright state of GX 339-4 were fitted with the Comptonization model of Sunyaev and Titarchuk (1980). For certain simplifying assumptions, this model predicts the gamma-ray spectrum that emerges from a spherical, hot plasma cloud, due to multiple Compton scatterings of soft ($\sim$ a few keV) photons injected in the central region of the cloud. The model is parameterized by the temperature of the cloud, kT, and by the cloud's optical depth, τ. Although the model does not provide a complete and detailed description of the source, it does appear to indicate that the spectra we observed are consistent with Comptonization.

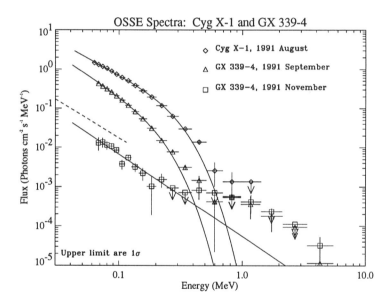

Figure 1: OSSE spectra of GX 339-4 and Cyg X-1. The Cyg X-1 (1991 August) and bright GX 339-4 spectra were fitted with the Comptonization model of Sunyaev and Titarchuk (1980); the dim GX 339-4 spectrum was fitted with a single power law model. The solid lines show the best-fit model photon spectrum; the data points represent the *model-dependent* deconvolved count spectra (see text). The symbols for the three obserations shown are given in the figure legend. The dashed line shows the extraplolation of the power law portion of the soft X-ray spectrum of GX 339-4 observed by Makishima *et al.* (1986).

The results of the fits are displayed in Figure 1, where the solid lines show the model photon spectra; for Cyg X-1, only the 1991 August observation is shown. The data points shown for each fit represent the "unfolded" count spectrum, determined by multiplying the measured count spectrum by the ratio of the best-fit model photon spectrum to the best-fit model count spectrum (the unfolded spectrum is also binned

Table 1: Comptonization Model Parameters

Observation	kT (keV)	τ	χ^2_ν	γ	$-\alpha$
Cyg X-1, 1991 May	60.06 ± 0.12	2.26 ± 0.05	10.8	3.27	-1.85
Cyg X-1, 1991 Aug.	59.05 ± 0.12	2.42 ± 0.06	1.16	2.99	-1.79
Cyg X-1, 1991 Dec.	52.16 ± 0.14	2.72 ± 0.09	18.6	2.81	-1.75
GX 339-4, 1991 Sept.	42.22 ± 0.11	2.39 ± 0.09	1.11	4.26	-2.05

in energy to improve the signal-to-noise). Figure 1 also shows the results of a power law model for the 1991 November observation of GX 339-4.

It appears that the Comptonization model provides a fairly good description of the spectra of Cyg X-1 and GX 339-4 during its gamma-ray bright state, at least out to about 400 keV. Beyond this energy, the spectra seem to be harder than the model would predict. At these energies the spectra are beginning to trace the OSSE sensitivity curve (c.f., Cameron et al., 1992), so the signal-to-noise is dropping off fairly quickly. But the discrepancy is undoubtedly also due in part to the inadequacy of the model and its assumptions, including the use of the classical Thompson scattering cross section and a single temperature for the entire plasma cloud.

With the understanding that the results are perhaps more qualitative than quantitative, we list the best-fit parameters (and reduced χ^2s) in Table 1, along with the derived quantities, γ and α. These are defined as:

$$\gamma = \frac{\pi^2}{3} \frac{mc^2}{(\tau + \frac{2}{3})^2 kT}$$

and

$$\alpha = -\frac{1}{2} + \sqrt{\frac{9}{4} + \gamma}.$$

γ gives the inverse of the average fractional energy gain of an emerging photon for the assumed geometery (i.e., the inverse of the Compton "y" parameter), and $-\alpha$ is the *photon* power law index which describes the spectrum at energies below kT of the scattering medium. As a caution that these results are less than decisive, we note that a simple thermal bremsstrahlung model with kT = 71.1 keV fits the bright spectrum of GX 339-4 nearly as well as the Comptonization model.

The spectrum of GX 339-4 from the second (November) observation was fitted with a simple power law photon model, because up to about 100 keV, where it "disappears" beneath the OSSE sensitivity curve, the spectrum clearly shows no signs of an exponential cutoff (Figure 1). The best-fit spectrum gives a photon index of $-\alpha = -2.09 \pm 0.20$, with a reduce χ^2 of 1.04.

DISCUSSION

GX 339-4 obviously underwent a transition during time between the two OSSE viewing periods when it was observed, and, as noted above, may even have been softening during the first viewing. High and low X-ray states, previously observed, are anticorrelated with the intensity of a hard tail, which is well-described by a power law with a photon index of approximately -2 (e.g., Nolan et al. 1982; Makishima et al. 1986). This agrees with the index derived from the power law fit to the second (dim) OSSE observation, as well as that for the extension of the Comptonization model to energies below kT. But it is not at all certain that the power law inferred from OSSE measurement is the continuation of the hard tail seen by lower energy experiments. As an example, the extrapolation of the high X-ray state observed by Makishima et al., shown in Figure 1, is seen to lie between the two gamma-ray states observed by OSSE. Similarly, the slopes of the power law extension of the Cyg X-1 spectra are in fair agreement with the results of lower energy observations.

The relatively good fits of the Sunyaev-Titarchuk model to the OSSE spectra of GX 339-4 and Cyg X-1 suggest that the spectra are shaped by Comptonization, despite the deficiency of the fits toward higher energies. By adding a second component to the model, with a distinct temperature and optical depth, a slight improvement to the fit can be obtained. A more realistic approach, however, probably requires proper treatment of the scattering cross sections, allowance for a temperature gradient in the scattering medium, and perhaps an alternative spatial distribution of the input soft photons. Time variability studies of the fluxes in these sources currently underway, including searches for QPO and cross correlation of different energy bands, may help provide further insights into the nature of GX 339-4 and Cyg X-1.

REFERENCES

Cameron, R.A. et al. 1992, in *The Compton Observatory Science Workshop*, eds., Chrader, C.R., Gehrels, N., and Dennis, B., NASA Conference Publication 3137, p. 3.
Liang, E.P., and Nolan, P.L. 1984, *Space Science Reviews*, **38**, 353.
Makishima, K. et al. 1986, *Ap. J.*, **308**, 635.
Nolan, P.L 1982, *Ap. J.*, **262**, 727.
Sunyaev, R.A., and Titarchuk, L.G. 1980 *Astron. Astrophys.*, **86**, 121.

OBSERVATION OF A HARD STATE OUTBURST IN THE GX339-4 SYSTEM

B. A. Harmon, C. A. Wilson, R. B. Wilson, G. J. Fishman,
C. A. Meegan, W. S. Paciesas[1], G. N. Pendleton[1],
B. C. Rubin[2], M. H. Finger[3]
NASA/Marshall Space Flight Center, ES-64,
Huntsville, AL 35812

W. A. Wheaton, J. C. Ling, R. T. Skelton
Jet Propulsion Laboratory, Pasadena, CA 91009

ABSTRACT

Monitoring of the x-ray binary system GX339-4 with the Burst and Transient Source Experiment (BATSE) on the Compton Gamma Ray Observatory indicated that the compact object in this system had entered its hard emission state in late June of 1991. X-ray flux between 20 keV and approximately 300 keV was observed to increase over a period of about 60 days, peaking at 300-400 mCrab. From late August to early October, the flux from GX339-4 dropped in intensity (with spectral softening) until it was not observable above 20 keV. We discuss this observation and compare it to other studies of GX339-4. We report results of a search for the 0.6 day orbital period[1], as well as for quasi-periodic oscillations that have been seen in the hard state.[2]

INTRODUCTION

The x-ray binary system GX339-4 has been studied for a number of years in order to establish the nature of the compact object emitting high energy photons. The following is a brief summary of observed properties of this system:

X-Ray Observations:
GX339-4 exhibits three x-ray states[2] (see figure 1): (1) A low or hard state with decreased 1-10 keV emission and a power-law-like spectrum extending to 200 keV or higher, (2) A high or soft state with a black-body-like spectrum with the bulk of the emission in the 1-10 keV band, and little to no emission above 20 keV (possibly a hard power law tail), and

---

[1] University of Alabama, Huntsville, AL
[2] Universities Space Research Association
[3] Computer Sciences Corporation

(3) An off state with spectrum similar to state (1) but at very low intensity.

Flickering on 10^{-3} - 10 secs timescales and quasi-periodic oscillations (QPO) have been seen mainly in the hard state (1) reminiscent of Cyg X-1.[2,3]

Optical Observations:
GX339-4 has been shown to be a binary system with an orbital period of 0.618 days with a probable F or G-type optical companion.[1]

Fast timescale variations are observed just as in x-rays, anticorrelated with soft x-rays, but proably correlated with hard x-rays[3]; 1.13 ms periodicity seen in one observation.[4]

Visual magnitude changes from 15 to 20th magnitude accompany x-ray state transitions (bright in low x-ray state).[3]

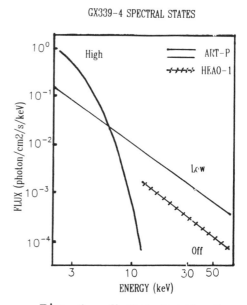

Fig. 1. X-ray spectral states of GX339-4. Data are from refs. 2 and 5.

Subsets of these properties are thought to be characteristic of black hole systems, such as the spectral pivoting and short timescale flickering. QPO have been observed in both suspected black hole and confirmed neutron star systems. A confirmation of the 1.13 msec periodicity[4] would provide an unambiguous determination of GX339-4 as a neutron star, but other properties suggest that its black hole candidacy be maintained.

RESULTS

In figure 2 we show a history of the flux for GX339-4 detected using the Earth occultation technique (see Harmon et al., these proceedings for details of this method) in the 20-320 keV energy band. Each point represents a three-day average of the flux measured in the BATSE large area detectors. The hard state outburst was observed from approximately June 23, 1991, to October 11, 1992. Continuous monitoring for two months prior to June 23 and from November 1991 to June 1992 indicate no detectable flux above 20 keV when GX339-4 was presumably either in the soft (2) or off (3) states.

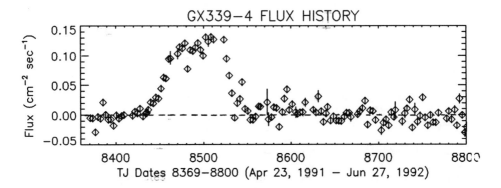

Fig. 2. Flux history for GX339-4 in the 20-320 keV band. Each point represents a three-day average of the flux.

Enhancement of the BATSE occultation technique to achieve greater sensitivity (see Ling et al., these proceedings) may eventually allow us to determine the spectral state before and after the outburst.

A number of spectral fits were performed throughout the hard state outburst. In figure 3, spectra for GX339-4 during the rising, middle and falling portions of the outburst are shown, along with fits using the Sunyaev-Titarchuk (S-T) Comptonized disk model[6]. Table I lists the fitting parameters for the S-T model (each time interval corresponds to one pointing period of the Compton Observatory). We see that there is considerable spectral evolution through the outburst from June to September 1991. The gamma parameter is inversely proportional to the product of kT_e and the square of the optical depth. Figure 4 shows the spectral index for a single power law fit for the same time intervals as in Table I. The power law fits also illustrate the spectral softening with time, although the S-T model, in general, gave the best chi-square over the entire energy range (20-320 keV).

Fluxes for single occultation edges throughout the 100 day outburst, as well as shorter time periods, were folded at the 0.618 day period reported from optical observations in the x-ray off state.[1] No variations indicating an orbital signature in hard x-rays were found. Searches for quasi-periodic oscillations are still in progress, but have not revealed significant results other than the presence of the low frequency noise (below 1 Hz) similar to Cyg X-1 and the recent outburst of GRO J0422+32 (see Kouveliotou et al., these proceedings.)

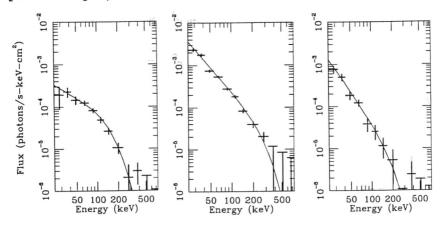

Fig. 3. Photon spectra from (1) early (TJ Dates 8435-8449, (2) near maximum (TJ Dates 8463-8476) and (3) late (TJ Dates 8532-8546) in the hard state outburst of GX339-4. Model fit is Sunyaev-Titarchuk[6] (See Table 1 for parameters.)

Table I.
SUNYAEV-TITARCHUK MODEL FITS

BATSE LARGE AREA DETECTORS (Channels 1-12 20-750 keV)

TJD	Amp (x 10e-5) (p/cm-2-keV)	kT (keV)	Gamma	Reduced Chi-2 (per 9 d.o.f.)
8435-8449	4.2 +- 1.8	38 +- 6	1.0 +- 1.0	5.7
8449-8463	5.2 +- 1.5	50 +- 5	2.3 +- 0.3	11.4
8463-8476	3.2 +- 1.8	60 +-11	3.7 +- 0.4	15.8
8476-8483	2.0 +- 4.0	33 +- 2	1.7 +- 0.4	20.3
8483-8490	8.9 +- 2.7	43 +- 4	3.1 +- 0.3	13.7
8490-8504	7.6 +- 2.6	40 +- 4	3.7 +- 0.3	27.3
8504-8511	14.0 +- 7.1	33 +- 5	3.6 +- 0.6	13.8
8511-8518	11.6 +- 9.3	32 +- 7	3.6 +- 0.9	20.9
8518-8532	2.3 +- 1.9	45 +-11	5.2 +- 0.6	26.6
8532-8546	0.6 +- 1.8	42 +-33	5.8 +- 2.2	8.1

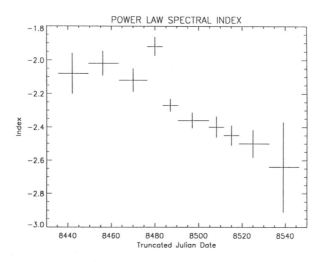

Fig. 4. Spectral index as a function of time for a single power law fit over the outburst interval of GX339-4.

REFERENCES

1. P. J. Callanan, et al., in Accretion Powered Compact Binaries, ed. by C. W. Mauche (Cambridge Press, Mass., 1989), p. 58.

2. S. A. Grebenev, et al., Sov. Astron. Lett. 17, 985 (1991).

3. C. Motch, et al., Astro. Ap., 119, 171 (1982).

4. J. N. Inamura, et al., Ap. J. 314, L11 (1987).

5. P. L. Nolan, et al., Ap. J., 262, 727 (1982).

6. R. A. Sunyaev and L. G. Titarchuk, Astro. Ap. 86, 121 (1980).

HARD X-RAY OBSERVATION OF GX339-4 WITH WELCOME-1

N. Y. Yamasaki, S. Gunji*, M. Hirayama, T. Kamae, S.Miyazaki,
Y. Sekimoto, T. Takahashi, T. Tamura, M. Tanaka

Dept. of Physics, Univ. of Tokyo, Bunkyo-ku, Tokyo, JAPAN 113

T. Yamagami

Inst. of Space Astronautical Science, Sagamihara, Kanagawa, JAPAN 229

M. Nomachi

Nat. Lab. for High Energy Physics (KEK), Tukuba, Ibaraki, JAPAN 305

H. Murakami

Dept. of Physics, Rikkyo University, Toshima-ku, Tokyo, JAPAN 171

J. Braga, J. A. Neri

Inst. Nacional de Pesquisas Espaciais, Sao Jose dos Campos, Sao Paulo, BRAZIL

ABSTRACT

The black hole candidate GX339−4 and the Galactic Center region were observed simultaneously in the hard X-ray region with the balloon-borne telescope, Welcome-1. The obtained energy spectrum of GX339−4 is harder than that of the Galactic Center and is consistent with a single power-law function of photon index $\alpha=1.52\pm0.34$ in the energy range 60–600 keV.

INTRODUCTION

GX339−4 is considered to be a black hole candidate (BHC) because its X-ray properties are similar to those of Cyg X-1. In the hard X/γ-ray spectrum of Cyg X-1, spectral features such as strong emission from 400 keV to 1.5 MeV and the broken spectrum at 10–1000 keV have been observed[1]. The broken spectrum of Cyg X-1 was explained by the optically thin accretion disk model with electron temperature $kT_e=20\sim100$ keV [1,2]. The energy spectrum in the hard X/γ-ray range gives direct information about the radiation process in the emission region. The physical condition in the accretion disk around a black hole such as the optical thickness, the electron temperature and e^+e^- production and annihilation rates can be obtained from studies of the hard X/γ-ray emission from BHCs like GX339−4.

The hard X-ray emission from GX339−4 was detected up to 200 keV [3,4,5]. In X-ray low states, the spectrum above 20 keV can be acceptably represented by a power-law spectrum with a typical photon index $\alpha=1.5\sim1.7$[4].

In order to obtain the spectrum of GX339-4 in hard X/γ-ray, the contamination of Galactic diffuse emission needs to be estimated because both GX339−4 and

---

*Now at Dept. of Physics, Yamagata Univ., Yamagata-shi, Yamagata, JAPAN 990

the hard X-ray sources near the Galactic Center are known to be variable. To resolve this possible confusion, we performed simultaneous observation of the Galactic Center and GX339−4 in the energy range 40–600 keV.

From our observations alone, we cannot determine whether GX339−4 was in a low or high state. BATSE reported that the hard X-ray flux from GX339−4 increased in summer 1991 and dropped below 0.1 Crab in September 1991[6]. This suggests that GX339−4 was in a low state in December 1991.

INSTRUMENT AND OBSERVATION

Welcome-1 is a low-background hard X/γ-ray telescope based on newly developed GSO(Ce)/CsI(Tl) well-type phoswich counters [7,8]. Welcome-1 was designed as a balloon-borne telescope in the energy range 60–800 keV. In the telescope, 64 GSO(Ce)/CsI(Tl) well-type phoswich counters and 36 CsI(Tl) guard anti-counters are assembled in 10×10 matrix configuration. Its total effective area is 740cm^2 at 122 keV and 220cm^2 at 511 keV. The field of view (FOV) defined by the depth of the CsI(Tl) well, is 15.4° (FWHM) at 122 keV and 18.2° (FWHM) at 511 keV. The pointing direction of the telescope is monitored by a geomagnetic sensor and a Sun camera with 0.2° accuracy.

The Welcome-1 telescope was launched in 1991 December 3, at Cachoera Paulista, Brazil (S22°39′44″, W45°00′44″). During the level flight, the balloon altitude was kept at 4.43±0.05g/cm^2. The telescope's elevation angle was fixed to 69.0°±0.05° by a technical problem in the control system.

The detector scanned an off-Galactic plane region from 13:41 to 14:08, and then scanned the Galactic Center region (hereafter referred to as the GC scan) from 14:08 (UT) to 14:30 as shown in Figure 1. After the GC scan, GX339−4 was observed from 14:35 (UT) to 16:05, alternately pointed at the source and background. During the observation, the atmospheric depth was kept constant and the telescope's fields of view are shown in Figure 2. The total observation time was 42 minutes for GX339−4 and 22 minutes for the GC scan.

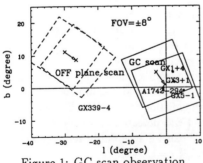

Figure 1: GC scan observation

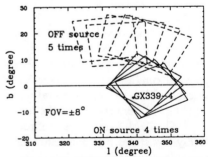

Figure 2: GX ON/OFF observation

ANALYSIS AND RESULTS

To derive the energy spectrum of GX339−4 shown in Figure 3, we subtracted the flux of the background observation from that of the on-source observation. Estimation of Galactic diffuse component will be discussed later. In the energy range 60–600 keV, the spectrum is consistent with a single power-law function with parameters shown in Table 1. Photon index α is 1.52±0.34 and intensity at 100 keV is about 60m Crab. The 3σ upper limit for 511 keV line emission is 7.7×10^{-4} photons/cm^2/sec.

For the GC scan, Welcome-1's FOV is too wide to separate point sources so we present the total sum of observed energy spectra from the GC scan in Figure 4. The spectrum probably is a complex sum of Galactic diffuse X/γ-ray emission and emission from many point sources. The parameters fitted to a single power-law function in the energy range 40–300 keV are shown in Table 2.

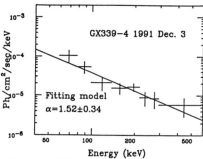
Figure 3: GX339−4:Flux & Fitting

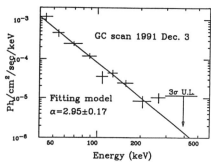
Figure 4: GC scan:Flux & Fitting

GX339−4 :1991 Dec.3 FOV=±8°		
Fitting range (keV)	60–600	
Photon index α	1.52	±0.34
Intensity at 100 keV (ph/cm^2/sec/keV)	3.71×10^{-5}	$\pm7.5\times10^{-6}$
χ^2	4.24	d.o.f =6

Table 1: GX339−4 fit parameters

GC scan :1991 Dec.3 FOV=±8°		
Fitting range (keV)	40–300	
Photon index α	2.96	±0.17
Intensity at 100 keV (ph/cm^2/sec/keV)	7.86×10^{-5}	±5.5×10^{-6}
χ^2	12.59	d.o.f =7

Table 2: GC scan fit parameters

DISCUSSION

To estimate the contamination of Galactic diffuse emission in the GX339−4 spectrum, we used our GC scan data and the data obtained by the LAC on board the Ginga satellite (1.2∼37 keV)[9]. The obtained intensity of GX339−4 at energy above 120 keV is greater than that of the GC scan. If all flux of the GC scan is the diffuse emission and independent of galactic longitude l, our GC scan spectrum limits the diffuse Galactic emission to be lower than the emission from GX339−4.

The spectrum obtained by the LAC at (l,b)∼(339°,4°), the opposite point of the Galactic plane, is represented by the thermal bremsstrahlung model with electron temperature kT_e=3.7 keV and intensity lower than 1m Crab[10]. If the emission is symmetric to the Galactic plane, the thermal component is expected to be less than 10^{-10} photons/cm^2/sec/keV at E≥60 keV. As for the non-thermal component, the intensity at (l,b)=(339°,−4°) is less than 5% of the intensity at (l,b)=(0°,0°) according to a Galactic bulge emission model determined by the LAC 6.4-17.9 keV band observation[11]. Thus we conclude that the contamination of diffuse Galactic emission is less than 10% of the obtained GX339-4 spectrum.

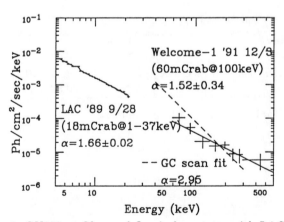

Figure 5: GX339−4:Observed flux in low states with LAC and Welcome−1

The emission spectrum from the hot accretion disk ($kT_e \geq$10 keV) is expected to break the power-law function in the hard X−ray range at an energy determined by the electron temperature kT_e and the optical thickness τ. To search for a break point or other spectral features, we compare our spectrum with the LAC spectrum and obtain the wide ranging spectrum of GX339−4. LAC observed GX339−4 four times in low states with the intensity 3m Crab∼160m Crab. Throughout these four observations, the photon index α in the 4∼20 keV range is 1.6∼1.7[10]. In Figure 5, we present the combined spectrum obtained with Welcome-1 and LAC (1989 September 2)[10]. Even though these were not simultaneous observations, the spectrum extends smoothly from 10 keV to 600 keV without a break. It implies that if the hard X/γ-ray emission is explained by an optically thin accretion disk model like Cyg X-1, the electron temperature kT_e should be greater than 200 keV[2].

REFERENCES

1. Ling, J.L., et al. Ap.J. 321, 1117 (1987)

2. Sunyaev, R.A. and Titarchuk, L.G., Astron.Astrophys. 86, 121 (1980)

3. Nolan, P.L., et al. Ap.J. 262, 727 (1982)

4. Dolan, J.F., et al. Ap.J. 322, 324 (1987)

5. Covault, C.E., et al. Ap.J. 388, 165 (1992)

6. IAUC 5327, 5352, 5395 (1991)

7. Kamae, T., et al. Adv. Space Res. Vol.13, No.2, 165 (1992)

8. Takahashi, T., et al. in SPIE meeting, San Diego (1992)

9. Makino, F. and ASTRO-C team ,Astrophys. Lett. Commun. 25, 233 (1987)

10. Ebisawa, K., et al. in preparation

11. Yamauchi, S., PhD. thesis ISAS RN 474 (1991)

IR and Optical observations of Be/X-ray binaries in conjunction with BATSE

M.J.Coe, C.Everall, A.J.Norton, P.Roche and S.J.Unger
Physics Department, The University, Southampton, SO9 5NH, U.K.

J.Fabregat and V.Reglero
Departmento de Matemática Aplicada y Astronomía,
Universidad de Valencia 46100 Burjassot (Valencia), Spain

J.M.Grunsfeld and T.A.Prince
Caltech, Pasadena, CA91125, U.S.A.

....

Abstract

IR and optical data are presented of a sample of 10-12 Be X-ray binaries. These sources are being monitored as part of a collaborative programme with BATSE. This paper deals with the ground-based side of the programme and addresses the role that the IR/optical data can play in complementing the X-ray data from CGRO. In particular, modelling of the IR-optical spectrum allows us to estimate the amount of material available for accretion onto the neutron star as well as the details of the temperature and density profiles of the circumstellar disk.

INTRODUCTION

Though the first Be star (Gamma Cas) was identified over a hundred years ago, and thirty years ago the basic model seemed clear for these objects, discoveries in the last ten years have demonstrated that we are far from being in such a secure position. There now exist several competing models for these systems, none of which adequately fit the observational data. In particular, the recently identified sub-group of Be stars in X-ray binaries pose particular problems in our understanding. In these systems, a neutron star is in an eccentric orbit ($P_{orb} \approx$ days–years) around the Be star, from which it accretes via an expanding circumstellar disk. The accretion powers strong, variable X-ray emission which is seen to be pulsed at the neutron star spin period ($P_{spin} \approx$ seconds–minutes). It is this group of systems that is the subject of this paper.
In order to help us explore the relationship between the X-ray and the optical/infrared emission we have established a long term programme to monitor a group of eleven of these targets in both the infrared and the optical. Waters[1] and others have shown that detailed information may be derived concerning the structure of the circumstellar disc provided good optical and infrared data are available. We are now also modelling this disc structure using an implementation of the Cloudy model[2] – a package which can

calculate the continuum and emission line fluxes of about 500 lines. This allows us to explore more detailed, and hence realistic, structures for the circumstellar disc and its interaction with the neutron star. Recent results using this model on a sample of four Be/X-ray systems are discussed elsewhere[3].

SUMMARY OF OBSERVATIONS

The programme began in the late 1980s using the 1m Nickel Telescope at the Lick Observatory for optical spectroscopy and a combination of UKIRT and the 1.5m Telescopio Carlos Sánchez (TCS), Tenerife, for infrared photometry. Since December 1990, the programme has been consolidated and we have been consistently collecting data approximately once a quarter on the targets listed in Table 1:

Source Name	Span of Observations	No. of sets of IR Obs.	No. of Hα Spectra
Gamma Cas	Sep. 1987 – Aug. 1992	22	10
4U0115+63	Dec. 1990 – Aug. 1992	13	5
LSI +61° 303	Sep. 1987 – Aug. 1992	20	8
V0332+53	Jul. 1984 – Aug. 1992	28	17
X Persei	Sep. 1987 – Aug. 1992	31	20
H0521+37	Sep. 1987 – Aug. 1992	25	18
A0535+26	Nov. 1987 – Aug. 1992	23	19
4U0728-25	Nov. 1987 – Apr. 1992	16	2
LSI +61° 235	Aug. 1991 – Aug. 1992	11	6
EXO2030+375	Jul. 1985 – Oct. 1992	8	3
4U2206+54	Sep. 1987 – Aug. 1992	24	8

To collect these data we have been using the TCS to obtain J, H and K photometry and the 2.5m Isaac Newton Telescope, La Palma and the 60" Palomar telescope to obtain high resolution (< 1 Å) Hα spectra. In addition, data on fainter targets have occasionally be taken as part of Service Programmes running on UKIRT and the William Herschel Telescope. Some collaborative work has also been carried out in conjunction with Penn. State.

DISCUSSION OF PARTICULAR SOURCES

In order to demonstrate the characteristic behaviour of these systems and the deductions that may be made from the data, we discuss in this section three systems taken from the list in Table 1.

V0332+53/BQ Cam

An excellent example of this group of sources is V0332+53. This transient X-ray source was first observed in 1973 during a bright X-ray outburst using the Vela 5B satellite[4,5]. During this outburst the source reached an intensity approximately 1.6 times that of the Crab. It was re-discovered ten years later using the Tenma satellite[6], though this time the X-ray flux was ten times lower than the peak activity in 1973. The system was found to consist of a neutron star with a spin period of 4.37s in a 34.25d orbit around a Be star companion [7-10]. There have been various studies of the optical continuum emission and the variability of its emission lines[11-12] as well as the infrared excess characteristic of Be stars in an active phase[13]. A third outburst was detected by Ginga in 1989 during which the X-ray flux peaked at 240mCrab[14].

In Fig.1 we show the long term infrared lightcurves for V0332+53, obtained from our monitoring programme. Two peaks in the J, H and K flux can clearly be seen occurring in 1983/4 and 1989. These infra-red "flares" occur simultaneously with the reported X-ray flares and indicate an increase in the amount of circumstellar material around the Be star, and hence fuel for accretion onto the neutron star. Currently (1992) the system is at its lowest brightness for the last 10 years and may be heading towards the complete loss of its circumstellar disk. If this happens, then we shall be able to use the opportunity to determine the "true" colours of the underlying B star in the same manner as we have recently accomplished with X Persei[15-16] – see next section.

X Persei/4U0352+30

The recent phase change from emission line object to a normal O9.5III star which we observed in X Per has been fully described elsewhere [17,15]. A further discussion of a similar episode in 1974–1977 is also discussed in Roche et al.[16]. Such events provide a great deal of information about the conditions required to form and disperse the circumstellar disc. Here we note that the detection and study of these events is only possible with a *regular* monitoring programme which allows us to examine the behaviour of the system before, during and after a phase change. In Figure 2 we present the optical V magnitude against the IR colours from data collected over the period 1987-1991 (for sources of data see Roche et al.[16] and references therein), during which the source has gone from a very active state to its current low state. In particular, we can see from this figure how the infrared colours provide an excellent indicator of the activity of the system, and hence also of X-ray activity levels.

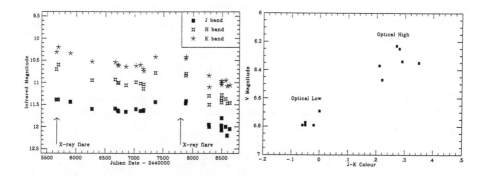

Fig 1 - *Infrared light curve for V0332+53 spanning the period 1983-present. The two arrows indicate the times of X-ray outburst.*

Fig 2 - *Optical magnitude versus IR colours for X Persei in various states of activity over the period 1987-1991.*

LSI +61° 235/RX J0146.9+6121

This new source was recently identified[18] from the ROSAT sky survey as a potential Be star system. A low resolution optical spectrum identified the presence of Hα emission – presumably from the circumstellar envelope or disk. We have since obtained detailed high resolution optical spectra confirming this result and have carried out infrared photometric measurements on four occasions, spread over eleven nights. We have found no evidence yet for any variability in the infrared, and the infrared excess, though present, is very weak in this object. Figure 3 shows the current location of this object on a IR excess/Hα plot in comparison to some of the other systems under observation in this programme. The very low state of this source is clearly visible from this diagram. These results are discussed in more detail elsewhere[19].

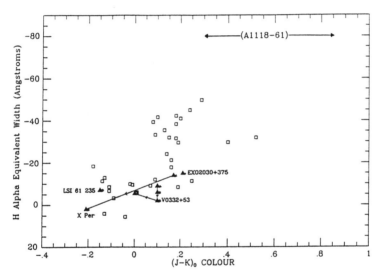

Fig 3 - Plot of IR excess versus Hα line strength for various Be/X-ray systems and for 33 isolated Be stars.

CGRO OBSERVATIONS

This programme was substantially enhanced from September 1992 by the addition of CGRO coverage of all these systems. As a result of the NASA Phase II selection, X-ray data from BATSE will be used to monitor all the sources in this programme. This instrument operates over the energy range 20 keV – 30 Mev and has proven to be very sensitive to detecting these Be X-ray transients. We know that many of these objects have strong, fairly hard spectra (e.g. V0332+53[20] and A0535+26[21]) and therefore should easily be detectable - at least in outburst. Since the beginning of 1992 two such objects (EXO2030+375 from our target list in Table 1, and A1118-61, in the southern sky) have been reported by the BATSE team[22] as having undergone a major outburst. We were able to obtain optical and infrared data on the southern object from the Anglo Australian Telescope courtesy of the Director as a Target of Opportunity, but the flare in EXO2030+375 occurred while the object was too close to the sun. However, it has recently become clear that the outbursts from this source are coincident with the periastron passage of the neutron star[23]. In October 1992 it proved possible to obtain an Hα spectrum from the WHT during such an outburst which showed the presence of the strong emission line. Future collaboration between this programme and the BATSE observations should greatly enhance our understanding of these complex, violent systems.

ACKNOWLEDGMENTS

The TCS is operated by the Instituto de Astrofísica de Canarias at the Teide Observatory (Tenerife, Spain). Thanks to the Service Programmes of UKIRT, INT, and WHT. The 60" telescope at Palomar Mt. is jointly owned by the California Institute of Technology and the Carnegie Institute of Washington.

REFERENCES

1. Waters 1986 A&A 162, 121 .
2. Ferland G. Ohio State Univ. Internal report 1991, no.1.
3. Everall C. et al. 1992 MNRAS (in press).
4. Terrell J. and Priedhorsky W.C. 1984 ApJ 285, L15.
5. Whitlock L 1989 ApJ 344, 371.
6. Tanaka Y 1983 IAU Circular no. 3891.
7. Argyle R.W. 1983 IAU Circular no. 3897.
8. Bernacca P.L., Ijima T. and Stagni R. 1983 IAU Circular no. 3897.
9. Kodaira K. 1983 IAU Circular 3897.
10. Stella L., White N.E. and Rosner R. 1986 ApJ 308, 669.
11. Bernacca P.L., Ijima T. and Stagni R. 1984 A and A 132, L8.
12. Corbet R.H.D., Charles D.A. and van der Klis M. 1986 A and A 162, 117.
13. Coe M.J., Longmore A.J., Payne B.J. and Hanson C.J. 1987 MNRAS 226, 455.
14. Makino F. et al 1989 IAU Circular no. 4858.
15. Fabregat et al 1992 A & A 259, 522.
16. Roche P et al 1992 A & A (In press).
17. Norton A.J., Coe M.J., Estela A. et al., 1991 MNRAS 253, 579.
18. Motch C et al 1991 A & A 246, L24.
19. Coe M.J. et al 1992 MNRAS (in press)
20. Unger S.J., Norton A.J., Coe M.J. and Lehto H.J. 1992 MNRAS 256, 725.
21. Coe M.J. et al. 1990 MNRAS 243, 475.
22. Wilson R.B. et al. 1992 IAU Circular 5454.
23. Stollberg M.T. et al. 1992 (this conference).

EGRET OBSERVATION OF THE CONSTELLATION CYGNUS

J.R. Mattox[a], D.L. Bertsch, B.L. Dingus[b], C.E. Fichtel
R.C. Hartman, S.D. Hunter, P. Sreekumar[b], D.J. Thompson, P.W. Kwok[c]
NASA/Goddard Space Flight Center, Code 662, Greenbelt, MD 20771
[a] GRO Science Support Center, Computer Science Corp.
[b] Universities Space Research Association
[c] NAS/NRC Postdoctoral Research Associate

K. Brazier, T. Halaczek, G. Kanbach, H.A. Mayer-Hasselwander,
C. von Montigny, K. Pinkau, H.D. Radecke, H. Rothermal, M. Sommer
Max-Planck Institut für extraterrestrische Physik, D-8046 Garching, Germany

J. Fierro, Y.C. Lin, P.F. Michelson, P.L. Nolan, J. Chiang
Hansen Experimental Physics Laboratory, Stanford University, Stanford, CA 94305

D.A. Kniffen
Dept. of Physics and Astronomy, Hampden-Sydney College, VA 23943

E. Schneid
Grumman Aerospace Corporation, Mail Stop A01-26, Bethpage, NY 11714-3580

ABSTRACT

The *Compton* Observatory was pointed at Cygnus twice during the full sky survey (5/30-6/8/91 & 8/8-8/15/91). The COS-B sources 2CG 075+00 and 2CG 078+01 are detected by EGRET, and smaller error boxes have been obtained[1], but the sources are not yet identified at other wavelengths. An excess of counts is noted at the position of Cygnus X-3 consistent with a point source. The corresponding flux is $8 \pm 2 \times 10^{-7} cm^{-2} s^{-1}$ (E>0.1 GeV) — a factor of 5 less than the flux reported by the SAS-2 group for Cygnus X-3. The observed intensities for the individual periods were consistent with no change in intensity between the first and second observations (the second observation began two weeks after the beginning of the July 1991 radio flare). More work is required to refine and verify the galactic diffuse model (which comprises at least 10 times more flux than Cygnus X-3 in this direction) before a definitive Cygnus X-3 point-source result will be obtained. The 4.8 hr X-ray periodicity of Cygnus X-3 is not detected in extant EGRET data with a χ^2 test with 10 bins[2], but because of the large background, a sinusoidal periodic flux of $8 \times 10^{-7} cm^{-2} s^{-1}$ (E>0.1 GeV) or less cannot be excluded. The possibility of GeV emission from other objects in this direction (Cygnus X, or the Cygnus OB2 association) cannot yet be dismissed. The additional 5 weeks of Cygnus exposure planned for phase II of the mission will provide improved sensitivity for periodicity, point-source detection, and localization.

The steep-spectrum, radio-galaxy Cygnus A was not detected (flux $< 2 \times 10^{-7} cm^{-2} s^{-1}$, E>0.1 GeV, with 95% confidence). An intense unidentified source (GRO J1837+59) was detected 26° off-axis during the first observation at α(J2000)=279.4, δ =59.2 (with an error circle of radius 0.4° at 68% confidence, and 0.6° at 95% confidence). The flux is $7.9 \pm 1.3 \times 10^{-7} cm^{-2} s^{-1}$, E>0.1 GeV. The photon spectral index is -1.9±0.3. GRO J1837+59 was not detected during the second observation (when it was 37° off-axis where the sensitivity is ≈0.1 of that for a source on-axis). The flux upper limit is $6 \times 10^{-7} cm^{-2} s^{-1}$, E>0.1 GeV, with 95% confidence. By analogy to other high-latitude sources seen by EGRET, this could be a blazar, although no known blazars are within 3° of this position. The 1987 Green Bank 4.85 GHz survey[3] detected a 179 mJy source (18 39 42.7 +58 50 22) within the GRO J1837+59 error circle which warrants further radio investigation.

REFERENCES

1. J.R. Mattox *et al.*, in preparation for submission to ApJ (1993).
2. P.F. Michelson *et al.*, ApJ, in press (1992).
3. P.C. Gregory and J.J. Condon, ApJ Suppl. **75**, 1011 (1991).

SIGMA OBSERVATIONS OF THE BURST SOURCE GX354-00

A. Goldwurm, A. Claret, B. Cordier, J. Paul
Service d'Astrophysique, Centre d'Etudes de Saclay,
91191 Gif-sur-Yvette Cedex, France

J.P. Roques, L. Bouchet, M.C. Schmitz-Fraysse, P. Mandrou
Centre d'Etude Spatiale des Rayonnements,
9 av. du Colonel Roche, BP 4346, 31029 Toulouse Cedex, France

R. Sunyaev, E. Churazov, M. Gilfanov, N. Khavenson,
A. Dyachkov, B. Novikov, R. Kremnev, V. Kovtunenko
Space Research Institute, Profsouznaya, 84/32, Moscow 117296
Russia

ABSTRACT

We report the first detection of the burst source GX 354-00 by the SIGMA telescope, occurred during a survey of the Galactic Center region performed between February and April 1992. During this $\approx$ 50 days survey the source underwent two high energy flares of about 15 days duration each. The spectrum can be described by a thermal Bremsstrahlung model with a temperature of 38 keV or by a broken power-law. The average integrated flux (30-200 keV) was $\approx 7 \times 10^{-10}$ ergs/cm^2/s.

INTRODUCTION

In the standard X-ray domain ($<$ 20 keV) GX 354-00 is a well-known persistent source, first detected by the Uhuru satellite[1] and successively observed to emit X-ray bursts[2]. The source is located in the Galactic Center (GC) region and is not optically identified. An association with a heavily reddened globular cluster[3] was not confirmed by later IR observations[4]. GX 354-00 therefore appears to be a typical low mass binary system (LMXB) of the low-luminosity burst source class[5] which very likely contains a neutron star.

As all other LMXB bursters GX354-00 has relatively soft spectrum and is not normally detected at energies higher than 30 keV. On the beginning of spring 1992 the coded aperture low gamma-ray (30-1300 keV) telescope SIGMA on the GRANAT satellite started a new survey of the Galactic Center region during which the source appeared to be very bright. The high quality data obtained during the observations of GX 354-00 provide the best known hard ($>$ 30 keV) X-ray spectrum of a bursting LMXB.

ANALYSIS AND RESULTS

Fifteen observations of the Galactic Center region were performed between February and April 1992 by the SIGMA telescope, spanning a total period of $\approx$ 50 days. Three sources were unambiguously detected during this survey[6]: 1E 1740.7-2942 which in its normal state dominates the hard X-ray emission of this region, the X-ray source in Terzan 2 globular cluster and, for the first time by SIGMA, GX 354-00. This source was always in the partially coded field of view of the instrument and observed with a sensitivity varying between 60 and 90 % of the on-axis telescope sensitivity. During the survey the source flux was seen to rise from the telescope sensitivity level to a maximum value in several days, to decrease on approximately the same time scale and then display another similar but weaker flare ≈ 10 days later.

Fig. 1 shows a contour plot image of the sky region around GX 354-00 in the 40-150 keV band obtained from the sum of images where the source was detected at more than 3 σ over the background level, for a total of 585205 s of dead-time corrected exposure. Crosses indicate the Einstein HRI position of GX 354-00[3] and the position of the nearby X-ray source X 1730-333, the Rapid Buster[7]. The 9 σ excess is well coincident with GX 354-00 and unambiguously separated from the Rapid Buster. The best fit position of the excess is at R.A.(1950) = 262.187° dec.(1950) = -33.810°, only 1.4 arcmin from the Einstein position, with an error circle of 2.9 arcmin radius (statistical uncertainty at 90 % confidence level in 4 parameters and satellite attitude errors).

The light curve of GX 354-00 in the 40-70 keV energy band is displayed in Fig.2, where is reported the detected source count rate versus the universal time. The source underwent two flares.

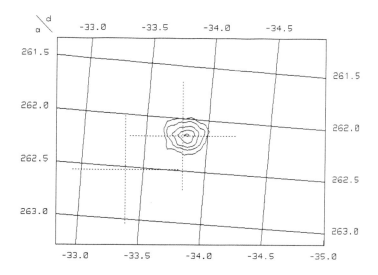

Fig. 1 Contour plot of the sky region around GX 354-00 in the 40-150 keV band obtained from the sum of the images of the 8 observations where the source was stronger. The contour levels are in units of standard deviations (σ) over the background mean, starting from 5 σ and spaced by 1 σ. GX 354-00 is here detected at 9.2 σ level. Crosses indicate the positions of GX 354-00 and the Rapid Buster.

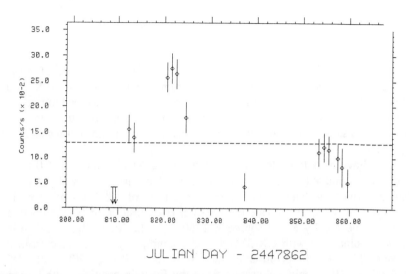

Fig. 2 Light curve of GX 354-00 in the 40-70 keV band measured by SIGMA between 17 Feb 1992 and 8 April 1992: detected count rate at the source position plotted versus the universal time (Julian Days). Errors and upper limits are at 1 σ level.

In the first one the flux increased at least of a factor of 10 on a time scale of 10 days and then decreased to the limit of the telescope sensitivity in $\approx$ 15 days. The second flare was weaker but the rise time and duration appear similar to the first one.

The average count spectrum obtained by the sum of the spectra measured when the source was visible at more than 1 σ in the 40-70 keV band is shown in Fig. 3. The source was detected up to $\approx$ 100-120 keV, with a relatively hard spectrum till 70 keV and very soft at higher energies. Power-law model gives a poor fit to the data points and better values of chi-square are obtained with thermal bremsstrahlung model or a broken power-law. Table I summarizes the parameters obtained from the fit of the data points between 30-200 keV with different models (errors are at 68 % confidence level in a single parameter). The best fit ($\chi_\nu^2 = 1.04$ for $\nu = 36$ d.o.f.) is in fact given by a broken power-law with photon indexes -1.4 and -4.6 and energy of break of $\approx$ 58 keV. The integrated photon flux in 30-200 keV assuming the best fit broken power-law is $7.0 \cdot 10^{-10}$ erg/cm^2/s. Using this spectral shape we estimate the upper limit of the source flux at 2σ level in the 40-70 keV band for the first 2 observations of $1.5 \cdot 10^{-10}$ erg/cm^2/s.

We searched for spectral variability by inspecting the variations of the hardness ratio between the different observations: the ratio of the count rate measured in the band 40-60 keV and the count rate in the 60-100 keV band did not vary significantly with the time. We also compared the average spectra of the source during the two flares (between Feb 21 - Mar 4 and Apr 2 - Apr 7) and we did not find significant differences between them.

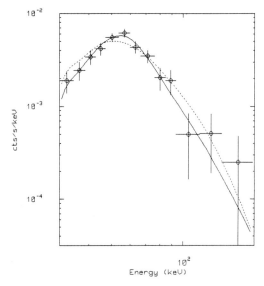

Fig. 3 Average count spectrum of GX354-00 obtained from the spectra of the observations where the source was detected by more than 1 σ in the 40-70 keV. The solid lines is the best-fit broken power-law model whereas the broken one shows the best-fit bremsstrahlung model (Table I).

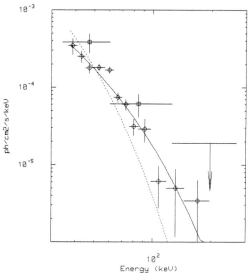

Fig. 4 Average photon spectrum of GX354-00 as measured by SIGMA (circles) with its best-fit bremsstrahlung thermal model (solid line) compared to the data points of GRIP experiment (squares) and the extrapolation of the MPC Einstein spectrum (thermal bremsstrahlung with kT = 17.7 keV) (dashed line).

CONCLUSIONS

In the standard X-ray energy band (< 30 keV) GX 354-00 is a persistent, though variable, source, but SIGMA detected it in hard X-rays only during about 12 out of a total 45 observations of the GC region (more than 700 hours of exposure) although it was always in the FOV of the instrument. GX354-00 is listed in the HEAO 1 A-4 catalogue[8] and was also detected in hard X-rays by the GRIP experiment[9].

Table I Spectral parameters for GX 354-00

Spectral model	Flux at 100 keV 10^{-5} ph/cm²/s/keV	Parameters	χ_ν^2 d.o.f $\nu = 38$
Power-Law	1.93 ± 0.20	$\alpha = -2.97 \pm 0.14$	1.59
Bremsstrahlung	1.66 ± 0.23	KT (keV) = 38 ± 4	1.31
Two Power-Laws	1.13 ± 0.35	$\alpha_1 = -1.40 \pm 0.70$	1.04
		$\alpha_2 = -4.64 \pm 0.65$	($\nu = 36$)
		E_0 (keV) = 57.6 ± 3.5	

Variations in the persistent flux at low energies up to a factor of 5 were observed previously[1,2,3] and the authors noted that the increase in intensity was correlated to an increase in the temperature. Fig. 4 show the SIGMA deconvolved spectrum of GX 354-00 compared to the GRIP data points[9] and the extrapolation of the spectrum observed by the Einstein observatory[3], a thermal bremsstrahlung with kT = 17.7 keV. The SIGMA spectrum is compatible with the GRIP points but not with the extrapolation of the Einstein spectrum. In particular the extrapolation at low energies of the thermal bremsstrahlung which describes the SIGMA points will clearly lie under the Einstein spectrum. Therefore in the hypothesis of a single thermal spectral component over the 1-100 keV band the SIGMA data are not consistent with the previous result of a correlation between intensity and temperature.

REFERENCES

1. Forman, W., et al., 1978, Ap. J. Suppl., 38:357.
2. Basinska, E.M., et al., 1984, Ap. J. 281:337.
3. Grindlay, J.E., and Hertz, P., 1981, Ap. J. 247:L17.
4. Van Paradijs, J., and Isaacman, R., 1989, Astr. Ap., 222:129.
5. White, N.E., et al., 1988, Ap. J., 324:363.
6. Cordier, B., et al., 1992, to be submitted to Astr. Astrop.
7. Bradt, H.V.D., and McClintock, J.E., 1983, Ann. Rev. A.A. 21:13.
8. Levine, A.M., et al., 1984, Ap. J. Supp. Ser. 54:581.
9. Cook, et al., 1991, Ap. J., 372:L75.

BATSE OBSERVATIONS OF EXO 2030+375

M. T. Stollberg, G. N. Pendleton, W. S. Paciesas
Department of Physics
University of Alabama in Huntsville

M. H. Finger
Computer Science Corporation

G. J. Fishman, R. B. Wilson, C. A. Meegan,
B. A. Harmon, C. A. Wilson
Marshall Space Flight Center, Huntsville

ABSTRACT

The transient x-ray pulsar EXO 2030+375, first detected by EXOSAT in 1985 May – August, has also been detected by the Large Area Detectors (LAD's) of BATSE. A major outburst occurred during 9 – 19 February 1992. BATSE has also seen this source on six other occasions at intervals of about 46 days. We present the pulse profile, spectrum, period history, and a refined orbital period for this binary pulsar.

INTRODUCTION

EXO 2030+375 was first detected by the EXOSAT satellite from May – August 1985 and again in October 1985[1]. Analysis of the data obtained from those viewings indicated EXO 2030+375 was an x-ray transient of the Be binary type with a binary period of 45.6 – 47.5 days and a pulse period of 41.8 sec.[2] The Burst And Transient Source Experiment (BATSE) aboard the Compton Gamma-Ray Observatory has seen EXO 2030+375 on more than six occasions with, the brightest outbursts being 9 – 19 February 1992 and 25 September – 9 October 1992. We present the first results of our observations.

PULSE PROFILE

BATSE detects many x-ray transients and pulsars using both pulsed and occultation analyses on its Continuous data. Figure 1 shows the pulse profiles of EXO 2030+375 for the two brightest periods seen by BATSE as a function of relative phase from epoch – folding analysis. The pulse phase origin is arbitrary. Two cycles are shown for clarity. In each case, the first profile is summed in energy from 20 – 120 keV, while each succeeding profile is over a smaller energy band. These bands are how BATSE divides up its observable energy range for the current configuration. A double peaked structure is seen in both profiles. By double peak we mean the peaks at 0.4 and 0.7 relative phase. These peaks are more pronounced than those of EXOSAT's[3], however, BATSE observes EXO 2030+375 in a higher energy range. BATSE has not detected the variations in pulse profile reported by EXOSAT[2] primarily due to our inability to observe this pulsar beyond a 14 day range centered about periastron.

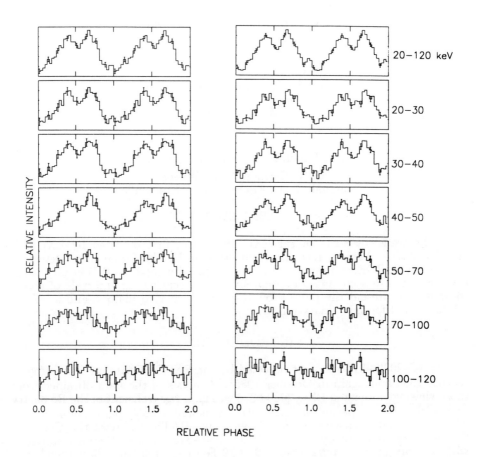

Figure 1. The pulse profile of EXO 2030+375 as seen by BATSE during the 9 − 19 February outburst (left) contrasted with the profile seen by BATSE during the 27 September − 9 October outburst (right). The zero phase point is arbitrary.

PHOTON SPECTRUM

The spectrum shown in figure 2 is from epoch - folded Continuous data for the 9 − 19 February viewing period. A power law fit to the data over the full viewing period gives a spectral index of $\gamma = -3.32 \pm 0.10$. Power law fits to each of the two peaks in the profile (relative phase 0.35 to 0.55 and 0.60 to 0.75) have indices consistent with each other and this result. Background for these spectral fits is obtained from the interval of relative phase 0.80 to 1.25.

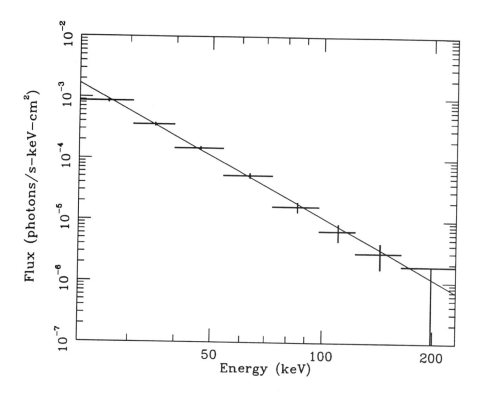

Figure 2. Photon spectra for the 9 – 19 February EXO 2030+375 pulse profile. A power law fit yields a spectral index of $\gamma = -3.32 \pm 0.10$.

Figure 3 shows the photon spectrum for the interval 25 September – 9 October 1992. A power law fit to the data yields a spectral index of $\gamma = -3.24 \pm 0.11$, consistent with the 9 – 19 February index. Inset to this is the photon spectrum produced using occultation data during the entire outburst period. The spectral index $\gamma = -4.19 \pm 1.12$ agrees with that obtained from the pulsed analysis, with reduced statistical significance due to the much shorter livetime available with this technique.

ORBITAL FIT

Figure 4 shows a preliminary orbital fit to the data from the two bright outbursts and one intermediate one. The results of this fit are presented below in Table 1. P_{orb} and the eccentricity are within 1σ of those values presented by EXOSAT.[2] $a_x sin(i)$ is 2σ from their lowest value and our ω is 5σ above EXOSAT's highest value. Work is in progress to further refine our orbital parameters.

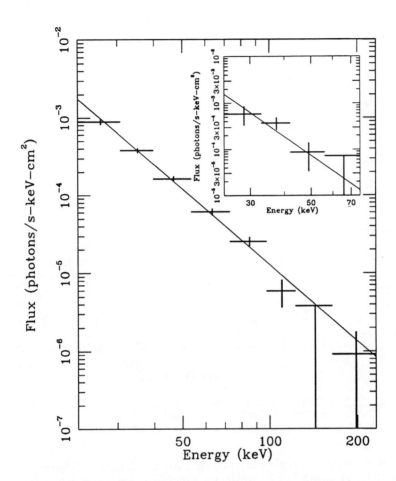

Figure 3. The photon spectrum for 25 September - 3 October. A power law fit gives $\gamma = -3.70 \pm 1.07$ for this data. Inset is the photon spectrum from the occultation analysis, for which $\gamma = -4.19 \pm 1.12$.

Table 1. Exo 2030+375 Parameters	
P_{orb}	46.02 ± 0.01 days
Eccentricity	0.37 ± 0.02
$a_x sin(i)$	206 ± 12 s
$T_{\pi/2}$	2448781.41 ± 0.03 JD
T_p	$2448800.38 \pm .03$ JD
ω	238 ± 3 deg

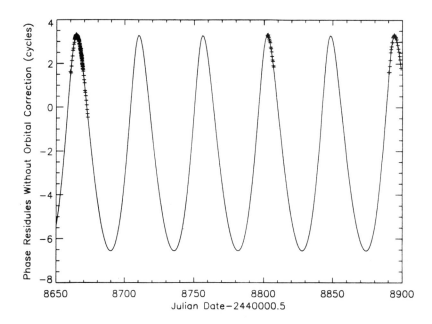

Figure 4. Orbital fit for EXO 2030+375 using BATSE data from 9 – 19 February, 29 June – 9 July, and 25 September – 9 October 1992. Data exists for the three other peaks shown and will be used for a better fit.

REFERENCES

[1] Parmar, A. N., Stella, L., Ferri, P., and White, N. E. 1985, IAU Circular No. 4066.
[2] Parmar, A. N., White, N. E., Stella, L., Izzo, C., and Ferri, P. 1989, Ap. J., 338, 359.
[3] Parmar, A. N., White, N. E., and Stella, L. 1989, Ap. J., 373 - 380.

BATSE OBSERVATIONS OF GS 0834-430

C.A. Wilson, B.A. Harmon, R.B. Wilson, G.J. Fishman, C.A. Meegan
NASA - Marshall Space Flight Center, ES-62, 35812

D. Chakrabarty, J.M. Grunsfeld, T.A. Prince
Caltech, 220-47, Pasadena, CA, 91125

M.H. Finger
CGRO Science Support Center

W.S. Paciesas and G.N. Pendleton
Dept. of Physics, University of Alabama, Huntsville, 35899

ABSTRACT

The X-ray pulsar GS 0834-430, first seen by GRANAT in 1990[1], and correctly identified as a new transient pulsar by GINGA[2], has been observed with the Large Area Detectors (LAD's) of the Burst and Transient Source Experiment (BATSE) on the Compton Gamma Ray Observatory. We present the history of the pulsed flux, the total flux, and the pulsed frequency as well as representative spectra and pulsed profiles.

INTRODUCTION

Using the LAD's, BATSE has observed GS 0834-430 essentially continuously since April of 1991 using Earth occultation, epoch-folding, and fast Fourier transform (FFT) analyses. Pulses from GS 0834-430 have been seen in the 20 - 70 keV range. With an orbital period of about 109 days and a spin period of about 12.31 seconds, GS 0834-430 is consistent with the distribution of Be/X-ray binaries on an orbital versus spin period diagram. Other characteristics that resemble other Be/X-ray binary systems are the bright outbursts and the observed 109 day period[3]. BATSE's nearly continuous coverage shows GS 0834-430 to have very regular outbursts. We also may be detecting weak x-ray flux between outbursts.

OCCULTATION AND FFT ANALYSES

Using Earth occultation analysis[4], BATSE has been able to observe GS 0834-430 continuously for the entire duration of the mission. Figure 1 is a flux history which shows frequent outbursts approximately every 109 days, lasting 40-50 days. Between the outbursts, there are interesting features in the occultation data. The measurements in these intervals suggest that the source has persistent emission between large outbursts. These fluxes are potentially due to an unmodeled source, which would produce a signature that repeats with the CGRO orbital precession period of 53 days. Analysis of intervals containing these features is in progress.

Observations of GS 0834-430 using FFT analysis have also shown 40-50 day outbursts every 109 days. Figure 2 shows the square root of the source power normalized to the local FFT noise power for GS 0834-430 on each day. Since the mean noise level probably varies from day to day, systematic trends in the normalized power may be present. Amplitudes below about 3 are not considered to be significant detections, though after TJD8700 a decrease in intensity due to data gaps must be taken into account.

The results of occultation and FFT analysis give similar time histories of GS 0834-430. Outbursts clearly correspond in both figures 1 and 2. The features between outbursts are not clearly evident in the FFT amplitude history. If these features are indeed from GS 0834-430 but are quite weak, the FFT analysis may not be sensitive enough to detect them, especially after the reduction in livetime after TJD 8700 due to the loss of the CGRO tape recorders.

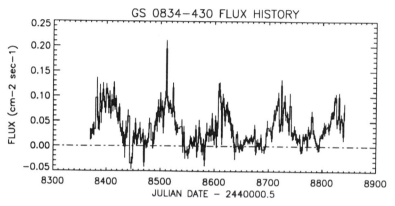

Figure 1. - The flux history (20-160 keV) obtained by Earth occultation for GS 0834-430.

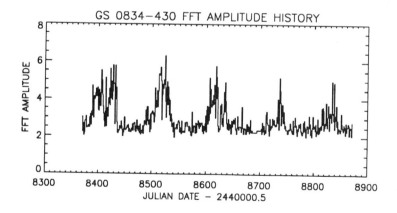

Figure 2. - The Fast Fourier Transform (FFT) amplitude time history for GS 0834-430.

GS 0834-430 PULSED PROFILES

Shown in figure 3 are representative pulsed profiles for GS 0834-430 from September 7 - 12, 1991, (TJD 8506-8511) the brighter portion of an outburst. Shown in figure 4 are pulsed profiles from August 22 to September 5, 1991, (TJD 8490-8504) for 30 - 40 keV and 40 - 50 keV that show a clear difference in pulse shape in the relative phase range 0.9 - 1.4. Due to the rebinning technique used, the data in figure 4 have neighboring bins that are correlated. The pulsed profiles in figure 3 were generated using a different technique that does not have correlated neighboring bins.

As in the other analyses, studies of the intervals between outbursts are in progress.

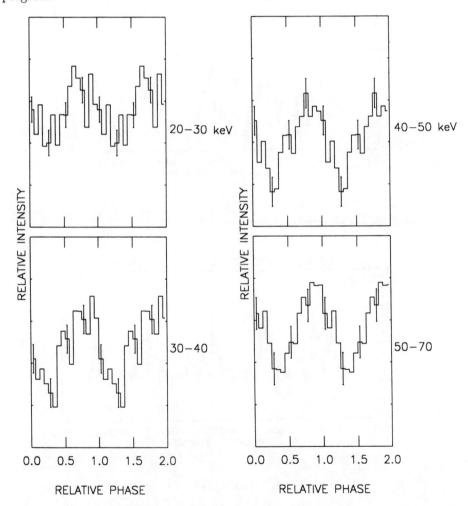

Figure 3. - Pulsed profiles for GS 0834-430 were generated by on ground epoch-folding of 2.048 second resolution continuous data, for four energy ranges (TJD 8506-8511).

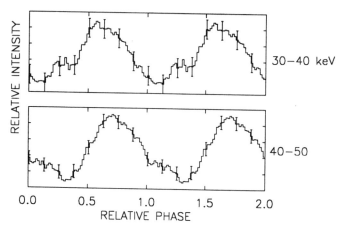

Figure 4. - A change in pulse shape occurs between the 30-40 keV and the 40-50 keV pulsed profiles (TJD 8490-8504).

SPECTRA FOR GS 0834-430

Several spectra for GS0834-430 were generated from occultation data and epoch-folded data. The occultation technique determines the total flux, while the epoch-folding technique measures only the pulsed component. Two representative spectra from TJD 8490-8504 are shown in figure 5. Spectra for both data types were fit with a simple power law. Although the fits are not extremely good, it appears that the spectrum generated from occultation data is steeper. The fits to the occultation data have spectral indices ranging from 3.7-4.1, while the fits to the epoch-folded data have spectral indices ranging from 2.9-3.3. The pulsed fraction at 40-50 keV is approximately 0.09.

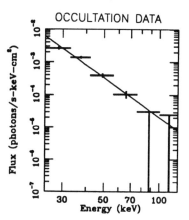
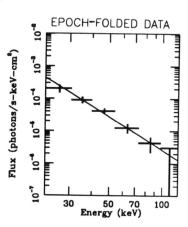

Figure 5. - Representative spectra for GS 0834-430, generated from Earth occultation and epoch-folding analyses (TJD 8490-8504).

GS 0834-430 PERIOD HISTORY AND ORBITAL FIT ATTEMPTS

Figure 6 shows the period history generated from epoch-folded data. Several orbital fits were applied to the phases from the epoch-folded data, using different models. The models ranged from ones with constant acceleration to ones assuming that acceleration occurred only during outbursts. These preliminary fits gave a wide range of possible parameters. The potential presence of torque noise in the data is causing considerable difficulty in determining an orbit. The data clearly show an approximately 109 day orbital period. Determination of the remaining orbital elements is in progress.

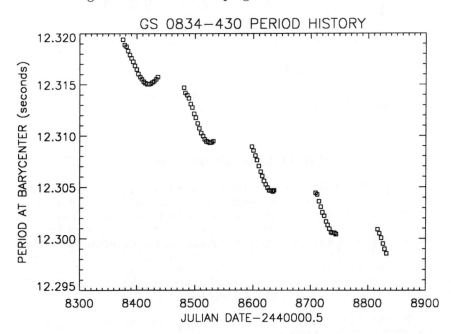

Figure 6. - The period history obtained for GS 0834-430 from epoch-folding analysis.

REFERENCES

1. R.A. Sunyaev, IAU Circ., No. 5122 (1990).
2. F. Makino, IAU Circ., No. 5142 (1990).
3. L.B.F.M. Waters, Proc. 23rd ESLAB Symp. on Two Topics in X-Ray Astronomy, Bologna, Italy, 1989 (ESA SP-296), p. 25, (1989).
4. B.A. Harmon et al., The Compton Observatory Science Workshop, eds. C.R. Shrader, N. Gehrels, and B. Dennis, (NASA CP-3137), p. 69, (1992).

BATSE OBSERVATIONS OF THE MASSIVE X-RAY BINARY 4U1700-37/HD153919

Brad C. Rubin

USRA/NASA – Marshall Space Flight Center, ES-62, 35812

B. A. Harmon, G. J. Fishman, C. A. Meegan, R. B. Wilson

NASA – Marshall Space Flight Center, ES-62, 35812

M. S. Briggs, W. S. Paciesas

Dept. of Physics, Univ. of Alabama, Huntsville, 35899

M. H. Finger

CSC/CGRO Science Support Center, ES-62, 35812

ABSTRACT

The Earth occultation technique has been used to monitor the massive x-ray binary (MXRB) 4U1700-37/HD 153919 with BATSE. This source has been important in the study of stellar winds from O-type stars and of radiation and flaring mechanisms of MXRB. We have detected several large flaring episodes, during which the 20 to 50 keV flux from this binary exceeds the flux from the Crab Nebula in the same energy range.

INTRODUCTION

The full sky coverage of the BATSE experiment allows nearly continuous monitoring of hard x-ray sources by the Earth occultation technique.[1,2] Most sufficiently strong sources can be observed to rise and set behind the Earth once per orbit (every 90 minutes). This capability should lead to a major advance in our understanding of the massive x-ray binary 4U1700-37. This source was first observed by Uhuru[3] and has been observed at x-ray energies by OSO-8[4], EXOSAT[5], and several other experiments. With BATSE, weak, quiescent emission is often detectable at times of Earth occultation and flaring episodes can be seen directly in the raw data.

4U1700-37 is known to be a neutron star (but not a pulsar) accreting material from the wind of a massive companion, the O6f star HD 153919. EXOSAT determined the period of the binary system to be 3.411652 ± 0.000026 days.[5]

LONG TERM BEHAVIOR

Figure 1 shows the flux history as measured at each Earth occultation over 130 days in the 20-160 keV energy band. The error bars in this plot show only the statistical errors. The quiescent emission from this source is quite variable and averages about 0.1 photons/cm^{2}sec. The Crab flux in this energy range is about 0.25 photons/cm^{2}sec. The light curve also contains evidence for the binary period.

Figure 2 shows the flux history folded into 20 phase bins at a 3.41 day period. A similar plot with 100 phase bins folded at a 3.41165 day period shows that eclipse of the neutron star by its companion lasts for at least 22% of the orbit and possibly slightly longer. This leads to an estimate of the eclipse semi-angle $\theta_E = 39.6° \pm 3.6°$. A quiescent spectrum computed by averaging over a two week interval (one GRO pointing interval) is shown in Figure 3. We fit this spectrum to the optically-thin-thermal-bremsstralung model $Ae^{-E/kT}/E$ and find an amplitude $A = 0.17 \pm 0.02$ photons/cm^2 sec keV and a temperature $kT = 29.6 \pm 2.3$ keV with chi-square per dof = 6.7/7, consistent with most earlier observations.

FLARING EPISODES

Several flaring episodes are evident in the light curve of Figure 1. Most episodes last fom a few hundred to a few thousand seconds, but occasionally the source will flare for a substantial fraction of a day. Many of these episodes are visible in the raw data returned by the spacecraft. We have analyzed a flaring episode which occured on truncated julian day 8552 (October 23, 1991). The episode lasted for at least 4000 seconds. Data from a portion of the episode, in the 20 to 50 keV band, are shown in Figure 4. Detectors 2 and 6 had nearly equal angles to the source. The setting edge near the end of the figure is an Earth occultation. A pulse train consisting of at least six pulses, separated by about 100 seconds, is evident in the figure. The spectrum during the flare, shown in Figure 5, fits well to an optically-thin-thermal-bremsstralung model. The temperature is consistent with the quiescent temperature. We find an amplitude $A = 1.00 \pm 0.23$ photons/cm^2 sec keV and a temperature $kT = 25.7 \pm 3.0$ keV with chi-square per dof = 7.4/7.

CONCLUSION

Near continuous monitoring of 4U1700-37 by BATSE provides important new data for the understanding of this source. The flux history reveals variable quiescent emission and significant flaring episodes every 10-20 days. We find an eclipse semi-angle of $39.6° \pm 3.6°$ and spectra of similar shape in quiescent and flaring states, consistent with earlier observations.

REFERENCES

1. Harmon, B. A., et al., "Earth Occultation Measurements of Galactic Hard X-Ray/Gamma-Ray Sources: A Survey of BATSE Results", these proceedings.
2. Skelton, R. T., et al., "Status of the Enhanced Earth Occultation Analysis Package for Studying Point Sources", these proceedings.
3. Jones, C., et al., Ap. J., 181 (L43) (1973).
4. Dolan, J. F. et al., Ap. J., 238 (238) (1980).
5. Haberl, F., White, N. E., Kallman, T. R., Ap. J. 343 (409) (1989).

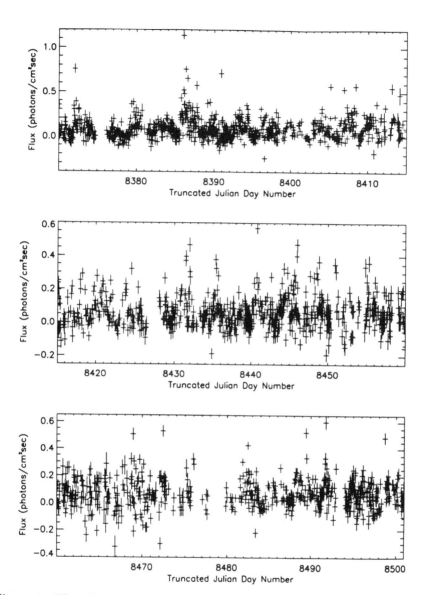

Figure 1. Flux Estimates for 4U1700-377 in the 20-150 keV energy band as a function of time. The vertical bars are statistical error bars. Each point occurs at a time of Earth occultation.

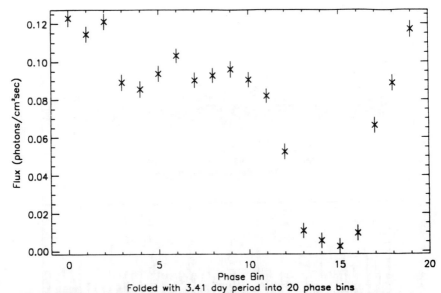
Figure 2. Flux estimates from TJD 8370-8500 folded with a 3.41 day period.

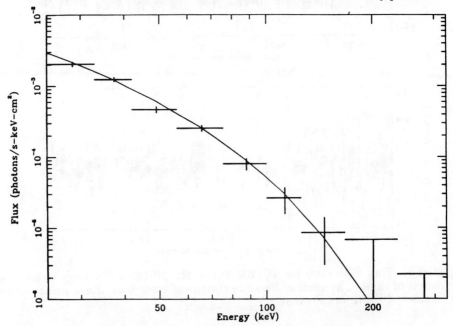
Figure 3. 4U1700-377 spectrum averaged over TJD 8422-8435.

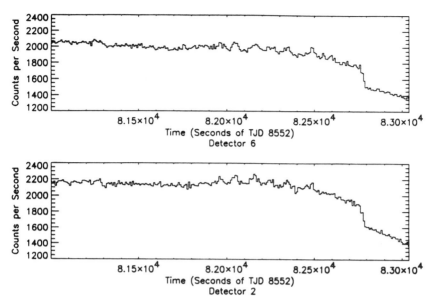

Figure 4. A portion of the flaring episode on TJD 8552.

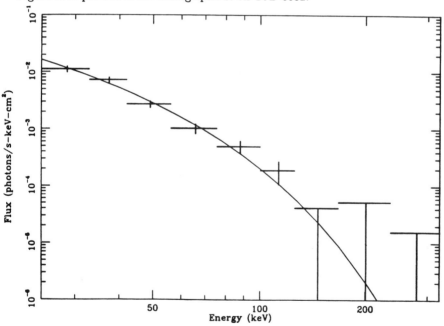

Figure 5. Spectrum at Earth occultation of flaring episode on TJD 8552.

BATSE OBSERVATIONS OF CEN X-3

M. H. Finger†, R. B. Wilson, B. A. Harmon
G. J. Fishman, C. A. Meegan
NASA Marshall Space Flight Center, AL 35812

W. S. Paciesas
Dept. of Physics, University of Alabama in Huntsville, AL 35899

ABSTRACT

We present results from 520 days of continuous hard x-ray observation of the x-ray binary pulsar Cen X-3. These observations were made with the Burst And Transient Source Experiment (BATSE) on the Compton Gamma Ray Observatory. Results of this analysis include a new determination of the binary orbital parameters, and histories of the pulse frequency and pulsed flux during the observation.

INTRODUCTION

Cen X-3 is an x-ray binary pulsar that has been observed sporadically over the last 20 years. The pulsar has a rotation period of 4.82 s which has been observed to decrease over the long term[1]. The pulsar is in a binary system with a massive O star. Optical observations of the companion have shown that the x-ray source has an accretion disk that is fed by Roche lobe overflow[2]. The binary orbit has a period of 2.087 days and is circular with a projected semimajor axis of 39.6 light seconds. Comparison of the precise orbit determinations made over the last 20 years reveal that the orbital period is gradually decreasing[3].

Hard x-ray observations of Cen X-3 have been performed continuously by the Burst And Transient Source Experiment (BATSE) on the Compton Gamma Ray Observatory since the spacecraft's launch. BATSE has eight uncollimated Large Area Detectors (LAD's) that allow pulsed sources like Cen X-3 to be monitored in the hard x-ray at all times when not occulted by the Earth. Each LAD consists of a .5 inch thick by 20 inch diameter NaI(Tl) scintillator with a frontal area of 2025 cm^2. Due to absorbing material the response of the detectors falls off rapidly below 20 keV[4].

We have performed an epoch folding analysis of the the pulse timing and pulsed flux of Cen X-3 over our 520 day observation period using the 20-50 keV energy channel of the BATSE LAD Discriminator (DISCLA) data. The 1.024 s resolution of this data is adequate for obtaining low resolution pulse light curves of Cen X-3. The 20-50 keV range of this channel is well matched to the detector's response to the spectrum of Cen X-3. The pulsed flux spectrum obtained from higher energy resolution BATSE data can be characterized by a photon power law index of -4.5 with a cutoff at approximately 70 keV.

After discussing our epoch folding procedure, we present the pulse frequency history and pulsed flux history obtained from the timing analysis of the DISCLA data. We then present a new determination of the Cen X-3 binary system orbital parameters.

† Astronomy Programs, Computer Sciences Corporation

EPOCH FOLDING ANALYSIS

In our analysis, data were rejected from periods when Cen X-3 was occulted by the Earth, or during periods when solar flares, gamma ray bursts, or other events caused rapid variations in the counting rates. The rates were then summed over the detectors exposed to Cen X-3, with a weighting for each detector given by the cosine of the angle between the detector's normal and the source direction.

Data in intervals of 10000 s were fit to a model. In this model the pulsed source flux was represented by a 3rd order harmonic expansion in the pulsar phase. Higher order terms were neglected because they cannot be measured with the 1.024 s resolution DISCLA data. To represent the background, the interval was divided into 300 s segments, and the background in each segment modeled as a quadratic in time. The background model value and slope were constrained to be continuous at segment boundaries. In the calculation of pulsar phases for the folding of the DISCLA data the planetary ephemeris DE200 was used in conjunction with a preliminary binary ephemeris and pulsar rest frame phase model.

The parameterized light curves obtained from the interval fits were then correlated with a pulse template to obtain a Cen X-3 pulse intensity and the phase offset from the preliminary ephemeris for each interval. The flux template resulted from folding the DISCLA data during a strong outburst, subtracting the mean, and normalizing to obtain a mean square value of one.

FREQUENCY AND PULSED FLUX HISTORY

Figure 1 shows an example of the pulse phase and intensity over a 25 day interval with 10000 s resolution. The phase has had a constant frequency model subtracted. The flux history shows several outbursts with durations of order 10 days, periodically interrupted by the eclipsing of the pulsar by its binary companion every 2.087 days. The phase history shows a smooth change in frequency with the angular acceleration reaching $6 \cdot 10^{-12}$ cycles/sec^2 and then reversing sign within 10 days.

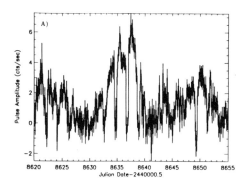 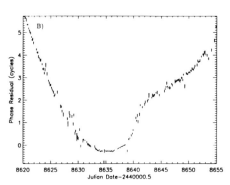

Figure 1. Example of the pulse phase and intensity obtained from the DISCLA timing analysis shown at 10000 s resolution. A) shows the pulse intensity, B) shows the phase minus a constant frequency (0.2075295 Hz) model.

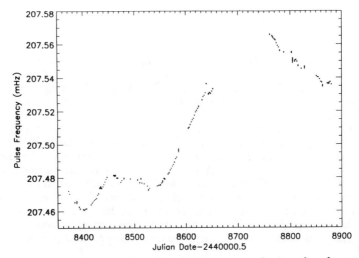

Figure 2. Cen X-3 rest frame pulse frequency during the observation.

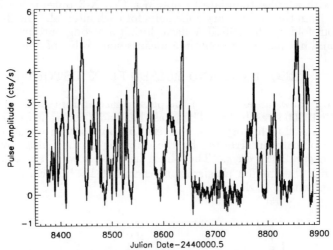

Figure 3. Cen X-3 pulsed flux during the observation.

Figure 2 shows the frequency history of Cen X-3 during our observation. This was obtained by fitting the corrected phases over single orbit periods. The missing intervals are periods with no positive detection. The source shows extended periods of both smooth spin-up and erratic spin-down.

The history of the pulsed flux with binary orbital period resolution is shown in Figure 3. A smooth fit was made to the phase history and eclipse-to-eclipse lightcurves accumulated with this phasing. The pulse amplitude was measured from these full orbit light curves. Cen X-3 is seen to episodically outburst in events with typical durations of 10-40 days. There is no obvious correlation between the outburst behavior and the spin-up or spin-down state.

For comparison, Figure 4 shows the total flux as derived by occultation analysis of the CONT data. There is a clear correspondence between many of the peaks of the pulse and occultation derived flux; however, there are also many discrepancies. This may be due to source confusion and other systematic effects as well as variation of Cen X-3's pulsed fraction.

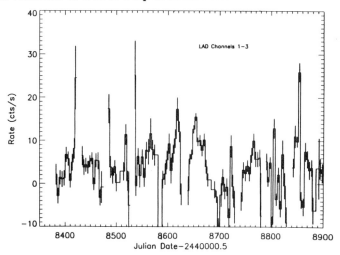

Figure 4. Total flux from Cen X-3 as obtained by occultation analysis of CONT data.

ORBIT ESTIMATION

Thirteen phase-connected segments of the phase history were selected and independent orbital parameters estimated from each. The mean of the estimated parameters was then obtained, and the scatter about the mean was used to critically evaluate the formal errors.

In each fit the orbital period was held fixed at a value of 2.0870686 days and all other orbital parameters estimated. In each fit the pulsar angular acceleration was modeled as an exponentially correlated noise process with a variance of $7.0 \cdot 10^{-12}$ cycles/sec^2 and a correlation time of 5 days. This is believed to be representative of the torque noise at orbital periods. The individual fits had reduced chi-squares in the range of 10-25. This is believed to be due to variations in pulse shape, which are currently under investigation.

The estimated orbital epochs were fit with a central epoch and an orbital period. The reduced chi-square of this fit was 1.83. The remaining parameters were fit to a constant value, with reduced chi-squares in the range of 0.8 to 1.4. From this we see that scatter in the estimated orbital parameters is nearly consistent with their formal errors, despite the fact that the reduced chi-squares of the fits were much to large. This indicates two things: First, the level of torque noise rather than the accuracy of our phase measurements is controlling the accuracy of the measurements. Second, our statistical model for the pulsar angular acceleration was reasonable at orbital time scales. The mean orbit, with errors scaled to the level of scatter, is given in Table I.

Epoch (mid Eclipse)	$T_{\pi/2}$	JD 2448562.156702(71) TDB
Period	P_{orb}	2.08706533(49) days
semi-major axis	$a_x \sin i$	39.627(18) sec
eccentricity vector	$e \cos \omega$	0.00013(35)
	$e \sin \omega$	-0.00090(53)

Table I. Cen X-3 mean orbital elements.

Figure 5 shows the individual orbital epochs estimates compared to previous measurements[3,5]. The ephemeris proposed by Nagase et al.[3],

$$T_{\pi/2} = T_0 + nP_{orb} + \frac{n^2}{2} P_{orb} \dot{P}_{orb} \qquad (1)$$

with n the orbit count, T_0 = JD 2440958.851128 TDB, P_{orb} = 2.08713845 days, and $\dot{P}_{orb} = -9.93 \cdot 10^{-9}$, has been subtracted to allow comparison.

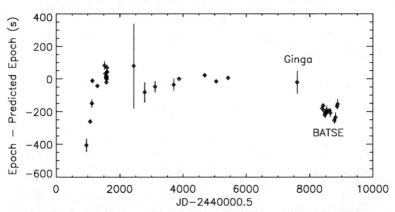

Figure 5. History of Cen X-3 orbital epoch. A quadratic ephemeris (see text) proposed by Nagase et al.[3] has been subtracted to allow comparison.

REFERENCES

1. F. Nagase, P.A.S.J. **41**, 1-79 (1989).
2. S. A. Tjemkes, E.J. Zuidewikjk, and J. van Paradijs, Astron. Astrophys. **154**, 77 (1986).
3. F. Nagase et al., Ap. J. **396**, 147-160 (1992).
4. G. J. Fishman et al., Proc. GRO Science Workshop (Goddard S.F.C, 1989), p. 2-39.
5. R. L. Kelly et al., Ap. J **268**, 790-799 (1983).

CORRELATED OPTICAL OBSERVATIONS OF SCO X-1 USING THE BATSE SPECTROSCOPY DETECTORS

B. McNamara, G. Fitzgibbons
New Mexico State University, Las Cruces, NM 88003

G.J. Fishman, C.A. Meegan, R.B. Wilson, B.A. Harmon
NASA - Marshall Space Flight Center, ES62, 35812

W.S. Paciesas
Department of Physics, University of Alabama, 35899

B.C. Rubin
USRA/NASA - Marshall Space Flight Center, ES62, 35812

M.H. Finger
CSC/CGRO - Marshall Space Flight Center, ES62, 35812

ABSTRACT

We have investigated the feasibility of using the 8-16 keV energy channel of the BATSE spectroscopy detectors to study the correlated short and long term optical and high energy behavior of Sco X-1. We have found that in this energy range, the spectroscopy detectors are well suited to these tasks. To demonstrate its short period monitoring potential, simultaneous BATSE 8-16 keV and optical Sco X-1 light curves are presented. Longer term activity cycles also appear to be present. BATSE spectroscopy data covering a 50 day period show distinct episodes of Sco X-1 activity.

INTRODUCTION

Sco X-1 is the brightest continuous X-ray source in the sky. Its high energy emission results from the transfer of mass from a low mass nondegenerate star onto a companion neutron star. It is representative of a class of similar objects referred to as low mass X-ray binaries (LMXB). Because Sco X-1 serves as the proto-type of an important class of X-ray emitters and since its detectable emission is so large, it warrants extended study.

One of the most fruitful techniques for studying a LMXB is by simultaneously monitoring its emission at a variety of different wavelengths. These correlated datasets can be used to probe the source environment, investigate emission mechanisms, and examine the mass transfer process itself. In principle, the BATSE spectroscopy detectors (SD) have the capability of providing a nearly continuous 8-16 keV record of Sco X-1 activity. The aim of this study is to investigate the feasibility of using BATSE SD data and simultaneous optical data to study this source. For a review of the spectroscopy detector properties see Fishman et al. 1989 and Schaefer et al. 1992.

OBSERVATIONAL CONSIDERATIONS

In order to obtain simultaneous ground-based data with BATSE, three satellite related constraints must be observed. Satellite observations can only be obtained when (1) Sco X-1 is unobscured by the earth, (2) when it is outside the South Atlantic Anomaly (SAA), and (3) when TDRSS is available to transmit data to ground facilities. Each of these constraints influences the extent of the joint optical/high energy observational windows. Currently the most severe constraint is the short notification period of the availability of TDRSS communication windows. This limits the planning time for joint satellite/ground-based observations to a few days prior to the actual observational date.

MEASURING THE BATSE SD 8-16 keV SCO X-1 SIGNAL

There is no doubt that a BATSE SD has sufficient sensitivity to monitor Sco X-1. Example signals, as seen in the raw SD data, are shown in Figure 1. The flux received from this source can be obtained in two fashions. The first procedure involves measuring the increase and decrease of detected flux as Sco X-1 rises or sets relative to the earth's limb as seen by CGRO. Given the celestial coordinates and spacecraft orientation, earth rise and set times can be calculated to within a few seconds. The technique, as applied to the BATSE large area detectors, has been discussed by Harmon et al. (1992). The same procedure can be applied to the spectroscopy detectors. The second technique yields the entire flux curve for the source while it is above the earth's limb provided the background signal can be successfully subtracted. Two methods were developed to estimate the 8-16 keV background.

Method 1: This method models the measured cosmic background radiation with a 3rd order polynomial fit in terms of the cosine of the angle between the detector and the center of the earth. Provided that flux measurements are available over a significant portion of the orbit prior to the rise of Sco X-1, this procedure works well. Because the background remains reasonably constant for a given earth angle over a few orbits, the background fit obtained during one orbit can be used for nearby orbits with little loss in accuracy. Because of the observational considerations mentioned above, this is sometimes necessary.

Method 2: This method also relies upon the assumption that over a few hours, the cosmic background 8-16 keV radiation remains constant with orbital position. An additional assumption is that during some orbits, the 8-16 keV emission from Sco X-1 is constant. This latter assumption has been tested using BATSE SD data and appears to be valid. Background subtraction is then accomplished by using an orbit which shows a constant Sco X-1 signal as a reference and subtracting it, point by point, from a nearby orbit. The computed rise or set times of Sco X-1 are used to determine the starting location of this process. The flux level of the subtracted signal is then adjusted using the measured Sco X-1 step size from the reference orbit.

RESULTS

Figure 2 shows the simultaneous optical and BATSE 8-16 keV SD signal for truncated Julian day 8791 (June 18 ,1992). The background was subtracted using Method 2. The top portion of the figure displays the optical data and the bottom portion displays the background subtracted BATSE SD data. During this orbit, Sco X-1 was in a flaring state. As can be seen, the BATSE 8-16 keV light curve closely tracks the optical data. There appears to be a slight time delay between the optical and high energy light curves in the sense that the optical activity follows the high energy signal. This time delay is presumably due to the reprocessing of the high energy signal in material located within the binary system. One would also expect the optical signal to appear broader than the high energy signal if it has been reprocessed over a significant area. At this point, data analysis is not sufficiently advanced to state whether this smearing is present or not. A second interesting feature is that the optical and high energy *continuum* radiation are not correlated. This may be due to inadequacies in the background subtraction model or may be due to the fact that optical continuum variations reflect a longer term system response to the high energy output.

The spectroscopy detectors give us a unique way of studying longer term, high energy activity cycles in Sco X-1. Figure 3 shows a nearly continuous 50 day record of 8-16 keV Sco X-1 activity. Each point represents the flux step observed either at an earth rise or earth set. There is an indication in this figure that Sco X-1 undergoes periods of significantly enhanced activity lasting about 2-3 days and separated by approximately 10 days of much more modest activity (ex: TJDs 8791-8793, 8810-8812, 8816-8819). A more extended monitoring time period is clearly needed to establish the reality of this cycle. The BATSE spectroscopy detectors provide a unique tool for examining this and other longer term variations in high energy emission from this source.

REFERENCES

Fishman, G.J. et al. (1989). BATSE: The Burst and Transient Source Experiment on the Gamma Ray Observatory *Proceedings of the GRO Science Workshop*, p2-39. Greenbelt, Maryland, April 10-12, 1989.

Harmon, B.A. et al. (1992) Occultation Analysis of BATSE Data - Operational Aspects,*The Compton Observatory Science Workshop*, p 69, Annapolis, Maryland, Sept 23-25, 1991.

Schaefer,B.E. et al. (1992). BATSE Spectroscopy Analysis System,*The Compton Observatory Science Workshop*, p53, Annapolis, Maryland, Sept 23-25, 1991.

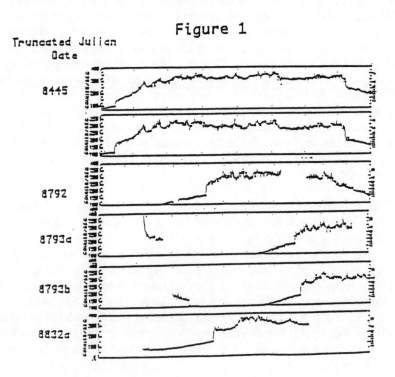

Figure 1

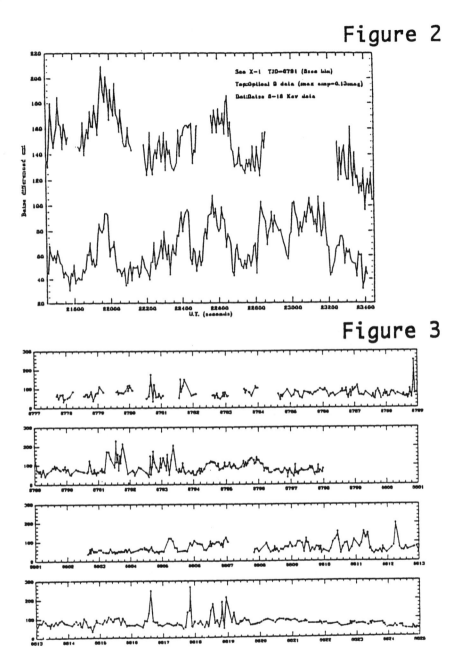

Figure 2

Figure 3

THERMAL HARD X-RAY AND GAMMA RAY EMISSION OF DISK-ACCRETING BLACK HOLES

Edison P. Liang
Rice University, Houston, TX 77251-1892

ABSTRACT

This paper reviews the observational data and thermal models of hard x-ray and soft gamma ray emissions of black hole candidates.

INTRODUCTION

Until recently it was very difficult to search for black hole candidates. For almost two decades the list of x-ray sources whose mass estimates based on binary orbital parameters comfortably exceed the canonical upper limit of ~ 3 $M_\odot$ for neutron stars consisted only of Cyg X-1 and LMC X-3, with LMC X-1 and A0620-00 as strong possibilities[1] (c.f. Table 1). This is because in the x-ray range black holes are greatly outnumbered by accreting neutron stars whose x-ray properties in most respects are very similar. This situation has now changed dramatically. Thanks to the discoveries of GRANAT and Compton Observatory (see contributions in this volume), we now have a whole new zoo of gamma-ray-selected black hole candidates listed in Table 1. The fact that several of these gamma-ray-selected candidates indeed turn out to have mass limits comfortably above the neutron star limit provide new support to the premise that enhanced gamma ray emissions (we define all photons > 100 keV as gamma rays here) is a good, though not necessarily conclusive, criterion for accreting black holes.

Table 1. New "Zoo" of Black Hole Candidates

	Mass-selected	(>100 keV) Gamma-ray-selected
quasi-steady sources	Cyg X-1 (>9 $M_\odot$)m+* LMC X-3 (>7$M_\odot$)m LMC X-1 (> 2.5 $M_\odot$)m	GX 339-4* 1E1740.7-2942+ GRS 1758-258
transients or XRN	A0620-00 (>3.2 $M_\odot$) V404 Cyg / GS 2023+338 (>6.3 $M_\odot$)[2] Nova Muscae / GRS 1124-684 (>3.7 $M_\odot$)+* [3]	Briggs Source+ GS 2000+25 GS 1826-24 GRS 1734-292 GRS 1915+105 Nova Per / GRO J0422+32+* EXS 1737.9-2952+

Footnotes: m = massive companion
   + = pair-related features (511 or 170 keV lines or MeV bumps)
   * = low frequency QPOs

A striking, almost universal, feature of all these candidates is the strong variability of their gamma ray emission, even among strong quiescent x-ray emitters such as Cyg X-1, GX339-4 and 1E1740.7-2942 (hereafter abbreviated as 1E). A large distinct subclass is the X-ray novae (XRN) which have no detectable quiescent x-ray emissions and have well-defined light curves during their outburst. These all seem to have low-mass companions and the outburst mechanisms have been likened to those of dwarf-novae: an accretion disk instability triggered by thermal instabilities at low temperatures[4]. Independent of the ultimate triggering mechanism, the duration of the outburst s(~ month) is much longer than any of the inner disk dynamic time scales (thermal, orbital or drift) so that we can treat the emissions as practically steady state. Similarity of their spectral behavior to those of Cyg X-1 and GX339-4 suggests that we should have a unified emission model for both classes of sources.

GAMMA RAY SIGNATURES OF BLACK HOLES

The following is a partial list of the high energy signatures of black hole candidates. Individually, none of them is sufficient as a telltale sign of a black hole. But taken together, they would strongly argue against a neutron star origin.
1. Power-law gamma ray tail (>100 keV) with photon number index $\lesssim 2$ and exponential cutoff above a few hundred keV (Fig.1).
2. Transient enhanced soft x-ray (~ keV) emission with blackbody-like spectrum (Figs.2 & 3).
3. Transient gamma-ray bumps above ~ 400 keV with thermal spectrum (Fig.4).
4. Transient broad or narrow line features near 500 keV and/or 170 keV, likely associated with annihilation of positrons (Figs 4 & 5).
5. Presence of low frequency QPOs (0.04 -10 Hz) in both soft and hard x-rays (Fig.6).

Historically, since the early dates of the discovery of Cyg X-1, much of the effort in the modeling of black hole emissions and their multi-state transitions was concerned with properties 1 and 2 above (see e.g. Liang and Nolan[12] for an early review). But now we have to also include properties 3, 4 and 5 into the framework of the standard disk models.

THERMAL DISK MODELS

Here we will concentrate on thermal disk models for the hard x-ray and gamma-ray emissions. While such models are clearly too naive and may even turn out to be conceptually wrong, at this point they provide the simplest and most predictive baseline framework to confront with observational data. Our prejudice is not to abandon them until they are convincingly proven wrong. Of course in the mean time we should continue to pursue nonthermal and nondisk models whenever possible.

The basic assumptions of the standard thermal disk models include[13-16]
a) viscosity $\alpha = \pi^{\phi r}/p$=constant; b) orbital velocity v_ϕ~ Keplerian; c) disk scale height h << r; d) electrons and ions are thermal; e) electron and ions couple via coulomb collisions; f) disk cools mainly via radiative loss. Under these assumptions the vertically-averaged steady disk structure can be solved as a function of the four parameters: M (mass of hole), $\dot{M}$ (accretion rate), α and r. For a given M and r, the ($\dot{M}$, α) domain of typical disks can be mapped onto the phase plane defined by electron temperature T and disk surface density Σ. In Fig.7 we illustrate such a phase diagram for a typical stellar mass black hole. The allowed (T, Σ) space is bounded from the right by the impossibility of heating balancing cooling at too high Σ and from the top by

the coulomb heating of electrons by thermal ions. Within these boundaries the phase plane can be divided into four main regions. The lower right region corresponds to the optically thick regime where the local emission spectrum is blackbody or modified blackbody[13]. The regime to the left of the boundary where Komponeet parameter $y \sim 1$ corresponds to optically thin emission with an abundance of soft photons so that Comptonization is unsaturated. The typical local spectrum is that of Sunyaev and Titarchuk[17]. The region with $y>1$ and $\tau_*<1$ corresponds to Comptonized bremsstrahlung or Wien spectral output[18]. Finally, at relativistic temperatures pair processes dominate[19-22] and the resultant spectrum is likely one of Comptonized bremsstrahlung together with a broad annihilation bump[23], perhaps including a pion-decay continuum if the ion temperature exceeds ~ 50 MeV and a narrow 511 keV line if a significant flux of positrons escape to annihilate in the ISM[24,25].

If the standard thermal disk model were optically thick and radiated like a blackbody for all radii, then the temperature would be too low and spectrum too soft to account even for the hard x-rays, not to mention the gamma rays. Historically, two scenarios were proposed to accommodate the hard x-rays. The radially zoned disk scenario assumes that the innermost part of the disk becomes optically thin and radiates the hard x-rays while the cool outer disk emits the blackbody soft x-rays[16]. The vertically stratified disk scenario assumes that the disk possesses a hot optically thin corona sandwiching the cool blackbody disk[26]. The spectral evolutions of many XRNs however show that the soft and hard components seem to evolve independently of each other (cf. Fig.3). This would argue slightly in favor of the radially zoned scenario since in the corona model all of the soft photons must pass through the corona and get reprocessed so the variability of the two components cannot be too independent. Theoretically, the radial-zoned model is also more predictive since the ratio of luminosities of the different spectral components are simply related to the radius separating their emission regions.

Within the framework of radially-zoned disks, Wandel and Liang[18] recently proposed a picture for the three emission states based on the different phases of Fig.7. The three states and their respective output spectrum are illustrated in Fig.8.

INSTABILITIES AND RADIAL OSCILLATIONS OF DISKS

One of the most exciting recent developments is the discovery of low frequency QPOs in the hard x-rays of black hole candidates[10,11]. Such QPOs are unlikely due to radiation-pressure driven instabilities in quasi-radial in-fall as proposed by Fortner et al[27] since the accretion rate is typically much below Eddington limit. This has motivated us to reexamine the dynamics of accretion disks and possible relation of disk oscillations to QPOs. Here we give a brief introduction to this topic. More details can be found in the paper by Luo and Liang[28] in this volume. To understand the various instability and oscillation regimes of thermal disk models it is instructive to plot the solution in the $\dot{M} - \Sigma$ plane for a given M, r and α. A typical solution is illustrated in Fig.9 Note that there is no solution above a certain maximum accreting rate $\dot{M}_{max}$ which can often be subEddington. Note also the existence of a triple-valued kink in the hot optically thin branch discovered by Bjornsson and Svensson[29] when pair effects are included. The three main branches have distinct thermal and secular instability properties. Only in the thermally and secularly unstable branch (b) do radial oscillations exist in the short wavelength limit where the thermal and secular modes merge into a single unstable mode. The longer wavelength perturbations which bifurcate into the two separate modes are purely growing, non-oscillatory modes. The wavelength-averaged oscillatory frequency lies right in the range of observed QPO frequencies. More importantly, the theory predicts that as the accretion rate increases towards $\dot{M}_{max}$ (Fig.9), the unstable disk gets hotter and optically thinner emitting harder spectra and the oscillatory frequency decreases, approaching zero when $\dot{M}$ reaches $\dot{M}_{max}$. This may explain why Cyg X-1 and Nova Per have such low QPO frequencies when the QPO is observed up to hundreds of keVs. At this point the possible connection between QPO and radial disk oscillation is still only suggestive. Much more work, especially in the areas of nonlinear saturation and global behaviors are needed before we can make contact with data. In particular, we need to know how disk oscillations translate into oscillations of the x-ray intensity and why the bandwidth is so narrow. However, if indeed QPOs are somehow proven to be related to disk radial oscillations, then QPOs will surely provide a new diagnostics of disks, including the measurement of the viscosity parameter.

PAIRS AND JETS

In addition to the increasing reports of pair related features in the form of broad or narrow annihilation features or MeV bumps from black hole candidates, the recent discovery of quasar-like radio jets associated with the 1E source[30], if confirmed, raises new questions about the origin of the pairs. Conventional scenarios for extragalactic **radio jets** tend to invoke **nonthermal** processes, such as electromagnetic acceleration processes powered by the rotation of the black hole rather than thermal accretion (see e.g. Blanford[31] in this volume). However, for Galactic sources such as 1E, there is no question that the bulk of the energy output, including the broad feature detected by SIGMA on Oct.13, 1990, appears to be of thermal nature. Hence it is an interesting and important question to ask if radio jets can be generated within the context of thermal disk models. Recall from the discussion of the previous section that within the standard thermal disk model, ultrahot pair-dominated plasma will be produced from first principles in the innermost part of the optically thin disk if the soft photon sources, whether internal or external, are quenched. Escaping pairs may come directly from the surface of the hot ion torus, or from gamma-gamma and x-gamma collisions above the disk. But they are most likely produced in the funnel region above the black hole by head-on gamma-gamma collision. Here there are no ions to inhibit the free escape of the pairs along the polar direction (Fig.10). Hence we believe that a hot thick ion torus is also a natural place to produce a collimated pair jet. The bulk Lorentz factor of the jet will depend on the amount of radiation pressure. Based on the radio data[30] of the 1E jet we estimate that the amount of escaped pair is a tiny fraction of the pairs present in the annihilation feature of this source on Oct.13, 1990. Hence the conventional assumption of local pair balance in calculating the structure, stability and emission spectrum of hot pair-dominated plasmas holds even in this case.

What may be causing the short duration and rather narrow width (compared with typical temperatures needed to produce copious amount of pairs) of the Oct 13, 1990 annihilation feature from 1E? Most **intriguing** is absence of enhanced gamma ray emission above pair producton threshold either during or prior to the emergence of the feature, such as the MeV bump observed in the gamma-1 state of Cyg X-1. We suspect that the emission of the 1E feature is due to some instability, in contrast to the steady stable emission of the Cyg X-1 MeV bump. A potential candidate is that found recently by Bjornsson and Svensson[29] (cf. Fig. 9). When pair process are included in the thermal disk solutions they find that in certain regimes the solution exhibits a kink in the $\dot{M}$ - Σ plane where two stable solutions are sandwiching an unstable solution, much like the thermal limit cycle near hydrogen ionization in the case of dwarf novae. Hence as $\dot{M}$ exceeds a certain value, it jumps from a hot ($T \sim mc^2$), pair-deficient branch to a cool ($T \sim 0.1mc^2$) pair-dominated branch, while the accretion rate jumps by a factor of a few. The spectral output of the cool pair-dominated branch would resemble that observed, and the jump in accretion rate would explain the dramatic increase in luminosity of the disk. Whether this may indeed be a viable model of the 1E flare on Oct 13, 1990 remains to be further investigated, but it is pointing in the right direction.

SUMMARY AND FUTURE WORK

While the thermal disk model may yet be proven wrong by observations, at present it is a useful paradigm for confrontation with observational data of galactic black hole emissions. We argue in this paper that most of the observed hard x-ray and gamma ray features, including the pair annihilation features, jets and QPOs, can in principle be accommodated within this framework. In addition, there are a number of useful checks with existing data that we can propose here:

1. Anti-correlated variations between different spectral components: If we roughly divide the luminosity into 3 bands: $L_s(<10\,\text{keV}) + L_h(10\text{-}200\,\text{keV}) + L_\gamma(>200\,\text{keV}) = L_0$(total luminosity), then according to the radial-zoning picture, the ratios $L_\gamma/L_0(r_1)$, $L_s/L_0(r_2)$ depend only on the separation radii r_1 and r_2. When L_s/L_0 goes down, for example, r_2 moves outward and we expect less soft photons cooling the middle region so that L_γ/L_0 should gain relative to L_h/L_0. It is important that we only use ratios since they are insensitive to changes in the overall accretion rate but only to the radial partition of the luminosity into different spectral components.

2. We can estimate α from the QPO frequency based on the disk oscillation model (cf. Ref.28). This value can then be checked against the α deduced from steady state models using detailed spectral fitting.

3. We can check the e^+ luminosity from the observed 511 keV flux and indirectly from the radio jet luminosity against the expected flux derived from first principles using x-gamma and gamma-gamma collisions of the observed spectrum. However, the observed 511 keV or radio flux usually have a time delay from the time of pair creation so it is hard to identify the episode when the gamma ray spectrum should be used.

4. The emergence of features such as that of 1E on Oct 13, 1990 should be checked against predictions based on the instability picture of Bjornsson and Svensson etc.

This work is partially supported by NASA NAG5-1547.

REFERENCES

1. J.E. McClintock and R.A. Remillard, Ap.J. 308, 110 (1986).
2. J. Casares et al., Nature 355, 614 (1992).
3. R.A. Remillard et al., Ap.J. Lett. to appear (1992).
4. S. Mineshige and J. Wheeler, Ap.J. 343, 241 (1989).
5. D.A. Grabelsky et al., contribution in this volume (1993).
6. S.A. Grebenev et al. & M. Gilfanov et al., in Proc. Workshop Nova Muscae 1991, ed. S. Brandt (DSRI, Denmark, 1991), p.19 & p.51.
7. J. Ling et al., Ap.J. Lett. 321, L117 (1987).
8. G. Riegler et al., Ap.J. Lett. 294, L13 (1985).
9. J. Paul et al., in AIP Conf. Proc. No. 232, ed. P. Durouchoux & N. Prantzos (AIP, NY, 1991).
10. Y. Tanake et al., in Proc. Workshop Nova Muscae 1991, ed. S. Brandt (DSRI, Denmark, 1991).

11. J.E. Grove et al., contribution in this volume (1993).
12. E.P. Liang and P.L.Nolan, Sp. Sci. Rev. 38, 353 (1994).
13. N. Shakura and R. Sunyaev, Ast. Ap. 24, 337 (1973).
14. J. Pringle and M. Rees, Ast. Ap. 21, 1 (1972).
15. I. Novikov and K.Thorne, in Black Holes, ed. B.and C. DeWitt (Gordon & Breach, NY, 1973).
16. S. Shapiro et al., Ap.J. 204, 187 (1976).
17. R. Sunyaev and L.G. Titarchuk, Ast. Ap. 86, 121 (1980).
18. A. Wandel and E. Liang, Ap.J. 380, 84 (1991).
19. R. Svensson, Mon.Not.Roy.Ast.Soc. 209, 175 (1984).
20. A. Zdziarski, Ap.J. 283, 842 (1984).
21. M. Kusunose and F. Takahara, PASJ 40, 435 (1988).
22. T. White and A. Lightman, Ap.J. 340, 1024 (1989).
23. E. Liang and C. Dermer, Ap.J. Lett. 325, L39 (1988).
24. C. Dermer and E. Liang, in AIP Conf. Proc. No. 170, ed. N. Gehrels & G. Share (AIP, NY, 1989), p.326.
25. R. Ramaty et al., Ap.J. Lett. 392, L63 (1992).
26. E. Liang and R. Price, Ap.J. 218, 247 (1977).
27. B. Fortner et al., Nature to appear (1990).
28. C. Luo and E. Liang, contribution in this volume (1993).
29. G. Bjornsson and R. Svensson, Stockholm Obs. preprint (1992).
30. I.F. Mirabel et al., Nature 358, 215 (1992).
31. R. Blandford, contribution in this volume (1993).

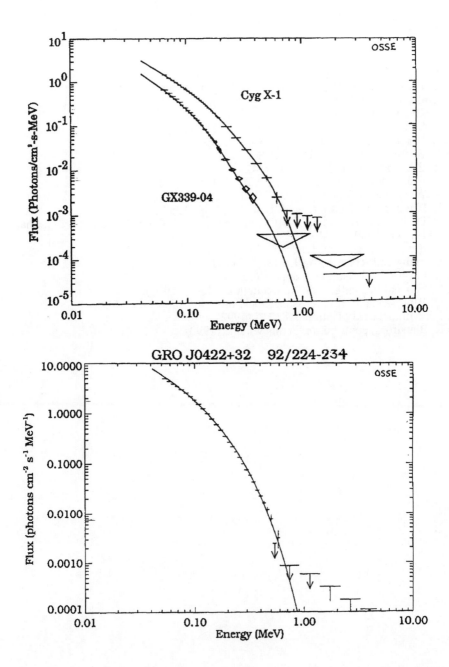

Fig.1 Sample hard x-ray / gamma ray spectra of black hole candidates (from Ref.5).

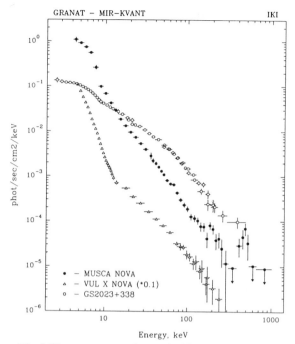

Fig.2 X-ray spectra of sample XRNs (from Ref.6).

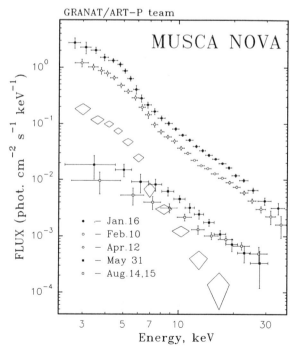

Fig.3 X-ray spectra of Nova Muscae showing the independent variations of the soft and hard components (from Ref.6).

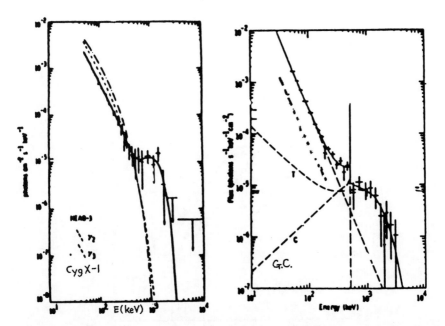

Fig.4 Gamma ray spectra of Cyg X-1 and the Galactic Center region (likely related to 1E) observed by HEAO3 in Fall 1979 showing the ~MeV bumps (from Refs. 7 & 8).

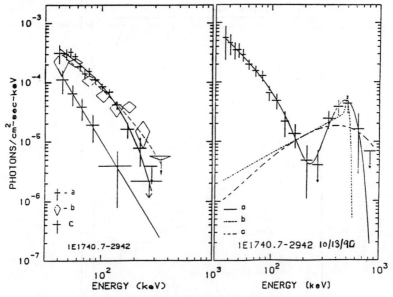

Fig.5 Broad feature around 500 keV detected by SIGME from 1E on Oct.13, 1990 strongly suggests a large positron flare (from Ref. 9).

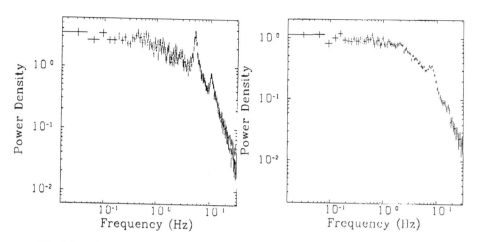

Fig.6 Samples of low frequency QPOs from Nova Muscae detected by GINGA (from Ref. 10).

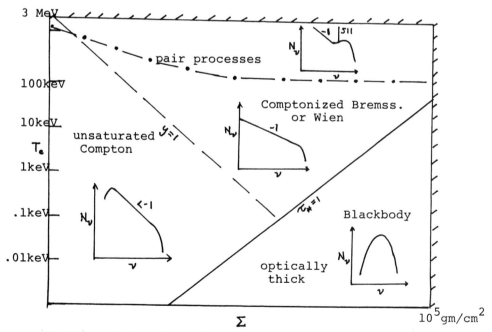

Fig.7 Schematic phase diagram for a thermal accretion disk model. Sketched also is the typical output spectrum expected from each phase.

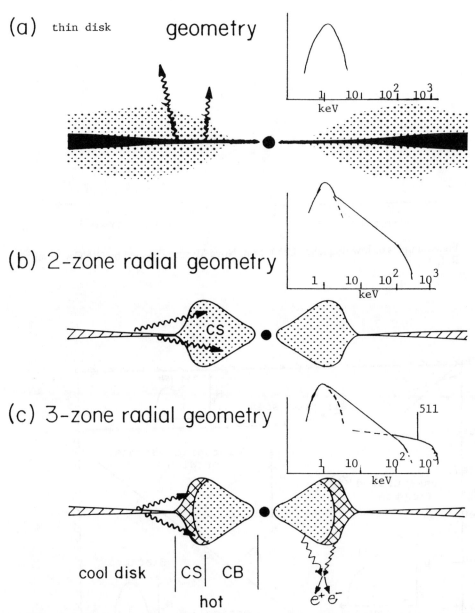

Fig.8 The three emission states of a radially zoned hybrid disk and their corresponding output spectra (from Ref. 18).

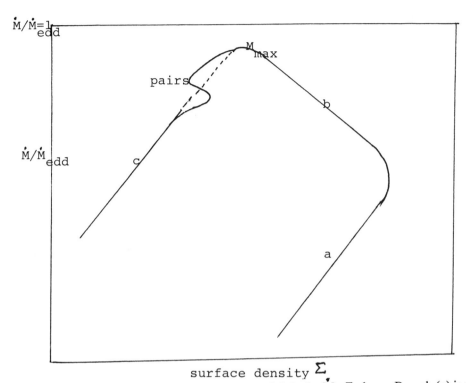

Fig.9 Schematic solutions of a steady-state disk in the $\dot{M}$ - Σ plane. Branch (a) is the optically thick gas-pressure-dominated blackbody branch (a) is thermally and secularly stable. The optically marginally thick radiation-pressure-dominated cool branch (b) is thermally and secularly unstable at long wavelengths. The optically thin branch (c) is secularly stable but thermally unstable at low $\dot{M}$. At high $\dot{M}$ pair processes dominate and the solution may exhibit a kink in which the middle solution is secularly unstable also (from Ref. 29).

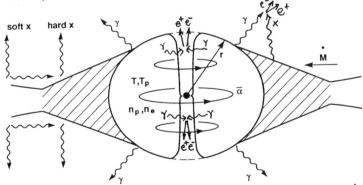

Fig.10 The funnel region of a hot gamma-ray emitting thick ion torus may be the natural birthplace of a collimated pair jet.

COMPTON BACKSCATTERED ANNIHILATION LINE FROM THE NOVA MUSCAE

XIN-MIN HUA

Space Astrophysics Laboratory, Institute for Space and Terrestrial Science,
Concord, Ontario, L4K 3C8 Canada

AND

RICHARD E. LINGENFELTER

Center for Astrophysics and Space Sciences, University of California, San Diego,
La Jolla CA 92093 USA

ABSTRACT

Assuming a simple dynamic model of an accretion disk around a compact source, we calculate the spectrum of Compton scattered annihilation line emission as a function of the velocity of disk material as well as the observing angles and disk opacities. We find that such spectral characters as the position and width of the line-like feature in the backscattered spectra of the annihilation line can provide new diagnostics of the innermost regions of accretion disks around compact objects. From the comparison of our calculations and the spectra observed from Nova Muscae, we also find that the inclination angle of the accretion disk is about 65°, implying a central black hole mass of about $5M_\odot$. We suggest that Compton backscattered annihilation line features may be a unique signature of accreting stellar mass black holes.

INTRODUCTION

Compton reflection of the gamma-ray continuum and lines has been studied in accretion disks around compact objects[1,2]. We have pointed out[3] that Compton backscattering of gamma-ray lines can produce *line-like reflection features* at lower energies. We did calculations based on the Compton scattering of 511 keV photons in cold, static spheres and disks with or without central holes. We showed that under a wide range of conditions, a line-like feature near 170 keV could be produced by Compton backscattering. Such features provided an explanation for previously unidentified 170 keV features observed[4-9] in the gamma-ray spectra of a variable compact source or sources near of the Galactic center observed over the last twenty years. Most recently, new observations also reveal an intense line-like feature accompanying the annihilation line emission from the black hole candidate Nova Muscae[10-11]. Other preliminary spectra from potential black hole candidates, GX339-4 and 4U1543-47, also show[12-13] suggestive features near 170-200 keV. Thus, we propose that Compton backscattered annihilation line features may be a unique signature of accreting stellar mass black holes.

In this study, we extend our previous calculations[2-3], using a simple but more realistic model of a warm, dynamic accretion disk around a compact object in order to explain the observed spectrum from Nova Muscae during the flaring on 20 January 1991. We find that the best fit to the spectral features observed from Nova Muscae requires reflection from the inner edge of an opaque disk viewed at an inclination angle of about 65°. Using the mass function determined from optical observations[14] this angle implies a central black hole mass of about $5M_\odot$.

COMPTON BACKSCATTERING FROM A DISK

The energy of the Compton-scattered photon relative to the initial photon energy is $E'_\gamma/E_\gamma = 1/(1 + \alpha - \alpha \cos\phi)$, where $\alpha = E_\gamma/m_e c^2$ and ϕ is the scattering angle. For a 511 keV source, $\alpha = 1$ and E'_γ/E_γ reaches its minimum 1/3 when $\phi = 180°$. Since photons from the first backscattering dominate the reflected spectra[2], the spectra have a sharp cutoff at $E' = 511/3$ keV so that a line-like feature forms at energies slightly greater than 170 keV. This is true for many geometries of reflecting media[3].

In particular, we have considered the spectra emerging from a cold ($kT \ll m_e c^2$), static gaseous disk with an isotropic 511 keV photon source in a central hole[3]. The disk has a thickness τ, defined in units of the Compton scattering optical depth at 511 keV, and a radius $\gg \tau$. The central cylindrical hole is intended to approximate the inner edge of an accretion disk around a compact object. The size of the hole is defined by an opening angle θ_o, which is the zenith angle of the inner edge of the disk measured from the center with respect to the axis of the disk. Thus, the radius of the hole is equivalent to $\frac{1}{2}\tau \tan\theta_o$.

Monte Carlo calculations[3] with this model showed that for large observing angles θ (measured from the disk axis), the 170 keV feature becomes more intense and narrower as the opening angle of the central hole increases. Because the reflecting matter behind the inner edge of the disk subtends a smaller solid angle from the source as the opening angle increases, the scattering angles for observable singly scattered photons are concentrated into a narrower band around 180° and the observable energy band above 170 keV is likewise narrower. Correspondingly at smaller observing angles, as θ becomes less than θ_o, an increasing opening angle concentrates the reflecting matter more nearly perpendicular to the line of sight and there is no reflecting matter directly behind the source. Thus, the peak energy of the reflection feature is shifted up toward 255 keV as scattering angles around 90° dominate the observable singly scattered photon emission, and there is a significant depression in the observable spectrum around 170 keV.

Concerning the effect of disk opacity, the calculation results show that for large observing angles the 170 keV feature also becomes narrower as the opacity of the disk increases. For small observing angles, increasing the disk opacity attenuates more severely the flux in the singly scattered feature around 255 keV from scattering around 90°.

In addition to the backscattered feature, at higher energies, between the 170 keV feature and 511 keV, there are photons from the scattering at angles between 0 and 180°. The calculation results show that this intermediate angle contribution is lower for thicker disks because there is more scattering material in the line of sight and the scattered photons are largely obscured.

GAMMA RAY LINES FROM NOVA MUSCAE

The black hole candidate GRS 1124-684, also known as the X-ray Nova Muscae 1991, was discovered in January 1991 by both the GRANAT and Ginga satellites[15–16]. Time dependent spectra have been measured by the gamma-ray telescope SIGMA on board GRANAT in the energy range 35 – 1300 keV. During an outburst on January 20 1991, two intense gamma-ray line features were observed[10–11] in the spectrum: an apparent redshifted annihilation line around 480 keV and an accompanying feature around 200 keV attributed to Compton backscattering (Figure 1). A best-fit[10] Gaussian line has a flux $(6.01^{+3.95}_{-2.76}) \times 10^{-3}$ photons cm^{-2} s^{-1} at a centroid of 481 ± 22 keV, with an intrinsic line width of 23 ± 23 keV. The feature around 200 keV has a flux of $\sim 1.8 \times 10^{-3}$ photons cm^{-2} s^{-1}, and it cannot be accounted for by backscattered photons in the instrument[10].

Nova Muscae is the first source outside the Galactic center region[4-5] that shows simultaneously the annihilation line and its reflection feature. These observations thus provide an new opportunity to use the reflection feature as a diagnostic tool to probe the innermost part of the accretion disk, which is assumed to surround the central compact object, possibly a black hole.

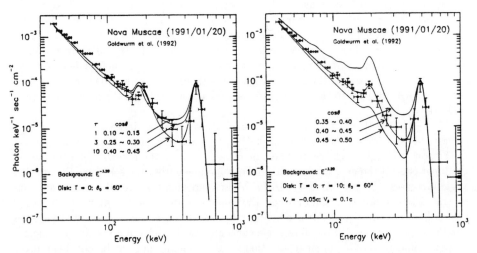

Figure 1. The observed[10] spectrum from Nova Muscae during the outburst on 20 January 1991, showing the blueshifted $\sim$170 keV feature plus a redshifted and broadened 511 keV line. The curves in the left panel are from calculations based on a static disk, and those in the right panel are from a dynamic disk with both rotational V_ϕ and infalling V_r velocities.

To compare with the observed spectrum of Nova Muscae, we consider not only models of static disks, such as we previously studied[2-3], but also models of more realistic dynamic disks, which include both rotational V_ϕ and infalling V_r velocities. In both models we calculate the angular dependent reflection spectrum from a central isotropically radiating source of annihilation line emission that is Gaussian broadened and gravitationally redshifted with a $z \approx 0.06 \pm 0.04$, consistent with the observed emission. In the comparisons shown here, we also assume that the Compton scattering region on the inner edge of the disk, where the reflection feature is produced, is far enough out of the gravitation potential well (i.e. $z \leq 0.01$) that the line photons are redshifted traveling from the source to the scattering region but that the scattered photons are not significantly redshifted on their way to the observer.

The calculated spectrum was first normalized so that the flux in the unscattered line is the same as observed. It is then superimposed upon a power law background with an index -3.2, normalized to the observed flux at 40 keV. This continuum spectrum is steeper than that determined[10] by the SIGMA team in the range from 35 – 300 keV, but the latter was flatter because the backscattered photons were included in the fit.

The width of the reflection feature, as discussed above, is dependent on both the opening angle of the inner edge of the disk and the opacity of the disk. As can be seen in Figure 1, we find that fits to the observed width of the reflection feature, require an accretion disk with a rather wide opening angle of $\sim 60°$, similar to that also found[3] for the compact source near the Galactic center. A smaller opening angle of the disk hole will produce a broader reflection feature than was observed.

The high ratio (~ 0.3) of the flux in the reflection feature to that in the annihilation line also requires[3] that the line photons were significantly attenuated by the disk material, so that the opacity must be relatively large. In the left panel of Figure 1, we show calculated spectra reflected from static disks with various opacities and $\theta_o = 60°$. As can be seen, the spectrum reflected from a disk with an opacity $\tau \sim 10$, observed at an inclination angle $\theta = 65°$, gives a better fit to the measured spectrum. In this direction, the line source is just obscured by the edge of the accretion disk and the emission is attenuated to such a degree that the flux of the reflected photons relative to that of the unscattered line matches the observed value. Disks with lower opacity ($\tau = 1$ and 3), observed at larger angles ($\sim 83°$ and $\sim 77°$ respectively) can still match the relative fluxes in the line and reflected feature close to 200 keV, but the flux is too high at intermediate energies between ~ 200 and 430 keV.

However, for these static disk calculations, no matter how the opacity opening angle or observing angle are varied, the calculated backscattered feature between ~ 150 to 200 keV is always displaced by at least ~ 10 keV to lower energies with respect to the observed peak. We believe this discrepancy between the calculation and observation can be explained as a blueshift of the backscattered feature by the infalling velocity of the material on the back inner edge of the disk, where most of the backscattering occurs.

To explore the possible blueshift of the backscattered feature quantitatively, we consider a more realistic dynamic disk with both rotational and infalling motion.

The cross section of Compton scattering between photons and moving electrons depends not only on the photon energy, but also on the velocity and direction of the electrons. However, a detailed calculation from the differential cross-sections[17] shows that as long as the electron velocity is much smaller than light velocity, the cross section differs from the Klein-Nishina cross section by a small fraction only. Also, if the electron velocity is perpendicular to the photon direction, which is true for radially emitted photons and rotating electrons in our model, the cross section is virtually unchanged even for larger velocities. Thus in the Monte-Carlo calculations with the disk material spiraling down to the central object, the Klein-Nishina formula is used for the cross-section. However the electron motion is always considered in the scattering kinematics.

The disk material at a radius r is assumed to have a Keplerian rotational velocity $V_\phi = (GM/r)^{1/2}$; and the infalling velocity is a small fraction of the free fall velocity. The curves in the right panel of Figure 1 are calculated spectra for such a dynamic disk in which the rotational $\beta_\phi = 0.1$ and the infalling $\beta_r = 0.05$ at the inner wall of the hole, an opening angle $\theta_o = 60°$ and an opacity $\tau = 10$. We see that the spectrum observed at an angle around 65° ($\cos\theta = 0.4 - 0.45$) from the axis of this dynamic disk not only shifts the backscattered reflection peak to ~ 190 keV, but also considerably improves the fit at energies above this peak. We also see from the calculated spectra at $\cos\theta = 0.35 - 0.4$ and $0.45 - 0.5$, that the reflected spectrum is quite sensitive to the inclination angle of the disk, significantly increasing or decreasing the reflected flux with respect to that in the line.

AN ESTIMATION OF THE PRIMARY MASS

Although we are still exploring the range of possible inclination angles, the values determined above can already give a good estimate of the mass of the central object. Optical photometry and spectroscopy of Nova Muscae in quiescence obtained recently[14] give a mass function, $(M \sin i)^3/(M + M_c)^2 = 3.07 \pm 0.4 M_\odot$, where M and M_c are the masses of the primary and companion respectively; and i is the orbital inclination angle of the system, which is believed to be the same as the accretion disk. Thus, an inclination angle of 65°, found above, gives a mass of the black hole primary $M = 5.3 M_\odot$, assuming a K0V - K4V companion mass[14] of $0.7 M_\odot$.

The best fit inclination angle in Figure 1 is not unique, however, because there is a range of opening angles of the disk hole, for which we can also find acceptable fit with inclination angles slightly larger than the opening angles. We have not yet fully explored this range. But as we pointed out above, if the opening angle is too small, the reflected feature will be too broad to agree with the observation. Our present calculations suggest that this limiting inclination angle is 40°. This limiting angle sets an upper limit of $13 M_\odot$ on the central black hole in Nova Muscae.

These inclination angles are much larger than that of $26 \pm 25°$ estimated by Chen and Gehrels[18] from rotational broadening in an *optically thin* disk, and the mass is much smaller than the mean of $\sim 38 M_\odot$ implied by such an angle.

Acknowledgements. We thank Reuven Ramaty for valuable discussions, and thank NASA for financial support under grant NAGW 1970 and the Province of Ontario for support at ISTS. The calculations were carried out on a SUN at ISTS.

REFERENCES

1. T. R. White, A. P. Lightman, and A. A. Zdziarski, Ap. J., **331**, 939 (1988).
2. X.-M. Hua, and R. E. Lingenfelter, Ap. J., **397**, 591 (1992).
3. R. E. Lingenfelter and X.-M. Hua, Ap. J., **381**, 426 (1991).
4. M. Leventhal, C. J. MacCallum and P. D. Stang, Ap. J. (Letters), **225**, L11 (1978).
5. M. Leventhal and C. J. MacCallum, Ann. N. Y. Acad. Sci., **336**, 248 (1980).
6. R.C. Haymes et al. Ap. J., **201**, 593 (1975).
7. W.S. Paciesas et al. Ap. J., **260**, L7 (1982).
8. J. L. Matteson, et al. in Gamma-Ray Line Astrophysics, ed. P. Durouchoux and N. Prantzos, (New York: Am. Inst. Phys., 1991), p. 45
9. D.M. Smith et al. Ap. J., in press (1992).
10. A. Goldwurm, et al., Ap. J., **389**, L79 (1992).
11. R. Sunyaev, et al., Ap. J., **389**, L75 (1992).
12. P. Mandrou et al. in this volume (1993).
13. R.A. Kroeger et al. in this volume (1993).
14. R.A. Remillard, J.E. McClintock and C.D. Bailyn, Ap. J., in press (1993).
15. N. Lund, and S. Brandt, IAUC 5161 (1991)
16. F. Makino, et al., IAUC 5161 (1991)
17. R. Ramaty, J.M. McKinley and F.C. Jones, Ap. J., **256**, 238 (1982).
18. W. Chen and N. Gehrels in this volume (1993).

TIME VARIABLE MULTIPLE BACKSCATTERED 511 KEV PHOTONS FROM BLACK HOLES

Ph. Durouchoux and P. Wallyn
Service d'Astrophysique, Centre d'Etudes Nucléaires de Saclay
91191 Gif sur Yvette Cedex FRANCE

ABSTRACT

The instabilities of the accretion flow in a disk surrounding a black hole heat the inner parts of this disk (T>10^9 K) where a pair plasma is formed. We suppose here that the outflows of matter during such supercritical states produce both annihilation of positrons and the subsequent scatterings (170 keV single backscattering.and 102 keV double backscattering).

We present in this paper Monte Carlo simulations which both spatially and temporally follow a burst of positrons and we compare these numerical results with recent observations of the galactic center region and Nova Muscae source.

INTRODUCTION

The balloon-borne experiments flown in May 1989 exhibit features in the spectra of the galactic center region. First, on May 9, EXITE, an imaging instrument working in the range 20-250 keV detected a "new" source: EXS1737.9-2952 (40' west of 1E1740.7-2942). This source exhibits significant flux in the 83-111 keV energy range, interpreted as a double backscattered annihilation line[1]. Then, on May 22, HEXAGONE, a high energy resolution spectrometer, detected a feature centered around 170 keV. We must point out the similarities between these results and the data concerning Nova Muscae obtained with the SIGMA telescope (presence of a 102 keV feature during the first subsession which lasted 7 hours and then a 170 keV feature present in the two following subsessions (7 hours and 6 hours). We want to investigate the possibility that the backscattered features can bring us new criteria to characterize black-holes in binary systems and to compare these observations with those of Nova Muscae[2].

BLACK-HOLE MODEL

We have proposed[3] a model involving a low mass x-ray binary containing a black hole; the black hole accretes matter from its companion and this accretion generates instabilities which periodically heat the inner part of the disk to relativistic temperatures. A pair plasma is formed and ejected outside the accretion disk (see Fig 1 in Ph. Durouchoux & P. Wallyn, these proceedings). The positrons of the plasma annihilate in the matter outflowings of the disk during supercritical states[4]. Single or double backscatterings also occur in the outflows pending the geometrical configuration, which varies versus time since these outflows are spiralling around the disk axis, due to the transfer of the angular momentum of the accreted matter.

MONTE CARLO SIMULATION

To verify this model, we ran extensive Monte Carlo simulations, based on a simple geometry (Fig. 1, see also Fig. 2 in Lingenfelter & Hua[5]) consisting of a spherical cloud with a radius R_1 and an opacity defined in terms of the Compton scattering optical depth τ at 511 keV, equivalent to 3.5×10^{24} electrons cm^{-2}. A monoenergetic and isotropic 511 keV point source is set to the surface (then $\theta_s = 180°$ in Lingenfelter & Hua geometry). An external sphere is defined and assumed to be at infinity (ratio of the radii of 10^5) and we then study the arrival time distribution and spectra of backscattered photons when they reach the external sphere. We follow the photons emitted by a burst at 511 keV and 11 equal and consecutive intervals of time (from t=0 to t=11) are defined. The twelfth interval regroups all the photons remaining after the eleventh. It is then possible to study in detail the evolution the spectra for a given observing angle θ (Fig. 1).

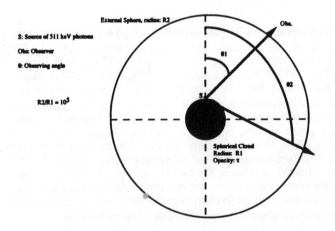

Fig 1: Geometry of a spherical cloud of matter with a point annihilation source (S). The external sphere is assumed to be at infinity. We study the spectrum evolution by varying both the observing angle θ and the time when scattered photons reach the external sphere. For $\theta=\theta_1$, the cloud is located behind both the observer and the annihilation region: the spectrum then contains a 170 keV feature and no

simultaneous forward scattered photons. For θ=θ₂, a part of the cloud is located between the annihilation region and the observer: no 170 feature is observed here.

We measure such evolution with time for nine different observing angles from 10° to 170° using 20° steps. For a given value of τ, the Monte Carlo simulation then gives 108 spectra binned in 491 channels of 1 keV (from 20 to 511 keV). We do not model here the slowing down and the annihilation of positrons in the dense cloud, and our simulation cannot reproduce the evolution of the annihilation line versus time. Only this simple configuration is studied to emphazise in a simple way the scenario proposed below; a detailed study of other configurations (like the geometries proposed in Lingenfelter & Hua[5]) are out of the range of the present analysis and will be developed in another paper. The evolution of the spectra versus time for two different observing angles (θ = 30°, which emphasizes the evolution of the single backscattered line and θ = 170°, which shows the evolution of the double backscattered line) is presented in Figure 2.

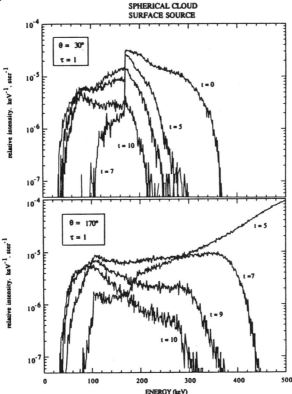

Fig 2: Time evolution of the spectrum for τ=1 and two different observing angles (θ=30° where the 170 keV feature is clearly present and no up-scattered photons appear, and θ=170° where the 170 keV feature is not seen but the 102 keV bump is prominent and up-scattered photons decrease very quickly).

This simulation does not take into account the duration of the interaction between a photon and an electron but we just only the time corresponding to the path between two interactions. The intervals of time are each 10^{-7} s long as we define a radius R_1 of 10^4 cm for the cloud.

Then, for a given value of t, it is possible to calibrate the duration of one interval, for a fixed value of τ, by in/decreasing by the same amount the radius of the cloud. This allows us to consider a time interval of typically 1 day in the case of Nova Muscae by considering a radius of approximately 8.6×10^{15} cm. Figure 2 emphasizes the large spectrum variations after a single 511 keV burst: (1), up-scattered photons quickly decrease with time (see t = 5 and t = 7) and do not appear for low observing angles (Fig. 3). (2) emerging photons are delayed for large values of θ, compared to small angles, and the whole phenomena duration is also decreased by a factor of 2 due to the increased scattering path for large values of θ (for example a spectrum appears at t = 5 only for θ = 170°, compared to t = 0 for θ = 30°); (3) for a given observing angle, it is not possible to have a bump around 102 keV *before* the 170 keV feature; (4) nevertheless, for a given value of t, it is possible to observe the correct sequence of the bumps (Fig. 4).

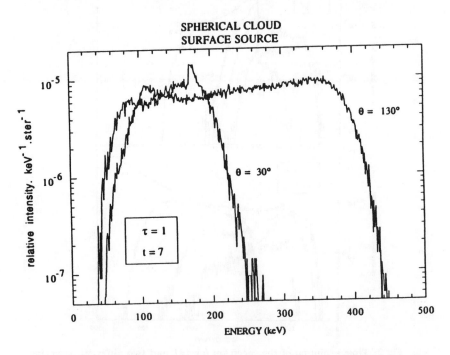

Fig 3: Comparison of the spectra at t=7, for two observing angles; it is possible to have a feature at 170 keV followed by a bump around 102 keV by decreasing θ.

Comparing the 102 and 170 keV fluxes in Nova Muscae, we found that for a given period, t = 7, a variation of the observing angle from 130° to 10° reproduces a correct evolution of the flux of both 102 and 170 keV features observed by SIGMA. It is not possible to perform a deeper analysis of the observations because of the lack of accuracy in the fluxes and energy resolution. In the framework of the cloud geometry, for t=7, as the observing angle is decreasing, the 170 keV line sharpens (Fig. 4) with a correlated increase of the flux. This trend was observed by SIGMA, but with a low confidence level.

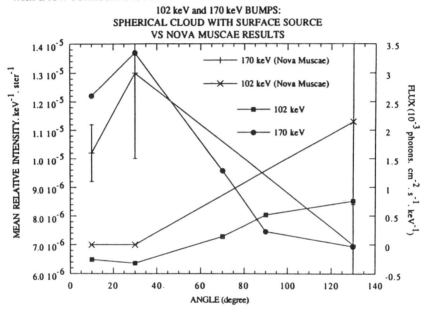

Fig 4: Time evolution of the intensities and widths of the 102 & 170 keV lines.

CONCLUSION

A spherical bubble allows the detection of an important flux around 102 keV for low scattering depth ($\tau \approx 1$) with simultaneous low flux above, and no feature at 170 keV if the cloud is located in front of the annihilation source. This model also explains for smaller observing angles, the presence of a feature at 170 KeV without the 102 keV feature when the cloud is located behind the annihilation region.

REFERENCES

1. J. Grindlay et al. 1992, Astr. Ap. Sup. Ser., in press
2. A. Goldwurm., et al. 1992, Astr. Ap. Sup. Ser., in press
3. Ph. Durouchoux & P. Wallyn, 1992, these proceedings
4. N. Shakura & R. Sunyaev., Astr. Ap., 1973, 24, 337
5. R. Lingenfelter & X. Hua., 1991, Ap. J., 381, 426

SPECTRAL MODELING OF GAMMA RAYS FROM BLACK HOLE CANDIDATES

Edison P. Liang
Rice University, Houston, TX 77251-1892

ABSTRACT

This paper gives sample model spectra generated with the Monte Carlo Radiation Code for Relativistic Thermal Plasmas developed by Canfield, Liang, Dermer, Howard and collaborators at LLNL. These spectra are applicable to the modeling of x-ray and gamma ray emission of many black hole candidates.

INTRODUCTION

Recent observations by GRANAT and the Compton Observatory have revealed strong hard x-ray / gamma ray emissions by many black hole candidates, both from quasi-steady sources and transient sources such as x-ray novae (see contributions in this volume). While details of their spectra show individuality, the overall continuum shapes show remarkable similarity, with a low-energy power-law terminated by an exponential cutoff above a few hundred keV. In addition there are episodic enhancements in the form of soft keV bumps and hard MeV bumps as well as annihilation-like features at ~ 500 keV and 170 keV. Detailed modeling of the continuum spectra in the context of thermal accretion disk models will allow us to constrain the model parameters and ultimately confront the black hole paradigm.

RESULTS

In this paper we illustrate the modeling process by presenting sample spectra generated with the Relativitstic Monte Carlo Thermal Radiation Code developed by Canfield, Liang, Howard, Dermer and collaborators at LLNL (cf. Ref. 7). Examples shown here include:
1. Comptonization of soft photons (Figs. 1 - 3).
2. Comptonization of self-bremsstrahlung (Figs. 3 - 5).
3. Comptonized bremsstrahlung and pair annihilation from pair-dominated plasmas (Figs. 6 - 8).
4. Reprocessing of narrow 511 keV line sources (Figs. 9 -13).

Some of the examples are from previous publications reproduced here for comparison.

REFERENCES

1. A. Wandel and E. Liang, Ap. J. 380, 84 (1991).
2. E. Liang, in Cosmic Rays, Neutrinos & Related Astrophysics, ed. M. Shapiro and J. Wefel (Kluwer, Dordrecht, 1989), p.73.
3. E. Liang and C.D. Dermer, submitted to Nature (1991).
4. P. Nolan and J. Matteson, Ap. J. 265, 389 (1983).
5. G. Riegler et al., Ap. J. Lett. 294, L13 (1985).
6. J. Paul el al., in AIP Conf. Proc. No. 232, ed. P. Durouchoux and N. Prantzos, (AIP, NY 1991), p.17.
7. E. Canfield et al., Ap. J. 323, 565 (1987).

E. P. Liang 419

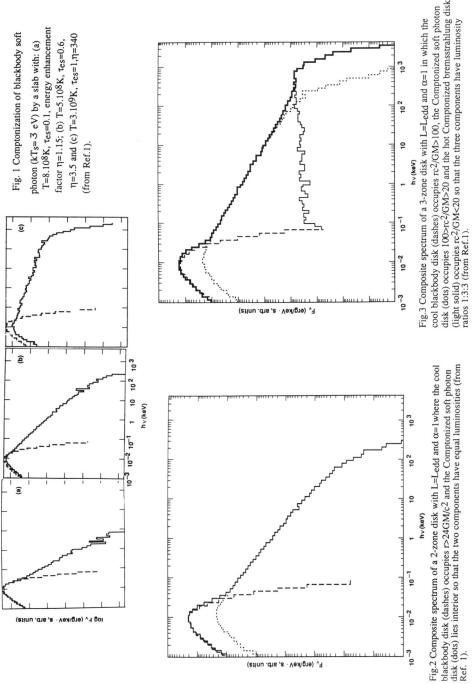

Fig. 1 Comptonization of blackbody soft photon ($kT_s = 3$ eV) by a slab with: (a) $T=8.10^8$K, $\tau_{es}=0.1$, energy enhancement factor $\eta=1.15$; (b) $T=5.10^8$K, $\tau_{es}=0.6$, $\eta=3.5$ and (c) $T=3.10^9$K, $\tau_{es}=1$, $\eta=340$ (from Ref.1).

Fig.2 Composite spectrum of a 2-zone disk with $L=L_{edd}$ and $\alpha=1$ where the cool blackbody disk (dashes) occupies $r>24GM/c^2$ and the Comptonized soft photon disk (dots) lies interior so that the two components have equal luminosities (from Ref. 1).

Fig.3 Composite spectrum of a 3-zone disk with $L=L_{edd}$ and $\alpha=1$ in which the cool blackbody disk (dashes) occupies $rc^2/GM>100$, the Comptonized soft photon disk (dots) occupies $100>rc^2/GM>20$ and the hot Comptonized bremsstrahlung disk (light solid) occupies $rc^2/GM<20$ so that the three components have luminosity ratios 1:3:3 (from Ref.1).

420 Spectral Modeling of Gamma Rays

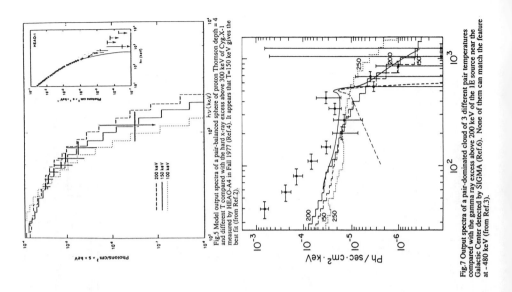

Fig.5 Model output spectra of a pair-balanced sphere of proton Thomson depth =4 and different T compared with the hard x-ray excess above 300 keV of Cyg.X-1 measured by HEAO-A4 in Fall 1977 (Ref.4). It appears that T=150 keV gives the best fit (from Ref.2).

Fig.7 Output spectra of a pair-dominated cloud of 3 different pair temperatures compared with the gamma ray excess above 200 keV of the 1E source near the Galactic Center detected by SIGMA (Ref.6). None of them can match the feature at ~480 keV (from Ref.3).

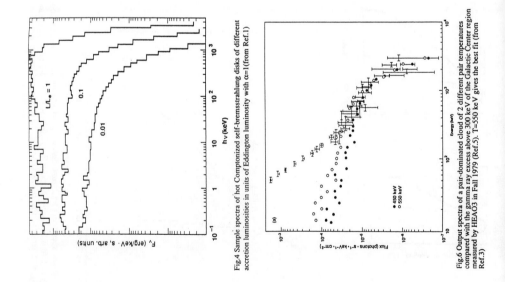

Fig.4 Sample spectra of hot Comptonized self-bremsstrahlung disks of different accretion luminosities in units of Eddington luminosity with α=1 (from Ref.1)

Fig.6 Output spectra of a pair-dominated cloud of 2 different pair temperatures compared with the gamma ray excess above 300 keV of the Galactic Center region measured by HEAO3 in Fall 1979 (Ref.5). T=550 keV gives the best fit (from Ref.3)

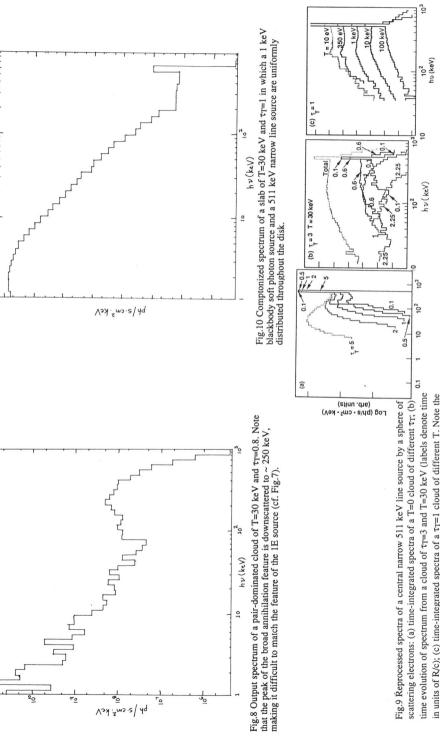

Fig.8 Output spectrum of a pair-dominated cloud of T=30 keV and τ_T=0.8. Note that the peak of the broad annihilation feature is downscattered to ~ 250 keV, making it difficult to match the feature of the 1E source (cf. Fig.7).

Fig.9 Reprocessed spectra of a central narrow 511 keV line source by a sphere of scattering electrons: (a) time-integrated spectra of a T=0 cloud of different τ_T; (b) time evolution of spectrum from a cloud of τ_T=3 and T=30 keV (labels denote time in units of R/c); (c) time-integrated spectra of a τ_T=1 cloud of different T. Note the gradual disappearance of the 170 keV feature with increasing T (from Ref.3).

Fig.10 Comptonized spectrum of a slab of T=30 keV and τ_T=1 in which a 1 keV blackbody soft photon source and a 511 keV narrow line source are uniformly distributed throughout the disk.

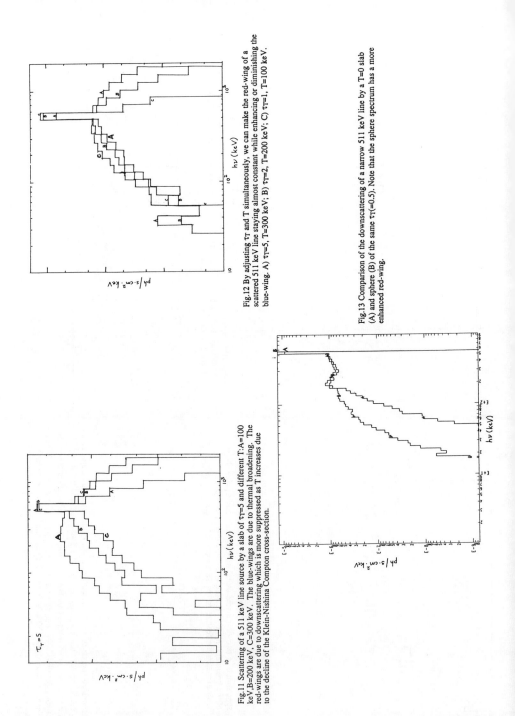

Fig.11 Scattering of a 511 keV line source by a slab of $\tau_T=5$ and different T: A=100 keV, B=200 keV, C=300 keV. The blue-wings are due to thermal broadening. The red-wings are due to downscattering which is more suppressed as T increases due to the decline of the Klein-Nishina Compton cross-section.

Fig.12 By adjusting τ_T and T simultaneously, we can make the red-wing of a scattered 511 keV line staying almost constant while enhancing or diminishing the blue-wing. A) $\tau_T=5$, T=300 keV; B) $\tau_T=2$, T=200 keV; C) $\tau_T=1$, T=100 keV.

Fig.13 Comparison of the downscattering of a narrow 511 keV line by a T=0 slab (A) and sphere (B) of the same $\tau_T(=0.5)$. Note that the sphere spectrum has a more enhanced red-wing.

THE BLACK HOLE MASS IN X-RAY NOVA MUSCAE: CONSTRAINT FROM SIGMA ANNIHILATION LINE DATA

Wan Chen[1] and Neil Gehrels
NASA/Goddard Space Flight Center, Greenbelt, MD 20771

ABSTRACT

We propose a new method to constrain the black hole mass in X-ray Nova Muscae 1991, using the annihilation line data from SIGMA observations. We discuss the possible mechanisms for the line redshift and broadening, and demonstrate that the annihilation radiation most likely originates in the inner region of an accretion disk around the black hole in Musca Nova. The line centroid and broadening can thereby be used to estimate the black hole mass in the system. The low statistics of the existing γ-ray data only allows the black hole mass in Nova Muscae to be constrained in a wide range. Future high-resolution γ-ray spectroscopy missions that monitor sources for transient line emission are strongly encouraged.

INTRODUCTION

The search for stellar black holes in the last few decades has produced more than a dozen candidates[1], all in close binary systems. For many of them the evidence is merely the spectral and temporal similarities to the best known and more firm candidates (e.g., Cyg X-1 and A0620-00). While such evidence is suggestive, the most reliable and indisputable proof of a black hole relies on the determination of a large compact object mass, $M > 3M_\odot$. A crucial step in the process is to obtain through optical spectroscopy the mass function of a binary system, $f = (M\,sin\,i)^3/(M + M_c)^2$, where M_c is the mass of the companion, and i is the inclination angle of the binary orbit to our line of sight ($i = 0$ is face-on). It is clear that if we know M_c and especially $sin\,i$, M can be determined from the mass function accordingly. Of course if f is much greater than $3M_\odot$, as it probably is in the case of V404 Cyg[2], the compact object will be most likely a black hole regardless of the value of M_c and i. However, if $f \leq 3M_\odot$, information on the inclination angle i becomes most important.

For (partial) eclipsing systems (e.g., A0620-00, ref. 3), one can deduce the orbital inclination with much confidence. For noneclipsing binaries with a low-mass companion, theoretically it is possible to infer the inclination from optical data alone by modeling the ellipsoidal variations in the orbital light curve since the cross-section of the Roche-lobe filling secondary is phase dependent[4]. In reality, however, this is difficult, especially for small inclination systems,

---

[1]Universities Space Research Association

because the light from the secondary is usually contaminated by disk radiation and the phase modulation diminishes as the inclination angle decreases.

We propose a new method to estimate the binary inclination from γ-ray line data. The basic idea is simple: if an emission line is produced in an *accretion disk* around the black hole and if one can determine the radius where the line is radiated, the disk (and so the orbital) inclination can then be inferred from comparison between the observed line width and the expected full line width caused by the Keplerian rotation of the disk at that radius. Of course one also has to take into account other line broadening effects. Unlike the optical method, this technique works well for small inclination systems in which the difference between the observed and expected line widths becomes most striking. A γ-ray line, e.g., a positron annihilation line, is best suited for this purpose, because it may be produced very *near* the black hole. The gravitational redshift of the line centroid is observable, which can tell precisely where it is emitted. The disadvantage of this method is that the chances of catching such a line, usually a transient phenomenon, are relatively small. Thus, the 1-day positron annihilation feature in X-ray Nova Muscae 1991 (GS1124-683) observed by GRANAT/SIGMA[5,6] is an important event which offers a unique opportunity to constrain the geometry of an accretion disk system using the γ-ray line data.

X-RAY NOVA MUSCAE 1991

The temporal and spectral behavior of the X-ray Nova Muscae[7] (XN Mus), which was first detected 8 January 1991, is typical of an ultrasoft X-ray transient[8]. Its optical and X-ray light curves[7] closely resemble those of the well-established black hole candidate A0620-00, indicating that it is probably also a black hole[9]. This conjecture was dramatically confirmed, a few months after XN Mus entered its quiescent stage, by optical observations[10] of its K dwarf secondary which revealed a mass function of $3.07 \pm 0.40\,M_\odot$. Because this is a noneclipsing system, a safe lower limit of the compact object mass is then $3.75\,M_\odot$[10].

Thanks to the persistent monitoring of XN Mus by the GRANAT/SIGMA telescope, a transient (13 hours$< \Delta t <$10 days) positron annihilation emission feature was detected 12 days after the outburst[5,6]. The line centroid is redshifted by about 6% to $E \sim 480$ keV and the intrinsic line width is only $\Delta E \sim 60$ keV, noticeably narrower than the transient annihilation lines observed in the only two other annihilation sources, the Galactic Center hard X-ray source 1E1740.7-2942[11,12] and a recently analyzed HEAO A-4 source[13], which have comparable line redshift but much greater line width, > 200 keV. In addition, no positronium continuum was detected below the annihilation line in XN Mus[5].

Another virtue of the XN Mus spectrum is that an additional emission feature near 200 keV appeared at the same time as the annihilation line was detected[5,6]. This can be attributed to the single back-scattered photons of the annihilation radiation. It is puzzling, however, that while the annihilation line is redshifted the back-scattering feature is apparently blueshifted from 170 keV to ~ 200 keV. This is significant because any model trying to explain the annihilation line must also successfully explain the back-scattering feature, and vice versa.

DISK INCLINATION AND BLACK HOLE MASS

The fact that the annihilation lines are redshifted in all three known sources argues favorably for their gravitational origin. Other alternatives like a relativistic jet or wind can be carefully ruled out[14]. Thus the redshift of the line centroid gives directly the distance of the line emitting region from the black hole, $r \sim 8r_s$ in case of XN Mus, where $r_s = 2GM/c^2$ is the Schwarzschild radius.

It is more troublesome to convince ourselves that the narrow line width is caused by annihilation taking place in an accretion disk with a small inclination angle, though we believe that this is the most natural and simple interpretation of the data. It is hard to imagine that a dense cloud or two near the black hole can do the job. Dynamically such a cloud has to move outward from or around the black hole. If it was moving outward even at a small speed of 100 kms^{-1} ($< 1\%$ of the Keplerian velocity at $10r_s$), it would be out of the observed annihilation region ($\leq 10r_s$) in less than 10 s. But the transient lasted at least 13 hours. If it was in the form of a continuous gas (or pair) stream, it would have created a broad annihilation line from the distribution of redshift alone, contradictory to the observed line width. If such a (pair) cloud was moving around the black hole (with some angle to the accretion disk), it would be hard to understand why there was only one rather than many, since just a few clouds moving in different angles (similar to a spherical turbulent cloud) would produce a much broader line than that observed. Futhermore, if some dense gas clouds somewhere above the disk were intercepting and annihilating pairs, the accretion disk would also be annihilating pairs, which certainly would not help reduce the observed line width. In our opinion, a tilted accretion disk offers a much simpler and cleaner option.

When $e^+ - e^-$ pairs are created via $\gamma - \gamma$ collisions inside the last stable orbit[15], they flow outward relativistically due to high radiation pressure. The high column density at the inner region of an accretion disk can intercept, quickly slow down, and then annihilate the high-energy positrons locally. This may explain the small range of the observed line redshift. The lack of positronium continuum also points to a warm ($> 10^5$ K) and high-density ($\gg 10^{14}$cm^{-3}) disk environment. On the other hand, the measured narrow line width in XN Mus places an upper limit on the temperature of the annihilation region to $<$ a few keV, consistent with the disk picture. Observationally, some direct (but weak) evidence exists indicating that the annihilation takes place in a disk. The narrowly binned line profile shows a double peak feature[7], characteristic of a disk emission profile, though the photon statistics are low, so this is not conclusive.

With the annihilation site located at the inner region of an accretion disk, it is straightforward to estimate the contribution to the observed line width from thermal broadening and disk rotation. The thermal width is $\Delta E_T = 36$ keV from a disk temperature of ~ 0.9 keV[16]. The Keplerian rotational width is then $\Delta E_K = [(\Delta E)^2 - \Delta E_T)^2]^{1/2} \sim 48$ keV. The disk inclination can be calculated from $\sin i = (2r/r_s)^{1/2} \Delta E_K / E \sim 0.4$. Thus the black hole mass in XN Mus is probably around 50 $M_\odot$ using the observed mass function (eq. 1) and a companion mass of 0.5 $M_\odot$. This is certainly a very large number. However, because of the large uncertainties (1 σ) in the observed line width, this proper value does not carry much statistical weight. If we follow the exact formulae to calculate the mean and standard deviation, assuming that the measured line width has a Gaussian error distribution[14], 1 σ lower limit of the black hole mass is 8 $M_\odot$.

TABLE 1   Annihilation line parameters and derived black hole mass

Parameters	Groups		
	Goldwurm et al.	Sunyaev et al.	Mean
Centroid E (keV)	481 ± 22	476 ± 15	479 ± 18
Redshift z	0.062 ± 0.049	0.074 ± 0.034	0.068 ± 0.042
Annih. radius (r_s)	8.8 ± 6.2	7.6 ± 3.1	8.2 ± 4.6
ΔE (keV)	54 ± 54	58 ± 34	56 ± 44
ΔE_K (keV)	48 ± 48	45 ± 37	46 ± 42
$\sin i$	0.39 ± 0.40	0.37 ± 0.30	0.38 ± 0.35
Inclination i	$28° \pm 30°$	$24° \pm 21°$	$26° \pm 25°$
BH mass ($M_\odot$)			
Lower limit (1 σ)	5	10	8

DISCUSSION

We have argued that the observed narrow annihilation line width in XN Mus can constrain the geometry of the accretion disk where the radiation originates. This new approach is useful in estimating the mass of black holes in X-ray binaries, though the low photon statistics in this particular source prevented us from a more precise conclusion.

However, the real situation may be more complicated since we have not yet tried to explain simultaneously the observed back-scattering feature in XN Mus. Hua, Lingenfelter, and Ramaty[17] have constructed a model for back-scattering which involves an accretion disk with a hollow central region inside which the annihilation photons are produced. Some of the photons are bounced back from the square inner wall of the hollow disk to our line of sight. This model requires a disk inclination angle of $\sim 60°$ to reproduce the observed direct/scattered annihilation line ratio and a radial speed of $0.05c$ of the inner wall to explain the apparent blueshift of the back-scattering feature. The narrow annihilation line width is explained cleverly by adjusting the size of the central hollow region so that only a small section of the annihilation source can be seen directly. It is clear that this model is very sensitive to the assumed geometrical structure of the accretion disk.

Finally, it is intriguing to note that a large fraction of the known black hole candidates belongs to the X-ray nova family[1]. The reason is that this subclass of X-ray transients usually has very low quiescent disk emission so the optical observation of the orbital modulation of the low-mass secondary is possible. We suggest that for current and future γ-ray missions high priority be given to monitor these sources while in outburst. A high-quality disk-related γ-ray emission line, when combined with an optically measured binary mass function, can be particularly valuable in determining the compact object mass in these X-ray binaries.

REFERENCES

1. A. P. Cowley, Ann. Rev. Astron. Astrophys. 30, 287 (1992).
2. J. Casares, P. A. Charles, & T. Naylor, Nature 355, 614 (1992).
3. C. Haswell, Ph.D. thesis, University of Texas, Austin (1992).
4. C. D. Bailyn, Astrophy. J. 391, 298 (1992).
5. R. Sunyaev, et al., Astrophys. J. 389, L75 (1992).
6. A. Goldwurm, et al., Astrophys. J. 389, L79 (1992).
7. S. Brandt (editor), Proceedings of the Workshop on Nova Muscae 1991 (DSRI, Lyngby, 1991).
8. N. E. White, J. L. Kalluzienski, & J. H. Swank, High Energy Transients in Astrophysics, ed. S. E. Woosley (AIP, N.Y., 1984), p.31.
9. M. Della Valle, B. Jarvis, & R. M. West, Nature 353, 53 (1991).
10. R. A. Remillard, J. E. McClintock, & C. D. Bailyn, Astrophys. J., 399, L145 (1992).
11. L. Bouchet, et al., Astrophys. J. 383, L45 (1991).
12. R. Sunyaev, et al., Astrophys. J. 383, L49 (1991).
13. M. Briggs, Ph.D. thesis, University of California, San Diego (1991).
14. W. Chen, N. Gehrels, & F. H. Cheng, Astrophys. J., 403, L71 (1993).
15. E. P. Liang, Astrophys. J. 367, 470 (1991).
16. J. Greiner, private communication (1992).
17. X. M. Hua, R. E. Lingenfelter, & R. Ramaty, this volume.

MAGNETIC FIELD RECONNECTION AND HIGH-ENERGY EMISSION FROM ACCRETING NEUTRON STARS

Marco Tavani
Joseph Henry Laboratories and Department of Physics
Princeton University, Princeton, NJ 08544

Edison Liang
Department of Space Physics
Rice University, P.O. Box 1892, Houston, TX 77251

ABSTRACT

We present preliminary results of a study aimed at clarifying the mechanism of particle acceleration caused by magnetic field reconnection in the inner region of accretion disks surrounding neutron stars. We consider systems containing weakly magnetized neutron stars (e.g., X-ray bursters). We discuss the relevant features of the magnetic configuration and of the accretion process influencing the emission of hard X-rays and γ-rays expected by reconnection events near the Alfven radius. Our results are relevant for ongoing high-energy missions such as SIGMA and the Compton Gamma-Ray Observatory (CGRO).

INTRODUCTION

Recent observations showed the existence of a galactic population of accreting compact stars which have a time variable hard X-ray emission[3,16]. Several of these sources are concentrated near the galactic center region and most of them are previously known X-ray sources. In particular, the fact that at least three sources detected by SIGMA with hard X-ray *power-law* tails are accreting neutron stars (the X-ray pulsar GX 1+4 and the X-ray bursters GX 354+0, KS 1731-260[1] indicate that particle acceleration and time variable hard photon emission occur near both strongly and weakly magnetized neutron stars. In this respect, the observational situation is very different compared to pre-SIGMA era. At that time there was no unambiguous detection of hard X-rays from accreting neutron stars. Furthermore, CGRO will provide a wealth of new data concerning galactic accreting neutron stars both in the all-sky coverage and in pointed observations.

The possibility of high energy emission from accreting neutron stars has been raised recently in the context of a model of the disk-magnetospheric boundary near a weakly magnetized neutron star (Ref. 10; hereafter KRST). We briefly report here preliminary results of a detailed study of the conditions near the disk-magnetospheric boundary which favor particle acceleration and the formation of hard X-ray and γ-ray tails in the spectrum.

PHYSICS OF THE DISK-MAGNETOSPHERIC BOUNDARY

Several processes occurring near the the disk-magnetospheric boundary need to be considered:

(1) the interaction of the neutron star magnetic field with the partially threaded inner disk which leads to flaring magnetic field reconnection;

(2) the effect of inverse Compton scattering on the accelerated particles by soft photons originating from the surface of the neutron star, magnetospheric flow and inner part of the accretion disk;

(3) the energization of a scattering corona around the neutron star;

(4) possible pair creation cascade and annihilation processes.

Based on the analogy with solar flare configuration of current loops emerging from a turbulent surface[11], we are studying a geometry of interaction similar to the geometry of Ref. 5. Previous studies of flaring inner parts of accretion disks have concentrated on thermal processes regulating the self-interaction and reconnection of magnetic field loops internally generated in the disk[4,5] and the threading of an external dipole disk[6,17]. A particularly promising acceleration mechanism is offered by reconnection of the ambient magnetic field with emerging loops of opposite polarity generated by differential rotation[5]. Particle acceleration by this mechanism becomes important when the magnetic field in the inner disk approaches the equipartition value.

A quantitative study of the acceleration process include:

(1) the determination of the maximum Lorentz factor of accelerated particles in different accretion regimes and soft-photons environments;

(2) the study of the role played by large cospatial reverse currents necessary for avoiding a catastrophic enhancement of ambient magnetic fields;

(3) the determination of the spectrum and intensity of high energy radiation from the accelerated electrons.

The efficiency and characteristics of acceleration depends ultimately on the magnitude of the neutron star magnetic field and the mass transfer rate producing the X-ray luminosity L_X.

A SIMPLE MODEL OF MAGNETIC FIELD RECONNECTION

In the following we use the convention of denoting the units chosen for a physical quantity Q by $Q_n = Q/(10^n)$. We assume a geometry of the inner edge of an accretion disk as in Ref. 5, with the neutron star magnetic field interacting with the differentially rotating disk of inner edge $R_A \simeq (3 \cdot 10^8 \text{ cm}) \dot{M}_{17}^{-2/7} B_{s,12}^{4/7} (M/M_\odot)^{-1/7}$, with $\dot{M}$ the mass accretion rate and M the neutron star mass. The corona over the inner edge of the disk can be thought as being perturbed by multiple and short-distance reconnection regions forming current sheets of length L, width w and thickness δr.

If we consider an initial configuration of the reconnecting magnetic field as in the tearing mode reconnection, we obtain that the energy available in a single reconnection episode is[2] $\mathcal{E}_c = e\, B_\parallel\, d\, (v_A/c)^{3/2}$ where d is the spatial acceleration length limited by cooling losses of the accelerated particles, $B_\parallel$ the accelerating field equal to the smallest of the two reconnecting fields, and v_A the Alfven velocity corresponding to the smallest of the reconnecting fields. The maximum energy per particle attainable in a reconnection event is ultimately limited by radiation and cooling processes. Furthermore, the total number of particles subjected to acceleration depends on complex phenomena occurring in the plasma. A limit imposed on the total number of current particles which ignores the role played by reverse currents can be simply derived by the constraint that the magnetic field generated by the current does not exceed the magnetic field of the reconnecting field. From the condition $4\pi \mathbf{j} \simeq c \nabla \times \mathbf{B} \sim B/d_B$ with d_B the radius of curvature of the reconnecting magnetic field, we obtain the B-limited number density $n_B \simeq (1.6 \cdot 10^9 \text{ cm}^{-3}) B_6 d_{B,5}^{-1}$. For comparison, the Goldreich-Julian number density is $n_{GJ} = \mathbf{\Omega_K} \cdot \mathbf{B}/(2\pi e c) \simeq (7 \cdot 10^7 \text{ cm}^{-3}) B_6/P_{-3}$ with P_{-3} the spin period of the neutron star in milliseconds comparable with the Keplerian period of the inner part of the accretion disk.

By equating the acceleration timescale defined by $\tau_a^{-1} = e E_r c/[\gamma m c^2]$ where E_r is the electric field associated with the reconnection event, with the cooling timescale due to inverse Compton losses $\tau_{IC} = m c/[(4/3) U_X \sigma_T \gamma]$ with γ the Lorentz factor and $U_X = L_X/(4\pi r^2 c)$ the energy density of soft X-rays (of total luminosity L_X) near the acceleration site, we get the general expression

$$\gamma_C \lesssim 10^4 \left(\frac{B_{*,8}}{L_{X,38} R_6} \right)^{1/2} \tag{1}$$

with B_* the surface magnetic field of the neutron star. Eq. (1) gives the limit Lorentz factor due to Compton scattering of the soft X-ray background permeating the acceleration region. The electrons quickly lose their energy through synchrotron radiation, pair creation and inverse Compton scattering. KRST showed that power-law emission is possible under favorable conditions with exponents depending on the large scale configuration of the magnetic field.

An estimate of the total high-energy power from the inner edge of the accretion disk for B-limited accelerated currents is

$$L_B = 4\pi (R_2^2 - R_A^2) n_B c (\gamma_C m c^2) f \tag{2}$$

where R_2 is the outer radius of the inner annular region of the accretion disk where most of the energetic reconnecting events take place, and $f \lesssim 1$ is a geometric filling factor. From previous equations we obtain the estimates

$$L_{GJ} \simeq (2.7 \cdot 10^{30} \text{ erg s}^{-1}) B_{s,12}^{-0.2} L_{X,38}^{-1/2} \dot{M}_{17}^{6/7} M^{3/7}$$

and

$$L_B \sim (3.7 \cdot 10^{33} \text{ erg s}^{-1}) f (R_7)^2 n_{B,10} [B_{*,8}/(L_{X,38} R_6)]^{1/2}$$

where we used Eq. (1). We notice that the high energy power depends inversely on the X-ray luminosity, suggesting that the hard X-ray and soft gamma-ray emission should be anticorrelated with the overall X-ray emission from the accreting neutron star.

High energy emission can become observable at the level of $L \sim 10^{36}$ erg s^{-1} if $R_7^2 n_{10} L_{X,38}^{-1/2} f \gtrsim 300$, a condition that it is not satisfied for B-limited currents. For a current sheet geometry, requiring that the induction magnetic field does not exceed the original ambient B gives a condition on the total number of electrons involved in a *single* reconnection event[8] $\dot{N}_e \lesssim c w B/(2\pi e) \sim (10^{31}\ \text{s}^{-1})\, w_5\, B_7$ and a total power per reconnection event $P_r \sim (3 \cdot 10^{28}\ \text{erg s}^{-1})\, w_5\, B_7^{3/2}\, L_{X,38}^{-1/2}$. An observable high energy emission requires the existence of $\gtrsim 3 \cdot 10^7$ simultaneous B-limited current channels (with each channel conforming to the upper limit on $\dot{N}_e$) near the disk-magnetospheric boundary of a weakly neutron star. As a comparison, the hard X-ray bursts ($E > 25$ keV) produced by solar flares require a number of at least 10^4 individual current channels[8].

The previous estimates neglect the possible role played by cospatial reverse currents both electrostatically and inductively driven by the electron beam accelerated in the reconnection events. The existence of reverse currents near the disk-magnetospheric boundary of weakly magnetized neutron stars has similarities with the situation occurring in solar flares[7,9,12,14]. As the accelerated beam propagates in the plasma, an induction electric field causes a plasma electron current to flow in a direction opposite to the beam current, reducing the net current and the self-field of the beam. The collective interaction of a beam in a plasma involves several complex effects yielding anomalous resistivity and plasma instabilities[15]. The return current eventually decays due to finite resistivity and contributes to heating the plasma electrons. If the beam pulse time is Δt and $\tau_D = 4\pi w^2/(\eta c^2)$ is the return current dissipation time with η the effective resistivity, the maximum resistive loss per each reconnection event can be estimated as $W \sim (\Delta t/\tau_D) L' I_b L_o$ ergs, where L' is the inductance per unit length of the beam, and L_o the spatial length of the reconnection event. Plasma instabilities reduce τ_D. An estimate based on the ion-acoustic instability has been found to correspond to a value[14] $\tau_D \sim 10 w_5^2 n_{10}^{1/2}$ s, which is larger than the Petschek-type reconnection time[5], $\Delta t =\sim 200 L/v_A \sim 10^{-2}$ s. It is therefore possible that reverse currents can alleviate the problem of extracting high-energy power from the inner region of an accretion disk by increasing the effective number density of current channels and by reducing their number. The plausibility of producing this situation and its stability require further study.

REFERENCES

1. Barret, D., et al., 1992, Ap.J., **394**, 615.
2. Bulanov, S.V., and Sasorov, P.V., 1976, Sov. Astron., **19**, 464.
3. Cook, W.R., et al., , 1991, Ap.L.(Letters), **372**, L75-L78.
4. Eardley, D.M., and Lightman, A.P., 1975, Ap.J., **200**, 187.
5. Galeev, A.A., Rosner, R., and Vaiana, G.S., 1979, Ap.J., **229**, 318.
6. Ghosh, P., and Lamb, F.K., 1979, Ap.J., **232**, 259.
7. Hoyng, P., Brown, J.C., and van Beek, H.F., 1976, Solar Phys., **48**, 197.
8. Holman, G.D., 1985, Ap.J., **293**, 584.
9. Knight, J.W., and Sturrock, P., 1977, Ap.J., **218**, 306.
10. Kluźniak, W., Ruderman, M., Shaham, J., and Tavani, M., 1988, Nature, **336**, 558.
11. Priest, E.R., 1987, Solar Magnetohydrodynamics (Dordrecht: Reidel).
12. Shapiro, P.R., and Knight, J.W., 1978, Ap.J., **224**, 1028.
13. Spicer, D.S., 1982, Space Sci. Reviews, **31**, 351.
14. Spicer, D.S. and Sudan, R.N., 1984, Ap.J., **280**, 448.
15. Sudan, R.N., 1984, in Handbook of Plasma Physics, vol. 2, eds. A.A. Galeev and R.N. Sudan (New York: North Holland), p. 337.
16. Sunyaev, R., et al.., 1991a, Astron. & Astrophys., **247**, L29.
17. Wang, A., 1987, A&A, **183**, 257.

GAMMA-RAY PRODUCTION BY NEUTRON STARS ACCRETING FROM A DISK

M. Coleman Miller, Frederick K. Lamb[†], and Russell J. Hamilton
Department of Physics, University of Illinois at Urbana-Champaign,
1110 W. Green St., Urbana, IL 61801-3080, U.S.A.

INTRODUCTION

In a system with a neutron star accreting from a disk, the electrical circuit formed by the star and disk has an electromotive force (EMF) $\sim 10^{15}$ Volts, even if the stellar rotation rate is slow and the plasma density in the magnetosphere is high (Lamb et al. 1992a,b). If the minimum density in the magnetosphere is less than $\sim 10^9 \, \text{cm}^{-3}$, this EMF may cause the formation of relativistic double layers with voltage drops $\sim 10^{12}$ Volts. If the minimum density is greater than $10^9 \, \text{cm}^{-3}$, magnetic reconnection events are likely to occur and could accelerate a small fraction of the charged particles to the full energy provided by the EMF. In either case, strong electric fields are probably generated in the magnetosphere above the disk (Hamilton et al. 1992). Electrons accelerated by these fields may reach energies $\sim 10 \, \text{TeV}$, while ions can be accelerated to the full energy provided by the EMF. Either electrons or ions may convert their energy (through bremsstrahlung and ion-ion collisions, respectively) into high-energy γ-rays. The precise high-energy radiation spectrum will depend on details such as the column depth of the disk, the magnetic field geometry, and the X-ray luminosity. Here we concentrate on the basic processes that will influence the shape of the γ-ray spectrum.

PHOTON ATTENUATION PROCESSES

The magnetospheres above neutron star accretion disks have magnetic fields $\sim 10^4 - 10^8$ G and X-ray photons of average energy ~ 1 keV and number density $\sim 10^{17} - 10^{18} \, \text{cm}^{-3}$. In this environment, γ-rays interact primarily via photon-photon pair production ($\gamma\gamma \to e^-e^+$), magnetic pair production ($\gamma B \to e^-e^+ B$), and photon splitting ($\gamma B \to \gamma\gamma B$). Where there is a high density of X-rays, energetic photons may produce pairs by photon-photon scattering if the product of the photon energies $\hbar\omega_1$ and $\hbar\omega_2$ exceeds $(m_e c^2)^2$. If this condition is satisfied, the cross section for the production of pairs is (Berestetskii et al. 1982)

$$\sigma_{\gamma\gamma}(v) = \frac{3}{16}\sigma_T(1-v^2)\left[(3-v^4)\ln\left(\frac{1+v}{1-v}\right) - 2v(2-v^2)\right],$$

where $v \equiv \left[1 - (m_e c^2)^2/(\hbar^2\omega_1\omega_2)\right]^{1/2}$ and $\sigma_T = 6.65 \times 10^{-25} \, \text{cm}^2$. This process may initiate a pair cascade.

[†] Also, Department of Astronomy.

In regions of strong magnetic field, photons moving at an angle to the field undergo magnetic pair production. The number of pairs created in a magnetic field by a flux of photons of energy $\hbar\omega$ in distance d is $n_{\text{pairs}} = n_{\text{photons}}(1 - \exp[-\alpha(\chi)d])$, where

$$\chi = (\hbar\omega/m_e c^2)(B_\perp/B_c),$$

$B_\perp$ is the magnetic field perpendicular to the direction of motion, $B_c \equiv m_e^2 c^3/(e\hbar) \approx 4.4 \times 10^{13}$ G, and (Erber 1966)

$$\alpha(\chi) = \frac{1}{2}\frac{\alpha_f}{\lambda_c}\frac{B_\perp}{B_c}T(\chi).$$

Here $\lambda_c = \hbar/(m_e c) \approx 3.8 \times 10^{-11}$ cm, $\alpha_f = e^2/\hbar c$, and in the limit $\chi \ll 1$, $T(\chi) \approx 0.46\exp[-4/(3\chi)]$. This process prevents the escape of photons of energy $\hbar\omega = \gamma m_e c^2$ from regions where the magnetic field B exceeds 2×10^{12} G/γ.

Photon splitting is generally less important than the other two processes. The fraction of photons that split in distance d is $1 - \exp[-Sd]$, where (Erber 1966)

$$S \approx 4 \times 10^{-19} \left(\frac{\hbar\omega}{m_e c^2}\right) \left(\frac{B_\perp}{10^5 \text{ G}}\right)^2 \text{ cm}^{-1}.$$

Unlike magnetic pair production, which is exponential in B, photon splitting depends only quadratically on B, so that it always operates at some level to decrease the mean photon energy.

Photon-photon pair production, magnetic pair production, and photon splitting are compared in Figure 1 for physical conditions typical of an accreting neutron star system. The dimension of the system is $\sim$ few $\times\, 10^8$ cm, so interaction probabilities less than $\sim 10^{-9}$ cm^{-1} are insignificant. For $n_\gamma = 10^{18}$ cm^{-3}, $B_\perp = 1$ G, and $\langle\hbar\omega\rangle = 1$ keV, photon-photon pair production cuts off the photon spectrum between ~ 200 MeV and ~ 1 TeV, while magnetic pair production cuts off the spectrum at $\hbar\omega \gtrsim 10$ TeV. Depending primarily on the average energy of X-ray photons, there will be two "windows" where high energy photons could escape the system. Photons with energies below the photon-photon pair production threshold escape, while higher energy photons escape if they have energies high enough to avoid efficient photon-photon pair production but below the threshold for magnetic pair production.

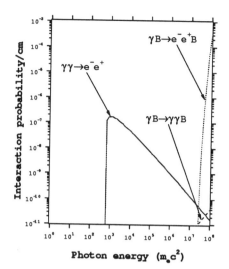

Fig. 1—Comparison of γ-ray attenuation mechanisms as a function of energy for a number density 10^{18} cm^{-3} of $\langle \hbar\omega \rangle = 1$ keV X-rays and a perpendicular magnetic field of 10^5 G.

ELECTRON ENERGY LOSSES

In the magnetospheres of accreting neutron stars, several processes prevent electrons from being accelerated by electric fields to arbitrarily high energies. Some of the possibilities include inverse Compton drag ($e\gamma \rightarrow e\gamma$), curvature and synchrotron radiation ($eB \rightarrow e\gamma B$), and triplet pair production ($e\gamma \rightarrow ee^-e^+$). If there is a high density of ambient X-rays, Compton drag can be important. An electron accelerated by an electric field $\mathcal{E}$ through a bath of photons of average energy $\langle \hbar\omega \rangle$ and number density n_γ has a probability of scattering off a photon of

$$p_s \approx 20 n_\gamma \sigma_T \frac{(m_e c^2)^2}{e\mathcal{E}\langle \hbar\omega \rangle} \lesssim 0.1 \left(\frac{n_\gamma}{10^{18}\,\text{cm}^{-3}} \right) \left(\frac{1\,\text{keV}}{\langle \hbar\omega \rangle} \right) \left(\frac{10^2\,\text{statvolt/cm}}{\mathcal{E}} \right).$$

This is less than unity for the conditions expected in magnetospheres.

The power emitted in curvature radiation by an electron with Lorentz factor γ moving along field lines with radius of curvature $R = 10^8\,R_8$ cm is $P \approx \frac{2}{3} e^2 c \gamma^4 / R^2$ (see, e.g., Jackson 1975). In an electric field $\mathcal{E} = 10^2\,\mathcal{E}_2$ statvolt/cm, this limits γ to $\gamma \lesssim 7 \times 10^6 R_8^{1/2} \mathcal{E}_2^{1/4}$.

The synchrotron loss rate for an electron with velocity βc and energy $E = \gamma m_e c^2$ travelling at an angle θ to a magnetic field in the limit $\gamma B/B_c \ll 1$ is $\dot{E}/E \approx 20 \gamma \beta^2 \sin^2\theta (B/10^5\,\text{G})^2$ s^{-1}.

For ultrarelativistic electrons, the cross section for triplet pair production increases logarithmically with electron energy. If the ambient photons have energy $\langle \hbar\omega \rangle = k m_e c^2$, the energy loss rate in the limit $2\gamma k > 100$ is (Mastichiadis et al. 1986)

$$d(E/m_e c^2)/dt \approx 6.6 \times 10^{-27} c n_\gamma k^{-1} (2\gamma k)^{0.77}\,\text{s}^{-1}.$$

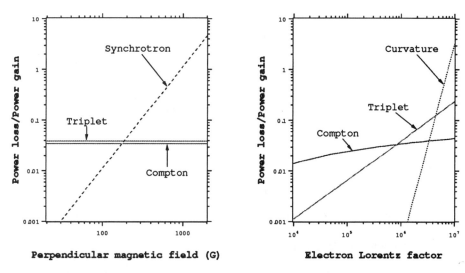

Fig. 2 *Left panel*—Comparison of electron energy loss processes as a function of perpendicular magnetic field, for an electron Lorentz factor $\gamma = 10^6$, a number density $n_\gamma = 10^{18}$ cm^{-3} of $\langle \hbar\omega \rangle$ =1 keV X-rays, and an electron number density $n_e = 10^{10}$ cm^{-3}. *Right panel*—Comparison of electron energy loss processes as a function of electron Lorentz factor, for a number density $n_\gamma = 10^{18}$ cm^{-3} of $\langle \hbar\omega \rangle$ =1 keV X-rays, a perpendicular magnetic field $B_\perp$ =1 G, and an electron number density $n_e = 10^{10}$ cm^{-3}.

These energy loss mechanisms are compared with each other and with the rate of energy gain from an electric field $\mathcal{E} = 10^2$ statvolt/cm in Figure 2. The left panel shows that synchrotron losses force particles in the magnetosphere to travel along field lines. Synchrotron radiation is the dominant energy loss mechanism for $B_\perp \gtrsim 2 \times 10^2$ G, and exceeds the rate of energy gain for $B_\perp \gtrsim 10^3$ G. The right panel of Figure 2 shows that, for particles moving along the field, curvature radiation sets the maximum Lorentz factor. Inverse Compton drag is most important for $\gamma \lesssim 10^6$, but is never significant compared to the rate of energy gain. For $10^6 \lesssim \gamma \lesssim 4 \times 10^6$, triplet pair production dominates, whereas for higher Lorentz factors curvature radiation is the most significant, exceeding the rate of energy gain for $\gamma \gtrsim 7 \times 10^6$.

PAIR CASCADES

Pair cascades may limit the energy of accelerated electrons. When the upper cutoff energy of curvature photons reaches $\hbar\omega_c \approx 500 m_e c^2$ (at $\gamma \approx 7 \times 10^6$), the curvature photons may produce pairs by scattering off of 1 keV X-rays. This Lorentz factor is close to that at which the energy loss from curvature radiation equals the energy gain from a 10^2 statvolt/cm electric field. If the electrons reach the pair threshold before the curvature limit, the number of curvature photons with $\hbar\omega > 500\, m_e c^2$ is likely to be at least several hundred per electron. Since the path length to pair production is only $\approx 4 \times 10^5$ cm, much smaller than the dimension of the system, an electron with $\gamma \gtrsim 7 \times 10^6$ will spawn hundreds of electron-positron pairs. Magnetic reconnection events do not last long enough for a cascade to develop, but double layers may be stable enough to generate a cascade.

Some authors have suggested that unless the magnetosphere is charge-separated, electric fields will be "shorted out" by the large currents induced by the EMF. However, the relativistic motion of the electrons reduces the field generated by these moving particles by a factor $\sim \gamma^2/\ln\gamma \sim 10^{10-11}$ parallel to the direction of motion. Thus, currents in the magnetosphere are not likely to be space-charge limited. However, the components of the electromagnetic field not aligned with the direction of motion are not affected, and hence the magnetic field generated by the large currents may halt cascades.

Acknowledgements: This work was supported in part by NSF grant PHY 91-00283 and NASA grant NAGW 1583 at the University of Illinois.

REFERENCES

Berestetskii, V. B., Lifshitz E. M., and Pitaevskii, L. P. 1982, *Quantum Electrodynamics*, 2nd ed., Pergamon Press.

Erber, T. 1966, *Rev. Mod. Phys.*, **38**, 4.

Hamilton, R. J., Lamb, F. K., and Miller, M. C., these proceedings.

Jackson, J. D. 1975, *Classical Electrodynamics*, 2nd ed. (New York: John Wiley).

Lamb, F. K., Hamilton, R. J., and Miller, M. C. 1992a, in *Isolated Pulsars*, ed. K. Van Riper, R. Epstein, and C. Ho (Cambridge U. Press), in press.

Lamb, F. K., Hamilton, R. J., and Miller, M. C. 1992b, these proceedings.

Mastichiadis, A., Marscher, A. P., and Brecher, K. 1986, *Astrophys. J.*, **300**, 178.

ACCELERATION OF PARTICLES IN THE MAGNETOSPHERES OF ACCRETING NEUTRON STARS

Russell J. Hamilton, Frederick K. Lamb[†], and M. Coleman Miller
Department of Physics, University of Illinois at Urbana-Champaign
1110 West Green Street, Urbana, IL 61801-3080, U.S.A.

INTRODUCTION

Interaction of an accretion disk with the magnetic field of a neutron star produces large electromotive forces (EMFs), which drive large conduction currents in the disk-magnetosphere-star circuit (Lamb, Hamilton, and Miller 1992a,b, and in preparation). Here, we argue that such large conduction currents will cause microscopic and macroscopic instabilities in the magnetosphere. If the minimum plasma density in the magnetosphere is relatively low ($\lesssim 10^9 \, \text{cm}^{-3}$), current-driven micro-instabilities may cause relativistic double layers to form, producing voltage differences in excess of 10^{12} V and accelerating charged particles to very high energies. If instead the plasma density is higher ($\gtrsim 10^9 \, \text{cm}^{-3}$), twisting of the stellar magnetic field is likely to cause magnetic field reconnection. This reconnection will be relativistic, accelerating plasma in the magnetosphere to relativistic speeds and a small fraction of particles to very high energies. Interaction of these high energy particles with X-rays, γ-rays, and accreting plasma may produce high energy radiation detectable by the *Compton Observatory*, *GRANAT*, or ground-based experiments.

MODEL

Consider a binary system in which a magnetic neutron star is accreting plasma from a geometrically thin Keplerian disk situated in the equatorial plane of the neutron star with its axis of rotation aligned with that of the star. We analyze this system using cylindrical coordinates (ϖ, ϕ, z) centered on the star, with the z-axis aligned with the stellar rotation axis. We assume the star has mass M_s, radius R_s, and rotates with angular velocity Ω_s, and that the stellar magnetic field is dipolar, with the dipole moment μ aligned with the rotation axis. We assume further that the mass flux $\dot{M}$ through the disk far from the star is constant.

The accreting plasma couples to the neutron star via its magnetosphere. Processes occurring within the magnetosphere mediate the transfer of mass, energy, and angular momentum between the accreting plasma and the star. Far from the star, where the magnetic field is unimportant, the accreting plasma orbits at the local Keplerian velocity $\Omega_K(\varpi)$. The Keplerian flow ends at the radius ϖ_0 where the magnetic stress is so large that it extracts the angular momentum of the disk plasma in a radial distance much less than ϖ (Ghosh and

---

[†] Also, Department of Astronomy.

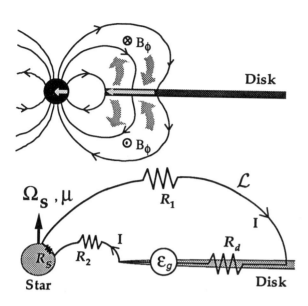

Fig. 1—*Top:* Orbiting motion of the disk plasma generates an EMF that drives cross-field currents (light shaded arrows) within the disk and the neutron star and field-aligned currents (shaded arrows) within the magnetosphere. *Bottom:* Equivalent circuit, showing the generated EMF $\mathcal{E}_g$, the circuit inductance $\mathcal{L}$, and the resistances $\mathcal{R}_s$ and $\mathcal{R}_d$ of the star and disk. Also shown are the resistances $\mathcal{R}_1$ and $\mathcal{R}_2$ and the current I in the two magnetospheric legs of the circuit.

Lamb 1991). The differential rotation between the star and accreting plasma generates an electromotive force (EMF) of magnitude (Lamb, Hamilton, and Miller 1992a,b)

$$\mathcal{E}_g \sim \varpi_0^2 \Omega_K(\varpi_0) B_p(\varpi_0)/c \sim 10^{15} \varpi_8^{-5/2} \mu_{30} (M_s/\dot{M})^{1/2} \text{ Volts} \qquad (1)$$

in the disk-magnetosphere-star circuit.

Although the magnitude of the *charge* density in the magnetosphere is expected to be comparable to the Goldreich-Julian value $n_{GJ} \approx |\mathbf{\Omega_s} \cdot \mathbf{B}|/(2\pi ec)$, the *number* density of ions (and electrons) in the magnetosphere is expected to be much larger. As a result, the EMF drives large conduction currents as shown in Figure 1 for a slowly rotating star. The azimuthal field B_ϕ produced by the conduction currents can be thought of as resulting from the twisting of the poloidal magnetic field B_p by the rotational motion of the accreting plasma relative to the star.

In the absence of other processes, the current asymptotically approaches the steady state value $I_{ss} = \mathcal{E}_g/\mathcal{R}$ with an e-folding time $\sim \mathcal{L}/\mathcal{R}$, where $\mathcal{R} = \mathcal{R}_1 + \mathcal{R}_2 + \mathcal{R}_d + \mathcal{R}_s$. The total resistance $\mathcal{R}$ in the circuit is very small, so that, in the absence of instabilities, a large asymptotic current I_{ss} would flow, generating an azimuthal magnetic field $\sim 10^{10}$ times larger than the poloidal field. Because of the low resistance, electrical currents become large enough to produce microscopic and/or magnetohydrodynamic (MHD) instabilities, which can lead to the acceleration of ions and electrons to energies greater than 1 TeV.

INSTABILITIES AND PARTICLE ACCELERATION

We consider the consequences of large electrical currents in the magnetosphere, with emphasis on processes that may lead to particle acceleration. Whether micro-instability or reconnection occurs first depends on the particle density in the most tenuous regions of the magnetosphere. If this particle density is moderately low, field-aligned currents are likely to trigger micro-instabilities and possible double layer formation before reconnection occurs. If instead the lowest particle density is higher, reconnection is likely to occur before the threshold for micro-instability is reached. In some circumstances, the current may continue to grow even after double layers have formed, eventually triggering a reconnection event.

Micro-instabilities will occur if the current density in the magnetosphere is high enough. If the current density exceeds

$$j_{\text{micro}} = en_e(kT_e/m_e)^{1/2} \sim 10^9 n_9 T_8^{1/2}, \qquad (2)$$

micro-instabilities will be produced. Here m_e, $n_e = n_9 10^9 \text{cm}^{-3}$, and $T_e = T_8 10^8$ K are the electron mass, density, and temperature. These instabilities can produce anomalous resistance, heating, or double layers (see Spicer 1982).

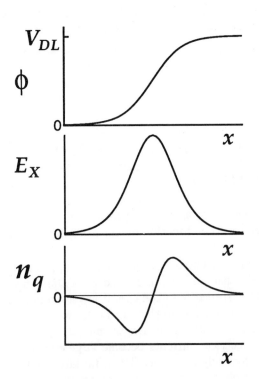

Fig. 2—Example of a stationary double layer solution, showing the electrical potential ϕ, normal electric field E_x, and charge density $n_q \equiv n_+ - n_-$ as functions of position x. Charge separation is maintained by a balance between inertial and electrostatic forces. Quasi-neutrality is violated at both edges of the layer, but the spatially integrated charge is almost zero, so the electric field outside the layer is small. Such a layer produces approximately monoenergetic electrons and protons (for $eV_{DL} \gg m_p c^2$), since in moving across the layer both attain energies $\sim eV_{DL}$.

Double layers, which are observed in both space and laboratory plasmas (see Block 1978; Carlqvist 1979), are related to the nonlinear BGK-wave solutions of the Vlasov-Maxwell equations and form when large electrical currents flow in response to applied voltages. These structures consist of two equal but oppositely charged, approximately parallel space charge regions. The electron number density n_e is many orders of magnitude larger in the disk than in the magnetosphere, while the temperature T_e is the same to within a factor of 10^2, so that j_{micro} is orders of magnitude smaller in the magnetosphere than in the disk. Thus, formation of double layers is most likely to occur in the magnetosphere. In these systems, we expect double layers to be strong (voltage $V_{DL} \gg kT_e$) and relativistic ($V_{DL} \gg m_p c^2 \gg m_e c^2$). Double layers are therefore effective particle accelerators.

Carlqvist (1969) found that for a relativistic double layer bordered by a thermal plasma consisting of ions and electrons with the ion mass $m_i \gg m_e$, the voltage drop is related to the current density by

$$V_{DL} \approx (\pi m_i c j d^2/e)^{\frac{1}{2}} , \qquad (3)$$

where $d = d_8 10^8 \text{cm}$ is the width of the double layer. For a current density equal to j_{micro}, the voltage drop is $\sim 10^{13} d_8 n_9^{1/2} T_8^{1/4}$ Volts.

The circuits linking the neutron star, magnetosphere, and disk are highly inductive. Therefore current disruption, which may occur if the current density greatly exceeds j_{micro}, would cause the energy stored in these circuits ($\sim \mathcal{L}I^2/2$) to be released rapidly via a discharge process in the region of the disruption. Such an event has been called an "exploding double layer" by Alfvén (1981) and would produce voltage drops across the double layer $\sim \mathcal{L}dI/dt$. Assuming that the current density is j_{micro} and that the current changes on a time scale d/c, the voltage drop across such an exploding layer is $\sim 10^{15}$ Volts.

Another possibility is that the continual twisting of the magnetic field by the differential rotation of the accreting plasma and the star causes the magnetic field configuration to become unstable to reconnection (see Parker 1979), since the field cannot be wound up indefinitely. We expect the magnetic field to become unstable to reconnection when twisting of the poloidal field produces an azimuthal field that exceeds the poloidal field by a factor $\gamma_{\text{mhd}} \sim 10$. The electrical current density corresponding to this value of γ_{mhd} is

$$j_{\text{mhd}} \sim (c\gamma_{\text{mhd}} B_p)/(4\pi \varpi_0) \sim 3 \times 10^9 \, (\gamma_{\text{mhd}}/10) B_6 \varpi_7^{-1} . \qquad (4)$$

If magnetic reconnection occurs steadily, it mainly produces heating of the plasma and bulk acceleration to the Alfvén velocity, $v_A = c[B^2/(4\pi n_e m_p c^2 + B^2)]^{1/2}$, with a small fraction of particles accelerated to higher energies. In the magnetosphere, $B^2 \gg 4\pi n_e m_p c^2$, so the resulting bulk plasma motion is relativistic, with Lorentz factor $\gamma_A \approx 10^2 \, n_9^{-1/2} B_6$. The corresponding electron and proton energies are $\sim 10^8$ eV and $\sim 10^{11}$ eV, respectively. If instead magnetic reconnection occurs in flaring processes, voltages comparable to the full EMF ($\sim 10^{15}$ Volts) may occur.

In summary, rotation of the disk plasma relative to the star continually twists the magnetic field, steadily increasing the field-aligned currents in the magnetosphere. Because the internal resistance of the disk-star generator is so low, these currents become large enough to trigger processes that accelerate particles. The detailed behavior of the system depends on what mechanism limits the current. Comparison of equations (2) and (5) shows that there are two regimes. If the particle density in the most tenuous regions of the magnetosphere is moderate ($n \lesssim 10^9 \text{cm}^{-3}$), field-aligned currents are likely to trigger microinstabilities that may produce relativistic double layers. If the particle density is higher ($n \gtrsim 10^9 \text{cm}^{-3}$), we expect magnetic reconnection to occur first. In either case, ions and electrons will be accelerated.

To determine the energies that particles can attain, their acceleration rates must be compared with their energy loss rates. In the magnetosphere, ions are freely accelerated to the full voltage drop. For electrons, energy losses are important. Emission of curvature radiation limits the electrons to energies $\lesssim 10^{13}$ eV. Emission of synchrotron radiation forces electrons to move along the field, but does not limit their acceleration along the field. Inverse Compton drag does not limit the energy here because electron-photon scattering is in the extreme Klein-Nishina regime. Gamma-rays are produced by interaction of the accelerated ions and electrons with X-rays and accreting plasma. A more detailed discussion of photon-electron interactions important for understanding the observable radiation expected from these accreting neutron stars is given elsewhere (Miller, Lamb, and Hamilton 1992, and in preparation).

Acknowledgements: This research was supported in part by NSF grant PHY 91-00283 and NASA grant NAGW 1583 at the University of Illinois.

REFERENCES

Alfvén, H. 1981, *Cosmic Plasma* (Dordrecht: Reidel).
Block, L. P. 1978, *Ap. Space Sci.*, **55**, 59.
Carlqvist, P. 1969, *Solar Phys.*, **7**, 377.
Carlqvist, P. 1979, in *Wave Instabilities in Space Plasmas*, ed. P. J. Palmadesso and K. Papadopoulos (Dordrecht: Reidel), p. 83.
Ghosh, P., and Lamb, F. K. 1991, in *Neutron Stars: Theory and Observation*, ed. J. Ventura and D. Pines (Dordrecht: Kluwer), p. 363.
Lamb, F. K., Hamilton, R. J., and Miller, M. C. 1992a, in *Isolated Pulsars*, ed. K. Van Riper, R. Epstein, and C. Ho (Cambridge U. Press), in press.
Lamb, F. K., Hamilton, R. J., and Miller, M. C. 1992b, these proceedings.
Miller, M. C., Lamb, F. K., and Hamilton, R. J. 1992, these proceedings.
Parker, E. N. 1979, *Cosmical Magnetic Fields*, (Clarendon: Oxford).
Spicer, D. S. 1982, *Space Sci. Rev.*, **31**, 351.

ELECTRODYNAMICS OF NEUTRON STARS ACCRETING FROM A DISK

Frederick K. Lamb[†], Russell J. Hamilton, and M. Coleman Miller
Department of Physics, University of Illinois at Urbana-Champaign
1110 West Green Street, Urbana, IL 61801-3080, U.S.A.

INTRODUCTION

The availability of MeV to GeV instruments onboard the *Compton Observatory* and *GRANAT* and the improving sensitivity of ground-based TeV and PeV detectors have focused attention on whether accreting neutron star systems may produce detectable fluxes of high energy electromagnetic radiation. Motivated by this question, we have investigated the electrodynamics of magnetic neutron stars accreting from Keplerian disks and the implications for particle acceleration and γ-ray emission by such systems. The summary of this work presented here is based on a more detailed account to be published elsewhere (Lamb, Hamilton, and Miller, in preparation; see also Lamb, Hamilton, and Miller 1992).

We argue that the particle density in the magnetosphere is likely to be many orders of magnitude larger than the Goldreich-Julian density, making acceleration by vacuum gaps unlikely. Interaction of the accretion disk with the magnetic field of the neutron star produces very large electromotive forces (EMFs), even if the star is slowly rotating. Since the resistivity of the disk-magnetosphere-star circuit is small, these EMFs drive very large conduction currents. Such large currents are likely to lead to magnetospheric instabilities, such as relativistic double layers and reconnection events, that can accelerate particles to very high energies.

DISK-STAR INTERACTION

Consider a system in which a neutron star with a strong dipolar magnetic field is accreting plasma from a Keplerian disk. For simplicity, assume that the neutron star magnetic and rotation axes are aligned and that the star is slowly rotating. (We expect the qualitative features of the system to be as described below, even if the axes are not aligned.) Assume also that the disk is geometrically thin and situated in the rotational equatorial plane of the neutron star. In analyzing this system, we use cylindrical coordinates (ϖ, ϕ, z) centered on the star. Our analysis focuses on the general features of the system rather than any particular model.

Plasma near the neutron star is highly ionized and has a high electrical conductivity. We therefore expect an electric field $\mathbf{E} \approx -\mathbf{v} \times \mathbf{B}/c$ in this region,

---

[†] Also, Department of Astronomy.

which causes the plasma to corotate with the star. The magnitude of the charge density near the star is then (see Goldreich and Julian 1969)

$$|n_+ - n_-| \approx n_{GJ} \approx |\mathbf{\Omega_s} \cdot \mathbf{B}|/2\pi ec = 7 \times 10^4 \mu_{30} r_8^{-3} P_s^{-1} \text{ cm}^{-3}. \quad (1)$$

where Ω_s is the stellar angular frequency, μ_{30} is the stellar magnetic moment in units of 10^{30} G cm^3, r_8 is the radius in units of 10^8 cm, and P_s is the stellar rotation period in seconds. We also expect the disk to have a magnetic field that causes the highly conducting plasma near the disk to rotate with an angular velocity close to that of the disk plasma to which it is connected by the magnetic field. The magnitude of the charge density in this region is given approximately by equation (1), with P_s replaced by the local rotation period.

We expect the charge density near the neutron star and the inner disk to be much smaller than the particle density there, for two reasons. First, the electrical potential energy difference between the inner disk and the star is $\sim 10^{15}$ eV (see below), whereas the gravitational potential energy confining protons to the disk is only $\sim 10^6$ eV. Hence, even small fluctuations in the electric field near the disk caused by irregularities in the accretion flow can readily pull protons as well as electrons out of the disk into the magnetosphere. Second, about 10^{40}–10^{42} particles per second flow from the disk to the neutron star. The differing rotation rates of the star and disk plasma twists the magnetic field in this region, creating MHD instabilities that inject plasma into the magnetosphere above the inner disk (see Ghosh and Lamb 1991). Even if only a fraction $\sim 10^{-10}$ of the accreting plasma is injected with a velocity comparable to the local Keplerian velocity, the particle density in the magnetosphere greatly exceeds n_{GJ}. Indeed, X-ray observations of Her X-1 (Becker et al. 1977; Bai 1980) indicate that the particle density above the inner disk is $\gtrsim 10^{11}$ cm^{-3}, far greater than n_{GJ}. Formation of a vacuum gap in the magnetosphere requires complete charge separation (see Cheng and Ruderman 1991) and therefore seems unlikely. Our treatment of the interaction between the disk and stellar magnetic field is based on the theory of disk-accreting neutron stars developed over the last fifteen years, which assumes that the particle density in the magnetosphere is large enough to support substantial conduction currents (see Ghosh and Lamb 1991).

The magnetic field of the star can interact with plasma in the disk in a variety of ways. The processes that transport angular momentum in the disk—such as turbulence and magnetic coupling—by their nature allow magnetic flux from the star to couple to the inner part of the disk. The differential rotation of the disk plasma and star generates an EMF that drives field-aligned currents within the magnetosphere. These currents create an azimuthal magnetic field component. The process can be thought of as twisting of the magnetic field due to the differing rotation rates of the disk plasma and star. The resulting magnetic stress on the disk extracts angular momentum from the disk plasma. The Keplerian flow ends at the radius ϖ_0 where the magnetic stress becomes so large that it extracts the angular momentum of the disk plasma in a radial distance much less than ϖ (see Ghosh and Lamb 1991). Inside ϖ_0, the angular

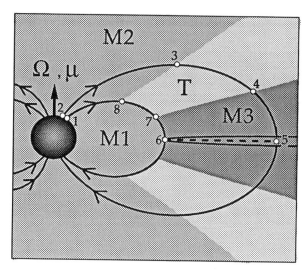

Fig. 1—Schematic diagram of the system analyzed in the text. Plasma on the magnetic field line that connects the inner edge of the disk to the stellar surface (points 6–7–8–1) corotates with the star, as does plasma in regions M1 and M2. Plasma in region M3 orbits the star with the angular velocity of the adjacent disk plasma. We expect velocities in the transition region T to be intermediate between the angular velocity of the star and that of the adjacent disk plasma. The size and shape of region T depends on the particular model.

velocity Ω_d of the disk plasma falls rapidly from the Keplerian value Ω_K, reaching the angular velocity of the star at ϖ_{co}.

SYSTEM ELECTRODYNAMICS

Consider now the electrodynamics of this system. We first calculate the EMF

$$\mathcal{E} \equiv \oint_C \left(\mathbf{E} + \frac{1}{c} \mathbf{v} \times \mathbf{B} \right) \cdot d\boldsymbol{\ell} \qquad (2)$$

around the closed path C denoted by the sequence of points 1–2–3 $\cdots$ 8–1 shown in Figure 1. As a first approximation, we assume that the electrical resistance of the matter along this path is negligible.

Noting that $\mathbf{v} \times \mathbf{B} \cdot d\boldsymbol{\ell} = 0$ wherever C follows the poloidal field, we find

$$\mathcal{E} = -\frac{1}{c} \int_A \frac{\partial B_\phi}{\partial t} dA + \frac{1}{c} \int_1^2 (\boldsymbol{\Omega}_s \times \mathbf{r}) \times \mathbf{B} \cdot d\boldsymbol{\ell} + \frac{1}{c} \int_5^6 (\boldsymbol{\Omega}_d \times \mathbf{r}) \times \mathbf{B} \cdot d\boldsymbol{\ell}. \qquad (3)$$

The first term on the right of this equation comes from the line integral of the electric field around C and describes the back-EMF produced by the inductance

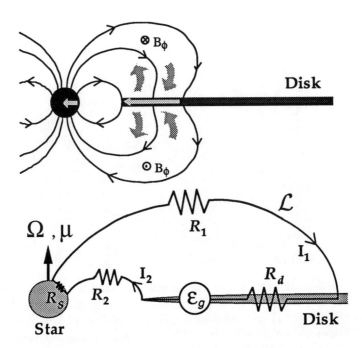

Fig. 2—*Top*: Orbiting motion of the disk plasma generates an EMF that drives cross-field currents (light shaded arrows) within the disk and the neutron star and field-aligned currents (shaded arrows) within the magnetosphere. *Bottom*: Equivalent circuit, showing the generated EMF $\mathcal{E}_g$, the circuit inductance $\mathcal{L}$, and the resistances $\mathcal{R}_s$ and $\mathcal{R}_d$ of the star and disk. Also shown are the resistances $\mathcal{R}_1$ and $\mathcal{R}_2$ and the currents I_1 and I_2 in the two magnetospheric legs of the circuit.

of the circuit. The second and third terms come from the line integral of $\mathbf{v} \times \mathbf{B}/c$. We call the sum of the latter two terms the "generated EMF" and denote it $\mathcal{E}_g$.

If the neutron star is slowly rotating, the third term in equation (3) dominates the second and

$$\mathcal{E}_g \sim \varpi_0^2 \, \Omega_K(\varpi_0) \, B_p(\varpi_0)/c \sim 10^{15} \, \varpi_8^{-5/2} \, \mu_{30} \, (M/M_\odot)^{1/2} \text{ Volts},$$

where $\Omega_K(\varpi_0)$ and $B_p(\varpi_0)$ are the Keplerian angular velocity and the poloidal magnetic field strength at the inner edge of the Keplerian flow. Thus, *even if the neutron star is slowly rotating, its interaction with the disk generates very large EMFs*.

CURRENTS AND INSTABILITIES

The large EMFs in the star-magnetosphere-disk circuit drive large electrical currents between the disk and the star. Although the details of the current distribution depend on the model, all produce current patterns qualitatively similar to the pattern shown in the top panel of Figure 2. The bottom panel shows an equivalent circuit that includes the electrical resistance of the disk, star, and magnetosphere. The time development of the current in this circuit is described by the equation

$$\mathcal{E}_g = \mathcal{L}\frac{dI}{dt} + I\mathcal{R}. \qquad (4)$$

The term on the left describes the EMF generated by the disk-star interaction while the first term on the right describes the effect of the circuit inductance $\mathcal{L} \sim r^2/c$. The second term on the right describes the effect of the total resistance $\mathcal{R}$ in the circuit.

Equation (4) shows that an initially small current grows linearly with time, corresponding to the linear increase of the azimuthal magnetic field B_ϕ as the field is twisted by the differential rotation of the disk and star. If no other processes intervened, the growth would slow after a time $\sim \mathcal{L}/\mathcal{R}$, and the current would asymptotically approach the value $\mathcal{E}_g/\mathcal{R}$. Our calculations of the resistance $\mathcal{R}$ of the circuit show that it is very small and hence that the asymptotic current would be very large, producing a magnetic pitch $\gamma_\phi \equiv B_\phi/B_p \sim 10^{10}$. We argue (Hamilton, Lamb, and Miller 1992 and in preparation) that long before the current reaches this value, microscopic and/or MHD instabilities will intervene, interrupting the growth of the current and the twisting of the magnetic field and accelerating particles to very high energies. Thus, instead of "shorting out" the EMF and thereby preventing acceleration of particles, as some authors have argued (see, e.g., Wang 1986; Katz and Smith 1988; Mitra 1991), *the low internal resistance of the disk-star generator circuit allows these currents to become large enough to trigger processes that accelerate particles.*

Acknowledgments: This research was supported in part by NSF grant PHY 91-00283 and NASA grant NAGW 1583 at the University of Illinois.

REFERENCES

Bai, T. 1980, *Ap. J.*, **239**, 328.
Becker, R. H., et al. 1977, *Ap. J.*, **214**, 879.
Cheng, K. S., and Ruderman, M. 1991, *Ap. J.*, **373**, 187.
Ghosh, P., and Lamb, F. K. 1991, in *Neutron Stars: Theory and Observation*, ed. J. Ventura and D. Pines (Dordrecht: Kluwer), p. 363.
Goldreich, P., and Julian, W. H. 1969, *Ap. J.*, **157**, 869.
Hamilton, R. J., Lamb, F. K., and Miller, M. C. 1992, these proceedings.
Katz, J. I., and Smith, I. A. 1988, *Ap. J.*, **326**, 733.
Lamb, F. K., Hamilton, R. J., and Miller, M. C. 1992, in *Isolated Pulsars*, ed. K. Van Riper, R. Epstein, and C. Ho, in press.
Mitra, A. 1991, *Ap. J.*, **370**, 345.
Wang, Y. M. 1986, *Ap. Space Sci.*, **121**, 193.

COMPTON SCATTERING POLARIZATION IN ACCRETION DISCS

Juri Poutanen and Osmi Vilhu
Observatory and Astrophysics Laboratory, University of Helsinki,
Tähtitorninmäki, SF-00130 Helsinki, Finland

ABSTRACT

We present the Compton scattering formalism for polarized light and compute the polarization of radiation from a two-phase accretion disc as a function of frequency and emission angle. The disk was assumed to be plane parallel and consisting of two parts: 1. optically thick and cool inner disc emitting black body radiation and 2. hot optically thin accretion disc corona (ADC). The central object was assumed not to contribute to the X-ray luminosity and we neglected the radiation of the corona. The radiative transfer equation of polarized light was solved in the approximation of single scattering taking into account the exact Compton matrix. It is shown that for high electron temperatures, or for large differences between the disc and coronal temperatures, the commonly used Rayleigh matrix predicts much larger polarization values than the Compton matrix.

INTRODUCTION

Compton scattering of polarized radiation (in the astrophysical context) has been considered in very few papers. The main reason for that situation is the lack of practical needs. Polarization in the region of spectrum (X-rays) where Compton scattering plays a major role is very difficult to observe. The launch of the Spectrum-X-Gamma satellite (1995) with the X-ray polarimeter[1] will change the situation. Information on the direction and magnitude of polarization would help to discriminate between various theoretical models of X-ray sources.

Among the first who drew attention to the importance of X-ray polarimetry for astrophysics was Rees.[2] The first model describing the polarization of radiation from accretion discs appeared in the work of Lightman & Shapiro.[3] They assumed that the disc consists of two parts. The polarization of the optically thin hot central region was calculated using the results of Angel.[4] The polarization properties of the optically thick outer region are well known from the classical works of Chandrasekhar[5] and Sobolev.[6] The signs of the polarization in these two regions are different.

Loskutov & Sobolev[7] considered multiple Thomson scattering in a slab where the temperature varied with optical depth. The linear integral equations for the emergent intensity were solved. Williams[8] calculated the polarization of radiation escaping from an infinite plane parallel atmosphere of cool electrons with high energy photon source. Sunyaev & Titarchuk[9] calculated theoretical spectra using the model of Comptonization of low-frequency radiation on hot electron gas. The results agree well with the observed spectra of many X-ray sources.

All the authors above used the Rayleigh matrix to describe the polarization. It is a good approximation if the electron temperature is comparable to or less than the photon energy. In the case of hot electrons ($kT_e > 50$ keV), where the photon frequency can increase several times after one scattering, the real polarization matrix of the Compton scattering is quite different from the Rayleigh matrix for Thomson scattering. To illustrate the difference between the Compton matrix and the Rayleigh matrix we calculate the polarization degree as a function of frequency and inclination angle for the simple two-phase accretion disc model: a cool disc emitting soft black body radiation surrounded by a hot ADC. We included only the disc luminosity and the scattered (Comptonized) radiation from the corona and studied only the single scattering case. In a forthcoming paper we plan to include multiple scattering in the treatment.

THE KINETIC EQUATION AND THE SCATTERING MATRIX

The relativistic kinetic equation describing propagation of polarized light through the isotropic nondegenerate electron gas (in the linear approximation) can be written in the following form[10,11]:

$$\frac{1}{c}\frac{\partial \tilde{I}}{\partial t} + \vec{\omega}\vec{\nabla}\tilde{I} = -n_e \sigma(x)\tilde{I} + \sigma_0 n_e x^2 \int_0^\infty \frac{dx_1}{x_1^2} \int d^2\omega_1 \hat{L}(-\chi)\hat{S}(x,x_1,\mu)\hat{L}(\chi_1)\tilde{I}', \quad (1)$$

where $\tilde{I}(x,\vec{\omega},\vec{r},t) = (I, Q, U, V)^T$ is the vector of the Stokes parameters, $\tilde{I}' = \tilde{I}(x_1,\vec{\omega}_1,\vec{r},t)$, σ_0 is the Thomson cross-section, $\sigma(x)$ is the Compton cross-section, n_e is the electron density, $\hat{L}(\chi)$ is the rotation matrix,[5] $x = h\nu/m_e c^2$ is the dimensionless photon energy, $\vec{\omega}$ is the unit vector of the photon propagation and μ — the cosine of the scattering angle.

The scattering matrix, which describes the redistribution of radiation in frequencies, angles and polarization degree due to the Compton scattering in the isotropic electron gas, can be presented as a single integral[11,12] over the electron distribution $f(\gamma)$:

$$\hat{S}(x,x_1,\mu) = \frac{3}{8}\int_{\gamma_*}^\infty f(\gamma)d\gamma \hat{R}(x,x_1,\mu,\gamma). \quad (2)$$

The matrix $\hat{R}$ has the same structure as the Rayleigh matrix.

Methods to compute the integral over the Maxwellian electron distribution are proposed by Kershaw et al.[13] and Nagirner & Poutanen.[14] Using these results it is possible to compute the integral in Eq. (2) with arbitrary values of x, x_1 and μ. Such calculations are performed by Poutanen.[12] If the electron temperature is not very high ($kT_e < 100$ keV) we can use the Gauss-Laguerre quadrature formula.

THE ADC MODEL AND NUMERICAL RESULTS

We consider a two-temperature model for a plane-parallel accretion disc. The cool inner disc with temperature T_{ph} is optically thick and emits black body radiation with the same temperature. The outer disc is a hot corona (ADC), with electron temperature T_e (Fig. 1). We assume that the optical depth of the ADC

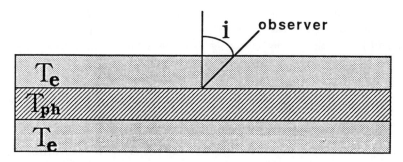

Figure 1: The structure of the accretion disc.

with respect to the electron scattering is small ($\tau \ll 1$). The X-ray luminosity was assumed to be produced by the disc with possible single Compton scattering in the corona. The direct coronal and central object luminosity were neglected. We did not take into account the frequency dependence of the scattering cross-section.

We have not considered the fourth Stokes parameter V, because the radiative transfer equation for V is independent of the other parameters (I, Q, U), and in our problem there are no sources of circular polarization.

We consider two cases of the initial radiation from the disc. In the first case **a** we assume that the disc emits unpolarized radiation isotropically. In the second case **b** we assume that the angular distribution of the initial radiation is the same as in the classical problem of the scattering of polarized light in the plane-parallel semi-infinite electron atmosphere.[5,6] The Chandrasekhar-Sobolev distribution can be approximated by the simple formula $1 + 2.06\eta$ (η is the cosine of the angle between the normal to the disc and photon momentum). The polarization degree in that case has a maximum 11.7% in the disc plane and is equal to zero along the normal due to symmetry.

As an example we give here results for $kT_{ph} = 0.05$ keV, $kT_e = 250$ keV and $\tau = 0.1$ (Fig. 2). This set of parameters can be applied to Seyfert galaxies.[15]

The polarization degree was calculated in two ways: in the first case we use the exact matrix $\hat{S}(x, x_1, \mu)$ and in the second one we use the exact redistribution function of the Compton scattering $S(x, x_1, \mu)$ to describe the intensity, but the polarization structure is computed using the Rayleigh matrix.

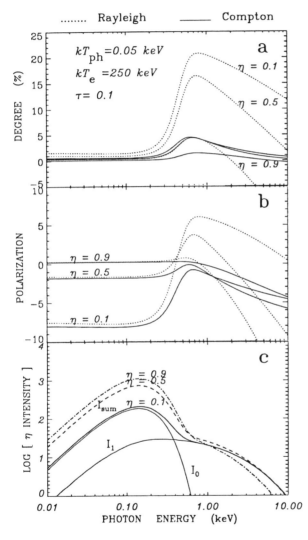

Fig. 2: **a** The polarization degree in the case of unpolarized radiation from the inner cool disc (Fig. 1) for three values of the inclination angle ($\eta = \cos i$). Dotted lines correspond to the Rayleigh scattering matrix; solid lines correspond to the exact Compton matrix. **b** Same as in **a**, but initial radiation is polarized in accordance with the Chandrasekhar-Sobolev distribution. **c** The frequency distribution of the observed flux $\eta I(\eta)$ for the case of polarized initial radiation. For $\eta = 0.1$ unscattered (I_0) and scattered (I_1) contributions to the total radiation (I_{sum}) are shown.

The radiation emitted from the disc, and scattered on the hot electron gas of the ADC, is shifted towards higher energies. Therefore with an increasing frequency the contribution of the scattered component increases and the degree of polarization changes quickly with frequency. The intensity depends weakly on $\eta = \cos i$, but the polarization degree is very sensitive to i.

CONCLUSIONS

We have calculated the polarization degree of X-rays as a function of the frequency and inclination angle in a model accretion disc with hot corona. We assumed that the optical depth of the hot ADC is rather small and solved the radiative transfer equation in a single scattering approximation. The calculations were made for two cases of the initial black body radiation from the relatively cool disc: unpolarized radiation and polarized in accordance with the Chandrasekhar law (optically thick plane parallel electron atmosphere). For high electron temperatures the Rayleigh matrix predicts much larger polarization values in the region of the first scattering than the Compton matrix. Future polarimetric observations need the theory of multiple Compton scattering of polarized light taking into account the exact scattering matrix, as formulated in the present paper.

REFERENCES

1. P. Kaaret, R. Novick, C. Martin et al., in: Observatories in Earth Orbit and Beyond (ed. Y. Kondo, Kluwer Acad. Publ., 1990), p.443.
2. M. J. Rees, MNRAS 171, 457 (1975).
3. A. P. Lightman, S. L. Shapiro, ApJ 203, 701 (1976).
4. I. R. P. Angel, ApJ 158, 219 (1969).
5. S. Chandrasekhar, Radiative Transfer (Dover Publ., N.Y., 1960).
6. V. V. Sobolev, A Treatise on Radiative Transfer (Van Nostrand, N.Y., 1956).
7. V. M. Loskutov, V. V. Sobolev, Afz 18, 81 (1982).
8. A. C. Williams, ApJ 279, 401 (1984).
9. R. A. Sunyaev, L. G. Titarchuk, A&A 143, 374 (1985).
10. D. I. Nagirner, J. J. Poutanen, Dokl. Acad. Nauk SSSR 318, 1125 (1991).
11. D. I. Nagirner, J. Poutanen, A&A, to be submitted.
12. J. Poutanen, these proceedings.
13. D. S. Kershaw, M. K. Prasad, J. D. Beason, JQSRT 36, 273 (1986).
14. D. I. Nagirner, J. J. Poutanen, AZh, in press.
15. F. Haard, L. Maraschi, in: Iron Line Diagnostics in X-ray Sources, (eds. A. Treves, G. C. Perola, L. Stella, Lecture Notes in Physics 385, Springer-Verlag, 1991), p.218.

COMPTON SCATTERING MATRIX FOR RELATIVISTIC MAXWELLIAN ELECTRONS

Juri Poutanen
Observatory and Astrophysics Laboratory, University of Helsinki,
Tähtitorninmäki, SF-00130 Helsinki, Finland

ABSTRACT

The methods of calculation of the Compton scattering matrix (CSM) describing redistribution of radiation in frequencies, directions and polarization degrees due to the scattering by relativistic Maxwellian electrons are proposed. Series expansions of CSM are found for three limiting cases: a) small frequencies ($h\nu \ll m_e c^2$); b) non-relativistic ($kT_e \ll m_e c^2$) and c) ultrarelativistic ($kT_e \gg m_e c^2$) electrons. A method of calculation of CSM for the general case is developed. The case of a monochromatic beam of photons is considered to show the differences between CSM and the Rayleigh scattering matrix.

INTRODUCTION

Compton scattering is one of the important mechanisms of the energy exchange between the electron and photon gases in the case of small electron density and sufficiently high temperature. The problems of radiative transfer taking into account Compton scattering are reduced in the general case to the integro-differential equation where the kernel is the Compton Scattering Kernel (CSK). The exact computations of the CSK are needed for understanding of different astrophysical, laser fusion and nuclear weapons effects. The Fokker-Planck approximation of kinetic equation (Kompaneets equation[1]) gives a possibility to solve many problems where the evolution of photon spectra due to the interaction with the electron gas was considered. This treatment is valid for small electron and photon energies ($h\nu, kT_e \ll m_e c^2$). The first investigations of CSK without any limits on the photon and electron energy gave its presentation in the form of a complicated multiple integral which then is calculated by means of Gaussian quadrature, but this gives incorrect results in a wide interval of scattering angles where the cross-section itself is large. Only a few years ago the first paper appeared where the correct computational scheme for CSK was presented.[2]

The first attempts to solve the radiative transfer equation for polarized light with the Compton scattering effect ran into the same difficulties as in the unpolarized case but also other problems arise connected with the Lorentz transformations of angles. Different limiting cases, for example, Fokker-Planck approximation including induced scattering[3] and a power-law distribution of relativistic isotropic electrons[4] were considered. One of the fruitful methods to solve this problem is the division into two parts: the frequency dependence of the intensity can be calculated by solving the Kompaneets equation and the angular and polarization structure of the radiation field can be obtained by iteration procedure based on an expansion in scattering orders using the Rayleigh matrix.[5]

The exact matrix for Compton scattering differs essentially from the Rayleigh one. The aim of this paper is to propose effective methods of calculating with high accuracy the Compton scattering matrix, averaged over the Maxwellian electron distribution. To show differences between the exact CSM and the Rayleigh matrix, the scattering of a monochromatic photon beam is considered and the polarization degree of scattered radiation is calculated for different kinds of initial photon polarization.

THE SCATTERING MATRIX
FOR MONOENERGETIC ISOTROPIC ELECTRONS

Let $x = h\nu/m_e c^2$, $x_1 = h\nu_1/m_e c^2$ be the dimensionless photon energies after and before the scattering, respectively, μ — the cosine of the scattering angle, γ — the electron energy in units of $m_e c^2$.

The Compton scattering matrix, averaged over the electron energy distribution $f(\gamma)$, can be presented as a single integral:

$$\hat{S}(x, x_1, \mu) = \frac{3}{8} \int_{\gamma_*}^{\infty} f(\gamma) d\gamma \hat{R}(x, x_1, \mu, \gamma) = \tag{1}$$

$$= \begin{pmatrix} S & S_I & 0 & 0 \\ S_I & S_Q & 0 & 0 \\ 0 & 0 & S_U & 0 \\ 0 & 0 & 0 & S_V \end{pmatrix} = \frac{3}{8} \int_{\gamma_*}^{\infty} f(\gamma) d\gamma \begin{pmatrix} R & R_I & 0 & 0 \\ R_I & R_Q & 0 & 0 \\ 0 & 0 & R_U & 0 \\ 0 & 0 & 0 & R_V \end{pmatrix},$$

where

$$\gamma_* = (x - x_1 + Qt_*)/2, \quad Q^2 = (x - x_1)^2 + 2q, \quad t_*^2 = 1 + 2/q, \quad q = xx_1(1 - \mu). \tag{2}$$

The matrix $\hat{R}(x, x_1, \mu, \gamma)$ is the Compton scattering matrix[6,7] of polarized radiation on isotropic electron gas with a fixed energy γ. The five functions R in (1) are the following:

$$\begin{aligned}
R &= R_g + R_f, \quad R_I = R_h + R_f, \\
R_U &= \frac{2}{Q} + \frac{4Q(x + x_1)^2}{r^2 q^2} + \frac{2}{q}\left(\frac{1}{a} - \frac{1}{a_1}\right) - \frac{4}{rq}(a - a_1) + 2R_h + \\
&\quad + \frac{4}{r^2 q^2}\left\{-3(x + x_1)[a(\gamma - x) + a_1(\gamma + x_1)] + 2(a_1^3 - a^3)\right\}, \\
R_Q &= R_U + R_f, \quad R_V = R_g - qR_f,
\end{aligned} \tag{3}$$

where

$$R_g = \frac{2}{Q} + \frac{q-2}{q}\left(\frac{1}{a} - \frac{1}{a_1}\right),$$

$$R_f = \frac{x+x_1}{q^2}\left(\frac{\gamma-x}{a^3} + \frac{\gamma+x_1}{a_1^3}\right) + \frac{2}{q}\left(\frac{1}{a^3} - \frac{1}{a_1^3}\right) - \frac{2}{q^2}\left(\frac{1}{a} - \frac{1}{a_1}\right), \quad (4)$$

$$R_h = \frac{2(x+x_1)}{rq^2}\left(\frac{\gamma-x}{a} + \frac{\gamma+x_1}{a_1}\right) + \frac{4}{rq^2}(a-a_1),$$

and

$$a^2 = (\gamma - x)^2 + r, \quad a_1^2 = (\gamma + x_1)^2 + r, \quad r = \omega^2 = \frac{1+\mu}{1-\mu}. \quad (5)$$

The expression for the first function R in (3) is well known.[2,8] Using the expressions (3)-(4) one can obtain different asymptotic and approximate formulae.

AVERAGING OVER A MAXWELLIAN ELECTRON DISTRIBUTION

This part of the work is based on the papers by Kershaw et al.[2] and Nagirner & Poutanen,[9] where the methods of integrating over a relativistic Maxwellian electron distribution

$$f(\gamma) = \frac{y}{4\pi K_2(y)} e^{-y\gamma}, \quad y = \frac{m_e c^2}{kT_e} \quad (6)$$

(here K_2 is the McDonald function) were proposed for the Compton scattering kernel (i.e., for the case of unpolarized radiation).

For example, in the general case we obtain the following presentation for the CSM[10]:

$$\hat{S}(x, x_1, \mu) = \frac{3y}{32\pi} \frac{e^{-y\gamma_*}}{K_2(y)} \left\{ \hat{\Omega} + e^{y\rho_+}\left(\hat{\Omega}^{0+}L_0^+ + \omega\hat{\Omega}^{1+}L_1^+\right) \right.$$
$$\left. + e^{y\rho_-}\left(\hat{\Omega}^{0-}L_0^- + \omega\hat{\Omega}^{1-}L_1^-\right)\right\}, \quad (7)$$

where quantities $L_n^\pm$ are the integrals

$$L_n^\pm = \int_{c_\pm}^\infty e^{-bp} \frac{p^n dp}{\sqrt{1+p^2}}, \quad n = 0, 1. \quad (8)$$

Here $c_\pm = \rho_\pm/\omega = [Qt_* \pm (x+x_1)]/2\omega$, $b = y\omega$. The matrices $\hat{\Omega}$ are the algebraic functions of y, x, x_1 and μ. The integrals $L_n^\pm$ can be evaluated in a variety of ways[2,9] with high accuracy.

In the same manner as in the papers above different series expansions[10] can be obtained for the following limiting cases:

1) Ultrarelativistic electron gas ($y\omega \ll 1$),
2) Small photon frequencies ($x, x_1 \ll 1$),
3) Non-relativistic electrons ($y \gg 1$).

A MONOCHROMATIC BEAM OF PHOTONS

We consider the scattering of the monochromatic photon beam on Maxwellian electrons. The Stokes vector of the scattered radiation is given by

$$\tilde{I}_1(x) = \begin{pmatrix} I_1 \\ Q_1 \\ U_1 \\ V_1 \end{pmatrix} \propto x^2 \hat{S}(x, x_0, \mu) \begin{pmatrix} i_0 \\ q_0 \\ u_0 \\ v_0 \end{pmatrix}, \qquad (9)$$

where the vector in the right hand side is the initial Stokes vector, x_0 is the initial frequency.

As an example, we consider the scattering of 1 keV photons on 100 keV electron gas and two cases of initial polarization. The first case is the unpolarized initial beam ($i_0 = 1$, $q_0 = u_0 = v_0 = 0$). The results for the polarization degree $P_1 = S_I/S$ are shown in Fig. 1. The degree of polarization is approximately two times smaller than for the Rayleigh scattering matrix ($P_1^R = (\mu^2 - 1)/(\mu^2 + 1)$).

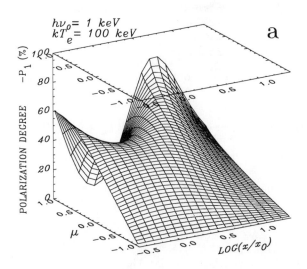

Figure 1: Polarization of scattered radiation. The case of unpolarized initial beam.

In the second case we consider the linear polarization normal to the plane of scattering ($i_0 = q_0 = 1$, $u_0 = v_0 = 0$). The degree of polarization is $P_2 = (S_Q + S_I)/(S + S_I)$ (Fig. 2) and it differs many times from the polarization which the Rayleigh matrix predicts ($P_2^R = 100\%$).

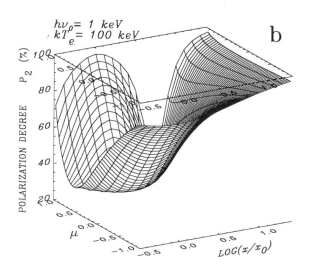

Figure 2: Polarization of scattered radiation. The case of the linear polarization normal to the plane of scattering.

In the case of the circular polarization the Rayleigh matrix gives nearly the same results as the exact CSM. Some astrophysical applications were made[11] using the proposed methods of calculation.

REFERENCES

1. A. C. Kompaneets, JETP 31, 876 (1956).
2. D. S. Kershaw, M. K. Prasad, J. D. Beason, JQSRT 36, 273 (1986).
3. R. F. Stark, MNRAS 195, 115 (1981).
4. S. Bonometto, P. Cazzola, A. Saggion, A&A 7, 292 (1970).
5. R. A. Sunyaev, L. G. Titarchuk, A&A 143, 374 (1985).
6. D. I. Nagirner, J. J. Poutanen, Dokl. Acad. Nauk SSSR 318, 1125 (1991).
7. D. I. Nagirner, J. Poutanen, A&A, to be submitted.
8. F. A. Aharonian, A.M. Atoyan, Ap&SS 79, 321 (1981).
9. D. I. Nagirner, J. J. Poutanen, AZh, in press.
10. J. Poutanen, A&A, to be submitted.
11. J. Poutanen, O. Vilhu, these proceedings.

ACTIVE GALAXIES

HIGH ENERGY GAMMA-RAY EMISSION FROM ACTIVE GALACTIC NUCLEI OBSERVED BY THE ENERGETIC GAMMA-RAY EXPERIMENT TELESCOPE (EGRET)

C. E. Fichtel, D. L.Bertsch, B. L. Dingus[1], R. C. Hartman, S. D. Hunter, P. W. Kwok[2], J. R. Mattox[3], P. Sreekumar[1], and D. J. Thompson
NASA/Goddard Space Flight Center, Code 662, Greenbelt, MD 20771, USA

D. A. Kniffen
Hampden-Sydney College, P.O. Box 862, Hampden-Sydney, VA 23943, USA

Y. C. Lin, P. L. Nolan, P. F. Michelson
Hansen-Experimental Physics Laboratory, Stanford University, Stanford, CA 94305, USA

G. Kanbach, H. A. Mayer-Hasselwander, C. von Montigny, K. Pinkau, H. Rothermel, M. Sommer,
Max-Planck Institut fur Extraterrestrische Physik, 8046 Garching, Munich, Germany

E. J. Schneid
Grumman Aerospace Corporation, Mail Stop A01-26, Bethpage, LI, NY 11714, USA

ABSTRACT

The Energetic Gamma Ray Experiment Telescope (EGRET) on the Compton Gamma Ray Observatory has detected 16 active galactic nuclei in high energy gamma rays thus far during the first year in orbit. Of these, the majority are quasars and four are usually classified as BL Lacertae objects. No Seyferts have been seen. Of the AGN's seen, 3C 279 had the highest flux when first observed, but it is variable and was at a significantly lower level when observed later. There is a wide range of distances among the observed set, with four quasars having known z values in excess of 1.0. One BL Lac has a z of 0.94. If the photon spectra are represented by power laws in energy, those which have been determined have spectral indices ranging from about -1.7 to -2.4. In some cases, there is a suggestion of a steepening at the highest energies observed.

I. INTRODUCTION

It was not long after the discovery of quasars in 1963 that Salpeter[1] and Zel'dovich and Novikov[2] proposed that quasars were very massive black holes accreting huge amounts of matter. Although neither consensus about the details of models based on this concept nor conclusive proof that they are massive black holes has come into being, the tremendous energy releases observed across many frequency bands and the observed time variations combined with the great distances to these objects seem to argue very strongly for very massive, quite compact objects. This deduction, in turn, seems to lead one in the direction of the supermassive black

1. Also Universities Space Research Association, Greenbelt, MD 20771,USA
2. Also NAS/NRC Research Associate, Greenbelt, MD 20771,USA
3. Compton GRO Science Support Center, CSC, Greenbelt, MD 20771, USA

hole explanation for these objects. The observations of large high energy gamma-ray emission described here add to the knowledge of these objects and hopefully ultimately provide a better understanding of them.

It seems appropriate to note here in the beginning that some of the more attractive current models for the emission of high energy gamma rays from active galaxies, specifically from jets associated with them, include a mechanism named for the physicist who is being honored in the title of this conference. The fact that the Compton effect might have astrophysical significance was suggested by two physicists at the host institution for this conference, namely Feenberg and Primakoff,[3] 44 years ago. That article was, in fact, the first paper known to us to be written on any aspect of the possible significance of gamma-ray astronomy.

Prior to the launch of the Compton Gamma Ray Observatory in April 1992, only one active galaxy, 3C 273, had been seen reasonably certainly in high energy gamma rays,[4] and it was just barely seen above the limit of detection. However, high energy gamma rays were not unexpected. The jets that play a central role in many of the models involve very energetic particles whose interactions lead to copious quantities of high energy gamma rays. In this paper, the general results on the 16 AGN's seen by EGRET thus far will be summarized. It should be recognized that this list is not the result of a systematic survey to the lowest detectable level; rather it represents the clearest detections in the fields of view studied through about July 1992.

II. THE EGRET OBSERVATIONS

EGRET has the standard elements of a high energy gamma-ray telescope, specifically an anticoincidence scintillator dome to discriminate against charged particles, a particle track detector consisting of spark chambers with interspersed high z material to convert the gamma rays into electron pairs, a trigger telescope which detects the presence of the pair and determines that the particles have the correct direction of motion, and an energy measurement device which in the case of EGRET is a NaI (Tl) crystal. A description of the instrument and its general capabilities is given by Kanbach et al.[5] The results of the instrument calibration, both before and after launch, are given by Thompson et al.[6] The telescope covers the energy range from about 20 MeV to 30 GeV. The effective area is about 1.5×10^3 cm^2 from 0.2 to 1.0 GeV, and lower outside of this range. The instrument is designed to be free of internal background, and the calibration tests have verified that it is at least an order of magnitude below the extragalactic gamma radiation. Hence the only significant radiation besides the sources themselves is the diffuse galactic and extragalactic radiation.

The 90% of the sky scanned for extragalactic gamma-ray sources contains about 550 AGN candidates from an a priori derived list. This gives much less than a 1% chance coincidence based on the typical positional uncertainties. Therefore the 16 detected sources should contain at most one chance association, and most likely none.

III. RESULTS

Before describing the whole set of observations, the results related to the first active galactic nucleus seen by EGRET will be described since, in many ways, they set the stage for the subsequent findings. The Compton Observatory had been in space just a little over 2 months when a source was seen far from the galactic plane that was approximately as strong as the three strongest sources seen by SAS-2 and subsequently COS-B, all of which were galactic. This source was consistent with the direction of 3C 279 within the 65% circle of direction uncertainty of 5 arcmin radius.[7] It was one of the AGN's thought to be a likely source of high energy gamma rays before the launch of the Gamma Ray Observatory.[5]

At the time of the first observation, the high-energy gamma-ray emission appears to have been the dominant energy output. If the emission from 3C 279 is isotropic, its gamma ray luminosity between 100 MeV and 10 GeV is about 1.6×10^{51} photons s^{-1}, or approximately 1.1×10^{48} ergs s^{-1} for $H_0 = 75$ km s^{-1} Mpc^{-1}. However, as discussed below, it is likely that the emission is beamed; for 0.01 sr, it would be about 1×10^{45} ergs s^{-1}, for example. For comparison, our own galaxy emits approximately 5×10^{38} ergs s^{-1} (E > 100 MeV).

Subsequently, 3C 279 has been observed again by EGRET and found to still be emitting high energy gamma rays, but at a lower rate. See Figure 1. Thus, 3C 279 is also variable--on time scales of the order of days in at least one case.

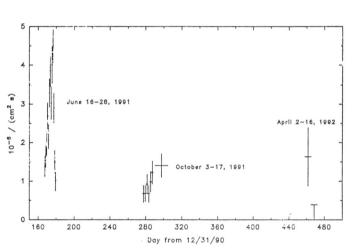

Fig. 1. The fluxes of gamma rays from 3C279 with energies above 100 MeV are given for the first 10 months of Compton /EGRET observations.[8] The object was only targeted in the June exposure and the first 2 weeks of the October exposure. For the remainder, there are only weak exposures at the edge of the field-of-view. The last April 1992 data point is a 95 percent confidence upper limit. Otherwise the errors given are 1 sigma.

Table I summarizes the positive EGRET AGN detections that exist. A very striking aspect of the results presented in this table is that nearly all of the observed AGN's are not the close ones; in fact, there is a broad distribution in z with four having values in excess of 1.0. Many are already known to be blazars. As noted earlier, these detections do not represent a complete or systematic survey, although most of the stronger sources at the time of the observation are probably in the list. Sources for which a range of flux is shown have exhibited significant variability. Note also that the spectra when represented by a power law in energy have spectral indices ranging from -1.7 to -2.4. Although harder spectra might seem more common, this may be due to their being more likely to be seen in the data or due to the limited statistical sample. Two examples of spectra are shown in Figures 2 and 3. Figure 4 shows the emission from an AGN over a broad range of frequencies.

Table 1: Characteristics of Active Galaxies detected by EGRET

source ID and characteristics	OVV	BL Lac	super lum.	radio loud	flat radio[1]	opt. pol.[1]	position diff.[2]	position uncert.[3]	flux (10^{-6} cm^{-2} s^{-1}) (E>100 MeV)	photon spectral index	z	relative luminosity[4]
0202+149 (4C+15.05)				✓	✓	✓	0.3°	0.4°	0.3±0.1		...	...
0208-512 PKS				✓	✓	✓	0.13°	0.13°	0.4 to 0.9	-1.7±0.1	1.00	2
0235+164 (OD+160)		✓		✓	✓	✓	0.10°	0.3°	0.8±0.1	-2.0±0.2	0.94	2.0
0420-014 (OA 129)	✓			✓	✓	✓	0.5°	0.4°	1.4±0.1		0.92	0.4
0454-463 PKS				✓	✓		0.27°	0.38°	0.25±0.1		0.86	0.3
0528+134 PKS	✓			✓	✓		0.13°	0.15°	0.4 to 1.6	-2.4±0.1	2.06	4 to 13
0537-441 PKS		✓		✓	✓	✓	0.4°	0.6°	0.3±0.1	-2.0±0.2	0.894	0.2
0716+714		✓	?	✓	✓		0.47°	0.4°	0.20±0.06	-1.8±0.2	...	...
0836+710 (4C+71.07)	✓		✓	✓	✓		0.58°	0.50°	0.15±0.04	-1.9±0.1	2.17	1.1
1101+384 (Mrk 421)		✓	✓	✓	✓	✓	0.3°	0.4°	0.14±0.03	-2.4±0.1	0.031	0.0002
1226+023 (3C 273)			✓	✓	✓		0.2°	0.5°	0.30±0.05	-2.4±0.1	0.158	0.008
1253-055 (3C 279)	✓		✓	✓	✓	✓	0.083°	0.08°	0.6 to 4.9	-2.0±0.1	0.54	0.3 to 2
1606+106 (4C+10.45)				✓	✓		0.42°	0.50°	0.5±0.2		1.23	1.6
1633+382 (4C+38.41)	✓		✓	✓	✓		0.08°	0.15°	0.4 to 1.4	-2.0±0.1	1.81	3 to 11
2230+114 (CTA 102)			✓	✓	?	✓	0.3°	0.4°	0.24±0.07	-2.4±0.1	1.037	0.5
2251+158 (3C 454.3)	✓		✓	✓	✓	✓	0.25°	0.22°	0.8±0.1	-2.0±0.1	0.859	0.5

1. Flat spectrum radio sources: $\alpha_r > -0.5$ (2-5 GHz band); Optical polaration: polarization > 3%
2. Difference between gamma-ray determined position and known position of idenfied source. Most are preliminary.
3. There is a 68% probability that the source is within a circle of this radius. Most are preliminary.
4. The source luminosity for the observed energy range (0.1 GeV $\leq \overline{E} \leq$ 5 GeV), computed using the known redshift assuming $H_0 = 75$ km s^{-1} Mpc^{-1} and $q_0 = 1/2$, is equal to the relative luminosity times (10^{48} erg s^{-1})f, where f is an unknown beaming factor. Typically f is thought to be in the range from 10^{-2} to 10^{-3}.

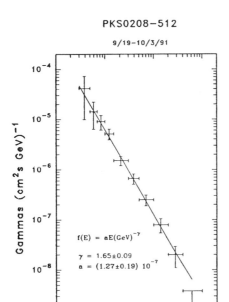

Fig. 2. The differential energy spectrum PKS 0208-512 seen by EGRET.[9]

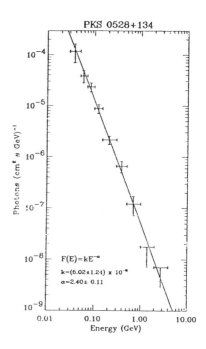

Fig. 3. The differential energy spectrum of PKS 0528+134 seen by EGRET.[10]

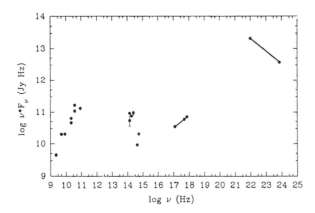

Fig. 4. The multifrequency spectrum of PKS 0528+134 seen by EGRET.[10]

Although there is difficulty in having observations at the same time and the domination of the gamma rays is particularly large here, in general when gamma rays are seen, their νF_ν is at least as large as that of other frequencies. However, it should be remembered that the gamma rays are likely to be beamed and the optical radiation, for example, may not be; so this representation may be misleading. No Seyfert galaxies have been detected.[11] Some details on upper limits to fluxes from other AGN's are given by von Montigny et al.[12] at this conference.

IV. DISCUSSION

As noted in the introduction, at this time a total of 16 active galactic nuclei have been identified as high energy gamma-ray emitters (E > 100 MeV) on the basis of detections by EGRET. A common characteristic of these sources is that they are all radio-loud, flat radio spectrum sources. Many of them are seen as superluminal sources as well. They all appear to be members of the blazar class of AGN (BL Lac objects, HPQ and OVV quasars), exhibiting one or more of the characteristics of this source class (flat radio spectrum, strong variability, optical polarization). The number of detected sources by itself is not particularly surprising in view of the theoretical expectations and the previously observed extragalactic diffuse radiation. However there are two generally unexpected results obtained from an analysis of the EGRET data. One is that several of these gamma-ray loud AGNs have large redshifts and hence are apparently emitting energy in gamma radiation at an extraordinary rate. At the same time, many active galaxies that are relatively close to the Earth, including Seyfert galaxies and some blazars, have not been detected in high energy gamma rays.

a. Implications for Emission Mechanisms

There are several arguments that suggest that the gamma-ray emission from the EGRET AGN sources is strongly beamed and not isotropic. If the gamma-ray emission is isotropic, then the observation of gamma-ray emission from the source places severe constraints on the size of the emission region. The usual argument is as follows. Typically, about 30% of the bolometric power of these sources is in X-rays and probably originates from the central engine, possibly from magnetic flares occurring on the surface of an accretion disk[13] or from either synchrotron or synchrotron self-Compton radiation occurring in the inner part of a relativistic jet.[14,15] Most of these sources also exhibit X-ray variability on relatively short timescales. As discussed by several authors,[16] from observations of continuing rapid variations in luminosity ΔL observed on timescales of Δt, several constraints can be placed on the internal properties of the source. Most important, as pointed out by McBreen[17], a compact x-ray source can become opaque to high energy gamma-rays through the process $\gamma_1 + \gamma_2 \rightarrow e^+ + e^-$. The optical depth of gamma-ray photons to photon-photon pair production can be expressed as

$$\tau_\gamma = (l/4\pi)(\sigma_{\gamma\gamma}/\sigma_T)(m_e c^2/E_x), \qquad (1)$$

where l is the so-called *compactness parameter*, defined as

$$l = (L_x/R)(\sigma_T/m_e c^3) \geq (L_x/\Delta t)/10^{39} \text{ erg s}^{-2}. \qquad (2)$$

L_x is the nonthermal source luminosity arising from within a region of radius R and E_x is the typical energy of a nonthermal source photon. σ_T is the Thompson cross section. An upper limit to R is given by the light crossing time $c\Delta t/(1+z)$. If most of L_x is emitted in the X-ray - soft gamma-ray range, the source will be optically thick for high energy gamma rays when l is greater than about 10.[18,19]

For the gamma-ray loud AGN observed by EGRET, the compactness parameter is typically greater than 10, suggesting that the gamma-ray emission is in fact beamed and associated with emission processes in a relativistic jet. Various emission models associated with jets have been proposed and are discussed below. Of course, the observational characteristics of a particular source will depend on the orientation of the jet to the observer's line-of-sight and the bulk velocity of matter in the jet. Beamed gamma-ray emission associated with a relativistic jet also implies that much less energy is required to be emitted at high energies compared to unbeamed models and that temporal variations in the flux can actually occur further from the central engine.

A unified picture of AGNs based on the theoretical idea of an accretion disk, surrounded by a thick torus of cooler gas, feeding a massive black hole, sometimes with a jet of particles emerging along the axis of the disk, had already emerged before the EGRET observations discussed here. This picture is supported by the observation of radio frequency jets, superluminal motion in some sources, and other observations (for a review, see Osterbrock.[20] A basic premise of this picture is that all AGN are different aspects of the same phenomena. Different classes of AGNs look different primarily because they are viewed at different angles on the sky. The hypothesized relationships are as follows: in the case of OVV quasars the jet is aimed almost directly at the observer while in Seyfert 1 and 2 galaxies the line-of-sight is oriented at a substantial angle with respect to the axis of the disk and torus. BL Lacs were originally proposed to be extreme examples of the quasar population with the jets seen almost exactly end-on.[21,22] The detection of high energy gamma emission from blazars and its non detection from Seyfert galaxies gives further support to this picture.

However, the situation is probably more complicated than indicated above. For one thing, there is probably an excess of nearby BL Lacs, and their space density is too large for them to be just a specially selected subset of quasars.[23] There are notable nearby BL Lac sources that have not been detected by EGRET. Also the redshift distribution of BL Lacs in flux-limited samples appears to be two-peaked, with maxima near $z \approx 0$ and $z \approx 1$[24]. It has recently been suggested that BL Lacs observed in elliptical galaxies at low redshift are members of a different class than weak-lined, luminous BL Lacs at high redshift.[23,25] An alternative picture has been proposed[23] in which very strong (FR II) radio galaxies are the parent population of themselves, normal quasars, HPQ/OVV quasars (as suggested by Barthel[26]), and strongly-variable BL Lacs. The less variable, relatively nearby BL Lacs are moderately strong (FR I) radio galaxies that are beamed towards us with generally larger viewing angles and smaller bulk Lorentz factors in their jets. This picture appears to account for the two-peaked redshift distribution of BL Lacs and a number of other observational characteristics of BL Lacs.[23] This may also account for the lack of EGRET detections of some relatively nearby BL Lac objects.

Within the relativistic jet picture of beamed gamma emission there are several different models of the emission mechanism. They can be broadly classified as either models involving Compton scattering[14,27,28,29] or nucleonic cascade models.[30,31] In the Compton scattering models, gamma-ray emission could arise either via the synchrotron-self-Compton process or from Compton scattering of UV flux from an accretion disk.[32,33]

All of the plausible gamma ray emission mechanisms discussed above require an extraordinary supply of relativistic particles that are likely beamed in a jet. The physical processes involved in accelerating particles, most likely electrons, to produce a relativistic jet are a very complex mixture of general relativity, plasma physics, hydrodynamics, and electrodynamics. While there seems to be a nearly universal consensus that the basic energy supply in AGN is provided by matter falling into the deep gravitational potential well around a supermassive black hole, little is known about how the accelerator in this central engine works. The standard arguments about the efficiency of disk accretion onto a black hole can easily account for the overall energetics of AGN, but accretion by itself does not provide the acceleration mechanism. It is clear that the accelerator must be capable of producing particles with energies extending at least to the several GeV range; at least to the TeV range in the case of Markarian 421.[34] At present the source of very high energy cosmic rays is unknown.

Can more familiar, and hopefully better understood, galactic sources of high energy gamma radiation provide any insight? Isolated, highly magnetized, rotating neutron stars are clearly the most solidly established examples of high energy gamma-ray point sources. The evidence for high energy gamma ray production from accreting neutron stars and stellar-mass black holes is not compelling. In the case of isolated pulsars it is clear that the dominant role in producing acceleration of charged particles is played by a rotating magnetic field configuration. The naive conclusion could be made that the most favorable conditions for producing efficient acceleration of charged particles and producing high energy gamma emission occur in systems not undergoing accretion. Extending this conclusion to AGNs is clearly at odds with the almost canonical picture of the central engine as an accreting, supermassive black hole. However, the accretion rates and magnetic field strengths in the case of AGN's, as compared with galactic cases, may lead to a configuration that is favorable for particle acceleration to very high energies.

There is a large class of theories of AGN beam formation in which magnetic fields do not have a dominant role (for a review of various models, see Wiita.[35] For example, models in which beams are accelerated by radiation pressure that fills a narrow funnel along the axis of a thick accretion disk have been developed by Lynden-Bell[36] and Paczynski and Wiita.[37] It appears that such hydrodynamical models can only produce mildly relativistic beams ($\gamma \leq 3$ or 4) with a moderate degree of collimation. Models in which magnetic fields play a dominant role appear necessary for accelerating particles to ultrarelativistic energies. The presence of significant magnetic fields are also consistent with observations of polarization. Indeed, the synchrotron self-Compton models mentioned above require such conditions close to the black hole. It is also possible in principle to extract rotational energy from the black hole via the Penrose process, but it is not clear how this mechanism could produce beaming.

Jet acceleration models in which magnetic fields are important generally fall into two categories; those in which energy is extracted from the black hole's rotational energy and those in which magnetic fields directly associated with an accretion disk are used to accelerate and collimate charged particles.

The Blandford-Znajek process[38] is a plausible mechanism by which rotational energy can be extracted from the black hole and produce relativistic beams. In this model a conducting accretion disk around a massive rotating black hole serves to confine a magnetic field that threads the hole's horizon ($B^2/8\pi \approx P_{disc}$). The horizon acts as a rotating conductor immersed in an external magnetic field; i.e., a unipolar generator that can deliver power to a magnetospheric load. Apparently if energy can be extracted in this way, then relativistic beams emerge along the rotation axis of the hole in a natural way. This power generation process is very similar to a process originally proposed for neutron stars by Goldreich and Julian.[39] A particularly illuminating analysis of the Blandford-Znajek process is given by Macdonald and Thorne.[40]

The potential drop available for particle acceleration is approximately

$$\Delta V = (10^{20} \text{ volts}) (a/M_{BH})(B/10^4 \text{ G}) (M_{BH}/10^9 M_O), \quad (3)$$

where B is the magnetic field threading the horizon and M_{BH} is the mass of the black hole and a is the angular momentum per unit mass of the black hole. For reasonable parameters, as much as 10^{20} to 10^{21} volts is available. If the charged particle existed in a vacuum then it could be accelerated up to a maximum energy of $e\Delta V$. But many effects can limit the energy to much lower values. These include synchrotron radiation losses and Compton cooling. In a plasma, collisions with other particles (ohmic dissipation) can limit the maximum attainable energy. Even if the plasma is tenuous and collisions are unimportant, a variety of plasma microinstabilities can also limit the energy. Various stochastic reacceleration mechanisms, such as Fermi-shock acceleration, may act to overcome some of these losses.

The total power in the relativistic beams is of order

$$L_b = (10^{45} \text{ergs/s}) (a/M_{BH})^2 (B/10^4 \text{ G})^2 (M_{BH}/10^9 M_O)^2 . \quad (4)$$

For reasonable astrophysical parameters the beam luminosity can approach 10^{45} to 10^{47} ergs/s.

Lovelace[41] proposed a different model in which magnetic fields generated in a differentially rotating disk can lead to acceleration of charged particles to ultrarelativistic energies and extraction of energy from the disk. The model apparently can produce either a proton beam or an electron-positron beam, depending on whether B_z is parallel or antiparallel to the angular momentum vector of the disk. The potential drop available for particle acceleration and the beam luminosity in this case are comparable to the above estimates for the Blandford-Znajek effect.

b. Relation to the extragalactic diffuse radiation

In addition to the galactic diffuse gamma radiation which varies strongly with position, past observations have shown that there appears to be a second component of the diffuse radiation which is isotropic at least on a very coarse scale. The fact that it is not galactic, and presumably extragalactic, is based on several analyses which show that the diffuse radiation over the measured part of the sky can be well explained by cosmic ray, matter interactions with a small Compton component only if a significant isotropic component is added (See, for example, Fichtel et al.,[42] Fichtel, Simpson, and Thompson[43]). The most current measurements, until the completion of the EGRET all sky survey and its analysis, are those of SAS-II (Fichtel et al.,[42] Fichtel et al.[44]).

One of the more promising explanations of this high energy gamma radiation has been that it is the sum of active galactic nuclei (See, e. g., early references such as Strong et al.;[45] Bignami et al.;[46] Fichtel and Trombka[47]). Based on rather limited information at the time, with the only quasar that had been seen in high energy gamma rays being 3C 273 (Bignami et al.[48]), it seemed possible that the extragalactic diffuse radiation was at about the right level and the spectral shape seemed reasonable.

There is now the new information on high energy gamma ray radiation from quasars described here. First, and most obvious, not all radio loud quasars can be emitting gamma rays isotropically at the level of the ones seen or the sky would be about a thousand times brighter in gamma rays than it is. Further, with the many BL Lac's and radio loud quasars which are not seen, it is known now that not all of these objects are so luminous in gamma rays, or they are highly beamed, or both.

If one supposes for the moment that the blazars are emitting high energy gamma rays at the high levels that are observed and that they are emitting them in a narrow cone of a few tenths to a hundredth of a steradian as is now believed, based in part on observed opening angles of nearby AGN's, then the luminosity of the blazars that have been seen and are listed in Table I would on the average be between 10^{45} to 10^{46} ergs s^{-1}. This is the range assumed in the original calculations based on the observation of 3C 273 and the assumption that it was isotropic and a typical quasar. Whereas the beaming means that this energy is concentrated in the beam direction to the degree of the beaming factor, only about one percent of quasars are blazars; so the two factors cancel. Assuming the same cosmology and number density as a function of look-back time as Fichtel and Trombka,[47] the integrated emission would be about what is needed at 100 MeV.

Another question that must be addressed is the spectrum. In terms of a power law in energy for the spectral shape, the slope of 3C 273 gave good agreement within errors with that of the observed extragalactic diffuse radiation, which is estimated to be -2.4+0.4/-0.3. (Note: In the cosmological integration, the slope of an individual source, if they are all the same, is reflected in the final spectrum in the range over which it applies.) As discussed above, the slopes measured thus far for the AGN's that have been seen in high energy gamma rays vary from about -2.4 to -1.7. This distribution is of concern since the harder ones will dominate at the higher energies. A better sample of AGN's will be needed as well as the final result for the extragalactic diffuse radiation before a definitive statement can be made.

Finally, there is the question of isotropy. If it is only beamed quasars that are significant emitters, and only one in a few hundred are beamed in this direction, the scale of observation will not have to be too fine before one sees an unevenness in the distribution of the diffuse gamma radiation. The early SAS-II measurements could certainly not have seen this level of deviation from isotropy, but, assuming that beamed AGN's do dominate, one might now be able to see variations in intensity and even spectral shape with position. Given the suggestion of time variability noted above, the uneven nature of the extragalactic component might even vary significantly with time.

c. Implication for extragalactic cosmic rays

As noted above, the intensity and spectral slopes of most of the observed quasars and BL Lac's suggest the presence of an extraordinary source of relativistic particles with spectra as hard or harder than the observed ultra high energy cosmic rays. The beaming into a very narrow cone combined with the knowledge gained from the study of high energy nuclear physics interactions suggests that the parent particles are extremely relativistic. Finally, one of the quasars already seen by EGRET[11] and disucussed here, namely Mk 421, has already been detected in gamma rays at 10^{12} eV.[34]

All of these considerations bring to mind the possibility suggested by Colgate[49] that blazar jets may be the source of the ultra high energy cosmic rays, those above about 10^{18} eV. There is, of course, a very long step from a few times 10^9 eV or even 10^{12} eV gamma rays to 10^{18} to 10^{21} eV cosmic rays, but the acceleration mechanism and the energy could be there for a supermassive black hole at the center of an AGN, as noted above. However, the answers to very critical questions are all at the moment unknown. For example: Are the particles actually accelerated to 10^{18} to 10^{21} eV? Do they escape from the AGN? What is their energy spectrum?

V. SUMMARY

Thus far, EGRET has detected 16 AGN's with a high degree of certainty. All are radio loud and probably blazars. There is a wide range of z values; hence, it is not the closest ones that are primarily being seen. The spectral slopes, determined for a power law in energy, range from -1.7 to -2.4. At least a few are seen to be time variable. There are several factors which suggest that the radiation is beamed. A very powerful source of extreme relativistic particles seems essential to explain the observations. This requirement raises again the possibility that AGN's may be the source of the extragalactic component of the cosmic rays. The question of whether AGN's are the source of the extragalactic diffuse high energy gamma radiation remains open, but results, which will be forthcoming over the next year, should help to answer the question.

REFERENCES

1. Salpeter, E. E., 1964, ApJ, 140, 796
2. Zel'dovich, Ya. B., and Novikov, I. D., 1964, Usp. Fiz. Nauk., 84, 377 [Sov. Phys., Usp., 7, 763 (1965)]
3. Feenberg, E., and Primakoff, H., 1948, Phys. Rev., 73, 449
4. Swanenburg, B. N., et al., 1978, Nature 275, 298
5. Kanbach, G., et al., 1988, Space Science Reviews, 49, 69
6. Thompson, D. J., et al., 1993, to be published in the Ap.J. Supp
7. Hartman, R. C., et al., 1992, ApJ, 385, L1
8. Kniffen, D. A., et al., 1993, submitted to the ApJ
9. Bertsch, D. L., et al., 1993, submitted to ApJ
10. Hunter, S., et al., 1993, submitted to ApJ
11. Lin et al., 1993, Submitted to Ap J.
12. von Montigny, C., et al., 1993, this proceedings
13. Galeev, A. A., Rosner, R., and Vaiana, G. S., 1979, ApJ, 229, 318
14. Ghisellini, Maraschi, and Treves 1985, A&A, 146, 204
15. Celotti, A., Maraschi, L., and Treves, A., 1991, ApJ, 377, 403
16. Done, C., and Fabian, A. C., 1989, MNRAS, 240, 81
17. McBreen, B., 1979, A&A, 71, L19
18. Guilbert, P. W., Fabian, A. C., and Rees, M., 1983, MNRAS, 205, 593
19. Zdziarski, A. A., 1991, in Extragalactic Radio Sources--From Beams to Jets, eds. J. Roland, H. Sol, and G. Pelletier, Cambridge University Press, Cambridge, p. 32
20. Osterbrock, D. E. 1991, Rep. Prog. Phys. 54, 579
21. Blandford, R. D., and Rees, M. J., 1978, in Pittsburgh Conference on BL Lac Objects, ed. A. M. Wolfe, Pittsburgh, p. 328
22. Blandford, R. D., and Konigl, A., 1979, ApJ., 232, 34
23. Valtaoja, E., Terasranta, Lainela, M., and Teerikorpi, P., 1992, in Variability of Blazars, eds. E. Valtaoja and M. Valtonen, Cambridge University Press, Cambridge, p. 70
24. Browne, I. W. A., 1989, in BL Lac Objects, eds. L. Maraschi, T. Maccacaro, M.-H. Ulrich, Springer, Berlin, p. 401
25. Impey, C. D., 1992, in Variability of Blazars, eds. E. Valtaoja and M. Valtonen, Cambridge University Press, Cambridge, p. 55

26. Barthel, P. D., 1989, ApJ, 336,606
27. Jones, O'Dell, and Stein 1974, ApJ, 188, 353
28. Gould 1979, A&A, 76, 306
29. Maraschi, Ghisellini, and Celotti 1992, in Testing the AGN Paradigm, AIP Conf. Proc. 254, eds., S. Holt, S. G. Neff, and C. M. Urry, AIP, New York
30. Kazanas, D., and Ellison, D. C., 1986, Nature, 319, 380
31. Marscher, A., and Bloom, S., 1992, in Proc. of the Compton Observatory Science Workshop, NASA Conf. Pub. 3137
32. Melia, F., and Königl, A., 1991, ApJ, 340, 162
33. Dermer, C. D., Schlickeiser, R., and Mastichiadis, A., 1992, A&A, 256, L27
34. Punch et al., 1992, Nature 358, 477
35. Wiita, P. J., 1991, in Beams and Jets in Astrophysics, ed. P. A. Hughes, Cambridge University Press, Cambridge, p. 379
36. Lynden-Bell, D., 1978, Phys. Scripta, 17, 185
37. Paczynski, B., and Wiita, P. J., 1980, A&A, 88, 23
38. Blandford, R. D., and Znajek, R., 1977, MNRAS, 179, 433
39. Goldreich, P., and Julian, W. H., 1969, ApJ, 157, 869
40. Macdonald, D., and Thorne, K. S., 1982, MNRAS, 198, 345
41. Lovelace, R. V. E., 1976, Nature, 262, 649
42. Fichtel, C. E., Simpson, G. A., and Thompson, D. J., 1978, Ap. J. 222, 597
43. Fichtel, C. E., Hartman, R. C., Kniffen, D. A., Thompson, D. J., Ogelman, H. B., Ozel, M. E., and Tumer, T., 1977. Ap. J. 217, L9
44. Fichtel, C. E., Hartman, R. C., Kniffen, D. A., Thompson, D. J., Ogelman, H. B., Tumer, T., and Ozel, M. E., 1978, NASA Tech. Memorandum 79650
45. Strong, A. W., Wolfendale, A. W., and Worrall, D. M., 1976, J. Phys. A: Math. Gen., 9,1553
46. Bignami, G. F., Fichel, C. E., Hartman, R. C., and Thompson, D. J., 1979, ApJ, 232, 649
47. Fichtel, C. E., and Trombka, J. I., 1981, Gamma Ray Astrophysics, New Insight into the Universe, NASA SP-453
48. Bignami et al., 1980, Astron. and Astrophys.
49. Colgate, S., 1990, American Astronomical Society Meeting, June 10-14,1990, 51.01

MONITORING THE LONG-TERM BEHAVIOR OF ACTIVE GALACTIC NUCLEI USING BATSE

W. S. Paciesas,[1] B. A. Harmon,[2] C. A. Wilson,[2] G. J. Fishman,[2]
C. A. Meegan,[2] R. B. Wilson,[2] G. N. Pendleton,[1] B. C. Rubin[2,3]

[1] Department of Physics, University of Alabama in Huntsville, AL 35899
[2] NASA/Marshall Space Flight Center, Huntsville, AL 35812
[3] Universities Space Research Association

ABSTRACT

The variability of hard x-ray/low-energy gamma-ray emission from active galactic nuclei (AGNs) has important implications for understanding the physical processes powering these sources. Use of the Earth occultation technique allows nearly continuous monitoring of sufficiently bright AGNs by BATSE. We have recently reprocessed a portion of the data base using an improved occultation algorithm and present preliminary results for a sample of eight AGNs known or suspected to emit above 20 keV. The sources Cen A, NGC 4151 and 3C 273 are routinely detected with $> 4\sigma$ statistical significance in 14-day accumulations. We present spectra of these sources for periods when they were observed by other *Compton* instruments. We investigate the time variability of Cen A and NGC 4151 and find marginal evidence for anti-correlation between spectral hardness and intensity.

INTRODUCTION

Prior to the launch of *Compton*, observations of AGNs were sparse and only a few sources had clearly been detected (see ref. 1 for a recent review). However, the observations were sufficient to show that at least some of the time AGNs emit most of their energy in gamma-rays. Similarities in the x-ray and gamma-ray properties of AGNs and Galactic black-hole candidates[2,3] may reinforce the arguments in favor of black-hole models for both types of sources, indicating that the physical mechanism powering these sources operates over a range of many orders of magnitude in black-hole mass. Furthermore, the "average" properties of AGNs are important for evaluating the contribution of unresolved point sources to the diffuse x-ray and gamma-ray backgrounds (e.g., ref. 4).

We are also studying AGNs in order to evaluate systematic effects in the BATSE occultation analysis. The AGN spectra are sufficiently hard that intercalibration with other instruments, particularly OSSE, will allow confirmation of BATSE's capability for measuring weak sources.

OBSERVATIONS

The Earth occultation technique which enables us to use BATSE as a near-all-sky monitor is continually being refined as our understanding of systematic effects advances. The basic technique and some of the systematics have been discussed elsewhere.[5,6] We use data from the large-area detectors (LADs) which provide 2.048 s time resolution in 16 energy channels. Rates near each source occultation are fitted with a model assuming a source immersion/emersion superimposed on a quadratically varying background, with additional terms for other potentially bright sources whose occultation times are sufficiently close. We have recently revised the software to use the energy-dependent atmospheric transmission in fitting each occultation step and to use more conservative criteria for rejection of interfering sources.

We have thus far reprocessed data from the beginning of the mission (TJD 8369) through TJD 8500. A catalog of 42 sources was used, including eight AGNs selected on the basis of previous measurements or extrapolations from lower energy data.[2,7] The reprocessed interval includes viewing periods 3 and 4 (TJD 8422–8449), during which 3C 273 and NGC 4151, respectively, were *Compton* primary (z-axis) targets. A portion of viewing period 8 (TJD 8490–8501), during which 3C 273 was an OSSE secondary (x-axis) target, is also covered. Since the *Compton* observation of Cen A did not occur until viewing period 12 (TJD 8546–8560), we reprocessed this interval specially for Cen A in order to provide a larger sample for intercalibration.

One complication with BATSE occultation analysis results from the satellite reorientations. Simple count rate histories are of limited use in long-term source monitoring because the effective exposure to a given source changes with each viewing period; the count rate spectra must be converted to incident flux in order to measure the true source variability. For this analysis, we deconvolved the 16-channel count rate spectra by folding a power-law spectrum $dN/dE = A_{80}(\frac{E}{80\,\text{keV}})^{-\alpha}$ through the appropriate detector response matrix and determining the best-fitting spectral index α and normalization A_{80} by χ^2 minimization. The flux was calculated by integration of the fitted power-law. We are not yet able to fit spectra from more than one viewing period simultaneously.

We derived a simple intercomparison of the sources in the sample by computing their fluxes during each viewing period assuming $\alpha = 2$. Table I summarizes the fluxes we measured for each source during viewing periods 3 and 6. The fluxes for Cen A, NGC 4151, and 3C 273 are typical of those derived at other times, with routine detections of $> 4\sigma$ statistical significance. The relatively strong signal seen from MCG−5-23-16 during viewing period 6 is atypical and may reflect a systematic error; we do not yet claim a detection of this source.

Table I. Intensities of Selected AGNs

Source	20–320 keV Flux (10^{-3} ph/cm^2-s)	
	TJD 8422–8435 (viewing period 3)	TJD 8463–8476 (viewing period 6)
* 3C 273	9.2 ± 1.4	4.7 ± 1.1
? 3C 279	1.4 ± 1.5	1.5 ± 1.1
* Cen A	36 ± 1.4	25 ± 1.8
? IC 4329a	2.4 ± 1.7	2.8 ± 1.4
? MCG−5-23-16	−1.0 ± 1.4	6.2 ± 1.2
* NGC 4151	12 ± 1.2	14 ± 1.1
? NGC 5506	−1.7 ± 1.5	3.3 ± 1.2
? NGC 7582	2.1 ± 1.0	1.9 ± 1.1

* positive detection   ? uncertain at present

For intercalibration we computed spectra for Cen A, NGC 4151, and 3C 273 at times coincident with *Compton* z-axis pointings. Figures 1a–c show

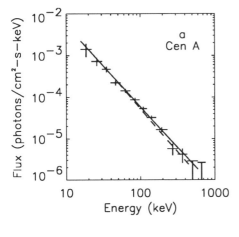
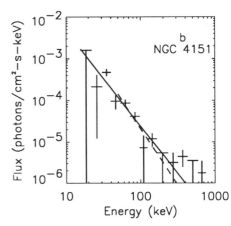
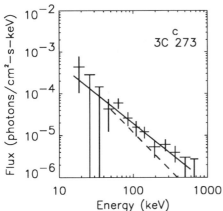

Figure 1. Comparison of BATSE occultation data with contemporaneous OSSE measurements. Solid lines are the best power-law fits to BATSE data. Dashed lines are the best power-law fits to OSSE data. a) Cen A (viewing period 12; TJD 8546–8560). b) NGC 4151 (viewing period 4; TJD 8435–8449). c) 3C 273 (viewing period 3; TJD 8422–8435).

the results together with best-fitting single power-law models. For Cen A, we find $\alpha = 1.96 \pm 0.07$ and $A_{80} = (9.1 \pm 0.4) \times 10^{-5}$ ph/cm^2-s-keV ($\chi^2 = 13.4$ for 11 d.o.f.). For NGC 4151, $\alpha = 2.3 \pm 0.2$ and $A_{80} = (4.0 \pm 0.5) \times 10^{-5}$ ph/cm^2-s-keV, although the fit is poor ($\chi^2 = 20.1$ for 11 d.o.f.). For 3C 273, $\alpha = 1.4 \pm 0.2$ and $A_{80} = (2.6 \pm 0.5) \times 10^{-5}$ ph/cm^2-s-keV ($\chi^2 = 10.0$ for 11 d.o.f.).

Cen A and NGC 4151 are sufficiently strong for us to be able to investigate their variability in more detail. For example, we summed the count rate spectra over four-day intervals for each source and fitted each interval separately with a power-law, allowing both A_{80} and α to vary. Figures 2a-b show the derived α as a function of the source flux. An anti-correlation of spectral hardness with intensity appears to occur in both sources. For Cen A, the effect appears stronger at low intensities and weaker or absent at high intensities. Again, systematic uncertainties require us to consider these results suggestive but preliminary. We plan further investigations of this sort after more of the data have been reprocessed.

The BATSE data for Cen A, NGC 4151, and 3C 273 agree well with more sensitive contemporaneous OSSE measurements[8-10] shown as dashed curves in

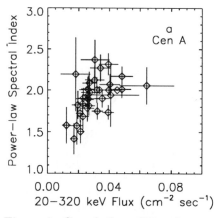 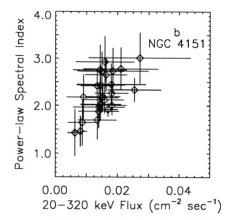

Figure 2. Correlation plots of power-law spectral index vs flux in the 20–320 keV energy range for a) Cen A and b) NGC 4151 in the interval TJD 8369–8498. Spectra were accumulated over four-day intervals. Points of less than 1.5σ significance have been excluded.

Figs. 1a-c. The spectral indices for all three sources are in reasonable agreement, considering the differences in energy coverage. The normalizations for Cen A and NGC 4151 are consistent to within ~20%, whereas a larger difference (~40%) occurs for 3C 273. We are investigating our systematics in more detail for the latter source.

DISCUSSION

Among our sample of AGNs, three objects (Cen A, NGC 4151, and 3C 273) are strong enough to be detected regularly with good statistical significance on timescales of 14 days or less. These sources have historically been the strongest AGNs in hard x-rays. The fluxes we observe for Cen A are well within the range of pre-*Compton* observations and those of NGC 4151 are typical of post-1977 measurements (cf. ref. 1).

Though widely spaced in time, pre-*Compton* observations have shown that the spectral index of Cen A is remarkably constant (~ 1.6) over a wide range of intensities,[1] whereas our four-day measurements show significant variations (~ 1.5 − 2.3). Although we cannot rule out all possible systematic effects in our data, one possible explanation for this is that our single power-law fit is too simplistic. We note that the OSSE spectra of Cen A during viewing period 12 are well fit by a broken power-law with a break energy around 180 keV.[8] Our data are also adequately fit by a broken power-law with $\alpha = 1.7 \pm 0.1$ below E_{break} (fixed at 180 keV). The steeper spectra in Fig. 2a may thus represent times when the break moves to a lower energy and/or the spectrum above the break becomes steeper, while the spectral index below the break remains constant.

The 2–20 keV photon spectra of Seyfert galaxies are known to be power-laws with spectral indexes clustering in the range 1.5–2.[11] We find that the average spectrum of NGC 4151 was softer than this above 20 keV during the period of our observations; if the 2–20 keV spectral index was normal at this time, then a single power-law could not fit the data over the 2–740 keV range. This would be consistent with recent GRANAT observations[12] which show a break in the spectrum of NGC 4151 around 40 keV. We cannot confirm the existence of such a break from our data alone.

Though our evidence for a hardness/intensity anti-correlation in NGC 4151 is rather weak, similar anti-correlations in the 2–10 keV band have been reported for a number of Seyfert galaxies (ref. 13 and refs. therein), including NGC 4151.[14,15] Comparable measurements in hard x-rays are generally not available; marginal evidence has been presented for a correlation of opposite sense in NGC 4151 at 100 keV.[16] Our preliminary results tend to support the lower energy measurements.

SUMMARY

With current algorithms, BATSE can monitor the hard x-ray/low-energy gamma-ray behavior of Cen A and NGC 4151 on timescales of several days or less. We find clear evidence for variability of Cen A and weak evidence for variability of NGC 4151 on these timescales during an interval of $\sim$ 130 days in 1991. Although some systematic errors may be present, our data suggest that the variability of both sources is such that softer spectra correlate with higher intensities. We also find that 3C 273 is regularly detectable during this same interval, although its lower intensity constrains our sensitivity for detecting source variability. Among the other AGNs in our sample, we find that several may have intermittent outbursts of significant flux, indicating that a modest improvement in our occultation analysis technique could allow monitoring of considerably more AGNs.

REFERENCES

1. N. Gehrels and C. Cheung, *Testing the AGN Paradigm*, eds. S. Holt, S. Ness and M. Urry (New York: AIP, 1992), p. 348.
2. H. Inoue, *Proc. 23rd ESLAB Symp. on two topics in X-Ray Astronomy*, eds. J. Hunt and B. Battrick (Noordwijk: ESA SP-296, 1989), p. 783.
3. R. A. Syunyaev *et al.*, Sov. Astron. Lett. **17**, 123 (1991).
4. D. Schwartz and W. Tucker, Ap. J. **332**, 157 (1988).
5. B. A. Harmon *et al.*, *The Compton Observatory Science Workshop*, eds. C. R. Shrader, N. Gehrels and B. Dennis (NASA CP-3137, 1992), p. 69.
6. W. S. Paciesas *et al.*, Astr. Astrophys. Suppl., in press (1992).
7. S. S. Holt *et al.*, *Proc. 23rd ESLAB Symp. on two topics in X-Ray Astronomy*, eds. J. Hunt and B. Battrick (Noordwijk: ESA SP-296, 1989), p. 1105.
8. R. L. Kinzer *et al.*, these proceedings.
9. R. A. Cameron *et al.*, these proceedings.
10. W. N. Johnson *et al.*, these proceedings.
11. R. F. Mushotzky, Adv. Space Res. **3**, 157 (1984).
12. E. Jourdain *et al.*, Proc. 22nd Intl. Cosmic Ray Conf., Dublin, v. 1, p. 177 (1991).
13. M. Matsuoka, M. Yamauchi and L. Piro, *Proc. 23rd ESLAB Symp. on two topics in X-Ray Astronomy*, eds. J. Hunt and B. Battrick (Noordwijk: ESA SP-296, 1989), p. 985.
14. G. C. Perola *et al.*, Ap. J. **306**, 508 (1986).
15. T. Yaqoob and R. S. Warwick, MNRAS **248**, 773 (1991).
16. W. A. Baity *et al.*, Ap. J. **279**, 555 (1984).

OSSE OBSERVATIONS OF ACTIVE GALAXIES AND QUASARS

R.A. Cameron[1], J.E. Grove[2], W.N. Johnson[2], J.D. Kurfess[2],
R.L. Kinzer[2], R.A. Kroeger[2], M.S. Strickman[2], M. Maisack[3,4], C.H. Starr[5],
G.V. Jung[1], D.A. Grabelsky[6], W.R. Purcell[6], M.P. Ulmer[6]

[1] Universities Space Research Association, Washington DC
[2] E.O. Hulburt Center for Space Research,
 Naval Research Laboratory, Washington DC
[3] George Mason University, Fairfax, VA
[4] Astronomisches Institut Tübingen, Germany
[5] Science Programs, Computer Sciences Corporation, Washington, DC
[6] Northwestern University, Evanston, IL

ABSTRACT

We present a summary of OSSE observations of galaxies and quasars that have been carried out during the Phase 1 all-sky survey by the *Compton* Observatory. The OSSE instrument has detected continuum emission from several Seyfert galaxies and quasars. Seyfert 1 galaxies make up the majority of the detections, typically at energies below 300 keV, with the measured spectra generally compatible with power-law continuum models with photon spectral indices around -2, or with thermal emission models with temperatures around 50 keV. The quasars generally have harder spectral indices than the Seyfert galaxies.

With the exception of Centaurus A and NGC 4151, there is little evidence of significant flux variability in the OSSE data sets for most of the Seyfert galaxies observed. In some cases, the OSSE detections are at flux levels significantly below those reported for previous observations.

While the analysis of the complete set of Phase 1 OSSE observations of active galaxies is still in progress, the OSSE data will clearly provide a major new database for the examination and testing of models of high-energy emission from active galactic nuclei.

INTRODUCTION

The OSSE instrument provides an excellent capability for hard X-ray and gamma-ray observations of active galaxies and quasars. Detailed descriptions of the capabilities and performance of the OSSE instrument are given elsewhere (Cameron *et al.*, 1992, Johnson *et al.*, 1993). OSSE typically observes 2 separate targets during an observation period, switching to a secondary target during the fraction of each orbit when the primary target is occulted by the Earth. The four

OSSE detectors are capable of observing a target over 90 degrees of detector positioning. Detector occultation occurs outside this range of detector position angles, though two unocculted detectors remain available for observations. Typical observations are of 2 or 3 weeks duration, although some observations may span only a few days, dictated by events such as targets of opportunity. The actual integration time that is useful for analysis on an object in any observation period is a function of several parameters, including the number of detectors used in the observation, and the relative priority of time allocation for the object during the observation period.

During the 18-month all-sky survey carried out in the first phase of the *Compton* Observatory mission, the OSSE instrument was able to perform pointed observations for more than 30 active galaxies and quasars. This paper presents a summary of these OSSE observations of galaxies and quasars, including indication of which objects have been detected in the analysis carried out to date. Several objects were observed more than once, which will provide an opportunity for examining source flux variability on timescales ranging from days to months.

THE OBSERVATIONS

A complete list of Phase 1 OSSE observations of galaxies and quasars is given in Table 1. Thirty-six objects are listed, with twenty-five Seyfert galaxies making up the majority of the objects. Detailed analysis has been carried out for about half of the objects. Work continues on refining the understanding of systematic errors in the datasets, which will allow final evaluation of model parameters.

Observations are listed only for those times when the object was the specific target for the observation. Observing periods when an object was in a background field or at low sensitivity in the field of another target are not included. The table gives the following information for each object: source name; position in J2000 equatorial coordinates; object type; redshift, or heliocentric velocity in km s^{-1}; the time interval(s) over which the object was observed; the gain of the OSSE instrument during the observation ($\times 1$ gain provides data up to 10 MeV, $\times 2$ gain provides data up to 5 MeV); the number of detectors used for the observation; and an indication of the detection of the object in OSSE data that has been analyzed to date. Object types, redshifts and heliocentric velocities were generally obtained from the NASA/IPAC Extragalactic Database.

Analysis has not yet been completed for all of the observations shown. However, more than 50% of the objects analyzed to date have been detected by OSSE. Marginal or doubtful detections are indicated in the table, for those objects with detections of between approximately 2σ and 5σ significance between 60 keV and 500 keV.

Analysis of OSSE observations of the starburst galaxies M 82 and NGC 253 are presented in these proceedings (The et al., 1993, Bhattacharya et al., 1993). For the Seyfert galaxies, only one Seyfert 2 has been detected, this being the

480 OSSE Observations of Active Galaxies

Table 1. Galaxies and Quasars Observed by OSSE

Object	RA	Dec	Type	$z(V_{hel})$	Obs. Dates	Gain	#Detectors	Detected?
MRK 335	00 06 19.4	+20 12 11	Sy 1	(7735)	92/114 – 92/119	×1	4+2	
					92/128 – 92/135	×1	4+2	
NGC 253	00 47 33.2	−25 17 19	SAB(s)c	(251)	92/233 – 92/240	×2	2	
					91/248 – 91/255	×1	4	
					91/311 – 91/318	×1	4	
					91/346 – 91/361	×2	2	
					92/037 – 92/051	×2	2	
					92/079 – 92/093	×1	4	
NGC 1068	02 42 40.1	−00 00 48	Sy 2	(1137)	92/051 – 92/065	×1	4+2	N
NGC 1275	03 19 48.1	+41 30 42	Sy 2	(5260)	91/332 – 91/346	×1/×2	4	N
3C 111	04 18 21.6	+38 01 37	Sy 1	0.049	91/179 – 91/193	×1	2	Y
3C 120	04 33 11.0	+05 21 16	Sy 1	0.033	92/135 – 92/156	×2	2	N
					92/135 – 92/156	×2	2	Y?
					92/156 – 92/198	×1	4	Y?
PKS 0528+134	05 30 56.4	+13 31 55	QSO	2.06	92/282 – 92/289	×1	2	N?
					92/308 – 92/322	×1	2	N?
PKS 0548−322	05 50 41.8	−32 16 11	BL Lac	0.069	92/163 – 92/177	×1	2	
MCG +8−11−11	05 54 55.2	+46 26 25	Sy 1	(6141)	92/163 – 92/177	×1	4	Y
QSO 0736+016	07 39 18.0	+01 37 05	QSO	0.191	91/167 – 91/179	×1	4	N
QSO 0834−201	08 36 39.1	−20 16 59	QSO	1.715	92/282 – 92/289	×1	2	
					92/308 – 92/322	×1	2	
MCG +5−23−16	09 34 15.4	+27 19 38	Sy 2	(3202)	92/225 – 92/261	×2	2	
NGC 2992	09 45 41.9	−14 19 35	Sy 2	(2314)	92/156 – 92/163	×1	4	
MCG −5−23−16	09 47 40.1	−30 56 55	Sy 2	(2498)	92/184 – 92/198	×1	4	
					92/219 – 92/225	×1	4	
					92/240 – 92/245	×1	4	
M. 82	09 55 53.9	+69 40 57	I0	(203)	91/221 – 91/227	×1	4	
					92/010 – 92/023	×1	4	
MRK 421	11 04 27.3	+38 12 32	BL Lac	(9234)	91/193 – 91/207	×1	2	N
					91/207 – 91/220	×2	2	N
					91/255 – 91/262	×2	4	N
NGC 3783	11 39 01.7	−37 44 19	Sy 1	(3033)	92/177 – 92/184	×1	4	Y?
NGC 4151	12 10 32.4	+39 24 20	Sy 1	(995)	91/179 – 91/193	×1	4	Y
					92/093 – 92/107	×2	2	Y

Table 1. Galaxies and Quasars Observed by OSSE (cont.)

Object	RA	Dec	Type	$z(V_{hel})$	Obs. Dates	Gain	#Detectors	Detected?
NGC 4388	12 25 46.6	+12 39 41	Sy 2	(2517)	92/261 – 92/282	×1	3	
3C 273	12 29 06.6	+02 03 09	QSO	0.158	91/166 – 91/179	×1	4	Y
					91/234 – 91/248	×1	4	Y
					91/276 – 91/290	×1	4	Y
					92/225 – 92/228	×2	2	Y
					92/251 – 92/257	×2	2	Y
M 87	12 30 49.4	+12 23 28	Sy	(1282)	92/261 – 92/282	×1	3	
NGC 4593	12 39 39.3	−05 20 39	Sy 1.9	(2492)	92/228 – 92/233	×2	2	N
3C 279	12 56 11.1	−05 47 22	QSO	0.538	92/257 – 92/261	×2	2	N
					91/262 – 91/276	×1	2	Y
					92/245 – 92/251	×2	2	Y
Centaurus A	13 25 28.9	−43 00 59	S0 pec, Sy 2	(562)	91/290 – 91/304	×1	4	Y
					92/303 – 92/308	×2	4	Y
MCG −6−30−15	13 35 50.8	−34 17 29	Sy 1	(2329)	92/282 – 92/289	×1	4	
IC 4329A	13 49 18.3	−30 18 34	Sy 1	(4813)	92/308 – 92/322	×1	4	
					92/282 – 92/289	×1	4	Y
					92/308 – 92/322	×1	4	Y
MRK 279	13 53 01.7	+69 18 29	Sy 1	(9144)	92/065 – 92/079	×1	4	Y
NGC 5548	14 17 59.4	+25 08 13	Sy 1.2	(5149)	91/227 – 91/234	×1	4	
					91/290 – 91/311	×1	2,4	
MRK 841	15 04 01.1	+10 26 19	Sy	0.036	92/107 – 92/114	×2	2	Y
3C 390.3	18 42 08.9	+79 46 17	Sy 1	0.056	91/290 – 91/304	×1	2	Y
					92/135 – 92/156	×2	2	
ESO 141−55	19 21 14.1	−58 40 15	Sy 1	0.037	92/219 – 92/224	×1	4	Y?
					92/240 – 92/245	×1	4	Y?
					92/289 – 92/303	×1	2	Y?
MRK 509	20 44 09.6	−10 43 25	Sy 1	0.034	92/303 – 92/308	×1	4	
PKS 2155−304	21 58 51.8	−30 13 31	BL Lac	0.17	92/289 – 92/303	×1	2	Y?
4C 04.77	22 04 17.5	+04 40 02	BL Lac	(8400)	92/233 – 92/240	×2	2	
NGC 7314	22 35 45.6	−26 03 03	Sy 1.9	(1422)	92/119 – 92/128	×2	4	
NGC 7582	23 18 23.1	−42 22 12	Sy 2	(1575)	91/346 – 91/361	×1	2	
					92/093 – 92/114	×2	2	

nearby galaxy Centaurus A. The other Seyfert 1 galaxies that have been detected generally show weak emission at hard x-ray energies, except for NGC 4151, which has a stronger flux than the other Seyferts by virtue of its proximity. Detailed analysis of the spectrum of NGC 4151 is described by Maisack *et al.* (1993). The detection rate for Seyfert 1 galaxies, including marginal detections, is close to 100%, indicating a promising class of objects for study with OSSE at hard x-ray energies. The quasars 3C 273 and 3C 279 show power-law spectra, with harder emission than a typical Seyfert 1 galaxy. Details of these detections will be given elsewhere as the final analysis is completed for each object.

Variability has been detected in three objects: Centaurus A (Kinzer *et al.*, 1993); NGC 4151, where a flux difference of $\sim 25\%$ was measured for the two observation epochs; and 3C 111, where the source was detected in its first observation by OSSE, but not in the second observation. Background subtraction systematics are still being investigated in the case of 3C 111.

SUMMARY

With the ability of the OSSE instrument to carry out high-sensitivity pointed observations of active galaxies at hard X-ray energies, and the detection rate demonstrated by the current analysis of OSSE data collected during Phase 1 observations, the OSSE instrument should provide a valuable dataset of high-energy emission measurements from extragalactic objects, which will refine the understanding of energetic processes in the nuclei of galaxies.

REFERENCES

Bhattacharya, D. *et al.* 1993, these proceedings.
Cameron, R.A. *et al.*, 1992, in *The Compton Observatory Science Workshop*, ed. C.R. Shrader, N. Gehrels, and B. Dennis, NASA Conference Publication 3137.
Johnson, W.N. *et al.*, 1993, Ap.J.Supp., accepted for publication.
Kinzer, R.L. *et al.*, 1993, Ap.J., in preparation.
Maisack, M., *et al.* 1993, these proceedings.
The, L.-S., *et al.* 1993, these proceedings.

SEARCH FOR GAMMA-RAY EMISSION FROM AGN WITH COMPTEL

W. Collmar, R. Diehl, G.G. Lichti, V. Schönfelder, H. Steinle, A.W. Strong
Max-Planck Institut für Extraterrestrische Physik, D/W-8046 Garching, Germany

H. Bloemen, J.W. den Herder, W. Hermsen, B.N. Swanenburg, C. de Vries
SRON-Leiden, P.B. 9504, NL-2300 RA Leiden, The Netherlands

M. McConnell, J. Ryan, G. Stacy
University of New Hampshire, Durham NH03824, USA

K. Bennett, O.R. Williams, C. Winkler
Astrophysics Division, ESTEC, 2200 AG Noordwijk, The Netherlands

ABSTRACT

The COMPTEL data (~0.7-30 MeV) were searched for emission from AGN. Four sources have been detected so far: the quasars 3C 273, 3C 279, PKS 0528+134, and the radio galaxy Centaurus A. 3C 273 and 3C 279 were detected in CGRO observation period 3 with quite different spectral shapes. There is also evidence for 3C 273 at a weak flux level in observation period 11. The quasar PKS 0528+134 was detected above 3 MeV as part of a search for AGN already observed by EGRET. Cen A was seen up to 3 MeV by combining data from different observation periods.

INTRODUCTION

The imaging Compton telescope COMPTEL is sensitive in the low energy γ-ray range from about 0.7 MeV to 30 MeV. A detailed description of the COMPTEL characteristics is given by Schönfelder et al[1].

COMPTEL's MeV energy range is particularly interesting for investigations into the physics of AGN. Measurements in the neighbouring energy ranges indicate that some AGN might have their peak luminosity in this range and that there might be breaks in the spectra at MeV energies. Moreover, before COMPTEL only three AGN were detected at MeV energies: NGC 4151[2], MCG 8-11-1[3], and Cen A[4]. Because each new detection increases our knowledge significantly, the COMPTEL data were searched for evidence of emission from AGN. The possibility of simultaneous measurements in neighbouring energy bands with CGRO would make AGN detections by COMPTEL even more valuable for the understanding of AGN physics.

In this paper we report on the results of the search for γ-ray emission from AGN in the COMPTEL data. In particular, the search was done for promising objects, with emphasis on AGN already detected by the EGRET experiment aboard CGRO. This paper represents a progress report with largely preliminary results. A more complete analysis for the different sources is underway.

THE QUASARS 3C 273 AND 3C 279

During the first year of its mission, the quasars 3C 273 and 3C 279 were in the COMPTEL field of view twice: in the observation period 3 (June 91) and 11 (October 91). In observation period 3, both quasars were detected by COMPTEL, as already reported[5]. 3C 273 was detected mainly in the lower part of the COMPTEL energy range (1 - 10 MeV), while at the highest energies (10-30 MeV) only an upper limit could be derived. In contrast, 3C 279 was detected at the higher COMPTEL energies (3-30 MeV) while at lower energies (1-3 MeV) only an upper limit could be derived. In figure 1 the COMPTEL fluxes for the three energy intervals together with the spectral shapes measured by EGRET and OSSE (simultaneously for 3C 273) are given for both quasars. The error bars on the COMPTEL spectral points are statistical only. For 3C 273 the OSSE spectrum (photon index -1.8[6]) is harder than the EGRET spectrum (photon index -2.4[7]) indicating a break somewhere in the MeV region. The COMPTEL data are consistent with a spectral steepening around 1 MeV. For 3C 279, EGRET and COMPTEL measurements together indicate a spectral break around 10 MeV[5].

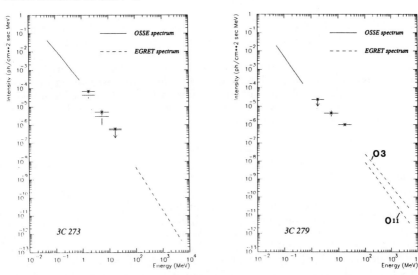

Figure 1. CGRO spectra of 3C 273 and 3C 279. a) Simultaneously (obs. 3) measured energy spectra of 3C 273 by OSSE[6], COMPTEL[5], and EGRET[7]. The lower COMPTEL spectral points are from October 91 (obs. 11). b) Spectra of 3C 279. The OSSE spectral shape was measured in September 91[6], COMPTEL in June 91[5], and EGRET in June and October 91[8].

In observation period 11, the COMPTEL data show evidence for 3C 273 at a weak flux level. The 1-3 MeV skymap of this region shows a feature with a significance of $\sim 4\sigma$ (figure 2). In the 3-10 MeV band a marginally significant feature ($\sim 2.4\sigma$) appears. Both excesses are consistent with the location of 3C 273. The derived flux level is $\sim 50\%$ of that observed in observation period 3 (same upper limit for the 10-30 MeV band). The spectral points including only the statistical errors are also given in figure 1. A more detailed analysis is underway.

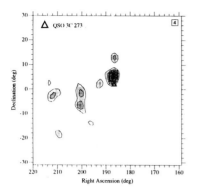

Figure 2. 1-3 MeV skymap of the Virgo region for observation period 11 (October 91). There is evidence ($\sim 4\sigma$) for 3C 273 at a weak flux level.

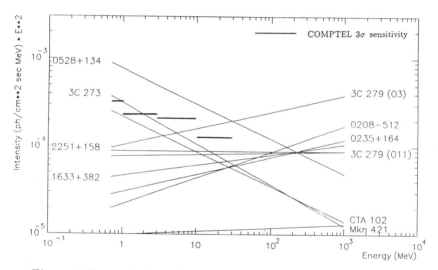

Figure 3. Extrapolation of the measured EGRET AGN spectra (≥ 100 MeV) into the COMPTEL energy range. For clarity the flux is multiplied by E^2. The horizontal lines indicate the conservative COMPTEL 3σ sensitivity limits[1] for the 4 standard energy intervals assuming an E^{-2} power law spectrum.

SEARCH FOR AGN DETECTED BY EGRET

Up to now EGRET has announced the detection of 16 AGN[9] mainly at energies above 100 MeV. For most of them spectral parameters are available. A search in the COMPTEL data for these AGN has been started. To select the promising candidates, the measured EGRET spectra were extrapolated into the COMPTEL energy range for comparison with the conservative COMPTEL sensitivity limits (figure 3).

Besides 3C 273 and 3C 279 in observation period 3, there is only one other promising candidate: the quasar PKS 0528+134. The extrapolation of the measured EGRET spectrum passes well above the sensitivity limits for all four COMPTEL standard energy intervals. A search for this quasar led to its detection during the CGRO validation period and also in the survey observation period 1. Skymaps for the energy ranges 3-10 MeV and 10-30 MeV from the galactic anticentre region during 9 days of the CGRO validation period (April 28 - May 7, 1991) are given in figure 4. A clear excess at the position of the quasar PKS 0528+134 is visible in both skymaps. Because of the very recent detection of PKS 0528+134 and its proximity to the Crab ($\sim 8°$), a precise flux value has not yet been determined. A first estimate leads to a flux level of the order of 10%-20% of the Crab flux for both energy ranges. No obvious signal from this source could be found so far for energies below 3 MeV, indicating a spectral break. A detailed analysis of this source is in progress.

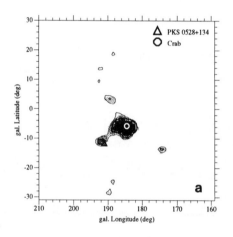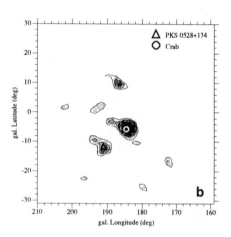

Figure 4. Skymaps of the galactic anticentre (April 28 - May 7, 91) for the energy ranges 3-10 MeV (a) and 10-30 MeV (b). In both images the Crab is saturated for better visibility of the image structure at lower intensity levels. In both energy ranges an excess at the position of the quasar PKS 0528+134 is visible.

The search for other AGN measured by EGRET is currently being pursued. No other AGN have been detected by COMPTEL so far. The recent EGRET result for the superluminal quasar CTA 102, having the softest spectrum[10] (photon index -2.6) of all AGN detected by EGRET, makes this source a candidate for COMPTEL.

THE RADIO GALAXY CENTAURUS A

The CGRO observation period 12 (October 1991) was devoted to the sky region containing the active radio galaxy Centaurus A (Cen A). Because Cen A has been observed before at MeV energies[4], it was a prime candidate for COMPTEL. Only an upper limit could be determined for observation period 12[11]. However, by combining the data of several pointings (Cen A was in the COMPTEL field of view several times) Cen A was detected up to 3 MeV in the COMPTEL data. Evidence for the detection can be found in [12]. The COMPTEL spectral points, together with spectral results of other experiments are shown in figure 5.

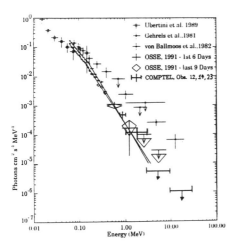

Figure 5. Spectral measurements of Cen A (from [13] with COMPTEL results added). The COMPTEL spectral points are derived from a combination of several pointings. The upper limits for the higher energies are 2σ.

SUMMARY

By searching the COMPTEL data for AGN in the energy range 0.7-30 MeV, 4 different objects have been detected so far: the quasars 3C 273, 3C 279, PKS 0528+134, and the radio galaxy Cen A.

The spectra of these objects are quite different. While two of them, 3C 273 and Cen A, were detected at lower energies, the other two, 3C 279 and PKS 0528+134 were only seen at higher energies, indicating a harder spectrum.

The presented findings are promising for the future of the CGRO mission, especially because of the possibility of simultaneous spectral measurements over about 5 decades in energy by the different CGRO experiments. This capability will surely improve the understanding of the still enigmatic AGN engine.

REFERENCES

1. V. Schönfelder et al., Ap.J. Supp. (in press) (1992)
2. F. Perotti et al.,, Ap.J. **247**, L63 (1981)
3. F. Perotti et al., Nature **292**, 133 (1981)
4. P. v. Ballmoos et al., Ap.J. **312**, 134 (1987)
5. W. Hermsen et al., Astron. Astrophys. Supp. (in press) (1992)
6. W.N. Johnson et al., these proceedings
7. C. v. Montigny et al., Astron. Astrophys. Supp. (in press) (1992)
8. D. Kniffen et al., Ap. J. submitted
9. C. Fichtel et al., these proceedings
10. P. Nolan et al., these proceedings
11. H. Steinle et al., Advances in Space Research (in press) (1993)
12. H. Bloemen et al., these proceedings
13. W.N. Johnson et al., Astron. Astrophys. Supp. (in press) (1992)

Detection of TeV photons from Markarian 421

M. Punch[a,b], C. W. Akerlof[c], M. F. Cawley[d], M. Chantell[a], D. J. Fegan[b], S. Fennell[a,b], J. A. Gaidos[e], J. Hagan[b], A. M. Hillas[f], Y. Jiang[a], A. D. Kerrick[g], R. C. Lamb[g], M. A. Lawrence[a], D. A. Lewis[g], D. I. Meyer[c], G. Mohanty[g], K. S. O'Flaherty[b], P. T. Reynolds[g], A. C. Rovero[a], M. S. Schubnell[c], G. Sembroski[e], T. C. Weekes[a], T. Whitaker[a], and C. Wilson[e].

a. Whipple Observatory, Harvard-Smithsonian CfA, Box 97, Amado, AZ 85645 USA
b. Physics Dept., University College Dublin, Belfield, Dublin 4, Ireland
c. Physics Dept., University of Michigan, Ann Arbor, MI 48109 USA
d. Physics Dept., St. Patrick's College, Maynooth, Co. Kildare, Ireland
e. Physics Dept., Purdue University, West Lafayette, IN 47907 USA
f. Physics Dept., University of Leeds, Leeds LS2 9JT, UK
g. Physics and Astronomy Dept., Iowa State University, Ames, IA 50011 USA

ABSTRACT

We report on the detection by the Whipple Observatory's γ-ray telescope of TeV γ-rays from the Markarian 421 at the 6.3σ level. The flux above 0.5 TeV is 0.3 that of the Crab Nebula. This is the most distant source detection at TeV γ-ray energies.

1. Source Characteristics

The successful detection of MeV - GeV emission from a number of active galactic nuclei (AGN's) by the EGRET instrument aboard the Compton Gamma-Ray Observatory[1] has given new impetus to the Whipple Observatory's programme of AGN observation. The detection of one of these "Egret AGN's" is presented here; observations on others are presented elsewhere in this symposium[2].

Markarian 421 is a giant elliptical galaxy with a BL Lacertae type nucleus[3-4]. BL Lacertae objects are characterized by absence of optical emission lines, polarization of radio and/or optical flux, and large and rapid variations in luminosity. It has been observed in the radio[5-7], optical[4,7], x-ray[7-10], and MeV - GeV[11] bands. It is the closest of the AGN's reported by EGRET ($z = 0.031$), but is also one of the weakest.

2. Instrumentation and Data Analysis

The Whipple Observatory's High-Resolution Imaging Cherenkov detector[12] consists of a 10 m aperture reflector which focuses Cherenkov light from air showers on a hexagonal array of 109 photo-multipliers, giving an image with a pixel size of $0.25°$ in the $4°$ field of view. Observations must be taken on moonless nights under clear skies. For d.c. source observations, the mode of operation is to view the source for a number of minutes ('on-source') and to view a comparison region which passes through the same range of azimuth and elevation ('off-source') for an equal amount of time.

The background for this technique results from the isotropic hadronic (cosmic-ray) air showers. The approximately elliptical images from both hadronic- and γ-induced showers are characterized by four parameters: the root-mean-square length and width of the ellipse; 'distance', the angular distance of the centroid of the shower image from the assumed source location in the image plane; and 'alpha', the angle between the major axis of the shower image and a line from its centroid to the source location, which gives the orientation of the image. The images of the γ-ray induced air showers are narrower in length and width than the hadron-induced air showers (shape criterion) and preferentially point to the source location, with smaller values of alpha (orientation criterion). The distance parameter is used to exclude events near the centre and edge of the field of view, where neither shape nor orientation are well-defined. The cut-off values of these parameters were optimized on a subset of Crab observations resulting in a set of criteria for γ-ray like images ("supercuts[13]"). This method was applied without modification to the data from Markarian 421.

Detections of the Crab Nebula using an earlier technique[14,15] based solely on Monte Carlos[16,17], and subsequently using the method above[13] show the efficacy of the instrument in locating sources of γ-rays. From Monte Carlos, the effective collection area of the detector is $\sim 3 \times 10^6 \, \text{m}^2$ and the threshold for the present observations is $\sim 0.5 \, \text{TeV}$. The angular resolution of the instrument is 6 arc minutes[18]. The 'supercuts' technique eliminates more than 99% of background events while retaining nearly half of the γ-rays. The data rate on the Crab Nebula is 90 photons/hour, but operational considerations limit the duty cycle of the instrument to $\sim 10\%$.

3. Results

Markarian 421 was observed at the Whipple Observatory during the period 24th March to 2nd June, 1992. The raw (uncut) on-source and off-source datasets contain 77,181 and 76,761 events respectively. After cuts are applied 302 events on-source and 166 events off-source remain, an excess with a 6.3σ significance. The excess corresponds to an average flux of 1.5×10^{-11} photons cm^{-2} s^{-1} above $0.5 \, \text{TeV}$, equivalent to 0.3 that of the Crab Nebula.[19]

We have investigated the possibility that such an excess might be due to systematic effects. Noise in a brighter off-source region might bias against the narrower γ-like events: for this source, however, the on-source region is negligibly brighter; reanalysis where noise is introduced to equalize the on and off-source regions result in a slight increase in the signal. The effect of a 5th magnitude star 2 arc-minutes north of Markarian 421 was considered: reanalysis with the central photomultipliers (which contain the star) excluded from the trigger and image have little effect on the signal; control observations on stars of similar magnitude show null results when subjected to supercuts analysis.

A visual comparison (figure 1) of the signals from the Crab Nebula (a) and Markarian 421 (b) may be made by plotting the distribution in alpha of those events which pass the shape criterion. As expected, the signal is concentrated in the range alpha < 15°, that is, the images point towards the source.

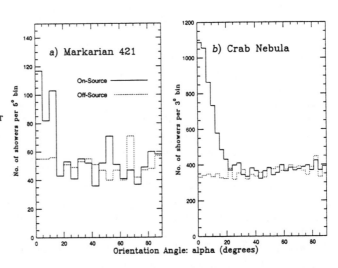

Figure 1. On and off-source orientation angle ('alpha') distributions for a, Mk 421 and b, Crab Nebula. The distributions are for those showers for which the other supercuts selection criteria[13] have been satisfied. The supercuts selection value for alpha is 15°.

From the observations a two-dimensional map of the source region[18] may be created. Figure 2 shows the map from the observations of Markarian 421. The centre of the field of view corresponds to the known direction of the source. The peak seen is within 0.1° of this direction.

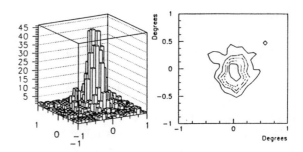

Figure 2. Maps of the on-source observations for Mk 421 made according to the prescription of ref. 18, figures 4 and 5. The peak intensity lies within 0.1° of the known location of Mk 421.

From the present observations, variability of the source on a time-scale of hours cannot be ruled out. At hard X-ray frequencies, variability has been observed on a time scale of 2 hours[10]. Correlated multi-frequency (optical, MeV - GeV, TeV) observations of Markarian 421 are scheduled for next May, when it will be in the field of view of the EGRET instrument.

Figure 3.
a) Integral photon fluxes for Mk 421 at 100 MeV[1] and 0.5 TeV[19]
b) Intergalactic transmission of high-energy photons according to ref. 21, for $z = 1.0$ and $z = 0.03$. The solid and dotted lines are upper and lower limits to the absorption.
(From ref. 20)

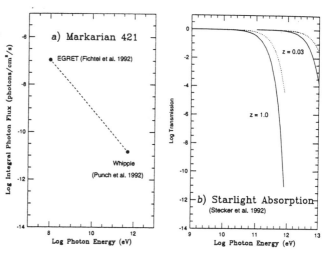

4. Conclusions

From an assumption of isotropic emission from this source, then the luminosity at a distance of 124 Mpc is $\sim 10^{43}$ erg s^{-1} at TeV energies. But as Markarian 421 is known to exhibit jet-like behaviour, the actual luminosity may be considerably less.

The fact that emission is seen from this source, but not from more distant AGN, many of which are brighter at GeV energies, may indicate absorption of the TeV photons by the IR background. Figure 3b shows the degree of absorption predicted by Stecker et al.[20] for objects at redshifts of $z = 1.0$ and 0.03, the distance of Markarian 421. For a $z = 1.0$ object the transmission at 0.5 TeV is less than 1%, whereas for a $z = 0.03$ object the transmission is 100%.

It is anticipated that with a larger database of observations on this and other AGN's that firmer conclusions on the TeV variability, luminosity, and intergalactic absorption from these objects may be obtained.

We thank Kevin Harris and Teresa Lappin for help in obtaining these observations. We acknowledge support from the US Department of Energy, NASA, the Smithsonian Scholarly Studies Research Fund, and EOLAS, the scientific funding agency of Ireland.

References

1. Fichtel, C. E. et al. *Astron. Ap.* in press (1992).
2. Fennell, S. et al. this symposium.
3. Ulrich, M.-H. et al. *Ap. J.* **198**, 261-266 (1975).
4. Maza, J., Martin, P. G. & Angel, J. R. P. *Ap. J.* **224**, 368-374 (1978).
5. Owen, F. N., Porcas, R. W., Mufson, S. L. & Moffett, T. J. *Astron. J.* **83**, 685-696 (1978).
6. Zhang, F. J. & Bååth, L. B. *Astro. Ap.* **236**, 47-52 (1990).
7. Mufson, S. L. et al. *Ap. J.* **354**, 116-123 (1990).
8. Mushotzky, R. F., Boldt, E. A., Holt, S. S. & Serlemitsos, P. J. *Ap. J. (Letters)* **232**, L17-L19 (1979).
9. George, I.M., Warwick, R.S. & Bromage, G.E. *M. N. R. A. S.* **232**, 793-808 (1988).
10. Giommi, P. et al. *Ap. J.* **356**, 432-455 (1990).
11. Michelson, P. F. et al. IAU Circular 5470, 1 (1992).
12. Cawley, M. F. et al. *Exper. Astr.* **1**, 173-193 (1990).
13. Punch, M. et al. in Proc. 22nd International Cosmic Ray Conf. (Dublin) **1**, 464-467 (1991).
14. Weekes, T. C. et al. *Ap. J.* **342**, 379-395 (1989).
15. Vacanti, G. et al. *Ap. J.* **377**, 467-479 (1991).
16. Hillas, A. M. in Proc. 19th International Cosmic Ray Conf. (La Jolla) **3**, 445-448 (1985).
17. Macomb, D. J. & Lamb, R. C. in Proc. 21st International Cosmic Ray Conf. (Adelaide) **2**, 435-438 (1990).
18. Akerlof, C. W. et al. *Ap. J. (Letters)* **377**, L97-L100 (1991).
19. Punch, M. et al. *Nature* **358**, 477-478 (1992).
20. Lamb, R. C. in Proc. Palaiseau Workshop, in press (1992).
21. Stecker, F. W., De Jager, O. C., & Salamon, M. H. *Ap. J. (Letters)* **390**, L49-L52 (1992).

OSSE OBSERVATIONS OF NGC 4151

Michael Maisack
George Mason University, Fairfax, VA
Astronomisches Institut Tübingen, Germany

W.N. Johnson, R.L. Kinzer, M.S. Strickman and J.D. Kurfess
E.O. Hulburt Center for Space Research, Naval Research Lab, Washington, DC

G.V. Jung
Universities Space Research Association, Washington, DC

D.A. Grabelsky, W.R. Purcell and M.P. Ulmer
Northwestern University, Evanston, IL

ABSTRACT

We report results of a two-week observation of NGC 4151 with the OSSE instrument. The source had a very soft spectrum which falls off exponentially with an e-folding energy of 39.2±1.5 keV in the energy range 65-800 keV, and had an intensity at 100 keV of 2.33±0.05 × 10^{-2} photons cm^{-2} sec^{-1} MeV^{-1}. Simple pair cascade models in a compact source, which have been suggested for Seyfert galaxies, cannot explain the steepness and shape of the spectrum above 100 keV. We suggest pair loading and reacceleration as the mechanism responsible for the observed X-ray and gamma-ray spectra and the previously reported luminosity - spectral index correlation between 2 − 20 keV.

INTRODUCTION

NGC 4151, the archetype of Seyfert 1 galaxies, has been the object of various campaigns at all wavelengths. It is also one of the 4 Active Galactic Nuclei (AGN) detected significantly prior to GRO in the energy range 100 keV to 10 MeV[1]. By studying a sample of Seyfert 1 galaxies with HEAO-A4, it was found[2] that Seyfert galaxies have a canonical spectral shape at energies between 20 and 100 keV, described by a power law with photon index $\alpha = 1.7$. Previous observations of NGC 4151 have detected power law spectra with photon indices $\alpha \approx 1.7$, sometimes up to MeV energies[3]. The current observations represent the most sensitive observation of NGC 4151 to date in this energy range, and provide important new information.

OBSERVATIONS AND RESULTS

OSSE observed NGC 4151 from 1991 June 28 to July 12. The total on-source observation time for the sum of four detectors was 8.6 × 10^5 seconds. A similar amount of time was spent on background observations. A significant signal between 65 and 300 keV was observed. The results of spectral fitting are shown in Table 1. The errors given in this table and the remainder of the paper are the 68% confidence limits for joint variation of the respective parameters of interest. The intensity at 100 keV is taken from the exponential fit. It has a value of $2.33^{+0.05}_{-0.05} \times 10^{-2}$ photons cm^{-2} s^{-1} MeV^{-1}, which is comparable to previous observations. A very soft spectrum is observed which cannot be described by a single power law. A spectral fit with this function yields a photon index of 2.72 ± 0.07. The value of χ^2 per degree of freedom is 1.45. The probability for this model is 5×10^{-4}. A broken power law fit gives a photon index of $\alpha_1 = 2.10^{+0.32}_{-0.27}$ below 103^{+12}_{-9} keV and $\alpha_2 = 3.35^{+0.28}_{-0.38}$ above that energy (χ^2_ν=1.09). The spectrum steepens exponentially with an e-folding energy of 39 ± 1.5 keV. A satisfactory fit can also be achieved by a more physical model like thermal Comptonization[4]. This model gives a plasma temperature of 33^{+9}_{-6} keV and optical depth of $2.7^{+1.0}_{-0.6}$ (χ^2_ν=1.15). Figure 1 shows the Sunyaev-Titarchuk spectral fit to the OSSE data.

Table 1: Spectral Fit Summary

Model	χ^2	I^a	α	α_2	E_{Br}	kT	τ
Power Law	1.45	2.11	2.72				
Broken PL	1.09	2.38	$2.10^{+0.32}_{-0.27}$	$3.35^{+0.28}_{-0.38}$	103^{+12}_{-9}		
exponential	1.19	$2.33^{+0.05}_{-0.05}$				$39^{+1.5}_{-1.5}$	
Comptoniz.	1.15	2.25				37^{+10}_{-6}	$2.7^{+1.0}_{-0.7}$
Pair Plasma	1.30	1.29	2.37				2.7

[a] flux at 100 keV in 10^{-2} photons cm^{-2} s^{-1} MeV^{-1}
[b] fixed parameter

The OSSE spectrum shows NGC 4151 in an extremely soft state with a thermal character which cannot be described by a single power law, contrary to previous, less sensitive observations. A spectrum as steep as the one observed here has been observed only once before: GRANAT recently found a hard spectrum from 3 − 30 keV[5] ($\alpha = 1.44 \pm 0.03$) together with a steeper spectrum from 50-300 keV[6] ($\alpha = 3.1^{+1.1}_{-0.9}$). In contrast, hard, non-thermal power law spectra ($\alpha \approx 1.5$) at energies above 100 keV have been reported for NGC 4151 on several occasions, sometimes extending to energies of several MeV[3]. More observations of NGC 4151 during the mission lifetime of GRO will show whether there are two distinct spectral states and which one is prevalent.

DISCUSSION

Previous observations have shown that the X-ray emission from NGC 4151 follows a luminosity-spectral index correlation in which the spectrum becomes steeper as the source brightens[7]. The spectral indices in the 2-20 keV band vary between 1.3 and 1.7. A model for the gamma-ray emission must also explain this correlation. The most successful models so far include pair cascades[7,8]. In such pair models, the primary X-ray spectrum is generated by Comptonization of soft photons from the "blue bump" by relativistic electrons. The photon density above 511 keV can be so high that electron-positron pair production by photon-photon collisions reprocesses a significant portion of the primary radiation. The effect of pair cascades is a steepening of the X-ray spectrum with increasing compactness. It reflects the depletion of the gamma ray spectrum above 511 keV by pair production and hence the spectral distribution of pairs, which by their emission contribute increasingly to the X-ray spectrum. A measure for the importance of this effect is the compactness parameter l defined as[9]

$$l = \frac{L_X \sigma_T}{R m_e c^3} \qquad (1)$$

where L_X is the X-ray luminosity, σ_T the Thomson cross section, R the source radius, m_e the electron rest mass and c the speed of light. The seed photon compactness l_s and the injected electron compactness l_e are defined accordingly. For NGC 4151, l_e is estimated to be in the range $1 - 10^{10}$.

The "canonical" index of $\alpha = 1.7$ is only compatible with the OSSE data up to an energy of 100 keV. Above 100 keV, the spectral slope derived from the broken power law fit is $\alpha = 3.35^{+0.28}_{-0.38}$. Such a steep high-energy spectrum cannot be explained by simple pair cascade models, in which a steepening due to Compton downscattering in a plasma of cooled pairs produces a spectrum which steepens by $\Delta \alpha = 1$ above an energy of $511/\tau^2$ keV, where τ is the optical depth of the pair plasma[11]. Fitting this spectral shape gives an unacceptable chi-square ($\chi^2_\nu = 1.30$ for 130 degrees of freedom).

It has been suggested[12] that a modification of pair models can reproduce a luminosity-spectral index correlation together with a steep spectrum as observed by OSSE. In this model, a fraction of the injected particles (electrons and positrons, collectively referred to as electrons) are continuously reaccelerated to high energies after they have radiated their primary energy. These reaccelerated particles then form a substantial part of the relativistic particle population. At the same time this reacceleration mechanism can only supply a limited amount of acceleration power. The number of particles eligible for acceleration therefore limits the average energy available per particle, i.e., the mean γ factor of accelerated electrons decreases when the number of accelerated particles increases once the accelerator operates at its maximum power output. Done, Ghisellini and Fabian[12] found that for a source in which reacceleration occurs the maximum γ factor can decrease to several tens (assuming γ factors of several 1000 for freshly injected particles) as the electron compactness increases to values of $l_e = 10\text{-}100$, if two conditions are met: The source must be photon starved and escape of particles from the source region is possible. Photon starvation occurs when there are more energetic particles than target photons, so that second-order Comptonization becomes important. This can result in spectra with a harder index than $\alpha = 1.5$. In a photon-starved source, the pair yield is higher than in the non-starved case. It thus provides more particles to load the accelerator and enhances the reduction in γ. Escape of particles from the source region is required to explain the behavior of NGC 4151. Escape suppresses the effect of loading at low compactness since it removes particles from the accelerator and makes injection more important, i.e., with escape, γ_{max} is larger for small l_e than in the purely photon-starved case. The low γ factors can explain why the spectrum does not extend to energies above 511 keV, since a first-order Comptonized spectrum extends[13] to energies of $\frac{4}{3}\gamma^2\epsilon$ where ϵ is the energy of the soft target photons in units of the electron rest mass energy. Assuming $\epsilon = 10^{-4}$ for the energy of the target photons (assuming a "blue bump" origin) and inserting the Sunyaev-Comptonization fit temperature of the OSSE data of 30-40 keV gives a maximum γ factor of $\approx$ 25-30. This suggests that l_e during the OSSE observation is on the order of several tens.

A source in which reacceleration occurs under the above assumptions therefore provides the observed index-luminosity correlation in the following way: at low l_e, the spectrum is hard ($\alpha < 1.5$) and may extend beyond 511 keV, while at the same time pair reprocessing is unimportant because of the low compactness. When the compactness l_e (i.e., number of injected particles) increases, the average electron energy decreases as long as l_e is on the order of several tens (see Fig.2b in Ref. 12). With more electrons being accelerated to lower energies, their energy spectrum becomes progressively softer as the importance of loading increases. Above a compactness of 100, injection dominates reacceleration as the main supplying mechanism of high-energy particles and the effect of loading diminishes.

SUMMARY

OSSE has observed NGC 4151 to have a spectrum with a thermal character which steepens exponentially with an e-folding energy of 39 ± 2 keV. Simple pair models cannot explain the observed behavior. An acceleration process which acts on both freshly injected and cooled particles and has a limited power output can explain the observations if the source is photon starved and if escape of particles from the source region is possible. The soft spectrum in this model reflects the lower particle energy spectrum, which is caused by the loading of the acceleration mechanism. Repeated observations of this object and Seyfert galaxies in general with the high sensitivity provided by OSSE are required to determine whether this spectral shape is common or even dominant in this class of source.

Acknowledgments: This work was supported under NASA grant DPR S-10987C.

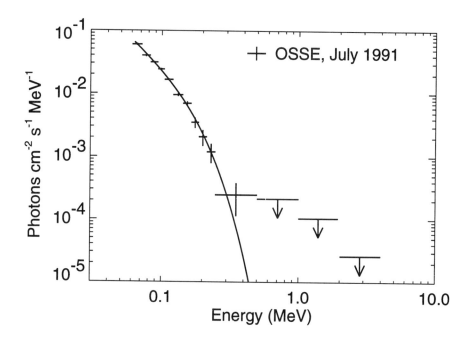

Figure 1: OSSE spectrum with thermal Comptonisation fit

REFERENCES

1. Bassani, L. and Dean, A. J., 1983, Space Science Reviews, 35, 367
2. Rothschild, R. E. et al., 1983, ApJ, 269, 423
3. Perotti, F. et al., 1990, ApJ, 356, 467
4. Sunyaev, R. A. and Titarchuk, L. G., 1980, A&A, 86,121
5. Apal'kov, Y. et al., A&A, in press
6. Jourdain, E. et al., A&A, in press
7. Yaqoob, T. and Warwick, R. S., 1991, MNRAS, 248,773
8. Lightman, A. P. and Zdziarski, A. A., 1987, ApJ, 319, 643
9. Guilbert, P. W., Fabian, A. C. and Rees, M. J., 1983, MNRAS, 233, 475
10. Maisack, M. and Yaqoob, T., 1991, A&A, 249, 25
11. Ghisellini, G., 1989, MNRAS, 238, 449
12. Done, C., Ghisellini, G. and Fabian, A. C., 1990, MNRAS, 245,1
13. Svensson, R., 1987, MNRAS, 227, 403

OBSERVATION OF THE STARBURST GALAXY NGC 253 WITH THE OSSE INSTRUMENT

Dipen Bhattacharya* and Neil Gehrels
Code 661, NASA/Goddard Space Flight Center, Greenbelt, MD 20771

J. D. Kurfess, W. N. Johnson, R. L. Kinzer, and M. S. Strickman
E. O. Hulbert Center for Space Research
Naval Research Laboratory, Mail Code 4150, Washington DC 20375-5320

Lih-Sin The
Dept. of Physics & Astronomy, Clemson University, Clemson, SC 29634-1911

G. V. Jung
Universities Space Research Association, Washington DC

D. A. Grabelsky, W. R. Purcell, and M. P. Ulmer
Dept. of Physics & Astronomy, Northwestern University, Evanston, IL 60208

ABSTRACT

Gamma-ray observations of the nearby starburst galaxy NGC 253 over the energy range 0.06–10 MeV have been obtained with the OSSE spectrometer. The source was detected up to 200 keV with a total significance of 4.2σ. When attributed to NGC 253 this corresponds to an estimated luminosity of 3×10^{40} ergs s^{-1}. The spectrum is fit by a power law of photon index ~ 2.5. A search for ^{56}Ni $\rightarrow$ ^{56}Co $\rightarrow$ ^{56}Fe supernova gamma-ray lines yielded no significant detection: the 3σ upper limits at 0.158, 0.847 and 1.238 MeV are 4×10^{-5}, 8×10^{-5} and 9×10^{-5} ph cm^{-2} s^{-1}, respectively. We find that inverse Compton scattering is insufficient to explain the observed continuum radiation. Bremsstrahlung and discrete sources may account for the flux. We also consider the possibility that the detected emission may result from low energy continuum from scattered gamma-ray lines produced by a very recent Type Ia or Ib supernova outburst in NGC 253.

INTRODUCTION

NGC 253 is the third brightest infrared galaxy with a luminosity of ~ 4×10^{10} $L_\odot$[1] in the far infrared band. This nearby (~ 3 Mpc) spiral Sc galaxy is undergoing extensive star formation within its central few kilo-parsecs where it is also very bright in X-rays and radio. The *Einstein* IPC images have revealed a plume of X-ray emission extending above the galactic plane[2] possibly bearing the evidence of nuclear outflows similar in nature to M82. The outflowing gas is believed to be heated by the energy released during supernova explosions[3]. The high rate of star formation in the starburst nuclei is biased towards producing massive stars. These stars, being luminous and massive (3–100$M_\odot$), have very short lifespans with most of them resulting in supernovae: massive C/O Wolf-Rayet stars undergo strong mass-loss and explode as Type Ib SN, whereas O and B stars generally evolve into Type II SN[4]. Supernova rates from starburst galaxies have been estimated from radio observations of M82[5] and NGC 253[6] and are in the range (0.1–0.3) supernova/yr within the inner 600 pc for those galaxies. However, more recently, Ulvestad and Antonucci suggest this rate may be too

---

*Resident Research Associate, Universities Space Research Association

optimistic[7]. A Type Ia or Ib supernova in the NGC 253 core may produce gamma-ray lines detectable to the OSSE instrument. In this paper, we present the results of the first gamma-ray observation of NGC 253 by the *Compton Gamma-Ray Observatory's (CGRO) Oriented Scintillation Spectrometer Experiment (OSSE)*.

OBSERVATIONS AND RESULTS

a) Observations: The OSSE instrument consists of four actively shielded NaI(Tl)-CsI(Na) phoswich detectors. Each of the detectors has a 3.8° × 11.4° (FWHM) field-of-view. The instrument observes each individual source in the "source–background" method over the energy range 0.06–10 MeV. A detailed description of the instrument can be found in Johnson et al.[8]. NGC 253 was observed with the OSSE in 5 separate viewing periods for a total observation time of ~ 10^6 s. The last two observations were carried out with a different scan mode and consequently had little source exposure. The observation schedule of the remaining 3 viewing periods are shown in Table 1 where col. 1 denotes the GRO viewing period (VP) number, col. 2 shows the dates of observation, col. 3 gives exposure times in seconds and col. 4 the total number of detectors used in the observation.

TABLE 1

Viewing Period	Date	Exposure (sec)	Number of Det.
9	9/5/91 16:13 – 9/11/91 23:34	3.31×10^5	4
13.5	11/7/91 18:20 – 11/14/91 17:03	9.76×10^4	4
16	12/14/91 01:40 – 12/28/91 18:06	2.71×10^5	2

NGC 253 is located very near to the galactic south pole (l = 24°.20 and b = –87°.86). This position was mostly observed as a secondary target whenever the primary target was the Galactic Center. The accumulation schedule for each of the four detectors alternated between source observations and background measurements. The background fields are checked for potential gamma-ray sources using the HEAO A2 all sky map[9]; no bright sources were found. A quadratic background estimation is performed by interpolating between two background observations accumulated before and after the source observation[10]. Background subtraction for each detector is carried out and the individual difference spectra for each detector is summed. A count spectrum, obtained by folding model photon spectrum through the instrument response matrices, is compared with the summed data and the model parameters are modified to achieve the best fit.

b) Results: We have used simple power law models to obtain the continuum photon fluxes. The emission spectrum of the full data set is fit by a power law with a photon index of 2.55 ± 0.79 (χ^2 = 415.07, ndof=445). In Table 2, the measured fluxes obtained during different viewing periods are shown. The significance of the detection is 4.2 σ. The flux over the 60–200 keV band is 3×10^{-11} ergs cm^{-2} s^{-1}. The derived photon spectrum is shown in Figure 1. At X-ray energies, fluxes measured by Ginga over the energy range 2–10 keV are also given. Ohashi et al.[11] find that both thermal bremsstrahlung and power law models give acceptable fits to Ginga results, although they prefer a thermal spectrum with kT ~ 6 keV because the derived N_H for power law model is much higher than the Galactic value and contradicts the *Einstein* IPC results. We estimate the Ginga photon fluxes from their power law spectral parameters and show that even a power law extrapolation of Ginga data into the OSSE regime yields fluxes an order of magnitude below the OSSE data points. Expected gamma-ray spectra for a Type Ia and a Type Ib supernova near their peak hard X-ray luminosity (see *Discussion* for details), scaled to the distance of NGC 253, are also shown.

TABLE 2

Energy Bins	VP 9 ph cm^{-2} s^{-1} MeV^{-1}	VP 13.5 + VP 16 ph cm^{-2} s^{-1} MeV^{-1}	Total ph cm^{-2} s^{-1} MeV^{-1}
60 – 72 keV	$(3.00 \pm 3.82) \times 10^{-3}$	$(3.44 \pm 3.81) \times 10^{-3}$	$(3.28 \pm 2.75) \times 10^{-3}$
72 – 88 keV	$(2.10 \pm 1.94) \times 10^{-3}$	$(1.95 \pm 1.92) \times 10^{-3}$	$(2.06 \pm 1.39) \times 10^{-3}$
88 – 110 keV	$(0.92 \pm 0.87) \times 10^{-3}$	$(2.60 \pm 0.85) \times 10^{-3}$	$(1.82 \pm 0.62) \times 10^{-3}$
110 – 133 keV	$(0.48 \pm 0.70) \times 10^{-3}$	$(1.02 \pm 0.67) \times 10^{-3}$	$(0.77 \pm 0.49) \times 10^{-3}$
133 – 166 keV	$(0.26 \pm 0.67) \times 10^{-3}$	$(1.11 \pm 0.65) \times 10^{-3}$	$(0.71 \pm 0.47) \times 10^{-3}$

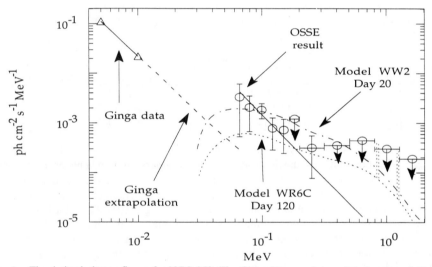

Fig. 1 – The derived photon fluxes for NGC 253. The Ginga X-ray points and their extrapolation to higher energies are also shown. The continuum emissions from a Type Ia (Model WW2) and a Type Ib SN (Model WR6C) are given. The 0.847 and 1.238 MeV lines of WW2 are not shown for clarity.

A line search was carried out at energies corresponding to lines from ^{56}Ni → ^{56}Co → ^{56}Fe decay at 0.158, 0.847 and 1.238 MeV. The line widths are taken as 3.5% of the rest energy following the calculation of Chan and Lingenfelter[12] for a Type Ia supernova. The search failed to show any significant flux. The 3σ upper limits on the flux for the 0.158, 0.847 and 1.238 MeV lines are 4.3×10^{-5}, 8×10^{-5} and 9×10^{-5} photons cm^{-2} s^{-1}, respectively.

DISCUSSION

The origin of diffuse hard X-rays or low energy gamma-rays in a starburst galaxy would include inverse Compton effects, electron bremsstrahlung or thermal emission from hot supernova gases. Discrete sources such as X-ray binaries or individual supernova outbursts may also substantially contribute to the emission. The emission from young, massive Population I binary X-ray sources (with $kT > 10$ keV) can be a component of the hard X-ray spectrum, but the number of such sources needed to explain the entire OSSE flux would be in the range of 10^3–10^4.

a) Inverse Compton Emission: The contribution of inverse Compton scattering of relativistic electrons off the far infrared photon field to the total X-ray emission in starburst galaxies has been considered with respect to M82[13,14]. The number density of the relativistic electrons can be assessed from the radio observations. The emission in the radio range is assumed to be primarily of synchrotron origin and coming from the same set of relativistic electrons responsible for the inverse Compton emission. The FIR photon density can be calculated from the total FIR luminosity and the spatial extent of the FIR emission. Our estimated IC luminosities for NGC 253, in the range 0.06–0.2 MeV, for magnetic field values of 1 and 5 μG, are 6×10^{39} and 4×10^{38} erg s^{-1}, respectively. The observed luminosity over the same energy range is ~ 3×10^{40} ergs s^{-1}. Even given the uncertainties in such calculations, we find that inverse Compton scattering accounts for only a small fraction of the emission seen.

b) Electron bremsstrahlung: At 100 keV, the bremsstrahlung contribution could be at least a factor of 2 higher than inverse Compton scattering in our galaxy[15]. Hence, it is not unlikely that the bremsstrahlung contribution could be large in NGC 253, but a proper estimation would require a better understanding of the cosmic ray and nucleon density in NGC 253 core.

c) Supernova continuum: We consider the possibility that the detected continuum emission arises from a Type Ia or Ib SN outburst in NGC 253 but with lines obscured by overlying material. The continuum emission is due to photons in the gamma-ray lines of the radioactive decay chain of ^{56}Ni which are degraded in energy through Compton down-scattering in the expanding envelope. We find that the Monte-Carlo calculations of Burrows and The[16] for a Type Ia SN (model WW2 of Woosley and Weaver[17]), when scaled to the distance of NGC 253, predict continuum flux values at 100 keV (at day 20–100) at the same level as the detected flux from NGC 253. However, the peak 847 keV flux for a Type Ia SN (WW2 model) occurs at day ~ 70 and is 3×10^{-4} ph cm^{-2} s^{-1}. The absence of this line in the observations makes it unlikely that the continuum emission is due to a Type Ia SN. Similar calculations (Model WR6C of The, Clayton and Burrows[18]) carried out for a Type Ib Wolf-Rayet SN (Ensman and Woosley[19]) yield peak continuum flux values (at day 80–150) at factors of 4–5 lower than what have been observed. Given the uncertainties involved in the estimation of the diffuse emission, discrete source contribution, parameters of the SN model and distance to NGC 253, we cannot rule out that a Type Ib explosion could account for much of the observed continuum flux. Furthermore, the peak flux in the 847 keV line for WR6C model ranges from 3.4×10^{-6} to 5×10^{-5} ph cm^{-2} s^{-1} and would not have been detected by the OSSE. The models WW2 and WR6C are used only for illustration purposes; the detailed structure of the spectrum depends on the initial mass, expansion velocity and Compton opacities of the supernova. The marginal variability seen in the 88–166 keV band over a period of 60 days (see Table 2) might be explained by the evolving Compton opacity of the supernova ejecta, although the low significance of the detection makes it difficult to choose a specific model to explain the observation.

A Type Ib event is more likely to occur than a Type Ia in a starburst core. The starburst phenomenon produces more massive stars, including Wolf-Rayet stars, which can undergo strong mass loss and explode as Type Ib SNe. What is the plausibility of

Type Ia supernovae in the starburst core? The conventional belief is that the progenitors of Type Ia's are old (~10^{10} yrs) and have low initial mass. But Oemler and Tinsley[20], using a sample of 178 supernovae in external galaxies, find that Type Ia SNe occur more frequently in spiral than in elliptical galaxies and the SNe rate is proportional to the current star formation rate. Although these authors do not distinguish between Type Ia and Ib SNe, Filippenko[21] argues that their sample contains relatively few Type Ib SNe and is representative of Type Ia. In star forming regions, possible evolution of massive stars (> $8M_\odot$) in close binaries into Type Ia SNe, within a short time scale (~ a few times 10^7 yrs), is not ruled out[22,23]. But the short timescale will not allow such systems to move very far from their birthplaces (HII regions) before explosion and as Type Ia SNe are not known to occur near HII regions[21] it would seem to exclude massive stars as progenitors of Type Ia SNe. Whether that is indeed the case, needs further observational verification. In this respect, gamma-ray observations, with the potential to detect supernovae signatures through the dense starburst nuclear regions, serve as a powerful diagnostic tool. The possibility, that we already have detected an extragalactic supernova signature, is extremely intriguing. If, indeed, the observed flux is due to a supernova outburst, future gamma-ray observations may not yield any detectable flux. Without a supernova explosion it is hard to envision rapid spectral or luminosity variability between observations for an otherwise normal galaxy such as NGC 253.

It is a pleasure to acknowledge the assistance of Chris Starr. We would like to thank Don Clayton and Mark Leising for many useful comments on the gamma-ray yield of supernovae. We thank the whole OSSE team for making the data available and providing supporting analysis tools. The work is supported under a OSSE Phase 1 Guest Investigation Program.

References
1. Soifer B. T. et al. 1987, ApJ, 320, 238.
2. Fabbiano, G. 1988, ApJ, 330, 672.
3. Kroenberg, P. P., Biermann, P., and Schwab, F. R. 1985, ApJ, 291, 693.
4. Melnick, J. and Terlevich, R. 1988, in High Energy Astrophysics, Ed. G. Borner, Berlin Springer-Verlag, 155.
5. Bartel et al. 1987, ApJ, 323, 505.
6. Antonucci, R. R. J. and Ulvestad, J. S. 1988, ApJ, 330, L97.
7. Ulvestad, J. S. and Antonucci, R. R. J. 1992, submitted to ApJ.
8. Johnson, W. N. et al. 1993, to appear in ApJ Suppl.
9. Jahoda, K. and Mushotzky, R. 1992, private communication.
10. Purcell, W. A. et al. 1991, in The Compton Observatory Science Workshop, ed. C. R. Shrader, N. Gehrels, and B. Dennis (Annapolis, NASA), 8.
11. Ohashi, T. et al. 1990, ApJ, 365, 180.
12. Chan, K. W. and Lingenfelter, R. E. 1991, ApJ, 368, 515.
13. Schaaf, R. et al. 1989, ApJ, 336, 722.
14. Seaquist, E. R. and Odegard, N. 1991, ApJ, 369, 320.
15. Skibo, J. G. and Ramaty, R. 1992, AA Suppl. Ser., 1, 1.
16. Burrows, A. and The, L.-S. 1990, ApJ, 360, 626.
17. Woosley, S. E. and Weaver, T. A. 1986, in Radiation Hydrodynamics in Stars and Compact Objects, ed. D. Mihalas and K.-H. A. Winkler (NY, Springer-Verlag), 91.
18. The, L.-S., Clayton, D. D., and Burrows, A. 1991, IAU Symp., in Wolf-Rayet Stars in Galaxies, ed. K. A. van der Hucht and B. Hidayat (Dordrecht, Kluwer), 537.
19. Ensman, L. and Woosley, S. E. 1988, ApJ, 333, 754.
20. Oemler, A., Jr. and Tinsley, B. M 1979, AJ, 84, 7.
21. Filippenko, A. V. 1989, in Supernovae and Stellar Evolution, ed. A. Ray and T. Velusamy, World Scientific, 58.
22. Nomoto, K. 1984, ApJ, 277, 791.
23. Iben, I., Jr. and Tutukov, A. 1984, ApJ, 54, 335.

OSSE Observations of Starburst Galaxy M82

Lih-Sin The, Donald D. Clayton, Mark D. Leising
Department of Physics & Astronomy, Clemson University, Clemson, SC 29634-1911

J.D. Kurfess, W.N. Johnson, R.L. Kinzer, M.S. Strickman
E.O. Hulburt Center for Space Research
Naval Research Laboratory, Mail Code 4150, Washington, DC 20375-5320

G.V. Jung
Universities Space Research Associates
Washington, DC 20375-5320

D.A. Grabelsky, W.R. Purcell, M.P. Ulmer
Department of Physics & Astronomy, Northwestern University, Evanston, IL 60208

ABSTRACT

OSSE observed the starburst galaxy M82 in two viewing periods of 8 and 14 days. M82's priority as a target had been established on the grounds that the average supernova rate may be very high there, so that a significant chance of ^{56}Co detection exists. If M82 is at 3.4 Mpc distance, normal Type II (e.g. SN1987A) are too dim in ^{56}Co lines, but the Wolf-Rayet derived Type Ib, which are also massive-star core implosion objects, might be detectable to OSSE.[1,2] Expected fluxes of the 847 keV γ-line of ^{56}Co would be near $(2\text{-}5)\times 10^{-5}$ γ cm^{-2} s^{-1}. A Type Ia in M82 would be very bright,[3] near $(3\text{-}6)\times 10^{-4}$ γ cm^{-2} s^{-1}. We present OSSE background subtracted spectra of the M82 region for viewing periods 7 and 18. These spectra show no significant excess at 847 keV or at 1238 keV, the two strongest ^{56}Co γ-lines. When we fit a smooth gamma continuum plus a feature having both the 847 keV and 1238 keV lines to the OSSE data, we obtain for the 847 keV line amplitude the values $(-4.18 \pm 3.45) \times 10^{-5}$ γ cm^{-2} s^{-1} for viewing period 7 and $(0.84 \pm 2.85) \times 10^{-5}$ γ cm^{-2} s^{-1} for viewing period 18. We discuss the implications on the supernova rate in M82.

INTRODUCTION

Evidence of a very high star formation rate in the starburst galaxy M82 is quite extensive. Its supernova rate[4,5] may be ~ 0.1 yr^{-1}, as suggested by Kronberg and Wilkinson.[6] Rieke et al.[4] analyzed IRAS infrared data to conclude that very high rates of star formation are occurring in M82. Moreover, this high rate of formation of O and B stars is also required to account for the radio luminosity of M82.[7] Detailed analysis of radio data has been interpreted as requiring the burst of star formation in M82 to produce only massive stars.[8] This implication has been revived recently by evolutionary models of starburst activity in M82.[9] The initial mass function of M82 requires stars to have mass ≥ 3 M$_\odot$ in order that the models can simultaneously match the observed M82 supernova rate and the dynamical mass.[4] Stars more massive than ~ 9 M$_\odot$ are believed to be the progenitors of Type II or Type Ib supernovae. Therefore these observations suggest that M82 probably has a high rate of core-collapse supernovae.

© 1993 American Institute of Physics

Clayton, Colgate, and Fishman[10] described an observation that could confirm such speculations. Radioactive ^{56}Ni is produced in the supernova event, and gamma radiation by its daughter nucleus ^{56}Co can be sought. This prediction has now been confirmed by the detection of ^{56}Co γ-lines from supernova 1987A (e.g., Leising and Share[11] and references therein). Gamma-ray observations have become a promising diagnostic tool of supernova structure. Gamma-ray line supernova are not obscured by dust, which is the condition in the inner nucleus of M82. The γ-line and x-ray fluxes of massive star supernovae have been calculated by many investigators.[12-17] Ensman and Woosley[18] have constructed detailed evolutionary models of Wolf-Rayet supernova explosion and find that only 4-6$M_\odot$ models produce acceptable fits to light curves of Type Ib supernovae. The γ-line fluxes from these models peak between 3.1×10^{-5} to 4.4×10^{-4} γ cm^{-2} sec^{-1} at distance of 1 Mpc.[1] Nomoto, Kumagai, and Shigeyama[2] in their calculations of helium star explosions[19] find maximum line fluxes to be 3.9×10^{-5} to 8.8×10^{-4} γ cm^{-2} sec^{-1} at distance of 1 Mpc. This indicates that OSSE has a potential range out to 4 Mpc for detecting Type Ib supernova. For Type Ia supernova the OSSE range is $\sim$ 10 Mpc.

OBSERVATIONS

The Oriented Scintillation Spectrometer Experiment (OSSE), which is sensitive to gamma rays in the range of 0.05 to 10 MeV, is designed to study nucleosynthesis by detecting γ-ray lines from radioactive elements. The characteristics and performance of the OSSE instrument have been described by Johnson et al.[20] OSSE observed the starburst galaxy M82 (NGC 3034) during its viewing periods 7 (8-15 August 1991) and 18 (10-23 January 1992). M82 is located at $\alpha=9^h56^m$ and $\delta=+69°41'$. The pointing strategy follows the simple OSSE pointing strategy shown in Figure 3 of Purcell et al.[21] with accumulation time for every pointing being $\sim$ 130 seconds. The total live time of four detectors for period 7 is 7.834×10^5 seconds and for period 18 is 1.651×10^6 seconds.

DATA ANALYSIS

The OSSE spectral analysis technique subtracts the background offset pointing of the detectors from the source spectrum.[21] The data analysis process presented here is similar to the one that has been reported in OSSE detection of ^{57}Co in SN1987A.[22] For each day of observation, initial data selection is performed to exclude poor data due to telemetry, environment, or positioning errors. The quadratically fitted background estimates are subtracted from the source spectrum to obtain a difference spectrum for each 2-minute integration. The sum of all spectra from four detectors over an entire observing period is obtained. See Figure 1 for period 7 (1 week) and period 18 (2 weeks) data from M82. We fit the spectrum of Figure 1 to power law, linear, or exponential continuum photon spectra plus a γ-line at energy 847 keV or 1238 keV. These photon spectra are folded through the OSSE instrument response and the resulting count spectra are least-squares fitted to the observed count spectra. Linear continuum photon spectra give the best fit for both observing periods for the count spectra. The results of a linear continuum plus a γ-line fitting are presented in Table 1 along with χ_ν^2, the χ^2 per degree of freedom (d.o.f), for the best-fit γ-line fluxes at 0.847 and 1.238 MeV.

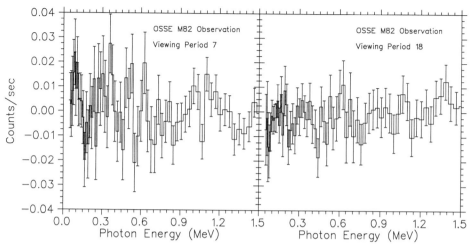

Figure 1: *Average count spectra of OSSE M82 observations for viewing periods 7 and 18. The spectrum is the average counts from four OSSE detectors of 8-day (VP 7) and 14-day (VP 18) observations.*

TABLE 1
RESULTS OF TWO M82 OBSERVATIONS BY OSSE

Period No.	847 keV γ-line Flux[a]		1238 keV γ-line Flux[a]		847 keV Combined Fit Flux[b]			
					$r_{1238/847}=0.68$		$r_{1238/847}=1.0$	
	χ^2_ν	d.o.f.	χ^2_ν	d.o.f.	χ^2_ν	d.o.f.	χ^2_ν	d.o.f.
7	-6.11 ± 3.77		+0.53 ± 4.32		-4.18 ± 3.45		-2.99 ± 3.03	
	0.86	23	1.00	19	1.01	39	1.02	39
18	+2.22 ± 3.43		-1.28 ± 3.61		+0.84 ± 2.85		+0.51 ± 2.53	
	0.68	23	0.56	19	0.63	39	0.63	39

[a] *When fitting the 847 (1238) keV γ-ray line, the linear continuum is fit from 0.6-1.2 (0.9-1.5) MeV; flux is in units of 10^{-5} γ cm^{-2} sec^{-1}.*
[b] *The 847 keV γ-line flux is obtained by fitting the 847 and 1238 keV γ-lines simultaneously with their ratio fixed to $r_{1238/847}$ together with a linear continuum from 0.5-1.5 MeV; flux is in units of 10^{-5} γ cm^{-2} sec^{-1}.*

Table 1 also shows the best fit of a linear continuum spectrum plus both γ-ray lines simultaneously for two values of the line flux ratio $r_{1238/847} = f_{1238}/f_{847}$. One flux ratio is chosen to be 0.68, the relative emission rate in ^{56}Co. We also fit the count spectra with the ratio fixed to 1.0, the approximate ratio of line fluxes at early times in the supernova. The results in Table 1 offer no evidence of the line features searched for. Fig. 2 we use the 3σ confidence limits for this simultaneous fit to test the probability of supernova recurrence rates. We also examined the eye-catching bump between 0.9-1.2 MeV of the count spectra of VP 7. The bump appears to be a statistical fluctuation. Surprisingly, it is dominated by the first day's counts, a curiosity for which we have no explanation.

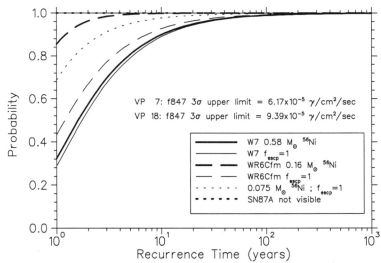

Figure 2: *Probability for the 847 keV γ-line fluxes from stochastically occurring Type Ia, Type Ib, and Type II supernovae to be less than both 3σ upper limits for the two M82 observations, which are separated by about 158 days.*

DISCUSSION

Monte Carlo simulations of randomly occurring supernova events in M82 were performed to evaluate the significance of our upper limit to the 0.847 and 1.238 MeV ^{56}Co γ-ray line fluxes on the supernova rate in M82. For this purpose we take the 3σ upper limit to mean the best-fit value plus three standard deviations from that value, even though this confidence limit may differ slightly from the expected 3σ sensitivity of OSSE to lines. The 3σ upper limit to the 0.847 MeV γ-line flux from both viewing periods (separated by 158 days) was compared with the line flux generated by the Monte Carlo histories. The fraction of galaxies in the simulations that would be fainter than both 3σ OSSE limits at 847 keV is shown in figure 2. It therefore represents the probability that that particular astrophysical simulation would not have been detected. Different types of supernova models are used in the simulation but are considered independently of each other. Model W7 is a Type Ia supernova deflagration model of carbon-oxygen white dwarf thermonuclear explosion.[23] WR6C is a Type Ib supernova model of Wolf-Rayet progenitor.[18] For Type II supernova we take the SN87A model that produces 0.075 M$_\odot$ ^{56}Ni.

Our simulations show that our observed upper limit to the γ-line fluxes does not give a strong constraint on the massive-star supernova rate of M82. For Type Ia SNe, even if the upper limit flux were as small as 1×10^{-6} γ cm^{-2} sec^{-1}, there would still be a 10% chance for the model flux to be less than that upper limit for a supernova recurrence time of 1 year. Therefore we conclude that γ-line flux upper limits from M82 cannot provide a strong constraint to the supernova rate of that galaxy unless the γ-ray detector is sensitive to fluxes of $\sim 1 \times 10^{-6}$ γ cm^{-2} sec^{-1} or better, or the recurrence time is much shorter than 1 year, hence much more frequent observations are needed. However, the search for γ-line supernova should continue, particularly in high-rate star formation dust-obscured galaxies such as M82 and NGC 253,[24] because a positive detection remains possible at that distance. Also, repeating the observations will somewhat improve the constraint on the recurrence rate of supernova.

L.-S.T would like to thank D. Bhattacharya and D. Hartmann for very useful discussions. This research was supported by NASA DPR S-10987C.

REFERENCES

1. The, L.-S., Clayton, D.D. & Burrows, A. 1991, in Wolf-Rayet Stars and Interrelationships with Other Massive Stars in Galaxies (Dordrecht: D. Reiedl) p.537
2. Nomoto, K., Kumagai, S., & Shigeyama, T. 1991, in Gamma-Ray Line Astrophysics, AIP Conf. Proc. 232, ed. N. Prantzos, & P. Durouchoux (New York, AIP), p.236
3. Burrows, A., & The, L.-S. 1990, ApJ, 360, 626
4. Rieke, G.H., Lebofsky, M.J., Thompson, R.I., Low, F.J., & Tokunaga, A.T. 1980, ApJ, 238, 24
5. Kronberg, P.P. & Sramek, R.A. 1985, Science, 227, 28
6. Kronberg, P.P. & Wilkinson, P.N. 1975, 200, 430
7. Kronberg, P.P., Biermann, P., & Schwab, F. 1981, 246, 751
8. Kronberg, P.P., Biermann, P., & Schwab, F. 1985, 291, 693
9. Doane, J.S., & Mathews, W.G. 1992, Lick Observatory Preprint
10. Clayton, D.D., Colgate, S.A., and Fishman, G. 1969, ApJ, 155, 75
11. Leising, M.D., & Share, G.H. 1990. ApJ 357, 638
12. Pinto, P.A., & Woosley, S.E. 1988. Nature, 333,534
13. Woosley, S.E., Pinto, P.A., & Hartmann, D. 1989. ApJ 346, 395
14. Nomoto, K., Shigeyama, T., Kumagai, S., & Hashimoto, M. 1988, Proc. Astr. Soc. Australia, 7, 490
15. Bussard, R.W., Burrows, A., & The, L.-S. 1989, ApJ, 341, 401
16. The, L.-S., Burrows, A., & Bussard, R.W. 1990, ApJ, 352, 731
17. Chan, K.W. & Lingenfelter, R.E. 1991, ApJ, 368, 515
18. Ensman, L., & Woosley, S.E. 1988, ApJ, 333, 754
19. Shigeyama, T., Nomoto, K., Tsujimoto, T., & Hashimoto, M. 1990, ApJ, 361, L23
20. Johnson, W.N. et al. 1993, to be published in ApJ Suppl., OSSE Preprint No. 6
21. Purcell, W.R. et al. 1991, in The Compton Observatory Science Workshop, ed. C.R. Shrader, N. Gehrels, & B. Dennis (Annapolis:NASA), 15
22. Kurfess, J.D. et al. 1992, ApJ, 399, L137
23. Nomoto, K., Thielemann, F.-K., and Yokoi, K. 1984, ApJ, 286, 644
24. Bhattacharya, D. et al. 1992, in this volume

SEARCH FOR TeV GAMMA-RAY EMISSION FROM AGN'S USING THE WHIPPLE IMAGING TELESCOPE

S. Fennell[1,2], C.W. Akerlof[3], M.F. Cawley[4], M. Chantell[2], D.J. Fegan[1], J.A. Gaidos[5], J. Hagan[1], A.M. Hillas[6], A.D. Kerrick[7], R.C. Lamb[7], M.A. Lawrence[2], D.A. Lewis[7], D.I. Meyer[3], G. Mohanty[7], K.S.O'Flaherty[1], P.T. Reynolds[1], A.C. Rovero[2], M.S. Schubnell[3], G. Sembroski[5], T.C. Weekes[2], T. Whitaker[2] & C. Wilson[5]

[1] Physics Dept., University College Dublin, Belfield, Dublin 4, Ireland
[2] Whipple Observatory, Harvard-Smithsonian CfA, Box 97, Amado, Arizona 85645, USA
[3] Physics Dept., University of Michigan, Ann Arbor, Michigan 48109, USA
[4] Physics Dept., St. Patrick's College, Maynooth, County Kildare, Ireland
[5] Physics Dept., Purdue Univ., West Lafayette, Indiana 47907, USA
[6] Physics Dept., University of Leeds, Leeds, LS2 9JT, UK
[7] Physics and Astronomy Dept., Iowa State University, Ames, Iowa 50011, USA

ABSTRACT

MeV-GeV emission has been reported by the EGRET team for a number of AGN's. We report the results of observations taken with the Whipple Gamma Ray imaging telescope during the spring of 1992, on QSO0836+710, PKS0528+134, 4C38.1, 3C273, 3C279, Mrk421, Mrk501, IZW187, and VRO42.22.01. The EGRET team also sees a short time-scale variation in luminosity for some of these objects. The results of a search for variability at TeV energies in these objects is reported.

DATA ANALYSIS

For all observations made on the AGN's the 10m telescope was operated in the standard ON/OFF tracking mode[1]. The Cherenkov light generated by both the gamma rays and hadronic particles is recorded, by an array of 109 photomultiplier tubes, positioned at the focal plane of the reflector.
In order to differentiate between gamma-ray induced showers and those generated by hadronic particles we apply a multi-parameter selection criteria to the shower images. Figure 1 illustrates the definition of the parameter as applied to a Cherenkov image, recorded by the telescope. The background events are rejected after we apply "Supercuts"[2], a predetermined selection criteria, to our parameterized data. One of the parameters we define, alpha, is a direct measure of the orientation of the shower. From Monte Carlo simulations we know that whilst background events will have a random distribution of orientation, the gamma-ray events from the source at the center of the field of view will be orientated towards the center of the camera to within 15°. By cutting events, which have previously passed the shape cut, at 15°

we can reject 99% of the background events.

OBSERVING PROGRAM

Many AGN's have been detected as MeV-GeV gamma ray emitters by the EGRET experiment. Since October, 1991 we have been looking at all the objects visible from the Whipple observatory in Arizona. Note that while MRK421 was not a very luminous object for the EGRET experiment, it is the only object visible at TeV energies to date. As this object has the lowest redshift (z=0.03) of the other EGRET AGN detections we expanded our search to other low redshift AGN's. The most suitable candidates for observation, in the spring, were the BL Lac objects IZW187, VRO42.22.01 and the galaxy MRK501, all of which have low redshifts. Table I summarizes our observations on these AGN's.

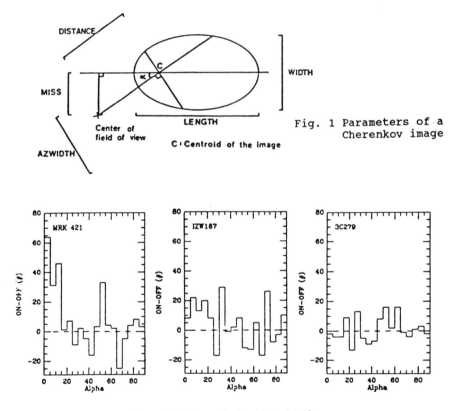

Fig. 1 Parameters of a Cherenkov image

Fig. 2 Alpha plots for AGN's

TABLE I
AGN Observation Summary

	RAW	SHAPE	ORIENTATION	COMBINED
MRK 421 (z=0.03)			Epoch: Mar - Jun '92	
ON	77355	1061	5976	307
OFF	76735	911	5757	168
Diff.	+620	+150	+219	+139
Sigma	+1.6	+3.4	+2.0	+6.4
IZW187 (z=0.055)			Epoch: Jun - Jul '92	
ON	159087	2309	12469	407
OFF	159225	2189	12081	362
Diff.	-138	+120	+388	+45
Sigma	-0.2	+1.8	+2.5	+1.6
MRK 501 (z=0.069)			Epoch: May - Jun '92	
ON	39689	427	2947	67
OFF	38883	377	2851	67
Diff.	+806	+50	+96	0
Sigma	+2.9	+1.8	+1.3	0
VRO42.22.01 (z=0.69)			Epoch: Jun - Jul '92	
ON	56511	786	4311	137
OFF	56312	736	4446	142
Diff.	+199	+50	-135	-5
Sigma	+0.6	+1.3	-1.4	-0.3
3C273 (z=0.158)			Epoch: Feb - Apr '92	
ON	13692	364	1188	63
OFF	13683	349	1160	71
Diff.	+9	+15	+28	-8
Sigma	+0.05	+0.6	+0.6	-0.7
3C279 (z=0.538)			Epoch: Dec '91 - Apr '92	
ON	94317	3917	8094	573
OFF	93652	3705	8174	562
Diff.	+665	+212	-80	+11
Sigma	+1.5	+2.5	-0.6	+0.3
4C38.41 (z=1.8)			Epoch: Apr '92	
ON	85246	834	6124	104
OFF	84786	806	6262	133
Diff.	+460	+28	-138	-29
Sigma	+1.1	+0.7	-1.2	-1.9
PKS0528+134 (z=1.9)			Epoch: Mar '92	
ON	9945	356	769	59
OFF	10198	443	837	75
Diff.	-253	-87	-68	-16
Sigma	-1.8	-3.1	-1.7	-1.4
QSO0836+710 (z=2.17)			Epoch: Mar '92	
ON	20019	1586	1846	260
OFF	20204	1476	1872	253
Diff.	-185	110	-26	7
Sigma	-0.9	+2.0	-0.4	+0.3

RESULTS

In the spring of 1992 we detected an excess of gamma-rays from the BL Lac object, MRK421[3]. This excess is clearly seen in figure 2 which shows the orientation of the ellipse (alpha) for both the on and off regions. On May 22nd 1992 we observed a threefold increase in the intensity of MRK421. This variability was not seen in any other AGN during the same viewing period (figure 3). For the rest of the AGN's we derive upper limits (Table II). The increase in threshold energy for 3C279 is attributed to the low elevation at which the source transitted.

TABLE II
UPPER LIMITS FOR AGN's

Source	Energy (TeV)	Flux $\times 10^{-11}$ (cm^{-2}s^{-1})	"Crabs"
MRK421	0.5	1.5 (Flux)	0.30
IZW187	0.5	< 4.4	< 0.09
MRK501	0.5	< 0.71	< 0.14
VRO42:22:01	0.5	< 0.73	< 0.15
3C273	0.55	< 2.0	< 0.44
3C279	0.6	< 0.76	< 0.19
4C38.41	0.5	< 0.45	< 0.09
PKS0528+134	0.5	< 3.0	< 0.60
QSO0836+710	0.6	< 2.3	< 0.57

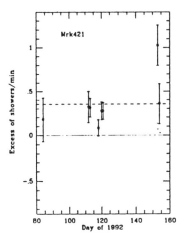

Fig. 3 Time variability of excesses from Mrk 421

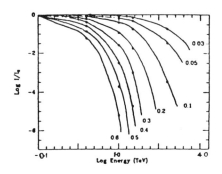

Fig. 4 Attenuation of gamma rays as a function of energy plotted for various values of redshift

FUTURE PLANS

According to a recent paper[4] the attenuation of TeV gamma-rays, by the extragalactic infrared field, may be a significant factor in the Whipple Collaboration's failure to detect the larger redshifted AGN's (figure 4). As a result of this we have decided to concentrate our efforts on the observation of low redshift AGN's. From GRO's list of 400 potential gamma-ray sources, derived prior to the mission, we have made a list of future observing targets (Table III). With GRANITE[5] coming online, the recoating of mirrors on the 10m reflector and the introduction of a new triggering system the energy threshold will be lowered by at least a factor of two from its present level.

TABLE III
OBSERVING TARGETS

SOURCE ID	OBJECT	REDSHIFT
4C31.04	Galaxy Pair	0.059
NGC 0985	Seyfert Galaxy	0.0432
0241+622	Quasar	0.044
3C111	Seyfert Galaxy (I)	0.0492
MRK142	Seyfert Galaxy (I)	0.0447
MRK180	BL Lac	0.0460
MRK464	Seyfert Galaxy (I)	0.0510
3C317	Radio Galaxy	0.0345
MRK506	Seyfert Galaxy (I)	0.0430
1746+676	Seyfert Galaxy	0.041
3C371	BL Lac	0.0512
3C390	Seyfert Galaxy (I)	0.0556
MRK509	Seyfert Galaxy (I)	0.0344
MCG-2-28-22	Seyfert Galaxy	0.048
2304+042	Seyfert Galaxy	0.042

ACKNOWLEDGEMENTS

We thank K. Harris and T. Lappin for help in obtaining these observations. We acknowledge support from the U.S. Department of Energy, NASA, the Smithsonian Scholarly Studies Fund and EOLAS, the scientific funding agency of Ireland.

REFERENCES

1. Vacanti et al., ApJ, **377**, 467 (1991)
2. Punch et al., Proc 22nd ICRC, **1**, 464 (1991)
3. Punch et al., Nature, **358**, 477 (1992)
4. Stecker et al., ApJ, **390**, L49-52 (1992)
5. Schubnell et al., 12.12, this conference

THE OPTICAL-UV SPECTRUM OF 3C 279 DURING OUTBURST

C.R. Shrader[1,6], J.R. Webb[2], T.J. Balonek[3], M.S. Brotherton[4],
B.J. Wills[4], D. Wills[4], S.D. Godlin[3] and B. McCollum[5]

1. Computer Sciences Corporation, Laboratory for High Energy Astrophysics, NASA Goddard Space Flight Center
2. Department of Physics and Astronomy, Florida International University
3. Department of Physics and Astronomy, Colgate University
4. Department of Astronomy and McDonald Observatory, University of Texas
5. Computer Sciences Corporation, IUE Observatory, NASA Goddard Space Flight Center
6. Guest Investigator, Compton Gamma Ray Observatory

ABSTRACT

Our ongoing program of photometric monitoring of *blazar* active galactic nuclei has revealed a significant outburst in the enigmatic high-energy source 3C 279. The event occurred during late May to early June of 1992, with a peak amplitude of 2.3 magnitudes in the R band. We obtained nearly simultaneous optical and ultraviolet spectroscopy ($\lambda\lambda 1150$-8600) of the source at the approximate midpoint of the outburst. The observations and data analysis procedures are described, and our the resulting photometric light curves and spectra are presented. Comparisons with previously published optical-UV outburst and quiescent spectra are made, including observations obtained contemporaneously with the major radio to X-ray outburst of 1988 and the 1991 gamma-ray outburst seen by EGRET.

INTRODUCTION

The *blazar* 3C 279 is well known as an enigmatic source of high energy gamma-radiation. It exhibited, during one observation, the highest apparent brightness of any quasar yet detected by EGRET and was the first such object detected by that experiment; additionally it is highly variable at high energies on timescales of days-months (Hartman et al. 1991; Kniffen et al. 1992). It also has a well documented history of optical outbursts, including flares of $\Delta V > 4$ magnitudes (Webb et al. 1990). There is not as yet any well established correlation between the optical and gamma-ray fluxes or variations for this or any other *blazar*, but observing these sources with EGRET during an optical flare offers the best chance of determining if such a relationship exists. Unlike at gamma-ray energies, sources such as 3C 279 can be effectively monitored at optical wavelengths, and once a flare is detected, GRO could be reoriented to monitor the gamma flux as well. Progress has been made in attempts to extend current theoretical models, which have been reasonably successful in accounting for the radio-UV spectra and moderately successful in accounting for the ≈ 1-10 keV X-ray emission, to incorporate the $\approx$GeV domain (e.g., Marscher et al. 1992; Dermer et al. 1992). All such attempts, however, are hampered by the current lack of a clear empirical multiwavelength variability picture. Only through extensive multiwavelength monitoring campaigns can this be accomplished unambiguously. Practical limitations on telescope and spacecraft time have and will likely continue to limit such efforts for some time to come. We report in this paper on an alternative approach which can, with greatly reduced investment in resources, begin to "scratch the surface" in establishing an empirical relationship between optical brightness and the global continuum energy distribution.

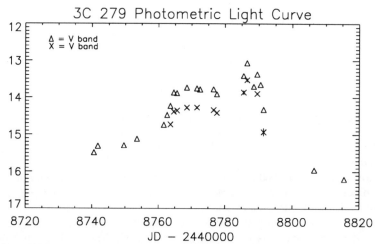

Figure 1. R- and V-Band photometric light curves for Spring 1992.

We have organized a cooperative program of photometric monitoring using small ground-based telescopes located at the Rosemary Hill Observatory (RHO) and at Colgate University. Monitoring by these instruments allows us to promptly identify periods of activity in a sample of *blazar*. This efficient flow of information then allows us to trigger our target of opportunity programs of follow-up spectroscopy in the optical-UV as well as with Compton GRO. For the latter we provide information from our other observations to the Science Working Team, who then, collectively, make a final decision as to whether or not GRO will be repointed.

Our photometric monitoring program led to the discovery of an outburst of 3C 279 during the spring of 1992. R-band CCD photometry revealed a brightening of the source by several magnitudes from late March to early June 1992 (figure 1). The March, preoutburst level represents the nominal quiescent level for this source. This brightening of ≈2.3 magnitudes can be characterized as an event of intermediate scale, referring to the long-baseline light-curve in figure 2 for a historical perspective. The mean $\Delta V/\Delta t$ over a 10-day interval shortly preceding our spectroscopic observations was 0.13 mag/day, which is small compared to the total amplitude of the event.

OBSERVATIONS

We obtained optical-UV spectra during the 1992 outburst at the epochs labeled on figure 1. The UV observations were made with the IUE short and long wavelength instruments in low dispersion mode. The source was not detected with the IUE Fine

Error Sensor (FES), thus no precisely simultaneous photometric measurement was possible. Optical spectra were subsequently obtained using the Large Cassegrain Spectrograph and the CRAF Cassini 1024x1024 CCD on the 2.7-m telescope of McDonald Observatory. Photometric coverage using the Colgate University 0.4-m telescope and Photometrics PM3000 CCD system continued over the duration of the event.

Observatory	Instrument	Exposure (min)	GMT (1992)
IUE	LWP	180	29.80 May
	SWP	180	29.67 May
McDonald	Cassegrain Spectrograph	10 20	01.25 Jun 03.18 Jun

Table 1 Log of Spectroscopic Observations

The *Compton* project was notified of the event and briefed on our observations. Serious consideration was given to reorienting the observatory, but ultimately the decision was made *not* to repoint. Table 1 is a log of our spectroscopic observations. We note that the UV and optical spectroscopic observations were not strictly simultaneous, but were obtained within a 4-day interval. Photometric coverage however, indicates that the source remained nearly constant during the interval.

DATA ANALYSIS

A small discontinuity was seen in the region of overlap of our ($\lambda\lambda$4620-7380) and

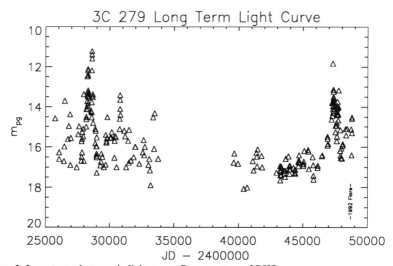

Figure 2. Long term photometric light curve. Data courtesy of RHO.

($\lambda\lambda$7138-8600) spectra. We interpret this as real variability, as our absolute flux calibration is expected to be reliable to ≈5%. For subsequent analysis, we have applied corrections for galactic reddening (E(B-V)=0.02, Seaton 1979) as well as a scaling factor to force the two optical spectra to connect smoothly. We detect no statistically significant emission lines, noting that our data are signal-to-noise limited in the UV and red bands. At a redshift of z=0.538, Ly-α is well situated on a sensitive portion of the SWP instrument; a very marginal (2σ) excess is seen, but this is consistent with being a noise spike. C IV (λ1550), typically the second strongest UV line seen in AGN spectra, is redshifted to a region of poor sensitivity on the LWP. We also examined each of the archival spectra and found no emission lines. It is likely that the line-to-continuum ratio decreases with increasing overall source brightness, however, the source in quiescence is generally undetectable by IUE. This is consistent with the outburst luminosity consisting of predominantly beamed radiation. We note, however that Webb et al. 1992 report no statistically significant line-continuum dependence on overall brightness for another OVV QSO 3C 345.

Another feature characteristic of QSO spectra is an excess above a power-law shortward of ~5000A; the so called "blue bump" (e.g., Malkan 1983), which for moderately red-shifted objects such as this, should be readily evident in the UV. Although this is typically less pronounced in OVV quasars than in non-OVV quasars, it is certainly seen in some cases (e.g., Webb et al 1993; Malkan and Moore 1986). No discernable blue bump component is present, the optical-UV light output being dominated by the *blazar* component.

Additionally, we searched for the presence of a Lyman discontinuity in the UV spectra, which would occur at 1402A. There is no statistically significant discontinuity seen at that position in the spectrum, thus an effective covering fraction of zero, which again is consistent with the beaming hypothesis.

Our spectrum is fitted by a power-law of α=1.74±0.03 (where $f_\nu \propto \nu^\alpha$; figure 3). This is significantly flatter than is seen in several quiescent cases, and similar to the 1988 outburst case (Makino et al. 1990). In attempt to establish a relationship between brightness and optical-UV spectral index we have analyzed all of the available UV archival data,

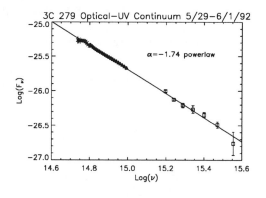

Figure 3. Power law fit to binned optical-UV continuum.

and identified cases where contemporaneous photometric measurement are available from our monitoring database or from the published literature; additionally, we derived powerlaw indices from published UBVRI photometric measurements (Webb et al. 1990; Mead et al. 1990; Kidger, et al. 1992). Figure 4 represents a compilation of powerlaw indices α (where $f_\nu \propto \nu^\alpha$) versus the 4400Å continuum fluxes derived in this manner. A weak statistical trend of flattening with increasing flux is evident from these observations. This is consistent with the analysis of the UV data alone by Bonnell et al. 1992, who also note a steep slope at an epoch approximately one month subsequent to the initial EGRET detection.

CONCLUSIONS

We have obtained a nearly simultaneous optical-UV spectrum of 3C 279 during the recent 2 magnitude optical flare. We find that the spectrum is best fitted by a power law with index 1.74+/-0.03. No lines were discernible in the spectra, nor was a Lyman discontinuity or a blue bump component. The absence of each of these features is consistent with the bulk of the outburst emission being beamed. Our analysis of the archival IUE data for this source, in conjunction with available contemporaneous photometric data suggest that the optical-UV slope decreases with increasing flux.

REFERENCES

Bonnell, J., Vestrand, T. and Stacy, J.G. 1992 (in preparation).
Dermer, C.D., Schlickeiser, R., and Mastichiadis, A. 1992, *Astr. & Astrophys.* **257**, L27.
Hartman, *et al.*, 1991, *Ap. J. Lett.*, **385**, L1.
Kidger, M., Garcia Lario, P. and de Diego, J.A. 1992, *A.&A. Supp.*, **93**, 391.
Kniffen, *et al.* 1992 preprint.
Makino, et al. 1989, *Ap. J.* **347**, L9.
Malkan, M.A. 1983. *Ap. J.*, **268**, 582.
Malkan, M.A., and Moore, R.L. 1986, *Ap. J.*, **300**, 216.
Mead, A.R.G., Ballard, K.R., Brand, P.W.J.L., Hough, J.H., Brindle, C., and Bailey, J.A., *Astron. & Astrophys. Suppl.* **83**, 183.
Seaton, M.J. 1979, *M.N.R.A.S.* **187**,73p.
Sitko, M. and Sitko, A. 1991, *P.A.S.P.* **103**, 160.
Webb, J.R., *et al.* 1990 *AJ.* **100**, 1452.
Webb, J.R. et al. 1993 (in press).

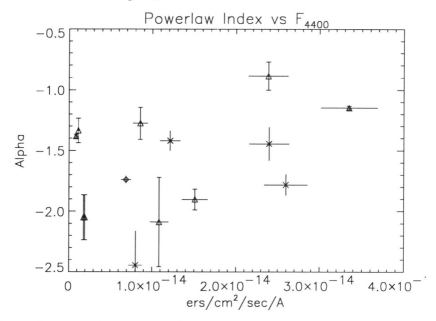

Figure 4. F_ν power-law index versus 4400Å flux; x = IUE + photometry, ▲ = UBVRI and ◊ = these observations.

EGRET UPPER LIMITS FOR THE BLAZARS 3C 345, BL LAC AND 3C 371

C. von Montigny, G. Kanbach, H. A. Mayer-Hasselwander
K. Pinkau, H. Rothermel, M. Sommer
Max-Planck-Institut für extraterrestrische Physik,
D-8046 Garching, Germany

D. L. Bertsch, C. E. Fichtel, R. C. Hartman, S. D. Hunter,
P.W. Kwok[†], J.R. Mattox[‡], P. Sreekumar, D. J. Thompson
NASA/Goddard Space Flight Center, Greenbelt, MD 20771

Y. C. Lin, P. F. Michelson, P. L. Nolan
Stanford University, Stanford, CA 94305

D. A. Kniffen
Hampden-Sydney College, Hampden-Sydney, VA 23943

E. Schneid
Grumman Aerospace Corporation, Bethpage, NY 11714

ABSTRACT

Most of the active galactic nuclei (AGN) detected by EGRET are known or suspected to be superluminal sources. We report here the results of the observations of the blazars 3C 345, BL Lac and 3C 371, which have been selected from the list of known and possible superluminal sources. For none of them we found a signal in high energy γ-rays. The 2σ upper limits for the γ- ray fluxes for energies above 100 MeV (equivalent to 95% confidence) are presented.

INTRODUCTION

EGRET has detected 16 AGN in the energy range 0.03 - 10 GeV up to now. Six of them are superluminal sources: 0235+164, 0836+710, 1226+023, 1253-055, 2230+114 and 2251+158 and one of them, the BL Lac object 0716+714, is a superluminal source if its redshift is larger than 0.28[1]. A recent report[2] indicates that also Mkn 421 shows evidence of superluminal flow.

Superluminal radio sources show properties similar to the so-called blazars: They are strong and compact radio sources with a flat or complex spectrum, their radio and optical emission is highly variable as well as polarized, and the polarization is also variable.

---

† NAS/NRC Research Associate
‡ Compton Observatory Science Support Center, operated by Astronomy Programs, Computer Sciences Corporation

Their individual appearance[3,4] can be very different: Some sources show one-sided inner jets (e.g. 3C 345, 3C 279, 3C 273), some are extended doubles (e.g. 3C 263, 1951+498, 3C 245). New radio components are ejected from a stationary core with time scales of a few months to a few years[3]. In general they are brightest as they emerge from the core and decay when moving away. The apparent superluminal velocities can be different for the components. Some sources show superluminal components and at the same time subluminal or quasi-stationary knots (e.g. 3C 454.3, 4C39.25 and 3C 395), sometimes components are accelerating and changing direction while moving out (3C 345, 3C 120, 3C 279) and some show decelerating knots (BL Lac).

We will report the result of the observation of three blazars (3C 345, 3C 371 and BL Lac) with the EGRET instrument. These three prominent sources were selected from the superluminal and possibly superluminal sources. Another report on the study of all known superluminal and suspected superluminal sources with the EGRET telescope will be given in a future publication.

ANALYSIS

The EGRET spark chamber instrument on the Compton Observatory has been described in detail by Hughes et al.[5] and Kanbach et al.[6,7] The results of the pre-flight calibration as well as the inflight verifications are given by Thompson et al.[8]

The analysis of the upper limits has been done using the counts and the exposure maps for photon energies greater than 100 MeV as well as the diffuse γ-ray maps predicted[9] by the standard EGRET analysis software from HI and CO distributions. The upper limits have been determined with a maximum likelihood method which simultaneously gives the best fit of the diffuse background to the data[10]. Prominent detected sources in the viewing period under consideration have been added to the diffuse background model.

For the determination of the upper limits, an assumed power law index of $\Gamma=2.0$ has been used for the spectral behaviour of the sources. All upper limits are given for the 95% confidence level.

RESULTS

3C 345 (l=63.46, b=40.95) is the best studied superluminal source. It is an optically violently variable source (z = .595) with fluxes and polarization varying on time scales of weeks[11]. At radio frequencies it is also a very strong and compact source and the total flux varies on time scales of years[12]. The jet is one-sided and shows superluminal motion of the components C4, C3 and C2 while component D is stationary [13]. There is also an indication of helical motion of the components in the innermost part of the jet. The observed total X-ray flux places a lower limit on the relativistic Doppler factor δ to be > 1. The largest limit ($\delta \geq 10.5$) is derived for the core[14]. From the observed proper motions of the components a lower limit for the Lorentz factor γ and an upper limit for the inclination angle of the jet to the line of sight θ can be determined and the following values are obtained[14]: $\gamma \geq 7$ and $\theta \leq 5°$.

3C 345 was in the field-of-view of EGRET in 1991 from 12 September, until 19 September. At that time the source 1633+382 (4C 38.41; l=61.09, b=42.34), about 2.25 degree from 3C 345, was a very strong gamma-ray source[15].

Although 1633+382 is very close, a simultaneous likelihood estimate of the flux of both 1633+382 and 3C 345 indicates a weak detection of 3C 345 on a $\sim 3\sigma$ level. The indicated 3C 345 flux is $(1.3 \pm 0.6) \cdot 10^{-7}$ cm^{-2}sec^{-1} (E>100 MeV). However, it might be that a statistical fluctuation in the 240 counts from 1633+382 has created this detection as an artifact. A more conservative interpretation of the data is that a 95% upper limit for 3C 345 is $2.4 \cdot 10^{-7}$cm^{-2}sec^{-1} (E>100 MeV).

Figure 1 compares this upper limit with the multifrequency power spectrum of 3C 345[16]. The upper limit derived from the EGRET instrument indicates that the γ-ray power does not dominate the spectrum the way it does in several of the other EGRET sources such as 3C 279[17]. However, it is important to note that the observations in the different energy bands were not simultaneous.

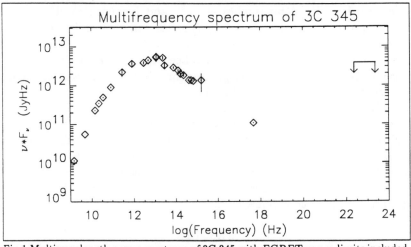

Fig.1 Multiwavelength power spectrum of 3C 345 with EGRET upper limits included. Data from radio to X-rays are from Landau et al.1986.

The elliptical galaxy **BL Lacertae** (l=92.59, b=-10.44; z = 0.069) showed two events (the emerging of knots from the core) with apparent deceleration when they reached a projected distance[18] of ~ 1.5-$2.0 h^{-1}$ pc. The ejection of new superluminal components is correlated with outbursts in the 10.6 GHz flux and the polarization [19].

BL Lac was observed by EGRET in 1991 during 30 May until 8 June (viewing period 2.0) and 8 - 15 August (viewing period 7.1). In both periods

BL Lac was at $\sim 22°$ from the pointing direction of EGRET. The derived upper limits for the flux are F(E>100 MeV)< $0.22 \cdot 10^{-6}$ cm^{-2}sec^{-1} and F(E>100 MeV)< $0.14 \cdot 10^{-6}$ cm^{-2}sec^{-1} for viewing periods 2.0 and 7.1, respectively.

A comparison of these limits with the multifrequency spectrum of BL Lac given in Impey and Neugebauer[20] shows that they are also consistent with the power in the gamma-ray regime not dominating the spectrum.

3C 371 (l=100.13, b=29.17; z = 0.05) has a one-sided jet which is surprisingly straight, although there might be an indication of helical motion[21]. It is considered to be a possible superluminal source.

3C 371 was observed by EGRET in 1991 during 12 - 19 September (viewing period 9.2) and in 1992 during 10 - 23 January (viewing period 18.0). In both periods it was at a very large angular distance ($\sim 33°$) from the pointing direction of EGRET. The derived upper limits are F(E>100 MeV)< $0.54 \cdot 10^{-6}$cm^{-2}sec^{-1} and F(E>100 MeV)< $0.14 \cdot 10^{-6}$cm^{-2}sec^{-1} for viewing periods 9.2 and 18.0, respectively.

DISCUSSION

There are no obvious differences between those superluminal sources detected by EGRET and those three reported here. In fact, we are dealing with sources which are highly variable and unsteady in all aspects. Therefore, it is very well possible that these superluminal sources will be γ-ray bright at some time.

Optical observations of the two EGRET sources 0716+714 and 0836+710 between September 1988 and February 1992 at Calar Alto/Spain showed that during the EGRET observations (January 10th - 23th, 1992) the activity in the optical range was very strong. At least one rapid optical flare was detected[22] on February 16th from 0836+710. Since there are indications that optical activity is related to the emergence of new superluminal components[23], it should once again be emphasized that only really simultaneous observations and monitoring over a broad energy range can help to improve our understanding of the processes going on in active galactic nuclei.

The upper limits derived for the three objects studied are consistent with the overall spectra predicted by the proton initiated cascade (PIC) model[24] and a proton/electron energy density ratio $\eta = u_p/u_e < 10$. For $\eta = 1$ the model predicts a gamma-ray power about a factor of 3 lower than the power in the infrared/optical while for $\eta = 10$ the power in the gamma-ray regime is already about a factor of 4 larger than in the infrared/optical range. A quite acceptable representation[24] of the multifrequency spectrum of 3C 279 during the "flare" state in June 1991 was obtained for $\eta = 10$.

This might suggest that during the "quiet" state the energy density of the protons and the electrons is about equal and for the "active" state the energy density of the protons is enhanced by some mechanism.

ACKNOWLEDGEMENTS

The EGRET team gratefully acknowledges support from the following: Bundesministerium für Forschung und Technologie, Grant 50 QV 9095 (MPE); NASA Grant NAG5-1742 (HSC); NASA Grant NAG5-1605 (SU) and NASA Contract NAS5-31210 (GAC).

REFERENCES

1. Witzel, A., Superluminal Radio Sources (eds.: Zensus, J.A. and Pearson, T.J., Cambridge University Press, 1987), p. 83.
2. Zhang, F.J. & Bååth, L.B., Astron. & Astrophys. 236, 47 (1990).
3. Porcas, R.W., Superluminal Radio Sources (eds.: Zensus, J.A. and Pearson, T.J., Cambridge University Press, 1987), p. 12.
4. Kellermann, K.I., IAU Symposium 121, 273 (1987).
5. Hughes, E.B. et al., IEEE Trans. Nucl. Sci. NS-27, 364 (1980).
6. Kanbach, G. et al., Space Science Reviews 49, 69 (1988).
7. Kanbach, G. et al., Proc. of Gamma Ray Observatory Science Workshop (ed. W.N. Johnson, GSFC, 1989), p. 2-1.
8. Thompson, D.J. et al., Ap. J. Suppl. submitted (1992).
9. Bertsch, D.L. et al., submitted to Ap. J. (1993).
10. Mattox, J.R. et al., in preparation (1993).
11. Moore, R.L. and Stockman, H.S., Ap. J. 243, 60 (1981).
12. Bregman, J.N. et al., Ap. J. 301, 708 (1986).
13. Bartel, N. et al., Nature 319, 733 (1986).
14. Biretta, J.A. and Cohen, M.H., Superluminal Radio Sources (eds.: Zensus, J.A. and Pearson, T.J., Cambridge University Press, 1987), p. 40.
15. Mattox, J.R. et al., submitted to Ap. J. (1993).
16. Landau, R. et al., Ap. J. 308, 78 (1986).
17. Hartman, R.C et al., Ap. J. Letters 385, L1 (1992).
18. Mutel, R.L. and Phillips, R.B., Superluminal Radio Sources (eds.: Zensus, J.A. and Pearson, T.J., Cambridge University Press, 1987), p. 60.
19. Aller, H.D., Aller, M.F. and Hughes, P.A, Ap. J. 298, 296 (1985).
20. Impey, C.D. and Neugebauer, G., Astronom. J. 95, 307 (1988).
21. Lind, K.R., Superluminal Radio Sources (eds.: Zensus, J.A. and Pearson, T.J., Cambridge University Press, 1987), p. 180.
22. v. Linde, J. et al., submitted to Astron. & Astrophys. (Letters) (1992).
23. Babadzhanyants, M.K. and Belokon, E.T., IAU Symposium 212, 305 (1987).
24. Mannheim, K., Astron. & Astrophys. in press (1992).

HARD X-RAY OBSERVATION OF CEN A AND A BREAK IN THE ENERGY SPECTRUM

T. Takahashi, S. Gunji, M. Hirayama, T. Kamae, S. Miyazaki,
Y. Sekimoto, M. Tanaka, T. Tamura, and N.Y. Yamasaki
Dept. of Physics, Univ. of Tokyo, Bunkyo-ku, Tokyo 113, Japan

H. Inoue, T. Kanou and T. Yamagami
Inst. of Space Astro. Science, Sagamihara, Kanagawa 229, Japan

M. Nomachi
Nat. Lab. High Energy Phys. (KEK), Tsukuba, Ibaraki 305, Japan

H. Murakami
Dept. of Physics, Rikkyo Univ., Toshima-ku, Tokyo 171, Japan

J. Braga and J. A. Neri
Instituto Nacional de Pesquisas Espaciais, Sao Jose dos Campos, Sao Paulo, BRAZIL

ABSTRACT

We studied hard X-ray/γ-ray emission from Cen A (NGC 5128) in a balloon experiment with a low background detector (Welcome-1) in Brazil. The energy spectrum of CenA is obtained from 40 keV to 600 keV. We combined the energy spectrum obtained by the Ginga satellite in a similar state. The combined spectrum indicates that there is a break at 185 ± 22 keV. The spectrum is fitted to the broken power law model with the photon index of $\alpha_1 = -1.79$ up to 188 keV, and $\alpha_2 = -3.7^{+0.9}_{-1.7}$ above the break.

INTRODUCTION

Radio galaxy Cen A is one of the brightest Active Galactic Nuclei (AGN) which is located at 5 Mpc. It is one of several extragalactic objects which are observed in a wide frequency range including the hard X-ray and γ-ray regions. VLBI observation revealed the radio structure of the nucleus of Cen A[1]. The nucleus consists of a compact self-absorbing core, a jet containing a set of three knots, and a very long narrow component elongated along the same position angle as the knots. A detailed X-ray map of Cen A was obtained with Einstein Observatory[2]: it consists of a variable compact nucleus, an X-ray jet almost aligned with the NE (north east) radio lobe, and a diffuse source extending several arcminutes around the nucleus.

The power-law component which extends to the hard X-ray region is an important characteristic of the continuum emission from AGNs. Photon indices being almost universal ($-1.6 \sim -1.8$) implies a common mechanism at the central engine of AGN, a black hole with mass in the range $10^6 - 10^8$ $M_\odot$. To explain observed results, many models have been proposed such as the Comptonization model[4,5], the Synchrotron Self Compton model[6], and the model involving pair creation and Compton reflection[7]. These models suggest some spectral feature yet unknown to emerge in the hard X-ray/γ-ray energy range (> 40keV). Thus, it is very important to carry out a high-sensitivity observation in the hard X-ray/γ-ray range.

Cen A is one of the most suitable for study at the hard X-ray/γ ray region because it is the nearest AGN. Below 100 keV, energy spectra have been studied by a number of experiments to find out that the spectrum is well described by a power law with the photon index around 1.6 [8]. Above 100 keV, the LEGS detector studied the spectrum at the high state but found no evidence of a break in the spectrum around a few hundred keV[9]. A balloon experiment reported emission in the MeV region[10]. SAS-2 [11] and COS-B [12] gave upper limits and this implies a spectral steepening between 100 keV and 10 MeV.

DETECTOR

The detector used in the observation is based on newly developed well-type phoswich counters [13,14]. In the first well-type phoswich counter, GSO(Ce) (Gd_2SiO_5 doped with Ce) is used as the detection part and CsI(Tl) as the shielding part. The detector consists of 64 GSO(Ce)/CsI(Tl) well-type phoswich counters and is called Welcome-1. The well-type configuration reduces background significantly, both externally and internally and allows us a high signal-to-noise ratio in balloon-borne experiments. The measured background level at an altitude of ~4.5g/cm^2 is 1×10^{-4}/cm^2/s/keV at 122 keV and is 4×10^{-5}/cm^2/s/keV at 511 keV. The detector characteristics relevant to the present observation are: energy range, 40 − 800 keV; geometrical area, 740cm^2; full peak efficiency, 98% at 122 keV and 30% at 511 keV; field of view (hwhm), 7.7° at 122 keV and 9.1° at 511 keV. The energy resolution of the GSO scintillator in the well-type phoswich counter is 28% at 122 keV and 12 % at 511 keV.

OBSERVATION

The detector (Welcome-1) was flown from the INPE (Instituto Nacional Pesquisas Espaciaus) balloon base in Cachoeira Paulista, San Paulo, Brazil (45.00'.34"W, 22.39'.44"S), on 19 November 1991. The detector reached the ceiling altitude of 36.7 km (4.6±0.1 g/cm^2). The observation of Cen-A started at 9:40AM, November 19, 1991 (UT:Universal Time). In total, 9 cycles of on-source and off-source observations (~10 minutes each) were performed until 13:00PM (UT). In the on-source observation, we followed Cen-A by redirecting the detector at 5-minute intervals. In order to minimize systematic errors introduced by the subtraction of the off-source spectrum from the on-source spectrum, the elevation angle (49−68 degrees) was kept the same in each set of on-off observation. During the off-source observations, the detector was pointed to the region where no strong hard X-ray source exists (RA 13^h28^m-15^h12^m DEC −14°- −19°). Pointing was done by using a geomagnetic sensor. A wide field CCD solar camera and an additional geomagnetic sensor are used to monitor the detector direction to about ± 0.2° accuracy. During the observation, all counters operated normally. In the flight, we observed the Crab nebula/pulsar, the Vela pulsar and PSR1509-58 together with Cen A.

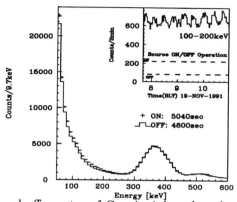

Figure 1: The on and off spectra of Cen-A. A broad peak at around 370 keV is mostly due to contamination of ^{152}Gd (2.14 MeV α decay, half-life 1.1×10^{14}y) in the GSO scintillator :

ANALYSIS AND DISCUSSION

In data analysis, first we rejected events within 2 ms of large energy deposits in the same counter or in a neighboring counter. This effectively eliminated unwanted baseline fluctuation after incidence of high-energy charged particles (E$\geq$60 MeV) in a CsI(Tl) crystal. We then rejected events with one or more hits in the neighboring counters to eliminate Compton-scattered γ-rays. As a result of the above event selection, the total on-source and off-source live times was reduced to 5040 s and 4800s for 64 counters (efficiency of about 75%).

The nine sets of on- and off-source spectra were then normalized according to the effective live time. Figure 1 shows the observed energy spectra during the on-source pointing and off-source pointing. The counting rate variation during the observation of Cen A for 100-200 keV is also shown in the figure. The observed signal-to-background ratio is about 15% in the energy range 100-200 keV. The off-source spectrum after the event selection agrees well with the measured internal background due to radioactive contamination in the detection part (GSO) and the simulated atmospheric and celestial diffuse background.

After correcting for the air-attenuation factor, nine sets of background subtracted spectra are summed to give the energy spectrum shown in Fig. 2. A possible break in the spectrum is visible at around 150-200 keV, although it is statistically not compelling by itself.

We compare the data with the observations in 1989 and 1990 by the Ginga satellite[3] to study the wide-band energy spectrum from 2 keV to 600 keV. The spectrum by the Ginga showed a power-law component with the absorption by $N_H = 1.6\times 10^{23} cm^{-2}$ below 10 keV. In the Ginga observation, the intensity varied by about 30% within a day but the photon index of the power-law spectrum remained stable at ~ -1.80 (1.79 - 1.84) throughout the observation periods both in 1989 and 1990. Among the seven sets of data, we find one set (Feb 1990)

Model	χ^2/dof	I^a	α_1	E^b(keV)	α_2
Single Power Law	17.1/11	6.6± 0.3	-1.79		
Broken Power Law	3.3/9	7.3± 0.4	-1.79	$185.2^{+22.3}_{-21.8}$	$-3.7^{+0.9}_{-1.7}$

Model	χ^2/dof	I^a	γ	kT_e(keV)
Comptonization	4.9/10	7.3±0.4	2.9941	$58.0^{+7.8}_{-5.8}$

a) Intensity at 100 keV ($\times 10^{-5}$ cm^{-2}s^{-1}keV^{-1}), b) Energy at break point

Table 1: The parameters obtained by spectrum fits to model functions

shown in Fig. 3 in which the absolute flux extrapolated to our energy range (40 keV to 150 keV) is 60 % of our data. From this we assume that the state of Cen A was the same for this set and our observation. The photon index obtained from this data set is $\alpha = -1.79 \pm 0.01$. In fact, the fit for our spectrum from 40 keV to 190 keV gives the photon index of $\alpha = -1.8 \pm 0.15$ which coincides with the photon index obtained by the Ginga observation. Under this assumption, we analyze the combined spectrum.

First, we fit our spectrum to the single power law with the photon index obtained from the Ginga ($\alpha = -1.79$). The fit gives $\chi^2 = 17.1$ for our part (40 keV - 600 keV) and it is inconsistent at a 10% confidence level (dof=11). Second we apply the fit to the broken power law. The break point in the spectrum is determined to be $E_b = 185.2^{+22.3}_{-21.8}$ keV when we fix the photon index below the break to be $\alpha = 1.79$. The photon index above the break is $\alpha = -3.7^{+0.9}_{-1.7}$ ($\chi^2 = 3.3$, dof=9 for our part).

The break in the power law spectrum can be explained by Comptonization of soft photons by mildly relativistic electrons in the hot accretion disk ($kT_e >$ a few 10 keV). Our spectrum is fitted with the Comptonization model by Sunyaev and Titarchuk [4,5]. According to the model, the spectrum is described in terms of the electron temperature (kT_e) and γ. An index of a power-law approximation to the data for $E_b < kT_e$ is given by $\alpha = (\gamma + 9/4)^{1/2} - 1/2$. We used $\gamma = 2.9941$ which is calculated from the photon index obtained by the Ginga. The best-fit parameter is the temperature of $kT_e = 58.0^{+7.8}_{-5.8}$. For a spherical geometry, an optical depth is given by

$$\tau = \sqrt{\frac{\pi^2 m_e c^2}{3kT\gamma} - \frac{2}{3}}$$

and becomes $\tau = 2.4 \pm 0.2$. Table 1 lists the results of the fits.

CONCLUSION

We measured the energy spectrum of Cen A from 40 keV to 600 keV by a balloon experiment. The data is compared with the Ginga observations in 1989 and 1990 from 2 keV to 25 keV. Out of seven Ginga data sets, we found one set

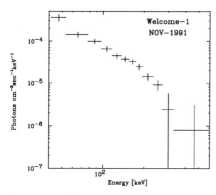

Figure 2: Energy spectrum of Cen-A:

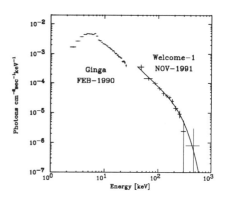

Figure 3: The spectrum of the combined Ginga/Welcome-1 data (Solid line: the fit to the S-T model)

(Feb. 1990) in which the state of Cen A seems to have been the same as that in our observation. When two spectra are combined, they indicate that there is a break at 185 ± 22 keV. The spectrum is fitted with the broken power-law model, one photon index is $\alpha_1 = -1.79$ up to 187 keV and the other is $\alpha_2 = -3.7^{+0.9}_{-1.7}$ above the break. The indices obtained by previous measurements in the 10-keV to a few hundred-keV range are harder than ones observed by the Ginga satellite[8]. If we combine our spectrum with these data, the break in the spectrum becomes statistically more significant. If the measurements mentioned above exhaust all possible states of Cen A, we can conclude that the spectrum has a break around 150-200 keV. With the Sunyaev-Titarchuk Comptonization model of the hot accretion disk, the electron temperature and the optical thickness have been determined to be $kT_e = 58.0^{+7.8}_{-5.8}$ keV and $\tau = 2.4 \pm 0.2$.

REFERENCES

1. Meier, M. et al. 1989, *Astron.J.*, **98**(1), 27.
2. Feigelson, E.D., 1981, *Ap. J.*, **251**, 31.
3. Kanou, T et al., Master Thesis, 1992.
4. Sunyaev, R.A., and Titarchuk, L. G. 1980, *Astr. Ap.*, **86**, 121.
5. Zdziarski, A.A., 1985, *Ap. J*, **289**, 514.
6. Jones, T.W., et al. 1974, *Ap. J.*, **188**, 353.
7. Zdziarski, A. et al., 1990, *Ap. J., Lett.*, **363**, L1.
8. Gehrels, N. and C. Cheung 1992, in AIP Conf. Proc. **254**, p.348.
9. Gehrels, N., et al., 1984, *Ap.J.*, **278**, 112.
10. von Ballmoos, P. et al., 1987, *Ap.J.*, **312**, 134.
11. Bignami, G.F. et al. 1979, *Ap. J.*, **232**, 649.
12. Bignami, G.F. et al. 1981, *Ap. J.*, **93**, 71.
13. Kamae, T. et al. 1992, *Adv. Space. Res.*, **2** 168.
14. Takahashi, T. et al, 1992, in *Gamma-ray Detector*, SPIE, **1734**, p.44.

AGN EMISSION ABOVE 100 KEV: THE HARD X-RAY DETECTION PROBLEM

Michael Maisack
Astronomisches Institut Tübingen, Germany
Kent S. Wood
Naval Research Lab, Washington DC, USA
Duane E. Gruber
CASS, University of California San Diego, La Jolla, CA, USA

ABSTRACT

We have searched the HEAO-A4 all-sky survey database for a class of AGN that is most likely to emit X-rays in the energy range covered by OSSE (50 keV - 10 MeV) on the Compton Gamma Ray Observatory (GRO). We have overlayed the survey signals of samples of AGN selected by their prominence and properties at other wavelengths, and in Einstein observations and the HEAO-A1 survey. In contrast to findings that radio and optical properties are correlated with the soft X-ray emission, we find no significant signal at hard X-rays from any of the chosen classes of QSOs. We find a signal from Seyfert galaxies up to 80 keV with the canonical index $\alpha = 0.7$.

1. INTRODUCTION

The OSSE instrument [1] on GRO is the first instrument that can perform high-sensitivity studies of a large group of AGN at energies of several 100 keV. The limited number of sources that can be observed by this non-imaging, collimated instrument – especially after the tape recorder failure – makes it important to identify suitable objects that are most likely to strongly emit in this energy range. Previous studies of correlations of properties of AGN at different wavelengths have shown that radio-loud, flat-spectrum AGN have flatter soft X-ray spectra than radio-quiet and steep spectrum sources [2]. This study attempts to investigate whether such or other properties at different wavelengths can be successfully used to predict emission at energies of several ten to ≈ 100 keV, the low-energy part of the OSSE energy range.

2. PREVIOUS OBSERVATIONS

2.1 Seyferts

Samples of Seyfert galaxies [3] have shown a canonical X-ray spectrum up to ≈ 100 keV with a power law index of $\alpha = 0.7$ with a small intrinsic spread. However, a roll-over at ≈ 50 keV cannot be ruled out. Such a roll-over has been observed for NGC 4151 on two occasions [4,5].

2.2 Quasars

Samples of quasars have not shown the spectral homogeneity of Seyferts. Apart from a few sources with very hard spectra ($\alpha < 0.5$), they have steeper spectra ($\alpha = 0.8\text{-}0.9$) and a higher intrinsic spread. It has been found that radio-loud quasars generally have harder spectra than radio-quiet, and core-dominated harder than extended sources [2,6].

2.3 GRO

GRO observations of AGN so far have produced some surprising results: EGRET has by now observed more than a dozen AGN at energies of several 100 MeV, the most prominent being the observations of a flare of the superluminal source 3C 279. It is not yet clear whether all these sources are superluminal, though. OSSE, on the other hand, has not observed many hard continua from AGN, with the exception of the prominent sources 3C 273 and Cen A. In NGC 4151, in contrast, a steep spectrum that falls off exponentially has been observed [5].

3. INTENTION

To identify a type of source with bright, hard X-ray emission, and to find out whether there are reliable indicators for such objects at other wavelength ranges, we have conducted a search of the HEAO-A4 all-sky survey data-base. We have selected samples of sources with certain characteristics which have not been identified as significant source detections in the survey [7] to overlay the signals to check whether a significant result could be obtained.

The samples included

- radio-loud, extended sources
- radio-loud, compact sources
- HEAO-A1 all sky survey QSOs
- HEAO-A1 all sky survey Seyferts
- HEAO-A1 all sky survey BL Lacs

The radio-loud sources were selected by X-ray luminosity and hard Einstein spectra from the sample of Worrall et al. (1987), the HEAO-A1 sources by flux and flat optical-to-X-ray spectra [8].

3.1 The database

HEAO-A4 was the high-energy instrument on the HEAO-1 satellite. The satellite performed both pointed observations and the first all-sky survey in this energy range. The Low Energy Detectors (LED) of the A4 instrument covered the energy range 13- 180 keV. We have analyzed data from the LED all-sky survey database in San Diego. The data are stored in four energy bands (12-25, 25-40, 40-80, 80-180 keV).

3.2 Data selection criteria

In all, there are 87 sources in our samples. The results of the flux search did not give qualitatively satisfactory results for all of these sources. For some, no flux could be evaluated due to the presence of a nearby bright source, for others, the background subtraction did not give a satisfactory result. Trying to fit a variable background was only accepted if the improvement was significant at the 90% confidence level applying an F-test. A minimum number of 5 pixels of the A4 skymap for evaluation of both estimates were required. This left 29 objects which satisfied all our criteria. Of these, 10 were Seyfert galaxies, 4 A1-selected QSOs and 15 Einstein-selected QSOs (see Tab. 1).

3.3 Adding technique

The A4 rates from the sources which were identified in the A1 survey were normalized by their relative A1 fluxes, to make sure that a single source could not account for the entire signal of a group. This procdure could not be applied to the Einstein-selected sources, since no A1 fluxes were reported for them.

4. RESULTS

4.1 Einstein-selected Quasars

We found no significant signal in any of the 4 A4 energy intervals from our samples of Einstein-selected quasars, nor any combination of them. The upper limit, though, are about as high as the fluxes from the Seyfert galaxies.

4.2 A1-selected quasars

We find no significant signal from the brightest quasars found in the A1 all-sky survey, which were in parts selected for their hard optical-to-X-ray spectra. However, since only 4 objects met the quality criteria for our selection, this is less significant than our findings for the Einstein selected QSOs.

4.3 A1-selected Seyferts

We selected two groups of Seyfert galaxies that were detected in the A1 survey, grouped by their A1 flux. These Seyfert galaxies are the only group of objects in our samples that give significant signals up to 80 keV. Adding up the signal from all 10 Seyferts with satisfactory data quality, we find that they roughly follow the canonical $\alpha = 0.7$ slope. A recent OSSE observation of NGC 4151 [5] has shown a different spectral shape, though. The spectrum above 60 keV is very soft, and can be best described by a Sunyaev-Titarchuk spectrum with plasma temperature 35 keV and optical depth 3.5. The spectrum steepens to a photon index of greater 3 above 100 keV. The spectrum above 100 keV will thus fall short of the expectations from the extrapolation of the canonical spectrum.

This has important implications for future OSSE observations:

- Seyfert galaxies will fall below the extrapolation of the canonical $\alpha = 0.7$ spectrum in the OSSE energy range. OSSE has detected the brightest Seyfert above 100 keV, NGC 4151, only up to ≈ 200 keV. Our samples of A1-bright Seyferts, on average, reaches only about 10% of the brightness of NGC 4151 in July 1991 and these soruces may therefore fall below the detection threshold at energies of 100-200 keV.

- The radio-properties of quasars seem to be good indicators only of the X-ray emission at ≈ 1 keV. Above that, their spectra seem to be significantly steeper. There are, however, individual exceptions like 3C 273. It may therefore be more promising to observe these outstanding members than a group of objects. Note, however, that the radio-loud quasars were chosen for their brightness during the Einstein survey, so simultaneity may be important for finding suitable candidates for hard X-ray emission.

- the flatter soft X-ray spectra of radio-loud, compact, flat spectrum QSOs don't seem to extend to energies above several 10 keV. This may indicate that the flat component is not different from high-energy components in other sources, but point to the lack of a soft excess in these sources.

- even simultaneous brightness in the X-ray range between 1-10 keV is not necessarily a good indicator of brightness around 100 keV, at least not for a class of objects. It seems that there are only a handful of individual objects emitting significantly in this energy. It may well be that emission at several 100 MeV, as detected by EGRET for more than 10 objects, is a more common phenomenon in AGN. Other than the synchrotron self-Compton model, in many models hard X-ray and gamma-ray emission arise from different populations of energetic particles, and the two types of emission may even be anti-correlated [9]. Therefore, the EGRET AGN may also not be suitable candidates for OSSE observations.

References

1. Johnson et al. 1989. Proceedings of the Gamma Ray Workshop
2. Worrall et al. 1987. ApJ 313, 596
3. Rothschild et al. 1983. ApJ 269, 423
4. Jourdain et al. 1992, Proceedings of the Cosmic Ray Conference
5. Maisack et al. 1992. in preparation
6. Williams et al. 1992. ApJ 389, 157
7. Levine et al. 1984. ApJ Supplememnt 1984
8. Remillard et al. 1990, preprint
9. Kafatos and Becker, these proceedings

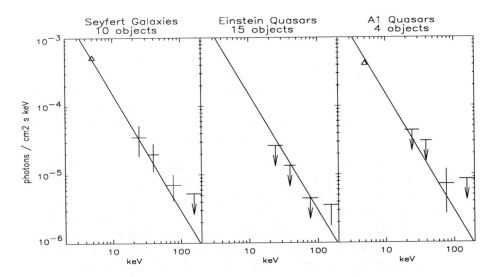

Figure 1: HEAO-A4 Spectra of Samples, triangles: Einstein fluxes, line is canonical Seyfert spectrum

Table 1: List of seleted AGN		
Seyfert Galaxies	A1-selected Quasars	Einstein-selected Quasars
NGC 4593	0016-257	3CR 48
MCG-6-30-15	PKS 0257-128	3CR 298
IC 4329A	0355-116	3CR 9
NGC 7582	PKS 0556-504	3CR 215
Fairall 9		PHL 1627
Fairall 303		PKS 0208-512
Mkn 279		Mrk 705
Akn 564		B2 1028+313
PG2304+042		CG0112-017
NGC 7674		4C-02.04
		0237-233
		3C 245
		Mkn 1383
		Mkn 841
		1548+114

GAMMA RAY JETS FROM ACTIVE GALACTIC NUCLEI

R. D. Blandford
Caltech, Pasadena, CA 91125.

ABSTRACT

Recent EGRET observations of core-dominated active galactic nuclei (AGN), can be interpreted as inverse Compton scattering of ambient nuclear X-rays by relativistic electrons and positrons accelerated behind shock fronts in a decelerating, relativistic jet. Specific application of this model to the observations of 3C279 suggests that the ~ 1 GeV γ-rays originate from a jet radius $r \sim 0.1$ pc, well inside the observed, superluminally expanding radio components. In general, it is suggested that there will be an approximate mapping between the γ-ray energy and the emission radius, which in turn implies that the variability timescale lengthens with increasing γ-ray energy. It is conjectured that jets originate as predominately electromagnetic conduits that decelerate as they interact with their surroundings and become increasingly particle-loaded and radiative. It is speculated that the jet power originates from a magnetized, spinning black hole and that a threshold spin angular frequency may be necessary for an AGN to become radio loud.

OBSERVATIONAL INTRODUCTION

It is simplest to start with low energy γ-ray observations.

μeV Photons: A gross oversimplification of ongoing research [1] [2] leads to the view that AGN are either *radio-loud* or *radio-quiet* and that the former class are observed to be (radio) *core-dominant* when viewed along their symmetry directions on account of the beaming of the emission from a pair of antiparallel, relativistic jets. Otherwise, they are *lobe-dominant*. This hypothesis is known as the *Unified Model*. Core-dominant sources (variously called BL Lac objects, OVV quasars, Blazars, flat spectrum quasars, etc) exhibit *superluminal expansion* with observed expansion speeds $\lesssim 12ch_{75}^{-1}$ (where h_{75} is the Hubble constant in units of 75 km s^{-1}Mpc^{-1}). In the jets with the most rapid observed expansion, the outflow Lorentz factors should be $\gamma_j \sim 12 h_{75}^{-1}$. Concomitant with this interpretation is the belief that jets are beamed in cones with opening angles $\theta = 0.1\theta_{-1} \sim \gamma_j^{-1}$ and that only a fraction $f \sim 1/2\gamma_j^2$ of all jets are beamed towards us. For the fastest jets, this fraction may be as small as $\sim 0.003 h_{75}^2$. Equivalently, the total luminosities of those sources that are beamed in our direction are smaller by a factor $\sim f$ than the luminosity of an isotropic source with similar observed flux. (It must be emphasized that most detailed models of compact radio sources, and surely the sources themselves, are much more complex than this.)

Now, the very fact that radio cores appear to be strongly relativistically beamed, while the radio structure on scales $\gtrsim 1$ kpc seems to move with no more than mildly relativistic speed, implies that jets are strongly decelerated as they propagate away from the AGN. This raises a very important question, namely "where does the associated loss in jet kinetic energy end up?" It is my contention that CGRO has effectively answered this question and that the jet kinetic energy is mostly radiated as high energy γ-rays.

meV-keV Photons: Core-dominant sources are also distinctive at intermediate energies. They show steep, variable, polarized, smooth continua which can overwhelm the atomic emission lines at eV energies. Although many qualifications ought, again, to be made, [1] the simple explanation is that this emission is a non-thermal infrared to X-ray continuum that, although only a minor contributor to the overall bolometric luminosity of the source, dominates the observed flux because of relativistic beaming. This continuum is believed to come from the inner regions of the same jet that is responsible for the radio emission. [3]

MeV-TeV Photons: There are now 16 detections by EGRET of γ-ray radio core-dominant sources in the 100 MeV-10 GeV range. [4] In addition, observations of AGN are being made by the OSSE [5] and COMPTEL [6] telescopes at lower energies. These sources range in redshift from $z = 0.03$ (MKN421) to $z = 2.16$ (0836+710). Their γ-ray fluxes per octave of energy are roughly energy-independent over the energy range $\sim$ 100MeV-$\sim$ 10GeV and their γ-ray spectral indices (in the radio astronomical usage) are $\alpha_\gamma \sim 1$. (In the case of MKN 421, [7] this spectrum can be extended to ~ 1 TeV.) It is highly significant that lobe-dominant and radio-quiet quasars are not represented on this list.

These observations are consistent with the Unified Model. If we suppose that the jets have a similar beaming factor f at γ-ray energies as at radio frequencies, Then the most powerful sources (*e.g.* 4C38.41, 0420-014), have luminosities $\sim 5 - 10 \times 10^{48} f h_{75}^{-2}$ erg s^{-1}. For $f \sim 3 \times 10^{-3} h_{75}^2$, as indicated by the fastest jets, the power is $\sim 1.5 - 3 \times 10^{46}$ erg s^{-1}, independent of the Hubble constant. This can be less than the UV power needed to account for the broad emission lines (when these are seen). Indeed, if the γ-ray emission were not beamed, then steep spectrum double radio sources would have been detected by EGRET and the totality of AGN would have produced a γ-ray background in excess of the measured value. [8] [2]

3C279: The source that has been best studied at γ-ray energies is 3C279. 3C279 is a radio loud quasar with $z = 0.54$. The 100MeV-10GeV luminosity is $L_\gamma \sim 1 - 6 \times 10^{48} f h_{75}^{-2}$ erg s^{-1}. [9] The six fold variation represents the observed variability which occurs over a timescale ~ 1 week. 3C279 was, historically, the first radio source to exhibit superluminal motion. In the 1970's, it expanded with a speed $\sim 12 c h_{75}^{-1}$; however now its speed has fallen to $\sim 3 c h_{75}^{-1}$. The deprojected radius of the expanding radio components is $r \sim 130 \theta_{-1}^{-1} h_{75}^{-1}$ pc, where the jet axis is supposed to make an angle $\theta \sim 0.1 \theta_{-1}$ to the line of sight. (The extended radio component extends out to $\sim 300 \theta_{-1}^{-1}$ kpc.) Optically, 3C279 has been as bright as 11^m, [10] equivalent to a lumnosity $L_{opt} \sim 3 \times 10^{48} f h_{75}^{-2}$ erg s^{-1}, though now it is only about 3 percent of this value. The total X-ray luminosity is $L_X \sim 0.3 L_{opt}$ but it has been reported to vary on hourly timescales, [11] suggesting a very compact origin close to the hypothetical central massive black hole.

RADIO JETS

Relativistic Jets: A simple model of relativistic radio jets has been developed. This comprises a collimated outflow within a cone of opening angle $\phi \sim \gamma_j^{-1}$. Near the base of this cone, synchrotron radiation at a given frequency is self-absorbed by its own emitting electrons. If, for simplicity, we treat the limiting brightness temperature and γ_j as constants, then the radio spectral index $\alpha_R \sim 0$ and the radius $r(\nu)$ of the radio photosphere will scale roughly inversely with radio frequency. In fact most self-absorbed cores cannot be resolved by earthbound telescopes. The actual moving features are believed to be instabilities, or more reasonably shock waves, formed at radii where the jet is optically thin. If we take the view that the jet is supersonic and must decelerate, then these shocks are most likely to come as forward-reverse pairs, just like in the solar wind. Most dissipation is associated with the reverse shocks.

Relativistic Shocks: The relativistic shock jump conditions are no harder to work with than their non-relativistic counterparts. [12] If, for example, the jet fluid mainly comprises relativistic particles, then the post-shock fluid speed, relative to the shock front is $c^2/3\beta_1$, where β_1 is the fluid speed ahead of the shock. The post-shock pressure is $p_2 = 2F/3c$ where F is the energy flux (again measured in the frame of the shock). A stationary shock in an ultrarelativistic jet will probably be very efficient at accelerating particles. [13] [14] However, the resulting radio synchrotron emission will only be subject to a modest Doppler boost factor (for privileged observers along the jet axis) of $[\gamma_2(1 - \boldsymbol{\beta}_2 \cdot \mathbf{n})]^{-(2+\alpha_R)} \lesssim 2.4$ where $\boldsymbol{\beta}_2$ is the post-shock velocity. By contrast, if we consider a weak traveling, or oblique shock, in which the post shock speed (in the jet frame) is similar to that ahead of the shock, then we have the opposite limit and there will be little dissipation and only a small radio emissivity in the post-shock frame. However, there will be a Doppler boost in observed flux by a factor that can be as large as 10^4 for on axis observers and $\gamma_j \sim 10$. The observed superluminal speed will be $|\boldsymbol{\beta}_s \times \mathbf{n}|/(1 - \boldsymbol{\beta}_s \cdot \mathbf{n})$, where $\boldsymbol{\beta}_s$ is the shock velocity.

Clearly, there is some intermediate strength of shock from which the observed flux is maximized. This depends upon the shock kinematics and the details of the particle acceleration. It is not special pleading or fine tuning to say that shocks with these characteristics are the ones that we will see; instead we should imagine that many shocks of all speeds and strengths form as the jet (which itself is likely to be quite non-uniform in speed and flux) is decelerated and that radio astronomers map only the brightest of these. As VLBI observations of M87 illustrate, several different speeds can be observed in a single jet. [15] (The kinematic possibilities are even richer if, as we shall argue below, the jet is magnetically dominated.)

Jet Speed: How fast are the jets? We know that they should have Lorentz factors at least as large as the observed superluminal speeds (in units of c). They may have to be even faster. Of considerable importance in this respect are recent reports of *intraday variability* [16] which, if interpreted naively, indicate that $\gamma_j \gtrsim 100$ in the most extreme cases. (This interpretation is not obligatory. Refractive scintillation in our interstellar medium may also be responsible for this variation. The matter will probably be settled fairly soon.)

The proposed deceleration of jets from ultra-relativistic speed is not just confined to the high power, FR2 sources. There is direct evidence for relativistic speeds in the lower power FR1 sources. [17] We may also see indirect evidence for deceleration in the form of *gaps*. [18] Observationally, these gaps are the regions of small or undetectable surface brightness at the base of low power jets. Physically, I suspect that they are caused by the deceleration of the flow which causes the emission cone to broaden until it includes the observer.

γ-RAY JETS

γ-ray photosphere: Relativistic expansion of compact radio sources was originally proposed to avoid the "inverse Compton problem". [19] However, it has never been clear whether the absence of catastrophic inverse Compton losses is a consequence of a natural feedback mechanism or a selection effect. I want to argue the former case on the basis of the γ-ray observations.

Let us first consider the escape of a beam of high energy γ-rays independent of their origin. If they are emitted too close to the nucleus they will be converted into electron-positron pairs. For the EGRET energies, most of the opacity is contributed by soft X-rays and far UV radiation. It is helpful to define a γ-ray photosphere beyond which the absorption optical depth to infinity is $\lesssim 1$. In several papers presented at this meeting, [20] [21] [22] it is supposed that the soft photons are also beamed, either because they originate in a compact region close to the base of the jet, or because they are either synchrotron or synchrotron self-Compton photons [8] [23] created in the frame of the outflowing plasma in the jet. Now, in the case of 3C279, we have direct evidence for the presence of a compact X-ray source. [11] In addition, inverse Compton radiation of X-rays by less energetic electrons than those responsible for the γ-rays in the outflowing jet seems inevitable. X-rays from both of these sources can catalyze pair production. However, I argue that a much larger pair production opacity is due to the minority of X-rays that originate close to the base of the jet and are scattered through large angles by intervening free electrons associated with the broad emission line cloud region through which the jet must propagate. Although only a minority of X-ray photons are likely to be scattered across the γ-ray beam, they will appear much more energetic to the γ-rays and will be encountered more frequently.

Let L_{Ei} be the isotropic luminosity per unit energy of soft UVX photons and let L_{Ej} be the corresponding beamed UVX luminosity per unit energy. The dominant pair production opacity is dominated by X-rays with near threshold energy

$$E_X \sim \frac{2(m_e c^2)^2}{(1-\cos\psi)E_\gamma} \sim 0.5 \left(\frac{E_\gamma}{1\text{GeV}}\right)^{-1} \text{keV}$$

where E_γ is the γ-ray energy and ψ is the angle between the photons in the AGN frame. We find that Thomson-scattered soft photons dominate the pair production opacity for γ-rays of energy E_γ produced at radius r if the direction-averaged local Thomson optical depth associated with an octave of radius r to $2r$ satisfies the inequality

$$<\tau_T>(r) \gtrsim \frac{L_{Ej}\left(4m_e^2 c^4/E_\gamma \theta^2\right)}{4L_{Ei}\left(2m_e^2 c^4/E_\gamma\right)}.$$

If we consider ~ 1 GeV observations of 3C279, and adopt a single X-ray spectrum, $L_{Ei}, L_{Ej} \propto E^{-\alpha_X}$, with $\alpha_X \sim 0.7$, $\theta \sim 0.1$, then we find that this condition becomes $\tau_T \gtrsim 0.01 L_{Ej}(1\text{keV})/L_{Ei}(1\text{keV})$. I shall assume, henceforth, that this condition (which is satisfied by models of broad emission line clouds) is also satisfied by the source and that, in general, the scattered ambient X-rays are of more importance than the beamed radiation.

The local Thomson depth $<\tau_T>$, associated with an octave of radius, is likely to decrease with radius. If we adopt a variation $<\tau_T>(r) \propto r^{-\beta}$, we find that the γ-ray photosphere is located at a radius r_γ satisfying

$$r_\gamma(E_\gamma) \sim \frac{\sigma_{PP} <\tau_T>(r_\gamma)}{\pi c} L_E(2m_e^2 c^4/E_\gamma)$$

$$\sim 10^{17} \left[L_{E55}(1\text{keV})\right]^{\frac{1}{(1+\beta)}} \left[\frac{<\tau_T>(10^{17}\text{cm})}{10^{-2}}\right]^{\frac{1}{(1+\beta)}} \left[\frac{E_\gamma}{1\text{GeV}}\right]^{\frac{\alpha_X}{(1+\beta)}} \text{cm}$$

where $\sigma_{PP} \sim 0.2\sigma_T$ is the pair production cross section near threshold and $L_E \equiv 10^{55} L_{E55} \text{s}^{-1}$ is the X-ray spectral luminosity. In the case of 3C279, we adopt for illustration, $L_{E55}(1\text{keV}) \sim 5, <\tau_T> \sim 0.01(r/0.1\text{pc})^{-0.5}$ to obtain $r_\gamma \sim 0.1(E_\gamma/1\text{Gev})^{0.5}$pc. Note that, in general, the γ-ray photospheric radius increases with E_γ. Therefore at a given radius r, there is a maximum energy $E_{\gamma m}(r)$ at which a γ-ray can escape.

Particle Acceleration and γ-ray emission: We have not yet considered the origin of the γ-rays. Again, there are many possibilities, most of which have already been considered in the literature. Consistent with the choice above, I shall assume that the dominant source is inverse Compton scattering of the ambient UVX radiation field by locally accelerated relativistic electrons and positrons. (Acceleration at reverse shocks formed in the decelerating flow provides one possible means of injecting these particles.) This process is complementary to γ-ray absorption by pair production. (This option is also explored at this meeting by Sikora, Begelman & Rees. [24]) In fact, if we believe that the pair production opacity is dominated by the unbeamed, ambient radiation, then we can be even more confident that this is true of the inverse Compton emission. This is because the ratio of the inverse Compton emission at energy E_γ of beamed soft photons to that of Thomson-scattered ambient photons is smaller than the corresponding ratio of the pair production opacities for γ-rays of similar energy by a factor $\sim \theta^2$.

Let us, for simplicity, continue to approximate the ambient UVX spectrum as a power law with $0.5 < \alpha_X < 1$. Particles that are accelerated to an energy at which the γ-ray opacity exceeds unity, will radiate most of their energy as high energy γ-rays. These γ-rays each create an electron and a positron, with roughly half the initial γ-ray energy, more or less on the spot. Subsequent pair production will result in an effective degradation of the original accelerated particle energy into many γ-rays near the threshold escape energy, $E_{\gamma m}(r)$. This is roughly what would have happened had the electrons (or positrons) been injected at the threshold energy.

What is the radiative efficiency of this inverse Compton emission? A straightforward estimate of the inverse Compton cooling length (measured in the AGN frame) of an electron of energy E gives

$$\frac{L_{\text{cool}}}{r} \sim \left(\frac{\sigma_{PP}}{\sigma_T}\right) \left(\frac{E}{E_{\gamma m}(r)}\right)^\alpha,$$

up to a numerical factor of order unity. As $\sigma_{PP} \sim 0.2\sigma_T$, we therefore expect that the electrons and positrons of energy $\lesssim 0.1 E_{\gamma m}$ will not be able to cool radiatively on the outflow timescale and, instead, will lose most of their energy through adiabatic expansion. The addition of, for example, a large ambient optical-infrared spectral component will enhance the cooling rate but reduce the γ-ray radiative efficiency.

This cooling criterion has two further consequences. Firstly, for the assumed UVX spectrum, most of the γ-ray emission from radius r will be at an energy close to $E_{\gamma m}(r)$. There is therefore an approximate *energy-radius mapping* that is monotonic as far the emergent radiation is concerned. Secondly for radii beyond r_{ann} where the escape energy is $E_{\gamma m}(r_{ann}) \sim m_e c^2$, the newly formed electrons and positrons will not be able to cool to non-relativistic energy. This means that they will not be able to annihilate.

If there is an energy-radius mapping, then this implies that the shape of the observed γ-ray spectrum will reflect the variation of the net dissipation rate, through particle acceleration, with radius. The reported γ-ray spectral indices (still adopting a radio astronomer's usage) are typically $\alpha_\gamma \sim 1 - 1.5$. If, for simplicity, we assume that the rate of dissipation of jet energy in form of the accelerated electrons and positrons varies as $dL_{e\pm}/d\ln r \propto r^{-\delta}$, then we expect that the γ-ray spectral index satisfies

$$\alpha_\gamma \sim 1 + \frac{\alpha_X \delta}{(1+\beta)}$$

For 3C279, $\alpha_\gamma \sim 1$ [9] suggesting that $\delta \sim 0$. In general, a quite detailed understanding of the flow conditions would be necessary to furnish an *a priori* estimate of the exponent δ.

γ-*ray variability*: Observed γ-rays must come from outside the photosphere. However, if the emission is variable, then they cannot come from too large a radius. Specifically, barring kinematic artifices which are unlikely to be generally appropriate, if the source region moves with a Lorentz factor γ_2, the γ-ray emission will come from within a radius $r \lesssim \gamma_2^2 t_{var}/c$, where t_{var} is the variability timescale. For 3C279, $t_{var} \sim 3 \times 10^5 s$, suggesting that $r \lesssim 10^{16} \gamma_2^2$cm. At an energy $E_\gamma \sim 1$ Gev, this implies that $\gamma_2 \gtrsim 3$, and as a consequence, the beaming fraction satisfies $f \lesssim 0.03$, consistent with the estimate above.

There is a second consequence of an energy-radius relation and this is that the typical variability timescales ought to *lengthen* with increasing γ-ray energy. As explained above, this is the opposite behavior from that expected (and generally observed) in the compact radio sources.

IMPLICATIONS FOR JETS

There are some further consequences of this (over) simple model. The first is that the absorption of γ-rays within the γ-ray photosphere leads to copious pair production. In fact, this will go on at such a rate that any baryons that may be present initially in the jet may become irrelevant. (Entrainment of baryons from the outflow surrounding the relativistic jet may be important, however.) Secondly, I have argued on the basis of the radio observations that the jet decelerates as it propagates outward from the smallest VLBI scales. It is natural to extrapolate to smaller radii and suppose that the Lorentz factor is even larger there. This is consistent with, though not required, by the γ-ray observations and the radio intraday variability. Thirdly, rapid jet deceleration is also an expected property of relativistic flows when they become radiatively efficient [13] because the photons remove significant momentum as well as energy from the jet flow. In fact the influence of radiative drag on a particle-dominated outflow close to the central black hole is so severe that it has been argued that the bulk Lorentz factor must be $\gamma_j \lesssim 10$.[25]

Finally, I have argued that there exists a radius r_{ann} in the outflowing jet within which electrons and positrons can cool to subrelativistic energy and so undergo two-photon annihilation. (Presuming that most of this annihilation happens within a decelerating relativistic jet and the pairs cool from ultrarelativistic energy there seems little prospect of observing the 0.5 MeV line.) For radii $r \lesssim r_{ann}$, we expect that the electron-positron density will be maintained at a value such that the annihilation time equals the outflow time, measured now in the jet frame. This limits the electron/positron density to $\sim 1/\sigma_T r \gamma_j$. Equivalently, the contribution to the jet thrust associated with the electrons and positrons is limited to

$$P_{je_\pm} \sim 10^{34} \left(\frac{r}{10^{17} \text{cm}}\right) \left(\frac{\gamma_j}{10}\right) \left(\frac{\theta}{0.1}\right)^2 \text{ dyne}.$$

This is generally small compared with total jet thrust inferred on the basis of large scale radio structure in the case of the powerful, FR2 radio sources.

However, beyond r_{ann}, the jet will become increasingly particle-loaded and we expect that an increasing fraction of its momentum flux will be carried mechanically. This will allow more and more fluid-decelerating and particle-accelerating shocks to form in the jet and further promote radiative deceleration and dissipation of the bulk kinetic energy.

These arguments for a large initial Lorentz factor and a small initial matter component both point to an alternative initial carrier of jet momentum. One possibility that has been advocated elsewhere for different reasons [26] is electromagnetic Poynting flux. I am therefore led to speculate that the jet energy is created in a fairly pure, electromagnetic form and that, as the jet propagates away from the nucleus, pairs are formed at an increasing rate and they interact with the surrounding radiation field so that further particle acceleration may occur. (These comments are also of relevance to cosmological models of γ-ray bursts.)

The source of this electromagnetic power may be the spin of the central black hole. [26] A more complete, qualitative interpretation is possible if we further suppose that black holes are surrounded by magnetized accretion disks from which gas is flung centrifugally along the poloidal field and collimated by the toroidal field with speeds related to the disk speed at the foot of the field lines. These jets may be responsible for extracting much of the angular momentum of the accreting gas. The relativistic electromagnetic energy derived from the spinning hole then constitutes a powerful, filamentary core of a much slower and more extensive outflow originating from the jet. Perhaps, in the lower power FR1 radio sources, the relativistic core contains less thrust than the subrelativistic sheath and the asymptotic speed is closer to that of the sheath, whereas the converse is true is the FR2 sources.

I will conclude by returning to the dichotomy with which I began this talk, the difference between radio-quiet and radio-loud AGN. It is clear that this distinction is maintained over a long period of time as the compact radio cores appear to survive as long as the extended components. One of the few sufficiently durable features of an AGN, that might be responsible for this difference, is the spin of the central black hole. Perhaps, only those holes that spin rapidly (relative to their maximum rates), can form relativistic, electromagnetic jets intense enough to overcome the inertia of their sub-relativistic, particle-dominated sheaths. (This explanation requires that the accretion be "demand-limited" instead of "supply-driven".) [27] Independent of the truth of these conjectures, it is clear that the CGRO γ-ray observations of core-dominant radio sources are of immense importance for the interpretation of VLBI jets and are forging a crucial link between the giant double radio sources and their putative prime movers some ten orders of magnitude smaller in size.

ACKNOWLEDGEMENTS

I thank Mitch Begelman, Paul Coppi and Martin Rees for helpful correspondence and Y. C. Lin, Peter Michelson and Pat Nolan for explaining the available observations. Support under NASA grants NAGW 2816 and 2372 is gratefully acknowledged.

REFERENCES

1. R. Antonucci, *Annu. Rev. Astron. Astrophys.*, (1992, in press).
2. C. D. Dermer & R. Schlickeiser, *Science* **157**, 1642 (1992).
3. G. Ghisellini & L. Maraschi, *Ap. J.* **340**, 181 (1989).
4. C. E. Fichtel et al., *These Proceedings*, (1993, in press).
5. W. N. Johnson et al., *These Proceedings*, (1993, in press).
6. W. Collmar et al., *These Proceedings*, (1993, in press).
7. C. W. Punch et al., *Nature* **358**, 477 (1992).
9. R. C. Hartman et al., *Ap. J.* **385**, L1 (1992).
8. A. Königl, *Ap. J.* **243**, 700 (1981).
10. L. J. Eachus & W. Liller, *Ap. J.* **200**, L61 (1975).
11. F. Makino et al., *Ap. J.* **347**, L9 (1989).
12. R. D. Blandford & C. F. McKee, *Phys. Fluids.* **19**, 1130 (1976).
13. R. D. Blandford & C. F. McKee, *MNRAS* **180**, 343 (1977).
14. M. Hoshino, J. Arons, Y. Gallant & A. B. Langdon, *Ap. J.*, (1992, in press).
15. J. Biretta, Astrophysical Jets. ed. M. Livio, C. O'Dea & D. Burgarella. Cambridge: Cambridge University Press (1992, in press).
16. A. Quirrenbach et al., *Astron. Astrophys.* **372**, L71 (1991).
17. R. Laing, Astrophysical Jets. ed. M. Livio, C. O'Dea & D. Burgarella. Cambridge: Cambridge University Press (1992, in press).
18. A. H. Bridle & R. Perley, *Annu. Rev. Astron. Astrophys.* **22**, 319 (1984).
19. M. J. Rees, *Nature* **211**, 468 (1966).
20. P. Coppi et al., *These Proceedings*, (1993, in press).
21. C. Dermer, *These Proceedings*, (1993, in press).
22. S. D. Bloom & A. P. Marscher, *These Proceedings*, (1993, in press).
23. L. Maraschi, G. Ghisellini, A. Celotti, *Ap. J.* **397**, L5 (1992).
24. M. Sikora, M. C. Begelman & M. J. Rees, *These Proceedings*, (1993, in press).
25. F. Melia & A. Königl, *Ap. J.* **340**, 162 (1989).
26. R. D. Blandford, Theory of Accretion Disks. ed. P. Meyer, W. Duschl, J. Frank & E. Meyer-Hofmeister, p.35 (1989).
27. T. A. Small & R. D. Blandford, *MNRAS*, (1992, in press).

GAMMA RAYS FROM ACTIVE GALACTIC NUCLEI

Charles D. Dermer

E. O. Hulburt Center for Space Research,
Naval Research Laboratory, Code 7653, Washington, DC 20375

ABSTRACT

We summarize the properties of active galactic nuclei (AGN) detected at hard X-ray and gamma ray energies. Two classes of gamma-ray AGN are evident. AGN in the first class display power-law X-ray spectra which soften considerably above ~ 100 keV, although periods of intense emission near 1 MeV may occur. These sources, generally associated with Seyfert 1 nuclei, have so far not been detected at $\gtrsim 10$ MeV photon energies. AGN in the second class are observed to radiate intensely at > 100 MeV energies. These objects are associated with extragalactic sources that display blazar properties. Models for the high energy radiation emitted by the two types of AGN are briefly reviewed. Thermal Comptonization models are in accord with the soft gamma-ray spectra observed from AGN of the first class. For the second class of AGN, we consider a model where nonthermal electrons and positrons in a relativistically outflowing jet Compton-scatter accretion disk photons. The power in gamma rays can dominate the bolometric luminosity due to beaming of the emission, as is observed in the multiwavelength spectra of these sources.

INTRODUCTION

Our present understanding of AGN is hindered by the restricted wavelength range and limited imaging capability of astronomical detectors. Several generalities have nevertheless emerged from the detailed phenomenology.[1,2,3] Radio-quiet AGN are generally associated with spiral galaxies, whereas radio-loud AGN are usually found in elliptical galaxies. Most of the radio-quiet AGN are classified as either Type 1 or 2 Seyfert (Sy) galaxies, a division that is widely understood to depend on the orientation of the observer with respect to the central engine. Sy 1 galaxies display broad permitted and narrow forbidden lines, a strong ultraviolet excess, and a fairly hard X-ray continuum. It is generally thought that in the case of Sy 1s, we are almost directly viewing the central nuclear source. On the other hand, Sy 2 galaxies display narrow emission lines and weak X-ray emission. It is generally thought that here the central source is obscured by a dense layer of dust or gas, which would most likely occur if we are observing the nucleus at shallow angles to the plane of the galaxy. QSOs may simply be scaled-up versions of Sy 1s, but at sufficiently large distances that the associated galaxies cannot be resolved.

Sy 1 nuclei are also often observed in radio galaxies. Indeed, radio-loud AGN may differ from Seyferts principally by the addition of radio jets. The radio jets introduce an additional orientation-dependent effect which could make radio-loud AGN appear, at different observing directions, as radio galaxies, steep-spectrum radio quasars, flat-spectrum radio quasars, or blazars. BL Lac objects apparently differ from quasars through the lack of emission line clouds, as well as having weaker radio jets.

By providing a detailed picture of AGN at gamma-ray energies, the Compton Observatory is significantly increasing our knowledge about the central engines of AGN. The observations show that there are two types of AGN so far detected at hard X-ray and gamma-ray energies. Those in the first class, which we call *gamma-ray soft AGN*, display hard X-ray spectra that show spectral softening at photon energies $\gtrsim$ 100 keV, and emit no measurable high energy ($\gtrsim$ 10 MeV) radiation. Sources in this class are primarily associated with radio-quiet Seyfert 1 galaxies, although this class also includes radio galaxies with Seyfert nuclei so oriented that we are not viewing along the jet axis. The emission from the Seyfert nucleus therefore dominates the observed gamma radiation, because the high energy emission from a radio galaxy is evidently strongly beamed along the radio axis. The soft gamma-ray spectra from Seyfert nuclei may be most simply explained with thermal Comptonization models.

Sources in the second class, which we call *gamma-ray hard AGN*, radiate intensely in the medium-energy ($\sim$ 100 MeV) gamma ray regime. These objects are associated with radio-loud AGN with blazar properties, implying that we are looking almost directly down the axis of a radio jet. The intense gamma radiation is anisotropically produced in an outflowing relativistic jet. In this paper, we describe a model where the electrons and positrons in the outflowing plasma Compton-scatter radiation from an accretion disk. This model can explain the time variability, the overall high energy spectral shape, and the dominance of the bolometric luminosity of gamma-ray hard AGN in the medium energy gamma-ray regime.

GAMMA-RAY OBSERVATIONS OF ACTIVE GALACTIC NUCLEI

Table 1 is a list of AGN that have been positively detected at photon energies > 20 keV. These include 11 of the 12 sources in the HEAO-1 sample[4] (the missing one, Mrk 335, was not observed at energies > 20 keV), the 16 AGN observed with the Energetic Gamma Ray Experiment Telescope (EGRET) in the medium energy gamma-ray regime[5], and three others. Note that this table is not meant to be complete. It omits, for example, some quasars that were weakly detected by Ginga[6] above 20 keV.

The Oriented Scintillation Spectrometer Experiment (OSSE) has reported positive detections of 11 AGN and upper limits for 6 AGN at photon energies $\gtrsim$ 80 keV.[7,8,9] One of the objects – 3C 111 – was positively detected during one viewing period but not during another, suggesting variability of this source at soft gamma-ray energies. Of the 11 sources detected in the OSSE energy range, 8 are classified as Sy 1s, 2 are quasars, and the remaining one is Cen A, the famous radio galaxy. OSSE upper limits have also been reported[7] for NGC 1068, a Sy 2 galaxy, and NGC 4593, a Sy 1.9 galaxy. By extrapolating HEAO-1 spectral data for Sy 1 galaxies into the OSSE energy range, many more, and stronger, detections of AGN were expected than have been observed. The implication is that the spectra of Sy 1 nuclei soften considerably when going from the hard X-ray to the soft gamma-ray regime. A quantitative measure of this softening will have important ramifications on source models, because the OSSE energy range covers the critical regime near 0.5 MeV where signatures of electron-positron annihilation are expected.

The quasars 3C 273 and 3C 279 have also been detected with OSSE.[9] These two sources are the most well-known representatives of the AGN detected at > 100 MeV energies with the EGRET instrument. Sixteen extragalactic AGN have been reported as sources of > 100 MeV gamma rays at the time of this writing.[5]

Table 1. Active Galactic Nuclei Detected at > 20 keV Photon Energies

Source	Type	20-165 keV	0.1-0.5 MeV*	> 100 MeV	> 0.5 TeV
ESO 141-55	Sy 1	Y	Y?	N	
NGC 6814	Sy 1	Y			
MCG-5-23-16	Sy 2	Y		N	
NGC 3783	Sy 1	Y	Y?	N	
Mrk 509	Sy 1	Y		N	
NGC 1275	RG, Sub, Sy 2	Y	N	N	
3C 120	Sy 1, RG, SL	Y	Y?		
Mrk 279	Sy 1	Y	Y	N	
3C 390.3	Sy 1, RG, SL?	Y	Y		
NGC 5548	Sy 1	Y			
NGC 4151	Sy 1-1.5	Y	Y	N	
MCG+8-11-11	Sy 1	Y	Y	N	
Cen A	RG, Sub	Y	Y	N	Y?
3C 111	Sy 1, RG, SL?		Y/N		
3C 273	FSRQ, SL	Y	Y	Y	N
3C 279	FSRQ, SL	Y	Y	Y	N
0454-463	FSRQ			Y	
3C 454.3	FSRQ, SL			Y	
PKS 0537-441	FSRQ, BL Lac			Y	
PKS 0420-014	FSRQ			Y	
PKS 0235+164	FSRQ, BL Lac, SL?			Y	
0208-512	FSRQ			Y	
CTA 102	SSRQ, SL			Y	
4C 15.05	RG?			Y	
4C 38.41	FSRQ			Y	
PKS 0528+134	FSRQ			Y	
0836+710	FSRQ, SL			Y	
0716+714	FSRQ, BL Lac, SL?			Y	
1606+106	FSRQ			Y	
Mrk 421	El, BL Lac, SL?		N	Y	Y

SL: superluminal; Sub: subluminal; RG: radio galaxy; El: elliptical galaxy

The identification of these sources with AGN that display blazar properties, which include high optical polarization, rapid optical variability, flat-spectrum radio emission from a compact core, and apparent superluminal motion, makes it virtually certain that the > 100 MeV gamma rays are produced in association with a radio jet.

The EGRET sources Mrk 421, a BL Lac object, and PKS 0528+134, a flat spectrum radio quasar (FRSQ), have also been viewed with OSSE but not positively detected.[10] Although the EGRET and OSSE observations were not contemporaneous, the lack of positive detections of these sources with OSSE implies that the spectra of Mrk 421 and 0528+134 soften between the hard X-ray regime and the medium energy gamma-ray regimes. Moreover, contemporaneous EGRET[11,12] and Compton Telescope[13] (Comptel) observations directly show spectral softening between $\sim$ 1 and 30 MeV in both 3C 273 and 3C 279, indicating that spectral softening may be characteristic of gamma-ray hard AGN as a class.

Variability of the > 100 MeV gamma ray flux from 3C 279 on a time scale $\lesssim$ 2 days is seen in the EGRET data.[14] Rapid variability is also observed from 0528+134 on the same time scale,[15] and may be a general feature of the gamma-ray hard AGN. This suggests that the emission region is within $\sim 10^{16}$ cm of the central source. Variability of 100 keV emission from Cen A is observed with OSSE[16] on an even shorter time scale of $\sim$ 6 hours. However, Cen A has not been reported as a source of > 100 MeV radiation, so it is not certain whether its radiation originates from the jet or the central source. In general, however, the variability observations imply that the high energy flaring emission is produced within a fraction of a pc from the central source, which may rule out models where the bulk of the high energy radiation is produced in the distant ($\gtrsim$ 1 pc) radio lobes.

TWO CLASSES OF GAMMA-RAY AGN

For the AGN that have been detected above 20 keV, Table 1 also lists the energy ranges in which detections (Y) and nondetections (N) took place. Blank entries mean that there have been no observations in that energy range. The 20-165 keV column refers to HEAO-1[4] and Ginga[17] observations, the 0.1-0.5 MeV column to OSSE observations, the > 100 MeV column to EGRET observations, and the > 0.5 TeV column to Whipple Observatory observations[18,19], although the ultra-high energy entry for Cen A refers to its unconfirmed detection by Grindlay et al.[20] at $\geq$ 0.3 TeV. Several of the sources in the EGRET column are listed as nondetections if they were z-axis pointing targets and have not been reported as sources of medium energy gamma rays. It is possible, though, that these AGN are > 100 MeV photon sources whose signal is weaker than the 5σ detection threshold criterion required of EGRET detections.

It is important to establish whether the probability of detection of an active galaxy is related to the energy range in which it is observed, and whether this probability is correlated with source type. For this purpose, we calculate the hardness ratios between various energy bands. Unfortunately, the HEAO-1 sources are mostly Sy 1 galaxies, and therefore do not provide a very diverse sample. It is possible to extend the data base by using the 10 keV fluxes for the sources listed in Table 1. We caution that the observations are almost all noncontemporaneous. Fig. 1 shows the ratios of the integrated > 100 MeV photon flux to the differential flux at 10 keV. The EGRET data is obtained from Ref. [5], and we use a value of 1.0×10^{-7} photons (> 100 MeV) cm^{-2} s^{-1} for the EGRET upper limits. We use the 10 keV flux data from the HEAO-1 measurements, except for 3C 279, 3C 454.3, and Mrk 421, which are from Ginga observations.[4,17] Note that the vertical extent of the hardness ratios reflects the range of flux values arising from source variability, not measurement error. Also, we use the larger upper limit when more than one 10 keV flux measurement is available.

Fig. 2 shows the hardness ratios of the integrated > 100 MeV photon flux measured with EGRET to the differential 100 keV flux measured with OSSE. An important point to note is that relatively few AGN detected with EGRET have also been observed with OSSE. Of the four sources in this category, two yielded positive OSSE detections, and the nondetection of PKS 0528+134 may reflect the limited exposure (3.2×10^4 seconds with only 2 detectors).[10] Thus we can expect that additional OSSE observations of the gamma-ray AGN detected with EGRET are likely to result in a substantial number of positive detections.

Figs. 1 and 2 show an interesting result. The Seyferts and radio galaxies do not radiate > 100 MeV gamma rays in the same proportion, compared to their hard X-ray emissions, as do the quasars and BL Lac objects. For the most extreme cases, namely NGC 4151, Cen A, 3C 273 and 3C 279, the hardness ratios differ by nearly two orders of magnitude. These results are consistent with the earlier COS-B and SAS-2 results which showed that a spectral softening must occur between the hard X-ray and medium energy gamma-ray regime.[21] But we now have a larger sample which is beginning to show that, as a class, Sy 1s and radio galaxies are qualitatively different in their high energy spectral properties than quasars and blazars.

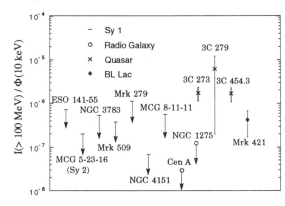

Fig. 1—Hardness ratios of gamma-ray AGN. Here we plot the ratio of integrated photon flux at energies > 100 MeV to the differential photon flux at 10 keV. No > 100 MeV emission has yet been detected from radio galaxies that are also classified as Seyferts.

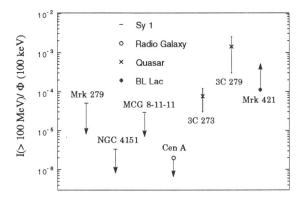

Fig. 2—Hardness ratios of gamma-ray AGN. Here we plot the ratio of integrated photon flux at energies > 100 MeV to the differential photon flux at 100 keV.

We also see from Figs. 1 and 2 that the gross spectral features of the high energy emission of radio galaxies and Sy 1s, as deduced from their hardness ratios, are similar. This suggests that to the extent that beamed emission from a jet is unimportant, the X-rays and gamma rays produced by radio galaxies originate from a Seyfert-type nucleus rather than from a jet. Observations of apparent superluminal (SL) motion in a source guarantee that directed relativistic outflow occurs and that we are viewing at a small angle to the jet axis. Thus we expect that the spectral properties of radio galaxies should be correlated with the appearance of SL motions: no SL or subluminal motion should reveal the Seyfert nucleus with virtually no jet emission, whereas sources with increasingly large SL velocities should be accompanied by the > 100 MeV jet emissions.

Cen A and NGC 1275 have so far not displayed SL motions, nor have they been reported as strong sources in the EGRET energy range, in accord with the interpretation proposed here. An interesting test is provided by the radio galaxies 3C 120, 3C 111, and 3C 390.3, each of which have been monitored for SL motions with VLBI techniques. Because 3C 120 has the firmest evidence for SL motion,[22] whereas the evidence is much weaker for 3C 111 and 3C 390.3,[23] we expect that 3C 120 is the most likely source of the three to have significant jet emission and be observable with EGRET. A spectrum more characteristic of a bare Sy nucleus is expected from the other two. Detailed spectral data from the OSSE, Comptel and EGRET detectors, in conjuction with lower energy measurements, have the potential to confirm or refute this classification scheme.

MODELS FOR GAMMA-RAY SOFT AGN

Explanations for the high energy emission from non-blazar AGN have divided into thermal and nonthermal models. Rather than attempt an in-depth review of the observations and models, I will briefly discuss the dominant modelling trends.

The rapid variability of the X-ray emission from AGN implies that the emission site is very close to the central powerhouse.[24] The HEAO-1 A-2 and A-4 experiments[4,25] showed that for 12 AGN at energies above 2 keV, the mean photon spectral index $\langle \alpha_x \rangle = 1.68$ with a dispersion $\sigma = 0.17$. Most of these sources were Sy 1s. The HEAO-1 results were confirmed by Exosat observations[26] in the 2-10 keV range, showing that $\langle \alpha_x \rangle = 1.70$ and $\sigma = 0.174$ in a sample of 42 objects, again consisting mostly of Sy 1s.

The power-law hard X-ray continuum is open to either a nonthermal or thermal interpretation. It was quickly noticed[25] that the mean X-ray spectral index is nearly equal to the average radio spectral index in radio jets where the radiating particles are thought to be energized by shock acceleration. An $\alpha = 1.7$ X-ray spectrum is produced when soft photons are Compton scattered by a power-law electron energy distribution with an index of 2.4. The formation of the electron spectrum could result from strong shocks accompanied by an energy-dependent leakage lifetime, as in the case of cosmic rays in our Galaxy, by somewhat weaker shocks, or as secondaries from hadrons with a 2.4 spectral index. Some work has been devoted to a shock acceleration scenario for the X-rays.[27,28] However, if the X-ray spectrum extends into the gamma-ray regime, the variability time scale, coupled with the observed source luminosity, implies that the upscattered gamma rays are optically thick to $\gamma - \gamma$ attenuation with the lower energy X-rays. Thus, a calculation of this model is incomplete unless the accompanying pair processes are treated.

The importance of pair attenuation of gamma rays is governed by the dimensionless compactness parameter

$$\ell \equiv \frac{L\sigma_T}{Rm_ec^3} = \frac{L}{R(3.7 \times 10^{28} \text{ergs s}^{-1} \text{ cm}^{-1})}, \qquad (1)$$

where R is the characteristic size of the emitting region, and L is the luminosity of the source above the pair production threshold (thus ℓ is technically dependent on photon energy). Nonthermal pair cascade models self-consistently treat the attenuation of upscattered photons, the reinjection and cooling of the produced pairs, and the subsequent rescattering and attenuation of the radiation. The driving force for the pair cascade models has been the attempt to find a natural explanation for the observed narrow range of spectral indices in Sy 1s. It was noted by Kazanas[29] that a cooling distribution of electrons injected with a sufficiently hard spectrum (he considered a monoenergetic injection source) produces an X-ray spectrum with $\alpha = 1.5$. In a very compact environment with $\ell \gg 1$, therefore, the upscattered photons are optically thick to pair production, constituting a new injection source of pairs. This would soften spectra to $1.5 \lesssim \alpha < 2$, providing a potential explanation for the HEAO results.

An immediate prediction of this model was an $\alpha - \ell$ correlation,[30] which was not observed. The lack of correlation could, however, be attributed to the uncertainty in calculating ℓ, which was originally obtained by setting $R = c\Delta t_X$, where Δt_x is the shortest X-ray variability time scale. It was also recognized that a softening in the spectra of Sy galaxies was required in order not to overproduce the observed gamma-ray background radiation. Attenuating the gamma rays requires large compactness parameters, but then $\alpha \to 2.0$, in conflict with the HEAO results. A possible way out of this difficulty was to invoke reflection from cold matter,[31] which is indeed inferred from observations of the fluorescence iron line emission. Because a reflection spectrum is harder than the incident spectrum at X-ray energies, the high compactness $\alpha \cong 2.0$ spectra could in principle be hardened to $\alpha \approx 1.7$.

Many variations of this basic model have followed, for example, by including magnetic fields[32] or reacceleration of the produced pairs.[33] The essential premise of the pair cascade models remains, however, that the injection of high energy electrons initiates a pair cascade. Thus at some level we should expect a pair annihilation feature from this class of models. The identification of this signature represents a principal goal of the OSSE instrument.

Thermal models, by contrast, do not necessarily involve prolific pair production and attenuation unless the electron temperature $\gtrsim 100$ keV. Thermal Comptonization of soft photons was considered as a model for black hole sources even earlier than the pair cascade models.[34,35] A simple spectral form was analytically derived for this system in the Thomson regime for a homogeneous sphere or uniform disk by Sunyaev and Titarchuk.[36] During the past decade, considerable effort was devoted to a self-consistent treatment of photon and pair processes when the temperature is sufficiently high that pair production is important.[37] More recently, the inclusion of pair processes in the accretion disk equations has led to the discovery of luminosity limits on the solutions.[38]

Earlier gamma-ray balloon and satellite experiments have shown that MeV bumps or annihilation features may represent a common spectral state of extragalactic AGN.[21] One of the surprising results from OSSE is that pair annihilation signatures have so far not been detected in the spectra of either galactic or extragalactic black-hole candidate sources.[8,39] The extremely soft spectrum of NGC 4151 measured by OSSE may be in conflict with all pair cascade models published to date.[9] Its spectrum is, however, accurately represented by the simple Sunyaev-Titarchuk spectral form. This is also the case for galactic black-hole sources, although a small hard excess may represent deviations due to the Thomson limit assumptions in the derivation of the Sunyaev-Titarchuk expression.[39] Fig. 3 shows a νF_ν diagram for NGC 4151. Scheduled Phase 2 OSSE observations will help show whether the soft gamma-ray spectrum shown here[8] represents the dominant spectral state of NGC 4151 and, indeed, of gamma-ray soft AGN in general.

Recent work on thermal models has been devoted to generating more realistic thermal Comptonization spectra in the context of accretion-disk theory. For example, a central X-ray emitting hot plasma may reprocess photons from an optically thick cool outer blackbody or modified blackbody accretion disk.[40] Variations in the level of soft photons may produce the observed X-ray spectral index/luminosity correlation.[41] In view of the Compton Observatory results, such an approach may hold the greatest promise for explaining the spectra of Seyfert nuclei from the optical through the gamma-ray regime.

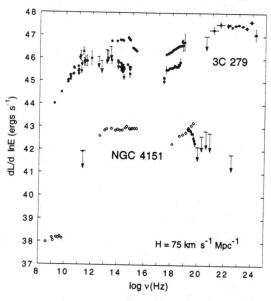

Fig. 3—Broadband power spectra of NGC 4151, a gamma-ray soft AGN, and 3C 279, a gamma-ray hard AGN. References to the gamma-ray data are given in the text. References to the lower-energy data for NGC 4151 and 3C 279 are found in Refs. [42] and [11], respectively. Note that the data are not contemporaneous.

MODELS FOR GAMMA-RAY HARD AGN

Gamma-ray hard AGN are associated with extragalactic sources that display blazar properties, implying that the emission originates in the outflowing plasma that forms radio jets. Fig. 3 shows the broadband power spectrum of 3C 279,[11,13] one of the brightest extragalactic sources of > 100 MeV radiation discovered with EGRET. Note that the luminosity in medium energy gamma rays dominates the bolometric luminosity in the observable wavebands.

Source properties and models of gamma-ray hard AGN have recently been reviewed.[43] Here we discuss a model in which nonthermal energetic electrons in the relativistically outflowing plasma Compton scatter accretion-disk radiation. The observational constraints that support this model are as follows: the gamma-ray emission can vary on < 2 day time scale,[14] implying that the emission site is fairly close to the central powerhouse. The detection of > 100 MeV gamma rays shows that the emission site cannot, however, be too close to the nucleus or else the high energy radiation would be severely attenuated. This implies the radiation originates between $\sim 10^{2-3}$ Schwarzschild radii and a few light-days from the central source. This constraint is stronger for Mrk 421, from which > 0.5 TeV gamma rays have been detected.[19] Another important constraint on source models is that the EGRET observations show that the emission between $\lesssim 100$ MeV and $\sim$ several GeV is well fit by single power laws.[5] In the case of 3C 279, a value of $\alpha_\gamma = 2.0 \pm 0.1$ is reported. Both softer and harder spectra have also been observed from other gamma-ray hard AGN. For example, $\alpha_\gamma = 1.8 \pm 0.2$ for 0235+164, and $\alpha_\gamma = 2.4 \pm 0.2$ for 3C 273.

The power-law spectra suggest that shock acceleration is energizing the particles in the jet. An important question to ask is whether hadrons or leptons are producing the medium energy gamma rays. It is commonly supposed that hadrons are more efficiently accelerated to high energies by shocks than electrons due to the greater energy losses suffered by the electrons. In the EGRET energy range, however, directly accelerated electrons are probably making the bulk of the emission. We argue in favor of this assertion by noting the importance of directly accelerated electrons in other astrophysical scenarios thought to involve shock acceleration. Note first of all that the EGRET detections of gamma-ray hard AGN extend only to several GeV, which are relatively low energies on the scale of the galactic cosmic radiation. And for cosmic rays, roughly equal energy is radiated at $\lesssim 1$ GeV energies by directly accelerated electrons and cosmic ray protons, as can be demonstrated from models of the diffuse galactic gamma radiation. Second, although shock acceleration undoubtedly occurs in the radio lobes of classical doubles such as Cygnus A, the evidence shows that the radiating particles are directly accelerated electrons. If the radio-emitting electrons were instead secondaries from proton-hadron interactions, there would be comparable energy emitted in > 100 MeV gamma rays as in 100 MHz - 10 GHz radio emission. Hence Cygnus A would be a fairly bright EGRET source, because its radio luminosity of 10^{45-46} ergs s^{-1} translates into an integrated $>$ 100 MeV gamma-ray flux of 10^{-10}–10^{-9} ergs cm^{-2} s^{-1}, a factor ~ 3 greater than the EGRET theshold of 3×10^{-11} ergs cm^{-2} s^{-1}. But Cygnus A has not been detected with EGRET,[44] which therefore discounts the hypothesis that its radio-emitting leptons are secondaries from hadronic interactions. Finally, we note that the absence of π^o features in the spectra of gamma-ray hard AGN, which are clearly observable in the EGRET spectra of the galactic gamma radiation[14] and the 11 June 1991 solar flare,[45] makes a secondary production model difficult to support.

If one accepts that directly accelerated electrons make the gamma rays in the EGRET energy range (the particles producing TeV emission from Mrk 421 could very well be hadrons), we can consider two scenarios for the high energy emission involving beamed electrons. In the beaming scenario treated by Königl and his collaborators,[46,47] and also recently by Protheroe, Mastichiadis, and myself,[48] electrons are accelerated directly outward along the jet axis. However, this model is not in accord with the interpretation that electrons are acclerated by shocks, since shock acceleration requires multiple traversals of the shock and a non-monodirectional angular distribution of the energetic particles. An electrodynamic accelerator injecting monoenergetic beamed relativistic electrons along the jet axis is more consistent with the assumptions of this model.

Are gamma-ray spectra calculated by injecting monenergetic high energy electrons outward from a radiating disk in agreement with the observations of gamma-ray hard AGN? Fig. 4 shows calculations of the high energy emission from beamed electrons that lose energy as they Compton scatter disk radiation and travel outward. Here we model the system by a disk of radius R that uniformly radiates thermal emission with temperature $\Theta = kT/m_e c^2$. The electrons are ejected outward along the disk's symmetry axis with initial Lorentz factor

$\gamma_i = 10^3/\Theta$, so that the scattering initially occurs in the Klein-Nishina regime. The values on the curves give the base 10 logarithm of the ratio of R to the Thomson mean free path of an electron at the surface of the disk. As can be seen, Klein-Nishina effects cause a break in the spectrum near $\Theta E \sim 0.2 - 1$, and the lower-energy regions of the spectra have spectral indices $\alpha \leq 1.5$. If this model is valid, spectral breaks near 3 MeV imply that the temperature of the disk is ~ 20 keV, which is not unreasonable in view of the temperatures derived for Sy nuclei with a Sunyaev-Titarchuk spectrum. However, it seems impossible to obtain X-ray spectra with spectral indices softer than 0.5 in the X-ray regime for a mono-energetic injection scenario, and the situation is not improved by calculating angle-dependent rather than the angle-integrated spectra shown here, since angle-dependent spectra that extend into the gamma-ray regime are even harder. The spectra are also rather curved in the high energy regime due to Klein-Nishina effects. Thus this model could be ruled out if the X-ray spectra of the beamed radiation from gamma-ray hard AGN have spectral indices softer than 1.5, although lack of contemporaneous X-ray/gamma-ray observations make this model currently viable.

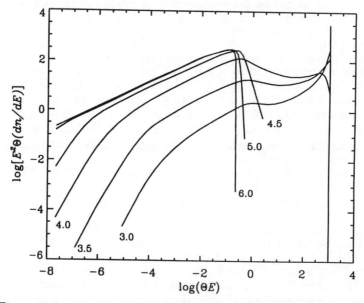

Figure 4. Power spectra as a function of photon energy E for a system where relativistic electrons are ejected outward along the axis of a uniform thermally emitting disk and lose energy by scattering disk radiation.[48] The values on the curve give the base 10 logarithm of the ratio of the disk radius to the Thomson mean free path at the surface of the disk. The injection Lorentz factor $\gamma_i = 10^3/\Theta$, where $\Theta = kT/m_e c^2$ is the dimensionless temperature of the disk.

A second model recently proposed by us,[49] which is consistent with a shock mechanism for electron acceleration, is a bulk outflow scenario in which relativistically outflowing shocks energize electrons that flow outward along the radio axis and lose energy mainly by scattering accretion disk radiation. Softer

X-ray spectra can be produced in this model than in the beaming scenario because the electron spectral index depends on the compression ratio of the shock which can be made arbitrarily weak.

The electrons will in general have a complicated angular distribution in the comoving fluid frame. For simplicity, we assume that this distribution is isotropic in a frame moving outward with bulk Lorentz factor Γ, and that the electrons are injected with a power-law electron energy distribution at height z_i Schwarzschild radii. As the electrons flow out from the system, they lose energy by Compton scattering the disk radiation, and through synchrotron and synchrotron-self Compton losses. The precise nature of the energy losses depends on the accretion disk model. We use a cool outer blackbody solution extending inward to the innermost stable orbit. A more realistic calculation must consider more complicated accretion disk spectra, particularly in view of the well-known instabilities to which the cool disk solutions are subject near the central black hole.

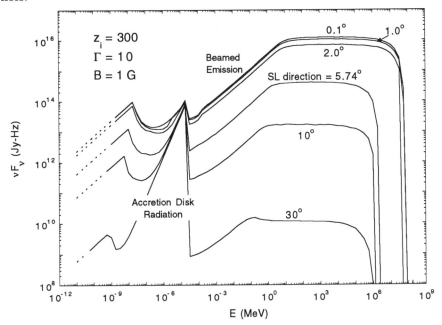

Figure 5. Calculations[50] of angle-dependent broadband spectra for a model of blazars where electrons, injected with a power-law distribution with spectral index = 2, lose energy by Compton scattering and synchrotron emission while flowing outward from the central source with bulk Lorentz factor $\Gamma = 10$. The electrons are assumed to have an isotropic angular distribution in the outflowing plasma frame, and Klein-Nishina effects are not included. The dotted curves show where the effects of synchrotron self-absorption, not treated in this calculation, are important. Other parameters are given in the text. The scale is normalized to a distance of 100 Mpc.

Figure 5 show some representative time-integrated spectra[50] calculated for this system. In the simulation shown here, the electrons are injected at

$z_i = 300$ from a 10^8 $M_\odot$ black hole radiating at 10% of the Eddington luminosity. Nonthermal electrons are injected with a luminosity of 1% of the Eddington luminosity in the rest frame of a relativistically outflowing blob moving with $\Gamma = 10$. The radius of the blob is set equal to 30 Schwarzschild radii. The magnetic field $B = 1$ Gauss, and we assume that neither B nor Γ vary with height z. We also neglect adiabatic losses, which may be important in more realistic calculations. We find that synchrotron and synchrotron self-Compton losses in the 1 Gauss B-field are small in comparison to Compton losses from the disk radiation until the electrons reach $z \gg 10^4$.

Several interesting features of this model should be noted. First, the bulk of the energy is radiated by photons with energies > 100 MeV, in agreement with the EGRET observations. Second, a spectral break of 0.5 is calculated, due to incomplete cooling of the electrons. This produces a spectral softening between the X-ray and medium energy gamma-ray regime, as is observed in these sources. Third, the variability time scale is $\approx z_i/c\Gamma^2 \sim 1\text{day}/\Gamma^2$, consistent with the EGRET observations. Note also that the accretion-disk soft photon source may be concealed by the radio emission when observations are made nearly along the jet axis. Moreover, although the power emitted in radio emission is much less than the power emitted in gamma rays for these parameters, a more complete calculation with a height-dependent B-field may be able to reproduce the entire radio through gamma-ray continuum of blazars. More details about this model can be found in Ref. [50].

SUMMARY

The data from the Compton Observatory offer a new perspective on the AGN unification scenario. The high energy results show that AGN come in two classes: gamma-ray soft AGN, where we are evidently looking at the emission from a Seyfert nucleus, and gamma-ray hard AGN, where we are viewing almost directly down the axis of a radio jet. The spectra of gamma-ray soft AGN soften at energies $\gtrsim 100$ keV, consistent with thermal Comptonization models, whereas the emission from gamma-ray hard AGN is clearly nonthermal and produced in association with a radio jet. Models where nonthermal electrons Compton scatter accretion-disk radiation are in accord with the observational constraints for the gamma-ray hard AGN.

Acknowledgements. This paper reviews work done in collaboration with Reinhard Schlickeiser, Apostolos Mastichiadis, and Ray Protheroe. I would also like to thank the OSSE team, especially Jim Kurfess, Bob Kinzer, Neil Johnson and Rob Cameron, for valuable discussions and for providing access to the OSSE AGN data.

REFERENCES

1. A. Lawrence, PASP **99**, 309 (1987).
2. J. N. Bregman, A&ApR **2**, 125 (1990).
3. M. C. Begelman, R. D. Blandford, and M. J. Rees, RMP **56**, 255 (1984).
4. R. E. Rothschild et al., ApJ **269**, 423 (1983).
5. C. E. Fichtel et al., these proceedings (1992).
6. T. Ohashi et al., Proceedings of the 23rd ESLAB Symposium, ed. N. White (ESA, Noordwijk, 1989), p. 837.
7. R. A. Cameron et al., these proceedings (1992).

8. M. Maisack et al., these proceedings (1992).
9. W. N. Johnson, private communication (1992).
10. J. D. Kurfess, private communication (1992).
11. R. C. Hartman et al., ApJ **385**, L1 (1992).
12. C. von Montigny et al., A&ApS, in press (1992).
13. W. Hermsen, et al., A&ApS, in press (1992).
14. C. E. Fichtel et al., Advances in Space Research (COSPAR), in press (1992).
15. S. D. Hunter et al., ApJ, in press (1992).
16. R. L. Kinzer et al., in preparation (1992).
17. F. Makino, Proceedings of the 23rd ESLAB Symposium, ed. N. White (ESA, Noordwijk, 1989), p. 803.
18. M. Punch et al., Nature **358**, 477 (1992).
19. R. C. Lamb, private communication (1992).
20. J. E. Grindlay et al., ApJ **197**, L9 (1975).
21. L. Bassani and A. J. Dean, SpSciR **35**, 367 (1983).
22. R. C. Walker, J. M. Benson, and S. C. Unwin, Superluminal Radio Sources, eds. J. A. Zensus and T. J. Pearson (Cambridge University Press, New York, 1987), p. 48.
23. C. Impey, ibid. (1987), p. 233.
24. H. R. Miller and P. J. Wiita, eds., Variability of Active Galactic Nuclei (Cambridge University Press, New York ,1991).
25. R. F. Mushotzky et al., ApJ **235**, 377 (1980).
26. T. J. Turner and K. Pounds, MNRAS **240**, 883 (1989).
27. R. J. Protheroe and D. Kazanas, ApJ **265**, 620 (1983).
28. A. A. Zdziarski, ApJ **305**, 45 (1986).
29. D. Kazanas, ApJ **287**, 112 (1984).
30. A. P. Lightman and A. A. Zdziarski, ApJ **319**, 643 (1987).
31. A. A. Zdziarski et al., ApJ **363**, L1 (1990).
32. G. Ghisellini, MNRAS **238**, 449 (1989).
33. C. Done, G. Ghisellini, and A. C. Fabian, MNRAS **245**, 1 (1990).
34. J. I. Katz, ApJ **206**, 910 (1976).
35. S. L. Shapiro, A. P. Lightman, and D. M. Eardley, ApJ **204**, 187 (1976).
36. R. A. Sunyaev and L. G. Titarchuk, A&Ap **86**, 121 (1980).
37. R. Svensson, MNRAS **209**, 175 (1984).
38. M. Kusunose and F. Takahara, PASJ **41**, 263 (1989).
39. D. A. Grabelsky et al., these proceedings (1992).
40. A. Wandel and E. P. Liang, ApJ **380**, 84 (1991).
41. C. D. Dermer, in Proceedings of the 23rd ESLAB Symposium, ed. N. White (ESA, Noordwijk, 1989), p. 925.
42. R. Ramaty and R. E. Lingenfelter, ARNuclPartSci **32**, 235 (1982).
43. C. D. Dermer and R. Schlickeiser, Science **257**, 1642 (1992).
44. J. R. Mattox et al., these proceedings (1992).
45. G. Kanbach et al., these proceedings (1992).
46. F. Melia and A. Königl, ApJ **340**, 162 (1989).
47. P. S. Coppi, J. F. Kartje, and A. Königl, these proceedings (1992).
48. R. J. Protheroe, A. Mastichiadis, and C. D. Dermer, Astroparticle Physics, in press (1992).
49. C. D. Dermer, R. Schlickeiser, and A. Mastichiadis, A&A **256**, L27 (1992).
50. C. D. Dermer and R. Schlickeiser, to be submitted to ApJ (1992).

GAMMA-RAYS FROM HOT ACCRETION DISKS IN AGN

M. Kafatos and P.A. Becker
CSI, and Physics Dept., George Mason University, Fairfax, VA 22030

ABSTRACT

The recent detection by EGRET of γ-rays from 3C 279 and several other blazars in the 100 MeV - several GeV imposes severe constraints on viable production mechanisms for high-energy radiation in AGN. One possibility that we explore here is that the high-energy emission is powered by proton-proton collisions occurring in two-temperature accretion disks around Kerr black holes. Neutral pions decay to γ-rays as high as several hundred MeV to $\lesssim$ 1 GeV as seen from infinity. Moreover, such hot disks produce large amounts of relativistic pairs via charged pion decays. These pairs comptonize the ambient soft photons to energies as high as several hundred MeV. In preliminary calculations, we show that relativistic inverse-Compton scattering significantly boosts the signal from hot disks in the OSSE/COMPTEL and EGRET spectral ranges, and that the resulting theoretical spectra agree closely with the 3C 279 observations by the CGRO. The hot disk model also predicts an anticorrelation between emission in the OSSE/COMPTEL and EGRET ranges due to γ-γ scattering.

INTRODUCTION

AGN were first identified as sources of γ-rays by previous missions[1]. Further clarification of the nature of the high-energy emission in AGN has awaited the acquisition of high-quality data at energies exceeding $\sim$ 1 GeV, which the EGRET instrument aboard the Compton Gamma-Ray Observatory (CGRO) is presently providing. The detection by EGRET of γ-rays from 3C 279 in the approximate range 100 MeV to 10 GeV[2,3] with an apparent luminosity in excess of 10^{48} erg/s has raised a number of questions regarding viable high-energy models in AGN. If unbeamed, this luminosity implies a central mass $\sim 10^{10}$ $M_\odot$ for near Eddington accretion rates. Moreover, the EGRET[2,3] and COMPTEL[4] detections imply that a major fraction of the bolometric luminosity of at least some AGN is emitted in the hard γ-ray region of the spectrum. Unless there is in situ acceleration, radiative losses restrict the Lorentz factor well below 10^4 which is the value required to make inverse Compton scenaria and synchrotron-self-Compton emission in beams plausible[5,6].

Even though all 16 sources detected so far by EGRET[3] belong to the class of astronomical objects known as blazars, which are radio loud quasars and BL Lacs, characterized by flat radio spectra, and other properties such as rapid variabilities, polarization and superluminar motions all of which are intimately connected to the beam phenomenon, it remains to be proven that the EGRET γ-rays are directly emitted in beams or jets. One should be searching for unique signatures of beamed emission in the high energy data, such as similar timescales in the radio (synchrotron), X-ray (first order Compton scattering) and γ-ray (second order Compton scattering) parts of the spectrum. In the absence of such evidence, a valid question remains; what is the contribution of accretion disks to the high-energy spectral properties of AGN? An attractive alternative hypothesis is that the high-energy emission arises in proton-proton collisions in hot accretion disks[7,8]. For example, the decay of neutral pions produced via proton-proton collisions gives rise to γ-rays of energies that can extend to several 100 MeV as seen from infinity. If the emission were due solely to beaming, then one would expect BL Lacs to dominate the list of extragalactic EGRET sources, and the remaining sources would be expected to show evidence for less precise alignment, such as superluminar motion. The nearest of these 3 BL Lacs (Mrk 421) is the closest AGN detected by EGRET. Combining this with the fact that the median redshift of all 16 EGRET AGN is ~ 1.0, we conclude that there is no significant tendency for EGRET sources to be associated with BL Lacs at large redshifts.

In the absence of a self-consistent theory of jet formation, it is important to study the role hot accretion disks may play as direct γ-ray emitters and as suppliers of relativistic electron-positron pairs which may serve as seed particles in the formation of jets. Furthermore, hot disks may provide the only plausible mechanism for producing EGRET-range emission in non-beamed AGN.

HOT TWO-TEMPERATURE DISKS

Whether beams or hot disks produce the observed high-energy emission, the ultimate energy source in AGN must be attributed to accretion onto supermassive black holes. As such, one may envisage a scenario where both beams and disks are required to explain the full spectrum of high-energy AGN emission. Since the physics of beams is not well understood and a number of ad-hoc assumptions need to be made, it is useful to explore the full potential of disk models. If accretion provides the luminous output observed in AGN and QSOs, then one would expect a high percentage of the bolometric luminosity to come out in X-rays and γ-rays. "Standard" α-disks, however, characterized by a single temperature, can at most reach values of that temperature as high as 10^{8-9} K, which are sufficient to produce copious amounts of X-rays[9,10] but fail to produce significant amounts of γ-rays above a few hundred keV and could not account for the OSSE/COMPTEL and, particularly, EGRET emission. Even hot corona models[11] with $T \sim 10^{11}$ K, do not produce γ-rays above ~ 100 keV.

Hot, optically thin regions embedded in standard disks can produce power-law comptonized spectra in X-rays to MeV γ-rays and have been proposed to explain the high-energy emission from the galactic black hole candidate Cygnus X-1[7,12]. The electrons cool efficiently and are, therefore, able to maintain a temperature of $T \sim 10^9$ K, while Coulomb collisions are not very efficient at cooling the ions, which become virialized[13] with a temperature $T \sim 10^{12-13}$.

The hot-disk spectra published before[8] were computed by summing components due to thermal X-ray and nonthermal γ-ray emission. The thermal X-ray emission arises when the 10^9 K electrons upscatter "blue bump" soft photons[14]. The comptonized X-ray spectrum has a nonthermal appearance with an energy spectral index $\alpha \sim 0.5 y^{-1}$, where the Comptonization parameter $y = (4kT_e/m_e c^2) \text{Max}(\tau_{es}, \tau^2_{es})$; typically $\alpha \sim 0.57$ for $y=1$ and $\alpha \sim 1.5$ for $y=0.3$. The two γ-rays emitted through the decay of neutral pions have energies peaked in the range 100 - 200 MeV in the rest frame of the disk. In previous work[8], gravitational focusing, Doppler boosts, and gravitational redshifts were incorporated in the spectra observed at infinity using the formalism of Cunningham[15]. These effects can boost the pion-decay γ-rays up to ≥ 300 - 500 MeV, depending on the viewing angle of the disk.

COMPTEL and EGRET observations of 3C 279 during observing period 3 (June 1991) when the quasar flared in high-energy γ-rays suggest a continuous spectrum extending between the hard X-rays and ~ 1 GeV, with a photon index $\alpha \sim 1.5$. Above ~ 1 GeV, the spectrum steepens to $\alpha \sim 1.8$. The observed emission in the 1 - 100 MeV band may represent the inverse-Compton cooling of relativistic pairs produced via charged pion decays in the hot disk.

We have calculated elsewhere[16] the energetic distribution of the relativistic pairs using a simple steady-state model including the effects of pair injection, inverse-Compton losses, and dynamical pair escape. These expressions are then used to calculate the steady-state population of relativistic pairs resulting from the injection of pion-decay pairs. The self-consistent inverse-Compton γ-ray component is then calculated using a simple homogeneous model for the upscattering of soft background photons. The complete high-energy spectrum as viewed at infinity is obtained by adding the inverse-Compton losses to the Eilek and Kafatos[8] original spectrum.

DISCUSSION AND CONCLUSIONS

We have obtained preliminary results for the Compton-cooled relativistic electron distribution and the associated X-ray and γ-ray spectra. We are able to fit the observed flare spectra of 3C 279 in the COMPTEL to EGRET ranges[4]. In general, we find that central black hole masses ~ 10^{10} $M_\odot$ are required and with accretion rates near the Eddington limit, or ~ 100 $M_\odot$/yr. We find that models which are optically thin to γ - γ scattering require a small y-parameter, $y \sim 0.3$. These are self-consistent models which can fit the results. The theoretical curves not only fit the COMPTEL/EGRET observations but also connect to the X-ray data observed by GINGA[17]. These models require high soft-photon compactness, ~ 170. Further analysis of OSSE data may provide additional constraints on the spectrum in the MeV range during October 1991 when 3C 279 was seen at a much lower level than in June. In particular, OSSE observations near ~ 1 MeV may help to determine the value of the compactness parameter. The GINGA X-ray data determine the value of the Compton y-parameter, which is a strong function of the X-ray energy spectral index, α.

Our hot disk models are mildly pair-dominated, $\dot{N}_e^{\pm} \sim \dot{M}/m_p$, near or super-Eddington accretion models, optically thin to γ-γ scattering and with small Comptonization parameter, y. They readily fit the γ-ray data up to $\sim$ 1 GeV. If Penrose processes operate above that energy range, natural exponential cutoffs are expected at $\sim$ (3 - 4) $m_p c^2$ or a few GeV[18].

The hot accretion disk model also predicts: i) an anti-correlation of keV X-rays (GINGA)/MeV γ-rays (OSSE) with EGRET γ-rays. This may explain the absence of EGRET emission from Seyfert galaxies. ii) Since γ-rays are emitted closer to the hole, we would expect (unlike beam models), that γ-rays would be varying faster than X-rays, by $\sim$ 2 - 3. iii) Since the same relativistic pairs produce radio waves via the synchrotron process, we expect both the spectral index in the radio part of the spectrum and the spectral index in the range of inverse-Compton scattering (1 - 100 MeV) to be the same and that the timescales of variability in the two regions of the spectrum should be similar.

Observations already indicate some trends in the data according to i) and iii).

Although this is a disk model, there is no inherent contradiction with beam models; in fact, the model produces $\sim 10^{49-51}$ pairs/s arising close to the hole. If instabilities are present, the disk would collapse over the free-fall timescale of $\sim$ a few days as observed for 3C 279. Electron-positron pairs escaping along the rotation axis of a two-temperature disk may form a relativistic jet, which is expected to carry a net positive charge since collisions between virialized protons in hot disks overproduce positrons relative to electrons. Acceleration in the electric field resulting from the charge separation can play an important role in mitigating the effects of the severe radiative losses suffered by particles in the luminous core of the nucleus. We are in the process of examining the boundary conditions close to the black hole that may allow the formation of a relativistic jet of e^+e^- pairs originating from a hot accretion disk.

If a geometrically thick disk is present, the hot disk would not be observable unless the line of sight was close to the axis of spin (and jet). One is justified to ask what is the contribution of an accretion disk to high-energy radiation? It is perhaps a quiescent γ-ray spectrum as seen in the October 1991 EGRET observations of 3C 279 that represents the disk contribution.

ACKNOWLEDGEMENTS

This work has been supported by a NASA CGRO Phase 2 grant.

REFERENCES

1. B.N.Swanenburg, et al. Nature, 275, 298 (1978).
2. R.C.Hartman, et al., Ap. J. (Letters) 385, L1 (1992).
3. C.E. Fichtel et al., present volume.
4. W.Hermsen et al., A. & A. Suppl. Ser. (1993).
5. A.P.Marscher, and S.D.Bloom, in Proc. Sec. CGRO Sci. Workshop, NASA (1992).
6. M.C.Begelman, R.D.Blandford, & M.J.Rees, Rev. Mod. Phys. 56, 253 (1984).
7. S.L.Shapiro, A.P.Lightman, & D.M.Eardley, Ap. J. 204, 187 (1976).
8. J.A.Eilek, & M.Kafatos, Ap. J. 271, 804 (1983).
9. N.I.Shakura, & R.A.Sunyaev, R. A., A. & A. 24, 337 (1973).
10. J.Novikov, & K.S.Thorne, in Black Holes, edit. C. de Witt and B. de Witt (New York: Gordon and Breach, 1973).
11. E.P.T.Liang, & R.H.Price, Ap. J. 218, 247 (1977).
12. C.D.Dermer, C. D., Ap. J. (Letters) 335, L5 (1988).
13. M.Rees, et al., Texas Symposium (1982).
14. M.A.Malkan, Ap. J. 268, 582 (1983).
15. C.T.Cunningham, Ap. J. 202, 788 (1975).
16. P.A.Becker, & M.Kafatos, Ap. J. (Letters), submitted.
17. F.Makino, et al., Ap. J. (Letters) 347, L9 (1989)
18. M.Kafatos, & D.Leiter, Ap. J. 229, 46 (1979).

HIGH ENERGY EMISSION FROM ULTRARELATIVISTIC JETS

P. S. Coppi, J. F. Kartje, A. Königl

Dept. of Astronomy and Astrophysics, Univ. of Chicago, Chicago, IL 60637

ABSTRACT

We consider the interaction of an initially ultrarelativistic beam of electrons and protons with the radiation field near the accretion disk of an Active Galactic Nucleus (AGN). The particles in such a beam suffer a strong radiative drag, photoproducing off and upscattering the ambient radiation field. The upscattered radiation is also beamed and can carry away a significant fraction of the luminosity initially in the particle beam. Such a mechanism may be responsible for the strong γ-ray emission seen by EGRET in objects like 3C279. We study the evolution of the beam in the anisotropic, inhomogeneous disk radiation field using an implicit kinetic code.

When the Lorentz factors of the initial beam particles exceed $\sim 10^4$, a photon-e^+e^- cascade develops in the beam. The cascade effectively converts the energy of the initial particles into a large number of e^+e^- pairs that cool to form an outflow with terminal Lorentz factor $\gamma_\infty \sim 10$. The characteristic radiation spectrum formed in the cascade has $\alpha \sim 0.5$ ($F_\nu \propto \nu^{-\alpha}$) and is insensitive to the details of the initial beam energy distribution. This spectrum will be steepened by photo-absorption of the cascade photons if an unbeamed, X-ray background field is present (e.g., from an X-ray disk corona). If the beam does not diverge rapidly, the cascade spectrum may also be modified by Compton downscattering on cooled electrons much farther down the beam. (The cascade photons are strongly beamed along the beam direction and can spend a long time in the jet.) For a fan jet geometry with an opening angle of a few degrees, the downscattered spectrum resembles the 3C279 EGRET spectrum. Radiation from a decelerating beam may thus provide a robust explanation of the EGRET quasar spectra. Note, however, that the photo-absorption optical depth for photons of energy $\gtrsim 50$ GeV is typically large within $\sim 10^{16}$ cm of the disk surface. Such a model, or any other which produces most γ-rays near a source of copious unbeamed UV photons (such as a disk or Broad Line Region), *cannot* explain the TeV emission claimed for Mkn 421.

MOTIVATIONS

In light of growing evidence for a moderately beamed X-ray component in AGN showing radio jets (e.g., in BL Lacs), it is worth pursuing further the possible contributions of a jet to AGN X- and γ-ray emission. They may be particularly relevant to hard or γ-ray bright sources such as 3C273. Current emission models based on isotropic pair plasmas have difficulty reproducing such spectra without requiring a fine tuning of the input parameters, overproducing the γ-ray background, or introducing unwanted spectral features through the accumulation of cooled e^+e^- pairs.

The observations by EGRET of time-variable emission ($\Delta T < 2$ weeks) up to energies ~ 10 GeV in 3C279 also cannot be explained by conventional models where the particle distributions and X-ray emission are isotropic. (The probability that a 10 GeV γ-ray will pair produce off one of the X-ray photons

© 1993 American Institute of Physics

before escaping the source region is very large.) Either the X-rays from 3C279 are beamed together with the energetic γ-rays, or energy is transported away from the X-ray emitting region before being converted into γ-rays. This suggests that a collimated jet is involved. The hypothesis is strengthened by the fact that the other strong EGRET extragalactic sources are all blazars, i.e., are thought to have a jet oriented at small angles to our line of sight. Even when plausible beaming corrections are taken into account, the amount of energy involved in an outburst like that seen from 3C279 is non-negligible. Transporting and releasing this energy at large distances from the central engine may be a problem. Hence, we consider a jet that starts out near the central engine and consequently interacts strongly with the ambient radiation there. Because protons interact much less strongly with the radiation than electrons, we will use them as the primary source of energy in the beam. Protons are in principle easier to accelerate and can travel sufficiently long distances to sample anisotropies in the radiation field due to geometric effects.

Such a scheme has two additional advantages. First, as noted, e.g., in Phinney[1], the radiative deceleration of a jet by Compton drag typically gives terminal jet Lorentz factors in the observed range, $\gamma_\infty \sim 10$, for a wide range of initial jet particle energy distributions. Second, if the initial beam particles are sufficiently energetic, photons Compton upscattered by the beam can pair produce. Such an energetic beam would quickly acquire an e^+e^- component if it did not already have one. The presence of such a component is supported by measurements of low Faraday rotation in AGN radio jets.

NUMERICAL CALCULATIONS

Three sets of calculations were carried out. The first assumed that the background radiation was isotropic and homogeneous and considered only the evolution of an electron-positron beam. The beam was taken to be perfectly collimated or cold in the beam frame (no transverse motion of beam particles allowed). The kinetic equations describing the beam evolution were solved using an implicit code. Because the radiation field was uniform, we were able to use the *exact* kinetic equations (e.g., the finite width of the photon energy distribution resulting from Compton upscattering by an electron is taken into account). The radiation processes considered were Compton scattering ($e\gamma \to e\gamma$), photon-photon pair production ($\gamma\gamma \to e^+e^-$), and triplet pair production ($e\gamma \to ee^+e^-$). Because of the high degree of beaming, interactions between high-energy particles (pairs, Compton upscattered photons) were neglected.

The second set of calculations took the radiation field to be that above an accretion disk (see Melia & Königl[2]). The anisotropy and inhomogeneity of this field were taken into account. However, because of computational constraints, we had to assume that the energy distributions of the particles produced in the various processes could be approximated by a delta-function, i.e., $\dot{N}(\epsilon) \propto \delta(\epsilon - <\epsilon_{final}>)$. The beam was started out with protons instead of electrons, and the additional processes of photon-proton (neutron) pion production and photon-proton pair production were considered (e.g., Begelman, Rudak & Sikora[3]). The effects of possible charge exchange (turning protons into neutrons and vice versa) during photo-pion production were also included, but triplet pair production was neglected. As the background radiation field was assumed to be constant, the calculations shown here are of *linear* cascades. Future calculations will try to relax this constraint (e.g., take into account the fact that the beam can "shield" itself from the external background photons if enough pairs are produced to make the beam Thomson thick perpendicular to the beam axis.) Sample results from the current calculations are shown in Fig. 1.

The third set of calculations involved taking the particle distributions computed via the kinetic code and computing the emergent (Compton upscattered) X- and γ-ray spectrum with a Monte Carlo code. This code allows us to perform a self-consistency check on the kinetic code, and provides much more accurate information on the angular distribution of the outgoing photons. In particular, the Monte Carlo code allows for multiple scatterings of photons off pairs. While energetically unimportant for the pairs, these scatterings can play a major role in determining the spectrum seen by an observer not looking straight down the beam.

EFFECTS OF COMPTON DOWNSCATTERING

Because photons Compton upscattered by the relativistic beam are directed very close to the beam axis, they can spend a long time inside the jet. Depending on details such as the rate of divergence and disruption of the cooled beam, there may be a reasonable probability that such a photon will Compton scatter off a cooled beam electron – even if that scattering is in the Klein-Nishina regime. The photon will lose energy in that scattering and will be deflected out of the jet. Typically, it will not undergo a second scattering before leaving the jet. It can be shown (see Fig. 2) that the downscattering of an incident power law gives rise to a steepened power law when the observer can see a range viewing angles (as in the case of a fan jet geometry). In particular, an incident power law of index 0.5 will be scattered to a power law at high energies with index $0.8-1$. At low energies, the scattered distribution preserves the original 0.5 spectral index. A spectrum of this type is what is needed to match the 3C279 GRO observations. (It is noteworthy that the cascades studied here almost always produce power law spectra with $\alpha = 0.5$.) We stress, however, that our calculations are still preliminary, and that we have not yet considered in detail the effects of finite jet optical depths or the effects of viewing a fan geometry from the side.

TEV EMISSION?

As noted by Stecker et al.[4], TeV photons with energies $\sim 10^5 - 10^6 m_e c^2$ can pair produce off photons with energies of $10^{-6} - 10^{-5} m_e c^2$ (i.e., photons at UV and optical frequencies). Since galaxies emit strongly at such frequencies, one might expect there to be a significant UV/optical intergalactic background radiation field. Sufficiently distant sources (e.g., 3C279) will thus not be visible at TeV energies due to intergalactic photoabsorption. Since the UV/optical background radiation is much larger inside a galaxy, particularly near the central engine of an AGN, one should also consider the possibility of *intrinsic* source absorption of TeV γ-rays. As an example, Figure 3 shows the optical depth for photon-photon pair production as a function of height above what is thought to be typical of an AGN accretion disk. The model described here requires a strong radiation field to decelerate the beam and achieve reasonable radiation efficiencies. Hence, the beam must be started near the disk and we do *not* expect to see significant emission above 50 GeV or so. If the intrinsic EGRET spectra do in fact continue to TeV energies as recently claimed for Mkn

421, the emitting region must be far from the central engine. The constraints may be worse than shown here if one allows for the existence of an extended region that scatters and reprocesses a significant fraction of the disk radiation (e.g., the Broad Line Region).

This research was supported in part by a GRO Fellowship and by NASA grants NAGW-830,-1284, and -1636.

REFERENCES

1. E.S. Phinney, in *Superluminal Radio Sources*, ed. J.A. Zensus and T.J. Pearson (Cambridge University Press, Cambridge, 1987), p. 301.
2. F. Melia and A. Königl, Ap. J. **340**, 162 (1989).
3. M.C. Begelman, B. Rudak, and M. Sikora, Ap. J. **362**, 38 (1990).
4. F.W. Stecker, O.C. De Jager, and M.H. Salamon, Ap. J. Lett. **390**, L49 (1992).

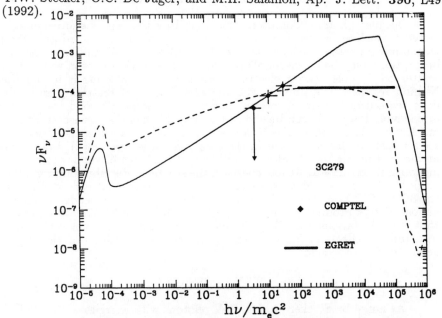

Figure 1. The *solid* line shows the spectrum produced by the radiative deceleration of an initially mono-energetic proton beam with $\gamma_i = 10^7$, starting $30 R_g$ above a flat accretion disk of maximum temperature 2×10^5 K. ($1 R_g = 3 \times 10^{13}$ cm.) Note the characteristic spectral index $\alpha = 0.5$ over most of the spectrum. The maximum energy at which there is significant emission in this model ($\sim 10^5 m_e c^2$) is about the highest possible in models of the type discussed here. The *dashed* line shows the effects of adding emission from a nonthermal X-ray corona ($L_x \sim 0.5 L_{disk}$, $\alpha_x \sim 0.5$) to the UV disk emission. The corona was mocked up as a flat, uniformly emitting disk of radius $40 R_g$. Note that the spectrum has been steepened to $\alpha \sim 0.7 - 1$. The fact that the primary beam particles are protons and lose their energy on a scale comparable to the anisotropy scale of the X-ray emission is crucial. If the primary particles had been electrons, the spectrum would have been exponentially attenuated starting at energies $\gtrsim 10$ MeV. The normalizations of the two spectra are arbitrary and were chosen to make the models go through the COMPTEL data points.

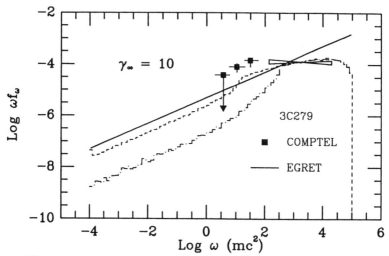

Figure 2. Spectrum produced by Compton downscattering of incident photons within the decelerated tail of an electron/positron jet. The incident spectrum (solid line) is taken to be a power law of spectral energy index -0.5; the photons are assumed beamed along the jet axis. The jet particles are monoenergetic with a Lorentz factor of 10. The *dashed* line shows the scattered spectrum integrated over a viewing angle range of 0° − 5°, measured with respect to the jet axis. The *dot-dashed* line shows the scattered spectrum integrated over a viewing angle range of 0° − 1°. This is what might be seen when looking down the axis of a fan jet with a corresponding opening angle of 5 or 1 degrees. Data points from COMPTEL and EGRET observations of 3C279 are also shown, normalized to our arbitrary flux units. We did not attempt a detailed fit to the data points (by varying the incident spectrum, viewing angle, beam geometry, etc.).

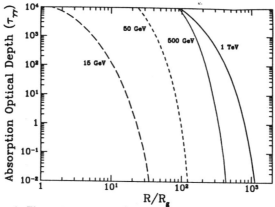

Figure 3. The optical depth for photon-photon pair production ($\tau_{\gamma\gamma}$) off disk photons as a function of height R above the disk and incident photon energy. The disk model used had an inner radius of $5R_g$ and outer radius of $100R_g$. The maximum temperature near the inner edge of the disk was 2×10^5 K. The black hole had a mass of 10^8 solar masses, for which $1R_g = 3 \times 10^{13}$ cm.

X-RAY AND GAMMA-RAY EMISSION FROM ACTIVE GALACTIC NUCLEI

K.Y. Ding, K.S. Cheng
Department of Physics, University of Hong Kong, Hong Kong

K.N. Yu
Department of Applied Science, City Polytechnic of Hong Kong, Hong Kong

ABSTRACT

We assume that the centers of some active galactic nuclei consist of a supermassive non-rotating black hole surrounded by a magnetized, cool Keplerean disk and a magnetosphere which is resembling that found surrounding a magnetic neutron star in an accreting binary system. Flux tubes passing through two differential rotation regions could result in a large potential drop along the B-field lines. Electrons are expected to accelerate and radiate curvature photons. These curvature photons are energetic enough to produce secondary $e^{\pm}$ pairs in collisions with the ambient infrared photons. These secondary pairs will radiate synchrotron photons in a wide energy range but only those photons with energies in the range of 10 GeV and below can escape from the ambient UV and soft X-ray photon field. The γ-ray spectral index is luminosity dependent, but the X-ray spectral index is very insensitive to the luminosity. We use 3C279 and 3C273 to illustrate our model.

I. INTRODUCTION

It seems that Active Galactic Nuclei (AGNs) have a universal X-ray spectral index $\alpha_x \sim 1.5-1.7$.[19-21,23] The recent GRO results[8] show that the γ-ray spectral indices of sixteen AGNs range from 1.7 to 2.4 but their luminosities range from 10^{44} erg s^{-1} to 6×10^{48} erg s^{-1}. Again, it appears that the γ-ray spectral indices are insensitive to, nevertheless depend on, the luminosity.

In this paper, we attempt to make use of models involving acceleration regions in the blackhole magnetosphere which result from the magnetic flux tubes passing through two differential rotation regions. Detailed models concerning charged particle acceleration mechanisms and radiation processes have also been applied to X-ray binaries[3,4] and proved to be successful in explaining TeV γ-ray emission from X-ray pulsars,[5] perhaps also in other energy ranges.[6,13]

II. THE MODEL

We assume that a supermassive non-rotating blackhole with mass M has formed in the centre of AGN and is surrounded by a magnetized, cool accretion disk (c.f. Fig. 1).

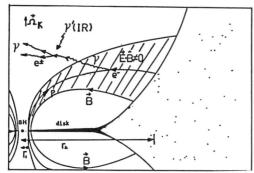

FIG. 1: Schematic representation of pair production and radiation from the assumed accelerator where $\vec{E} \cdot \vec{B} \neq 0$ in the supermassive blackhole magnetosphere.

It was argued that the blackhole magnetosphere[27,26] should be similar to that found surrounding a magnetic neutron star in an accreting binary system.[25] It is not clear what the strength of the magnetic field in the blackhole should be, nevertheless, over sufficient time, the field should reach equipartition. The density $\rho(r_1)$ of the disk at the inner edge, r_1, is given by the standard accretion disk model for the case where the disk is gas pressure dominated with the opacity dominated by electron scattering.[22] For simplicity, we choose a fiducial radius with r_1 equals $10 r_g$, where r_g is the gravitational radius. Hence, we have[14]

$$B(r) \simeq B_\circ \simeq 4.5 \times 10^7 M_8^{-11/20} \dot{M}_{25}^{1/5} \text{ Gauss} \quad (r \leq r_1)$$
$$B(r) = B_\circ (\frac{r_1}{r})^2 \quad (r > r_1). \qquad (1)$$

Since the accretion material surrounding the blackhole may be assumed highly conducting, an electric field perpendicular to the magnetic field line will be produced whenever a magnetic flux tube is sweeping through the disk. If the charge density inside this flux tube is charge separated and corotating with that part of the disk linked by the magnetic field, we have $\vec{E} \cdot \vec{B} \neq 0$ whenever the magnetic flux tube passes through two differential rotation regions labeled by radii r_1 and r_2 respectively.[3,4] The maximum potential difference can be achieved along the field line and is given by

$$\Delta V_{max} \simeq \frac{\Omega_k(r_1) r_1^2 B(r_1)}{c} = \frac{\Omega_k(r_1) \Phi}{c} \quad \text{if } r_2 \gg r_1 \qquad (2)$$

where Φ is the magnetic flux enclosed inside the flux tube.[7]

III. PRODUCTION OF X-RAYS AND GAMMA-RAYS

Inside the accelerators, both protons and (primary) electrons can be accelerated. In this paper, we shall restrict ourselves to the acceleration of electrons. The primary electrons are moving in a curved field line and their energies are limited by curvature radiation. The size of the accelerator is approximated by r_2. Most of these curvature photons will be attenuated to form secondary $e^\pm$ pairs because numerous radio to optical photons are around the AGN. We shall assume that for the 1st generation, all curvature photons will become pairs. These secondary $e^{\pm}$ pairs will then lose their energy through synchrotron radiation. Of course, part of these secondary photons (1st generation of synchrotron photons) are also energetic enough to produce a third generation of $e^{\pm}$ pairs which can also emit the third generation of γ-ray photons via the synchrotron radiation mechanism[7] (2nd generation of synchrotron photons). Finally, the total output spectrum is given by summing over the ith generation surviving spectrum.

IV. APPLICATION TO 3C279 AND 3C273

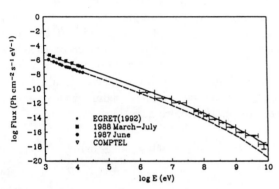

FIG. 2: Comparison between the model X-ray and γ-ray spectra and the observed data of 3C279.

It is not easy to determine exactly where the other end of the accelerator (r_2) should be as the real situation of AGNs is far more complicated. According to the results of GRO, the spectrum is extended at least up to 10 GeV for 3C279. We require the attenuation depth of the 10 GeV photons to be less than unity. Using the typical values of 3C279 in the outburst state,[17] we obtain r_2 approximately equals 10^{20} cm. We further assume that $\frac{r_2}{r_1}$ equals a constant.

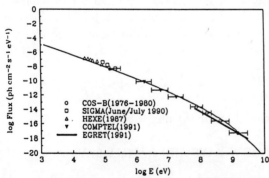

FIG. 3: Comparison between the model X-ray and γ-ray spectra and the observed data of 3C273. $\dot{M} = 10^{26}$ g s^{-1} is used in both model curves in order to match the observed L_γ of COS-B.

(a) 3C279 — Before calculating the spectrum, there are two parameters we need to specify, namely, M and $\dot{M}$. First the mass accretion rate is related to the luminosity by $L_\gamma = \frac{GM\dot{M}}{r_1}$. Secondly, we assume that the observed luminosity of 3C279 equals the Eddington luminosity. Taking $L_\gamma \simeq 10^{48}$ erg s^{-1}, we obtain $M = 10^{10} M_\odot$ and $\dot{M} = 10^{28}$ g s^{-1}. Fig. 2 shows the comparison between model curves (the solid line refers to $\dot{M} = 10^{28}$ g s^{-1} for the out-

TABLE I: X-ray and γ-ray Spectral Indices for Various parameters.

log L_γ log $\dot M$†	44	45	46	47	48	49
8	1.58	1.57	1.55	*	*	*
	3.10	2.60	2.30	*	*	*
9	1.60	1.58	1.56	1.55	*	*
	3.50	2.90	2.60	2.20	*	*
10	1.62	1.58	1.57	1.55	1.54	*
	4.60	3.30	2.80	2.40	2.10	*
11	1.64	1.61	1.58	1.56	1.55	1.54
	5.10	4.10	3.20	2.70	2.30	1.99

†where M is in units of $M_\odot$ and L_γ is in units of erg s^{-1}. The first number and the second number in each cell correspond to the mean X-ray spectral index from 10keV – 300 keV and to the mean γ-ray spectral index from 100MeV – 3GeV respectively.
*we consider that these are unphysical because L_γ is larger than that implied by Eddington limit.

burst state, and the dashed line refers to $\dot M = 10^{27}$ g s^{-1} for another data set).

The γ-ray data are from EGRET[9] and COMPTEL[11] and the X-ray data are from Ref.[17] We remark that the X-ray spectral indices ($\alpha_x \sim 1.5$) are roughly unchanged for the two sets of fit and the γ-ray spectral indices change only slightly ($\alpha_\gamma \sim 1.8$ for the higher intensity one and $\alpha_\gamma \sim 2.0$ for the lower one).
(b) 3C273 — In calculating the spectrum of 3C273, $\dot M$ is somewhat undetermined because it is difficult to decide whether the observed luminosity equals the Eddington luminosity. Fig. 3 shows that $M = 10^8 M_\odot$ seems to fit the observed data better. The γ-ray data come from COS-B,[10] EGRET[18] and COMPTEL[11] and the X-ray data come from HEXE[15] and SIGMA.[1]

V. DISCUSSION

According to our model, most of the low energy pairs are cascaded from high energy pairs via synchrotron radiation. Therefore, the energy distribution of low energy pairs is always $\frac{dN_e}{dE_e} \propto \frac{1}{E_e^2}$, hence the corresponding low energy photon spectrum[2] $F(E_\gamma) \propto E_\gamma^{-(2+1)/2} = E_\gamma^{-1.5}$. On the other hand, the γ-ray spectrum changes rapidly around the typical energy of the synchrotron photon. In general, when $E_\gamma > E_{syn}$, the spectral index will become steeper than 2, and when $E_\gamma < E_{syn}$, the spectral index will become less than 2 and eventually approach 1.5. Since E_{syn} is proportional to $\dot M^{1/2}$, we predict that α_γ will be inversely proportional to $\dot M$. In Table I, we list the model X-ray and γ-ray spectral indices for various M and $\dot M(L_\gamma)$. We see that $\alpha_x \simeq 1.54 - 1.64$, while α_γ varies from 2 – 5. These seem to be consistent with the observed x-ray and γ-ray spectra. Finally, we remark that although our model spectra fit well with the observed data, the existence of accelerators in the magnetosphere of blackholes remains an open question.

REFERENCES

[1] Bassani *et al.*, Ap. J., in press (1992).
[2] Blumenthal, G.R., and Gould, R.J., Rev.Mod.Phys., **42**, 237, (1970).
[3] Cheng, K.S., and Ruderman, M.A., Ap.J.Lett., **337**, L77, (1989).
[4] Cheng, K.S., and Ruderman, M.A., Ap.J., **373**, 189, (1991).
[5] Cheng, K.S., Yu, K.N., Cheung, T., and Lau, M.M., Ap.J., **379**, 290, (1991).
[6] Cheng, K.S., *et al.*., Ap.J., in press (1992).
[7] Cheng, K.S., Yu, K.N., and Ding, K.Y., submitted (1992).
[8] Fichtel, C.E., *et al.*, in these proceeding (1992).
[9] Hartman, R.C., *et al.*., Ap.J.Lett., **385**, L1 (1992).
[10] Hermsen W. *et al.*, Proc. 17th ICRC, 1,230, (1981).
[11] Hermsen W. *et al.*, Astron. Astrophys. Suppl., in press (1992).
[12] Jauch, J.M., and Rohrlich, F., The Theory of Photons and Electrons, 2nd ed. (New York: Springer, 1976).
[13] Leung, P.P., Cheng, K.S., and Fung, P.C.W., submitted (1992).
[14] Lovelace, R.V.E., Nature, **262**, 649, (1976).
[15] Maisack M. *et al.*, Astron. Astrophys., in press (1992).
[16] Misner, C.W, Thorne, K.S., and Wheeler, J.A., Gravitation (Freeman: San Francisco, 1973).
[17] Makino, F., *et al.*., Ap.J.Lett., **218**, L113 (1989).
[18] von Montigny C. *et al.*, Astron. Astrophys. Suppl., in press (1992).
[19] Mushotzky, R.F., and Marshall, F.E., Ap.J.Lett., **239**, L5 (1980).
[20] Mushotzky, R.F., *et al.*., Ap.J., **235**, 377 (1980).
[21] Mushotzky, R.F., Proceedings COSPAR meeting on High Energy Astrophysics and Cosmology, 157 (1984).
[22] Novikov, I.D., and Thorne, K.S., in Black Holes, ed. C. DeWitt and B. DeWitt (New York: Gordan and Breach, 1972), p.343.
[23] Petre, R.,*et al.*., Ap.J., **280**, 499 (1984).
[24] Rees, M.J., Ann.Rev.Astron.Astrophys., **22**, 471 (1984).
[25] Slane, P., and Wagh, S.M., Ap.J., **364**, 198 (1990).
[26] Suen, W.M., Price, R.H., and Redmount, I.H., Phys.Rev.D., **37**, 2761 (1988).
[27] Thorne, K.S., Price, R.H., and MacDonald, D.A., (eds) Black Holes: The Membrane Paradigm (Yale University Press: New Haven, 1986).

GAMMA RAYS FROM THE STRONG MAGNETIC FIELD AGN CENTRAL ENGINE

Howard D. Greyber
Greyber Associates, 10123 Falls Road, Potomac, MD 20854, U.S.A.

It is reasonable to assume that the mechanism for production of gamma rays in "blazar" QSOs and BL Lacs (16 observed so far) and for production of gamma ray bursts (GRB) are essentially the same (Occam's Razor). Therefore it is logical to examine physical models for **the Central Engine of AGN and Quasars**. The failed fast rotating accretion disk model is ignored.

Rather we analyze the **Strong Magnetic Field model (SMF)**, created originally in 1961 to explain spiral arms, then extended for radio galaxies, quasars and their jets and AGN in general.[1-12] SMF postulates that a new physical construct, a gravitationally bound current loop **(GBCL)**, is produced when the galaxy is formed by gravitational collapse of a giant cloud. Since in the pregalactic, precollapse giant cloud, the Debye length is small compared to other lengths of interest (probably less than ten kilometers), the cloud is a *plasma cloud* and *electromagnetic effects dominate*.

The introduction of a new constituent such as **GBCL** shocks many astronomers because theorists have arbitrarily ruled that there are **only four** large, basic components of the Universe, conventional stars, white dwarfs, neutron stars and black holes. But *does* the Universe (over 99% plasma) *obey* **the current theorists' dogma?** For several decades a minority view has argued against the galactic dynamo model, and for a primordial magnetic field existing before the galaxies were formed. SMF agrees with this minority view. SMF argues that before gravitational collapse, large-scale turbulence in an isolated giant plasma cloud will *naturally* over time, by the processes of dissipation, produce a toroidal volume current and an approximately poloidal magnetic field, suitable for amplification.[8] Spiral galaxies (found either isolated or in loose clusters) follow this path, but the ordinary ellipticals (found in *rich* clusters) do not, having interacted and cancelled out poloidal effects.[3] The obvious exceptions are the *most massive* galaxies in a *tight* cluster, the "giant ellipticals", since their poloidal development is too strong and well developed to be cancelled out by interaction.[3]

As described in 1966, one examines the dissipation term in a basic equation from the theory of turbulence,

$$\frac{\partial}{\partial t} u(k) = \int u(k)u(k'-k)dk' + \gamma k^2 u(k)$$

where $u(k)$ is the Fourier velocity component whose square is proportional to the energy in wave number k.[3] The first term is called the Kolmogoroff term. However, over long times in an isolated system for moderate viscosity, γ, the second (dissipation) term determines the energy transfer. Clearly the small wavelengths dissipate energy *before* the large wavelengths. Then, after some time far less than 10^9 years, the non-poloidal components will die out in this bounded pregalactic cloud volume, leaving only the largest wavelength, rotation around an axis.[8]

A similar viscous term is found in the Navier-Stokes hydrodynamic equations. Again, a similar dissipation term, $\eta \nabla^2 H$, is found in electromagnetic theory, where η is the electrical resistivity and H is the magnetic field intensity.[8] Thus the precollapse cloud is naturally prepared over time with a poloidal magneticfield suitable for amplification under gravitational contraction, whose axis is the angular momentum axis and axis of the poloidal field.

A dipole magnetic field is *not* "special" or "artificial". The simple truth is that there is no simpler organized field conceivable in astrophysics than a dipole magnetic field. This follows from the fact that a uniform magnetic field over galactic dimensions D would imply gigantic solenoidal currents organized over distances much larger than D, while a dipole's current loop is anchored in a distance far smaller than D. A fundamental result of Maxwell's Equations is that, viewed from far away, the magnetic field of even a tangle of currents, confined to a small volume, looks like a dipole field.

The toroidal volume current is made up of myriads of individual filamentary current loops. Gravitational collapse means that the myriads of individual current loops are forced to coalesce into a **GBCL** anchored around the central object, presumably a black hole, producing an extremely strong dipole magnetic field.[9] The coherent current loop is extremely intense and highly relativistic. The bursting force of this strong unified magnetic field system is in equilibrium with the gravitational attraction between the slender toroidal plasma, bound to the loop by the Maxwell "frozen-in" magnetic field condition, and the black hole.

Particles in the completely coherent relativistic current loop (ccrcl) **store a significant fraction of the huge energy of gravitational collapse**, and, upon interaction with incoming matter, cause much of the phenomena observed in AGN and quasars. **One can see** such a current loop forming in the figures in an old, but very fine paper by Leon Mestel and Peter Strittmatter, although they did not recognize this.[13] Their interest was in the formation of X points in the magnetic field. A crude underestimate of the magnetic field amplification is to assume flux conservation.[8] Another rough estimate, with a circuit analogy, done together with Donald H. Menzel, showed enormous amplification.[3] However, a large computer calculation is needed, using particle-in-cell (PIC) simulation of gravitational collapse, starting with a toroidal volume current, in a large three-dimensional electromagnetic particle code such as TRISTAN.

GAMMA RAY PRODUCTION

It is important to note that in faint very young galaxies still forming under gravitational contraction, ($z > 5$), (a) the toroidal **GBCL** is obviously larger and fatter than after the collapse finishes, (b) the relatively rigid toroidal plasma bound to and protecting the current loop is more easily penetrated than later in the collapse, and (c) star formation has not really begun (except perhaps for the beginnings of globular clusters) and so the luminosity in the optical range is very small. If these faint young forming galaxies are the source of GRB, then we understand why it has been difficult to find optical counterparts for GRB! Also one understands that a "rock" (piece of neutral matter) falling in towards the center of the galaxy sees a larger target and much more easily penetrated **GBCL** than at later stages of galaxy formation.

The obvious concept from **SMF** is that when a large "rock" (perhaps of asteroid size) races across the primordial relativistic current loop associated with young forming galaxies at cosmological distances, gamma ray bursts of roughly 0.1 to 1000 seconds are produced, similar to when one passes a foil target across the beam of a synchrotron accelerator. One figure illustrates this. Obviously, there will be **strong relativistic beaming of the burst,** drastically reducing the energy requirements. The other figure illustrates the **SMF** model for an AGN after gravitational collapse is essentially complete. Clearly for the same GBCL and "rock", a **long** burst comes when the "rock" passes roughly along a GBCL diameter, while a **short** burst comes when it passes along a short chord. <u>**The timing fits the observations.**</u> For instance, with reasonable values, a "rock" (or "flying mountain") moving at 1000 km/second crossing a 100,000 km <u>small diameter</u> of the current loop beam would produce a burst lasting 100 seconds.

In this concept from **SMF**, the less than millisecond rise times observed in GRB are describing the size of the very slender filaments of current that make up the beam. It is well known from solar system space plasma studies that currents in space are made up of slender filaments due to the pinch effect.[14] The intensity-time pattern of the burst gives the beam current structure in that particular source. Cyclotron resonance lines observed previously in GRB are consistent with the extremely intense magnetic field adjacent to the **GBCL**.

From the table in reference (9), which is deduced from a classic paper by Julian Schwinger, the beam energy is deduced to reach as high as or higher than 100 GeV for a young forming galaxy.[15] If only one in 10^{10} of the galactic mass makes up the **GBCL**, the stored energy in the GBCL is 10^{58} ergs, and a significant fraction of this can be converted into gamma rays by a variety of well-known reactions, including ionization, nuclear excitation, brehmsstrahlung, Compton scattering, pair production, electromagnetic showers, etc. A stream of matter, passing close to or crossing the beam of the current loop in an AGN for several weeks, can explain the outbursts from "gamma-ray loud" AGN such as 3C279.

An elaborate computer calculation of the electromagnetic shower produced when the rock passes through the intense relativistic current beam, using the cross-sections and formulas such as in the book by W.S.C. Williams, is necessary for proper comparison of theory with observations.[16] However, a crude estimate, to demonstrate that the energies observed in gamma ray bursts are produced by the model, is as follows: Using reasonable values, one estimates $n_1 n_2 \sigma v$ to be 10^{31} reactions per cm^2 per second. If the "rock" is 3000 km in radius, and the burst lasts 100 seconds, then the total number of reactions during the burst is 3×10^{50}. Now if the particles are 100 GeV in energy, the total energy of the burst is 3×10^{49} ergs, which is comparable with the observations.

DISCUSSION AND CONCLUSIONS

From the viewpoint of the **Strong Magnetic Field model (SMF)**, which was created in 1961 before the discovery of AGN and quasars, a simple, direct explanation for gamma ray production is the rapid passage of matter through, or very near, the intense relativistic gravitationally bound current loop **(GBCL)** in the central engine (nucleus) of the galactic object.

For GRB, the model suggests that one may be surveying a new cosmological population of **optically faint, young forming galaxies**, mostly from around z = 5 and beyond! This opens up the possibility that one may now have a means of studying a population of objects **even earlier in the Universe than the quasars**, a subject of great importance to cosmology.

As I pointed out one year ago, the V/V_{max} of about 0.34 observed from GRO for gamma ray bursts in the BATSE experiment resembles that observed for quasars whose cosmological distance is greater than 3.0.[11] This is confirmed from a result by Maarten Schmidt (private communication) that for quasars from z = 2.75 to 4.75, the V/V_{max} is 0.38. Of course the isotropic distribution of GRB suggests a cosmological origin, as many astrophysicists have pointed out. The temporal profile of bursts is affected by time dilation as discussed by J. P. Norris et al.[19] **SMF agrees with the insightful remark by Martin Rees,** that "the phenomena of quasars and radio galaxies cannot be understood until they are placed in the general context of galactic evolution".[17] The connection of **SMF** with the interesting large scale structures of galaxies in the Universe is made in a recent paper by the author.[18]

It is a pleasure to acknowledge the late Donald H. Menzel, and John A. Wheeler, for comments, and Gart Westerhout for permission to use the U.S. Naval Observatory library.

REFERENCES

1. H.D. Greyber, Air Force Office of Scientific Research, Report No. 2958, June 1, 1962.
2. H.D. Greyber, Quasistellar Sources and Gravitational Collapse, ed. I. Robinson et al, U. of Chicago Press, Chap.31, 1964.
3. H.D. Greyber, Instabilite Gravitationelle et Formation des Etoiles, des Galaxies et de Leurs Structures Caracteristiques, Memoirs Royal Society of Sciences of Liege, **XV**, 189 (1967).
4. H.D. Greyber, ibid. **XV**, 197 (1967).
5. H.D. Greyber, Publications Astronomical Society of the Pacific, **79**, 341 (1967).
6. H.D. Greyber, 11th Texas Symposium., Annals New York Academy of Sciences, **422**, 353 (1984).
7. H.D. Greyber, Supermassive Black Holes, ed. M. Kafatos, Cambridge University Press, p. 360 (1988).
8. H.D. Greyber, Comments on Astrophysics, **13**, 201 (1989).
9. H.D. Greyber, The Center of the Galaxy, ed. Mark Morris, Kluwer Academic Press, p. 335 (1989).
10. H.D. Greyber, 14th Texas Symposium, Annals New York Academy of Sciences, **571**, 239 (1990).
11. H.D. Greyber, Testing the AGN Paradigm, eds. S. Holt et al, A.I.P. Conf. Proc. **254**, 467 (1992).
12. H.D. Greyber, in STScI Symposium Proceedings Report on "Astrophysical Jets", December 1992.

13. L. Mestel and P. Strittmatter, Monthly Notices Royal Astronomical Society, **137**, 95 (1967).
14. H. Alfven, Cosmic Plasma, Chapter II, D. Reidel Publishers, (1981).
15. J. Schwinger, Physical Review, **75**, 1912 (1949).
16. W.S.C. Williams, Nuclear and Particle Physics, Clarendon Press, Oxford, Chapter 11 (1991).
17. M.J. Rees, Highlights of Modern Astrophysics, eds. S. Shapiro and S. Teukolsky, Wiley, p.163 (1986).
18. H. D. Greyber, "The Formation of Sheets of Galaxies and Voids", to be published.
19. J. P. Norris et al, in this Compton Symposium volume.

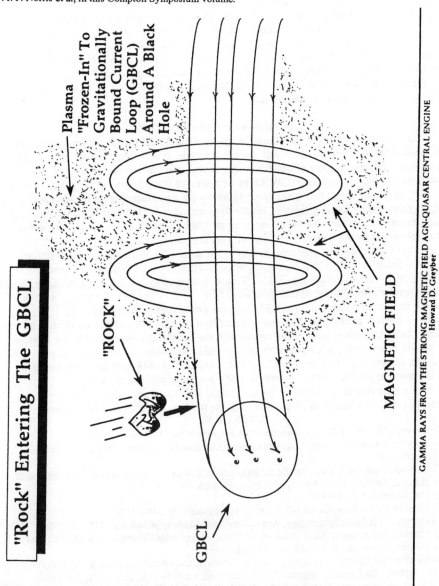

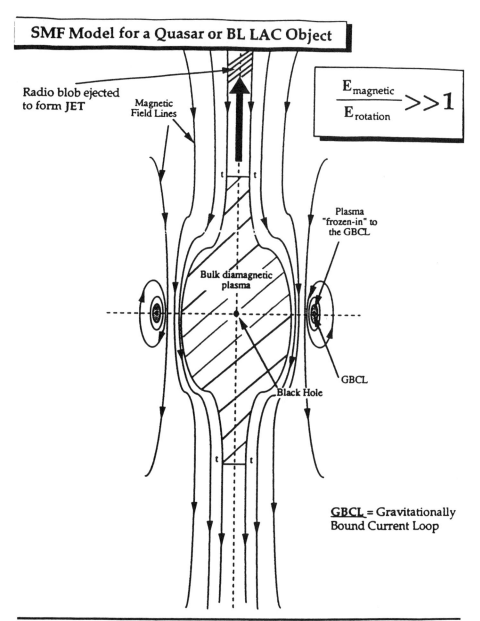

The Strong Magnetic Field AGN-*Central Engine* Quasar-Galaxy Formation Paradigm
H. D. Greyber

GAMMA- RAY-EMISSION FROM STAR - ACCRETION DISK COLLISIONS IN ACTIVE GALACTIC NUCLEI

O. Heinrich [+], G. Shaviv [*] & R. Wehrse [+]

[+] Institut für Theoretische Astrophysik
Heidelberg, Germany

[*] Department of Physics and Asher Space Research Institute
Israel Institute of Technology
Haifa, Israel

It is shown that star- disk encounters in active galaxies lead to energetic X- and gamma- ray transients. We estimate the total energy and the rising time of such transients. The rate of star- disk collisions may be as high as ~ 100 per year.

There is strong evidence that the main mechanism for producing the luminosity of active galactic nuclei (AGN) is viscous heating in an accretion disk [1]. However, the required supply of a considerable amount (1...10 solar masses per year) of matter to the vicinity of the central black hole poses a severe problem [2]. For black holes smaller than $10^8 M_\odot$ tidal disruption of stars can provide the disk material [3,4]. For more massive black holes direct star- disk encounters have been proposed as an efficient fueling mechanism [5,6,7]. The aim of this paper is to show that star- disk collisions are inevitably connected with energetic X- ray and γ - ray transients. An observational identification of such events should be possible since several predictions about the properties of such transients can be made.

We begin with an estimate of the relative velocity v_{rel} at which an encounter typically takes place. v_{rel} is given by the angular velocity in the Keplerian accretion flow and by the velocity of the star. The latter we assume to be randomly distributed. Thus

$$v_{rel} \approx \sqrt{\frac{GM}{r}} = c\sqrt{\frac{r_{grav}}{r}}$$

where r_{grav} is the gravitational radius. At this relative velocity the protons of the accretion disk will collide with the protons in the surface layers of the star. The energy in the center of mass system of the p-p- collision is

$$\epsilon_{cms} = m_p c^2 \cdot \frac{r_{grav}}{r}.$$

If we assume a disk extending from $r_{in} = 3r_{grav}$ (nonrotating black hole) to $r_{out} \approx 1000 r_{grav}$ we obtain a range

$$\epsilon_{cms} = 1 \ MeV...330 \ MeV.$$

For most cases this is clearly above the threshold for electron- positron pair production. Since the threshold for pion production in p-p collisions is 290 MeV, one also has to expect $\gamma-$ rays from pion decays for the most energetic collisions in the inner region . For a detailed description of pion reactions in accretion disks see the paper of Eilek [8]. For a rotating black hole the upper limit for ϵ_{cms} is even higher but then a fully relativistic treatment of the collision is necesssary. In any case, the maximum available cms- energy provides a cutoff E_c for the $\gamma-$ spectrum of the transient.

Several authors (e. g. [9,10,11]) have discussed the possibility of optically thin disks which due to the lack of an efficient cooling mechanism reach very high temperatures. It seems unlikely that the optically thin regime can be reached in a standard AGN- disk, but it is quite possible that a star- disk collision evaporates part of the disk material and changes the remaining disk locally to a state corresponding to the hot disk models. Therefore the decay time for the high- energy activity may be significantly longer than the passage time of the star.

The next point is to estimate the total amount of energy E which can be released in a star- disk encounter. It is approximately given by the kinetic energy of the matter in the Keplerian disk flow that collides with the star during its passage time

$$\Delta t \sim \frac{h_d}{v_{star}},$$

(h_d is the height of the disc). If one adopts the standard thin Keplerian disk model [12], the disk thickness h_d should be smaller than $0.1 r_g$, or $h_d \sim 10^{-2} r_g$. If, however, the disk changes to a thick torus [13] in its innermost parts, one may have $h_d \approx r_g$. Assuming a star on a nearly parabolic orbit one has

$$v_{star} = \sqrt{\frac{2GM}{r}} \approx 10^3...10^5 \, km s^{-1}$$

leading to passage times in the range

$$\Delta t \approx 10^2...10^5 s$$

(i.e. between about one minute and one day).

The volume ΔV directly affected by the star is therefore

$$\Delta V \sim v_{rel} \cdot \Delta t \cdot \pi R_{star}^2$$

and the energy involved is

$$E \sim \frac{1}{2}\rho v_{rel}^2 \cdot \Delta V = \frac{1}{2}\rho v_{rel}^3 \cdot \frac{h_d}{v_{star}} \cdot \pi R_{star}^2.$$

For the mean density in the disk one can reasonably assume (see the recent calculations of AGN- disk structure by Stoerzer [14]) $\rho \approx 10^{-7} g cm^{-3}$. Thus one has

$$E \approx 10^{46}...10^{50} erg.$$

The star- disk model discussed here makes a definite prediction about the rising time for a X-ray/gamma- ray outburst connected with a star- disk encounter. The rising time τ is roughly given by the stellar radius divided by its velocity perpendicular to the disk. That gives approximately

$$\tau = 10s...10^3 s.$$

A further prediction can be made about the distribution of the total energy output among the transients if the disk model is specified. Let us assume, that the density varies as $\rho \sim r^{-\alpha}$ and the disk height as $h_d \sim r^\beta$. Then one has roughly

$$E \sim r^{\beta-\alpha-1},$$

while for the number of events within dr one has $dN \sim rdr$. Thus

$$\frac{dN}{dE} \sim E^{-\gamma}$$

with

$$\gamma = \frac{3+\alpha-\beta}{1+\alpha-\beta}.$$

Considering the distribution of events with respect to the spectral cutoff E_c one has completely independent from the disk structure $E_c \sim r^{-1}$ and therefore

$$\frac{dN}{dE_c} \sim E_c^{-3}.$$

Let us assume now that direct star- disk encounters are indeed the dominant feeding mechanism for the accretion phenomenon. What is the rate at which star- disk encounters take place ? In any case, one has to assume a high mass density for the central region of quasars in order to fuel the central engine. Most of it will be contained in stars. Following the work of Hills [15] we consider

$$n_s \approx 10^9 pc^{-3}$$

to be a reasonable value. The star- disk collision rate is given by

$$\Lambda = n_s \cdot \pi r_{out}^2 \cdot \bar{v}$$

where $\bar{v}$ is obtained from the velocity dispersion of stars in the central region of galaxies and is of the order of magnitude

$$\bar{v} \approx 250 kms^{-1}.$$

Thus one has

$$\Lambda \approx 82 y^{-1}.$$

It should be stressed that the rate Λ increases if one takes into account the possibility that stars suffer an orbital decay and are captured in bound orbits around the black hole [6,7].

In conclusion, one can say that star- disk collisions should strongly contribute to the variability of AGNs up to $\gamma-$ ray energies.

Acknowledgments: This work was supported in part by the Deutsche Forschungsgemeinschaft , the DARA and the GIF .

REFERENCES
1. Ulrich,M.-H. in "Theory of Accretion Disks"
(ed. Meyer,F. et. al.) 3-18 (Kluwer,Dordrecht, 1989)
2. Shlosman,I., Begelman,M.C. & Frank,J., Nature 345,679-686 (1990)
3. Rees,M.J., Nature 333, 523-528 (1988)
4. Cannizzo,J.M.,Lee,H.M. & Goodman,J., Astrophys. J. 351,38-46(1990)
5. Ostriker,J.P., Astrophys. J. 273 (1983) 99
6. Allen, A.J. and Hughes,P.A., Astrophys. J. 313 (1987) 152
7. Syer,D.,Clarke,C.J. & Rees,M.J. Mon. Not. R. Astr. Soc. 250 (1991) 505
8. Eilek,J.A., Astrophys. J. 236 (1980) 664
9. Shapiro S.L.,Lightman,A.P. & Eardley,D.M.,
Astrophys. J. 204,187- 199(1976)
10. Liang,E.P.T., Astrophys. J. 234 (1979) 1105
11. White,T.R. and Lightman,A.P., Astrophys. J. 352 (1990) 495
12. Shakura,N.I. & Sunyaev,R.A., Astron. Astrophys. 24,337-355(1973)
13. Abramowicz,M.A., Calvani,M. & Nobili,L.,
Astrophys. J. 242,772-778(1980)
14. Stoerzer,H., submitted to Astron. Astrophys.
15. Hills,J.G., Mon. Not. R. Astr. Soc. 182 (1978) 517

EXAMINING THE SYNCHROTRON SELF-COMPTON MODEL FOR BLAZARS

Steven D. Bloom and Alan P. Marscher

Department of Astronomy, Boston University

ABSTRACT

We examine the synchrotron self-Compton (SSC) process as the primary γ-ray emission mechanism in blazars. Contrary to recent published criticisms, we find that the SSC model conforms with γ-ray observations of blazars. In a follow-up of a previous paper, we demonstrate through numerical calculations how the shape and level of the high energy spectrum changes with certain crucial parameters of blazars, such as $F_{\nu x}/F_{\nu s}$, optically thin synchrotron spectral index, and the maximum and minimum relativistic electron energies. The hard γ-ray log νF_ν vs. ν spectrum tends to be flat, corresponding to an energy index $\alpha_\gamma \sim 1$, as observed. The TeV γ-ray flux from Mkn 421 is consistent with first order scattering. We discuss future observations that can test the SSC model.

1. INTRODUCTION

The EGRET and COMPTEL instruments on CGRO have detected hard and medium-energy (3 MeV–10 GeV) γ-rays from a number of blazars. The γ-ray emission from the BL Lac object Mkn 421 extends to photon energies up to 0.5 TeV (Punch et al. 1992). Most blazars have also been detected as keV X-ray sources. All of the blazars detected at γ-ray energies are bright, flat-spectrum radio sources. For blazars observed with the Einstein Observatory, the X-ray luminosities are also strongly correlated with the luminosities of the compact radio emission (Worrall et al. 1987). These observations show that there is a strong relation between the emission at both very high and very low photon energies. One emission mechanism that produces such a relationship is synchrotron self-Compton (SSC) radiation. Several authors have discussed the SSC process in the context of blazars (eg., Jones et al. 1974; Jones 1979; Band and Grindlay 1986). In a previous paper (Bloom and Marscher 1992; hereafter called Paper I) we presented formulae for the flux densities from synchrotron radiation plus 1st and 2nd order scattering in a homogeneous spherical source. We showed that if the X-ray emission mechanism is assumed to be first order self-Compton, the ratio of X-ray to optically thin synchrotron (submm–IR) flux density, $F_{\nu x}/F_{\nu s}$, can be used to predict the second order self-Compton γ-ray flux density. In this paper, we show more explicitly, in terms of νF_ν spectra, how the radio–IR, X-ray, and γ-ray spectra are related.

2. FIRST AND SECOND ORDER SSC SPECTRA

We use the analytic formulae of Marscher (1987) to calculate the physical parameters B (magnetic field), N_0 (relativistic electron density), and γ_{min} and γ_{max} (lower and upper cutoff of electron energy spectrum) given a set of observed parameters (angular size, turnover frequency, radio flux density, synchrotron cutoff frequency, relativistic Doppler factor inferred from superluminal motion of VLBI components, and spectral index) for a hypothetical blazar. We then calculate the first and second order flux densities using expressions similar to those presented in Paper I (e.g., equation 4). The full Klein-Nishina cross-section is used. For calculational purposes we assume that the synchrotron source is uniform and spherical and also moving relativistically at a narrow angle to the line-of-sight. Even if the emission originates in a jet, this approximation is appropriate if most of the scattering occurs over a small portion of the jet (Königl 1981).

We present a number of calculated SSC spectra to illustrate how changing certain key parameters affects the self-Compton emission. Of particular interest is the detailed dependence of the hard γ-ray spectrum on the optically thin radio spectral index and electron energy cutoffs. We display the spectra in terms of $\log \nu F_\nu$ vs. $\log \nu$, as these show more explicitly how the emitted energy is distributed per logarithmic frequency interval. We note, however, that sharp curvature in such plots may correspond to only very gradual curvature in plots of $\log F_\nu$ vs. $\log \nu$, which is more closely coupled with the observational data. In general the spectra have similar overall shape. Over restricted frequency ranges, both the first and second order spectra tend to mimic the synchrotron spectrum, as expected from basic theory. However, at the lowest photon energies the scattered spectrum gradually curves downward toward lower frequency due to the lower electron energy cutoff. This effect is even more exaggerated for second order scattering. At the highest photon energies, the spectra also depart from the shape of the synchrotron spectrum. The first order spectrum steepens and then cuts off at frequencies $\gtrsim \gamma^2 \nu_{cutoff}$. Beyond frequencies $\sim \frac{\gamma_{min} mc^2}{h}$, the slope of the second order spectrum α_γ increases owing to Klein-Nishina effects. Beyond $\nu = \frac{\gamma_{max} mc^2}{h}$ there is a sharp cutoff, since no photon can have more energy than the highest energy electron.

Of primary interest in terms of comparison with observations, is the detailed dependence of the high energy spectrum on the ratio of X-ray to radio flux densities, optically thin radio spectral index, and electron energy cutoffs. The results are summed up in Figs. 1-6. In Figs. 1 and 2 we examine the dependence of the γ-ray flux on the ratio $F_{\nu x}/F_{\nu s}$. When $F_{\nu x}/F_{\nu s}$ is decreased (moving from Fig. 1 to 2), the γ-ray flux is significantly reduced. In these figures, the first-order scattered emission dominates up to frequency $\nu \sim \frac{\gamma_{min} mc^2}{h}$ (about 10 MeV). In the example shown in Fig. 2, νF_ν decreases considerably beyond about 10 Mev. In this case, a source could be detected by COMPTEL, yet remain undetected by EGRET. In Fig. 3 we increase the upper synchrotron cutoff (relative to Fig. 1). With all other parameters remaining constant, the higher cutoff implies a greater upper electron energy cutoff, and hence a much larger 1st order contribution to the observed γ-ray flux. Fig. 4 shows the effect of varying the lower energy cutoff to the relativistic electron distribution, γ_{min}. For lower γ_{min} the X-ray flux at several

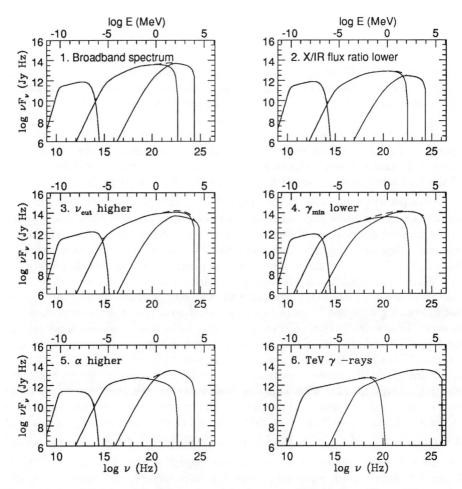

Figures 1-6. Multi-waveband spectra of a hypothetical blazar. The solid curves correspond to (left to right) synchrotron radiation, 1st order, and 2nd order self-Compton scattering, and the dashed curve is the combined emission from all three processes. Figs. 2–5 show the effects of varying the parameter noted. Fig. 6 corresponds to another hypothetical blazar with a spectrum roughly similar to that observed for Mkn 421.

keV is about the same; however, the soft γ-ray flux is greater since the region of strong emission from second order scattering extends to lower frequencies. In addition, νF_ν is flatter at γ-ray energies. At the lower energies we find that the curvature of the 1st order spectrum is less dramatic, causing the optical flux to be greater. In fact, the requirement that the first-order inverse Compton flux not exceed that observed places a lower limit on γ_{min} for any given source model. In Fig. 5 we examine the effects of varying the spectral index.

Fig. 6 shows the spectrum of a model source, with a very high value of γ_{max}, that roughly corresponds to Mkn 421. The multifrequency spectrum of this source (Makino 1989) is consistent with the synchrotron emission extending all the way to X-ray frequencies, which indeed implies that γ_{max} is quite high. Obviously, a value $\gamma_{max} \sim 10^6$ is needed to produce TeV γ-rays. In this model, the TeV γ-rays are due primarily to first order self-Compton scattering.

3. VIABILITY OF THE SSC MODEL FOR BLAZAR γ-RAYS

Jones (1979) discusses the virtues of the SSC process in explaining the γ-ray emission from 3C 273. Indeed, compact incoherent synchrotron sources must produce self-Compton scattered high-energy photons. Nevertheless, with the detection of hard γ-rays from a number of flat-spectrum radio sources, the SSC model has been criticized on the grounds that the γ-ray spectrum should show more curvature than is observed (Dermer and Schlickeiser 1992) and that Klein-Nishina effects preclude the detection of high fluxes of γ-rays at 0.5 TeV (Zdziarski & Krolik, these proceedings), as observed from Mkn 421. Our calculations show that both criticisms are unwarranted. The νF_ν vs. ν spectra are quite flat, as observed, with only slight curvature. In addition, Mattox et al. (1993) report that the γ-ray spectrum of 1633+382 *does* show downward curvature.

One advantage of the SSC process is that it can occur far from the central engine, perhaps near the radio core at a distance of parsecs, thereby avoiding high opacities to pair production. At infrared and longer wavelengths, flares in compact nonthermal sources are well modeled as shock waves propagating down a turbulent relativistic jet (Marscher, Gear, and Travis 1992). The emission region behind such shocks can be very thin owing to radiative losses. The resultant timescale of variability can be quite short, of order days, despite a distance of parsecs from the central engine.

The SSC model makes certain strong predictions that can be used to test its application to γ-ray emission in AGN. A source whose hard γ-ray and submillimeter–infrared fluxes vary simultaneously or with short delay must have an X-ray luminosity that is intermediate between the luminosities in the other two wavebands. That is, if the γ-rays result from 2nd order scattering of the synchrotron radiation from the dominant submm-IR component, the X-rays must also have high luminosity at the same time (cf., e.g., Fig. 5). A second prediction is that any source with strong TeV emission should have a synchrotron spectrum that continues up to X-ray frequencies.

It is theoretically possible for a small region of a source to produce copious γ-rays but a low luminosity of X-rays and even lower luminosity of synchrotron radiation such that the latter two are "buried" beneath the stronger emission from other portions of the source. Such a "γ-ray only" component would need to be a compact site of extraordinarily efficient particle acceleration. However, since the "Compton catastrophe" is mitigated by Klein-Nishina effects beyond 2nd order scattering, there are no fundamental arguments against such an explanation for luminous emission of hard γ-rays. In the shock model for submm–IR flares (Marscher and Gear 1985), rapidly variable self-Compton emission is expected to dominate the luminosity prior to the onset of the synchrotron outburst.

4. FUTURE OBSERVATIONS FOR TESTING THE SSC MODEL

Simultaneous observations in the radio through γ-ray wavebands for a well selected sample are necessary to determine whether $F_{\nu x}/F_{\nu s}$ is a good indicator of $F_{\nu \gamma}$. The parameters of the synchrotron emission needed to predict the self-Compton flux can be determined through observations at radio–submm and IR–optical wavelengths. With the essential parameters determined by observation the SSC model (in its simplest form) will be subject to stringent tests. In more general terms, repeated simultaneous multi-frequency observations will reveal the connections between the emission at different wavebands. Each model for the γ-ray emission predicts specific multi-waveband behavior. It is only through such (massive!) efforts that the origin of hard γ-rays from blazars can be determined.

This research is funded by the NASA Gamma Ray Observatory Guest Investigator Program under grant NAG5-1566.

REFERENCES

Band, D. L., & Grindlay, J. E. 1986, ApJ, 308, 576

Bloom, S. D., & Marscher, A. P. 1992, in The Compton Observatory Science Workshop, ed. C. Shrader, N. Gehrels, & B. Dennis (NASA), 339

Dermer, C. D., & Schlickeiser, R. 1992, Science, 257, 1642

Jones, T. W. 1979, ApJ, 233, 796

Jones, T. W., O'Dell, S .L., & Stein, W. A. 1974, ApJ, 188, 353

Königl, A. 1981, ApJ, 243, 700

Makino, F. 1989, in Proc. 23rd ESLAB Symposium on Two Topics in X-Ray Astronomy (ESA SP-296, Nov. 1989), 803

Marscher, A. P. 1987, in Superluminal Radio Sources, ed. J. A. Zensus, & T. J. Pearson (Cambridge: Cambridge University Press), 280

Marscher, A.P., & Gear, W.K. 1985, ApJ, 298, 114

Marscher, A. P., Gear, W. K., & Travis, J. P. 1992, in Variability of Blazars, ed. E. Valtaoja & M. J. Valtonen (Cambridge University Press), 85

Mattox, J.R., et al. 1993, ApJ, submitted

Punch, M., et al. 1992, Nature, 358, 477

Worrall, D. M., Giommi, P., Tananbaum, H., & Zamorani, G. 1987, ApJ, 313, 596

A MODEL OF THE COSMIC X- AND γ-RAY BACKGROUND

M.Matsuoka, N.Terasawa and M.Hattori
The Institute of Physical and Chemical Research (RIKEN),
Wako, Saitama 351-01, Japan

ABSTRACT

The origin of wide band cosmic X- and γ-ray background has not been well understood, but recently it has been considered that the cosmic X-ray background in the energy range below several 100 keV is well reproduced by the superposition of AGN X-ray spectra with a hard bump (Morisawa et al. 1990; Fabian et al. 1990; Rogers & Field 1991; Terasawa 1991). It should be noted that this interpretation is based on the recent *Ginga* observations of the X-ray spectrum of AGN (Morisawa et al. 1990; Matsuoka et al. 1990).

On the other hand, the cosmic γ-ray background above 0.5 MeV which shows a broad bump around a few MeV seems to require an additional component to the superposition spectrum of AGN X-ray radiation. In order to explain this cosmic γ-ray background we investigate the model in which π^0–γ–rays are produced by p-p and/or α–p interaction of cosmological cosmic rays. The cosmological cosmic rays could be provided from explosions of Population III objects. It is known that this basic theory has been proposed by Stecker (1971).

COSMIC X-RAY BACKGROUND IN THE ENERGY BAND OF 1-500 KEV

Recent discovery of a hard bump around 10-30 keV in AGN (Matsuoka et al. 1990) led to great progress in the cosmic X-ray background theory. The hard bump is commonly interpreted as created by the Compton reflection of X-rays on the dense matter near a non-thermal X-ray emitter (Lightman & White 1988; Guilbert & Rees 1988; Matsuoka et al. 1990; Pounds et al. 1990; Piro et al. 1990).

The hard bump of AGN spectra could resolve the difficult problem of superposition of canonical power law spectra ($\alpha=0.7$) of AGN, because generally AGN spectra with a hard bump do not show a canonical spectrum. Especially distant faint sources would show a fairly flat spectra although this has not been confirmed by the z-dependence of the spectral index in the hard X-ray band concerned.

Now we have several works to interpret the cosmic X-ray background along the above idea based on the observational results (Morisawa et al. 1990; Fabian et al. 1990; Roger & Field 1991; Terasawa 1991; Zdziarski et al. 1992 and so on). In Figure 1 we show the recent result based on the observational results of AGN spectra. In this theory it is noted that we need the three key points: (1) X-ray spectra of most AGN have the hard bump around 10-30 keV which would be created by Compton reflection of a power law spectrum incident on cold matter. (2) There is a certain luminosity evolution in AGN. (3) We need some tuning factor in AGN spectra such as a uniform absorption component (Terasawa 1991) or a constraint on an incident power law index (Zdziarski et al. 1992).

COSMIC γ-RAY BACKGROUND IN THE ENERGY BAND OF 0.5-10 MEV

On the other hand, the cosmic γ-ray background (CGB) in the energy band of 0.5-10 MeV is observed with fairly reliable results by several authors, as shown in Figure 2 (Fabian 1981 and references therein). The CGB is approximately flat between 500 keV and 3 MeV, and then drops off as a power law with energy index ~1.5 above 5 MeV up

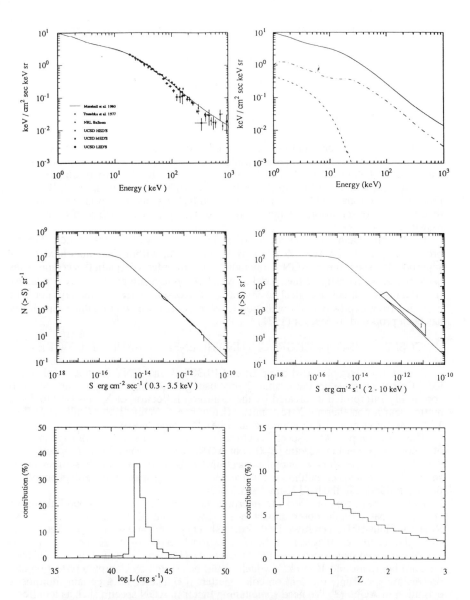

Figure 1. The CXB spectrum reproduced by superposition of AGN spectra which have the reflection component and uniform absorption component in addition to the power law component (see Terasawa (1991) for more details).
Top left: predicted CXB spectrum compared with observations. Top right: predicted spectrum (solid line) and cluster contribution (broken line) with a model spectrum of AGN (dash-dotted line). Middle left: the predicted log N - log S (0.3-3.5 keV) superposed with observed log N - log S by EMSS. Middle right: the predicted log N - log S (2-10 keV) superposed with observed log N - log S by *Ginga*. Bottom left: contribution to CXB at 10 keV from AGN with different luminosity. Bottom right: contribution to CXB at 10 keV from AGN with different red shift.

to ~10 MeV. This spectral shape seems to require a somewhat broad band emission mechanism with a peak around 2 MeV, but may not require two spectral indices. The interpretation of this "MeV bump" origin has not been established, although a number of models have been proposed so far (Stecker 1971; Brown & Stecker; Olive & Silk 1985; Daly 1988; Strong et al. 1976; Bignami et al. 1978; Grindlay 1978; Lichti et al. 1978; Schönfelder 1978; Bignami et al. 1979; Mereghetti 1990; Rogers & Field 1991).

Recently a smart idea for the integrated emission of AGN has been proposed by Rogers & Field (1991). They consider that the spectrum of AGN is composed of three parts; an X-ray power law, a broad hump from ~10 to 100 keV due to reprocessing of the power-law X-ray spectrum by cold gas, and γ-rays which are due to the inverse-Compton scattering of X-rays. They also have derived constraints on the physical conditions in AGN to fit the observed CGB to their model. One of the essential points is to assume that generally the γ-ray spectrum of AGN has an MeV bump. Consequently they expect that future observations of γ-ray spectra of AGN will provide a useful test of their model.

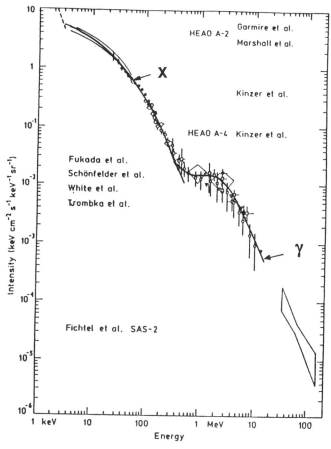

Figure 2. Observed cosmic X- and γ-ray background (taken from compilation by Fabian (1981)). The "X" marked curve is a model of superposed spectra of AGN (see Fig.1), while the "γ" marked curve is a tentative model spectrum consisting of red shifted π^0–γ–rays and bremsstrahlung γ-rays by cosmological cosmic rays.

In this paper we investigate the model in which π^0-γ-rays are produced by p-p and/or α-p interaction of cosmological cosmic rays. The peak energy of π^0-γ-rays forms at ~70 MeV (Dermer 1986 and other references therein). Figure 3 shows the expected γ-ray spectrum which is estimated assuming current cosmic-proton and electron spectra near our solar system. An integrated γ-ray spectrum creates a 60-100 MeV bump structure as a result of contributions of primarily bremsstralung γ-ray emission from cosmic ray electrons in addition to π^0-γ-ray production. On the contrary, COS B result of the galactic diffuse γ-ray spectrum did not show a clear 60-100 MeV bump (Strong et al. 1982). This can be attributed to greater contribution of bremsstrahlung γ-ray emission from primarily accelerate cosmic ray electrons (Schlickeiser 1982). Generally it is difficult to estimate primarily accelerated comic ray electrons, because this flux depends on the acceleration mechanism and condition.

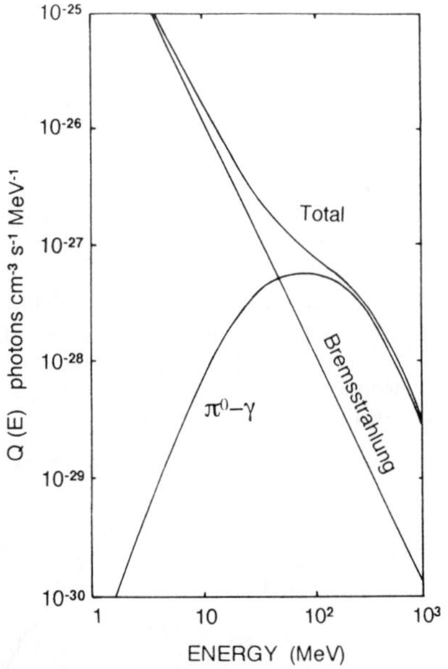

Nevertheless we consider a possibility of redshifted contribution of π^0-γ rays due to cosmological cosmic rays. This idea was already proposed by Stecker (1971) long ago. A problem of heavy element creation in primordial gas is one of the difficult problems in current cosmology.

Figure 3. γ-ray production rate in our Galaxy, where the intensity of cosmic ray electrons is assumed to be the observed one in the solar system (see Dermer (1986) and Stecker (1977)) for more details).

If some cosmological objects such as Population III could provide heavy elements before galaxy formation, this would be resolved. Recently de Araujo & Opher (1991) have studied a possible formation of Population III objects that can be formed taking account of cold collisionless dark matter. They show that the minimum mass of Population III to collapse is $\sim 10^3$ $M_\odot$ when z=100~300, although Jeans mass depends on a mass of dark matter.

Once Population III objects are formed, super nova explosions would occur in a final phase of envolution of these objects. Consequently cosmic rays also are produced accompanying the explosions. The cosmic rays would interact with interstellar matter and produce π^0-γ-rays. The phenomena to be observed in this process would be red-shifted X- and γ-rays. Red-shifted π^0-γ-rays at z~100 could form the cosmic γ-ray background which is possible to contribute to the observed CGB. In this model we can derive constraints on the cosmic ray spectrum and primarily accelerated cosmic ray electrons. Comparing the observed CGB with this γ-ray background, we can derive a result that the power-law index (differential index) of cosmic rays (protons) is gradually changing from 2 to 2.5 in the energy of the GeV region and that the primarily accelerated cosmic ray electrons are negligibly small.

A possible formation of Population III objects is supported from the following order estimation regarding the flux of CGB and an expected flux of primordial cosmic rays.

The upper limit of metal abundance of Population II is about 1% of that of solar abundances; $z_{II} \geq 0.01\ z_\odot$. If this metal is provided by Population III objects, the metal density of Population III objects is estimated to be $\rho_{zIII} \sim z_{III}\ m_{Fe}\ \rho_b = (3.5-7) \times 10^{-36}$ g cm^{-3}, where we take the upper limit of Population II for z_{III} and the barion density must be $\rho_b = (2-4) \times 10^{-31}$ g cm^{-3} from a current nuclear synthesis theory. Assuming $M_z \sim 0.07 M_\odot$ and $E_{SN} \sim 10^{51}$ erg for the respective amounts of metal and energy by a super nova explosion, a production rate of the energy of cosmic rays concerned due to all Population III objects is estimated to be $q_{HE} \sim (\rho_{zIII}/M_z)\ E_{SN}\ f_{HE} \geq (0.5-1)\ 10^{-16}\ (f_{HE}/1)\ (z_{III}/0.01 z_\odot)$ erg cm^{-3}. A factor of f_{HE} is an efficiency of cosmic ray production. From this production rate the π^0-γ-ray flux is estimated to be $F_\gamma = q_{HE}\ (c/4\pi) = (1.3-2.5) \times 10^{-7}\ \xi$ erg cm^{-2} s^{-1} sr^{-1}., where ξ is the efficiency of γ-ray production from cosmic rays. This last value is compared with the observed γ-ray flux around 0.5-3 MeV, which is $(6-8) \times 10^{-8}$ erg cm^{-2} s^{-1} sr^{-1} from Figure 2. Both the flux values are comparable to each other. Although a fairy efficient γ-ray production is required in this scenario, it is noted that this kind of order of estimations has further uncertainty. Thus it is concluded that the present model gives one possibility to explain the CGB around 0.5-10 MeV.

The "γ" marked line in Figure 2 shows the present model which is fitted to the observed CGB.

REFERENCES

Bignami,G.F., Lichti,G.G. & Paul,J.A. 1978, Astron. Astrophys. **68**, L15.
Bignami,G.F., Fichtel,C.E., Hartman,R.C. & Thompson,D.J., 1979, Ap.J. **232**, 649.
Brown,R.W. & Stecker,F.W., 1979, Phys. Rev. Letters **43**, 315.
Daly,R.A., 1988, Ap. J. Letters **324**, L47.
De Araujo,J.C. and Opher,R., 1991, Ap. J. **379**, 461.
Dermer,C.D., 1986, Astron. Astrophys. **157**, 223.
Fabian,A.C., George,L.M., Miyoshi,S. & Rees,M.J., 1990, MNRAS **242**, 14p.
Fabian,A.C., 1981, Annals New York Academy of Sciences **375**, p235.
Grindlay,J.E., 1978, Nature **273**, 211.
Lightman,A.P. & White,T.R., 1988, Ap.J. **335**, 57.
Guilbert,P. & Rees,M.J., 1988, MNRAS, **233**, 475.
Matsuoka,M., Piro,L., Yamauchi,M. & Murakami,T., 1990, Ap. J. **361**, 440.
Mereghetti,S., 1990, Ap.J. **354**, 58.
Morisawa,K. Matsuoka,M., Takahara,F. & Piro,L., 1990, Astron. Astrophys. **236**, 299.
Olive,K.A. & Silk,J., 1985, Phys. Rev. Letters, **55** 2362.
Piro,L., Yamauchi,M. & Matsuoka,M., 1990, Ap. J. Letters, **360** L35.
Pounds,K.A., Nandr,K., Stewart,G.C., George,I.M. & Fabian,A.C., 1990, Nature, **44**, 132.
RogersR.D. & Field,G.B., 1991a, Ap.J. Letters 370, L57.
Rogers,R.D. & Field,G.B., 1991b, Ap.J. Letters 378, L17.
Schlickeiser,R., 1982, Astron. Astrophys. **106**, L5.
Schönfelder,V., 1987, Nature 275, 719.
Stecker,F.W., "Comic Gamma Rays" Mono Book Co. Baltimore (1971).
Stecker,F.W. , 1977, Ap. J. **212**, 60.
Strong,A.W., Wolfendale,A.W. & Worral,D.M., 1976, MNRAS **179**, 23p.
Strong,A.W. et al., 1982, Astron. Astrophys. **115**, 404.
Terasawa,N., 1991, Ap.J. Letters **378**, L11.
Zdziarski,A.A., Zycki,P.T., Svensson,R. & Boldt,E. 1992, Ap. J. in press.

A TEST OF THE UNIFIED MODEL FOR ACTIVE GALACTIC NUCLEI

Alan Owens and S. Sembay
Department of Physics & Astronomy, University of Leicester, Leicester LE1 7RH, UK

Paul Nandra
Institute of Astronomy, Madingley Road, Cambridge CB3 0HA, UK

Ian George
NASA/Goddard Space Flight Center, Greenbelt, Maryland 20771, USA

INTRODUCTION

Unified models for active galactic nuclei (AGN) propose that some classes of objects with apparently different observational properties are, in fact, physically similar and are simply orientated at different angles to the observer. One such unified model is illustrated in Fig.1. It has been suggested that Seyfert 1 and 2 galaxies may be unified if obscuring matter distributed as a torus lies between the broad and narrow line regions; if the torus obscures the line of sight to the broad line region and central engine then the object is classified as a Seyfert 2 galaxy whereas if there is a direct view into the inner nucleus the object is classified as a Seyfert 1 galaxy.

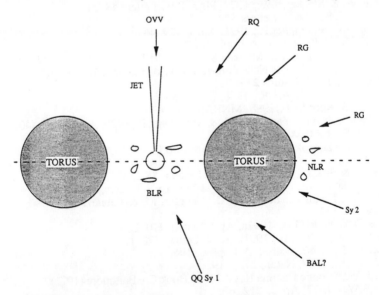

Fig. 1: A Unified Model for AGN (adapted from [13]). An absorbing torus of a few parsecs in extent surrounds the central engine and BLR clouds. Sources containing relativistic jets are represented in the upper half of the figure whilst those with weak or no jets by the lower half. The viewing angle of the observer defines the classification of the source.

Continuum γ-rays are produced by processes associated with the innermost regions of the central engine and, due to their extreme penetrative powers, provide a useful probe of the nucleus, free from most of the effects of absorption. If Seyfert 2s are simply obscured Seyfert 1s then there should be no systematic difference in the range of their γ-ray luminosities. We have calculated the expected spectra of two representative Seyfert 2 galaxies, NGC 4507 and MKN 78. The hard X-ray continuum of NGC 4507 has been measured by the Ginga Large Area Counter and has been shown to be highly absorbed with a column density of Log N(h) = 23.3. If the central engine of NGC4507 was not obscured by this column it would be one of the brightest extragalactic objects in the X-ray sky. MKN 78 has not been detected at hard X-ray energies, yet has a soft X-ray flux comparable to that of the soft excess in NGC 4507. These observations may be explained if the central engine in MKN 78 is obscured by a column density larger than observed in NGC 4507 and the soft X-ray flux arises as a scattered component, similar to that thought to contribute to the soft X-ray flux in the archetype "obscured Seyfert 2", NGC 1068. The results of our calculations show that both objects should be easily detectable by the OSSE instrument on GRO if they do indeed have γ-ray luminosities comparable to that observed in Seyfert 1 galaxies.

THE X- AND γ-RAY PROPERTIES OF AGN

Spectral surveys of low luminosity quasars and Seyfert 1s by *HEAO-1* in the 2-20 keV band [1] and *EXOSAT* in the 2-10 keV band [2], found that the data were well represented by a single power-law model with differing amounts of low energy absorption. The range of spectral indices was found to be relatively narrow, with a mean photon index of ~1.7, and not correlated with the source luminosity. The uniformity of this so-called 'canonical' power law slope imposes strong constraints on the range of possible medium to hard X-ray emission mechanisms (for a summary see [3]). In the soft X-ray band (0.1-2.0 keV) the *EXOSAT* observations revealed a significant excess above the extrapolation of the hard X-ray power law in ~30% of the objects surveyed. This excess has been attributed to thermal emission from optically thick material which may be the accretion disk in the standard central engine model.

More recent observations by the *Ginga* satellite performed at greater sensitivity than the previous generation of instruments have shown that the hard X-ray emission in a number of objects can be resolved into at least three components. Superimposed on the underlying nonthermal power law is an Fe K_α emission line (*e.g.* [4]) and 'hard tail' above 10 keV [5]. Illumination of the optically thick material responsible for the soft excess by the nonthermal radiation can, in most cases, account for the iron line *via* fluorescence and the hard tail by continuum reflection. The presence of the hard tail has important consequences for emission models as the underlying power law is steeper ($\Delta\Gamma \sim 0.2$) than the canonical value. In particular some electron-positron pair models can be revived [6].

As a class Seyfert 2s are between 10^2-10^3 times fainter in hard X-rays than Seyfert 1s but are also typically more highly absorbed with measured hydrogen column densities often as great as 10^{23}-10^{24} cm^{-2}. It is important to realize that, in consequence, the inferred range of X-ray luminosities are similar in both Seyfert classes. For example, if the Seyfert 2 galaxy NGC 4507 did not have a column of 5×10^{23} cm^{-2} it would have an observed flux in the 2-10 keV band of $\sim 6.6 \times 10^{-11}$ erg cm^{-2} which would make it one of the brightest extragalactic objects in the X-ray sky

[7]. The X-ray continuum of the archetype obscured Seyfert 2 NGC 1068 has an apparent low energy absorption of less than 5×10^{21} cm^{-2} and a derived 2 - 10 keV luminosity of only 4×10^{41} erg s^{-1}. However, strong evidence for a much higher *intrinsic* luminosity comes from the observation of an intense iron emission line at 6.5 keV with an equivalent width of $\approx$ 1300 eV [8]. The most likely origin for the line is that is arises from the fluorescence yield of the continuum X-rays scattering off a partially ionized cool gas. This picture is consistent with the obscuration model suggested by the observation of the polarized broad components to the optical/UV emission line spectrum; the direct line-of-sight to the central engine is obscured and we are detecting some small fraction of the continuum luminosity *via* scattering. Hence, the intrinsic X-ray luminosity is probably in the range 10^{43}-10^{44} erg s^{-1}, or a factor of 10^2 - 10^3 larger than observed.

Some Seyfert 2s (*e.g.* Mkn 78, Mkn 507 and Mkn 573) which have been detected with the *Einstein* IPC in soft X-rays have been found to lie below the detection limit of the *Ginga* satellite in hard X-rays. If these objects are similar to NGC 1068 then the soft component can be attributed to a scattered continuum which is too weak to be detected by *Ginga*. One problem with this picture, in the case of the highly polarized Seyfert Mkn 78, is that no polarized broad lines have so far been detected [9]. The other possibility, of course, is that these objects do not have a central engine as luminous as a Seyfert 1.

NGC 4507

The 2 to 20 keV *Ginga* spectrum of this source can be represented by a flat power law with photon index, $\Gamma = 1.39^{+0.13}_{-0.19}$, on which is superimposed a strong iron emission line with $EW = 400 \pm 50$ eV [3]. The hydrogen column density has been measured at $(5.0 \pm 0.5) \times 10^{23}$ cm^{-2}. To be able to predict the expected γ-ray flux we need to extrapolate the hard X-ray continuum to the higher energy regime. We have done this for two different continuum models. In the first model we simply assume that the observed X-ray slope extends up to higher energies until it cuts off at 1 MeV. If, however, the origin of the iron line is fluorescence from the accretion disk then the hard X-ray continuum will consist of a combination of a steeper underlying nonthermal power law and a higher energy reflection component (unresolved in this observation). Using the reflection model of George and Fabian [10] we can estimate the slope of the underlying continuum from the strength of the observed iron line. The slope is not particularly well constrained because the model is dependent on unknown parameters such as the disk inclination and the relative element abundances. However, because of the strength of the iron line, we can say with reasonable confidence that the slope is unlikely to be steeper than $\Gamma = 1.7$. We have assumed this value and calculated the reflection component using a Monte-Carlo simulation. For simplicity the input nonthermal spectrum was assumed to extend down to 1 MeV and then sharply cut-off, as seems typical of most AGN observed in the γ-ray regime [11,12]. As the reflected photons are Compton down-scattered from higher energies this approximation leads to a spectral break at ~ 0.2 MeV which is sharper than would be observed.

In Fig. 2 we have plotted the two continuum models in the range 0.01 to 1 MeV and the OSSE 3σ continuum sensitivity curve (for a 5×10^5 s observation) in the range 0.05 to 10 MeV. In both models the source is easily detectable up to ~ 0.5 MeV in a standard observation period of two weeks and with sufficient strength that the

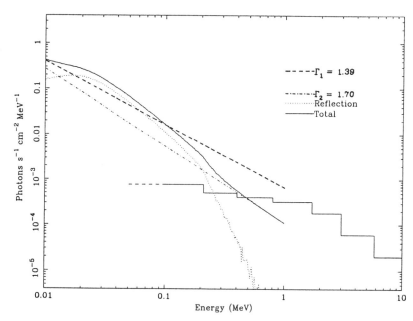

Fig. 2: Predicted γ-ray spectrum for NGC 4507. Two continuum models are shown. The dashed line is a power law slope of index, Γ=1.39. The solid line is the total continuum for the sum of a power law slope of index, Γ=1.70 (dot-dashed line) and the reflected component (dotted line). For comparison, the 3σ OSSE continuum sensitivity curve is also shown.

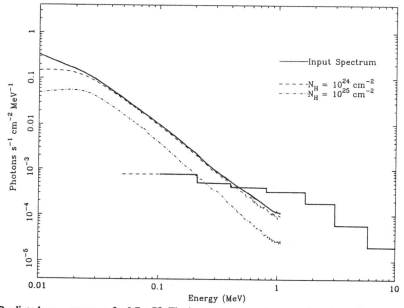

Fig. 3: Predicted γ-ray spectrum for Mkn 78. The input spectrum of a power law slope of index, Γ=1.70 is shown by the solid line. The two dashed lines show the predicted observed spectra for column densities of 10^{24} and 10^{25} cm^{-2}. For comparison, the 3σ OSSE continuum sensitivity curve is also shown.

continuum slope can be well defined. Hence, in addition to extending the γ-ray database on Seyfert 2 galaxies, OSSE can provide an important test of the reflection model. The predicted spectrum for this model has the appearance of a power law with slope $\Gamma \simeq 2.0$ in the OSSE band and is steeper by $\Delta\Gamma \approx 0.6$ than the observed X-ray slope. This difference is larger than that expected from intrinsic spectral variability.

MKN 78

The expected γ-ray flux from Mkn 78 is obviously more uncertain than for NGC 4507 as we have no direct observation of the hard X-ray continuum (*Ginga* set a 1σ upper limit of 3.4×10^{12} erg s^{-1}cm^{-2} in the 2-10 keV band [7]. If, however, the obscuration model is correct we can use the observed flux in the 0.2-4.0 keV band measured with the *Einstein* IPC to get an estimate of the hidden intrinsic X-ray luminosity of the central engine. The IPC fluxes of NGC 4507 and the similar source, Mkn 3 were 5.3×10^{-13} and 7.8×10^{-13} erg s^{-1} cm^{-2} respectively. These compare with values derived from the extrapolated *Ginga* spectra of 1.65×10^{-13} and 0.4×10^{-13} erg s^{-1} cm^{-2} respectively. While source variability between the observations could perhaps account for the differences, a more plausible explanation is that the excess soft X-rays are the scattered component as is the case in NGC 1068. If the same is true for Mkn 78 and the scattering efficiency is similar in these sources, then the ratio of the soft to hard X-ray luminosities should also be similar.

Mkn 78 has a flux of 3.5×10^{-13} erg cm^{-2}s^{-1} in the IPC band which is almost identical to the excess soft X-ray flux of NGC 4507 and, therefore, we will assume that the hard X-ray luminosities are the same. The predicted γ-ray spectrum was then calculated with a Monte-Carlo simulation of a point source surrounded by an absorbing torus. For the shape of the input spectrum we have assumed a typical nonthermal power law of slope $\Gamma = 1.8$ plus reflection component appropriate for Seyfert 1s [5]. A number of calculations were performed with various values of the hydrogen column density. The results are shown in Fig. 3. Above column densities of 10^{25} cm^{-2} only the scattered component is observed so the predicted γ-ray flux remains roughly constant. In this case the source is still detectable in the 0.05 - 0.2 MeV band given the assumed input spectrum.

REFERENCES

[1] Mushotzky, R.F., 1984. *Adv. Space. Res.*, **3**, 10.
[2] Turner, T.J., & Pounds, K.A., 1989. *Mon. Not. R. astr. Soc.*, **240**, 833.
[3] Zdziarski, A.A., 1986, *Astrophys. J*, **305**, 45.
[4] Nandra, K., et al., 1989. *Mon. Not. R. astr. Soc.*, **236**, 36.
[5] Nandra, K., et al., 1991. *Mon. Not. R. astr. Soc.*, **248**, 760.
[6] Zdziarski, et al., 1990, *Astrophys. J.*, **363**, L1.
[7] Awaki, H., 1991. *ISAS Resrarch Note 473*, ISAS, Kanagawa, 229, Japan.
[8] Koyama, K., et al., 1989. *Publ. Astron. Soc. Japan*, **41**, 731.
[9] Miller, J.S. & Goodrich, R.W., 1990. *Astrophys. J.*, **355**, 456.
[10] George, I.M. & Fabian, A.C., 1991. *Mon. Not. R. astr. Soc.*, **249**, 352.
[11] Perotti, F., et al., 1981. *Astrophys. J.*, **247**, L63.
[12] Ubertini, P., et al., 1984. *Astrophys. J.*, **284**, 54.
[13] Woltjer, J., 1990. In: *Active Galactive Nuclei*, Springer Verlag publ. Co., 1.

STARBURST AND REFLECTION-DOMINATED AGN CONTRIBUTIONS TO THE DIFFUSE X-RAY BACKGROUND

P. M. Ricker
University of Chicago, Chicago, IL 60637

P. Mészáros
Pennsylvania State University, University Park, PA 16802

ABSTRACT

Massive X-ray binaries (MXB) in starburst galaxies and reflection-dominated AGN have been proposed as models of the cosmic X-ray background (CXB). We calculate the CXB spectrum to be expected from an evolving population of MXB in starburst galaxies and find that starbursts cannot produce much of the background at low energies ($\lesssim 30$ keV) without significant evolution. We find their contribution at higher energies to be negligible, even with strong evolution. We also investigate the consistency of the MXB and AGN models. If MXB in starburst galaxies do produce much of the low-energy CXB, some other type of source must be responsible for the high-energy background. If reflecting AGN are this other type of source, up to 92% of their emission must be reprocessed. Even without a starburst contribution, the AGN spectrum must still be about 85% reprocessed.

INTRODUCTION

Recently two potentially important contributions to the CXB have been discussed: that of starburst galaxies[1-5], and that of AGN in which a cold gas component reflects an X-ray power-law[6-10]. There have not been detailed observational or theoretical investigations of the hard X-ray spectrum of starbursts, although at energies $\lesssim 10$ keV their spectrum based on that of X-ray binaries may be comparable to that of the residual CXB. Reflection-dominated AGN, on the other hand, are theoretically expected to produce a spectrum which appears rather similar to that of the CXB between a few to at least 100 keV. Here we investigate in some detail the X-ray spectrum that may be expected from starbursts and consider its implications for the CXB both alone and in conjunction with that of the reflection-dominated AGN at higher energies.

X-RAY BINARIES AS SOURCES OF X-RAYS IN STARBURSTS

Few starburst X-ray spectra have actually been observed[11,12], even fewer at energies above about 10 keV[13]. The spectrum must therefore be modeled on the basis of extrapolations of our knowledge of the spectrum of X-ray binaries, complemented by plausible assumptions about scaling with metallicity and the luminosity evolution. Here we take the starburst galaxy X-ray spectrum to be a composite of the spectra of massive X-ray binaries (MXB) and low-mass X-ray binaries (LMXB).

The canonical MXB spectrum consists of a power-law component of energy index α_M at low energies ($E < E_B$) and a combination of the power-law with an exponential tail of folding energy E_F at high energies[14,15]. We have used average values of $\alpha_M = 0.25$, $E_B = 16.3$ keV, and $E_F = 15.7$ keV. In addition, MXB generally produce emission lines with an average equivalent width $W_{Fe} \approx 200$ eV at ~ 6.4 keV due to iron fluorescence[16]. The MXB spectrum includes the effect of intrinsic absorption at low energies, with an average hydrogen column density $N_H \approx 3.5 \times 10^{22}$ cm^{-2}.

The canonical LMXB spectrum consists of a blackbody component of temperature $E_{BB} \approx 1.2$ keV, which provides about 28% of the luminosity, and a hard component, which is thought to be due to inverse Comptonization of blackbody photons from the neutron star within the accretion disk[17-19]. The hard component is characterized by a power-law of index $\alpha_L \approx 0.2$ and an e-folding energy $E_L \approx 5.0$ keV[19]. LMXB have a weaker iron line than MXB ($W_{Fe} \approx 50$ eV) at 6.7 keV. The effect of an intrinsic absorbing column of average density $N_H \approx 1.3 \times 10^{22}$ cm^{-2} is included in the LMXB spectrum[19].

The model X-ray spectrum of a starburst galaxy is a composite which is parametrized by the fractional number of LMXB, $\beta \equiv N_{LMXB}/(N_{MXB}+N_{LMXB})$. Composite spectra for three values of β are shown in Figure 1. The addition of LMXB clearly softens the composite spectrum; therefore we take the value of β to be ~ 0.1 as a representative of the best-case starburst model. The power-law slope of this starburst galaxy spectrum below 10 keV with $\beta = 0.1$ is about 0.35, close to the value assumed by Griffiths and Padovani[4]. The estimated background intensity depends only on the observed luminosity density in starburst galaxies, the spectral shape, and the assumed evolution law, leaving the luminosities and numbers of X-ray binaries undetermined.

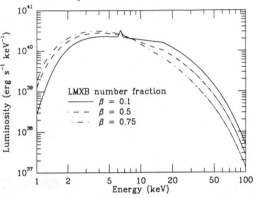

Figure 1. Composite Starburst Rest-Frame Spectrum.

CXB SPECTRUM FROM STARBURST GALAXIES

The CXB intensity is obtained by summing the product of the starburst luminosity function with the intensity due to a single galaxy over all luminosities and redshifts:

$$I_\nu(E) = \frac{\rho_x(0)c}{4\pi H_0} \int_0^{z_{max}} \frac{e^{\gamma z/(1+z)}}{(1+z)^2\sqrt{1+2q_0 z}} \left[(1-\beta)\xi_\nu^M(E(1+z)) + \beta\xi_\nu^L(E(1+z))\right] dz, \quad (1)$$

where $\xi_\nu^M(E)$ and $\xi_\nu^L(E)$ are the normalized MXB and LMXB spectral shapes. Here $\rho_x(0) \approx 7.7 \times 10^{37}$ erg s^{-1} Mpc^{-3} is the integrated 0.5 – 3.0 keV luminosity density due to starbursts at low redshift[4]. Source counts indicate that starbursts undergo strong luminosity evolution in the radio and infrared out to large redshifts[20], so if we assume a positive correlation between X-ray and IR luminosities, we may parametrize $\rho_x(z)$ as

$$\rho_x(z) = \rho_x(0)\exp[\gamma z/(1+z)], \quad (2)$$

where $z/(1+z)$ is approximately the lookback time $\tau(z)$ multiplied by the Hubble constant H_0, and γ is an evolutionary parameter. The infrared source counts are consistent with a γ as large as 5. We have taken H_0 to be 50 km s^{-1} Mpc^{-1} and the deceleration parameter q_0 to be 0.5.

We numerically evaluated the integral in equation (1) for γ equal to 0.0, 2.5, and 5.0, checking the energy dependence of the fractional starburst contribution to the residual CXB for $z_{max} = 3.0$ (Figure 2) and the dependence of the 3 keV starburst contribution on z_{max} (Figure 3). The residual CXB intensity is given by Boldt[21]:

$$I_\nu^{res}(E) = 3.65 e^{-E/23 \text{ keV}} \text{ keV s}^{-1} \text{ cm}^{-2} \text{ keV}^{-1} \text{ sr}^{-1}. \qquad (3)$$

At 3 keV, starbursts can produce up to 100% of the residual CXB with strong evolution, provided they are found up to a redshift of 4.5. With no evolution, they contribute almost 10% of the 3 keV residual if they are found up to a redshift of 1.5. However, the energy dependence of the starburst contribution does not match that of the residual. For $z_{max} = 3.0$, the fractional starburst contribution to the residual drops below 10% above 30 keV, even for maximal starburst evolution. The redshifted starburst spectrum is steeper than the residual at high energy. As the maximum redshift for starbursts increases, this spectral discrepancy worsens; as it decreases, the magnitude of the starburst contribution decreases. Above 100 keV, starbursts contribute almost nothing to the residual CXB. Thus while it appears that MXB-dominated starburst galaxies can produce a significant fraction of the residual CXB at low to intermediate energies, at high energies some other kind of source must be responsible for the large majority of the background.

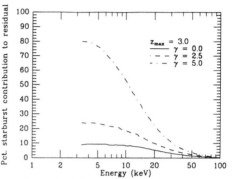

Figure 2. Starburst Contribution to CXB Spectrum.

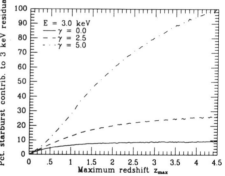

Figure 3. Starburst Contribution to 3 keV CXB.

REFLECTION-DOMINATED AGN AND THE CXB

Recently, AGN observations with EXOSAT[22] and Ginga[23,24] have indicated that some AGN have spectra which are significantly flatter than the canonical slope of ~ 0.7. The reflection of an X-ray power-law spectrum by a semi-infinite cold gas leads to a spectrum which shows a hump at energies above ~ 10 keV due to the effect of scattering above and absorption below that energy[6,7]. This hump mimics the flattening seen in these AGN, and in fact the spectrum resembles that of the CXB[8-10,25], provided one assumes that a fraction $\gtrsim 0.85$ of the input spectrum is reflected. The fits to the CXB presented by these authors cover from upward of 100 keV down to 2 keV. The question that arises is then, if this is the explanation for the hard X-ray background, what constraints does this impose on the starburst contribution to the 2-10 keV region? And conversely, if starbursts do contribute to this soft region, what constraints does this imply for the AGN reflection model?

To answer these questions, we fitted the CXB contribution from reflection-dominated AGN like those considered by Rogers and Field[9,10] to the spectrum remaining after the starburst contribution has been subtracted from the total CXB. Included is a contribution due to galaxy clusters amounting to 5% of the total CXB at 2 keV. The AGN source spectrum is described by

$$\xi_\nu(E) = E^{-\alpha}[(1-f) + fR(E)], \qquad (4)$$

where a fraction f of the emission is reflected, and the remainder is direct. The energy-dependent reflectivity $R(E)$ is found using the Green's function calculated by Lightman and White[6], which includes the effects of photoelectric absorption below about 50 keV and Compton scattering at higher energies. The CXB contribution is given by an integral similar to equation (1), with the AGN spectrum (4), properly normalized, taking the place of the composite starburst galaxy spectrum. The 2 − 10 keV AGN luminosity evolution law is assumed to be a power law, as in Rogers and Field[9,10]:

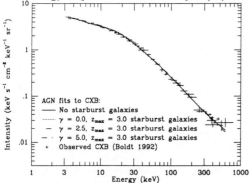

Figure 4. AGN Fits to CXB Spectrum.

$$\rho_x(z) = \rho_x(0)(1+z)^\delta. \qquad (5)$$

We used a chi-square minimization procedure to fit the AGN background contribution to the starburst residual between 3 and 500 keV, with z_{max} (for the AGN distribution), δ, α, f, and the present-day AGN luminosity density $\rho_x(0)$ as free parameters.

Table I Best-fit Parameters for Reflection-Dominated AGN Models

Starburst parameters ($z_{max} = 3.0$)	z_{max} (AGN)	α	δ	f	$\rho_x(0)^b$	Reduced χ^2
none	$5.95^{+0.41}_{-0.83}$	$1.02^{+0.01}_{-0.01}$	$1.35^{+0.10}_{-0.08}$	$0.85^{+0.01}_{-0.01}$	$7.3^{+0.7}_{-0.7}$	3.99
$\gamma = 0.0$	$6.00^{+1.79}_{-0.69}$	$1.02^{+0.01}_{-0.01}$	$1.10^{+0.09}_{-0.10}$	$0.84^{+0.01}_{-0.01}$	$8.1^{+0.7}_{-0.6}$	4.26
$\gamma = 2.5$	$5.20^{+0.86}_{-1.98}$	$1.00^{+0.02}_{-0.01}$	$1.05^{+0.06}_{-0.05}$	$0.85^{+0.01}_{-0.02}$	$7.6^{+0.7}_{-0.5}$	4.55
$\gamma = 5.0$	$4.45^{+0.42}_{-2.98}$	$0.86^{+0.07}_{-0.01}$	$1.05^{+0.15}_{-0.05}$	$0.92^{+0.01}_{-0.01}$	$4.1^{+2.4}_{-0.1}$	8.16

[a]Parameter ranges are 90% confidence limits. [b]Luminosity density in units of 10^{39} erg s^{-1} Mpc^{-3}.

In Figure 4 we have plotted the best-fit combined (starburst plus reflecting AGN) spectrum against that of the total measured CXB (Boldt, private communication) for different starburst evolution laws, as well as for a model in which only reflecting AGN contribute. Table I gives the best-fit AGN parameters for each case. The reduced χ^2, given by the best-fit value of χ^2 divided by the number of degrees of freedom (number of data points minus number of fitting parameters), is a measure of the relative probability of each fit occurring by chance (goodness-of-fit). These values are quite high, apparently due to a bump in the observed spectrum at ~ 40 keV which the reflected AGN model does not reproduce[26]. Moreover, the fits confirm the high reflected fractions found by earlier authors. For the pure AGN model, the required fraction is 0.85, and for the maximal starburst evolution law, the reflected fraction from AGN must be $\gtrsim 0.92$. The pure AGN (lower f) model gives a far better goodness-of-fit, but none of the models can be said to be probable. In any case, it does not appear possible to reduce the reflected fraction below ~ 0.85, which is still rather high, given that the reflecting medium considered here is optically thick. Rogers[27] has discussed a physical model which could produce a high reflected fraction, but it differs geometrically from the reflection model used to produce the CXB spectral fit.

It appears that if the X-ray emission of starburst galaxies is dominated by MXB, starbursts cannot be responsible for the hard X-ray background, regardless of their rate of evolution. Starbursts are an observed component at low energies, whose numbers at high redshift are uncertain, but at high energies we are faced with the need to invoke a second class of sources. The reflection-dominated AGN models currently under consideration provide a qualitative fit whose statistical significance must be improved, particularly when starbursts are a significant component at low energies.

Acknowledgements: We would like to thank E. Boldt for permission to use, and D. Gruber for supplying the CXB data. This research was supported in part through an NSF REU grant (AST 88-15266), NASA NAGW 1522, NAGW 830, and NAGW 1284.

REFERENCES

1. J. Bookbinder et al., Ap. J. **237**, 647 (1980).
2. D. W. Weedman, in *Star Formation in Galaxies*, ed. Carol J. Lonsdale (NASA CP-2466, 1986), p. 351.
3. L. L. Cowie, in *Proc. 23rd ESLAB Symposium on Two Topics in X-Ray Astronomy* (ESA SP-296, 1989), p. 707.
4. R. Griffiths and P. Padovani, Ap. J. **360**, 483 (1990).
5. L. P. David, C. Jones, and W. Forman, Ap. J. **388**, 82 (1992).
6. A. P. Lightman and T. R. White, Ap. J. **335**, 57 (1988).
7. P. Guilbert and M. Rees, MNRAS **233**, 475 (1988).
8. A. C. Fabian et al., MNRAS **242**, 14P (1990).
9. R. D. Rogers and G. B. Field, Ap. J. **370**, L57 (1991a).
10. R. D. Rogers and G. B. Field, Ap. J. **378**, L17 (1991b).
11. G. Fabbiano, Ann. Rev. Astr. Ap. **27**, 87 (1989).
12. E. J. A. Meurs and T. Boller, in *X-Ray Emission from Active Galactic Nuclei and the Cosmic X-Ray Background*, eds. W. Brinkmann and J. Trümper (MPE 235, 1992), p. 338.
13. T. Ohashi et al., Ap. J. **365**, 180 (1990).
14. F. Nagase, Publ. Astron. Soc. Japan **41**, 1 (1989).
15. N. White, J. Swank, and S. Holt, Ap. J **270**, 711 (1983).
16. S. H. Pravdo, in *(COSPAR) X-Ray Astronomy*, eds. W. A. Baity and L. E. Peterson (New York: Pergamon, 1979), p. 169.
17. K. Mitsuda et al., Publ. Astron. Soc. Japan **36**, 741 (1984).
18. N. White, in *Variability of Galactic and Extragalactic X-Ray Sources*, ed. Aldo Treves (Milan: Associazione per L'avanzamento Dell'astronomia, 1986), p. 165.
19. N. White and K. Mason, Space Sci. Rev. **40**, 167 (1985).
20. L. Danese et al., Ap. J. **318**, L15 (1987).
21. E. Boldt, Phys. Rep. **146**, 215 (1987).
22. K. Pounds et al., Nature **344**, 132 (1990).
23. K. Morisawa et al., Astr. Ap. **236**, 299 (1990).
24. M. Matsuoka et al., Ap. J. **361**, 440 (1990).
25. A. C. Fabian, in *Proc. 23rd ESLAB Symposium on Two Topics in X-Ray Astronomy* (ESA SP-296, 1989), p. 1097.
26. A. Zdziarski et al.., Ap. J. preprint, 1992.
27. R. D. Rogers, Ap. J. **383**, 550 (1991).

COMPTONIZATION OF EXTERNAL RADIATION IN BLAZARS

Marek Sikora[1,2], Mitchell C. Begelman[1] and Martin J. Rees[3]

[1] Joint Institute for Laboratory Astrophysics,
University of Colorado and National Institute of Standards and Technology,
Boulder, Colorado 80309

[2] Copernicus Astronomical Center, Polish Academy of Sciences, Warsaw, Poland

[3] Institute of Astronomy, Cambridge, England

ABSTRACT

The observed luminosities of many blazars are dominated by gamma-rays which, supposedly, are produced by Compton cooling of ultrarelativistic electrons in relativistic jets. Several puzzling properties of blazar radiation spectra can be understood if the high energy spectra are produced by Comptonization of an external radiation field, rather than of internally produced synchrotron radiation.

The best candidate for the "seed" radiation field is diffuse UV radiation, produced in the central engine of the AGN and scattered by gas surrounding the jet. As measured in the frame comoving with the jet, this radiation dominates easily over the direct radiation from the central source. Up to distances ~ 1 pc and for bulk Lorentz factors of the radiating plasma $\Gamma \geq 10$, it also dominates over synchrotron and synchrotron-self-Compton radiation produced in the jet.

According to our scenario, the break between the X-ray and gamma-ray portions of blazar spectra results from inefficient Compton cooling of electrons with energies below a certain value. The main advantage of the model is that it is consistent with the assumption that both the X-to-gamma-ray Compton radiation and the IR–UV synchrotron radiation are produced by the same source component, and by a single relativistic electron population.

ASSUMPTIONS AND INPUT PARAMETERS

We assume the following:
1. Geometry of the source is as shown in Fig. 1;
2. Production of X-rays and gamma-rays is dominated by Comptonization of diffuse external UV radiation;
3. Low-energy break in gamma-ray spectrum ($\alpha_\gamma \simeq 1 \Rightarrow \alpha_X \simeq 0.5$) is caused by inefficient radiation of electrons/positrons below a certain value of random Lorentz factor;
4. A single source component with a single electron population produces both IR–UV synchrotron radiation and X–gamma Compton radiation;
5. Most observed radiation is produced over a region $\Delta r \sim r$, where r is the distance from the central engine.

The input parameters are: time scale of γ-ray flare, Δt_f; time scale of sharp drops in flux, Δt_m; "observed" γ-ray luminosity per log energy bin, $\partial \tilde{L}_C/\partial \ln E$, IR-UV luminosity per log energy bin, $\partial \tilde{L}_s/\partial \ln E$, and soft X-ray ($\sim 1$ keV) luminosity per log energy bin, $\partial \tilde{L}_{SX}/\partial \ln E$; energy of the "$\alpha_\gamma \Rightarrow \alpha_X$" spectral break, E_b; and maximum energy of the synchrotron component, $E_{syn,max}$.

DISTANCE AND SOURCE DIMENSIONS

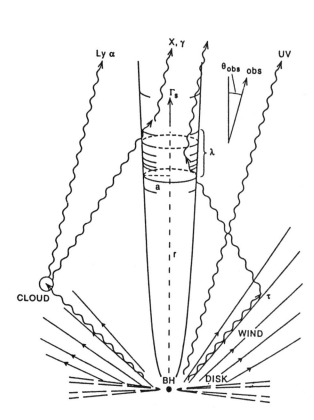

Fig. 1 Geometry of the source.

High luminosity states of blazars are represented by flares with duration $\Delta t_f \sim$ weeks but, possibly, with much shorter drops, with doubling time scale $\Delta t_m \sim 1$ day.[1,2] These time scales together with assumption #5 impose constraints on r and source sizes, λ and a (see Fig. 1). For the observer located at $\theta_{obs} \sim 1/\Gamma$ we have[3]: $r \simeq c\Delta t_f \Gamma_s^2$, $\lambda_{max} \simeq c\Delta t_m$, $2a_{max} \simeq \Gamma c\Delta t_m$ for a case in which the flare time scale is determined by the source travel time (hereafter case A); and $r_{max} \simeq c\Delta t_m \Gamma_s^2$, $\lambda \simeq c\Delta t_f$, $2a_{max} \simeq \Gamma c\Delta t_m$, for a case in which the flare time scale is determined by the source length (hereafter case B). In these expressions Γ is the bulk Lorentz factor of the radiating plasma and Γ_s is the Lorentz factor of the source (e.g. a group/pattern of shocks within which particle acceleration takes place[4]). Note that for a fixed distance the constraint on r_{max} in case B provides $\Gamma_{s,min}$.

CONSTRAINTS ON THE FLOW LORENTZ FACTOR

Electron radiation efficiency is depressed if the time scale of $e^\pm$ energy losses due to radiation processes is longer than the time scale of $e^\pm$ energy losses due to adiabatic expansion. The critical energy, γ_b', below which this happens is given by

$$t_e(\gamma_b') = \Gamma t_e'(\gamma_b') \sim r/c \qquad (1)$$

where t_e is the time scale of $e^{\pm}$ energy losses due to Comptonization of external UV radiation which, in our model, dominates radiation energy losses. Primed quantities are measured in the plasma flow frame.

Taking into account that the energy around which the gamma-ray spectrum has a low-energy break is $E_b \sim \Gamma E_b' \sim \Gamma \gamma_b'^2 E_{UV}'$ and parametrizing the average Doppler shift of seed photons between the external frame and the source frame by $\xi\Gamma = E_{UV}'/E_{UV}$ we can find the energy density of the external UV radiation field, u_{UV}'. Then, using the inequality

$$u_{UV}' > \left(\frac{\partial \tilde{L}_C / \partial \ln E}{\partial \tilde{L}_{SX} / \partial \ln E}\right) u_{syn}' \qquad (2)$$

which guarantees domination of Comptonization of an external radiation field over Comptonization of a synchrotron radiation field, we obtain Γ_{min}.

CONSTRAINTS ON GEOMETRY OF EXTERNAL MEDIUM

Scattered by a relativistic flow of cold electrons/positrons, the seed UV photons are received by the observer at $\theta_{obs} \sim 1/\Gamma$ with average energy $E_{C,min} = \xi\Gamma^2 E_{UV}$. The lack of any significant X-ray excesses in blazars down to ~ 1 keV yields $\xi \leq (1/3)/(\Gamma/16)^2$. This means that the radiation energy density, as seen in the flow frame, should be dominated by photons scattered at radii somewhat less than r.

On the other hand, ξ cannot be much smaller than this upper limit, since this would require such a large L_{UV} that a prominent UV bump should be observed in blazars. Even with the assumption that the UV co-varies with the nonthermal source, a lower limit imposed by the lack of any significant UV excess during high states of blazars is still rather restrictive and for 3C 279 provides $\xi_{min} \sim 0.1$.

An independent constraint yielding ξ_{min} comes from the transparency of the external radiation field to GeV photons. If the soft X-ray luminosity of the central source is scaled with L_{UV} with the factor ~ 0.1 and the temperature of the medium which scatters the radiation to the jet is higher than $\sim 10^7$K, then already at $\xi = 1/3$ the optical thickness for absorption of 10 GeV photons by soft X-rays due to pair production, $\tau_{\gamma\gamma}^{(ext)}$, is larger than 1.

We conclude that the model provides a strong prediction about the presence of soft X-ray excesses, which eventually can be checked against data from ROSAT.

VERIFICATION OF ASSUMPTION # 4

If the same component produces both synchrotron radiation and Compton radiation, then $\tilde{L}_C/\tilde{L}_{syn} = u_{UV}'/u_B'$, from which the magnetic field can be obtained. Knowing B' and $E_{syn,max}$, one can find γ_{max}' and $E_{C,max}$, and the

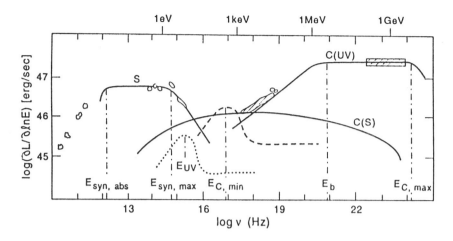

Fig. 2. Spectrum of 3C 279: patchy fragments – observations; S - synchrotron radiation; dotted curve – central source radiation (CSR); dashed curve – Comptonization of scattered CSR by cold matter in jet; C(UV) – Comptonization of scattered CSR by relativistic matter in jet; C(S) – Comptonization of synchrotron radiation by relativistic matter in jet.

latter can be compared to the highest observed energies. For 3C 279 the model predicts $E_{C,max} > 3\,\text{GeV}$, provided that the source is located at $r > 3 \times 10^{16}\,\text{cm}$ from the nucleus. This is enough to explain EGRET observations, but is much below the energies detected in Mkn 421.[5] However, the soft X-ray spectrum of Mkn 421 seems to be a continuation of the synchrotron component, and since $E_{C,max} \propto \gamma'^2_{max} \propto E_{syn,max}$, TeV photons can be easily produced in this object according to our phenomenological model.

OTHER OUTPUT PARAMETERS

The list of output parameters of the model can be extended to include: the energy density, u'_e, and total energy, $E'_{tot,e}$, of the relativistic $e^\pm$ plasma; internal source luminosity $L' = L'_C + L'_{syn}$; the energy injected to the source through acceleration of electrons/positrons, $E'_{tot,inj}$; the frequency below which the spectrum breaks due to synchrotron absorption, $E_{syn,abs}$; and the optical thickness for pair production on X-rays produced in the source by Comptonization of external UV radiation, $\tau^{(int)}_{\gamma\gamma}$.

The full list of output parameters of the model is specified in Table I as a function of r, Γ, ξ and η, where $\eta(A) = \Delta t_m/\Delta t_f$ and $\eta(B) = 1$ [ref. 3], for the following 3C 279 input parameters: $\Delta t_f \simeq 12$ days; $\Delta t_m \simeq 2$ days; $\partial \tilde{L}_C/\partial \ln E \simeq 2.9 \times 10^{47}$ erg/s; $\partial \tilde{L}_s/\partial \ln E \simeq 6.7 \times 10^{46}$ erg/s; $\partial \tilde{L}_{SX}/\partial \ln E \simeq 2.8 \times 10^{46}$ erg/s; $E_b \simeq 2.5$ MeV; and $E_{syn,max} \simeq 10^{15}$ Hz.

TABLE I. Output parameters for 3C 279

Γ_{min}	$16(3\xi)^{1/12}\eta^{-1/6}$	$\tau_{\gamma\gamma}^{(ext)}$	$1.5(E/10\text{GeV})(3\xi)^{-3/2}$
$\Gamma_{s,min}^{(B)}$	$4.4r_{17}^{1/2}$	$\tau_{\gamma\gamma}^{(int)}$	$1.6 \times 10^{-2}(E/10\text{GeV})^{1/2}r_{17}^{-1}(\Gamma/16)^{-3}\eta^{-1}$
$r^{(A)}$	$3.6 \times 10^{18}(\Gamma/16)^2$ cm	u'_{UV}	$2.6r_{17}^{-1}(3\xi)^{1/2}(\Gamma/16)^2$ erg/cm^3
$\lambda^{(B)}$	1.8×10^{16} cm	u'_{syn}	$0.29r_{17}^{-1}(\Gamma/16)^{-4}\eta^{-1}$ erg/cm^3
a_{max}	$2.1 \times 10^{16}(\Gamma/16)$ cm	u'_B	$0.70r_{17}^{-1}(3\xi)^{1/2}(\Gamma/16)^2$ erg/cm^3
γ'_b	$54(3\xi)^{-1/2}(\Gamma/16)^{-1}$	u'_e	$0.27(\Gamma/16)^{-2}\eta^{-1}$ erg/cm^3
γ'_{max}	$2.4 \times 10^3 r_{17}^{1/4}(3\xi)^{-1/8}(\Gamma/16)^{-1}$	τL_{UV}	$3.6 \times 10^{44}r_{17}(3\xi)^{-3/2}$ erg/s
$E_{C,max}$	$5.1r_{17}^{1/2}(3\xi)^{3/4}$ GeV	L'	$2.1 \times 10^{45}r_{17}^{-1}(\Gamma/16)^2$ erg/s
$E_{C,min}$	$0.65(3\xi)(\Gamma/16)^2$ keV	$E'_{tot,e}$	$5.3 \times 10^{50}(\Gamma/16)^{-3}$ erg
$E_{syn,abs}$	$1.9 \times 10^{12}r_{17}^{-5/14}(3\xi)^{1/2}\eta^{-2/7}$ Hz	$E'_{tot,inj}$	$8.8 \times 10^{50}(\Gamma/16)^{-3}$ erg

and $r_{max}^{(B)} = \eta r^{(A)}$, $\lambda_{max}^{(A)} = \eta\lambda^{(B)}$.

MORE ABOUT SOURCE DISTANCE

As can be deduced from Table I, the source distance in 3C 279 should lie between 3×10^{16} cm and 3×10^{17} cm. Otherwise one gets too high UV luminosities or too low maximum photon energies. At such distances, sources with $\lambda = c\Delta t_f$ are favored (case B). For sources with $\lambda \leq c\Delta t_m$ (case A) shocks with $\Gamma_s < 3 \ll \Gamma_{min}$ would be required, which should impose thermalization of the plasma in the shock up to relativistic temperatures. Then $\gamma'_{min} \gg 1$ and the power-law X-ray spectrum should start from $E_{C,min} > 1$ keV, contrary to observations.

This research has been supported in part by NSF grants AST 88-16140, AST 91-20599, INT 90-17207, and the Polish State Committee for Scientific Research grant 221129102.

REFERENCES

1. C. E. Fitchel et al. 1992, A&A, in press
2. R. Edelson 1992, ApJ, in press
3. M. Sikora, M. C. Begelman, and M. J. Rees, in preparation
4. K. R. Lind and R. D. Blandford, R. D. 1985, ApJ, 295, 358
5. M. Punch et al. 1992, Nature, 358, 477

GAMMA RAYS AND NEUTRINOS FROM POINT SOURCES

YUKIO TOMOZAWA

Randall Lab of Physics, University of Michigan, Ann Arbor, MI 48109-1120

ABSTRACT

We examine the prediction that the energy spectrum from point sources such as AGN (Active Galactic Nuclei) and GBHC (Galactic Black Hole Candidates) is universal, irrespective of the nature of the emitted particles. γ-rays from quasars observed by CGRO (Compton Gamma Ray Observatory) are reported in this symposium. The average of the photon indices from 11 quasars is compared with that of cosmic rays at the source. The former is 2.1 ± 0.1 while the latter is 2.2 ± 0.1. The compilation of the observed neutrino data by Kamiokande and IMB yields 1.2 for the ratio of ν_μ/ν_e, instead of 2 as expected from atmospheric neutrinos. If the observed neutrino data is interpreted as cosmic neutrinos, this result is consistent with the prediction of the present model. It has an index approximately equal to 2.2.

I. Introduction

The observed γ-rays from quasars by CGRO have power law spectra with photon index ranging[1] from 1.7 to 2.4. Recent data on a few TeV γ-rays from MRK421 observed by the Whipple Observatory[2] seems to indicate that the power law spectrum of γ-rays from quasars is of a general nature and clearly calls for an explanation. Understanding the γ-ray emission mechanism from compact sources is the key element for resolving the secret of quasars. In this article, the author analyzes the recent CGRO data for γ-rays from quasars and compares it with that of cosmic rays and neutrinos. It will be shown that the data are consistent with the model[3] for particle emission from quasars proposed by the author in 1985.

II. The Model

We summarize the pertinent features of the theory.[3]

1. The rotation of black holes represented by the Kerr metric as well as quantum effects on the Einstein equation yield repulsive gravitational potential at short distances.

2. As a result, one concludes that black holes pulsate and lead to the emission of particles.

3. The spectrum of particles emitted during pulsation is a reflection of the expansion rate of the system. This suggests a universal energy spectrum for different particles from an individual pulsating compact object. This model allows variation of the spectral index for various objects and also variation for an individual object at different times. [It depends on the environment at the time of pulsation.]

4. The prediction of universality can be tested by the data for cosmic rays (at the source), γ-rays as well as neutrinos.

III. Cosmic Rays

It is known[4] that shock wave acceleration in a supernova explosion does not explain the high energy component of cosmic rays (above the so called knee energy, 10^{16} eV). It was then suggested that strong magnetic fields around pulsars may be responsible for the acceleration of high energy cosmic rays[5] above the knee energy. In this case, however, drastically different mechanisms are responsible for cosmic ray acceleration below and above the knee energy. This makes it difficult to explain the continuous spectrum observed in the experimental data. An alternative scheme is to invoke shock wave acceleration in the galactic wind,[6] the existence of which is yet to be established by observation.

The application of BHP (black hole pulsation) leads naturally to the emission of particles. Since the temperature decreases with the expansion of quantum mechanical black holes or rotating black holes, the energy spectrum of the emitted particles can be computed, provided the expansion rate is known.

The number of particles of type x emitted with energy E is given by

$$f_x(E) = \frac{(2s+1)}{2\pi^2} \int \eta_x(E/kT) \frac{E^2 4\pi R^2 dt}{e^{E/kT - \mu/kT} \pm 1}, \tag{1}$$

where R is the radius of the system, $\eta_x(E/kT)dt$ is the fraction of particles x emitted in time interval dt and μ and s are the chemical potential and the spin for particles of type x. The $+(-)$ sign in the denominator is for fermions (bosons). Assuming the relationship $R = a/kT$ and an expansion rate

$$t = bR^\alpha, \tag{2}$$

we obtain

$$f_x(E) = \frac{A_{x,\alpha}}{E^\alpha}, \tag{3}$$

where

$$A_{x,\alpha} = \frac{2(2s+1)\alpha b(a)^{2+\alpha}}{\pi} \int_0^\infty \frac{\eta_x(s)s^{\alpha+1}ds}{e^{s-\mu_0} \pm 1} \tag{4}$$

is a constant, and $\mu_0 = \mu/kT$.

Some discussion is in order. First, how can one explain the observed nuclei in cosmic rays ($\sim$10% of primary cosmic rays which are mostly protons below the knee energy). Nuclei can be emitted from BHP, since the density is extremely high so that the Fermi temperature is also extremely high. As a result, the situation can be like a low temperature state even if the temperature T is high. Thus, due to Boson condensation, even nuclei can be emitted in this system.[7]

Secondly, the cosmic ray energy spectrum observed at the earth, $E^{-2.7}$, may not be the same as that at the sources. Using leakage, spallation and information on the chemical abundance of cosmic rays, the Chicago group has derived the power index λ of the cosmic ray energy spectrum $E^{-\lambda}$ at the sources. The most elaborate analysis[8] gives

$$\lambda_{\text{source}} = 2.2 \pm 0.1. \tag{5}$$

This index should be compared with the index for the γ-ray energy spectrum.

Finally, in our model the power law spectrum for cosmic rays is a reflection of the power law expansion rate. The knee energy is caused by the difference between the expansion rates in the nonrelativistic and relativistic regime. Then this would require the existence of an energy scale of $\sim$ several hundred TeV which differentiates the two regimes. However, modification of the low energy spectrum by Eq. (5) brings the knee energy to a much lower energy scale, around 1 TeV (instead of several hundred TeV). This may signal a Higgs particle at 1 TeV.

IV. Gamma-Rays

It is clear that the energy spectrum derived in the last section applies for any particles emitted from BHP. The important prediction is, then

[Proposition] Any particles emitted from point sources such as GBHC or AGN should have identical energy spectra, a universal spectrum for each source, when they left the sources. The spectral index must be 2.2 ± 0.1 on the average.

The photon spectral indices from 11 AGN have been determined by the EGRET group. [See the list in reference (1).] The average of these indices is

$$\lambda_{\text{av}} = 2.1 \pm 0.1. \tag{6}$$

The average of the photon spectral index weighted with the flux intensity quoted in ref. (1) also gives a result similar to eq. (6).

The proximity of Eqs. (5) and (6) seems to support the concept of universality proposed in ref. (3) and (9). It should be emphasized, however, that universality of the energy spectrum is valid only in the sense of the average so that a certain amount of fluctuation is inevitable. As a matter of fact, the spectrum of cosmic rays itself should be the outcome of an average of the spectrum from many sources whose power index has a fluctuation.

V. Muon Neutrino Deficiency and Cosmic Neutrinos

Neutrinos are emitted from pulsating black holes with an intensity 3/8 that of γ-rays, with the same spectral index (with the same variation for individual sources, of course). Moreover, the intensity is the same for all species $\nu_e, \bar{\nu}_e, \nu_\mu, \bar{\nu}_\mu, \nu_\tau$ and $\bar{\nu}_\tau$ and they have the advantage that their flux is hardly modified once it leaves the sources, in contrast to γ-rays.

Recently, the underground neutrino detectors at Kamiokande II and IMB compiled the neutrino flux between 100 MeV and 1.5 GeV and concluded[10] that the ratio of ν_μ and ν_e is 1.2 instead of 2. The latter value of 2 is expected if the observed neutrinos are produced in the atmosphere, since pions and kaons are the neutrino source. This riddle, called the muon neutrino deficiency problem, is solved if neutrino oscillation ($\nu_\mu \to \nu_x$) ensued after production in the atmosphere.[11,12] However, study of the up-going muon suggests that such oscillation does not take place.[12] Also, it can be shown that the calculated atmospheric neutrino flux[13] tends to be an overestimate. From these considerations, it is very likely that the neutrinos observed by the underground detectors are mainly not atmospheric but cosmic, *i.e.* most neutrinos observed are coming from outside the atmosphere of the earth. But, of course, an ordinary mechanism for neutrino production ends up with $\nu_\mu/\nu_e \simeq 2$, by the same reason as for atmospheric neutrinos. The model proposed in this project is the only one which predicts $\nu_\mu/\nu_e = 1$. Moreover, it is worthwhile to mention that the neutrino flux spectrum inferred from the underground detector data[10] is close to $E^{-2.2}$. Further observation of the neutrino flux will decide the validity of the model.

VI. Summary and Further Predictions

The prediction of universality for the energy spectrum from point sources (with variation of the spectral index for individual sources) is dramatically borne out by the GRO data and the cosmic energy spectrum at the source. Also the approximate equality of ν_μ and ν_e in the underground detector seems to lend support to this model.

The author is indebted to David Williams for reading the manuscript. The work is supported in part by the U.S. Department of Energy.

REFERENCES

1. C. Fichtel, EGRET data report in this symposium.
2. M. Punch et al., Nature **358**, 477 (1992).
3. Y. Tomozawa, Proceedings of the INS Internationsl Symposium on Composite Models of Quarks and Leptons (ed. H. Terazawa and M. Yasue, 1985), p. 386; in Quantum Field Theory (ed. F. Mancini, Elsevier Pub., 1986), p. 241; Lectures at the Second Workshop on Fundamental Physics, Univ. of PUerto Rico, Humacao (ed. E. Estaban, 1986), p. 144.

4. A.R. Bell, M.N.R.S., 182, 147, 443 (1978); P.O. Lagage and C.J. Casarsky, Astr. Ap. 118, 223 (1983); R.D. Blandford and J.P. Ostriker, Ap. J. 221, L29 (1978).
5. J.E. Gunn and J.P. Ostriker, Phys. Rev. Lett. 22, 728 (1969).
6. J.P. Jokipii and G.E. Morfill, Ap. J. 290, L1 (1985).
7. A. Majumder and Y. Tomozawa, Prog. Theoret. Phys. 82, 555 (1989).
8. S.P. Swordy et al., Ap. J. 349 (1990) 625; D. Müller et al., Ap. J. 374 (1991) 356; P. Meyer et al., ICRC (1991) OG6.1.11.
9. Proceedings of the Compton Observatory Science Workshop (ed. C. Shrader, 1991), p. 335;
AIP Conference Proceedings 254, Testing the AGN Paradigm (ed. S.S. Holt et al., 1991), p. 345;
Gamma Rays from Point Sources and a Universal Energy Spectrum, to appear in Astr. & Astrophys., 1992.
10. K.S. Hirata et al., Phys. Lett. B280, 146 (1992).
11. E.W. Beier et al., Phys. Lett. B283, 440 (1992).
12. R. Becker-Szendy et al., Phys. Rev. Letters, to appear (1992).
13. e.g., E.V. Bugaev and V.A. Naumov, 20th International Cosmic Ray Conference, HE 4.1-18 (1987); Sov. J. Nucl. Phys. 45, 857 (1987); G. Barr, T.K. Gaisser and T. Stanev, Phys. Rev. D39, 3532 (1989).

COMPTONIZATION OF A SOFT PHOTON DISK SPECTRUM BY MOVING BLOBS AS A SOURCE OF GAMMA-RAYS IN AGN

Magda Zbyszewska
Department of Astronomy and Astrophysics and Enrico Fermi Institute
University of Chicago, Chicago, IL 60637

ABSTRACT

We investigate the production of gamma-rays in AGNs through the Comptonization of soft photons by relativistic electrons. Soft photons are produced in an accretion disk and illuminate a spherical cloud of relativistic electrons which moves in the direction of the observer. This model applies to AGNs that are similar to 3C 279. It has been shown that gamma-rays observed from 3C 279 by EGRET are produced by an object whose beaming factor is greater than 7. Because the existence of an accretion disk is expected in AGNs, interaction between the moving object and the disk radiation should occur. We assume that the moving object has a Lorentz factor $\Gamma \sim 10$, and the Lorentz factors of the electrons which form this object reach $\gamma \sim 10^3$. Models for AGNs predict that the inner parts of the disk radiate in optical/UV bands, so the Comptonized soft photons originally have energies about 5 eV in our model. We compare our results with the luminosity observed from 3C 279 in the energy range 100 MeV – 5 GeV.

DESCRIPTION OF MODEL AND RESULTS

Several previous works (*e.g.* Dermer 1992, Zbyszewska 1992) concerning gamma-ray radiation from quasar 3C 279 considered models in which the gamma-rays are produced via the interaction between a moving spherical cloud of relativistic electrons and soft photons radiated by an accretion disk. The soft photons which interact with the relativistic electrons are Comptonized and emitted as gamma-rays by the moving cloud. The cloud (blob) moves in the direction of the observer. The soft photons illuminate the blob from the direction opposite to that of the observer. In our calculations we assume that:

1. The intensity of the soft photons is described by a monoenergetic distribution $I(\epsilon, \Omega) = F_0 \epsilon_s \delta(\epsilon - \epsilon_s)$. F_0 is the proportionality coefficient, $\epsilon_s = h\nu_s/m_e c^2 = 10^{-5}$ is the energy of the soft photons. The intensity is constant

for $0 < \Theta < \Theta_{max}$, and is absent for $\Theta_{max} < \Theta < \pi$, where Θ is the angle between the velocity of the incoming photon and the velocity of the blob in the observer rest frame.

2. The medium of the blob is optically thin, so that a photon escapes after the first scattering.

3. The density of electrons is uniform and isotropic in the blob rest frame: $N_j(\gamma_j, \varphi_{aj}, \cos\alpha_j) = N_j(\gamma_j) = C_e \gamma_j^{-p}$, where γ_j is the Lorentz factor of an electron in the blob rest frame, $p = 2\alpha_{ph} - 1$, α_{ph} is the spectral index of photons in the gamma-rays range (in this case $\alpha_{ph} = 2$), and C_e is the proportionality coefficient.

4. Scattering is isotropic in the rest frame of the electron. This is a qualitatively good approximation when scatterings occur in the Thomson limit i.e. $\epsilon_s \gamma \ll m_e c^2$, where γ is the Lorentz factor of an electron in the rest frame of observer (Rybicki and Lightman 1979).

5. Gamma-ray luminosity in the range 100 MeV – 5 GeV is produced by electrons with Lorentz factors $(100\,\mathrm{MeV}/\epsilon_s)^{1/2} \lesssim \gamma m_e c^2 \lesssim (5\,\mathrm{GeV}/\epsilon_s)^{1/2}$ (*e.g.* Rybicki and Lightman 1979).

Following the procedure from Rybicki and Lightman 1979, we obtained emissivity in the blob rest frame:

$$j_j(\epsilon_1, \mu_j) = \int_{\gamma_{j\,min}}^{\gamma_{j\,max}} \int_{-1}^{1} \int_{0}^{2\pi} \frac{\epsilon_1^2}{\epsilon_s^2} \frac{\sigma_T}{4\pi} \frac{N_j}{\gamma_j} \frac{F_0}{\gamma\beta} 2\varphi_{max}\, d\varphi_{aj}\, d\cos\alpha_j\, d\gamma_j, \qquad (1)$$

where ϵ_1 is the energy of the Comptonized photon, φ_j, $\mu_j = \cos\theta_j$ are the angular coordinates of the Comptonized photon with respect to the direction of the velocity of the blob. All variables are in the rest frame of the blob. φ_{max} follows from the integration of the number of incoming photons over solid angle. For a given ϵ_1, φ_{max} depends on the angular coordinates and the Lorentz factor of the electron and energy of the photon in the electron rest frame. Photons in the electron rest frame may have energies between some limits (Fig. 1) which result from the range of allowed angles (*e.g.* Ghissellini *et al.* 1991).

According to Lind and Blandford 1985, we obtain the observed flux:

$$S\left(\frac{\epsilon_1 \delta}{1+z}\right) = \frac{\delta^3}{(1+z)^3} D_A^{-2} V_j j_j(\epsilon_1, \mu_j) = \frac{\delta^3}{(1+z)^3} D_A^{-2} V_j \frac{\sigma_T}{4\pi} C_e F_0\, \eta(\epsilon, \theta_o) \qquad (2)$$

and total observed flux:

$$S = \int_{\epsilon_{min}}^{\epsilon_{max}} S\left(\frac{\epsilon_1 \delta}{1+z}\right) d\frac{\epsilon_1 \delta}{1+z} = \frac{\delta^4}{(1+z)^4} D_A^{-2} V_j \frac{\sigma_T}{4\pi} C_e F_0 \int_{(\gamma_{j\,min}\Gamma)^2 \epsilon_s}^{\epsilon_{max}} \eta(\epsilon_1 \delta, \theta_o)\, d\epsilon_1 \qquad (3)$$

$$= \frac{\delta^4}{(1+z)^4} D_A^{-2} V_j \frac{\sigma_T}{4\pi} C_e F_0\, H(\theta_o),$$

where D_A is the angular distance to the source, z is the cosmological redshift V_j is the volume of the blob in its rest frame, $\epsilon = \epsilon_1 \delta$, $\delta = \Gamma^{-1}(1 - B\cos\theta_o)^{-1}$ is the beaming factor, and B is the velocity of the blob. μ_j corresponds to the μ_o transformed to the blob rest frame, where μ_o is the cosine of the angle, θ_o, between the velocity of the blob and the direction to the observer in the observer rest frame.

We obtained results for the blob Lorentz factor $\Gamma = 10$, and two values of Θ_{max}: $\pi/2$ and $\pi/4$. In Fig. 3 $\delta^4 H(\theta_o)$ is plotted as a function of θ_o for $\Theta_{max} = \pi/2$ and $\Theta_{max} = \pi/4$. As already noticed by Dermer et al. 1992, in the limiting case $\Theta_{max} \to 0$, the maximum of the observed flux is achieved for an angle $\theta_o > 0$, because of the anisotropic illumination of the blob.

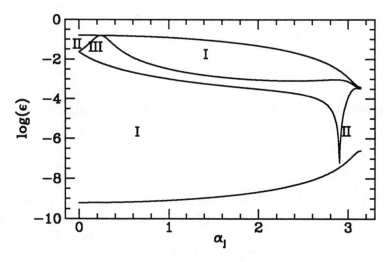

Fig. 1. The limits on the energy of the incoming photon in the rest frame of the electron as a functions of angle α_j (α_j is the angle between the blob velocity and the electron velocity in the rest frame of the blob), for $\gamma_j = 4 \times 10^2$. For energies in regions I, $\varphi_{max} = \pi$; for II, $\varphi_{max} = 0$; and for III, φ_{max} has intermediate values: $\varphi_{max} = \arccos((\cos\Theta_{maxj} - \mu_j \cos\alpha_j)/\sin\alpha_j \sin\theta_j)$ (Θ_{maxj} — Θ_{max} transformed to the blob rest frame).

DISCUSSION

We compare our results with the spectrum observed for 3C 279 in the optical/UV/X-ray/gamma-ray range. If we assume that the disk radiates approximately a blackbody spectrum with a temperature 3.7×10^4 K (for which the blackbody spectrum gives most photons with energies about $\epsilon_s = 10^{-5}$), we can compare the observed flux with the one produced in our model. The

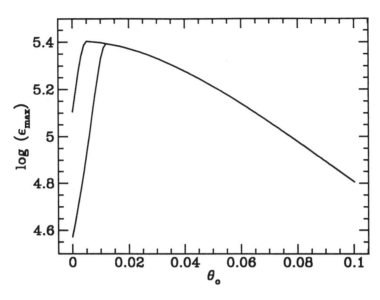

Fig. 2. The maximum observed photon energy as a function of θ_o for $\Theta_{max} = \pi/2$ (the first curve from the left), and $\Theta_{max} = \pi/4$ (the second curve from the left).

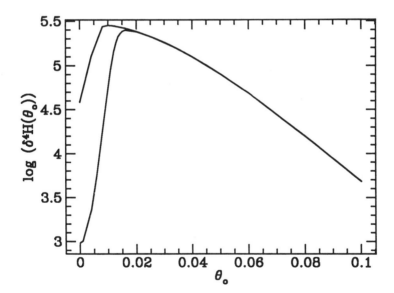

Fig. 3. The function $\delta^4 H(\theta_o)$ for $\Theta_{max} = \pi/2$ (the first curve from the left), and $\Theta_{max} = \pi/4$ (the second curve from the left).

flux measured by Makino et al. 1989 at this energy is $F_{\nu\rm UV} \simeq 3\times 10^{-26} - 3\times 10^{-27}$ erg cm^{-2}s^{-1}Hz^{-1}. To fulfill this limit, the region of the disk which emits optical/UV flux should have a radius $R_d \lesssim 2 \times 10^{15} - 7 \times 10^{15}$ cm. This means that the central black hole must have a radius $r_{bh} \lesssim 10^{14}$ cm. We assume that the blob has a size comparable to the black hole so that $R = 10^{14}$ cm. This is consistent with the limit given by the variability time scale for gamma-ray radiation $R < ct_{var}\delta/(1+z) = 5.2 \times 10^{15}\delta/(1+z)$ cm, where $t_{var} = 2d$ is the variability time scale Kanbach et al. 1992.

We assume that the X-rays and gamma-rays are produced in the same blob. As was shown by Maraschi et al. 1992, this gives a constraint on δ from the requirement that the optical depth for photon-photon pair production, $\tau_{\gamma\gamma} < 1$. For the X-ray flux observed by Makino et al. 1989 we obtained $\tau_{\gamma\gamma} \simeq 2(\delta/(1+z))^{-4.2}(10^{18}$ cm$/R)$. For $R = 10^{14}$ cm the constraint gives $\delta \gtrsim 7$, which is consistent with $\Gamma = 10$; for $R = ct_{var}\delta/(1+z)$ we obtained $\delta \gtrsim 4.5$.

Comparing our results and the observed luminosity in gamma-rays, we can estimate the number density of the electrons in the blob required to produce such a luminosity. In our model we have $F_o \simeq 3\times 10^{18}$ erg cm^{-2}rad^{-2}s^{-1} which gives, for $\delta^4 H(\theta_o) = 10^5$, $C_e \simeq 5 \times 10^6$ cm^{-3} and the number density of high energy electrons is about 10^2 cm^{-3}. If we include electrons that produce soft gamma and X radiation, the number density of electrons becomes 10^6 cm^{-3}. With this density the blob is still optically thin to Compton scattering. The broken power law spectrum in the X-ray/gamma-ray range results from a broken power law in the electron energy distribution.

This research was supported in part by NASA grants NAGW-830 and Compton GRO Fellowship (GRO/PFP-91-28).

REFERENCES

Dermer, C. D., Schlickeiser, R., Mastichiadis, A., 1992. *A.&A.*, **256**, L27.

Ghisellini, G., 1991. *M.N.R.A.S.*, **248**, 449.

Hartman, R.C., et al., 1991. *Ap. J. (Letters)*, **385**, L1.

Kanbach, G., et al., 1992. *IAU Circ.* 5431.

Lind, K. R. & Blandford, R. D., 1985. *Ap. J.*, **295**, 358.

Makino, F., et al., 1991. *Ap. J., (Letters)*, **347**, L9.

Maraschi, L., Ghisellini, G., and Celotti, A., 1992. *preprint*.

Rybicki, G. B. & Lightman, A. P., 1979. *Radiation Processes in Astrophysics*, Wiley, Interscience, New York.

Zbyszewska, M., 1992. *B.A.A.S.*, **24**, 730.

Why so many (so few) EGRET AGN's, or, which ones next?

G.F. Bignami (*), P.A. Caraveo, P. Ciliegi

Istituto di Fisica Cosmica del CNR, Milano, ITALY

(*) University of Cassino, ITALY

Abstract

Data on the 16 AGN's (mostly blazars) seen recently by EGRET to emit the majority of their power in gamma-rays have been collected, in an effort to understand which, if any, common property renders them a γ-ray sample. A striking correlation is found between the beaming Doppler factor δ and the gamma-ray emitting property of blazar. Pending confirmation for a few cases where the data are as yet insufficient (and the relevant proposed observations will be discussed), this appears as a promising method to predict which AGNs will next be observable by EGRET.

1. INTRODUCTION

The discovery of a γ-ray flux (>100 MeV) from 16 AGNs, by the EGRET instrument aboard GRO (Hartman et al. 1991), tightly constrains the physical conditions of the sources, possibly throwing light on the mechanism operating in these objects.

In order to unveil possible common characteristics in the EGRET extragalactic dataset, we have collected data on the radio optical and x-ray fluxes of the objects, as well as on their redshift and Doppler factor. These data are shown in Table 1.

It is interesting to note that most of these objects belong to the *blazar* family, a particular class of AGNs containing highly polarized quasars (HPQ), optically violent variable quasars (OVV) and BL Lac objects.

Blazar are characterized by strong and compact radio emission, variability and high polarization at optical and infrared wavelengths, and a smooth continuum over many decades in frequency. Some blazars have strong emission lines typical of quasar (HPQ), but most (BL LAC) have weak emission and absorption lines.

Table 1 : γ-ray sources detected by EGRET

Source	Other name	Type	z	S_{5GHz} (Jy)	V (mag.)	S_{2keV} (μJy)	δ
0202+149	4C+15.05	QSO		2.43	20.9	0.02	
0208-512	PKS	HPQ	1.003	3.31	17.5		
0235+164	PKS	BLLAC	0.940	1.85	15.5	0.17	6.19[1]
0420-014	PKS	HPQ	0.915	3.50	17.8	0.38	2.57[1]
0454-463	PKS	QSO	0.858	1.58*	1.74		
0528+134	PKS	QSO		3.98	20.0	0.12	
0537-441	PKS	BLLAC	0.894	4.00	15.5	0.10	>3.64[2]
0716+714		BLLAC		1.12	15.5	0.22	2.28[1]
0836+710	4C71.07	HPQ	2.170	2.59	16.5		6.99
1101+384	MKN 421	BLLAC	0.031	0.54	13.4	1.59	0.49[1]
1226+023	3C273	QSO	0.158	42.9	13.0	10.7	3.48-8.14[2]
1253-055	3C279	HPQ	0.538	14.9	16.8	0.54	4.15[2]
1606+106	4C10.45	QSO	1.240	1.14	18.5	<0.10**	3.04
1633+382	4C38.41	OVV	1.814	2.93	18.0	0.04	6.38
2230+114	CTA102	HPQ	1.037	3.69	17.3	0.21	1.85[1]
2251+158	3C454.3	HPQ	0.859	17.4	16.1	0.29	5.93[1]

* 8.4 GHz   ** 1 KeV   (1) Madau et al. 1987   (2) Bassani and Dean 1986

For the three sources (4C71.07, 4C10.45 and 4C3841) with unpublished Doppler factor, we have calculated δ following Marsher (1987) and Ghisellini et al. (1992) as :

$$\delta = f(\alpha) F_m \left(\frac{\ln(\nu_b/\nu_m)}{F_x \, \theta^{(6+4\alpha)} \, \nu_x^{\alpha} \, \nu_m^{5+3\alpha}} \right)^{1/(2(2+\alpha))} (1+z)$$

where F_m (in Jy) is the flux at the self-absorption frequency ν_m (in GHz), F_x (in Jy) is the X-ray flux at ν_x (in KeV), α is the spectral index of the thin synchrotron emission (we have assumed α=0.75 for all sources), θ (in mas) is the angular size of the source, ν_b is the synchrotron high frequency cut-off (10^{14} Hz) and $f(\alpha) \simeq 0.08\alpha + 0.14$ (Ghisellini, 1987). For 4C71.07 and 4C10.45 we have replaced F_x and ν_x with the optical flux and the corresponding optical frequency.

2. DISCUSSION

The region where γ-rays are produced cannot be very compact. Otherwise photon-photon collisions, leading to pair creation, would prevent γ-rays from escaping the source. The compactness l of a spherical source with radius R and luminosity L is defined by $l = L\sigma_T/R\, m_e\, c^3$, where σ_T is the Thomson cross section. The observed luminosity L_{obs} is connected to the intrinsic luminosity L_{int} by $L_{int} = \delta^{-n}L_{obs}$, where n is between 3 and 4 (Lind and Blandford 1985, Maraschi et al. 1992) and the intrinsic size R of the emitting region is estimated from the variability timescales, $R = ct_{var}\delta$, where $\delta = [\Gamma(1-\beta\cos\theta)]^{-1}$ is the beaming Doppler factor. For angle θ (the angle between the bulk velocity of the emetting plasma and the line of sight) smaller than $\theta_l = \sin^{-1}[2/[\Gamma-1)]^{1/2}$, δ is greater than unity and correspondingly the source appears brighter than in its rest frame and the intrinsic size R is greater than the value estimated by ct_{var}. For compactness $l <1$ (i.e. fairly low to allow escape of the γ-ray emission) we must have low luminosity and values of R fairly high. Therefore, because the AGNs with γ-ray emission are characterized by high luminosity and small variability timescales, the most probable AGN candidates to become γ-ray sources are those with $\delta >1$.

This is easly see in fig.1, where we plotted δ versus intrinsic x-ray fluxes ($F_{x\ 1keV}/\delta^n$, assuming n=3) for the newly detected γ-ray sources and a sample of 41 blazars (Madau et al. 1987).

The first AGNs detected by GRO (excluding the BL Lac MKN 421, the source with lower redshift) are all consistent with the prediction of $\delta >1$.

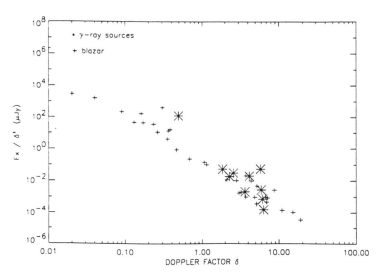

Figure 1. Doppler factor δ versus $F_{x\ 1\ KeV}/\delta^n$ (for n=3).

Then the condition $\delta > 1$ can be used as a selection criterion to obtain a list of probable γ-ray AGNs.

Fig. 1 contains 26 blazar from the sample of Madau et al. with $\delta > 1$. Of these, 7 (PKS 0235+164, PKS 0420-014, 0716+714, MKN421, 3C279, CTA102 and 3C454.3) are already detected by EGRET. The remaining 19 sources are OB081, 0212+73, PKS 0306+102, 0454+84, 0I158, OI090, 0J287, 4C14.60, 4C09.57, 1803+78, 2007+776, 2200+420, PKS 0106+ 013, CTA 26, 1308+326, 3C 345, 3C 446, 2234+282 and PKS 2345-167.

Thanks to the sky survey done by the GRO EGRET instrument these 19 sources have all been within the instrument FOV and on the basis of our "high δ" selection criterion, they should be considered promising candidates deserving closer look in the near future.

REFERENCES

Bassani,L. and Dean 1986, A&A, 161, 85
Ghisellini, G., 1987. *Ph. D. Thesis*, SISSA, Trieste
Ghisellini, G., Padovani, P., Celotti, A. and Maraschi, L. 1992, Ap.J., in press
Hartman,R.C., et al., ApJ, 385, L1
Lind, K.R. and Blandford, R.D., 1985, Ap.J., 295, 358
Madau,P., Ghisellini,G. & Persic,M. 1987, MNRAS, 224, 257
Maraschi, L., Ghisellini, G. and Celotti, A. 1992, Ap.J., in press
Marscher, A.P., 1987. In *Superluminal Radio Sources*, eds. Zensus, A. and
    Pearson, T.J. (Cambridge: Cambridge University Press), p. 280

SOLAR FLARES

OSSE OBSERVATIONS OF SOLAR FLARES

R.J. Murphy, G.H. Share, J.E. Grove, W.N. Johnson, R.L. Kinzer
R.A. Kroeger, J.D. Kurfess, M.S. Strickman
E.O. Hulburt Center for Space Research
Naval Research Laboratory, Washington, DC 20375

S.M. Matz, D.A. Grabelsky, W.R. Purcell, M.P. Ulmer
Dept. Physics & Astron., Northwestern University, IL 60208

R.A. Cameron, G.V. Jung
USRA, Naval Research Laboratory, Washington, DC 20375

C.M. Jensen
George Mason University, Fairfax, VA 22030

W.T. Vestrand, D.J. Forrest
University of New Hampshire, Durham, NH 03824

ABSTRACT

The Oriented Scintillation Spectrometer Experiment (OSSE) on the *Compton Gamma Ray Observatory* (*CGRO*) provides solar-flare observation capabilities that represent a significant improvement over previous experiments. In this paper, we discuss the OSSE solar observation modes, give an overview of the solar-flare observations to date, and discuss in some detail the observations obtained during the June 1991 *CGRO* solar Target of Opportunity. We show evidence for time structure of flare hard X-ray (>100 keV) emission as short as tens of msec. Several nuclear deexcitation lines along with emission >10 MeV were seen from the four *GOES* X10+ flares observed by OSSE. Neutrons were observed from two of the flares. Preliminary results from a search of the OSSE flare database for nuclear emission from weak flares are presented and compared to results from a similar search of the *Solar Maximum Mission* Gamma-Ray Spectrometer database.

INTRODUCTION

The Oriented Scintillation Spectrometer Experiment (OSSE) on the *Compton Gamma Ray Observatory* (*CGRO*) incorporates a variety of detection and measurement capabilities for solar flares (Johnson et al., 1993). OSSE is similar in construction to the *Solar Maximum Mission* Gamma-Ray Spectrometer (*SMM/GRS*, Forrest et al., 1980) but represents a significant step forward in sensitivity and background rejection. While solar-flare observations are not the primary mission of *CGRO*, excellent observations of a number of solar flares have been accomplished. In this paper, we give an overview of the OSSE solar-flare database in addition to preliminary results of analyses of some specific flares. In the next section, we describe the OSSE instrument and the various solar data-collection modes used. We then describe the OSSE solar observations to date and give some results from the analyses of flares that occurred during the June 1991 *CGRO* solar Target of Opportunity. We provide context for the OSSE observations by discussing results from a broad analysis of the *SMM/GRS* database.

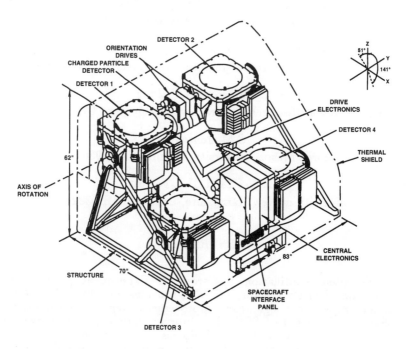

Figure 1. The OSSE instrument.

INSTRUMENT AND DATA-COLLECTION DESCRIPTION

OSSE consists of four independently oriented phoswich scintillation detectors with both passive and active shielding for reducing background and defining its aperture. The OSSE experiment is shown in Figure 1. Its principal detector element is a 33-cm-diameter phoswich (effective area 480 cm^2 at 511 keV), consisting of 10.2-cm-thick NaI and 7.6-cm-thick CsI crystals, optically coupled to each other and viewed by seven photomultiplier tubes. The CsI and NaI pulses are electronically separated by pulse-shape discrimination, thus providing a compact anticoincidence system for charged particles and background γ-rays. The energy resolution is 3.8% FWHM at 6.1 MeV, increasing to 8.2% at 662 keV. The combined sensitivity for narrow-line detection from a 1000-sec observation is 1×10^{-3} photons cm^{-2} sec^{-1}.

Photon spectra covering the range from 0.05 to >200 MeV in 528 channels are typically accumulated with a spectral accumulation time (t_{sac}) of ~16.4 sec; event-by-event data covering limited spectral ranges with integration times

≥0.125 ms and rates in broad energy ranges with integration times ≥4 ms may also be acquired. Neutron spectra above ~10 MeV are accumulated at t_{sac} resolution. A tungsten collimator above the phoswich defines a 3.8° × 11.4° FWHM field of view (FOV). Surrounding the phoswich and collimator is a NaI annular shield, made up of four segments and having the capability of providing 0.1- to 8-MeV spectra in 256 channels at ~ 10% FWHM energy resolution. The projected area of the shields for each detector ranges between 1000 and 1750 cm^2, depending on incident angle.

Each detector is mounted in an independent elevation-angle gimbal which provides 192 degrees of rotation about the spacecraft Y-axis. Targets near the extremes of rotation may be accessible by only two of the four detectors due to detector shadowing. The ability to slew allows OSSE to view a secondary target when the primary target is occulted by the Earth. It also allows some or all of the OSSE detectors to slew to the Sun for observations of solar flares when the position of the Sun is sufficiently near the OSSE scan plane. In general, *CGRO* is placed into a fixed orientation for a 2-week viewing period and the OSSE detectors alternate between viewing the primary and secondary targets.

Since the primary mission of *CGRO* is celestial, the Sun is generally not an explicit OSSE target. Nevertheless, excellent observations of a number of solar flares have been accomplished via solar-flare triggers provided by BATSE. BATSE continuously examines count rates from 50 to 300 keV in each of the eight Large Area Detectors (LADs) for statistically-significant increases above background. When such increases are detected (and when some additional criteria are met), a signal is sent to the other *CGRO* instruments indicating that a transient has been detected. After approximately 2 seconds of data accumulation, the BATSE on-board electronics determine if the relative count rates from the eight LADs are consistent within upper and lower limits for a burst coming from the direction of the Sun. If they are not, the signal is terminated at 2 seconds. If they are, and if certain other criteria (which may or not be invoked, such as the most sunward detector must be triggered, that the hardness ratio be between certain limits, and that the maximum count rate be above a specified lower limit) are met, then the burst is declared to be solar in origin and the signal being sent to the other instruments is extended. OSSE on-board electronics examine the signal 3 seconds after the initial trigger and, if it is still present, OSSE is placed in one of the solar data-collection modes described below.

In addition to the flare observations obtained via BATSE triggers, the Sun has been an OSSE target during four 1-to-2-week periods since launch: during May 1991 before the start of Phase I (the activation phase), during the June 1991 solar Target of Opportunity, during December 1991, and during September 1992. During these periods, the OSSE detectors directly viewed the Sun for varying periods of time. In general, OSSE was placed in a solar data-collection mode identical to that of the full, Sun-accessible mode discussed below.

Several OSSE solar data-collection modes are utilized depending on the trigger type and the observational conditions. If BATSE fails to identify the source of the burst as solar, OSSE obtains standard burst readout data. These are 4096 16-msec shield count rates above threshold (~93 keV), summed over all shields. Five hundred of these 16-msec rates are accumulated prior to the trigger. If BATSE successfully identifies the burst as solar, the standard burst readout is again obtained along with an additional 1000-sec OSSE response, depending on whether the Sun is accessible to the FOV via a slew.

If the Sun is not accessible, OSSE collects spectra from the two sunward-facing shields at 4-t_{sac} temporal resolution and 256-channel energy resolution (from $\sim$0.1 to $\sim$8 MeV). In addition, one 96-t_{sac} accumulation of high-energy, 2-dimensional (pulse-duration vs. energy, see Johnson et al., 1993) spectra is collected in the phoswich nearest the Sun. Such spectra can be used to help distinguish high-energy photon events from neutron events. If the Sun is accessible to the OSSE FOV, either two or four detectors rotate (at $\sim 2°$ sec^{-1}) to the solar direction, depending upon how many can acquire it. The time required to slew and acquire the Sun depends on the relative position of the current OSSE celestial target and that of the Sun.

If all four detectors acquire the Sun, two maintain a fixed orientation with the Sun in the FOV while the other two alternately point on and off the Sun at 2-minute intervals to facilitate background determination and to provide non-saturated spectra for intense flares. If only two detectors can acquire the Sun, they typically alternate between on- and off-pointing. Full energy-resolution photon spectra (from $\sim$60 keV to $\sim$200 MeV) and 16 channels of neutron spectra (>10 MeV) are obtained from the central detectors at 8-sec t_{sac} resolution along with 16-msec count rates in four broad energy windows (200–450 keV, 570–750 keV, 4–7 MeV and >10 MeV).

OSSE SOLAR OBSERVATIONS

Since the launch of *CGRO* in April of 1991 to the present (28 October 1992), 4550 solar flares have been given soft X-ray flare classifications by the *GOES* satellite solar-monitoring network. Thirteen percent have been B-class, 74% have been C-class, 12% have been M-class, and 1% have been X-class. Of these flares, 3081 occurred both during the sunlit portion of the *CGRO* orbit and not during a South Atlantic Anomaly (SAA) passage when all detectors are shut down. Since launch, BATSE has issued 553 burst triggers due to solar activity with emission above 60 keV and 377 of those were successfully identified on-board as solar in origin, thus providing a BATSE "solar" trigger. (It should be noted that, since the *CGRO* tape recorders have been disabled, solar coverage has been reduced and so the BATSE solar-triggering rate has been correspondingly reduced.) For 162 of these events, the Sun was accessible to the OSSE FOV and OSSE slewed to view the Sun.

The OSSE data obtained through 1991 December as a result of BATSE triggers due to solar activity (whether or not the transient was identified on-board as solar) has been searched for detected emission. Sixty-nine of the triggered events resulted in emission detectable in the central detectors up to energies of at least 300 keV. Twenty-six of the events were detectable up to at least 800 keV, 22 up to 3 MeV and 20 up to or greater than 10 MeV. Detailed, high time-resolution shield count-rate data were obtained for 65 of the flares. During those times when the Sun was an explicit OSSE target (such as during the June 1991 period), additional data were obtained on flares for which BATSE failed to provide a solar trigger, but which nevertheless were detected since OSSE was directly viewing the Sun.

OSSE JUNE 1991 SOLAR OBSERVATIONS

In June of 1991, solar active region 6659 produced some of the largest *GOES* flare events ever recorded by the satellites. Fortunately, *CGRO* Phase I Viewing Period 2 began on 30 May with the Sun accessible to the OSSE FOV (although not within the COMPTEL or EGRET FOVs). On 1 June, AR 6659 appeared at the East limb and produced an X12+ flare. At the time of flare onset, OSSE was shut down due to an SAA transit and no BATSE flare trigger was sent. Upon exit from SAA, OSSE slewed to the primary celestial target (Cygnus) and no significant solar data were obtained. As a result of the high probability for intense flare production, the Sun was declared an OSSE Target of Opportunity and replaced the existing secondary celestial target. On 4 June, AR 6659 produced another X12+ flare while OSSE was viewing the Sun. Excellent observations were obtained of the rise, peak and decay of the event. The decay was interrupted by spacecraft night but observations were resumed at daylight of the next orbit and additional observations were obtained. On 6 June, another X12+ flare occurred while OSSE detectors were pointed at the Sun. High-quality data were obtained until the detectors slewed to the primary celestial target. Because of this unprecedented solar activity, a *CGRO*-wide Target of Opportunity was declared on 8 June and the spacecraft was reoriented so that COMPTEL and EGRET could also view the Sun. On 9 June, AR 6659 produced an X10 flare. At the time, OSSE was viewing Cygnus (which had been given observing priority) but immediately slewed to the Sun in response to the BATSE trigger. Solar observations continued for over 3000 seconds until the detectors slewed back to Cygnus and were then resumed during the next orbit. Another X12+ flare occurred on 11 June while OSSE was again viewing Cygnus. OSSE slewed to the Sun and observed it for approximately 3000 seconds and during the following orbits. Finally, an X12+ flare occurred on 15 June during a *CGRO* SAA transit. Observations beginning late in the decay phase were obtained upon SAA exit. Observations of a number of weaker flares were also obtained during the 1 June – 15 June period since OSSE was often viewing the Sun.

Figure 2 shows 16-msec temporal resolution total-shield count rates (>100 keV) for up to 60 seconds following the BATSE triggers for the flares of 4, 6, 9 and 11 June. Temporal variations as short as tens of msec can be observed in some of the flares. (The data for the 11 June flare is from a second BATSE trigger that occurred after the primary trigger for that flare.) Figure 3 shows preliminary, time-integrated spectra at energies above 1.5 MeV from one of the sunward-viewing OSSE detectors for the same 4 flares. Clearly observed are the 2.223-MeV line from neutron capture and the deexcitation lines from carbon, oxygen and nitrogen at 4.44, 6.13 and ~7 MeV. Figure 4 shows a spectrum obtained late in the 4 June flare, when the contribution from primary electron bremsstrahlung is small. The delayed positron-annihilation line at 511 keV is easily seen.

We emphasize that these data are preliminary. Electronic saturation due to the intense fluxes produced during the peaks of these exceptionally large flares caused a number of problems with data obtained during those times. For example, distortion of pulse shapes due to pulse pile-up effects broadened the pulse shape of on-board ^{241}Am calibration-source events, causing the line to be

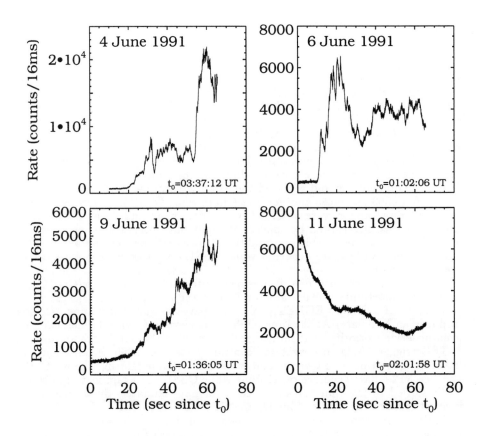

Figure 2. OSSE 16-msec total-shield count rates for several solar flares from 1991 June.

included in "valid" γ-ray spectra. The apparent line near 2.7 MeV in the spectral plots of Figure 3 is due to ^{241}Am. Similar pulse pile-up effects produce an energy-dependent change in the effective area of the detectors since pulse shapes of true γ-ray events can be sufficiently distorted to have their measured durations fall outside the valid time window. This is particularly apparent at energies less than $\sim$1.5 MeV. Also, the large count rates at low energies filled data buffers of the low channels resulting in "wrapping," sometimes several times. We have attempted to "dewrap" these channels with a statistical technique. Finally, the fractional livetimes during the most intense portions of the flares fell below 1%, making livetime corrections suspect. We are currently working to model the above effects and hope to recover or correct much of the data during the intense portions of such large flares.

We have fit the 2.223-MeV neutron-capture line and obtained integrated fluences over the OSSE accumulation times for the flares of 4, 6, 9 and 11 June. No attempt to correct the results for data gaps caused by spacecraft night has been made. For this reason, and for the data-problem reasons discussed above, these fluences are to be considered lower limits. We obtained about 300, 200, 30 and 100 photons cm^{-2}, respectively. These fluences are to be compared with a fluence of $\sim$300 photons cm^{-2} for the 3 June 1982 flare observed by *SMM/GRS*. Gamma-ray emission >10 MeV was observed from all four of these flares and neutrons were directly observed from the flares of 4 and 6 June.

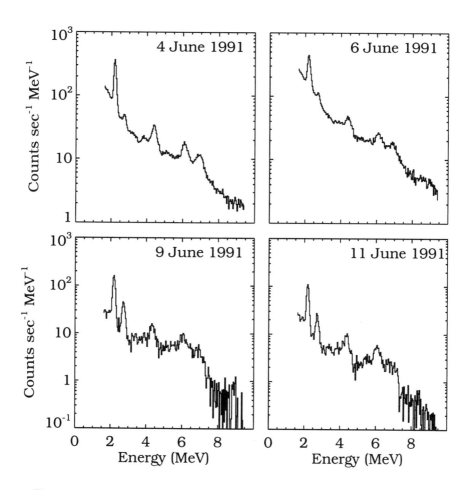

Figure 3. OSSE count spectra for several solar flares from 1991 June. The 2.223-MeV neutron-capture line and the deexcitation lines at 4.44, 6.13 and $\sim$ 7 MeV can be clearly seen. The line at $\sim$ 2.7 MeV is a calibration line from the on-board ^{241}Am source and appears in the spectra because of pulse pile-up effects due the large fluxes from these intense flares.

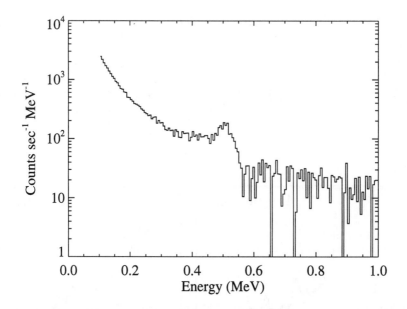

Figure 4. OSSE count spectrum accumulated late in the 4 June 1991 solar flare showing the presence of the 511-keV positron-annihilation line.

The 11 June flare is unique in that EGRET has reported observing high-energy (>50 MeV) emission many hours after the peak of the X- and γ-ray emissions from this flare. Mandzhavidze and Ramaty (1992) have analyzed the EGRET data and found that this late emission can be fit with a combination of pion-decay radiation and primary electron bremsstrahlung. They argue for a model in which the bulk of the particles responsible for the late, high-energy emission are accelerated early during the impulsive phase and are subsequently trapped in coronal magnetic loops. Since lower-energy particles responsible for nuclear-line emission are expected to lose their energy faster than the higher-energy particles responsible for the pion-decay emission, a consequence of such a model is that the nuclear-line emission should decay faster than the high-energy emission. If the high-energy emission is due to continuously-accelerated particles, the two types of emission should have comparable time profiles.

The most reliable measure of nuclear-line emission would be a fit to a narrow nuclear line, such as the 4.44-MeV ^{12}C line. At the late times of interest in the 11 June flare, the flux in such a line is expected to be quite weak and a reliable measure requires a good determination of the background contribution to the line. Such a determination is, at present, not available. An alternative would be to fit the broad, nuclear-dominated energy range from $\sim$4 to 7 MeV with a model of the narrow-line nuclear spectrum. As of this writing, the OSSE data analysis tools needed to support such an approach were not available but are expected to be soon. For the results presented here, we will use the 2.223-MeV neutron-capture line as an estimator of the nuclear-emission time profile. Because of the time required for the neutrons produced in the nuclear reactions to slow down and combine with ambient hydrogen to produce the 2.223-MeV γ ray, the time profile of this line is delayed relative to that of the nuclear reactions themselves. However, this delay is on the order of ~100 seconds, which is much less than the timescales of interest here (hours). The neutron-capture line time profile, therefore, should provide an adequate measure of the nuclear-line production time profile.

We have only used OSSE data obtained after the peak of the 11 June flare to avoid the data problems near the flux maximum discussed above. The data from each of the two detectors which were continuously pointed at the Sun were summed into 4 time bins: two bins after the peak but during the same orbit as that of the peak, and two bins during the following orbit. The count spectra near the region of the neutron-capture line were then fit with a narrow Gaussian line centered at 2.223 MeV.

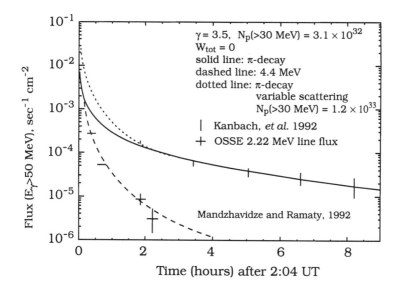

Figure 5. Measured and calculated time-dependent fluxes of the >50-MeV γ rays and predicted 4.44-MeV γ rays from the 11 June 1991 solar flare (from Mandzhavidze and Ramaty, 1991). Also shown are the renormalized OSSE measurements of the 2.223-MeV neutron-capture line from this flare.

Figure 5 (from Mandzhavidze and Ramaty, 1992) shows the EGRET data from the 11 June flare (the peak of the *GOES* soft X-ray emission occurred at 2:09 UT). The solid and dotted curves shown in the figure are fits by Mandzhavidze and Ramaty to the high-energy data assuming impulsive acceleration. The dashed curve is their prediction for the 4.44-MeV nuclear line. The parameters listed in the figure relate to their loop model. We have overplotted the OSSE measurements of the 2.223-MeV line, renormalized such that the earliest OSSE measurement falls on the predicted 4.44-MeV line curve. The predicted curve is consistent with the OSSE observations providing support for the conclusion by Mandzhavidze and Ramaty that the particles are accelerated impulsively. It must be noted, however, that high-energy particles trapped in magnetic loops and producing pions will also produce neutrons, although not as efficiently as the lower-energy particles. It would be expected, then, that even under the trapped-particle scenario, the neutron-capture line time profile will be similar to that of the high-energy emission at late times in the flare. Comparison of the OSSE data with the high-energy curve shows that this is not the case for the time period covered by the OSSE data, implying that the bulk of the emission at these times is produced by particles accelerated impulsively.

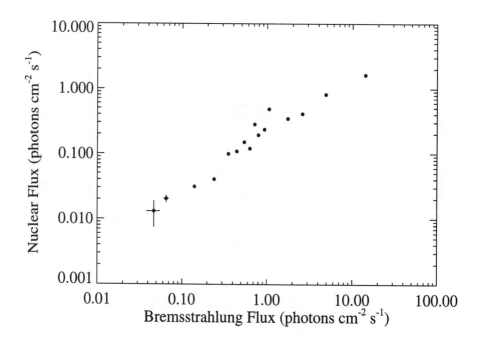

Figure 6. Total >0.3-MeV nuclear flux versus total >0.3-MeV bremsstrahlung flux of flare spectra obtained by summing individual *SMM/GRS* flare spectra into 16 spectra according their >0.3-MeV bremsstrahlung fluxes and fitting.

SMM/GRS OBSERVATIONS

While OSSE has observed the Sun for a total of about 2 weeks since its launch, the Gamma Ray Spectrometer on board the Solar Maximum Mission satellite was continuously pointed at the Sun over its full operational period of nearly 10 years. During that time, *SMM/GRS* observed almost 200 flares with emission above 300 keV and with analyzable spectra. These flares have been analyzed for contributions from primary electron bremsstrahlung and nuclear interactions. They were fit with a model composed of a power law to represent the bremsstrahlung contribution and a nuclear-emission template derived from the detailed analysis (Murphy et al., 1990) of the 27 April 1981 flare observed by *SMM/GRS*. Two additional narrow lines at 0.511 and 2.223 MeV were also included. The results of these analyses were summarized (Share et al., 1992) at the Spring 1992 AAS meeting and will be published in detail elsewhere (Murphy et al., 1993). Close to 50% of the flares exhibit evidence for ion acceleration either by the detection of specific lines (e.g., 0.511 or 2.223 MeV) or of the nuclear template. On average, the nuclear and bremsstrahlung fluences are strongly correlated but do exhibit large flare-to-flare variations in addition to variations within individual flares. The flare-averaged, integrated bremsstrahlung fluxes >0.3 MeV are generally about two to four times the averaged total prompt nuclear fluxes >0.3 MeV. No evidence has been found for flares exhibiting nuclear lines without a bremsstrahlung continuum.

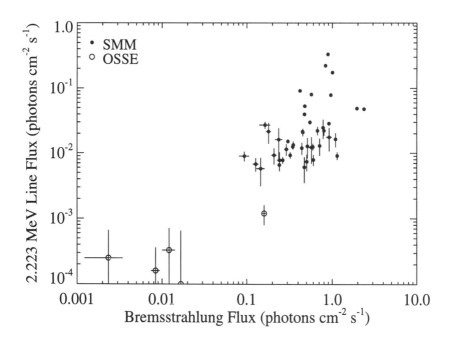

Figure 7. Measured 2.223 MeV line fluxes versus the electron bremsstrahlung fluxes (>0.3 MeV) for *SMM/GRS* and OSSE flare data.

To show the bremsstrahlung-nuclear correlation clearly, we have divided the range of measured >0.3-MeV bremsstrahlung fluxes into 16 bins and summed the spectra of those flares whose fluxes fall within each bin. We have analyzed the resulting spectra for bremsstrahlung and nuclear contributions using the fitting procedure discussed above. The results are shown in Figure 6 where we have plotted the total nuclear flux above 0.3 MeV versus the total bremsstrahlung flux above 0.3 MeV. Down to the limit of the *SMM/GRS* sensitivity, no evidence is found for a threshold below which only electrons are accelerated without acceleration of ions.

Although OSSE has observed the Sun for a much shorter time than has *SMM/GRS*, it is significantly more sensitive (effective areas of $\sim$150 and $\sim$37 cm^2 for *SMM/GRS* and $\sim$480 and $\sim$190 cm^2 for each of the OSSE detectors at 0.511 and 4.44 MeV, respectively). This should allow the extension of the ion-electron acceleration study to flares weaker than those detectable by *SMM/GRS*. Here we use the 2.223-MeV neutron-capture line as a measure of the nuclear emission. In Figure 7 we show the measured *SMM/GRS* 2.223-MeV line fluxes for those flares with at least a 2-σ measurement of the line plotted versus the integrated bremsstrahlung fluxes >0.3 MeV. We have selected 5 relatively weak flares from the OSSE database and fit them with a narrow line centered at 2.223 MeV and with a power law above 0.3 MeV. The results are also shown in Figure 7. We see that OSSE has detected flares with an order of magnitude weaker bremsstrahlung flux than any detected by *SMM/GRS* but that the 2.223-MeV line measurements of 4 of the 5 OSSE flares are only upper limits. We expect that a thorough analysis of all OSSE flares will provide data with well-measured 2.223-MeV line and bremsstrahlung fluxes.

ACKNOWLEDGMENTS

We would like to acknowledge the technical assistance of R. J. Mackinnon.

REFERENCES

1. Forrest, D.J., *et al.* , Sol. Phys. **65**, 15 (1980).

2. Johnson, W.N., *et al.* , Ap. J. , (submitted).

3. Kanbach, G., *et al.* , A&AS , (in press).

4. Mandzhavidze, N., and Ramaty, R., Ap. J. (Letters) **396**, L111 (1992).

5. Murphy, R.J., Share, G.H., Letaw, J.R., and Forrest, D.J., Ap. J. **358**, 298 (1990).

6. Murphy, R.J., *et al.* , Ap. J. , (to be submitted).

7. Share, G.H., Murphy, R.J., and Mackinnon, R.J., Bull. AAS **24**, 755 (1992).

COMPTEL GAMMA RAY AND NEUTRON MEASUREMENTS OF SOLAR FLARES

J. Ryan, D. Forrest, J. Lockwood, M. Loomis, M. McConnell, D, Morris, W. Webber
Space Science Center, Morse Hall, University of New Hampshire,
Durham, NH 03824

G. Rank, V. Schönfelder
Max-Planck Institut für Extraterrestrische Physik, 8046 Garching, Germany

B.N. Swanenburg
SRON-Leiden, 2300 RALeiden, The Netherlands

K. Bennett, L. Hanlon, C. Winkler
Astrophysics Divsion, European Space Research and Technology Center, Noordwijk,
The Netherlands

H. Debrunner
University of Bern, Bern, Switzerland

ABSTRACT

COMPTEL on the Compton Gamma Ray Observatory has measured the flux of γ-rays and neutrons from several solar flares. These data have also been used to image the Sun in both forms of radiation. Unusually intense flares occurred during June 1991 yielding data sets that offer some new insight into of how energetic protons and electrons are accelerated and behave in the solar environment. We summarize here some of the essential features in the solar flare data as obtained by COMPTEL during June 1991.

INTRODUCTION

COMPTEL on the Compton Gamma Ray Observatory (CGRO) was designed to measure the flux of cosmic gamma ray sources down to a level of $\sim 10^{-5}$ γ-cm^{-2}-s^{-1}-MeV^{-1} in the energy range of 1 - 30 MeV. However, it is also capable of measuring the γ-ray and neutron flux from solar flares. Although the intensity of large solar flares exceeds the dynamic range of the telescope, useful measurements still can be performed even for the most intense events. Such was the case in June 1991 when active region 6659 was declared to be a CGRO Target of Opportunity and the capabilities of the observatory were directed toward making the highest quality measurements of flares during the transit of the active region across the solar disk.

From the emergence of the region on the east limb on 1 June until 8 June 1991 the Sun was outside the field-of-view of COMPTEL. From 8 June until passage of the region around the west limb on 15 June, the Sun was in the telescope field-of-view at a zenith angle of ~ 10° to 15°. Other periods of solar activity have produced solar flares that COMPTEL measured; however, we report here on the more spectacular events seen in June 1991, including those from active region 6659. The flare data summarized in this paper are from events on 4, 9, 11, 15 and 30 June 1991. Only the flares of 9, 11 and 15 June 1991 were within the COMPTEL field-of-view.

The design of COMPTEL is illustrated schematically in Figure 1. COMPTEL has two modes of measuring γ-rays from cosmic bursts and solar flares[1]. We only present here the salient features of the telescope operation as they relate to solar measurements. The two methods of γ-ray detection are the so-called burst mode and telescope mode.

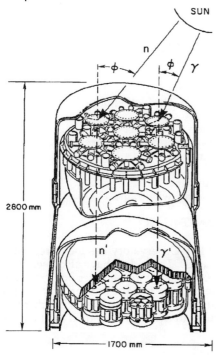

COMPTEL
IMAGING COMPTON TELESCOPE

Figure 1. Schematic of COMPTEL with typical γ-ray and neutron interactions.

In the telescope mode an incoming γ-ray scatters off an electron in one of seven D1 detectors and proceeds down to one of fourteen D2 detectors scattering again. Events fully absorbed in D2 constitute the ideal type of γ-ray interaction. The material in D1 is a liquid organic scintillator NE213A with the properties of low density and low Z (H/C ratio = 1.286). The material scatters both γ-rays and neutrons elastically, off electrons and hydrogen nuclei respectively. The scatters take place according to the Compton kinematic formula

$$\phi = \cos^{-1}\{(1 - \varepsilon/E_2 + \varepsilon/(E_1 + E_2))\}, \qquad (1)$$

where ε is the electron rest mass energy, E_1 is the energy deposit in D1, E_2 is the energy deposit in D2 and ϕ is the Compton scatter angle provided $E_1 + E_2$ is the full incident γ-ray energy.

As an imaging solar γ-ray telescope in the energy range of 0.8 - 30 MeV, COMPTEL relies on the full energy deposit of the γ-ray to correctly estimate the scattering angle ϕ of the photon within the instrument. For a solar flare γ-ray interacting in COMPTEL, the inferred scatter angle ϕ about the vector of the scattered γ-ray must be such that the photon is assigned a solar origin as indicated schematically in Figure 1. Hence, we know that the photon deposited its full energy in the detector. The response of the telescope to such events is simple. The energy or pulse height distribution is Gaussian in shape, but with a heavily suppressed Compton tail at low energies. Since the solar γ-ray spectra are rich in lines from C, N, O, Ne, Mg etc., a simple instrumental response function facilitates correct de-convolution of the pulse height spectra.

The liquid organic scintillator in D1 possesses pulse shape discrimination (PSD) properties, in that energetic protons produce light pulses with longer rise times than those of electrons (and other minimum ionizing particles). This capability allows for efficient identification of signals from recoil protons produced by fast neutrons scattering elastically off hydrogen in D1.

The D1 and D2 subsystems of the telescope are each completely surrounded by charged particle detectors (see Fig. 1). These 4 domes of plastic scintillator NE110 are 1.5 cm thick and do not significantly attenuate the incident energetic γ-ray or neutron fluxes, yet are virtually 100% efficient in identifying charged cosmic rays. The charged particle shields and other intervening material heavily attenuate the solar flare hard X-ray flux, minimizing pulse pile-up effects in the D1 and D2 detectors. They do, however, saturate under the intense soft thermal X-ray flux associated with flares. The effect of this is seen below when we discuss the behavior of the telescope during intense events.

Neutrons are also measured in the telescope mode. The ideal type of neutron interaction in COMPTEL occurs when the incoming neutron scatters elastically off a hydrogen nucleus in the D1 detector. The scattered neutron then proceeds to the D2 detector where it may interact, depositing some of its energy to produce a trigger signal as indicated in Figure 1. The energy of the incident neutron is computed by summing the proton recoil energy E_1 in the D1 detector with the energy of the scattered neutron E_s deduced from the time-of-flight (TOF) from the D1 to the D2 detector. The scatter angle for non-relativistic neutrons (< 150 MeV) can be computed by the formula:

$$\tan^2 \phi = E_1 / E_s. \qquad (2)$$

As with γ-rays, neutrons can be traced backwards from D2 to D1 through the angle ϕ to a cone mantle restricting the incident direction to include the Sun. This is a geometrical constraint identical to that of the γ-ray measurements. The pulse shape from recoil protons is sufficiently different from that of electrons to reject more than 95% of electron-recoil events greater than about 1 MeV, the energy threshold in D1 for neutron detection. The PSD and TOF criteria in this channel are set such that solar neutrons incident on D1 in the energy range from about 10 MeV to 150 MeV are recorded. In this energy interval COMPTEL can observe neutrons from about 14.5 to 55 minutes after release from the Sun. This corresponds to a minimum observed delay time of 6 to 47 minutes after the onset of the γ-ray flash (assuming neutrons are not produced without accompanying γ-rays).

The burst mode of the COMPTEL instrument can also be used to detect solar γ-rays[2]. One D2 detector module covers the energy interval from 0.1 to 1 MeV and another the interval from ~ 1 to 10 MeV. Each detector module has an unobscured field-of-view of about 2.5 sr and a physical area of ~ 600 cm^2. Outside this field-of-view varying amounts of intervening material exist attenuating the solar γ-ray flux.

OBSERVATIONS

4 June 1991

This X12+/3B flare occurred at coordinates N30E70. The X-ray flare started at 0337 UT. At this time the Sun was ~ 105° from the COMPTEL axis, well outside the telescope field-of-view. The following data come only from the high range burst detector operating from ~ 600 keV to ~ 10 MeV. The low energy range burst detector was not operating at this time. The intensity-time profile of the burst detector count rate > 600 keV is shown in Figure 2.

The burst spectrometer was activated by the trigger signal provided by BATSE at 0337 UT. The peak intensity occurred at 0341:15 UT as seen in the large reduction in the count rate produced by dead time effects. The flare continued up to at least 0350 UT at which time the burst spectrometer reverted back to its low time resolution (100 s) background cadence.

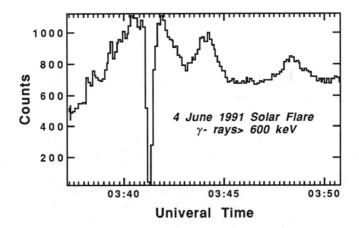

Figure 2. The uncorrected high range COMPTEL burst detector count rate from the 4 June 1991 solar flare.

9 June 1991

The γ-ray emission from this flare was measured by COMPTEL[3]. The Sun was ~ 15° off the telescope axis, well within the field-of-view. Figure 3 displays selected housekeeping data from the flare on 9 June 1991.

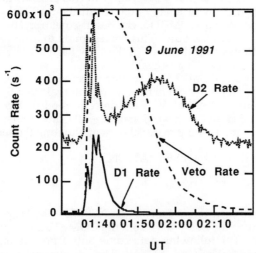

Figure 3. COMPTEL housekeeping data of the flare of 9 June 1991.

This X10/3B flare was also impulsive in nature. The GOES X-ray flare started at 0134 UT peaking at 0143 UT. The γ-ray onset occurred at 0136 UT and the impulsive phase lasted until ~ 0142 UT. Satellite sunrise precedes the flare only by a few minutes, thereby allowing a long observation of any extended flare emissions, such as neutrons. The count rates in the D1 and D2 detecting systems with thresholds of ~ 60 and ~ 400 keV, respectively, show how the flare behaved in hard X-rays and γ-rays. The slow rise and fall of the D2 counts rate after the impulsive phase arises from excursions in geomagnetic latitude and is not related to the flare. Also shown is the count rate in the Anticoincidence detection system. These detectors reject charged cosmic ray particles but are also sensitive to the thermal X-ray flux. At the X10 level, the X-ray flux falling upon the large Anticoincidence detectors results in pulse pile-up and large dead time effects for the whole instrument. The large dead time effects persist at least until 0155 UT, where the dead time from the Anticoincidence system drops to the ~ 50% level. From the γ-ray events in the telescope mode, although not shown here, an image of the sun can be constructed[3].

During the impulsive phase the γ-ray emission spectrum is obtained by selecting the telescope events consistent with the Sun's location, as shown in Figure 4. The strong line from deuterium formation is seen at 2.2 MeV. Other lines are also present. No background spectrum has been subtracted, but few photons are expected, since the background is suppressed by the large dead time factor making the signal to noise ratio large.

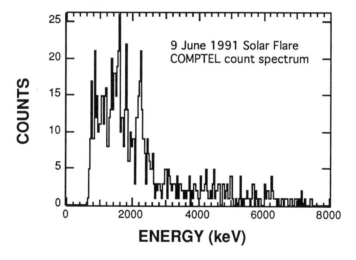

Figure 4. The raw telescope mode count spectrum from the 9 June 1991 solar flare.

Solar neutron events from 0155 UT to 0222 UT were selected from the data and reprocessed. This time selection avoids the troublesome period around the impulsive phase with large dead times and other instrumental effects. The time 0155 UT corresponds to a 50 MeV neutron produced at the flare start 0136 UT or a 60 MeV neutron from the flare maximum at 0139 UT. Some selection bias, therefore, exists for neutrons detected shortly after 0155 UT (real time). The selected data were subjected to similar geometric constraints as were the γ-rays, i.e., that the origination direction of the neutron event must be consistent with the solar direction (± 10°). The scatter angle (φ) of the individual neutron events was restricted to 20° to improve the signal-to-noise ratio. The energy of each neutron is used to compute its production time at the Sun. Shown in Figure 5 is the live time corrected and background subtracted intensity-time profile of the neutrons as produced at the Sun but plotted at the production time plus the light travel time over 1 AU (507 s).

Therefore, photons and neutrons produced simultaneously are plotted at the same time value even though the neutrons arrive later.

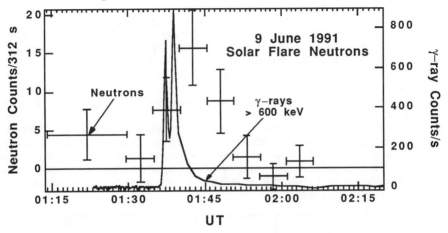

Figure 5. Neutron and γ-ray emission-time profiles for the 9 June 1991 solar flare, plotted at the time corresponding to a photon arrival time.

The background was estimated from the measured neutron flux ~ 24 hours later when the same orbital-geophysical conditions were reproduced. The on-board instrument software, however, was configured differently, being less restrictive in accepting neutron events. This resulted in a uniform 36% increase in the neutron background count rate. The background rate was scaled downward by this factor in order to provide a representative background for the flare orbit on 9 June. The background corrections were successfully tested on similar background orbits to ensure that a null result is obtained. Identical data cuts then were made on both the flare data and the prescaled background data. Live time fractions of 60% to 100% were applied to the data as a function of real time from 0155 - 0220 UT.

The most important feature to note in Figure 5 is the time coincidence of the onsets of neutron and the γ-ray production. The neutron intensity-time profile is expected to be smoother and somewhat broader than that of γ-rays because of the neutron energy resolution which maps into an error (FWHM) in production time at the Sun (~ 1-3 minutes, dependent on energy). Although the γ-rays and neutrons are observed to start simultaneously, there is evidence that the neutron emission persists for ~ 10 minutes after the > 600 keV γ-ray flux has subsided.

This extended emission is evidence for a significant evolution (hardening) of the progenitor proton spectrum, arising from either additional acceleration or differential trapping and precipitation of the protons. As seen in Figure 4 a large fraction of the γ-rays > 1 MeV are of nuclear origin. With no evolutin of the proton spectrum, we would expect the > 600 keV γ-ray flux to follow the neutron production profile. However, a hardening of the spectrum would enhance the neutron emissivity with respect to that of the nuclear γ-rays. It should be noted, though, that the primary electron bremsstrahlung component has not yet been separated from the nuclear component, so that part of the decay of the flux > 600 keV could result from the decay of the pure electron component of the γ-ray flux.

The trapping scenario is consistent with the > 50 MeV γ-ray flux detected by EGRET after 0145 UT[4]. The net neutron count rate from 0136 UT to 0150 UT is positive at the 4.2 σ significance level.

11 June 1991

This flare also occurred in region 6659 at heliographic coordinates N31W17. As with the 9 June solar flare this flare occurred shortly after satellite sunrise within the COMPTEL field-of-view giving the longest available period of γ-ray and neutron measurements. This event like the others produced intense emission of γ-rays. The impulsive phase extended from 0158 UT until approximately 0210 UT, after which there was emission lasting until at least 0225 UT. The background subtracted burst mode intensity-time profile is shown in Figure 6a. In Figure 6b is the integrated count rate (not photon) spectrum from the telescope mode clearly showing a strong 2.2 MeV emission line and detection of photons up to ~ 7 MeV.

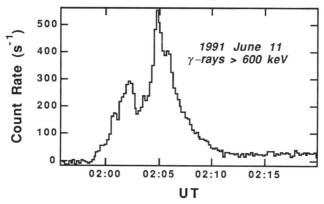

Figure 6a. The intensity-time profile of the 11 June 1991 solar flare measured by the COMPTEL burst spectrometer (> 600 keV).

Kanbach et al.[5] reported that this flare emitted γ-radiation > 50 MeV for a period of ~ 10 hours with the radiation decaying away with a time constant of ~ 25 minutes for 2 hours following the flare, after which the flux then decayed more slowly with a time constant of 4.5 hours. As with the EGRET observations, COMPTEL measures prolonged γ-ray emission as evidenced by the image of the Sun during the first post-flare orbit using all data in the narrow 2.2 MeV band (Figure 7).

In Figure 8 we show the intensity-time profile of the 2.2 MeV emission during the flare orbit having selected data consistent with the solar position[6]. The impulsive phase is excluded from this diagram. The thermal X-ray flux and the intense γ-ray flux make the estimates of the 2.2 MeV count rate highly uncertain before 0213 UT. The emission at 2.2 MeV falls off with a decay time of ~ 10 minutes. The > 50 MeV flux decay constant of 25 minutes has a large uncertainty and should be considered an upper limit[7]. The 10 minute decay constant measured here for 2.2 MeV γ-rays may, therefore, be consistent with this number.

The high energy radiation seen by EGRET > 50 MeV and the 2.2 MeV emission seen by COMPTEL both falling off with similar time constants point toward a common origin, i.e. a nuclear origin for both forms of the γ-radiation. If this is so, then the > 50 MeV radiation stems from pion decay (charged and neutral) produced by nuclear interactions and not by primary electron bremsstrahlung as concluded by Mandzhavidze and Ramaty[8]. The neutrons which also arise from these interactions manifest in the deuterium formation line at 2.2 MeV. It should be noted that the time scale for the thermalization and capture of the neutrons (~ 100 s) is short compared to the time constants of the γ-ray emission (~ 600 s), so that the 2.2 MeV flux represents the "instantaneous" neutron production rate.

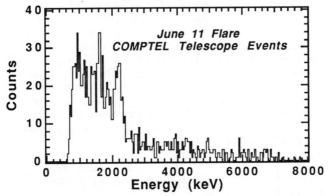

Figure 6b. The time integrated count spectrum measured in the telescope mode.

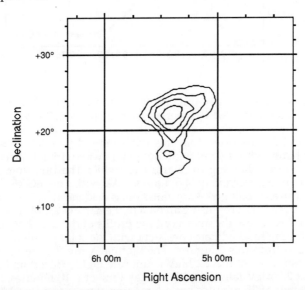

Figure 7. The COMPTEL image of the Sun in the emission around the 2.2 MeV line, from data later than ~ 0315 UT.

15 June 1991

The X12+ solar flare on 15 June was located at N33W69, peaking in X-rays around 0820 UT. During this time the Compton Observatory was behind the earth unable to measure the solar γ-ray flux. Upon entering daylight, the spacecraft found itself in the South Atlantic Anomaly (SAA). Only after exiting the SAA was COMPTEL able to acquire data on the remaining flare emission. This flare was also measured by the GAMMA-1 instrument[9] and was reported to have emission > 30 MeV after ~ 0837 UT lasting for at least 20 minutes after the X-ray maximum. The COMPTEL measurements began at ~ 0859 UT, while the GAMMA-1 measurements ended at 0902 UT. The COMPTEL instrument continued making measurements until 0930 UT. During this time there was significant emission in the MeV energy range[3] declining in intensity with a decay time of 830 ± 100 s as shown in Figure 9.

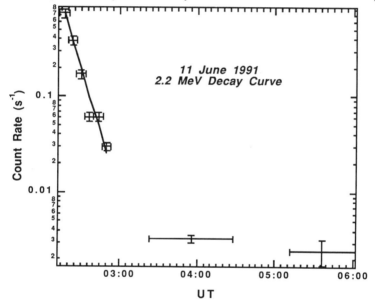

Figure 8. The count rate of the background subtracted 2.2 MeV emission from the 11 June 1991 solar flare. Also shown is the best exponential fit to the decay of the line in the flare orbit (10.2 minutes).

The γ-ray spectrum (or 2.2 MeV/4 - 7 MeV emission ratio) is typical of many solar flares in which the impulsive phase was observed indicating that, at least up to 40 minutes after the flare, there was no appreciable change in the shape of accelerated proton spectrum. Neutrons were also measured during this time[10] and reflect nuclear activity at times before the γ-ray emission measured by COMPTEL, but coincident with the GAMMA-1 measurements. In Figure 10 the measured intensity-production time profile of the detected neutrons is background corrected.

As with Figure 5, the points are plotted at the arrival time of photons created simultaneously with the neutrons. The curve suffers from velocity dispersion effects, in that the counts at the earliest times in the plot are the slowest neutrons, while the counts at the later times come from higher energy neutrons. The time intervals do not sample uniformly the available neutron energy spectrum. The plot does show, however, that neutron emission occurred at times near the X-ray maximum and extended for over an hour. This coupled with the detection of γ-rays up to 0930 UT means that proton acceleration or precipitation persisted for at least 70 minutes.

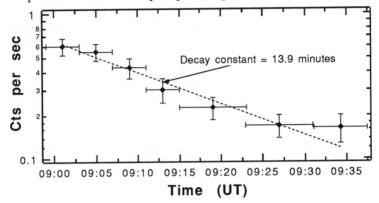

Figure 9. The γ-ray intensity-time profile of the 15 June 1991 solar flare for energies from 1 - 10 MeV. The decay is fit to a single exponential decay with time constant 830 ± 100 s.

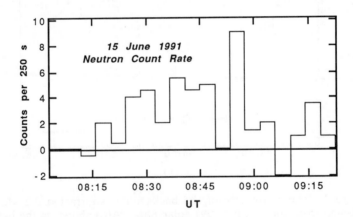

Figure 10. The background subtracted neutron production-time profile from the 15 June 1991 solar flare.

The validity of the neutron data is demonstrated by the neutron image of the Sun produced from the full neutron data set (0859 - 0930 UT)[10]. The neutron events consistent with the solar location in the image are the ones used to construct the production-time profile in Figure 10.

30 June 1991

This flare occurred well outside the field-of-view of COMPTEL (~ 70° off axis), but was successfully measured in the burst mode. The background subtracted intensity-time profile is shown in Figure 11.

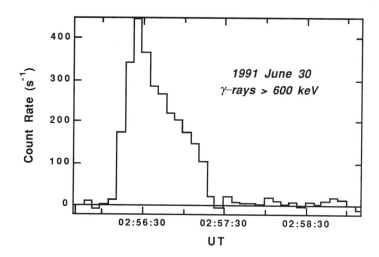

Figure 11. The background subtracted intensity-time profile for the 30 June 1991 solar flare.

The background subtracted-integrated count rate spectrum is plotted in Figure 12, indicating emission beyond the nuclear part of the spectrum. We suspect that this indicates there was a significant number of primary electrons > 10 MeV radiating via bremsstrahlung to produce the intensity at the higher end of the spectrum.

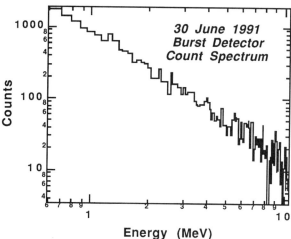

Figure 12. The count rate spectrum from the solar flare of 30 June 1991.

CONCLUSIONS

Several large and intense events were observed by COMPTEL during June 1991 in both the telescope and burst modes. We have evidence of extended emissions of γ-rays and neutrons on a time scale of hours after the X-ray maximum. From the 9 June 1991 flare we have neutron evidence for prolonged high energy emission from a flare which would otherwise have been considered impulsive. We have evidence of significant proton spectral evolution (hardening) in two flares (9, 11 June 1991) while the data of the 15 June 1991 flare shows no such evolution. We have produced the first γ-ray and neutron images of the Sun, demonstrating the high signal-to-noise ratio of the measurements. The intensity of the flares, the location of the Sun far away from the telescope axis, and the incomplete knowledge of the neutron response function mean that there is much work to be done in removing instrumental effects from the data. The final results should illuminate the phenomena reported above, yielding insights into the nature of proton acceleration and storage in the solar atmosphere.

REFERENCES

[1] V. Schönfelder et al., Astrophys. J. (Suppl.) accepted for publication (1992).
[2] C. Winkler et al., Adv. Space Res. 6 113-117 (1986).
[3] M. McConnell et al., Adv. Sp. Res., to be published (1992).
[4] G. Kanbach (private communication).
[5] G. Kanbach et al., Astron. Astrophys. in press (1992).
[6] G. Rank et al., Compton Centennial Symposium Proc., to be published (1992).
[7] G. Kanbach (private communication).
[8] N. Mandzhavidze and R. Ramaty, Astrophys. J. (Lett.) (1992).
[9] V. V. Akimov et al., in Proceedings of the 22nd International Cosmic Ray Conference, (Dublin, Ireland, 1991), 3, 73.
[10] J. Ryan et al., Adv. Sp. Res., to be published (1992).

HIGH ENERGY PROCESSES IN SOLAR FLARES

Reuven Ramaty and Natalie Mandzhavidze[1]
Laboratory for High Energy Astrophysics
NASA/GSFC, Greenbelt, MD 20771

ABSTRACT

We review the highlights of gamma ray and neutron observations from solar flares and their relevance to particle acceleration at the Sun. We also discuss the overall energetics of solar flares and present a possible scenario for the origin of the accelerated particles in impulsive and gradual flares.

OVERVIEW

Unlike cosmic sources of high energy emission whose luminosities peak in the hard X-ray or gamma ray bands (e.g. gamma ray bursts, Galactic black hole candidates, AGNs and some quasars), the radiative output of solar flares is maximal in a band ranging from optical to EUV energies (Figure 1). Nevertheless, there are compelling arguments, based on the overall flare energetics and the impulsiveness of the hard X-ray and gamma ray emissions, which indicate that flares are fundamentally high energy phenomena. The transition from thermal flare radiation to nonthermal high energy emission is thought to occur around 20 keV. While below this energy the emission is produced by hot flare plasma, at higher energies the emissions are nonthermal, resulting from interactions of flare accelerated particles with the ambient solar atmosphere. The energy deposited in the solar atmosphere by the accelerated particles which produce these hard X-rays and gamma rays is quite large. The energy contained in ions of energies greater than several MeV, determined[3] from gamma ray line observations, amounts to several percent of the total flare energy. More energy could be contained in ions of energies less than 1 MeV, but since in this energy range the ions do not produce radiations directly, their energy content cannot be reliably calculated. The nonrelativistic electrons which produce the observed hard X-rays via bremsstrahlung could contain an even larger amount of energy, again depending on the low energy cutoff of the nonthermal electron distribution. The energy content in these electrons could approach the total available flare energy, depending on whether the hard X-rays around 20 keV are mostly thermal or nonthermal[4,5].

The study of solar flares can thus provide basic information on impulsive energy release and particle acceleration in astrophysics. The aim of the present article is to provide an overview of the information that has already been learned on high energy particles in solar flares, emphasizing observations of gamma rays, hard X-rays and neutrons. Complementary to these data are observations of the accelerated particles which escape from the Sun and are detected in interplanetary space. Particle acceleration is a wide spread process occurring at many astrophysical sites. But only in the case of the Sun can both the high energy emissions and the particles which produce them be simultaneously observed.

---

[1] NAS/NRC Res. Research Assoc.; also at the Inst. of Geophysics, Tbilisi, Georgia

Gamma rays from solar flares (at ~ 500 keV) were first detected in 1959 with a detector flown on a balloon[6] and gamma ray lines were first observed[7] in 1972 with a NaI(Tl) scintillator on OSO-7. But it was not until 1980 that routine observations of gamma ray lines and continuum became possible with the much more sensitive NaI(Tl) Gamma Ray Spectrometer (GRS) on SMM[8,9]. The SMM/GRS detector operated successfully until 1989, making important observations during both the declining portion of solar cycle 21 (1980-1984) and the rising portion of cycle 22 (1988-1989). Additional gamma ray observations were carried out with a CsI(Tl) spectrometer on HINOTORI[10]. Neutrons from solar flares were also detected with SMM/GRS[11]. In addition, solar flare neutrons were observed indirectly via the protons resulting from their decay in interplanetary space[12] and the secondary cascades produced by the neutrons in the atmosphere (see[13]). Currently, solar flare gamma ray emission is detected with all four CGRO instruments, as well as with instruments on GRANAT[14] and YOHKOH[15]. Solar flare gamma rays were also detected[16] in 1991 with GAMMA-1, which is no longer operational. Among the highlights of these new observations are the detection of gamma rays of energies up to a GeV with both GAMMA-1 and EGRET[17], and the direct measurement of neutron energy spectra with COMPTEL[18].

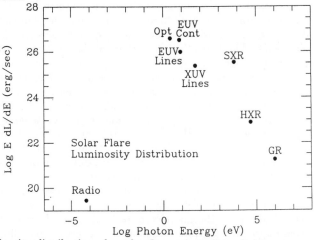

Fig. 1. Luminosity distribution of a solar flare, where E is photon energy and dL/dE is differential luminosity. The radio to soft X-ray points represent luminosities at flare maximum for the 5 September 1973 flare[1]. There are no hard X-ray and gamma ray observations from this flare. The hard X-ray luminosity, plotted at 50 keV, was normalized to the lower energy data such that the ratio of peak hard X-ray luminosity at energies greater than 20 keV to the total peak luminosity equals $Y(t_{ther}/t_{nt})$, where Y is the neutral gas bremsstrahlung yield at 30 keV and $t_{ther}/t_{nt} = 10$ is a typical ratio of the durations of the thermal and nonthermal emissions (e.g. soft and hard X-rays) in an impulsive flare. The gamma ray luminosity, plotted at 1 MeV, was normalized to the X-ray luminosity at 50 keV using observations of the 27 April, 1981 flare[2].

The first evidence for particle acceleration in solar flares came from observations with ground based cosmic ray monitors[19,20] which recorded increased particle fluxes in association with large solar flares in 1942 and 1946. Subsequently, many direct observations of flare accelerated particles determined their properties[21-25]. Their energy spectrum has been measured over a broad energy range, from below 1 MeV to several

GeV and their elemental composition has been determined. The observed composition is highly variable, changing from flare to flare, and with time within a flare. It is thought that these variations are caused mostly by the acceleration process; however, the variations could also reflect variations in the composition of the ambient solar atmosphere from which the particles are accelerated. The observations have also determined the isotopic composition of He, showing very large variations of ^{3}He/^{4}He from one flare to another. It has been shown (see[25]) that enhanced abundances of elements heavier than O and enhanced abundance ratios of ^{3}He to ^{4}He distinguish acceleration in compact, impulsive flares from acceleration in spatially extended, gradual flares. The charge states of some of the ions have also been measured[26], indicating that in impulsive flares the particles are accelerated from flare plasma at $\sim 10^7$K, as opposed to acceleration in gradual flares, where the charge states are consistent with coronal gas at only $\sim 10^6$K. Electrons of energies extending to tens of MeV have been observed, showing[22,27-29] that the ratio of electrons to protons (e/p) is much higher in impulsive flares than in gradual flares.

Several of the characteristics that distinguish impulsive flares from gradual flares, (e/p), heavy element abundances, and α particle to proton (α/p) and ^{3}He to ^{4}He ratios, can also be derived from gamma ray observations for the particles which interact at the Sun. The analysis[30] of the gamma ray spectrum of the 27 April 1981 flare revealed very significant enhancements of the abundances of heavy elements, particularly Fe. This analysis also showed that α/p, and probably also ^{3}He/^{4}He are enhanced. Values of e/p can be determined from the observed gamma ray continuum and prompt line fluences, as the former is bremsstrahlung from relativistic electrons and the latter is nuclear deexcitation emission produced by ion interactions. We have shown[31] that the values of e/p derived from gamma ray observations of both impulsive and gradual flares are at least as high, and probably higher than the e/p's of the accelerated particles observed in interplanetary space from impulsive flares. We have therefore suggested[31] that the particles which produce gamma rays in both impulsive and gradual flares are accelerated by the same mechanism as that which accelerates the particles observed in interplanetary space from impulsive flares. This mechanism is probably stochastic acceleration by plasma turbulence, as such acceleration is the only one that so far has been shown to be capable of producing the observed abundance anomalies[32,33].

Based on these and other arguments, we have presented[34] the following schematic picture for particle acceleration in solar flares (Figure 2). Flares are probably triggered by magnetic reconnections caused by the rearrangement of existing magnetic fields and the emergence of new magnetic fluxes. Reconnection produces electric fields which heat the ambient plasma and accelerate electrons (possibly also ions) up to energies of few tens of keV. These electrons, as well as the heated plasma, are partially responsible for the observed hard X-ray and microwave emissions. Because of the low efficiency of nonthermal bremsstrahlung relative to collisional losses, the number of energetic electrons required to produce the observed hard X-rays is very high. The direct electric fields, however, are not expected to accelerate particles to energies much higher than about 100 keV, although this possibility can certainly not be ruled out. On the other hand, the streams of runaway electrons produced by the electric fields could generate plasma turbulence, which further stochastically accelerates particles to higher energies. The resonant interactions of the plasma waves with the particles could, in principle, account for the ^{3}He and heavy element enhancements observed in impulsive flares[32,33,35]. The majority of the particles remain trapped in closed magnetic structures (loops) where they produce hard X-rays, gamma rays and neutrons. The small fraction of particles that escapes into interplanetary space, along with the particles directly accelerated on open field lines, constitute the particles detected in space. This is a scenario for the purely impulsive flare. There are other particle acceleration phenomena as well. Coronal mass ejections[36] (CME) could drive shocks through the corona and interplanetary space, and these shocks will accelerate particles. The associated gradual particle events exhibit typical coronal abundances and charge states, do not show ^{3}He enhancements, and are characterized by large proton fluxes which are the primary reason for the observed low electron-to-proton ratios. The two scenarios correspond to the extremes of purely impulsive and purely gradual flares. The majority of the observed events probably are mixtures of these two. Depending on which one prevails, the flare is classified as impulsive or gradual.

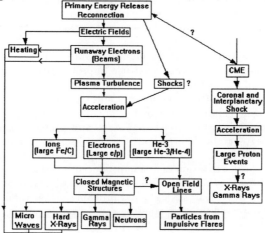

Fig. 2. Schematic particle acceleration in solar flares.

In the next section we present the fundamentals of solar flare gamma ray and neutron production, along with the key conclusions obtained from the analysis of these emissions.

SOLAR FLARE GAMMA RAYS AND NEUTRONS

<u>Continuum.</u> Interactions of the electrons with ambient gas in the flare region produce continuum X-ray and gamma ray emission via nonthermal bremsstrahlung. This continuum extends from about 20 keV to over 100 MeV. At the low energy end it merges into the thermal bremsstrahlung produced by hot flare plasma. There is no known high energy cutoff; the highest energy observed[16] bremsstrahlung is around several hundreds of MeV. Bremsstrahlung produced by ultrarelativistic electrons is strongly beamed along the direction of motion of the electrons, and thus reflects the angular distribution of the interacting electrons. At lower energies, the angular spread of the bremsstrahlung is broader. Moreover, the interacting electrons are expected to be much less anisotropic because Coulomb scattering will tend to isotropize the electrons before they lose much energy to bremsstrahlung. In addition, below about 300 keV there is a significant contribution to the observed emission from Compton backscattering in the photosphere[37].

SMM/GRS observations of gamma rays above 300 keV in solar cycles 21 and 22 revealed that the distribution of the flare positions on the Sun shows limb brightening over the distribution expected for isotropic emission[38,39]. This limb brightening effect is even more pronounced[9,38,39] at energies above 10 MeV, especially in the cycle 21. In cycle 22, however, >10 MeV gamma ray emission was also observed from several disk flares. In Figure 3 we show the heliocentric angle distribution of 28 flares for which >10 MeV gamma ray emission was observed with SMM/GRS. Also shown is the calculated[34] angular distribution expected for isotropically emitting sources. In this calculation, we assumed that the integral flare size distribution is inversely proportional to the flare size and that the flares are uniformly distributed on the Sun between heliolatitudes ±40°. The probability that the observed distribution is consistent with isotropic emission is less than 10^{-3} for the cycle 21 and is 0.24 for cycle 22. Thus, the hypothesis of isotropic emission in cycle 22 cannot be ruled out. Recently, more gamma ray flares with emission above 10 MeV were detected with instruments on GRANAT, GAMMA-1 and CGRO (see Table 2 in[34]). Because the sensitivities of these instruments are different from that of SMM/GRS, we did not include these flares into the distribution given in Figure 3. However, the fraction of disk events is also quite high among these flares. We should also note that during the rising portion of cycle 21, stereoscopic observations[40] with PVO and ISEE-3 in the 0.1-1 MeV range revealed essentially no anisotropy.

It has been shown that the angular pattern of bremsstrahlung produced in magnetic loops[41–45] is sufficiently anisotropic to account for the observed flare distribution in cycle 21. On the other hand, for certain loop geometries the emission could be much less anisotropic. This can occur in either low lying dense loops, in very large loops with high mirror ratios, or in tilted loops. Thus, the difference between the heliocentric angle distributions of flares with >10 MeV emission in cycles 21 and 22 could reflect changes in the conditions of particle accelerations and/or trapping, most likely related to the changes in the magnetic field structure.

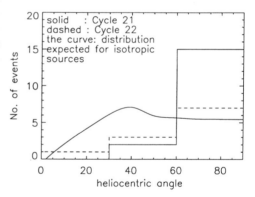

Fig. 3. Heliocentric angle distributions of flares with >10 MeV gamma rays detected with SMM/GRS.

The observed continuum spectra of solar flares show considerable structure. Two examples[15] are shown in Figure 4. The solid curves are isotropic thick target bremsstrahlung spectra[31] calculated to fit the observations in the 0.3-1 MeV range. While this isotropic interaction model is obviously an over simplification, the comparison of the calculated spectra with the data does bring out the essential features of the structure of the observed spectrum. The most obvious one is the nuclear emission seen in panel (a) in the range 1-8 MeV. This will be discussed in the next section. However, as can be seen in panel (b), there are flares for which the nuclear lines cannot be seen because of strong bremsstrahlung emission in the MeV region. This 'electron rich' phenomenon was first indentified[46,47] in 3 emission episodes during the 6 March 1989 flare. We have derived[31] the ratio of the electron flux at 0.5 MeV to the proton flux at 10 MeV ($J_e(0.5)/J_p(10)$) for 10 flares and for 6 individual emission episodes during the 6 March flare, including the 3 electron rich episodes. We found that the lower limit on the average $J_e(0.5)/J_p(10)$ for the events with no nuclear lines exceeds the average $J_e(0.5)/J_p(10)$ for the events with lines by about a factor of 3. There is currently not enough data to decide whether the distribution of $J_e(0.5)/J_p(10)$ is bimodal, indicating different acceleration mechanism for flares with and without nuclear lines, or whether the high values of $J_e(0.5)/J_p(10)$ in electron rich events simply represent extreme cases in a continuous distribution of electron to proton flux ratios.

The other prominent feature seen in the spectra of Figure 4 is the break at about 300-400 keV, representing a flattening of the continuum with increasing energy. Similar breaks have also been reported for the 4 June 1980 flare[48] and the 27 April 1981 flare[2]. While possibly a common feature of many flares, the nature of the transition between the hard X-ray regime below about 300 keV to the soft gamma ray regime above 400 keV is poorly studied because it represents the transition between instruments optimized for hard X-ray studies (e.g. SMM/HXRBS) and those optimized for gamma ray observations (e.g. SMM/GRS). The spectrum in panel (b) shows no further structure, while that in panel (a) suggests that the continuum around 50 keV is simply the continuation of the higher energy continuum with a possible 'bump' in the range 70-300 keV. Only one solar flare (29 June 1981) has so far been observed[49] with a high resolution Ge detector. The observed spectrum of this flare showed a strong steepening around 60 keV, which would not be inconsistent with the data shown in Figure 4 (a) obtained with much coarser energy resolution.

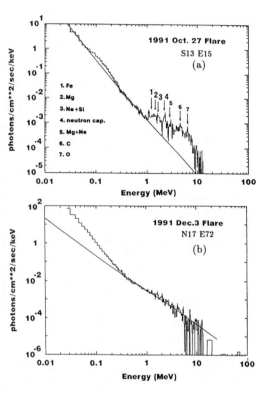

Fig. 4. Hard X-ray/gamma ray spectra observed with HXS-GRS on YOHKOH[15]

For the only two flares (21 June 1980 and 3 June 1982) for which there are published data[38,50] on the continuum below 1 MeV and above 10 MeV, the extension of the low energy continuum to energies greater than 10 MeV shows that the spectrum either maintains its slope or steepens somewhere between 1 to 10 MeV. This is in contrast with the only reported[46] case of flattening around 1 MeV. However, since the spectrum which showed the flattening was a count rate spectrum, we assume, more conservatively, that there are no further flattenings in the solar flare bremsstrahlung spectrum above 400 keV. The only flattening above 10 MeV is caused by pion decay emission, which is discussed below.

Line Emission. Nuclear deexcitation lines result from the bombardment of ambient C and heavier nuclei by accelerated protons and α particles, and from the inverse reactions in which ambient hydrogen and helium are bombarded by accelerated carbon and heavier nuclei[51]. Because of their low relative abundances, interactions between accelerated and ambient heavy nuclei are not particularly important and were ignored in past research. Furthermore, since H and He have no bound excited states, p-p and p-He interactions can also be ignored. However, interactions of α particles with ambient He produce two strong lines, at 478 keV from ^{7}Li and at 429 keV from ^{7}Be.

Neutron production in solar flares was studied in detail[52-56]. The neutrons can penetrate into the photosphere where the density is high enough for the neutrons to be captured before they decay. Most of the captures are on H and ^{3}He. Capture on H produces the 2.223 MeV line, which is the strongest line from solar flares, except for flares very close to the limb of the Sun. For limb flares the 2.223 MeV line is strongly attenuated by scattering in the photosphere[55,57]. The other nuclear lines originate higher in the atmosphere and are attenuated only for flares behind the solar limb[58].

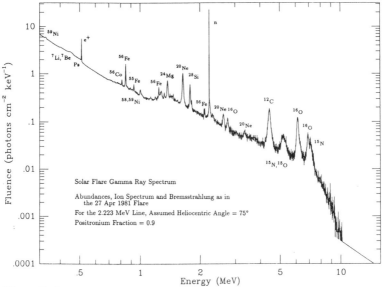

Fig. 5. Theoretical gamma ray spectrum.

Positrons in solar flares result from the decay of radioactive nuclei and charged pions[59]. The 511 keV line resulting from positron annihilation has been observed from several flares, and its delayed nature, due to the finite lifetimes of the radioactive nuclei, has been confirmed[60]. Some of the positrons form positronium before annihilation, which, in addition to annihilating into 511 keV line photons, also produces a characteristic 3-photon continuum[61]. This continuum has not yet been observed from solar flares, even though it has been seen with high resolution gamma ray spectrometers from the direction of the Galactic center. There are two reasons for this: (i) solar flares have not yet been studied in detail with Ge spectrometers; (ii) the ratio of the solar flare 511 keV line flux to the underlying continuum flux is much smaller than the corresponding ratio for radiation from the direction of the Galactic center. This second reason would render the observability of positronium continuum from solar flares quite difficult, except at late times in a flare, when the strong bremsstrahlung continuum has disappeared but positrons are still generated from the longer lived radioactive emitters.

Observations of gamma ray line emission from solar flares have been used in the following investigations: (i) abundance determinations from the analysis of nuclear deexcitation line spectra[30]; (ii) determinations of ion energy spectra from line fluence ratios[31,55]; (iii) determination of the accelerated electron to proton ratios from line to continuum fluence ratios[31]; (iv) determination of the number of interacting ions and the energy contained in them[3,55]; (v) determination of the photospheric ^{3}He/H ratio from the time dependence of the 2.223 MeV line flux[62]; and (vi) studies of the angular distribution of the interacting ions from gamma ray line shapes[63].

Here we briefly summarize the highlights of the ambient gas abundance determination. Normalized to O, the abundances of Mg, Si and Fe are consistent with coronal but not photospheric abundances; C is consistent with both photosphere and corona; and Ne is enhanced relative to both the photosphere and the corona. The coronal enrichment of Mg, Si and Fe, elements of low FIP (first ionization potential), is thought to be due to a charge dependent transport process which favors the collisionally ionized, low FIP elements in the photosphere. The enrichment of Ne (a high FIP element) could be due to photoionization by soft X-rays[64]. Many unresolved questions, however, remain on the nature of the processes responsible for these abundance variations.

We show in Figure 5 a theoretical solar flare gamma ray spectrum derived for the parameters of the 27 April 1981 flare, except for the 2.223 MeV line whose fluence was obtained by calculating[31] neutron production with the parameters of this flare and assuming that the heliocentric angle of the flare is 75°. This latter assumption is for illustration purposes only; the actual heliocentric angle of the 27 April flare was around 91°. The rich gamma ray spectrum embodied in this calculation awaits study with high resolution, high sensitivity detectors.

<u>Pion Decay Radiation.</u> In the energy range above 10 MeV, along with the bremsstrahlung from primary electrons, there can also be a significant contribution from pion decay radiation. Neutral and charged pions are produced mostly in high energy p-p, p-He and He-He interactions. Neutral pions decay directly into two photons, while charged pions decay (via muons) into secondary electrons and positrons which produce gamma rays via bremsstrahlung and annihilation in flight. The radiation from these secondary electrons dominates the pion decay emission at energies below 30-40 MeV, while the contribution of the neutral pions is dominant above about 100 MeV. Pion decay radiation carries information on the high energy part of the accelerated ion spectrum, above the pion production threshold($\sim$ 200-300 MeV). The combination of this information with that derived from nuclear line emission allows us to determine[54,56,65] the energy spectrum of the interacting ions over a broad band, from about 10 MeV up to few GeV. The mechanisms of producing pion decay emission have been considered in detail[54]. The problem was further studied[66] recently, taking into account the effects of particle transport in solar flare magnetic loops and radiative transfer. It was shown that, although the emission is slightly anisotropic, the degree of anisotropy is not enough to account for the heliocentric angle distribution of flares with >10 MeV emission observed in cycle 21, indicating that for the majority of these flares the emission was dominated by primary electron bremsstrahlung.

Pion decay emission was first observed with SMM/GRS during the 3 June 1982 flare[67]. The characteristic feature that distinguishes pion decay radiation from primary electron bremsstrahlung is the flattening of the spectrum around $\sim$ 70-100 MeV due to the contribution of the π° component. Pion decay radiation was subsequently observed with SMM/GRS from 3 more flares (see Table 1). While the SMM observations were limited to energies below ~ 140 MeV, the recent EGRET and GAMMA-1 observations revealed for the first time emission at energies above 1 GeV during the 11 June 1991 and 15 June 1991 flares[16,17]. In both cases, the measured energy spectrum shows clear evidence of pion decay emission. The 11 June 1991 event was studied in detail[65]. The striking feature of this event was the extremely long duration of the GeV emission, lasting for up to 8 hours after the beginning of the flare. The observed time profile and energy spectrum of the 11 June 1991 event in the 30-2000 GeV energy range were modeled[65] with the combination of the pion decay emission and primary electron bremsstrahlung, assuming that the particles were accelerated impulsively at the beginning of the flare and subsequently trapped in magnetic loops. The required conditions for long term trapping are matter density not higher than 5×10^{11} cm^{-3} in the coronal portion of the loop (to prevent Coulomb and nuclear losses), low magnetic field (to prevent synchrotron losses for the primary electrons), relatively high mirror ratios (to prevent the bulk of the particles from penetrating into the photosphere), low plasma turbulence (to prevent the scattering of the particles into the loss cones), and large and/or twisted loops (to prevent the particles from drifting out of the loops). The model predicts that the primary electrons, as well as the ions that produce nuclear lines, lose their energy faster than the high energy ions that produce pions. Therefore, it is expected that the time profiles of the nuclear line emission and primary electron bremsstrahlung will decay faster than that of the pion decay radiation. This was in fact confirmed by OSSE[73] and EGRET[17] observations of nuclear lines and 50-70 MeV continuum, providing strong support for trapping. The power law spectral index of the accelerated ions that provides[65] a good fit to the EGRET pion decay emission data is essentially the same as that derived[14] from the fluence ratios of the deexcitation and 2.223 MeV lines measured with SIGMA during the impulsive phase of the 11 June 1991 flare. This means that the ion spectrum for this flare can be described by a single power law with an exponent of 3.5 in the energy range from about 10 MeV up to at least 5 GeV.

TABLE 1
Flares for which Pion Decay Emission and Neutrons were Detected

Date	Position	Pions	Neutrons
21 Jun 1980	W90N20		SMM/GRS[11]
			ISEE-3[12] (NDP)
3 Jun 1982	E72S09	SMM/GRS[11]	SMM/GRS[11]
			ISEE-3[12] (NDP)
24 Apr 1984	E43S12	SMM/GRS[68]	SMM/GRS[68]
			ISEE-3[12] (NDP)
16 Dec 1988	E37N26	SMM/GRS[69]	SMM/GRS[71]
6 Mar 1989	E69N35	SMM/GRS[70]	SMM/GRS[70]
4 Jun 1991	E70N30		CGRO/OSSE[73]
6 Jun 1991	E44N33		CGRO/OSSE[73]
9 Jun 1991	E04N34		CGRO/COMPTEL[18]
11 Jun 1991	W17N31	CGRO/EGRET[17]	
15 Jun 1991	W69N33	GAMMA 1[16]	CGRO/COMPTEL[72]

We are currently analyzing the 30-4000 MeV GAMMA-1 data for the 15 June 1991 flare. Similar to the 11 June 1991 flare, this emission was also quite extended, lasting for at least 2 hours. The measured energy spectrum can be well modeled with pion decay emission, although in this case the ion power law spectrum is steeper than that found[65] for the 11 June event. Nuclear lines were also observed from this event with COMPTEL[75]. Incorporation of these data in the analysis will allow us to constrain the parameters of the accelerated particles and the loop model. In particular, we should be able to distinguish between a single power law ion spectrum, and more complex spectra such as a power law with an exponential cutoff at high energies or a Bessel function.

Neutrons. We have already mentioned the general studies of neutron production. The effects of accelerated particle transport in the magnetic loops on the time profiles, energy spectra and angular distributions of the high (>50 MeV) neutrons were also studied[56]. Similar to the ultrarelativistic electron bremsstrahlung, the high energy neutron emission can also be significantly anisotropic[56]. However, because of the relatively small number of flares from which neutrons were detected, at this point it is not possible to arrive at conclusions regarding the anisotropy of the neutron emission.

Neutron production in solar flares was first indicated by observations of the 2.223 MeV line which we have already discussed. The neutrons that contribute to this line have mostly energies below ~ 100 MeV. There are 5 flares from which neutrons were directly detected with SMM/GRS which was sensitive to neutrons with energies from about 50 to a few hundreds of MeV. During 3 out of these 5 flares, neutron decay protons (NDP) in the energy range of 20-200 MeV were also observed on ISEE-3 (see Table 1).

Recently neutrons were directly detected with OSSE and COMPTEL during the 4,6,9 and 15 June 1991 flares, although quantitative information on the neutron fluxes is currently available[18] only for one of these flares (9 June 1991). Compared to SMM/GRS, OSSE and COMPTEL are sensitive to the lower energy neutrons (>15 MeV). The advantage of COMPTEL is that so far it is the only instrument in space that can directly measure neutron energy. The construction of neutron energy spectra from SMM/GRS data was based on the Sun-Earth transit time method. This method can introduce a certain ambiguity, unless the time of neutron production at the Sun is well known, as in the case of the 3 June 1982 flare when it was possible to tie high energy neutron production rates to the measured time profile of the pion decay emission[67]. This may have been the reason that, while the energy spectrum of neutrons obtained for the 3 June 1982 flare is in good agreement with theoretical calculations[56], discrepancies with theory were found[69,70] for the 16 December 1988 and 6 March 1989 flares. Consideration of particle transport effects, including possible differences between the time profiles of nuclear deexcitation lines, neutrons and pion decay emission, may help to remove these discrepancies.

There are a number of studies dedicated to the search for relativistic solar neutrons in neutron monitor (NM) data (for example[76]). The two strongest ground level neutron events have been observed recently at several North American NMs during the 24 May 1990 flare[77] and at the Haleakala NM during the 22 March 1991 flare[78]. Some of the events identified with NMs were undoubtedly due to the solar neutrons, for example the 3 June 1982 event [67,79,80] and the 24 April 1984 event[76,81]. During these flares the neutrons were simultaneously observed by several sufficiently well located NMs, and neutrons, NDPs, as well as gamma rays, were also directly observed with detectors on spacecraft. However, we must note that in general, a certain ambiguity is always present in the identification of solar neutrons from the NM data.

Recently, a new type of ground based neutron detector, a neutron telescope, was installed at Mt. Norikura[82]. Compared to NMs, this telescope has the advantages of better (10 s) time resolution, directivity and the ability to measure neutron energy.

The first detection of neutrons with this instrument was reported for the 4,6,9 and 11 June 1991 flares[83], and neutron fluxes were given for the 4 June 1991 flare[84]. Solar neutron signals were also detected from the 4 June 1991 flare with the Mt. Norikura NM[85]. The authors attempted to derive a neutron energy spectrum from this observation, assuming that the neutrons were produced instantaneously. The resulting spectrum shows a hardening at energies above a few hundreds of MeV, which is in contradiction with theoretical calculations. A possible reason of this discrepancy can be due to the time extended production of neutrons, as is expected in the magnetic loop model[56]. Time extended production of neutrons in this flare is also consistent with the Akeno Giant Air Shower observations, which provide evidence for the possible detection[86] of neutrons with energies above 10 GeV. We also mention the first report[87] of the possible detection of relativistic neutron decay protons during the 19 October 1989 flare with Canadian NMs.

CONCLUSION

As we have seen, much can be learned from gamma and neutron observations on accelerated particles in solar flares. However, many basic questions still remain. The most important of these is the nature of the acceleration mechanism and its relationship to the basic energy release process. Other questions concern the magnetic configurations which trap the particles, or, conversely, allow them to escape. Magnetic loops most likely play a dominant role in both the acceleration and transport. Particles trapped in closed magnetic loops produce the bulk of the observed gamma rays and neutrons by interacting with gas throughout the loop. What fraction of the particles escape from the loops to interplanetary space, and the dependence of this escape fraction on particle energy and type (e.g. differences between electrons and protons) are not known. Simultaneous observations of gamma rays, neutrons and interplanetary particles from more flares are required to answer these questions. And finally, we should emphasize again, that the information obtained from the solar observations should be incorporated into the modeling of the cosmic sources of high energy emission.

REFERENCES

1. R. C. Canfield, et al., in Solar Flares, ed. P. A. Sturrock (Boulder:Colorado Associated University Press), p. 451 (1980).
2. M. Yoshimori, J. Phys. Soc. Japan, 54, 4462 (1985).
3. R. Ramaty, in Physics of the Sun, ed. P. A. Sturrock (Dordrecht:Reidel), p. 291 (1986).
4. R. P. Lin and H. S. Hudson, Solar Physics, 50, 153 (1976).
5. G. D. Holman and S. G. Benka, Ap. J., 400, L79 (1992).
6. L. E. Peterson and J. R. Winckler, J. Geophys. Res., 64, 697 (1959).
7. E. L. Chupp et al., Nature, 241, 333 (1973).
8. E. L. Chupp, Physica Scripta, T18, 15 (1987).
9. E. Rieger, Solar Physics, 121, 323 (1989).
10. M. Yoshimori, Ap. J. Supp., 73, 227 (1990).
11. E. L. Chupp, Ap. J. Supp., 73, 213 (1990).
12. P. Evenson, R. Kroeger, P. Meyer, and D. Reames, Ap. J. Supp.73, 273 (1990).
13. K. Takahashi, Space Science Reviews, 51, 123 (1989).
14. F. Pelaez et al., Solar Physics, 140, 121 (1992).

15. M. Yoshimori, Y. Takai, K. Morimoto, K. Suga, and K. Ohki, Publ. Astron. Soc. Japan, 44, in press (1992).
16. V. V. Akimov et al., 22nd Internat. Cosmic Ray Conf. Papers, 3, 73 (1991).
17. G. Kanbach et al., Astr. and Ap. Suppl., in press (1992).
18. J. M. Ryan et al., in The Compton Observatory Science Workshop, p. 470 (1992).
19. I. Lange and S. E. Forbush, Terr. Magnetism and Atm. Elect., 47, 185 (1942).
20. S. E. Forbush, 1946, Phys. Rev., 70, 771.
21. F. B. McDonald, C. E. Fichtel, L. A. Fisk, in High Energy Particles and Quanta in Astrophysics, ed. F. B. McDonald and C. E. Fichtel (Cambridge:MIT), p. 212 (1974).
22. H. V. Cane, R. E. McGuire, and T. T. von Rosenvinge, Ap. J., 310, 448 (1986).
23. R. P. Lin, Rev. Geophys., 25, 676 (1987).
24. D. Moses, W. Dröge, P. Meyer, and P. Evenson, Ap. J., 346, 523 (1989).
25. D. V. Reames, Ap. J. Supp., 73, 235 (1990).
26. A. Luhn, B. Klecker, D. Hovestadt, and E. Möbius, Ap. J., 317, 951 (1987).
27. P. Evenson, P. Meyer, S. Yanagita, and D. Forrest, Ap. J., 283, 439 (1984).
28. E. I. Daibog, 20th Internat. Cosmic Ray Conf. Papers, 3, 45 (1987).
29. M. B. Kallenrode, E. W. Cliver, and G. Wibberenz, Ap. J., 391, 370 (1992).
30. R. J. Murphy, R. Ramaty, B. Kozlovsky, and D. V. Reames, Ap. J., 371, 793 (1991).
31. R. Ramaty, N. Mandzhavidze, B. Kozlovsky, and J. Skibo, Adv. Space. Res.(COSPAR), in press (1993).
32. A. Temerin and I. Roth, Ap. J., 391, L105 (1992).
33. J. A. Miller and A. Viñas, Ap. J., in press (1993).
34. N. Mandzhavidze and R. Ramaty, Nuclear Physics B, Proc. Suppl., in press (1993).
35. L. A. Fisk, Ap. J., 224, 1048 (1978).
36. S. W. Kahler, Ann. Rev. Astr. and Ap., 30, 113 (1992).
37. T. Bai and R. Ramaty, Ap. J., 219, 705 (1978).
38. W. T. Vestrand, D. J. Forrest, E. L. Chupp, E. Rieger, and G. H. Share, Ap. J., 322, 1010 (1987).
39. W. T. Vestrand, D. J. Forrest, and E. Rieger, 22nd Internat. Cosmic Ray Conf. Papers, 3, 69 (1991).
40. S. R. Kane et al., Ap. J., 326, 1017 (1988).
41. P. E. Semukhin and G. A. Kovaltsov, 19th Internat. Cosmic Ray Conf. Papers, 4, 106 (1985).
42. J. A. Miller and R. Ramaty, Ap. J., 344, 973 (1989).
43. A. L. MacKinnon and J. C. Brown, Astr. and Ap., 232, 544 (1990).
44. L. G. Kocharov and G. A. Kovaltsov, Solar Physics125, 67 (1990).
45. J. M. McTiernan and V. Petrosian, Ap. J., 379, 381 (1991).
46. E. Rieger and H. Marschhäuser, in Proc. of the 3rd MAX'91/SMM Workshop on Solar Flares, eds. R. M. Winglee and A.L. Kiplinger (Univ. of Colorado:Boulder) p. 68 (1990).
47. H. Marschhäuser, E. Rieger, and G. Kanbach, 22nd Internat. Cosmic Ray Conf. Papers, 3, 61 (1991).
48. B. R. Dennis, Solar Physics, 118, 19 (1988).
49. R. P. Lin and R. A. Schwartz, Ap. J., 312, 462 (1987).
50. D. J. Forrest et al., 19th Internat. Cosmic Ray Conf. Papers, 4, 146 (1985).
51. R. Ramaty, B. Kozlovsky, and R. E. Lingenfelter, Ap. J. Supp., 40, 487 (1979).
52. R. E. Lingenfelter and R. Ramaty, in High Energy Nuclear Reactions in Astrophysics, ed B. S. P. Shen, (Benjamin:New York), p. 99 (1967).

53. R. Ramaty, B. Kozlovsky, and R. E. Lingenfelter, Space Science Reviews, 18, 341 (1975).
54. R. J. Murphy, C. D. Dermer, and R. Ramaty, Ap. J. Supp., 63, 721 (1987).
55. X. M. Hua and R.E. Lingenfelter, Solar Physics, 107, 351 (1987).
56. V. G. Gueglenko, G. E. Kocharov, G. A., Kovaltsov, L. G. Kocharov, and N. Z. Mandzhavidze, Solar Physics, 125, 91 (1990).
57. H. T. Wang and R. Ramaty, Solar Physics, 36, 129 (1974).
58. X. M. Hua, R. Ramaty and R. E. Lingenfelter, Ap. J., 341, 516 (1989).
59. B. Kozlovsky, R. E. Lingenfelter, and R. Ramaty, Ap. J., 316, 801 (1987).
60. G. H. Share, E. L. Chupp, D. J. Forrest, and E. Rieger in Positron Electron Pairs in Astrophysics, eds. M. L. Burns, A. K. Harding, and R. Ramaty (New York:AIP), p. 15 (1983).
61. C. J. Crannell, G. Joyce, R. Ramaty, and C. Werntz, Ap. J., 210, 582 (1976).
62. X. M. Hua and R. E. Lingenfelter, Ap. J., 319, 555 (1987).
63. R. J. Murphy, B. Kozlovsky, and R. Ramaty, Ap. J., 331, 1029 (1988).
64. A. Shemi, Mon. Not. Royal Astr. Soc., 251, 221 (1991).
65. N. Mandzhavidze and R. Ramaty, Ap. J., 396, L111 (1992).
66. N. Mandzhavidze and R. Ramaty, Ap. J., 389, 739 (1992).
67. E. L. Chupp et al., Ap. J., 318, 913 (1987).
68. D. J. Forrest, private communications (1988).
69. P. P. Dunphy and E. L. Chupp, in: Particle Acceleration in Cosmic Plasmas, eds. G.P. Zank and T.K. Gaiser (AIP:New York) p. 253 (1992).
70. P. P. Dunphy and E. L. Chupp, 22nd Internat. Cosmic Ray Conf. Papers, 3, 65 (1991).
71. P. P. Dunphy, E. L. Chupp, and E. Rieger, 21st Internat. Cosmic Ray Conf. Papers, 5, 75 (1990).
72. J. M. Ryan et al., Adv. Space. Res.(COSPAR), in press (1993).
73. R. J. Murphy et al., this volume.
74. G. Trottet et al., Astr. and Ap. Suppl., in press (1992).
75. M. McConnel et al., Adv. Space. Res.(COSPAR), in press (1993).
76. K. Takahashi, M. Wada, M. Yoshimori, M. Kusunose, and I. Kondo, 20th Internat. Cosmic Ray Conf. Papers, 3, 82 (1987).
77. M. A. Shea, D. F. Smart, and K. R. Pyle, Geophys. Res. Letters, 18, 1655 (1991).
78. K. R. Pyle and J. A. Simpson, 22nd Internat. Cosmic Ray Conf. Papers, 3, 53 (1991).
79. Yu. Efimov, G. Kocharov, and K. Kudela, 18th Internat. Cosmic Ray Conf. Papers, 10, 276 (1983).
80. N. Iucci, M. Parisi, C. Signorini, M. Storini, and C. Villoresi, 19th Internat. Cosmic Ray Conf. Papers, 4, 134 (1985).
81. M. A. Shea, D. F. Smart, and E. O. Flückiger, 20th Internat. Cosmic Ray Conf. Papers, 3, 86 (1987).
82. S. Shibata et al., 22nd Internat. Cosmic Ray Conf. Papers, 3, 788 (1991).
83. Y. Muraki et al., 22nd Internat. Cosmic Ray Conf. Papers, 3, 49 (1991).
84. Y. Muraki et al., Ap. J., 400, L75 (1992).
85. K. Takahashi et al., 22nd Internat. Cosmic Ray Conf. Papers, 3, 37 (1991).
86. N. Chiba et al., Astroparticle Physics (1992), to be published.
87. M. A. Shea, D. F. Smart, and E. O. Flückiger, Geophys. Res. Letters, 18, 829 (1991).

SOLAR FLARE NEUTRON SPECTRA AND ACCELERATED ION PITCH-ANGLE SCATTERING

RICHARD E. LINGENFELTER
Center for Astrophysics and Space Sciences,
University of California, San Diego, La Jolla, CA 92093
XIN-MIN HUA
Institute for Space and Terrestrial Science,
Concord, Ontario, L4K 3C8 Canada
BENZION KOZLOVSKY
Physics & Astronomy Department, Tel Aviv University, Tel Aviv, Israel
REUVEN RAMATY
Laboratory for High Energy Astrophysics,
Goddard Space Flight Center, Greenbelt, MD 20771

ABSTRACT

We have included accelerated ion pitch angle scattering and magnetic mirroring in our Monte Carlo programs for neutron production in solar flare magnetic loops, and we have thoroughly updated the neutron production kinematics and cross sections used in these programs. We are now making new calculations to explore the effects of flare-accelerated ion pitch angle scattering on the angular and temporal dependence of the escaping neutron spectra. These calculated spectra can be directly compared with the new direct measurements by COMPTEL of the time dependent spectra of flare neutrons to determine both the energy spectrum and the effective pitch angle scattering of the accelerated ions in the flares.

INTRODUCTION

Our understanding of solar flare ion acceleration, interactions and propagation has been greatly advanced by comparisons of solar flare neutron and gamma-ray line observations with theoretical calculations of the fluxes expected from flare-accelerated ion interactions in the solar atmosphere. Because they are produced directly by nuclear interactions of the flare-accelerated protons and heavier ions with ambient gas in the solar atmosphere, these neutrons and gamma-rays give us the most direct information available on the total number, energy spectrum, time dependence, and angular distribution of ion acceleration and propagation in flares. They can also provide unique information on the composition, scale height, magnetic field convergence and magnetohydrodynamic turbulence in the flare region. Here we focus on what can be learned from the new direct measurements of the time dependent spectra of flare neutrons.

Neutrons were observed[1,2] from five solar flares between 1980 and 1989 with the SMM detector. Relativistic neutrons were also observed[3,4] from several flares with ground based cosmic-ray neutron monitors, and protons from the decay of solar flare neutrons were detected[5,6] in interplanetary space by ISEE-3. Most recently the neutron and gamma-ray spectrometers on Compton GRO have measured[7-9] gamma-ray spectra from several flares and neutron spectra from at least 3 of them. These measurements not only roughly double our data base, they also give us the first direct measurements of the neutron spectra and provide gamma-ray spectra over a much broader energy range with much finer time resolution and a wide range of heliocentric observing angles.

Detailed theoretical calculations of neutron production by flare accelerated ion interactions were carried out[10-13] in order to interpret the earlier measurements. Now we have expanded our neutron and gamma-ray line production Monte Carlo simulation programs[14-15] to investigate the effects of pitch angle diffusion and magnetic mirroring of the flare accelerated ions in the solar atmosphere. This will enable us to better understand the angular dependence of the neutron and gamma-ray line emission observed from flares at different heliocentric angles.

We have also thoroughly re-evaluated and revised the neutron kinematics and cross sections assumed in our calculations. These revisions have focused primarily on the neutron production kinematics used for the proton-α and α-α interactions, which have a dominant effect on the neutron production spectrum. We have developed new kinematic approximations for the differential angular and energy dependent neutron yield from multiparticle breakup that are in very good agreement with the available laboratory measurements and theoretical simulations. We have also gained a more quantitative measure of the systematic uncertainties in the calculated spectra. Some preliminary results of these calculations have already been presented[16].

NEUTRON PRODUCTION IN FLARE LOOPS

Using these new cross sections and kinematics in our Monte Carlo simulation program, we calculate the angle, time and depth dependence of the neutron production by accelerated ions solar flare loops. The models for solar atmosphere and the flare magnetic loop used in this calculation are the same as those which we have used in calculating the gamma-ray line production[14]. Basically, the loop consists of a semicircular coronal segment and two straight segments extending from the ends of the coronal segment through the chromosphere and into the photosphere. In the corona, the gas is assumed to be completely ionized with a constant density and the magnetic field is also taken to be constant. In the chromosphere and photosphere, the gas is neutral with a pressure profile given by Avrett[17] and a magnetic field B proportional to a power δ of the pressure[18]. The transition between the corona and the chromosphere is 1500 km above the photosphere.

The acceleration of ions is assumed to take place in the corona. The accelerated ion spectra are assumed to be either a modified Bessel function in momentum or a power-law in energy. The modified Bessel function spectrum, expected for stochastic acceleration[19], is characterized by the parameter αT, such that harder spectra correspond to larger values of αT. The power law spectrum in kinetic energy per nucleon, more appropriate[20] to shock acceleration, is defined by the power-law index, S.

The propagation and interaction of the accelerated ions in the magnetized solar atmosphere is also similar to that used in our γ-ray line production program[14], where the ions are subject to the following interactions with the ambient gas and magnetic field: 1) ion energy loss due to Coulomb scattering, 2) neutron and gamma-ray production by ion in inelastic nuclear reactions, 3) ion removal by various nuclear reactions, 4) ion reflection by the mirroring force of the convergent magnetic field at the feet of the loop, and 5) ion pitch-angle scattering by the MHD turbulence in the coronal ionized gas. The pitch-angle scattering is characterized by its mean free path length Λ, which is related to the energy density in MHD turbulence[21-22].

The propagation and interaction of the neutrons produced in the ion interactions is then simulated in our Monte-Carlo program by following each neutron through many scatterings, until it either escapes from the solar atmosphere, decays, or slows down and is captured either radiatively on ^{1}H to form ^{2}H with the emission of a 2.223 MeV gamma ray, or nonradiatively on ^{3}He to form ^{3}H with the emission of a proton. Finally we calculate the time and angle dependent spectrum of neutrons surviving to a distance of 1 AU from the Sun for comparison with the directly measured neutrons.

EFFECTS OF ION PITCH ANGLE SCATTERING

In the present calculations, we consider two limiting cases of pitch-angle scattering of the ions: 1) the case of negligible pitch-angle scattering with the scattering mean free path $\Lambda = \infty$, and 2) the case of nearly saturated pitch-angle scattering, with $\Lambda = 40$ times the assumed magnetic loop length of 11,500 km.

Ion pitch angle scattering directly affects the depth and angular distributions of the neutron production in the solar atmosphere. We find that most of the neutrons are produced at depths corresponding to an overlying column density ≤ 10 g cm^{-2}. At these depths, neutrons initially directed upward have a good chance of escaping from the solar atmosphere without scattering. But if they are initially headed downward, they can lose most of their energy in the solar atmosphere and have little chance of escaping from it. Thus the neutron escape probability depends strongly on the initial angular distribution of neutrons. We find that in a typical case, the escape probability for neutrons with energies > 1 MeV is 55% without pitch-angle scattering of the ions, while it is only 25% for nearly saturated pitch-angle scattering.

The angular dependences of the escaping neutrons produced by ions without pitch-angle scattering and by those with nearly saturated pitch-angle scattering, are also quite different. We find that without ion pitch-angle scattering, the escaping high-energy neutrons have a rather strong limb-brightened distribution, with the peak fluence at the zenith angle close to 90°. This closely reflects the neutron production distribution which peaks at $\sim 90°$, resulting in turn from the pancake-like distribution of the mirroring ions. In this case, the neutrons are mostly produced in the corona and upper chromosphere, and they escape with very little scattering.

On the other hand, in case of nearly saturated pitch-angle scattering, the limb-brightening is much weaker because the neutrons are produced deeper in the atmosphere and preferentially in downward directions. As a result, the neutrons suffer greater attenuation as they escape in the directions nearly tangential to the solar surface. This limb-darkening effect is more striking for neutrons at lower energies, because the scattering cross section is larger. These effects shift the peak fluence of the escaping neutrons to zenith angles around $\sim 77°$.

The limb-brightening of the escaping neutrons is further enhanced for those neutrons that survive at a distance 1 AU from the Sun, since the fluence of the limb-darkened low-energy neutrons becomes negligible because of decay. As a result the energy spectrum of the escaping neutrons also depends strongly on the angular distribution of the neutrons. Our calculations of the time dependences of the neutron flux at a distance 1 AU from the Sun also show: 1) how the different accelerated ion spectral forms (Bessel function and power law) can be distinguished and how the accelerated ion spectrum can thus be determined from the neutron flux measurements, independent of the degree of pitch-angle scattering experienced by the ions, and 2) how the amount of pitch-angle scattering can also be determined from the same neutron flux measurements, independent of the accelerated ion spectra.

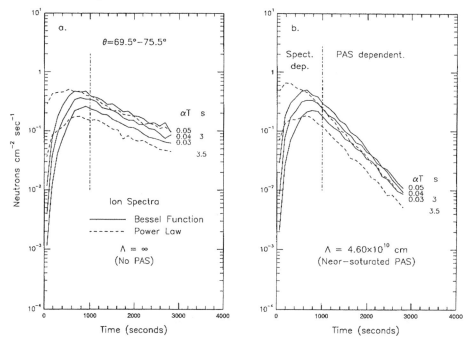

Fig. 1. Calculated time dependences of the neutron flux at a distance 1 AU from the Sun resulting from interactions of accelerated ions with either Bessel function (αT) or power law (s) spectra in the bracketing cases: a without pitch-angle scattering (pas) of the ions, and b with nearly saturated pitch-angle scattering. These fluxes are from a flare at a heliocentric angle of $\sim 72°$ and they are normalized to a constant 4–7 MeV nuclear deexcitation line fluence.

As we see in Figure 1, in the first 1000 seconds after the impulsive phase of a flare, the shape of the time dependence of the neutron flux at a distance 1 AU depends very strongly on the form of the ion spectrum and only very weakly on whether or not there is any significant pitch-angle scattering of the ions. Independent of the degree of ion pitch-angle scattering, ions with power-law spectra produce much harder neutron spectra than ions with more exponential-like Bessel function spectra, for the same nuclear deexcitation gamma-ray line fluence. Therefore, we see that the neutron flux at 1 AU from ions with a power-law spectrum (Figure 1 dashed lines), rises very quickly with the first 100 seconds to within a factor of 2 of the peak flux, while the neutron flux from ions with a Bessel function spectrum (solid lines), rises much more slowly to the peak flux. Thus, from measurements of the neutron flux time dependence in the first 1000 seconds, we should be able to distinguish between power-law and Bessel function spectral forms and clearly determine the ion spectrum, independent of the degree of ion pitch angle scattering.

Moreover, as we also see in Figure 1, after the first 1000 seconds of a flare, the time dependence of the neutron flux at 1 AU no longer depends on either the form or the index of the ion spectrum and instead depends very strongly on the degree pitch-angle scattering experienced by the ions. Quite independent of the ion spectrum, we see (Figure 1a) that without ion pitch-angle scattering, all of the neutron fluxes after the first 1000 seconds fall off with essentially the same relatively slow exponential ($\sim$ 1200 sec) time decay, determined by the extended production of neutrons by unscattered ions mirroring and slowly losing their energy in the upper atmosphere of the Sun. On the other, with nearly saturated pitch-angle scattering, we see (Figure 1b) that irrespective of the ion spectrum, all of the neutron fluxes after the first 1000 seconds fall off with a very similar but much more rapid exponential ($\sim$ 600 sec) time decay, as the ions are more quickly flushed out of magnetic loop by pitch-angle scattering which causes them to mirror deeper in the solar atmosphere where they lose their energy more rapidly. Thus, from measurements of the neutron flux time dependence after the first 1000 seconds, we should be able to determine the degree of ion pitch angle scattering quite independent of the ion spectrum.

Acknowledgements. The calculations in this study were carried out on the computers at ISTS and GSFC/NASA. XMH thanks ISTS for the support, and REL thanks NASA for support under grants NAG5-1597 and NSF under grant ATM-8921658.

REFERENCES

1. E.L. Chupp, AnnRevA&A, 22, 359 (1984).
2. P.P. Dunphy, & E.L. Chupp, Proc. 22nd Int. Cosmic Ray Conf., 3, 65 (1991).
3. E.L. Chupp, et al. ApJ, 318, 913 (1987).
4. K.R. Pyle & J.A. Simpson, 22th ICRCPapers, 3, 53 (1991).
5. P. Evenson, R. Kroeger, & P. Meyer, 19th ICRCPapers, 4, 130 (1985).
6. P. Evenson, P. Meyer, & K.R. Pyle, ApJ, 274, 875 (1983).
7. J. Ryan, et al. (1992), in this volume.
8. G. Share & R. Murphy (1992), in this volume.
9. G. Kanbach et al. (1992), in this volume.
10. X.-M. Hua, & R.E. Lingenfelter, ApJ, 323, 779 (1987).
11. R.J. Murphy, C.D. Dermer & R. Ramaty, ApJSupp, 63, 721 (1987).
12. V.G. Guglenko, et al. Solar Phys, 125, 91 (1990a).
13. V.G. Guglenko, et al. ApJSupp, 73, 209 (1990b).
14. X.-M. Hua, R. Ramaty, & R.E. Lingenfelter, ApJ, 341, 516 (1989).
15. X.-M. Hua, B. Kozlovsky, R. Ramaty, & R.E. Lingenfelter, in preparation, (1992).
16. R. Ramaty, X.-M. Hua, B. Kozlovsky, R.E. Lingenfelter & N. Mandzhavidze, Compton Observatory Science Workshop (NASA CP 3137) 480 (1992).
17. E.H. Avrett, in The Physics of Sunspots ed. L.E. Cram and J.H. Thomas (Sacramento Peak Obs., AURA, 1981), p. 235.
18. E.G. Zweibel & D. Haber, ApJ, 264, 648 (1983).
19. R. Ramaty, in Particle Acceleration Mechanisms in Astrophysics, ed. J. Arons, C. Max & C. McKee (New York: AIP, 1979), 135
20. M.A. Forman, R. Ramaty, & E.G. Zweibel, in The Physics of the Sun, ed. P.A. Sturrock (Dordrecht: Reidel, 1986), v2, 249
21. D.B Melrose, Solar Phys, 37, 353 (1974).
22. R. Ramaty, J.A. Miller, X-M. Hua, & R.E. Lingenfelter, Nuclear Spectroscopy of Astrophysical Sources, ed. N. Gehrels and G.H. Share, (New York: AIP, 1988), p. 217.

Observations of the 1991 June 11 solar flare with COMPTEL

G. Rank, R. Diehl, G. G. Lichti, V. Schönfelder, M. Varendorff*
Max-Planck Institut für extraterrestrische Physik, D-8046 Garching, Germany

B. N. Swanenburg
SRON-Leiden, P.B. 9504, NL-2300 RA Leiden,The Netherlands

D. Forrest, J. Macri, M. McConnell, J. Ryan
University of New Hampshire, Institute for the Study of Earth, Oceans and Space,
Durham NH 03824, U.S.A.

K. Bennett, L. Hanlon, C. Winkler
Astrophysics Division, Space Science Department of ESA/ESTEC, NL-2200
AG Noordwijk, The Netherlands

ABSTRACT

The COMPTEL instrument onboard the Compton Gamma Ray Observatory is well suited for the observation of solar flares. It is sensitive to γ-rays in the energy range from 0.75 to 30 MeV. In addition COMPTEL has the novel capability to detect individual neutrons from solar flares.
During the period of unexpectedly high solar activity in June 1991 several flares from active region 6659 were observed by COMPTEL. For one of these - that of June 11th - we present the latest results from the analysis of COMPTEL data. This includes the time history of the γ-ray emission extending for at least two hours after the impulsive phase.

INTRODUCTION

On June 11th the active solar region 6659 produced a huge flare, being accompanied by a loop prominence. It was classified as a 3B event optically and as a X-12 GOES event. The X-ray emission has started at 1:56 UT, had a maximum at 2:09 UT and faded away at about 2:20 UT. Orbital sunrise of Compton GRO has taken place at 1:48 UT, a few minutes before the flare onset. So the evolution of the flare can be studied for a whole orbital observation period. In its telescope mode COMPTEL uses a double scattering of the γ-rays and provides energy and incoming direction of the radiation. To avoid background of charged particles, the whole instrument is shielded by anti-coincidence domes. The remaining γ-ray background originating in the earth's atmosphere and the instrument itself can be supressed using the imaging capability of COMPTEL. Fig. 1 demonstrates these imaging qualities for a time interval during the flare.

---

* Present adress: University of New Hampshire, Institute for the Study of Earth, Oceans and Space, Durham NH 03824, U.S.A.

662 Observations of the 1991 June 11 Solar Flare

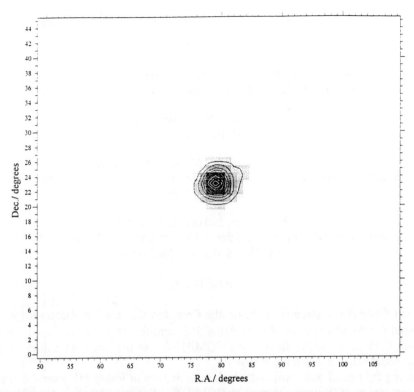

Fig 1 Image of the solar flare during the same time period as the spectrum in fig.3a. To perform this Maximum Entropy picture γ-ray data in the spectral range from 0.8 to 8 MeV were used.

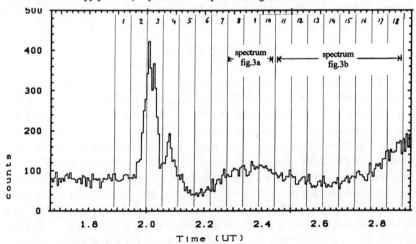

Fig 2 Time profile of raw telescope events in the energy range from 0.75 to 30 MeV. The indicated intervals are explained in the text.

So only events really coming from the sun can be selected. The remaining γ-ray background of the telescope shows orbital variations due to changes in rigidity and spacecraft orientation relative to the earth. After every 15 or 16 orbits the orbital parameters are reproduced quite precisely. Therefore, data from 15 and 16 orbits apart from the flare can be used to get a model background.

Because of the enormous γ- and X-ray flux during the flare the data interpretation is very difficult due to lifetime effects, the occurence of multi-hit processes in the telescope and restrictions by the event buffers and the telemetry rate. For the impulsive phase these effects are not yet completely understood. Therefore, the data from to the impulsive phase cannot be analyzed in a reliable manner and will not be shown here.

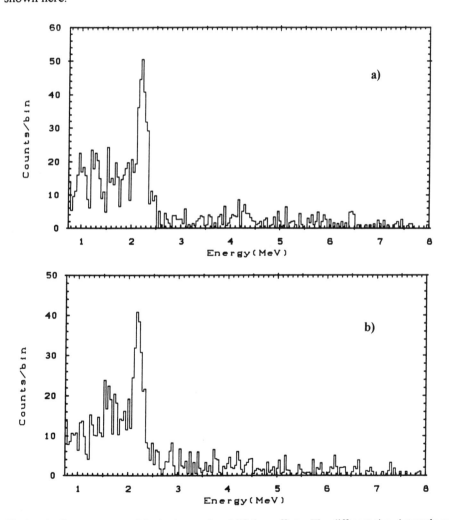

Fig 3 a, b Spectra corrected for background and lifetime effects. The different time intervals are illustrated in fig. 2. The used energy binning is 40 keV, the integration time is 800s for the first and 1200s for the second spectrum.

RESULTS

Our analysis concentrates on the flare spectrum and its evolution throughout the satellite orbit in which the flare occured. To study the time history of the event, eighteen time intervals of 200s each were defined, beginning at the onset of the flare. Fig. 2 shows the raw uncorrected count rate of γ-rays and illustrates the position of the intervals. As established above the first 7 intervals are excluded from the studies. For the remaining intervals raw spectra in the energy range from 0.75 MeV to 15 MeV were obtained and corrected for background and lifetime effects.

Emission from the solar event can also be measured in the next two satellite orbits following the flare (see McConnell et al., 1992).

The spectra in fig.3 show the energy loss in the telescope for two different time intervals as indicated in the raw event plot (fig.2). The spectrum after the impulsive phase is dominated by a strong 2.2 MeV emission line of the neutron capture. This line feature declines exponentially with a time constant of 11.8 minutes (see fig.5). At energies below 2.2 MeV there is a continuum which rises slowly in time relatively to the line. The continuum measured at higher energies above 4 MeV is low and also the line fluxes at 4.4 and 6.3 MeV are not remarkable. The time history for the γ-ray emission from 0.75 to 15 MeV is displayed in fig.4.

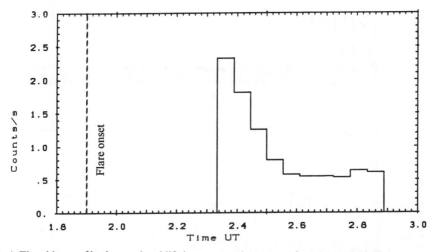

Fig 4 Time history of background and lifetime corrected count rate for 0.75 to 15 MeV.

CONCLUSIONS

We could obtain a history of spectra of the 1991 June 11th flare from COMPTEL data. A clear evolution of the spectrum as demonstrated in fig. 2 was found. Interesting values for theoretical models are expected from flux ratios as (2.2MeV line)/(4-7MeV) and (1-2MeV)/(2.2MeV line) but there are still uncertainties remaining in the present analysis that makes more precise studies necessary.

In addition we found evidence for the detection of flare neutrons in the raw data, but the analysis may be difficult and remains to be done.

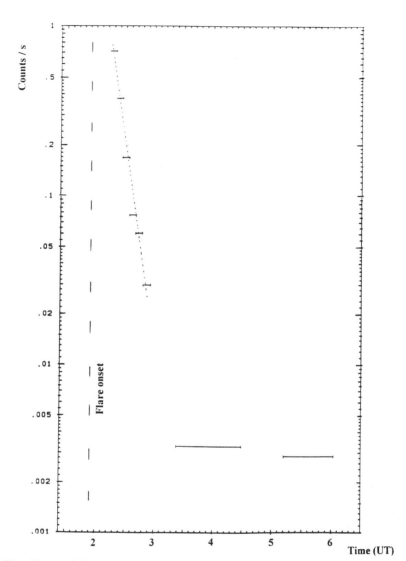

Fig 5 Time history of the γ-ray count rate 2.2 MeV line for the flare orbit and the next two orbits. The data are obtained from the background and lifetime corrected count rates. No line fitting has been done.

REFERENCE

M. McConnell, K. Bennett, H. Bloemen, H. de Boer, M. Busetta, W. Collmar, A. Connors, R. Diehl, J.W. den Herder, W. Hermsen, L. Kuiper, G. G. Lichti, J. Lockwood, J. Macri, D. Morris, R. Much, G. Rank, J. Ryan, V. Schönfelder, G. Stacy, H. Steinle, A. W. Strong, B. N. Swanenburg, B. G. Taylor, M. Varendorff, C. de Vries, W. Webber, C. Winkler (1992): COMPTEL observations of solar flare gamma-rays, COSPAR paper E.3-S.5.06

GAMMA-RAY BURSTS: DISTRIBUTIONS

BATSE OBSERVATIONS OF GAMMA-RAY BURSTS

G. J. Fishman
NASA/Marshall Space Flight Center

ABSTRACT

The BATSE experiment on the Compton Observatory has provided unprecedented, detailed observations of gamma-ray bursts during its 18 months of operation. Some of the key observations are: the celestial distribution of almost 500 gamma-ray bursts, the intensity distribution of bursts, high-energy spectral observations, measurements of rapid spectral changes, the form of the continua, a search for cyclotron lines, rapid search for counterparts, high time resolution measurements, and detailed time profiles of a large number of bursts. Numerous papers from BATSE are presented at this Symposium; a selection of highlights from some of them are given here. Samples of time profiles from the first BATSE burst catalog are shown; one is immediately struck by their diversity.

In spite of the recent observational advances and the renewed, intense theoretical work that they have inspired, the answers to the fundamental questions of the distance, central source, and emission mechanism of gamma-ray bursts remain elusive.

INTRODUCTION

Gamma-ray bursts remain one of the longest-standing problems of all time in astronomy, as described in a recent article on astronomical puzzles.[1] There is a now growing feeling that, based on the current spatial distribution observations, a satisfactory resolution to this puzzle will not be found until new and/or more detailed types of analyses are performed, guided by theoretical work, or until a verified counterpart to gamma-ray bursts is observed. Several new types of burst analyses, burst counterpart searches, and new theoretical directions can be found in these Proceedings. It is obvious that research in this field is now quite active.

The BATSE experiment on the Compton Observatory continues to provide sensitive observations of gamma-ray bursts as never before possible. At this Symposium, over 40 papers describe various details of these burst observations. My purpose here is not to summarize these observations, but rather to offer some of my personal impressions of some aspects of these data with an emphasis on the diverse time histories (morphology) of gamma-ray bursts. I will also mention the status of the first BATSE burst catalog and some future directions in BATSE burst observations.

BATSE is currently observing an average of 0.85 bursts per day out of a total of 4.0 onboard triggers per day. The determination and separation of the true cosmic gamma-ray bursts from the other sources of onboard triggers is described by Mallozzi et al.[2] In addition, a significant number of untriggered cosmic gamma-ray bursts are expected to be found as archived data are processed on the ground by new software.[3] The all-sky gamma-ray bursts rate at the BATSE sensitivity, corrected for sky exposure and including untriggered bursts, is about 1000 bursts per year.

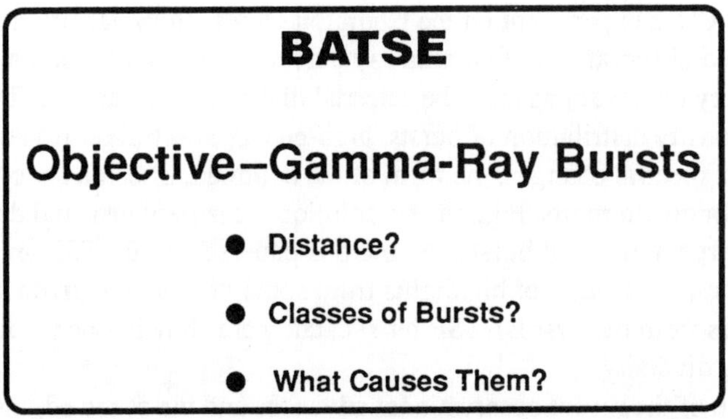

BATSE

Objective — Gamma-Ray Bursts

- Distance?
- Classes of Bursts?
- What Causes Them?

Fig. 1. Viewgraph of BATSE primary objectives

Several weeks ago, I came across an old, pre-launch viewgraph (Fig. 1) that I had used for briefing senior NASA officials and that was also used by me at a pre-launch NASA press conference. As I will describe in this paper, and it is obvious from these Proceedings, the answer to the first two bullets (in NASA parlance) is far from being realized. And, of course, the last question cannot be answered until the first two are well in hand.

BURST DISTANCE SCALE

Prior to the launch of the Compton Observatory, most workers in the field expected that the weak bursts would be seen to be concentrated in the plane of the Galaxy and/or toward the Galactic center. This expectation was based not only on the well-developed paradigm of the Galactic neutron star origin of gamma-ray bursts,[4-6] that was developed in the preceding decade, but also on the observation of a deficiency of weak gamma-ray bursts that had been measured from balloons by our group[7] and others.[4] [However, it should be pointed out that B. Paczynski, for one, had maintained a minority viewpoint[8] (a cosmological origin). It was also noted by K. Hurley, the year before the launch of the Compton Observatory, that several weak, well-localized gamma-ray bursts did not appear to cluster in the galactic plane.[9]]

The isotropy of gamma-ray bursts observed by BATSE, together with their observed inhomogeneity, as measured by the deficiency of weak gamma-ray bursts,[10] has re-opened the distance question, allowing anything from Oort cloud distributions[11] to cosmological models.[12] Arguments for and against various distribution models and distance scales are made in these Proceedings. The enormous range of distance scales and associated energies are given in Table 1.

Model	Presumed Objects	Typical Distance	Typical Burst Energy
I. Solar System (Oort Cloud)	?	~0.001 pc	~10^{28} ergs
II. Galactic Halo or Corona	Neutron Stars	~30–100 kpc	~10^{43} ergs
III. Cosmological	?	~1 Gpc	~10^{52} ergs

Table 1. Burst distribution models.

Galactic disc models can almost certainly be ruled out now due to the observed absence of a galactic quadrupole moment.[13-15] The galactic halo or coronal models are now also severely constrained by the lack of an observed dipole moment.[13,14] Any such distribution would require an extremely large, homogeneous, spherical core radius. This is unlike any other known galactic component. Also, the lack of clustering around the LMC may provide difficulties for such models. If the isotropy holds up as hundreds or thousands of gamma-ray bursts continue to be detected, these galactic halo models will become implausible.

CLASSIFICATION OF GAMMA-RAY BURSTS

It is generally recognized that in any new branch of science, classification (taxonomy) must first be well developed before the science can successfully enter the explanatory phase. Prior to the launch of the Compton Observatory, several classes of gamma-ray transients were known and it was anticipated that the detailed temporal and spectral observations by BATSE would delineate additional sub-classes of gamma-ray bursts.[16] In Table 2 are listed three types of gamma-ray transients that are easily separated in their properties. Theorists would be hard pressed (and could mislead themselves and others) if they were to attempt to explain all three gamma-ray

Gamma Ray Transients 1972–1992

	Approximate Number
Gamma Ray Bursts	~1,000
Soft Gamma Repeaters	2–3
March 5 Event (1979)	1

Table 2. Classes of transients

transient types with a single theory. The March 5, 1979, gamma-ray event remains a unique event because of its intense initial spike, soft spectrum, periodic after-pulses, repeated outbursts for several months afterward, and its identification with the SNR N49 in the LMC. BATSE has seen no comparable events. Likewise, soft gamma repeaters (SGRs) are certainly a different class of gamma-ray transients, easily identifiable by their soft spectra, repeating nature, and association with the galactic plane. (The term "gamma-ray" in SGR is actually a misnomer; they are better characterized as hard x-ray events.) BATSE has detected a recurrence of one of these: SGR1900+14.[17] Some consider the March 5 event as a special type of SGR.

By far the most populous class of gamma-ray transients are the classical gamma-ray bursts, the subject of this paper. Prior to the launch of the Compton Observatory, there were about 450 gamma-ray bursts detected. The directions were known for about 180 of them, although less than 50 had precise, unambiguous locations determined. BATSE has now detected a comparable number (500 as of December 1992). Thus there are now roughly 1000 gamma-ray bursts known, of which about 600 have had their locations determined.

DIVERSITY OF GAMMA-RAY BURSTS

In the study of gamma-ray bursts, one is immediately struck by the diversity of the durations, spectral characteristics, and time profiles. The duration distribution is roughly known; it spans over five decades, from a few milliseconds to a few hundred seconds.[18,19] There is most likely an observational bias against the shortest and the longest bursts. The spectral diversity of gamma-ray bursts observed by BATSE is described in recent papers by Band, et al.[20,21]

Gamma-ray burst temporal profiles, or time histories, have a great potential for illuminating the size, geometry, and beaming characteristics of the burst-emitting region. While some quantitative work in characterizing BATSE burst time profiles is underway,[22,23] I will describe here some qualitative aspects of BATSE burst profiles.

In Figures 2-7 are shown some time profiles, selected from the 260 gamma-ray bursts in the first BATSE burst catalog now being prepared. This does NOT imply that all bursts are easily categorized into these few groups. In fact, just the opposite is true - gradations of morphological types, intermediate to those shown, are commonly seen and mixtures of morphologies are seen within the same burst (Fig. 7).

Single Peak, Smooth Profile

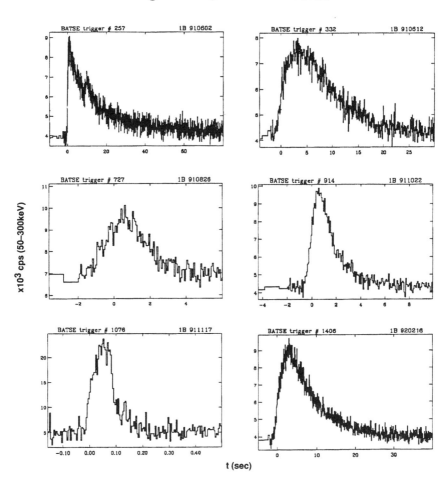

Fig. 2.

Few Peaks, Smooth Profile

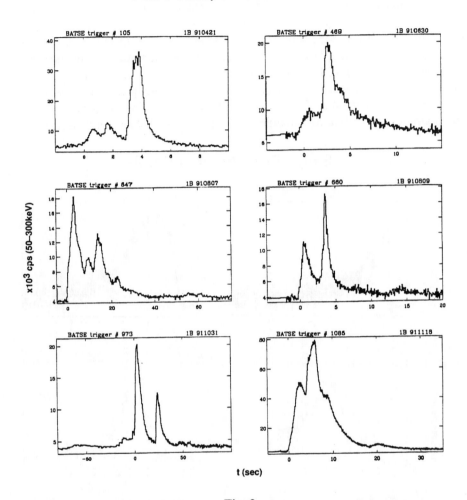

Fig. 3.

This latter mixture of morphologies, and the fact that the different morphologies all have an isotropic distribution seems to indicate that the central object and emission mechanism of all gamma-ray bursts are similar. Further complicating the gamma-ray burst problem is the fact that the diversity in durations, rise-times, and minimum time structures spans four or five orders of magnitude, yet the intrinsic peak luminosity spread is only perhaps two or three orders of magnitude, at most. This is evidenced by the fact that the bend-over in the slope of the $\ln N$-$\ln P$ distribution occurs over this limited range of P.[24] Models of the burst process are expected to have difficulty explaining this.

Complex, Well-Separated Peaks

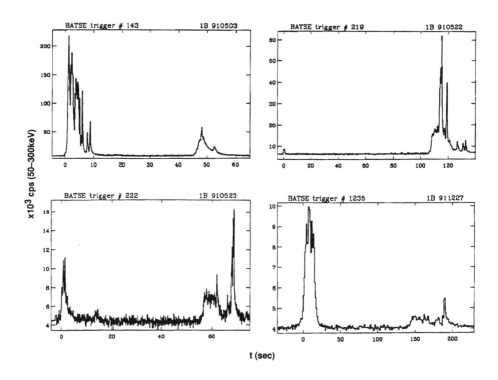

Fig. 4.

FUTURE DIRECTION IN BATSE BURST OBSERVATIONS

As solar activity continues to decrease over the next few years, we expect that the number of non-burst triggers will be reduced and that the livetime for burst observations will increase. Portions of data from many bursts are currently lost due to the failure of the tape recorders on the Observatory. New flight software is being developed that will allow us to receive more complete data on most bursts. The detectors all continue to operate with their expected performance; no degradation has been observed.

The first BATSE catalog is in the process of being released in electronic form.[25] The BATSE science team will issue updates of the burst catalog on approximately a yearly basis. The delay time between the bursts and the issuance of catalogs is expected to decrease as the burst data processing becomes more routine

Complex, Many Overlaping Spikes & Peaks

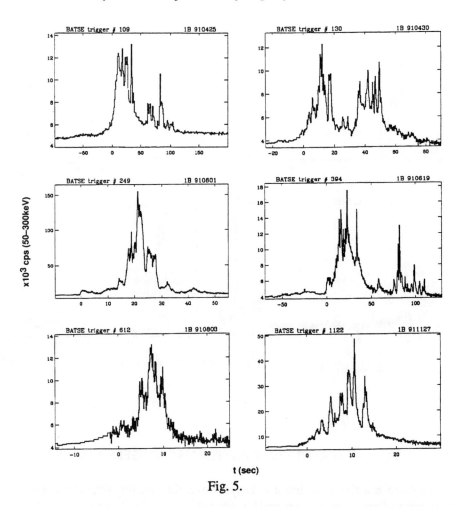

Fig. 5.

and the analysis software matures. Considerable improvements in the burst location accuracy have occurred over the past 18 months,[26] as on-orbit experience improves and more burst locations become available for comparison with the interplanetary network (IPN).[27,28] However, the very precise locations from the three spacecraft (Ulysses-BATSE-Mars Observer) IPN will not become available routinely for another year, when the Mars Observer spacecraft is in orbit around Mars.

The BATSE team is participating in more than 20 collaborative investigations, both as guest investigators and as informal collaborators, to search for coincident objects or phenomena with gamma-ray bursts detected by BATSE. The wide range of

Very Complex, Chaotic

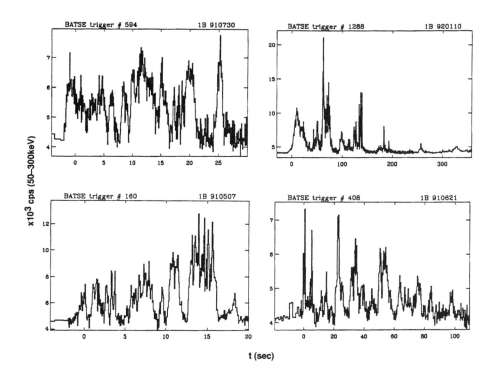

Fig. 6.

these investigations include: rapid response optical searches, archival searches through ground-based and spacecraft data, targets-of-opportunity with orbiting spacecraft, and time coincidences with air-shower arrays, neutrino detectors, and gravitational wave detectors. So far, all searches have been negative. Some of these efforts are described in these Proceedings and the proceedings of the Huntsville Gamma-Ray Burst Workshop.[29] Continued attempts to correlate BATSE gamma-ray bursts with distributions of known objects will be made by the BATSE team[30] and soon by others outside of the team, as the catalog becomes released to the scientific community.

This paper does not address the extensive efforts in progress to search for spectral features (cyclotron lines) in BATSE gamma-ray bursts; that research is presented in other papers of these Proceedings.[31-33] The BATSE spectroscopy detector exposure is now comparable to that of Ginga and other experiments that have

Blend of Profiles – Difficult to Classify

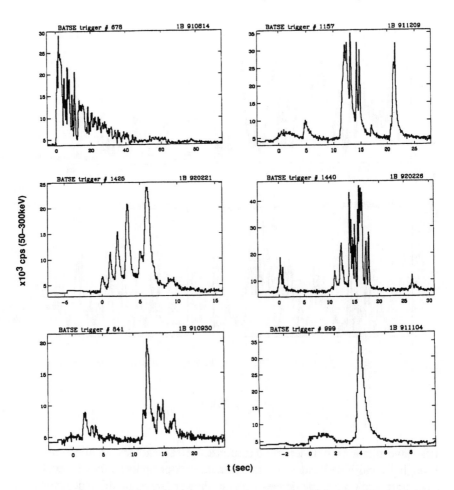

Fig. 7.

observed spectral line features.[31] Thus, BATSE should soon be able to make a more definitive statement regarding the existence of line features in the BATSE spectroscopy data. We are also beginning to utilize the large area detector (LAD) data for burst spectral analysis.[34] Although these detectors have coarse spectral resolution, their large effective area can be used for selected spectral analyses above 30 keV.

As more coordinated observations become available in the coming years, BATSE is expected to continue its productive studies of gamma-ray bursts.

ACKNOWLEDGEMENTS

As Principal Investigator and the primary spokesman for BATSE, I am merely reflecting on the work of the BATSE science team. The names of the members of this team are on the numerous BATSE contributions throughout these Proceedings.

The development of BATSE was made possible through the capabilities of a large number of technical personnel, primarily at the NASA/Marshall Space Flight Center (MSFC). The continuing mission operations are being carried out by a dedicated staff of operations personnel, headed by Dr. William Henze. The outstanding support of the BATSE programming staff at MSFC, GSFC, and UCSD is gratefully acknowledged.

REFERENCES

1. V. Trimble, "Gamma-Ray Bursts" (Proc. Taos Workshop, 1990) p. 481, eds. C. Ho, R. Epstein, and E. Fenimore (Cambridge U. Press) (1992).
2. R. Mallozzi et al., these proceedings.
3. B. Rubin et al., these proceedings.
4. J. Higdon and R. Lingenfelder, Ann. Rev. Astron. & Ap., 28, 40 (1990).
5. B. Schaefer, Scientific American, 252, 52 (1985).
6. C. Ho, R. Epstein, and E. Fenimore, "Gamma-Ray Bursts" (Proc. Taos Workshop, 1990) (Cambridge U. Press) (1992).
7. C. Meegan, G. Fishman, and R. Wilson, Ap. J., 291, 479 (1985).
8. B. Paczynski, Ap. J., 308, p. 143 (1986).
9. K. Hurley, "Gamma-Ray Bursts" (Proc. Taos Workshop, 1990) p. 183, eds. C. Ho, R. Epstein, and E. Fenimore (Cambridge U. Press) (1992).
10. C. Meegan et al., Nature, 355, 143 (1992).
11. J. Horack et al., these proceedings.
12. D. Hartmann, these proceedings.
13. J. Hakkila and C. A. Meegan, "Gamma Ray Bursts" Proc. Huntsville Workshop, AIP Conf. Proc. 265, p. 70, eds. W. S. Paciesas and G. J. Fishman (AIP Press: New York) (1992).
14. J. Hakkila et al. these proceedings.
15. D. Hartmann, these proceedings.
16. G. Fishman et al., Proc. GRO Science Workshop, p. 3-47, ed. W. N. Johnson (1989).
17. C. Kouveliotou et al., these proceedings.
18. K. Hurley, Annals. N.Y. Acad. Sci., 571, 442 (1989).
19. N. Bhat et al., Nature, 359, 217 (1992).
20. D. Band et al., these proceedings.
21. D. Band et al., Ap. J., submitted (1992).
22. J. Lestrade et al., these proceedings.
23. J. Norris et al., these proceedings.

24. C. Meegan et al., these proceedings.
25. S. Howard et al., these proceedings.
26. M. Stollberg et al., these proceedings.
27. T. Cline et al., these proceedings.
28. K. Hurley et al., these proceedings.
29. W. S. Paciesas and G. J. Fishman, eds. "Gamma Ray Bursts," Proc. Huntsville Workshop, AIP Conf. Proc. 265 (AIP: New York) (1992).
30. S. Howard et al., these proceedings.
31. B. Teegarden, these proceedings.
32. D. Palmer, these proceedings.
33. L. Ford, these proceedings.
34. R. Preece, these proceedings.

THE SPATIAL DISTRIBUTION OF GAMMA-RAY BURSTS OBSERVED BY BATSE

Charles Meegan, Gerald Fishman, Robert Wilson, Martin Brock, John Horack
NASA/Marshall Space Flight Center

William Paciesas, Geoffrey Pendleton
University of Alabama, Huntsville

Chryssa Kouveliotou
USRA (on leave from University of Athens, Greece)

ABSTRACT

BATSE has now detected over 400 cosmic gamma-ray bursts. The angular distribution is isotropic to within the statistical limits. The intensity distribution exhibits fewer weak bursts than would be expected for a spatially homogeneous distribution of sources. The measured $\langle V/V_{max} \rangle = 0.324 \pm 0.016$. The strongest bursts do appear to follow the $-3/2$ power law expected for a homogeneous distribution. These observations imply that we are near the center of an isotropic distribution of burst sources whose space density decreases with distance, or that the sources are at cosmological distances.

INTRODUCTION

The origin of gamma-ray bursts remains unknown after two decades of study. The mystery deepened with the first BATSE observations[1], which showed that the burst sources were distributed isotropically in the sky, but with a space density decreasing with distance. No known galactic objects have such a distribution.

The BATSE instrument is described by Fishman et al.[2] It consists of 8 detector modules situated at the corners of the Compton Observatory. A burst trigger occurs when the count rate on two or more detectors exceeds the background level by at least 5.5σ. The background is recomputed every 17 s, and the rates are tested independently on 64 ms, 256 ms, and 1024 ms timescales.

A comprehensive sky exposure map has been maintained[3]. It is a compilation of total observing time as a function of right ascension, declination, and burst peak flux on each of the three trigger timescales. It is used to correct both the intensity and angular distributions.

INTENSITY DISTRIBUTION

The intensity distribution of GRB s provides indirect information on the radial distribution of sources. If the sources are distributed homogeneously in space out to the limit of detection, then the integral number $N(>P)$ of bursts brighter than peak flux P follows a $-3/2$ power law, for any luminosity distribution.

Figure 1 shows the integral number of bursts as a function of peak flux in the energy band 50 keV to 300 keV. The integration time for the computation of the peak flux is 1024 ms. This value was chosen because it has the lowest

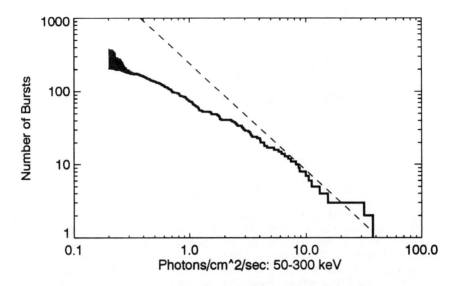

Figure 1. The integral $\log N(>P) - \log P$ distribution.

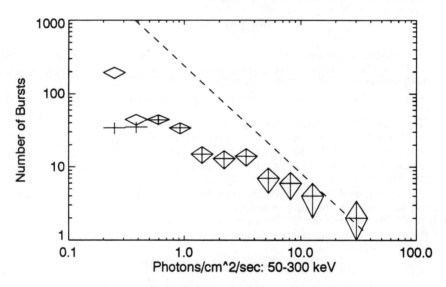

Figure 2. The differential $\log N - \log P$ distribution.

threshold of the three BATSE trigger timescales, thereby showing the deviation from homogeneity most dramatically.

At peak fluxes below about 0.4 photons/cm^{2}s, the detection efficiency decreases. The observations in this flux range are shown in Figure 1 as a solid region. The lower limit is derived from the uncorrected number of bursts observed. The upper limit is derived using the sky exposure map to determine the correction. This is an upper limit because the exposure correction does not yet include the effects of atmospheric scattering, which increases the instrument's sensitivity near the threshold. It is clear that the deviation from the $-3/2$ power law is not a result of trigger threshold effects. Figure 2 shows a differential peak flux distribution. The crosses represent the observations; the diamonds represent the upper limits after correcting for exposure. Here the corrections are more obvious in the two lowest flux bins. Note that the vertical axis is number of bursts per bin, where the bins are equally spaced logarithmically. The criterion for homogeneity again requires a $-3/2$ power law in this formulation. After correcting for instrument efficiency and sky blockage, the full sky burst rate for BATSE is ~800 bursts per year.

The V/V_{max} test for homogeneity[4] removes biases due to detector threshold effects by measuring intensity relative to trigger threshold. The value of V/V_{max} for a burst is calculated as $(C_{max}/C_{min})^{-3/2}$, where C_{max} is the maximum count rate during the burst and C_{min} is the trigger threshold count rate. The average $\langle V/V_{max} \rangle$ is 1/2 for a homogeneous distribution of sources. For bursts detected by BATSE, $\langle V/V_{max} \rangle = 0.324 \pm 0.016$, indicating a highly significant deviation from homogeneity. Other recent observations[5,6,7] have also found statistically significant deviations from $\langle V/V_{max} \rangle = 1/2$.

If the BATSE threshold were a factor of 20 higher, 26 bursts would have been detected with $\langle V/V_{max} \rangle = 0.47 \pm 0.05$. This result implies a homogeneous distribution for the brighter bursts, which is consistent with PVO results[8].

ANGULAR DISTRIBUTION

Each burst detected by BATSE will directly illuminate four of the eight detector modules. Each module contains a Large-Area Detector whose angular response approximates a cosine function. The relative count rates are used to determine the direction to the burst source.

The distribution in galactic coordinates of 447 bursts detected by BATSE is shown in Figure 3. The distributions are characterized[9] by computing $\langle \cos \theta \rangle$, where θ is the angle between the direction to the burst and the galactic center, and $\langle \sin^2 b \rangle$, where b is the galactic latitude. These functions are measures of the dipole and quadrupole moments of the distribution. To properly test for anisotropy in the observed distribution, these moments must be compared to the moments of the sky exposure map. Errors in the moments of the distribution are dominated by the statistical error due to the finite number of bursts observed. There is a smaller contribution due to instrumental error in locating each burst[10]. This location error consists of a systematic error of about 2.5 degrees[11], and a statistical error that depends on the number of photons detected. The error in the computation of the sky exposure map is negligible. A summary of the moments of the BATSE burst distribution is presented in Table 1. The first error on the measured moments represents the sampling error; the second represents the instrumental error. The anisotropy is the observed moment minus the moment of the sky exposure map, divided by the total error.

The first two entries in Table 1 are the moments in galactic coordinates. The last two entries are moments in equatorial coordinates, where the anisotropy in the exposure is now becoming significant. Briggs et al.[12] have shown that there is no significant dipole or quadrupole moment in any coordinate system.

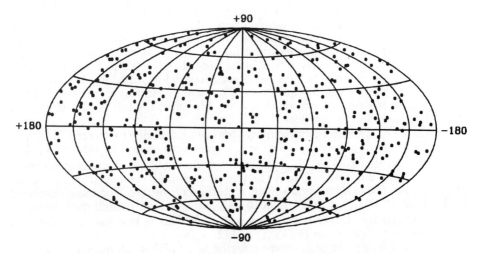

Figure 3. The angular distribution of 447 BATSE bursts in galactic coordinates.

Statistic	Observed Value	Sky Exposure Map	Anisotropy
$\langle \cos\theta \rangle$	$0.034 \pm 0.027 \pm 0.004$	-0.014	1.8σ
$\langle \sin^2 b - 1/3 \rangle$	$-0.017 \pm 0.014 \pm 0.003$	-0.004	0.9σ
$\langle \sin\delta \rangle$	$-0.012 \pm .028 \pm 0.004$	0.026	1.4σ
$\langle \sin^2\delta - 1/3 \rangle$	$0.025 \pm 0.015 \pm 0.003$	0.026	0.1σ

Table 1. Summary of BATSE burst angular distributions

DISCUSSION

The high degree of isotropy in the burst distribution, combined with the observation of non-homogeneity, is difficult to reconcile with galactic distributions of burst sources[13]. Disc populations are either homogeneous, if nearby, or have significant quadrupole moments, if more distant. Scenarios have been recently proposed[14,15] to place the sources in an extended galactic corona. These would require a minimum core radius of $\sim$30 kpc[16]. If the sources are at cosmological distances, the isotropy is required and the apparent inhomogeneity is due to redshift effects.

REFERENCES

1. C. Meegan et al., Nature **355**, 143 (1992).
2. G. Fishman et al., Proceedings of the GRO Science Workshop , 2-39 (1989).
3. M. Brock et al., AIP Conference Proceedings **265**, 399 (1992).
4. M. Schmidt, J. Higdon, & G. Heuter, Ap.J.(Letters) **329**, 85 (1988).
5. Y. Ogasaka et al., Ap.J.(Letters) **383**, L61 (1991).
6. J.-L. Atteia et al., Nature **351**, 296 (1991).
7. Higdon et al., AIP Conference Proceedings **265**, 89 (1992).
8. K. Chuang et al., Ap.J. **391**, 242 (1992).
9. D. Hartmann & R. Epstein, Ap.J. **346**, 960 (1989).
10. J. Horack et al.,"Effects of Location Uncertainties in the Observed Distribution of Gamma-Ray Bursts Detected by BATSE", these proceedings.
11. M. Brock et al.,"Improvements in Measuring the Direction to Gamma-Ray Bursts with BATSE", these proceedings.
12. M. Briggs et al.,"Coordinate-System Independent Dipole and Quadrupole Tests of Isotropy Applied to the BATSE Gamma-Ray Burst Locations", these proceedings.
13. Bohdan Paczyński, Ap. J. **348**, 485 (1990).
14. H. Li & C. Dermer, Nature **359**, 514 (1992).
15. D. Eichler & J. Silk, Science **257**, 937 (1992).
16. J. Hakkila et al.,"Constraints on Galactic Gamma-Ray Burst Models from BATSE Angular and Intensity Distributions", these proceedings.

LIMITATIONS ON DETECTING DIPOLE AND QUADRUPOLE ANISOTROPIES IN BATSE'S GAMMA-RAY BURST LOCATIONS

Michael S. Briggs, William S. Paciesas

Dept. of Physics, University of Alabama in Huntsville, Huntsville, AL 35899

Martin N. Brock, Gerald J. Fishman, Charles A. Meegan, Robert B. Wilson

NASA Marshall Space Flight Center, Code ES-64, Huntsville, AL 35812

ABSTRACT

Dipole and quadrupole tests of isotropy have been and will continue to be applied to the locations of the gamma-ray bursts observed by BATSE. To date, these tests show the locations to be statistically consistent with isotropy. This paper examines three effects that limit BATSE's ability to detect a small dipole or quadrupole anisotropy: 1) the nonzero errors on the locations, 2) the finite number of bursts observed, and 3) the nonuniform sky exposure. The dominant limitation is, and will continue to be, the finite number of bursts observed.

INTRODUCTION

An important constraint upon theories of the origin of gamma-ray bursts is the previously announced finding that the locations of the gamma-ray bursts observed by BATSE are consistent with isotropy[1,2]. This finding is a statistical statement: the values of various statistics calculated from the burst locations are consistent with the hypothesis that the sample is drawn from an isotropic population. The purpose of this paper is to discuss the limitations on BATSE's ability to detect dipole and quadrupole anisotropies.

For any experiment, there is a threshold below which the sought for effect cannot be found. Three factors that limit BATSE's ability to detect dipole and quadrupole anisotropies are discussed herein: 1) the nonzero errors in the locations determined for the bursts, 2) the finite number of bursts observed, and 3) the nonuniform sky exposure caused by earth blockage and other effects.

THE STATISTICS

Dipole and quadrupole statistics and the corresponding tests are used because they are sensitive to large-angular-scale anisotropies: dipole tests are sensitive to concentrations towards one location while quadrupole tests are sensitive to concentrations in a plane or towards two poles or to more complicated patterns. Neither are sensitive to clustering that averages to isotropy on large scales. Six different dipole and quadrupole tests are used herein–they are listed in Table 1 and graphically depicted in Figure 1. Dipole and quadrupole statistics are discussed in more detail in reference 3 and references therein.

Coordinate-system-based tests are used because they are the most sensitive tests to anisotropies that originate in the particular coordinate system: galactic-based tests are used because the most likely cause of an anisotropy is a galactic origin for the bursts, while equatorial-based tests are used because they are the most sensitive tests to the artificial anisotropies induced by BATSE's nonuniform sky exposure. Coordinate-system-independent tests are used because they search for anisotropies anywhere on the sky in a model-independent manner[3,4].

Table 1. Location Statistics of BATSE's first 447 GRBs					
Statistic	Moment Tested	Coord. System	Value Expected for Isotropy with Uniform Sky Exposure	Value Expected for Isotropy with BATSE's Sky Exposure	Observed Value (with propagation of location errors)
$\langle \cos\theta \rangle$	Dipole	Galactic	0 ± 0.027	-0.014 ± 0.027	0.034 ± 0.004
$\langle \sin^2 - \frac{1}{3} \rangle$	Quad.	Galactic	0 ± 0.014	-0.004 ± 0.014	-0.017 ± 0.003
$\mathcal{W}$	Dipole	Indepen.	3 ± 2.45	3.90 ± 3.14	1.98 ± 0.36
$\mathcal{B}$	Quad.	Indepen.	5 ± 3.16	8.40 ± 4.90	9.45 ± 1.29
$\langle \sin\delta \rangle$	Dipole	Equatorial	0 ± 0.027	0.026 ± 0.028	-0.012 ± 0.004
$\langle \sin^2\delta - \frac{1}{3} \rangle$	Quad.	Equatorial	0 ± 0.014	0.026 ± 0.015	0.025 ± 0.003

THE NONZERO LOCATION ERRORS

BATSE determines the locations of bursts by comparing the rates in its 8 Large Area Detectors[5,6]. The locations are not exactly determined, which causes statistics calculated from the locations to be not exactly determined. The locations have both statistical errors, which are primarily determined by the burst's fluence and which are estimated by the burst location software, and systematic errors, which are caused by factors such as the imperfect modeling of the instrument and which are estimated to be typically 2.5°[6]. For each burst, the 1σ location error is estimated as the root-mean-square sum of the statistical error and 2.5° systematic error. These location errors, assumed to be independent and Gaussian, are propagated to the values of the statistics via Monte Carlo simulations[7]. The locations determined by BATSE for the 447 gamma-ray bursts observed from 21 April 1991 to 7 October 1992 are shown elsewhere[2]. The values of the statistics for these locations and the Monte Carlo error estimates on those values are shown in the last column of Table 1 and in Figure 1 as dots with vertical error bars.

THE FINITE NUMBER OF LOCATIONS

The values of the statistics for a given sample obtained by a perfect instrument observing an isotropic population will fluctuate from sample to sample; e.g. each sample will have a small but nonzero dipole moment. The expected means of the statistics for an isotropic population observed with uniform sky exposure are listed in the fourth column of Table 1 and shown in Figure 1 with bold, dashed lines. Similarly, the standard deviations σ are listed in the fourth column of Table 1 and the $\pm 1\sigma$ envelopes of the distributions of the statistics are shown in Figure 1 with the non-bold dashed lines. The widths of these distributions limit any instrument's ability to detect a small anisotropy because, for a given sample, the observed statistics are expected to differ from their means by an amount comparable to the standard deviations.

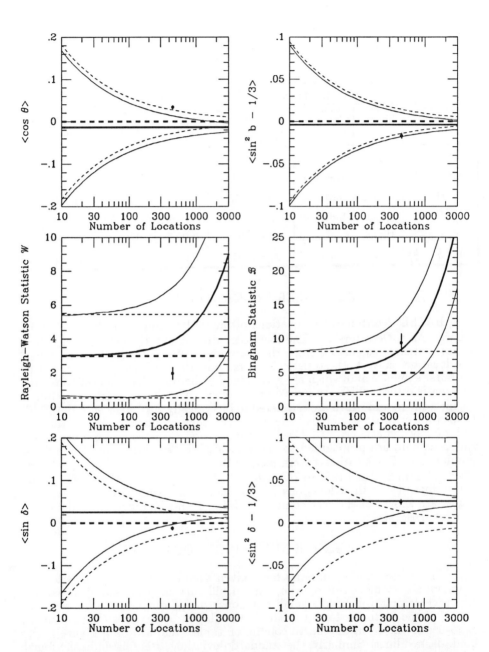

Figure 1. For each of 6 statistics, the value and estimated error of the statistic for the locations of BATSE's first 447 gamma-ray bursts are shown with a dot and vertical error bars. The expected mean values and $\pm 1\sigma$ envelopes of the distributions of the statistics for an isotropic population observed with uniform sky exposure are shown with dashed lines; the same quantities for an isotropic population observed with BATSE's sky exposure are shown with solid lines.

BATSE'S NONUNIFORM SKY EXPOSURE

Because of the anisotropic response of BATSE's detectors (the effect used to determine the locations), BATSE's intensity threshold for detecting a burst varies across the sky. This intensity-dependent effect is best described in GRO coordinates. Because of the many orientations of GRO for the various observation periods, this effect averages away and will not be further considered herein.

Because GRO is in a low-earth orbit, a considerable fraction of the sky is blocked by the earth. Averaged over the precession period of GRO's orbit, this causes a region of low declination to be observed with less exposure than the regions of high declination. Additionally, because the instrument HV is turned off for the South Atlantic Anomaly (SAA), Southern declinations receive less exposure than Northern declinations. Both of these effects are intensity independent and best described in equatorial coordinates. BATSE's sky exposure is monitored[8] and is shown as a function of declination (the dependence on right ascension is very small) for the first 165.5 days of the mission in Figure 2.

The effects of this nonuniform sky exposure have been found by Monte Carlo simulations: a location is drawn from the isotropic distribution and is then included in a simulated sample with a probability obtained from the sky exposure map. This process is repeated until a sample of N locations has been obtained and then the values of the statistics are calculated. Many simulated samples are created in order to obtain the means and 1σ widths of the distributions of the statistics. These sky-exposure-modified means and 1σ widths are listed in the 5th column of Table 1 and are depicted with solid lines in Figure 1.

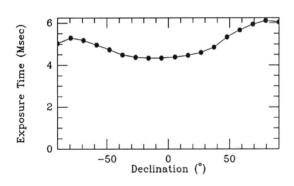

Figure 2. BATSE's sky exposure for the first 165.5 days of observations. For each point on the sky, the exposure time is the accumulated time during which a burst could have been seen.

CONCLUSIONS

Three limitations of BATSE's ability to detect dipole and quadrupole anisotropies in the locations of gamma-ray bursts have been discussed. These limitations are shown graphically in Figure 1.

The first effect is the nonzero errors on the locations of the bursts, which propagates into errors on the observed values of the statistics: these errors are shown as error bars on the observed values (dots). The second effect is the expected fluctuations in the values of the statistics about their means for a finite sample: the $\pm 1\sigma$ envelope of the distributions are the outer lines on Figure 1. The Figure shows that for all of the statistics the effect of the nonzero location errors is less important than the effect of the finite size of the observed sample.

This will continue to be the case; e.g. for the coordinate-system-based statistics both the error on the observed values and the fluctuations expected for a finite sample of size N are inversely proportional to $\sqrt{N}$.

The third effect is BATSE's nonuniform sky exposure, which if ignored, would cause an artificial anisotropy. The expected means of the statistics and their $\pm 1\sigma$ envelopes for a uniform sky exposure are depicted with dashed lines in Figure 1, while the same quantities for BATSE's actual sky exposure are shown with solid lines. The nonuniform sky exposure is an important effect when the expected mean for BATSE's sky exposure (middle solid line) lies outside of the $\pm 1\sigma$ envelope for a uniform sky exposure (outer dashed lines). The statistic $\langle \sin^2 \delta - \frac{1}{3} \rangle$ detects the nonuniform sky exposure at $N \gtrsim 200$: it is the most sensitive statistic to nonuniform sky exposure because it detects the earth blockage by measuring the quadrupole moment about the equatorial plane. In contrast, the effect of the nonuniform sky exposure does not become important for the galactic-coordinate based statistics, $\langle \cos \theta \rangle$ and $\langle \sin^2 b - \frac{1}{3} \rangle$, until $N \gtrsim 2000$: this is because the equatorial plane is highly inclined to the galactic plane. However, the nonuniform sky exposure does not limit BATSE's ability to detect an anisotropy because we know the magnitude of the effect for each statistic based upon monitoring BATSE's sky exposure.

On Figure 1, a significant anisotropy would manifest itself as one of the observed values of the statistics, including vertical error bars, lying well-outside the $\pm 1\sigma$ envelope of the distribution of the statistic expected for BATSE's sky exposure, which is shown with the outer solid lines. All of observed values are within 2σ and most are within 1σ of the expected values, consequently the locations of BATSE's gamma-ray bursts are consistent with isotropy. As the number of bursts observed increases, the effects due to BATSE's nonuniform sky exposure will increase in relative importance, but these effects can be taken into account since BATSE's sky exposure is carefully monitored. The dominant limitation on BATSE's ability to detect an anisotropy is and will remain the finite number of bursts observed.

REFERENCES

1. C. A. Meegan et al., Nature, **355**, 143 (1992).

2. C. A. Meegan et al., "The Spatial Distribution of Gamma-Ray Bursts Observed by BATSE", these proceedings.

3. Michael S. Briggs, Ap. J., to appear on 10 April 1993.

4. Michael S. Briggs et al., "Coordinate-System-Independent Tests of Isotropy Applied to BATSE's Gamma-Ray Burst Locations", these proceedings.

5. M. N. Brock et al., pp. 383-387 in AIP Conference Proceedings 265: Gamma-Ray Bursts, ed. W. S. Paciesas & G. J. Fishman (1992).

6. M. N. Brock et al., "Improvements in Measuring the Direction to Gamma-Ray Bursts with BATSE", these proceedings.

7. W. H. Press et al., Numerical Recipes (Cambridge, Cambridge, U.K., 1986), pp. 529-531.

8. M. N. Brock et al., pp. 399-403 in AIP Conference Proceedings 265: Gamma-Ray Bursts, ed. W. S. Paciesas & G. J. Fishman (1992).

COORDINATE-SYSTEM-INDEPENDENT TESTS OF ISOTROPY APPLIED TO BATSE'S GAMMA-RAY BURST LOCATIONS

Michael S. Briggs, William S. Paciesas

Dept. of Physics, University of Alabama in Huntsville, Huntsville, AL 35899

Martin N. Brock, Gerald J. Fishman, Charles A. Meegan, Robert B. Wilson

NASA Marshall Space Flight Center, Code ES-64, Huntsville, AL 35812

ABSTRACT

Two statistics and the corresponding tests of isotropy are discussed. These tests are coordinate-system independent and thus model-independent, and they yield an analytic significance for any detected anisotropy. The Rayleigh-Watson statistic W measures the size of the dipole moment of the locations, while the Bingham statistic B measures the deviation of the quadrupole moments from the values expected for isotropy. These tests are applied to the locations of the first 447 BATSE gamma-ray bursts.

INTRODUCTION

We discuss two coordinate-system-independent statistics and apply the corresponding tests of isotropy to the locations of the first 447 gamma-ray bursts observed by BATSE. Although the coordinate-system-independent statistics are 'new' to the field of gamma-ray bursts, an extensive mathematical literature exists[1,2]. Further discussion of both coordinate-system-independent and coordinate-system-dependent dipole and quadrupole statistics is given in reference 3.

The determination by BATSE that the gamma-ray burst locations are consistent with isotropy is a strong constraint on theories of the origin of gamma-ray bursts[4,5]. This finding is, of course, a statistical statement: statistics calculated from the observed locations have values consistent with the hypothesis that the locations are a sample of an isotropic population. Statistical measures of the dipole and quadrupole moments of the locations are powerful tests for large-scale anisotropies[6]. Galactic coordinate-system based statistics commonly used are $\langle \cos \theta \rangle$ (where θ is the angle between a location and the Galactic Center), the dipole moment to the Galactic Center, and $\langle \sin^2 b - \frac{1}{3} \rangle$, the quadrupole moment about the Galactic Plane[7]. The coordinate-system independent statistics discussed herein are the Rayleigh-Watson statistic W, a function of the dipole moment of the sample, and the Bingham statistic B, which measures the deviation of the quadrupole moments from the values expected for isotropy.

All of these statistics have the advantages that they are backed by mathematical theory and have simple physical interpretations in terms of dipole and quadrupole moments. The statistics W and B have the advantage that because they are coordinate-system-independent they are model-independent and thus will find an anisotropy anywhere on the sky. The disadvantage of W and B is that they are less sensitive than the galactic-coordinate-based statistics, $\langle \cos \theta \rangle$ and $\langle \sin^2 b - \frac{1}{3} \rangle$, to anisotropies that originate in galactic coordinates.

THE RAYLEIGH-WATSON DIPOLE STATISTIC $\mathcal{W}$

Represent the N locations as unit vectors: $\vec{r}_i = (x_i, y_i, z_i)$. Then the length of the dipole vector is

$$\mathcal{R} = \left| \sum_{i=1}^{N} \vec{r}_i \right| \qquad (1)$$

and the Rayleigh-Watson statistic is defined to be

$$\mathcal{W} = \frac{3}{N} \mathcal{R}^2. \qquad (2)$$

Instead of thinking of $\mathcal{R}$ as the dipole moment of the locations, it is more fruitful to consider $\mathcal{R}$ as the distance from the origin achieved by an N-step unit-step random walk. Random-walk theory then tells us that for isotropic locations $\mathcal{W}$ is asymptotically distributed as χ_3^2 and hence the expected value of $\mathcal{W}$ is 3 with a standard deviation of $\sqrt{6} = 2.45$. The asymptotic distribution is a useful approximation to the true distribution for $N \gtrsim 50$. Hence, the procedure to use the Rayleigh-Watson statistic $\mathcal{W}$ to test the isotropy of a sample of 50 or more gamma-ray bursts locations is to: 1) calculate $\mathcal{W}_{\text{obs}}$ for the sample and 2) calculate the chance probability $\alpha_\mathcal{W}$ of obtaining $\mathcal{W} \geq \mathcal{W}_{\text{obs}}$ under the hypothesis that the locations are isotropic using the χ_3^2 distribution: $\alpha_\mathcal{W} = P(\chi_3^2 \geq \mathcal{W}_{\text{obs}})$. If $\alpha_\mathcal{W}$ is very small then the hypothesis of isotropy is contradicted and the locations have a significant dipole moment.

THE BINGHAM QUADRUPOLE STATISTIC $\mathcal{B}$

The orientation matrix $\mathbf{M}_N$ is defined somewhat differently from the definition of the quadrupole matrix customarily used in physics:

$$\mathbf{M}_N = \frac{1}{N} \sum_{i=1}^{N} \begin{bmatrix} x_i x_i & x_i y_i & x_i z_i \\ y_i x_i & y_i y_i & y_i z_i \\ z_i x_i & z_i y_i & z_i z_i \end{bmatrix}. \qquad (3)$$

Since $\mathbf{M}_N$ is real and symmetric, it has three real eigenvalues λ_k and since the diagonal elements of $\mathbf{M}_N$ are the sums of squares, the λ_k are non-negative. Since the $\vec{r}_i$ are unit vectors, $\text{trace}(\mathbf{M}_N) = \sum \lambda_k = 1$. For isotropic locations the symmetry of the matrix implies that the λ_k should be equal within statistical fluctuations and thus $\lambda_k \approx 1/3$. The Bingham statistic $\mathcal{B}$ measures the deviation of the λ_k from the values $1/3$ expected for isotropy:

$$\mathcal{B} = \frac{15N}{2} \sum_{k=1}^{3} \left(\lambda_k - \frac{1}{3} \right)^2. \qquad (4)$$

For isotropic locations, $\mathcal{B}$ is asymptotically distributed as χ_5^2 and hence the expected value of $\mathcal{B}$ is 5 with a standard deviation of $\sqrt{10} = 3.16$. The asymptotic distribution is a good approximation for $N \gtrsim 40$. Hence, the procedure to use the Bingham statistic $\mathcal{B}$ to test the isotropy of a sample of 40 or more gamma-ray burst locations is to: 1) calculate $\mathcal{B}_{\text{obs}}$ for the sample and 2) calculate the chance probability $\alpha_\mathcal{B}$ of obtaining $\mathcal{B} \geq \mathcal{B}_{\text{obs}}$ under the hypothesis that the locations are isotropic using the χ_5^2 distribution: $\alpha_\mathcal{B} = P(\chi_5^2 \geq \mathcal{B}_{\text{obs}})$. If $\alpha_\mathcal{B}$ is very small then the hypothesis of isotropy is contradicted and the locations have a significant quadrupole moment.

RESULTS AND CONCLUSIONS

A map of the locations of the first 447 gamma-ray bursts observed by BATSE is presented in these proceedings[5]. The value of the Rayleigh-Watson dipole statistic calculated for these locations is $\mathcal{W}_{\text{obs}} = 1.98$ and the value of the Bingham quadrupole statistic is $\mathcal{B}_{\text{obs}} = 9.45$. (A more detailed discussion of the practicalities of applying these statistics to the BATSE data is given in reference 8). Assuming that the locations are a sample of an isotropic population, the probabilities of obtaining values this large or larger are $\alpha_\mathcal{W} = 0.58$ and $\alpha_\mathcal{B} = 0.09$. Both of these probabilities are reasonable and thus the observed locations are consistent with the hypothesis of isotropy. The coordinate-system independent tests for dipole and quadrupole anisotropies show no anisotropy in BATSE's gamma-ray burst locations.

REFERENCES

1. G. S. Watson, Statistics on Spheres (John Wiley & Sons, N. Y., 1983).

2. N. I. Fisher, T. Lewis & B. J. J. Embleton, Statistical Analysis of Spherical Data (Cambridge, Cambridge, U.K., 1987).

3. M. S. Briggs, Ap. J., to appear on 10 April 1993.

4. C. A. Meegan, G. J. Fishman, R. B. Wilson, W. S. Paciesas, G. N. Pendleton, J. M. Horack, M. N. Brock & C. Kouveliotou, Nature, **355**, 143 (1992)

5. C. A. Meegan, G. Fishman, R. Wilson, M. Brock, J. Horack, W. Paciesas, G. Pendleton & C. Kouveliotou, "The Spatial Distribution of Gamma-Ray Bursts Observed by BATSE", these proceedings.

6. D. Hartmann & R. I. Epstein, Ap. J., **346**, 960 (1989).

7. B. Paczyński, Ap. J., **348**, 485 (1990).

8. M. S. Briggs, W. S. Paciesas, M. N. Brock, G. J. Fishman, C. A. Meegan & R. B. Wilson, "Limitations on Detecting Dipole and Quadrupole Anisotropies in BATSE's Gamma-Ray Burst Locations", these proceedings.

BATSE OBSERVATIONS OF GAMMA–RAY BURSTS IN SUN–REFERENCED COORDINATE SYSTEMS

J. M. Horack, S. D. Storey, G. J. Fishman, C. A. Meegan, R. B. Wilson,
NASA – Marshall Space Flight Center, ES–64, Alabama, 35812

T. M. Koshut, R. S. Mallozzi, W. S. Paciesas
Dept. of Physics, University of Alabama, Huntsville, 35899

ABSTRACT

The results of the BATSE experiment have shown that the gamma–ray bursts possess an angular distribution that is consistent with isotropy, and a spatial distribution that is inhomogeneous.[1,2] The distance scale to the bursts, however, still remains uncertain. Several possible models have been suggested to explain the observed distribution of the bursts: heliocentric distributions such as the Oort Cloud, large galactic halos, and cosmological models. We report on an investigation into the distribution of bursts detected by BATSE in Sun–referenced coordinate systems. We find no statistically significant anisotropy in the angular distribution of the bursts in these systems; there is no dipole moment in the direction of the Sun, and no quadrupole moment associated with the ecliptic plane. Monte Carlo simulations constrain possible heliocentric burst distributions, and provide limits to burst energy in the range $\sim 10^{28} - 10^{29}$ ergs.

THE OBSERVED BURST DISTRIBUTIONS IN SUN–REFERENCED COORDINATES

The possible association of bursts with the solar-system might best be uncovered through an analysis of the angular distribution in some Sun-based coordinate system. In ecliptic coordinates, one may investigate the distribution for possible concentration towards the ecliptic plane. However, the annual motion of the Earth around the Sun will tend to remove any dipole moment in the direction of the Sun that may actually be present, resulting in an observed distribution more consistent with isotropy. To counter the motion of the Earth, and attempt to enhance any possible dipole moment with respect to the Sun, we have transformed the BATSE burst locations into a coordinate system that is centered on the Earth, but maintains one axis directed at the position of the Sun on the sky, and another directed at the ecliptic pole. This system has the advantage that it maintains a constant direction for the 1 AU offset of the Earth with respect to the Sun, thereby enhancing any existing dipole moment that may be present, and otherwise masked by the Earth's motion around the Sun.

Figure 1 shows the distribution of 377 triggered bursts seen by BATSE in this Sun–referenced system. The Sun is located at the center of the figure. In this system, we use the parameters $\langle \cos \phi \rangle$, where ϕ is the angle between the burst and the Sun, and $\langle \sin^2 d \rangle$, where d is the angle out of the ecliptic plane to investigate any deviation from isotropy. In this system, $\langle \cos \phi \rangle = -0.004 \pm 0.03$, and $\langle \sin^2 d \rangle = 0.359 \pm 0.016$. Both of these values are consistent at the $< 2\sigma$ level with what is expected from an isotropic distribution, namely 0.0 and 1/3 respectively. There is no evidence of a dipole offset towards the Sun.

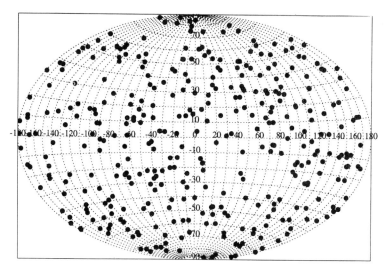

Figure 1 – Locations of 377 gamma–ray bursts in the Sun–referenced coordinate system.

Similar analyses were performed in the Sun–referenced coordinate system on sub–sets of the BATSE bursts. Distributions were analyzed based on morphology, intensity, and time of year in order to identify any possible seasonal effects. Using the parameters $\langle \sin^2 d \rangle$ and $\langle \cos \phi \rangle$, each of these sub–sets was found to be consistent with isotropy. The sub–set of the strongest bursts, with intensities greater than 2050 counts/second as measured in 0.256 second intervals, shows a slight ($\sim 3\sigma$) tendency to avoid the ecliptic plane. At this time, this deviation is not thought to be statistically significant in light of the number of sub–sets analyzed, and the rather slight deviation from isotropy. No sub–set contains a significant dipole moment in the direction of the Sun, or a concentration of sources towards the ecliptic plane. Table 1 summarizes the values of $\langle \sin^2 d \rangle$ and $\langle \cos \phi \rangle$ obtained for each subset.

Distribution	# of Bursts	$\langle \sin^2 d \rangle$	$\langle \cos \phi \rangle$
Smooth Profile	120	0.354 ± 0.026	-0.059 ± 0.052
Spike Profile	61	0.382 ± 0.039	0.039 ± 0.075
Complex Profile	168	0.353 ± 0.023	0.033 ± 0.043
1st 92 Days	80	0.329 ± 0.033	-0.036 ± 0.065
2nd 92 Days	63	0.377 ± 0.036	0.028 ± 0.072
3rd 92 Days	76	0.373 ± 0.033	0.060 ± 0.062
4th 92 Days	88	0.350 ± 0.032	0.027 ± 0.062
Weakest	95	0.363 ± 0.032	0.058 ± 0.058
Mid–Intensity	94	0.319 ± 0.029	-0.024 ± 0.062
Strongest	94	0.423 ± 0.031	-0.062 ± 0.052

Table 1 – Values of $\langle \sin^2 d \rangle$ and $\langle \cos \phi \rangle$ obtained from sub–sets of the BATSE detected bursts

If one assumes a large spherical halo of bursts around the Sun, similar halos around other stars may be detectable. Figure 2 shows the distribution of angles between α Centauri and the BATSE burst locations. One expects approximately 5 bursts within 12° of the star if the bursts are distributed uniformly on the sky. BATSE does not detect a statistically-significant excess of bursts from the direction of α Centauri.

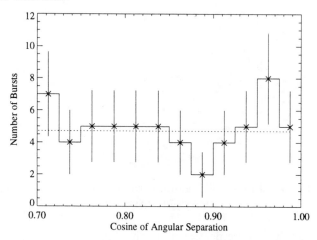

Figure 2 – Angular separation distribution of 377 gamma–ray bursts with respect to α Centauri.

MONTE CARLO SIMULATION OF THE OORT CLOUD

We have constructed a Monte Carlo simulation based on assumed parameters of the Oort Cloud to determine whether a heliocentric distribution of burst sources is consistent with the BATSE observations. After 45 years of study, the concept of a large nearly spherical cloud of comets proposed by Jan Oort[3] remains essentially correct. Weissman[4] provides an excellent review of the Oort Cloud. For the simulation, a cometary number density distribution calculated by Bailey[5] is assumed. Based on analysis of Oort's original model, the number density of cometary bodies is computed to be

$$n(r) = (R_o/r - 1)^\alpha, \qquad (2)$$

where R_o is the maximum extent of the Oort Cloud, approximately 10^5 AU.

We assume a mono–luminous burst population. Under this assumption, there is some distance, R_{vis}, at which a burst will no longer be detectable. The simulation also requires a distance d_o, interior to which, no bursts are found. The values of R_{vis} and d_o are parameters in the Monte Carlo simulation.

The relatively small offset of the Earth from the Sun, the annual motion of the Earth around the Sun, and the positioning of the burst sources at relatively large distances does not allow for an observation of a significant dipole moment directed towards the Sun. With values of d_o many times larger than the offset of the Earth from the Sun, we have assumed that the detector is at the center of the distribution. This assumption is further validated by the observation of no statistically significant dipole moment in the direction of the Sun in the BATSE data, either in equatorial or Sun–referenced coordinates. Consequently, all of the Monte Carlo distributions will satisfy the angular isotropy required by the BATSE observations. Changing the parameters will, however, change the observed value of $\langle V/V_{\max}\rangle$ in the simulation.

RESULTS OF THE MONTE CARLO SIMULATIONS

The simulation was constructed to provide $\langle \cos\theta \rangle$, $\langle \sin^2 b \rangle$, and $\langle V/V_{\max} \rangle$ for a series of values of R_{vis} when a value of α, and d_o were chosen. The parameter α was assigned two values; 3/2, as prescribed by the work of Bailey,[5] and 1/2, to investigate the dependence of the results on the exponent used. For each of these values of α, four different values of d_o were used. The smallest value of d_o was 100 AU, and is representative of the minimum distance to a gamma–ray burst allowed under an absence of parallax effects in the locations of bursts obtained by BATSE and interplanetary timing analyses. Two values of d_o, 10^4 AU and 2×10^4 AU, were chosen to be representative of the distance at which the inner Oort Cloud transitions into the more dynamically active outer cloud region.[4] A fourth value of 10^3 AU provides an intermediate radius. For all simulations, R_o was fixed at a value of 10^5 AU.

Figure 3 contains the results of the simulations performed with the parameter $\alpha = 3/2$. The 90% confidence intervals obtained for $\langle V/V_{\max}\rangle$ are plotted as a function of R_{vis}/R_o. The values of $\langle \cos\theta \rangle$ and $\langle \sin^2 b \rangle$ from the simulation are consistent with the BATSE measured value in all cases because of the inherent angular isotropy of the simulation. For 262 bursts, BATSE observes $\langle V/V_{\max}\rangle = 0.33 \pm 0.02$.[2] Acceptable values of $\langle V/V_{\max}\rangle$ are dependent on the values of d_o and R_{vis} used in the simulation.

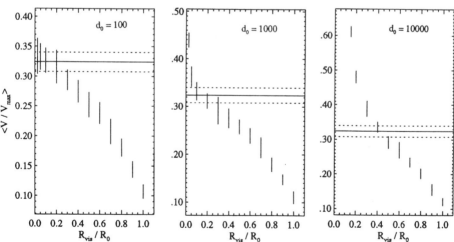

Figure 3 – Results of the Monte Carlo simulation of bursts in the Oort Cloud with $\alpha = 3/2$.

For small values of d_o (100 – 1000 AU), values of $\langle V/V_{\max}\rangle$ consistent with the BATSE data are obtained, provided that R_{vis} is less than $\sim 0.4 R_o$, or approximately 40,000 AU. This would be the distance to a burst just barely detectable by BATSE. The weakest bursts seen by BATSE have a fluence of $\simeq 10^{-8}$erg/cm^2, while the strongest bursts have a fluence of about $\simeq 10^{-5}$erg/cm^2. If located at a distance of 40,000 AU, the fluence of the weakest burst detectable by BATSE corresponds to an intrinsic energy output of $\sim 4.5 \times 10^{28}$ ergs. The strongest bursts observed by BATSE, if located at a distance of $\sim d_o$, imply an energy release of $\sim 10^{26} - 10^{28}$ ergs, depending on whether the value of d_o used is 100 AU or 1,000 AU. The average cometary nucleus is estimated to have a mass of $\sim 3.8 \times 10^{16}$ g.[4] A gamma-ray burst of 10^{28} ergs represents a release of energy corresponding to $\sim 10^{-9}$ of the total rest mass energy of the comet.

Larger values of d_o (10,000 – 20,000 AU), corresponding to the regime of the dynamically active outer Oort Cloud, can produce values of $\langle V/V_{max}\rangle$ consistent with the BATSE data as well. However, this requires that the parameter R_{vis} lie somewhere between $\sim 0.4 R_o$ and $0.7 R_o$, depending on the exact value of d_o. The weakest bursts detected by BATSE, placed at a distance of 60,000 AU, correspond to an energy output of $\sim 10^{29}$ ergs.

The simulations run for $\alpha = 1/2$ show a markedly different $\langle V/V_{max}\rangle$ profile than for the 3/2 case. Regardless of the value of d_o used in the $\alpha = 1/2$ case, in order to obtain values of $\langle V/V_{max}\rangle$ consistent with the BATSE data, one must require that the R_{vis} parameter be larger than $\sim 0.9 R_o$. In other words, BATSE is capable of detecting bursts out to at least 90,000 AU in this scenario, and is effectively sampling the entire Oort Cloud.

CONCLUSIONS

We have analyzed the angular distribution of the BATSE gamma-ray bursts in a coordinate system designed to heighten any anisotropy that may otherwise be masked by the motion of the Earth around the Sun. We find no evidence for a dipole moment towards the Sun in the bursts detected by BATSE.

We have constructed a Monte Carlo simulation of burst sources in the Oort Cloud to determine whether a distribution of bursts consistent with the BATSE data can be obtained, and to place quantitative constraints on its parameters. We find sets of parameters consistent with BATSE observations where the closest bursts are approximately 100 – 1000 AU from the Sun, and the BATSE experiment can detect bursts out to a distance of approximately 40,000 AU. Consistent parameters are also found if the bursts are confined to the dynamically active outer cloud, with BATSE detecting bursts as remote as 70,000 AU.

Future work will compare the results of these simulations to the observed C_{max}/C_{min} BATSE distribution. Fitting these scenarios to the observed intensity distribution may prove to be difficult because of the observed homogeneity of the strong nearby bursts.[1,2]

REFERENCES

1. Meegan, C. A., Fishman, G. J., Wilson, R. B., Paciesas, W. S., Pendleton, G. N., Horack, J. M., Brock, M. N., Kouveliotou, C., *Nature*, **355**, 143 – 145, (1992)
2. Meegan, C. A., et al., IAU Circular #5478, (1992)
3. Oort, J., *Bull. Astron. Inst. Neth.*, **11**, 91, (1950)
4. Weissman, P. R., *Nature*, **344**, 825 – 830, (1990)
5. Bailey, M. E., *Mon. Not. R. Astr. Soc.*, **204**, 603 – 633, (1983)

PRELIMINARY ANGULAR CORRELATION ANALYSES OF GAMMA-RAY BURSTS DETECTED BY BATSE

J. M. Horack, G. J. Fishman, C. A. Meegan, R. B. Wilson, M. N. Brock
NASA – Marshall Space Flight Center, ES-64, 35812

J. Hakkila
Dept. of Mathem., Astron., and Stat., Mankato St. Univ., MN 56002

W. S. Paciesas, G. N. Pendleton, M. S. Briggs
Dept. of Physics, University of Alabama, Huntsville, 35899

ABSTRACT

Angular correlation analyses are standard tools used to understand the distributions of numerous celestial objects, including gamma–ray bursts. Three useful methods of testing for clustering are binning analyses, nearest neighbor tests, and the correlation function. We present the results of a preliminary investigation into the clustering properties of the bursts detected by BATSE. We find no statistically significant evidence of bursts clustering around the directions of M31, the LMC, or the Virgo Cluster, and there is no evidence for bursts tending to cluster among themselves.

INTRODUCTION

Clustering analyses have been widely used in the study of the distributions of distant galaxies,[1,2,3] quasars,[4,5] and gamma–ray bursts.[6] Evidence of the bursts tending to cluster near one or more known objects would be a great clue to their distance and energy output. We investigate the possibility that bursts may tend to cluster in the direction of three specific celestial objects: the LMC, M31, and the Virgo Cluster of galaxies. In addition to clustering around a specific object or direction, one might also suspect that bursts themselves may tend to cluster or clump together in a statistically significant manner. Evidence of bursts tending to cluster in pairs or triplets, for example, would provide new constraints for theories explaining the origins of the gamma–ray bursts. We present an analysis of the bursts' self–clustering properties.

We employ three different methods of analysis to test the BATSE gamma–ray bursts, each of which yields no statistically significant evidence for clustering of any type. The distribution of angles between the BATSE bursts is examined, a nearest–neighbor test is employed, and we compute the correlation function for the bursts seen by BATSE, including corrections for non–uniform sky-coverage and uncertainties in the burst locations. These three methods of analysis are standard tests for clustering, and have been used, for example, by Osmer[4] to search for clustering among distributions of quasars.

CLUSTERING OF BURSTS IN THE DIRECTION OF KNOWN OBJECTS

If one is investigating the possibility of bursts in a large galactic halo,[7,8,9] one may ask if a concentration of bursts is seen in the direction of other galaxies, presumed to also have a large halo of bursts. We investigate the possibility of gamma–ray bursts clustering in the directions of the Large Magellanic Cloud, M31, and the Virgo Cluster of galaxies.

To quantify the amount of clustering that may exist around these objects, we use a binning analysis technique. Beginning at the location of the object in question, we divide the sky into equal–area regions and determine the number of bursts in each region at an angle between θ and $\theta + \Delta\theta$ from the object in question. For an isotropic distribution covering the entire sky, the number of bursts expected, N_{exp}, in a bin of solid angle Ω is simply given by

$$N_{\text{exp}} = N_{\text{total}}(\Omega/4\pi), \qquad (1)$$

where N_{total} is the total number of bursts in the distribution. Clustering is evidenced by a statistically significant excess of bursts in the bin or bins nearest the object in question when compared to the number of bursts expected from an isotropic distribution of the same size.

Figure 1 shows the distribution of angles from the LMC, M31, and the Virgo Cluster. The dotted line represents the number of bursts expected in a given bin from an isotropic distribution. These data have not been corrected for the non–uniform exposure of BATSE to the celestial sphere, measured to be a small effect.[10] The non–uniformity in the BATSE sky exposure is a smoothly varying function, and does not change substantially over angles of $< 30°$.[10] Consequently, regions near a certain object will be sampled by BATSE with about the same efficiency as BATSE samples the object itself. For each of these objects, no statistically significant excess of bursts is found in the bins nearest to the object in question.

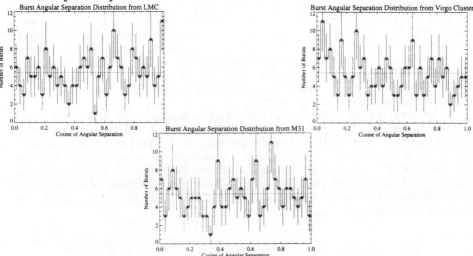

Figure 1 – Angular Separation Distributions for the BATSE gamma–ray bursts with respect to the LMC, M31, and the Virgo Cluster.

The binning–analysis method has some disadvantages. The first bin in Figure 1 ($\cos\theta \simeq 1$) corresponds to a distance of $\simeq 12°$ from the object. Small-scale clustering, on the order of $\sim 1°$ for example, may not be found through this method, unless the bin size were decreased. With burst location uncertainties that are on the order of $\sim 5°$,[10] the bin size was chosen to provide accounting for all bursts actually within $\sim 5°$ of a given object, but possibly localized an additional $5° - 7°$ away. Tests for clustering on scales smaller than the location uncertainty using this method may be somewhat suspect. Decreasing the bin size substantially would also result in poorer counting statistics. Despite these disadvantages, this method provides a quantitative test for clustering in the direction of specific objects. Using this method, we find no evidence for an excess of bursts from the directions of M31, the LMC, or the Virgo Cluster.

SELF–CLUSTERING PROPERTIES OF THE BATSE GAMMA–RAY BURSTS

In the second part of this work, we investigate the possibility that gamma–ray bursts may tend to cluster among themselves. We utilize a binning analysis technique similar to the one described in the previous section, in addition to a nearest neighbor test, and calculation of the auto–correlation function.

The binning analysis used in this section is similar to the one described above. However, instead of using a fixed direction in space as the point of comparison, each burst in turn is used as the center, and the angles to the remaining bursts are binned accordingly. In this manner, one obtains the distribution of angles between the burst pairs. Self–clustering among the bursts would be evidenced by a statistically significant excess in the number of burst pairs in a bin or bins, compared to what one expects for an isotropic distribution of bursts. Figure 2 contains the results of this comparison. These data, like those of Figure 1, are not corrected for the non–uniform exposure of BATSE to the celestial sphere. No statistically significant excess is found for any angle separation angle between burst pairs.

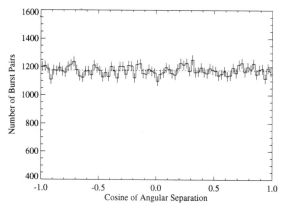

Figure 2 – Angular Separation Distribution of all pairs of bursts detected by BATSE.

We have also employed a nearest neighbor analysis to search for self–clustering among the BATSE gamma–ray bursts. The analysis proceeds by selecting the first burst from a set of n bursts, and measuring the angles to the $(n-1)$ remaining bursts in the order of their distance from the first burst used. This measurement is repeated using the $(n-1)$ remaining bursts as the center. One then obtains the distribution of mean distances to the n_{th} nearest neighbor. The result is compared to what one obtains from an isotropic distribution. This type of test is sensitive to weak clustering on a small–scale, but is rather insensitive to larger–scale clustering.[11]

Figure 3 contains the results of the nearest neighbor test on the BATSE burst distribution. The average angular separation between a burst and its n_{th} nearest neighbor is plotted against the neighbor in question. The triangles represent the 90% confidence intervals obtained from Monte Carlo simulations of isotropic distributions. The filled circles are the BATSE data. These data are consistent with an isotropic distribution, and show no evidence of clustering.

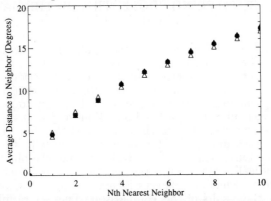

Figure 3 – Nearest Neighbor Distribution of the BATSE gamma–ray bursts.

The two–point angular correlation function, $w(\theta)$, is perhaps the most widely used tool to investigate distributions for clustering. In principle, $w(\theta)$ is simple to calculate, and is given by[12]

$$\frac{\delta P}{\delta \Omega} = N[1 + w(\theta)]. \qquad (2)$$

In Equation 2, N is the average number of bursts per steradian over the sample, and δP is the probability of finding a burst in a region of solid angle $\delta \Omega$ a distance θ from the point in question.

Non–uniformities in sky coverage and burst location uncertainties are important effects to be accounted for in computing $w(\theta)$. If these are not accounted for, bursts found in less frequently sampled parts of the sky will have fewer counterparts, and correlations that may be found on scales smaller than the location uncertainty are suspect. In light of these selection effects, $w(\theta)$ was calculated using Gaussian positional errors and weighting the bursts according to the sky-exposure of BATSE[10]. Details of the inclusion of these selection effects can be found in Hakkila et al.[8]

The resulting $w(\theta)$, computed for the first 285 gamma–ray bursts detected by BATSE, is shown in Figure 4. The crossed symbols represent the 90% confidence intervals obtained by computation of $w(\theta)$ from isotropic distributions of bursts. The filled circles are values computed from the BATSE data. The large value of $w(\theta)$ at the extreme left of the figure occurs because the separation angle is smaller than the size of the location uncertainty. This "overlapping" of the uncertainty regions causes an artificial correlation. The BATSE data agree with the results obtained from isotropic distributions of bursts. No statistically significant evidence for clustering on any scale at the $> 2\sigma$ level is apparent in Figure 4.

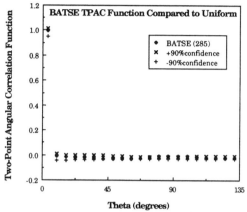

Figure 4 – The BATSE two–point angular correlation function $w(\theta)$, incorporating non–uniform sky exposure and location uncertainties.

CONCLUSIONS

We have investigated the angular distribution of the gamma–ray bursts detected by BATSE in an attempt to find self–clustering, or clustering in the direction of the LMC, M31, and Virgo Cluster. We find no statistically significant evidence for clustering in the angular distribution of the gamma–ray bursts detected by BATSE using three different measurement techniques: binning analyses, nearest neighbor distributions, and the auto–correlation function.

REFERENCES

1. Abell, G. O., in *Galaxies and the Universe*, (eds. A. Sandage, M. Sandage, and J. Kristian), (Univ. of Chicago Press: Chicago), 601, (1975)
2. Bachall, N., *Ann. Rev. Astron. and Astrophys.*, **15**, 505, (1977)
3. Fall, S. M., *Rev. Mod. Phys.*, **51**, 21, (1979)
4. Osmer, P. S., *Ap. J*, **247**, 762 – 773, (1981)
5. Hartwick, F. D. A., & Schade, D., in *Ann. Rev. Astr. and Astrophys.*, **28**, 437, (1990)
6. Hartmann, D., Linder, E., and Blumenthal, G., *Ap. J*, **367**, 186, (1991)
7. Brainerd, J. J., *Nature*, **355**, 522–524, (1992)
8. Hakkila, J., et al., these proceedings.
9. Horack, J. M., et al., *Ap. J*, (submitted), (1992)
10. Brock, et al., in *Proc. Gamma Ray Bursts*, (W. S. Paciesas & G. J. Fishman, eds.), 399 – 403, (AIP: New York), (1992)
11. Webster, A., *MNRAS*, **175**, 61, (1976)
12. Peebles, P. J. E., *Ap. J*, **185**, 413 – 440, (1973)

CONSTRAINTS ON GALACTIC GAMMA-RAY BURST MODELS FROM BATSE ANGULAR AND INTENSITY DISTRIBUTIONS

J. Hakkila
Mankato State University, Mankato, MN 56002-8400

C. A. Meegan, G. J. Fishman, R. B. Wilson, M. N. Brock, J. M. Horack
NASA/Marshall Space Flight Center, ES-62, Huntsville, AL 35812

G. N. Pendleton, W. S. Paciesas
University of Alabama in Huntsville, Huntsville, AL 35899

ABSTRACT

The paradigm that gamma-ray bursts originate from Galactic populations is studied in detail using the BATSE angular and intensity distributions. Monte Carlo models of gamma-ray burst spatial distributions and luminosity functions are created, then folded through mathematical models of BATSE selection effects. The BATSE Burst Catalog and the Monte Carlo Burst Catalogs are analyzed using modified statistical homogeneity and isotropy tests. Numerous catalog results are studied so that errors can be established.

Analysis of BATSE angular and intensity distributions greatly constrains origins and luminosities of bursts. No single population of Galactic Disk, Halo, or Local Spiral Arm sources satisfactorily explains BATSE observations (and burst luminosity functions are of secondary importance when considering such models). One family of models still satisfying BATSE data is comprised of sources in a spherical Galactic Corona. Coronal models are limited to small ranges of burst luminosity and core radius. Multiple-population models work only if (1) the primary population accounts for the general isotropy and inhomogeneity seen by BATSE, and (2) secondary populations either have characteristics similar to the primary population or contain small numbers.

INTRODUCTION

The intensity and angular distributions of gamma-ray bursts have been measured[1] by the Burst And Transient Source Experiment (BATSE) on NASA's Compton Gamma-Ray Observatory. These distributions are used to identify burst sources statistically because no burst has been positively identified at other wavelengths. BATSE results are well-known: (1) few faint bursts exist, indicating that any Euclidean distribution must be strongly heterogeneous, and (2) bursts are located isotropically on the sky, suggesting that a bounded distribution is being observed from near its center.

Until recently, the accepted astrophysical model for the bursts had been that they originated on Galactic neutron stars. Other observations have been supportive of this model: bursts exhibit short timescale variability (indicative of compact sources), and some bursts exhibit what appear to be emission and/or absorption features. In order for the Galactic neutron star model to be correct, spatial locations and luminosity functions of some Galactic population(s) must explain BATSE angular and intensity distributions. Constraints cannot be placed on Galactic models without first correcting for observational selection effects introduced by BATSE.

METHODOLOGY

The method used is to compare Monte Carlo angular and intensity distributions to those observed by BATSE. It is encoded within three computer programs: the first generates Monte Carlo models of GRB spatial positions and luminosity functions then incorporates BATSE selection biases, the second characterizes Monte Carlo burst catalogs in terms of homogeneity and isotropy, and the third compares Monte Carlo analyses to BATSE results, obtaining errors via significance levels.

Spatial models currently generated by this program include uniform, Galactic Disk, Galactic Halo, Local Spiral Arm, and Galactic Coronal distributions. The parameters defining these distributions as well as their functional forms are alterable. Flexibility exists for constructing multiple population models (either generic or specific, such as inclusion of M31). Luminosity functions that can be generated include standard candle, Gaussian, log-normal Gaussian, and power-law distributions. Monte Carlo sources are sifted through subroutines that mimic BATSE effects of (1) inhomogeneous, anisotropic sky sampling, and (2) burst localization errors. These give the Monte Carlo catalogs the same appearance as the BATSE catalog.

The Monte Carlo catalogues are analyzed in terms of standardized homogeneity and isotropy tests, including BATSE selection effects. Homogeneity analyses used are $<V/V_{max}>$, the V/V_{max} histogram, and various size-frequency distributions. Isotropy analyses of Monte Carlo catalogues currently include calculation of multipole moments (rigorously and in Galactic coordinates), the two-point angular correlation function, and an angular excess test.

Errors are generated through many runs of each Monte Carlo model (this study uses the 90% significance level). Formal errors are not as accurate as these because anisotropic, inhomogeneous sky sampling and burst localization error effects alter results in complex ways.

GALACTIC MODELS

Analysis is made of Galactic gamma-ray burst models using the 260 bursts listed in the first BATSE Burst Catalog[2] as well as a larger set (400 bursts) of as yet unpublished data. The analysis (discussed in more detail elsewhere[3]) is directed at Disk, Halo, Spiral Arm, Corona, Extended Corona, and two-population models. It is convenient when discussing results to refer to the model's sampling distance, which is the distance to which the most luminous bursts in the distribution could be observed above the minimum BATSE detection threshold.

No single population of sources located in the Galactic Disk, Galactic Halo, or Local Spiral Arm can account for the angular and intensity distributions observed by BATSE. Parameters for these models are "best case" scenarios in which the distributions conform to BATSE observations without violating boundary conditions imposed by the structure of the Milky Way. Disk and Halo distributions can be sampled in zones representing roughly three sampling distance scales: (1) local zones, in which the distributions are homogeneous and isotropic, (2) mid-distance zones, in which the distributions are heterogeneous and anisotropic, and (3) distant zones, in which the distributions are less anisotropic and extremely heterogeneous (essentially all bursts have V/V_{max} near zero). When sampling Disk or Halo distributions in mid-distance zones, heterogeneities are always coupled to measurable anisotropies. The effects of applying broad luminosity functions are found to be small when compared to those introduced by the spatial structure, in agreement with other findings[4]. Spiral arm models do not satisfy BATSE observations, and attempts to vary the structure of the Arm and the luminosity function cannot hide the fact that heterogeneities in such models are accompanied by strong anisotropies.

Sources lying in a spherical dark matter Corona (with a radial density of the form $n(r)=n_c[1+(r/r_c)^\alpha]^{-1}$ in which $\alpha=2$) explain BATSE data[5], but only for a narrow range of model parameters. Although no angular dependence is assumed, the Galactic Corona is possibly non-spherical (spherical coronae are best-case scenarios). An additional constraint is that a similar corona surrounds M31 (it is assumed that other Local Group galaxies have coronae insignificant when compared to those of the Milky Way and M31). The corona of M31 is modeled in identical fashion to that of the Milky Way and both coronae terminate at their respective tidal radii. Coronal models that satisfy BATSE observations (assuming standard-candle luminosities) are presented in figure 1 in a plot of core radius vs. sampling distance. Acceptable models occupy a bounded region because (1) the distribution is too homogeneous when sampled only slightly beyond the core radius and is too heterogeneous when sampled too far beyond it, and (2) excessive numbers of bursts are observed from the Galactic Core when the core radius is too small and from M31 if the sampling distance is too large. This allowed phase space shrinks with additional observations (statistical errors), which is shown by results of 100, 200, 400, 800, and 1600 bursts (the latter two values are projected). At the present time this implies a core radius in excess of 34 kpc and a distribution sampled to at least 140 kpc, with a luminosity range of a factor of six only. If no excess is observed by BATSE in the directions of either the Galactic Center or towards M31, then the parameter space of Coronal models will be eliminated after roughly 2200 bursts.

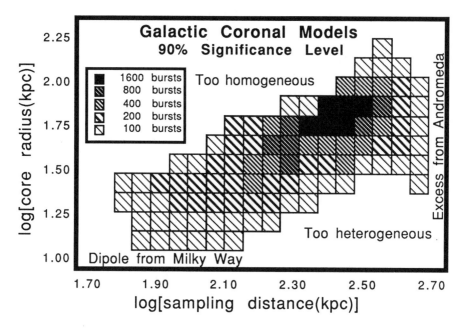

Figure 1. Spherical Galactic Coronal models satisfying BATSE data.

Another model studied is that in which an extended corona of Local Group galaxies is sampled beyond its "edge." The best-case scenario for this distribution is a spherical one, which is unlikely considering the bimodal distribution of Local Group material. The radial density gradient for this scenario is assumed to be that used in Coronal models. A core radius is needed to counteract the off-center location of the Milky Way in the Local Group (assumed to be 0.35 Mpc). This off-center location and the Sculptor and the M81 groups of galaxies (roughly 2.5 Mpc away and with masses assumed equal to that of the Local Group) are the greatest constraints on this model. The extended corona can be modeled by simply increasing the scale of Coronal models by a factor of $350/8.5 \approx 40$. Figure 1 then indicates that (after 200 bursts) the minimum sampling distance needed to not observe a dipole moment from the Local Group is 1.4 Mpc, which should overlap coronae of both the Sculptor and M81 groups. This model is therefore unacceptable.

Two-population models have also been studied, with results that are remarkably constraining to Galactic populations. Disk and Halo populations are mixed with isotropic populations in varying degrees to determine when the combined distribution shows measurable anisotropies. Two solutions are allowed. The first occurs when the Disk or Halo population is sampled in the local or mid-distance zone. Any strong anisotropy produced by this population can be balanced by decreasing the number of these sources. All cases with marginally acceptable anisotropies are fairly homogeneous, forcing the isotropic population to be heterogeneous (the best fit occurs when no Disk or Halo sources are present, it is quite unlikely that more than 30% of such sources are present, and it is physically impossible that more than 65% of such sources are present). The second solution (suggested previously[6]) occurs when the Disk or Halo distribution has been sampled in the distant zone. In this zone both Disk and Halo distributions are roughly isotropic and have $<V/V_{max}>=0$. Although they can be combined with any isotropic distribution to yield a reasonable value of $<V/V_{max}>$, the size-frequency distribution of this mix is wrong because its slope bends in the opposite sense as that observed by BATSE[7]. Nonetheless, it is possible to mix a small number of these sources into the sample (generally < 5 to 10%) such that their presence is disguised by statistical errors. The inescapable conclusion is that Disk or Halo sources (or Spiral Arm sources, for similar reasons) are essentially contaminants to the isotropic population and must therefore be rare. Other sources *may* be mixed with a Galactic Coronal population, but only because the Coronal distribution works by itself.

CONCLUSIONS

Tight constraints exist on Galactic gamma-ray burst models based on angular and intensity analyses that include BATSE selection effects. These constraints arise from the confined distribution of light-producing material, the Sun's off-center location in the Milky Way, and the presence of a relatively nearby massive neighbor (M31). Galactic Disk, Halo, and Spiral Arm sources cannot be common among BATSE bursts; otherwise the anisotropies introduced by them would be observed. Although Galactic Coronal models are still viable, statistical errors in BATSE data have reduced their allowed spatial properties to a small range of values and their luminosities to essentially one order of magnitude.

REFERENCES

1. C. A. Meegan et al., Nature 355, 143 (1991).
2. G. J. Fishman et al. (in preparation, 1992).
3. J. Hakkila et al. (in preparation, 1992).
4. S. Mao and B. Paczynski, preprint (1991).
5. B. Paczynski, Acta Astronomica 41, 157 (1991).
6. R. E. Lingenfelter and J. C. Higdon, Nature 356, 132 (1992).
7. B. Paczynski, Acta Astronomica 42, 1 (1992).

IMPROVEMENTS IN MEASURING THE DIRECTION TO GAMMA-RAY BURSTS WITH BATSE

M. N. Brock, G. J. Fishman, C. A. Meegan, R. W. Wilson
Marshall Space Flight Center, AL., 35812

G. N. Pendleton, M. T. Stollberg, W. S. Paciesas
University of Alabama in Huntsville, AL., 35899

ABSTRACT

To understand the source of gammy-ray bursts, we must distinguish between theories placing the sources within and around our galaxy and theories placing them well beyond our galaxy. To make that distinction, we must accurately determine the direction to bursts. Previous directions determined by the Burst and Transient Source Experiment on the Compton Gamma ray Observatory included *systematic* errors of around 7 degrees in addition to *statistical* errors between 1 and 20 degrees resulting from the Poisson uncertainty in photon flux measurements. Although these errors did not significantly affect our reported measurement of moments of the γ-ray burst distribution, we made a concerted effort to understand and reduce the systematic errors in BATSE's burst directions. We report here on the result of our efforts.

SOURCES OF LOCATION ERROR ADDRESSED

We have examined six potential sources of error in the direction to bursts detected by BATSE, the response of BATSE's large area detectors to flux nearly parallel to a detector's face, removal of background γ radiation from our measurement of the burst, burst flux reflected from the Earth's atmosphere, deadtime during the accumulation of burst photon counts, calibration of the thresholds of BATSE's energy channels, and the selection of detectors for use in the direction determination. Our analysis focused on relatively hard solar flares and bursts with directions determined by the Interplanetary Network. The bursts in our sample are more intense than typical γ-ray bursts. Weaker bursts may be more vulnerable to certain systematic effects, particularly incorrect detector selection, but we believe we have removed most of the systematic error from previously reported directions.

IMPROVED CALIBRATION OF BATSE'S RESPONSE

A correction to the response of BATSE's large area detectors significantly reduced our direction errors.[1] The correction removed 36% of the average error in the sample presented here. The correction of detector response was calibrated with a sample of solar flares collected early in the CGRO's flight (not the sample presented here), and flare locations improved more than burst locations. Flares have softer spectra than bursts and a detector's response varies most with angle of incidence when the source flux is soft. In time, a larger sample of IPN located bursts may permit us to further improve our knowledge of the location dependence of BATSE's response to harder sources.

REMOVING BURST FLUX REFLECTED FROM THE ATMOSPHERE

Because the Earth's atmosphere is an efficient reflector of γ-rays, we must remove burst flux reflected from the atmosphere from our burst measurement before computing the burst's location. After subtracting the interpolation of background flux before and after a burst, we attempt to remove reflected flux by first computing a location which ignores reflection, then predicting the reflection from the first order location, subtracting the reflected rates from the observed rates, and repeating this procedure until a stable location is reached. The model atmosphere we initially used to predict reflected flux assumed a flat atmosphere, simple absorption interactions, and single Compton scattering of reflected flux. After launch, we performed a more complete monte carlo analysis of atmospheric reflection and incorporated the results into our burst location software. The effect of removing the new correction for atmospheric reflection is shown in **Figure 3**. Correcting for reflection improves the error in burst locations by an average of about 50%, but the correction does not improve the location of solar flares. The softer spectrum of a flare suffers much less reflection than the spectrum of a γ-ray burst.

SUBTRACTING BACKGROUND FLUX FROM BURST FLUX

To obtain the γ-ray flux from a burst, we must subtract from BATSE's measurement of flux during a burst the background flux which BATSE would have measured in the absence of the burst. Background subtraction is straightforward for the majority of bursts since the background flux is unchanging or changing at a constant rate around the time of the burst. For previously reported directions, our burst location software computed the level of background flux during a burst using a two-point interpolation between flux measurements before and after a burst. However, when a burst is very long or occurs during a period of high particle activity in the atmosphere, the slope of the background flux as a function of time may change significantly around the time of the burst. To improve the background correction of these bursts, we modified our software to sample the background and burst flux in several different time intervals, up to four intervals before and after each burst and up to eight intervals during each burst.

In addition to allowing second order variations in our model of background flux during a burst, we carefully reevaluated the selection of burst and background time intervals to avoid unusual background rates, and we removed intervals during a burst when the flux was very weak and very intense to increase the significance of our measurement and to avoid known limits of our deadtime correction. Those corrections produced substantial improvements in the direction to some events, and we believe that careful attention to the selection of burst and background measurements significantly reduced the direction errors in our catalog.

LOCATIONS WITH ALL CORRECTIONS TO DATE

After making a number of improvements to BATSE's burst location software and detector calibration, we selected a sample of solar flares with relatively hard spectra and bursts located by the Interplanetary Network[2] to evaluate the result of our efforts. The computed locations of the flares and bursts are presented in **Figure 1** as offsets from known sun locations and IPN locations, respectively. Azimuth is measured perpendicular to the line of constant right ascension through each known location, and elevation is measured along the line of constant right ascension through each known location. The largest error on the chart is 15 degrees. The relatively weak flare with this error apparently suffered a background subtraction error which led the location software to locate the flare with a detector which should have detected no flux and to ignore a detector which should have detected flux. This detector selection problem is complicated by many different systematic effects, and the problem continues to elude a complete solution. Our location software now includes up to six detectors in each direction calculation. The response of a detector to flux arriving from behind is not well known, but a more thorough model of BATSE'S response may in the future make detector selection less critical. Locations computed without the corrections to detector response cited earlier are illustrated in **Figure 2**.

COMPARISON WITH PREVIOUSLY PUBLISHED LOCATIONS

Most of the burst locations computed with our improved software and calibration changed no more than we expected from previously published locations.[3] Some locations moved more than expected. In most cases, we have located the source of the larger than expected error and convinced ourselves of the superiority of the new location. By far the largest of these errors (102 degrees) afflicted the event of April 23 1991 when the burst location software used an incorrect spacecraft orientation to transform a burst location into the equatorial coordinate system. As we had hoped, the changes in location did not significantly affect our initial finding of little or no correlation of burst directions with the galactic disk, galactic center, etc. The distribution of burst directions observed by BATSE remains highly isotropic.[4]

A small class of bursts suffered from errors in the calibration of the thresholds of the energy channels of BATSE's large area detectors. We corrected the error by searching our energy calibration for instances of the error and by averaging the calibrated parameters over longer time intervals.

THE SYSTEMATIC COMPONENT OF LOCATION ERRORS

In order to determine the extent to which our location errors are not due to the inherent uncertainty in our detectors' measurement of photon arrival rates, we defined the *systematic error*, σ_s, in our sample of bursts and flares as follows.

$$\sigma_s^2 = \frac{1}{n} \sum_i (e_i^2 - \sigma_i^2),$$

where n is the number of events in our sample, e_i is the observed error in the ith burst, σ_i is the error resulting from uncertainty in the Poisson counting of photons, and σ_s is the average systematic component of location error. σ_s is a simple measure of the systematic error in our burst direction and assumes that statistical and systematic errors vary independently. For the locations illustrated in **Figure 1**, σ_s is 2.5 degrees.

While we have not exhausted all the improvements we might make to BATSE's burst directions, we have approached a limit beyond which more improvement will require much greater effort affecting only small classes of bursts. Since BATSE determines the direction to weak bursts with an accuracy which is unavoidably greater than most of the systematic errors we have observed, we will likely concentrate in the future on ensuring that weak bursts, which are numerous in BATSE's catalog, do not suffer from systematic errors larger than those we have seen in stronger bursts. Analysis already conducted on the low intensity portions of well located, strong bursts indicates that errors in our directions to weak bursts increase as expected with decreasing burst intensity.[5]

REFERENCES

1. Pendleton, G. N., et al., AIP Conference Proceedings 265 (AIP, 1992), p. 395.
2. Hurley, K., et al., these proceedings.
3. Brock, M. N., et al., AIP Conference Proceedings 265 (AIP, 1992), p. 383.
4. Briggs, M. S., et al., these proceedings.
5. Horack, J. M., et al., these proceedings.

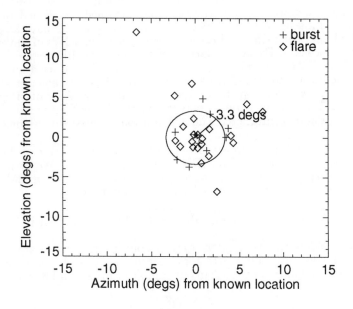

Figure 1. BATSE locations of 11 IPN located γ-ray bursts and 22 hard solar flares.

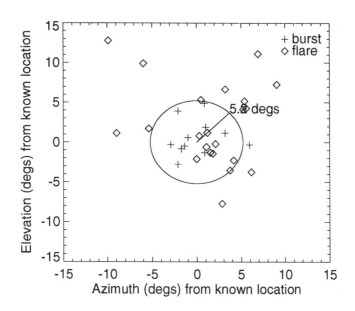

Figure 2. BATSE locations of 33 events computed without corrections to the response of BATSE's large area detectors.

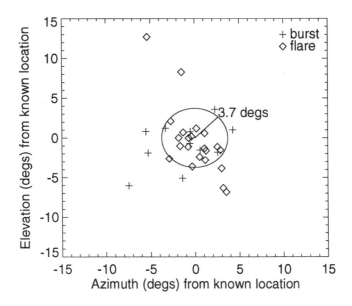

Figure 3. BATSE locations of 33 events computed without removing burst flux reflected by the Earth's atmosphere.

EFFECTS OF LOCATION UNCERTAINTIES ON THE OBSERVED DISTRIBUTION OF GAMMA-RAY BURSTS DETECTED BY BATSE

J. M. Horack, C. A. Meegan, G. J. Fishman, R. B. Wilson, M. N. Brock
NASA – Marshall Space Flight Center, ES-64, 35812

W. S. Paciesas, A. G. Emslie, G. N. Pendleton
Dept. of Physics, University of Alabama, Huntsville, 35899

ABSTRACT

Three scenarios of burst distributions seem geometrically compatible with the data obtained by BATSE[1,2]: association of bursts with a heliocentric distribution, bursts in an extended galactic halo, and bursts at cosmological distances. Detection of a slight deviation from isotropy in the angular distribution would provide an important clue to determining the distances to the bursts, by discrimination among the three scenarios. The ability to tightly constrain the amount of anisotropy present in the distribution is limited, in part, by uncertainties in the individual burst locations. Using bursts whose locations have been determined accurately and independently by the Third Interplanetary Network, we derive an estimate of the BATSE location uncertainties. We then incorporate these into 3–dimensional Monte Carlo simulations of burst distributions to show that despite individual location uncertainties of ~5°, the ability of BATSE to place stringent limits on the anisotropy present in the angular distribution of the gamma–ray bursts is not significantly degraded. Once the amount of allowable anisotropy is known, constraints can be placed on the possible parent distributions of the burst sources. We conclude with a brief discussion of these constraints for galactic halo models.

QUANTIFICATION OF LOCATION UNCERTAINTIES

Several factors complicate the simple cosine–like response of the BATSE detectors used to determine the direction to a burst. If not properly accounted for, these factors can introduce errors as large as ~10° in the computed position.[3] First–order corrections in the BATSE location algorithms reduce these errors substantially, however do not remove all sources of error in determining the location.

Scattered radiation from the atmosphere and the spacecraft is a major component of systematic error in the burst location. The scattered photons can produce significant counting rates in detectors not exposed to the burst direction, and substantially increase the rates in detectors with small projected areas to the burst. The BATSE location algorithm contains corrections for scattering from the atmosphere, however does not employ a correction for spacecraft scattering. Errors in the atmospheric scattering correction, and from the uncorrected spacecraft scattering cause the computed location to differ from the actual location.

An additional class of errors is statistical in nature, dependent on the number of counts observed. The magnitude of these errors is primarily a function of the intensity of the burst, but also depends on the geometry involved in detecting an event. The location methodology[4] used for the initial analyses of the BATSE distribution[1,2] is sensitive to the correct determination of the third brightest detector, which uniquely defines the burst location. If the event is faint enough that the counting rate in the third brightest is barely above background, statistical fluctuations in the rate may lead to erroneous identification of the third brightest detector. This causes the derived location to be incorrect, by an amount larger than expected from simple propagation of the statistical errors.

We have separated and quantified the scattering–related and counting–related uncertainties present in the BATSE burst locations by employing the positions for seven bright bursts as determined by the Third Interplanetary Network[5]. These bursts were bright enough to be seen by three or more experiments (including BATSE), and could be independently localized to an accuracy of better than $1°$ using timing analyses. A comparison of this independent location with the BATSE location gives a measure of BATSE's total error, both systematic and statistical, in the location for that burst. The brightness and the duration of these strong bursts allow for a separation of the error into two parts, one of which is predominantly statistical, related to the number of counts, and the other systematic, dominated by the scattered photons produced in the detection environment.

An intense burst with an appreciable duration can be artificially divided into many weaker and shorter bursts, all of which have the same celestial location and geometrical detection environment. The geometric factors of the scattering environment that are the principal contributors to the systematic portion of the location error are therefore the same for each sub–interval. Any variation in the derived locations for these sub–intervals is then, to first order, due to the varying number of counts seen by the detectors in the interval. Errors associated with the sub–interval are thus representative of a burst with a similar intensity which, because of its inherent faintness, could not be detected by other less sensitive instruments. Most of the bursts detected by BATSE fall into this category. Having obtained a large set of sub–intervals and their associated uncertainties, the systematic and statistical contributions to the uncertainty in any burst location can be accurately estimated through comparison to a similar intensity sub–interval.

Figure 1a shows the result of dividing data from GRB910522 into 2.048–second sub–intervals, and determining their locations. These locations are shown (•), along with their centroid (⋆), and the location of the event as determined independently by the Third Interplanetary Network. The scatter of the locations around the centroid is representative of the statistical (i.e., counting–related) uncertainties inherent in bursts of comparable intensity to the sub–interval under examination. The two out–lying circles at the top and left of the figure are sub–intervals with relatively low intensities, and are degraded by poor counting statistics in the third and fourth brightest detectors. Figure 1b shows a plot of angular distance from the centroid as a function of intensity in the third brightest detector for sub–intervals of four bursts seen by BATSE. The solid line is an empirical fit to the data that has been over–estimated at the lower end ($< 10\ \sigma$) to incorporate large offsets due to identification of the wrong third brightest detector. The gradual increase in the separation of the sub–interval location from the centroid at large intensities is caused by increased dead–time that is greater than modeled for the detectors most directly facing the burst. From this figure, an estimate of the count–rate related uncertainty in the location can be obtained for any burst, provided the intensity is known.

The offset of the centroid location from that of the interplanetary location is representative of the systematic errors associated with the geometry of the detection environment at the time of the burst. For the sample of seven bursts, the average offset of the centroids from the interplanetary network locations was found to be 5.8° ± 1.4°.

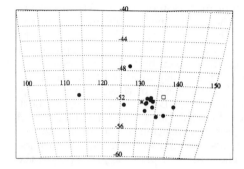
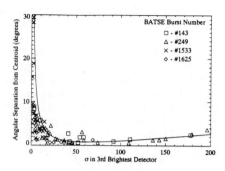

Figure 1 – (a) Sub–interval locations in RA and DEC for GRB910522.
(b) Offset of Sub–interval location as a function of intensity in the 3rd brightest detector for four BATSE bursts.

MONTE CARLO SIMULATIONS OF GALACTIC GRB DISTRIBUTIONS

In order to determine the effect of the location uncertainties obtained in the previous section, we designed a Monte Carlo simulation to generate distributions of bursts with a known anisotropy. The subsequent introduction of location uncertainties causes the distribution to 'relax' into a more isotropic set of burst locations. The amount of relaxation caused by the introduction of uncertainties is then a measure of the effect of the location uncertainties on the ability to discern the anisotropy present in the initial distribution.

Before starting the simulations, the intensities of the first 300 BATSE bursts were measured. From these intensities, a counting–related uncertainty was determined using the empirical fit in Figure 1b. These uncertainties were used as the statistical uncertainties to be imparted to each of the computer-generated burst locations.

We concentrate here on galactic halo distributions, adopting a standard density function[6,7]

$$n(r) = \frac{1}{1 + (r/r_o)^2} , \qquad (1)$$

where r_o is the core–radius, and r is the distance from the galactic center. Galactic halo models are used because of the simple form of the distribution, and the relevance to current discussions on the origins of the bursts. However, any distribution with a known anisotropy is acceptable for determining the effect of the location uncertainties.

We assume for the purposes of the Monte Carlo simulation that all bursts have the same intrinsic brightness. Consequently, there is some distance from the detector, R_{vis}, at which a burst will not be detectable.

In the simulations, 1000 sets of 300 bursts were generated with the density profile of Equation 1. We then measured the values of $\langle\cos\theta\rangle$ and $\langle\sin^2 b\rangle$ for each set of bursts. Each location was then perturbed twice. These small displacements represented the statistical and systematic uncertainties in each burst location.

Having previously measured the intensities and computed a statistical uncertainty using the curve in Figure 1b for the first 300 BATSE bursts, we perturbed the location of each Monte Carlo burst by an amount equivalent to the statistical uncertainty computed for that location's companion in the BATSE flight data. The perturbation direction is random.

The second perturbation represents the systematic error inherent in the BATSE locations. The sizes of these perturbations were obtained by sampling a Gaussian, centered at 5.8°, with a σ of 1.4°. These parameters were chosen to match the computed BATSE centroid offsets from the seven interplanetary locations. The actual distribution of systematic errors is a complex function of detection geometry, spacecraft orientation, and other parameters. A Gaussian representation, however, is an acceptable first-order approximation. The systematic offsets of the BATSE locations for the seven bursts seen by the interplanetary network were examined to see if there was a preferred direction for the systematic error, e.g. towards the Earth-center. No preferred direction was found for the systematic offset. Consequently, the direction of the second perturbation, representing the systematic error, is random.

After the two perturbations, we again measured $\langle\cos\theta\rangle$ and $\langle\sin^2 b\rangle$ for the smeared distribution. The differences in the values of $\langle\cos\theta\rangle$ and $\langle\sin^2 b\rangle$ for the original and the smeared distributions are measurements of the degradation inherent in BATSE's ability to constrain the amount of isotropy in the burst distribution because of location uncertainties.

RESULTS AND DISCUSSION

Figure 2 depicts the results from two sets of Monte Carlo runs. The values of $\langle\cos\theta\rangle$ are plotted against of r_o, the core-radius, in units of kpc. The upper data set corresponds to $R_{\rm vis} = 100$ kpc, with the lower data representing $R_{\rm vis} = 125$ kpc. For both of these sets, $R_{\rm max}$, the distance beyond which the density of sources is negligible, is set at 150 kpc. The BATSE measured value[2] of $\langle\cos\theta\rangle$ for 262 bursts, 0.008 ± 0.035, is plotted as a horizontal solid line. The dotted line represents the 1σ uncertainty associated with this value.

The filled symbols represent the value of $\langle\cos\theta\rangle$ for the unperturbed distributions. After introduction of the location uncertainties, the measured value of $\langle\cos\theta\rangle$ decreases slightly to the position of the open symbols. Because of the large numbers of bursts, the uncertainties on $\langle\cos\theta\rangle$ from the Monte Carlo distributions are smaller than the symbols used to plot the data. In each case, the addition of location uncertainties makes the distribution appear slightly more isotropic than the original one actually is; however, the effect is very small. For example, the location uncertainties have their largest effect for $\langle\cos\theta\rangle \simeq 0.10$ – 0.12. For the two values of $R_{\rm vis}$ shown, $\langle\cos\theta\rangle$ decreases by ~0.005 after the introduction of the location uncertainties. This shift is ~7 times smaller than the statistical uncertainty on the BATSE value[2] of $\langle\cos\theta\rangle$ due to the finite number of bursts in the sample. When $\langle\cos\theta\rangle \simeq 0.05$, near the 1σ upper-limit on the BATSE measured value, the change to $\langle\cos\theta\rangle$ imparted by the location uncertainties is reduced from the previous amount by nearly a factor of two. The small differences between $\langle\cos\theta\rangle$ for the smeared and unsmeared distribution when compared to the size of the BATSE measurement uncertainty indicate that the effect of individual location uncertainties on the ability to place constraints on the amount of anisotropy present in the burst distribution is negligible.

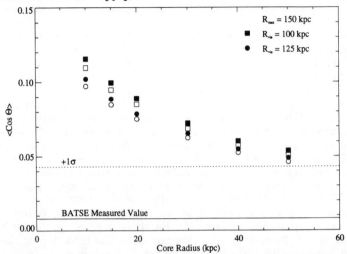

Figure 2 – $\langle \cos\theta \rangle$ vs. Core Radius for two galactic halo models.

Distributions with core radii of 30 – 50 kpc produce values of $\langle \cos\theta \rangle$ consistent with the BATSE value at a level of ~1 –2 σ. Recently postulated distributions of high–velocity pulsars in a large galactic halo[8] as the source of bursts are compatible with this result. Smaller halos are consistent with the BATSE measured value of $\langle \cos\theta \rangle$ at approximately the 2 – 3 σ level. Hakkila et al.[9], provide more detail on the currently allowed parameters for halo models. Constraints on the values of core radii and the amount of allowable anisotropy consistent with the BATSE observations are much more sensitive to the choice of R_{vis} and to the number of bursts in the BATSE sample than to location uncertainties in individual bursts. The ability of BATSE to impose constraints on the amount of anisotropy present in the angular distribution is therefore not substantially degraded by the uncertainties in the individual locations.

REFERENCES

1. Meegan, C., Fishman, G., Wilson, R., Paciesas, W., Pendleton, G., Horack, J., Brock, M., Kouveliotou, C., *Nature*, **355**, 143 –145, (1992)
2. Meegan, C. A., et al., IAU Circular #5478, (1992)
3. Pendleton, G. N., private communication, (1992)
4. Brock, M. N., et al., in *Proc. Gamma Ray Bursts 1991*, (eds. Paciesas, W. S., & Fishman, G. J.), 383, (Huntsville, 1991)
5. Hurley, K., private communication, (1992)
6. Brainerd, J. J., *Nature*, **355**, 522–524, (1992)
7. Paczynski, B. *Acta Astronomica*, **41**, 157 – 166, (1991)
8. Eichler, D., & Silk, J., *Science*, **257**, 937 – 942, (1992)
9. Hakkila, J., et al., These Proceedings, (1992)

A SEARCH FOR UNTRIGGERED GAMMA-RAY BURSTS IN THE BATSE DATA

Brad C. Rubin

USRA/NASA – Marshall Space Flight Center, ES-62, 35812

J. M. Horack, M. N. Brock, C. A. Meegan, G. J. Fishman, R. B. Wilson

NASA – Marshall Space Flight Center, ES-62, 35812

W. S. Paciesas

Dept. of Physics, Univ. of Alabama, Huntsville, 35899

J. van Paradijs

Univ. of Amsterdam & Center for High-Energy Astrophysics, The Netherlands

ABSTRACT

In the BATSE data there exist cosmic gamma-ray bursts which do not meet the on-board trigger criteria. We report here the results of an off-line, *ex-post-facto*, search for bursts which would trigger on 1.024 s or 4.096 s timescale integrations. These bursts did not set the on-board trigger.

INTRODUCTION

The BATSE experiment triggered on about 450 cosmic gamma-ray bursts in almost one and a half years of operation[1]. One obvious question regards whether there exist bursts which could be identified in the data, but which do not set the on-board trigger. There are two reasons why a burst would fail to set the trigger. Either it fails to meet the nominal trigger criteria (too weak, too soft, or rising too slowly), or other conditions exist which prevent trigger activation despite the presence of a burst. We have searched BATSE data with an off-line program which works almost exactly like the on-board trigger, but which allows the user to change parameters of that trigger.

There are at least three reasons to search for untriggered bursts. First, if there are a significant number of bursts which meet the nominal trigger criteria, but do not trigger, then these bursts should be included in searches for gravitational lensing. Secondly, the longest duration burst reported by BATSE lasts about 600 seconds.[1] However, if emission from a triggered burst lapsed for a period between several minutes and 90 minutes, and then resumed, existing BATSE procedures could fail to recognize resumption of emission, and hence a burst of much longer duration. Finally, some bursts from soft gamma repeaters and slow rising bursts do not meet the nominal trigger criteria. Detection of some SGRs requires searching for untriggered events at low energies.

PROCEDURE

On-board, BATSE LAD discriminator data are tested for trigger activation on separate 64 ms, 256 ms, and 1.024 s timescales. Only the 1.024 s data is continuously telemetered to the ground. The on-board trigger criteria are listed in Table I.

Table I. BATSE On-Board Trigger Criteria.
1) LADs tested on timescales of 64 ms, 256 ms, and 1.024 s.
2) Background rate averaged for 17.408 s segments of data (not sliding).
3) Trigger threshold 5.5 σ above background on any timescale.
4) Energy range 50-300 kev (Sum of DISCLA channels 2 and 3).
5) Two or more detectors above threshold required.
6) Burst overwrite by stronger event on 1.024 s timescale only.

Off-line, the nominal criteria dispense with item 6 in table I. We changed items 1, 3, and 4 and searched for bursts in three different modes:

1. SGR mode: Channel 1 (20-50 keV), 1.024 s time scale, trigger threshold at 3.5 σ.
2. Burst mode in Channel 1: Separate 1.024 s and 4.096 s time scales, separate trigger thresholds at 4.5, 5.0, and 5.5 σ.
3. Burst mode (50-300 keV): Separate 1.024 s and 4.096 s time scales, separate trigger thresholds at 4.5, 5.0, and 5.5 σ.

On the 4.096 s timescale the background is averaged over 20.480 s. Results from modes 1 and 2 are described in Kouveliotou et al.[2] and van Paradijs et al.[3] and will not be further discussed in this paper. Here we will concentrate on the results obtained from mode 3.

The off-line trigger program outputs a list of triggers for each timescale and each threshold level used. To prevent many false triggers from long magnetospheric events, the trigger test is disabled for 240 seconds following a trigger. Scanners compare the trigger list to plots of data showing all 8 detectors at high time resolution. Figure 1 shows a plot containing a burst. A typical day yields 25 triggers at 5.5 σ and 50 triggers at 4.5 σ, including both timescales. The great majority of these could easily be ascribed to solar flares, magnetospheric electron precipitations, Cygnus X-1 flares, on-board triggers, strong source rises, SAA entries, and rising background. However, there are events we classified as gamma-ray bursts. Table II shows the breakdown of events for a typical day.

RESULTS

In 40 days we identify 12 off-line triggers as additional gamma-ray bursts. In the same period there were 172 on-board triggers of which 39 were bursts. Details of the untriggered bursts are listed in Table III. The errors in RA and DEC are typically about 10 degrees. The last column lists the reason each burst failed the on-board trigger. We see that one burst rose too slowly, one was too weak, and 7 occured during readout of a stronger event. Note that since there are 4 triggers per day and trigger readout takes 90 minutes, there is an effective 25% trigger deadtime for weak bursts which meet the nominal trigger requirements. Three bursts failed to trigger probably because of a roundoff effect of the trigger near threshold.

Table II. 1 Day Off-line Trigger Breakdown.				
Timescale —	1.024 s		4.096 s	
Classification	4.5 σ	5.5 σ	4.5 σ	5.5 σ
On-Board Trigger	4	4	4	4
Cygnus X-1 Flare	0	0	10	3
SAA	1	1	6	3
Source Rise	0	0	5	1
Magnetospheric Event	1	0	3	2
Rising Background	1	0	1	1
Solar Flare	3	2	3	3
Gamma-Ray Burst	1	1	1	0
Total	11	8	33	17

Table III. List of Untriggered Bursts.						
No.	Scale	Sigmas	Duration	RA	DEC	Reason
1	1,4	5.7, 6.6	40 s	194°	-3°	roundoff
2	1,4	6.9, 12.1	25	240	27	roundoff
3	1,4	6.6, 5.0	4	232	49	readout
4	1,4	8.7, 4.9	3	25	-41	readout
5	1,4	8.9, 7.2	35	298	6	readout
6	1,4	6.1, 6.3	5	15	-12	readout
7	4	7.4	19	261	40	slow rise
8	1,4	4.9, 7.0	8	211	34	too weak
9	1	6.4	15	60	-47	readout
10	1	6.1	30	20	-23	readout
11	1	5.6	3	9	-54	roundoff
12	1,4	8.5, 12.7	130	8	-39	readout

Burst 3 is a 200 second early precursor to a triggered burst, but none of the other events are associated with triggered bursts. Figure 2 shows a closer view of the burst found in Figure 1. Figure 3 shows the longest burst found in this study, 130 seconds duration. Figure 4 shows the burst which triggered only on the 4.096 s timescale. In Figures 2, 3, and 4, count rates are summed over detectors viewing the burst. The lowest estimated fluence, for burst 11, is 5×10^{-8} ergs/cm^2, and the highest, for burst 12, is 4×10^{-6} ergs/cm^2, in the 50-300 keV energy band.

It has been estimated that the probability of a gravitationally lensed burst is between 0.04% and 0.4%[4]. More than half of the untriggered events found in this study have time profiles distinctive enough to compare with other bursts in a lensing search. Though it is more likely that a lensed event would trigger at least twice, the probability that a lensed event will not trigger is not negligible.

The slowest rise times known to exist for bursts are a few seconds.[4,5] About 15% of BATSE bursts would trigger only on the 1.024 s timescale. This search does not find evidence for a comparable number of weak bursts triggering only on the 4.096 s timescale, which would hint at an extension of the rise time distribution. However, it is still possible that rare events with long rise times exist. Finding these would require searching more data, and searching longer timescales in conjunction with more sophisticated methods for subtracting the background.

CONCLUSIONS

Our preliminary search reveals that more than 20% of gamma-ray bursts detectable by BATSE do not activate the on-board trigger. This fraction will be smaller, about 10%, for data with gaps due to tape recorder malfunction. This sample should not be neglected in searches for lensed events. We find no preliminary evidence for bursts with rise times longer than a few seconds. This type of search can check for the possibility of long duration bursts with weak reemission after a quiescent interval.

ACKNOWLEDGEMENTS

The authors acknowledge useful discussions with Dr. Michael Briggs and Dr. Chryssa Kouveliotou.

REFERENCES

1. Fishman, G. J., "BATSE/Compton Observations of Gamma-Ray Bursts", these proceedings.
2. Kouveliotou, C. K., et al., "BATSE Observations of a Soft Gamma Repeater (SGR)", these proceedings.
3. van Paradijs, J., et al., "A Search for Untriggered Low-Energy Events in the BATSE Data Base", these proceedings.
4. B. Paczyński, private communication.
5. Hurley, K., (AIP Conference Proceedings 265, 1990), p. 3.
6. Barat, C. et al., Ap. J. 200 (791) (1984).

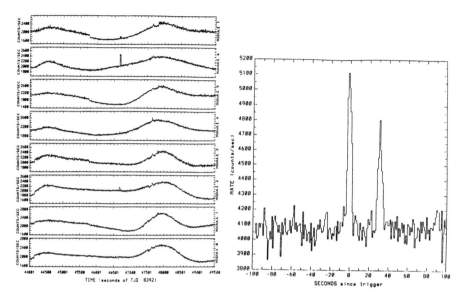

Figure 1. A plot of BATSE burst channel discriminator data for all 8 detectors. Burst 1 triggered off-line at 46658 s in detectors 2 and 6.

Figure 2. Burst 1 summed over detectors 2 and 6.

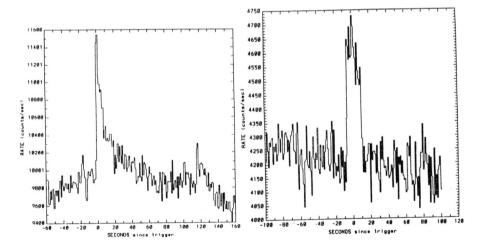

Figure 3. Burst 12 summed over detectors 1, 3, 5, and 7.

Figure 4. Burst 7 summed over detectors 3 and 7.

SOME STATISTICAL PROPERTIES OF 153 BATSE GAMMA-RAY BURSTS

W. Kluźniak

Physics Department, University Of Wisconsin, Madison, WI 53706

ABSTRACT

Disk models are inconsistent at the 2.5σ level with the mean value of C_{min}/C_{max} determined for 153 BATSE gamma-ray bursts. The significance of this result is increased to 5.7 sigma if the published limits on the quadrupole moment of the celestial distribution of 153 BATSE bursts are also taken into account. Conclusions about the isotropy of GRB sources are somewhat weakened by the unusual distribution (a strong octupole) in galactic longitude of the published samples of BATSE. A random distribution in longitude can be formally ruled out at the .995 confidence level for samples of 153 and of 256 bursts.

COUNT STATISTICS

It has already been noted in Meegan *et al.*, 1992, that the distribution of the first BATSE sample of 153 gamma-ray bursts (GRBs) excludes Galactic disk populations, but no statistical significance was quoted for this result, which relied simultaneously (Paczyński, 1991) on the apparent isotropy of the observed bursts and on the slope of the $\log N - \log C$ curve, as determined by the V/V_{max} test.

The Meegan *et al.*, 1992, value for 141 gamma-ray bursts of $< V/V_{max} >= 0.348 \pm 0.024$ seems[1] to exclude the asymptotic value of plane-parallel disk models, ($V/V_{max} = 0.4$), at the 2σ level. A more stringent limit can be obtained by computing the harmonic mean of C_{max}/C_{min}, i.e., performing the "A/A_{max}" test (Kluźniak, 1992). For this asymptotic disk model, the expected distribution of the maximum (background subtracted) count rate (in the second[2] most brightly illuminated detector) of the burst divided by the threshold count rate value (of the same detector) is expected to be flat, with $\langle C_{min}/C_{max}\rangle = 0.5 \pm (12N)^{-1/2}$.

The first 153 BATSE bursts (of those with known thresholds) yield the mean $\langle C_{min}/C_{max}\rangle = 0.44$, which is 2.5$\sigma$ below the asymptotic disk value of 0.5 (± 0.023 for 153 bursts). The use of this harmonic mean of the count rate has a few practical advantages over that of the mean of $V/V_{max} = (C_{max}/C_{min})^{-3/2}$. The main one is that the harmonic mean is less corrupted by systematic effects. For example, a systematic error of 20% in the computation of C_{max} (perhaps caused by conspiring errors in the background subtraction procedure) could raise the mean value of V/V_{max} from the reported 0.35 to 0.50, which would be in full agreement with the value expected for a uniform three-dimensional distribution of sources. The same 20% error would elevate the value of $\langle C_{min}/C_{max}\rangle$ reported here only to 0.53, which is clearly far below the value 0.67 expected for the harmonic mean in the uniform three-dimensional

---

[1] In fact, the quoted error $\pm 0.024 = (12N)^{-1/2}$ can only be used to rule out a uniform three-dimensional (Euclidean) distribution. The expected dispersion for other models is different.

[2] As appropriate for the case of a two-detector trigger. I thank Dr. Meegan for clarifying the experimental procedure used by the BATSE team.

model. Note that while this elevated value of $\langle C_{min}/C_{max} \rangle = 0.53$ is allowed for some disk models, the sampling depth required is incompatible with BATSE sensitivity, and is ruled out by the reported limits on $\langle \sin^2 b \rangle$, see Table 1 of Kluźniak, 1992. Thus, even if the value of $\langle C_{min}/C_{max} \rangle$ is corrupted by a 20% systematic error, disk models are still ruled out.

Note that the result reported so far (disk models excluded at the 2.5σ level) made no use of directional information and applies equally well to all disk-like distributions with no reference to the orientation of the disk(s) with respect to the observer. To give an (outrageous) example, an observer situated at the intersection of two planes (e.g., Zel'dovitch pancakes) may well report a zero dipole and quadrupole moment of the distribution of sources on the sky (and a non-zero octupole), in agreement with observations, but that situation is excluded by the observed value of $\langle C_{min}/C_{max} \rangle$ just as strongly as is a distribution of sources in the ecliptic or Galactic planes.

THE ANGULAR DISTRIBUTION

The 2.5σ result reported here can be strengthened if, following Paczyński, 1991, we make use of limits on the quadrupole moment of the distribution of gamma-ray bursts on the sky. Assuming that the sources of gamma-ray bursts are distributed in a plane parallel fashion (i.e., in one and not two or more disks) with a spatial density that does not increase with distance from the plane of symmetry (in which the observer is assumed to reside), I find the best-fit model to be excluded at the 5.7 sigma level.

Loosely speaking, the quadrupole moment can be used to place an upper limit on the sampling depth, while the count statistics yield a lower limit, as in the previous section. More precisely, the departure from the model is computed by adding the respective dispersions in quadrature. Table 1, of Kluźniak 1992, gives the predicted values of $< \sin^2 b >$ and of $< C_{min}/C_{max} >$ for a uniform disk model with intrinsic luminosity function of slope 2. That model was chosen because it gives the largest value of $< \sin^2 b >$ for a given value of $\langle C_{min}/C_{max} \rangle \equiv \langle A/A_{max} \rangle$. If that model is ruled out, so are *a fortiori* all other disk models satisfying the assumptions just mentioned above.

The observed values for 153 BATSE bursts are: $< \sin^2 b > = 0.310 \pm .029$ (Meegan *et al.*, 1992), and $< C_{min}/C_{max} > = 0.443 \pm 0.023$ (see above). The errors correspond to the one-sigma errors expected from an isotropic distribution or from a two-dimensional, uniform surface density distribution respectively. Comparison with Table 1 of Kluźniak, 1992, shows that for disk distributions aligned with the Galactic plane, $\sqrt{\{[(< \sin^2 b >_t -0.310)/.03]^2 + [(< C_{min}/C_{max} >_t -0.443)/.023]^2\}} \geq 5.7$ where the subscript t indicates theoretical values. As the limits on the quadrupole have been found not to depend on the coordinates used (Briggs, 1992), I conclude that any plane parallel distribution (satisfying the assumptions made in the opening paragraph of this section) is ruled out at the 5.7σ level by the properties of the BATSE sample of 153 bursts.

Unfortunately, the same sample of 153 bursts cannot be used to demonstrate conclusively that only spherically symmetric models are allowed by the data. This is because the observed GRB positions on the sky are formally inconsistent at the 0.995 confidence level with a random distribution. If the underlying distribution is in fact isotropic, the most likely explanation of this departure from randomness (Figure 2) is a systematic effect, perhaps connected with the octahedral symmetry of the detector

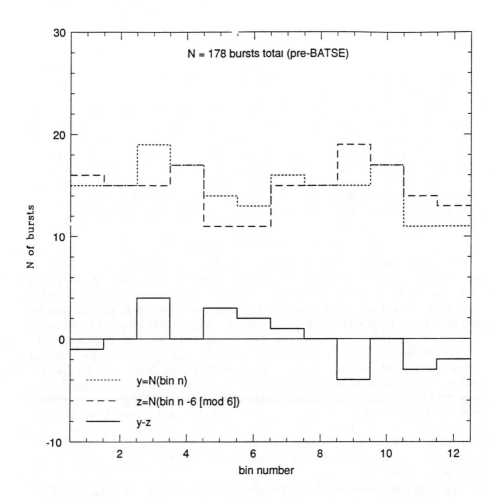

Figure 1. The distribution in Galactic longitude of 178 gamma-ray bursts detected before the launch of CGRO. The burst positions on the sky are binned into twelve equal area, contiguous strips bounded by successive lines of constant longitude separated by 30°. To exhibit the lack of large-scale correlations, the distribution of bursts obtained after a rotation of the sky by 180° is also shown, as is the difference between the original and the rotated distributions.

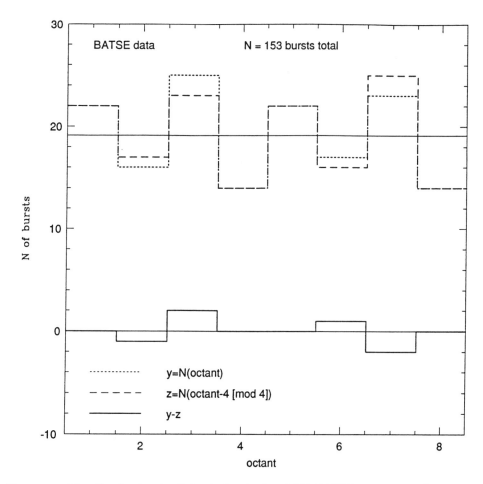

Figure 2. The distribution in Galactic longitude of 153 BATSE gamma-ray bursts. The histogram shows the number of bursts in each of eight contiguous strips in the sky bounded by successive lines of constant longitude separated by 45°, as drawn in Figure 2 of Meegan *et al.*, 1992. Note that this distribution is nearly invariant under a rotation of 180°: the difference between the actual distribution and that rotated by 180° nearly vanishes.

(a statistical fluke is by definition an unlikely possibility).

To check whether the 153 bursts are randomly distributed on the sky, I have binned the GRB positions in Figure 2 of Meegan et al., 1992, in eight equal areas in the sky. The bins were uniquely defined by the coordinate grid given in that figure, specifically by the lines of constant Galactic longitude. Upon dividing the binned sky into two hemispheres (there are many ways of doing that, each gives the same answer), one can test the hypothesis, H_0, that the burst positions are drawn randomly from the same parent distribution in each hemisphere. If H_0 is true, the quantity $\chi_4^2 = \Sigma_{i=1}^4 (y_i - z_i)^2/(y_i + z_i)$ should be distributed according to the chi square distribution for four degrees of freedom. "It is customary to set $Q(\chi^2) = 0.05$ as the level at which to consider rejecting a set" (Meyer, 1975). Not being interested in 2σ results, I set a higher threshold of rejection at the confidence level of 0.995, corresponding to $\chi_4^2 < 0.21$ or $\chi_4^2 > 14.86$. From the data (Figure 2), $\chi_4^2 = 0.11 < 0.21$; i.e., hypothesis H_0 can be rejected.

A comparison of Fig. 2 with Fig. 1, which shows the pre-BATSE data, makes the apparent non-randomness in the BATSE sample quite obvious. Monte Carlo results, typified by the observed distribution in Figure 1, confirm that the BATSE sample is very unusual.

To test whether the detected correlation on scales of 90° (and 180°) in longitude persists, I have also tested the largest published BATSE sample, Figure 1 of Fishman et al., 1992. Folding 256 GRB positions into two equal bins spaced at 45° in Galactic longitude (modulo 90°) I find 153 bursts positioned in one bin and 103 in the other. The hypothesis, H_1, that the bursts are uniformly distributed in Galactic longitude gives $\chi^2 = 9.77$ for one degree of freedom. Again $Q(\chi_1^2 = 9.7) < 0.005$ and therefore the hypothesis H_1 can be rejected at the 0.995 confidence level.

CONCLUSIONS

In spite of the apparently non-random distribution on the sky of the initial 256 gamma-ray bursts detected by BATSE, planar distributions, and combinations thereof, are excluded at the 2.5σ level by the value of $\langle C_{min}/C_{max} \rangle = 0.44$ obtained for the first 153 BATSE bursts. Single-disk distributions of sources are excluded at the 5.7σ level.

I thank Dr. Hakki Ögelman and Dr. John Finley for sound advice.

REFERENCES

1. Briggs, M.S. 1992, *Astron. Astroph.*, in press.
2. Fishman, G.J. et al. 1992, *Astron. Astroph.*, in press.
3. Hurley, K. 1992, in *Gamma-Ray Bursts*, W.S. Paciesas & G.J. Fishman eds., (New York: AIP), pp. 3–12.
4. Kluźniak, W. 1992, in *Gamma-Ray Bursts*, W.S. Paciesas & G.J. Fishman eds., (New York: AIP), pp. 105–107.
5. Meegan, C. et al. 1992, *Nature* **355**, 143–145.
6. Meyer, S.L. 1975, *Data Analysis for Scientists and Engineers* (New York: Wiley).
7. Paczyński, B. 1991, *Acta Astronomica* **41**, 157.

GAMMA RAY BURST REDSHIFTS: ZERO OR ONE?
IN SEARCH OF A BEND IN LOG N–LOG S

Ralph A.M.J. Wijers[†] and Lori M. Lubin
Princeton University Observatory, Peyton Hall, Princeton, NJ 08544–1001, USA

ABSTRACT

We have compared the observed distribution of peak count rates, $C_{\max}$, and $C_{\max}/C_{\min}$ values of gamma-ray bursts with simulated distributions. We find that some observed samples for which the data were available to us show significant deviations from simple power-law distributions. We also find that presently available samples of gamma-ray bursts are too small to distinguish significantly between galactic and extra-galactic models for their origin on the basis of the $\log N - \log S$ distribution. The BATSE sample will have to increase to 1000–2000 bursts before this becomes feasible.

INTRODUCTION

There can hardly be a problem in astronomy that is more classic and characteristic of the field than the determination of distance, especially in the case where we know little or nothing about the physics of the object that we are seeking the distance of. In fact, our present task in establishing the gamma-ray burst distance scale is more difficult than that of finding the distances to the stars. After all, at the time when astronomers were establishing stellar distances, they could compare the stars on the night sky with their well-studied cousin of the daytime sky. No such Rosetta stone exists as yet for gamma-ray bursts.

The combination of an isotropic distribution on the sky with a value of $V/V_{\max}$ significantly less than 0.5 allows only two popular distance scales: (1) the galaxy. In this case, the galactic disk is already ruled out, as are halos with core radii less than 18 kpc[1]. (2) extragalactic distances. In this case, a lower limit to the typical distance of about 100 Mpc can be inferred from the absence of a detectable angular correlation between gamma-ray bursts on angular scales greater than 1 degree[2]. Of course, the small BATSE value for $V/V_{\max}$ (e.g. ref. 3) implies a much greater typical distance of a few Gpc, since the inhomogeneity is explained in this case by the fact that we are seeing beyond redshift 1.

One of the ways of finding clues to the nature of gamma-ray bursts is identifying a characteristic flux. This flux can then be associated with a characteristic distance or luminosity in each of a set of possible models for their origin. Because power laws are scale-free, this implies that one must find a deviation of the $\log N - \log S$ curve from a simple power law. That the overall flux distribution of all gamma-ray bursts deviates from a power law seems quite clear, because the slope of $\log N - \log S$ varies from -1.5 for bright PVO bursts[4] to about -0.8

[†]Compton Fellow

for the weaker BATSE bursts[3]. However, since it is not easy to exactly connect observations done with instruments that differ in spectral response and overall sensitivity, we shall explore whether this trend is visible in any of the individual data sets. Also, we shall demonstrate that with a sample size of 400 bursts it is not possible to tell the difference between the simplest allowed galactic halo and cosmological models.

DEVIATIONS FROM POWER LAWS IN $\log N - \log S$

To find out whether the peak count rates C_{max} of gamma-ray bursts or their values of C_{max}/C_{min} are distributed as a power law, many methods may be applied. The most straightforward one is to fit a power law probability distribution of form

$$f(S) = \begin{array}{ll} \frac{1-\alpha}{S_{min}} \left(\frac{S}{S_{min}}\right)^{-\alpha} & S > S_{min} \\ 0 & S \leq S_{min} \end{array} \qquad (1)$$

to the observed values S_i in a data set. Then one can investigate whether the best fit is good enough, and if it is not the observed distribution apparently deviates from a power law. We shall use this technique, with maximum-likehood estimation as the method of fitting.

We shall also apply a second technique, which we name the 'bend statistic' B. It is defined as folllows: consider a flux distribution $N(> S)$ (we shall henceforth refer to both C_{max} and C_{max}/C_{min} as 'flux' for brevity). Define S_q by the relation $N(> S_q) = 1 - q$. It is easy to show that the statistic

$$B \equiv \frac{\log(S_{1/3}/S_0)}{\log(S_{2/3}/S_0)} - \frac{\log(2/3)}{\log(1/3)} \qquad (2)$$

satisfies the property that it has zero expectation value if the distribution is a pure power law. It has a positive (negative) expectation value if the slope of $\log N(> S)$ steepens (flattens) with increasing S. Another way of deciding whether a sample of fluxes is distributed as a power law therefore is to get estimators s_q of the various S_q involved, and therefrom compute an estimator b of B. In this paper, we compute s_q for a given sample by constucting its cumulative distribution $F_{obs}(S)$, and then interpolating linearly to find s_q. (s_0 is set to the lowest flux in the sample.)

Application of the above techniques was done as follows: first the values of b_{obs}, the maximum-likelihood estimate $\hat{\alpha}_{obs}$ of the slope of the observed distribution, and the likelihood L_{obs} were computed. Then we generated 10000 Monte Carlo samples with the same number of elements as the observed sample, but drawn from a pure power-law distribution with slope $\hat{\alpha}_{obs}$. The values of b, $\hat{\alpha}$, and L were computed for each. It turned out that the average value of L was always very close to L_{obs}, indicating that on the basis of likelihood-analysis no difference could be found between the observed distribution and a pure power law. This was

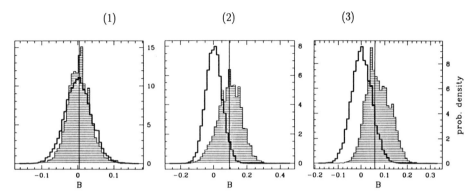

Figure 1: The bending test applied to three observed samples. (1) the BATSE sample of 241 $C_{\max}/C_{\min}$ values from three different trigger modes. (2) $C_{\max}$ values of a complete BATSE subsample consisting of all 133 bursts triggered in the 256 ms bin mode, which satisfied the constraint that $C_{\max} > 147.8$ and $C_{\min} < 147.8$. (3) The $C_{\max}/C_{\min}$ values of all 166 bursts detected by the French-Soviet SIGNE experiment. The solid vertical line marks the measured value of b, and the heavy unshaded histogram is the distribution of b in the simulated samples. The shaded histogram indicates the distribution of b for 10000 bootstrap samples drawn from the observed sample. It is seen that samples (2) and (3) show significant evidence for a steepening slope of $\log N - \log S$ within the sample, because only a small fraction of the heavy histogram lies to the right of the observed b.

not the case with b, however. Most samples showed a clearly positive value of b. One can assign a significance to the detection of the steepening as indicated by a positive b by computing what fraction of the simulated samples have a value of b exceeding the measured value. This is the probability that the sample is a power law and yielded a positive value of b by chance.

In Fig. 1, we show the result of the bending test for three observed samples. The probability that there is no steepening in the samples is 46% in the full BATSE sample, 3.5% in the complete sample constructed from the BATSE 256 ms triggers, and 10% in the full SIGNE sample (a steepening of the slope in the SIGNE sample has been noted previously[5]). It is not unlikely that a change of slope would be washed out in the full BATSE sample, because we computed the $C_{\max}/C_{\min}$ values in that sample as the maximum of $C_{\max}/C_{\min}$ over the three trigger modes, so the sample is a bit of a mixed bag. We may therefore conclude that there is clear, albeit not overwhelming, evidence for a steepening of the $\log N - \log S$ diagram of gamma-ray bursts with increasing S, as was conjectured previously based on the difference in slope between PVO and BATSE.

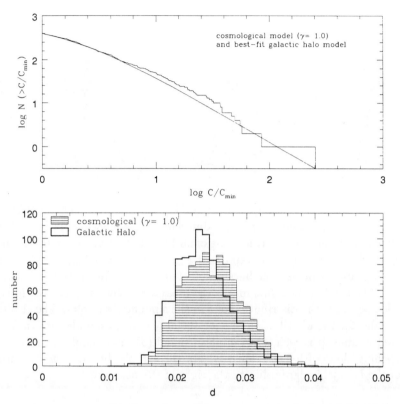

Figure 2: top: a sample of 400 bursts from a cosmological model with its best-fit halo model. bottom: comparison of the goodness of fit of 1000 fits of a halo model to a cosmological sample (shaded histogram) and 1000 fits of a halo model to a halo sample. The difference between the two is too small to be detected in a typical sample.

GALACTIC OR EXTRA-GALACTIC ORIGIN?

Now that we have found some reasonable evidence for a characteristic flux of gamma-ray bursts, we may attempt to address the question of the distance scale corresponding to that characteristic flux: is there enough information in the present BATSE sample of over 400 bursts to differentiate between simple galactic and extra-galactic models for the distribution of gamma-ray burst peak count rates? To investigate this, we performed the following experiment: we took a simple cosmological model and a simple halo model (using standard-candle bursts). For the cosmological model, we chose standard-candle bursts with a uniform rate per unit comoving time and volume[6]. For the galactic-halo model, we took a rate of standard-candle bursts that depends on galactocentric distance R as $n(R) = n_0/[1 + (\frac{R}{R_c})^\beta]$.

We then generated samples by drawing 400 burst fluxes from the cosmological distribution. The lower flux limit was chosen such that the $V/V_{\max}$ value of the samples averaged to the observed BATSE value of 0.34. These samples were then fitted with the halo model. This procedure was repeated 1000 times, and the resulting Kolmogorov-Smirnov d values were recorded. (d is a measure of goodness of fit: the smaller it is, the better the fit.) The shaded histogram in Fig. 2 is the distribution of d values thus obtained. For comparison, we then generated another 1000 samples of 400 bursts, now from a galactic-halo model with parameters equal to the average best-fit parameters in the fits to cosmological samples. The distribution of d values for these fits is given by the heavy-lined histogram in Fig. 2. It is seen that these fits are hardly better than the fits to the cosmological model. We therefore conclude that the present BATSE sample contains too few bursts to distinguish between galactic and extra-galactic models on the basis of the flux distribution. We estimate that the BATSE sample will have to increase to 1000–2000 bursts before a significant difference between the models could show up.

CONCLUSION

We have shown that a significant steepening of the slope of $\log N - \log S$ with increasing S is present in some observed samples of gamma-ray bursts, and that we may therefore consider the fact that we see the edge of the distribution of gamma-ray bursts with our detectors as firmly established. We also show that considerably larger samples of gamma-ray bursts than those presently available are needed to detect the difference between galactic and extra-galactic models on the basis of the $\log N - \log S$ distribution.

ACKNOWLEDGEMENTS

It is a pleasure to thank Bohdan Paczyński and Robert H. Lupton for stimulating discussions, and Dr. J.-L. Atteia for making the SIGNE data available to us. This work was supported in part by NASA grant NAG5-1901. RW is supported by a Compton Fellowship (grant NAG5-1816).

REFERENCES

1. S. Mao and B. Paczyński, ApJ 389, L13 (1992)
2. D. Hartmann and G.R. Blumenthal, ApJ 342, 521 (1989)
3. C.A. Meegan, G.J. Fishman, R.B. Wilson, W.S. Paciesas, G.N. Pendleton, J.M. Horack, M.N. Brock, and C. Kouveliotou, Nature 355, 143 (1992)
4. K.W. Chuang, R.S. White, R.W. Klebesadel, and J.G. Laros, ApJ 391, 242 (1992)
5. J.-L. Atteia, C. Barat, E. Jourdain, M. Niel, G. Vedrenne, A. Chernenko, V. Dolidze, A. Kozlenkov, A. Kuznetsov, I.G. Mitrofanov, A. Pozanenko, R. Sunyaev, and O. Terekhov, Proc. Huntsville Conference on Gamma-Ray Bursts (1992)
6. S. Mao and B. Paczyński, ApJ 388, L45 (1992)

THE USE OF V/V_{max} IN THE STUDY OF GAMMA RAY BURSTS

D. Band

CASS, UC San Diego, La Jolla, CA 92093

ABSTRACT

V/V_{max} is a powerful test of burst homogeneity. However, V/V_{max} is not the proper distribution for the study of an inhomogeneous spatial distribution since it involves a convolution of the intensity and detection threshold distributions. In particular a variable detection threshold can produce much of the structure of the V/V_{max} distribution observed by BATSE. The peak count rate distribution corrected for detection efficiency or maximum likelihood techniques should be used to study the spatial inhomogeneity.

INTRODUCTION

The Burst and Transient Source Experiment (BATSE) on board the Compton Gamma Ray Observatory (GRO) has shown definitively that the sources of gamma ray bursts are not distributed homogeneously.[1] Now the question is what radial distributions are consistent with observation. Many modelers and theorists[2-9] have attempted to learn about the radial source distribution by reproducing the BATSE V/V_{max} distribution with spatial models.

It is my contention that the V/V_{max} distribution is inappropriate for studies of an inhomogeneous distribution. The variable detection threshold modifies the shape of the BATSE V/V_{max} distribution, obscuring the true nature of the underlying source density. A broad burst luminosity function will also smear the V/V_{max} distribution, perhaps more than the detection thresholds; however modeling the effects of various luminosity functions is within the purview of the modeler. Thus V/V_{max} adds an additional instrumental effect which at best must be removed, and at worst will mislead the modeler.

A fuller discussion of my results will be published elsewhere.[10] Other studies have concluded that the V/V_{max} distribution and its average are biased[11,12], and that the spatial distribution is best studied by other methods.[13,14]

DEFINITIONS AND FORMALISM

Here I define homogeneous as distributed uniformly in three-dimensional Euclidean space. Since isotropy is observed,[1] I assume the density of burst sources, $n[r]$, is a function of r alone. If L is the burst *photon* luminosity, then the intrinsic luminosity function is $\phi[L]$, normalized to unity. Since I do not distinguish between the instrumental count rate and the incident photon flux, every burst is characterized by the peak count rate C (sometimes written as C_{max}); in Euclidean space $L = 4\pi r^2 C$. The threshold count rate C_{min} at which the detector would have triggered has a distribution $g[C_{min}]$, again normalized to unity.

Without independent measures of a burst's distance or intrinsic luminosity, the most fundamental information available at the Earth is the count rate distribution $\psi[C]$ (a convolution of the spatial source and the intrinsic luminosity distributions) or the related cumulative distribution $\Psi[> C]$. When $n[r] = n_0$ and the source population is homogeneous (at least out to the maximum distance observable), $\psi[C] \propto C^{-5/2}$ and $\Psi[> C] \propto C^{-3/2}$.

Unfortunately, a variable detection threshold distorts $\psi[C]$, particularly at the faint end. The observed count rate distribution $\psi_o[C]$ is

$$\psi_o[C] = \int \psi[C]\theta[C - C_{min}]g[C_{min}]\,dC_{min} \tag{1}$$

where $\theta[x] = 1$ for $x \geq 0$ and $\theta[x] = 0$ for $x < 0$; the related observed cumulative distribution is $\Psi_0[> \bar{C}]$. The distribution $\Xi[\xi]$ of $\xi = C/C_{min}$ removes the effects of the variable detection thresholds for power law distributions $\psi[C]$:

$$\Xi[\xi] = \int \psi[C]\theta[C - C_{min}]g[C_{min}]\delta\left[\xi - \frac{C}{C_{min}}\right]\,dC_{min}\,dC \quad . \tag{2}$$

Note that $\Xi[\xi]$ will have the same index as $\psi[C]$ if $\psi[C]$ is a power law.

Finally, the distribution $\chi[v]$ of $v = \xi^{-3/2}$ is derived from $\Xi[\xi]$ by a change of variables:

$$\chi[v] = -\frac{2}{3}\Xi[\xi]\xi^{5/2} \quad . \tag{3}$$

For a homogeneous source distribution $\psi[C] \propto C^{-5/2}$ and therefore $\Xi[\xi] \propto \xi^{-5/2}$. Consequently, $\chi[v]$ will be constant for a homogeneous population of sources.

THE V/V$_{max}$ TEST

The V/V$_{max}$ test determines whether the observed intensities are consistent with a homogeneous source distribution by compensating for the variable threshold intensity. Originally created to test homogeneity of AGN,[15] the V/V$_{max}$ test was subsequently applied to bursts[16] as a consequence of the difficulty in interpreting the observed deviation of $\Psi_0[> C]$ from the $C^{-3/2}$ power law expected for homogeneity. In brief, the V/V$_{max}$ test is based on the expectation that $\chi[v]$ (see equation [3]) will be constant for a homogeneous source distribution. The quantity $v =$V/V$_{max}$ can be understood as the ratio of the volume on the surface of which the burst was found to the volume within which the burst could have been found. Note that actual distances and volumes are not required. Therefore, if the burst sources are distributed homogeneously then V/V$_{max}$ should be distributed uniformly between V/V$_{max}$=0 and 1, and $\langle$V/V$_{max}\rangle = 1/2 \pm [12N]^{-1/2}$.

The detector-independent property of $\chi[v]$ is predicated on the homogeneity of the source population. For an inhomogeneous source distribution, the distribution of detection thresholds affects $\chi[v]$: $\Xi[\xi]$ (from which $\chi[v]$ is derived by equation [3]) is the convolution of $\psi[C]$ with the detection threshold distribution $g[C_{min}]$. Features in $\psi[C]$ will be broader in $\Xi[\xi]$. Not only is $g[C_{min}]$ necessary to recover the more fundamental $\psi[C]$ from $\Xi[\xi]$ and hence $\chi[v]$, but information may be lost in the convolution forming these distributions, making the recovery of $\psi[C]$ more difficult. More significantly, using $\chi[v]$ in place of $\psi[C]$ while assuming a constant detection threshold will lead to quantitatively incorrect results.

To see how a variable detection threshold smears out features in the V/V_{max} distribution, consider a source density which is homogeneous out to a radius r_0 and zero beyond, and assume standard candle bursts with photon luminosity L_0. For a constant detection threshold C_1, the $v = V/V_{max}$ distribution $\chi[v]$ will be constant with a sharp upper cutoff at $v_0 = L_0/4\pi r_0^2 C_1$ if $v_0 < 1$. However, if the detection threshold is distributed uniformly between C_1 and C_2, then various $\chi[v]$ are possible, characterized by the ratios $L_0/4\pi r_0^2 C_1$ and $L_0/4\pi r_0^2 C_2$. In many cases the sharp spatial cutoff is smeared out, and some distributions bear a resemblance to the $\chi[v]$ observed by BATSE.[10] It is extremely unlikely that the true source density has a sharp spatial cutoff, but this example demonstrates how a variable threshold can obscure steep density gradients in forming the V/V_{max} distribution.

VARIABLE BATSE THRESHOLDS

A variable detection threshold is an unavoidable consequence of the way BATSE operates and detects bursts:[17]

1. BATSE triggers when the count rate in two or more detectors increases by 5.5σ over the background; σ is proportional to the square root of the background count rate, which fluctuates because GRO is in a 28.5° inclination low Earth orbit.

2. Each detector is built around a large, thin NaI(Tl) crystal which has an angular response approximately proportional to the cosine of the angle to the detector normal. A burst's count rate changes with the detector-burst angle. Thus at some points in the sky the threshold is higher than at others.

3. The count rates are integrated on three different timescales (0.064 s, 0.256 s and 1.024 s), each of which has an associated threshold.[18] Consequently, the threshold relevant to a given burst depends on the burst morphology. In addition, the occurrence of the burst relative to BATSE's clock may also determine whether or not a burst triggers: the burst's peak may fall entirely within one integration time bin, or may be split between two bins.

4. During the orbit ($\sim$ 90 minutes) after a triggering event (not all triggers are bursts) BATSE transmits the data accumulated immediately after the trigger. During this readout period the trigger threshold for a subsequent event is raised to the triggering event's peak count rate. This effect can be removed by omitting those bursts which occurred during a readout period from the determination of $\psi[C]$ or $\chi[v]$. Indeed such an editing of the burst list will probably be necessary because the distribution of detection thresholds $g[C_{min}]$ which will be published will probably not include this effect.[18]

Brock et al.[19] find significant detection threshold variations of at least a factor of two. While this may not seem significant, note that V/V_{max} is proportional to $C_{min}^{3/2}$, and only spans the range 0 to 1.

WHAT SHOULD BE USED?

The distribution of burst intensities $\psi[C]$ observable at the Earth is the most fundamental quantity obtainable in the absence of additional information about the burst luminosity or distance. However, instrumental information is needed to use the observed $\psi_o[C]$ as a probe of the source distribution. As equation (1) shows, the factors dependent on C_{min} in the integrand are independent of the rest of the integrand. If

$$G[C] = \int_C^\infty g[C_{min}]dC_{min} \qquad (4)$$

is the fractional sensitivity of a detector to C then observers can calculate and present the instrument-independent $\psi[C] = \psi_o[C]/G[C]$. Although this requires $g[C_{min}]$ from the detector's entire operational lifetime while V/V_{max} requires C_{min} only when each burst was detected, the BATSE $g[C_{min}]$ is being calculated and will be published.[18] Petrosian[13] provides a non-parametric, maximum likelihood method to determine both $\psi[C]$ and $g[C_{min}]$ directly from the observed bivariate distribution of C and C_{min}.

While $\psi[C]$ can be used to gain a qualitative sense of the burst distribution, most modelers wish to evaluate quantitatively the agreement of a model $\psi'[C]$ with the observed bursts, or to optimize the parameters of such a model. One could compare $\psi'[C]$ to $\psi[C]$, or even $\chi'[v]$ calculated for a given $\psi'[C]$ to the observed $\chi[v]$. However, it is better to use the observed data directly, as in the following maximum likelihood method[20] which also provides a means of comparing models and computing parameter uncertainties.

The expected number of observed bursts in dC is $dN = \psi'[C]G[C]dC$. The likelihood is the product of the Poisson probabilities of finding the observed bursts at the observed C_i, and not finding bursts at other values of C:

$$L = \prod_i \psi'[C_i]G[C_i]dC_i e^{-\psi'[C_i]G[C_i]dC_i} \prod_{j \neq i} e^{-\psi'[C_j]G[C_j]dC_j} \qquad (5)$$

With the usual manipulations (i.e., $S = -2 \ln L$), and dropping terms which are not dependent on ψ':

$$S = -2 \sum_i \ln \psi'[C_i] + 2 \int_0^\infty \psi'[C]G[C]dC \qquad (6)$$

S should be minimized for the best model ψ'. This methodology uses the information $G[C]$ to account not only for the observed bursts, but also for the absence of observable bursts. Note that C_{min} at the time of the observed bursts is subsumed within $G[C]$, and is thus irrelevant! Confidence contours for the parameters of ψ' can determined by decreasing L or increasing S away from the extremum by a given amount. This method has been applied to the BATSE bursts.[14]

Normalized distributions such as $\Xi[\xi]$ and V/V_{max} can test whether $\psi[C]$ is a power law of any index since $\Xi[\xi]$ will also be a power law of the same index as $\psi[C]$. If the burst sources are found in a volume of a different dimension (e.g., on a plane, which is ruled out by the observed isotropy) or on a fractal surface then $\psi[C]$ will also be a power law, but with an index other than -5/2. The A/A_{max} test[21] for a uniform distribution on a plane (e.g., the Galactic disk) is based on this concept.

IMPLICATIONS FOR GAMMA RAY BURSTS

Since $g[C_{min}]$ smears out sharp features in the V/V_{max} distribution, the true radial decrease of the density distribution $n[r]$ is more extreme than in the spatial models fit to the V/V_{max} distribution.[2-9] In addition, bursts are undoubtedly not standard candles and the true luminosity function also smooths the V/V_{max} distribution (and ψ). Consequently, the radial decrease in the source density $n[r]$ is more extreme than found by modeling thus far.

ACKNOWLEDGEMENTS

It is a pleasure to thank G. Fishman, D. Gruber, D. Hartmann, R. Lingenfelter, T. Loredo, C. Meegan, R. Rothschild, and B. Schaefer for incisive discussions and encouragement. I also thank the other members of the BATSE team for their detailed explanations of the detectors' operation. This work was supported by NASA contract NAS8-36081.

REFERENCES

1. C. A. Meegan, et al., Nature **355**, 143 (1992).
2. C. D. Dermer, Phys. Rev. Lett. **68**, 1799 (1992).
3. C. D. Dermer and H. Li, Gamma Ray Bursts, AIP Conference Proceedings 265, eds. W. S. Paciesas and G. J. Fishman (AIP, New York, 1992), p. 115.
4. D. Eichler and J. Silk, Science **257**, 937 (1992).
5. S. Mao and B. Paczyński, Ap. J. Lett. **388**, L45 (1992).
6. B. Paczyński, Acta Astronomica **42**, 1 (1992).
7. T. Piran, Ap. J. Lett. **389**, L45 (1992).
8. M. Schmidt, Gamma Ray Bursts, AIP Conference Proceedings 265, eds. W. S. Paciesas and G. J. Fishman (AIP, New York, 1992), p. 94.
9. R. A. Shafer and C. M. Urry, unpublished work (1992).
10. D. Band, Ap. J. Lett., in press (1992).
11. D. H. Hartmann, et al., Gamma Ray Bursts, AIP Conference Proceedings 265, eds. W. S. Paciesas and G. J. Fishman (AIP, New York, 1992), p. 120.
12. D. H. Hartmann and L.-S. The, Ap. Sp. Sci., in press (1992).
13. V. Petrosian, Ap. J. Lett., in press (1992).
14. T. Loredo and I. Wasserman, these proceedings (1993).
15. M. Schmidt, Ap. J. **333**, 694 (1968).
16. M. Schmidt, J. C. Hidgon, and G. Hueter, Ap. J. Lett. **329**, L85 (1988).
17. G. J. Fishman et al., Proceedings of the Gamma Ray Observatory Science Workshop, ed. W. N. Johnson (1989), p. 2-39.
18. C. Meegan, private communication (1992).
19. M. N. Brock, et al., Gamma Ray Bursts, AIP Conference Proceedings 265, eds. W. S. Paciesas and G. J. Fishman (AIP, New York, 1992), p. 399.
20. H. L. Marshall, Y. Avni, H. Tannabaum, and G. Zamorani, Ap. J. **269**, 35 (1983).
21. W. Kluźniak, Gamma Ray Bursts, AIP Conference Proceedings 265, eds. W. S. Paciesas and G. J. Fishman (AIP, New York, 1992), p. 105.

THE CONSISTENCY OF STANDARD COSMOLOGY AND THE BATSE NUMBER VS. BRIGHTNESS RELATION

W.A.D.T. Wickramasinghe[1], R.J. Nemiroff[2,3],
C. Kouveliotou[2,4], J. P. Norris[3],
G.J. Fishman[4], C.A. Meegan[4], R.B. Wilson[4] and W.S. Paciesas[4,5]

ABSTRACT

The integrated number - peak flux relation measured by BATSE is compared with different standard cosmological distributions for GRBs to check for consistency, which if any cosmological model is particularly implied, and hence what brightness - redshift relation is implied. The standard Friedmann Robertson Walker models were used with the assumption that the bursts are standard candles and have no number or luminosity evolution. Given a specific cosmology we used a free parameter, essentially the co-moving coordinate number density of bursts, for a given Ω and spectral shape, to generate a best fit between the cosmology and the measured relation. Preliminary results are shown for the first 155 BATSE bursts. The best fits show a maximum burst redshift of the order of unity; however, significantly higher redshifts are possible for other fits.

INTRODUCTION

Understanding gamma-ray bursts (hereafter GRBs) has been a long standing problem in astronomy. The BATSE on the Compton Gamma Ray Observatory has been detecting GRBs at the rate of almost one a day. The data show an isotropic angular distribution of GRBs. The integrated number counts (N) versus C_{max}/C_{min} curve does not obey the -1.5 law expected for a spatially flat infinite homogeneous distribution of sources (Meegan et al. 1992). To solve this enigma, there have been many suggestions that GRBs in fact do originate at cosmological distances (van den Bergh 1983, Paczynski 1986). There have also been suggestions that a cosmological distribution of GRBs cannot be reconciled with data if evolutionary effects are neglected (Fenimore et al. 1992). As for an idealized case, assuming that they are cosmological and evolutionary effects are negligible, there have been a number of studies of their distribution. Assuming GRBs are cosmological and $\Omega = 1$ and $\Lambda = 0$, statistical properties of GRBs have been analyzed by Mao and Paczynski (1992). They concluded that the calculated distribution of burst intensities is consistent with that found by the BATSE for the weak bursts and by the PVO for the strong ones. Thus they arrived at the values of $z_{BATSE} \approx 1.5$ and $z_{PVO} \approx 0.2$. Making similar assumptions, Dermer (1992) has extensively studied the statistics of GRBs. His analysis has also been for a $\Omega = 1$ flat universe. He also concludes that the results are in good agreement with BATSE data. Deriving C_{max}/C_{min} ratio as

---

[1] University of Pennsylvania, Philadelphia, PA 19104
[2] University Space Research Association
[3] NASA Goddard Space Flight Center, Greenbelt, MD 20771
[4] NASA Marshall Space Flight Center, Huntsville, AL 35812
[5] University of Alabama, Huntsville, AL 35899

a function of burst redshift, hence a cosmological C_{max}/C_{min} test, Piran (1992) in his extensive analysis showed that a cosmological GRB population would be a natural explanation for the observed N versus C_{max}/C_{min} relationship. There has also been a suggestion (Paczynski 1992) to measure the redshift of GRBs making a color - luminosity diagram for GRBs. In this analysis he has used a Friedmann cosmology with $\Omega = 1$ and $\Lambda = 0$. Thus there exists the necessity of investigating cosmological populations of GRBs. In order to do that we will consider the Friedmann models only with $\Lambda = 0$.

THEORY

We investigated several cosmologies assuming GRBs to be unevolving standard candles with constant co-moving density. We also assumed that their spectra can be represented by $J(\nu) \sim \nu^\alpha$ identical power law, where $J(\nu)$ is the intensity distribution of a burst. The α can be regarded as a free parameter of the theory. Let L be the intrinsic luminosity of the burst and $\Delta\nu_0$ be the wavelength band in which the burst is observed at the present epoch, then using usual notations we have

$$C_{max} \propto \frac{LJ(\nu_0.1+z)\,\Delta\nu_0}{(1+z)\,4\pi r_1^2\,S^2(t_0)}, \quad (1)$$

where C_{max} is the maximum peak intensity observed, r_1 is the coordinate distance of the burst, S is the scale factor and t_0 represents the present epoch. Thus, writing a similar equation for C_{min} and taking the ratio between the two, we get

$$\frac{C_{max}}{C_{min}} = \frac{J(\nu_0.1+z)}{J(\nu_0.1+z_{min})} \frac{1+z_{min}}{1+z} \left(\frac{r_{min}}{r_1}\right)^2 \quad (2)$$

where C_{min} would be the peak intensity at which the burst would disappear were it taken away to z_{min}. For Friedmann universes the integrated number counts N up to a redshift of z' can be calculated as follows. For $k = 1$

$$N(z') = \frac{4\pi n}{q_0^4}(2q_0-1)^{\frac{3}{2}} \int_0^{z'} \frac{[zq_0 + (q_0-1)(\sqrt{1+2zq_0}-1)]^2}{(1+z)^3\sqrt{1+2zq_0}} dz \quad (3)$$

and

$$r_{(k=\pm 1)} = \frac{g(q_0)}{(1+z)}\left[zq_0 + (1-q_0)(1-\sqrt{1+2zq_0})\right], \quad (4)$$

where n is the co-moving number density and g is a function of q_0 only. For $k = -1$

$$N(z') = \frac{4\pi n}{q_0^4}(1-2q_0)^{\frac{3}{2}} \int_0^{z'} \frac{[zq_0 + (q_0-1)(\sqrt{1+2zq_0}-1)]^2}{(1+z)^3\sqrt{1+2zq_0}} dz. \quad (5)$$

Eliminating coordinate distance factors in Eqn. (2) for $k = \pm 1$ models using (4) and the fact that $J(\nu) \sim \nu^\alpha$, one gets after some manipulations

$$\frac{C_{\max}}{C_{\min}} = F(q_0, z_{\min}) \frac{(1+z)^{\alpha+1}}{\left[zq_0 + (1-q_0)(1-\sqrt{1+2zq_0})\right]^2}, \quad (6)$$

where $F(q_0 = \frac{\Omega_0}{2}, z_{\min})$ is a function of q_0 and $z_{\min}$ only. Now using (3), (5) and (6) and remembering for $k = +1$, $q_0 > \frac{1}{2}$ and for $k = -1$, $0 \leq q_0 < \frac{1}{2}$, it is easy to plot the curve N versus $\frac{C_{\max}}{C_{\min}}$. In doing so we chose a statistical sample where $C_{\min}$ remains approximately constant. The normalization was done by demanding that the population have the appropriate average value of $\frac{C_{\max}}{C_{\min}}^{-3/2}$. The behavior of the curve is not to be confused with an edge of a Euclidian geometry of a burst population - it can be attributed completely to cosmology. For $k = 0$ model, however, we have

$$\frac{N}{N_{\min}} = \left(\frac{r}{r_{\min}}\right)^3 \quad (7)$$

and

$$r \propto \left[1 - \frac{1}{\sqrt{1+z}}\right]. \quad (8)$$

As before, eliminating the coordinate distance in (7) and (2) using (8) one gets

$$N(z) = G(N_{\min}, z_{\min})(1+z)^{\frac{3}{2}(\alpha-1)} \left(\frac{C_{\max}}{C_{\min}}\right)^{-\frac{3}{2}} \quad (9)$$

and

$$\frac{C_{\max}}{C_{\min}} = H(z_{\min}) \frac{(1+z)^\alpha}{(\sqrt{1+z}-1)^2}, \quad (10)$$

where G and H are functions that can be determined from the normalization process. Again the normalization was done when the curves approach a Euclidian line with the slope of $-\frac{3}{2}$. Equations (9) and (10) were used to make the plots N versus $C_{\max}/C_{\min}$ in this model ($k = 0$).

The complete analysis was done for the first 155 BATSE bursts. We tried to fit the theoretical curves with the real plot of N versus $C_{\max}/C_{\min}$ for a class of values of Ω and α. For α, 0.0, -1.0 and -2.0 values were chosen (note that the usual definition of α differs from ours). Since n, the coordinate number density, is a free parameter of the theory, another parameter, N_{free}, was defined for convenience such that

$$N_{\text{free}} \propto n. \quad (11)$$

The physical interpretation of N_{free} in a flat universe is that it is just a measure of the total number of bursts that the detector sees up to infinite redshift($z \to \infty$). For a flat model N_{free} can be calculated to be

$$N_{\text{free}} = \frac{N_{\min}}{\left(1 - \frac{1}{\sqrt{1+z_{\min}}}\right)^3}. \quad (12)$$

For the case $\Omega \neq 1$, it turns out that $N_{\text{free}} = 4\pi n$. Thus it is in the same order as the coordinate number density of the bursts.

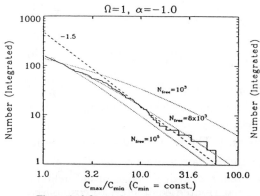

Theoretical fits to the measured number (N) versus brightness (C_{max}/C_{min}) distribution. We demanded that $< (C_{max}/C_{min})^{-3/2} > = 0.367$, the measure value of the data used. Fits are shown for an $\Omega = 1$ universe and $\alpha = -1$ where $L \sim \nu^{\alpha}$. The free parameter N_{free}, related to the number density of bursts, was allowed to vary in order to find a good fit.

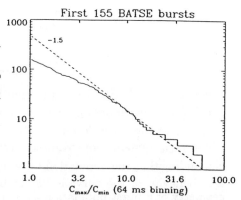

A plot of peak flux in a 64 ms time bin (C_{max}) divided by the triggering flux (C_{min}) versus the integrated number count for the first 155 BATSE triggered bursts. Only bursts with $C_{max}/C_{min} > 1$ which have C_{min} within 2 standard deviations of the mode of the C_{min} distribution have been selected. The dotted line represents the power law expected from a Euclidean distribution of standard candles.

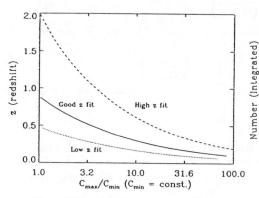

The resulting redshift (z) versus C_{max}/C_{min} for several fits to the BATSE data. A good fit implies a maximum burst redshift of less than one, however redshifts of greater than one are allowed by somewhat less good fits.

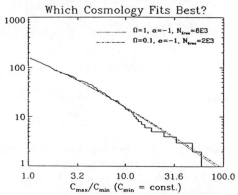

It is seen that good fits can be found to the data in both an open and critical universe. Hence, it is not possible to distinguish between cosmologies from theoretical fits to the measured N versus C_{max}/C_{min} distribution.

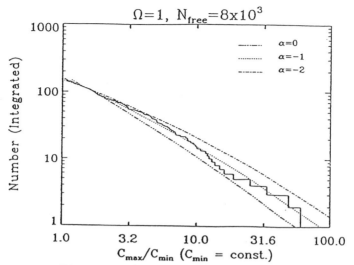

Theoretical fits to the measured number (N) versus brightness (C_{max}/C_{min}) distribution. Here we held Ω and N_{free} constant and varied the spectral index α in $L \sim \nu^\alpha$. It can be seen that the fit is indeed somewhat sensitive to the value of α used.

DISCUSSION AND CONCLUSIONS

In the present analysis, it was assumed that GRBs were cosmological in origin and they were standard candles with identical power law spectra with no break at higher energies. It was also assumed that there were no luminosity or number density evolution of bursts. We tried to fit cosmological models with the N versus C_{max}/C_{min} relation (the number versus brightness relation) measured by BATSE.

We find that quite acceptable fits to standard cosmology can be found without number or luminosity evolution. However, one cannot distinguish between currently popular cosmologies from the goodness of these fits. Best fits are obtained for a maximum redshift of the order of unity, in contrast to limits from a lack of optical identifications. Marginally acceptable fits do allow the maximum redshifts to exceed unity, however.

REFERENCES

1. Dermer, C. D. 1992, Phy.Rev.Lett., **68**, 1799
2. Fenimore, E. E., et al. 1992, *Nature*, **357**, 140
3. Meegan, C. A., et al. 1992, *Nature*, **355**, 143
4. Mao, S., & Paczynski, B. 1992, ApJ(*Letters*), **388**, L45
5. Paczynski, B. 1992, *Nature*, **355**, 521
6. Paczynski, B. 1986, ApJ, **308**, L51
7. Piran, T. 1992, ApJ(*Letters*), **389**, L45
8. van den Berg, S. 1983, Ap& SS, **97**, 385

THE NORMALIZATION OF THE PVO LOG N-LOG P DISTRIBUTION

E. E. Fenimore, R. W. Klebesadel, J. Laros, C. Lacey
C. Madras, M. Meier, and G. Schwarz
Los Alamos National Laboratory, MS D436, Los Alamos, NM 87545

ABSTRACT

The long duration of the Pioneer Venus Orbitor (PVO) mission has provided coverage of the infrequent, intense Gamma-Ray Bursts (GRB) that occupy the -3/2 power law region of the log N–log P distribution. The combination of the PVO events with the BATSE results on weaker events[1] could provide the frequency distribution of GRBs over a range of 10^4 in intensity. Earlier work[2] combined the BATSE and PVO results and concluded that they are not consistent with simple (i.e., no evolution) cosmological models involving standard candles. However, to utilize both databases, careful attention must be paid to the normalization of each data set. This paper presents a new normalization of the PVO log N–log P distribution which, when compared with the BATSE results, reduces the disagreement with the simple cosmological models from a greater than 3σ result to one between 2 and 3σ.

Pioneer Venus Orbitor (PVO) observed 273 events that triggered the on-board high-resolution memory. Since the trigger provides a standard selection criterion, these "Memory Read Out" (MRO) events have been the basis for giving the shape of the log N–log P distribution. To obtain a normalized rate one must include the effects of background variations and determine what fraction of the time the instrument was capable of recording a trigger. The threshold of 2.2×10^{-5} erg/sec/cm^2 was selected because usually the background does not affect the ability to trigger at that level. Of the triggered events, 197 had peak intensities greater than 2.2×10^{-5} erg/sec/cm^2 during the period 78/9/13 to 91/8/14.

The key correction factor, thus, is the time the memory is unavailable. There are two components to this off-time. The first component is caused by the instrument being turned off either because of solar conjunction (when it cannot be tracked) or to conserve electrical power. During the 4718 days between 78/9/13 and 91/8/14, PVO was powered off about 6.6% of the time. The second component of the memory off-time is the time that the memory is completely filled so a new burst will not be recorded regardless of its intensity. This factor is more difficult to estimate since the only available observable it is related to is the number of bursts missed by the trigger.

There are two methods which can estimate the number of events missed by the trigger because the MRO was filled. The first was used in our earlier work[2] and used the number of events seen in the "real-time" data to correct the number seen in the MROs. For $\sim 57\%$ of the time, PVO had a "real-time" telemetry link with earth and events missed by the trigger can be observed if the sample duration is comparable to the duration of the burst. Chuang[3] studied the PVO realtime data for the period 78/9/13 to 88/7/31 and identified 217 GRBs.

By determining the number that exceeded 2.2×10^{-5} erg/sec/cm^2, Fenimore et al.[2] arrived at a normalization of the PVO $\log N$–$\log P$.

The second method to estimate the number of events missed by the trigger is to sum the time between when each trigger filled the memory and when the ground controllers read and reset the memory. This information was not retained in the initial analysis of the database. There were $\sim$1325 triggers, not all of which filled the memory. Note that the 273 triggered GRBs make up a small fraction of the total number of triggers. The rest are solar flares, proton events, and some noise spikes caused by particles or showers of particles depositing very large energies.

We have finished a complete reanalysis of the entire PVO database that has determined the cumulative time between triggers that filled the memory and the time the memory was reset. This determination of the MRO-unavailable time should have been consistent with that determined from the number of realtime events seen by Chuang[3] but, in fact, disagreed by a large factor. Thus, we also reanalyzed all of the realtime events and have concluded that errors in the analysis by Chuang led to the difference.

From Table 1 of Chuang, one finds 14 triggered GRB with peak intensities of $\sim 2.2 \times 10^{-5}$ erg/sec/cm^2 and they have an average count rate of 256 cts/sec. From Chuang's Table 6, we find 73 events as intense as 256 cts/sec. In our Table I, we reproduce Chuang's values for the duration of the events (in seconds), the risetime of the events (sec), the peak count rate (cts/sec), and the hardness ratio defined to be the counts between 200 and 2000 keV divided by the counts between 100 and 200 keV. The data for two events (GB811102 and GB811104) was not available and we could not locate one event (GB861108) at the time specified.

Chuang used an algorithm based on the hardness ratio and the duration to determine if an event was a GRB or a solar event. Long duration (> 18.8sec) events with a hardness ratio greater than 1.38 were identified as GRBs and events shorter than 18.8 sec, events were deemed to be GRBs if their hardness ratio was greater than $(1.38/18.8)T_w$ where T_w is the duration of the event. Although we suspect this criterion may not be sufficient to distinguish solar events from GRBs, we will accept it.

There are two errors that we have found when reanalyzing the events in Table 1. First, the background was not removed from the peak intensity listed in Chuang's Table 6 (although the background was removed from the events listed in Chuang's Table 1). Second, a casual inspection of the durations shows them to be wrong. For some events (e.g., GB830804) a duration of 1.14 sec is reported but with a risetime of 60.57 sec. We have extracted each of the realtime events identified by Chuang and estimated the duration by a visual inspection of time history plots. Our values are reported in column 6 of our Table I. There are other problems among the realtime events. Although all events in Table 1 were claimed to be non-triggered events, we found that 8 of them had triggers (noted by an "M" is the last column). One event (GB800501) was listed as being confirmed as a GRB by the ICE experiment but a reanalysis of the PVO and ICE data has convinced us that it is a solar event.

We have reanalyzed each realtime event to determine its hardness ratio (see column 8 in Table I). Many of these disagree strongly with the values from Chuang. Of the 69 events we found and reanalyzed, we classified 20 of these as solar events, 35 as probable GRBs, 12 as GRBs confirmed by another spacecraft, and 2 were of uncertain origin (see last column of Table I).

Table I
Bright Real Time Events in PVO

Event	Chuang Analysis				This Analysis			Class
	Dur	Rise	Peak	HR	Dur	Peak	HR	
790101	1.35	1.92	392	3.35	15	259	2.11	CGRB
790113	0.39	2.69	360	2.24	8	223	1.76	GRB
790218	29.23	6.15	269	1.46	30	56	1.66	GRB
790403	5.39	11.92	304	0.92	20	181	0.58	Solar(M)
790514	0.39	0.58	344	3.87	8	223	3.16	CGRB
790610	2.86	0.86	256	1.50	5	109	1.24	CGRB
790719	3.8	5.24	294	1.44	10	181	1.47	GRB
780919	1.35	0.96	304	1.68	5	159	0.88	GRB
780930	2.12	5.39	280	1.14	20	143	1.13	CGRB
781006	0.58	0.19	336	1.47	10	198	1.64	CGRB
781025	0.39	0.10	320	2.52	0.5	187	2.38	CGRB
781026	7.31	1.73	344	1.94	10	212	1.32	CGRB
781119	0.58	0.19	304	2.75	1	166	2.71	GRB
781217	12.89	1.73	488	1.06	12	354	1.34	CGRB
800130	0.19	0.39	328	2.98	1	218	3.05	GRB
800222	27.69	1.54	608	1.50	50	496	1.67	GRB
800222	2.69	3.46	376	0.77	300	264	0.53	Solar
800223	9.33	1.36	1077	0.79	100	967	0.73	Solar(M)
800501	164.8	10.84	1281	1.76	200	1174	0.45	Solar
800522	9.5	2.86	340	4.09	20	202	1.46	GRB
800528	5.71	4.76	1507	5.48	120	1273	0.99	Solar
800602	0.77	0.67	304	4.40	1	198	4.38	GRB
800621	71.43	60.95	1936	1.50	130	1832	3.03	GRB
800710	1.15	0.77	352	4.40	5	252	5.24	GRB
800908	7.14	9.43	400	0.61	15	304	0.41	Solar(M)
800924	3.14	1.14	280	3.06	15	186	2.20	GRB
801016	3.43	2.29	304	2.04	5	179	1.79	CGRB
801108	91.67	13.3	1024	1.69	140	1020	0.75	Solar
810104	125.3	53	380	33.05	150	291	0.22	Solar
810123	23.08	1.73	336	1.89	20	244	1.89	GRB
810127	1.43	0.86	264	1.73	4	152	1.28	GRB
810714	211.9	4.76	397	4.80	100	668	0.41	Solar
810717	149.4	43.37	674	79.49	>300	576	0.64	Solar
810719	113.3	69.9	380	3.29	>300	270	0.58	Solar
810721	4.76	2.38	258	2.09	16	121	2.13	CGRB
810824	4.53	8.3	1800	0.56	20	1702	0.57	Solar(M)
811102	3.14	0.57	304	5.54		183		
811104	27.69	3.08	332	2.10		228		
820720	13.08	7.69	400	1.01	30	291	0.80	Solar
820822	2.89	1.45	258	1.71	10	163	2.03	CGRB
830105	45	8.33	919	18.79	>200	821	1.03	Solar
830106	138.1	38.1	806	4.10	180	710	0.37	Solar
830301	25.14	1.74	304	1.71	40	204	1.34	CGRB
830420	4	2.29	680	0.43	10	565	0.41	Solar(M)
830420	1.91	2.86	1871	1.57	30	1757	0.85	Solar
830427	2.29	2.29	312	0.59	200	211	0.56	Solar(M)
830804	1.14	60.57	776	0.35	80	665	0.30	Solar(M)
830928	0.29	0.14	288	1.20	2	179	1.20	GRB

831119	12.57	5.14	272	2.41	12	157	2.10	GRB
840205	123.10	15.39	974	1.87	150	850	0.19	Solar
840317	3.81	0.48	257	0.50	10	136	0.52	Solar
840503	1.91	7.62	264	1.70	10	123	1.60	GRB
840518	0.95	1.43	342	0.59	4	233	0.58	GRB
841221	0.95	0.24	512	0.47				
850115	4.76	2.38	280	2.18	5	148	3.30	GRB
850618	5.70	7.69	331	2.80	40	192	2.46	GRB(M)
850729	0.48	4.29	288	4.00	40	145	6.18	GRB
860401	11.91	5.24	268	8.24	15	117	18.27	GRB
860929	1.92	0.96	432	0.86	2	267	1.70	GRB
861012	0.2	0.1	400	3.12	0.25	232	3.11	GRB
861017	0.1	0.19	608	0.57	0.25	442	0.53	GRB
861023	0.1	0.2	560	3.42	0.25	392	3.24	GRB
861108	0.1	0.39	496	0.67				?
861123	0.19	0.1	768	4.86	0.25	607	4.78	GRB
861128	0.39	0.58	432	1.76	0.50	267	3.14	GRB
861217	0.1	0.19	384	3.95				?
861229	1.91	1.91	288	2.02	10	118	1.60	GRB
870114	2.86	0.48	277	1.14	8	104	1.18	GRB
870127	0.48	0.95	400	7.36	2	224	7.80	GRB
870212	4.76	6.67	272	3.20	130	95	3.79	GRB
870707	4.76	0.49	388	2.60	8	166	2.50	GRB
880128	0.24	1.45	328	0.80	1	178	0.80	GRB(?)
880128	7.71	4.82	1192	4.37	15	1042	3.61	GRB

Table II shows parameters from the realtime events that were used to normalize the PVO log N–log P distribution. The first row is from our previous analysis[2] and the second row is from the analysis presented here. The first issue is to correct for the variations in the telemetry rate. PVO triggers on either 0.25, 1.0, or 4.0 sec so when the realtime has at least one sample per 2 seconds the detection ability of the realtime data is comparable to that of the triggered system. $N_{\Delta T<2}$ is the number of realtime events found when PVO had samples of 2 sec or less and $N_{\Delta T>2}$ is the number of events with greater than 2 sec samples (c.f. Table 9 of Chuang). In both the previous analysis and this analysis, $N_{\Delta T<2} = 166$ and $N_{\Delta T>2} = 51$. $T_{\Delta T<2}$ is the cumulative time (in days) spent in 2 sec or better mode and $T_{\Delta T>2}$ is the time spent in greater than 2 sec mode. An estimate of the number of realtime events that are missed because of poor time resolution is:

$$N'_{\Delta T>2} = N_{\Delta T<2}\frac{T_{\Delta T>2}}{T_{\Delta T<2}} - N_{\Delta T>2}. \quad (1)$$

Then the efficiency of modes with samples greater than 2 sec is

$$\epsilon_{\Delta T>2} = \frac{N_{\Delta T>2}}{N'_{\Delta T>2} + N_{\Delta T<2}}. \quad (2)$$

Let N_{RT} be the total number of non-triggered GRBs with intensities greater than 2.2×10^{-5} erg/sec/cm^2 seen in the realtime data over the duration $T_{\Delta T>2} + T_{\Delta T<2}$. The inefficiency of the MRO, ϵ_{MRO}, is defined to be the ratio of events missed by the MRO to those seen by the MRO and is related to the above parameters by

$$N_{RT} = \epsilon_{MRO} \frac{(T_{\Delta T<2} + \epsilon_{\Delta T>2} T_{\Delta T>2})}{T_{MRO}} N_{MRO} \qquad (3)$$

Here, N_{MRO} (=197) is the total number of triggered GRBs more intense than 2.2×10^{-5} erg/sec/cm^2 in time T_{MRO} (which is $[1 - 0.066]4718$ days).

Table II
Parameters for Normalization of PVO

	$T_{\Delta T<2}$	$T_{\Delta T>2}$	$N'_{\Delta T>2}$	$\epsilon_{\Delta T>2}$	N_{RT}	ϵ_{MRO}	R_{PVO}
Previous Analysis	1426	709	31.5	0.62	73	0.89	36
This Analysis	1429	655	25	0.67	7	0.11	21

The total GRB rate for events seen above 2.0×10^{-5} (a more convenient number than 2.2×10^{-5}) is found by correcting for the MRO inefficiency and scaling by a $-3/2$ power law:

$$R_{PVO} = \frac{(1 + \epsilon_{MRO}) N_{MRO}}{T_{MRO}} \left(\frac{2.0}{2.2}\right)^{-3/2} \qquad (4)$$

The earlier analysis used $N_{RT} = 73$ in Eq. 3 which gave $\epsilon_{MRO} = 0.89$ resulting in a rate near 36 events per year. The reanalysis provided ϵ_{MRO} directly by finding the cumulative time between triggers and resets (T_{off}) which is related to the MRO inefficiency by $\epsilon_{MRO} = T_{off}/T_{MRO}$. We found T_{off} to be 400.1 days yielding a ϵ_{MRO} of 0.11. Using 0.11 for ϵ_{MRO} in Eq. 4 gives a PVO rate of 21 events per year. Using 0.11 in Eq. 3 predicts that one should have expected 8.1 events in the realtime search and, indeed, our reanalysis shows that only 7 of the events were likely to be in fact GRBs more intense than 2.2×10^{-5} erg/sec/cm^2.

We have compared cosmological models assuming GRB peak intensities are standard candles with the newly normalized PVO log N–log P distribution. The model and method of fitting are identical to Fenimore et al[2]. The best fit luminosity is unchanged since that depends on the shape of the curve, not the normalization. The normalization affects only the density of standard candles. The new best fit value is 1.7×10^4 events within a volume of $(4\pi/3)(c/H)^3$ where c is the speed of light and H is Hubble's constant. The extrapolation of the PVO best fit curve predicts more events than seen by BATSE. However, the BATSE point falls between the 2σ and 3σ fits to the PVO distribution.

REFERENCES

1. Meegan et al., Proceedings of the Huntsville Gamma-Ray Burst Workshop (AIP, New York, 1991).
2. E. E. Fenimore R. I. Epstein, C. Ho, R. W. Klebesadel, and J. Laros, Nature 357, 140 (1992).
3. K. Chuang, Ph. D. Thesis (UC Riverside, 1989).

INFERRING THE SPATIAL AND ENERGY DISTRIBUTION OF BURST SOURCES FROM PEAK COUNT RATE DATA

Tom Loredo and Ira Wasserman
Center for Radiophysics and Space Research, Cornell University

ABSTRACT

We analyze the distribution of gamma-ray burst peak count rates reported by BATSE in the context of a variety of isotropic phenomenological and physical models using Bayesian methods. These methods fit the *differential* count distribution, but do so without binning or averaging data (in contrast to, say, χ^2 or $\langle V/V_{\max}\rangle$ analyses), and take known selection effects and biases into account. Using phenomenological broken power-law models, we find significant evidence for steepening in the $\log R(>\Phi)$–$\log \Phi$ distribution from $R(>\Phi) \propto \Phi^{-1}$ below $\Phi_b \approx 4 - 40$ cm^{-2}s^{-1}, to $R(>\Phi) \propto \Phi^{-3/2}$ above Φ_b. We then study three families of cosmological models: standard candle models, standard candle models with power-law density evolution, and models with power-law luminosity functions. We find that although the data require curvature in the $\log R(>\Phi)$–$\log \Phi$ distribution, they are too sparse to distinguish among the cosmological alternatives.

In the absence of direct measurement of the distances to burst sources, or association of bursts with well-localized counterparts, the spatial and energy distribution of bursters must be inferred from the distribution of burst strengths and directions. The methods currently used to analyze the burst distribution all ignore or destroy some of the information in the data in various ways.

Here we follow the rules of Bayesian inference to draw conclusions about phenomenological and physical models for the burst distribution. Prominent in the Bayesian treatment is the *likelihood function*, which overcomes the weaknesses of other methods. Likelihood functions can be derived for the "rawest" data available, avoiding the information loss that inevitably accompanies processing performed to remove selection effects. The likelihood function is the joint probability for all available data, and thus can explicitly use information provided by nondetection data (*e.g.*, specification of the detector threshold over time and direction), and any correlations in jointly analyzed strength and directional data; it can also be used to combine data from different experiments that have different known selection effects and biases. It depends on the values of each individual datum, and does not unnecessarily bin or average the data. Finally, since it is the probability for the data, it takes known selection effects into account by construction.

In the remainder of this paper we briefly describe the likelihood function for the recently reported BATSE peak count rate data, and the results of analyses of isotropic models for this data. Further details will appear elsewhere, with descriptions of likelihood functions for more complete data. A brief review of the Bayesian methods and notation we use is available in §2 of Gregory and Loredo (1992).

The likelihood function is a function of the model parameters, which we denote collectively by Θ, and the data, D. Of course, it also depends on the model, which we denote by M; the model includes specification of both the burst distribution model, and any other needed assumptions (*e.g.*, use of the

Poisson distribution). We assume that the model allows one to calculate the differential burst rate per unit flux and direction, $dR/d\Phi d\Omega$, as a function of observed peak flux, Φ, and direction, Ω. The differential rate also depends on the values of the model parameters. We use the Poisson distribution based on the rate to model burst occurrence as a function of time, direction, and strength.

To facilitate calculation of the likelihood, we separate the data into two parts: detection data, $\{d_i\}$, denoting the data associated with burst events detected at times t_i; and nondetection data, $\{\bar{d}_j\}$ denoting the data associated with nondetections of bursts at times t_j. For the available BATSE data, each detection datum simply specifies c_i, the peak count rate (in the second most brightly illuminated detector) for burst i. (In the absence of a model for burst time histories, we can analyze data from only one trigger timescale; we use the 165 bursts detected with a 64ms trigger time.) Nondetection data are specified by the peak flux threshold, $\Phi_{\rm th}(t,\Omega)$, as a function of time and direction. In the course of the analysis we find that only certain integrals of this function are needed; these can be inferred (at least approximately) from sky exposure data published by the BATSE team (Brock et al. 1992).

The Poisson assumption implies that the probabilities for each datum are independent of the others, since each refers to a distinct time interval. For isotropic models, the resulting likelihood function, up to a Θ-independent factor, is

$$\mathcal{L}(\Theta) = \exp\left[-T\int d\Phi\,\eta(\Phi)\frac{dR}{d\Phi}(\Phi)\right]\prod_i \int \frac{d\Phi}{\Phi}\,w\left(\frac{c_i}{A_0\Phi\delta t}\right)\frac{dR}{d\Phi}(\Phi).$$

Here T is the duration of the observations; $\eta(\Phi)$ is the time-averaged fraction of the sky from which bursts with a flux of Φ could be detected (sky exposure); A_0 is the effective area of a single detector for incident fluxes normal to the detector; and $w(\mu)$ is the probability density for the cosine of the angle between the burst direction and the detector normal (we assume purely geometric dependence of effective area on incident angle). The tabulated data are for the second most brightly illuminated detector; we use the corresponding form for $w(\mu)$.

The first exponential factor comes from the product of nondetection probabilities. The value of the detection threshold affects this factor only. The second factor is simply the product of the event rate for each burst, but "blurred" by $w(\mu)$ since the burst directions are unreported, so the precise burst fluxes are unknown.

Most models for $dR/d\Phi$ can be written in the form $dR/d\Phi = A f(\Phi)$, where A is an amplitude parameter, and the remaining parameters, θ, define the "shape" of the distribution, $f(\Phi)$. We will concentrate here on the shape parameters alone. Eliminating A using standard methods (marginalization) leaves a quasilikelihood for the shape parameters,

$$\mathcal{L}_q(\theta) = \prod_i \frac{\int \frac{d\Phi}{\Phi}\,w\left(\frac{c_i}{A_0\Phi\delta t}\right)f(\Phi)}{\int d\Phi\,\eta(\Phi)\,f(\Phi)}.$$

We use flat prior densities for shape parameters throughout this work; thus the posterior distribution for the shape parameters is simply the quasilikelihood, normalized with respect to its arguments.

We begin by analyzing two phenomenological models to get a sense of what information is in the data, independent of a particular physical model for the burst distribution.

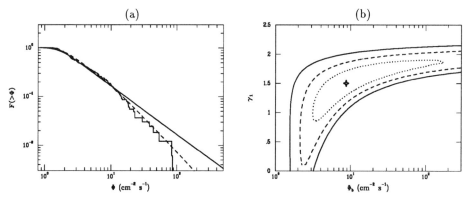

Figure 1. (a) Best-fit single power law (solid) and smooth broken power law (dashed) predicted cumulative distributions, compared with histogram of estimated burst fluxes. (b) Contours of $p(\gamma_1, \Phi_b \mid D, M_2)$, for smooth broken power law model; dotted, dashed, and solid contours bound approximate 68%, 95%, and 99.7% credible regions, respectively. Cross marks the mode.

First, we consider a simple power-law model, M_1, with $dR/d\Phi = A\,\Phi^{-\gamma}$. This has one shape parameter, the power-law index, γ. The posterior for γ is nearly Gaussian, indicating that $\gamma = 2 \pm 0.2$ ("2σ"). The index corresponding to an isotropic distribution ($\gamma = 2.5$) is outside even the "5σ" range. The solid curve in Figure 1a gives a sense of how this model, with γ set to its best-fit value, compares with the data. Plotted are the normalized effective cumulative rate distribution, $F(>\Phi) = [\int_\Phi^\infty d\Phi'\, \eta(\Phi')\, f(\Phi')]/\int_0^\infty d\Phi'\, \eta(\Phi')\, f(\Phi')$, as a solid curve, and a histogram of estimated burst fluxes, $\Phi_i = c_i/(A_0 \bar{\mu})$, where $\bar{\mu} = 0.62$ is the expectation value of μ for a burst.

Figure 1a seems to indicate that the logarithmic slope of the rate is not constant over the observed range. Also, PVO observations imply that at large fluxes ($\Phi > 20$–50) $dR/d\Phi \propto \Phi^{-2.5}$ (Chuang et al. 1992). Accordingly, we investigate a smooth broken power law model, M_2, with $dR/d\Phi = A\,(\Phi/\Phi_b)^{-\gamma_1}/[1 + (\Phi/\Phi_b)^{\gamma_2 - \gamma_1}]$, with $\gamma_2 \equiv 2.5$. This model has two shape parameters, γ_1 and Φ_b.

The best-fit parameter values are $\gamma_1 = 1.5$ and $\Phi_b = 8.7$. The best-fit model is ≈ 10 times more likely than the single power law model (which is the $\Phi_b \to \infty$ limit of this model); this represents fairly significant but not definitive evidence for curvature.

Figure 1b shows joint credible regions for Φ_b and γ_1. Values of $\Phi_b \sim 10$ are favored, but significantly larger values cannot be ruled out. Thus although the data are better fitted with a break, the location of the break is not well determined. We interpret this as implying that the data require curvature in $\log R$–$\log \Phi$, but not necessarily a sharp break. The allowed values of γ_1 are systematically smaller than the best-fit γ for a single power law, revealing that the data favor significant flattening at low Φ. The dashed curve in Figure 1a is the best-fit model, illustrating how allowing curvature improves the fit.

We now consider cosmological models. The differential burst rate for these models can be calculated according to,

$$\frac{dR}{d\Phi}(\Phi) = \int d\dot{N} \int dV\, \frac{\dot{n}_c(z, \dot{N})}{(1+z)^4}\, \delta(\Phi - \Phi_{\rm obs}[z, \dot{N}]),$$

where $\dot{n}_c$ is the burst rate per comoving volume element for bursts at redshift

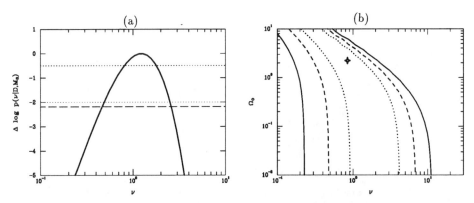

Figure 2. (a) Log posterior for dimensionless photon luminosity, ν, for standard candle model with $\Omega_o = 1$. Dotted lines mark values bounding approximate 68% and 95% credible regions; dashed line marks the likelihood of the single power law model. (b) Contours of $p(\Omega_o, \nu \mid D, M_3)$, for standard candle model; dotted, dashed, and solid contours bound approximate 68%, 95%, and 99.7% credible regions, respectively. Cross marks the mode.

z with photon luminosity $\dot{N}$, and the observed flux from a burst at redshift z of photon luminosity $\dot{N}$ is $\Phi_{\rm obs}(z, \dot{N}) = K(z)\dot{N}/[4\pi(1+z)d^2(z)]$, where $d(z)$ is the proper distance to a redshift z, and $K(z)$ is a "color correction" (similar to the optical "K-correction") that depends on the shape of the burst spectrum (we assume photon number spectra inversely proportional to photon energy).

The volume element, dV, depends on the cosmology adopted. We study $\Lambda = 0$ cosmologies, for which dV is specified by the Hubble constant, H_0, and the density in terms of the critical density, Ω_0. The integrals over luminosity and direction (in dV) can be performed analytically, leaving a single integral over redshift, which must be done numerically.

The models we investigate differ with respect to the choice of functional form for the comoving burst rate, $\dot{n}_c(z, \dot{N})$. The shape parameters of the models are the parameters defining $\dot{n}_c$, and Ω_0. Hubble's constant is absorbed into the amplitude parameter.

First we consider a standard candle model, M_3, with comoving burst rate $\dot{n}_c(z, \dot{N}) = n_0\, \delta(\dot{N} - \dot{N}_c)$. The shape parameters for this model are Ω_0 and $\dot{N}_c$. The photon luminosity is best written in terms of a dimensionless photon luminosity, ν, according to $\dot{N}_c = \nu(4\pi c^2/H_0^2)\Phi_{\rm fid}K(0)$, where $\Phi_{\rm fid}$ is a fiducial value of the observed flux, which we set equal to 1 cm^{-2}s^{-1}. For $\nu = 1$, $\dot{N} \approx 10^{57}$s^{-1}, corresponding to a luminosity of approximately 10^{51} erg s^{-1} for $H_0 = 100$ km s^{-1}.

For a flat universe ($\Omega_0 = 1$), Figure 2a shows the constraints placed on the photon luminosity. The best-fit value is $\nu = 1.2$, with $\nu = 0.5$ to 2 at the "2σ" level. These parameters imply typical burst redshifts ≈ 1. This model has a best-fit likelihood very nearly equal to that of the broken power-law model.

Allowing Ω_0 to vary, however, broadens the allowed range of ν, as shown in Figure 2b. This figure also shows that we can not make any useful inferences about Ω_0 from the available data.

Next we consider a standard candle model with power-law density evolution, M_4, for which $\dot{n}_c(z, \dot{N}) = n_0\,(1+z)^{-\beta}\,\delta(\dot{N} - \dot{N}_c)$. The shape parameters are now Ω_0, ν, and β. The previous model corresponds to $\beta = 0$.

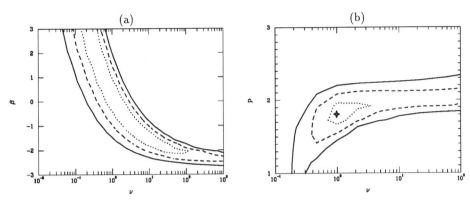

Figure 3. (a) Contours of $p(\beta, \nu \mid D, M_4)$, for standard candle model with density evolution. (b) Contours of $p(p, \nu \mid D, M_5)$, for power-law luminosity function model. Dotted, dashed, and solid contours bound approximate 68%, 95%, and 99.7% credible regions, respectively. Cross marks the mode.

Figure 3a shows joint credible regions for β and ν, for $\Omega_0 = 1$. Allowing density evolution only slightly increases the likelihood, but it greatly weakens the ability of the data to constrain the burst luminosity: the allowed range of ν now spans over four decades, with the typical burst redshift varying from $z \sim 0.1$ to $z \gg 10$. This emphasizes how critically inferences about burst characteristics can depend on theoretical assumptions.

Finally, we consider a model with no density evolution, but with a power-law photon number luminosity function (Wasserman 1992). For this model, M_5, we set $\dot{n}_c(z, \dot{N}) = A \dot{N}^{-p}$ over a finite range, $\dot{N}_l$ to $\dot{N}_u$. The shape parameters for M_5 are Ω_0, p, $\dot{N}_u$, and the dynamic range, $\rho = \dot{N}_u / \dot{N}_l$.

If ρ is smaller than the dynamic range of the data (≈ 100), these models resemble standard candle models; also, the likelihood function is essentially independent of ρ for values larger than the dynamic range of the data. Accordingly, we simply fix ρ at a large value (10^5) to explore this model. Here we have also fixed $\Omega_0 = 1$. Two shape parameters remain, p and $\dot{N}_u \equiv \nu(4\pi c^2 / H_0^2) \Phi_{\rm fid} K(0)$.

Figure 3b shows the allowed range for p and ν. The best-fit point is at $p = 1.8$, $\nu = 0.98$; but ν is largely unconstrained. Typical burst redshifts are 0.1 to a few. This model has the largest likelihood of those we have studied (≈ 13 times as likely as a simple powerlaw), but it is only marginally likelier than its cosmological competitors. The data cannot distinguish among the cosmological alternatives.

This work was supported in part by NSF grant AST-89-13112 and NASA grants NAGW-666 and NAG 5-1758.

Brock, M.N, Meegan, C.A., Fishman, G.J., Wilson, R.B., Paciesas, W.S., and Pendleton, G.N. 1992, in Gamma-Ray Bursts, Huntsville, AL 1991, eds. W.S. Paciesas & G.J. Fishman (NY: AIP), 399

Chuang, K.W., White, R.S., Klebesadel, R.W., and Laros, J.G. 1992, ApJ, 391, 242

Gregory, P.C., and Loredo, T.J. 1992, ApJ, 398, 146

Wasserman, I. 1992, ApJ, 394, 565

DISTRIBUTION OF PEAK COUNTS OF GAMMA-RAY BURSTS

Vahé Petrosian and Walid J. Azzam
Center for Space Science and Astrophysics
and
Bradley Efron
Department of Statistics

Stanford University, Stanford, CA 94305

Abstract

We describe a new method for the direct determination of the so-called $\log N$-$\log S$ relation of gamma-ray bursts and discuss its relation and advantages compared to other commonly used methods. We present results from the application of this method to three data sets and discuss their significance.

1. Introduction

There has been considerable discussion of the analysis of the so-called $\log N$-$\log S$ distribution of gamma-ray bursts (GRBs) and its implications for their spatial distributions. As is well known for a homogeneous, isotropic, static and Euclidean situation (HISE for short) the differential distribution of peak photon count rates C_p, obeys the relation $n(C_p) \propto C_p^{-5/2}$ and the cumulative distribution function (CDF) $N(C_p) = \int_{C_p}^{\infty} n(C_p) dC_p \propto C_p^{-3/2}$. The extent to which the observations show such a distribution and the value of C_p at which the distribution deviates from HISE together with the actual form of the diviations is the distinguishing factor among various possible scenarios for the spatial distribution of GRBs.

The derivation of the above distributions from GRB data is not straightforward because, unlike other situations, the threshold or the limiting photon count rate C_{lim} is not constant or easily definable. The usual practice has been to analyze the source count data in terms of the distribution of the ratio (C_p/C_{lim}) or equivalently the ratio $(V/V_{max}) = (C_p/C_{lim})^{-3/2}$. When HISE is valid the differential distribution $f(C_p/C_{lim})$ or its CDF, $F(C_p/C_{lim})$, are also expected to have power law forms with indices $-5/2$ and $-3/2$, respectively. (Equivalently, the distribution of V/V_{max} is expected to be uniform, see Petrosian 1992a). However, as pointed out recently (Band 1992; Hartmann and The 1992, and Petrosian 1992b), the interpretation of the deviations of these distributions from such power laws is not straightforward. The last of the above three works (denoted as paper I), in addition, presents a method for obtaining the distributions $n(C_p)$ and $N(C_p)$ directly from the observed bivariate distribution $\psi(C_p, C_{lim})$ such as the one shown in Figure 1.

We first note that it seems unlikely that the threshold C_{lim} prior to a burst will have any correlation with its peak count C_p. As we will briefly describe below, using a new test of independence (Efron and Petrosian 1992a), one can show that C_p and C_{lim} are indeed independent. Because of this independence, we can write $\psi(C_p, C_{lim}) = n(C_p)\phi(C_{lim})$ and obtain the following cumulative and differential distributions for $c \equiv (C_p/C_{lim})$:

$$F(c) = c^{-1} \int_0^{\infty} N(C_p)\phi(C_p/c)dC_p, \qquad f(c) = c^{-2} \int_0^{\infty} C_p n(C_p)\phi(C_p/c)dC_p. \qquad (1)$$

In the usual case when there is a single threshold, $\psi(C_p, C_{lim}) = n(C_p)\delta(C_{lim} - C_{l_o})$, these equations show that the distributions of C_p and C_p/C_{lim} are identical. But in the opposite limit when the $\phi(C_{lim})$ distribution is uniform and broad (at least as broad as the observed range of C_ps), then $F(c) \propto c^{-1}$ [or $f(c) \propto c^{-2}$] independent of the intrinsic distribution $n(C_p)$. If this were true, then the fact that most observations (see e.g. Meegan et al, 1992) show roughly $F(c) \propto c^{-1}$ would have no bearing on the homogeneity, isotropy or any other geometrical aspect of the distribution of GRBs. This, of course, is an extreme condition and has been used here to emphasize that care is necessary in the interpretation of the distributions of C_p/C_{lim} or V/V_{max} vis-a-vis the actual counts $n(C_p)$ and $N(C_p)$.

In general, the distribution of C_p/C_{lim} is a convolution of the distribution $n(C_p)$ with the distribution of instrumental threshold, so that the current approach would require some kind of "deconvolution" of equation (1) which, if not impossible, would magnify the noise considerably rendering it useless for detection of subtle effects. In the next section we describe a new method proposed in paper I specifically for the analysis of GRB data. This method yields the required distributions $n(C_p)$ and $\phi(C_{lim})$ directly from the data. In section 2 we also present some results from the application of the method to various data sets. In section 3 we discuss the significance of the results.

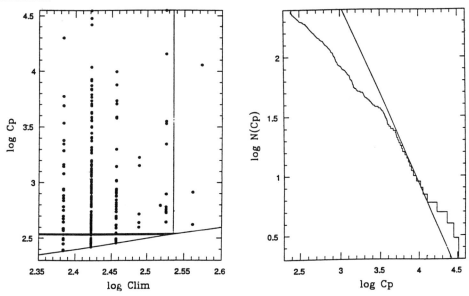

Figure 1. Gamma-ray burst data obtained by BATSE instrument of Compton Observatory. *Left panel*: Distribution of observed peak and threshold photon count rates C_p and C_{lim} for 223 bursts triggered by 1024 milisecond integration time. The horizontal lines show the strip on C_p containing one data point i and the left hand square shows the box containing the $\tilde{N}_i$ untruncated sources associated with this point. *Right panel*: The cumulative counts obtained from equation (2) for this data set. The straight line has slope $-3/2$.

2. Direct Determination of Distribution of Peak Counts

The problem is to determine the distribution $n(c_p)$ or $N(C_p)$ from the observed joint distribution of data on the C_p, C_{lim} plane, with $C_p \geq C_{lim}$ (see Figure 1). This problem occurs in the determination of bivariate (or multivariate) distributions whenever observational biases produce *truncated* data in the sense that points with certain values of the two (or more) variables are not accessible. It is difficult, if not impossible, to obtain the bivariate distribution in the general case when the two variables have arbitrary relation to each other. This problem simplifies considerably if the two variables are stochastically independent. In the next section we show that C_p and C_{lim}, as observed by various instruments, are independent, in which case the best method for obtaining the distribution of the variables is the maximum likelihood method first described by Lynden-Bell (1971) and elaborated and utilized for obtaining the evolution of the luminosity function of quasars by Jackson (1974), Petrosian (1986) and Caditz and Petrosian (1990). This method avoids binning of the data and correctly accounts for all observational selection biases. In fact, as shown by Petrosian (1992), all other correct methods, in the limit of single data per bin, reduce to this simple and elegant method which we briefly describe below.

2.1. Description of the Method

Assuming C_p and C_{lim} are stochastically independent we can separate the bivariate distribution; $\psi(C_p, C_{lim}) = n(C_p)\phi(C_{lim})$. Then the method, in its simplest form, gives a histogram of CDF's $N(C_p)$ and $\Phi(C_{lim}) = \int_0^{C_{lim}} \phi(y)dy$. Whether the fixed integration limit is taken to be infinite or zero (or, for that matter, any other fixed number) is immaterial. Then referring to Figure 1, it is easy to show that the logarithmic increment $d\ln N(C_p)$ around the ith data point is equal to the ratio of the number of data points in strip i to that in the Box i ($C_p > C_{p,i}, C_{lim} < C_{p,i}$). If we choose the strip narrow enough to include only the data point i, then it can be shown that $\delta \ln N(C_{p,i}) = \ln(1 + \tilde{N}_i^{-1})$, where $\tilde{N}_i$ is the number of objects in the Box i. Now repeated applications of this expansion across strips containing the data points (in order of decreasing C_p), gives the histogram

$$N(C_{p,i}) = N_1 \prod_{j=2}^{i}(1 + \tilde{N}_j^{-1}). \tag{2}$$

where N_1 is the assumed CDF for the largest observed C_p. It is obvious then that by the same procedure we can obtain the CDF $\Phi(C_{lim,i}) = \Phi_1 \prod_{j=2}^{i}(1 + \tilde{M}_j^{-1})$, where Φ_1 is the value of Φ for the smallest C_{lim}, and $\tilde{M}_j$ is the number of data points in the box with $C_{lim} < C_{lim,j}$ and $C_p > C_{lim,j}$. For more details see paper I and the references cited above.

The right hand panel of Figure 1 shows the cumulative distribution $N(C_p)$ obtained by this method for the subset of BATSE data shown in the left hand panel. A small portion of this distribution at high values of C_p appears to obey a power law with index $-3/2$ as expected from HISE. But for the most part the index is closer to -1. As shown in paper I, similar results are obtained for data from SIGNE instruments on Venera 13 and 14 (Attiea et al, 1991) and for GRBs observed by the GRS instrument on SMM (Matz et al, 1992) and Ginga (Ogasaka et al, 1991).

2.2. Independence Test

There exist many statistical tests of independence if there are no observational selection biases (or truncations) in the data. A non-parametric method which avoids binning of the data involves the comparison of the rank distribution of the actual data points with that expected from randomly distributed data obtained, for example, from random permutation of the two variables. Such a procedure cannot be naively applied to truncated data. We use a method developed recently by Efron and Petrosian (1992a) which can account for truncation and is closely related to the method of the determination of the CDFs described above. Without going into further detail, we mention in passing that the application of this method to the GRB data shows that, C_p and C_{lim} appear to be stochastically independent except for an almost insignificant correlation in the SIGNE data. For more details on this the reader is referred to Efron and Petrosian (1992b).

2.3. Differential Distributions

It is also possible to obtain the differential distribution $n(C_p)$. It is easy to show that differentiation of equation (2) yields (see Petrosian, 1986, 1992)

$$n(C_p) = \sum \frac{\delta(C_p - C_{p,i}) N(C_{p,i})}{\tilde{N}_i + \Theta(C_p - C_{p,i})} \qquad (3)$$

where $\Theta(x) = 0$, for $x < 0$, and $= 1$ for $x > 0$. This distribution is equal to a series of delta functions of different weights at the data points $C_{p,i}$. This form is not convenient for graphical presentation of the results. A histogram for $n(C_p)$ can be obtained by binning the data, or by various smoothing procedures (see, e.g., Caditz and Petrosian, 1992). Care is needed in the selection of the degree of smoothing. Too little smoothing or too few data points leads to a noisy distribution. Larger smoothing and fewer bins can cause loss of relevant information. Figure 2 shows the binned differential distributions for the SMM and BATSE data with ten data points per bin. In both these figures there is a hint of a slope of $-5/2$ at large values of C_p. We now examine the significance of these slopes and the deviation from it.

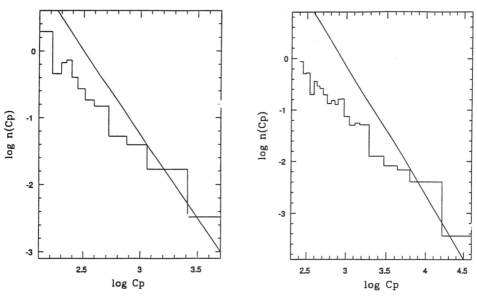

Figure 2. Differential counts obtained from equation (3) for binned SMM (left panel) and BATSE (right panel) data. Each bin contains 10 data points. The straight lines have a slope of $-5/2$.

2.4. Variation of Power Law Index

From equations (2) and (3) one can show that the logarithmic slope $s_n = -d\log N/d\log C_p$ can also be obtained directly from the data. This slope, like the differential counts, is not well defined at the data points. Its inverse changes from $\tilde{N}_i$ to $\tilde{N}_i + 1$ across the ith data point. An excellent approximation for the slope is $s_{n,i} = (\tilde{N}_i + \frac{1}{2})^{-1}$. The unbinned slopes obtained this way will also be a series of spikes. A smoother version can be obtained by averaging the slopes over several data points as shown in Figure 3 (see Efron and Petrosian, 1992b).

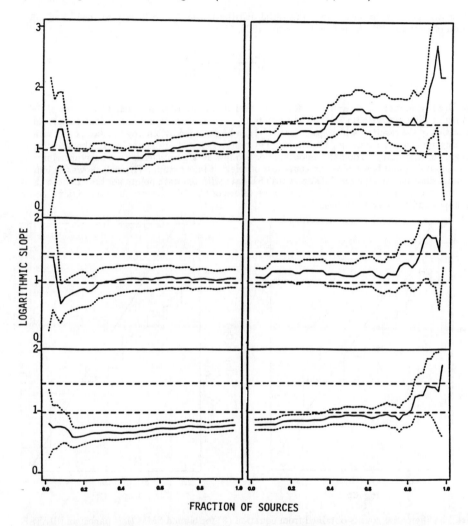

Figure 3. Variation with normalized peak count, C_p, of the cumulative average slope for SIGNE, SMM and BATSE data (from top to bottom, respectively) starting from right (right panel) or from left (left panel). The two horizontal lines are drawn for slopes of 3/2 and 1. The dashed lines show the 90% confidence range. Vertical axis shows $50[1 - \exp(.02 s_n)] \simeq s_n$ and the horizontal axis shows the fraction of sources above the minimum observed C_p.

Note that only SIGNE data show a slope of 3/2 over a significant portion ($\simeq$ 60%) of the high photon count rates. SMM and BATSE indicate the presence of such a slope only over the highest 10% range where the statistical fluctuations are large. At low values of C_p (the left portion of the Figures) the slopes are all smaller than 3/2. For most of the BATSE range, the slope is even smaller than one which is the value expected for a disklike distribution of sources.

3. Summary and Conclusions

We have indicated the difficulties in the interpretation of the log$N$-log$S$ relation for GRBs when presented in the commonly used manner in terms of the ratio C_p/C_{lim}. We also point out that it is not necessary to analyze the data in this manner and that we can obtain the distribution of C_p and C_{lim} directly from their observed bivariate distribution. We have described procedures to obtain the cumulative and differential distributions and their logarithmic slopes. This procedure relies on the fact that C_p and C_{lim} are independent, a hypothesis which can be tested using a new method described by Efron and Petrosian (1992a and 1992b). Some results from the application of our methods to various data sets on GRBs are presented in paper I. Here we show the results for about 220 bursts observed by BATSE and compare the differential distributions from this data to that obtained by SMM and compare the logarithmic slopes obtained from the above two data sets with that from SIGNE. For further details on the methods the reader is referred to the preceeding references, to paper I and to a review paper by Petrosian (1992c).

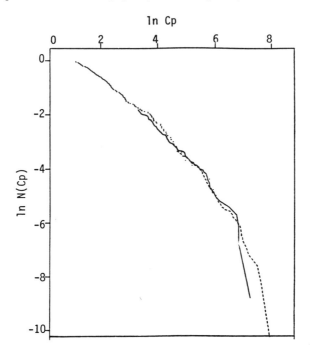

Figure 4. Cumulative distribution $N(C_p)$ for the combined data from SIGNE (dashed), SMM (solid) and BATSE (dotted) lines. The counts are shifted relative to each other to obatin the best fit in the overlap region. Unlike individual data sets this combined data shows distinct variation of slope s_n.

Our main results are that in each data set it is difficult to ascertain with a high degree of confidence a systematic and clear flattening of the distribution from that expected for HISE as we proceed from high to low values of C_p. The combined data set, however, can clearly show this trend. It is difficult to combine the various data sets because a given value of C_p refers to bursts of different peak flux density for different instruments. One could have an approximate normalization if there exist bursts observed simultaneously by two instruments. A third possibility is to normalize the distribution according to the rate of observations of the bursts which, except for some variation due to differences in the spectral responses of the instrument, should be the same for a given peak flux density (not C_p). This normalization requires comparison of the duration of the observations and their duty cycles (or their live times) which we currently do not know.

We hope to be able to carry out these procedures in the future. At the moment we have carried out the following pure mathematical exercise for the normalization of the SIGNE, SMM and BATSE data sets. This method attempts to identify the portions of the distributions for each data sets which have the most similar shape. For details the reader is referred to Efron and Petrosian (1992b). Figure 4 shows the outcome of this exercise. As evident in the overlapping region the cumulative distributions are very close to each other. We emphasize that this normalization is purely statistical and not based on known characteristics of the instruments and should not be taken literally. Nevertheless, we point out that the shifting among the C_p values which produced this normalization agrees with what is expected from a preliminary consideration of the characteristics of the three data sets.

The combined distribution shown in Figure 4 clearly shows the gradual variation of slope from its HISE value of 3/2 to one and even values less than this. There are various conclusions which can be drawn from this. These are tentative conclusions and are listed here only to indicate the usefulness of direct and exact knowledge of the distribution $n(C_p)$ or $N(C_p)$ over a larger dynamical range than is possible for a single data set. We can, for example, conclude that the presence of a significant range with slope less than one, as shown in Figure 4, further strengthens the already strong argument against the disklike distribution obtained from the isotropy of the GRBs (Meegan et al, 1992). The shape of the combined curve can constrain the halo models as well as the cosmological models, considering that this distribution does not look like the $\log N$-$\log S$ curve of galaxies, quasars, radio sources or any other known extragalactic population.

Acknowledgement: This work is supported by NASA grants NAG5 2036, NAGW 2290 and NAGW 3223.

References

Atteia, J. L. et al 1991, Nature 351, 296.
Band, D. 1992, ApJ Letters 400, L63.
Caditz, D. M. and Petrosian, V. 1990, ApJ 357, 326.
Caditz, D. M. and Petrosian, V. 1992, submitted to ApJ.
Efron, B. and Petrosian, V. 1992a, ApJ 399, 345.
Efron, B. and Petrosian, V. 1992b, submitted to J. of Am. Statistical Assoc.
Hartmann, D. and The, L.-S. 1992, Ap and Space Sci, in press.
Jackson, J. C. 1974, N.N.R.A.S., 166, 281.
Lynden-Bell, D. 1971, M.N.R.A.S., 155, 95.
Matz, S. M. et al 1992, *Gamma-ray Bursts*, eds. C. Ho, R. I. Epstein, page 175.
Meegan et al 1992, Nature, 355, 143.
Ogasaka, Y. 1991, ApJ Letters, 383, L61.
Petrosian, V. 1986, Structure and Evolution of Active Galactic Nuclei, ed., G. Giunian (Dordrecht, Rdidel), page 355.
Petrosian, V. 1992a, *Gamma-Ray Bursts*, eds W. S. Paciesas and G. J. Fishman, AIP Conference Proceedings, 265, page 97.
Petrosian, V. 1992b, ApJ Letters (in press), Paper I.
Petrosian, V. 1992c, *Statistical Challenges in Modern Astronomy*, eds. E. Feigelson and G. J. Babu.

STATISTICS OF COSMIC GAMMA-RAY BURSTS: EVOLUTION OF DIFFERENT POPULATION OF GALACTIC SOURCES

I.G. Mitrofanov, A.A. Kozlenkov A.A., A.M. Chernenko,
V.Sh. Dolidze, A.M. Pozanenko, and D.A. Ushakov

Space Research Institute, Moscow, Russia

J.-L.Atteia, C.Barat, M.Niel, G.Vedrenne

Centre d'Etude Spatiale des Rayonnements, Toulouse Cedex, France

ABSTRACT

Below new results are presented of statistical studies of cosmic Gamma-Ray Bursts (GRBs) detected with the Russian-French APEX experiment onboard the PHOBOS-2 spacecraft in 1988–89. For 48 APEX events, the averaged time history and hardness ratio profiles are provided with one second time resolution. For the two subsets of events (19 'strong' GRBs with $0 < V/V_{max} < 0.3$, and 29 'weak' ones with $0.3 < V/V_{max} < 1$), the above-mentioned profiles compare as follows: (i) the time histories differ before the bursts reach their maxima, but their tails are almost identical; (ii) the hardness ratio profiles differ significantly close to the bursts' maxima. Several implications of these observations are drawn for cosmological and galactic models of GRBs.

INTRODUCTION

The APEX experiment has been done jointly by Space Research Institute (Russia), Instiute of Geochemistry and Analitic Chemistry (Russia) and Centre d'Etude Spatiale des Rayonnements (France). From July 1988 till March 1989, 62 classical GRBs had been detected (e.g., see [1,2]), 48 of them have been loaded to the APEX-database[3]. This database with direct access to any measurements of each event, permits to study averaged characteristics for all bursts as well as for any particular subsets of GRBs, selected by specific values of their parameters. In the following, the APEX GRBs are divided into subsets corresponding to the 2 subranges of V/V_{max} ($0 < V/V_{max} < 0.3$ and $0.3 < V/V_{max} < 1$). We call these subsets 'strong' and 'weak' bursts, respectively.

AVERAGED TIME-HISTORIES OF APEX GRB'S

The counting rate F in the broad range from 100 to 1000 keV has been used for the study of the time histories. The universal time scale has been used for averaging, so that the sampling intervals contain the whole number of 1/8 second. When the time scale for averaging is chosen (further it will be 1 sec), the interval of maximum rate is found for each event (t_{max} with F_{max}). Then on each side of the maximum, relative intensities are calculated and stored for all 1 sec time intervals

($f = F/F_{max}$). For any chosen subset of bursts, the averaged time history is calculated by aligning the maxima at $t = 0$ and averaging all relative intensities for the corresponding time intervals before and after t_{max}.

Figure 1 shows the averaged time history for all events from the APEX database. It has one peak shape with smooth back slope. The main pulse has duration of about 1 second. The total duration of averaged time history is more than 20 seconds.

Averaged time histories of 'strong' and 'weak' events are given in Fig. 2. They are different only in the rising parts, while the back slopes are identical to all statistical criteria. It is interesting to note that the tails are characterized by the power law rather than exponential decay.

AVERAGED SPECTRAL EVOLUTION OF APEX GRB'S

For each time interval of averaged time history t_i, the mean energy spectra could be calculated. To separate average spectral evolution from variations of bursts' intensities, spectra for interval of each burst have been normalized to the number of counts N_i in the energy range 100–1000 keV. After averaging the normalized spectra corresponding to the same time interval t_i, the mean hardness ratio H.R. is evaluated for each of them (H.R. is defined as the ratio between the number of counts in the 300–1000 keV band to that in 100–300 keV).

The evolution of H.R. is clearly seen along the averaged time history of all bursts (Fig. 1). The H.R. is larger on the rise front and at the top than on the back slope. It is in agreement with the well-known results based on the measurement of individual bursts[4]. On the other hand, the evolution of H.R. is rather different for 'strong' and 'weak' events (see Table 1). For the t_{max} interval the difference between the averaged H.R. is 0.15 ± 0.02.

Table 1 Comparison of average hardness ratios for 'strong' and 'weak' APEX

Time intervals	Relative intensity		Hardness ratio	
	'strong' events	'weak' events	'strong' events	'weak' events
1 s before maximum	0.53 ± 0.01	0.49 ± 0.03	0.37 ± 0.02	0.25 ± 0.03
1 s at the maximum	1.0	1.0	0.45 ± 0.01	0.30 ± 0.02
1 s after maximum	0.53 ± 0.02	0.46 ± 0.02	0.29 ± 0.02	0.27 ± 0.03

ASTROPHYSICAL CONCLUSIONS

The observations lead to the following astrophysical implications:

1) The similarity of averaged time histories of 'strong' and 'weak' GRBs presents serious problem for the hypothesis of their cosmological origin. Indeed, to explain the mean value of V/V_{max} about 0.33 ([5]), one has to assume the range of redshifts from $z = 0$ for the 'strong' events to $z = 1-2$ for the 'weak' ones. For the cosmological model, one would have expected all timescales of 'weak' events to be $1 + z \sim 2-3$ times larger than those of 'strong' bursts.

2) The similarity of averaged time histories of 'strong' and 'weak' GRBs suggests that their sources are both of the similar nature and the mechanisms of their emission are comparable. Two-population models of sources are known to address statistical properties of 'strong' and 'weak' GRBs. It is hardly acceptable that sources of different astronomical origin have very similar time histories.

3) On the other hand, the difference in the hardness ratios between 'strong' and 'weak' GRBs points out that sources of the same astronomical population have different hardnesses in their maxima depending on how close or distant they are located. It could be an evolutionary effect if we assume that closer objects are younger than the more distant ones. Moreover, in this case the burst sources seem to be characterized by the outward expansion and evolution.

4) Provided the galactic origin of GRBs is postulated, neutron stars are the most probable candidates for their sources. This population has to have the isotropic distribution on the Sky, but with a decreasing number of sources at larger distances). There are only two possibilities, when the Observer is located in the Center of such population: (i) the outbursting neutron stars belong to Extended Galactic Corona with scales ~ 100 kpc[6], and (ii) the outbursting neutron stars belong to a local population with scales about 1-3 kpc, born $\sim 10-30$ Myrs ago in strong 'burst' of star formation in Solar Galactic Vicinity[7].

Both possibilities agree with the idea of outward evolution. In the first case the evolution is continued during the total lifetime of NS:

$$\sim 300 \text{ Myrs} \left(\frac{d_{max}}{100 \text{ kpc}}\right) \left(\frac{300 \text{ km/s}}{V_{NS}}\right), \tag{1}$$

in the second, the evolution lasts shorter than

$$\sim 30 \text{ Myrs} \left(\frac{d_{max}}{1 \text{ kpc}}\right) \left(\frac{30 \text{ km/s}}{V_{NS}}\right). \tag{2}$$

Here d_{max} is the distance scale for outside sources, and V_{NS} is the mean velocity of outgoing NS's. In the second case the outbursting activity of NS's should eventuallty come to end after 30 Myrs because otherwise old NS's from the Thick Disk would dominate in the Solar Galactic Vicinity, and neither isotropy, nor evolution of sources could be observed.

CONCLUSIONS

Statistical analysis, as presented above, suggests that galactic origin of GRBs is more compatible with the APEX observational data, than the cosmological one. In this case neutron stars either from Extended Galactic Corona, or from Local Galactic Vicinity, should be accepted as sources of GRBs. However, it should be emphasized that the above statistical results are based on 48 APEX events only, so they should be verified using larger statistics. Hopefully, BATSE events will provide such possibility.

ACKNOWLEDGEMENT

The author is grateful to the GRO Conference Organizing Committee for the kindful invitation and appropriate financial support.

REFERENCES

1. C.Barat et al. Instruments and Methods for Space Studies, ed. V.M.Balebanov (Nauka, Moscow, 1989), p.213.
2. I.G.Mitrofanov et al. Pis'ma v Asron.Zhurnal, 16, 302 (1990).
3. I.G.Mitrofanov, A.A.Kozlenkov, V.Sh.Dolidze et al., Astronom.Zh. (in russian), 69, 5, 1052 (1992).
4. S.M.Matz et al. Ap.J. Lett., 288, L37 (1985).
5. G.J.Fishman. These Proceedings.
6. I.S.Shklowskii and I.G.Mitrofanov MNRAS, 212, 545 (1985).
7. I.G.Mitrofanov Proceedings of Taos Workshop "Isolated Pulsars" (Cambridge University Press, 1992)

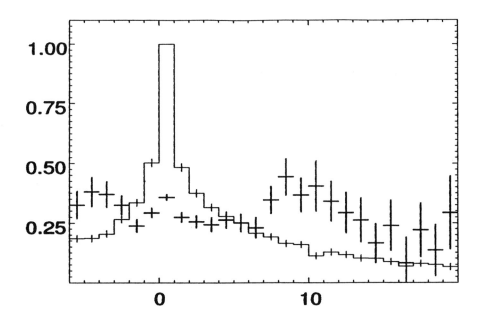

1. One second resolution averaged time history (thin line) and hardness ratio (thick line) for the total sample of APEX bursts stored in database (48 events).

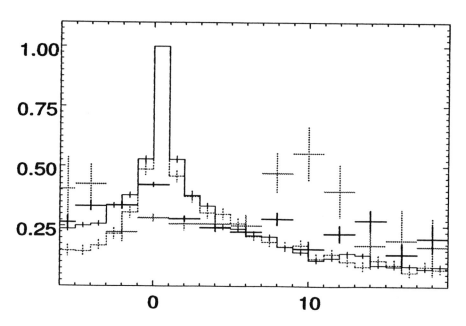

2. Averaged time histories (thin lines) and hardness ratios (thick lines) for the 2 subsets of bursts: $0 < V/V_{max} < 0.3$ — solid line (19 events), and $0.3 < V/V_{max} < 1$ — dotted line (29 events).

GAMMA-RAY BURSTS:
COUNTERPARTS

GAMMA-RAY BURST LOCATIONS WHILE-YOU-WAIT: FIRST RESULTS OF THE THIRD INTERPLANETARY NETWORK BURST WATCH

K. Hurley
University of California, Space Sciences Laboratory, Berkeley, CA 94720

M. Sommer
Max-Planck Institut für Extraterrestrische Physik, 8046 Garching-bei-München, Germany

T. Cline
NASA - Goddard Space Flight Center, Greenbelt, MD 20771

M. Boer, M. Niel
Centre d'Etude Spatiale des Rayonnements, 31029 Toulouse Cedex, France

E. Fenimore, R. Klebesadel, J. Laros
Los Alamos National Laboratory, Los Alamos, NM 87545

G. Fishman, C. Kouveliotou, C. Meegan, R. Wilson
NASA - Marshall Space Flight Center, Huntsville, AL 35812

ABSTRACT

The gamma-ray bursts of May 1, 17, 25, and July 11 and 20, 1992 were rapidly localized to regions with dimensions of several arcminutes by the spacecraft of the third interplanetary network. Radio, optical, and soft X-ray follow-up observations were carried out, and a soft X-ray source was detected in one error box. Although the Pioneer Venus Orbiter spacecraft is no longer operating, its place in the network will be taken by Mars Observer, and locations will continue to be produced rapidly.

INTRODUCTION

Multiwavelength searches for counterparts to gamma-ray burst sources have been carried out for over a decade, but the vast majority of these searches took place at least months, and usually years, after the burst. Starting in May 1992 it became possible to access all or part of the data from the spacecraft of the third interplanetary network within a day or so of the arrival of a gamma-ray burst, and subsequently process it within another half day or so to obtain relatively accurate burst locations by the method of triangulation. (Until the recent entry of Pioneer Venus Orbiter into the atmosphere of Venus, this network was composed primarily of the Ulysses, PVO, and Compton Observatory

(BATSE) spacecraft, as well as other near-earth missions.) We have dubbed this mode of operation a "burst watch". The typical preliminary error boxes obtained in this way had dimensions of several to ten arcminutes, small enough to encourage rapid follow-up observations by the VLA, by optical observers, and by ROSAT. Because the data from Ulysses and PVO were often incomplete at this stage, later processing of the data reduces the error box dimensions by a factor of at least two. When all three spacecraft were operating continuously in their optimum modes, a burst was detected about once every two weeks. However, as PVO was nearing the end of its lifetime in June 1992, the May - September period was punctuated by stretches when the instrument was off. A total of five Ulysses/PVO/BATSE triggers were recorded during this period, as well as several events involving Ulysses, PVO, and another near-earth spacecraft (i.e., not Compton).

FIVE RAPID LOCALIZATIONS

Table 1 lists the five rapidly localized bursts, their preliminary error box dimensions, and the dates of VLA, optical, and ROSAT observations that we are aware of. As the locations were given in IAU circulars, there may be other observations as well. Figures 1 - 3 show some triangulated positions in relation to the BATSE error boxes.

Table I. Rapidly localized gamma-ray bursts

Event Date (1992)	Error Box Dimensions	VLA Observation	Optical Observation	ROSAT Observation
May 1	2' x 2'	May 15	May 13	May 19
May 17	2' x 2'	No	May 20	No
May 25	2' x 2'	No	May 30	No
July 11	12' x 12'	No	July 22	TBD
July 20	3' x 5'	No	July 29	No

The optical observations listed were carried out at the European Southern Observatory (Schmidt camera), Lick Observatory (double astrograph), and the South African Astronomical Observatory (1 m telescope). The ROSAT observation of the May 1 event was done as a target of opportunity and lasted $\approx$ 2700 seconds; a similar target of opportunity observation of the July 11 event is being scheduled.

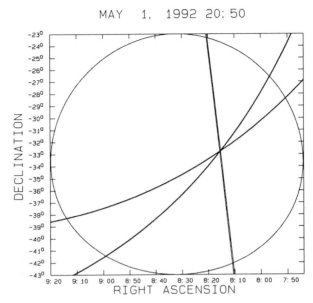

Figure 1. The BATSE (10° radius error circle) and triangulated positions (Ulysses-BATSE, Ulysses-PVO, and PVO-BATSE annuli) of the May 1, 1992 event.

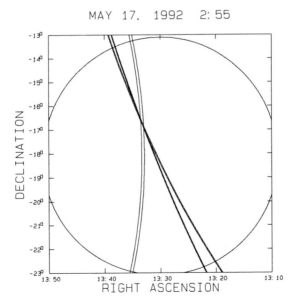

Figure 2. The BATSE (5° error radius circle) and triangulated positions (Ulysses-BATSE, Ulysses-PVO, and PVO-BATSE annuli) of the May 17, 1992 event.

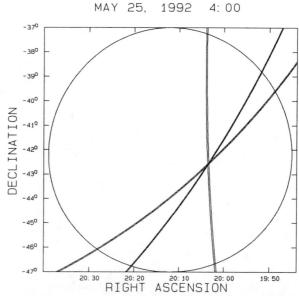

Figure 3. The BATSE (5° radius error circle) and triangulated positions (Ulysses-BATSE, Ulysses-PVO, and PVO-BATSE annuli) of the May 25, 1992 event.

DISCUSSION

The most interesting set of observations to date is for the May 1 event. A very weak ROSAT source (strength $\lesssim 10^{-13}$ erg/cm^2 s) was detected in the error box of the burster. Given the number of sources in this ROSAT field, the chance probability of finding a source in this error box is about 10^{-3}. The error box lies in the galactic plane, however, and this estimate does not take into account the more general statistics of ROSAT sources in the plane. The ESO J plate for this field reveals a faint, diffuse object, at the plate limit, probably an anonymous galaxy, whose position appears to be consistent with that of the X-ray source. This object appears to be below the POSS plate limit. The VLA observation yielded two weak sources, both outside the error box (R. Hjellming, private communication).

CONCLUSIONS

Deeper optical observations have already been carried out for the May 1 event, and a longer ROSAT observation will be requested in the AO4 period.

While it is certainly too early to conclude anything about the relation between the ROSAT source, the diffuse object, and the gamma-ray burst, the feasibility of carrying out relatively deep multiwavelength counterpart searches days after a gamma-ray burst has been demonstrated. Although most past searches have been fruitless, the locations being produced by the third interplanetary network offer two distinct advantages with respect to them, namely decreased delay and smaller error boxes.

With the demise of PVO, the network has temporarily collapsed to a Ulysses/Earth baseline. However, Mars Observer is currently on course, and the gamma-ray burst detector will be turned on shortly, returning the network to a 3 point configuration. We intend to continue working in the burst watch mode, with some possible improvements. One is to extend VLA coverage to more events. Another is to attempt to overcome the day/night, weather, and dark time limitations of ground-based observatories by using the Hubble Space Telescope Wide Field Camera for selected targets of opportunity; proposals have been submitted for these purposes. (A fundamental limitation exists for ROSAT, since a source position must lie in a band $90\pm15°$ from the sun, comprising 26% of the sky; thus rapid follow-up opportunities with ROSAT will remain relatively rare.) Yet another is to use EUVE for targets of opportunity. Finally, we will work towards the goal of reducing the localization delays to their absolute minimum, which in some cases will be less than one day.

ACKNOWLEDGEMENTS

The Ulysses GRB experiment was constructed at the CESR (Toulouse, France) with a grant from CNES, and at the Max-Planck Institut für Extraterrestrische Physik with support from FRG Contracts 01 ON 088 ZA/WRK 275/4-7.12 and 01 ON 88014. We are indebted to R. Williams, C.-H. Yu, J. Callet, S. Davies, S. Vilke, and W. Wartelle for assistance in data reduction operations. KH acknowledges support from NASA under JPL Contract 958056.

GAMMA-RAY BURST OBSERVATIONS
WITH THE *COMPTON / ULYSSES / PIONEER-VENUS* NETWORK

T. L. Cline
NASA / Goddard Space Flight Center, Code 661, Greenbelt, MD 20771

K. C. Hurley
University of California, Space Sciences Laboratory, Berkeley, CA 94720

M. Sommer
Max-Planck Institut fur Extraterrestrische Physik, Garching, Germany

M. Boer and M. Niel
Centre d'Etude Spatiale des Rayonnements, Toulouse, France

G. J. Fishman, C. Kouveliotou*, C. A. Meegan, W. S. Paciesas** and R. B. Wilson
NASA / Marshall Space Flight Center, ES-62, Huntsville, AL 35812

E. E. Fenimore, J. G. Laros and R. W. Klebesadel
Los Alamos National Laboratory, MS D436, Los Alamos, NM 87545

ABSTRACT

The third and latest interplanetary network for the precise directional analysis of gamma ray bursts consists of the Burst and Transient Source Experiment on *Compton Gamma Ray Observatory* and instruments on *Pioneer-Venus Orbiter* and the deep-space mission *Ulysses*. The unsurpassed resolution of the BATSE instrument, the use of refined analysis techniques, and *Ulysses'* distance of up to 6 AU all contribute to a potential for greater precision than had been achieved with former networks. Also, the departure of *Ulysses* from the ecliptic plane in 1992 avoids any positional alignment of the three instruments that would lessen the source directional accuracy.

The 1.5-year life of this network has now been adequate to provide at least 15 precisely determined source fields and 30 narrow, elongated source loci. The results as of the time of writing, like the results (other than N49) of former networks, show no positional agreement with any previously catalogued candidate objects. An exposure of a preliminary source field using the *Rosat* mission did, however, produce a candidate x-ray counterpart[1,2]. Detailed scrutiny of this agreement is in progress. Additional source counterpart searches, both in the x-ray and in other domains, are also underway.

Given the conflicting clues regarding the nature of gamma-ray bursts, along with BATSE's maintenance of source isotropy, counterpart searches in fields from this network may yet provide the best hopes for a resolution of the gamma-ray burst puzzle.

*   also, on leave from the University of Athens, Greece.
**   also, Physics Department, University of Alabama, Huntsville, AL 35899.

THE NETWORKS

The latest generation of gamma-ray burst interplanetary networks is based on the primary baseline from the *Compton Gamma Ray Observatory* to the *Ulysses* deep space probe. The *Pioneer-Venus Orbiter* (the sole survivor of both the first two interplanetary burst networks of the late 1970s to early 1980s) operated successfully until recently, in September, 1992, with its termination and Venusian atmospheric entry. Its 14-year survival made possible this initial interplanetary network with *Compton-GRO*.

The *Mars Observer* was launched in October 1992, with its burst detector being operated perhaps 30% of the time during the 1-year trajectory, to be on full-time after the *MO* is established in planetary orbit. The *CGRO/Ulysses/MO* network (given that Mars is typically at a greater distance than Venus) should then provide precision equivalent to, or slightly better than, that of the *PVO* network it replaces. The types of existing and/or expected results from the *Ulysses-CGRO* connection are as follows:

Dates	Missions participating	Observations
11/1990 to 04/1991	*Ulysses + PVO*	> 10 2-sc. events
04/1991 to 10/1992	*CGRO + Ulysses + PVO*	~ 20 3-sc. events
04/1991 to - -	*Ulysses* or *CGRO* or *PVO*	> 30 2-sc.events
-?-/1993 to - -	*Ulysses + MO + (CGRO* or *Wind)*	-?- 3-sc. events
-?-/1993 to - -	*Ulysses* or *MO* or *CGRO* or *Wind*	-?- 2-sc. events

Other burst detectors, e.g., on *Granat* or *Ginga*, can provide added capabilities. In particular, *GGS-Wind* will provide a survey burst instrument and a high-resolution burst spectrometer, after next year. Also, *HETE* will later search for optical transients associated with bursts. *Ulysses* may enjoy an extended mission to 2001, and other interplanetary probes, now being considered, may also contribute to future networks.

OBSERVATIONS

Burst source locations are expected to scatter over a considerable dynamic range of precision; thus, depending on circumstances, results can be categorized as follows:

1) Highly precise source fields, up to tens of arc seconds in diameter:
Celestial areas considerably smaller than 1 square arc minute will result from events with the structure and variability that provide the best timing correlations, observed with the interplanetary geometries that give the optimum, large-angle source locus intersections, when the instruments are operating in modes with the highest resolution.

2) Precise source fields, from under one to a few arc minutes in size:
Areas in the vicinity of 1 square arc minute are expected for 3-spacecraft events detected in various, more typical circumstances of instrument operation, or in the cases of events having a smoother or less complicated profile.

3) Elongated source fields, as thin as arc seconds, by a few degrees in length:
Resulting either from the superposition of a 2-spacecraft source annulus onto a several-degree BATSE source region or from the grazing intersection of two sources into a crescent, these areas can be small enough, in the arc-minute-squared region, to be of practical use for statistical studies, but are generally too irregular for deep searches.

4) Low-utility source fields, of much larger areas or of unrestricted lengths:
These are most useful to show an absence of consistency with a trial source direction.

The list of all 3-spacecraft burst events from the *CGRO/Ulysses/PVO* network is given below in preliminary form. Some of the circumstantial limitations that affect source precision are listed (locus intersection geometry and rate-only data recovery), but not all (e.g., Earth-*PVO* distance or time-history resolution mode). It is evident that about one third of the source fields to be finalized should be both of relatively high precision and found at higher than average galactic latitude, best suited for counterpart searches. (However, the first candidate x-ray source thus far located in any of these fields was found for the event of 1992 May 01, at only ~ 1 degree galactic latitude[1,2].)

The numbers listed in this table are based on presently incomplete recoveries of burst data and of trajectory information; also, the event analyses are in various stages of development. Thus, the results shown are in the preliminary state, varying up to the 'IAU-telegram' stage for a few events; none are near the final status at this time: thus, these values should be conservatively taken to be ± several degrees in accuracy.

Date	UT (s)	RA, dec	b_{II}	loci x angles	remarks
91 Apr 21	33248.	270, 26	22	23, 52, 105	low-res. rate
91 Apr 30	61719.	135, 01	28	24, 31, 125	low-res. rate
91 May 22	44029.	137, -51	-02	5, 15, 160	-
91 Jun 27	16162.	200, -04	18	0.5, 3, 176	(grazing xn)
91 Jul 17	16385.	247, -58	-08	0.9, 8, 171	(grazing xn)
91 Aug 14	69273.	339, 31	-24	0.2, 5, 175	(grazing xn)
91 Nov 04	54285.	212, 35	71	2, 15, 163	(moderate xn)
91 Nov 09	12460.	111, -22	-03	2, 14, 164	(moderate xn)
91 Nov 18	68259.	166, -23	33	8, 50, 122	-
91 Nov 26	46127.	-?-	-?-	-?-	-?-
92 Mar 11	08424.	132, -36	03	3, 9, 168	(moderate xn)
92 Mar 25	62256.	350, 13	-44	5, 20, 155	-
92 Apr 06	09929.	289, -59	-27	6, 16, 158	-
92 May 01	76731.	124, -33	01	2, 4, 174	(moderate xn)
92 May 17	11882.	203, -16	44	5, 6, 169	-
92 May 25	12426.	301, -42	-31	7, 8, 165	-
92 Jul 11	58157.	282, 73.	26	11, 33, 136	-
92 Jul 20	11524.	203, 37	77	11, 28, 141	-

The observations, not listed here, from the additional 30 or more events with only one timing baseline will result in thin annular-segment fields, each limited (by the corresponding BATSE source region definitions) to a few degrees in length. For those event source loci as narrow as a few seconds of arc, the total celestial field areas will thus be only a few square arc minutes. These fields, though irregular, can be expected to be statistically useful in searches for confirmation of any promising link that might be inferred between the precise source locations and some counterpart population.

ANALYSES

The analytic techniques for the derivation of precise source fields are well-defined in their simplest form, and have been refined over the years, both to optimize precision and to minimize the possibility of systematic error. Calculational methods were originally derived independently by each of the international groups participating in the first network. This work has been and is still being done independently by each group for each event, cross checking all known assumptions with as many differing derivations as possible, both for redundancy and for the elimination of 'clerical' error.

Improvements on the early techniques have evolved to fit the circumstances. One modification, used first for the late-1979 events, incorporates all the data from any number of interplanetary spacecraft with the statistically weighted combination of all possible 2-spacecraft timing comparisons into one final source field. This result can be compared with the each of several fields individually obtained from the 3-spacecraft analyses. We do not have, and will not soon have, the luxury of four spacecraft mutually separated with interplanetary baselines. If one did, the optimum result would be a redundancy of mutually consistent fields; note that if the results were to disagree, however, a determination of which one of the spacecraft is 'wrong' is not possible.

Advances in instrumentation incorporated in *CGRO*-BATSE and *Ulysses* both improve event resolution and inspire new analytic approaches. For example, since the limitation on a timing comparison depends, in part, on the binning times, as well as on counting statistics and profile complexity (with errors formerly in the > tens of ms, except for the 79 March 5 event, with its anomalous < 1-ms rise), improving this aspect is critical. The BATSE, on occasion, time tags each photon individually, in a portion of the event history, permitting a certain improvement in the comparison accuracy.

The *Ulysses* data, in turn, can be artificially time tagged, introducing little error in events with a low counting rate in a narrow time sample. This permits observations from one detector to be rebinned using the timing period of the other, with the phase varied as a vernier; the profiles are then directly capable of correlation, in order to find the best-fit relative time delay. Since any two spacecraft are likely to be asynchronous, if there is enough structure in the burst profile and enough definition of the variability, then the total timing error may become narrower than the wider of the binning times.

Clearly, a primary issue in all such analyses is the assignment of the timing error, which fixes the source field extent. For 3-spacecraft events, one definitive check is possible: any two time delays determined by the correlations from spacecraft a to b and from b to c defines the third as the sum; this value can be compared with that found independently from the a to c correlation. Experience with such self-checking events provides confidence in 2-spacecraft event studies. The total errors in this network may become as precise as a few ms. Given that *Ulysses* can be at 3000 light-s distance, source locus accuracies could result down to 2 arc seconds, almost 'optical' in quality.

RESULTS

Preliminary analyses of some *CGRO/Ulysses/PVO* events have been released[1], some with precision to a few arc minutes, prompting the detection of one *Rosat* x-ray source[1,2]. Various possibilities (of the survival of this association with additional refinement of the burst source location, of additional *Rosat* x-ray associations, of the nature of their composite source pattern, if any, and also of counterpart associations in other wavelengths) are now all being researched. This effort may yet surface one definitive clue, locked in the existing data, as to the nature of gamma-ray burst sources.

Acknowledgements: The *Ulysses* instrument was constructed at the CESR in Toulouse, France, and at Max Planck Institut fur Extraterrestrische Physik with support from FGR contracts 01 ON 088 ZA/WRK 275/4-7.12 and 01 ON 88014.

References:
1. K. Hurley et al., Proceedings of this Conference.
2. M. Boer et al., Proceedings of this Conference.

A SEARCH FOR LONG-LIVED EMISSION FROM WELL-LOCALIZED GAMMA-RAY BURSTS USING THE BATSE OCCULTATION TECHNIQUE

J. M. Horack, B. A. Harmon, G. J. Fishman, C. A. Meegan, R. B. Wilson,
NASA – Marshall Space Flight Center, ES-64, 35812

W. S. Paciesas, G. N. Pendleton,
Dept. of Physics, University of Alabama, Huntsville, 35899

C. Kouveliotou
USRA / on leave from the University of Athens, Greece

ABSTRACT

We present the results from a search for discrete gamma–ray emission from the positions of 14 well-localized strong bursts, using the Earth occultation technique. These bursts have been precisely localized by timing analyses from several spacecraft in the solar-system (the Third Interplanetary Network). Their locations are used as input to the standard BATSE source occultation processing software. We have searched for emission in a five–day interval centered on the time of the gamma–ray burst, with a typical 3σ sensitivity of ∼0.15 Crab.

INTRODUCTION

In addition to the primary objective of detecting and localizing gamma-ray bursts, BATSE has several secondary scientific objectives, one of which is to monitor known discrete sources in the sky.[1] The experiment utilizes the Earth limb as an occulting disk to measure the intensities of various discrete sources as they rise or set at the horizon, and to discover new sources that appear in the sky. This technique was used, for example, in the discovery of the transient source GRO J0422+32.[2] With an orbital period of approximately 1.5 hours, the BATSE experiment can perform in excess of 150 measurements of the discrete flux from a known celestial location in a five day period.

The discovery of long–lived emission before or after a burst would provide an important diagnostic in understanding the burst environment. Fencl et al.,[3] for example, have discussed the usefulness of secondary signals, or afterglows, that may arise from interaction of primary burst photons with some surrounding environment, as a means to investigate the physics of the burst region. The spectrum of any detectable long–lived emission may also greatly constrain the physics of gamma–ray bursts. Murakami[4] has shown, for example, that soft X–ray emission from GRB 870303 at late times can be fit with a blackbody spectrum and a temperature of ∼1 keV. The assumption of blackbody emission from the surface of a neutron star mandates that this particular burst be of a galactic origin, based on the required ratio of the emitting area radius to the distance. The assumption of blackbody emission in this case, however, is uncertain because other models of emission also give good fits to the data.[4]

METHOD OF ANALYSIS

The method used to search for discrete emission is rather straightforward. The locations of intense gamma–ray bursts are obtained from the Third Interplanetary Network.[5,6] Because these strong bursts were detected by at least three experiments, their locations can be determined very precisely using timing analyses. Typical error boxes for these bursts are on the order of a few arc–minutes in size.[6] With precise locations established, the times when BATSE views the source in transit across the Earth's limb are computed for a 5–day window centered on the time of the burst's detection. At the time of each occultation, 14 energy channels from the BATSE continuous (CONT) data are analyzed to search for evidence of discrete emission from the burst source.

For this search, we utilized 120 s intervals of data centered on the times of each occultation, and employed a 10 s occultation width, to account for attenuation from the atmosphere. In each of the CONT data channels, a linear fit is made to the background prior to and after the time of the occultation step. By comparing the background levels obtained before and after the transit, a measurement of the source intensity over a particular energy channel range can be obtained, and a significance can be computed. The bursts used in the analysis are shown in Table 1.

GRB	Trigger #	TJD of Detection
GRB910421	105	8367
GRB910430	130	8376
GRB910522	219	8398
GRB910627	451	8434
GRB910717	543	8454
GRB911104	999	8564
GRB911109	1025	8569
GRB911118	1085	8578
GRB920311	1473	8692
GRB920325	1519	8706
GRB920406	1541	8718
GRB920501	1576	8743
GRB920711	1695	8814
GRB920720	1711	8823

Table 1 – Bursts Used in Search for Long–Lived Emission

The sensitivity of BATSE to long–lived emission using this technique is dependent on several parameters. The primary factor is the spacecraft orientation with respect to the source location. As the orientation of the spacecraft changes with respect to a given location on the sky, the projected areas to detectors observing the position change. A secondary contributor to the sensitivity is angle β, the angle of the source out of the orbital plane of the satellite. Sources near the poles of the orbit are not occulted by the Earth, and consequently cannot be detected using this technique. Sources that lie near the orbital plane have the most rapid transition through the atmosphere limb, producing the sharpest occultation edges that are more easily identified. Figure 1 shows the estimated

3σ sensitivity of BATSE for a one–orbit step. The Crab is also shown for comparison.

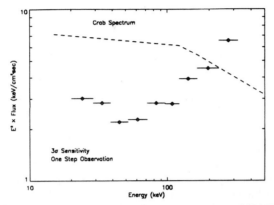

Figure 1 – The estimated 3σ sensitivity of BATSE to discrete emission as measured with one occultation edge.

RESULTS OF THE SEARCH

For each burst in Table 1, we examined 5 days of data centered on the time of the detection by BATSE. In total, over 2,000 occultation times were measured in this survey. Each occultation measurement consisted of 14 individual energy channel measurements.

Figure 2 shows the typical result obtained from the search: no significant emission at the level of ~ 0.15 Crab is identified for approximately 2.5 days prior to or after the burst. The sources of gamma–ray bursts appear to be rather quiet unless, of course, they are bursting. This particular figure is from a measurement of GRB010522 in CONT channel 11, which would include any 511 keV emission, but the result is typical. Combining energy channels to search for broad–band emission and combining occultation steps to improve sensitivity yielded no statistically significant detections of emission.

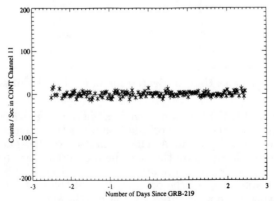

Figure 2 – Time History of flux measurements in CONT channel 11 from the location of GRB910522.

BATSE trigger #1695, GRB920711, was a very intense, complex burst lasting over 100 seconds. The profile shows an extensive amount of time structure, with significant fluctuations on the order of ~100 ms. This burst was detected by BATSE on TJD 8814 at 58140.573 seconds UT, and provided a significant occultation edge. The location of this event and the orientation of the sky combined to produce a less-than-favorable geometry for detection in multiple detectors. BATSE detector B7 was nearly face-on to the source, producing minimal projected areas in other detectors.

The location of GRB920711 was calculated to rise above the Earth's limb at 7,510.61 seconds of TJD 8814. This particular rise was ~ 50,000 seconds *before* the burst was detected by BATSE. Figure 3 contains a plot of this occultation edge as seen by detector B7 in CONT channels 1-7. The corresponding energy range is approximately 20 – 150 keV. At approximately 7,510 seconds, the time of the predicted rise, the data from this detector show a clear occultation-like feature. The dotted lines to the left and right of Figure 3 indicate the computed background levels in the regions before and after the rise of the location above the horizon.

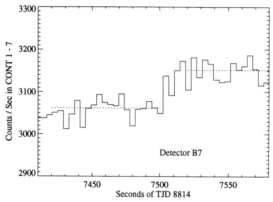

Figure 3 – Measurement of CONT channels 1–7 in detector B7 at the time of a rise of the location determined for GRB920711.

The other seven BATSE detectors were also examined at this time. The detectors facing away from the source showed no evidence of an occultation step. Detector B3, with the second-largest projected area to the source, showed a hint of an occultation step. The configuration of detectors displaying occultation steps is in agreement with the orientation of the satellite with respect to the source at the time of detection. The location of the burst is in a region of the sky that is void of discrete sources observed regularly by BATSE. The occultation software checks for possible interfering sources at other points on the Earth's limb, and found none at this time.

The BATSE occultation software computed an intensity of 134.9 ± 25.0 counts/second over this channel range, a significance of ~5 σ.

Figure 4 displays a count-spectrum of this occultation step as measured by the BATSE software. The intensity in counts/second is plotted as a function of BATSE CONT channel. Each of the energy channels 1–7 show a positive detection, with channel 3, approximately 40 – 50 keV, showing a ~3 σ detection by itself. No other occultation edge measured for any of the 14 bursts displayed this type of consistent positive detection over a large range of consecutive channels.

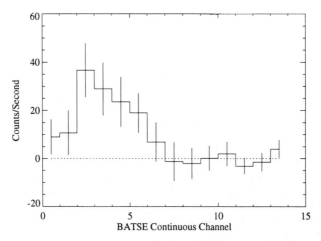

Figure 4 – A count spectrum of the occultation step shown in Figure 3.

The significance of the step is somewhat difficult to assess. Searches for un–triggered transients in the BATSE data have revealed a moderate number of significant events.[7] We note that the occultation feature shown in Figure 3 occurs at an expected time, and has the proper direction (i.e. a rise in the data with a predicted rise of the source). It occurs only in the illuminated detectors facing the source location. The step occurs ∼50,000 seconds *before* the detection of the burst event. The shape of the count spectrum is not unusual, peaking near a corresponding energy of ∼50 keV, and falling below the sensitivity of BATSE above ∼150 keV.

CONCLUSIONS

We have examined occultation steps from 14 well–localized bursts. We find no significant emission from 13 of the 14 bursts in five day intervals centered on the time of detection. One measurement from the location of GRB920711 displays a feature that is consistent with an occultation step. If this detection is real, the peak of the detected emission is near 50 keV, and falls below the BATSE sensitivity at approximately 150 keV. Further investigation into this event is clearly warranted, and additional bursts need to be examined.

REFERENCES

1. Fishman, G. J. , et al., in Proc. Gamma Ray Obs. Science Wkshp., (ed. Johnson, W. N.), 3–47, (NASA/GSFC Greenbelt, Md.), (1989)
2. Paciesas, W. S., IAU Circular #5580, (1992)
3. Fencl, H. S., et al., in *Proc. Gamma Ray Bursts*, (eds. Paciesas, W. S. & Fishman, G. J.), 267 – 271, (AIP: New York), (1992)
4. Murakami, T., in *Physics of Neutron Stars and Black Holes*, (ed. Y. Tanaka), 405 – 412, (Univ. Acad. Press: Tokyo), (1988)
5. Cline, T. L., et al., in *Proc. Gamma Ray Bursts*, (eds. Paciesas, W. S. & Fishman, G. J.), 72 – 76, (AIP: New York), (1992)
6. Hurley, K., private communication, (1992)
7. van Paradijs, J., et al., these proceedings

THE E_{max} DISTRIBUTION OF GAMMA-RAY BURSTS OBSERVED BY THE PHEBUS EXPERIMENT

J.-P. Dezalay, C. Barat, R. Talon
CESR, BP 4346, 31029 Toulouse Cedex, France

R. Sunyaev, O. Terekhov, A. Kuznetsov
Space Research Institut, Profsoyuznaya 84/32, 117810, Moscow, Russia

ABSTRACT

We present the E_{max} distribution of 78 cosmic gamma-ray bursts (GRB) recorded by the PHEBUS experiment during the first two years of operation. This distribution is compared with that obtained from the Gamma-Ray Spectrometer[1] (on SMM) cosmic event data set. We observe that the proportion of events displaying high-energy emission above 1 MeV (hereafter $P_{(>1)}$) is lower in the PHEBUS population of GRBs. The possible reasons for this difference are discussed and several tests are applied to understand the influence of statistical effects on the value of $P_{(>1)}$.

INTRODUCTION

The existence of high-energy emission by GRBs has been established mainly from the analysis of the GRS/SMM data set[1]. These observations have revealed that emission above 1 MeV is a common and energetically important component of classical cosmic γ-ray bursts. About 60% of the 72 events detected by the GRS instrument exhibit a 3σ excess above the measured background for energies above 1 MeV. The problem is knowing if this percentage can be extrapolated to the entire burst population, i.e., if some fraction of the events have no high-energy emission. This is the purpose of the analysis reported here from a study of the PHEBUS data set.

PHEBUS EXPERIMENT

The PHEBUS experiment onboard GRANAT (launched in December, 1989) was designed to record high-energy transient events in the 100 keV–100 MeV energy range. It consists of six independent detectors with their associated electronics. Each sensor includes a BGO crystal (78 mm in diameter, 120 mm in height) surrounded by a plastic anti-coincidence jacket. These detectors are arranged on the spacecraft in order to cover a nearly 4π steradian field of view. The total efficiency of the BGO crystal is close to unity over the entire energy range. The burst mode is triggered when the count-rate in the 0.1–1.5 MeV energy range exceeds the background level by 8σ in 0.25 or 1s. In this case, 176 time-to-spill spectra are recorded for each detector with 116 energy channels over the 100 keV–100 MeV energy range. The observation average duty cycle of PHEBUS was about 35% up to the end of 1991.

© 1993 American Institute of Physics

Between December 1989 and December 1991, 78 GRB candidates were recorded by PHEBUS. During these first two years of operation, 40% have been confirmed by other experiments and now, with the launches of ULYSSES and CGRO, the confirmation rate is greater than 60%. This percentage is consistent with the duty cycle and the sky coverage of the BATSE/CGRO experiment. The PHEBUS event detection rate is about 115 bursts per year for full sky coverage, with an average low-energy threshold of 110 keV.

DATA ANALYSIS AND RESULTS

We have analyzed the maximum energy (called E_{max} hereafter) observed in the spectrum integrated over the total duration of PHEBUS events. The E_{max} value has been estimated using information only from the detector having the highest C_{max}/C_{min} on a 1s or 0.25s time interval[2]. As in the SMM analysis, E_{max} was defined to be the highest energy such that the integrated count-rate above that energy is 3σ above the background level. To perform this calculation, we first make a cumulative spectrum. Each channel of this spectrum contains all counts of the primary spectrum recorded above its lower energy. Then, the significance (number of σ) is calculated[3] for each channel of this cumulative spectrum. In order to reduce statistical fluctuations, the significance is fitted as a function of energy and the maximum energy is deduced from the intersection with the 3σ line. To estimate the uncertainty in the E_{max} value, we have made one hundred simulations. In each simulation, we have done a Poissonian trial around the observed count-rate in each channel and calculated the maximum energy using the method described above. The final E_{max} is defined as the most probable value observed in the maximum energy distribution from the simulations. Asymmetrical 1σ error bars are derived from this set of simulations by taking the smallest energy range that contains at least 68% of the simulated E_{max}. The proportion of events displaying a high-energy emission ($P_{(>1)}$) and its 1σ error bars are derived from a bootstrap calculation[4] (with 10,000 trials) using the 78 simulated E_{max} distributions.

The E_{max} versus C_{max}/C_{min} distribution is shown in Figure 1. The proportion of events displaying a high-energy emission above 1 MeV is $34.2 \pm 3.4\%$ for PHEBUS, to be compared with the value of 60% found by Matz et al. (1985) from the GRS/SMM data set. This apparent discrepancy can be explained by the differences in detector efficiency, burst triggering criterion, and low-energy threshold which yield different burst detection rates for the two experiments. The rate is nearly 50 events per year for GRS and about 115 per year for PHEBUS.

In order to test for the existence of an instrumental effect, we have selected events from the PHEBUS data set that would have been seen by GRS according to its detection rate (using the C_{max}/C_{min} criterion). This divides our data into two subsets of 32 intense events and 46 weak events. For these new samples of GRBs, the proportion $P_{(>1)}$ is $54.9 \pm 4.5\%$ for the subset of strong bursts and falls off to $19.7 \pm 4.7\%$ for the subset of weaker events. The difference in the value of $P_{(>1)}$ between the two samples is greater than 5σ indicating that the ratio C_{max}/C_{min} is correlated with the maximum energy.

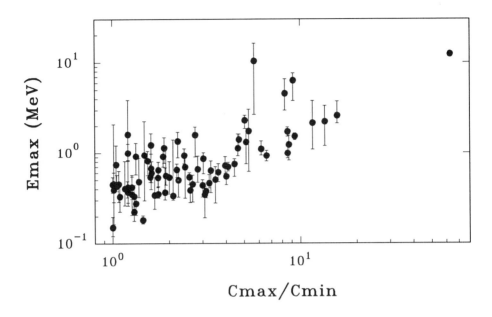

Figure 1. Maximum Energy vs C_{max}/C_{min}.

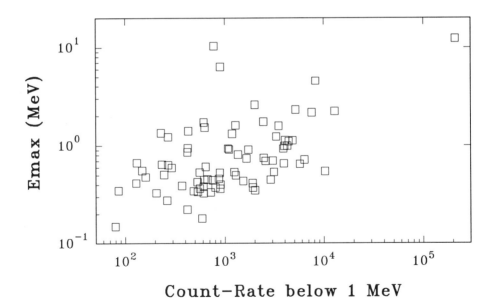

Figure 2. Maximum Energy vs. Integrated Count-Rate in the 150 – 1000 keV energy range.

If we select the most intense events of the PHEBUS data set, we find that the SMM and the PHEBUS value of $P_{(>1)}$ are fully consistent. This most likely indicates that the method used to estimate $P_{(>1)}$ is not responsible for the observed difference between the two experiments.

The observed E_{max} value for a typical GRB spectrum is dependent upon the continuum shape and the total count-rate. The proportion $P_{(>1)}$ is therefore affected by instrumental limitations (effective area, background, etc.) that could explain the difference between 'strong' and 'weak' events. Figure 2 shows the maximum energy plotted versus the total count-rate in the 150 keV –1 MeV energy range. In order to determine the influence of the total count-rate on the $P_{(>1)}$ value, we have created a third sample of events. This new sample is composed of strong bursts which have been weakened to have on average the C_{max}/C_{min} mean value of the 'weak' events. This operation was achieved by dividing the signal count-rate in a channel by a factor $k = 3.74$ (which is the ratio of the C_{max}/C_{min} mean value of the two subsets). Then, the original background has been added to the new signal. An additional background has also been added to obtain the expected amplitude of Poissonian fluctuations of the quantity $S/k + B$ (where S and B are the expected signal and background values in the channel). Events which have a C_{max}/C_{min} value lower than k are not taken into account in the new sample since they would not have been detected by PHEBUS. This third sample of 22 bursts has been treated like the two others and the derived value of $P_{(>1)}$ is 24.7±5.2%. According to this result, in terms of E_{max}, there is no difference between 'strong' and 'weak' events.

DISCUSSION AND CONCLUSION

The difference between GRS and PHEBUS in the proportion of events displaying a high-energy emission above 1 MeV, arises from the fact that PHEBUS has a higher detection rate and therefore records more weak bursts than GRS. As demonstrated in the previous section, these 'weak' events that would not have been detected by GRS have a lower value for $P_{(>1)}$. On the contrary, the sample of 'strong' bursts, detectable by the GRS instrument, has a value for $P_{(>1)}$ in good agreement with those found by Matz et al. 1985. Therefore, this result indicates that, for bursts that are intense enough, the proportion, $P_{(>1)}$, is not strongly affected by the existing differences in energy threshold or spectral accumulation time between the two experiments. The PHEBUS population of short bursts (<4 s) represents nearly 30% of events[5] and the time-to-spill mode used to record spectra allows an accurate calculation of the maximum energy on short time scales in comparison with GRS. It should be noted that these short events do not introduce any discrepancy in the value of $P_{(>1)}$. In order to determine if the correlation between the maximum energy and the burst intensity can be explained by statistics, we have simulated the detection of strong events with a smaller value of the signal to noise ratio. In this case, the 'weakened strong' and the true 'weak' events become indistinguishable. The difference in the value of $P_{(>1)}$ for these two samples is not significant (0.7σ). This result has several consequences. First, this test confirms that, in terms of E_{max}, there is no difference between strong and weak bursts. The observed correlation between the 'hardness' (value of $E_{\max}$) and the intensity is mainly due to instrumental limitations. In this case, one can conclude that the unbiased $P_{(>1)}$ value for the subset of 'weak' events is near the one obtained for the sample of 'intense' bursts. Therefore, the SMM results are confirmed by the entire PHEBUS GRB population. According to these results, the proportion $P_{(>1)}$ of 55% can be seen as a manifestation of some spectral shape distribution in classical GRBs.

Second, no physical mechanism is required to describe the correlation between the maximum energy and the burst intensity. However, if a cosmological effect makes 'weak' events softer than 'intense' ones (Paczynski, 1992[6]), then it cannot be observed with this method due to the small number of GRB's and the large dispersion in spectral shapes.

REFERENCES

1. S.M. Matz, D.J. Forrest, W.T. Vestrand, E.L. Chupp, G.H. Share and E. Rieger, Ap. J. **288**, L37-L40 (1985).
2. M. Schmidt, J.C. Higdon, and G. Hueter, Ap. J. **329**, L85-L87 (1988).
3. T. Li and Y. Ma, Ap. J. **272**, 317-324 (1983).
4. G. Simpson and H. Mayer-Hasselwander, Astron. and Astrophys. **162**, 340-348 (1986).
5. J-P. Dezalay, C. Barat, R. Talon, R. Sunyaev, O. Terekhov and A. Kuznetsov, Gamma-Ray Bursts (AIP Eds W.S. Paciesas and G.J. Fishman, 1991), p. 304.
6. B. Paczyński, Nature **355**, 521 (1992).

Gamma-Ray Burst Results from DMSP Satellites

J. Terrell, P. Lee, and R. W. Klebesadel
Los Alamos National Laboratory, Los Alamos, NM 87545
J. W. Griffee
Sandia National Laboratory, Albuquerque, NM 87185

ABSTRACT

Gamma-ray burst detectors are aboard three U.S. Air Force Defense Meteorological Satellite Program (DMSP) spacecraft, in orbit at 800 km altitude. A large number of bursts have been detected by DMSP, often confirming and supplementing data from GRO and other spacecraft, sometimes detecting bursts not otherwise known. The position of an unknown source may be considerably restricted by knowledge of the several DMSP spacecraft locations and fields of view. These data may be of considerable assistance in understanding the gamma-ray burst phenomenon.

INTRODUCTION

Three U.S. Air Force DMSP spacecraft are currently in use, monitoring the gamma-ray environment; the most recent launch (DMSP 11) took place on 28 November 1991. They are in near-polar (99°) sun-synchronous orbits, coming within 9° of the North and South poles. A considerable number of gamma-ray bursts have been detected by these spacecraft, which have two gamma-ray detectors, each with 100cm^2 area of NaI, plus charged-particle detectors. A fuller description has been published elsewhere[1]. The relatively high orbits (800 km) allow detection of gamma-ray burst events occurring within 117° of the spacecraft zenith, and the presence of three birds means that there is a good chance of detection of any given event.

RESULTS

A typical example of DMSP data for a 100-minute orbit is shown in Figure 1. There are 4 high counting-rate peaks in each orbit, corresponding to passage through the horns of the outer Van Allen radiation belts, plus in this case a long passage through the South Atlantic Anomaly. During this particular orbit a strong gamma-ray burst was detected by DMSP 10, with the first peak occurring at 46925.7 sec UT, as seen in

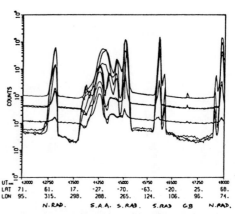

Fig. 1: Data for 3/10/91 (100 min).

Figure 2. The time has been corrected for delay in transmission of data, and is the center of the 2-sec counting interval with maximum counting rate. Five channels of gamma-ray pulse height data are shown in Figure 2, corresponding to thresholds of 20, 100, 200, 430, and 550 keV and upper limits of 1000 keV, for each of the two counters on DMSP10.

This gamma-ray burst on 10 March 1991 was also detected by DMSP 8 and DMSP 9, as well as by PVO, Granat, and Ulysses; GRO was not launched until 7 April 1991. The overlapping 117° fields of view of the DMSP spacecraft for this event are shown in Figure 3. The intersection of these fields considerably restricts the area of the sky in which the burst necessarily occurred. The location derived[2] for GB910310 (GBS1217+06.6) is also shown in Figure 3, and is, as expected, within the overlapping fields of view.

Other examples of 1991 DMSP burst data have been presented previously[1]. A few 1992 events will be discussed here, as examples of the field-of-view data from DMSP satellites, as well as of the time history and spectral information.

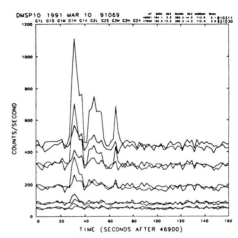

Fig. 2: DMSP 10 data for 3/10/91.

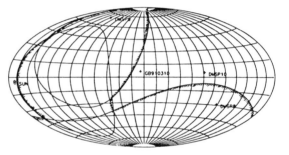

Fig. 3: DMSP fields of view, 3/10/91.

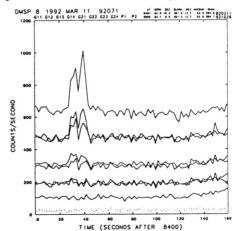

Fig. 4: DMSP 8 data for 3/11/92.

Figure 4 shows the detection of a very intense gamma-ray burst on 11 March 1992 by DMSP 8, which has thresholds of 50, 100, 200, and 420 keV. Similar time histories were measured by DMSP 9 (which had thresholds of 50 and 100 keV), as seen in Figure 5, and by DMSP 11 (which has the same thresholds as DMSP 10). As seen in Figure 6, the published location[3] of this gamma-ray burst falls within the fraction of the gamma-ray sky observed by these three DMSP spacecraft.

Another very intense gamma-ray burst was detected by GRO on 6 April 1992, but was not detected by any DMSP spacecraft. Figure 7 shows the positions of DMSP 8 and DMSP 11 at the announced burst time; DMSP 10 was in the south radiation belt and thus not useful. The announced position of the burst[4], as seen in the figure, falls within the relatively small portion of the sky not visible to the two DMSP spacecraft which would otherwise have been able to observe the event. Thus the non-detection by two DMSP spacecraft would have given considerable positional information even if no more than the time had been known.

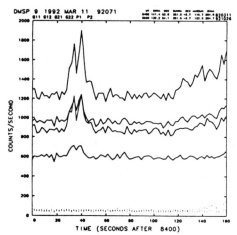

Fig. 5: DMSP 9 data for 3/11/92.

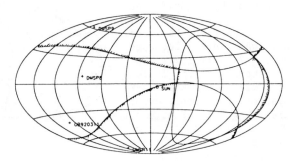

Fig. 6: DMSP fields of view, 3/11/92.

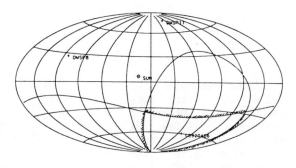

Fig. 7: DMSP fields of view, 4/6/92.

A final example of the information contributed by the DMSP spacecraft occurred on 29 June 1992. DMSP 10 and 11 each detected a complex event, apparently a soft gamma-ray burst, peaking at 75897 and 75907 UT, as seen in Figures 8 and 9. No BATSE data were recovered[5] for this time period. This burst could not have been of solar origin, since the sun was 124° away from DMSP 11. Furthermore, the sun was 77° away from DMSP 8, which did not observe the event although it was otherwise able to. Thus the source of this event must have been in that portion of the sky more than 117° away from the DMSP 8 zenith, as seen in Figure 10.

Much more data on gamma-ray bursts is available from the DMSP spacecraft, but has not yet been fully analyzed. It is hoped that data with much better time resolution will soon be available, including trigger times. Thus the three DMSP spacecraft may soon be able to contribute much more positional and timing data to the problem of understanding gamma-ray bursts.

This work was supported by the U.S. Department of Energy, by NASA, and by the U.S. Department of Defense.

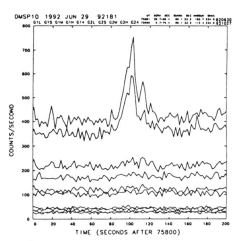

Fig. 8: DMSP 10 data for 6/29/92.

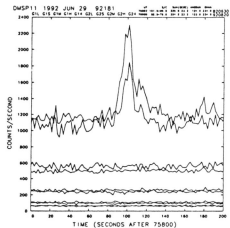

Fig. 9: DMSP 11 data for 6/29/92.

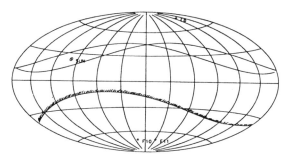

DMSP POSITIONS+VIEW, 92/06/29 75896 UT
EQUATORIAL COORDINATES

Fig. 10: DMSP fields of view, 6/29/92.

REFERENCES

1. J. Terrell, R. W. Klebesadel, P. Lee, and J. W. Griffee, Gamma-Ray Burst Data from DMSP Satellites, pp. 48-52 in Gamma-Ray Bursts, AIP Conference Proceedings 265, ed. W. S. Paciesas and G. J. Fishman (American Institute of Physics, New York 1992).

2. N. Lund, S. Brandt, and A. Castro-Tirado, IAU Circular 5207 (1991).

3. BNEWS (Batse Bulletin Board) No. 17 (1992).

4. BNEWS No. 19 (1992).

5. W. S. Paciesas, private communicatiobn.

Search for correlations of BATSE GRBs with known objects

S. Howard
USRA/Marshall Space Flight Center

G.F. Fishman, C.A. Meegan, R.B. Wilson
NASA/Marshall Space Flight Center

W.S. Paciesas
UAH/Marshall Space Flight Center

ABSTRACT

Correlations are presented between BATSE GRB locations and the following sets of objects: UV Ceti Flare stars, 2048 nearby galaxies, Abell Clusters, BL Lacs, Seyfert galaxies, globular cluster, members of the Local Group, quasars, H II regions and AGNs, all from the NSSDCA. These range from sources in the solar neighborhood to sources at cosmological distances. The result to date is null; no significant correlations are found. Presented are the $\langle \cos \theta \rangle$ and the $\langle \sin^2 b \rangle$ for each distribution considered as well as a measure of the minimum distance between pairs.

DESCRIPTION OF TECHNIQUE

To date there has been no identification of a gamma-ray burst with an observed catalog object. Although the uncertainty in angular location for BATSE's gamma-ray bursts is about 3° (Stollberg, this conference), there are a sufficient number of them to warrant coarse correlations with catalog objects. The following catalogs were used: UV Ceti Flare stars, 2048 nearby galaxies, Abell Clusters, BL Lacs, Seyfert galaxies, globular cluster, members of the Local Group, quasars, HII regions and AGNs, all from the NSSDCA. These range over sources in the solar neighborhood to sources at cosmological distances. The result to date is null; no significant correlations are found.

Table I presents the $\langle \cos \theta \rangle$ and the $\langle \sin^2 b \rangle$ for each distribution considered. The BATSE value is quite different from these, an effect of the data coverage in the catalogs. No close correlation with the BATSE values can be expected. Many of the catalogs have no coverage in the Galactic plane. When the gamma-ray bursts with galactic latitudes less than 15° are removed from the sample, the $\langle \cos \theta \rangle = 0.03$ and $\langle \sin^2 b \rangle = 0.427$.

The angular location of each catalog object was compared with the first 403 BATSE gamma-ray bursts. Two sets of angles are computed. One is set of the smallest angles between each catalog member and all the bursts. The other is the set of smallest angles between each burst and all members of each catalog.

TABLE I

catalog	$\langle \cos\theta \rangle$	$\langle \sin^2 b \rangle$
GRB	0.030	0.330
UV Ceti stars	0.024	0.317
AGNs	-0.058	0.579
near galaxies	-0.064	0.621
Local Group	-0.097	0.366
BL Lacs	-0.137	0.424
Abell clusters	-0.165	0.597
quasars	-0.189	0.731
Seyferts	-0.345	0.511
globulars	0.665	0.179

Table II presents the count of objects that are near a gamma-ray burst, and also the count of gamma-ray bursts that are near a catalog member. The angles were binned into $< 1°$, $< 5°$, $< 10°$, $< 20°$ circles and the number in each bin computed. The column marked $max\theta$ gives the maximum value for the separation angle. Clearly, the near-by objects show a larger "maximum" value than most of the extragalactic objects.

TABLE II
DOES A GRB HAVE AN OBJECT NEARBY?
ex: 4 GRBs have a BL Lac within 1 degree

catalog	1° away	5° away	10° away	20° away	$max\theta$
BL Lacs	4 (1%)	60 (15%)	185 (46%)	322 (80%)	45.6
AGNs	26 (6%)	245 (61%)	355 (88%)	403 (100%)	20.6
UV Ceti	1 (0.2%)	40 (10%)	154 (38%)	352 (87%)	38.1
HII regions	6 (1%)	64 (16%)	123 (30%)	214 (53%)	65.8
near galaxies	55 (14%)	285 (71%)	371 (92%)	397 (99%)	26.5
Abell clusters	63 (16%)	283 (70%)	291 (72%)	343 (85%)	47.5
quasars	31 (8%)	166 (41%)	224 (55%)	271 (68%)	70.3
globulars	4 (1%)	47 (12%)	154 (38%)	316 (78%)	39.8
Seyferts	2 (0.5%)	72 (18%)	184 (45%)	310 (77%)	39.7
local group	1 (0.2%)	24 (6%)	71 (18%)	189 (47%)	60.0

DOES AN OBJECT HAVE A GRB NEARBY?
ex: 4 BL Lacs have a GRB within 1 degree

catalog	1° away	5° away	10° away	20° away	$max\theta$
BL Lacs	4 (3%)	57 (49%)	109 (93%)	117	14.2
AGNs	39 (3%)	627 (56%)	1057 (95%)	1108	14.3
UV Ceti	1 (1%)	40 (43%)	88 (94%)	93	12.6
HII regions	5 (4%)	73 (62%)	115 (98%)	117	11.4
near galaxies	116 (4%)	1683 (60%)	2690 (96%)	2809	15.0
Abell clusters	86 (3%)	1441 (53%)	2592 (96%)	2712	16.3
quasars	91 (3%)	1792 (60%)	2829 (94%)	3000	15.2
globulars	4 (3%)	52 (43%)	111 (92%)	120	13.4
Seyferts	2 (16%)	61 (50%)	115 (95%)	121	11.6
local group	1 (3%)	15 (15%)	24 (83%)	29	13.4

Figure 1 shows these results for the AGN catalog. The top figure shows the locations in galactic coordinates of the BATSE GRBs and the AGNs. The middle figure shows the distribution of the smallest angle each GRB and all the members. The bottom figure shows the distribution of the smallest angle between each member and all the GRBs. Since the error radius for the position of the BATSE GRBs is on the order of 35°, no closer correlation can be made at this time. The data indicate that there is no "clustering" of objects around any GRB (and vice versa). No class of objects has a large percentage of its membership within 1° of a GRB. Several classes had obvious data gaps. In those cases where the Galactic plane was empty, the GRBs within ±15° of the plane were eliminated from the comparison.

A general comment can be made. There is a break in the "clustering" between "local objects" (H II regions, UV Ceti stars, globulars, local group members) and the members of extragalactic catalogs. The close objects have a larger scatter in their minimum angles than the distant objects do.

The distribution of AGNs covers the map with no gaps except the Galactic plane. So a further comparison was made with this data set to investigate whether the AGNs that fall within 1° of any GRB have brighter magnitudes than the rest of the sample. Figure 2 shows the magnitudes of all the AGNs in the sample. Those AGNs that fall within 1° of any GRB are marked with the rectangle. There appears to be no correlation with magnitude. The result for BL Lacs is the same.

One hundred data sets of randomly located points (in galactic coordinates) were created (with the Galactic place removed). Each set had the same number as the AGN population. The average of the minimum angle between each member of the simulated data sets and all the GRBs was computed. These numbers fell within 1σ of the average for the actual AGN data set. The averages of the minimum angle between each GRB and all the members of each of the 100 simulated data sets were computed. The actual average fell several σ *above* the average for the random distributions. These results are shown in Figure 3. One concludes that the BATSE GRBs will correlate with any random distribution as well as they do with the sample of AGNs.

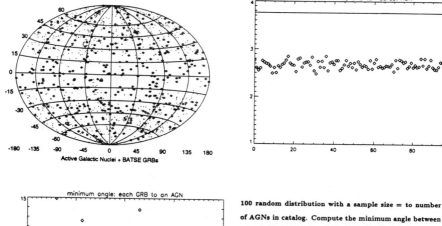

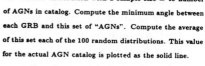

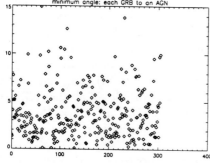

100 random distribution with a sample size = to number of AGNs in catalog. Compute the minimum angle between each GRB and this set of "AGNs". Compute the average of this set each of the 100 random distributions. This value for the actual AGN catalog is plotted as the solid line.

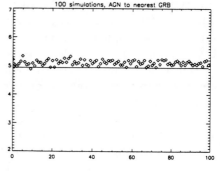

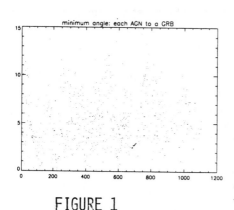

100 random distributions with a sample size = to number of AGNs in catalog. Compute the minimum angle between each "AGN" and the set of GRBs. Compute the average of this set for each of the 100 random distributions. This value for the actual AGN catalog is plotted as the solid line.

FIGURE 1

FIGURE 3

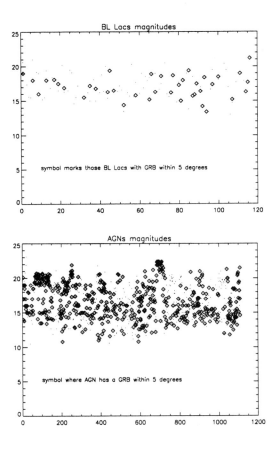

FIGURE 2

ROSAT WIDE FIELD CAMERA SEARCH FOR XUV COUNTERPARTS OF GRB AND OPTICAL TRANSIENTS

Alan Owens, S. Sembay, Mark Sims and Alan Wells
Department of Physics and Astronomy, University of Leicester, Leicester LE1 7RH, UK

Bradley E. Schaefer
Laboratory for High Energy Astrophysics, NASA/Goddard Space Flight Center, Greenbelt, MD 20771, USA

ABSTRACT

We have searched the ROSAT Wide Field Camera (WFC) all sky survey for possible steady state XUV counterparts within well located (< 5 arc-min) GRB and Optical Transient (OT) error boxes. In a search of the error boxes of 12 GRB and 9 OT, we found no evidence for such emission down to a limiting flux of ~ 10^{-11} ergs cm^{-2}s^{-1} (2×10^{-4} counts s^{-1}arcmin^{-2}). Combining our GRB results and assuming a simple neutron star black body model, our results suggest a minimum population distance of 30 pc for a temperature of 10^5 K and 100 pc for a 10^6 K temperature.

INTRODUCTION

Cosmic Gamma-Ray Bursts (GRB) were discovered accidentally two decades ago by the VELA satellites [1] whilst attempting to verify the nuclear test ban treaty. Although over 300 events are reported each year, it has still not been possible to identify counterparts at other wavelengths or even establish a convincing correlation with a specific class of astronomical object, despite deep searches at specific GRB locations (see Hurley [2] for a review). This result is at variance with most emission models which, not only predict bursting X-ray, UV and optical fluxes, but also observable quiescent emission. Soft X-ray observations of GRB error boxes have only been carried out for a few specific GRB [3,4]. In this paper, we report the results of a search for XUV quiescent counterparts of well located GRB and Optical Transients in the (60 - 200 Å) wavelength range (62 - 206 eV) using the ROSAT Wide Field Camera (WFC) all sky survey.

OBSERVATIONS AND ANALYSIS

The Wide Field Camera [5] is one of the core instruments of the ROSAT spacecraft [6] which was launched into a 53°, 580 km circular orbit on June 1, 1990. It is centered on a three-fold nested mirror system based on the Wolter-Schwarzschild Type I geometry. The focal plane detector is an XUV sensitized microchannel plate. Spectral information is obtained using a filter wheel to position one of eight X-ray filters into the focal path. For the data described here only two filters were used, each being moved into the prime focus on alternate days. These are the S1 (90 - 206 eV) and S2 (62 - 110 eV) filters. After an initial check-out and calibration period, ROSAT began its planned 6 month All Sky Survey (ASS) on July 30.

The primary data products of the WFC ASS are two maps of the XUV sky in the S1 and S2 energy regions. Each map is divided along lines of ecliptic latitude and longitude into 13,560 - 2° × 2° sky cells. The exposure is dependent on ecliptic latitude and ranges from ~ 1500 sec at the equator to ~ 80,000 secs at the poles, the average exposure per cell being approximately 2500 secs.

TABLE I WFC Gamma-Ray Burst Counterpart Search

Source	Exposure Times, secs		3σ Upper Flux Limits†, cts s^{-1} × 10^{-3}		Max Column
GRB	S1	S2	S1	S2	H cm^{-2}
6 Oct. 78b	1256.73	2133.48	12.40	10.80	5.0×10^{20}
4 Nov. 78	1336.27	1268.17	13.40	20.30	7.8×10^{20}
19 Nov. 78	1442.99	1482.99	6.03	12.30	1.7×10^{20}
13 Jan. 79	781.06	692.35	28.00	25.30	$*6.8 \times 10^{20}$
5 Mar. 79	13557.87	13107.90	3.16	3.91	$*6.0 \times 10^{20}$
25 Mar. 79	4227.34	2569.22	4.76	14.00	5.7×10^{20}
6 Apr. 79	1097.64	1108.06	11.00	12.50	$*2.0 \times 10^{20}$
18 Apr. 79	2328.46	2431.80	3.35	7.11	1.9×10^{21}
13 Jun. 79	2866.12	2800.23	5.20	6.14	2.9×10^{20}
1 Nov. 79	2808.84	2169.11	7.65	10.60	1.7×10^{21}
5 Nov. 79	1482.08	1109.81	5.60	9.46	5.1×10^{20}
16 Nov. 79	1140.20	1686.02	6.75	11.20	4.1×10^{20}

TABLE II WFC Optical Transient Counterpart Search

Source	Exposure Times, secs		3σ Upper Flux Limits†, cts s^{-1} × 10^{-3}		Max Column
OT	S1	S2	S1	S2	H cm^{-2}
1901	1440.02	1102.29	6.458	11.091	5.1×10^{20}
1928	1613.44	1966.52	5.318	8.406	1.7×10^{20}
1944	720.98	702.05	26.838	26.067	1.7×10^{20}
1959	2808.61	2168.48	7.666	10.643	1.7×10^{21}
1963	2792.60	2892.27	3.549	5.179	7.9×10^{21}
1964	2713.86	2546.45	4.043	9.229	4.3×10^{21}
1965	2412.15	2444.01	8.262	8.738	4.9×10^{21}
1966	1256.70	2133.46	12.429	10.767	5.0×10^{20}
1979	2170.22	1994.71	3.993	10.848	4.9×10^{21}

† A count of 3×10^{-3} cts s^{-1} corresponds to $\sim 10^{-11}$ ergs cm^{-2} s^{-1}.
* Columns in these directions are not known and were estimated by using values corresponding to equivalent directions in the opposite galactic hemisphere.

Given the large exposure to every region of sky, we have searched the integrated ASS at individual GRB error boxes for evidence of XUV steady state emission. For completeness we have also searched the error boxes of optical transients which, at the present time, are the only credible evidence of GRB at other wavelengths [7]. In order to minimize the number of false associations, we have chosen only those reported error boxes whose extent is well located and comparable to the WFC point spread function (*i.e.*, < 5 arc min). Based on the WFC source list [8] the expected number of sources per error box is « 10^{-6}. The results are listed in Tables I and II along with the observation parameters appropriate to each source. The total exposure per source per filter ranged from 700 secs (13 Jan 1979) to 13,500 secs (5 March 1979) - the average being ~ 2500 secs. No significant emission was detected from either GRB or OT's and so the last column of the tables lists the 3σ upper flux limits for steady state emission in the 62-110 eV and 90-206 eV energy bands. Assuming a simple black body source spectrum (described below), a count of 3×10^{-3} cts s^{-1} corresponds to a limiting flux at the Earth of ~ 10^{-11} ergs cm^{-2}s^{-1}.

RESULTS AND DISCUSSION

Our null results for steady-state counterparts are consistent with those of the other soft X-ray searches [3,4]. Assuming a simple black body neutron star model, we have used the measured fluxes to derive a distance limit based on the following assumptions: a neutron star radius of 10 km and two representative black body temperatures. The results are shown in figures 1 (T=10^5 K) and 2 (T=10^6 K) which show the minimum derived source distances as a function of column density for both the S1 (*a*) and S2 (*b*) filters. The minimum distance curves for all 12 candidate GRB are encompassed within the shaded band. The upper left hand portion of the graphs bounded by the solid diagonal lines can be excluded because the lines correspond to an integration of the column out in that direction, assuming a hydrogen density of 0.1 atom cm^{-3} out to the local super-bubble radius and a value representative of the interstellar medium thereafter (1 atom cm^{-3}). The allowed source distances are shown by the shaded regions from which we deduce that the minimum distance the GRB population can have is ~ 30 pc for a 10^5 K and ~ 100 pc for a 10^6 K neutron star temperature (otherwise the Wide Field Camera would have detected them).

Work is underway to search for general transient phenomena on all time scales ranging from 1 sec to several hundred sec. This work was carried out using the Leicester Starlink computer facilities and was funded by the SERC.

REFERENCES

[1] Klebasadel, R.W., Strong, I.B. & Olsen, R.A., 1973, *Astrophys. J. Lett.*, **182**, L85.
[2] Hurley, K., 1989, In: *Proc. 14th Texas Symposium on Relativistic Astrophysics*, Annals of the New York Academy of Sciences, **571**, 442.
[3] Boer, M., et al., 1988, *Astron. & Astrophys.*, **202**, 117.
[4] Pizzichini, G., et al., 1986, Astrophys. J., **301**, 641.
[5] Sims, M.R., et al., 1990, *Opt. Eng.*, **26**, 649.
[6] Trumper, J., 1984, *Physica Scripta*, **T7**, 209.
[7] Schaefer, B.E., 1986, In *Gamma-Ray Bursts*, eds. E.T. Liang and V. Petrosian, New York, AIP, 47.
[8] Pounds, K.A., et al., 1992, *Mon. Not. R. Astr. Soc.*, in press.

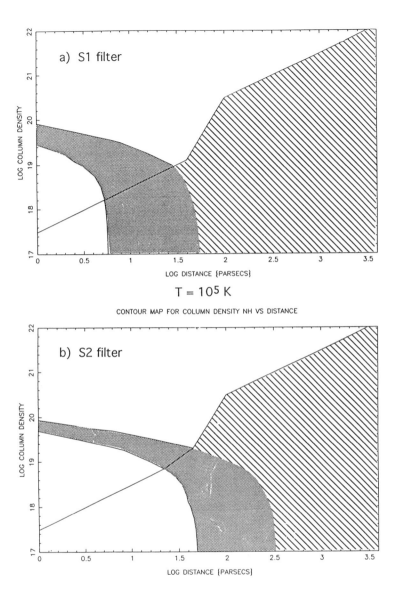

Figure 1. The allowed GRB distances (sectioned regions) for both the 90-206 eV (S1 filter) and 62-110 eV (S2 filter) energy regions, based on the WFC upper flux limits and assuming a simple black body neutron star model of temperature 10^5K. The shaded band encompasses the derived minimum distances as a function of column for all 12 candidate GRB (see text for details).

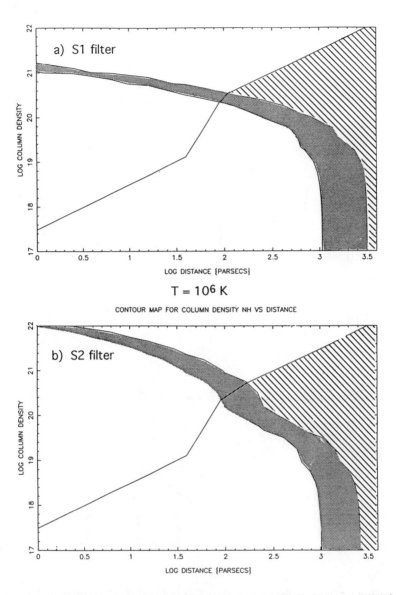

Figure 2. The allowed GRB distances (sectioned regions) for both the 90-206 eV (S1 filter) and 62-110 eV (S2 filter) energy regions, based on the WFC upper flux limits and assuming a simple black body neutron star model of temperature 10^6K. The shaded band encompasses the derived minimum distances as a function of column for all 12 candidate GRB (see text for details).

GAMMA-RAY BURSTER DISTANCES FROM SOFT X-RAY OBSERVATIONS

Bradley E. Schaefer
NASA/Goddard Space Flight Center, Code 661, Greenbelt, MD 20771

ABSTRACT

If soft x-ray spectra can be obtained for a small set of bursts, then the burster distances can be reliably measured by the variation of absorption from the intervening neutral gas. The variation of the measured column density with burst brightness and galactic latitude provides a unique signature of the distance scale to bursts. For example, an extragalactic hypothesis would be proved if the observed column density is greatly larger than can be contributed by our own Milky Way, while a Solar System origin would be proved if the observed column density is zero. To achieve this goal, a wide field (>2 sr) detector with area greater than roughly 70 cm^2 and reasonable spectral resolution down to 0.1 keV needs to be flown for longer than roughly a year. The primary goal for this wide-field soft x-ray spectrometer would be to establish a distance scale to bursters - and this one result would by itself be the most important discovery about the biggest mystery in modern astrophysics. For other gamma-ray burst studies, the spectrometer would also measure the soft precursors and tails, search for x-ray cyclotron lines and iron lines, and be ideally suited for the discovery of Soft Gamma Repeaters. For non-burst studies, the instrument would provide long-term spectra and light curves for bright persistent sources, search for new phenomena such as the ultrasoft transients seen by Einstein, and possibly detect the shock break-out of a supernova.

THE PROBLEM

Over the past decade, Gamma-Ray Bursters were usually thought to have distances of a kiloparsec or so. With the startling isotropy seen by BATSE[1], the latest shift of opinion has been to consider distances of megaparsecs. Nevertheless, Oort Cloud models, heliosphere models, and galactic halo models are still as reasonable as extragalactic models. Many theorists still hope for a local galactic disk origin and at least one model has been proposed that reconciles this distance scale with the BATSE data[2]. At this time, there is no reasonable evidence to set the distance scale of bursters even to within many orders of magnitude. Physics Today[3] has stated the problem that the uncertainty in the burst energy is 26 orders of magnitude! Until, the distance scale is known, there is no possibility of having an understanding of bursts that is any better than one of many speculations. As such, the distance scale is one of the most important questions to be addressed by future experiments.

So how can we hope to measure the distance to bursters with either the current generation of burst detectors or with the generation of detectors currently under construction? The confident identification of a quiescent counterpart at any wavelength is likely to allow for a distance scale measurement by any of the many techniques developed for this purpose by optical, infrared, and x-ray astronomers. Unfortunately, counterpart searches to the limits of modern technology in radio[4], infrared[5], optical[6], soft x-ray[7], and x-ray[8,9] have all turned up no candidates. If bursts are really cosmological, then perhaps they will show gravitational lensing or red shift effects. Unfortunately, detailed analysis shows that both possibilities are either very unlikely or impossible to measure with even the next generation of burst experiments[10,11].

So the problem can be restated as "What instrumental concept for a new burst experiment can establish the distance scale?"

A SOLUTION

If soft x-ray spectra (above 0.1 keV) can be obtained for a small set of bursts, then the burster distances can be reliably measured by the variation of photoelectric absorption due to unionized gas along the sight line to the bursts. The goal is to measure the column density causing the absorption for some bright and faint bursts over a range of galactic latitudes. The pattern of column densities will distinctively select the distance scale:

A.) If bursts are close to our Solar System (perhaps associated with the heliosphere or the Oort cloud) then the column density will be zero for all bursts. So if a non-zero column density is discovered for any burst, then this class of models can be strongly rejected. Similarly, the lack of any absorption will be proof that bursters are within a few parsecs of the Sun.

B.) If bursts have a typical distance scale of under 100 parsecs, then the column density will (1) scale roughly as the -0.5 power of the burst brightness, (2) range roughly up to $10^{20.5}$ cm^{-2}, and (3) not be a function of galactic latitude.

C.) If bursts are a galactic disk population with a typical distance scale of perhaps 1000 parsecs, then the column densities will be a simple function of both galactic latitude and the brightness of the burst and range up to perhaps 10^{22} cm^{-2}.

D.) If bursts are a galactic halo population, then most bursts will have a column density consistent with the line-of-sight that passes through the galactic disk. However, some fraction of a halo population will be immersed within the galactic disk, so that the column density for some bursts (such as bright or low galactic latitude events) will be significantly smaller than appropriate for a line-of-sight that passes outside our galaxy.

E.) If bursts are at cosmological distances, then the observed column densities will be consistent with the line-of-sight through the Milky Way plus absorption in the burster's host galaxy. So for example, if even one burst at high galactic latitude is seen with a column density much greater than 10^{21} cm^{-2}, then an extragalactic origin is proven.

DETAILS

This solution requires that bursts emit flux down to 0.1 keV. While it is unknown whether bursts have soft x-ray flux, its existence seems likely since many satellites[12-14] have found a spectral shape of roughly $E^{-0.5}$ down to 2 keV with no sign of a turn over.

The shape of the absorption cutoff due to photoelectric ionization of neutral gas is distinctive and relatively sharp (see Figure 1), so that the value of the column density does not depend greatly on the shape of the intrinsic burst spectrum. This is illustrated in Figure 2, where variations in the intrinsic spectrum from E^{-1} to E^0 below 2 keV appear as only a small variation in column density.

What about the possibility of a column density intrinsic to the burster itself? Fortunately, photoelectric absorption requires the material to be unionized, whereas the burst will ionize all matter to great distances. For material at even greater distances from the burst (for example, from a shell of gas ejected by a supernova in the burst progenitor), the column density will be negligible.

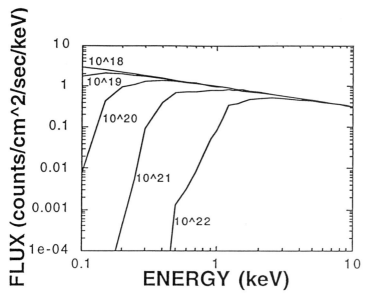

Figure 1. The soft x-ray and x-ray spectra for a bright burst (with an intrinsic spectrum that is an $E^{-0.5}$ power law) observed through column densities of 10^{18}, 10^{19}, 10^{20}, 10^{21}, and 10^{22} cm^{-2}. The photoelectric absorption is quite sharp and distinctive, with the column density being easily measured.

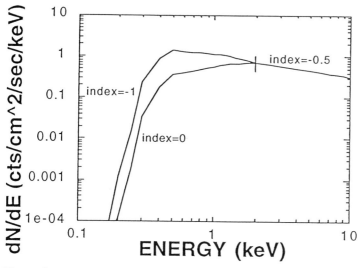

Figure 2. The cutoff from photoelectric absorption does not strongly depend on the the intrinsic shape of burst's spectrum. The diagram shows the observed spectrum where the power law below 2 keV is either E^{-1} or E^0. Such strong differences in the intrinsic spectrum will appear only as a ~20% change in the deduced column density.

The cosmological red shift for any extragalactic source will not affect the absorption from our own galaxy, but the cutoff energy from the host galaxy will depend on the host's red shift. For extragalactic models, bursts of moderate brightness have z equal to roughly 0.5, so the apparent absorption from the host galaxy will have an error in the column density of only 30%.

INSTRUMENTATION

This idea will not work for any current or planned burst instrument (BATSE, Ulysses, Mars Observer GRS, SAX, and HETE), since they do not go below ~2 keV. At this energy, the very small photoelectric cross section will not yield any significant observable absorption except for a source viewed through the galactic center (with a column density around 10^{23} cm^{-2}, see Figure 1). The probability that a distant burst will appear at a position with this minimum detectable column density is roughly 0.5%, so that the instruments that go to ~2 keV (SAX and HETE) are unlikely to see even one case of photoelectric absorption in their lifetime.

It may be possible for various soft x-ray experiments (such as ROSAT or EUVE) to spot a burst. If they even catch one event, then they will measure the absorption and perhaps this one observation will be sufficient to settle the distance scale problem. Unfortunately, these experiments have a small field of view and are unlikely to detect a burst. For example, the ROSAT Wide Field Camera[15] has a circular field with a diameter of 5° (or 0.05% of the sky) and a duty cycle of roughly 40%, so that for 800 bursts per year over the entire sky[1] (since the ROSAT detection threshold is comparable to the BATSE threshold for typical burst spectra) it will be lucky to detect even one event during its lifetime.

The detector needs to spot bursts from all over the sky with enough photons to measure a column density. Let us translate this into the requirements that (1) at least 100 counts be detected from 0.1 to 1.0 keV, (2) the background noise be less than 10 counts, (3) 20 bursts be detected within one year, (4) the bursts be over a range of a factor of ten in brightness, and (5) the bursts be uniformly sampled in galactic latitude. Let us adopt a typical duration of ten seconds for a burst, the BATSE LogN-LogS curve, the average x-ray background of McCammon & Sanders[16], the $E^{-0.5}$ spectral shape of Laros et al.[12], and the brightness of the brightest BATSE bursts. To achieve the requirements, I calculate that the detector should have a total field of view larger than 2 sr, an angular resolution better than roughly 2°, and an effective area of over 70 cm^2.

The instrument might be made with solid-state detectors or with proportional counters. The solid-state detectors have high spectral resolution ($E/\Delta E \sim 30$) all the way down to ~0.1 keV. The instrument should be flown on a spacecraft with a traditional burst detector (sensitive above 30 keV or so) so that the gamma-ray burst nature of any event can be confirmed.

Several configurations are reasonable to achieve the necessary sensitivity and solid angle. Possible configurations include an array of many medium field of view (perhaps 5° to 45° in size) detectors, an array of pinhole cameras or an array of coded mask detectors. Another possibility is to mount a sensitive detector with a small field of view (several degrees in size) onto a movable platform (much like the SUNFLOWER experiment[17]).

ADDITIONAL SCIENCE TASKS

A wide-field soft x-ray transient spectrometer can perform additional science tasks beyond measuring the column density to gamma-ray bursts:

For bursts studies, the spectrometer will measure the soft precursors and tails as seen by GINGA[13]. An important question that will be answered is whether the spectra are really of a blackbody, since if so then distances and sizes can then be measured. Another important question is whether low-energy line features (e.g., the iron line near 6 keV or cyclotron lines for magnetic fields stronger than 10^{10} Gauss) appear in the soft x-ray regime.

The spectrometer would be ideal for the study of short-duration transients. For example, large numbers of x-ray bursters could be simultaneously monitored for spectra and burst rates. Another example is that the Soft Gamma Repeaters could be discovered and spectrally analyzed. Finally, there is always the possibility of discovering an unknown class of soft x-ray transients, much as the ultrasoft transients possibly detected by Einstein[18,19].

If the spectrometer is placed in low Earth orbit, then it could act as an all-sky monitor for slow transients, much like the BATSE is currently being used[20]. Thus, the spectrometer could provide long-term nearly continuous light curves for bright persistent sources. Also, slow transients might be detected, with one possibility being due to the shock break-out of a supernova explosion.

REFERENCES

1. C. A. Meegan et al., Nature, 355, 143 (1992).
2. W. Kundt and H.-K. Chang, Ap. Space Sci., submitted (1992).
3. B. Schwarzschild, Physics Today, 45, no. 2, 21 (1992).
4. B. E. Schaefer et al., Ap. J., 340, 455 (1989).
5. B. E. Schaefer et al., Ap. J., 313, 226 (1987).
6. B. E. Schaefer, Gamma-Ray Bursts (Cambridge Univ. Press, Cambridge, 1992), p. 107.
7. A. Owens, S. Sembay, M. Sims, A. Wells, and B. E. Schaefer, this conference.
8. G. Pizzichini et al., Ap. J., 301, 641 (1986).
9. M. Boer et al., Astron. Ap., 202, 117 (1988).
10. S. Mao, Ap. J. (Letters), 389, L41 (1992).
11. B. E. Schaefer, Ap. J. (Letters), submitted (1992).
12. J. G. Laros et al., High Energy Transients in Astrophysics (AIP, NY, 1984), p. 378.
13. A. Yoshida et al., PASJ, 41, 509 (1989).
14. M. Yoshimori, K. Okudaira, Y. Hirasima, and I. Kondo, 18th ICRC in San Diego, (XG2-1), p. 39.
15. J. P. Pye, Adv. Space Res., 6, 139 (1986).
16. D. McCammon and W. T. Sanders, Ann. Rev. Astron. Ap., 28, 657 (1990).
17. R. L. Aptekar et al., Gamma-Ray Bursts (AIP, NY, 1992), p. 317.
18. D. J. Helfand and S. D. Vrtilek, Nature, 304, 41 (1983).
19. D. J. Helfand, E. Gotthelf, and T. T. Hamilton, Compton Observatory Science Workshop, NASA Conf. Proc. No. 3137 (NASA, Annapolis, 1992), p. 317.
20. R. B. Wilson et al., Compton Observatory Science Workshop, NASA Conf. Proc. No. 3137 (NASA, Annapolis, 1992), p. 35.

AN X-RAY COUNTERPART TO THE 5 MARCH 1979 GAMMA RAY BURST?

R. E. Rothschild, R. E. Lingenfelter
CASS, University of California, San Diego, La Jolla, CA 92093

F. D. Seward
Smithsonian Astrophysical Observatory, Cambridge, MA 02138

and

O. Vancura
Johns Hopkins University, Baltimore, MD 21218

ABSTRACT

We have searched the Einstein HRI images of the supernova remnant N49 for evidence of any point-like x-ray enhancement within the error box of the repeating gamma ray burst source GBS 0525-66, the source of the 5 March 1979 and other bursts. Although nothing stands out strongly from the diffuse supernova emission, a most likely location exists, centered within an unusual, bright, x-ray emitting region. Emission is brightest at location $5^h\ 25^m\ 56.^s5$ and $-66°\ 07'\ 05"(1950)$. We suggest that a 5" radius circle, centered here, provides a more precise error box for study at other wavelengths. The x-ray flux from a possible point-like source at this position is $1.0\pm0.3 \times 10^{-12}$ ergs cm^{-2} s^{-1} in the range 0.5 – 4.5 keV. We discuss models for the emission, and conclude that thermonuclear runaway on a nearby (<1 kpc) neutron star or a starquake on a neutron star at the distance of the LMC (55 kpc) is consistent with the data.

INTRODUCTION

The failure to find a quiescent source of any gamma ray burst at any wavelength has severely limited our understanding of these objects and the physical processes involved. Identification with objects at other wavelengths is crucial to our understanding of gamma ray burst sources. For the few bursts with small positional error boxes, only upper limits[1] exist for optical counterparts down to $m_V \sim 25$. The smallest positional error box to date[2] is for the 5 March 1979 event from the repeating burst source GBS0525-66. This 0.05 arcmin^2 region is unique in that it lies within the 2 arcmin^2 supernova remnant N49 in the Large Magellanic Cloud[3]. The coincidence of this error box with the remnant of a supernova that may have created a neutron star is quite consistent with a neutron star origin for the burst. Such an origin is also strongly suggested[4] by observations of a 430 keV emission line in the burst spectrum, attributed to gravitationally redshifted (z = 0.19) annihilation radiation, and the rapid (0.2 ms) rise time of the burst emission, implying a source size < 60 km.

We report here the coincidence of an X-ray emission enhancement within the gamma ray error box for both of the Einstein HRI images of N49. We suggest that this may be a likely location for the quiescent source of the gamma ray bursts.

OBSERVATIONS AND ANALYSIS

The first Einstein/HRI observation of the supernova remnant N49 was made on 19 April 1979, some 45 days after the 5 March 1979 burst (and its first weak repeat burst that followed it by 0.6 days) and just 15 days after the second of its weak repeating bursts. Subsequently, Cline et al.[2] reported a reduced and refined 0.09 arcmin^2 gamma ray burst error box, and Mathewson et al.[5] published Einstein HRI X-ray isophotes from the April 1979 observation showing the remnant N49 in greater detail (Figure 1). A second Einstein HRI exposure was taken on 3 February 1981.

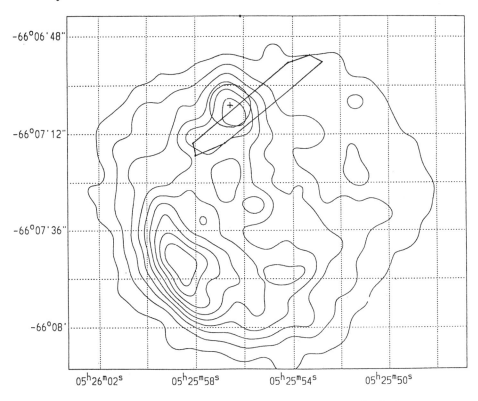

Figure 1. 19 April 1979 Einstein HRI X-ray contours[5] of the supernova remnant N49 with gamma ray burst source GBS0525-66 error box[2] superimposed.

We combined the two HRI images (see Figure 2), and the enhancement stands out clearly above any fluctuations seen elsewhere in the nebula, except for the broad diffuse nebular flux in the southeast. Since the optical nebulosity that correlates with the bright flux in the southeast is significantly reduced in the region of the enhancement, the possibility that the hot spot is not associated with the nebular flux is increased.

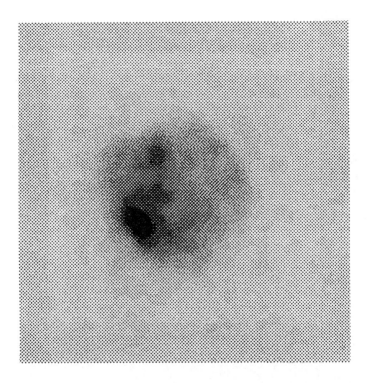

Figure 2. Halftone rendering of the x-ray flux from the combined HRI images of N49. The region of enhanced flux in the northeast stands out clearly above any fluctuations seen elsewhere in the nebula, except for the broad diffuse nebular flux in the southeast.

We have searched for signs of a point-like source within the enhancement. The shape of the emission maximum is compatible with a compact source, if a general background for emission from the remnant is subtracted. Up to 50% of the emission could be from a compact object within a 5" radius circle centered at a 1950 position $5^h\ 25^m\ 56.^s5$ and $-66°\ 07'\ 05"$. The point source flux is calculated to be $1.0 \pm 0.3 \times 10^{-12}$ ergs cm^{-2} s^{-1} in the range 0.5–4.5 keV, assuming a Crab-like power law spectrum. This is consistent with the previous upper limits[3,6]. At the distance of the LMC, this possible point source has a luminosity $L_x = 3.6 \pm 1.1 \times 10^{35}$ erg s^{-1} for a Crab-like spectrum, or $7.2 \pm 2.2 \times 10^{35}$ erg s^{-1} for a blackbody at 10^6K.

POSSIBLE SOURCES OF X-RAY EMISSION

If the enhanced X-ray emission does come from an unresolved "point source", the most likely candidate is a neutron star. As mentioned, several observations strongly suggest[4] that the source of the 5 March 1979 burst is a neutron star. Blackbody radiation from the surface of such a neutron star could produce the enhanced X-ray emission observed by the HRI, if the star has a high enough surface temperature.

For the case of the location within N49 being accidental and the neutron star source actually lying within the Galactic disk at a distance less than 1 kpc, the required blackbody surface temperature would be less than $\sim 1.6 \times 10^5$ K, assuming a neutral hydrogen column density of 5×10^{20} H cm^{-3} along the line of sight[6], and the corresponding blackbody luminosity would be less than $\sim 3 \times 10^{32}$ erg s^{-1}. Temperatures and luminosities of this order are consistent with those expected from current cooling models[7-8] for nonaccreting neutron stars of ages $> 10^6$ yr.

However, if the neutron star source is, in fact, in the supernova remnant N49 in the LMC at a distance of ~ 55 kpc, then the blackbody surface temperature would have to be $\sim 1.4 \times 10^6$ K and the luminosity $\sim 7 \times 10^{35}$ erg s^{-1}. This temperature is a factor of 2 to 3 times that expected for a neutron star with an age of 10^4 yr, comparable to that of the N49 remnant[9] and the blackbody luminosity is consequentially 1 to 2 orders of magnitude larger. Thus, in this case, additional heating is required to account for the higher temperature. Such heating could be caused either by the gamma ray bursts or by subsequent accretion.

Assuming a simple cooling model in which the luminosity decreases exponentially with a cooling time, we can estimate the total thermal energy, that would have to have been deposited as heat during a burst. For the 19 April 1979 observation the blackbody luminosity of $\sim 7 \times 10^{35}$ erg s^{-1} (15 days after the repeater burst of 4 April 1979) requires a minimum total thermal energy deposition of $\sim 8 \times 10^{41}$ erg. Since, this energy is 30 times that radiated at photon energies >30 keV during the 4 April 1979 burst, it would appear to be a rather unlikely heating source.

For the much more luminous 5 March 1979 burst (45 days before the X-ray observation) the required minimum thermal energy deposition would be $\sim 2 \times 10^{42}$ erg which is only 4×10^{-3} of the $\sim 5 \times 10^{44}$ erg radiated at energies >30 keV during that burst. Moreover, if a larger fraction, e.g. 5%, of the 5 March 1979 burst energy were deposited in thermal energy with an even longer cooling time, e.g. 700 days, such heating could also account for the x-ray flux in the 3 February 1981 HRI observation.

A cooling time of 45 days, or more, is much longer than that of hours expected[10] from thermonuclear runaway models of gamma ray bursts, and the total energy release in the burst is a factor of ~ 300 larger. Thus, a much deeper and much larger energy source, such as a star quake[11-12] would be required for the 5 March 1979 burst, if the source is in the LMC.

Alternatively, the candidate source could be a binary system[13] and the X-ray enhancement could result from the emission from accreting matter falling onto a neutron star. A luminosity of $\sim 7 \times 10^{35}$ erg s^{-1} would require a minimum accretion rate of $\sim 6 \times 10^{-11}$ M$_\odot$ yr^{-1} onto a neutron star with a surface z of 0.19, assuming that all of the gravitational energy is radiated. This minimum accretion rate is a factor 30 less than the average rate of $\sim 2.2 \times 10^{-9}$ M$_\odot$ yr^{-1} which would be required to accumulate the critical mass needed for a He flash thermonuclear burst on 24 April 1979 in the 20 days following the 4 April 1979 burst, and thus this scenario is not acceptable. However, either a variable accretion rate or an anisotropic, nonthermal emission which radiates only a small fraction (3%) of the energy in the HRI band could allow consistency with the thermonuclear model.

CONCLUSIONS

We report that a region of enhanced x-ray emission coincident with the refined error box for the 5 March 1979 gamma ray burst (and the subsequent repeat bursts) exists in two Einstein HRI images of N49 taken afterwards. Furthermore, this enhancement is consistent with a point source of emission. If this X-ray enhancement is the quiescent source of these bursts, we have further refined the position to $5^h\ 25^m\ 56^s.5$ and $-66°\ 07'\ 05"$ with a 5" radius uncertainty.

We find that 1) a solitary, cooling neutron star without subsequent heating would have to be nearby (<1 kpc) to account for the emission by current cooling models 10^6 yr or more after formation, 2) with additional heating, a neutron star source could be at the distance of the supernova remnant in the LMC and the 5 March 1979 burst would have deposited at least 5% of its energy in the star mantle, 3) cooling times for thermonuclear runaway models are inconsistent with the data under these assumptions, and 4) a model, such as a starquake event deep within the neutron star, does reproduce the characteristics of the 5 March 1979 event and subsequent soft x-ray emission.

Acknowledgements. We gratefully acknowledge NASA Contract NAS5-30720 and grant NAGW 1970. A longer version of this paper has been submitted to The Astrophysical Journal.

REFERENCES

1. C. Motch et al. Astron. & Ap. Supp., 145, 201 (1985).
2. T.L. Cline, et al. Ap. J., 255, L45 (1982).
3. D.J. Helfand and K.S. Long, Nature, 282, 589 (1979).
4. J.C. Higdon and R.E. Lingenfelter, Ann. Rev. Astron. & Ap., 28, 401 (1990).
5. D.S. Mathewson, et al. Ap. J. Supp., 51, 345 (1983).
6. G. Pizzichini et al. Ap. J., 301, 641 (1986).
7. K. Nomoto and S. Tsuruta, in Supernova Remnants and their X-ray Emission, J. Danziger and P. Gorenstein, eds. (Dordrecht: Riedel), 509 (1983).
8. K.A. van Riper, in Supernova Remnants and their X-ray Emission, J. Danziger and P. Gorenstein, eds. (Dordrecht: Riedel), 517 (1983).
9. P. Shull, Ap. J., 275, 611 (1983).
10. S.E. Woosley, in High Energy Transients in Astrophysics, S. E. Woosley, ed., (New York: AIP), 485 (1984).
11. R. Ramaty, et al. Nature, 287, 122 (1980).
12. D. Kazanas, Nature, 331, 320 (1988).
13. R.E. Rothschild and R.E. Lingenfelter, Nature, 312, 737 (1984).

ROSAT X-ray observations of GBS J0815-3245 = GRB 920501

M. Boër*

Max Planck Institut für Extraterrestrische Physik,
D 8046 Garching bei München, Germany

K. Hurley

Space Sciences Laboratory, University of California Berkeley

J. Greiner, M. Sommer

Max Planck Institut für Extraterrestrische Physik,
D 8046 Garching bei München, Germany

M. Niel

CESR, BP 4346, F 31029 Toulouse Cedex, France

G. Fishman, C Kouveliotou, C. Meegan, W. Paciesas, R. Wilson

NASA Marshall Space Flight Center, Huntsville, Alabama

E. Fenimore, R. Klebesadel, J. Laros

Los Alamos National Laboratory

T. Cline

NASA Goddard Space Flight Center

---

*Present address: CESR, BP 4346, F 31029 Toulouse Cedex (France)

ABSTRACT

We report the detection of a soft X-ray source in the error box of GBS J0815-3245 = GRB 920501, during a TOO observation by the ROSAT X-ray satellite. Some preliminary results of this observation are presented.

1. INTRODUCTION

The search for gamma-ray burst (hereafter GRB) quiescent counterparts has not resulted in any unambiguous detection at any wavelength, including soft X-rays[1,2,3,4]. In one case[1] there was a marginally significant detection (3.5σ) of a point like Einstein source with an energy flux of about 1.5×10^{-13} erg $\cdot$ s$^{-1} \cdot$ cm^{-2}. Most galactic GRB models predict quiescent emission[5,6,7,8] or a thermal afterglow[9].

We present here some preliminary results of an observation of the GRB 920501 GRB source, observed by the ROSAT X-ray telescope shortly after the burst itself.

2. OBSERVATIONS

GBS J0815-3245 was observed as a target of opportunity by the ROSAT X-ray telescope on May 19, 1992, for about 2680s. Since GBS J0815-3245 is the source of GRB 920501, the observation took place only 19 days after the burst, the shortest delay between a burst and an associated X-ray observation to our knowledge. Optical data on this source have been also obtained at the ESO Schmidt telescope and are reported elsewhere in these proceedings[12]. The X-ray observation was processed by the authors using the standard ROSAT EXSAS software develloped at MPE as well as the ROSAT PROS package in the US. Because of its low galactic latitude ($\simeq 1 \deg$), the hydrogen column density is fairly high, $n_H = 6 \times 10^{21}$ cm^{-2}. In order to map adequately the X-ray telescope (hereafter XRT) sensitivity curve the observation was analysed in several broad energy bands. 5 sources were detected at the 4σ level in the full ROSAT energy band (0.07 - 2.4 keV) and 8 in the "hard" energy band (0.4 - 2.4 keV). As shown in figure 1, one of these sources lies inside the error box of GBS J0815-3245. This source is present in both the hard and full energy bands.

Figure 2 presents a count spectrum of the 18 detected photons. The bins are constructed so that each bin represents a 3 sigma count rate above the background. The blackbody flux, for a distance of 1 kpc and a temperature of 7×10^5K, is 10^{-13} erg $\cdot$ s$^{-1} \cdot$ cm^{-2}. Obviously, the source is not very hard and there is a clear cutoff at low energies due to interstellar absorption. We were not able to detect any temporal variation (or clustering) in the source flux.

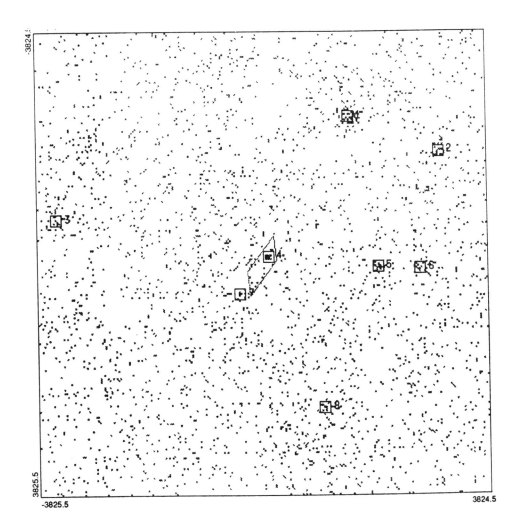

figure 1. 0.4 - 2.4 keV image of the inner 1 deg of the ROSAT X-ray telescope field of view. The GRB 920501 error box is also plotted

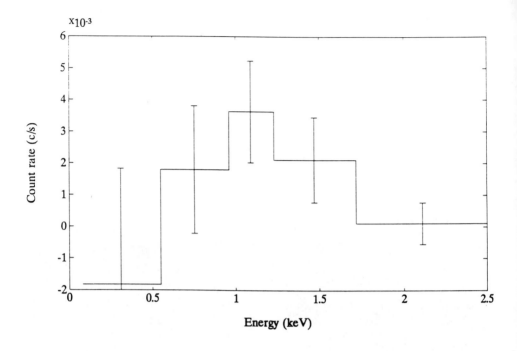

figure 2. Count spectrum of source 4 from figure 1. The bins are constructed so that each bin represent a 3 sigma count rate above the local background

3. DISCUSSION

One may question the reality of the association of the source with the gamma-ray burst in this densely populated region near the galactic plane. The local X-ray source density is about 8 sources per square degree; using the error box size, 1.4 arcmin^2 the probability of a chance coincidence is at the several times 10^{-3} level, although this will be reduced as the error box size is refined. The small number of detected photons does not allow a detailed study of the source X-ray properties. Given the position, a late-type star may not be excluded, and is in fact a good candidate. The spectral shape we observe is consistent with this hypothesis. Further studies, both at optical and X-ray wavelengths, are needed to determine the nature of this source, i.e. galactic vs. extragalactic. This is in progress and will be reported in a later publication.

ACKNOWLEDGEMENTS

MB wishes to thank the kind hospitality of the Max Planck Institut für Extraterrestrische Physik, in the framework of a CNRS/MPG agreement. We are grateful to D. Eichler for suggesting this project, and to J. Trümper for granting target of opportunity time on ROSAT. The Ulysses project is supported at the CESR by the Centre National d'Etude Spatiales, at the Max-Planck Institute by FRG contracts 01 ON 088 ZA/WRK 275/4 - 7.12 and 01 ON 88014, and at UC Berkeley by JPL contract 958056.

REFERENCES

1. Pizzichini et al., *Ap. J.*, **301**, 641 (1986)
2. Boër et al., A&A, **202**, 117 (1988)
3. Boer et al., A&A, **249**, 118 (1991)
4. Murakami et al., A&A, **227**, 451 (1989)
5. Hameury et al., A&A, **111**, 142 (1982)
6. Woosley, S.E., and Wallace, R.K., *Ap. J.*, **258**, 716 (1982)
7. Michel, F.C., *Ap. J.*, **290**, 721 (1985)
8. Epstein, R.I., *Ap. J.*, **291**, 822 (1985)
9. Eichler, D., and Chang, A.F., *Ap. J.*, **336**, 360 (1989)
10. Meegan et al., IAUC 5358 (1991)
11. Fishman et al., *Nature*, **355**, 143 (1992)
12. Hurley, K., et al., these proceedings

GAMMA-RAY BURST ASTROMETRY IV:
THE ATTEIA ET AL. CATALOG

L. G. Taff, J. H. Scott,* and S. T. Holfeltz
Space Telescope Science Institute, Baltimore, MD 21218

ABSTRACT

We continue to make progress in understanding the analytical nature of the triangulation problem and in computing gamma-ray burst positions. In particular, the crucial role of the area occupied by the detectors is illustrated and the Atteia et al.[1] catalog has been completely re-reduced. There are some significant discrepancies, even for $N = 3$, and the ambiguity with regard to the burst source direction in the $N = 3$ case is successfully resolved.

INTRODUCTION

An alternative to the customary "time difference of arrival" method of gamma-ray burst source location was presented by Taff.[2] In contrast to the time difference of arrival method this technique predicts a unique position for the source of the burst. The conventional method only defines a (circular) locus of points on the sky, on which the burst source location would reside were there no observational errors, one circle for each pair of detectors. Three detections yields two possible places. When there are more than three detections of the same burst a finite area on the celestial sphere is, in practice, delineated via a pair-wise analysis. With real data—and a difficult problem of time registration, differing thresholds to be exceeded before recording is initiated, and so forth—the geometrically pure version of the problem is degraded into one whose best possible outcome is that the intersection points lie near each other. This area is used to define an 'error-box' in which the burst source is believed to lie.

The region deduced in this fashion will, in general, not include the most likely position for the burst. Furthermore, the supposition of a uniform region of probability surrounding the predicted source location circle is simplistic. The forcing of a Gaussian probability distribution is unconvincing.[3] Finally, random errors in detector locations or the spacecraft-to-spacecraft timing become systematic errors in the "time difference of arrival" method circle.[4]

In the next section we briefly review the fundamental ideas. The third section describes a new procedure for providing a starting point to the general problem. It uses the largest area triplet subset of the observing spacecraft and the exact, three-dimensional solution to the $N = 3$ problem. In addition, geometrically intuitive explanations of the numerical solvability and stability of the co-planar

---

* Research supported in part by ST ScI Director's Research Funds.

are provided which can be generalized to the non-co-planar case. A summary of new computations, for the Atteia et al.[1] catalog, is also given.

REVIEW AND EXTENSION OF THE BASIC CONCEPT

The essential concept is to use the one piece of information that we know; that the phase of the burst at a particular wavelength (whether planar or spherical as would be the case for a solar burst) is an invariant. The phase is given by $\phi = \mathbf{k} \cdot \mathbf{r} - \omega t$ where $\mathbf{r}$ is the solar system barycentric location of the spacecraft, t is the time of arrival of the burst at that spacecraft, $\mathbf{k}$ is the wave vector of the burst wave front, and ω is the angular frequency of the burst ($= 2\pi\nu$ where ν is the frequency of the photon). Rewriting $\mathbf{k}$ as $k\mathbf{u}$, where $\mathbf{u}$ is the wave front unit normal, the pseudo-invariant Φ can be defined, $\Phi = \mathbf{u} \cdot \mathbf{r} - ct$. Although neither ϕ nor Φ can be directly measured, they do include all the observational data at our disposal and the quantity we want to determine, the source direction, $-\mathbf{u}$. Taff's hypothesis[2] was enforcing the constraint that each sensor's value of Φ be the same would lead to a mathematically well-posed problem for the computation of $\mathbf{u}$. This is now known to be the case.[4,5]

In particular, minimize their pair-wise phase discrepancies by forming, and then minimizing, the quantity T,

$$T = \frac{1}{2} \sum_{n,m=1}^{N} (\Phi_n - \Phi_m)^2, \quad \mathbf{u} \cdot \mathbf{u} = 1 \qquad (1)$$

The physical basis for this choice is in Taff.[6]

One deficiency of the above formulation is that it is monochromatic and, as some of the difficulties we face when attempting to isolate the positions of gamma-ray bursts have to do with the different responsivities and sensitivities of the various detectors, this is apparently a substantial shortcoming. We can not be certain that different times represent the same phenomena within the burst. However, this difficulty is an inherent part of gamma-ray burst observing and the novel statistical estimation problem formulated in Eq. (1) is explicitly designed to help overcome it. Moreover, this complication, for one energy band, is not severe for we can always imagine that the quantity T is part of an integral over frequency. The other components of the integrand would be the spectral energy distribution of the source multiplied by the spectral response of the detector. Because the latter function is relatively sharply peaked, the integral approximately reduces to the product of the other two terms, evaluated at the effective wavelength, multiplied by the energy passband width. These factors are phase independent and only result in an unimportant re-normalization of T.

This generalization can not be easily extended to cover widely separated wavelength bands both because the above mathematical approximations rapidly degrade and, more importantly, without an explicit model for the mechanism of the gamma-ray burst, one has to make a strong assumption regarding the temporal evolution of the burst over all observed frequencies. This compounds, when it is not the origin of, the trying problem with regard to systematic errors in the timing and no merely statistical adjustment technique is going to overcome it.

ANALYTICAL PROGRESS

The Co-planar Case

If one writes the unit normal of the burst wave front **u** as (α, β, γ) and uses two of the direction cosines as the independent variables, then we can regard T as a function of α and β. In other words,

$$\gamma = \gamma(\alpha, \beta) = \pm(1 - \alpha^2 - \beta^2)^{1/2} \qquad (2)$$

whence $T = T(\alpha, \beta)$. The ambiguity in the square root for gamma in Eq. (2) is the cause of the bi-directionality uncertainty associated with the case of only co-planar detectors. Because we search for a minimum of T we can resolve it as was suggested.[5]

Now let the plane $z = 0$ be the plane of the ecliptic. If all the detectors are in this plane, then all the gamma dependent terms disappear because they are all multiplied by the z coordinate of one of the sensors and these are all now equal to zero. In other words, the problem to minimize T is just

$$\partial T/\partial \alpha|_{z=0} = \sum_{n,m=1}^{N} (\Phi_n - \Phi_m)(x_n - x_m) = 0$$
$$\partial T/\partial \beta|_{z=0} = \sum_{n,m=1}^{N} (\Phi_n - \Phi_m)(y_n - y_m) = 0. \qquad (3)$$

Equations (3) are a system of two linear, inhomogeneous equations in the two unknowns α and β. Hence, their solution is trivial and is given by

$$\alpha = D^{-1} \begin{vmatrix} \sum X_{nm}\tau_{nm} & \sum X_{nm}Y_{nm} \\ \sum Y_{nm}\tau_{nm} & \sum Y_{nm}^2 \end{vmatrix}, \quad \beta = D^{-1} \begin{vmatrix} \sum X_{nm}^2 & \sum X_{nm}\tau_{nm} \\ \sum X_{nm}Y_{nm} & \sum Y_{nm}\tau_{nm} \end{vmatrix} \qquad (4a)$$

$$D = \begin{vmatrix} \sum X_{nm}^2 & \sum X_{nm}Y_{nm} \\ \sum X_{nm}Y_{nm} & \sum Y_{nm}^2 \end{vmatrix} \qquad (4b)$$

where $\mathbf{R}_{nm} = \mathbf{r}_n - \mathbf{r}_m$ and $\tau_{nm} = c(t_n - t_m)$. The areal distribution of the detectors plays a crucial role in the solvability of the problem (i.e., D must not vanish) and in the stability of the solution (e.g., $|D|$ should be of non-negligible norm). We shall illustrate this, in a geometrically clear fashion, immediately below. The ability to interpret D in this way exemplifies why Taff's method has the intrinsic capability to accurately characterize its output.

Consider the case of $N = 3$ detectors. Using the determinant form of the area of a triangle with vertices at (x_1,y_1), (x_2,y_2), and (x_3,y_3), viz.

$$\text{area} = A = (1/2) \begin{vmatrix} x_1 & y_1 & 1 \\ x_2 & y_2 & 1 \\ x_3 & y_3 & 1 \end{vmatrix}, \qquad (5)$$

one can show that D in Eq. (4b) is exactly equal to $48A^2$. Thus, we have an explicit proof that our geometrical intuition is a good guide; the larger the area of the triangle occupied by the three detectors the larger the determinant of the system in Eq. (4) and, therefore, the more stable its numerical solution. Clearly this holds for the case of only 2 sensors also.

If we try to generalize this result, by for instance exploiting the fact that the area of any quadrilateral is the sum of the areas of the two triangles which comprise it, one can also explicitly show that when N is four D is no longer equal to *any* quadratic function of the area of the quadrilateral. Accordingly we see that the case of $N = 4$ represents a real bifurcation between Taff's procedure and the pairwise treatment of the customary time of arrival method.

Lastly, there **is** an interpretation of D in the general (i.e., $N > 3$) co-planar case which does simply reflect on both the solvability for the burst wave front direction and the numerical stability of the solution. Imagine that instead of dealing with D in the ecliptic coordinate system as above, first rotate to principal axis coordinates (say u and v). In this coordinate system the off-diagonal elements of D will be zero (by construction) and $|D|$ itself will just be the product of the eigenvalues. Moreover, in the principal axis coordinate system the eigenvalues of this matrix are the variances of the rectangular coordinates u and v. Thus, $|D|$ will be large, the solution for the unit normal to the burst wave front well-determined, and numerically stable, precisely when the distribution of the spacecraft is such to maximize their 'spread'. So no matter how large N, a nearly collinear constellation of spacecraft (i.e., one principal axis variance nearly equal to zero) would not be able to produce a good estimate for the source of the gamma-ray burst.

An Algorithm for the Non-Co-planar Case

In case the three dimensionality of the detector configuration is important an alternative starting point exists but to use it we have to compute our initial guess based on the restriction that the number of sensors is only three. However, for most gamma-ray bursts detected by more than three sensors, the majority of the satellites will be in cislunar space rather than interplanetary space. As it is the area occupied by the sensors which is most important to precisely determining **u**, rather than their number (remember that $D = 48A^2$ for $N = 3$) we first find the subset of the $N > 3$ detectors which occupy the most area and then compute an estimate for **u** based on the timings from these three. This can be exactly performed.[3] In particular, if we use Heron's formula for the area of a triangle, viz. $A = [s(s-a)(s-b)(s-c)]^{1/2}$ [where the semi-perimeter s is given by $(a+b+c)/2$ in terms of the lengths of the sides a, b, and c] and compute the lengths of the sides from the Pythagorean theorem, then determining the maximum area subset from $N > 3$ sensors is simply an enumeration problem.

As there is an analog of Heron's theorem for quadrilaterals, one might conjecture that expanding this to polygons of more sides would be beneficial. Unfortunately, four points need not be co-planar so that finding the first value for **u** would be as difficult as finding the ultimate value for **u**.

RESULTS

We have upgraded our Monte Carlo software to work on the real data behind the Atteia et al.[1] catalog kindly supplied by K. Hurley. In general, as predicted,[2] we obtain the same results for $N = 3$ although there are some significant discrepancies. In addition, as mentioned above, we can resolve the sign ambiguity in Eq. (2) because we are searching for a minimum of T. The wrong sign either leads to an improperly normalized wave front unit normal or to a larger value of T. (Analytically, T vanishes for the correct value of $\mathbf{u}$ when N is three.) Finally, there are also several degree differences in some of the $N > 3$ positions. For 25 of the 37 $N = 3$ cases the mean positional difference between us and one of the Atteia et al.[1] positions was $0.21° \pm 0.41°$. For the entire sample it was $16.3° \pm 34.8°$ showing that for about 1/3 of the cases there is a sizeable disagreement. We are currently trying to better understand these instances as well as the few $N > 3$ examples we have. A complete list of the new positions was provided in the accompanying poster.

REFERENCES

1. J. L. Atteia et al. ApJ Suppl **64**, 305 (1987).
2. L. G. Taff, ApJ **326**, 1032 (1988a).
3. G. Pizzichini, Adv. Space Res. **1**, 227 (1981).
4. L. G. Taff and S. T. Holfeltz, Proceedings of the Compton Observatory Science Workshop, eds. C. R. Shrader, N. Gehrels, and B. Dennis, NASA Conf. Publ. 3137, pg. 301–308, (1992b).
5. L. G. Taff and S. T. Holfeltz, Proceedings of the Huntsville Gamma-Ray Workshop, eds. W. S. Paciesas and G. J. Fishman, AIP Conf. Proc. 265, pg. 378–382, (1992a).
6. L. G. Taff, Phys. Rev. **37A**, 3943 (1988b).

THE SEARCH FOR GAMMA-RAY BURST COUNTERPARTS: A COORDINATED COMPTEL/BATSE RAPID RESPONSE APPROACH

R. M. Kippen, J. Macri, J. Ryan
University of New Hampshire, Durham, NH 03824

B. McNamara
New Mexico State University, Las Cruces, NM 88003

C. Meegan
NASA/Marshall Space Flight Center, ES-62, Huntsville, AL 35182

ABSTRACT

The COMPTEL experiment onboard CGRO can image cosmic gamma-ray sources in the energy range 0.7-30 MeV with an accuracy of about 1 degree. Within its 1 steradian field of view, COMPTEL detects several strong cosmic gamma-ray bursts per year. When combined with burst trigger notification from the BASTE experiment on CGRO, the COMPTEL data processing and imaging analysis can be accelerated to yield burst position localizations within hours of onset. This rapid response, coupled with good localization, can be exploited in searching for burst counterparts at other wavelengths. We present here a rapid response plan which will allow scans of COMPTEL gamma-ray burst localization regions at other wavelengths soon after detection. The majority of this plan has been successfully implemented and preliminary work is discussed.

INTRODUCTION

With few rare and questionable exceptions, attempts to identify gamma-ray burst (GRB) counterparts at other wavelengths in the past twenty-five years have failed. Most of these searches relied upon localizations determined through timing analyses between widely separated spacecraft; taking at least several days to obtain a localization precise enough to search for counterparts. No doubt, part of the problem is that there have been no deep searches of GRB localizations concurrent or soon after their detection in gamma-rays.

The Imaging Compton Telescope (COMPTEL) on the Compton Gamma Ray Observatory (CGRO) has the ability to directly image MeV gamma-rays[1], eliminating the need for coordination between multiple spacecraft, while preserving good spatial resolution. However, COMPTEL has no on-board transient event triggering system to quickly identify GRB events. Fortunately, the BATSE experiment on CGRO does incorporate such a system which can be used to notify COMPTEL when a GRB occurs in it's ~1 steradian field of view. Thus far, BATSE burst trigger notification has only been used to search existing COMPTEL data well after the events. During its first year of operation, COMPTEL images with ~1° positional accuracy have been generated for several of the most intense of these events[2,3]. This subset of 'imagable' bursts represents only a small fraction of all the bursts occurring within COMPTEL's field of view and an even smaller fraction of all BATSE GRB triggers. Thus BATSE GRB triggers indicating COMPTEL 'imagability' must be filtered with care.

If only the strongest events occurring near the COMPTEL field of view are filtered from the BATSE triggers a rapid response action can be initiated to accelerate COMPTEL data processing and analysis, yielding as its end result a localization which can then be scanned at other wavelengths. This rapid response approach will, under optimum circumstances, allow multi-wavelength scans of COMPTEL localization regions for several bursts per year within ~hours of GRB onset.

RAPID RESPONSE SCENARIO

A scenario of the COMPTEL/BATSE rapid response plan is schematically illustrated in figure 1. The scenario begins with the detection of a gamma-ray burst event which activates the BATSE transient event triggering system. BATSE trigger signals are continuously monitored in real-time at the CGRO Payload Operations Control Center (POCC) by the Flight Operations Team (FOT). Relevant BATSE data are automatically checked against thresholds to monitor data and instrument operation. One set of thresholds has been implemented to determine the COMPTEL detectability of any burst trigger based on BATSE data parameters. When this set of thresholds is exceeded, a limit violation message appears and the FOT immediately contacts the BATSE team. The event must be of cosmic origin, sufficiently strong with a hard spectrum from a direction near the COMPTEL axis to be acceptable. In addition, basic engineering checks such as whether COMPTEL was activated and transmitting real-time data at the time of the burst must also be made. The ultimate acceptability of events which trigger the POCC automatic thresholds is determined by the BATSE team.

If the event is acceptable, the BATSE team notifies the COMPTEL team at the University of New Hampshire (UNH), and then continues to analyze the BATSE data to compute a source direction. The BATSE-derived source direction is valuable in the process since it can be used later to aide in analysis of COMPTEL data. Ever since CGRO started transmitting all data during 'real-time passes' of the spacecraft (no longer relying on tape recorders to store data), approximately 60% of COMPTEL data can, or does go directly to UNH electronically in real-time. If the COMPTEL data containing the burst are

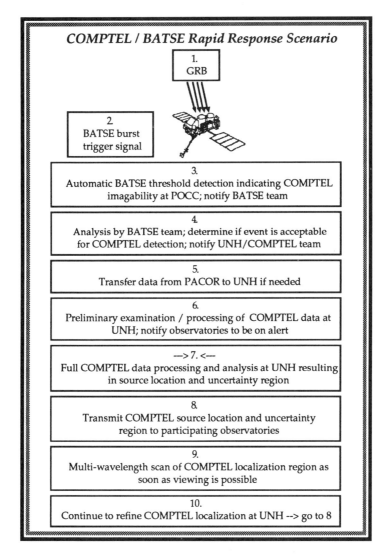

Fig. 1. Outline of the COMPTEL/BATSE GRB rapid response plan from burst trigger to multi-wavelength search. See text for details.

not already at UNH a request is made that they be transmitted from the Packet Processor (PACOR). Upon receipt of the data at UNH, raw count rates in the individual COMPTEL detectors (including COMPTEL's burst detectors[4]) and VETO domes are examined. This early check helps to identify obvious burst events in the COMPTEL data, however lack of a clear signal here is not sufficient to reject the possibility of a detection. Thus, for each BATSE notification, data is processed at least to a point where detection can be absolutely rejected. When acceptable detection is seen in the COMPTEL data, UNH notifies all participating observatories, such as the New Mexico State University team[5].

At this point, the COMPTEL team performs full data processing and imaging analysis at UNH[6]. This is by far the most time and resource consuming step in the entire plan. The amount of time required for full processing and analysis is being optimized, but will require at least a few hours. The end result of the COMPTEL data processing and analysis is a maximum likelihood ratio skymap[2,7] of the burst region containing the GRB position and uncertainty contours. The GRB localization from the COMPTEL skymap is then communicated to all participating observatories where the region can be scanned for counterparts if and when observing is possible. In the mean time, the COMPTEL imaging analysis is iterated to refine the localization.

PRESENT AND FUTURE STATUS

The implementation of the rapid response plan is now complete both with BATSE and with COMPTEL. One effort was to determine characteristics in BATSE data of 'COMPTEL imagable' bursts. This was done by examining BATSE data for the several GRB events which were successfully imaged by COMPTEL in its first year of operation. This allowed a set of POCC thresholds to be defined; the first, crucial step which filters out the majority of BATSE burst triggers. Other efforts included developing reliable communications between all participants and implementing all COMPTEL processing and analysis software at UNH. The COMPTEL portion of the rapid response procedure has been tested on several of the previously imaged GRB's; yielding source positions within ~hours. These tests served to measure/improve overall processing time, determine optimum default software settings, identify and correct software problems, optimize source location uncertainty and provide overall practice with the system. The full system was exercised on August 30, 1992 when the automatic thresholds were activated by GRB920830. Communication problems between BATSE and COMPTEL, coupled with the lack of an evident signal in raw COMPTEL data resulted in a source position ~days rather than ~hours after the burst. This 'first try' was, however, a valuable experience with the full rapid response process since it revealed several problem areas and verified that the automatic thresholds are appropriately set.

The ultimate success of the rapid response approach depends not only on the refinement of the plan outlined here (resulting in consistent, rapid and accurate GRB source locations), but also on the involvement of multi-wavelength observatories. There is currently only one dedicated observatory[5] and we therefore request more support from the multi-wavelength community.

SUMMARY

The lack of any conclusive observations of gamma-ray bursts at other wavelengths may be tied to the fact that no observations have yet been attempted soon after the detection in gamma-rays. The COMPTEL/BATSE rapid response approach to the search for GRB counterparts will allow us to search several COMPTEL GRB localizations per year at other wavelengths within hours of burst onset. The identification or non-identification of GRB counterparts or any lingering effects will provide invaluable new clues about the nature of GRB's. The success of this approach is contingent both upon the refinement of this plan and the active involvement of as many multi-wavelength observatories as possible.

REFERENCES

1. R. Diehl et al., Proceedings 4th International Workshop on Data Analysis in Astronomy (Plenum Press, Erice, Italy, 1991).
2. A. Connors et al., Astron. Astrophys. Suppl. Ser. in press (1993).
3. C. Winkler et al., Astron. Astrophys. **255**, L9 (1992).
4. C. Winkler et al., Adv. Space Res. **6**, 113 (1986).
5. B. McNamara, Burns, J. and Webber, W., Proceedings Compton Observatory Science Workshop, Annapolis, **NASA CP-3137**, 158 (1992).
6. J. W. den Herder et al., Proceedings 4th International Workshop on Data Analysis in Astronomy (Plenum Press, Erice, Italy, 1991).
7. H. de Boer et al., Proceedings 4th International Workshop on Data Analysis in Astronomy (Plenum Press, Erice, Italy, 1991).

SIMULTANEOUS OPTICAL/γ-RAY OBSERVATIONS OF GRBs

J. Greiner
Max-Planck-Institut für Extraterrestrische Physik, 8046 Garching, FRG

W. Wenzel
Sternwarte Sonneberg, 6400 Sonneberg, FRG

R. Hudec, P. Pravec, T. Rezek, M. Nesvara
Astronomical Institute Ondřejov, 25165 Ondřejov, CSFR

E.I. Moskalenko
Sternberg Astronomical Institute, 119899 Moscow, Russia

A. Karnashov
Astronomical Observatory, Odessa University, 270014 Odessa, Ukraine

V. Metlov
Sternberg Astronomical Institute, Crimean Observatory,
334413 Crimea, Ukraine

N.S. Chernych
Crimean Astrophysical Observatory, 334413 Crimea, Ukraine

V.S. Getman
Institute of Astrophysics, Tadshikistan Academy of Sciences,
743042 Dushanbe, Tadshikistan

C. Kouveliotou[1], G.J. Fishman, C.A. Meegan, W.S. Paciesas[2], R.B. Wilson
Marshall Space Flight Center, Huntsville, AL 35812, U.S.A.

ABSTRACT

The photographic sky patrols of the observatories Sonneberg (FRG), Ondřejov (CSFR), Odessa (Ukraine), Dushanbe (Tadshikistan) and Crimea (Ukraine) are used to look for patrol plates which have been exposed simultaneously with a γ-ray burst detected by GRO. We present the results for the first year of data.

For several GRBs we have plates which contain the GRB location, and the exposure time of which contains the time of the GRB. We will discuss the results of the search for optical flashes on these simultaneous plates.

INTRODUCTION

The search for optical counterparts of γ-ray bursters in the past has primarily focused on the photoelectric monitoring of known γ-ray burst (GRB) error boxes[1] (i.e. several years after the burst event), and on the examination

---

[1] Also Universities Space Research Association

[2] Also University of Alabama in Huntsville

of archival plates taken during the last 60-90 years[2-6]. Both these approaches rely on the assumption that GRBs repeat. With the exception of three bursters which are widely believed to form a separate class of bursts, such repetitions have never been observed.

In spite of differing opinions on the reality of unique archival optical objects there is observational evidence that optical flashes really exist. However, up to now the relation to GRBs is still open. Therefore, simultaneous optical and γ-ray observations are needed to clarify this possible relation.

There are several photographic sky patrols still active in the world. In each cloud-less night the available sky or specific regions are regularly photographed. The idea of our collaborative project is to wait for accidental coincidences of the completely independent sky surveys, i.e. BATSE of the γ-ray sky and the optical sky patrols. That is, we search for sky patrol plates, the exposure time of which sometimes covers the time of a GRB.

Previous results with this approach are rather sparse due to the lack of a large number of localized GRBs. The first attempt was made by Grindlay et al.[7] who found time-correlated plates for 2 events at SAO out of a small set of localized bursts. Of $\approx$ 60 localized GRBs up to 1982 at $\delta \geq -20°$ and occuring during night in Middle Europe time-correlated plates have been found for 5 events at the Ondřejov plate collection[8], and for 1 event at Sonneberg Observatory[9].

OBSERVATIONS

The instruments which are used for this search fall into three categories (for details see Greiner et al.[10]): i) usual sky patrol with guided wide field cameras of about $25°\times 25°$ (or more) field of view and typical exposure times of 30 min, ii) meteor patrol with primarily unguided fish-eye cameras and typical exposure times of several hours, and iii) astrographic cameras of about $10°\times 10°$ field of view and typical exposure times of 30 min. The limiting magnitudes (for a 1s flash) of the instruments in these three categories are of the order of 6-7 mag : 2-4 mag : 11 mag and the probability of finding simultaneous plates (which mainly scales with exposure time $\times$ sky area) is roughly $5\times 10^{-3} : 5\times 10^{-2} : 1\times 10^{-5}$.

For the sky patrol plates taken at Sonneberg and Odessa, two identical sets of several (5-7) of these cameras are operated at the same time in two colours at each of the observatories (independently of each other). The astrographic cameras at Sonneberg and Crimea are regularly used for specific monitoring tasks in variable star research and minor planet search. With the aim of meteor patrol a total of nine fish-eye cameras are operating independently at 7 different stations throughout Czechoslovakia and three mountings with 6 wide field cameras ($40° \times 40°$) each are operating at Dushanbe.

In order to minimize the amount of data to be exchanged we have established the following search procedure: In the first step we take the times of BATSE detected GRBs and check in our observatories whether plates have been taken at these times or not. In the second step each observatory gets the position of the GRBs for which there are simultaneous plates, and checks whether these GRB positions (including the whole error box) are located on one (or more) simultaneously exposed plate(s).

During the first year of operation we have looked through 140 GRBs (between 1991 April 23 and 1992 May 8) which have been detected by BATSE on the northern hemisphere and which have earth crossing times between 14 UT and 6 UT (night time span on the observatories). These 140 GRBs have been completely checked at all observatories involved. For 17 of these 140 GRBs we have simultaneously exposed plates of the burst location (Table I). With the exception of GRB 910814 for which the simultaneous plate was exposed at Dushanbe, all other plates stem from the meteor network in Czechoslovakia. For most of these hits, several simultaneously exposed plates from different observational stations are available. No optical flash could be found on any of these plates at the position of the bursts including a 7 degree error radius. That is, if there was any optical emission together with the high-energy burst then the optical flux (erg/cm^2/s) must have been less than the minimum detectable optical flux of the corresponding plate. The resulting ratio of the γ-ray flux at the peak (erg/cm^2/s) to the optical flux limit is given for each burst in Table I.

There are also a number of near-simultaneous plates, i.e. plates with its exposure time a few hours before or after the GRB. Again, no optical flash or a considerable brightening of a star could be found within the 7 degree error radius. Since these findings come out of step two of our search procedure, we can identify such near-simultaneous plates only for the about 15% of GRBs for which the coordinates are communicated to the observatories. However, already the few examples given in Table II show that we can give limits for the optical emission shortly before and after the GRB which are lower by a factor of 10-200 than the simultaneous limits.

DISCUSSION

Due to the detection frequency and location ability of BATSE we were able to double within one year the number of simultaneous optical and γ-ray observations of GRBs. From the present statistics it seems to be obvious that typical GRBs have (1) optical emission at the time of the burst at a level at least below $(F_\gamma/F_{opt})^{-1} \approx 0.5^{-1}$ and (2) optical emission a few hours after the burst is lower by a factor of 10-200 than the simultaneous emission. Given the fact that GRB spectra are rather broad over the observed energy range of, say 20 keV up to 100 MeV, our observations indicate that the broad spectral shape may not continue into the optical range.

In the future (with improving statistics) we expect to find more simultaneous plates with deeper limiting magnitudes (as for the GRB 910814) and possibly also for brighter GRBs. This will either reveal optical emission or allow us to set more stringent limits on the low-energy behaviour of GRB spectra.

Acknowledgement

This project is supported by the Deutsche Agentur für Raumfahrtangelegenheiten (DARA) GmbH under contract 50 OR 9104 3 and by the Deutsches Elektronen-Synchrotron (DESY-PH) under contract 05 5S0414(4).

Table I: Simultaneous plates for BATSE GRBs of the first year

GRB	Time (UT)	Location (RA, DEC)	No. of simultaneous plates	limiting magnitude mag_{pg} (for 1s flash)	Limit[1] for F_γ/F_{opt}
910505	20:15	187°,44°	4	2-3	1.3
910809	0:58	238°,39°	3	2.5	0.7
910814	19:14	344°,29°	1	4.5	42
910829	22:40	57°,22°	1	1	0.04
910902	22:56	300°,-2°	8	2-3	0.8
911004	1:27	351°,59°	8	3-4	6.7
911025	21:25	329°,26°	8	0.5	0.4
911027	1:45	360°,71°	8	1.0	0.05
911127	4:22	268°,51°	2	-0.5	0.5
911129	18:00	44°,1°	4	2-3	0.6
911209	0:57	79°,17°	3	1.5	0.4
911228	17:22	281°,40°	3	2	0.3
920209	21:41	291°,57°	5	2-3	0.5
920224	21:38	170°,14°	9	3	0.4
920227	20:50	198°,46°	7	3-4	4.9
920305	1:01	294°,57°	5	1-2	0.6
920505	21:55	42°,74°	6	1-4	[2]

[1] In the case of several simultaneous plates the ratio F_γ/F_{opt} is calculated for the best (deepest) plate. The value for F_γ refers to the energy range 20 keV - 2 MeV. Note that F_γ/F_{opt} is a true flux ratio (peak γ-ray flux in erg/cm^2/s divided by the optical flux limit in the same units) rather than the usually given[2] "flux" ratio L_γ/L_{opt}, which is really a fluence ratio, i.e. it depends on the burst duration (still to be defined in a general way for these bursts) and the exposure time of the plate. F_γ/F_{opt} is always lower than L_γ/L_{opt} if the burst duration is larger than 1s (cf. for example $L_\gamma/L_{opt} = 1600$ for GB 910814).

[2] Flux F_γ not yet available

Table II: Examples of near-simultaneous plates for BATSE GRBs

GRB	Time (UT)	Location (RA, DEC)	Plate center (RA, DEC)	Exposure time	mag$_{lim}$ (1s flash)
910601	19:22	309°,32°	300°,40°	21.52-22.32	6
				21.52-22.32	6.5
911209	0:57	79°,17°	75°,20°	20.12-20.42	6.5
			90°,20°	22.29-22.49	7
920121	21:57	31°,34°	30°,40°	18.03-18.33	6
911129	18:00	44°,1°	45°,0°	21.23-21.53	5
			30°,0°	20.50-21.20	5
920227	20:50	198°,46°	195°,40°	0.05-0.35^a	6
			180°,40°	21.43-22.13	6
			210°,40°	22.57-23.27	6
920305	1:01	294°,57°	270°,60°	2.38-3.08	5.5

^{a} Time of next day.

REFERENCES

1. Pedersen, H., Danziger, J., Hurley, K. et al., Nature **312**,46 (1984)
2. Schaefer, B.E., Nature **294**, 722 (1981)
3. Greiner, J., Flohrer, J., Wenzel, W. & Lehmann, T., Astrophys. Space Sci. **138**, 155 (1987)
4. Hudec, R., Borovička, J., Wenzel, W. et al., Astron. Astrophys. **175**, 71 (1987)
5. Moskalenko, E.I., Karnashov, A.N., Kramer, E.N. et al., in *Gamma-Ray Bursts*, ed. C. Ho et al, (1992), p. 127
6. Schaefer, B.E., Astrophys. J. **364**, 590 (1990)
7. Grindlay, J.E., Wright, E.L., & McCrisky, R.E., Astrophys. J. **192**, L113 (1974)
8. Hudec, R., Prep. Astr. Inst. Ondrejov, No. 120, 1991 (Astrophys. Lett. and Comm., in press)
9. Greiner, J., Laros, J.G., Fenimore, E.E. & Klebesadel, R.W., Astron. Astrophys. **227**, 115 (1990)
10. Greiner, J., Wenzel, W., Hudec, R., Moskalenko, E.I., Fishman, G.J., Kouveliotou, C., Meegan, C.A., Paciesas, W.S. & Wilson, R.B., in *Gamma-Ray Bursts*, ed. W.S. Paciesas & G.J. Fishman, AIP Conf. Proc. 265 (1992), p. 327
11. Schneid, E.J., Bertsch, D.L., Fichtel, C.E., Hartman, R.C. et al. , Astron. Astrophys. **255**, L13 (1991)

A METHOD FOR SEARCHING THE WHIPPLE OBSERVATORY GAMMA RAY DATA BASE FOR EVIDENCE OF GRB'S

M. Chantell[2], C.W. Akerlof[3], M.F. Cawley[4], V. Connaughton[1], D.J. Fegan[1]s., Fennell[1,2], J.A. Gaidos[5], J. Hagan[1], A.M. Hillas[6], A.D. Kerrick[7], R.C. Lamb[7], M.A. Lawrence[2], D.A. Lewis[7], D.I. Meyer[3], G. Mohanty[7], K.S.O'Flaherty[1], P.T. Reynolds[1], A.C. Rovero[2], M.S. Schubnell[3], G Sembroski[5], T.C. Weekes[2], T. Whitaker[2] & C. Wilson[5]

[1] Physics Dept., University College, Dublin, Belfield, Dublin 4, Ireland
[2] Whipple Observatory, Harvard-Smithsonian CfA, Box 97, Amado, Arizona 85645, USA
[3] Physics Dept., University of Michigan, Ann Arbor, Michigan 48109, USA
[4] Physics Dept., St. Patrick's College, Maynooth, County Kildare, Ireland
[5] Physics Dept., Purdue University, West Lafayette, Indiana 47907, USA
[6] Physics Dept., University of Leeds, Leeds, LS2 9JT, UK
[7] Physics and Astronomy Dept., Iowa State University, Ames, Iowa 50011, USA

ABSTRACT

In light of recent Gamma Ray Burst (GRB) data from GRO, it is desirable to have a method for searching data from the Fred Lawrence Whipple Observatory 10 meter Cherenkov telescope for evidence of GRB's at TeV energies. In particular, this instrument's narrow field of view and high angular resolution make it well suited for observations of M31 as a test of the galactic halo model of GRB origin. We present here such a method, and give the results of its application to data taken on M31. Also reported here is an upper limit for the detection of a DC signal from M31.

INTRODUCTION

The Whipple Observatory 10 meter reflector images the Cherenkov radiation produced in air showers from TeV gamma rays and cosmic rays. The light from these showers is focused onto an array of 109 photomultipliers which are packed in a hexagonal pattern. Images are recorded by computer for later analysis. Gamma-ray images are separated from cosmic ray images using the differences in the shape of the images (Fig. 1)[1]. The images are parameterized using moment-fitting software developed from Monte Carlo simulations. Cuts on these parameters have recently been optimized on data taken on the Crab Nebula[1,2]. These techniques have been shown to be effective in rejecting 99% of hadrons while accepting 50% of the photon signal. The instrument has a field of view of 3.5 degrees, collection area ~5×10^8 cm^2, and an energy range of 0.5 to 5 TeV. Background rates for gamma-rays vary with elevation and are ~3-5 per minute. In its normal mode of operation, the 10 meter telescope is used to search for point sources of gamma-rays. While the instrument is not well suited to looking for GRB's in general, due to the small FOV, it is well suited for testing the galactic halo model by observing M31.

GRB SEARCH TECHNIQUE

The GRB search method presented here consists of three parts. First, gamma-ray events are selected from the data based on their shape. Second, the selected gamma-ray events are scanned for groups of events arriving in time statistically in excess of the background rates. Third, the groups of events selected as potential bursts are examined to determine if their constituent events originate from a common point in the sky.

The first step of the analysis is performed using standard parameterization and selection software that yielded detections of the Crab Nebula and Markarian 421[2]. Groups of gamma rays were selected in time by passing a 'time window' through the data and counting the number of events arriving within the boundaries of the window. Whenever the number of events within the window exceed a predefined threshold, the events are tagged as a group. The duration of the window and the threshold can be adjusted to look for bursts on time scales ranging from milliseconds to minutes. The final step is to determine if the events in a potential burst originate from a common point in the sky. This is equivalent to asking if the major axes of the images in the group intersect at a common point on the face of the camera (fig. 2 & 3)[3]. This is determined by superimposing a grid of points with a spacing of 0.1 degrees on the face of the camera. The impact parameter of the major axis of each event about each point is calculated. If the events in a group originate from a given point on the camera, the events should have impact parameters about that point of less than 0.08 degrees. The grid size and impact parameter are chosen to conform to the pointing accuracy of the telescope.

RESULTS

This method was applied to 11.37 hours of data taken in 1991 on M31. There were a total of 112,189 raw events which reduced to 3,232 candidate gamma-ray events corresponding to an average rate of 4.7 gamma rays per minute. This data was searched for groups on time scales of 0.1, 1.0, and 10 seconds. The following is a listing of the number of groups found along with the size of the groups in parentheses.

Time scale	10s	1.0s	0.1s
Threshold	5	3	3
	23 (5 events)	11 (3 events)	no groups found
	4 (6 events)	1 (4 events)	

None of these groups are statistically significant, based on the expected background rates. The analysis of the events within the groups also showed no statistically significant clustering of events within any of the groups. Hence we see no GRB's from M31 with the following upper limits on the fluence: For a time scale of 10s we see no GRB's with more than 8 events corresponding to $\sim 1.3 \times 10^{-8}$ ergs/cm^2; at 1.0s no GRB's with more than 5 events corresponding to $\sim 5 \times 10^{-8}$ ergs/cm^2; and for 0.1s no GRB's with more than 3 events corresponding to $\sim 5 \times 10^{-9}$ ergs/cm^2. Since we see no GRB's in 11 hours, we can place an upper limit on the rate of GRB's from M31 at 2595 per year.

SEARCH FOR DC POINT SOURCE IN M31

The 11.37 hours of data on M31 were also analyzed to see if there is a steady TeV gamma-ray point source within the galaxy. Since only ON source data is available, the usual ON minus OFF source analysis technique cannot be applied. However, a method exists for looking for a dc signal from a potential source using ON source data only. This method can also be applied to situations where the source may be located away from the center of the FOV of the camera; this is an important consideration since M31 is an extended object that nearly fills FOV and the source could be anywhere in the galaxy. The method involves first selecting gamma-ray events as previously described, then calculating the angle alpha as defined in Fig. 4 for each gamma-ray event. If there is a source in the center of the FOV, the distribution of alpha (called an alpha plot)[2] will show an excess for values of alpha less than 15 degrees, as shown in Fig. 5. If there is no source present, the distribution will be flat except for a slight rise at large angles because distortion effects at the edge of the camera. To check for a source away from the center of the camera, alpha plots are generated separately for points covering the whole surface of the camera. For this analysis, the center of each PM tube in the camera was considered as the location of a potential source. The results of this analysis showed no source in the FOV of the camera. A typical alpha plot for M31 is shown in Fig. 5. From this we can assign an upper limit to the flux from

M31 of 1.7E-11 photons/cm² sec at an energy of 0.55 TeV.

CONCLUSIONS

After searching 11.37 hours of data on M31, we see no evidence for TeV energy GRB's nor for a point source of TeV gamma-rays within the galaxy. The limited amount of data used in this analysis precludes making any strong statement about GRB's emanating from M31; however, the methodology used demonstrates what could be done on this source and others given sufficient amounts of data. This same technique could be used to test the local hypothesis (Oort Cloud model) by observing a local star with similar properties to the sun.

ACKNOWLEDGEMENTS

We thank K. Harris and T. Lappin for help in obtaining these observations. We acknowledge support from the U.S. Department of Energy, NASA, the Smithsonian Scholarly Studies Fund and EOLAS, the scientific funding agency of Ireland.

FIGURES

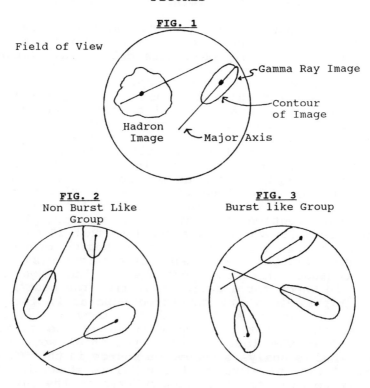

FIG. 1

FIG. 2
Non Burst Like Group

FIG. 3
Burst like Group

FIG. 4

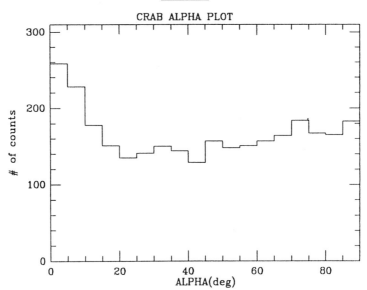

FIG. 5

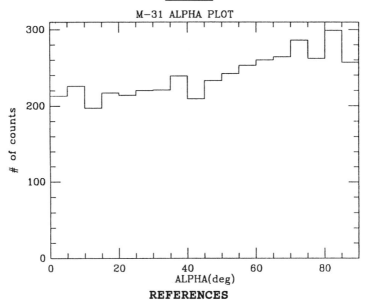

REFERENCES

1. Vacanti et al, ApJ , 377, 467 (1991)
2. Punch et al, PROC 22nd ICRC 1, 464 (1991)
3. Connaughton et al, PROC 22nd ICRC 1, 69 (1991)

A SEARCH FOR NEUTRINO AND GAMMA RAY BURST TEMPORAL CORRELATIONS WITH THE IMB DETECTOR

R. Becker-Szendy[4,*], C. B. Bratton[5], J. Breault[1], D. Casper[6,†], S. T. Dye[6], W. Gajewski[1],
M. Goldhaber[3], T. J. Haines[7], P. G. Halverson[1], D. Kielczewska[1,8], W. R. Kropp[1],
J. G. Learned[4], J. LoSecco[9], S. Matsuno[4], J. Matthews[2], G. McGrath[4], C. McGrew[1],
R. S. Miller[10], L. Price[1], F. Reines[1], J. Schultz[1], D. Sinclair[2], H. W. Sobel[1], J. L. Stone[6],
L. R. Sulak[6], R. Svoboda[10]

[1] University of California, Irvine, California 92717
[2] University of Michigan, Ann Arbor, Michigan 48109
[3] Brookhaven National Laboratory, Upton, New York 11973
[4] University of Hawaii, Honolulu, Hawaii 96822
[5] Cleveland State University, Cleveland, Ohio 44115
[6] Boston University, Boston, Massachusetts 02215
[7] University of Maryland, College Park, Maryland 20742
[8] Warsaw University, Warsaw, Poland
[9] University of Notre Dame, Notre Dame, Indiana 46556
[10] Louisiana State University, Baton Rouge, Louisiana 70803
* Now at SLAC, Stanford, California 94309
† Now at CERN, CH-1211 Geneva, Switzerland

Abstract

If Gamma Ray Bursts (GRBs) are associated with a stellar collapse-like phenomenon then it is reasonable to expect neutrino production to occur at the source. We have performed a temporal correlation analysis with GRBs using the IMB low-energy neutrino dataset during 809 days of livetime between 1986 and 1990. No correlations were observed placing a 90% C.L. limit of 0.046 ν interactions per GRB. The dependence of the GRB distances to neutrino yield using volume and shell distribution models is discussed. Lower limits are derived which exclude galactic stellar collapse-like models.

Introduction

Determination of the sources capable of producing GRBs is one of the great mysteries of astrophysics. There is no well-accepted theory as to the mechanism capable of producing the gamma ray spectra or temporal histories observed in GRBs since their discovery [1]. Many astrophysicists are searching for counterparts to GRBs at different electromagnetic wavelengths in an attempt to understand the source objects and deduce the gamma ray production mechanism [2]. We have taken a different approach by searching for neutrino emission from GRBs. Whether through accretion processes, stellar collapse or fundamental particle emission and decay it is thought that neutrino and gamma ray production may be related in many astrophysical processes [3]. Therefore it is not unreasonable to expect neutrino emission to be correlated with a GRB.

Neutrino and GRB Datasets

The GRBs and neutrino events used in the search have been detected during the same period of time. The neutrino dataset is a subset of the contained interactions within the volume of the Irvine-Michigan-Brookhaven (IMB) detector [4]. Specifically, the neutrino dataset consists of 223 contained neutrino events recorded over 809 livedays between 1986 and 1990 firing between 30 and 70 photomultiplier tubes. This corresponds to neutrino energies of roughly $25-60 MeV$. The energetics of mechanisms expected to have correlated neutrino and gamma ray emission suggest that this low-energy neutrino sample is the most reasonable for comparison with the gamma ray energies associated with most GRBs. All 8 neutrino events recorded from SN1987a are in this range.

During the IMB livetime two satellites, GINGA and the Solar Maximum Mission (SMM), were operational and recorded a total of 50 GRBs. Detailed information on the satellites can be found elsewhere [5]. The detection times of the neutrinos and GRBs form the basis of the coincidence search.

Analysis

The search for coincidences between the detected neutrinos and GRBs is performed by counting the number of neutrinos arriving within a given temporal window around a GRB detection time. The appropriate window is model dependent. If GRBs are the result of stellar collapse-like events then neutrino emission is expected to take place over a period on the order of $\leq 1\,minute$, as our experience with SN1987a has shown [6]. If, on the other hand, we are dealing with a particle or nuclear decay phenomena, or a diffusion process through some type of stellar atmosphere then neutrino emission, relative to gamma ray production, may take place over many minutes or hours. Since IMB records, on average, a cosmic-ray induced neutrino every $3.6\,days$ in the energy range of interest, correlation times longer than a day would not be detectable. These arguments suggest a range of temporal windows: 60 *seconds*, 1 *hour* and 1 *day*.

In order to compute the significance of any observed coincidences, the expected background due to random neutrino arrival times must be determined. The expected background, due to cosmic-ray induced neutrinos in the Earth's upper atmosphere, can be estimated using two methods which rely only on the recorded neutrino data sample and not on atmospheric neutrino production models.

The first background estimation method (*M1*) is Poisson statistics. This estimate is only accurate if the neutrino detection rate is constant. In actuality the threshold of IMB was systematically lowered during its 6 year operation.

A better method ($M2$) estimates the background for n-fold coincidences by randomly selecting a GRB detection time during the IMB detector livetime, and computing the number of coincidences. A series of background trials was performed, with the estimated background being the average n-fold coincidence of all trials. If the neutrino and GRB arrival times are correlated, then this randomization process should destroy any observed coincidence signal. However, if uncorrelated, this background estimate should give approximately the number of n-fold coincidences as the actual GRB detection times.

Results

Table 1 gives the number of n-fold coincidences along with the two background estimates for the three different temporal windows. There are no significant excesses in any multiplicity bin. Based on this non-observation of ν/GRB coincidences we can place a 90% C.L. limit of 0.046 detected ν interactions per GRB.

Using the neutrino emission limit derived for GRBs, the mean inverse square distance is given by

$$< \frac{1}{d^2} > = \frac{0.046}{\eta \cdot d_0^2} = 1.9 Mpc^{-2} \qquad (1)$$

where η is the neutrino yield in IMB and $d_0 = 0.055 Mpc$, the distance to SN1987a. The relationship of the mean inverse square distance to the source distance depends on the spatial distribution. It is not clear whether the spatial distribution of GRBs is that of a spherical shell or volume distribution. For distributions with a maximum radius r the mean inverse square distance is given by

$$< \frac{1}{d^2} > = \frac{3}{r^2} \qquad (2)$$

for a volume distribution, and

$$< \frac{1}{d^2} > = \frac{-1}{2Rr} \ln \frac{1 - \frac{R}{r}}{1 + \frac{R}{r}} \qquad (3)$$

for a shell distribution. R is the offset of the observer from the center of the shell. *Figure 1* gives the lower limit to the GRB distance as a function of neutrino yield in IMB for both distrubutions. For the shell distribution the center of the galaxy is assumed to be the center of the distribution (R=0.0075 Mpc).

If GRBs are assumed to be due to a stellar collapse-like phenomena a lower limit on the GRB distance can be computed using the limit derived above. The 8-neutrino events observed in coincidence with SN1987a become 12 neutrino interactions within the volume of the IMB detector when corrected for triggering

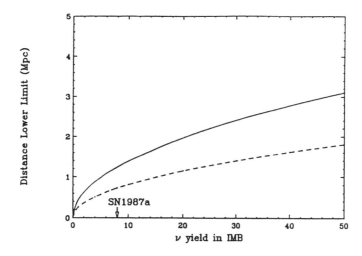

Figure 1: GRB Distance vs ν yield for volume (solid) and shell (dashed) distributions. Regions below the curves are excluded at the 90% C.L.

efficiency due to $\frac{1}{4}$ of the detector being inoperative during the supernova. Without prior knowledge of a supernova, or close-in-time correlation of several events, the data reduction efficiency for finding the low-energy events used in this neutrino dataset is 70% based on comparisons between independent data analysis chains. Thus the neutrino yield is 8 interactions for a supernova at a distance of $55 kpc$. After 1988 modifications lowered the detector threshold thus increasing the neutrino yield per supernova; however using the experimentally tested pre-1988 neutrino yield for the entire detector livetime gives the most conservative distance limit.

Summary

No temporal correlations of GRBs and neutrinos were found during 809 days of livetime between 1986 and 1990 thus setting a 90% C.L. limit of 0.046 ν interactions per GRB. This limit when used with a model dependent neutrino yield places constraints on the GRB distances. We set a distance lower limit of 0.73 (1.26) Mpc on stellar collapse-like models of GRBs in a shell (volume) distribution. This limit excludes galactic stellar collapse-like models for GRBs.

This work was supported in part by the U.S. Department of Energy. We also wish to thank the Morton Salt Company, which operates the Fairport Mine. One author (R.S.Miller) is currently a NASA Fellow.

References

1. P.V.Ramana Murthy and A.W.Wolfendale, Gamma-ray Astronomy, Cambride Press, Cambridge (1986).

Table 1: n-fold ν/GRB Coincidences

n	$\Delta t = 1\,minute$			$\Delta t = 1\,hour$			$\Delta t = 1\,day$		
	Signal	M1	M2	Signal	M1	M2	Signal	M1	M2
0	50.0	49.9	50.0	50.0	48.9	49.5	32.0	28.8	35.8
1	0.0	0.1	0.0	0.0	1.1	0.5	15.0	15.9	9.4
2	0.0	0.0	0.0	0.0	0.0	0.0	2.0	4.4	4.0
3	0.0	0.0	0.0	0.0	0.0	0.0	0.0	0.8	0.5
4	0.0	0.0	0.0	0.0	0.0	0.0	1.0	0.1	0.3
5	0.0	0.0	0.0	0.0	0.0	0.0	0.0	0.0	0.0

2. W.Neil Johnson ed., Proceedings of the Gamma Ray Observatory Science Workshop (1989); C.R. Schrader, *et al.* ed., The Compton Observatory Science Workshop (1991).

3. T.K. Gaisser, Cosmic Rays and Particle Physics, Cambridge Press, Cambridge (1990).

4. R.M. Bionta, *et al.*, *Phys. Rev. Lett.* **51** (1983) 27; R. Claus *et al.*, *Nucl. Instr. and Meth.* **A261** (1987) 540; R. Becker-Szendy *et al.*, *Phys. Rev.* **D42** (1990) 2974.

5. B.R.Dennis *et al.*, NASA Technical Memorandum 4036 (1988); Y. Ogasaka *et al.*, *Ap. J.* **383** (1991) L61.

6. R.Bionta, *et al.*, *Phys. Rev. Lett.* **58** (1987) 1494; K.Hirata *et al.*, *Phys. Rev. Lett.* **58** (1987) 1490.

**GAMMA-RAY BURSTS:
SPECTROSCOPY**

GAMMA-RAY BURST STUDIES BY COMPTEL DURING ITS FIRST YEAR OF OPERATION

C. Winkler, K. Bennett, L. Hanlon, O.R. Williams
Astrophysics Division of ESA, ESTEC, 2200 AG Noordwijk, The Netherlands

W.Collmar, R. Diehl, V. Schönfelder, H. Steinle, M. Varendorff
Max-Planck Institut für extraterrestrische Physik, 8046 Garching, F.R.G.

J.W. den Herder, W. Hermsen, L. Kuiper, B.N. Swanenburg, C. de Vries
SRON - Leiden, 2300 RA Leiden, The Netherlands

A. Connors, D. Forrest, M. Kippen, M. McConnell, J. Ryan
University of New Hampshire, Durham, NH 03824, U.S.A.

ABSTRACT

During the first year of Compton GRO operations, more than 20 cosmic gamma-ray bursts - detected by the BATSE instrument - occurred inside the 1 sr field of view of the imaging gamma-ray telescope COMPTEL[1]. Using COMPTEL's primary mode of operation (the telescope mode) direct images (with ~ 1° GRB location accuracy) and event spectra (0.7 MeV - 30 MeV) with spectral resolution better than 10% FWHM have been obtained. In its secondary mode of burst operations, COMPTEL has recorded time resolved spectra (0.1 MeV - 10 MeV) from its large NaI detectors. This paper summarises the results on cosmic GRB sources obtained by COMPTEL during its first year of operation.

INSTRUMENT AND OPERATING MODES

Gamma-ray bursts can be observed by COMPTEL[1] using two independent operating modes, the "Double Scatter Mode" and the "Single Detector Mode". These modes are described in detail elsewhere[2,3]. Summarizing, the "Double Scatter Mode", which is the normal imaging mode, is used to obtain direct locations of gamma-ray bursts within a field of view of ~ 1 sr and to obtain telescope spectra with an energy resolution of better than 10% FWHM. The operating range in this mode is 0.7 MeV to 30 MeV.

In the "Single Detector Mode"(see[2,3] for detailed description) COMPTEL uses 2 of the lower 14 NaI detectors to accumulate burst spectra upon receipt of a trigger signal from BATSE. The detectors are, in principle, 4π sensitive. However, their on-axis field of view is largely obstructed by the upper D1 detector array. At larger zenith angles (> 45°), obstruction is due to other CGRO instruments, electronics boxes, spacecraft structure etc. The two detectors measure different energy regions: a low range (apx. 0.1 MeV - 1.1 MeV, binwidth ~ 9.8 keV) and a high range (apx. 1 MeV - 10 MeV, binwidth ~ 84.7 keV).

GAMMA-RAY BURST OBSERVATIONS

During its first year of operation, beginning on 25 April 1991, COMPTEL received in total 1358 trigger messages from BATSE identifying 305 cosmic

© 1993 American Institute of Physics

gamma-ray bursts, 432 solar flares, 456 particle events, 81 triggers on SAA entry/exit, and 84 others. Out of the 305 isotropically distributed[4] cosmic gamma-ray bursts, 29 events occurred inside or close to the COMPTEL field-of-view and they are therefore candidate objects for direct imaging and spectral analysis (Table I). The data include the strong bursts GRB 910425, GRB 910503, GRB 910601 and GRB 910814. Six bursts could not be observed due to instrument switch-off and telemetry gaps. The low range burst detector D2-14 was out of commission from 25 May 1991 until 13 May 1992.

BATSE Trigger	Burst ID Date	Burst ID Seconds	COMPTEL gal. long.	COMPTEL gal. lat.
109	910425	2268	228.1	-21.1
143	910503	25455	172.6	5.2
249	910601	69737	74.4	-5.0
298	910609	2909		
451	910627	16160	314.2	58.4
503	910709	41604		
537	910714	74779		
678	910814	69275	93.7	-25.9
692	910818	49487		
856	911002	31974		
1051	911113	49306		
1073	911117	16543		
1085	911118	68260	271.4	33.1
1125	911127	83316		
1154	911209	3410		
1197	911219	79040		
1211	911224	21946		
1221	911225	61720		
1297	920113	75141		
1298	920114	62628		
1318	920127	77219		
1365	920207	6263		
1551	920413	82534		

Bursts detected by COMPTEL

BATSE Trigger	Burst ID Date	Burst ID Seconds	COMPTEL Status
829	910927	84415	COMPTEL off
1469	920308	63226	Bit error gaps
1493	920318	54420	Bit error gaps
1510	920321	84902	Telemetry gap
1517	920324	75269	Telemetry gap
1550	920412	72126	Telemetry gap

Bursts not detected by COMPTEL

Table I. Journal of gamma-ray burst observations by COMPTEL during one year following 25 April 1991.

All bursts listed in Table I are in the process of being analysed, i.e. for the stronger bursts, locations by direct imaging and telescope spectra in the 0.7 MeV to 30 MeV range, as well as single detector spectra up to 10 MeV are obtained. Some of the bursts listed in Table I are either too weak or too soft for direct imaging. The final results of the spectral analysis for all of these bursts will be given in a separate paper in preparation. Preliminary results have been reported earlier [3,5,6,7,8]. Here we concentrate on updated results obtained for GRB 910503, GRB 910601 and GRB 910814.

LOCATIONS

Locations of strong gamma-ray bursts (S (> 1 MeV) > 10^{-6} erg cm^{-2}) can be obtained using the Maximum Likelihood and the Maximum Entropy methods. The former is used in this paper because it yields accurate error estimates on the source location. The locations of 6 bursts are listed in Table I. Location maps of two bursts are shown in Figure 1 together with a triangulation annulus using independent timing information from the ULYSSES spacecraft. All bursts localized by COMPTEL have a source origin consistent with the position derived from the independent triangulation method. The statistical error on source location depends on the burst fluence and the inclination angle (relative to the instrument line of sight). A strong burst like GRB 910503 (S (> 1 MeV) = 1.2×10^{-4} erg cm^{-2}) has an error radius of 1° for the 2σ confidence interval, whereas the weaker event GRB 910425 (with a flux 3 times less than GRB 910503) has a 2σ error radius of $\sim 3°$. Systematic errors are estimated to be of the order of 0.5°.

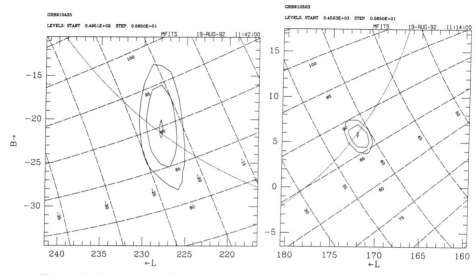

Figure 1. Gamma-ray burst locations obtained by COMPTEL using the Maximum Likelihood method for GRB 910425 (left panel) and GRB 910503 (right panel). A (R.A., Decl.) grid for epoch 2000.0 is shown. Inner contour: most likely source position; two outer contours: 2σ and 3σ confidence intervals; dotted line: triangulation annulus using ULYSSES timing information.

SPECTRA

We obtain gamma-ray burst spectra using telescope event data ("Double Scatter Mode") and single detector data (" Single Detector Mode"). In fact, theses constitute two independent instruments using different response information and deconvolution techniques. Photon spectra have been constructed using the traditional method by assuming a model photon spectrum which is folded by the response after which comparison is performed with the observed data in count space. Best fit results are obtained using χ^2 or similar statistics. Detailed results obtained on all bursts (Table I) will be published in a forthcoming paper. For the purpose of this workshop it is of benefit to compare our results with those from other CGRO experiments. We report therefore updated spectral analysis results for GRB 910503, GRB 910601 and GRB 910814.

GRB 910503: One of the strongest bursts observed so far resulting in significant deadtime effects in the COMPTEL event data stream. So far it is the only event with a significant spectral hard-to-soft evolution. Preliminary results for this event are published elsewhere[9]. Figure 2 shows a spectrum accumulated during a 2-second period starting 1 second after the onset of the burst. The COMPTEL best fit photon spectrum can be described by a hard single power-law $a \cdot E^\alpha = (7.92 \pm 0.42) \cdot (E/\text{MeV})^{(-2.03\pm 0.08)}$ ph/(cm² s MeV). Comparison with EGRET[10] and BATSE[11] results shows good agreement across the CGRO instruments in both normalization and slope. The EGRET photon spectrum has been determined using data from 1 MeV up to 150 MeV[10].

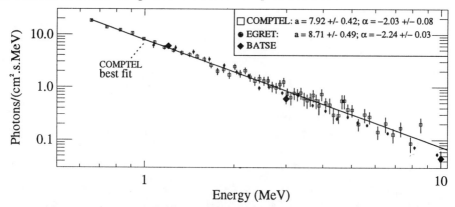

Figure 2. Photon spectrum of GRB 910503 during 2 seconds starting one second after the BATSE trigger.

We note that at the end of the first pulse of this event (i.e. during a 6 second time interval starting 3 seconds after burst onset) we have to reject a single power-law for the input spectrum and favour a broken power-law similar to findings by EGRET[10].

GRB 910601: We have performed a spectral analysis covering the full burst duration (33 s). The data can be described by a single power-law with, however, a much softer spectral index ($\alpha = -2.82$) compared to GRB 910503. No spectral (hard-to-soft) evolution has been found. The photon spectrum which was derived using the high range COMPTEL burst detector data is shown in Figure 3, together with the photon spectrum obtained from COMPTEL telescope double scatter events. The comparison shows excellent agreement between these two datasets which have been derived using different deconvolution techniques and response matrices.

GRB 910814: Using single detector data, spectral analysis of this event which is characterized by a sharp rise followed by an exponential decay with a total duration of about 40 seconds indicates, that the individual burst spectra and time integrated spectra cannot be fitted by single power-law models. A broken power-law provides acceptable fit results. Break energies are found between 1 MeV and 3 MeV with power-law indices α in the range of -1 to -2 below, and -2.5 to -3.5 above the break energy. Independent spectral analysis of the COMPTEL telescope data accumulated over the full burst duration confirm both the single detector results and BATSE observations published recently[13]. In addition, EGRET data analysis[14] shows a clear break at around 2 MeV. The break at high energies may indicate the presence of photon annihilation processes, e.g. due to interaction of single photons with strong magnetic fields[12].

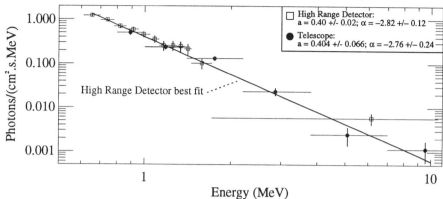

Figure 3. GRB 910601 photon spectrum of (full burst, 33 s).

CONCLUSIONS

COMPTEL can locate gamma-ray bursts (by direct imaging) to an accuracy of typically 1° (error radius of the statistical 2σ confidence interval) for 10^{-4} erg cm^{-2} bursts with correspondingly larger radii for weaker events. Systematic errors are estimated to be of the order of 0.5°. All locations derived so far are fully consistent with independent methods using triangulation with other spacecraft. Spectral analysis of COMPTEL burst data shows that many bursts have photon spectra which can be described by single power-laws with slopes in the range of $\alpha \sim -1.8$ to ~ -3. Only GRB 910503 shows significant spectral evolution (hard-to-soft). GRB 910814 and the last 6 s of the first main pulse (9 s) of GRB 910503 deviate from single power-law and need a broken power-law model. Comparison of data from EGRET and BATSE with the COMPTEL burst detector and the COMPTEL telescope shows a good agreement across the instruments using different deconvolution methods and response matrices. Up to now we have not detected any lines, periodicities or any pre- or post-cursors in the COMPTEL data.

REFERENCES

1. V. Schönfelder et al., IEEE Trans. Nucl. Sci. **31**, 766 (1984).
2. V. Schönfelder et al., Ap.J.Suppl. in press (1993).
3. C. Winkler et al., AIP Proceedings **265**, 22 (1992).
4. C. Meegan et al., Nature **355**, 143 (1992).
5. A. Connors et al., A& A Suppl. in press (1993).
6. W. Collmar et al., A& A Suppl. in press (1993).
7. A. Connors et al., Proc COSPAR Washington in press (1993).
8. M. Varendorff et al., AIP Proceedings **265**, 77 (1992).
9. C. Winkler et al., A&A **255**, L9 (1992).
10. E. Schneid et al., A&A **255**, L13 (1992).
11. B. Schaefer, priv. comm.
12. M. Baring, MNRAS **244**, 49 (1990).
13. B. Schaefer et al., Ap.J. **393**, L51 (1992).
14. P. Kwok et al., these proceedings

EGRET OBSERVATIONS OF GAMMA-RAY BURSTS

E.J. Schneid[1], D.L. Bertsch[2], B. L. Dingus[3], C.E. Fichtel[2], R.C. Hartman[2], S.D. Hunter[2], G. Kanbach[4], D.A. Kniffen[5], P.W. Kwok[6], Y.C. Lin[7], J.R. Mattox[8], H.A. Mayer-Hasselwander[4], P.F. Michelson[7], C. von Montigny[4], P.L. Nolan[7], K. Pinkau[4], H. Rothermel[4], M. Sommer[4], P. Sreekumar[2], D.J. Thompson[2]

1. Grumman Corporate Research Center, M/S: A01-26, Bethpage, NY 11714
2. NASA/Goddard Space Flight Center, Code 662, Greenbelt, MD 20771
3. Universities Space Research Association, NASA/GSFC, Code 662, Greenbelt, MD 20771
4. Max-Plank Institut fur Extraterrestrische Physik, 8046, Garching, Munich, Germany
5. Hampden-Sydney College, P.O. Box 862, Hampden-Sydney, VA 23943
6. NAS/NRC Postdoctoral Research Associate, NASA/GSFC, Code 662, Greenbelt, MD 20771
7. Hansen Experimental Physics Laboratory, Stanford University, Stanford, CA 94305
8. GRO Science Support Center, Computer Sciences Corporation, NASA/GSFC, Code 668.1, Greenbelt, MD 20771

ABSTRACT

The EGRET instrument has detected gamma-ray bursts in its large-volume NaI spectrometer, anticoincidence shield and imaging spark chamber. Its spectral range from 1 MeV to greater than 200 MeV overlaps with and extends beyond those of other instruments on the Compton Gamma Ray Observatory. For intense high-energy bursts in the telescope field of view, detection of gamma rays in the spark chamber telescope allows the location of a burst to be determined. Only six bursts, three of which were strong bursts with gamma-ray energies greater than 10 MeV, have been observed to date as a result of BATSE triggers. Interest in spectral breaks at low energies led to a careful examination of the EGRET sensitivity in the overlap region with BATSE. The low-energy sensitivity was studied using the detailed CGRO mass model; the resulting sensitivity values are presented as a function of burst direction relative to EGRET.

INTRODUCTION

The high-energy emission of gamma-ray bursts is important for understanding the emission mechanism nature of the bursts. Gamma rays above 1 MeV can provide upper limits on the magnetic field strength in the region of gamma-ray emission[1] and the distance to the source[2]. Spectral breaks in the MeV region[3,4] can be caused by electron-photon interaction, pair production by intense magnetic fields, or degradation of photon energy by Compton scattering near the burst source.

The EGRET spectral range overlaps with the BATSE in the MeV range but only six of the BATSE-detected bursts with MeV gamma-ray emission were observed by EGRET. Three of these bursts had spectra extending beyond the BATSE range and were measured by EGRET.

Two sigma upper limits for the MeV range were studied for the EGRET instrument as a function of burst direction using the detailed Compton Gamma Ray Observatory (CGRO) mass model[5]. These limits will establish EGRET capability to contribute to the BATSE data and extend the spectra to higher energies.

INSTRUMENT DESCRIPTION

The EGRET instrument is a typical spark chamber gamma-ray telescope with substantially greater sensitivity than its predecessors, SAS2 and COSB. It consists of a spark chamber with interleaved tantalum plates to convert gamma rays to electron position pairs, and to image the trajectories of the pair. A NaI scintillation spectrometer measures the energy of the pair as it emerges from the bottom of the spark chamber. A large anticoincidence shield to reject charged particle events surrounds the spark chamber but does not extend to cover the NaI spectrometer. Details of the instrument and calibration can be found in Hughes, et al.[6], Kanbach, et al.[7], and Thompson, et al.[8]

The NaI spectrometer (76x76x20 cms) has a special burst/flare mode for recording gamma-ray bursts and solar flares. In this mode a pulse height spectrum is accumulated for all events from 1 MeV to approximately 200 MeV. Spectra are routinely accumulated every 32.76 seconds but when activated by a BATSE trigger, EGRET accumulates 4 sequential spectra with integration times variable from 1 to 16 seconds. For high-energy bursts coming within the telescope field of view, the spark chamber telescope, with an energy range from 30 MeV to 20 GeV, can provide valuable spectral and directional information.

BURST RESULTS

EGRET has observed six BATSE triggered bursts. Table I contains information on the burst direction, measured peak low-energy flux or two sigma upper limits at 1.3 and 2.0 MeV, and the EGRET component in which detection occurred. Three of these bursts, GRB910503, GRB910601, and GRB910814-2 (the second BATSE burst of the day), are very strong and were easily observed by EGRET in the NaI spectrometer and the anticoincidence shield. The other three bursts were observed with the anticoincidence shield but were near or below the sensitivity threshold for the NaI spectrometer.

Table I EGRET-DETECTED GAMMA-RAY BURSTS

| Burst | Direction | | Flux (Photons/Mev-cm^2-sec) | | A | Spark |
	Zen	Azimuth	@ 1.3 MeV	@ 2.0 MeV	Shield	Chamber
GRB910503	28	231	8.0	1.9	Yes	Yes
GRB910503-t			0.35	<0.1	Yes	No
GRB910601	12	305	0.55	0.10	Yes	No
GRB910814-2	29	279	4.5	1.6	Yes	-
GRB920508	22	327	<0.11	<0.06	Yes	No
GRB920622	44	113	<0.15	<0.08	Yes	No
GRB920830	20	1.4	<0.09	<0.05	Yes	No

The time curves for the GB910503 burst[9] are shown in figure 1. The initial burst is easily observed as well as a second pulse, GRB910503-t, occurring approximately 45 seconds later. This burst has high statistical significance in both the

NaI spectrometer and the anticoincidence shield count rates. The anticoincidence shield count rate data can resolve the burst time structure to 0.256 seconds. For this burst, high energy gamma rays were observed in the spark chamber. The occurrences of associated spark chamber events are identified by x's in this figure. The location derived from these spark chamber gamma-ray events is shown in figure 2 along with the COMPTEL source location. The spectrum for the second pulse was soft, a spectral index of 3.8±0.1, similar to the soft component needed for a two-component fit of the third spectrum.

Bursts, GRB910601 and GRB910814-2, are discussed in detail by Kwok, et al.[10] in another paper at this conference.

Schaefer, et al.[4] has identified 18 BATSE gamma bursts with signals near or above 1 MeV that occurred before January 1992. EGRET's first three bursts are in this list but the remaining 15 bursts have not been observed by EGRET. Most of these bursts were at a very large angle (>90°) to the EGRET axis so that there was substantial shielding by the spacecraft.

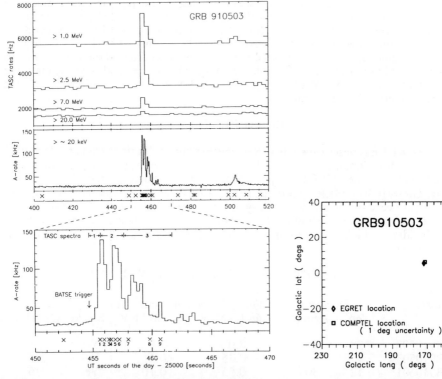

Fig. 1 The observed light curves for the GRB910503 burst in the NaI spectrometer and the anticoincidence shield. The x's indicate the time of related events in the EGRET spark chamber used to locate the burst direction.

Fig. 2 The GRB910503 source location plot.

LOW-ENERGY BURST DETECTION SENSITIVITY

To establish the EGRET low-energy burst detection sensitivities, the transport of low-energy gamma rays through the spacecraft and entering the NaI spectrometer was calculated as a function of direction and energy. For each direction, a parallel beam of monoenergetic gamma rays, directed at the NaI spectrometer, impinged on the detailed CGRO mass model over (Jensen, et al.[5]) an area larger than the projected area of the NaI spectrometer. The calculated energy loss spectrum in the NaI contains the directly transmitted gamma rays as well as scattered radiation from the surrounding spacecraft. Knowing the NaI projected area, the number of incident gamma rays, and the number of events in the spectrum, an effective area efficiency can be determined for each direction and energy.

The geometry of the calculation is straightforward. The center of the NaI spectrometer is at (-229, 0, 12) cms in the CGRO coordinate system. The +x direction (azimuth angle = 0.0) looks through the bulk of the CGRO and other instruments, the +y direction (azimuth angle = 90.0) looks out the side of the CGRO and the +z direction looks along the EGRET principal axis.

To generate two sigma detection limits certain assumptions were made. First, the bursts are highly peaked and integration times of 2 seconds can be used. Second, a wide energy interval of 0.6 MeV was selected to enhance sensitivity. Third, while the low-energy background rate changes as a position in orbit, an average rate was used for the sensitivity calculations.

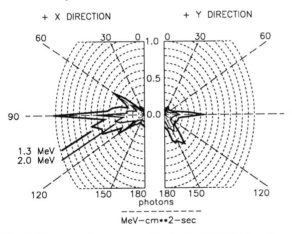

Fig. 3 The two sigma upper limits for 1.3 MeV and 2.0 MeV gamma rays as a function of the zenith angle. The +x direction looks along the spacecraft and +y looks through the side of EGRET and the spacecraft.

Two sigma detection limits at 1.3 and 2.0 MeV are plotted in figure 3 as a function of zenith angle from the z axis for the +x direction and the +y direction. EGRET has its greatest sensitivity for zenith angles less than 45 degrees but degrades rapidly as more matter comes between the burst and the NaI detector. The two sigma upper limits along the z axis are 5.0 and 3.8 x 10^{-2} photons/MeV-cm^2-sec for 1.3 and 2.0 MeV gamma rays, respectively. Substantial differences can be observed between the +x and +y directions reflecting the differences in mass distribution.

At large zenith angles, it is possible for a hard burst with a spectral index less than 1.6 to get a positive measurement at 2.0 MeV and have only a two sigma upper limit value at 1.3 MeV.

Four of the bursts observed by BATSE[4] and not by EGRET, GRB910717, GRB911202, GRB910430, and GRB910814-1, have 1.3 MeV flux values of 0.01, 0.03, 0.003, and 0.005 gamma rays/MeV-cm^2-second, respectively. These bursts occurred at zenith angles >90° and are below the two sigma limits for the NaI spectrometer.

CONCLUSIONS

The EGRET NaI has its best sensitivity over a large field of view of within 45.0 degrees from the EGRET axis. This sensitivity decreases as spacecraft mass intervenes between the NaI and the gamma-ray burst. For bursts below the EGRET two sigma limits, EGRET misses many bursts that BATSE can still measure. For intermediate and strong bursts, EGRET with its extended energy range will provide useful information on the high-energy emission of bursts and the confirmation of possible high-energy spectral breaks.

ACKNOWLEDGMENTS

The EGRET team acknowledges support from the following grants: Bundesministerium fur Forschung and Technologie, Grant 50 QV 9095 (MPE); NASA Grant NAG5-1742 (HSC); NASA Grant NAG5-1605 (SU); and NASA Contract NAS5-31210 (GAC).

REFERENCES

1. Matz, S.M., et al., 1986, Ap. J., 288, L37
2. Schmidt, W.K.H., 1978, Nature, 271, 525
3. Share, G., et al., 1986, Adv. Spac Res., 6, 15
4. Schaefer, B.E., et al., 1992, Ap. J., 393, L51
5. Jensen, C., et al., this conference
6. Hughes, E.B., et al., 1988, IEEE Trans. Nucl. Sci, NS-27, p 364.
7. Kanbach, G., et al., 1989, Proc of the Gamma Ray Observatory Science Workshop, p 2-1
8. Thompson, D.J., et al., 1992, to be pub. in Ap. J. Suppl.
9. Schneid, E.J., et al., 1992, A&A, 255, L13
10. Kwok, P.W., et al., this conference

EGRET OBSERVATIONS OF GAMMA-RAY BURSTS ON JUNE 1, 1991 AND AUGUST 14, 1991

P.W. Kwok[a], D.L. Bertsch, B.L. Dingus[b], C.E. Fichtel, R.C. Hartman,
S.D. Hunter, J.R. Mattox[c], P. Sreekumar[b], D.J. Thompson
NASA/Goddard Space Flight Center, Code 662, Greenbelt, MD 20771
[a] NAS/NRC Postdoctoral Research Associate
[b] Universities Space Research Association
[c] GRO Science Support Center, Computer Science Corp.

E.J. Schneid
Grumman Aerospace Corporation, Mail Stop A01-26, Bethpage, NY 11714-3580

G. Kanbach, H.A. Mayer-Hasselwander, C. von Montigny, K. Pinkau,
H. Rothermel, M. Sommer
Max-Planck Institut für Extraterrestrische Physik, 8046 Garching, Munich, Germany

Y.C. Lin, P.F. Michelson, P.L. Nolan
Hansen Experimental Physics Laboratory, Stanford Univ., Stanford, CA 94305

D.A. Kniffen
Hampden-Sydney College, P.O. Box 862, Hampden-Sydney, VA 23943

ABSTRACT

The Energetic Gamma Ray Experiment Telescope (EGRET) instrument detected the gamma ray bursts on June 1, 1991 (GRB 910601) and August 14, 1991 (GRB 910814)*. Significant time structures down to 0.256 s were resolved by the anti-coincidence dome. Spectra accumulated by the NaI(Tl) crystal are consistent with a power law model with evidence of spectral evolution. Four gamma ray events within 5 degrees from the direction of GRB 910601 were found in the spark chamber, when 1.5 could be expected from the detection rate, but they arrived only after the main burst emission had decayed. When GRB 910814 arrived, the spark chamber was disabled because the earth's atmosphere was in its field of view.

INTRODUCTION

The high-energy gamma ray emission of gamma ray bursts is important to understand the emission mechanism and the nature of gamma ray bursts. High-energy measurements can provide constraints on the magnetic field strength at the site of radiation[1] and the distance of the source assuming isotropic emission[2].

---

*This gamma ray burst, which was triggered at 69275 s (19:14:35) U.T., corresponds to the second burst (GRB 910814_69275) detected by BATSE on August 14, 1991.

Spectral breaks in the range 0.511-2 MeV may indicate conditions at the source[3] such as photon-photon interactions, photon-magnetic field interactions[4] or degradation of photon energy by Compton scattering.

Although the EGRET instrument is not optimized to detect gamma ray bursts, its high-energy capabilities (1-200 MeV in the NaI(Tl) crystal spectrometer and 30 MeV – 30 GeV in the spark chamber) can nevertheless provide valuable information at high energies if the burst is coming from a favorable direction with respect to the instrument axis. Several bursts have been observed by the EGRET instrument, including the highest energy gamma ray burst[5] on May 3, 1991. In this paper we present the analysis of two additional bursts, on June 1, 1991 and August 14, 1991.

OBSERVATIONS AND RESULTS

The EGRET instrument on the Compton Observatory is a typical spark chamber type gamma ray detector, similar to the predecessors SAS-2 and COS-B in the 1970s but with better sensitivity. It consists of a spark chamber stack with interleaved high Z foils to convert incoming photons to electron-positron pairs. The energy of the pair is measured by a NaI(Tl) crystal at the bottom of the spark chamber. Charged particle triggers are vetoed by a plastic scintillator surrounding the spark chamber. The NaI(Tl) crystal itself can also function as an omnidirectional counter, but it is shielded by various materials in different directions. Details of the instrument and its calibration can be found in Kanbach et al.[6,7] and Thompson et al.[8].

On June 1, 1991 at 19:22:16 U.T., the Burst and Transient Source Experiment (BATSE) instrument detected a burst and sent a trigger signal to EGRET. The NaI(Tl) crystal counter of the EGRET responded by activating the BURST mode with integration times of 1, 2, 4, 8 s sequentially in addition to the normal synchronous mode of 32.768 s. Fig. 1 shows the time history of the burst at different energy thresholds. In the hard x-ray energy range seen by the anti-coincidence dome, it is clear that BATSE triggered at a precursor about 15 s before the main phase of emission. The main phase lasted for about 12 s followed by a weak tail of 16-20 s. The time profiles at higher energies are less informative because of poorer time resolution and loss of precision in count rate digitization.

Four spectra were accumulated during the BURST mode. However the main phase of the emission only began near the end of the BURST mode and the main peak was recorded in the normal mode of 32.768 s integration which included the weak emission that followed.

Background was established by interpolating the data in each channel typically 2 minutes before and after the burst. Using the mass model of the Compton Observatory, response functions were calculated and the spectra deconvolved. Because the four spectra obtained in the BURST mode were before the main phase of emission and of weak intensity, the spectra were statistically limited. Fig. 2 shows the spectrum of the main phase of emission. The spectrum might start turning over at about 4 MeV. However, a power law, $(0.98 \pm 0.08)E^{-(3.67 \pm 0.20)}$

photons cm^{-2} s^{-1} MeV^{-1}, gives a good fit to all the points, with a χ^2 of 4.5 (dof = 5). Alternatively, a thermal bremsstrahlung fit of kT = 0.676 MeV to all the points has a χ^2 of 3.1 (dof = 5).

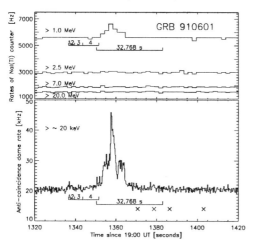 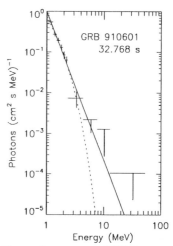

Fig. 1 Time history of GRB 910601 at different energy thresholds. Also indicated are the time intervals of the accumulated spectra: BATSE triggered spectrum 1, 2, 3, and 4, and normal mode of 32.768 s integration. The crosses are the arrival times of the gamma ray events in the spark chamber.

Fig. 2 Spectrum of GRB 910601 main phase of emission. Solid line is the power law model fit. Dotted line is the thermal bremsstrahlung fit. Upper limits are one standard deviation.

Four gamma ray events (shown as crosses in Fig. 1), which were consistent with the direction of the burst [9,10], were recorded in the spark chamber after the main peak. The energies of the 4 events are $314\pm32, 46\pm22, 268\pm28, 200\pm22$ MeV and the arrival times are 69770.798 s, 69778.804 s, 69786.485 s, and 69803.048 s U.T. respectively. The probability that 4 events were found when 1.5 events were expected in 100 s around the burst is 6%. It should be noted that the spark chamber was not paralyzed by the anti-coincidence dome during the main peak. The flux represented by the 4 events is inconsistent (high) with the power law extension even if they occurred near the peak of the burst, making the association of the events with the burst somewhat questionable.

The August 14, 1991 burst occurred at 19:14:35 U.T.. Unlike that on June 1, this burst rose quickly with no precursors (Fig. 3). It started with an intense peak for about 4 s after trigger, followed by some short time scale (0.256 s) emissions for about 9 s and then a second peak of about 4 s. Four spectra (1, 2, 4, 15.875 s) were sequentially accumulated during the BURST mode which essentially covered the whole burst. The four spectra can be fitted with power laws (Table 1) except for the low-energy ends, which roll over and suggest a spectral break which is also seen by BATSE[3] . The first three spectra are associated with the first peak, indicating soft-hard-soft changes (-2.94, -2.72, -2.92) (Table 1) while the second peak is covered by the last spectrum (-2.49).

This burst came from a direction 38.7 degrees[9,10] from the instrument axis at a time when the spark chamber was disabled because the earth was in the field of view of EGRET. Therefore, no events were recorded in the spark chamber.

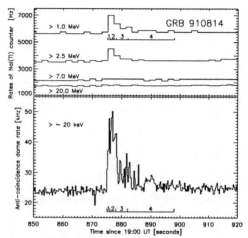

Fig. 3 Time history of GRB 910814 and the time intervals of the four spectra accumulated during the BURST mode.

Table 1. Power law fit, A $E^{-\alpha}$, to spectra of GRB 910814

	1†	2†	3†	4
A	14.36 ± 1.44	12.69 ± 0.86	4.79 ± 0.73	1.22 ± 0.14
α	2.94 ± 0.10	2.72 ± 0.06	2.92 ± 0.13	2.49 ± 0.10
χ^2 (dof)	12.1 (14)	15.9 (18)	4.8 (13)	5.4 (14)

† First 2 points in spectrum 1 and 2, and first 3 points in spectrum 3 are not used in the fit.

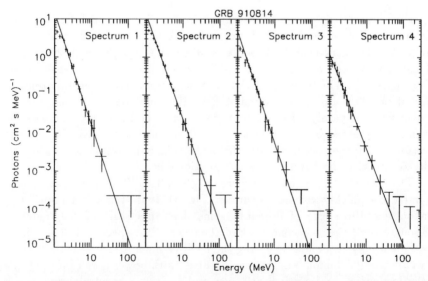

Fig. 4 Spectra of GRB 910814. See Table 1 for fit parameters. Upper limits are one standard deviation.

DISCUSSION

Previous observations of gamma ray bursts suggested that most of the spectra are power laws at MeV energies[11] although there are cases where a thermal synchrotron shape gives a better fit[12]. The present observations seem to further support power laws as common features of gamma ray bursts at MeV energies, but with possible cutoffs in some cases.

For GRB 910814, the power law model gives very good fit to the data except the roll over at around 1 to 2 MeV indicating a spectral break which may be caused by single-photon pair production[4]. For GRB 910601 the limited energy range makes it difficult to infer the spectral change near 1 MeV. The fact that the optically thin thermal bremsstrahlung gives a good fit does not necessarily mean that it is the relevant radiation process, because of the inefficiency of bremsstrahlung emission and the required extreme source geometry aspect ratio[13].

The emission of GRB 910814 up to about 50 MeV may suggest some anisotropy of the radiating electrons[4], because photons emitted closer to magnetic field lines can escape from an intense magnetic field at higher energies.

ACKNOWLEDGMENTS

The EGRET team gratefully acknowledges support from the following: Bundesministerium für Forschung und Technologie, Grant 50 QV 9095 (MPE); NASA Grant NAG5-1742(HSC); NASA Grant NAG5-1605 (SU) and NASA Contract NAS5-31210(GAC).

REFERENCES

1. S.M. Matz, et al., ApJ, **288**, L37 (1985).
2. W.K.H. Schmidt, Nature, **271**, 525 (1978).
3. B.E. Schaefer, et al., ApJ, **393**, L51 (1992).
4. M.G. Baring, MNRAS, **244**, 49. (1990).
5. E.J. Schneid, et al., A & A, **255**, L13 (1992).
6. G. Kanbach, et al., Space Sci. Rev., **49**, 69 (1988).
7. G. Kanbach, et al., Proc. of the Gamma Ray Observatory Science Workshop, p.2-1 (1989).
8. D.J. Thompson, et al., to be published in ApJ Suppl. (1992).
9. V. Schönfelder, IAU circular no.5369 (1991).
10. BATSE private communication.
11. S.M. Matz, Ph.D. Thesis, University of New Hampshire (1986).
12. P.L. Nolan, et al., High Energy Transients in Astrophysics, ed. S.E. Woosley (New York: American Institute of Physics), p.399 (1984).
13. R.W. Bussard & F.K. Lamb, Gamma Ray Transients and Related Astrophysical Phenomena, ed. R.E. Lingenfelter, H.S. Hudson & D.M. Worrall, p.189 (1982).

Search for Gamma-Ray Burst Spectral Features in the Compton GRO BATSE Data

B. J. Teegarden, T. L. Cline, D. Palmer, B. E. Schaefer (NASA/GSFC), G. J. Fishman, C. Meegan, R. B. Wilson (NASA/MSFC), M. Briggs, W. S. Paciesas, G. Pendleton (UAH), P. Lestrade (MSU), D. L. Band, L. Ford and J. L. Matteson (UCSD)

INTRODUCTION

The presence of line-like features in the spectra of gamma-ray bursts has stood as one of the strongest pieces of evidence that these events originate at neutron stars. Such features have been interpreted as cyclotron absorption in the $> 10^{12}$-gauss magnetic fields that are believed to be characteristic of neutron stars. First reported by Mazets and co-workers (1981), these features were subsequently confirmed by HEAO-A4 (Hueter, 1982) and Ginga (Murakami *et al.*, 1988; Fenimore *et al.*, 1988). The line energies ranged between 18 and 70 keV, and the presence of both a first and second harmonic was reported by Ginga (Murakami *et al., 1988*; Yoshida *et al.*, 1992; Graziani *et al.*, 1992). Typically such features were found in ~ 10% of bursts that were spectroscopically examined.

In this paper we report on a search of the BATSE Spectroscopy Detector (SD) database for such "cyclotron" features. These detectors were added to the BATSE experiment specifically to search for and study such features in gamma-ray burst spectra.

OBSERVATIONS

The BATSE experiment consists of eight each of two types of detectors, the Large Area Detectors (LADs) and the Spectroscopy Detectors (SDs). The LADs are optimized for the detection of weak gamma-ray bursts, whereas the SDs are optimized for spectroscopy of the brighter bursts. The SDs (and LADs) are placed in pairs at the four corners of the spacecraft. The eight detectors are oriented such that their front surfaces have a complete and nearly uniform view of the sky. At least two, and normally four, detector faces view a burst.

Each SD consists of a 12.7 cm diam. x 7.6 cm height NaI scintillator, which has an aluminum housing with a 7.6 cm diam. beryllium window on the front face. This geometry results in an effective area to burst photons which is a function of energy and angle. Below ~ 20 keV the aluminum has low transmission. Thus the effective area is dominated by that of the Be window, which has a cosine (viewing angle) dependence and cuts off below ~ 5 keV due to photoelectric absorption. Above ~ 20 keV all housing elements have significant transmission, and the effective area varies more slowly with angle. The lowest energy analyzed by the electronics depends on the commandable settings of the electronic threshold and detector gain. However, some electronic nonlinearities near threshold were not fully corrected in all the data analysis. Therefore, the spectral fitting used here was limited to energies above 17 keV. This restriction has limited the effectiveness of the SDs in the search for lines at ~ 20 keV.

The data presented in this paper cover the period from 20 Apr 1991 to 8 Sept. 1992, or 1.4 years. During this period, the BATSE experiment operated essentially continuously with an average duty cycle of ~ 75 %. Even after the loss of the CGRO's on-board tape recorders the BATSE burst duty cycle for burst detection and analysis was not significantly degraded due to the use of a large buffer memory for burst storage.

When a burst is detected by the LADs the BATSE burst mode is initiated and data are stored in a number of different formats. The data type primarily used in this search is the so-called Spectroscopy High Energy Resolution/Burst (SHERB) data. This data type consists of 192 spectra accumulated for each burst. The duration of each spectrum is keyed to the accumulation of a fixed number of counts in the LAD detectors, so that the brighter the burst the shorter the accumulation time. Accumulation times for SD spectra typically vary between a few tenths of a second and ~ 5 seconds. Each spectrum consists of 256 quasi-logarithmically spaced channels. The energy range covered varies according to the gain of the individual detector. For most bursts there were at least two SDs that provided high-quality spectra.

The database used in the analysis reported here comprises the brightest ~ 17% of the bursts detected by the BATSE LADs. Only spectra from high-gain detectors (4x or 7x) were included in the database since at lower gains the low-energy thresholds were too high for cyclotron line detection. The distribution of bursts in the spectroscopy database is given as a function of burst fluence in Fig. 1. For comparison, the distribution of the fluences of the 23 Ginga bursts that were spectroscopically analyzed is also plotted (E. Fenimore, private communication). It is important to note that there are more bright bursts in the Ginga sample than in the BATSE sample.

The entire spectroscopy database has been visually searched for spectral features in the 17 - 1000 keV range by three independent observers, and a list of candidate features has been compiled. These visually identified features have been fit to a model photon spectrum consisting of a broken power-law continuum and one or two gaussian lines. The model photon spectrum is multiplied by the BATSE response matrix for the particular burst to create a model counts spectrum and χ^2 is minimized using the procedure of Bevington (1969). This procedure was repeated for models with and without lines. $\Delta\chi^2$ and $F = (\Delta\chi^2/\Delta v)/\chi_v^2$ were then calculated, and the probability, P_F, that F could have been greater than the measured value was determined (the classical F-test). In Fig. 2 the calculated P_F values are plotted for 16 visually identified spectral features. The most significant feature had $P_F \sim 10^{-3}$. An important question is, what significance is required in order to claim a positive detection? To be able to detect an absorption feature at a certain significance, the continuum spectrum must have a certain minimum significance or signal-to-noise ratio. Our data sample contained 1063 independent spectra, but most of these had insufficient signal-to-noise ratios to permit line detections. The true size of the sample is therefore dependent on the threshold criterion. We find that, of the 1063 examined spectra, 48 had sufficient statistics to allow the detection of a 40-keV line (using the Ginga 880205 line parameters) at a significance of 10^{-3}. Since the sample size is about 48, the critical significance level for detecting a line should be decreased from our single spectrum critical level of 10^{-3} to at least 10^{-4}. By this criterion, BATSE has not yet detected any cyclotron lines.

For comparison, the F-test significances of the three Ginga events are also shown in Fig. 2. Two time periods, S1 and S2, were reported to contain line features in the Ginga burst 870303 (Graziani et al., 1992). The former was best fit with a single line at 21.1 keV, whereas the latter was best fit with a 2-line model with line energies of 21.4 and 42.8 keV. The authors argued that the F-test was not a good measure of the significance of the line features since the reduced χ^2 value for the fit was anomalously low (~0.5). Nonetheless, for the sake of completeness, we have included 870303 in our plot. Fig. 2 shows that BATSE has not yet detected any line features as significant (using the F-test criterion) as those in 870303 and 880205. Given our criterion of 10^{-4}, we would not be able to claim a BATSE detection for a line feature of the significance of 890929.

In Figs. 3 and 4, the two most significant BATSE "lines" are shown. In both cases, spectra are shown from two independent detectors. For BATSE burst 920210, the detector with the smallest viewing angle to the burst (detector 4) can be well fit by a model with two absorption lines at 24.5 and 52.1 keV yielding an F-test probability of 2.0×10^{-3}. The detector with the second smallest viewing angle (detector 7) shows no evidence of lines, but because it has lower statistics than detector 4, there is no inconsistency. For BATSE burst 920718 (Fig. 4), detector 6, which has the second smallest viewing angle, is a best fit model with a single emission line at 34.7 keV. This line has a significance of 9.6×10^{-4}. However, the spectrum of detector 7 has no significant lines, which is inconsistent with detector 6, since detector 7 is the most favorable detector because it has the smallest viewing angle.

Table 1. BATSE/Ginga Comparison

	BATSE	GINGA
No. of Detectors	8 (SDs)	2[†]
Energy Range (this analysis)	17 - 2800 keV	1.5 - 375 keV
Detector Area	127 cm^2	~ 60 cm^2
Dates of Spectral Survey	4/91 - 9/92	3/87 - 11/91
	(1.4 yr)	(4.7 yr)
Exposure Factor (solid angle × live time)	~ 6.6 sr yr	~ 2.8 sr yr*

[†] Scintillator + proportional counter.
* E. Fenimore, private communication. This is an approximate preliminary result.

BATSE/GINGA COMPARISON

To decide whether there is any discrepancy between the BATSE and Ginga results, it is necessary to evaluate the relative sensitivities of the two experiments for cyclotron line detection. The basic characteristics of the two experiments are summarized in Table 1. BATSE has greater collecting area and a larger effective solid angle for burst detection (due to the fact that there are 8 SDs covering all of the visible sky). Furthermore, the BATSE orbit with its lower altitude and inclination has lower particle fluxes and therefore is more favorable than the Ginga orbit. On the other hand, Ginga's spectral coverage extends to lower energies, which is of crucial importance to the detection of first harmonic cyclotron lines at ~ 20 keV. The limit of 17 keV on our spectral fitting is insufficient to strongly anchor the continuum below a possible line at ~ 20 keV, which leads to a lower significance for such lines.

In order to assess the overall importance of these effects, we have simulated the response of the BATSE experiment to the Ginga bursts. Best-fit models of the Ginga photon spectra were convolved through the BATSE response matrices and Poisson noise was added. A typical BATSE background spectrum was then added. Continuum and continuum + line models were fit in the usual way and the F-test probabilities were determined. Fig. 5 shows a realization of the counts spectrum of Ginga burst 880205 in two of the BATSE detectors having viewing angles of 32° and 62° respectively. The 20- and 38-keV absorption features are obvious in the 32° detector, but the 20 keV feature is less apparent in the 62° detector. This is because angle-dependent absorption in the housing of the detector is important at 20 keV, but not at 38 keV. Table 2 summarizes the results of averaging the logarithm of the significances over 12 trials for each of the cases and compares them to Ginga. For 880205, BATSE is considerably more sensitive at 32° and marginally more sensitive at 62°. For the S1 interval of 870303, BATSE is less sensitive at both angles, but there may be a problem with the Ginga F value for this burst (Graziani et al., 1992).

Table 2. Comparison of F-Test Significances

Burst	BATSE		GINGA
	32°	62°	
870303 (S1 interval, 1 line at 21 keV)	1.5 × 10^{-4}	.19	1.5 × 10^{-7}
880205 (2 lines at 20 and 38 keV)	2.2 × 10^{-7}	1.1 × 10^{-5}	2 × 10^{-5}

CONCLUSIONS

We have reported on a search for spectral features in the BATSE spectroscopy database over the energy range 17 to 1000 keV. In the data taken between launch and early Sept. 1992, we have found no statistically significant features. Our criterion is that the F-test probability $P_F < 10^{-4}$. The most significant features found in the data had $P_F \approx 10^{-3}$. We have evaluated the relative sensitivities of the BATSE and Ginga experiments. We find they are roughly comparable, but vary according to the energy of the line feature and the angle of incidence of the burst. Similar conclusions have been reached by Freeman, Lamb and Fenimore (1992) using a Bayesian analysis. More detailed information on this subject is also contained in papers by Ford *et al.* (1992) and Palmer *et al.* (1992). It appears that BATSE would have unambiguously detected lines in only one of the three Ginga bursts. According to Table 1, the BATSE exposure factor was 2.4 x greater than Ginga's. However, Fig. 1 shows that Ginga has seen more bright bursts than BATSE (15 for Ginga vs. 11 for BATSE for fluences > 2.5 x 10^{-5} erg cm^{-2}). If we take the exposure factors as a guide, then, based on the Ginga detection rate, BATSE would be expected to detect line(s) in ~ 2-3 bursts. If we use the actual detected bright bursts as the guide, then BATSE should have detected line(s) in ~ 1 burst. In either case, we are dealing with the statistics of small numbers, and there is not yet any discrepancy between the BATSE and Ginga results.

REFERENCES

Bevington, P. R., 1969, Data Reduction and Error Analysis for the Physical Sciences, (McGraw-Hill, New York).

Graziani, C, et al., 1992, in Gamma-Ray Bursts, ed. C. Ho, et al., (Cambridge Univ. Press, Cambridge), 407.

Ford, L. et al., 1992, these proceedings.

Freeman, P., Lamb, D. Q. and Fenimore, E. E., 1992, these proceedings.

Hueter, G. J., 1984, in High Energy Transients in Astrophysics, ed. S. Woosley, (AIP, New York), 373.

Mazets, E. P., et al., 1981, Nature, 290, 378.

Murakami, et al., 1988, Nature, 335,234.

Palmer, D., et al., 1992, these proceedings.

Yoshida, A., et al., 1992, in Gamma-Ray Bursts, ed. C. Ho, et al., (Cambridge Univ. Press, Cambridge), 399.

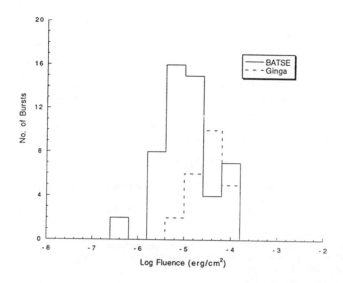

Fig. 1. Distribution of fluences of bursts in BATSE spectroscopy data base compared with fluences of Ginga bursts.

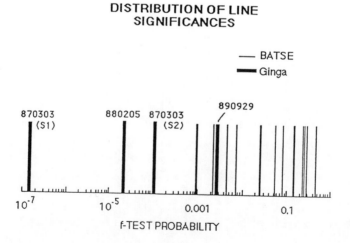

Fig. 2. Distribution of F-test significances of BATSE features and Ginga lines. Note that these significances are from two different instruments.

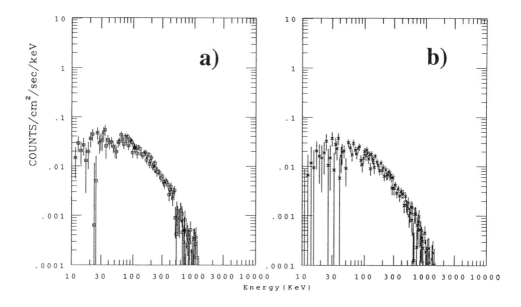

Fig. 3. Spectra from BATSE burst 920210. a) detector 4 with viewing angle = 39°, b) detector 7 with viewing angle = 72°.

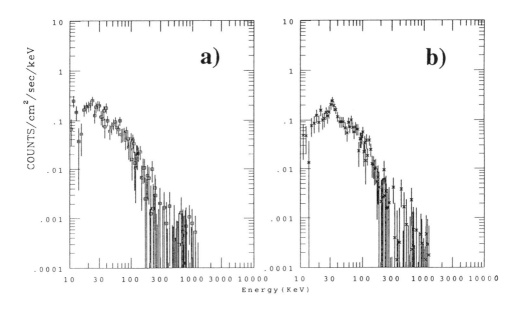

Fig. 4. Spectra from BATSE burst 920718. a) detector 7 with viewing angle = 20°, b) detector 6 with viewing angle = 49°.

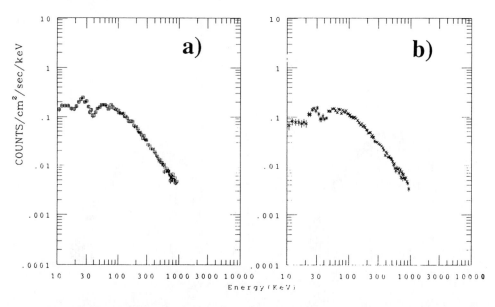

Fig 5. Simulated spectrum of Ginga burst 880205. a) viewing angle = 32°, b) viewing angle = 62°.

SEARCHING FOR GRB SPECTRAL FEATURES IN THE BATSE LARGE AREA DETECTOR DATA

R. D. Preece[†], G. J. Fishman, C. A. Meegan, R. B. Wilson, M. N. Brock
NASA, ES64, Marshall Space Flight Center, AL 35812

W. S. Paciesas, G. N. Pendleton, M. S. Briggs
University of Alabama at Huntsville, Dept. of Physics, Huntsville, AL 35899

B. Teegarden
NASA, Code 661, Goddard Space Flight Center, Greenbelt, MD 20771

ABSTRACT

The Large Area Detectors of the BATSE instrument combine the benefits of large collecting area for good statistics and good energy resolution, thus they should be a very useful tool for the detection of spectral features in gamma ray bursts. The best candidate line features would have to be moderately broad (≥ 20 keV at around 50 keV) to be able to be resolved. The BATSE Spectroscopy Detectors have been at the focus of efforts to confirm previous observations of line-like spectral features, particularly those observed by the GINGA instrument. Here, we have made a study to characterize the spectroscopic capabilities of the Large Area Detectors in their response to GINGA-like features. With two essentially distinct all-sky monitoring spectroscopy instruments, the Large Area Detectors and the Spectroscopy Detectors, BATSE has the potential to confirm its own observations of spectral features.

INTRODUCTION

Much effort has been put into comparisons between the BATSE Spectroscopy Detectors (SDs) and the gamma ray spectroscopy instruments on GINGA, in order to determine whether or not both are capable of detecting spectral features of comparable equivalent widths in gamma ray bursts (GRBs) of comparable intensities. In particular, several papers in these proceedings are devoted to resolving this question [1,2]. What has been largely left out in these discussions is that the BATSE Large Area Detectors (LADs) also have the capability to resolve spectral features in GRBs and other cosmic sources. While the BATSE SDs have better energy resolution than the LADs, the LADs have a considerable advantage in counting statistics, due to their larger collecting area. What remains to be determined are the LADs' response to spectral features with equivalent widths such as have been reported in the observations by GINGA [3,4]. We

---

[†] NAS/NRC Resident Research Associate

have begun a study with this goal in mind, prior to a survey of the LAD dataset itself for bursts containing spectral features.

PROCEDURE

An ensemble of synthetic spectra was created by multiplying a simple model of a GRB by the LAD response matrices calculated for several different angles. For this study, we used values of 30° and 60° between the normal to the plane of the detector and the vector directed at the source. Since the LADs are thin in profile, the detector energy response changes considerably as a function of the source's zenith angle after about 30° (it is roughly a cosine function of angle) [5]. In the worst case, where a source is placed on one of the spacecraft's axes, it would be seen by four detectors with equal angles to the source of $\tan^{-1} \sqrt{2} \approx 55°$. Thus, the poorer response of the 60° case serves as an indication of what one might expect if the confirmation of a line-like spectral feature would require detection by a second detector. The photon model consists of a -2 power-law index continuum with a Gaussian absorption line. The strength of the line is expressed in terms of a fraction of the continuum flux at the line centroid, with the constraint that the strength never exceeds 1, representing 100% absorption. After multiplication by the response matrix, random noise is added to the resulting count spectra, which are then fit with and without an absorption line component. For the purpose of model fitting, the MFIT program, developed at MSFC, was used. The resulting values for χ^2 are subjected to the F-test to obtain the probability that an 'observed' apparent line feature would occur if no line were present in the photon model. 'Detection' of an absorption line was predetermined to be a probability $< 1.0 \times 10^{-4}$, in line with the SD survey [6].

As the BATSE LAD energy response is quite a bit poorer than the SD's at lower energies, there would be little gained from investigating the response of the LADs for the lower-energy GINGA-like features, so the energy chosen for the centroid of the Gaussian 'absorption' feature in the model was that of the second harmonic of GB880205, 40 keV [4]. The energy resolution of the detectors at 40 keV is, on the average, 33%, or ~13 keV. The equivalent width reported for the GINGA second harmonic was 9.1 keV, thus the possibility for a LAD detection of a similar feature hinges upon the trade-off between the large collecting area and poorer resolution of the LADs with respect to GINGA. In our model, the absolute value of the equivalent width, $w = -EW$, is calculated as a function of the fraction of absorption of the Gaussian line, f, the Gaussian width, ΔE_g, and the detector resolution, ΔE_s:

$$w = f\sqrt{\pi}\sqrt{2\Delta E_g^2 + \Delta E_s^2}. \tag{1}$$

A similar model is described more fully elsewhere in these proceedings [1]. The reported FWHM for the GINGA second harmonic of 16.0 keV translates to a Gaussian width of ~11.3 keV, so the parameter f is constrained in our model, for comparison with the

GINGA results, to be 0.25. As the individual probabilities resulting from the fits in our ensemble varied over a large range in values, it makes sense to use the geometric mean (equivalently, the mean of the logs) in presenting the results. For a sample of four model spectra (admittedly small, so the error will be quite high), from a source at a zenith angle of 30°, and with $f = 0.25$ absorption at 40 keV, the geometric mean of the probabilities was 5.7×10^{-4}, which does not meet the detection criterion (*e. g.*: Fig. 1). However, an increase in absorption by 12% to $f = 0.28$ gives a mean probability of 7.4×10^{-5} (Fig. 2). To better understand these results, a positive detection ($P < 10^{-4}$) of the model line feature was made in 2 out of the 6 sample spectra created with the deeper line, while none were positively detected out of the four spectra containing a shallower line. While these results will have to await a larger statistical sample, they are indicative of what can be expected from the LADs. With a zenith angle to the source of 60°, the results become somewhat worse. The mean probability (from a sample of four spectra) is 6.2×10^{-3} for a GINGA-like e. w., and decreases by about an order of magnitude to 7.0×10^{-4} for $f = 0.30$, which is 20% more absorption. One positive detection was made out of each ensemble of four spectra.

CONCLUSION

This preliminary study indicates that, while not as sensitive as GINGA to the detection of line-like spectral features in the 40 keV region, the BATSE LADs still have good spectral capabilities. In particular, Konus-like lines, which are typically broader than the reported GINGA features, and at higher energies, would be excellent candidates to look for in a search through the LAD dataset. A spectral feature which has been reported by several observers is annihilation-like emission, for which a LAD detection may be very possible. At the higher end of the spectrum, above ~300 keV, the LAD energy resolution becomes quite good, less than 20%. Any feature detected independently by either the SDs or the LADs can possibly be confirmed or shown to be inconsistent with the other instrument's data. With a larger collecting area than the BATSE SDs, the sample of bursts collected by the LADs is considerably greater, and represents one of the most extensive catalogs of GRBs which exists today.

REFERENCES

1. Ford, L. A., Band, D. L., Matteson, J. L., Palmer, D., Schaefer, B. E., Teegarden, B. J., Cline, T. L., Preece, R. D., Fishman, G. J., Meegan, C. A., Wilson, R. B., Briggs, M. S., Paciesas, W. C. & Pendleton, G. 1992, these proceedings.
2. Freeman, P. E., Lamb, D. Q. & Fenimore, E. E. 1992, these proceedings.
3. Murakami, T. *et al.* 1988, *Nature*, **290**, 378.
4. Fenimore, E. E. *et al.* 1988, *Ap. J.*, **335**, L71.

5. Pendleton, G. N., Paciesas, W. S., Lestrade, J. P., Fishman, G. J., Wilson, R. B., Meegan, C. A., Roberts, F. E., Horack, J. M. & Brock, M. N. 1992, in "Gamma-Ray Bursts", *AIP Conference Proceedings 265*, W. S. Paciesas & G. J. Fishman, eds. (AIP).
6. Teegarden, B. J., Palmer, D., Schaefer, B. E., Cline, T. L., Fishman, G. J., Meegan, C. A., Paciesas, W. C., Lestrade, P., Pendleton, G., Preece, R., Band, D. L., & Matteson, J. L. 1992, these proceedings.

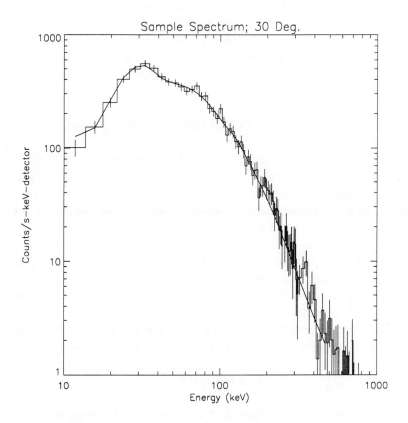

Fig. 1 Count spectrum created by multiplying a -2 power-law photon model with a 40 keV line, 25% absorption, by a BATSE LAD response matrix for a source angle of 30° from the zenith. Random noise has been added to the count spectrum, which is plotted here with the result of a model fit. The selection of the model parameters results in an equivalent width of 9.1 keV.

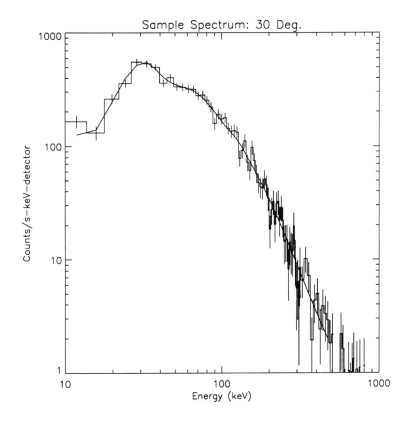

Fig. 2 Same as Fig. 1, for 28% absorption.

BATSE OBSERVATIONS OF GAMMA-RAY BURST SPECTRAL DIVERSITY

D. Band, J. Matteson, L. Ford
CASS, UC San Diego, La Jolla, CA 92093

B. Schaefer, D. Palmer, B. Teegarden, T. Cline
Goddard Space Flight Center, Greenbelt, MD 20771

G. Fishman, C. Meegan, R. Wilson
Marshall Space Flight Center, Huntsville, AL 35812

W. Paciesas, G. Pendleton
University of Alabama at Huntsville, AL 35899

P. Lestrade
Mississippi State University, MS 39762

ABSTRACT

The spectra of 58 bursts observed by the BATSE Spectroscopy Detectors were fit satisfactorily from less than 30 keV to more than an MeV by a 4 parameter model consisting of two power laws with a smooth transition. We find that a distribution of spectral indices is required instead of the low energy E^{-1} dependence found in previous work. The break energy ranges from less than 100 keV to more than an MeV, peaking below 200 keV; this suggests that the break is not tied directly to pair processes in all bursts. We did not find any conclusive correlations among the spectral parameters; our sample is probably inappropriate for the correlations predicted by the cosmological models.

INTRODUCTION

In the current study we investigate burst-averaged spectra from the Spectroscopy Detectors (SDs) of the Burst and Transient Source Experiment (BATSE) on board the Compton GRO. We search for global properties such as the continuum shape, as well as correlations with other burst properties. Since the spectral hardness changes significantly during a typical burst,[1] the spectra studied here are averages over widely varying instantaneous spectra, and thus reflect both the underlying physical processes and the temporal evolution.

The characterization of burst spectra has evolved over time as instruments have become more sensitive. Burst spectra typically have been described[2] as $N_E \propto E^\alpha \exp[-E/E_0]$ flattening out smoothly to $N_E \propto E^\beta$, with $0 > \alpha > \beta$. Early studies[3] described burst spectra with $\alpha = -1$ and $E_0 = 150$ keV, and also observed evidence of a high energy component $E^{-2.5}$; a later study[4] found the E_0 distribution ranged between ~ 20 keV to ~ 2 MeV, peaking at 250 keV. High energy tails have been detected by a number of detectors.[5,6] The SDs with their larger effective area, greater energy resolution, and increased spectral range (typically two decades) can better determine the true shape of burst continua.

A more complete discussion will be presented elsewhere.[7]

METHODOLOGY

Each of the 8 BATSE modules includes an SD. The detector's 5" diameter by 3" thick NaI(Tl) crystal provides an energy resolution of $\sim 7\%$ FWHM at 662 keV. The energy deposited by a photon in the crystal is first analyzed into 2782 linear pulse height channels which are then compressed into 256 pseudo-logarithmic channels. For the spectra used in this study these compressed channels are finer than the detectors' energy resolution. The BATSE detectors and their scientific capabilities are described by Fishman et al.;[8] and the calculation of the detector response is outlined by Pendleton et al.[9]

After a burst is detected, BATSE collects a prodigious quantity of data during a ~ 4 minute burst mode. The data type used for this study (known by the acronym SHERB) consists of a series of 192 spectra from the four most brightly illuminated SD detectors. A time-to-spill criterion determines each spectrum's accumulation time. In this study we averaged the SD spectra over the duration of each burst, although in a few cases the SHERB ran out of spectra before the end of the burst.

Our sample of 58 bursts comprises almost all the strong bursts until the end of May 1992: data processing difficulties excluded some strong bursts; a number of weaker events with interesting time histories were included; and the SHERB data type was not appropriate for extremely short duration bursts. Thus the sample is not rigorously statistically complete, but is most probably effectively complete. Because our sample's intensity distribution has the same intensity dependence as the entire BATSE database, in this study we may be sampling bursts out to distances where the source distribution is inhomogeneous. We also find that the bursts are spread uniformly across the sky. We cannot draw firm conclusions concerning differences between the different morphological classes because 47 of the 58 bursts are complex, multipeaked events.

To include the flat low energy component, the spectra were required to have a low energy cutoff of 10-30 keV. Depending on the gain of the detector, the high energy cutoff was 1200-3000 keV. We present fits to spectra from only one detector for each burst, avoiding the difficulty of relative energy and flux calibration. Joint fits from multiple detectors are generally consistent with the single detector fits if one spectrum extends to low energies since the greatest spectral curvature is at low energy.

The spectra were analyzed using the Burst Spectral Analysis Software (BSAS) package.[10] The spectra were rebinned so that each spectrum we fit generally contained ~ 100 nearly logarithmic channels. The background spectra for background-subtraction were interpolated from observed background spectra accumulated ~ 1000-2000 seconds before and after the burst. As judged by the size of the uncertainties and the channel-to-channel variations, the calculated background spectra were usually acceptable.

Based on previous work we modeled the background-subtracted continuum by the functional form

$$N_E = AE^\alpha \exp(-E/E_0) \qquad (\alpha - \beta)E_0 \geq E$$
$$= A\left[(\alpha - \beta)E_0\right]^{(\alpha-\beta)} \exp(\beta - \alpha)E^\beta \qquad (\alpha - \beta)E_0 \leq E \qquad (1)$$

which we call the GRB model. This functional form characterizes the spectrum without any direct relation to the underlying physical processes. Since bursts are unlikely to involve thermal processes,[11] E_0 probably does not correspond to a temperature. We define the hardness H as the maximum in $E^2N(E)$ (proportional to νF_ν used in other astrophysical subfields), indicating the energy band where most of the energy is radiated. Because the high energy spectral index is less certain than the low energy index, we calculate the maximum of $E^2N(E)$ from the low energy component alone, giving $H = (2+\alpha)E_0$.

In order to understand and improve our spectral analysis procedures, we ran simulations in which we created and then fit model SD spectra. Our conclusions will be summarized in detail elsewhere.[7] One noteworthy result is that *a priori* information about the existence of a high energy tail[5,6] should be used since neglecting the high energy component in the fit biases the α and E_0 parameters to be too hard.

The results presented here are not final fits to the burst spectra because the analysis software and methodology were improved in parallel with this study. In particular, the detector response we used did not include scattering off the Earth's atmosphere. For a small fraction of the bursts atmospheric scattering steepened the count spectrum; note that we found flatter spectra than previous studies. In addition, a new channel-energy calibration[12] was introduced after the completion of this study. Comparisons of fits using the two calibration methods demonstrate that the small quantitative differences should not affect our conclusions.

AVERAGE SPECTRAL SHAPE

We compared fits of the spectra in our sample with $\alpha = -1$ and $\beta = -2$ to fits with all four parameters free to vary. The fits with all four parameter free to vary are generally statistically acceptable, while those with α and β fixed often are not. Using the F-test we found that fixing $\alpha = -1$ and $\beta = -2$ was acceptable ($P > 0.1$) for only 17.2% (13/58) of the bursts in our sample, marginally acceptable ($0.01 < P < 0.1$) for 15.5% (9/58), and unacceptable ($P < 0.01$) for 62.1% (36/58).

We therefore conclude that the BATSE burst spectra in our sample can be characterized by the simple four parameter GRB form in equation (1), but without a universal set of burst parameters, contrary to past suggestions. The spectral indices populate a region of parameter space: β is typically found between -2 and -2.5, and α between 0 and -1.5. In general, the continuum gives the impression of continuous curvature below a few 100 keV which the GRB functional form is able to approximate over the observed energy range with good statistics. In the future our four parameter model may be inadequate to describe spectra observed by detectors with better spectral resolution.

Because the current emission theories are not very constraining, the spectral diversity we find should not be difficult for these theories to explain. Varying the physical conditions (e.g., electron distribution or magnetic field) may provide the necessary diversity in Galactic theories, while no successful burst spectrum has been calculated yet for the cosmological models.

DISTRIBUTION OF SPECTRAL BREAKS

We find the spectral break ranges from less than 100 keV to over an MeV. It peaks below 200 keV, and shows few breaks above 400 keV. Figure 1 presents the distribution of three different measures of the spectral break.

Our earlier study, Schaefer *et al.*[13], fit spectra from a sample of bursts

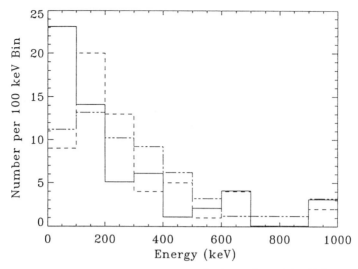

Figure 1. Distribution of the break energy E_0 from fits with all four parameters free to vary (solid curve) and from fits with $\alpha = -1$ and $\beta = -2$ (dot-dot-dashed curve). Also shown is the hardness distribution, the peak of $E^2 N(E)$, for the fits with α and β free (dashed curve).

observed by the BATSE SDs, but over the energy range 100 or 300 keV to ~ 27 MeV. Five spectra required a break in the model spectrum: a broken power law gave a better fit than a single power law. Our new fits are consistent with our earlier Schaefer et al. spectral analysis: we find approximately the same break energy for four of the five Schaefer et al. breaks. The break energy calculated for the fifth burst, 1B 910503, depends on the energy range fit as the model attempts to fit a continuously curving spectrum; the high energy spectrum in this burst is known from other GRO observations[6] to be a power law with $\beta \sim -2$, while the low energy spectrum is much harder. Therefore we suspect our earlier Schaefer et al. analysis did not find a larger number of spectral breaks, or any breaks below 400 keV, because the previous study fit spectra above 100 keV, whereas the spectra in our current study extend below 30 keV; the high energy breaks that the Schaefer et al. study observed are the tail of the break distribution.

The broad hardness and break energy distributions may indicate that the spectrum is sculpted by a complex series of physical processes such as evolving electron distributions and varying magnetic fields, as opposed to processes tied to specific fundamental energies such as the electron rest mass (e.g., by photon-photon or photon-magnetic field pair opacity[13]), or an atomic or nuclear line.

CORRELATIONS

If bursts occur at cosmological distance, relativistic effects should produce correlations among the hardness, duration, fluence and peak photon flux. We would thus expect harder bursts to be shorter, brighter and more energetic. However, we did not find the expected correlations between the hardness and the fluence, duration or peak flux in our sample. Unfortunately the broad distribution of intrinsic burst properties may obscure the cosmological signature. Moreover, our burst sample is not statistically complete, and may suffer selection effects.[14] In addition, the cosmological signature may be fairly weak since the sample includes the strong, predominantly nearby burst sources. Thus the absence of a cosmological signature is not conclusive.

We searched for differences in the hardness distributions of a number of spatial subsamples. The subsamples were: Galactic plane—$|b^{II}| < 30°|$ vs. $|b^{II}| > 30°|$; Galactic hemispheres—$|b^{II} > 0°|$ vs. $|b^{II} < 0°|$; equatorial hemispheres—declination greater than vs. less than 0°; Galactic Center—$l^{II} < 90°$ or $l^{II} > 270°$ vs. $90° < l^{II} < 270°$; and right ascension—less than vs. greater than 12^h. In no case were the differences within any set of spatial subsamples significant by the Kolmogorov-Smirnov test.

Unfortunately our sample consists predominantly of complex, multipeak bursts, and the other burst morphologies are not sampled sufficiently. Of the 11 bursts with different morphologies, 4 have steep high energy spectra, but more definitive conclusions await a larger, more complete sample.

ACKNOWLEDGEMENTS

We thank D. Gruber, R. Lingenfelter, and R. Rothschild for stimulating discussions on GRB spectra. This work was supported in part by NASA contract NAS8-36081 (UCSD group).

REFERENCES

1. J. P. Norris, et al., Ap. J. **301**, 213 (1986); D. L. Band, et al., in Gamma-Ray Bursts, AIP Conference Proceedings 265, eds. W. S. Paciesas and G. J. Fishman (AIP, New York, 1992), p. 169.
2. K. Hurley, Cosmic Gamma Rays, Neutrinos, and Related Astrophysics, NATO ASI Series C, vol. 270, eds. M. M. Shapiro and J. P. Wefel (Kluwer Academic Publishers, Dordrecht, 1989), p. 337; J. C. Higdon and R. E. Lingenfelter, Ann. Rev. Astron. Astrophys. **28**, p. 401 (1990).
3. T. L. Cline, et al., Ap. J. Lett. **185**, L1 (1973); T. L. Cline and U. D. Desai, Ap. J. Lett. **196**, L43 (1975).
4. E. P. Mazets, et al., Astrophys. Space Sci. **82**, 261 (1982).
5. S. M. Matz, et al., Ap. J. Lett. **288**, L37 (1985).
6. E. J. Schneid, et al., Astron. Astrophys. **255**, L13 (1992); C. Winkler, et al., Astron. Astrophys. **255**, L9 (1992).
7. D. L. Band, et al., in preparation for submission to Ap. J. (1993).
8. G. J. Fishman, et al., Proceedings of the Gamma Ray Observatory Science Workshop (1989), p. 2-39, 3-47.
9. G. N. Pendleton, et al., Proceedings of the Gamma Ray Observatory Science Workshop(1989), p. 4-547; G. N. Pendleton, et al., in Gamma-Ray Bursts, AIP Conference Proceedings 265, eds. W. S. Paciesas and G. J. Fishman (AIP, New York, 1992), p. 395.
10. B. E. Schaefer, BATSE Spectral Analysis Software: User's Guide (Internal Memo, 1991).
11. A. K. Harding, Physics Reports **206**, 327 (1991).
12. D. L. Band, et al., Experimental Astronomy, submitted (1992).
13. B. E. Schaefer, et al., Ap. J. Lett. **393**, L51 (1992).
14. B. E. Schaefer, submitted to Ap. J. Lett. (1993).

A SEARCH FOR UNTRIGGERED LOW-ENERGY EVENTS IN THE BATSE DATA BASE

Jan van Paradijs
Astronomical Institute "Anton Pannekoek" & CHEAF, Amsterdam

Bradley C. Rubin, Chryssa Kouveliotou
Universities Space Research Association, Huntsville, AL 35812

John M. Horack, Gerald J. Fishman, Charles A. Meegan, Robert B. Wilson
NASA/MSFC, Huntsville, AL 35812

William S. Paciesas
University of Alabama in Huntsville, AL 35899

Walter H. G. Lewin
Center for Space Research, MIT 37-627, Cambridge, MA 02139

Michiel van der Klis
Astronomical Institute "Anton Pannekoek" & CHEAF, Amsterdam

ABSTRACT

We have started a search in the BATSE data base for low-energy ($\sim 25 - 55$ keV) events, which cannot (under the current configuration) cause an onboard trigger. We present here some preliminary results, in particular the detection of several very weak events temporally clustered in a period when the experiment was triggered by emission attributed to the Soft Gamma-ray Repeater SGR 1900+14 (Mazets et al.[1]). We suggest that these events most likely originate from the same source.

INTRODUCTION

BATSE consists of eight large-area detectors (LADs) of ~ 2000 cm^2 each, and eight smaller spectroscopic detectors. The combined surfaces of the LADs form a regular octahedron to provide all-sky monitoring in the 25 keV – 1 Mev photon energy range. On-board processing of the data stream leads to a trigger when the signal in at least two detectors increases by 5.5 σ or more simultaneously, in the 55–320 keV energy range. The rates are measured and tested independently against the background during time intervals of 64, 256, and 1024 ms. After a trigger, a burst acquisition mode starts in which high temporal and spectral resolution data are collected. A detailed description of the experiment has been given by Fishman et al.[2].

Soft-spectrum transient events which occur mainly in the low-energy channel (25–55 keV) of the LADs cannot lead to a burst trigger. These events may comprise known astrophysical phenomena, e.g. solar flares, magnetospheric disturbances, or soft gamma repeaters (see Norris et al.[3] for a recent review). They may also include new hitherto undetected phenomena. We have started a project to search the BATSE LAD low-energy data base for such transient events; we describe here our search strategy, and present some preliminary results.

© 1993 American Institute of Physics

THE SEARCH STRATEGY

Several types of data are continuously recorded with BATSE. For this project we used the discriminator data recorded in four energy channels (25 – 55 keV, 55 – 120 keV, 120 – 320 keV, and 320 keV – 1 MeV) with a time resolution of 1.024 s, which we occasionally rebinned into 4 s bins. Our search technique simulates the on-board trigger procedure; we calculate a running-mean value of the count rate over 17 time bins ("background") and its r.m.s. scatter σ. An offline "trigger" occurs when the signal in a particular time bin exceeds the background by a preselected number (n_{tr}) times σ, in at least two of the eight detectors simultaneously. Any combination of the time resolution, the value of n_{tr} and the energy channel to be searched, can be selected. So far we have used only the 25 – 55 keV data (i.e., channel 1). The number of offline triggers depends strongly on the selected parameter values, and on the activity of the Sun and a limited number of bright hard X-ray sources, in particular Cyg X-1. For a 4.5 σ trigger level the number of triggers is typically a few dozen per day, but values as high as a hundred per day have been occasionally measured.

As a second step, we visually inspected the channel-1 intensity curves of the eight detectors, which are daily plotted with 8.192 s time resolution as part of the BATSE standard data analysis. We were thus able to reject offline triggers for which there was an obvious cause that was not interesting in the context of our search. We mention here solar flares, the steep rise above the Earth's horizon of strong sources, triggered or untriggered gamma-ray bursts, magnetospheric disturbances, and intense fluctuations of bright hard X-ray sources, particularly Cyg X-1 (see Meegan et al.[4]). This step turned out to be a very efficient filter: it left an average of only about one trigger per day that could not be explained by immediately obvious causes.

As a final step we calculated the locations on the sky of these "surviving" events from their relative intensities as registered in the different BATSE detectors (Brock et al.[5]). This led to a further reduction of the number of unexplained events, e.g., by the rejection of events that originated from within $\sim$ 15 degrees from the Sun.

PRELIMINARY RESULTS

We have applied the above search procedure to four time intervals of approximately 30 days each varying the input parameters, to determine the best combination of time resolution and trigger levels. The first two intervals, which occur early in the mission, are TJD 8392–8421 and TJD 8438–8471, where TJD (Truncated Julian Day) = JD - 2,440,000.5. This search was made with time resolutions of 1.024 and 4.096 s, and a trigger level of 4.5 σ.

During these 64 days we detected a total of 16 events which did not lead to an on-board trigger. Twelve of these are gamma-ray bursts which were also detected in the channel 2 and 3 data; some of these events are described in some detail by Rubin et al.[6]. Most of these gamma-ray bursts did not lead to an on-board trigger because they occurred during the readout of a previously triggered event which was stronger. In two cases, the events were too weak to allow a unique solution for the location; the difference between the rates of the remaining 6 detectors was very small and barely above the background. Thus the selection of the brightest 4 detectors viewed by the "source" became ambiguous and the location could not be determined. The remaining two events were not detected in channels 3 and 4, i.e., above 120 keV. They both lasted ~10 s and they come from the same general region of the sky with $(\alpha, \delta) = (312,+54)$ and $(306,+66)$, respectively.

WEAK EVENTS FROM SGR 1900+14?

After the discovery (Kouveliotou et al.[7,8]) of repeated on-board triggered events from the Soft Gamma Repeater SGR1900+14 we searched two time intervals of BATSE data for possible fainter (non-triggering) events from this source. The intervals searched are TJD 8787–8821, and TJD 8722–8751. The first interval covers the period just after the source activity was discovered (the three triggered events occurred on TJD 8792, 8811, and 8853). The second interval is ~ 2 months prior to this. In view of the short duration of SGR events (typically 0.1 s), the search was executed on channel-1 data with a time resolution of 1 s only. We used a 3.5 σ offline trigger level, because we found that in channel 1 the three triggered events gave rise to a barely 3.5 σ signal when integrated over 1 s.

A summary of this search is given in Fig. 1, which shows, separately for the two time intervals, the sky distribution of the detected events which lasted 1 s or less (i.e., the count rate increase occurred in only 1 time bin). It is clear from the figure that during the period when SGR1900+14 events led to on-board triggers there is an excess of brief low-energy events from the general area of the sky where SGR 1900+14 is located. Most of these events triggered at the 3.5–3.8 σ level. Their fluence is typically $\sim 10^{-8}$ ergs cm^{-2}; if the durations of these events are similar to those of the events from SGR1900+14 this corresponds to a peak flux of $\sim 5 \times 10^{-7}$ ergs cm^{-2} s^{-1}. These values are a factor of ~ 5 lower than the corresponding values for the events recently reported by Kouveliotou et al.[7,8] and a factor $\sim 10^2$ below those of the events detected with KONUS (Mazets et al.[1]).

We have estimated a location for a possible single source of these excess events by taking the average of the (unit) direction vectors of the events in the upper-right quadrant in Fig. 1b. This average is at $\alpha = 297°$, $\delta = +36°$. The statistical accuracy of this location is $\sim 10°$, as estimated by dividing the r.m.s. scatter of the events ($\sim 30°$) by the square root of their number. (If we would have taken all points in Fig. 1b the average direction would have shifted by less than 3°).

This average direction is offset from the location of SGR 1900+14 (Mazets et al.[1]) by $\sim 24°$. Can this be considered a convincing argument against the assumption that the weak untriggered events originate from SGR 1900+14? We think not: apart from the similarity of these events to the triggered events from this source (with respect to their short duration, their soft spectra, and the times of their occurrence), which strongly argues for a common origin, there is also evidence that the locations of the untriggered events may contain a substantial systematic error. This is caused by the fact that all these events are brightest in one particular BATSE detector (no. 7) whose low-energy (channel 1) response appears to differ from that of the other detectors but has not yet been fully calibrated (Pendleton[9]). (Note that this effect is not present in the locations of events that are based on channel 2 and 3 data, such as gamma-ray bursts and the on-board triggered events from SGR 1900+14). The resulting systematic error on the souce location would be expected to depend on the satellite orientation,

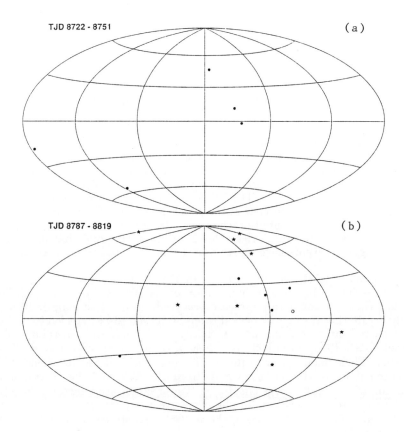

Figure 1. a) Sky-map (in galactic coordinates) of the brief low-energy events detected in the interval TJD 8722–8751. b) Same as Figure 1a for the interval TJD 8787–8819. Different symbols indicate different satellite orientations as follows: filled circles TJD 8787–8798; open circles TJD 8798–8805; asterisks TJD 8805–8819.

i.e., to vary with the pointing periods of GRO. The sky map of the events shown in Fig. 1b suggests that such a dependence of the event locations on the pointing period may actually be present, although the number of events seems too small for firm statements.

We note that the average direction of the untriggered events is close to that of Cyg X-1 ($\alpha = 299°$, $\delta = +35°$). We cannot on a priori grounds exclude Cyg X-1 as the source of the brief low-energy events. If the events last of order of 1 s, then their peak flux of $\sim 10^{-8}$ ergs cm^{-2} s^{-1} is within the range of the well known shot noise in the 1–50 keV X-rays from Cyg X-1. These shots have a soft spectrum (Negoro et al.[10]). All emission from Cyg X-1 detected so far with BATSE in the > 25 keV region, be it steady emission observed via Earth occultation intensity steps, or rapid fluctuations that give rise to on-board triggers, has a much harder spectrum than that of the untriggered events discussed here. If these untriggered events would nonetheless originate from Cyg X-1 our results would indicate that rapid increases in the flux of Cyg X-1 have quite variable spectral properties.

ACKNOWLEDGEMENTS

JvP thanks the BATSE group at the Marshall Space Flight Center where most of this work was done, and the University of Alabama at Huntsville, for their hospitality. He acknowledges financial support from the Leids Kerkhoven Bosscha Fonds. WHGL acknowledges support from NASA under contract NAG 5 - 2030. This work was supported in part by the Netherlands Organization for Scientific Research (NWO) through grand PGS 78 - 277 to MvdK.

REFERENCES

1. E. P. Mazets, S. V. Golenetskii and Yu. A. Gur'yan, Sov. Astron. Lett. **5**, 343 (1979).
2. G. J. Fishman, et al., Proc. GRO Science Workshop (ed. W. N. Johnson, NASA/Goddard Space Flight Center, 1989), p. 39.
3. J. P. Norris et al., Ap. J. **366**, 240 (1991).
4. C. A. Meegan et al., these proceedings (1993).
5. M. Brock et al., Gamma Ray Bursts: Huntsville, AL 1991, eds W. S. Paciesas & G. J. Fishman (AIP Conference Proceedings 265, 1992), p. 383.
6. B. C. Rubin et al., these proceedings (1993).
7. C. Kouveliotou et al., IAU Circular 5567 (1992).
8. C. Kouveliotou et al., these proceedings (1993).
9. G. Pendleton, private communication (1992).
10. H. Negoro et al., Frontiers of X-Ray Astronomy (Nagoya Proceedings, 1992), p. 313.

BATSE OBSERVATIONS OF A SOFT GAMMA REPEATER (SGR)

C. Kouveliotou
Universities Space Research Association, Huntsville, AL 35812

G. J. Fishman, C. A. Meegan, R. B. Wilson, R. D. Preece[†], J. M. Horack
NASA/MSFC, Huntsville, AL 35812

W. S. Paciesas, M. S. Briggs, T. M. Koshut
University of Alabama in Huntsville, AL 35899

J. van Paradijs
Astronomical Institute "Anton Pannekoek" & CHEAF, Amsterdam

ABSTRACT

We have detected repeated soft gamma-ray emission from a region centered on $\alpha = 290°$ and $\delta = 11°$ with an average error radius of $\sim 5°$. The source has produced three intense outbursts that have triggered BATSE on June 19, July 8 and August 19, 1992, respectively. The emissions are very short (of the order of 0.1 s) and their spectra are among the softest of all cosmic events detected by BATSE; they are consistent with a typical SGR spectrum with an upper energy cut off of ~ 120 keV. Our results indicate that SGR 1900+14[1] has become active again.

INTRODUCTION

The three SGRs so far known are characterized by repetitive intense bursts in the energy range from 25–100 keV. With one exception (the famous March 5, 1979 initial outburst of SGR 0526-66[2]) all emissions from these sources have unique and rather uniform characteristics[3]. They exhibit simple light-curves with very short durations (typically ~ 0.1 s) and fast (< 5 ms) rise times[4]. Their spectra are much softer than a classical Gamma Ray Burst (GRB) spectrum[5] with no evidence of spectral evolution[4,6]; they can be well fitted with an optically thin thermal bremsstrahlung (OTTB) function with temperatures between 30–40 keV[4,7]. The total numbers of events observed from SGR 0526-66, SGR 1900+14, and SGR 1806-20 were 16, 3, and over 100 bursts, respectively, with time intervals between emissions that vary stochastically from hours to years.

All three SGRs were discovered with the KONUS experiment onboard the Venera 11–14 spacecraft[1,7] and with the first and the second Interplanetary Network (IPN) of satellites [4,8,9,10]. The Burst and Transient Source Experiment (BATSE) onboard the Compton Gamma Ray Observatory (CGRO) is unique in combining unprecedented sensitivity to weak gamma and SGR-like bursts and the capability of determining their locations in a single experiment. A detailed description of BATSE is given elsewhere[11]. The experiment consists of 8 identical Large Area Detectors (LADs) of 2000 cm^2 each, pairwise arranged with 8

[†] NAS/NRC Resident Research Associate

smaller Spectroscopy Detectors. The surfaces of the LADs form a regular octahedron to provide nearly all-sky monitoring in the 20 keV–2 MeV energy range. On-board processing of the count rates collected by the 8 LADs leads to a trigger when the signals in the 50–300 keV energy range increase by more than 5.5 σ simultaneously in at least two detectors. The rates are tested independently against the background in 64, 256, and 1024 ms intervals. Burst locations are obtained from the relative strength of their signals in different detectors. The position determination includes the detailed spectral dependence of the angular sensitivity of the detectors, and corrections for photon back-scattering by the Earth's atmosphere[12].

I. TEMPORAL ANALYSIS

The first SGR event detected with BATSE (Fig 1a,b) lasted ~ 0.6 s and consisted of two pulses separated by ~ 0.5 s. The first pulse (Fig 1a) has a total duration of 40 ms and a trapezoidal shape. It initially rises gradually for ~ 10 ms and then increases sharply in ~ 0.5 ms to a plateau lasting ~ 20 ms; it then decays to near the background level in less than 5 ms with some residual weak emission continuing above background for the following 30 ms. These rise and decay times are the shortest ever resolved for an SGR event (they are not limited by the instrument resolution of 0.5 μs but by the counting statistics). The second pulse (Fig 1b) lasts also 40 ms and has a triangular shape. It rises gradually, in 15 ms, and then decays to the background level in 25 ms. Neither of the two pulses shows significant substructure. The second SGR burst (Fig 1c) lasts ~ 500 ms with most of the emission concentrated in an unresolved initial spike lasting less than 64 ms. The third SGR event (Fig 1d) lasts ~ 80 ms and has a trapezoidal (flat-topped) shape with a plateau lasting ~ 36 ms. It has a fast rise of < 5 ms and a more gradual decay of ~ 40 ms.

II. SPECTRAL ANALYSIS

We have sufficient data for deriving a spectrum only for the second pulse of the first event. Figure 2 shows the best fit of an optically-thin thermal bremsstrahlung function of the form $A\exp(-E/kT)/E$, to the time-integrated spectrum of this pulse. The gap in the spectrum between ~ 20–40 keV is due to a telemetry outage during transmission of the data. We find for the best fit temperature $kT = 39 \pm 3$ keV($\chi^2 = 10.7$ for 11 d.o.f.). A power-law fit results in a spectral index of -3.0 ± 0.1 but has a significantly worse χ^2 (29.8 for 12 d.o.f.). Spectral fits for two independent 16 ms time bins show no spectral evolution during the pulse.

Some spectral information for the remaining three pulses can be obtained from their hardness ratios R_{21} and R_{32}, i.e., the ratios of their background-corrected count rates in the 50–100 to 20–50 keV band and 100–300 to 50–100 keV band, respectively (see Table 1). The hardness ratios of all four pulses are the same within statistics (with the first pulse of the 19 June event being slightly harder). The ratio R_{21} is much larger than R_{32}; this is exactly opposite in the case of the classical GRBs[13]. In addition, the time histories (in 1 ms time bins) of the hardness ratios of three of the pulses (Fig 1a, b and d) do not reveal a significant spectral change during any of the events.

The fluences of the four pulses range from $(4.5$–$6.6)\times 10^{-8}$ ergs/cm^2 (Table 1). These values are ~ 5 times weaker than those of the weakest event from SGR

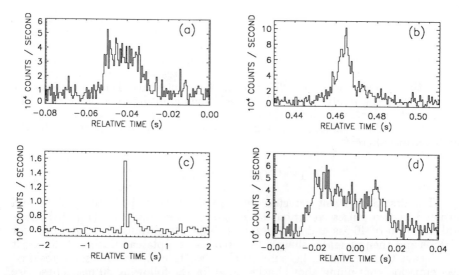

Figure 1. Light curves of the three SGR bursts detected with BATSE. a) and b) First and second pulse, respectively, of the June 19, 1992 trigger. The counts are integrated between 20 and 100 keV. Time resolution is 0.512 msec. c) Event of July 8, 1992 recorded with 64 msec time resolution (the only available). The initial spike of < 64 msec appears only up to 120 keV. d) Event of August 19, 1992 recorded with 0.512 msec between 20 and 100 keV. Notice the similarity in structure between 1a and 1d.

1900+14 previously reported[1].

Table 1. Summary of SGR 1900+14 temporal and spectral properties

Date	UT (sec)	RA (°)	Dec (°)	Duration (ms)	Hardness Ratios R_{21}	R_{32}	Fluence (10^{-8} erg cm^{-2})
920619a	64666	288	11	40	0.94 ± 0.02	0.16 ± 0.010	4.5 ± 0.03
920619b				40	0.64 ± 0.01	0.15 ± 0.010	4.3 ± 0.06
920708	18957	286	10	< 64	0.72 ± 0.02	0.14 ± 0.010	3.0 ± 0.04
920819	17589	297	13	80	0.64 ± 0.01	0.13 ± 0.004	6.6 ± 0.08

III. LOCATIONS

We have used the spectral index of -3.0 as an initial estimate of the input power-law spectrum required to compute the locations of the events. All three locations are consistent with a single positional error circle in the sky centered at $\alpha = 290°$, $\delta = 11°$ with an error radius of $\sim 5°$. The size of this circle reflects mainly the scatter of the three measured positions (the statistical error of each position is of the order of 8°). Unfortunately, no other spacecraft detected these

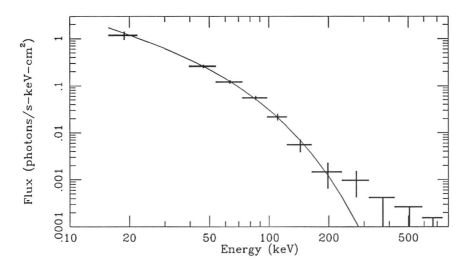

Figure 2. Photon spectrum of the pulse of Fig 1b. The spectrum is integrated over 32 msec and corresponds to the peak of the pulse.

emissions, which precludes the possibility of obtaining a much more accurate location from triangulation[14]. This error box is consistent with the location of SGR 1900+14, which was obtained by a combination of triangulation and cosine response of the KONUS experiments; three bursts were detected from this source in the period 24–27 March, 1979[1]. In view of the evidence of temporal and spectral similarity between the BATSE events and the previously reported SGR 1900+14 emissions, and the positional coincidence of the recent bursts with this source location, we conclude that the BATSE events originate from SGR 1900+14.

A search of the 1 s time resolution BATSE data for additional, untriggered emissions from this source shows that during the intervals between the SGR triggers there is an excess of short (< 1 s), low-energy events from the general region of the sky near SGR 1900+14[15]. This indicates that the three triggered events are accompanied by a possibly much larger number of weaker (and softer) events.

Finally, we would like to point out an intriguing temporal and spatial coincidence: the newly discovered Aquila x-ray transient GRS 1915+105[16] is located at $\alpha = 288.2°$, $\delta = 10.9°$[17] and thus included in the BATSE error box of SGR 1900+14. In addition, inspection of the BATSE flux history of the source, indicates that the transient started rising one day before the first SGR trigger and it remained at the maximum during the subsequent two triggers. Are these SGR-like emissions related to the x-ray transient? The positions for both SGR 1900+14 and GRS 1915+105 are known with high accuracy (of the order of arcmin) and they differ by more than $\sim 2°$. This result alone excludes that the KONUS events could be related to the Aquila transient. Moreover, recent results[18] indicate that there was no persistent x-ray emission from either region in the sky (SGR 1900+14, GRS 1915+105) during the 1979 SGR bursts.

We believe that an origin of the BATSE events from GRS 1915+105 is a most improbable option.

DISCUSSION

In conclusion, our results indicate that the "old" SGR 1900+14 has been detected again ~13 years after its discovery. We observed the fastest rise times (< 1 ms) ever seen from an SGR emission and confirmed the absence of spectral evolution on time scales longer than 1 msec. The source reactivation stregthens the argument[3,4] that the SGR sources are related to galactic (Population I) objects, quite plausibly neutron stars. This is very different from recent results on classical gamma ray bursts, for which a galactic origin is excluded[19]. Recurrent SGR emissions do not signify a unique (catastrophic) event in the life cycle of the source, as is the case in the cosmological models currently favoured for the classical GRBs[20,21]. The long term monitoring of the source by BATSE gives hope of obtaining valuable information on its recurrence timescale and (combined with other spacecraft) an improved source position, which may lead to a better understanding of the nature of these objects.

ACKNOWLEDGEMENTS

JvP thanks the BATSE group at the Marshall Space Flight Center and the University of Alabama in Huntsville, for their hospitality. He acknowledges financial support from the Leids Kerkhoven Bosscha Fonds.

REFERENCES

1. E. P. Mazets et al., Sov. Astron. Lett. **5**, 343 (1979).
2. T. L. Cline, Ann. NY Acad. Sci. **375**, 314 (1981).
3. J. P. Norris et al., Ap. J. **366**, 240 (1991).
4. C. Kouveliotou et al., Ap. J. Lett. **322**, L21 (1987).
5. D. Band et al., submitted to Ap. J (1993).
6. S. V. Golenetskii et al., Nature **307**, 41 (1984).
7. E. P. Mazets et al., Ap. & S. S. **80**, 1 (1981).
8. K. Hurley, talk presented at the Taos Gamma-Ray Stars Conference (1986).
9. J.-L. Atteia et al., Ap. J. Lett **320**, L105 (1987).
10. J. G. Laros et al., Ap. J. Lett. **320**, L111 (1987).
11. G. J. Fishman et al., Proc. GRO Science Workshop, ed. W. N. Johnson (NASA/Goddard Space Flight Center, 1989), p. 39.
12. M. Brock et al., Gamma Ray Bursts: Huntsville, AL 1991, eds. W. S. Paciesas & G. J. Fishman (AIP Conference Proceedings 265, 1992), p. 383.
13. C. Kouveliotou et al., The Compton Observatory Science Workshop, ed. C. R. Shrader, N. Gehrels & B. Dennis (NASA CP-3137, 1991), p. 61.
14. K. Hurley, private communication (1992).
15. J. van Paradijs et al., these proceedings (1993).
16. A. J. Castro-Tirado et al., IAU Circular No. 5590 (1992).
17. B. Cordier et al., these proceedings (1993).
18. J. Lochner & L. Whitlock, IAU Circular No. 5658 (1992).
19. C. A. Meegan et al., Nature **355**, 143 (1992).
20. B. Paczynski, Acta Astr. **41**, 157 (1991).
21. B. Paczynski, Acta Astr. **41**, 257 (1991).

OBSERVABILITY OF LINE FEATURES IN BATSE SPECTROSCOPY DETECTOR OBSERVATIONS OF GAMMA RAY BURSTS

L.A. Ford, D.L. Band, J.L. Matteson
CASS, UC San Diego, La Jolla, CA 92093

D.M. Palmer, B.E. Schaefer, B.J. Teegarden, T.L. Cline
Goddard Space Flight Center, Greenbelt, MD 20771

R.D. Preece, G.J. Fishman, C.A. Meegan, R.B. Wilson
Marshal Space Flight Center, Huntsville, AL 35812

M.S. Briggs, W.S. Paciesas, G.N. Pendleton
University of Alabama at Huntsville, Huntsville, AL 35899

ABSTRACT

We have performed a series of numerical simulations and developed an analytic model of how the significance of absorption features as seen by the BATSE Spectroscopy Detectors (SDs) varies as the continuum and line parameters are changed. Both the simulations and model predict that the significance of lines depends approximately on the square of the signal-to-noise, the square of the equivalent width of the line, and the inverse of the natural width of the line.

INTRODUCTION

A great deal of effort has gone into the search for spectral features in gamma ray burst data returned by the BATSE Spectroscopy Detectors (SDs). Absorption features in the 20-70 keV range have been observed[1-3] and were interpreted as cyclotron lines, implying a neutron star origin for bursts. To this point, however, significant lines have not been found in the BATSE data[4,5] which prompts the question of how sensitive the BATSE SDs are to lines of given strength and width. To answer this question, we have performed a series of numerical simulations to investigate how the statistical significance of lines varies as the signal-to-noise (S/N), the equivalent width, and the natural width of the lines vary. Guided by these simulations, we constructed a model to predict the significance of a line given the aforementioned parameters.

In both the numerical and analytic work, we created simulated spectra containing lines. Significance was determined by comparing the quality of fits of a continuum model and a continuum with a line to the simulated spectrum. This comparison yielded a probability that the line was random fluctuation as determined by the F-test, which compares the difference in χ^2 between the models[6]. For numerical simulations, finding $\Delta\chi^2$ was straightforward. In analytic studies however, noise must be modeled. To examine the effect of noise on χ^2, consider a spectrum y_{d_i} which can be expressed as some smooth function y_{s_i} with noise σ_i added to it: $y_{d_i} = y_{s_i} \pm \sigma_i$. Then writing out χ^2 for some model y_{m_i},

$$\chi^2 = \sum_{i=1}^{n} \frac{1}{\sigma_i^2} \left[y_{m_i}^2 + y_{s_i}^2 - 2y_{m_i} y_{s_i} + \sigma_i^2 \pm 2y_{s_i}\sigma_i \mp 2y_m \sigma_i \right]. \qquad (1)$$

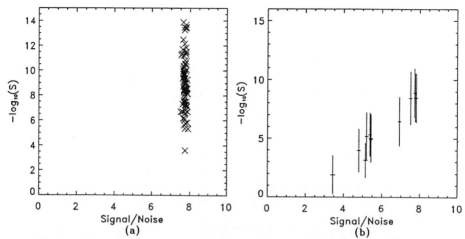

Figure 1. Results of the simulations. Frame (a) shows the significance of lines for the 100 spectra viewed at 3.6°, 1 second accumulation time, $EW = -9.6$ keV, and $\Delta E_l = 9.6$ keV. Frame (b) shows median values for all spectral sets with $EW = -9.6$ keV and $\Delta E_l = 9.6$ keV. The individual points represent different viewing angles and accumulation times. The error bars are the standard deviations about the medians.

The first three terms in the sum represent what would be seen if there were no noise in the data. Assuming the last two terms cancel for large n, noise adds n to the total χ^2 of a given model. For fits of continuum models, χ^2 will, on average, be the first three terms in equation (1) plus n. If the continuum plus line model is the same function used for the simulated spectrum, then its χ^2 will be n since $y_{m_i} = y_{s_i}$. Therefore, $\Delta\chi^2$ between the two models is just χ^2 of a continuum fitted to a noiseless simulated spectrum.

NUMERICAL SIMULATIONS

To generate a data set appropriate for model testing, we simulated bursts by creating a photon spectrum and folding it through real response matrices[7]. The photon spectrum consisted of a power law continuum of index -0.9 and a pair of Gaussian lines at 20 and 40 keV. From these and earlier simulations, we found that the sensitivity (as measured by significance) of the SDs to lines at 20 keV is at least an order of magnitude worse than for lines at 40 keV (see also Palmer et al.[5]). This is because absorption by the detector housing at low energies reduces the effective area of the detector. To avoid this issue, we limit our discussion here to the 40 keV line, which was isolated by fitting the spectrum from 25 to 100 keV. The continuum was chosen so that the count rate was ten times a typical background rate at 50 keV. The equivalent width (EW) of the 40 keV line was varied from -2.4 to -14.4 keV and its natural width ($\Delta E_l = 0.6 \times$FWHM) ranged from 2 to 15 keV. Spectra were accumulated for 0.5 and 1 seconds with no dead time and the viewing angles relative to the detector normal were 3.6°, 17.0°, 32.2°, 44.2°, 62.2°, and 80.2°. For every

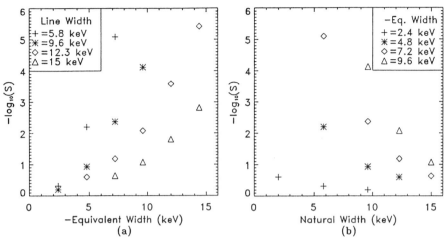

Figure 2. Variation in median significance from numerical simulations as a function of (a) equivalent width and (b) line width. $S/N = 5$ for both figures.

combination of angle, accumulation time, and line parameters, 100 spectra were made with Poisson-distributed noise. The significance of the line was evaluated by comparing the fit of a power law continuum to that of a continuum with a three parameter Gaussian line.

The significances were plotted as a function of S/N for each set of parameters and all showed large significance variations (see figure 1a). The definition of S/N used here was the arithmetic average of the ratio of signal to standard deviation of the data in 1 keV bins for the region between 25 and 50 keV. We assumed variations in the viewing angle affected only S/N and plotted together the medians of spectra with the same line parameters (figure 1b). These plots showed a correlation between S/N and significance although the rms variance allowed a range of likely significances of about four orders of magnitude. Because of this large spread, more detailed models of line detectability will require the distribution of significance. For the moment though, we are only interested in general proportionalities. To find the dependence of significance on line parameters, a linear fit was made to the data in figure 1b and the value of the significance at $S/N = 5$ was plotted against both EW and ΔE_l (figure 2). These figures suggest that the logarithm of the line significance varies as the square of EW and is inversely proportional to ΔE_l. This clear correspondence suggests that an analytic model can be made which can predict the significance of lines observed by the SDs.

AN ANALYTIC MODEL FOR $\Delta \chi^2$

Observed spectra depend not only on conditions at the source, but are also affected by the instrument performing the measurement. An analytic model must include effects such as smearing of the signal by the SD's NaI(Tl) crystal, binning of the data, and Compton downscatter filling of the line. We take these effects into account in our model and determine $\Delta \chi^2$ in count space. The signal was assumed to be a power law continuum and a Gaussian line. The line was broadened by convolving it with a Gaussian of width equal to the detector resolution at the line center. Since we considered strong bursts, the background was negligible.

Mathematically, the spectrum in count space can be expressed as

$$y_s = y_c \left(\frac{E}{E_0}\right)^\alpha + \frac{y_c EW_{\text{eff}}}{\sqrt{\pi}\sqrt{\Delta E_s^2 + \Delta E_l^2}} \left(\frac{E_l}{E_0}\right)^\alpha \exp\left[-\frac{(E-E_l)}{\sqrt{\Delta E_s^2 + \Delta E_l^2}}\right]^2, \quad (2)$$

where $E_0 = 50$ keV, E_l is the line energy and ΔE_s is the detector resolution. Downscattering changes the effective equivalent width of the line although this depends both on the location of the line and its natural width. Since the effect of downscattering was difficult to model analytically, it was determined empirically by examining the spectra of the numerical simulations (without added noise) and finding EW in count space. Comparing the observed EW to the known input value yielded an effective equivalent width formula at 40 keV,

$$EW_{\text{eff}} = \frac{EW_{\text{true}}}{(1.24 + 0.0176\Delta E_l)}. \quad (3)$$

The value of y_{s_i} for a binned spectrum is y_s averaged over the energy of the bin, which corresponds to the noiseless simulated spectrum used in equation (1). The continuum model y_{m_i} fit to y_{s_i} is a power law of index α but with normalization to be determined by minimization of χ^2. Performing the averaging and applying the result of equation (1),

$$\Delta\chi^2 = \sum_{i=1}^{r} \left(\frac{S}{N}\right)_i^2 \frac{1}{\Delta E_{\text{bin}}^2} \left\{ \frac{E_0}{\alpha+1}\left(1 - \frac{y_f}{y_c}\right)\left[\left(\frac{E_{\text{hb}_i}}{E_0}\right)^{\alpha+1} - \left(\frac{E_{\text{lb}_i}}{E_0}\right)^{\alpha+1}\right] - \frac{EW_{\text{eff}}}{2}\left(\frac{E_l}{E_0}\right)^\alpha \text{nerf}\left(\frac{E_{\text{lb}_i} - E_l}{\sqrt{\Delta E_s^2 + \Delta E_l^2}}, \frac{E_{\text{hb}_i} - E_l}{\sqrt{\Delta E_s^2 + \Delta E_l^2}}\right) \right\}^2. \quad (4)$$

where $\text{nerf}(a,b) \equiv \text{erf}(b) - \text{erf}(a)$, E_{hb_i} and E_{lb_i} refer to the upper and lower energy boundries of bin i, and ΔE_{bin} is the bin width. The normalization of the continuum fit (y_f) is found by minimizing equation (4) with respect to y_f. Since the logarithm of the significance is nearly proportional to $\Delta\chi^2$, this quantity varies as the square of both S/N and EW and noting that for small values of x, $\text{erf}(x) \approx x$, the dependence on natural width is essentially $1/\Delta E_l$.

The predictions of this model for the parameters of the numerical simulations are plotted in figure 3. The model curves show the same dependencies as the numerical results but deviations indicate that the model does not include all detector effects. The cause of this discrepancy is probably the assumption that viewing bursts at various angles affected only S/N. Hua & Lingenfelter[8] have shown that scattering off the BATSE structure (which is angle dependent) makes an important contribution to the signal, some of which was probably included in the empirical correction for downscattering. Making a similar correction but including angular variation should produce an improved model. Despite the small disagreement with the numerical results, the analytic model of $\Delta\chi^2$ does exhibit the right dependencies on S/N and line parameters and can therefore be used to understand the minimum requirements for line observability.

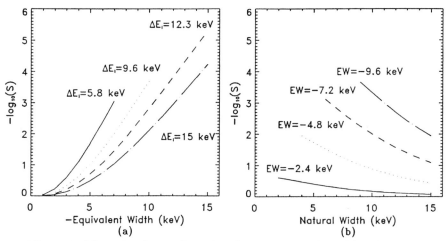

Figure 3. Variation in median significance as predicted by the analytic model as a function of (a) equivalent width and (b) line width. $S/N = 5$ for both figures.

CONCLUSION

We have presented a model of how the statistical significance of lines in gamma ray burst spectra varies as the continuum and line parameters change. This model shows that the relevant quantities are the equivalent width and natural width of the line, and the signal-to-noise of the spectrum in the region around the line. Since the model is analytic, it can be applied to map out the parameter space (equivalent width and natural width) of lines in terms of significance which can be used to determine the observability of a given line for an observed spectrum.

BATSE research at UCSD is supported by NASA contract NAS8-36081.

REFERENCES

1. E.P. Mazets, et al., Nature **290**, 378-382 (1981).
2. G.J. Hueter, Ph.D. Thesis, UCSD (1988).
3. T. Murikami, et al., Nature **335**, 335-336 (1988).
4. B.J. Teegarden, et al., these proceedings (1992).
5. D.M. Palmer, et al., these proceedings (1992).
6. B.R. Martin, Statistics for Physicists (Academic Press, London, 1971).
7. G.N. Pendleton, et al., Proceedings of the Gamma Ray Observatory Science Workshop (1989), p. 4-547.
8. X.-M. Hua, & R.E. Lingenfelter, these proceedings (1992).

THE SENSITIVITY OF THE BATSE SPECTROSCOPY DETECTORS TO CYCLOTRON LINES IN GAMMA-RAY BURSTS

D. M. Palmer* , B. Teegarden, B. Schaefer, T. Cline
Goddard SFC, Greenbelt, MD 20771

G. Fishman, C. Meegan, R. Wilson
Marshall SFC, Huntsville, AL 35812

W. Paciesas, G. Pendleton, M. Briggs
UA–Huntsville, Huntsville, AL 35899

D. Band, L. Ford, J. Matteson
CASS, UC San Diego, La Jolla, CA 92093

J. P. Lestrade
Mississippi State University, MS 39762

Abstract

As of October 16, 1992, we have examined 72 gamma-ray bursts (GRBs) using the BATSE Spectroscopy Detectors (SDs) and have found no convincing evidence for cyclotron absorption lines. This contrasts with the results of Ginga, which found lines in 3 out of 23 examined GRBs.[1,2,3] These results are not necessarily in conflict, as many of the 72 SD bursts were so weak that lines with the relative strengths measured by Ginga would not have been detectable. A comparison of these results therefore requires a determination of the sensitivity of each instrument to cyclotron lines in the observed bursts.

Simulations based on the cyclotron line spectra found by Ginga were used to estimate our sensitivity. We find that 12-18 bursts (depending on the assumed duration of the lines) were bright enough that typical Ginga second harmonic cyclotron lines would have been detectable at the $P<10^{-4}$ (3.8σ) level in the most favorable detector. Comparison to a similar analysis by Fenimore et al.[4] shows that the Ginga and SD results are not significantly in conflict.

Line Search Status

Currently, the brightest ~15% of the GRBs detected by the BATSE Large Area Detectors (LADs) are selected for further study using the higher energy-resolution Spectroscopy Detectors (SDs). As part of the standard analysis of these bursts, detectors with high gain (low energy threshold $\leq$30 keV) are used to search for absorption lines in the burst spectra.

Such absorption lines were found by Ginga to be in the ~20-50 keV energy range, to last for a short period (2 – 9 seconds) during a longer burst, and, in the most convincing cases, to be multiple with energy ratios of 2:1, suggesting that they are the first and higher harmonics of cyclotron absorption by electrons in teragauss magnetic fields.[1,2,3]

To search for lines with the SD data, background-subtracted count spectra from the high gain detectors are produced over the whole burst, and over consecutive shorter time periods during the burst. These short time periods were determined by

---

* This work was done while D. Palmer held a National Research Council/Goddard Space Flight Center Research Associateship.

circuitry on BATSE on a Time-To-Spill basis, and are typically 0.5-2 s long during the brightest portions of the bursts.

These count spectra were plotted and visually examined for features over the energy range of ~15–1000 keV. If matching features were found in consecutive periods, the spectra were summed to give the strongest feature.

The spectra which show features were then fit by χ^2 minimization to photon spectrum models which either do or do not have line features like the ones identified visually. From the difference in χ^2 between the featured and featureless spectra, an F-test probability[5] can be calculated which gives the statistical significance of the feature.

Because of the large number of GRB spectra measured by the SDs, a feature is considered significant only if its F-test probability $P_F < 10^{-4}$, which corresponds to a significance of $>3.8\sigma$. In addition, the spectra for all detectors must be consistent.

The most significant line feature found by this method so far[6] has $P_F \sim 10^{-3}$. Thus, no significant line features have yet been detected. To find an upper limit on the fraction of GRBs which have absorption lines, it is necessary to determine in how many of the GRBs lines would have been seen if present.

Simulations

To measure the sensitivity of the SDs to lines in GRBs, we did a series of simulations, based on the SD response function and the line spectra found by Ginga. As far as possible, we used the BATSE Spectral Analysis Software (BSAS)[7] system used in the analysis of real bursts for these simulations. We used the results of these simulations to estimate the line sensitivity in the GRBs actually seen by the SDs.

To generate simulated count spectra, we first produced count spectra of data taken from the period following a very short GRB, and subtracted a background derived from data taken over longer time periods surrounding the burst. This mimics the background subtraction procedure used in the standard burst analysis, and produces 'null spectra' which have the same statistics and systematics as would be associated with the background subtraction in real burst spectra. The durations of these null spectra are approximately the same as the durations of the lines found by Ginga.

BSAS was used to model the expected detector response to models based on three photon line spectra from Ginga[8] at incidence angles of 30° and 60° to the axes of the detectors. Poisson-distributed counts based on the predicted count spectra were added to the null spectra to get simulated background-subtracted count spectra. Approximately a dozen of these simulated count spectra, each using a different null spectrum, were produced for each model and at each angle. Three samples of these count spectra, one of each model, are shown in Figures 1a-c for each of the 30° and 60° cases. The parameters of the original models are also shown.

The simulated spectra were then fit by χ^2 minimization to models consisting of a continuum and 0, 1, or 2 line features. The continuum models were power laws, or broken power laws, based on the continuum used in the generation of the simulated spectrum. For single line fits to two-line spectra, the higher energy line was modelled. F-test probabilities were then calculated for $\Delta\chi^2$ to determine the statistical probability that the lines were required. For each model and each incidence angle, the geometric mean of the probabilities of the different simulation runs was calculated.

At 30°, the GB880205 model fits required the ~40 keV line at the $P_F=2.2\times10^{-7}$ level, while the further addition of the ~20 keV line was only required at the $P_F=0.024$ level. For a 60° incidence angle, the probabilities are 1.2×10^{-5} and 0.26. For the GB870303 (S1) simulations, the single ~20 keV line is required at the 1.5×10^{-4} and 0.19 level for 30° and 60°, respectively. The GB870303 (S2) 30°

894   Cyclotron Lines in Gamma-Ray Bursts

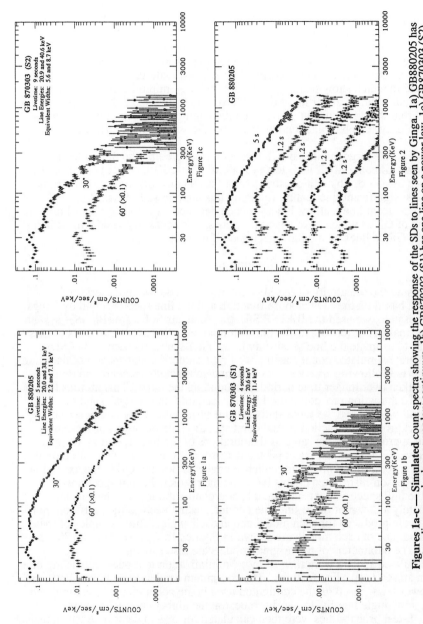

Figures 1a-c — Simulated count spectra showing the response of the SDs to lines seen by Ginga. 1a) GB880205 has two lines on a broken power law continuum; 1b) GB870303 (S1) has one line on a power law, 1c) GB870303 (S2) (after period S1 of the same burst) has two lines on a broken power law.

Figure 2 — Simulated count spectra for short periods (~1.2 s), corresponding to the spectra seen in the standard line search, and over a longer period (~5 s) corresponding to the full duration of the absorption lines.

simulations require the ~40 keV and the further ~20 keV line at the 1.6×10^{-5} and 7×10^{-5} levels. For the 60° simulations of GB870303 (S2), the ~40 keV line is required at the 4×10^{-5} level, but the addition of the ~20 keV line does not reliably and significantly improve the goodness of fit.

Note that the ~40 keV lines are strongly detected in the simulations, while the ~20 keV lines are not. This is generally consistent with what you would expect from examining the samples in Figures 1a-c. The ~40 keV lines are visible to the eye even in shorter segments of time corresponding to the durations used in the standard line search, as is shown in Figure 2. These shorter segments must be summed, however, to produce features strong enough to result in a significant line detection. Note also that if two or more detectors view the burst, the significance of the detection goes as the product of the significances from the individual detectors.

Implications of Simulation Results

The results of these simulations can be scaled to estimate the visibility that lines would have if they existed in the bursts which have been observed by the BATSE SDs. For a line of a given energy, shape, and equivalent width, $-\log(P_F)$ is approximately proportional to the square of the Signal-Noise Ratio (SNR) of the continuum, where the SNR can be defined as the statistical significance of the continuum measurement over an appropriate energy band.[9]

Using the results of the GB 880205 simulation, we can obtain an estimate of the detectability of a 'standard' 40 keV line with an equivalent width of 7.1 keV, using an SNR based on the continuum flux in the 35-45 keV range. This standard line has values typical of or conservative for the second harmonic cyclotron lines found by Ginga in GB880205, GB870303 (S2), and GB890929. The continuum flux (simulated without lines) of GB880205 gives SNRs of 29.5σ and 24.8σ for the 30° and 60° detectors in 5 seconds. This yields:

$$\mathrm{Log_{10}}(P_F) \approx -7.6\times10^{-3}\ \mathrm{SNR}^2$$

for the significance of a standard line. To meet our detection criterion of $P_F < 10^{-4}$ therefore requires an SNR ≥ 22.9σ.

Figure 3 shows the number of bursts in the BATSE SD data set which exceed a given SNR over intervals no more than 5 or 10 seconds long. 12 and 18 bursts have periods ≤5 seconds and ≤10 seconds long, respectively, over which a 'standard' line would be detected with $P_F < 10^{-4}$. It is in these bursts that lines similar to those seen by Ginga could be detected, if they existed.

A similar analysis by Fenimore et al.[4] finds that, if all bursts seen by Ginga had cyclotron line complexes of the strength seen in GB880205, the expected number of line detections at a level of $P=10^{-3}$ would have been 7.6, compared to 2 actual line detections at this level. (The lines in GB890929 are at the $P=2.7\times10^{-3}$ level,[3] and are not included in this count.) These line detection rates (2 out of 7.6 for Ginga, 0 out of 12 for SD) have a statistical probability of $P=14\%$[†], which is not a significant disagreement.

Conclusions

The BATSE SDs would have clearly seen lines at 40 keV in bursts similar to GB880205. The first harmonic cyclotron line at 20 keV would be much less visible, especially in detectors at large angles to the burst.

[†] This probability is calculated as follows: There are 2 line detections in 12+7.6=19.6 'samples'. The probability that both detections are among the 7.6 Ginga samples is $(7.6/19.6)\times(6.6/18.6)=0.138$

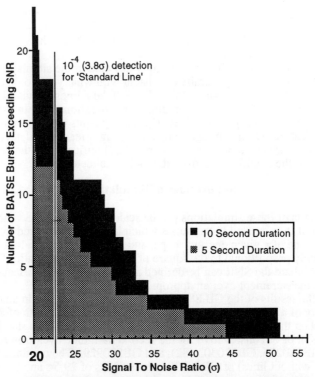

Figure 3 — Number of BATSE bursts exceeding a specific Signal to Noise Ratio in a single detector over the brightest 5 and 10 second period. The vertical line marks the threshold for a line detection at the $P_F = 10^{-4}$ level.

No bursts have yet been found in the BATSE SD data which meet our criteria for line detection. The maximum significance found so far is $P_F \sim 10^{-3}$.

Of the 72 bursts examined so far, 12–18 of them were bright enough that lines similar to the 40 keV lines seen in GB 880205 and GB 870303 would be visible during their 4–9 second durations, if the lines occurred during the intense periods of the burst.

These results do not significantly contradict those of Ginga.

References

1. T. Murakami *et al.*, *Nature*, **335** 234 (1988).
2. E.E. Fenimore *et al.*, *ApJ (Letters)*, **335**, L71, (1988).
3. A. Yoshida *et al.*, *PASJ* **43**(6) (1992)
4. E.E. Fenimore *et al.* (these proceedings) (1992)
5. M. Briggs, Ph.D. Thesis (Appendix), UCSD (1991)
6. B. Teegarden *et al.*, (these proceedings) (1992)
7. B. Schaefer, *BATSE Spectral Analysis Software User's Guide* (1991)
8. P. Freeman, University of Chicago, *private communication.*
9. L. Ford *et al.* (these proceedings) (1992)

ESTABLISHING THE EXISTENCE OF HARMONICALLY-SPACED LINES IN GAMMA-RAY BURST SPECTRA USING BAYESIAN INFERENCE

C. Graziani and D. Q. Lamb
Department of Astronomy and Astrophysics
University of Chicago, Chicago, IL 60637

T. J. Loredo
CRSR, Cornell University, NY 14853

E. E. Fenimore
Los Alamos National Laboratory, NM 87545

T. Murakami
Institute of Space and Astronautical Research, Japan

A. Yoshida
Institute of Physical and Chemical Research, Japan

ABSTRACT

We use a rigorous method derived from Bayesian inference to establish the existence of lines in the spectra of γ-ray bursts. Line detection involves a comparison of nested models. The method amounts to the calculation of the odds O favoring models with lines over models without lines. O is given by the product of the maximum likelihood ratio and a second factor which includes the ratio of the posterior uncertainty of the line parameters to their prior uncertainty. The maximum likelihood ratio always favors the more complex model, since the likelihood of the more complex model can never be larger than that of the simpler model. The second factor penalizes the more complex model, since the posterior uncertainty for the extra parameters is generally smaller than their prior uncertainty. Thus an "Ockham's Razor" automatically appears in Bayesian model comparison.

We fit the count spectrum for time interval S1 of GB870303 with a power-law continuum model and with a power-law continuum multiplied by an exponentiated Gaussian line. We also fit the count spectrum for time interval S2 of GB870303 with a two-segment power-law continuum model and a two-segment power-law continuum multiplied by two exponentiated, harmonically constrained Gaussian lines. In each case we parametrize the exponentiated lines in terms of their equivalent widths and their full widths at half-maximum. We use the results of the fits to compute the odds favoring the model with lines in each case. We find that the odds in favor of the line model are $O \approx 110:1$ for S1, and $O \approx 2.8:1$ for S2. We repeat the line fit to S2, fixing the line center energy at the value determined by the line fit to S1, and find that the resulting odds in favor of this line model are $O \approx 660:1$.

We extend the analysis to include having searched different time intervals for lines by considering a model in which the lines appear at time t_1, disappear at time t_2, and are constant in between. We find that the uncertainties σ_{t_1} and σ_{t_2} in the times at which the lines appear and disappear in GB870303 are large for the *Ginga* data. Indeed, we cannot exclude the possibility that lines are present throughout the burst. Thus the corresponding Ockham factors are of order unity for this burst.

INTRODUCTION

Observations using the Los Alamos/ISAS burst detector on the *Ginga* satellite have provided compelling evidence of the existence of line features in γ-ray bursts. Fenimore et al.[1] analyzed the spectrum of GB880205 reported by Murakami et al.[2], and reported that the line features at ≈ 20 and 40 keV are statistically significant. Graziani et al.[3] analyzed spectra from two time periods during GB870303, referred to as S1 and S2 respectively. They reported that S1 had a statistically significant line feature at $\approx$ 20 keV, and that S2 had line features at $\approx$ 20 and 40 keV.

Establishing unambiguously the existence of line features in burst spectra has long been a nettlesome problem. Fenimore et al.[1] used the F-test[4] to calculate significance levels of the line features in GB880205. Graziani et al.[3] advocated a test suggested by qualitative Bayesian arguments, according to which the significance is determined by evaluating the difference in the best-fit χ^2 between the no-line fit and the line fit, and looking up the significance by consulting the χ^2 distribution with degrees of freedom equal to the total number of parameters in the line fit.

The problem with these approaches is that "significance" is a characteristically "frequentist" concept, and is subject to the well known ambiguities that afflict frequentist statistics.[5,6] In the case of the F-test, there is more than one possible F-like statistic that can be formed, and different choices yield different significance levels. As far as the χ^2 test is concerned, there is ambiguity in the choice of number of degrees of freedom.

Recently, Loredo[7,8] used Bayesian inference to develop a method that is rigorous, intuitive, and unique. The key element of the method is the calculation of the *odds* O favoring the models with lines over models without lines.

In this work we apply this method to the analysis of the spectra of time intervals S1 and S2 in GB870303 (see Figure 1). By computing odds, we establish that in both cases the lines are, with high probability, real. We also extend the analysis to account for the many time periods searched for lines during GB870303, and conclude that the resulting modification of the odds is of order unity.

BAYESIAN MODEL COMPARISON

Spectral line verification requires the comparison of nested models. This is because models with continua but no spectral lines are special cases of models with continua and spectral lines. The Bayesian inference approach to comparison of such classes of models involves computation of the odds favoring the more complicated model (with lines) over the simpler model.[7,8] The approximate expression given in Loredo & Lamb[8] assumes that the prior parameter ranges are uncorrelated, and that there are no correlations between the parameters in the posterior probability distribution. Both of these assumptions fail in the case at hand. Equation (2) of Loredo & Lamb[8] must therefore be modified. The suitably generalized expression for the odds is

$$O \approx \frac{\mathcal{L}_{\max}^{(\text{Lines})}}{\mathcal{L}_{\max}^{(\text{No Lines})}} \times (2\pi)^{L/2} Z(\langle A_\alpha \rangle) \left(\frac{\det V^{(\text{Lines})}}{\det V^{(\text{No Lines})}} \right)^{1/2}. \quad (1)$$

$\mathcal{L}_{\max}^{(\text{Lines})}$ is the maximum value of the likelihood found for the fit of the model with lines to the data, while $\mathcal{L}_{\max}^{(\text{No Lines})}$ is the maximum likelihood found for the fit of the pure continuum model; L is the number of line-related model parameters A_α, $\alpha = 1, \ldots, L$; $Z(\langle A_\alpha \rangle)$ is the prior probability distribution for the line-related parameters evaluated at the maximum likelihood parameter values; $\det V^{(\text{Lines})}$ is the determinant of the covariance matrix of the posterior probability distribution for the line model; and $\det V^{(\text{No Lines})}$ is the determinant of the covariance matrix for the pure continuum model. Equation (1) assumes

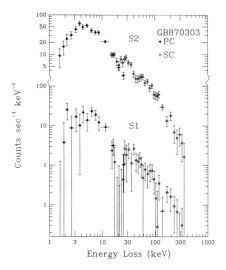 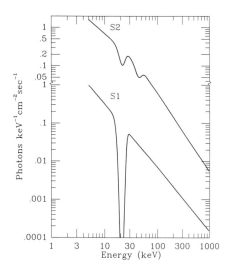

Figure 1. Time intervals S1 and S2 of GB870303. Left: Count spectra. Right: Photon spectra.

that the posterior distributions are approximately Gaussian in the vicinity of the maximum likelihood points, and that there are no correlations between the line parameters and the continuum parameters in the prior probability distribution for the model with lines. We have verified that both assumptions are valid in the present case (see below).

The first factor in equation (1) is the likelihood ratio. It rewards models with more parameters over models with fewer by boosting the odds, since models with more parameters fit the data better and yield larger likelihoods. The remaining factor is the "Ockham factor." It *penalizes* models with more parameters over models with fewer by decreasing the odds, since it is the ratio between posterior and prior uncertainty in the parameters — a small number, if the experiment has added any new information. Equation (1) may be viewed as a quantitative expression of the principle of "Ockham's Razor": A more complicated model is favored only if its greater ability to describe the data (represented by the likelihood ratio) outweighs its additional complexity (represented by the Ockham factor).

SPECTRAL LINE MODELS

The models that we fit to the various spectra are segmented power-law continua and segmented power law continua multiplied by harmonically constrained exponentiated Gaussian lines.[1,3] The spectrum is given by

$$\Phi(E) = \phi_l(E) \exp\left[-\sum_{n=1}^{m} \beta_n G(E; E_n, \Delta E_n)\right], \quad (2)$$

where $\phi_l(E)$ is an l-segment broken power-law, and

$$G(E; E_n, \Delta E_n) = \frac{1}{\Delta E_n \sqrt{\pi}} \exp\left[-\frac{(E - E_n)^2}{\Delta E_n^2}\right]. \qquad (3)$$

Extensive tests using simulated spectra show that the optimal search strategy for harmonically spaced lines is to use a model in which the E_n and ΔE_n are harmonically constrained: $E_n = nE_1$, and $\Delta E_n = n\Delta E_1$. No-line models are obtained by setting $\beta_n = 0$. The values of l and m in equation (2) are $l = 1$, $m = 1$ for S1, and $l = 2$, $m = 2$ for S2.

The parametrization of the lines in terms of β and ΔE is inconvenient for two reasons. First, it is difficult to formulate the allowed parameter ranges in such a way that the prior probability distributions for the line parameters and for the continuum parameters are independent. This is an essential condition for the applicability of equation (1). Second, this parametrization becomes degenerate for "saturated" lines (lines that are black near the line center). For such lines, vast ranges of (very large values of) β and of (very small values of) ΔE result in lines which are essentially indistiguishable from each other using current detectors. This effect distorts the posterior distribution from the Gaussian shape assumed in the derivation of equation (1).

A better parametrization of the lines is the line equivalent widths (EW) and the full widths at half-maximum ($FWHM$) of the lines themselves (*not* the full widths of the Gauusians in equation [3]). For the exponentiated Gaussian model above, we have shown that $EW/FWHM \leq 1.015$, and that almost all models (except some extremely saturated ones) have $EW/FWHM \leq 1$. We therefore parametrize the lines by EW and $R \equiv EW/FWHM$ rather than by β and ΔE, and confine R to the range $0 \leq R \leq 1$. We fit two harmonically constrained exponentiated Gaussian lines to S2. As a result, this fit has four line parameters, E_1, EW_1, R_1, and R_2 (which can be inverted to yield the old parametrization E_1, ΔE_1, β_1, and β_2). We also fit a single exponentiated Gaussian line to S1, with the R parameter fixed to the value $R = 1$, as is appropriate for a saturated line. This fit therefore has two line parameters, E_1 and EW_1. Finally, we fit two harmonically constrained exponentiated Gaussian lines to S2, with E_1 fixed to the value determined by the fit of the line model to S1. This fit has three line-related parameters, EW_1, R_1, and R_2.

OCKHAM FACTORS

We assume uniform prior probability distributions for all the line parameters. The constant values of these uniform priors are the reciprocals of the volumes of the allowed regions of parameter space. These regions are determined by the conditions

$$0 \leq R_i \leq 1 \quad ; \quad 0 \leq FWHM_1 \leq 2E_1 \quad ; \quad E_{\min} \leq E_1 \leq E_{\max}/m, \qquad (4)$$

where $E_{\min}$ and $E_{\max}$ are the lower and upper energy limits of the detector response function, and m is the number of lines in the fit. The upper limit on the $FWHM$ is set to the value such that a line cannot be distinguished from a low-energy rollover. The range of allowed $FWHM$ translates simply to the range $0 \leq EW_1 \leq 2R_1E_1$.

RESULTS

BAYESIAN ANALYSIS OF SPECTRAL LINES

Fit	Likelihood Ratio	Ockham Factor	Odds
S1	8.5×10^5	1.3×10^{-4}	110 : 1
S2	1.4×10^6	2.0×10^{-6}	2.8 : 1
S2 (E_1 fixed)	1.2×10^6	5.5×10^{-4}	660 : 1

Table 1 shows that it is highly probable that the line in S1 is real. The lines in S2 are less clearly present, and taken alone are not really convincing. However, when information from S1 is brought to bear on the analysis of S2 (by fixing E_1 to the value determined by the line fit to S1) the odds in favor of the line model become substantial. It is then highly probable that the lines in S2 are real.

We have extended this analysis to include having searched different time intervals for lines by considering a model in which the lines appear at time t_1, disappear at time t_2, and are constant in between. We find that the uncertainties σ_{t_1} and σ_{t_2} in the times at which the lines appear and disappear in GB870303 are large for the *Ginga* data. Indeed, we cannot exclude the possibility that lines are present throughout the burst. Thus the corresponding Ockham factors are of order unity, and the results for GB870303 cited above are not greatly affected.

REFERENCES

1. Fenimore, E. E., *et al.* 1988, *Ap. J. Lett.*, **335**, L71.
2. Murakami, T. *et al.* 1988, *Nature*, **335**, 234.
3. Graziani, C. *et al.* , in Gamma-Ray Bursts: Observations, Analyses, and Theories, ed. C. Ho, R. I. Epstein, and E. E. Fenimore (Cambridge University Press, 1992), p. 407.
4. Bevington, P. R. 1969, Data Reduction and Error Analysis for the Physical Sciences (New York: McGraw-Hill).
5. Loredo, T. J. 1990, in Maximum Entropy and Bayesian Methods, (Dordrecht: Kluwer Academic Publishers), p. 81.
6. Loredo, T. J., in Statistical Challenges in Modern Astrophysics, ed. E. Feigelson and G. Babu, in press (1992).
7. Loredo, T. J. 1991, private communication.
8. Loredo, T. J., and Lamb, D. Q., in Gamma-Ray Bursts, ed. W. Paciesas and G. Fishman (AIP Press, 1992), p. 414.

SPECTRAL VARIABILITY IN GAMMA-RAY BURSTS ON MILLISECOND TIMESCALES

T. M. Koshut, W. S. Paciesas, G. N. Pendleton
Dept. of Physics, University of Alabama in Huntsville, Huntsville, AL 35899

C. Kouveliotou
Universities Space Research Association, Hunstville, AL 35812

G. J. Fishman, C. A. Meegan, R. B. Wilson
Marshall Space Flight Center, Huntsville, AL 35812

ABSTRACT

We report on preliminary results of an investigation of spectral variability on millisecond timsescales within short duration (≤ 0.5 s) gamma-ray bursts using data obtained with the Burst and Transient Source Experiment onboard the Compton Gamma Ray Observatory. We use hardness ratios to examine short duration bursts for trends in spectral evolution, as well as to identify any association of spectral evolution with burst intensity. Previous studies of spectral variability within bursts have shown a tendency for hard-to-soft evolution throughout the burst as well as throughout individual pulses.[1,2] The capability to examine short duration bursts has been limited in previous experiments by the lack of significant source counts on timescales sufficient to resolve the temporal variability characterizing these events.

DATA ANALYSIS

BATSE consists of eight uncollimated Large Area Detectors, each with a frontal area of 2025 cm^2, spanning an energy range of $\sim$ 20–1800 keV. Details of BATSE instrumentation can be found elsewhere.[3,4] Upon satisfaction of a number of trigger criteria, the instrument begins to accumulate various burst data types. Time-Tagged Event (TTE) data are chosen for this study. TTE data consist of individual photon arrival times in four discriminator energy channels, with a time resolution of 2μs. The TTE onboard memory is organized as a continuously running ring-buffer. Once a burst trigger occurs, the accumulation of TTE data halts after three-fourths of the memory fills. The remaining memory contains data immediately prior to the trigger. Therefore, the interval of TTE data coverage depends upon the intensity of the triggered event. Accumulation of TTE data is not restarted until a burst data readout is completed ($\sim$ 5600 seconds after trigger time); therefore, TTE data are not available for bursts that trigger during a readout. For these cases we use a combination of Preburst (PREB) and Discriminator Science (DISCSC) data, which consist of count rates in four discriminator channels, with a time resolution of 64 ms. PREB data are available for all eight detectors, and cover the 2 seconds immediately prior to the trigger time. DISCSC data are summed onboard over the triggered detectors, and span the 240 seconds following the trigger time.

The TTE and PREB data are summed over the triggered detectors to improve the statistical significance of our results. The TTE data are rebinned to obtain a time resolution between 1 and 32 ms. There is no need to rebin the

PREB or DISCSC data into a coarser time resolution. For each burst analyzed, we use a linear fit to model background in each of the four discriminator channels. These fits are subtracted from the data to obtain background subtracted count rates R_i, where R_1 corresponds to the count rate in discriminator channel 1 (20–50 keV), R_2 corresponds to channel 2 (50–100 keV), R_3 corresponds to channel 3 (100–300 keV), and R_4 corresponds to channel 4 (> 300 keV). A linear fit is sufficient over the short timescales involved here. We define two parameters, a hardness ratio HR and a softness ratio SR:

$$HR = \frac{R_3 + R_4}{R_1 + R_2} \quad \text{and} \quad SR = \frac{R_1}{R_2 + R_3} \quad (1)$$

These ratios are calculated over a time interval containing significant rates above background. We define intensity as the sum over the four energy channels of the background-subtracted count rates. To measure the association of HR and SR with intensity, we calculate the linear correlation coefficient r and $P_c(r,N)$[5]. $P_c(r,N)$ is the probability of measuring a linear correlation coefficient greater than or equal to r from two completely uncorrelated data sets, each having N data points. A small value of $P_c(r,N)$ indicates a significant r.

RESULTS

A total of 23 gamma-ray bursts were analyzed. Examples of the evolution of the parameters HR and SR through time are shown in Figures 1–4. 1σ error bars are plotted on the data; when a data point drops to zero, or negative (possible because rates are background subtracted), 2σ upper limits are plotted. The burst intensity time history is the bottom curve of each figure. Figure 1 shows an example of a burst exhibiting indications of correlation between HR and intensity, calculated on a 4 ms timescale. Figure 2 shows a burst exhibiting indications of a correlation between HR and the intensity, as well as indications of anticorrelation between SR and intensity, both calculated on a 64 ms timescale. Figure 3 shows an example of a burst characterized by hard-to-soft evolution throughout the event, indicated by the evolution of both HR and SR, calculated on a 16 ms timescale. Figure 4 shows an example of hard-to-soft spectral evolution, on a 1 ms timescale, through an intense single pulse (FWHM ≤ 3 ms) occuring within a longer burst (FWHM ~ 150 ms).

Table 1 gives the results of the search for correlations between either HR or SR and burst intensity. The first two columns give the BATSE trigger number and the corresponding BATSE Burst Catalog name for each burst analyzed. The third column indicates the time resolution of the data used in the calculation of HR and SR. The fourth and fifth columns give the correlation coefficient r and significance $P_c(r,N)$ resulting from the comparison of HR and burst intensity. The last two columns give r and $P_c(r,N)$ resulting from the comparison of SR and burst intensity.

CONCLUSIONS

There seems to be a significant correlation between HR and the burst intensity for those bursts analyzed on a 64 ms timescale. In addition, these same events seem to show a significant anticorrelation between SR and burst intensity on the same timescale. When one looks at the results for those bursts analyzed on timescales less than 64 ms, it seems that a correlation between

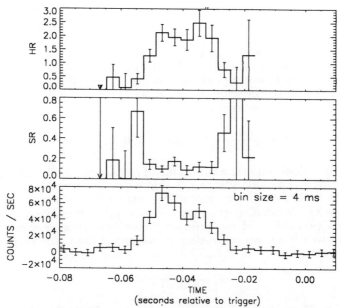

Figure 1. Spectral variations in Burst 432.

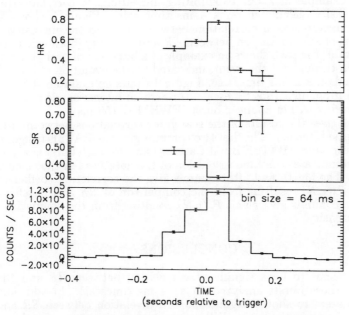

Figure 2. Spectral variations in Burst 444.

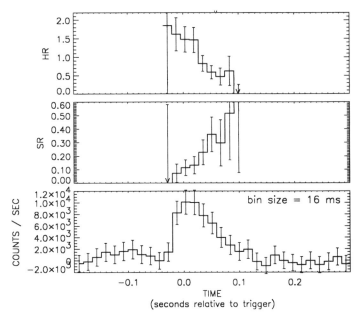

Figure 3. Spectral variations in Burst 1308.

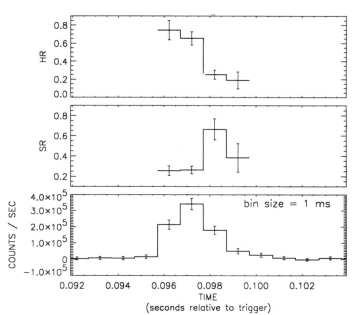

Figure 4. Spectral variations in Burst 1453.

HR and burst intensity is an occasional case, rather than a rule. A significant correlation between SR and burst intensity seems to be much less common on these shorter timescales; in fact many of these bursts give values for r and $P_c(r,N)$ that are consistent with zero correlation.

Table I. Spectral/Intensity Correlations						
TRIGGER NUMBER	BURST NAME	BINSIZE (ms)	HARDNESS RATIO		SOFTNESS RATIO	
			r	$P_c(r,N)$	r	$P_c(r,N)$
138	910502B	32	0.689	0.009	0.307	0.307
207	910518	4	0.715	0.007	0.507	0.245
298	910609	64	0.654	0.078	-0.246	0.557
353	910614	16	0.759	0.003	0.530	0.176
432	910625	4	0.842	0.0003	0.171	0.576
444	910626	64	0.969	0.006	-0.953	0.012
547	910717B	64	0.826	0.006	-0.698	0.036
551	910718	64	0.985	0.002	-0.720	0.170
575	910725	64	0.816	0.014	-0.808	0.015
729	910827	16	0.090	0.791	0.191	0.574
788	910912	16	0.859	0.0007	0.225	0.505
799	910916	8	0.454	0.138	0.129	0.690
830	910928	8	0.378	0.080	-0.102	0.648
1073	911117B	64	0.784	0.215	-0.965	0.034
1088	911119	64	0.980	0.126	-0.898	0.290
1096	911120C	64	0.964	0.171	-0.990	0.090
1097	911120B	16	-0.664	0.104	0.698	0.081
1112	911123B	16	-0.104	0.576	0.005	0.787
1289	920110C	64	0.547	0.053	-0.121	0.694
1308	920121	16	0.228	0.586	-0.263	0.529
1453	920229	8	0.008	0.970	0.097	0.657
1463	920305B	8	-0.386	0.125	-0.227	0.380
1482		4	0.709	0.00007	-0.451	0.023

REFERENCES

1. J. Norris et al., Ap. J. **301**, 213 (1986).
2. D. Band et al., Proc. Huntsville Gamma-Ray Burst Workshop (AIP, N. Y., 1991), p. 174.
3. G. Fishman et al., Proceedings of the Gamma Ray Observatory Science Workshop , 2-39 (1990).
4. J. Horack, NASA Reference Publication No. 1268 , (1991).
5. W. Press, B. Flannery, S. Teukolsky, W. Vetterling, Numerical Recipes (Cambridge University Press, 1986), p. 484.

RAPID SPECTRAL EVOLUTION ANALYSIS OF BATSE GRBs

Vincent E. Kargatis, Edison P. Liang
Dept. of Space Physics and Astronomy, Rice Univ. , Houston, TX 77251

G. Fishman, C. Meegan, R. Wilson
NASA/Marshall Space Flight Center, Huntsville, AL 35812

W. Paciesas
University of Alabama in Huntsville, AL 35899

B. Schaefer, B. Teegarden
NASA/Goddard Space Flight Center, Greenbelt, MD 20771

J. Matteson, D. Band
CASS, UC San Diego, La Jolla, CA 92093

ABSTRACT

We analyze the evolution of BATSE GRB continuum spectra using the BATSE Spectral Analysis Software. We check the consistency of various methods used to characterize spectral hardness: hardness ratio, single power law index, OTTB temperature, and break energy of a broken power law fit. Time evolution of spectral parameters is compared with derived energy fluxes. We also search for correlations between the different spectral parameters and derived quantities, such as break energy, power law indices above and below the break, and energy flux.

INTRODUCTION

The launch of Compton Observatory and BATSE has provided the best available tool to investigate the ongoing mystery of gamma-ray bursts. Due to the lack of observed counterparts and the large variety of time profiles characteristics, much work has focussed on spectroscopy. Although spectral line features would probably provide the most information about the burster environment, the rarity of such features motivates study of the continuum and its spectral variability.

DATA AND ANALYSIS

This study employs the BATSE Spectroscopy Detector (SD) database and BATSE Spectral Analysis Software (BSAS). The detectors used nominally span an energy range of $\sim 50 - 10,000$ keV. Typical time resolution for the spectra are $0.128 - 1$ s, depending on the source intensity observed by the large area detectors.

We analyze spectra from three long, bright bursts in the SD catalog. The spectral fits were usually done at full or half time resolution, to investigate rapid changes in the spectrum. We are interested in characterizing the spectrum with various hardness parameters, one of which is the characteristic energy of the spectrum's peak power. As a preliminary analysis, we use simple 1-3 parameter models to model the spectrum. Following previous work on the SIGNE and BATSE databases[1,2], we use simple power law (PWER), optically thin thermal bremsstrahlung (OTTB), and single broken power law (BRPWR).

Due to the simplicity of the models and the frequency of spectra with low-energy turnovers in flux, we set the lower energy limit to 100 keV. When the limit is lowered to 50, these turnovers will often dominate and force the spectral fit to describe the low energy shape, discarding any information about the spectral hardness at higher energies, where the power peaks. Future analyses including a double broken power law analysis may prove insightful in describing the low-energy turnovers and revealing information about absorption processes at the site creating the x-ray deficit.

This preliminary analysis includes 1B 910503, 1B 910814, and 1B 911118.

RESULTS

As seen in Schaefer et al.[2] for integrated spectra for a larger number of bursts, BRPWR fits are consistent with the three bursts studied here at higher time resolution. PWER gave poorer fits for each. 910814 shows significantly better fits with BRPWR than with OTTB. 910503 and 911118 were roughly equally well fit by OTTB and BRPWR. (However, 911118 is sufficiently soft so that BRPWR high-energy power indices are very steep and often try to exceed the constraint of -7 slope. We therefore use OTTB results to characterize the spectrum.)

Results from the analyses of the SIGNE database and hard X-ray solar flares motivate the present study. High time resolution (0.5 s) spectral fitting of tens of GRBs showed no consistent pattern of spectral evolution: hard-to-soft, soft-to-hard, intensity-hardness tracking, and chaotic patterns are all observed[3,4]. Eight bursts show a trend of increasing luminosity with spectral hardness, a pattern first suggested in 1983[5]. The trend is most notable on the decay side of peaks. The trend was characterized with a power law form of $L \propto kT^\alpha$; the correlation index α was found to have a large range: 1.4 ± 0.6.

Fig. 1 show spectral evolution plots in energy flux - spectral hardness space. Hardness is characterized by either OTTB kT or BRPWR break energy, E_b. Each point is an individual interval within the burst, fitted by OTTB or BRPWR. Energy flux is estimated by integrating each best-fit model spectrum over the fit energy range, then dividing by interval duration (flux error bars are not available for this preliminary analysis). We believe energy flux to be a more meaningful parameter than other traditional measures of intensity, like counts or fit normalizations, which have no physical significance.

The 910503 OTTB intervals cover $0 - 10$ s. The burst begins at a hard, low-intensity state and quickly moves to hard, high intensity. The rest of the burst follows a rough correlation of hardness and intensity as the spectrum softens. The approximate value of the correlation index for the decaying part of the spectrum is ~ 2.2. The BRPWR spectra for 910814 cover the period $0 - 16$s. There is a general progression from hard, high intensity to softer, low intensity intervals. The correlation index here is ~ 2.3. Note that this burst, although spiky, is basically decaying after the first two intervals. 911118 OTTB fits cover $0 - 13$ s. Again, the spectral evolution is largely hard-to-soft, with the flux increasing to the second peak and then falling linearly (in log space) with kT on the decay side of the burst. The correlation slope for the decay side is

~ 2.4. We observe that for these three bursts, <u>the decay phases</u>, spiky for 910503 and 910814 and smooth for 911118, <u>all show a degree of correlation, with similar correlation indices around ~ 2.3</u>. This value is similar to those found for solar hard x-ray flares[6,7]. We also note that the behavior at the beginning of 910503 and in the two peaks of 911118 resemble Type II spectral evolution suggested by Liang[1].

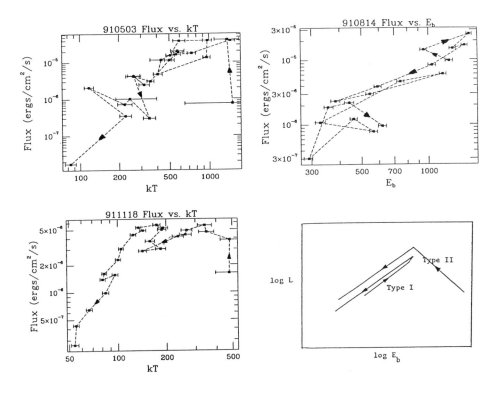

Fig. 1. Energy flux vs. spectral hardness for the three BATSE bursts analyzed here. Each point corresponds to an interval in the time profile. The dashed lines trace the time evolution. Flux error bars are missing because the software does not produce them at this time. Bottom right shows two types of evolutionary trajectories in the L - E_b plane (from ref. 1).

A correlation between the upper and lower power law indices of a series of broken power law fits has been observed in a solar hard x-ray flare[7] (fig. 2). For comparison, we plot the same parameters γ_1 (lower index) and γ_2 (upper index) from BRPWR fits to 910503 and 910814 (fig. 3). Errors and scatter are large enough to preclude any conclusive correlation, though 910503 appears consistent with the solar flare result. γ_2 for 910814 stays roughly constant regardless of γ_1. (These comparisons aren't made in reference to any physical model as of yet - we are simply looking for similarities between two impulsive high-energy phenomena).

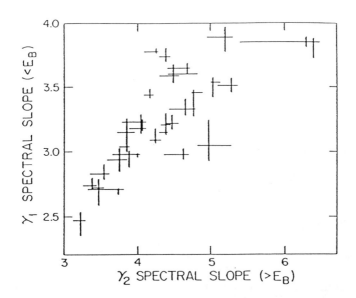

Fig. 2. A plot upper and lower power law indices from a sequence of broken power law fits to a solar hard x-ray flare. They seem to be roughly correlated. (From ref. 7)

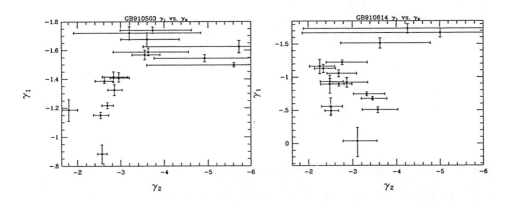

Fig. 3. The same plots for a sequence of BRPWR fits to GB910503 and 910814. There is no definite correlation present, though GB910503 is consistent with fig. 2.

A disturbing result arose when we checked the consistency of two different measures of spectral hardness. Both OTTB kT and BRPWR E_b have been suggested and used (even in this work) as two possible parameters to characterize hardness. Recall that 910503 showed no significant preference for either OTTB or BRPWR fits. Because, for the two spectral forms, kT and E_b are the energies at which the spectrum peaks in power, we expect that for the same set of spectra, kT and E_b would be highly correlated if both models fit equally well. Fig. 4

shows kT plotted against E_b for the same set of spectra in 910503. Instead of a linear correlation expected for consistent measures of hardness, we see essentially a scatter plot.

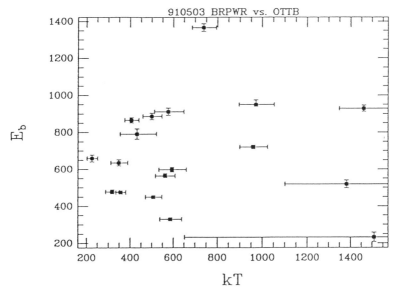

Fig. 4. E_b vs. kT for a sequence of BRPWR and OTTB fits to the same spectra for GB910503. Since both parameters indicate the peak power energy of their respective model spectra, one might expect them to be tightly correlated. They are not.

It is not difficult to understand the specific instances where kT and E_b significantly differ. In cases of $E_b > kT$, the spectrum is usually soft with steeper-than-average power law indices, and the lower index extends to higher energies before diverging from the data. For $E_b < kT$, the lower index is usually flatter than -1. Since OTTB can only approach -1 at low energies, the temperature is "dragged out" to higher energies to fit the data. Since neither model is statistically preferred overall, the wisdom of using these parameters to characterize hardness is called into question. We would like to warn the community (ourselves included) that interpretations of traditional spectral characterizations may not be reliable. Care should be taken in future discussions when using hardness parameters to describe spectral evolution.

This research was supported in part by NASA grant NGT 50924.

REFERENCES

1. E. P. Liang, Proc. of the Hunstville GRB Workshop (AIP, NY, 1992).
2. Schaefer et al., Ap. J. **393**, L51 (1992).
3. K. Hurley et al., this proceedings (AIP, NY, 1992).
4. V. Kargatis, E. Liang, K. Hurley, Ap. J. , to be submitted (1992).
5. S. V. Golenetskii et al., Nature **306**, 451 (1983).
6. E. Priest, Solar Flare Magnetohydrodynamics (Reidel, Holland, 1987).
7. P. Lin and R. Schwartz, Ap. J. **312**, 462 (1987).

SPECTRAL EVOLUTION STUDIES OF A SUB-CLASS OF GAMMA RAY BURSTS OBSERVED BY BATSE

P.N. Bhat[†], G.J. Fishman, C.A. Meegan, R.B. Wilson, C. Kouveliotou[‡]
ES-62, Space Science Laboratory
NASA/Marshall Space Flight Center, Huntsville, AL 35812

W.S. Paciesas and G.N. Pendleton
Department of Physics, University of Alabama in Huntsville
Huntsville, AL 35899

B.E. Schaefer
Code 661, NASA/Goddard Space Flight Center
Greenbelt, MD 20771

ABSTRACT

Among the gamma-ray bursts (GRB) observed by BATSE, we define a sub-class of bursts characterized by fast rise times (< 0.5 sec) and a relatively slow decay times. Their time histories exhibit diverse variations. GRB s with such temporal features were selected and divided into shorter time segments. Spectral features were parameterized for each of the time segments using the BATSE Spectral Analysis Software (BSAS) from the large area detector data. The evolution of these parameters during the burst were studied in relation to the other observed variables like the intensity and temporal structure. A possible correlation between the time histories and spectral characteristics is explored.

INTRODUCTION

One of the characteristics of a "classical" gamma ray burst (GRB) is its hard energy spectrum, often with a very high energy tail extending up to tens of MeV[1]. These spectra are remarkable in that nearly all of the emission is at γ-ray energies. The power per logarithmic energy interval rises steeply in all burst spectra at low energies and often peaks around 100 keV. A study of the evolution of the spectral characteristics is expected to lead to an understanding of the possible emission mechanisms. A hard-to-soft spectral evolution in the 50 keV- 2 MeV energy range has been observed in many GRB s with durations greater than 1 s[2]. The hardness ratio decreases monotonically from the burst rise through the decay phase. However there are several exceptions to this general behaviour. Continuum spectra of GRBs seem to be variable on time scales as short as the detector time resolution and there is some indication of a correlation between temporal variability of luminosity (derived from the count spectra) and a parameter measuring the hardness, like the temperature[3,4]. Also, some bursts begin with hard spectra while in others hardness seem to be correlated with intensity[5] and do not show an unambiguous correlation of luminosity with temperature[6]. In 1990 Jourdain[7] analyzed the count spectra of several bursts

---

[†] NASA/NRC Senior Resident Research Associate; On leave from Tata Institute of Fundamental Research, Bombay 400 005, India

[‡] USRA; On leave from the University of Athens, Greece

observed by the APEX experiment and concluded that spectral evolution has no correlation with the time history. Thus spectral evolution varies significantly over the distribution of observed GRB s.

There are several problems connected with the determination of GRB spectra. The observed counts spectrum is a convolution of the incident photon spectrum with the detector response. The detector response in turn is a function of the photon energy, source angle, relative contribution from the earth scattered photons, background estimation etc. Thus, it is absolutely necessary to understand this function in any spectral evolution study to ensure that the results obtained are not detector dependent.

We chose a sample of bursts which have short rise time (≤ 0.5 s) and a nearly exponential decay time. Some of the GRBs chosen have a smooth profile while others are highly structured.

OBSERVATIONS

Each of the eight BATSE Large Area Detectors (LAD s) consist of 1.27 cm thick NaI(Tℓ) crystal of size 50.8 cm in diameter. The large diameter-to-thickness ratio of the crystal produces a detector angular response similar to that of a cosine function at low energies. The eight planes of the LAD s are parallel to the eight faces of a regular octahedron, thus providing nearly uniform and complete sky coverage[8]. A simultaneous 5.5 σ increase in the background counting rate (in any of the 3 time scales: 64 ms, 256 ms and 1.024 s) in any 2 of the 8 detectors constitutes a burst trigger. The average rate is about one burst per day.

ANALYSIS

A background spectrum is generated by a polynomial fit to the spectra before and after each burst and then interpolating to the burst interval. Each burst belonging to this sample is subdivided into segments such that a background subtracted spectrum for this segment has statistically significant signal in the energy range of 40 - 2000 keV. Each spectrum is then fitted to either an optically thin thermal Bremsstrahlung spectrum (OTTB) or a power-law with an exponential cut-off function (COMP) in order to parameterize the spectral hardness of each segment. The choice of spectral function is merely to parameterize the spectrum rather than to study the physical process responsible for the photon emission at the burst source. In some cases like the burst 1B910814, the spectrum is not well represented by the OTTB function while either a COMP function or combination of a black-body spectrum and a power-law function fit the spectra best. In cases where more than one function fits the spectrum, that function with lesser number of parameters is chosen. We used the BATSE Spectral Analysis Software (BSAS) for deriving the spectra and the functional fits. The detailed simulation studies[9] of the detector response at various energies and source angles have been used in deconvolving the count spectra before fitting to standard functions.

To study the spectral evolution over short time scale ($\sim$ 64 ms) we used the hardness ratio (defined as the ratio of the number of counts above 100 keV to those below 100 keV) computed using 4-channel spectral data. Thus we could study the spectral evolution over different time-scales during each burst.

RESULTS AND DISCUSSION

Table 1 summarizes the results of the spectral fits. Column 3 lists the range of integration times while column 4 lists the range of temperatures resulting from the fits to functions listed in column 2. The last column lists the correlation coefficient of T with the burst intensity. It may be noted that 3 bursts show a rather good correlation while one of them shows a weaker correlation and one shows no spectral evolution during the entire burst. Band et al.[10] also discovered wide variations in the spectral evolution of GRB s, based on the bursts detected by the BATSE spectroscopic detectors. Figures 1 and 2 show two extreme examples of spectral evolution of GRB s with similar temporal characteristics, viz: 1B910717 and 1B910602.

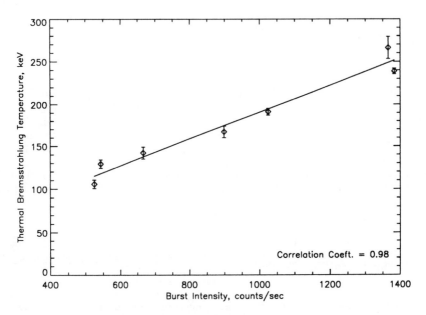

Figure 1. Burst 1B910717 showing evidence for strong spectral evolution. The straight line is a least square fit to the data points

Figure 3 shows the time history of 1B910814 with a time resolution of 128 ms. The variation of the hardness ratio with time during the burst is shown in dotted lines (suitably scaled). The correlation coefficient of the hardness ratio with the burst intensity is 0.84 consistent with that of the fitted temperatures. It was found that the correlation coefficient remains unchanged even at a time resolution of 64 ms showing that the spectral parameters change even at short time scales consistent with the variation in the count. A similar analysis carried out for bursts 1B910717 and 1B910718 show that the correlation coefficients improve when the integration time is increased from 64 ms to 1.28 s (0.72 to 0.92 and 0.12 to 0.64 respectively).

To summarize the results of this study we find that the spectral evolution is an important property of some bursts while it is not a general feature of the sub-class chosen here. Similarity in temporal characteristics does not require similarity in spectral properties. Some bursts show spectral evolution at tens of millisecond time-scales while others show a much slower variation. The burst spectra could be fitted to a variety of spectral functions. Burst 1B910814 was also fitted to a combination of a black-body spectrum and a power-law. It was

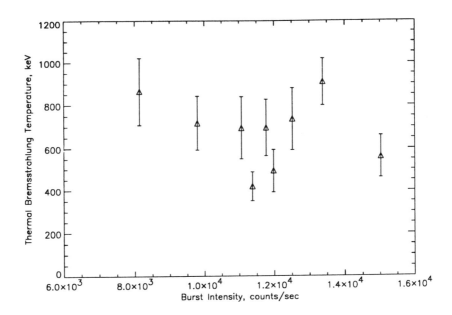

Figure 2. Burst 1B910602 which does not show any spectral evolution.

found that both the black-body temperatures and the power-law indices show good correlation with burst intensity. The reason why some spectra do not fit a given spectral function may be due to spectral evolution during the integration time. This is particularly true of those bursts which show fast evolution over tens of milliseconds.

Table I. Summary of Spectral Fits						
Burst Name	Spectral Function	Int. Time min. max.	T (keV) min max	No. of Spectra	Duration (sec)	Corr.Coeft. with Inten.
1B910602	OTTB	0.896 7.42	418.4 908.6	9	23.1	-0.11
1B910629	OTTB	0.384 6.33	228.4 389.2	6	11.0	0.92
1B910717	OTTB	0.64 1.34	105.6 266.2	7	5.2	0.98
1B910718	COMP	0.832 4.42	83.3 156.7	8	11.4	0.64
1B910814	COMP	0.768 2.56	185.5 811.2	16	20.4	0.86

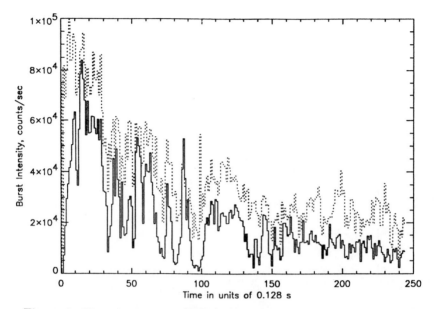

Figure 3. Time history of 1B910814 with a time resolution of 128 ms. Also shown is the variation of the hardness ratio suitably scaled showing a good correlation of the hardness ratio with burst intensity even on very small time scales.

REFERENCES

1. S.M. Matz et al., Ap. J. Letters **288**, L37 (1985).
2. J.P. Norris et al., Ap. J. **301**, 213 (1986).
3. E.P. Mazets et al., Astrophys. and Space Sci. **82**, 261 (1982).
4. S.V. Golenetskii et al., Nature **306**, 451 (1983).
5. K. Hurley et al., AIP Conf. Proc. - Gamma Ray Bursts; Huntsville (Ed.: W.S. Paciesas and G.J. Fishman, 1992), p. 195.
6. V.E. Kargatis et al., AIP Conf. Proc. - Gamma Ray Bursts; Huntsville (Ed.: W.S. Paciesas and G.J. Fishman, 1992), p. 201.
7. Jourdain, E., Ph. D. Thesis (C.E.S.R., Paul Sabatier University, Tolouse, France, 1990).
8. G.J. Fishman et al., Proc. Gamma Ray Observatory Science Workshop; Greenbelt (Ed.:W.N. Johnson, 1989), p. 2-39.
9. G.N. Pendleton et al., Proc. Gamma Ray Observatory Science Workshop; Greenbelt (Ed.:W.N. Johnson, 1989), p. 4-547.
10. D. Band et al., AIP Conf. Proc. - Gamma Ray Bursts; Huntsville (Ed.: W.S. Paciesas and G.J. Fishman, 1992), p. 169.

THE GINGA RATE OF OBSERVING CYCLOTRON LINES

E. E. Fenimore and G. Schwarz
Los Alamos National Laboratory, MS D436, Los Alamos, NM 87545

D. Q. Lamb and P. Freeman
The University of Chicago, Chicago Il 606037

T. Murakami
ISAS, Tokyo Japan

ABSTRACT

The Ginga Gamma-Ray burst detector has observed three gamma-ray bursts with absorption lines yet BATSE has not reported any lines. If the lines are cyclotron lines then gamma-ray bursts are almost certainly not at cosmological distances. Based on the Ginga data, we estimate the fraction of bursts that should contain lines if all bursts had lines like GB880205, GB870303, or GB890929 is consistent with roughly 35%. Using estimates of the relative steradian-sec factors for Ginga and BATSE implies that BATSE should be seeing about 5.5 events with lines per year although that number is rather uncertain. The features Ginga detects are most certainly not statistical flukes yet it is not clear why Ginga has such strong lines and BATSE does not.

Ginga has detected three Gamma-Ray Bursts (GRBs) with lines: GB880205[1,2], GB870303[1,3], and GB890926[4]. Although 23 of the Ginga events were searched, the fraction of bursts with lines should not be taken as 3 out of 23 because that does not take into account whether or not we had the sensitivity in each of the 23 events to detect lines. Whether we would have the sensitivity to observe lines in a burst, of course, depends on the strength of the lines and, thus, all bursts that one can reasonably analyze were checked. Through simulations we can determine which events we had the sensitivity to see lines at a particular strength. We have done extensive simulations to determine the expected number of events Ginga would see lines if all GRBs had lines similar to GB880205. Figure 1a shows simulations of GB880205 scaled to various intensities, Consider the curve labelled "1.0" which corresponds to GB880205 at its observed strength. We simulated 128 examples with lines at the strength reported by Fenimore et al[2] using the background and deadtime characteristic of Ginga. The response matrix was evaluated for an incident angle of 45 degrees. For each simulation we first fit with a broken power law but no lines (4 free parameters) and then with a broken power law plus two lines whose positions are linked by a harmonic relationship but with separate free parameters for the widths and intensities (9 total free parameters). For each simulation the difference in χ^2 between the fit without lines and one with lines is found. Such a value should be distributed as χ^2 with 5 degrees of freedom. The probabilities for the 128 simulations were formed into integral probability distributions shown in Fig. 1a. The probability of observing lines at a specified significance can be determined from Fig. 1a. For example, a significance of 10^{-5} or better occurred for 73%

of the simulations when the event was scaled by 1.0. This means that, given a sample like GB880205 that actually contains lines, those lines could be found with a 10^{-5} significance 73% of the time. In Fig. 1a we have also simulated GB880205 scaled by 2.0, 1.5, 1.25, 0.75, and 0.5. The sample in GB880205 that we originally detected lines had a signal-to-noise (SNR) of about 60 (the third sample under GB880205 in Table I). Given the background, it would have had a SNR of 38, 52, 69, and 77 at scalings of 0.5, 0.75, 1.25, and 1.5, respectively. The significance scales as the square of SNR (Band et al[5]). In Fig. 1b, we show the probability of detection of a GB880205-like line with Ginga as a function of the square of the SNR of a sample.

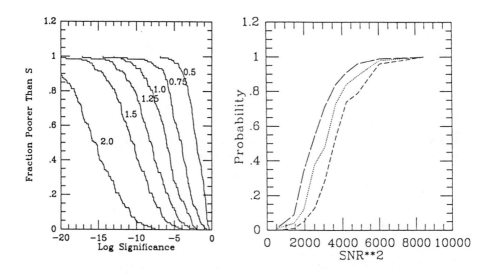

Figure 1. a (left): Simulations of GB880205 into Ginga scaled by the factors that label the curves. b (right): The probability of detecting lines like GB880205 in Ginga at various significance as a function of the signal-to-noise squared of a sample. The long dashed, dotted, and short dashed curves are for $10^{-6}, 10-5$, and 10^{-4} significance, respectively.

Table I shows 23 Ginga events that had relatively flat backgrounds and reasonable temporal samples such that line searches could be attempted. Bursts were analyzed as they were received from the satellite and the method of analysis varied over the four years of the Ginga flight operations to take into account improvements in the analysis methods and software. Here we will simulate a single method based on our recent assessments of techniques[6,7]. Our goal is to estimate the number of lines that Ginga could detect if all events had lines similar to those observed. The variation in the methods used to initially find the lines should not substantially change our conclusions. Since each burst was initially analyzed by different teams using different techniques, the way bursts were divided up into individual samples in Table I is not necessarily the way they were divided in the initial analyses but they have similar characteristics (number and duration of samples).

The first column of Table I is the event date. The second column is a rough estimate of the total fluence in erg/cm^2 and the third column is the peak erg/sec/cm^2. The forth column list the integral background corrected counts within temporal samples that could be analyzed. The fifth column gives the duration of the sample. In triggered mode, the temporal resolution is 0.5 sec and typically 10 such samples were summed before inspection. For untriggered events, the temporal resolution of useful samples varied from 2 to 16 seconds. In one case (GB870902) the burst had substantial emission after the triggered memory with 64 sec resolution and those samples were included in this analysis. The sixth column gives the SNR using the background at the time of the event. The seventh column gives the probability that we could detect lines in the sample assuming the sample has lines like GB880205 (cf. Fig. 1b). Some samples with a zero probability were omitted from the table. The eighth column gives the probability that the event would show at least one sample with lines.

The sum of the seventh column gives the expected number of samples that would show lines if all samples had lines like GB880205 and the sum of the last column gives the expected number of events that would have at least one sample with lines. If all bursts were like GB880205, we would expect to find 5.4 events with lines. We observed 3 events with lines but at strengths and significance different than GB880205. We have done preliminary simulations of the S2 region[3] of GB870303 and roughly 10 events would show lines if all bursts were like GB870303. The S1 region of GB870303 occurred in a portion of the burst that was barely above background so we expect virtually all events would show such lines if they were present in all bursts. Thus, depending on which line is consider, the fraction of bursts that would show lines if they were present in each burst varies from 1/23 to 1/5.4. Event GB890929[4] has not been simulated so its fraction is unknown but is probably similar to GB870303. We have assumed it is $\sim 1/10$. The joint probability that an event would show lines like either GB880205, GB870303, or GB890929 is, therefore, $\sim 35\%$.

To compare Ginga and BATSE rates of detecting bursts, one must determine the relative ster-sec factors. Ginga has four modes: a triggered mode which can operate independent of the other modes and which provides the bulk of the useful data for line searches. The other threes modes take a spectral sample either every 2 sec, 16 sec, or 64 sec. The samples taken at 2 seconds are just as effective for searching for lines as is the triggered system. The samples taken with 16 sec resolution are probably 1/8 as effective as the triggered system. We have analyzed the period 87/03/03 to 90/10/10 (1.08×10^8 sec). We estimate that 55% of the time Ginga was in a low background with the trigger system operating. However, the triggered memory was filled 36% of the time. Thus, the triggered memory had a livetime of 3.8×10^7 sec. During the period of study, we had 2 sec spectral samples for 3.9×10^5 sec and 16 sec samples for 3.0×10^6 sec. Combined with the memory time gives a total livetime of 3.9×10^7 sec. Thus, the average fraction of the time we could search for lines is 0.36. In contrast, given the better telemetry coverage of GRO, BATSE's memory is rarely filled and its livetime is roughly $\sim 75\%$ and now has 1.4 yr of operations[8].

We have simulated both the Ginga and BATSE instruments to estimate their steradian coverage. The simulation found the number of events that had an average SNR per energy channel of at least 4 between 15 and 95 keV assuming a typical log N–log P distribution, background level, and spectral shape (250

Table I
Probability of Detecting GB880205 Lines in GINGA Events

Event	S	P	counts	time	SNR	Prob	Expected # Events
870303	6.3E-05	3.8E-06	1660	5	24.6	0.01	0.01
870414a	2.0E-05	1.8E-06	1058	5	18.6	0.00	0.00
870521	7.3E-05	2.0E-06	2720	16	21.6	0.00	0.00
870707	3.1E-05	1.0E-05	3742	5	44.8	0.15	0.15
870902	1.0E-04	2.8E-06	1661	5	24.3	0.01	
			8256	64	38.2	0.04	0.05
880128b	9.4E-05	1.5E-05	2163	5	35.5	0.03	
			6093	5	69.7	0.89	
			4166	5	55.1	0.48	0.95
880205	1.5E-04	9.3E-06	2069	5	32.8	0.03	
			3826	5	50.6	0.39	
			5035	5	60.5	0.73	
			4736	5	58.1	0.62	
			2408	5	36.7	0.04	
			1754	5	29.0	0.02	0.94
880725	4.0E-05	9.1E-06	2681	5	35.3	0.03	0.03
880830b	2.4E-05	3.1E-06	1485	5	25.0	0.01	0.01
881009	1.8E-05	1.8E-06	1344	16	13.8	0.00	0.00
881130	1.7E-05	5.5E-06	1112	5	18.4	0.00	0.00
890704	2.2E-05	7.3E-06	2083	5	30.6	0.02	0.02
890915	1.6E-05	2.7E-06	1070	5	18.4	0.00	0.00
890929	6.2E-05	1.1E-05	5770	5	63.2	0.80	
			2521	5	35.3	0.03	0.80
900126	5.4E-05	1.5E-05	1525	5	26.3	0.02	
			5373	5	63.2	0.80	0.80
900129	4.4E-05	9.6E-06	4445	5	52.4	0.42	0.42
900221	5.8E-05	5.4E-06	2256	5	32.5	0.03	
			1984	5	29.5	0.02	
			1744	5	26.6	0.02	0.06
900322b	1.7E-05	1.9E-06	1043	5	19.1	0.00	0.00
900623	2.0E-05	5.6E-06	2106	5	29.5	0.02	0.02
900709	6.4E-06	3.9E-06	1188	5	18.7	0.00	0.00
900906	1.2E-04	1.2E-05	5160	16	49.2	0.34	
			9918	16	79.0	0.98	0.99
900928	1.3E-05	5.6E-06	1875	5	33.4	0.03	0.03
901001	5.3E-05	6.7E-06	2040	5	34.7	0.03	
			2692	5	42.0	0.09	0.12
						7.1	5.4

KeV thermal bremstrahlung). The BATSE simulation used eight detectors distributed like those on GRO with dimensions and materials similar to the BATSE SD detectors (Pendleton, private communication). The front window was taken to be solely a beryllium window although the actual design has some of the front area covered by aluminum. The Ginga detector used the actual dimensions and materials of the Ginga design which had substantial shielding on the sides and is only 1 cm thick so reacts much like a planar detector. Simulations were run with an isotropic distribution for the locations and with all events at normal incidence into the detector. Events blocked by the earth from low-earth orbit were, of course, not detected. For Ginga, the ratio of normal-incidence events detected to isotropic events detected was about 0.3 and, thus, Ginga's steradian coverage is about $2\pi/3$. (The earth blocks 2π). For BATSE, effectively all events that would be detected at normal incidence were also detected if they were had an isotropic distribution. Thus, the BATSE steradian coverage is 2π, as expected.

In conclusion, the BATSE ster-sec coverage is 6.25 times larger than Ginga, that is, $(0.75 \times 2\pi)/(0.36 \times 2\pi/3)$. Ginga saw lines at a rate of about 0.9 per year (i. e., 3/3.4) so BATSE should be seeing lines at a rate of $\sim$5.5 per year. This rate should actually be larger since BATSE is usually more sensitive than Ginga[6] and, thus, should see to a lower point on the $\log N$–$\log P$ distribution. After 1.4 years of operation BATSE is yet to observe lines. The probability that BATSE would see no events with lines while Ginga sees 3 has a maximum when the rate of events with lines is 0.5 events/yr/4π. The maximum probability is only 1.5×10^{-3}. However, the Ginga lines, particularly those in GB880205 were very significant and certainly are not statistical flukes[7].

REFERENCES

1. T. Murakami et al., Nature **335**, 234 (1988).
2. E. E. Fenimore et al., Ap. J. **335**, L71 (1988).
3. C. Graziani et al., Gamma-Ray Bursts: Observations, Analyses and Theories (Cambridge Press, eds Ho, Epstein, and Fenimore, 1992a), p. 407.
4. A. Yoshida et al., Gamma-Ray Bursts: Observations, Analyses and Theories (Cambridge Press, eds Ho, Epstein, and Fenimore, 1992), p. 399.
5. D. Band et al., these proceedings, 1992.
6. P. Freeman et al., these proceedings, 1992.
7. C. Graziani et al., these proceedings, 1992.
8. B. Teegarden et al., these proceedings, 1992.

SENSITIVITY OF THE BATSE SPECTROSCOPY DETECTOR TO GAMMA-RAY BURST SPECTRAL LINES LIKE THOSE SEEN IN GINGA

P. E. Freeman and D. Q. Lamb
Dept. of Astronomy and Astrophysics, University of Chicago, IL, 60637
E. E. Fenimore
Los Alamos National Laboratory, NM, 87545
T. J. Loredo
Dept. of Astronomy, Cornell University, NY, 14853

ABSTRACT

We use simulations to explore the sensitivity of the Burst and Transient Source Experiment (BATSE) Spectroscopy Detector (SD) on the *Compton* Gamma-Ray Observatory to harmonically-spaced lines like those seen by the *Ginga* Burst Detector. We simulate bursts using the best-fit spectral parameters derived from the *Ginga* data for time intervals S1 and S2 of GB870303 and time interval b of GB880205. These bursts represent the weakest and strongest bursts with lines seen by Ginga.

Our results show that, within the context of the Gaussian line model, comparison of a power-law continuum model and a power-law continuum plus "constrained" harmonic lines model ($E_2 = 2E_1$, $\sigma_2 = 2\sigma_1$, and $\beta_2 \neq \beta_1$, where E_i is the centroid energy, σ_i is the width, and β_i is the strength of the line) constitutes the optimal search strategy for harmonically-spaced lines in the SD data.

We find that the ability of the SD to detect a line is a sensitive function of the line energy, strength, and width. The SD is able to detect strong lines like that at ≈ 20 keV in interval S1 of GB870303 only at angles $\lesssim 30°$, due to the Al shielding. The SD can detect the weaker, narrower, harmonically-spaced lines like those at ≈ 20 and 40 keV in interval S2 of GB870303 at angles $\lesssim 60°$ because the SD is able to detect the second harmonic at larger incident angles. Similarly, the SD is able to detect lines like those at ≈ 20 and 40 keV in interval b of GB880205 at angles $\lesssim 60°$.

INTRODUCTION

The Burst and Transient Source Experiment (BATSE) on the *Compton* Gamma-Ray Observatory has discovered that even faint γ-ray bursts are distributed isotropically on the sky.[1] This discovery has intensified debate about whether some, or even all, γ-ray bursts are cosmological in origin. The existence of the harmonically-spaced low-energy lines seen by *Ginga*,[2,3] and the success of the cyclotron scattering model in explaining them[4,5] provides compelling evidence that these particular γ-ray bursts originate from strongly magnetic neutron stars, and is the strongest evidence that some γ-ray bursts are galactic in origin.

Given the importance of this implication, it is imperative to search for spectral lines in the BATSE SD data and to assess the consistency of the *Ginga* and *Compton* observations. Here we report the results of extensive simulations to determine the sensitivity of the BATSE SD to lines like those seen by *Ginga*.

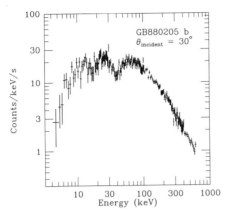 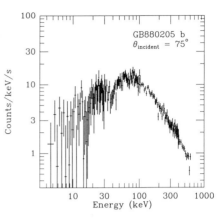

Figure 1. Simulated spectra like that of GB880205 b, as seen by the BATSE SD. (Left panel) Incident angle 30°, where the harmonically-spaced lines at $\approx$ 20 and 40 keV are clearly seen. (Right panel) Incident angle 75°, where the first harmonic at $\approx$ 20 keV is lost, due to the Al shielding, and the feature at $\approx$ 40 keV is no longer significant.

THE SIMULATIONS

To investigate the sensitivity of the BATSE SD to lines like those seen by *Ginga*, we simulate bursts using the best-fit spectral parameters derived from the *Ginga* data. To investigate a relatively weak burst, we use periods S1 and S2 of GB870303,[3] which extend for 4 and 9 seconds and have fluences 1.6 x 10^{-6} and 9 x 10^{-6} erg cm^{-2}, respectively. To investigate a relatively strong burst, we use period b of GB880205,[2] which extends 5.5 seconds and has a fluence 2.2 x 10^{-5} erg cm^{-2}.

We fit the *Ginga* data for period S1 of GB870303 using a single power-law continuum and a truncated Gaussian line (truncated meaning that if the model flux, dN/dE, at energy E is less than zero, we reset the flux to zero). We use a truncated Gaussian line because the maximum likelihood method, unlike χ^2, does not permit a negative value for the burst flux. We fit the *Ginga* data for periods S2 of GB870303 and b of GB880205 using a two-component power-law continuum plus harmonic Gaussian line model, in which $E_2 = 2E_1$ but the line strengths and widths are all independent.

In our simulations, we assume that bursts strike the BATSE SD detector from four angles: 30°, 45°, 60°, and 75°, at a gain setting of 7X nominal. We then fold the bursts through the response matrices to derive a counts spectrum. We add typical BATSE SD backgrounds that have been provided to us by the BATSE team.[6] We then apply Poisson statistics to generate a simulated spectrum. We assume no dead time. Figure 1 shows sample count spectra for interval b of GB880205.

We fit the simulated spectra for period S1 of GB870303 using two models: (a) a one-component power law; and (b) a one-component power law plus truncated Gaussian line.

In order to determine the optimum search strategy for harmonically-spaced lines, we fit the simulated spectra for period S2 of GB870303 and for period b of GB880205 using four models: (a) a two-component power law; (b) a two-component power law plus one Gaussian line at the second harmonic; (c) a two-component power law plus two "constrained" harmonic Gaussian lines ($E_2 = 2E_1$, $\sigma_2 = 2\sigma_1$, and $\beta_2 \neq \beta_1$, where E_i is the centroid energy, σ_i is the width, and β_i is the strength of the line); and (d) a two-component power law plus two harmonic Gaussian lines ($E_2 = 2E_1$, $\sigma_2 \neq 2\sigma_1$, $\beta_2 \neq \beta_1$).

Furthermore, we fit the data using three different methods: (a) maximum likelihood; (b) χ^2 with variances derived from the model; and (c) χ^2 with variances derived from the data. We expect that, for absorption-like features, χ^2 with variances derived from the model gives results closer to those found using the rigorous maximum likelihood method than does χ^2 with variances derived from the data, and that the difference becomes larger as the line becomes stronger.

RESULTS OF THE SIMULATIONS

We examine how well the BATSE SD can detect lines like those seen by *Ginga* using fits to 25 simulated spectra at each of four angles: 30°, 45°, 60°, and 75°. We derive the significance by computing $\Delta\chi^2$, the difference between the χ^2 for the power-law continuum model and the power-law continuum plus lines model, and finding its Q-value from the χ^2 distribution with the number of degrees of freedom equal to the number of line parameters.[7] We adopt a significance (Q-value) $< 10^{-5}$ as a rough criterion for detection of a line or lines. Figure 2 shows the cumulative distributions of the fraction of fits to these simulations with significance poorer than S.

GB870303 Period S1

We carried out fits to simulated spectra at only three angles: 30°, 45°, and 60°; attempts to fit simulations at 75° failed. We find that the BATSE SD can marginally detect the line at 30° but not at larger angles (see Figure 3).

In the case of the saturated line, the three different fitting methods give disparate results, as expected. Using variances derived from the model gives significances which are similar to the rigorous maximum likelihood method at incident angles of 45° and 60°, but which are three times more optimistic at an incident angle of 30°. In contrast, using variances derived from the data gives significances which are more optimistic than the rigorous maximum likelihood method by a factor of nearly 20 at incident angles of 30° and 45°, and by a factor of more than 4 at an incident angle of 60°. These differences are not small, which indicates that, for saturated lines, χ^2 with variances derived from the model is far better than χ^2 with variances derived from the data, but that the rigorous maximum likelihood method is clearly best.

GB870303 Period S2

We find that the BATSE SD can unambiguously detect the lines for angles $\lesssim 60°$ (see Figure 3) when we fit the data using the constrained harmonically-spaced Gaussian lines model, with four parameters. Adding additional parameters by unlinking first the widths, and then the energies, decreases the significance of detection.

In this case, we find that χ^2 with variances derived from the data gives slightly optimistic results, while χ^2 with variances derived from the model and the rigorous maximum likelihood method give similar results. Thus, the use of χ^2 with variances derived from the model is apparently adequate for establishing the existence of weak lines, although use of the rigorous maximum likelihood method is still preferable.

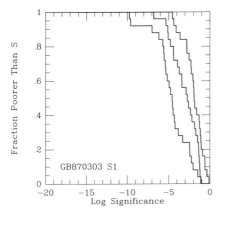
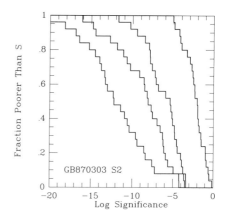
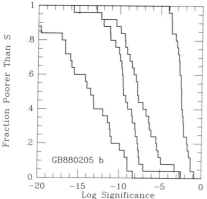

Figure 2. Cumulative distributions of the fraction of fits with significance poorer than S for simulated BATSE SD spectra with lines like those seen by *Ginga*. (Upper left panel) GB870303 S1. (Upper right panel) GB870303 S2. (Lower left panel) GB880205 b. In each case, the histograms correspond (from left to right) to incident angles of 30°, 45°, 60°, and 75°, except in the case of S1 of GB870303 where there is no histogram for 75°.

GB880205 Period b

We find a result similar to that for GB870303 S2; namely, that the BATSE SD can unambiguously detect harmonically-spaced lines at angles ≲ 60° (see Figure 3). Unlike the case of GB870303 S2, however, we find that we can always fit a single Gaussian line to the second harmonic and that most of the significance of the detection comes from this line. Thus at 30°, the ratio of the significance of the constrained harmonic Gaussian line fit and that of a single Gaussian line fit to the second harmonic is $\approx 10^{-2}$. This is due to the fact that the second harmonic has three to four times the equivalent width of the first harmonic in interval b of GB880205.

CONCLUSIONS

We find that the BATSE SD cannot detect a saturated Gaussian line like that seen in GB870303 S1, in spectra with fluence $\sim 10^{-6}$ erg cm^{-2}, except at incident angles ≲ 30°. We find that the BATSE SD should be able to detect harmonically-spaced lines like those seen in GB870303 S2 and GB880205 b in spectra with fluence $\sim 10^{-5}$ erg cm^{-2} for burst incidence angles ≲ 60°.

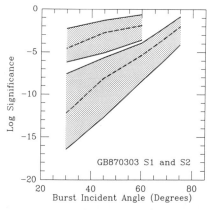 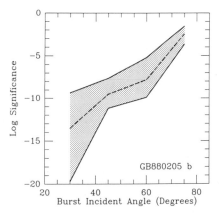

Figure 3. (Left panel) Significances of the detection by the BATSE SD of lines like those seen by *Ginga* in GB870303 S1 (upper shaded region) and S2 (lower shaded region). (Right panel) Same as the left panel, except for GB880205 b.

Detection at larger angles is unlikely due to the Al shielding.

We reach two other important conclusions in this work. First, we find that the detectability of lines is greatest using a constrained harmonic Gaussian line model with $E_2 = 2E_1$ and $\sigma_2 = 2\sigma_1$. While the model is not physically realistic, it provides the optimal search strategy for harmonically-spaced lines in the BATSE SD data. Second, we find that for saturated lines, χ^2 fits using variances derived from the model and from the data give significances that are markedly more optimistic than those found using the rigorous maximum likelihood method; for weaker lines, χ^2 fits using variances derived from the model apparently suffices.

We thank Carlo Graziani for numerous discussions about fitting to the *Ginga* data. PEF acknowledges the support of NASA Graduate Traineeship NGT-50778. We also acknowledge NASA grants NAGW-830, NAGW-1284, and NAG5-1454. Work at LANL was done under the auspices of the DOE.

REFERENCES

1. Meegan, C., et al. 1992, *Nature*, **355**, 143.

2. Fenimore, E. E., et al. 1988, *Ap. J. Lett.*, **335**, L71.

3. Graziani, C. et al. , in Gamma-Ray Bursts: Observations, Analyses, and Theories, ed. C. Ho, R. I. Epstein, and E. E. Fenimore (Cambridge University Press, 1992), p. 407.

4. Lamb, D. Q., Wang, J. C. L., Loredo, T. J., Wasserman, I., Salpeter, E. E., and Fenimore, E. E. 1989, *Ann. N.Y. Acad. Sci.*, **571**, 460.

5. Wang, J. C. L., et al. 1989, *Phys. Rev. Letters*, **63**, 1550.

6. Schaefer, B., 1992, private communication.

7. Lampton, M., Margon, B., and Bowyer, S. 1976, *Ap. J.*, **208**, 177.

COMPTON SCATTERING OF GAMMA-RAY BURST SPECTRA

Xin-Min Hua
Space Astrophysics Laboratory,
Institute for Space and Terrestrial Science,
Concord, Ontario, L4K 3C8 Canada

AND

Richard E. Lingenfelter
Center for Astrophysics and Space Sciences, C-0111
University of California, San Diego, La Jolla, CA 92093

ABSTRACT

We investigate the effects on the measured spectrum of gamma-ray bursts of Compton scattering of the gamma-ray burst emission both in the Earth's atmosphere and in the Compton Observatory. We have developed a Monte Carlo program to calculate the spectra of Compton scattered burst photons incident upon each of the eight BATSE Spectroscopy Detectors, as a function of energy and angle with respect to both the burst source and the geocentric directions. From these calculations, we find that the effects of the Compton reflected spectra depend strongly on the orientation of each detector with respect to the burst and the Earth. Thus, the effects differ greatly from detector to detector. We also find that, in general, the reflected spectra are significantly steeper than the incident spectra of the bursts, so that the effects of reflected photons are much more important at lower energies than they are at higher energies. Finally, we show that the Compton reflected photons can greatly reduce the detectability of spectral features at energies less than ~ 100 keV.

INTRODUCTION

Because each of the unshielded, BATSE gamma-ray burst Spectroscopy Detectors is effectively an all-sky detector, the gamma-ray spectrum measured by each detector can be significantly affected by Compton scattering of the burst emission in the atmosphere of the Earth, which fills roughly 33% of the sky, and in the Compton Observatory spacecraft, which fills the backward hemisphere of each detector. In order to investigate the effects of such Compton scattering on the measured spectrum of gamma-ray bursts, we have developed a Monte Carlo program to calculate the spectra of Compton scattered photons incident upon each of the eight Spectroscopy Detectors, as a function of energy and angle with respect to both the burst source and the geocentric directions.

Here, we briefly describe the Monte Carlo program and then discuss our preliminary calculations of the Compton reflected flux, with particular emphasis on the detection of absorption features in the gamma-ray burst spectra.

MONTE CARLO SIMULATION PROGRAM

At the time of each gamma-ray burst, we know the pointing directions of the three axes of the Compton Observatory spacecraft, the spacecraft altitude, the direction of the geocenter and the approximate direction of the burst. For the purpose of the Monte Carlo simulations of the Compton scattering from the Earth's atmosphere, we transform these directions into an Earth centered coordinate system which has the burst direction as one of its axes. We approximate the Earth's atmosphere as an optically thick, sphere with a radius of 6439 km corresponding to the limb of the atmosphere as seen by $\sim$ 100 keV photons; this is 68 km above the mean surface of the Earth. We assume a uniform density for the atmosphere, taking the value at the limb altitude, and from the mean atmospheric composition[1] and the appropriate Compton scattering and photoelectric cross sections[2], we determine the energy-dependent scattering and absorption mean free paths.

In simulating the Compton scattering, we randomly select incident photons from an assumed burst spectral distribution and randomly chose the point at which each is incident on the spherical surface of the atmosphere which is visible to both the spacecraft and the burst source. We then follow the photon as it travels in the atmosphere until it either escapes from the atmosphere, is absorbed, or is scattered to a low enough energy (< 10 keV) that it is no longer of interest.

In doing so, we determine the path length between two scatterings from the scattering mean free path at the photon energy and calculate the direction and energy of the scattered photon by usual Monte Carlo method. At each scattering point, however, we also calculate the the probability that the photon would be scattered in the direction of the spacecraft and the corresponding energy of the scattered photon. We then calculate the survival probability determined from the column depth of the atmosphere along a line of sight from the scattering point to the spacecraft and the photon scattering and absorption mean free paths at the scattered energy. The accumulation of products of these two probabilities at all scattering points along the path of the photon consists the contribution to the escape spectrum of that photon. We then calculate the effective weight, or probability of detection of the photon at each of the detectors that could see it. This is determined by the energy-dependent effective area of that detector for a photon of this energy and arrival direction, which have calculated from the detector design specifications[3,4] and composition. We thus count the scattered photons in each detector weighted by the product of all of their respective weights. In this way we calculate the contributions of the Compton scattering of some 10^5 to 10^6 incident photons.

For the Monte Carlo simulations of the Compton scattering of the burst emission from the Compton Observatory spacecraft, we use only a very simple approximation at the present to show the comparative effects. We plan to make much more detailed calculations using the spacecraft mass model. At present, we assume a cylindrically symmetric coordinate system along the axis of each Spectroscopy Detector, and determine the direction of the burst with respect to that axis. We approximate the spacecraft simply as an optically thick disk in the backward hemisphere of each detector with a disk radius of 10 times the detector radius.

We simulate only the Compton scattering in the spacecraft of photons coming directly from the burst and ignore the second order effects of Compton scattering in the spacecraft of Compton scattered burst photons from the Earth. In the spacecraft simulations, we follow the same procedure with the Compton scattering of incident burst photons as we used in the simulations in the Earth's atmosphere.

Finally, we calculate for comparison the unscattered burst spectrum seen directly by each detector, weighted again by the energy-dependent effective area of the detector as a function of burst direction. The sum of these three weighted spectra, the direct emission, and the Compton scattered reflections from the spacecraft and the Earth's atmosphere, approximate the total spectrum actually seen by each detector.

COMPTON SCATTERED SPECTRA

Carrying out Monte Carlo simulations for a number of configurations, we find that the effects of such reflected spectra depend strongly on the directions of the burst and Earth with respect to the axis of each detector, and thus the effects differ greatly from detector to detector. We also find that, in general, the reflected spectra are significantly steeper than the incident spectra of the bursts, so that the effects of reflected photons are much more important at lower energies than they are at higher energies.

These reflected photons can greatly reduce the detectability of spectral features at energies less than ~100 keV. This can be seen in Figure 1, in sample calculations of the direct and Compton scattered photon spectra observed by the two adjacent, low-threshold Spectroscopy Detectors 6 and 7, looking closest to the burst direction. This corresponds to the conditions for an actual burst GB910803. In this simulation, we have assumed an incident burst continuum spectrum of the form, $(E + E_o)^{-2}$ with $E_o = 45$ keV and a 10 keV wide square-wave absorption feature centered at 45 keV with an equivalent width of 5 keV.

We see that even for the Spectroscopy Detectors that were looking closest to the burst direction, the Compton reflected flux from the adjacent parts of the Compton Observatory spacecraft is expected to be comparable to the directly measured flux from the burst at energies < 100 keV. Moreover, although the total flux seen by the two detectors is nearly the same, the direct flux from the burst is roughly 25% lower in detector 6 (pointed 64° from the burst) than in detector 7 (pointed only 38° away), while the Compton reflected flux is roughly

33% higher in detector 6 (pointed 94° from the geocenter and only about 23° above the atmospheric limb) than in detector 7 (pointed 117° away from the geocenter and twice as far above the limb). As a result, the ratio of the signal to noise in the line differs by a factor of ~ 1.8 between that in the detector pointed closest to the burst and the adjacent detector pointed farther away. Thus, a 3σ, or 0.1% significance, line detection in the closer detector would be seen as only a 1.7σ, or barely 5% significance, line in the adjacent detector. Even a 5σ, or 10^{-5} significance, line detection in the first detector would be barely a 2.4σ, or 10^{-2} significance, line detection in the other. This makes it very difficult to confirm the detection of an absorption line by detecting it in two separate Spectroscopy Detectors.

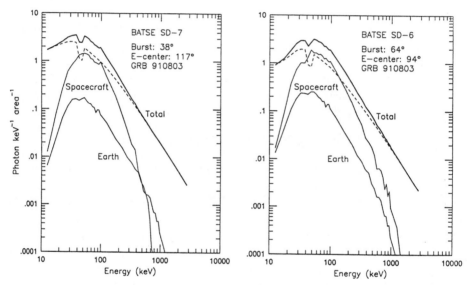

Figure 1. The effect of Compton scattered burst photons on the detectability of absorption lines in gamma-ray burst spectra, shown by a comparison of calculated direct burst fluxes (dashed line) and Monte Carlo simulated Compton scattered fluxes (solid lines) from the spacecraft and the Earth, seen by two adjacent Spectroscopy Detectors 6 and 7, looking closest to the burst direction.

We also see that in this case (Figure 1) the Compton reflected flux from the Earth is small compared to that from the spacecraft.

The effects of Compton scattering in the spacecraft have been considered in the detector response matrices developed[5] for the burst spectral analysis. The program that we have developed, when modified to include a realistic spacecraft mass model, will allow us to make an independent check of the spectra determined from these matrices, so that the reliability of the burst spectra can be confirmed.

Acknowledgements. We thank D. Band, D. Gruber, J. Matteson and G. Pendleton for helpful discussions. This work was supported in part by NASA under contract NAS8-36081 (UCSD group).

REFERENCES

1. U.S. Standard Atmosphere, (Washington: GPO, 1976).
2. E.F. Plechaty, D.E. Cullen and R.J. Howerton, UCRL-50400 (1975).
3. J.M. Horack, NASA RP1268 (1991).
4. J.L. Matteson, private communication.
5. G.N. Pendleton, private communication.

GAMMA-RAY BURSTS: GALACTIC OR COSMOLOGICAL IN ORIGIN?

D. Q. Lamb

Dept. of Astronomy and Astrophysics, Univ. of Chicago, Chicago, IL 60637

ABSTRACT

We describe our work that addresses the question of whether γ-ray bursts are galactic or cosmological in origin. We consider first the brightness and sky distributions of the bursts, and second the lines that have been observed in the spectra of a few bursts.

INTRODUCTION

Gamma-ray bursts continue to confound astrophysicists a quarter of a century after their discovery.[1] The bursts span a wide range of durations and characteristic spectral energies, and exhibit a bewildering variety of time histories (see, e.g., Fishman et al.[2]). The challenge of deciphering the nature of the bursts is exacerbated by the fact that one cannot predict when or from where on the sky the bursts will come, and the fact that it has been impossible to find quiescent counterparts of the bursts at radio, infrared, optical, ultraviolet, X-ray, or γ-ray energies.

The lack of quiescent counterparts puts a premium on garnering knowledge from the bursts themselves. The brightness and sky distributions of the bursts[3] constrain, e.g., the spatial distribution of burst sources, and suggest an astronomical distance scale. The narrow low-energy lines seen in the spectra of a few bursts[4-6] constrain the nature of the burst sources (see, e.g., Wang et al.[7]), and provide (indirectly) a physical distance scale.[8] Below we describe some of our work on each approach.

BRIGHTNESS AND SKY DISTRIBUTIONS

It is difficult to extract information from the brightness distribution of the bursts due to the complexities introduced by correlations among the intrinsic properties of the bursts and their sources, and among the properties of the bursts and those of the detector.[9] Only when the cumulative C_{max}/C_{min} distribution for the bursts has a -3/2 slope can a clear implication be extracted; namely, that the burst brightness distribution is *consistent* with the burst sources being uniformly distributed in space. But it can never be proven that the sources are uniformly distributed in space; indeed, if even one experiment yields a slope different from -3/2, this is sufficient to rule out a uniform source distribution.

Two-population Galactic models. Prior to the launch of the *Compton Gamma Ray Observatory*, it was generally believed that the γ-ray burst sources were neutron stars located at distances $\lesssim 300$ parsecs in the Galactic disk with luminosities $\sim 10^{37}$ ergs s^{-1} (see, e.g., Higdon and Lingenfelter;[10] Harding[11]). A Galactic disk source distribution was consistent with the -3/2 slope of the cumulative C_{max}/C_{min} distribution for bright bursts and their isotropic distribution on the sky.

The confirmation of a roll-over in the brightness distribution of the bursts

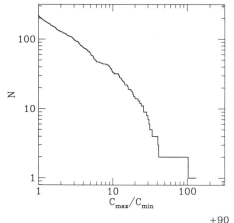

Figure 1. BATSE cumulative C_{max}/C_{min} distribution for 217 γ-ray bursts. (From Meegan et al.[12])

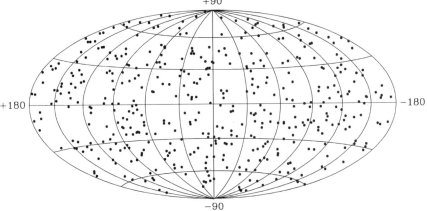

Figure 2. BATSE sky distribution for 447 γ-ray bursts. (From Meegan et al.[13])

(see Figure 1) and the discovery that even faint γ-ray bursts are distributed isotropically on the sky (see Figure 2),[12] suggest that we are at or near the center of a roughly spherical distribution of burst sources which is finite in extent. Consequently, it is now thought that, if γ-ray bursts originate in the Galaxy, the burst sources cannot lie in the Galactic disk. Rather, they must lie in an extended Galactic halo at distances $\sim 50-100$ kpc, implying luminosities $\sim 10^{41}$ ergs s^{-1}, or at cosmological distances $\sim 1-3$ Gpc, implying luminosities $\sim 10^{51}$ ergs s^{-1}.

However, we have shown that the angular and brightness distributions seen by BATSE are consistent with as much as 70% of the burst sources lying nearby in the Galactic disk, with the remainder in an extended Galactic halo.[14] The fraction of sources that can lie in the disk depends on the type of halo: No more than $\sim 20\%$ of the sources can lie nearby in the Galactic disk, if the rest reside in a standard "dark matter" halo; while $\sim 2/3$ of the burst sources can lie in the Galactic disk, if the rest reside in a Gaussian shell halo (see Figure 3). (A lack of distant halo sources is the key feature that allows a large fraction of the sources to be in the Galactic disk.) Nevertheless, the startling conclusion is that many, and perhaps even the majority, of observed bursts can originate

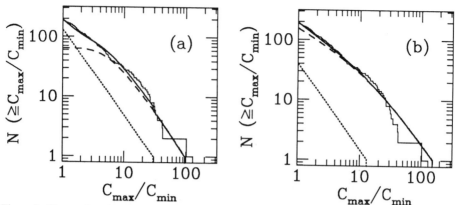

Figure 3. Comparison between disk-halo models and the BATSE cumulative C_{max}/C_{min} distribution for the 193 bursts detected by the 1024 msec trigger for which C_{max}/C_{min} is available. The dotted line shows the contribution from the disk, the dashed line shows the contribution from the halo, and the solid line shows the total. (a) Disk plus Gaussian halo. (b) Disk plus "dark matter" halo. (From Smith and Lamb.[14])

from local neutron stars in the Galactic disk.

Gamma-ray bursts as a probe of large-scale structure in the universe. If γ-ray bursts are cosmological in origin, the burst sources are expected to trace the large-scale structure of luminous matter in the universe. The use of γ-ray bursts to probe large-scale structure offers several advantages. Existing galaxy and quasar surveys have studied the two-point correlation function on angles $\lesssim 1°$ (see Luo and Schramm[15], and references therein), but such angular scales are too small to reflect the spectrum of primordial density fluctuations.[16] While future galaxy and quasar surveys will provide information on the behavior of the two-point correlation function on larger angular scales, they will face difficulties due to absorption of light by dust and gas clouds in the Galaxy. In contrast, the Galaxy is transparent to γ-rays and studies of γ-ray bursts can easily probe large angular scales. The BATSE catalogue of γ-ray bursts constitutes a homogeneous sample covering the entire sky, unlike existing galaxy and quasar surveys. Because γ-ray bursts are rare events ($R_b \sim 10 \text{ Gpc}^{-3} \text{ yr}^{-1}$), the burst sources form a sparse sample. Their two-point angular correlation function $w(\theta)$ is therefore diluted very little by projection effects, unlike that of galaxies. Finally, the brightness distribution of the bursts seen by BATSE suggests that we are seeing beyond the edge of the burst source distribution.[12] This contrasts with galaxy surveys where the observer function decreases markedly before the edge of the galaxy distribution is reached, complicating attempts to detect structure on the largest scales.

The use of γ-ray bursts to probe large-scale structure suffers from two disadvantages. Foremost is the limited number of bursts ($N_b \approx 1000$ for 3 years of BATSE observations and ≈ 3000 for 10 years of BATSE observations). Second, no estimates of the distances to burst sources exist, restricting studies of source clustering to the burst two-point angular correlation function $w(\theta)$.

We have shown that it may well be possible to use γ-ray bursts to probe the structure of luminous matter on the largest scales known,[17] consistent with recent determinations from studies of superclusters[18-21] and pencil-beam

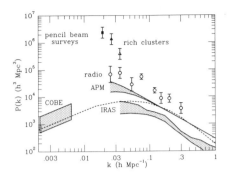
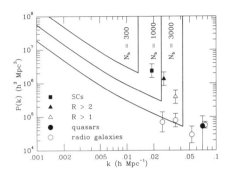

Figure 4. Power spectrum $P(k)$ of density fluctuations as a function of wavenumber k corresponding to distance scales of order $10h^{-1}$ Mpc to $3h^{-1}$ Gpc. The shaded regions show the power spectrum of total mass-energy density fluctuations found from analysis of the Cosmic Background Explorer (COBE) data, and the power spectra of luminous matter density fluctuations found from analysis of the APM optical and the IRAS infrared survey data, respectively. The data points shown are from analysis of surveys of faint radio galaxies, quasars, clusters of galaxies and from pencil beam surveys and studies of superclusters. The dashed line shows a typical power spectrum for the Cold Dark Matter model.

Figure 5. Power spectrum $P(k)$ of density fluctuations as a function of wavenumber k corresponding to distance scales of order $10h^{-1}$ Mpc to $3h^{-1}$ Gpc. The data points shown are from analysis of surveys of faint radio galaxies, quasars, clusters of galaxies and from pencil beam surveys and studies of superclusters. Also shown are the regions of the $(k,P[k])$-plane that can be probed with 300, 1000, and 3000 γ-ray bursts, taking $D = 1h^{-1}$ Gpc ($z \approx 0.3$) for illustrative purposes. (From Lamb and Quashnock.[17])

surveys[22-23] (see Figures 4 and 5). Hartmann and Blumenthal[24] earlier searched for correlations in the angular distributions of the γ-ray bursts in the Konus and interplanetary network catalogues. They detected none, but this is not surprising given the small numbers (160 and 54, respectively) of bursts in the catalogues and the very large uncertainties in the locations of the Konus bursts.

Figure 5 shows the power spectrum $P(k)$ of density fluctuations as a function of wavenumber k corresponding to distance scales of order $10h^{-1}$ Mpc to $3h^{-1}$ Gpc. The data points shown are from analysis of surveys of faint radio galaxies,[25] quasars,[26] clusters of galaxies,[27,28] and from pencil beam surveys[22,23] and studies of superclusters.[18-21] Also shown are the regions of the $(k,P[k])$-plane that can be probed with 300, 1000, and 3000 γ-ray bursts, choosing $D = 1h^{-1}$ Gpc ($z \approx 0.3$) for illustrative purposes. The righthand boundary of the region that can be probed scales in k like D^{-1}, while the lower boundary scales in $P(k)$ like D^2. Therefore, if $D = 0.3h^{-1}$ Gpc ($z \approx 0.1$), the righthand boundary shifts rightward by a factor of 3, while the lower boundary shifts downward by a factor of 9. Conversely, if $D = 3h^{-1}$ Gpc ($z \approx 3$), the righthand boundary shifts leftward by a factor of 3, while the lower boundary shifts upward by a factor of 9.

Figure 5 shows that, if luminous matter is clustered on large scales with a density contrast of order unity and γ-ray bursts trace this matter, the resultant clustering of bursts will be detectable in the two-point angular correlation function $w(\theta)$ of the bursts if $D \lesssim 1h^{-1}$ Gpc. If the clustering of luminous matter continues like $P(k) \sim P(k_0)k_0/k$ to somewhat larger scales, the resultant clustering of bursts will be detectable even for $D = 3h^{-1}$ Gpc.

A positive result would provide compelling evidence that most γ-ray bursts are cosmological in origin, and would allow comparison between the distribution of luminous matter and the distribution of dark matter on large scales. Conversely, a negative result might cast doubt on the cosmological origin of the bursts, provide evidence that the power on large scales is less than that expected from studies of superclusters and pencil beam surveys, or indicate that γ-ray bursts have some more exotic origin.

CYCLOTRON LINES

The continuum spectrum of γ-ray bursts can provide important constraints on theory, but inverting it uniquely to determine the radiation mechanism, let alone physical parameters like the density and temperature, is exceedingly difficult. In contrast, the power of spectral lines is well known: Analyses of atomic lines transformed astronomy into astrophysics.

Some time ago Mazets et al.[29,30] reported the observation of single, low-energy dips in the spectra of approximately 15% of the γ-ray bursts detected by the Konus experiment on Venera 11-14. Heuter[31] reported the observation of single, low-energy lines in the spectra of several bursts detected by the HEAO-1 A-4 experiment.

Recently, Murakami et al.,[4] Graziani et al.,[5] and Yoshida et al.[6] reported the observation of well-resolved, harmonically-spaced, low-energy lines in the spectra of three γ-ray bursts: GB870303, GB880205, and GB890929. We fit the observed photon count rate spectrum for GB880205 by folding Monte Carlo spectra through the *Ginga* Burst Detector response functions.[7] The results showed that cyclotron resonant scattering in a strong magnetic field can account *quantitatively* for the positions, strengths, and widths of the observed lines (see Figure 6).

The existence of cyclotron lines in the spectra of some bursts and the success of the cyclotron scattering model in explaining them provides compelling evidence that these particular γ-ray bursts originate from strongly magnetic neutron stars, and is the strongest evidence that some γ-ray bursts are galactic, rather than cosmological, in origin.

Given the importance of this implication, it is imperative to test the cyclotron line hypothesis further. This we are doing. First, we are continuing our development of rigorous statistical methods based on Bayesian inference that address the crucial problems of establishing the existence of spectral lines and of determining the best-fit values and credible regions for the physical parameters describing them. Second, we are undertaking a search for spectral lines in the BATSE Spectroscopy Detector (SD) data and Large Area Detector (LAD) data, using these methods. Third, we are assessing the consistency of the *Ginga* and *Compton* observations, using extensive simulations to determine the sensitivities of the *Ginga* Burst Detector and the *Compton* BATSE SD and LAD. Below we describe our work in each of these areas.

Model Comparison. One of the most important, yet nettlesome, issues in the study of spectral lines in γ-ray bursts is establishing the existence of the lines themselves. Here we describe a rigorous method[32,33] derived from Bayesian inference (for discussions of the conceptual and methodological advantages of the Bayesian approach, see Loredo;[34,35] for a discussion of Bayesian model comparison, see Gregory and Loredo[36]).

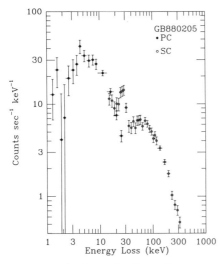
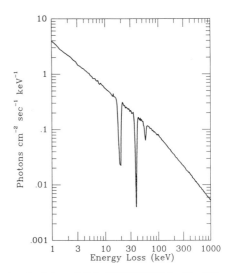

Figure 6—(Left panel) Observed count-rate spectra for interval (b) of GB880205 for the PC and SC data on *Ginga*. (Right panel) Best-fit photon-number spectra for Monte Carlo radiation transfer model. (After Wang et al.[7])

In Bayesian inference, the probability for a model as a whole is the product of a prior probability and a global likelihood. In the absence of any information suggesting otherwise, we take the prior probabilities of competing models to be equal. Then the odds ratio in favor of one model over another is given by the ratio of their global likelihoods. Suppose that model 1 has M_1 parameters, denoted A_α, and has a minimum χ^2 equal to $\chi^2_{1,\min}$. Suppose further that model 2 has M_2 parameters, denoted A'_α, and has a minimum χ^2 equal to $\chi^2_{2,\min}$. Assuming Gaussian errors and uniform priors, the odds favoring model 2 over model 1 is,

$$O_{21} = \frac{\mathcal{L}_2}{\mathcal{L}_1}(2\pi)^{(M_2-M_1)/2}\sqrt{\frac{\det V_2}{\det V_1}}\frac{\prod_{\alpha=1}^{M_1}\Delta A_\alpha}{\prod_{\alpha=1}^{M_2}\Delta A'_\alpha}, \quad (1)$$

where $\mathcal{L}_2/\mathcal{L}_1 = e^{\frac{1}{2}\Delta\chi^2}$ is the ratio of maximum likelihoods, $\Delta\chi^2 = \chi^2_{1,\min} - \chi^2_{2,\min}$, V_1 and V_2 are the covariance matrices for the estimated parameters, and ΔA_α and $\Delta A'_\alpha$ are the prior uncertainties for the parameters A_α and A'_α.

Equation (1) reveals the odds ratio to be the product of the maximum likelihood ratio and a factor which includes the ratio of the posterior uncertainty of the extra parameters to their prior uncertainty. The maximum likelihood ratio will always favor the more complex model, since χ^2 of the more complex model can never be larger than that of the simpler model and therefore $\Delta\chi^2 \geq 0$. But the second factor penalizes the more complex model, since the posterior uncertainty for the extra parameters will generally be smaller than their prior uncertainty. Thus an "Ockham's Razor" automatically appears in Bayesian model comparison (the dependence of this factor on the prior range superficially resembles correcting a frequentist statistic for the number of parameter values examined, but the horrendous problems associated with choosing the number

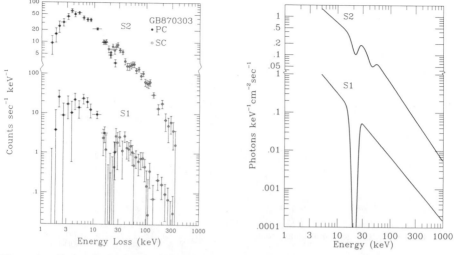

Figure 7—(Left panel) Observed count-rate spectra for intervals S1 and S2 of GB870303 for the PC and SC data on *Ginga*. (Right panel) Best-fit photon-number spectra for exponentiated Gaussian line model. (From Graziani et al.[5])

and location of the examined values are absent in the Bayesian approach). Thus the ratio of maximum likelihoods $\mathcal{L}_2/\mathcal{L}_1$ must exceed some critical value before the more complex model is favored.

An interesting special case of model comparison is the case of *nested models*, where one model is a special case of a more complicated model when the additional parameters in the more complicated model take on some default value (often zero). Line detection is an example of just this kind of comparison: We want to compare a model consisting only of a continuum spectrum to a model with an additional feature in it. Typical examples are the Gaussian line or the exponentiated Gaussian line, where the extra parameters are the centroid energy E, the strength β, and the width ΔE of the line. Figure 7 shows the results of fits we have made to the *Ginga* data for time intervals S1 and S2 of the burst GB870303 using an exponentiated Gaussian model.[5]

However, the correlations between line parameters are generally non-linear. The assumption of Gaussian errors is then invalid. Guided by the knowledge that present detectors are sensitive primarily to the centroid energy E and the equivalent width EW of narrow lines, we have re-parameterized the Gaussian line and the exponentiated Gaussian line in terms of E, EW, and the ratio $R \equiv FWHM/EW$, where $FWHM$ is the full width at half maximum of the line. We find that the resulting errors are Gaussian, due to a felicitous relation between EW and $FWHM$,[37] permitting us to evaluate the covariance matrices by inverting the Hessian at the best fit values of the line parameters. Further, we find that parameterization in terms of equivalent width allows a straightforward treatment of saturated lines and largely decouples the line and continuum parameters. Thus parameterization of a Gaussian line in terms of equivalent width has enormous advantages over the the usual parameterization. It is an important breakthrough, which allows the existence of a line to be established rigorously.

Table 1 shows the results of such a rigorous evaluation.[38] The odds favoring the existence of a line in GB870303 S1 are substantial. Taken alone, the harmonically-spaced lines in GB870303 S2 do not yield convincing odds; however, when information from S1 is brought to bear on the analysis of S2 (by fixing the line centroid energy in S2 to the best-fit value found for the fit to S1), the odds favoring the existence of harmonically-spaced lines in S2 become very large. We conclude that it is highly probably that the lines in S1 and S2 are real.

We extend this analysis of GB870303 to include having searched different time intervals for lines by considering a model in which the lines appear at time t_1, disappear at time t_2, and are constant in between. We find that the uncertainties σ_{t_1} and σ_{t_2} in the times at which the lines appear and disappear are large for the *Ginga* data. Indeed, we cannot exclude the possibility that lines are present throughout the burst. Thus the corresponding Ockham factors are of order unity, and the results cited above are little altered.

TABLE 1. BAYESIAN ANALYSIS OF SPECTRAL LINES

Fit	Likelihood Ratio	Ockham Factor	Odds
S1	8.5×10^5	1.3×10^{-4}	110 : 1
S2	1.4×10^6	2.0×10^{-6}	2.8 : 1
S2 (E_1 fixed)	1.2×10^6	5.5×10^{-4}	660 : 1

Parameter Estimation. The use of Bayes' theorem to determine what one can learn about a set of M model parameters A_α from data is called *parameter estimation*, although strictly speaking, Bayesian inference does not provide estimates for parameters. Rather, the Bayesian solution to the parameter estimation problem is the full posterior *distribution*. Of course, it is often useful to summarize this distribution in terms of a "best-fit" value and "credible" intervals or regions.

Frequently a model will have more than one parameter, but we want to focus attention on a subset of the parameters. For example, we may want to focus on the implications of spectral data for the parameters of a line (*e.g.*, its centroid energy E, strength β, and width ΔE), independent of the shape of the continuum spectrum. If the parameters have non-Gaussian errors (as is usually the case), one must project the full multi-dimensional confidence volume onto one or two dimensions to obtain confidence intervals or regions within frequentist statistics.[39] The resulting confidence intervals or regions are often excessively large. For this reason, the parameters which one is not focusing attention on have come to be called "nuisance parameters."

In contrast, Bayesian inference offers a powerful and straight-forward method: Integration of the posterior density over the parameters one is not focusing on.[34,40,41] The resulting credible intervals or regions are far smaller, and correctly give the probability that the true value of a parameter lies within the credible interval or region. The case of fitting an exponential decay model of neutrino emission to the data from supernova SN 1987A illustrates the power of the Bayesian method.[40,41] In this particular example, the 99.8% credible region found from Bayesian statistics lies well within the projected 68% confidence region found from frequentist statistics!

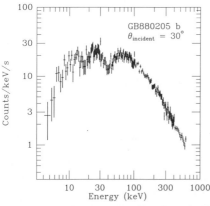 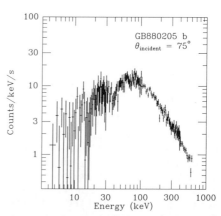

Figure 8. Simulated spectra like that of GB880205 b, as seen by the BATSE SD. (Left panel) Incident angle 30°, where the harmonically-spaced lines at $\approx$ 20 and 40 keV are clearly seen. (Right panel) Incident angle 75°, where the first harmonic at $\approx$ 20 keV is lost, due to the Al shielding, and the feature at $\approx$ 40 keV is no longer significant. (From Freeman et al.[42])

The *EW* parameterization of the line is again advantageous because the credible volume in parameter space is ellipsoidal, and the credible intervals of the individual parameters can then be well-characterized in tabular form.

Sensitivity of the BATSE SD to lines like those seen by *Ginga*. We use simulations to explore the sensitivity of the BATSE Spectroscopy Detector (SD) to harmonically-spaced lines like those seen by the *Ginga* Burst Detector[42] (see also Band et al.[43] and Palmer et al.[44]). We simulate bursts using the best-fit spectral parameters derived from the *Ginga* data for time intervals S1 and S2 of GB870303 and time interval b of GB880205 (see Figure 8). These bursts represent the weakest and strongest bursts with lines seen by Ginga.

Our results show that, within the context of the Gaussian line model, comparison of a power-law continuum model and a power-law continuum plus "constrained" harmonic lines model ($E_2 = 2E_1$, $\sigma_2 = 2\sigma_1$, and $\beta_2 \neq \beta_1$, where E_i is the centroid energy, σ_i is the width, and β_i is the strength of the line) constitutes the optimal search strategy for harmonically-spaced lines in the SD data. They also show that for saturated lines, χ^2 fits using variances derived from the model and from the data give significances that are markedly more optimistic than those found using the rigorous maximum likelihood method; for weaker lines, χ^2 fits using variances derived from the model, but not from the data, apparently suffice.

We find that the ability of the SD to detect a line is a sensitive function of the line energy, strength, and width (see Figure 9). The SD is able to detect strong lines like that at $\approx$ 20 keV in interval S1 of GB870303 only at angles $\lesssim$ 30°, due to the Al shielding. The SD can detect the weaker, narrower, harmonically-spaced lines like those at $\approx$ 20 and 40 keV in interval S2 of GB870303 at angles $\lesssim$ 60° because the SD is able to detect the second harmonic at larger incident angles. Similarly, the SD is able to detect lines like those at $\approx$ 20 and 40 keV in interval b of GB880205 at angles $\lesssim$ 60°.

Assessing the Consistency of the Ginga and Compton Observations To date no statistically significant lines have been found in any burst by the BATSE team. Given this situation, it is imperative to assess the consistency

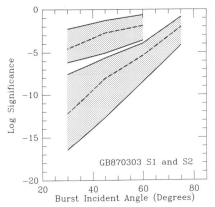 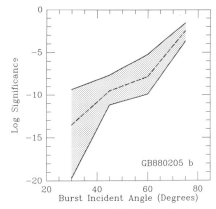

Figure 9. (Left panel) Significances of the detection by the BATSE SD of lines like those seen by *Ginga* in GB870303 S1 (upper shaded region) and S2 (lower shaded region). (Right panel) Same as the left panel, except for GB880205 b. (From Freeman et al.[42])

of the *Ginga* and *Compton* observations. We are doing so[45] (see also Teegarden et al.[46]).

Using Bayesian statistical inference and more than 10^3 simulations, we have evaluated the sensitivity of the *Ginga* Burst Detector to harmonically-spaced lines like those in GB880205 in a fashion similar to that described above for the BATSE SD. These preliminary results suggest that roughly 35% of bursts should contain lines if all lines were like those in GB880205. A preliminary comparison of the *Ginga* and *Compton* sensitivities and angular and time exposures then implies that the BATSE SD should be seeing roughly 6 events with lines per year, although this number is highly uncertain. We are now evaluating the sensitivity of the *Ginga* Burst Detector and the *Compton* BATSE SD to harmonically-spaced lines like those seen in GB870303 and GB890929 in order to refine this number.

THE FUTURE

The central question about γ-ray bursts is the location of the burst sources: Are the bursts Galactic or cosmological in origin? Below we assess the ways in which this question might be answered.

Distinct morphological classes. To date, no clear correlations between burst properties or distinct morphological classes of bursts have been found. However, the Burst and Transient Source Experiment (BATSE) on the *Compton* Gamma-Ray Observatory has now detected more than 600 γ-ray bursts.[13] The BATSE catalogue of bursts is the largest homogeneous sample of γ-ray bursts ever assembled and, as such, constitutes a unique resource for studying burst properties. It is therefore possible that correlations between burst properties or distinct morphological classes of bursts which were previously not visible will now become apparent.

Brightness distribution. Little further knowledge of bursts is likely to come from the burst brightness distribution because of the great complexities potentially introduced by correlations among the intrinsic properties of the bursts and between the properties of the bursts and those of the BATSE instrument.

Sky distribution. Information may come from the angular distribution of bursts if, with larger numbers of bursts detected, $\langle \cos\theta \rangle \neq 0$ or $\langle \sin^2 b \rangle \neq 0.33$, and/or clustering of the bursts is found. The former would support a Galactic origin, while the latter would likely support a cosmological origin.

Continuum and line spectra. Study of burst continuum spectra is not likely to lead to a definite conclusion, unless low-energy ($E \lesssim 3$ keV) X-ray emission is seen from bursts lying almost directly in the Galactic plane. The latter would imply that these sources lie nearby in the Galactic disk.

Line spectra may further constrain possibilities, if cyclotron lines are observed in some bursts detected by BATSE or by the High Energy Transient Experiment.

Quiescent counterparts. If quiescent counterparts to the bursts were found, they would likely provide the most immediate and powerful information about the distance to the burst sources, but the chances of finding them are unclear. The most accurate locations for this purpose will come from the Interplanetary Burst Network and from the High Energy Transient Experiment.

Serendipitous discovery. The discovery of γ-ray bursts was serendipitous and many of the most important advances in the field have come from unexpected discoveries (witness the discovery by Meegan et al.[12] that faint bursts are isotropically distributed on the sky). Thus, if history is a guide, the most likely way in which the distance to the burst sources will be established is through some entirely unexpected or serendipitous discovery.

This research was supported in part by NASA grants NAGW-830 and NAGW-1284, and NASA contract NASW-4690.

REFERENCES

1. R. W. Klebesadel, I. B. Strong, & R. A. Olson, Ap. J. (Letters) **182**, L85 (1973).
2. G. J. Fishman, et al., in Gamma-Ray Bursts (W. S. Paciesas and G. J. Fishman, eds., AIP, New York, 1992), p. 13.
3. C. A. Meegan, et al., in Gamma-Ray Bursts (AIP: New York, 1992), p. 13.
4. T. Murakami, et al., Nature **335**, 234 (1988).
5. C. Graziani, et al., in Gamma-Ray Bursts: Observations, Analyses, and Theories (C. Ho, R. I. Epstein, & E. E. Fenimore, eds., Cambridge University Press, Cambridge, 1991), p. 315.
6. A. Yoshida, T. Murakami, J. Nishimura, I. Kondo, & E. E. Fenimore, Pub. Astron. Soc. Japan **43**, L69 (1991).
7. J. C. L. Wang, et al., Phys. Rev. Letters **63**, 1550 (1989).
8. D. Q. Lamb, J. C. L. Wang, & I. Wasserman, Ap. J. **363**, 670 (1990).
9. T. J. Loredo & I. Wasserman, these proceedings (1993).
10. J. C. Higdon & R. E. Lingenfelter, Ann. Rev. Astron. Ap. **28**, 401 (1990).
11. A. Harding, Phys. Rep. **206**, 327 (1991).
12. C. A. Meegan, et al., Nature **355**, 143 (1992).
13. C. A. Meegan et al., these proceedings (1993).
14. I. A. Smith & D. Q. Lamb, these proceedings (1993).
15. X. Luo and D. N. Schramm, Science **256**, 513 (1992).
16. E. W. Kolb and M. S. Turner, The Early Universe (Addison-Wesley, Reading, MA, 1990).
17. D. Q. Lamb & J. M. Quashnock, these proceedings (1993).

18. N. Bahcall and R. Soneira, Ap. J. **270**, 20 (1983).
19. N. Bahcall and W. Burgett, Ap. J. (Letters) **300**, L35 (1986).
20. M. Geller & J. P. Huchra, Science **246**, 897 (1989).
21. L. Guzzo, et al., Ap.J. **393**, L5 (1992).
22. D. C. Koo and R. Kron, in Observational Cosmology (A. Hewitt, G. R. Burbidge, and L. Z. Fang, eds., Reidel, Dordrecht, 1987), p. 383.
23. T. J. Broadhurst, et al., Nature **343**, 726 (1990).
24. D. Hartmann and G. R. Blumenthal, Ap. J. **342**, 521 (1989).
25. J. A. Peacock and D. Nicholson, M.N.R.A.S. **253**, 307 (1991).
26. N. Bahcall and A. Chokshi, Ap. J. **385**, L33 (1992).
27. S. A. Schectman, Ap. J. (Supplements) **57**, 77 (1985).
28. O. Lahav, et al., M.N.R.A.S. **238**, 881 (1989).
29. E. P. Mazets, et al., Nature **290**, 378 (1981).
30. E. P. Mazets, et al., Ap. Space Sci. **82**, 261 (1982).
31. G. J. Heuter, Ph.D. Thesis (University of California, San Diego). 1988
32. T. J. Loredo, private communication (1991).
33. T. J. Loredo & D. Q. Lamb, in Gamma-Ray Bursts (W. S. Paciesas and G. J. Fishman, eds., AIP, New York, 1992), p. 414.
34. T. J. Loredo, in Maximum-Entropy and Bayesian Methods (P. Fougère, ed., Kluwer Academic Publishers, Dordrecht, 1990), p. 81.
35. T. J. Loredo, in Statistical Challenges in Modern Astrophysics (E. Feigelson and G. Babu, eds., Springer-Verlag, New York, 1992), p. 275.
36. P. C. Gregory & T. J. Loredo, Ap. J. **398**, 146 (1992).
37. C. Graziani, P. Freeman, & D. Q. Lamb, Ap. J., to be submitted (1993).
38. C. Graziani, et al., these proceedings (1993).
39. M. Lampton, B. Margon, B., and & S. Bowyer, S., AP. J. **208**, 177 (1976).
40. T. J. Loredo, & D. Q. Lamb, in Proceedings of the 14th Texas Symposium on Relativistic Astrophysics, N.Y. Acad. Sci. **571**, 1989 (601).
41. T. J. Loredo & D. Q. Lamb, Phys. Rev. **D**, to be submitted (1993).
42. P. E. Freeman, et al., these proceedings (1993).
43. D. Band, et al., these proceedings (1993).
44. D. Palmer, et al., these proceedings (1993).
45. E. E. Fenimore, G. Schwarz, D. Q. Lamb, P. Freeman, and T. Murakami, these proceedings (1993).
46. B. Teegarden, et al., these proceedings (1993).

GAMMA-RAY BURSTS: PULSE PROFILES

POSSIBLE DETECTION OF SIGNATURE CONSISTENT WITH TIME DILATION IN GAMMA-RAY BURSTS

J. P. Norris[1], R.J. Nemiroff[1,2], C. Kouveliotou[2,3], G. J. Fishman[3], C. A. Meegan[3], R. B. Wilson[3], W. S. Paciesas[3,4]

[1]NASA/Goddard Space Flight Center, Greenbelt, MD 20771
[2]Universities Space Research Association
[3]NASA/Marshall Space Flight Center, Huntsville, AL 35812
[4]University of Alabama in Huntsville, AL 35899

ABSTRACT

If gamma-ray bursters are at cosmological distances – a possibility suggested by their isotropic distribution and radial spatial inhomogeneity – then the temporal profiles of more distant sources will be time dilated compared to those of relatively nearby sources. We apply two brightness-independent tests for time dilation. Selection effects arising from intensity differences are removed by rescaling all bursts and associated noise levels to a canonical dim peak intensity. The first test measures the total normalized flux above background and indicates that dim bursts have approximately twice as much temporal structure as do bright bursts, suggesting that dim bursts are longer. Using wavelet transforms we show that, on average, dim bursts have significantly more structure than bright bursts on all time scales from 128 ms to 64 s. Simulations that approximate actual burst profiles indicate that time dilation should be detectable, given the observed distribution of burst durations and the assumption of a modest range in intrinsic luminosity. The results from analysis of about 100 BATSE bursts are consistent with cosmological distances, i.e., with a relative time dilation factor, between brightest and dimmest bursts, of $(Z+1) \sim 1.5 - 3$.

INTRODUCTION

The practically isotropic celestial distribution and the differential V/V_{max} distribution for gamma-ray bursts detected and localized[1] by *Compton's* Burst and Transient Source Experiment (BATSE) are most naturally interpreted in terms of either a cosmological[2] or heliospheric distribution of sources. However, admixtures of galactic disk and halo populations cannot be definitively ruled out with the present statistics[3]. Without (near) real-time coincident detections from other wavebands accompanied by constraining interpretations, there are few methods to discriminate among these possibilities (detection of a gravitationally lensed burst would be definitive).

One promising program is a search for time dilation: If the bursters are at cosmological distances, then the time profiles of more distant sources will be dilated relative to those of nearer sources. Since burst durations range over more than four orders of magnitude, dilation can be detected only in a statistical sense. A positive finding would not prove outright the cosmological hypothesis, but would increase the burden on any galactic scenario, requiring additional ad hoc theories to explain an anti-correlation of intensity with duration. We employ two tests for time dilation – total normalized flux and the wavelet transform – and describe measures taken to neutralize selection effects and to address effects arising from the wide variety of time profiles. Using simulations we show that the wavelet transform test is capable of making a definitive detection. Results are consistent with the expected signature of time dilation.

© 1993 American Institute of Physics

ANALYSIS

Some measure of luminosity must be assumed in order to arrive at a working measure of redshift, which is necessary for interpretative and calibration purposes. We use the burst peak intensity, determined at 256-ms resolution, ignoring for this analysis any intrinsic luminosity function. The counts are summed over the 4 channels of the LAD's (Large Area Detectors) DISCSC data type, for which the energy range is ~ 25 keV to > 1 MeV, and the temporal resolution is 64 ms. Approximately 100 bursts were selected, distributed in three brightness groups. All bursts longer than 1 s were included; shorter dim bursts will tend not to activate the 64-ms and 256-ms triggers. The selected brightest and dimmest bursts differ by up to more than two orders of magnitude according to peak intensity, C_p:

brightest	18,000	<	C_p	<	250,000	cts s^{-1}
dim	2,100	<	C_p	<	4,500	
dimmest	1,400	<	C_p	<	2,100	

The burst peak intensities and backgrounds must be rendered uniform in order to realize a satisfactory treatment of noise for some tests, e.g., almost any kind of test which defines an interval within a burst or which measures flux or structure. For each burst a background interval was defined, fitted with a first-order polynomial and subtracted. The remaining "signal" profile was then diminished to a canonical peak intensity (that of the dimmest burst used, 1400 cts s^{-1}), and a canonical flat background was added (the average background for dimmest bursts). Over the fixed length of interval analyzed, 1024 64-ms bins, the actual backgrounds are relatively free of curvature and fairly constant among dimmer bursts. The requisite amount of empirical variance was then added to compensate for diminishing the burst profile. These compensating noise intervals were taken from the ends of (~ 3700-bin) records for those dim bursts which appear to have subsided soon enough to afford a flat 1024-bin background interval. The judgement to utilize a particular background interval as substitute noise was made by fitting and subtracting a first-order polynomial and then examining the trend of residuals expressed in standard deviations. Because noise in the diminished burst and the added substitute noise are uncorrelated on all time scales, for the wavelet analysis small corrections for the respective fractional noncorrelations were applied to the three brightness groups. Each burst was then required to activate at least one trigger (256-ms or 1.024-s) to be retained. This last measure coarsely approximates the actual instrument criterion and avoids, for example, including a bright, short burst which would not have triggered the instrument had the source been situated at a much larger distance.

Concomitant with time dilation is redshift, which will affect temporal measurements. Because pulses tend to be narrower at higher energy, redshifted bursts would have these narrower structures shifted to lower energies. At lower energies, the disparity between pulse widths as a function of energy is not as large and therefore redshift is less of a complicating factor. From fits to about 45 isolated pulses in BATSE bursts we find the average pulse width ratios, channels 2:1 and 3:2, to be roughly 0.85 in both cases. This is not enough of a departure from unity to mask a dilation factor of order 2. Therefore, we used DISCSC data summed over channels 1 and 2 (~ 25 to 50 keV, and 50 to 100 keV, respectively) for all analyses reported here.

The total structure in each prepared time profile was then measured by subtracting background and summing over the 1024-bin interval. Empirical one-sigma error bars were generated and sample errors determined for each brightness group.

Table I: Total Normalized Flux above Background
Per Brightness Group

Brightness Group	# of bursts	<flux>	+ 1 sigma	- 1 sigma
Bright	33	4590	1300	610
Dim	35	8790	1350	1125
Dimmest	28	9450	1780	1330

Table I presents results of the flux test. The average flux can be translated into an average "equivalent duration": divide by the canonical peak intensity of 1400 cts s^{-1}, and then divide the result by a filling factor – for specificity, choose 0.3. The average equivalent duration for the bright group is then ~ 10 s (approximately the average duration of actual bursts), whereas the equivalent durations for the two dim groups are about twice as long. This difference is significant at about the 3 sigma level. One-sigma fluctuations in the average flux measurements would translate into a range in the ratio of equivalent durations of about 1.5 – 3. A more robust test would utilize all of the temporal information to measure the structure on various time scales.

The second test, employing wavelet transforms, does characterize the temporal information more fully, and provides a structure decomposition method that is tailored to the gamma-ray burst problem. A wavelet transform illustrates the amount of "activity" on all (discretely spaced) time scales within a time series; our measure of activity is the sum of absolute values of differences between pairs of flux values on a given time scale. This measure is implemented using the first order Daubechies wavelet representation, otherwise known as Harr wavelets[4]. Essentially, a Harr wavelet transform is a decomposition of the profile into localized, orthonormal "building blocks" which span a temporal range (128 ms to 64 s in this analysis) in factor-of-two steps. Since wavelet transforms are indexed in position *and* frequency, they are superior to Fourier tranforms for characterizing the spectrum of a time series which contains a variety of heterogeneous structures – often the case in gamma-ray bursts. If dim bursts are dilated relative to bright bursts, then the *average* activity on any given time scale will be higher by approximately the average dilation factor of the dim bursts. (Average activity is not merely "shifted" to longer time scales by dilation because the activity in each burst scales differently with time scale. If all bursts had the same time profile and if activity scaled in a self-similar manner on all time scales, then a continuous wavelet transform would not show a distinction between dilated and nondilated bursts.)

Figure 1 illustrates the average wavelet transforms for each of the three brightness groups. The average transform for the bright bursts (long dashed line) lies below that for the other two groups (dim: dashed, dimmest: dotted) on all time scales. Empirical one sigma errors for each group are shown as envelopes, solid lines in each case. For our normalization, the appearance of Poisson noise is flat, or white. In fact, the actual average noise level is close to white, with a value of about 10 units. This level is sub-Poisson; simulations of Poisson-distributed noise yield a value of 15 units. Each time the analysis is run, different associations of substitute noise intervals and burst profiles are made, however, the resulting curves for each realization have similar loci. Thus, the error envelopes are mostly determined by the ranges of durations and profile shapes within the three brightness groups. For each time scale between 2 s and 64 s, the

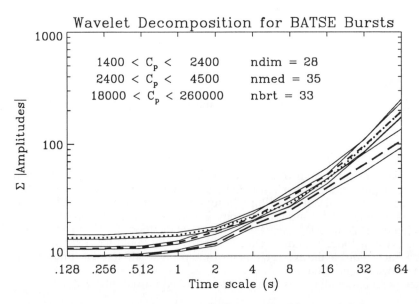

Fig. 1. Average wavelet transforms of BATSE bursts for three brightness groups: dotted = dimmest, dashed = dim, long dash = bright. Solid lines indicate empirical one-sigma errors for respective samples.

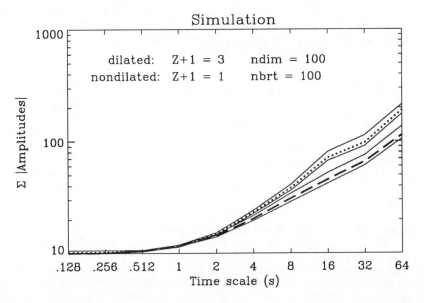

Fig. 2. Similar to Figure 1, for simulated bursts: dotted = dilated bursts; long dash = nondilated bursts. Simulations approximate temporal characteristics of bright BATSE bursts on time scales longer than about 1 s.

averages for the two dim groups lie essentially on top of each other and above the bright group's average by one or more sigma, except at 8 s and 16 s, where the envelopes of the dimmest and bright groups overlap.

All three curves are even more significantly separated on time scales less than about 2 s, even though noise then begins to dominate. We have investigated this effect thoroughly in terms of background variations within the three brightness groups and found such effects to be negligible in our treatment: Although bright bursts trigger over a wider range of backgrounds, their signals are reduced by at least 90%, hence any noise effect is proportionately diminished. Also, a larger (positive) variation in background would introduce more variance into the average wavelet transform of the bright bursts, thus having an effect opposite from that observed. Since the bright average is only marginally above the expected noise level on short time scales, we suspect that the same would be true for the dimmer bursts. A possible explanation for the observed difference is that the dimmer bursts are in fact redshifted, bringing the spikier structure of time profiles at higher energy into the lower energy band which we analyze. This hypothesis will be investigated by redshifting the bright bursts, but results are necessarily model dependent (see below).

To calibrate the wavelet transform technique we simulated time profiles using attributes derived from fitting most of the bright bursts (a few were too long to be accommodated with present fitting programs). We fitted 220 pulses in 30 bursts[5]. Most of these pulses overlap substantially, nevertheless the sum of the fitted pulses per burst represent the total signal on time scales longer than about 1 s. Higher frequency components comprise the fit residuals, not well modelled by our simulations. The simulations approximate distributions of several fundamental properties of bursts: duration, number of pulses per burst, pulse rise and decay times and pulse peakedness. The tendency for pulses to cluster was not simulated for results discussed here. We utilized empirical noise (discussed above) for the simulations. Bright bursts were simulated from exactly the parent population of fitted pulses, preserving all fitted parameters per burst except for the positions of pulses, which were randomized for each burst realization. Dim bursts were simulated in the same manner with these additional modifications: rise and decay times and burst durations were dilated by a constant factor. Figure 2 illustrates the average wavelet transforms for 100 dilated and 100 nondilated burst simulations. The relative dilation factor is 3.0. On time scales longer than about 2 s, the simulated curves follow loci similar to those of the bright and dim/dimmest groups. This result would suggest that the dimmer bursts are at redshifts of order 2, if indeed the difference in activity between the brightness groups is mostly due to time dilation. However, some refinements are in order – especially pulse clustering and redshifting – which will modify the appearance of the observed and simulated curves.

Spectral redshifting is a problem because burst pulse shapes are energy dependent – at higher energy, pulses tend to be narrower. Thus, when attempting an unbiased measure of, for example, pulse widths, there are two competing effects, and both increase with distance: time dilation which broadens a structure, and spectral redshifting which moves the narrow structures at high-energy into lower energy bands. Both time dilation tests described here were performed without compensation for redshifting. Properly, when attempting an accurate measure of time dilation, redshift must be taken into account. However, the analysis is then cosmology dependent. We will report model-dependent results in detail elsewhere.

Pulse clustering is the tendency for the distribution of intervals between pulses to be peaked towards shorter values than would expected if pulse placement is random. That pulses are clustered can be demonstrated by calculating $\chi = <(x_i - <x>)^2>$, where the x_i are count samples, for actual bursts and for simulated bursts with the same pulses randomly distributed; actual bursts will have a larger average value for χ. Apparently, clustering will tend to shift some activity to longer time scales.

CONCLUSIONS

A brightness-independent measure of the total flux in a given burst for three brightness groups indicates that dimmer bursts have approximately twice as much flux – or structure – as do bright bursts, and therefore would appear to have "equivalent" durations approximately twice as long as bright bursts. The result is significant at about the 3 sigma level. The uncertainties would easily admit time dilation factors for the dimmer bursts in the range of ~ 1.5 – 3.

A wavelet transform test, which more fully exploits the information in each time profile, indicates that dimmer bursts have more structure than bright bursts *on all time scales*. The results of the wavelet transform test appear to be significant at the several sigma level for a sample of about 100 bursts. Simulations patterned after bright BATSE bursts yield a quantitatively similar picture, giving us confidence that the wavelet transform algorithm is sensitive enough to discern the sought-after signature of time dilation. Given the uncertainties incurred by neglecting redshift effects on temporal structure in the energy band which we analyze, the simulations indicate an acceptable range of time dilation factors similar to those found using the total flux test.

Clearly, the result will become more quantitative as more bursts are included in the study. For improved measurements and predictions, the simulation program will be modified to include the effect of pulse clustering, and cosmology-dependent redshift and intrinsic luminosity functions will be modelled.

REFERENCES

1. C.A. Meegan, G.J. Fishman, R.B. Wilson, W.S. Paciesas, G.N. Pendleton, J.M. Horack, M.N. Brock, & C. Kouveliotou, Nature, 355, 143 (1992).
2. S. Mao & B. Paczinski, Ap. J., 388, L45 (1992).
3. D. Lamb, these proceedings.
4. I. Daubechies, Ten Lectures on Wavelets (Capital City Press, Philadelphia, 1992).
5. J.P. Norris, S.P. Davis, C. Kouveliotou, G.J. Fishman, C.A. Meegan, R.B. Wilson, W.S. Paciesas, these proceedings.

MORPHOLOGICAL STUDY OF SHORT GAMMA RAY BURSTS OBSERVED BY BATSE

P.N. Bhat[†], G.J. Fishman, C.A. Meegan, R.B. Wilson
ES-62, Space Science Laboratory
NASA/Marshall Space Flight Center, Huntsville, AL 35812

W.S. Paciesas
Department of Physics, University of Alabama in Huntsville
Huntsville, AL 35899

ABSTRACT

Gamma-ray bursts (GRB) of durations less than about 0.6 s constitute nearly 20% of the BATSE bursts. These bursts are distinct from the soft gamma-ray repeaters due to their hard spectra and may form a separate class by themselves. The time histories of these bursts exhibit a variety of shapes. The implications of short spikes in understanding the GRB origin are enormous. Using high time resolution (2 μsec) data from the BATSE large area detectors, we carry out a systematic search for sub-millisecond structures in the GRB time histories from the triggered detectors. We also test the predictions of Mitrofanov [1] who suggested that GRB light curves may consist of superposed microsecond flares. Three new parameters are introduced to parameterize the temporal features. A systematic search for a possible correlation between the temporal characteristics with spectral features is carried out for a sample of 16 short bursts.

INTRODUCTION

Gamma Ray Burst (GRB) durations vary greatly, spanning at least 5 orders of magnitude, from the shortest of < 12 ms (FWHM) for GB820405[2] to the longest of 1000 s for GB840304[3], with the range limited by possible instrumental effects on either extreme. The median burst duration is about 10-15 s. Soft γ- repeaters generally have durations less than 1 s and are distinct from the 'classical' short GRB's considered here. Some bursts exhibit very short time structures[4,5] which are expected to constrain the parameters characterizing the GRB emission models. Burst duration is a detector and distance dependent parameter[6]. However there are reports of a break in the burst duration distribution at $\sim$ 600 ms based on the bursts detected by SIGNE[7,8]. Based on 66 bursts detected by the PHEBUS experiment on GRANAT, Dezalay et al.[9] find that short rise times and hard spectra are common features of bursts lasting < 2 s and hence may form a separate class of GRB's. However, attempts[10] to search for possible correlations of spectral properties with temporal structures of longer bursts were not successful. Hence short bursts may form an interesting sub-set of GRB's.

In this paper we address the following questions:

(a) What fraction of short bursts contain very short time (millisecond / sub-millisecond) structures in the time histories?

[†] NASA/NRC Senior Resident Research Associate; On leave from Tata Institute of Fundamental Research, Bombay 400 005, India

(b) Is there any evidence for a coherent emission of γ-rays in a burst?
(c) Is there a spectral evolution during short bursts?
(d) Are any of the spectral characteristics related to any of the temporal features of the burst?

OBSERVATIONS

BATSE large area detectors are described elsewhere[11]. Among the various data types available we use the time tagged event (TTE) data which have the highest time resolution (2 μs) available in any GRB experiment so far. TTE data have 4 channel spectral information; the approximate photon energies at the channel boundaries are: 20, 50, 100 and > 300 keV respectively. The photon arrival times are read out for each short burst and the time histories are generated for each triggered detector for each of the 4 energy channels.

TERM DEFINITIONS

The temporal and spectral features for each of the bursts analyzed here are quantified using the following parameters:
The spectral characteristics are defined by:
(i) **Hardness Ratio (HR)** which is defined as the ratio of the number of signal photons above 100 keV to that below 100 keV. In general, HR of the burst derived from the time history summed over triggered detectors, is affected by the earth reflected photons[5]. However the contribution from the scattered photons is different for different bursts. Hence, for the sake of uniformity, the average HR is used in this study.
(ii) **Correlation Coefficient (CC)** is the simple correlation coefficient between HR and the burst intensity.
(iii) **Energy Evolution Parameter (EEP):** In a majority of the bursts the higher energy photons lead the lower energy counter-parts. As a result, the centroid of the burst light curve at higher energies leads in time with respect to that of the lower energy light curve. We compute the slope of the straight line fitted to the centroid positions in the first 3 channels as a function of the channel number, for the 2 brightest detectors. We define the mean of the two slopes as the EEP of the burst.
The temporal characteristics are defined by:
(i) **Burst duration (D)** is the apparent quantity derived from the summed time history of all the triggered detectors. The burst boundaries are defined as the interval from the earliest to the latest significant source excess in the energy range 20 to 2000 keV.
(ii) **Burst Complexity Index (CI):** This is a measure of the total number of peaks in the burst time history. For this purpose the burst time history of duration D and the background of same duration & time resolution (*viz:* 0.5 ms) are simultaneously analyzed. The number of peaks are counted by counting the number of zero crossings (in one direction) of the first differential of the time history. Shot noise is removed by applying a one dimensional unweighted low pass filter of kernel size 3 to the time histories successively until the background shows 0 or 1 peak. The resulting number of peaks in the burst interval is CI. CI is independent of the burst intensity as the differential is expressed in units of standard deviation. The process is speeded up by using finite threshold for a peak to be counted. CI is conceptually similar but technically different from the burst structure parameter introduced by Lestrade et al.[12].
(iii) **Burst Rapidity Index (RI):** CI is insensitive to sharp peaks with very low rise and decay times within burst time history because of the filtering tech-

nique employed to determine CI, as described above. Hence RI is introduced to quantify the sharpness of the peaks in the time history. This is given by the ratio of the variances of the first differential of the time histories of burst and background. An advantage of RI is that one needs to search only those bursts with higher than average RI for short time structures. This will improve the statistical significance of such structures if and when one finds them. A frequency distribution of RI shows that a majority of the bursts have RI around 0.9 and very few of them have a significantly higher RI, thus suggesting that bursts with higher RI may form a separate class. A plot of CI as a function of RI shows that they are uncorrelated and hence independent (figure 1).

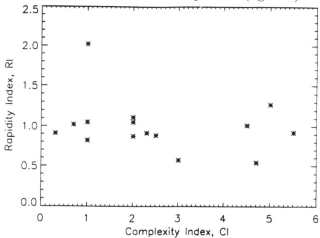

Figure 1. A plot of Complexity Index, CI versus Rapidity Index, RI, showing no functional relationship between them.

(iv) Channel energy at Maximum RI (CM): It is an observed feature of most of the bursts that the time histories become sharper at higher energies. The RI computed for different energy channels increases with energy and in some cases shows a maximum in channels 2-4. The channel energy where the RI is maximum is defined as CM.

RESULTS AND CONCLUSIONS

Only one case (1B910711, RI=2.0) of a burst showing a significant sub-millisecond structure has been found which is already reported[5]. A plot of the cross-correlation function between first 2 brightest detectors (figure 2a) shows a significant peak at zero lag showing the genuineness of the structure. A similar plot for another burst (1B920229) which also shows sub-millisecond structure (RI=10.9) shows no such peak at zero time lag (figure 2b) confirming that the structure is only a statistical fluctuation.

Using the statistical distribution of event arrival time differences in pairs of bright detectors we find no evidence for coherent emission of γ-ray photons (on time scales as short as 5 μs) in any of the bursts analyzed, a result consistent with a similar search carried out on longer, stronger bursts[13].

Some bursts show spectral evolution even at ms/sub-ms time scales (*e.g.*,

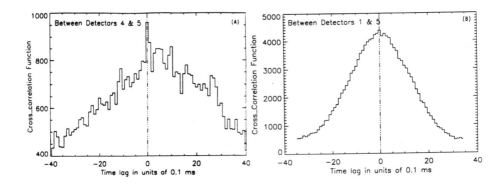

Figure 2. A plot of the cross-correlation function for the light curves of the two detectors as a function of time lag in (a) 1B910711 showing a significant sharp peak at zero lag and (b) 1B920229 showing no significant peak at zero lag.

1B910814B) and some do not (1B920229). This result is in agreement with the spectral evolution studies carried out on longer bursts which have similar time histories[14]. The frequency distribution of CI is flat showing that different bursts show different levels of spectral evolution consistent with the results of Koshut et al.[15] Evidence shows that spectral evolution is independent of D. However there is a weak suggestion that harder bursts show stronger spectral evolution than softer bursts.

In a limited sample of 16 bursts analysed so far, we find that longer bursts show more time lag between the softer and harder photons (figure 3). This is consistent with conventional picture of X-ray reprocessing being responsible for γ- ray emission. Also there is convincing evidence to show that bursts with higher CM have a higher HR. (figure 4). However, CI is seen to be independent of HR, consistent with earlier results[10]. An inverse correlation of spectral hardness with D is not ruled out.

These positive correlations have a strong bearing on the burst source models and emission mechanisms. However the sample size of 16 bursts is too small to draw any firm conclusions at the present time.

REFERENCES

1. I.G. Mitrofanov, Astrophys. Space Sci. **155**, 141 (1983).
2. E.P. Mazets et al., Positron Electron Pairs in Astrophysics (Ed.:M.L.Burns, A.K. Harding and R. Ramaty, New York: AIP, 1983), p. 36.
3. Klebesadel, R.W., Laros, J.G. and Fenimore, E.E., Bull. Am. Astron. Soc. **16**, 1016 (1984).
4. J.G. Laros et al., Nature **318**, 448 (1985).
5. P.N. Bhat et al., Nature **359**, 217 (1992).
6. E.P. Mazets and S.V. Golenetskii, Astronomia **32**, 16 (1987).
7. E.P. Mazets and S.V. Golenetskii, Ap. Space Sci. **75**, 47 (1981).

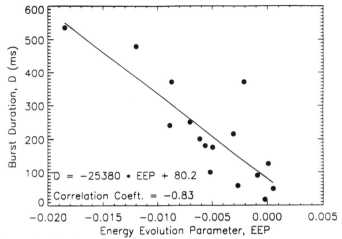

Figure 3. A plot of D as a function of EEP showing a significant correlation. The straight line is a least square fit to the points.

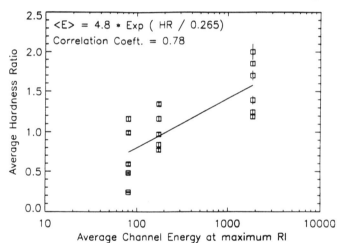

Figure 4. A plot of HR as a function of CM, again showing a strong correlation. The straight line is a least square fit to the points.

8. C. Barat et al., Ap. J. **285**, 791 (1984).
9. J-P. Dezalay et al., AIP Conf. Proc. - Gamma Ray Bursts; Huntsville (Ed.: W.S. Paciesas and G.J. Fishman, 1992), p. 304.
10. E. Jourdain, Ph. D. Thesis (C.E.S.R., Paul Sabatier Univ., Tolouse, France, 1990).
11. G.J. Fishman et al., Proc. GRO Science Workshop; Greenbelt (Ed.:W.N. Johnson, 1989), p. 2-39.
12. J.P. Lestrade et al., these proceedings (Ed.: N. Gehrels, 1992).
13. B.E. Schaefer et al., Ap. J. , in press (1992).
14. P.N. Bhat et al., these proceedings (Ed.: N. Gehrels, 1992).
15. T.M. Koshut et al., these proceedings (Ed.: N. Gehrels, 1992).

FRACTAL ANALYSIS OF THE GRB LIGHT CURVES

N.I.Shakura, E.I.Moskalenko, K.A.Postnov, M.E.Prokhorov, N.N.Shakura

Sternberg Astronomical Institute, 119899 Moscow, Russia

The profiles of gamma-ray bursts (GRB) are observed to be of great diversity (see e.g. Ref.1). The observed gamma emission might actually be generated by different mechanisms. Hence the need for classifying GRBs by their shapes arises. Here we propose a simple independent method for measuring the observed jagged profiles.

For a given GRB profile we do the following iterations:
1. choose a sample consisting of 2^n bins;
2. take a sum (length) L_i of absolute values of count rate differences in each consecutive pair of bins;
3. average each pair of bins, thus two times reducing their numbers;
4. do the same as in item [2] $n-1$ times so that two wide bins remain.

If the GRB profile assigns to a fractal-type curve, then a quantity $\alpha_i = \log(L_{i+1}/L_i)/\log 2$ does not depend on i.

We have analyzed some GRBs from a catalogue[1] (see Fig. 1-3). In the central and right panels we present the calculated "length" L as a function of iteration # expressed in number of the initial bins taken for this step (scale of smoothing). α_i is a local value of the inclination of this broken line. The fractal dimension of the curve is $D = 1+ <\alpha_i>$. We estimate errors of α_i (marked with vertical arrows in Fig. 1-3) by taking different bins as the burst beginning.

For comparison we have calculated the slope α for a random number sequence (see Fig. 4). The obtained value $\alpha_{rnd} = 1.5$ proved to be the same as in the post-burst record in GRB 060479 (bottom of Fig. 3).

Many authors believe that GRB can originate as a result of starquakes in the degenerate crystallized stars. In this connection it would be of great interest to compare the structural properties of GRB profiles with those of the earthquake seismograms making use of the proposed method.

REFERENCE

1. J.-L.Atteia, C.Barat, K.Hurley, M.Niel, G.Vedrenne, W.D.Evans, E.E.Fenimore, R.W.Klebesadel, J.C.Laros, T.Cline, U.Desai, I.V.Estulin, V.M.Zenchenko, A.V.Kuznetsov, V.G.Kurt, Astrophys. J. Suppl. **64**, 305 (1987).

DECONVOLUTION OF PULSE SHAPES IN BRIGHT GAMMA-RAY BURSTS

J. P. Norris[1], S. P. Davis[2], C. Kouveliotou[3], G. J. Fishman[4],
C. A. Meegan[4], R. B. Wilson[4], W. S. Paciesas[4,5]

[1]NASA/Goddard Space Flight Center, Greenbelt, MD 20771
[2]Catholic University of America, Washington, DC
[3]Universities Space Research Association, Huntsville, AL 35812
[4]NASA/Marshall Space Flight Center, Huntsville, AL 35812
[5]University of Alabama in Huntsville, AL 35899

ABSTRACT

We deconvolve gamma-ray burst time profiles into pulses using a model-dependent, least-squares pulse-fitting algorithm. The algorithm has been applied to 25 bright gamma-ray bursts detected by BATSE. Profiles with 64-ms resolution are fitted in four energy bands covering ~ 25 keV to > 1 MeV. A successful pulse model employs separate rise and decay time constants and a peakedness parameter. Some trends in pulse shape with energy and time of occurrence are apparent: within a given burst, pulse widths are often comparable; decay times almost invariably are longer than rise times, and pulses sometimes peak earlier in the higher energy bands. The dominant trend of spectral softening seen in the large majority of pulses arises mostly from faster onsets at higher energy and longer decays at lower energies. Broader temporal structures at low energies result from overlapping pulses and the appearance of smaller subsidiary pulses.

INTRODUCTION

Most gamma-ray burst pulse structures appear to be comprised of several pulses, often with considerable overlap. In some portions of burst profiles the pulses are sufficiently widely spaced to allow accurate fits of pulse shape parameters. To achieve an acceptable uncertainty, we apply the criterion that the peaks of two overlapping pulses be separated such that the interjacent minimum is at most half the intensity of the smaller pulse; then perhaps one-sixth of all major pulses may be deconvolved. We quantify the effects of pulse overlap elsewhere.

There are several reasons for performing quantitative measurements of pulse shapes. From a phenomenological point of view, it has not been completely clear what the fundamental "event" is in bursts, i.e., whether bursts are comprised of monolithic pulses, or noise processes on a continuum of time scales, or both. Indeed there may be more than one paradigm for burst emission processes. Our work sheds some light on these questions for a wide variety of burst profiles. With the much improved signal to noise afforded by BATSE's LADs, measurement of temporal development in bright bursts throughout the event is possible on time scales down to ~ 16 ms. These measurements will translate into constraints on possible emission mechanisms. Also, the fitted pulse shapes can be utilized in simulations where it is necessary to reproduce closely the aspects of temporal variability. We present partial results here, including fitted shapes in 4 energy channels and rise vs. decay times.

© 1993 American Institute of Physics

FITTING PROCEDURE

We have developed and applied an interactive graphical IDL routine for fitting pulses in gamma-ray bursts. The routine iteratively spawns a Fortran program, Superfit, that performs a least-squares fit to burst profile for a given model pulse. Superfit employs two methods for seeking convergence, the Marquardt technique far from convergence, and Gauss-Newton near convergence. Pulse shapes as a function of energy and position within the burst are fitted following a rule-based approach for pulse placement, initial pulse shape estimates, and chronology of parameter constraint.

After consideration of the problem using simulations and real BATSE data, we chose a pulse model which well represents many pulse shapes occurring in real bursts. The selected model is:

$$F(t) = A \exp(-(|t - t_{max}|/\sigma_{r,d})^\nu)$$

where t_{max} is the time of the pulse's maximum intensity, A; σ_r and σ_d are the rise ($t < t_{max}$) and decay time constants ($t > t_{max}$), respectively; and ν is the pulse "peakedness" (lower values => more peaked). Combined, the parameters σ_r, σ_d, and ν afford a wide variation in pulse shape.

Pulses were added to profiles until acceptable visual fits were obtained and reduced chi square was near unity. Fits were performed by one investigator and reviewed by a second investigator. The distribution of rise vs. decay times, at 50% of pulse peak, is plotted in Figure 1. The fitted pulses illustrated in Figure 2 were fairly well separated – we required the condition described in the introduction for retention. In many cases the fit to the pulse is not shown for all four energy bands, either because pulses were too crowded at lower energy, or the pulse was not present at higher energy.

CONCLUSIONS

Several trends are readily apparent from inspection of the fitted pulse shapes. At lower energies, rise and decay times tend to be longer, and thus valleys between pulses are more filled in. Pulses tend to be more peaked at higher energies. Pulses often peak nearly simultaneously in the 4 energy bands, within 64-ms resolution. In about 80% of the cases shown in Figure 2 the pulse either peaks earlier at higher energy and/or (usually) decays more quickly. Thus, evolution of many individual pulse shapes and peak position migration as a function of energy account for the general softening trend often observed in bursts. Most contrary occurrences were found in narrow pulses where the 64-ms resolution may not be sufficient to resolve fast fluctuations (subpulses?). In these 20% of pulses, little or no spectral evolution is discernible.

From two points of view our work supports the conclusion that, to a large extent, monolithic pulses – rather than a continuum of noise structures – do comprise burst profiles. First, the pulse model we employ usually results in a parsimonious representation of burst profiles in that a relatively small number of pulses is required to achieve an acceptable fit. This is especially apparent in the highest energy band, whereas at lower energy it often appears that additional pulses have "spawned," and pulses are broader, making the fitting procedure more difficult. Also, the large majority of fitted pulses exhibit the familiar softening trend that has been widely observed for conglomerate pulse structures – an indication that the primary energy generation mechanism is manifested and temporally resolved on the pulse time scale.

There may be other unresolvable components in bursts, including long smooth structures at lower energies and very short spiky features at higher energies. However, we note that this broadening trend with decrease in energy band is exactly what is seen for deconvolved pulses. Thus from one point of view, the unresolvable components may be just overlapping pulses. Yet, as noted above, sometimes broad structures appear at lower energy unaccompanied by clear counterparts at higher energy. Since pulses do evolve spectrally on a wide range of time scales, it is quite likely that superposed pulses can occasionally give rise to depressions in the resultant continuum which resemble cyclotron absorption lines.

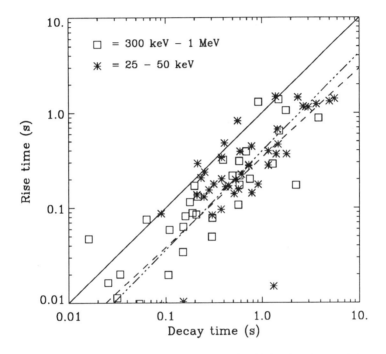

Fig. 1. Distribution of fitted rise vs. decay times (at 50% of pulse peak intensity) for the two extreme energy bands, for pulses of Figure 2. Over a range of approximately two decades, the average ratio of decay time to rise time, with large scatter, is about 2.5 for both channels. Dashed = chan 1, dash-dotted = chan 4.

Legend, Figures 2a & 2b: Fitted pulses from BATSE temporal profiles at 64-ms resolution (except burst #143 – 16-ms resolution) in 4 energy bands. Pulse amplitudes are normalized to unity. Relative peak positions per energy band are fitted values. Pulses are grouped and labeled by burst trigger number. An attempt has been made to arrange pulses (left to right, top to bottom) in order of increasing *burst* temporal complexity. As can be seen, pulses of comparable duration often occur in a given burst. First 4 pulses, Figure 2a, are single spike events, not well resolved at 64 ms. Pulses in bottom row, Figure 2b, are from bursts with wide structures. Approximate energy bands: dotted, 25 – 50 keV; dash-dots, 50 – 100 keV; dash, 100 – 300 keV; solid, 300 keV – > 1 MeV. NOTE that time scale changes between plots.

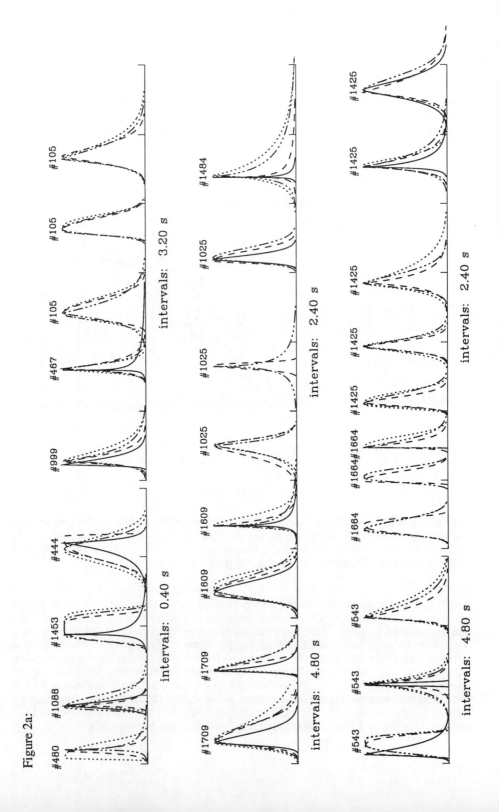

Figure 2a:

Figure 2b:

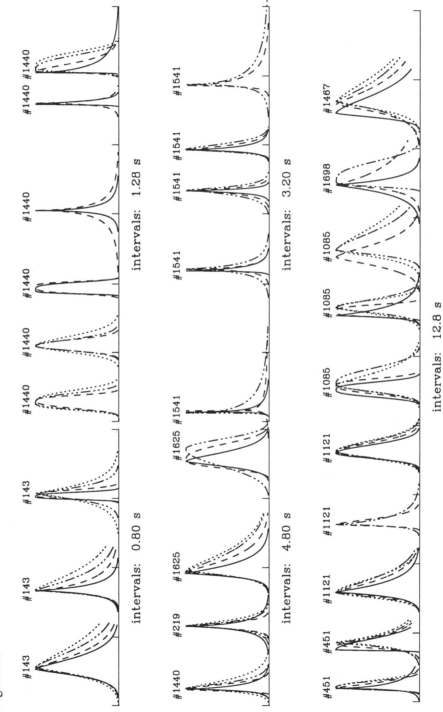

CALIBRATION OF AN ALGORITHM FOR DECONVOLUTION OF OVERLAPPING PULSES IN GAMMA-RAY BURSTS

S.P. Davis[1,2], J.P. Norris[1]

[1]NASA/Goddard Space Flight Center, Greenbelt, MD 20771
[2]The Catholic University of America, Washington, DC 20064

ABSTRACT

We present results of temporal deconvolution into constituent pulses of simulated gamma-ray bursts, using a rule-based least-squares fitting algorithm that accommodates varying pulse shapes. We examine limitations of the pulse deconvolution method which arise from pulse overlap. We employ a pulse model that includes rise and decay time constants, and pulse peakedness. This model often fits real gamma-ray burst profiles parsimoniously in that relatively few pulses are required to achieve an acceptable fit. The simulations include sets with amplitude variations, with width variations, and with variations in peak-to-peak separation. We quantify the limits to which simulated pulse parameters can be recovered.

INTRODUCTION

Real gamma-ray burst pulse structures appear to be composed of many overlapping primary pulses, whose shapes are often effectively masked by considerable pulse overlap. We have developed a program that fits pulses in bursts, assuming a particular pulse model. The program is a multi-parameter least-squares approach, which uses a system of rules governing placement of pulses, initial conditions, and parameter constraint. We quantify contributions to the error budget which arise from overlapping pulses by performing simulations of burst profiles and then fitting them.

The simulations approximate pulse shapes and pulse intensities in actual gamma-ray bursts. Each simulated profile is deconvolved into its constituent pulses and the fitted pulse parameters are compared with the corresponding input parameters, thus quantifying the degree to which pulse parameters can be recovered for a given arrangement of pulses. We performed three groups of simulations, in which pulse amplitude, width, and separation were varied, sometimes in combination. In each case, the pulses have some degree of overlap. We report some of the results here.

Understanding the errors associated with fitting overlapping pulses in bursts is important not only for measuring pulses themselves, but also for quantifying the fidelity of simulations that may be used to evaluate temporal analysis algorithms.

PROCEDURE

We have developed and applied an interactive graphical IDL interface routine, Pulsefit, for performing variable-shape pulse fits to pulses in real and simulated gamma-ray burst time profiles. The routine iteratively spawns a Fortran program ("Superfit") that performs a least-squares fit to the simulated dataset for a model pulse. Superfit employs two methods for seeking convergence: the Marquardt technique far from convergence, and the Gauss-Newton near convergence.

Three different sets of simulations were performed, each reflecting an aspect of ambiguity resulting from pulse overlap: pulse amplitude variations combined with variation of pulse separation; variation of pulse separation only; and pulse amplitude and width variations.

These simulations were done such that pairs of pulses overlapped. We generated 10 simulated profiles per group, each comprised of 4 pairs of overlapping pulses plus background (a first-order polynomial), and each of a set of ten differing only in that a different Poissonization was performed. The simulated profiles were fitted by employing a rule-based approach for initial estimates of pulse peak position and pulse shape, and for order of parameter constraint. Fits to the simulations were performed using the model function

$$F(t) = A \exp(-(|t - t_{max}|/\sigma_{r,d})^v)$$

where t_{max} is the time of the pulse's maximum intensity, A; σ_r and σ_d are the rise ($t < t_{max}$) and decay time constants ($t > t_{max}$) espectively; and v is the pulse "peakedness" (lower values => more peaked). Combined, the parameters σ_r, σ_d, and v afford a wide variation in pulse shape. All parameters are initially fixed at some canonical values.

The profiles, each comprised of 8 pulses, were fitted using Pulsefit, iterating until all parameters were unconstrained, a reduced χ^2 of order unity was obtained, and the fit was visually acceptable. That is, if a fit to one pulse appeared to "bloom" into another clearly separate pulse, then the profile was refitted using different initial conditions. One investigator generated the simulations, and a second investigator performed the fits. The fits were then reviewed and uncertainties propagated for derived quantities; e.g., the decay time – half-width post-maximum – is a function of both σ_d and v. For all pulses the value for v was set at 1.45, roughly the average peakedness found for many pulses in real bursts.

In simulations of pulses with varying amplitudes and interval separation within pulse pairs, the rise and decay times, and peakedness remained constant. In the second set of simulations, only the separations between pulses were varied. We show one such simulation in Figure 1. In simulations of pulses with varying pulse widths, the values for interval separation, rise time and peakedness were held constant, whereas the decay times and amplitudes were varied.

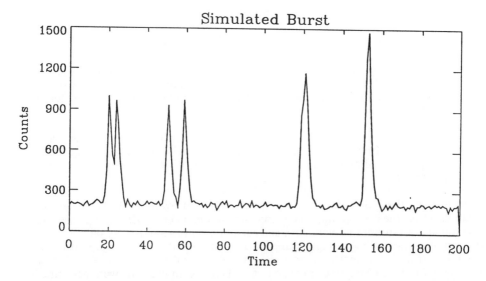

Fig. 1. A simulation where only the pulse separations varied. Pulse amplitudes, rise and decay times, and peakedness are constant for all pulses. See text for values.

RESULTS

Results for the first two groups of simulations are reported here. Analysis of the third group – in which decay time and amplitude both varied – is complex and will be reported elsewhere. Table I summarizes the results of deconvolution of pulse profiles with varying amplitudes, pulse separation and, therefore, ratio of total width to separation. These parameters are major contributors to the uncertainty of recovered input parameters.

The columns of the table (left to right) refer to pulse number, pulse separation, amplitude, and average fractional errors (rms) in decay time and amplitude, and error in peak position compared to (fixed) rise time. The peakedness (1.45), rise and decay times (0.8, and 1.5, respectively) are constant. The two values for pulse amplitude are 200 and 800. The ratio of total width to separation between pulses increases from about 0.6 for pulse pairs 1 & 2 and 3 & 4, to 0.7 for pulse pairs 5 & 6 and 7 & 8.

The large error associated with the low-amplitude pulse number 7 arises from its being almost completely emerged in the long leading trail of pulse number 8, as contrasted with pulse pairs 1 & 2 and 3 & 4. This is another example of the effect that pulse ordering has on the amount of error in the recovered input parameters. For an amplitude of 200, errors are unacceptably large regardless of the separation between pulses. Only for amplitudes of about 800 and a width that is approximately twice the separation do we obtain errors that are acceptably small (~ 20%). Thus, we have probed the minima for pulse separation and amplitude necessary for acceptable recovery of input parameters.

TABLE I: Dependence of Fitted Parameter Errors on
Amplitude Variation and Pulse Separation

Pulse #	Interval	A	$<\varepsilon_{\tau dec}>/\tau_{dec}$	$<\varepsilon_A>/A$	$<\varepsilon_{tmax}>/\tau_{rise}$
1	4.2	800	0.2	0.1	0.2
2		200	1.6	1.1	3.2
3	4.2	200	7.4	5.3	13.6
4		800	0.2	0.1	0.5
5	3.1	800	1.6	8.1	2.2
6		200	3.2	4.9	5.0
7	3.1	200	31.1	10.5	38.8
8		800	1.9	3.0	5.2

TABLE II: Dependence of Fitted Parameter Errors on
Pulse Separation at Constant Amplitude

Pulse #	Interval	$<\varepsilon_{\tau dec}>/\tau_{dec}$	$<\varepsilon_A>/A$	$<\varepsilon_{tmax}>/\tau_{rise}$
1	8.2	0.1	0.3	0.1
2		0.1	0.2	0.1
3	4.2	0.2	0.3	0.2
4		0.4	0.5	0.7
5	2.1	14.4	46.8	15.9
6		1.1	25.1	4.9
7	1.1	9.4	171.7	30.0
8		1.1	15.0	6.0

Table II summarizes the results of deconvolution of pairs of pulses that possess only variable pulse separation between pairs of pulses; all other parameters are held constant. We investigate our ability to recover pulse input parameters with acceptable uncertainty.

The simulations for Table II featured a constant peakedness value of 1.45, constant rise and decay times of 0.8 and 1.5, respectively, and a constant amplitude of 800. The ratios of total width to separation between pulses were 0.3, 0.6, 1.1, and 2.1, for the four pairs of pulses.

The table shows a significant increase in error with decrease in separation for pulse pairs 5 & 6 and 7 & 8. These pairs have a second pulse partially immersed in the first pulse's decay. The resulting errors in the recovered parameters are unacceptably large.

CONCLUSIONS

Pulses with low ratio of total width to separation (~0.6), and high amplitudes preceding those with a high ratio (~0.9), and low amplitude (~one-half its lagging neighbor) cannot be deconvolved successfully, therefore resulting in higher uncertainty in the recovered parameters. Pulses containing a low-amplitude pulse that is less than or equal to one-fourth its high-amplitude neighbor and having a ratio of total width to separation greater than ~0.6 cannot be deconvolved successfully. This is especially true if the low-amplitude pulse precedes its high-amplitude neighbor. Pulses with a low ratio of total width to separation (~0.3) can be deconvolved successfully.

We would like to thank the members of the BATSE team for informative discussions and comments.

An Analysis of the Structure of Gamma-Ray Burst Time Histories

John Patrick Lestrade
Dept. of Physics, Mississippi State University, MS 39762

G. J. Fishman, C. A. Meegan, and R. B. Wilson
Marshall Space Flight Center, ES-64, Alabama 35812

W. S. Paciesas and G. N. Pendleton
University of Alabama at Huntsville, Huntsville, AL, 35803

P. Moore
Boeing Computer Support Services, MSFC, Alabama, 35812

H. E. Cody
CS-GSD22, Kennedy Space Center, FL 32899

ABSTRACT

If gamma-ray bursts arise from a small number of distinctly different physical phenomena, then this might be revealed by a clustering of time profile characteristics into a small number of groups. We have applied a "spike" counting algorithm to 107 GRB profiles. Here we present graphs of spike frequency and spike amplitude versus burst intensity and duration. So far, we see no evidence of grouping.

INTRODUCTION

One of the challenges for both theorists and experimentalists in the field of gamma-ray bursts is to explain the seemingly disparate results of recent observations. One of the pressing questions in the ongoing debate is whether the majority of bursts originate from a single- or multi-component source distribution.

If bursts arise from a small number of distinctly different physical phenomena, then this might be revealed by a clustering of time profile characteristics into a small number of groups. In this paper, we analyze the amplitude and frequency of occurrence of spikes in GRB time profiles. Our goal is to see if these characteristics divide the GRB population into groups which could indicate the presence of a multi-component source population.

Previous attempts to characterize gamma-ray bursts by their time histories have met with very limited success. Barat et al.[1] calculated a wide range of burst profile characteristics with no apparent groupings or trends. They also found no correlation of time profiles with spectral character.

Belli,[2] analyzing five Venera 11 and 12 GRB time profiles from the Mazets catalog, found that the non-Poissonian "noise" could be explained by shot noise processes with characteristic shot times (shot-averaged amplitude) of about 1 s, but with relatively large uncertainty in the shot frequencies due to poor counting rate statistics.

ALGORITHM

In a previous paper[3] we presented the idea that a reasonable statistic that measures the structure or "spikiness" of a time profile is the number of times the profile shows a run of monotonically increasing (or decreasing) bin-to-bin differences of a predetermined length. The algorithm that we use permits a definition of a "spike" as a monotonic increase of at least N bin differences followed by at least M monotonically decreasing bin differences. (The mirror image can also be counted, if desired, i.e., N decreasing followed by M increasing.)

In that paper, we showed that runs of 7 or more bin differences were effective in differentiating profiles with different amounts of structure and also separating the weakest bursts from background profiles. This algorithm is also sensitive, as it should be, to spikes riding on top of a broad emission feature and is insensitive to the broad feature itself. In order for this method to work, we found it necessary to first smooth the profile with a 5-point moving average. This does not alter the number of spikes in the post-burst background but it does enhance the spikes in the burst.

In addition to the width of the spike (length of the "run"), we have added another variable to the algorithm. This second variable is the peak height, equal to the difference in the number of counts between the last bin in the run and the first bin. This variable, called the "peak height filter" or filter, is measured as a number of σ above the post-burst background.

In this paper, we apply these criteria to 107 burst time histories. The data that we use are the BATSE 64-ms DISCSC data ($E > 20$ keV). Thus, a monotonic run of 8 bins spans a time of 1/2 second. The 107 profiles were chosen because their durations were greater than 12 seconds.

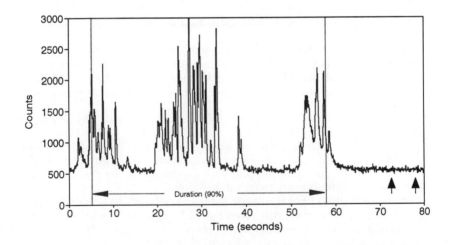

Figure 1. Time history of BATSE GRB #1676 (1B 920627, at 46956 sec). Indicated are the duration, T_5^{95}, and the region of the post-burst profile used to calculate the average background.

Figure 1 presents a sample time profile. Superimposed on the figure is a measure of duration, T_5^{95}. An average background level is calculated in the post-burst region between the small arrows on the right side of the figure. This average background is subtracted and the T_5^{95} duration begins after 5% of the burst counts are recorded and ends after 95%.

Figure 2 shows histograms of the average spike frequency (spikes/sec) seen in the 107 profiles for three different values of spike size, N. Notice that as the spike size is decreased, the number of spikes increases. In this paper, we use a spike size of $N = 7$. The average frequency in this case is about 0.2 spike/sec.

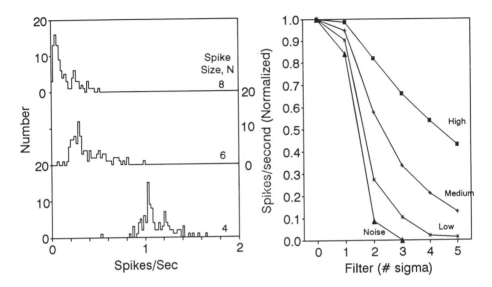

Figure 2. Spike frequencies for all 107 profiles for 3 different spike sizes.

Figure 3. Spike frequency versus filter for 3 profile types (high, medium, and low structure and background.)

To demonstrate the effect of the filter on the measured spike frequency, Figure 3 presents the frequency as a function of filter (σ). We have divided the 107 bursts into three groups according to their structure. In addition, we include the results of measurements on ten theoretically-generated noise profiles, each one with a duration of 250 seconds. Notice that the spike frequency in the profiles with high structure do not decrease as rapidly with increasing filter as do the others.

RESULTS

Here we present graphs of profile structure and several other parameters in the hope of finding unexpected correlations or groupings of data points. The presence of groups may indicate the presence of different causal phenomena and/or a multi-component source distribution.

Figures 4, 5, and 6 present i) the values of $C_{\max}$ (i.e., the largest count in a 64-msec bin) versus the T_5^{95}, ii) the spike frequency vs. T_5^{95}, and iii) the spike frequency vs. $C_{\max}$, respectively. As might be expected, there is no correlation in the first two graphs. In Figure 4, long bursts have the same maximum intensity as short bursts[†]. In Figure 5, the frequency of spikes evidently does not depend on burst duration. Figure 6 does show an expected correlation. Brighter bursts have small peaks that stand out above the background and are therefore counted, thus increasing the spike frequency. However, dimmer bursts lose spike counts because of the background noise and therefore have lower frequencies.

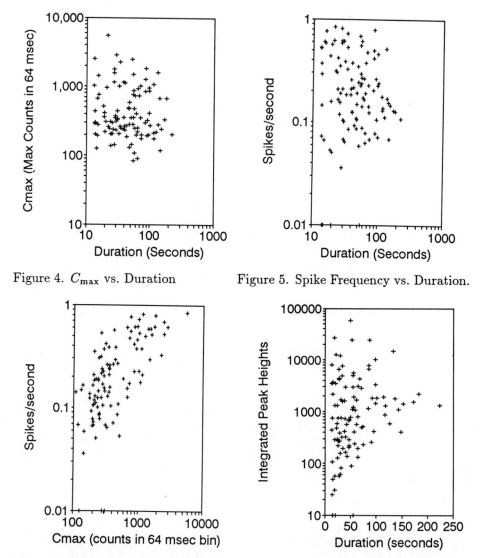

Figure 4. C_{max} vs. Duration

Figure 5. Spike Frequency vs. Duration.

Figure 6. Spike Frequency vs. C_{max}

Figure 7. Integrated peak heights vs Duration.

[†] A cosmological effect, if present, would at most change the durations by a factor of 2 or 3. In Figure 4, the ratio of maximum to minimum duration is 25.

Figure 7 presents the integrated peak heights vs. duration. The integrated peak height is the sum of all heights for qualified spikes in a profile. It is interesting that the longest bursts do not show the same spread in this parameter as do the shorter bursts. However, we see no evidence of separate components in this figure.

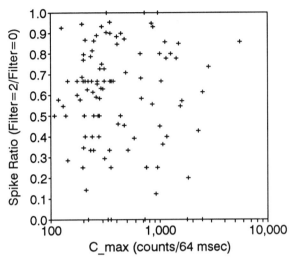

Figure 8. Ratio of spike frequency at a filter of 2 σ to that at 0 σ vs. C_{max}.

Finally, we present in Figure 8 the graph of the ratio of spikes seen with a filter equal to 2 σ to that at 0 σ versus C_{max}. The abscissa here reflects the absence of smaller spikes in a profile. If a profile has no small peaks, this ratio will be unity; if most of its spikes are small (i.e., less than 2 σ) then its ratio will be close to zero. As before, there is no evidence for grouping nor the presence of subsets in these data.

CONCLUSIONS

At this stage we see no evidence in burst time profiles that would suggest that gamma-ray bursts originate from a small number of distinctly different phenomenological sources. The distributions of profile characteristics that we measure, such as spike frequency and spike amplitude, are homogeneous and show no unexpected correlations. We are continuing this line of investigation and will include correlations with other parameters such as spectral hardness. Furthermore, since our algorithm provides us with not only the number of spikes in a profile but also the time between spikes, we will extend the analysis to investigate spike clustering.

REFERENCES

1. C. Barat, R. I. Hayles, K. Hurley, and G. Vedrenne, Ap. J. **285**, 791 (1984).
2. B. M. Belli, Ap. J. **393**, 266 (1992).
3. J. P. Lestrade, et. al., Gamma-Ray Bursts (AIP Conf. Proc. 265, 1992).

SEARCH FOR GRAVITATIONAL LENS ECHOES IN GAMMA-RAY BURSTS

R.J. Nemiroff[1,2], J.M. Horack[3], J. P. Norris[2],
W.A.D.T. Wickramasinghe[4], C. Kouveliotou[1,3], G.J. Fishman[3],
C.A. Meegan[3], R.B. Wilson[3] and W.S. Paciesas[3,5]

ABSTRACT

All BATSE triggered gamma-ray bursts before 1992 July 21 with recorded positions within 20 degrees of each other have been compared in a search for echoes. The ~2300 burst pairs were compared to see if one could possibly be a dimmer replica of the other and hence a gravitational lens echo. No significant echoes have yet been found, but more detailed comparisons involving weak, noisy, and extremely short bursts are ongoing. Candidate coincidences are discussed. Assuming the bursts lie at cosmological distances, new limits on the cosmological abundance of compact dark matter in the range near 10^6 $M_\odot$ have been derived. Based on these constraints, we expect future observations to be able to either test currently popular cosmological dark matter paradigms, or indicate that GRBs do not lie at cosmological distances.

INTRODUCTION

That gamma-ray bursts might be affected by gravitational lensing was first suggested by Paczynski (1986, 1987). Lensed bursts at cosmological distances can be used to trace the matter between the observer and the burst (Webster and Fitchett 1986, Nemiroff 1991, Blaes and Webster 1992), including potential diffraction effects (Bliokh and Minakov 1975, Gould 1992). It has even been suggested that lensing causes gamma-ray bursts (McBreen and Metcalfe 1988) but this idea has been criticized (Kovner 1990).

The primary effect of a gravitational lens on a gamma-ray burst would be to create many images of the burst instead of one. These images could not be spatially resolved with present technology, however, but they could be temporally resolved. A typical result would be that some bursts would have "echoes" of themselves arriving at later times. These echoes would have identical spectral and temporal profiles, although they would appear either dimmer or brighter than the previously recorded burst. Detecting lensing by temporal analysis was suggested by Paczynski (1986) and has been discussed by Krauss and Small (1991), who showed that the mass of the lens can be determined uniquely, by Mao (1992), who estimated the frequency of this effect due to the known galaxy distribution, and by Blaes and Webster (1992), who estimated what dark matter mass scales could be explored with GRBs. Narayan and Wallington (1992) have determined what lens characteristics can be determined from such a temporal detection.

---

[1] University Space Research Association
[2] NASA Goddard Space Flight Center, Greenbelt, MD 20771
[3] NASA Marshall Space Flight Center, Huntsville, AL 35812
[4] University of Pennsylvania, Dept. Astron. & Ap., Philadelphia, PA 19104
[5] University of Alabama, Huntsville, AL 35899

In this work we report on an active search for gravitational lens induced echoes in BATSE data. We also report on the limits this search has been able to provide on the cosmological distribution of dark matter.

THE SEARCH FOR AN ECHO

Were a bright echo to be measured by BATSE it would probably be noticed immediately by the MSFC team (GJF, CAM, RBW, JMH, CK, WSP). JMH in particular plots the burst location and checks conspicuous nearby bursts for similarity in temporal profiles. RJN systematically checked all the bursts within 20° of each other that occurred before 1992 July 21 for similarity in discriminator channel 3 temporal profiles, noted any two bursts that were even vaguely similar, and recorded them for subsequent close comparison. No compelling burst echo was found in the original comparison. These detailed comparisons are not yet complete, however, on dim and extremely short bursts.

Statistical background. There are several reasons why two temporal profiles would appear similar even if they are not lens-induced images of the same burst. The first is that the bursts are both "spikes" and not time resolved. These bursts might be distinguished from lens events by their relative hardness. The second is that the bursts are both "FRED" (Fast Rise Exponential Decay) bursts or pulse structures that are quite common to GRBs. Again, FREDs might be distinguished from gravitational lensing by relative hardness. Lastly, two very dim bursts would be dominated by noise and might appear statistically similar. These events constitute a background against which a real gravitational lensing event must be measured. Presently it is hard to quantify statistically how this background affects lens detection statistics.

Suggested Echoes. Previously there has been a suggestion that burst trigger 143 consists of a main image and its gravitationally lensed echo. Specifically the "second" pulse has been suggested as a gravitational lens echo of the first pulse. These two pulses are shown in detail below. Clearly, the two pulses have temporal profiles that are significantly different. This has been born out in a detailed statistical comparison. Unless a mechanism is proposed that can change the temporal profiles of images relative to each other, BATSE trigger 143 cannot be considered an example of gravitational lensing.

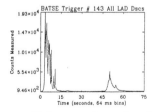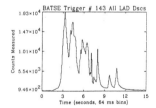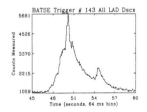

Only once have two bursts been recorded where the statistical comparison (by RJN) came out significant: between BATSE triggers 235 and 1447 (see figures below). These bursts had derived positions 19° apart, so their candidacy as lensed images was immediately greeted with skepticism. One of the bursts was quite weak, however, and hence its position was subject to significant uncertainty. In fact, the positions of the two bursts were consistent to within about 2 standard deviations. Burst 235, it turned out, the weaker burst, was found (by CAM, JMH, and CK) to have a position-consistent precursor, while no such precursor was found for the brighter burst. Hence, these two bursts are now not thought to be an example of gravitational lensing.

976 Search for Gravitational Lens Echoes in Gamma-Ray Bursts

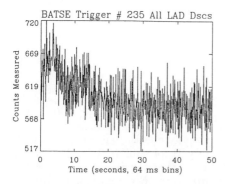
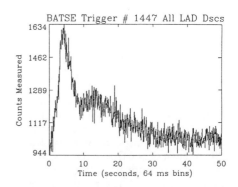

A time profile coincidence between two GRBs much closer together exists between triggers 493 and 1430. These bursts are located a mere 2.5 degrees apart - well within the statistical errors of their positions. Here two similar time profiles are seen (see figures below). However, when compared statistically in detail, we find that these two bursts are significantly different (see figure next page: f is the ratio of the brightness of the two images). The statistical procedure used was based on a scaled χ^2 comparison. A similar but different statistic has been discussed by Wambsganss (1992).

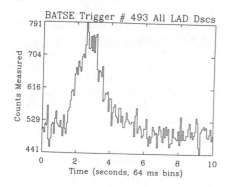
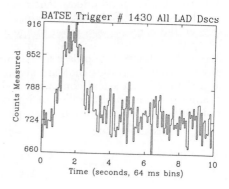

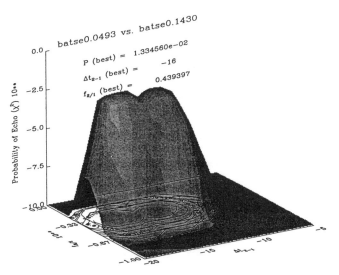

LIMITS ON DARK MATTER

Weaker bursts make better gravitational lens candidates for echoes than stronger bursts. This is because the probability of being lensed goes as $P \sim D^2 \Phi$ where Φ is related to the magnification of lensing such that $\Phi \sim \sqrt{C_{\max}}$. However $D^2 \sim C_{\max}^{-1}$, so that $P \sim C_{\max}^{-1/2}$. Therefore the most likely to be gravitationally lensed are the weak bursts, and the lack of lensing of weak bursts will provide the most constraining limits on the abundance of dark matter in the universe.

We wished to use the BATSE data and its lack of gravitational echoes to set limits on the abundance of cosmological dark matter in the universe. We have assumed the best fit redshift distribution of GRBs consistent with the observed N versus $C_{\max}/C_{\min}$ computed by Wickramasinghe et al. (1993), such that a burst with $C_{\max}/C_{\min} = 1$ would have a redshift less than unity. We assume an $\Omega = 1$ filled beam cosmology where $H_o = 50$ km sec^{-1} Mpc^{-1}. Were we to use a larger H_o, the excluded range would be larger. One of us (RJN) inspected the first 50 BATSE bursts for echoes internal to the burst. Not finding any, conservative limits in the dynamic range between images and the minimum and maximum delay times were set, past which it was clear no echo existed. In cases where a candidate echo signal existed, this signal was compared to the primary burst in both time profile and hardness to determine if it was, in fact, an artifact of gravitational lensing. No cases of gravitational lensing were found. The null probability gravitational lens detection volumes (Nemiroff 1991) were then added until there was a greater than a 75 % chance of detecting an echo were the universe at the hypothesized lens fraction (Ω_L) and lens mass M_L. This (Ω_L, M_L) pair was then excluded.

Following this procedure, we were able to find that one lensed image was to be expected for a universe composed of 10^6 solar masses (see figure next page). Since no image was found, this universe can be considered excluded. However, we are only beginning now to be able to exclude interesting regions of the (Ω_L, M_L) plane. With the full current BATSE data, which includes a factor of 10 more GRBs, the limit on Ω_L can be reduced by a factor of 10, testing currently popular dark matter paradigms.

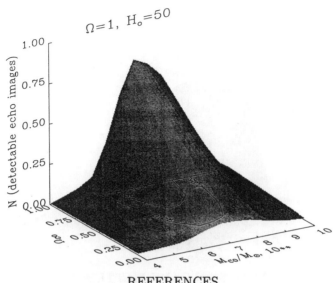

REFERENCES

Blaes, O.M. and Webster, R.L. 1992, ApJ 391, L63

Bliokh, P.V. and Minakov, A.A. 1975, ApSpSci 34, L7

Gould, A. 1992, ApJ 386, L5

Kovner, I. 1990, ApJ 351, 114

Krauss, L. M. and Small, T.A. 1991, ApJ 378, 22

Mao, S. 1992, ApJ 389, L41

McBreen, B., and Metcalfe, L. 1988, Nature 332, 234

McBreen, B., Plunkett, S., and Metcalfe, L. 1992, preprint

Narayan, R. and Blandford, R. 1992, AnnRevAstronAp 30, 311 (eds: G. Burbidge, D. Layzer, and A. Sandage; Annual Reviews: Palo Alto)

Narayan, R., and Wallington, S. 1992, ApJ, in press

Nemiroff, R.J. 1991, PhysRevLett 66, 538

Nemiroff, R.J. 1991, CommAp 15, 139

Paczynski, B. 1991, ActaAstron 41, 257

Paczynski, B. 1987, ApJ 317, L51

Paczynski, B. 1986, ApJ 308, L42

Wambsganss, J. 1992, ApJ, in press

Webster, R., and Fitchett, M. 1986, Nature 324, 617

Wickramasinghe, W.A.D.T. et al. 1993, The consistency of standard cosmology and the BATSE number versus brightness relation, in these conference proceedings

GAMMA-RAY BURSTS: MODELS

GAMMA-RAY BURSTS

Bohdan Paczyński

Princeton University Observatory, Princeton, NJ 08544-1001

ABSTRACT

The distribution of γ-ray bursts as observed by BATSE rules out the galactic disk origin, and it is consistent with the cosmological origin. Much work is currently done to reconcile galactic halo distribution with the absence of dipole and quadrupole moments in the positions of BATSE's bursts. A relativistic expansion of the source is implied by both the halo as well as cosmological origin.

While the correct distance scale is likely to be established observationally, a lot of work is done on a variety of possible models. The most promising is a neutron disk or torus accreting onto a stellar mass black hole. Such objects can form as a result of a merger/collision between two neutron stars or a neutron star and a black hole in a close binary system, or more directly from a massive stellar core collapse, a "failed supernova". Up to $\sim 10^{53} - 10^{54}$ erg is released while the disk accretes, but there is no quantitative model yet for the conversion of some of this energy into γ-rays.

THE DISTRIBUTION

There are two central questions related to γ-ray bursts: how far away are they, and what are they? These are two separate issues, even though they are often treated as one. I consider the distance to γ-ray bursts to be the most important at this time; it can be settled even if we have no understanding of the bursts. Currently, the only clue about the distance is the observed bursts' distribution.

The distribution of burst intensities, as measured with the peak count rate C_{max}, can be approximated as a power law:

$$N_b(> C_{max}) \sim C_{max}^{-\beta}, \tag{1}$$

where C_{min} is the minimum (threshold) count rate. For every burst a parameter V/V_{max} is defined as

$$V/V_{max} \equiv (C_{min}/C_{max})^{1.5}, \tag{2}$$

The power law exponent β and the average value $\langle V/V_{max} \rangle$ are related as[1]

$$\langle V/V_{max} \rangle = \frac{\beta}{\beta + 1.5}, \qquad \beta = \frac{1.5 \langle V/V_{max} \rangle}{1 - \langle V/V_{max} \rangle}. \tag{3}$$

Notice that for any power law distribution the value of $\langle V/V_{max}\rangle$ is not affected by the variations of the detection threshold C_{min}, as long as C_{min} is not correlated with C_{max}.

$\beta = 1.5$ and $\langle V/V_{max}\rangle = 0.5$ correspond to a uniform distribution of sources in the Euclidean space, while $\beta < 1.5$ and $\langle V/V_{max}\rangle < 0.5$ imply that the source number density decreases with distance, or the space is not Euclidean, i.e. the uniform distribution in the Euclidean space is bound.

Pioneer Venus Orbiter (PVO) worked for thirteen years, longer than any other γ-ray burst experiment, and it provided the best statistics for ~ 200 strong events. Two recent papers[2,3] report $\beta_{PVO} \approx 1.5$ and $\langle V/V_{max,PVO}\rangle = 0.46 \pm 0.02$, both consistent with a uniform distribution. The strong events are also known to be distributed isotropically in angle[4]. The Burst and Transient Source Experiment (BATSE) on the Compton Gamma-Ray Observatory is the most sensitive instrument to date. Its 447 bursts[5] are distributed isotropically over the sky, with no statistically significant dipole or quadrupole moment, while their radial distribution is bound, with $\beta_{BATSE} \approx 0.7$, and $\langle V/V_{max,BATSE}\rangle = 0.324 \pm 0.014$.

The relative calibration of the two instruments has not been done yet, but there is no doubt that BATSE bursts (the effective rate ~ 800 per year) are much weaker than PVO bursts (the effective rate ~ 38 per year). All other instruments have intermediate sensitivity as measured with the effective burst rate per year, and intermediate values of $\langle V/V_{max}\rangle$. As BATSE improves its statistics of strong and rare bursts a gradual change from $\beta \approx 1.5$ at the bright end towards $\beta \approx 0.7$ at the faint end becomes apparent.

THE DISTANCE SCALE

All observations of γ-ray bursts imply that we are at the center, or close to the center, of a spherically symmetric and bound distribution, and the number density of sources is uniform out to some distance. The most important question is: what is that distance? Among the many astronomical objects there are examples of many different distributions, but there is only one type that fits the γ-ray bursts: we are at the center of the whole observable universe. All sources that can be seen out to cosmological distances, galaxies and their clusters, active galactic nuclei, appear to be distributed more or less uniformly out to a few Giga-parsecs, and their distribution is observed to be bound and spherically symmetric. No other distribution is **observed** to have such properties. Therefore, the observed distribution of γ-ray bursts implies cosmological distances[6-10].

There are two other competing distance scales: the extended galactic halo[6,11-19] and the Oort cloud[6,15,20,21]. Any halo ever observed or proposed for any class of objects other than γ-ray bursts had a strong concentration towards the galactic center, and was not strictly spherical[6]. In particular, none had a core radius of ~ 25 kpc or more, as required to explain the lack of observable dipole and quadrupole moments. As for the Oort cloud, it has never been demonstrated that it can accommodate an inner constant density core. Therefore, the non-cosmological distance scales require new, unprecedented distributions, with the sole purpose of being **not inconsistent** with the BATSE and PVO data. Notice, there is **no positive evidence** yet for any of those alternative distance scales.

THE NATURE OF THE BURSTS

What can we learn from other properties of γ-ray bursts?

The diversity of γ-ray bursts' durations, time variability and spectra is enormous[22-27]. It is clear that their spectra are non-thermal. Some experiments, like KONUS, GINGA, and others, found lines in the spectra of some bursts, while SMM and BATSE were not able to detect any lines so far. The situation is somewhat confusing, but one point is clear: the majority of bursts do not have detectable lines in their spectra. It is also clear that whenever any burst parameter is plotted against any other, there is no clear correlation present, and no clear clustering of events into distinct sub-types. The only exception are the three soft gamma repeaters, which are very distinct from all the rest, and will not be discussed here. The classical hard γ-ray bursts are not observed to repeat. They seem to form a very diverse group which cannot be broken up into clearly separable sub-groups. This suggests that there is only one population of sources with a large number of degrees of freedom.

Notice that there is no model, and there has never been one, that would cover in a quantitative way the ultimate energy source, the energy conversion to the observed γ-ray spectrum with or without spectral lines, and the types of variability observed among hard γ-ray bursts. Theoretical models developed for neutron stars in the galactic disk covered a variety of separate physical processes, but the many pieces of the puzzle have never been put together. It is very sobering to compare the so called models of γ-ray bursts with the truly successful models of X-ray bursts[28].

The two most popular distance scales currently proposed for γ-ray bursts are cosmological and the galactic halo. It is important to recognize that while the previously popular galactic disk distance scale implied burst luminosity of $\sim 10^{37}\ erg\ s^{-1}$, the galactic halo and the cosmological distance scales imply $\sim 10^{42}\ erg\ s^{-1}$ and $\sim 10^{51}\ erg\ s^{-1}$, respectively. This means that while all disk models were sub-Eddington, and hydrostatic equilibrium could be assumed for the bursting region, any halo or cosmological model must be highly super-Eddington, and hence the bursting region must expand relativistically. This means that the "technology" developed in the past for the disk bursts has no direct application to any of the current models.

Given the diversity of γ-ray bursts, and the energetics implied by the galactic halo and the cosmological distance scales, can we find any objects or events that are known to exist, and which might be relevant? Neutron stars or related objects are the prime candidates. We know very little about them in the extended galactic halo, whereas we may assume that objects known to exist in the disk of our galaxy also exist in galaxies at cosmological distances.

A cosmological distance scale would imply that γ-ray bursters are the most luminous objects in the universe, with a peak power of $\sim 2 \times 10^{51}\ h_{100}^{-2}\ (\Omega_b/4\pi)\ erg\ s^{-1}$ where h_{100} is the Hubble constant in units of $100\ km\ s^{-1}\ Mpc^{-1}$, and Ω_b is the solid angle into which the burster's emission is beamed. The rate of events would have to be $\sim 2 \times 10^{-6}\ (4\pi/\Omega_b)\ yr^{-1}$ per galaxy like ours, or $\sim 30\ (4\pi/\Omega_b)\ h_{100}^3\ Gpc^{-3}\ yr^{-1}$ in our neighborhood. In the cosmological scenario the PVO bursts would be at a distance corresponding to a redshift of

$z \leq 0.2$. All objects known at $z = 0.1 - 0.2$, i.e. at distances $\sim 500\ Mpc$, are randomly distributed, as the largest scale structures known to exist in the universe are $\leq 100\ h_{100}\ Mpc$ across. And indeed, there is no correlation between the positions of various γ-ray bursts in the pre-BATSE data[29].

It is well established that an iron core is formed at the end of evolution of a massive star. When that core exceeds the effective Chandrasekhar mass limit it collapses to form a neutron star or a black hole. Of the order of $\sim 10^{53} - 10^{54}\ ergs$ is released in a neutrino burst that lasts ~ 10 seconds. Many massive stars are in binary systems. Hence, at the end of their evolution pairs of neutron stars, and presumably neutron star and black hole pairs, are formed. Some of those are known to have binary orbits so tight that the two compact components will merge within Hubble time. The merger is as violent as a collision, with up to $10^{53} - 10^{54}\ ergs$ released in a neutrino burst[30]. These are not hypothetical events, they must happen in our galaxy once every $10^5 - 10^6$ years. There is no doubt they release a lot of energy, but it is not clear how to convert enough energy into hard γ-rays.

Can the end products of massive star evolution meet the demands imposed by γ-ray bursts at $\sim 1\ Gpc$? I think there is a chance they can. A merger/collision of a pair of neutron stars, or a neutron stars and a stellar mass black hole lasts for a few milliseconds only, after which most mass is likely to end up in a black hole, but some may remain outside as a neutron disk or torus[6,30]. A similar configuration can be formed more directly and efficiently as a product of a "failed supernova"[31]. Whatever its origin, a neutron disk or torus spinning around a stellar mass black hole with a period of $\sim 10^{-3}\ s$, can release $\sim 10^{53} - 10^{54}\ erg$ while accreting on a time scale which may be much longer than $\sim 1\ s$. This means the system may be capable to provide a fine time structure required by γ-ray bursts' variability, and supply energy for the long lasting bursts. There is a number of degrees of freedom in the system: the mass and the spin of a black hole, the angle between the spin axis and the line of sight, the mass and angular momentum of the neutron disk, and finally the magnetic fields which are known to exist in neutron stars. Therefore, the general properties required for γ-ray bursts are in place. What is missing is a quantitative theory of γ-ray emission.

THE DISTANCE SCALE AGAIN

Even though the cosmological distance scale to γ-ray bursts is currently more plausible than any other, it is too early to claim that is has been established. A major breakthrough may be expected if optical or X-ray counterparts can be identified in small error boxes to be obtained for the strongest bursts using the interplanetary network. If the positions were well correlated with galaxies the cosmological distance scale would become evident[6,30]. If soft thermal X-ray counterparts were found with the intensity consistent with neutron stars at $\sim 30\ kpc$, the case for the galactic halo would be strong[19].

A discovery of the cosmological redshifts in bursts' durations or spectra[32], or a clear dipole or quadrupole anisotropy related to our galaxy would also provide a decisive evidence for one or another distance scale. In a few years, when the number of BATSE bursts increases to $\sim 1,000$ or so, some excess of γ-ray bursts towards M31 may become apparent, or the halo origin will be ruled out[33,34].

A major step towards understanding γ-ray bursts will be made when any correlations are found between various bursts' properties and/or a reproducible classification scheme for γ-ray bursts is developed. This might prove to be equivalent of the H-R diagram for stars or the Hubble diagram for galaxies.

This work was supported with NASA grants NAG5-1901 and NAGW-2448, and NSF grant AST-9023775.

REFERENCES

1. B. Paczyński, *Acta Astronomica*, **42**, 1 (1992).
2. E. E. Fenimore, R. I. Epstein, C. Ho, R. W. Klebesadel and J. Laros, *Nature*, **357**, 140 (1992).
3. K. W. Chuang, R. S. White, R. W. Klebesadel and J. G. Laros, *Astrophys. J.*, **391**, 242 (1992).
4. J. L. Atteia *et al.*, *Astrophys. J. Suppl.*, **64**, 305 (1987).
5. C. A. Meegan *et al. IAU Circ. No.* 5641 (1992).
6. B. Paczyński, *Acta Astronomica*, **41**, 257 (1991).
7. S. Mao and B. Paczyński, *Astrophys. J.*, **388**, L45 (1992).
9. T. Piran, *Astrophys. J.*, **389**, L45 (1992).
9. C. D. Dermer, *Phys. Rev. Lett.*, **68**, 1799 (1992).
10. E. E. Fenimore, R. I. Epstein, C. Ho, R. W. Klebesadel and J. Laros, *Nature*, **357**, 140 (1992).
11. B. Paczyński, *Acta Astronomica*, **41**, 157 (1991).
12. S. Mao and B. Paczyński, *Astrophys. J.*, **389**, L13 (1992).
13. J. Shaham, *Nature*, **354**, 439 (1991).
14. J. J. Breinerd, *Nature*, **355**, 522 (1992).
15. T. Nakamura, N. Shibazaki, T. Murakami and A. Yoshida, *Prog. Theoret. Phys.*, **87**, 879 (1992).
16. D. Eichler and J. Silk, *Science*, **257**, 937 (1992).
17. E. E. Salpeter and I. M. Wasserman, presented at *Planets Around Pulsars* held in Pasadena, California (1992).
18. H. Li and C. D. Dermer, *Nature*, **359**, 514 (1992).
19. H. Li, C. Thompson and R. C. Duncan, presented at the *Compton Symposium*, St. Louis, Oct. 15-17 (1992).
20. A. V. Kuznetsov, *Cosmic Res.*, **20**, 72 (1982).
21. C. M. Gaskell, preprint (1992).
22. E. P. Liang and V. Petrosian, eds, *Gamma Ray Bursts*, AIP Conf. Proc. 141, New

York (1986).
23. W. N. Johnson, Ed., *Proc. of Gamma Ray Obs. Sci. Workshop*, (NASA, 1989).
24. J. J. Higdon and R. E. Lingenfelter, *Annu. Rev. Astron. Astrophys.*, **28**, 401 (1990).
25. C. Ho, R. I. Epstein and E. E. Fenimore, eds., *Gamma Ray Bursts Observations, Analyses and Theories*, (Cambridge Univ. Press, 1992).
26. C. R. Shrader, N. Gehrels and B. Dennis, eds., *Compton Obs. Sci. Workshop*, (NASA, 1992).
27. W. S. Paciesas and G. J. Fishman, eds., *Gamma Ray Bursts*, AIP Conf. Proc. 265, New York (1992).
28. F. Melia, *Fundamentals of Cosmic Physics*, **12**, 97 (1987).
29. D. Hartmann and G. R. Blumenthal, *Astrophys. J.*, **342**, 521 (1989).
30. R. Narayan, B. Paczyński and T. Piran, 1992, *Astrophys. J.*, **395**, L83 (1992).
31. S. E. Woosley, *Astrophys. J.*, submitted (1992).
32. B. Paczyński, *Nature*, *355*, 521 (1992).
33. J. Hakkila and C. A. Meegan, in *Gamma-Ray Bursts*, AIP Conf. Proc. 265, eds. W. S. Paciesas and G. J. Fishman, New York (1992), p70.
34. J. Hakkila and C. A. Meegan, presented at the *Compton Symposium*, St. Louis, Oct. 15-17 (1992).

THE IMPACT OF RELATIVISTIC FIREBALLS ON AN EXTERNAL MEDIUM: A NEW MODEL FOR "COSMOLOGICAL" GAMMA-RAY BURST EMISSION

M.J. Rees
Institute of Astronomy, Madingley Road,
Cambridge, CB3 0HA, UK

P. Mészáros
525 Davey Laboratory, Pennsylvania State University,
University Park, PA 16802, USA

ABSTRACT

The expansion energy of a relativistic fireball can be reconverted into radiation when it interacts with an external medium. For expansion with Lorentz factors $\gtrsim 10^3$ into a typical galactic environment, the corresponding time-scale in the frame of the observer is of the order of seconds. This mechanism would operate in any cosmological scenario of gamma-ray bursts invoking initial energies of order a per cent of a stellar rest mass, and implies photon energies and time-scales compatible with those observed in gamma-ray bursts.

INTRODUCTION

Catastrophic events involving compact stars – binary neutron star mergers, for instance – occur with the appropriate frequency to account for gamma-ray bursts if we are sampling the entire volume out to a redshift $z \simeq 1$; and may produce a sufficient *initial* radiation energy. However, two major problems arise. One is that it seems almost impossible to release this much energy in radiation without the fireball being to some degree contaminated by baryons; the associated opacity would then trap the bulk of the energy until it had been converted into (expansion) kinetic energy. This would be the case even in the relatively "clean" fireballs needed to break out from a dense precursor wind (Mészáros and Rees, 1992b), for a contamination mass $\gtrsim 10^{-9} M_\odot$. The other problem which plagues cosmological GRB models is that the typical timescales computed for radiative fireballs are far shorter than the observed timescales of up to ~ 100s.

In this paper we point out that interaction with even very tenuous external matter (e.g. an ordinary interstellar medium) can efficiently reconvert the kinetic energy of a baryon-loaded relativistic fireball into a burst of high energy photons, which would be observed on timescales comparable to those of GRB. The only requirement is that a substantial part of the fireball should have acquired a Lorentz factor $\Gamma \gtrsim 10^3$ – this is a much less demanding requirement than is imposed by other models for GRBs at cosmological distances. Further details are given in Rees and Mészáros (1992) and Mészáros and Rees (1992c).

HOMOGENEOUS FIREBALLS

Consider a simple model for a homogeneous fireball with energy E_f. With respect to a comoving frame, the contents of the fireball would cool as it expanded. After the internal thermal motions have become sub-relativistic, the boundary Lorentz factor Γ_f saturates, and remains constant for as long as external material can be neglected. The internal density n_b decreases as t_{com}^{-3}, where t_{com} is the time that would be measured by a clock comoving with the expanding material. We can then ask what happens if the fireball runs into a stationary external medium of negligible thermal energy which may also be approximated as homogeneous. The behavior resembles an ordinary supernova remnant expanding into an undisturbed medium: a contact discontinuity advances with speed characterized by a Lorentz factor Γ_f; a shock wave moves ahead of this into the external medium characterized by $2^{1/2}\Gamma_f$ (Blandford and McKee, 1976); a reverse shock builds up and propagates back into the fireball, after the latter starts to decelerate (McKee, 1974). Deceleration becomes important when the boundary r attains a value given by

$$E_f \sim \frac{4}{3}\pi \rho_{ext} \Gamma_f^2 r_{dec}^3 . \tag{1}$$

This occurs when the fireball has swept up a mass $\Gamma_f^{-2} E_f/c^2$, i.e. not more than Γ_f^{-1} of the rest mass of polluting baryons carried in the fireball.

In the earliest adiabatic stage of the expansion, the thermal energy of the fireball is converted into bulk kinetic energy of expansion, but when deceleration becomes significant the process reverses itself and the fireball starts to reconvert its bulk kinetic energy into thermal form, which can be radiated away if cooling times are short enough.

The kinematics of a fireball as viewed by a distant observer, taking retarded-time effects into account, are similar to the simplest models of relativistically expanding radio sources (Rees, 1966): the part of the shell coming directly towards the observer has an apparent speed $\sim 2c\Gamma_f^2$, so the effects of the shell's deceleration are observed on a timescale of only $\sim r_{dec}/2\Gamma_f^2 c$. The bulk of the observed radiation comes from parts of the fireball whose doppler factor is $\gtrsim \Gamma_f$. To take an illustrative example, suppose that

$$E_f = 10^{51} \text{ erg}, \quad \Gamma_f = 10^3, \quad n_{ext} = 1 \text{ cm}^{-3}, \tag{2}$$

implying a baryon rest mass $\sim 10^{-6} M_\odot$ in the fireball and an external particle density typical of the interstellar medium. Reconversion of kinetic energy occurs when the swept-up baryon rest mass is $\sim 10^{-9} M_\odot$, the corresponding radius being $r_{dec} \sim 10^{16}$ cm.

As it continues to expand to radii exceeding r_{dec}, the fireball bulk Lorentz factor decreases as $\Gamma_f \propto r^{-3}$, if radiative cooling carries away all the re-randomized energy, or as $r^{-3/2}$ if radiative losses are unimportant. The observed timescale ($\propto r\Gamma_f^{-2}$) would then increase as $t \propto r^7$ or r^4 respectively. If radiative processes were prompt and efficient, the fireball would have a bolometric luminosity that first rose as $L \propto t^2$ until a peak of order $L \sim E_f c \Gamma_f^2 / r_{dec}$ was reached at $t \sim r_{dec}/c\Gamma_f^2$; thereafter the luminosity would decay rapidly. In the first stage of the deceleration, a reduction from $\Gamma_f = 10^3$ to 5×10^2 in the expansion Lorentz factor re-randomizes about half of the fireball's total energy during an expansion by only a factor of 1.3 in the radius; if this energy could then be radiated, we would receive it in a "burst" with an observer-frame timescale of a few seconds.

The Lorentz factors of the external shocked particles would be $\sim \Gamma_f$ in the frame of the contact discontinuity, and therefore of order $\gamma \sim \Gamma_f^2$ in the observer's frame (c.f., reflection or bouncing off an expanding mirror or wall). The internal reverse shock (which starts off weak but strengthens as the acceleration builds up) would become relativistic when $r \gtrsim r_{dec}$, with $\gamma_{rs} - 1 \geq 1$ in its frame, and this would be the initial thermal Lorentz factor of the fireball particles between the discontinuity and the reverse shock.

If radiative processes were efficient both within the fireball itself and in the (high-gamma) shocked external material, then all the shocked material (both inside and outside the contact discontinuity) would cool and pile up in a dense shell. Roughly equal amounts of energy are thermalized inside and outside the contact discontinuity, as can be seen from the fact that the post-shock pressures are the same on both sides of the discontinuity.

To check the radiative efficiency, we use the parameter values in our numerical example, and assume that the fireball rest mass of $10^{-6} M_\odot$ is in ordinary baryonic matter. What would be the conditions within the fireball at the stage when it starts to "feel" the external medium? If the fireball were uniform (i.e. such that, away from the boundary, the expansion and density appeared isotropic to any comoving observer, as in Milne's zero-deceleration cosmology) then 75 per cent of the baryons would have bulk Lorentz factors $\Gamma > 0.5\Gamma_f$ with respect to the centre. If $\Gamma_f = 10^3$, these particles would all be expanding outward at speeds differing from c by $\lesssim 1$ part in 10^6, and would therefore, in our example, be concentrated in a shell of thickness 10^{10} cm when $r \simeq r_{dec} \simeq 10^{16}$ cm. In the frame comoving with this shell the thickness is larger by Γ_f – i.e. $\sim 10^{13}$ cm. The corresponding density is then $n_b \simeq 10^6$ cm^{-3}; the expansion timescale in the comoving frame is $t_{com} \simeq 10^3$ secs.

The pressure of marginally relativistic plasma of density 10^6 cm^{-3} inside the contact discontinuity that has been reheated by the reverse shock is balanced by ultrarelativistic plasma with mean particle energy 10^{12} eV and density 10^3 cm^{-3} just outside. The equipartition magnetic field corresponding to this pressure is 100 G. In this field, the synchrotron lifetime of any electrons with energies $> 5 \times 10^7$ eV is less than t_{com}. The contact discontinuity would be Rayleigh-Taylor unstable, so the internal and external regions would probably mix. The processes that determine the particle energy distribution in an ultrarelativistic plasma of this type are uncertain. However, synchrotron emission of photons which would be observed as ~ 1 MeV gamma rays (i.e. with energies ~ 1 keV in the frame of the contact discontinuity) requires electrons with energies 3×10^{10} eV. These electrons would not be the extreme part of some high energy tail – indeed, the required energy is actually *lower* than the *mean* energy of particles in the external shocked shell. Such electrons would, moreover, radiate all their energy within a time $< t_{com}$. So it at least seems rather likely that a fireball with properties resembling (2) could generate $> 10^{51}$ ergs of gamma rays, which would reach a distant observer in a burst lasting a few seconds.

The radiation comes, in effect, from a region $\sim 10^{16}$ cm in extent. Its escape would not be impeded by opacity. Nor would the region be compact enough for photon-photon collisions to yield $e^- - e^+$ pairs (and in any case, because of the Doppler boost factor, the threshold for this process is 10^3 MeV in the observer frame). If the fireball contained pairs, then thermal processes (which are ineffective on the expansion timescale in our illustrative example) could be important in the reverse-shocked region.

INHOMOGENEOUS FIREBALLS

In a dynamically realistic fireball, we would expect a wider range of Lorentz factors than in the homogeneous case discussed above. We cannot model the detailed formation and dynamics of a pair-dominated fireball, (of the kind discussed by, for instance, Shemi and Piran 1990), even for an idealized simple geometry. However, the simpler "fluid" case is more tractable, and has been extensively discussed, particularly by Colgate and his collaborators (e.g., Colgate and Noerdlinger, 1971, Johnson and McKee, 1971 and earlier work cited therein). A shock strengthens as it propagates down a density gradient, the fraction of the kinetic energy going into shells of matter which attain Lorentz factors $> \Gamma$ being

$$f(> \Gamma) \propto \Gamma^{-\alpha} . \tag{3}$$

Detailed calculations show that $\alpha \simeq \frac{1}{2}$ is realistic (e.g. McKee and Colgate, 1973). Some kind of a power law is probably generic whenever material is blown off a compact object by some violent hydrodynamical disturbance.

Material ejected in this way would eventually be braked by the external medium. In contrast to the homogeneous fireball discussed in section 2, the particles cannot now be characterized by one "typical" Γ; instead, the fraction in shells with a particular energy will be determined by some expression such as (3). Essentially, the energy that is received within the first $\sim 1-10$ seconds is equivalent to the energy that goes into material with $\Gamma \gtrsim 10^3$. If $\alpha = \frac{1}{2}$ (eqn. (3)), this is 3% of the energy of relativistic ejecta. Therefore any process that could convert most of a neutron star's binding energy ($\sim 10^{53}$ ergs) into dynamical motions could plausibly generate a $\sim 10^{51}$ erg burst.

DISCUSSION

We have addressed here two of the major difficulties which are generic to almost all cosmological (and many of the halo) gamma-ray burst models: the recovery of the expansion kinetic energy and the relevant burst timescales. It is clearly possible to achieve photon energies comparable to the initial fireball total energy ($\sim 10^{51}$ erg) over timescales of a few seconds or more, in the observer frame. The scaling laws for energy, Lorentz factor, fireball loading, external density, etc., are quite straightforward, based on the expansion dynamics of fireballs that would be produced, e.g., by neutron star binary mergers. Even if the bursts involve the kind of mechanisms discussed by Narayan, Paczyński and Piran (1992), Usov (1992) or Mészáros and Rees (1992a,b), which can generate a short γ-ray pulse, those same models ought to produce high energy radiation *by the present mechanisms too*, and with a more appropriate timescale for fitting the observations. The timescale over which the peak power is received (given by the time in the observer frame corresponding to $r = r_{dec}$) is sensitive primarily to Γ_f, scaling as $\Gamma_f^{-8/3} E_f^{1/3} n_{ext}^{-1/3}$. The key requirement is bulk expansion with $\Gamma \gtrsim 10^3$: it makes little difference whether the fireball arises from $\nu + \bar{\nu} \rightarrow e^- + e^+$, from electromagnetic effects, or from purely hydrodynamical processes.

Depending on the particular model, the fireball will expand either over a broad range of angles, or within a narrower range of angles, forming one or two jets. However, even in this case it is likely that the opening half-angle α_o exceeds Γ^{-1}, so that the isotropic results apply with minor modifications. As a result of the beaming, the frequency of detection would go down as α^2, which is compatible with the expected frequency of compact binary mergers, while the total energy needed goes down by α^{-2}.

Whatever the initial fireball production mechanism, the burst occurring when it first becomes optically thin is unlikely to be the main "archetypal" γ-ray burst observed: the thinning burst is far too short on average, and usually does not have enough energy to be the most prominent at cosmological distances. It also tends to be rather softer and thermal-like. Instead, we propose that the usual GRB main bursts occur when the contaminating baryon kinetic energy is reconverted into radiation by the deceleration shock as it piles up against external matter. This produces predominantly non-thermal bursts of energy comparable to the initial fireball energy ($\sim 10^{51}$ erg) extending over a timescale $t \sim 10$ s. The burst properties would depend on the external environment; time substructure could arise from shock instabilities, or from inhomogeneities in the external medium. The thinning burst, when at all observable, may be identified with the precursor burst. Alternatively, if deceleration occurs in a very dense surrounding medium such as a nebula or a precursor wind, a stronger deceleration burst may occur first, followed by a weaker and longer thinning afterburst.

Acknowledgements: This research has been partially supported through NASA NAGW-1522.

REFERENCES

Blandford, R.D. and McKee, C.F., 1976, Phys. Fluids, **19**, 1130

Colgate, S.A. and Noerdlinger, P.D., 1971, Ap.J., **165**, 509

Johnson, M.H. and McKee, C.F., 1971, Phys.Rev, **D3**, 858

McKee, C.R. and Colgate, S.A. 1973. Ap.J. **181**, 903.

Meegan, L.A., Fishman, G.J., Wilson, R.B., Paciesas, W.S., Brock, M.N., Horack, J.M., Pendleton, G.N. and Kouveliotou, C., 1992, Nature, **335**, 143.

McKee, C.F., 1974, Ap.J., 188, 335

Mészáros, P. and Rees, M.J., 1992a, Ap.J., **397**, 570

Mészáros, P. and Rees, M.J., 1992b, M.N.R.A.S., **257**, 29P.

Mészáros, P. and Rees, M.J., 1992c, Ap.J., in press

Narayan, R., Paczyński, B. and Piran, T., 1992, Ap.J. (Lett.), **395**, L83

Rees, M.J. 1966. Nature **211**, 468.

Rees, M.J. and Mészáros, P., 1992a, M.N.R.A.S., **258**, 419.

Shemi, A. and Piran, T. 1990, Ap.J. (Lett.), **365**, L55.

Usov, V.V., 1992, Nature, **357**, 472

CONTRIBUTION TO PANEL DISCUSSION ON GAMMA RAY BURSTS

R. D. Blandford
Caltech, Pasadena, CA 91125.

ABSTRACT

Taken at face value, the BATSE observations of γ-ray bursts favor (but do not yet verify) extragalactic source models. The observations pose some general challenges that must be addressed by such models. Explanations invoking relativistic blast waves raise additional questions.

PHENOMENOLOGICAL CHALLENGES

The apparent isotropy and radial inhomogeneity of the BATSE γ-ray burst distribution [1] came as a considerable surprise to aficionados of the local neutron star interpretation. As one of these, I now have to concede that the extragalactic source interpretation seems more probable. [2] Nevertheless, several features of the observations raise concerns, independent of the detailed source model. These include:

(i) If we naively translate the C_{max}/C_{min} data to the language of source counts, then $N(>S) \propto S^{-0.8}$. Equivalently $<V/V_{max}>= 0.32$. [1] This is slightly surprising for an extragalactic neutron star population. It might have been expected that the rate of formation of neutron stars would have been much greater in the past and that the γ-ray burst population evolved just like quasars with $<V/V_{max} >> 0.5$. Curiously, the one extragalactic population that they do mimic quite accurately is that of the faintest, "blue galaxies" which show similar source counts. [3] If γ-ray bursts do come from these enigmatic objects, then there is almost no chance of identifying them from accurate positions, because they are only $\sim 7"$ apart on the average.

(ii) As many contributors have emphasized, quantum electrodynamical imperatives demand that cosmologically distant sources expand ultrarelativistically. This allows the sources to vary on timescales short compared with the light crossing time and results in tremendous Doppler boosts of the observed γ-rays. Now samples of γ-ray bursts are famously heterogeneous, [4] especially in their burst durations which range from milliseconds to minutes. This is usually attributed to inhomogeneity in the sources. It is then very surprising that this variation is not reflected in the burst spectra. One might expect that the median burst energy, typically ~ 1 MeV, would instead range from UV to UHE γ-rays. This does not appear to be the case, but how sure are we of this?

(iii) This same relativistic expansion surely dooms any explanation of the hard X-ray, (supposed cyclotron), lines, [5] however ingenious. Averaging over a relativistic flow generally produces broader lines than reported in the GINGA observations.

(iv) The physical processes that were associated with slowly accreting local neutron stars ought still to be important. Although the nuclear physics details remain uncertain, [6] it did not seem likely that bursts of sufficient energy and frequency to account for the observations would occur. Instead, the accumulation of up to 10^{45} erg of latent nuclear energy on the surface of an old, slow, long-period pulsar might permit outbursts that could be seen from nearby galaxies. Such bursts might be a minority of the total population.

(v) In addition, the recently discovered ROSAT soft X-ray sources might be associated with steady accretion onto local neutron stars. [7] These sources are found to be isotropic, just like the γ-ray bursts. However, in this case, we must draw the opposite conclusion. It seems unavoidable that a population of sources with luminosity function depending solely on height above the Galactic plane and in compliance with the inverse square law, cannot simultaneously be isotropic and exhibit a deficit of faint sources. A general, formal demonstration of this statement is contained in the Appendix to Blaes et al.(1992). [8] It is therefore claimed that the γ-ray bursts are of cosmological origin. In the case of the ROSAT soft X-ray sources, interstellar absorption is unavoidable and presumably is compensated by an anisotropy in the source distribution. This rules out a cosmological origin.)

THEORETICAL CHALLENGES

Eighteen years ago, in an article that still repays reading, [9] Ruderman surveyed theoretical explanations of γ-ray bursts and their attendant problems. (The most *speculative* model that he could devise involved antimatter comets falling onto white holes; but we have come a long way since then!) The most *promising* model to date for cosmological bursts invokes a very high entropy explosion that drives a blast wave into the surrounding medium. [10] The particle acceleration and γ-ray emission come mainly from the bounding shock front. (It is ironic that theoretical studies of relativistic blast waves [11] were originally motivated by observations of low frequency radio variability, not UHE γ-rays!) Three aspects of this particular model seem to deserve attention.

(i) Relativistic explosions ought to decelerate dramatically if the shock becomes radiative. This is because the radial momentum is actually lost to the escaping photons rather than merely shared with increasing mass of swept-up gas as in non-relativistic, radiative blast waves. Although the outcome is fairly model-dependent, I would expect to see much more pronounced spectral softening and "Fast Rise, Exponential Decline" temporal evolution than is reported.

(ii) In order for the bounding shock fronts to become radiatively efficient, it is necessary for the magnetic field strength to be raised from the pre-shock microgauss levels to post-shock equipartition values ~ 100 G. This is somewhat unusual.

(iii) Much of the physics of relativistic blast wave models is independent of the nature of the explosion. Nevertheless, the requirements on the explosion mechanism are so constraining that any viable model developed in moderate detail would be of immense value even if not ultimately correct. Reluctant as I am to augment the burgeoning list of ingenious, though surely ephemeral, suggestions that already exist in the literature, I can't resist adding two more. Both of these are motivated by the belief that the best way to avoid baryonic contamination of the outflow is to extract the spin energy of a Kerr black hole using strong electromagnetic fields. In the first scenario, it is suggested that the source is a coalescing binary black hole. If, as is qualitatively correct, [12] we treat both holes as spinning, (and orbiting), conductors with $\sim 100\Omega$ resistance strong electric fields can be induced in the region between the holes. These can cause sufficient numbers of electron-positron pairs to be created to carry electric current and motivate the relativistic MHD approximation, without providing significant inertia. Now, in non-relativistic MHD, two sphere, kinematic dynamos have been studied. [13] These allow feedback between the two spinning spheres to cause the magnetic field strength to grow with time. It is possible that a relativistic generalization of this mechanism involving black holes might allow a seed interstellar field to grow exponentially with time constant of order the spin period after the orbit has shrunk to a certain limiting radius. If enough e-foldings can occur prior to coalescence, then it may be possible for the field to become dynamically important, ($i.e. B \gtrsim 10^{15}$ G) and extract the spin energy in $\lesssim 1$ s. Efforts by the author and B. Ratra to understand this in more detail have been unsuccessful. In the second scenario, which has been explored in collaboration with J. Ostriker, it is hypothesized that galactic halos are bound by $\lesssim 10^6 M_\odot$ black holes and that γ-ray bursts are associated with their stellar captures. The star forms an orbiting torus linked by an increasing magnetic flux, which can, again, extract spin energy from the hole. Although the incidence of capture is compatible with the burst rate, it is far from clear that the field will grow fast enough and that mass loss from the torus can be ignored. Also the initial hypothesis raises some extra questions.

ACKNOWLEDGEMENTS

I thank my fellow panel members, Dieter Hartmann, Don Lamb, Bohdan Paczyński and Virginia Trimble for helpful discussions. Support under NASA grant NAGW 2920 is gratefully acknowledged.

REFERENCES

1. C. A. Meegan et al., These Proceedings (1993, in press).
2. B. Paczyński, These Proceedings (1993, in press).
3. J. A. Tyson, *Astronom. J.* **96**, 1 (1988).
4. G. J. Fishman, These Proceedings (1993, in press).
5. D. Q. Lamb, These Proceedings (1993, in press).
6. P. C. Mock & P. C. Joss, These Proceedings (1993, in press).
7. O. M. Blaes & P. Madau, *Ap. J.*, (1993, in press).
8. O. M. Blaes, R. D. Blandford, P. Madau, & L. Yan, *Ap. J.*, (1992, in press).
9. M. Ruderman, *Ann. N. Y. Acad. Sci.* **262**, 164 (1975).
10. M. J. Rees & P. Mészáros, These Proceedings (1993, in press).
11. R. D. Blandford & C. F. McKee, *MNRAS* **180**, 343 (1977).
12. Thorne, K. S., R. M. Price & D. MacDonald, Black Holes: The Membrane Paradigm. New Haven: Yale University Press (1986).
13. H. K. Moffatt, Magnetic Field Generation in Electrically Conducting Fluids. Cambridge: Cambridge University Press (1978).

STELLAR COLLAPSE AND GAMMA-RAY BURSTS

S. E. Woosley

Board of Studies in Astronomy and Astrophysics, UCO/Lick Observatory
University of California at Santa Cruz, Santa Cruz, CA 95064

Abstract. The energy requirements for a cosmologically situated gamma-ray burst lead naturally to the consideration of neutron star formation and black hole accretion as possible models. It is shown that all spherically symmetric models are doomed to failure because of baryonic overloading of the ejected plasma. In two or three dimensions however, merging neutron stars, or better still, the collapse to a black hole of the iron core of a massive helium star endowed with rotation (failed Type Ib supernova), can in principle set up conditions where energetic plasma is ejected in jets that contain a much smaller fraction of baryonic matter. These models resemble those that have been discussed for active galactic nuclei and it is suggested that similar physics may be at work.

INTRODUCTION

We live in interesting times. I will not take up space lamenting the ignorance of theorists when it comes to determining the nature of gamma-ray bursts. I am one of them and have demonstrated my lack of character by proposing three totally different scenarios - the thermonuclear model, in its original incarnation a thick disk model (Woosley & Wallace 1982); a planetesimal accretion hypothesis which functions in an extended halo (Woosley 1992a); and a model based upon stellar collapse at cosmological distances (Woosley 1993). Here I will be reviewing models of a cosmological nature. My studies and those of others have convinced me that such models are, in principle, possible. This should not be construed to mean that other possibilities have been abandoned. With regards to the distance scale, I have followed a long path to agnosticism.

The questions to be briefly discussed are the general physical requirements of a cosmological model and the demonstration of a plausible example.

THE FAILURE OF SPHERICALLY SYMMETRIC MODELS

A viable cosmological model must not only produce in excess of 10^{51} erg, but make it either in the form of gamma-radiation or extremely relativistic particles whose energy can be converted to gamma-radiation with high efficiency (e.g., Rees & Meszaros 1992). In practice, it proves very difficult if not impossible to craft such a model provided that one is restricted to spherical symmetry.

A relevant example is shock break out from a supernova (Colgate 1968, 1975; Colgate & Petschek 1979, and references therein). The shock wave could originate from iron core collapse in a massive star, which may itself be a red or blue supergiant (Type IIp and 87A prototypes respectively), or in certain detonation models for Type Ia. The radiation would be produced either as the surface of the star is heated to high temperature by shock passage or as relativistic matter (i.e., cosmic rays, Colgate & Johnson 1960) impacts circumstellar material.

Unfortunately modern calculations of shock break out in all varieties of supernovae give neither gamma-rays nor appreciable relativistic matter (Woosley 1992b). For Type II and Ib supernovae the hardest radiation is around 100 eV, certainly less than 1 keV. The radiation from a Type Ia would be extremely brief ($\lesssim 1$ ms) and probably no harder than 1 keV. Though these calculations have all been carried out using non-relativistic hydrodynamics, it is still possible to put upper limits on the amount of mass and energy that could be in relativistic matter. For a typical Type II-p supernova (15 $M_\odot$, R = 2 AU, total kinetic energy 10^{51} erg), only 10^{48} erg is concentrated in the outer 10^{-4} $M_\odot$ of ejecta and this matter, for the most part, has velocity near 0.1 c. A much smaller amount of energy and mass would be concentrated in relativistic matter. For Type Ib and Ia the numbers are even smaller. Less than 10^{-6} $M_\odot$ has speeds over 1/3 c and the total energy is less than 10^{47} erg. The rapid time variation observed in gamma-ray bursts would also be difficult to understand.

One possible exception, so far unexplored within the context of cosmological models, is a phase transition in a neutron star. The transition would have to occur many years after the supernova explosion when the nebula was thin to gamma-rays and the readjustment would need to be both abrupt and major, releasing a substantial fraction of the binding energy of the neutron star. This might give rise to a shock of 10^{51} erg or more that would propagate to the surface without much dissipation and might deposit a significant fraction of its energy in accelerating relativistic matter. Whether the spectrum and time scale could be made to work would depend upon the uncertain circumstellar interaction – the duration of the electromagnetic pulse itself would be very brief. The nature of the instability that leads to such a major structural readjustment in an old neutron star would also need elucidating.

Because of these problems with direct shock acceleration, several groups (e.g, Dar et al. 1992) have proposed tapping the binding energy of the neutron star indirectly by using some fraction of the neutrino energy released during core collapse. The neutrinos encounter their antiparticles in a region of presumed low mass concentration, annihilate to make electrons and positrons, and a relativistically expanding plasma of pairs is produced. Clever as this idea may be, it has been shown repeatedly not to work in the case of spherically symmetric gravitational collapse (Woosley & Baron 1992; Woosley 1992b). The problem is best illustrated by considering a neutron star forming from the accretion-induced collapse of a white dwarf, though similar arguments would apply to a spherically symmetric neutron star formed any other way, for example by the merger of two neutron stars.

The basic problem is that the "neutrinosphere" (surface of last probable interaction for a neutrino) is always clothed with matter. In fact, this emitting surface typically lies at a density of 10^{11} g cm^{-3}. As the transition is made from optically thick (to neutrinos) to thin, some small fraction of the neutrino energy will deposit. Indeed this small deposition is thought to be the powerhouse behind most supernova explosions. It is easy to demonstrate the unavoidable existence of an energy gaining region in the protoneutron star atmosphere (Woosley 1992b). Near the base of the atmosphere an equilibrium exists between energy producing and energy losing weak interactions. Energy is deposited by $p(\bar{\nu}_e, e^+)n$ and $n(\nu_e, e^-)p$; energy is lost by $p(e^-, \nu_e)n$ and $n(e^+, \bar{\nu}_e)p$. Detailed balance then gives a condition on the temperature that it scale as $r^{-1/3}$. However the pressure scale height is very short for a neutron star, typically 20 m times the radius squared in units of 10 km. The density must therefore decline rapidly to the point where radiation pressure dominates. Once that is the case, hydrostatic equilibrium (a very accurate assumption here) demands that T scale as r^{-1}. Since hydrostatic equilibrium is a stronger condition than weak equilibrium, the temperature declines abruptly and the energy depositing reactions (neutrino capture) dominate the energy losing reactions (pair capture).

The physics of this neutrino driven wind was first studied in detail by Duncan, Shapiro, & Wasserman (1986) and more recently by Woosley & Baron (1992) and Woosley (1992b). Both analytically and numerically the mass loss rate is $\sim 0.0015 \, L_{\nu 53}^{5/2}$ M$_\odot$ s^{-1} where $L_{\nu 53}$ is the *total* neutrino luminosity in units of 10^{53} erg s^{-1}. The power deposited in this wind is $\sim 10^{50} \, L_{\nu 53}^{5/2} R_7^{-1}$ erg s^{-1} with R_7 the neutrinosphere radius in units of 100 km. The matter is not relativistic. Moreover the ratio of mass to radiation is so large that the energy deposited by neutrino interactions, including neutrino annihilation outside the "gain radius", is degraded by adiabatic expansion to values far too low to make a gamma-ray burst. No tinkering with circumstellar matter distribution, precollapse model, or energy transport can alter this fundamental conclusion.

THE SUCCESS OF AXIALLY SYMMETRIC MODELS

The arguments of the preceding section had led me to conclude in early 1992 that cosmological models for gamma-ray bursts were untenable. That conclusion was altered by the preprint of Meszaros & Rees (1992a) who presented a scenario which, while still utilizing neutrino annihilation, invoked a situation where the spherical symmetry was broken. Two merging neutron stars heated by tidal interaction emit neutrinos which meet and partially annihilate along the revolutionary axis. Pairs are produced which escape as jets along the axis. In the original model the radiation was released as the expanding pair plasma became optically thin. More recently these same authors have invoked shock interaction with circumstellar matter to produce the gamma-rays (Rees & Meszaros 1992; Meszaros & Rees 1992b; 1993).

A potential problem with such models is that the time scale for coalescence of two neutron stars by gravitational radiation is short, only a few milliseconds at the time when substantial neutrino emission would occur. While this period can be lengthened by relativistic effects (Rees & Meszaros 1992), the burst duration would then be keyed to specific values for the Lorentz factor and the circumstellar density - values that while reasonable are not yet demonstrated to be natural. There may also be a problem with getting enough total energy out of merging neutron stars. This relates to the fact that in models where the neutrinos are not only the incident particle but the target as well (i.e., neutrino annihilation), the efficiency for conversion to pairs is quadratic in the neutrino luminosity. Unless the tidal heating leads to the emission of a substantial fraction of the neutron star binding energy as neutrinos before the stars physically merge (or unless the gamma-ray burst is highly beamed), the efficiency for converting neutrinos to pairs may be too small. Once the neutron stars have merged there may still be matter left behind in an accretion disk, but the total amount of matter and efficiency for neutrino annihilation would both be small (Shibata, Nakamura, & Oohara 1992; Woosley 1993).

The successful cosmological model must produce about 10^{51} ergs in gamma-radiation times the beaming solid angle over 4 π. Scaled to a beaming fraction of 10% (there must be enough sources!) and an efficiency for converting gravitational energy into gamma-rays of 0.1%, this means that in accretion models we are dealing with solar masses of material, even if we are able to extract a substantial fraction of $\dot{M}c^2$. Part of the appeal of the merging compact object scenarios is that they provide these amounts of mass in the requisite short times. But there may be other ways of doing this.

Elsewhere (Woosley 1993), I have presented a cosmological gamma-ray burst model based on black hole accretion. Briefly, it begins with a massive star that loses its envelope (either owing to a wind or binary interaction) and evolves to iron core collapse. Unlike the usual Type Ib supernova that this would ordinarily produce, it is presumed that for some reason, presumably too large an iron core mass, the supernova "fails" and a black hole forms (see also Bodenheimer & Woosley 1983). The remaining core of helium and heavy elements, which is rotating with specific angular momentum $\sim 10^{16} - 10^{17}$ cm^2 s^{-1}, then accretes onto this black hole. Without rotation, the collapse time is about 10 to 100 seconds. With rotation, an accretion disk is formed that, with much greater uncertainty, also has a time scale for viscous transport in the same range. As a result several solar masses of material accrete into a black hole which also has initial mass several M$_\odot$. The inner edge of the accretion disk is optically thick to neutrinos and a thermal emission surface forms. Characteristic numbers are 30 km for the dimension, 4 MeV for the neutrino temperature, and several tenths M$_\odot$ s^{-1} for the accretion rate, very similar to those characterizing a Type II supernova explosion. With similar efficiency, $\sim 0.1\%$, the neutrinos meet along the relatively evacuated region of the rotation axis, annihilate, and produce a pair gas with temperature about 1 MeV. The pairs expand relativistically, eventually annihilating and, in the simplest case, produce a beamed gamma-ray burst directly. Following Goodman (1986) and Paczynski (1986), one can estimate that recombination would occur at at temperature of about 15 keV, a radius of a few thousand km, and a Lorentz factor of about 100, the blue shift moving the radiation back up into the gamma-ray domain. The event rate for such occurrences could be as much as one per thousand years per galaxy (about 10% of the Type Ib supernova rate), considerably more frequent than neutron star mergers.

However, one must still be concerned with baryonic contamination of the pairs. Following Meszaros & Rees (1992b; 1993) one can calculate that the ideal case, one where all the energy in the pair gas comes out as gamma-rays, occurs only for baryonic contaminations less than 10^{-9} M$_\odot$. Between 10^{-9} and 10^{-4} M$_\odot$ most of the energy is contained in baryons with a decreasing Lorentz factor starting at $\Gamma \sim 10^5$ for 10^{-9} M$_\odot$ and decreasing inversely with mass to one at 10^{-4} M$_\odot$. Above 10^{-4} M$_\odot$, neither hard radiation nor relativistic baryons are produced.

Lacking detailed models, one is inclined to guess that even 10^{-4} M$_\odot$ of contamination may be an underestimate, in which case neither this nor any other cosmological model with which I am familiar will work. But it might be reasonable to have a smaller mass for which the mechanism proposed by Rees & Meszaros (1992; circumstellar interaction) would function to convert the streaming kinetic energy back into hard radiation. The quest for a successful model has seemingly become a quest for ever higher entropies ($S_{rad} \sim T^3/\rho$).

A MINIATURE QUASAR?

Discussions so far have centered on a pair jet energized by neutrino annihilation, but there may be other ways to achieve the same desired goal, perhaps with greater efficiency and entropy. The specific angular momentum in the collapsing stellar core described above is $\sim 10^{16} - 10^{17}$ cm^2 s^{-1} (Woosley 1993). The specific angular momentum of an extreme Kerr black hole is $\sim cR_s$, i.e., for black holes of a few solar masses, in the same range. The black hole may be born rotating rapidly, but in any case will acquire near critical angular momentum after accreting its own mass. Blandford & Znajek (1977) have discussed electrodynamic means for extracting the angular momentum of a rapidly rotating black hole and converting it with high efficiency into beamed jets of relativistic pairs. The motivation for their model was to explain the central power source of an active galactic nucleus, but similar physics may operate in a gamma-ray burst. Blandford & Znajek estimate a luminosity in the form of relativistic particles of $10^{52}(a/a_{crit})^2 \dot{M}$ erg s^{-1} where $\dot{M}$ is the accretion rate measured here in M$_\odot$ per *second* (typically 0.3) and a is the specific angular momentum. An appropriate description of the geometry is given in their Fig. 1. Since the jet is beamed, even angular momenta of less than 10% critical would suffice to produce a powerful gamma-ray burst or perhaps the efficiency for converting this energy to gamma-rays is less than one. While a collapsing stellar core would provide the appropriate conditions, similar conditions might characterize a black hole merging with a compact object or a black hole surrounded by an accretion disk following the merger of two neutron stars (though see Shibata et al. 1992).

One interesting aspect of these models is that a non-trivial fraction of the entire rest mass of the system comes out in jets over the space of a few seconds. 10^{51} erg extracted from 5 M$_\odot$ is 0.01% of the mass. If the two jets are not precisely aligned and balanced, the emitting system may both recoil and precess. Thus the orientation of the beam may not be constant. Depending on the solid angle, the Earth may be in and out of the beam more than once resulting in quasiperiodic recurrences of the burst. On the less imaginative and more problematic side, the wandering of the beam may make it more difficult to keep the baryons out.

Naryan, Paczynski, & Piran (1992) have suggested a different sort of model where the magnetic field also plays a critical role. There the field, a fossil in the accretion disk, is wrapped up and amplified by differential rotation, eventually experiencing a Parker instability. The gamma-rays are produced in an explosive reconnection event similar to a solar flare.

CONCLUSIONS

The riddle of gamma-ray bursts is such a deep and enduring one that any solution is likely to be very exciting, all the more so if bursts are at cosmological distances. The idea that we may be witnessing, every day, the birth of a black hole in a process whose violence dwarfs even the central engines of supernovae and galactic nuclei is a terribly exciting one, but, fortunately, not everything that can be imagined happens. It is the observers who must ultimately solve this problem. Will the bursts continue to be isotropic on the sky after 5 years of BATSE data? Will an optical flash accompany any of the bursts observed by the High Energy Transient Experiment? A single optical flash would have enormous implications, not the least for source localization. None of the previously proposed models can produce an optical flash for a halo population of neutron stars; no cosmological models have yet addressed the issue. On the other hand, further observations of line features may strengthen the case for magnetic neutron stars. The observation of a single recurrence (other than a soft repeater) would also have dramatic implications. The cosmological models discussed here do not repeat (though they may be lensed or the observer may move in and out of a narrow beam); most halo models do, though the time scale may be long. Cosmological models may be associated with galaxies and, in the case of the stellar collapse model, star forming regions of certain classes of galaxies (i.e., not ellipticals). And of course the confirmed identification of just one quiescent counterpart at any wavelength would lead to a quick resolution of the problem.

The solution will almost certainly come during the next decade and I, for one, can hardly wait.

This work has been supported by NASA (NAGW 2525).

REFERENCES

Blandford, R. D., & Znajek, R. L. 1977, MNRAS, 179, 433.

Bodenheimer, P., & Woosley, S. E. 1983, ApJ, 269, 281.

Colgate, S. A. 1968, Can J Phys, 46, S476.

Colgate, S. A. 1975, ApJ, 198, 439.

Colgate, S. A., & Johnson, M. H. 1960, Phys Rev Lett, 5, 235.

Colgate, S. A., & Petschek, A. G. 1979, ApJ, 229, 682.

Dar, A., Kozlovsky, B. Z., Nussinov, S., & Ramaty, R. 1992, ApJ, 388, 164.

Duncan, R. C., Shapiro, S. L., & Wasserman, I. 1986, ApJ 309, 141.

Goodman, J. 1986, 308, ApJL, L47.

Meszaros, P., & Rees, M. J. 1992a, ApJ, 397, 570.

Meszaros, P., & Rees, M. J. 1992b, MNRAS, in press.

Meszaros, P., & Rees, M. J. 1993, ApJ, in press.

Narayan, R., Paczynski, B., & Piran, T. 1992, Harvard-Smithsonian CfA Preprint 3396, submitted to ApJ Lett.

Paczynski, B. 1986, ApJL, 308, L43.

Rees, M. J., & Meszaros, P. 1992, MNRAS, in press.

Shibata, M., Nakamura, T., & Oohara, K. 1992, Kyoto Univ preprint "Coalescence of Spinning Binary Neutron Stars - 3D Numerical Simulations", preprint YITP/K 985 and 989, submitted to Prog. Theor. Phys.

Woosley, S. E. 1992a, in *Planets Around Pulsars*, ed. J. A. Phillips, S. E. Thorsett, & S. R. Kulkarni, to be published by the Astron. Soc. Pac. Conference Series.

Woosley, S. E. 1992b, in Recent Advances in High Energy Astronomy, Proceedings of the Toulouse Conference, A&A Suppl., in press

Woosley, S. E. 1993, ApJ, 405, 000.

Woosley, S. E., & Wallace, R. K. 1982, ApJ, 258, 716.

Woosley, S. E., and Baron, E. 1992, ApJ, 391, 228.

On the Extended Halo Origin of Gamma-Ray Bursts

Dieter H. Hartmann, Eric V. Linder, & Lih-Sin The
Department of Physics & Astronomy, Clemson University, Clemson, SC 29634-1911

ABSTRACT

The γ-ray burst brightness distribution is inhomogeneous and their distribution on the sky is nearly isotropic. These features argue against an association of γ-ray bursts with Galactic objects whose spatial distribution is similar to that of Population I tracers. Circumstantial evidence suggests that neutron stars are involved in the burst phenomenon. We consider Population II neutron stars in an extended Galactic halo (EGH) as an alternative to cosmological scenarios. The BATSE data presented at this conference suggest deviations from isotropy slightly below the 2 σ level of statistical significance. If confirmed by additional data, the anisotropies would rule out cosmological scenarios. Extended halo models predict small anisotropies like those observed by BATSE. We consider simple EGH models to determine the generic properties such halos must have to be consistent with the observational constraints.

INTRODUCTION

The BATSE experiment aboard Compton Observatory has detected 447 cosmic γ-ray bursts (GRB's) from 910424 through 921007. The observed burster brightness distribution shows that the source distribution in space must be non-uniform, while the localizations suggest that their sky distribution is nearly isotropic.[1-5] If the burster sky were indeed isotropic, cosmological models would provide the simplest framework for their origin.[6] Deviations from isotropy would rule out a cosmological origin, but one could resort to multi-component models to satisfy the observational constraints. Strong anisotropies expected from Galactic disk populations rule out the possibility that most bursters are associated with the Galactic disk, but a small "contamination" of an otherwise isotropic background is conceivable.[7] Similar arguments apply to the nearly, but not perfectly, isotropic distribution expected from sources in an extended halo, although contamination by disk sources might be expected in this context.

Halo distributions similar to that inferred for the Galactic dark matter corona do not yield sky distributions that have the required statistical properties,[8-10] but it was argued[11-14] that neutron stars in an extended galactic halo could satisfy all observational constraints, if one considers Population II stars. An extended halo created by injection of high velocity neutron stars from the Galactic plane[15] still causes unacceptably large anisotropies, unless early bursting activity is suppressed.[16] The recent discovery[17] of high velocity radio pulsars well above the Galactic plane lends some support to the notion that neutron star formation in the halo could provide the GRB sources.

The extent of such a bursting halo population is constrained by the BATSE statistics. While coordinate independent techniques are superior for the task of detecting arbitrary

© 1993 American Institute of Physics

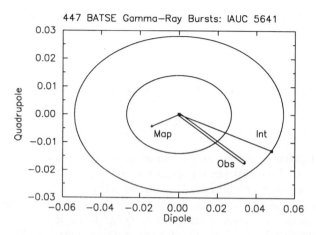

Figure 1: *Galactic gamma-ray burst multipoles for 447 events observed with BATSE. The data suggest a 2σ detection of sky distribution anisotropies. The intrinsic (INT) source moments are obtained from combining the observed values (Obs) with those expected for an isotropic distribution convolved with the non-uniform sky sensitivity map of BATSE. 1 and 2 σ statistical error contours are for 447 events.*

anisotropies on the sky,[18] EGH models can be tested more efficiently using low order multipoles in galactic coordinates.[19] The BATSE statistics of 447 events[5] shows a dipole moment of

$$D = \langle \cos\theta \rangle = 0.034 \pm 0.027 \;,$$

where the error is statistical only. The quadrupole moment is

$$Q = \langle \sin^2(b) \rangle - 1/3 = -0.017 \pm 0.014 \;.$$

These values are very close to those expected from isotropy, $(D,Q)_{\text{iso}} = (0,0)$, so that small effects due to uneven sampling of the sky may become important. Therefore, the BATSE team spent great efforts to derive the sky exposure map for their instrument.[20] Based on this map, a perfectly isotropic burst distribution would have resulted in the measured moments $(D,Q)_{\text{map}} = (-0.014, -0.004)$.[5] Because errors in (D,Q) due to the localization uncertainties of individual bursts are expected to be small,[21] we can subtract the map-moments from the observed moments to obtain an estimate of the intrinsic moments. Figure 1 shows the results; the current BATSE data suggest anisotropies of the intrinsic burst distribution at the 2σ level. While this may not be significant enough to rule out a cosmological origin of gamma-ray bursts, we note the fact that the moments point in the direction expected from EGH models.

NEUTRON STARS IN THE GALACTIC HALO

The high degree of isotropy clearly requires a very extended halo of bursting objects. So if one retains the neutron star paradigm, the crucial question is how to populate such an extended region and how to avoid the highly anisotropic contributions from neutron stars residing in the disk. A clue to the answer might be contained in a recent interferometric pulsar proper motion survey.[17] The radio data suggest a significant extension of neutron star space velocities over that inferred from previous surveys.[22] In several cases, velocities of ~ 1000 km s^{-1} were detected. A significant fraction of pulsars in the high velocity tail of the distribution are not gravitationally bound to the Galaxy. An interesting aspect of these observations is the presence of several high velocity pulsars that appear to travel toward the Galactic plane, while most pulsars are known to migrate away from the plane.[22] Although there are currently less than a handful of these pulsars, their observed properties cast some doubt over the traditionally held views about pulsar birthplaces. A possible interpretation of these observations is, that neutron star formation is an ongoing process high above the Galactic plane.[12,14] We will show below that the formation of high velocity neutron stars in an already extended halo produces a spatial distribution that can satisfy the BATSE isotropy constraint. In contrast to injection from the disk, halo injection produces a much more isotropic distribution.

Eichler & Silk[14] suggested a white dwarf merger scenario in which the Galactic halo could be a prolific producer of pulsars. To make this scenario work, the formation of white dwarfs, neutron stars, or black holes, must avoid the problems of overproducing light and/or metals. Eichler & Silk argue that white dwarf binaries are promising candidates. These systems coalesce on a time scale of $\sim 10^{10}$ yrs. The present day merger rate could be as high as $R_{II} \sim 0.5$ yr^{-1}, depending on the mass distribution of progenitor stars, and out of these mergers neutron stars with velocities of order 10^3 km s^{-1} could be formed. We call these remnants Pop II neutron stars and suggest that the observed GRB rate is predominantly due to a Galactic halo population formed in this or a similar way. The total number of Pop II pulsars is $N_{II} = \eta\, R_{II}\, T_H \sim 5 \times 10^9\, \eta$, where η is constrained by the lack of recent Galactic neutrino bursts due to mergers and also by their associated nucleosynthesis. Since about 1985 a small network of neutrino detectors monitored the sky for transient neutrino bursts. These detectors would have detected a halo merger event to a distance of ~ 20 kpc during the full period, and a sampling depth of ~ 100 kpc was achieved for an effective time of 2-3 years.[23] Thus, the absence of observed neutrino bursts constrains the halo merger rate parameter η to less than unity. The nucleosynthesis for these mergers can be estimated from those of accretion-induced collapse of white dwarfs.[24] Although the total amount of ejected material and the fraction of very neutron-rich material depends on the detailed dynamic history of the merger event, production of isotopes such as ^{62}Ni, ^{66}Zn, ^{68}Zn, ^{87}Rb, and ^{88}Sr provide stringent upper limits on an average Galactic merger rate. If about 0.005 M$_\odot$ of the ejecta has a neutron excess greater than 8%, the most stringent limit gives a recurrence time of $\sim 3,000$ yrs for halo mergers,[24,25] thus η less than $\sim 10^{-3}$. To alleviate this constraint, one must assume that mergers either produce much less neutron-rich material, that it does not escape the explosion site, or that mixing of the ejecta into the Galactic disk is incomplete. Ejection of material with small neutron excess will be dominated by ^{56}Ni,[26] which leads to additional limits through the γ-ray emission in the decay of nickel to cobalt and iron. The ejecta of merger events are likely to become optically thin before the half-life of ^{56}Ni (6.1 days)[24], so that even as little as $10^{-4} M_\odot$ of ^{56}Ni would produce a γ-ray flux of ~ 0.01 photons cm^{-2} s^{-1} at a typical halo distance of 10 kpc. This is much brighter than SN1987a, so that even small γ-ray detectors can monitor merger events in the halo. The absence of transient γ-ray line emission from ^{56}Ni in about 3 years of SMM data[27] (covering about 1/3 of the sky) limits the merger rate to ~ 1 yr^{-1}, consistent with the neutrino limits. In summary, the current merger rate may be as high as 0.5/yr, but the nucleosynthesis limit on the average Galactic rate suggests much lower values. We may also consider Population III neutron stars, injected during Galaxy formation. Velocities typical of the bulk of present-day pulsars would have sufficed then to unbind these stars from the Galaxy. An initial burst of neutron stars with velocities of order 200 km s^{-1} would at present reside at distances of a few Mpc. Models of this kind would have to include the halos of nearby galaxies and the dynamic effects of local group galaxies.

COMPARISON WITH BATSE DATA

We may calculate the spatial distribution of these stars from orbit integrations in a realistic Galactic potential, but at the birth sites and velocities considered here the orbits are close to straight lines. Isotropic injection at some location r_0 with velocity v_0 leads to expanding galactocentric shells of widths $\Delta R = 2 R_\odot$, with $R_\odot = 8.5$ kpc. The burster brightness distribution is a function of many (unknown) properties; radial distribution of birthplaces and birthdates, initial velocities, burst rate as a function of age, and the intrinsic luminosity function. Obviously, it won't be too hard to find a set of parameters that satisfies the observed V/Vmax statistics. One has less freedom with the angular distribution. In the case of spherical, galactocentric shells, the observed anisotropies depend on the sampling depth from the Earth and the radial "occupation probabilities" of the shells sampled. For simplicity, we ignore effects due to varying detector thresholds[28,29] and the dependency on the (unknown) luminosity function, and consider the appearance of bursts uniformly located on shells of radius $x = R/R_\odot$. A single shell contributes a dipole moment

$$D = \frac{2}{3} x^{-1}$$

and a quadrupole moment

$$Q = \frac{1}{8} x^{-1} \left\{ x^3 + x - \frac{1}{2}(1-x^2)^2 \ln\left(\frac{1+x}{|x-1|}\right) \right\} - \frac{1}{3}.$$

Obviously, the dipole component provides the most stringent constraints for this class of models, but because of the relationship between the two moments (Q = const D^2 for large x), it is advantageous to consider both of them simultaneously. Figure 2 shows a series of shells with x between 1 and 100. The quadratic relationship is evident. Any weighted combination of two shells would create a (D,Q) point somewhere along a curved path like that shown in the figure for the (x1,x2)=(1,100) case. Summing many shells over some dynamic range (x1,x2) would yield a point inside the segment

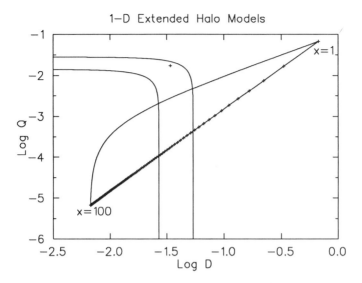

Figure 2: *Crosses mark the dipole (D) and quadrupole (Q) moments of single shells with radii between x=1 and x=100. A linear combination of any two shells yields curves like the one shown for the (1,100) combination. One and two sigma statistical deviations from isotropy are shown for 447 bursts. The BATSE average (cross) is located close to the two sigma contour.*

shown in the figure. For 447 events the one and two sigma statistical errors expected for an isotropic distribution would be consistent with an extended halo that includes observable sources out to $\sim$ 200 kpc (1 sigma) or $\sim$ 100 kpc (2 sigma). However, the BATSE data point (cross in Fig 2), located close to the 2 sigma contour, cannot be matched by these simple EGH models unless one allows a much larger dynamic range or an additional contribution from a second population. These considerations demonstrate the importance of considering both moments simultaneously to constrain EGH models. One can proceed by integrating over shells using a parametrized density profile,[30,31] but in doing so one constrains only particular models. The discussion given here shows that the geometric properties of EGH models imply relations between observable moments that provide powerful constraints, independent of the specifics of how many burst events originate on the various shells.

DISCUSSION AND CONCLUSIONS

We have argued that the ratio N_{II}/N_I could be much less than unity, so that the absence of "contamination" of BATSE data with bursts from Pop I stars either argues against the model considerd here, or suggests that Pop II neutron stars are much more prolific bursters. On the other hand, some disk contamination appears to be required by the observations (Figure 2), and a fraction as large as $\sim$ 0.1 might be tolerable. [7]

The total event rate observed with BATSE is ~ 1000 per year,[1] implying a recurrence time of less than 10^5 years for the conservative estimate $N_{II} = 10^8$. If a typical burst fluence is 10^{-6} ergs cm^{-2}, this implies a lifetime storage requirement of $\sim 10^{46}$ ergs. If this energy is provided by accretion or thermonuclear explosions, less than $\sim 10^{-5}$ M$_\odot$ have to be stored in orbit around the neutron star. The low metallicity of population II objects could be responsible for a much higher efficiency of fall-back of material onto the neutron star,[32] leading to the build-up of a planetesimal accretion disk. Lin et al.[33] estimate that the total mass of the disk is $\sim 10^{-5}$ M$_\odot$ and that the mass range of planetesimals is 10^{-12} M$_\odot$ to 10^{-6} M$_\odot$. With these parameters one can match rate, brightness-, and sky distribution of GRBs. Slowly accreting neutron stars in the halo undergoing rare Rayleigh-Taylor instabilities due to pycnonuclear reactions don't satisfy the energy requirements.[34] Of course, neutron stars formed in the halo could have, for some reason, more energy stored in their crusts than those formed in the disk.

Our calculations show that the Galactic neutron star origin of GRBs does not yet have to be abandoned. Because disk distributions do not satisfy the observational constraints, one is forced to consider very extended halos, perhaps with a small contamination by anisotropic sources that have an enhanced quadrupole moment relative to their dipole moment (see Figure 2). The observed angular distribution suggests that such halos were created by injection from a pre-existing halo (either during Galaxy formation or an ongoing process). Neutrino-, γ-ray, and nucleosynthesis limits suggest that the halo/disk ratio of neutron stars is small, so that the observed isotropy implies enhanced bursting activity for halo stars and/or suppressed activity for disk stars. Cosmological scenarios predict that continued GRB observations should show an increasing degree of isotropy. On the other hand, the EGH model predicts a small but measurable anisotropy. Current BATSE data favor the EGH hypothesis at the 2σ level of statistical significance.

This work was supported in part by NASA grant NAG 5-1578.

REFERENCES

[1] Meegan, C. A., et al., *Nature*, **355**, 143 (1992).
[2] Meegan, C. A., et al. *Proc. of the Huntsville Gamma-Ray Burst Workshop*, American Institute of Physics, New York, (in press) (1992).
[3] Meegan, C. A., et al., IAU Circular No. 5358 (1991)
[4] Meegan, C. A., et al., IAU Circular No. 5478 (1992)
[5] Meegan, C. A., et al., IAU Circular No. 5641 (1992)
[6] Paczynsky, B. 1992, these proceedings
[7] Lamb, D. Q. et al. 1992, these proceedings
[8] Paczynski, B., Acta Astr., 41, 157 (1991)
[9] Hakkila, J. & Meegan, C. A., in *Gamma-Ray Bursts*, (AIP : N. Y.), ed. W. S. Paciesas & G. J. Fishman, p. 120 (1992)
[10] Hakkila, J., et al., these proceedings
[11] Hartmann, D. H., The, L.-S., Clayton, D. D., Schnepf, N. G., & Linder, E. V., in *Gamma-Ray Bursts*, (AIP : N. Y.), ed. W. S. Paciesas & G. J. Fishman, p. 120 (1992)

[12] Hartmann, D. H. 1992, Comm. in Astrophys., in press
[13] Brainerd, J. J., Nature, 355, 522 (1992)
[14] Eichler, D. & Silk, J. 1992, Science, 257, 937
[15] Shklovskii, I. S., & Mitrofanov, I. G. MNRAS, 212, 545 (1985)
[16] Li, H. & Dermer, C. D., Nature, 359, 514 (1992)
[17] Harrison, P. A., Lyne, A. G., & Anderson, B., MNRAS, in press, (1992)
[18] Hartmann, D. & Epstein, R. I., ApJ, 346, 960 (1989)
[19] Paczynski, B., ApJ, 348, 485 (1990)
[20] Brock, M. N., Meegan, C. A., Fishman, G. J., Wilson, R. B., Paciesas, W. S., & Pendleton, G. N., in *Gamma-Ray Bursts*, (AIP : N. Y.), ed. W. S. Paciesas & G. J. Fishman, p. 399 (1992)
[21] Horack, J. M., Meegan, C. A., Fishman, G. J., Wilson, R. B., Paciesas, W. S., Emslie, A. G., Pendleton, G. N., & Brock, M. N., ApJ, in press (1992)
[22] Lyne, A. G., Anderson, B., & Salter, M. J. MNRAS, 201, 503 (1982)
[23] Cline, D. B., (private communication) (1992)
[24] Woosley, S. E. & Baron, E., ApJ, 391, 228 (1992)
[25] Woosley, S. E. & Hoffman, R. D. ApJ, (in press), (1992)
[26] Hartmann, D., Woosley, S. E., & El Eid, M. F. , ApJ, 297, 837, (1985)
[27] Leising, M. (private communication) (1992)
[28] Band, D., ApJ, in press (1992)
[29] Hartmann, D. & The, L.-S., Ap. Space Sci., in press, (1992)
[30] Wasserman, I., ApJ, 394, 565 (1992)
[31] Hakkila, J. et al. , these proceedings
[32] Woosley, S. E., *Astr. Soc. Pacific*, in press (1992)
[33] Lin, D. N. C., Woosley, S. E., & Bodenheimer, P., Nature, 353, 827 (1991)
[34] Blaes, O., Blandford, R., Madau, P., & Koonin, S., ApJ, 363, 612 (1990)
[35] Silk, J. *Phys. Rep.*, (in press) (1992)
[36] Paczynski, B., ApJ, 304, 1 (1986)

A POSSIBLE CONTRIBUTION OF LOCAL (< kpc) NEUTRON STARS TO THE γ-RAY BURSTS

J. C. Higdon
Joint Science Department, Claremont McKenna College, Claremont, CA 91711

ABSTRACT

I conjecture that a large number of low-velocity, nearby neutron stars were produced in a single, giant OB association by multiple, type II supernovae, $\sim 5 \times 10^7$ years ago. Such correlated supernova explosions have been suggested as a possible mechanism, for the production of a local expanding superstructure, called Gould's belt. I performed a simple Monte Carlo simulation to judge the efficacy of my conjecture. First, I assumed that $\sim$ half the nearby, active bursters were created in a single event, $\sim 5 \times 10^7$ years ago, located at what was then the center of Gould's belt, and their initial velocity distribution was Gaussian. Then the trajectories of the individual, candidate bursters were followed to the present via numerical integration for a range of Gaussian dispersions. I find anisotropy limits can be significantly less than those of previous galactic-disk, source models when low-speed, Gould's belt neutron stars are included. To reproduce the values of $\langle V/V_{max} \rangle$ determined from spacecraft measurements of ≤ 0.4, this composite model for galactic disk neutron stars reduces the dipole anisotropy by a factor as large as ten, and the quadrupole anisotropy by a factor as large as three compared to the anisotropies of simple disk population models. If such a composite model is viable, it has critical implications on the fraction of low-velocity neutron stars created in type II supernovae, the effective lifetime of the burst phase, and/or the burst repetition rate.

INTRODUCTION

Significant structure exists in the local (< kpc) density distribution of OB stars, the stellar precursors of neutron stars, the most likely sources of γ-ray bursts. Superimposed on the local galactic disk population is an expanding complex[1] of O-B5 stars and molecular clouds, inclined $\approx 20°$ to galactic plane. This expanding superstructure is called Gould's belt in honor[2] of the first astronomer, who recognized its importance. Gould's belt contributes a significant number of OB stars to the local galactic disk; in fact the bulk[1] of the most massive, young stars, O-B2, in the solar neighborhood, belong to this system.

Gould's belt is an ellipse, ≈ 1000 by 800 pc. From the analysis of the positions of ≈ 1300 OB stars, Stothers and Frogel[1] found that the center of Gould's belt is in the outer galaxy, at a distance of 210 pc, at a galactic longitude, l, of $\sim 180°$ and a latitude, b, of $\sim -16°$. However, the location of Gould's belt center is somewhat uncertain. From analyses of the distortions in the velocity field of HI, Olano[3] found the center somewhat closer, at a distance of ~ 170 pc, at $l \sim 130°$. From a similar analysis of HI profiles, Linblad[4] et al. found the center closer still, at ~ 140 pc, at $l \sim 150°$. The age of Gould's belt is uncertain by a factor of two. From the main sequence turn up in the number of massive stars with spectral type near B2.5, Frogel and Stothers[1] estimated an age of $\sim 2 \times 10^7$ years. Later, Frogel and Stothers[5] found kinematical stellar ages of either 2×10^7 or 6×10^7 years. Finally, from the expansion of HI gas Linblad[4] et al. estimated an expansion age of 6×10^7 years.

The origin for such large-scale, expanding structures seems to be multiple, type II supernovae in a single, large OB association (see review[5]). However, the required number of supernovae is very uncertain, since thermal energy in such expanding plasma shells can be vented into the galactic corona. Simulations of such shell expansion suggests that ~ 10 to 10^3 supernovae explosions are required to power such expansions (see review[5]).

I conjecture that such correlated type II supernovae explosions are a source of local neutron stars in addition to those generated in the galactic disk.

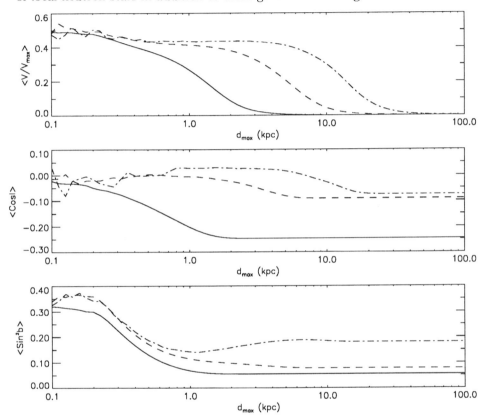

FIG. 1 The top panel shows the mean$\langle V/V_{\max}\rangle$ as a function of distance, d_{max} from Earth. The middle panel illustrates the dipole moment,$\langle \mathrm{Cos}l \rangle$, as a function of distance from Earth. The bottom panel shows $\langle \mathrm{Sin}^2 b \rangle$. The dot-dashed curve indicates the model fits for a population with an isotropic, one-dimensional velocity dispersion of 100 kms^{-1}. The dashed curve indicates the model fits for a population with an isotropic, one-dimensional velocity dispersion of 40 kms^{-1}. The dashed curve indicates the model fits for a population with an isotropic, one-dimensional velocity dispersion of 15 kms^{-1}.

MONTE CARLO SIMULATIONS

I investigated the contribution of such galactic neutron stars via a simple Monte Carlo simulation. I assumed all neutron stars associated with Gould's belt were created in a single event 5x10^7 years ago. I identified their birth site with the center[1] of Gould's belt, propagated back 5x10^7 years. I assumed that these neutron stars were formed with a Gaussian velocity distribution. The trajectories of individual neutron stars in the galactic gravity field were determined following the approach of Paczynski[6]. The effect of three velocity dispersions, 15, 40, and 100 kms^{-1} were investigated. In each simulation the trajectories of 10^6 Gould's belt neutron stars were modeled.

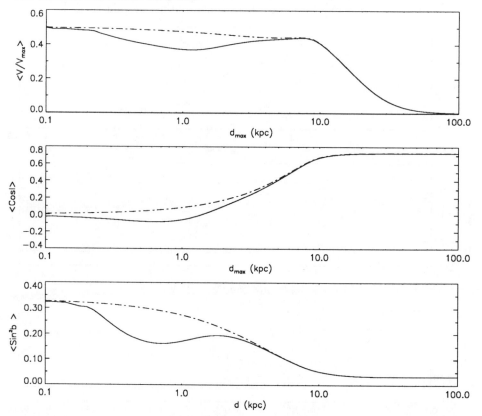

FIG. 2 The top panel shows the mean $\langle V/V_{\max}\rangle$ as a function of distance, d_{max} from Earth. The middle panel illustrates the dipole moment, $\langle \text{Cos}l\rangle$, as a function of distance from Earth. The bottom panel shows $\langle \text{Sin}^2 b\rangle$. The dot-dashed curve indicates the model fits for a simple, galactic disk population with an exponential scale height, transverse to the galactic plane, of 1 kpc.

For these three velocity dispersions I determined $\langle V/V_{\max}\rangle, \langle \text{Cos}l\rangle$, and $\langle \text{Sin}^2 b\rangle$, as function of maximum distance from the Earth, where l and b are galactic longitude and latitude respectively. The dipole moment[6,7] is proportional to $\langle \text{Cos}l\rangle$, and the quadrupole moment is essentially proportional to $\langle \text{Sin}^2 b\rangle - 1/3$, for large-scale galactic distributions. These distributions are plotted in Fig. 1. Note the irregularities at the smaller distances which are caused by low-sample statistics. For the low-velocity stars $\langle V/V_{\max}\rangle$ decreases continuously, and anisotropy increases, with increasing distance. Note that for such low velocities the mean dipole is always negative. Note also, the concentration toward the galactic plane, as implied by the low values of $\langle \sin^2 b \rangle$. For the moderate and higher-velocity stars $\langle V/V_{\max} \rangle$ decreases much more slowly with distance, while anisotropy, particularly the dipole term, is reduced. In all cases a significant quadrupole term remains at large distances. However, if the bulk of Gould's belt neutron stars were created with velocities ≥ 50 kms^{-1}, the contribution of such a modest number ($\sim 10^2$ to 10^3), would be overwhelmed by the general galactic disk population, where the typical type II supernova birthrate[8] is expected to be $\geq 10^{-2}$ yr^{-1}. However, if the Gould's belt neutron stars possess significantly lower speeds, ≤ 20 kms^{-1}, then most of them would reside with a kpc of the Earth. Such a nearby concentration could contribute then as many neutron stars as the galactic disk at least within the last $\sim 5 \times 10^7$ years.

There exists independent evidence that a significant fraction of the local neutron stars possess low velocities. From an analysis of radio pulsar statistics, Tutukov et al.[9] found that the sample of nearby (≤ 500 pc) pulsars was significantly slower and less luminous than the full pulsar sample. For example, they found that 70% of the nearby pulsars possessed velocities, transverse the line of sight, ≤ 50 kms^{-1}. Consequently, the remainder of this investigation will concentrate on the contribution of low-speed neutron stars.

To visualize the contribution of such low-speed neutron stars, I added the contribution of a model of Gould's belt neutron stars, created with a velocity dispersion, σ_{1D}, of 15 kms^{-1} to the contribution of a typical galactic disk population with an exponential scale height transverse the plane of 1 kpc. I normalized both source models to contribute equally at d_{max} of 1 kpc. The resulting distributions are shown in Fig. 2. I find at a distance of 1 kpc: $\langle V/V_{\max} \rangle = 0.37$, $\langle \cos l \rangle = -0.06$, and $\langle \sin^2 b \rangle = 0.18$. Where the modeled Gould's belt neutron stars contribute significantly, the anisotropy of the composite model is less than that of a simple galactic disk model for the same $\langle V/V_{\max} \rangle$. For example, a value of $\langle V/V_{\max} \rangle = 0.37$ in a disk model with an exponential scale height of 1 kpc generates corresponding $\langle \cos l \rangle$ of 0.7 and $\langle \sin^2 b \rangle$ of 0.05. Thus to reproduce the values of $\langle V/V_{\max} \rangle$, determined from spacecraft measurements of ≤ 0.4, this composite model for galactic disk neutron stars can reduce the dipole anisotropy by a factor as great as ten, and the quadrupole anisotropy by a factor as great as 3 compared to those values of simple disk population models.

Introducing more parameters reduces the model anisotropies. For example, the contribution of low-velocity, neutron stars born in the expanding Gould's belt after its creation, greatly reduces the expected dipole anisotropies. Modeled quadrupole anisotropies are decreased, if anisotropic velocity dispersions are introduced.

CONCLUSION

The contribution of this conjectured Gould's belt population rests on a very uncertain value for the fraction of low-speed, galactic neutron stars. However, there seems to be evidence for the existence of multiple populations[10] in radio pulsars, with slower neutron stars rotating less rapidly. Biases[11] also exist in the determination of the velocity distribution of pulsars. Whether there exists a significant number of low-luminosity, low-speed neutron stars, as found by Tututkov el al.[9], and required by the present model, needs further study.

If this model has any validity, it has several major implications. First, the lifetime of a burster in the galactic population must be $\leq 10^8$ years. Otherwise the contribution of Gould's belt neutron stars would be overwhelmed by the galactic disk population. Second, it implies that recurrence times must be ~ year for this class of γ-ray bursts. This implied because, in the most optimistic case, ~ 700 low-velocity neutron stars would be generated by Gould's belt formation. In a dual population model[10] ≈ 25 % of bursters could be expected to be generated by such local neutron stars. However, BATSE detections[11] correspond to a full sky rate of ≈ 800 yr^1.

REFERENCES

1. R. Stothers, and J. A. Frogel, A. J., **79**, 456 (1974).
2. B. A. Gould, Am. J. Sci. (2), **8**, 324 (1874).
3. C. A. Olano, A. & Ap., **112**, 195 (1982).
4. P. O. Linblad et al., A. & Ap., **24**, 309 (1973).
5. G. Tenorio-Tagle and P. Bodenheimer, Ann. Rev. A. & Ap., **26**, 145 (1988).
6. B. Paczynski, Ap. J., **348**, 485 (1990).
7. D. Hartmann, R. I. Epstein, and S. E. Woosley, Ap. J., **348**, 625 (1990).
8. S. van den Bergh and G. A. Tammann, Ann. Rev. A. & Ap., **29**, 363 (1991).
9. A. V. Tutukov, N. N. Chugai, and L. R. Yungelson, Sov. A. J., **10**, 244 (1984).
10. R. Narayan, and J. P. Ostriker, Ap. J., **352**, 222 (1990).
11. J, Greiner, and H. J. Wiebcke, A. Lett. & Comm., **27**, 381 (1990).
12. R. E. Lingenfelter, and J. C. Higdon, Nature, **356**, 132 (1992).
13. C. A. Meegan et al., Nature, **355**, 143 (1992).

HIGH LORENTZ-FACTOR $e^{\pm}$ JETS IN GAMMA-RAY BURST SOURCES

P. Mészáros
525 Davey Lab., Pennsylvania State University, University Park, PA 17802

M.J. Rees
Institute of Astronomy, Madingley Road, Cambridge, CB3 0HA, U.K.

ABSTRACT

We discuss the conditions under which a high-entropy pair jet or fireball can result from the violent merger of compact binaries consisting of a neutron star and another neutron star or a black hole. We estimate the wind mass-loss, and consider scenarios in which the pair-photon plasma produced by $\nu\bar{\nu}$ annihilation can attain highly relativistic expansion velocities, resulting in gamma-ray emission of sufficient energy to be observable at cosmological distances.

INTRODUCTION

If gamma-ray bursts (GRB) are located at cosmological distances, the typical photon energies at the source needed for them to be detectable are in excess of 10^{50} erg, or $\sim 10^{-3} M_\odot c^2$. Such energies are typical of supernovae, where however the bulk of the energy appears in kinetic energy form and at wavelengths much softer than that of γ-rays. Models have been considered based on white dwarf collapse involving a superstrong magnetic field analogue of the pulsar mechanism (Usov, 1992, Duncan and Thompson, 1992), and failed type Ib supernova (Woosley, 1992). An alternative source of comparable total energy is the merger of a compact binary made up of two neutron stars (NS-NS) or a neutron star and a black hole (NS-BH) possibly leading to a disk and magnetic flares (Narayan, et al., 1992). A general (and recognized) problem of all GRB scenarios is that a large energy density is required, implying a minimum mass density and a significant opacity, so that a super-Eddington wind appears unavoidable. Under these circumstances, it is not obvious whether most of the γ-ray photons avoid degradation by scattering and absorption, nor why most of the energy is not simply converted into kinetic energy of the matter being pushed by the intense radiation field. This difficulty is encountered in galactic scenarios as well, and becomes extreme in the case where the GRB are at cosmological distances. If the energy is released in a region populated by the stellar baryonic matter, the resulting $e^\pm, \gamma$ and baryon fluid will have a low specific entropy per baryon and a high opacity. As a result, the photons remain trapped and are adiabatically cooled in their own rest frame as the heated fluid expands, while the expansion bulk velocity remains subrelativistic and thus unable to provide a large enough Doppler boost to shift these photons back up to high energies in the observer frame.

These problems can be avoided if the pair plasma is created in a region which is relatively free of baryons at the time of the burst. This situation is most likely to occur in a compact binary scenario, where there exist regions in which the baryon wind density is expected to be very low. This leads to a high-entropy relativistic pair fireball, which is the basic requisite for producing observable photon bursts in the γ-ray energy range at cosmological distances. Furthermore, since the fireball is likely to be strongly anisotropic or jet-like, the total energy required in photons is significantly smaller than for isotropic bursts, while the detection probability of beamed events provides a better fit between expected mergers and GRB statistics. A more detailed discussion is given in Mészáros and Rees (1992a;1992b).

PRE-EJECTED BARYONIC WIND

For observed GRB flux levels $F \lesssim 10^{-4}$ erg cm^{-2} s^{-1} the required luminosity $L \sim 10^{38}(F/10^{-4}$ erg cm^{-2} s$^{-1})(M/M_\odot)^{-1}(D/0.1$ kpc$)^2$ erg s^{-1} exceeds the (spherical) Eddington limit for all distances beyond the galactic disk, which would lead to an optically thick wind.

In a compact binary, a wind will begin even before tidal disruption of one or both members, fed by the tidal heating rate. The latter can be estimated as (Mészáros and Rees, 1992a)

$$\dot{E}_t \sim (GM_*^2/R_*)(R_*/D)^6(qt_{orb})^{-1}, \qquad (1)$$

where D is the separation. The parameter q is itself a function of D, being likely to increase as D decreases. When $D \gg R_*$ only the crust will dissipate, and the (small) tidal distortions change slowly compared with the characteristic internal timescale of the star; q would consequently then be $\ll 1$. However, as the stars approach coalescence (on the gravitational radiation timescale), the tidal distortions become violent enough to tear the crust, and perhaps even to heat the core enough to destroy its superfluid properties; q may then become $\gtrsim 10$. A rough numerical estimate of the dissipation rate is $\sim 10^{56} q_1 D_6^{-15/2}$ erg s^{-1} (similar expressions are given by Bildsten and Cutler, 1992, and Kochanek, 1992). The maximum rate is roughly the energy of two stars thermalizing their gravitational binding energy in a free-fall time multiplied by a q factor (even after disruption, the gas blobs will collide, and all the mass is heated to temperatures above the Fermi value). Equating the tidal heating rate to the neutrino energy losses one obtains an estimate of the temperature, and consequently of the radiation pressure, giving a probable mass loss of

$$\Delta M \gtrsim 10^{-3} q_1^{-4/3} D_6^{-8} \, M_\odot \qquad (2)$$

in a baryonic wind before the final coalescence. Single white dwarf collapse models will achieve a comparable heating rate from the thermalization of a fraction of the kinetic energy of the infalling and bouncing envelope over a free-fall timescale, so a radiation-driven baryonic wind of comparable mass is expected also in this case. A similar situation would arise in a NS-BH binary *after* the NS is disrupted and forms a disk orbiting the BH: the maximum heating and mass loss may be somewhat smaller than above, since one would integrate only down to the last stable orbit, but nonetheless, as soon as it is formed, the disk will produce an extremely super-Eddington wind leading to an opaque baryonic wind.

The crucial question in the various models is whether there are regions which are sufficiently free of baryons, even temporarily, where pairs are being formed. This problem of making "photon bubbles" which can push and escape from the denser baryonic environment has been considered also in other physical situations (e.g., Hsieh and Spiegel 1976, Colgate, 1991). In the specific GRB case, any significant intermixing with baryons will result in an optically thick, slowly expanding single-fluid observable only at lower energies; this is the "baryon pollution" problem described, e.g. by Paczynski (1990). In the single star collapse, the formation of an unpolluted fireball appears very unlikely: pairs can be formed by the nearly head-on collision of $\nu\bar{\nu}$, e.g. as in supernovae (e.g., Goodman, et al., 1987), but only very close to the star, i.e., in the wind. Similarly, $e^{\pm}, \gamma$ produced by a superstrong field pulsar-type mechanism will occur in or inside the wind. Also the $e^{\pm}, \gamma$ produced by magnetic flares in a disk (formed *after* disruption of the NS) will have been preceded by a baryonic wind. Any γ-rays thus produced will be smothered and degraded by the drag of the outer envelope and/or the wind. In addition the maximum bulk Lorentz factor of the wind is $\Gamma \lesssim E_o/\Delta M c^2 \sim 1\ E_{51} \Delta M_{30}^{-1}$, so the adiabatically cooled photons that do escape are not appreciably blue-shifted, and would appear below the γ-ray range.

"CLEAN" FIREBALL SITES

In GRB models based on single star scenarios or binary collapse scenarios, the companion is too far to induce a significant asymmetry of the collapse and the baryons are distributed fairly symmetrically. In compact binary (NS-NS or BH-NS) *merger* scenarios, on the other hand, the flow symmetry is of lower order and there are regions of space where (especially before complete disruption of the NS) the baryon density is expected to be much less than elsewhere.

In the NS-NS merger, such a baryon-minimum is expected in the last stages before disruption and contact, along the binary rotation axis between the two stars, due to centrifugal barrier effects. Given enough time, viscosity could fill in this region, but before that the two stars are tidally heated to temperatures where copious $\nu\bar{\nu}$ emission occurs. The $\nu\bar{\nu}$ of each star will collide with those of the other star and annihilate into $e^{\pm}, \gamma$ preferentially near the midplane between the stars (the cross section being $\propto [1 - \cos\theta_\nu]^2$), which contains the binary rotation axis. A relatively "clean" fireball may therefore be possible here, which will have a very high specific entropy per baryon. This (highly relativistic) fireball will tend to break out from the surrounding low-entropy baryonic medium preferentially along the least-resistance direction, which is the binary rotation axis.

In the BH-NS merger scenario, an even cleaner fireball is possible, just before disruption of the NS. Near the tidal radius the star will release of the order of its binding energy in $\nu\bar{\nu}$, which are emitted in all directions; the fraction emitted within an angle $\sim 2R_S/R_t$ of the BH (where R_S, R_t are the Schwarzschild and tidal radii, respectively) are focused on the other side of the BH. Thus, behind the black hole, the $\nu\bar{\nu}$ collide in the focal region (of volume $\sim R_*^3$) almost head-on, producing $e^{\pm}, \gamma$. Since neutrinos travel faster than any baryons, this region is expected to be quite free of baryons for a time large enough to allow a high entropy fireball to develop. The fireball energy is (Mészáros and Rees, 1992b)

$$E_f \sim 10^{50}(M_H/M_N)^{-1/3}\tau_{-3}\ \text{erg}, \qquad (3)$$

where $\sim 10^{-3}\tau_{-3}$ s is a typical neutrino burst duration, and M_H, M_N are the BH and NS masses. This is a lower limit to the pair luminosity, since once some pairs have formed the higher cross-section $\nu + e \to \nu' + e^+e^-$ process will start producing even more pairs. For a Kerr hole, the plunge orbits, tidal shear and gravitational focussing differ in details (and may be even more favorable), but the essential features of producing copious pairs in a baryon-free region will remain the same. The fireball will tend to break out possibly along the axis joining the two mass centers, opposite to the NS, or else along the binary rotation axis.

RELATIVISTIC PAIR-PLASMA EXPANSION

Once the high-entropy pair-plasma fireball has broken through, it will expand in conventional fashion (e.g. Cavallo and Rees, 1978, Goodman, 1986, Shemi and Piran, 1990). Its thermal energy is gradually converted into kinetic energy of expansion, with mean bulk Lorentz factor growing as $\Gamma \sim r/R_o$. This goes on either until the plasma becomes optically thin (and a burst occurs), or until the plasma becomes non-relativistic and Γ saturates to a value comparable to the initial specific entropy $\eta = E_o/M_f c^2$, where M_f is the mass of polluting baryons carried with the fireball.

The under-loaded fireball ($M_f \lesssim 10^{-9} E_{o,51} M_\odot$, or $\eta \gtrsim 4 \times 10^5 E_{o,51}$) becomes optically thin while still in the acceleration stage. The emitted photons are blueshifted into the observer frame by a factor $\Gamma \sim r/R_o$ which just compensates the adiabatic energy loss of the fireball radiation. If the fireball is beamed into a half-angle $\alpha > \Gamma^{-1}$ (a condition which is likely to be met) the jet behaves for an observer within α of its axis just as a spherical fireball would, except that its total energy can be smaller. For a given fluence, the energy inferred under the assumption of isotropy is $E_{obs} \sim 10^{51} E_{o,51} \alpha^{-2}$ erg, i.e. for $\alpha \sim 0.1$ it could be $\sim 10^{49}$ erg. At the same time, the detection probability goes down by a factor $\alpha^{-2} \sim 10^2$, which brings the expected number of mergers within observable distances (e.g. Narayan, et al., 1991, Phinney, 1991) into much closer agreement with the observed rate of $\sim 10^3$ yr^{-1}. On the other hand, the typical burst timescale is $t_b \sim r_{th} c^{-1} \Gamma_{th}^{-2} \lesssim 10^{-4}$ s, where quantities are evaluated at the point where optical thinness is reached. This is rather short, and has led to efforts aimed at lengthening it, e.g. by considering a slower energy release in a disk that evolves on a viscosity timescale. However, disks should self-pollute themselves through their super-Eddington wind, and it is not clear that such a solution would lead to γ-rays. Another concern is that the spectrum would be close to thermal, with an observer-frame effective photon temperature which is in the soft γ-ray range, instead of the observed power-law behavior. The other case is that of the high-load fireballs ($M_f \gtrsim 10^{-9} M_\odot$, $\eta \lesssim 10^5$), which become thin after saturation; these expand by a larger factor and the Doppler blueshift boost $\gamma \sim \eta$ now no longer compensates the adiabatic cooling factor $R_o/r_{th} > \eta$, so the spectrum is softer and the observable photon energy when it becomes optically thin is lower than the initial total burst energy.

DISCUSSION

The radiation burst arising when a pair fireball or jet becomes optically thin provides a known mechanism for γ-ray bursts. In this paper we have concentrated on showing that high entropy fireballs can arise in compact binary mergers, and that there is a high probability of their being able to break through the optically thin wind which is the unavoidable consequence of such mergers. The fluid dynamics of relativistic fireballs is difficult to investigate in its details. Thus, one cannot be sure of the details of the spectrum of the thinning fireball; while deep inside the spectrum may be quasi-thermal, instabilities could develop at the contact discontinuity which could lead to non-thermal effects. However, the timescales over which the thinning process occurs are unlikely to differ significantly from the expansion timescale, which in the laboratory frame is rather shorter than the observed average durations ~ 10 s of classical GRBs. Thinning bursts may represent a sub-class of GRBs, or may also be associated with brief precursor bursts or afterbursts (Mészáros and Rees, 1992c). In fact, in the case of high-load fireballs ($M_f \gtrsim 10^{-9} M_\odot$, most of the initial energy $E_o \sim 10^{51} E_{o,51}$ erg is *not* converted into radiation as the fireball thins out, due to the relatively lower Doppler boost: most of the energy is carried on as kinetic energy of the baryons, with bulk Lorentz factor $\Gamma \sim \eta \sim 10^2 (M_f/0.5 \times 10^{-5} M_\odot)^{-1}$. Elsewhere in these Proceedings (Rees and Mészáros, 1992b; further details are in Rees and Mészáros, 1992a, Mészáros and Rees, 1992c) we argue that it is the recovery and thermalization of this kinetic energy in a blast wave which gives rise to the main burst, whose duration and energy is comparable to the observed values.

Acknowledgements: We are grateful to G. Fishman, B. Paczyński and S. Woosley for useful discussions. This research has been partially supported through NASA NAGW-1522.

REFERENCES

Bildsten, L. and Cutler, C., 1992, Ap.J.(Letters), in press

Cavallo, G. and Rees, M.J., 1978, M.N.R.A.S., 183, 359

Colgate, S., 1991, in *Supernovae*, ed. S. Woosley

Duncan, R. and Thompson, C., 1992, Ap.J.(Letters), in press

Goodman, J., 1986, Ap.J.(Lett.), 1986, 308, L47

Goodman, J., Dar, A. and Nussinov, S., 1987, Ap.J.(Lett.), 314, L7

Hsieh, S.H. and Spiegel, E.A. 1976. Ap.J., **207**, 244.

Kochanek, C., 1992, Ap.J., in press

Mészáros, P. and Rees, M.J., 1992a, Ap.J., 397, 570.

Mészáros, P. and Rees, M.J., 1992b, M.N.R.A.S., 257, 29P.

Mészáros, P. and Rees, M.J., 1992c, Ap.J., in press

Narayan, R., Piran, T. and Shemi, A., 1991, Ap.J.(Lett.), 379, L17

Narayan, R., Paczyński, B. and Piran, T., 1992, preprint

Paczyński, B., 1986, Ap.J.(Lett.), 308, L43

Paczyński, B., 1990, Ap.J., 363, 218

Phinney, E.S., 1991, Ap.J.(Lett.), 380, L17

Rees, M.J. 1966. Nature 211, 468.

Rees, M.J. and Mészáros, P., 1992a, M.N.R.A.S., 258, 41P.

Rees, M.J. and Mészáros, P., 1992b, these Proceedings

Shemi, A. and Piran, T., 1990, Ap.J.(Lett.), 365, L55

Usov, V.V., 1992, Nature, 357, 472

Woosley, S., 1992, Ap.J., in press

DISK PLUS HALO MODELS OF GAMMA-RAY BURST SOURCES

I. A. Smith, D. Q. Lamb
Dept. of Astronomy and Astrophysics, Univ. of Chicago, Chicago, IL 60637

ABSTRACT

We show that, in principle, $\approx 70\%$ of the γ-ray bursts observed by BATSE can come from local ($\lesssim 1$ kpc), galactic disk neutron stars, with the rest in an extended galactic halo. We consider three possible forms for the distribution of the galactic halo sources: a Gaussian halo, an exponential halo, and a standard "dark matter" halo. We find that for the Gaussian halo the fraction of bursts that can come from the galactic disk can be $\approx 2/3$, close to the maximum possible value; for exponential and dark matter halos the fraction can be $\approx 1/2$ and $\approx 1/5$ respectively. In each case, the values of $< V/V_{max} >$, $< \sin^2 b >$, $< \cos \theta >$, and the C_{max}/C_{min} distribution are all easily consistent with the BATSE observations. Dividing the bursts into three, equal-sized groups of the brightest, intermediate, and weakest, there is little difference in the values of $< \sin^2 b >$ and $< \cos \theta >$, agreeing with the BATSE observations. The disk sources have luminosities $\sim 10^{36-37}$ ergs s^{-1}, while those in the halo have luminosities $\sim 10^{41-42}$ ergs s^{-1}. The brightest observed bursts must come from the halo; therefore, given current neutron star γ-ray burst models, one might not expect to see cyclotron line features in the brightest γ-ray bursts.

DISK PLUS HALO MODELS

The BATSE observations[1] cannot be explained if all the γ-ray burst sources are local in the galactic disk[2], but they can be explained if all the sources are in an extended galactic halo. Two population models were considered by Lingenfelter and Higdon[3], but were criticized by Paczyński[4]. We show here that up to $\sim 2/3$ of the bursts can be local in the galactic disk, with the other $\sim 1/3$ in a galactic halo, and still match the BATSE observations. Such models make it easier to understand the existence of cyclotron lines in some bursts; however, the existence of two separate populations has yet to be proven.

For the disk, we assume the sources are standard candles with luminosity L_d and that the disk number density is $n(z) = n_d e^{-|z|/z_0}$; the disk has infinite extent in the galactic plane, z_0 is the disk scale-height, and the distance to the faintest disk source that could be observed by BATSE will be called D_d. Given D_d/z_0, the disk $< V/V_{max} >_d$ and $< \sin^2 b >_d$ can be calculated.

For the halo, we assume that the sources are standard candles with luminosity $L_h \neq L_d$, that they are distributed spherically symmetrically about the Galactic Center, and that the solar system is displaced a distance $R_0 = 8.5$ kpc from the Galactic Center. We consider three possible forms for the distribution of halo sources: a Gaussian halo in which the number density of sources has the form $n(R) = n_h e^{-\frac{1}{2}(R-R_s)^2/\sigma^2}$, where R is the distance of the source from the Galactic Center; an exponential halo in which $n(R) = n_h e^{-R/\bar{r}}$; and a standard "dark matter" halo[5] in which $n(R) = n_h/(1 + (R/R_c)^2)$. We take $n_h \neq n_d$. The halo $< V/V_{max} >_h$, $< \cos \theta >_h$, and $< \sin^2 b >_h$ can be calculated given the distance D_h to the faintest halo burst that BATSE could detect, and given (depending on the halo chosen) R_s, σ, $\bar{r}$, or R_c.

Combining the disk and halo sources, to obtain an observed value of $<\sin^2 b>$, the fraction of disk sources allowed is:

$$\frac{N_{disk}}{N_{total}} = \frac{<\sin^2 b>_h - <\sin^2 b>}{<\sin^2 b>_h - <\sin^2 b>_d}. \quad (1)$$

Similarly, to obtain an observed value of $<V/V_{max}>$,

$$\frac{N_{disk}}{N_{total}} = \frac{<V/V_{max}> - <V/V_{max}>_h}{<V/V_{max}>_d - <V/V_{max}>_h}. \quad (2)$$

Figures 1 and 2 assume an "ideal" halo for which $<V/V_{max}>_h = 0$ and $<\sin^2 b>_h = 1/3$. The solid curves on Figure 1 show N_{disk}/N_{total} as a function of D_d/z_0 for (from top to bottom) $<\sin^2 b> = 0.25, 0.27, 0.29, 0.31, 0.33$. The dashed curves on Figure 1 show N_{disk}/N_{total} as a function of D_d/z_0 for (from top to bottom) $<V/V_{max}> = 0.40, 0.37, 0.34, 0.31, 0.28$. At each value of D_d/z_0, the smaller value of N_{disk}/N_{total} must be used to satisfy the $<\sin^2 b>$ and $<V/V_{max}>$ constraints simultaneously. Figure 1 shows that for the current BATSE observations of $<\sin^2 b> \approx 0.32$ and $<V/V_{max}> \approx 0.32$, N_{disk}/N_{total} can be as large as $\approx 70\%$, in principle, for a disk with $D_d/z_0 \sim 1$.

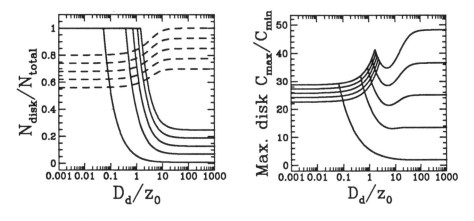

Figure 1. Fraction of disk sources for "ideal" halo.
Figure 2. Maximum value of C_{max}/C_{min} for the disk sources.

Figure 2 shows the maximum value of C_{max}/C_{min} that the disk sources can have as a function of D_d/z_0. The "ideal" halo is used as in Figure 1, and $N_{total} = 193$ is used to compare with the BATSE results we show later. The $<V/V_{max}>$ constraint leads to the curves starting from the left; from top to bottom, $<V/V_{max}> = 0.40, 0.37, 0.34, 0.31, 0.28$. The $<\sin^2 b>$ constraint leads to the curves starting from the right; from top to bottom, $<\sin^2 b> =$

0.25, 0.27, 0.29, 0.31, 0.33. Figure 2 shows it is not possible for the disk sources to be the brightest ones observed by BATSE, whose C_{max}/C_{min} distribution extends beyond 100.

It is also crucial to get the shape of the C_{max}/C_{min} distribution correct, and this limits the number of sources that can be in the disk. However, in the next section we show an example for a Gaussian halo that has 2/3 of the sources in the disk; this is very close to the "ideal" case. For an exponential halo, we have found that up to $\approx 1/2$ of the sources can be in the disk. Finally, for a standard "dark matter" halo, it is only possible for $\approx 1/5$ of the sources to be in the disk; we give an example of this in the final section. Further details will be published elsewhere[6].

GAUSSIAN HALO PLUS DISK

For the disk we use $D_d/z_0 = 2/3$, and for the Gaussian halo we use $R_0 = 8.5$ kpc, $R_s = 25$ kpc, $\sigma = 38$ kpc, and $D_h = 200$ kpc. We take 129 disk sources and 64 halo sources (i.e. 66.8% of the sources are in the disk). The values of $< V/V_{max} >$, $< \cos \theta >$, and $< \sin^2 b >$ for the disk, halo, and combined disk plus halo are given in Table 1. The values for the combined model are all easily consistent with the BATSE observations; at this meeting, $< V/V_{max} > = 0.324 \pm 0.016$ was quoted for 336 bursts, and $< \cos \theta > = 0.048 \pm 0.027$, and $< \sin^2 b > = 0.320 \pm 0.014$ were quoted for 447 bursts.

Table 1. Average quantities for Gaussian halo plus disk.

	Disk	Halo	Disk + Halo
$< V/V_{max} >$	0.4833	0.0734	0.347
$< \cos \theta >$	0.0	0.0944	0.031
$< \sin^2 b >$	0.2932	0.3314	0.306

The publicly accessible BATSE C_{max}/C_{min} catalog currently contains the 241 bursts seen from launch up to March 5, 1992. C_{max}/C_{min} is determined on three different timescales; 64 ms, 256 ms, and 1024 ms. We have chosen to use the values of C_{max}/C_{min} from only the 1024 ms timescale, since the $< V/V_{max} >$ for these bursts is smaller than those using the 64 ms or 256 ms timescales; the predictions of the two-component model are the most conservative if this timescale is used. From the 241 bursts, we delete those whose $C_{max}/C_{min} < 1$ and those whose peak rate is "undetermined" on the 1024 ms timescale (thus all the "overwrite" bursts are removed). This leaves 193 bursts; performing the above cuts on the 64 ms or 256 ms timescales results in fewer bursts than this, which is another advantage of using the 1024 ms timescale. C_{max}/C_{min} is not the most appropriate distribution to use for fitting population models, but it is the only data that is currently available.

Figure 3(a) shows the C_{max}/C_{min} distribution for the combined model (solid curve) and the BATSE data for the 193 bursts, showing that this gives an adequate fit. The model curve shown does not include the temporal variation in C_{min}; model curves including this will be published elsewhere[6]. The model gives a slight steepening of the C_{max}/C_{min} distribution at small C_{max}/C_{min};

however, this is not nearly as pronounced as predicted by Paczyński[4], removing a main objection to two-component models.

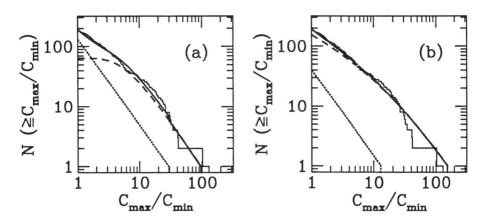

Figure 3. C_{max}/C_{min} distribution for (a) Gaussian halo plus disk, (b) "Dark matter" halo plus disk.

The dashed line in Figure 3(a) shows the C_{max}/C_{min} distribution for just the halo sources, and the dotted line is for just the disk sources. It can be seen that the halo curve is almost flat at low C_{max}/C_{min}. The reason is that BATSE is not seeing any distant halo bursts, because the halo number density drops rapidly with distance; all but 1 of the 64 halo bursts are within a distance ~ 140 kpc, well inside $D_h = 200$ kpc. The lack of distant halo bursts is the key feature that allows a large fraction of the sources to be in the disk.

It can be seen from Figure 3(a) that the largest value of C_{max}/C_{min} for the disk sources is 29, and there are 6 halo sources that are brighter than this disk source. Current models for the formation of cyclotron features in γ-ray bursts[7] have usually assumed the sources are local neutron stars; if this is correct, one might not expect to see cyclotron line features in the brightest γ-ray bursts.

The scale-height for the galactic disk neutron stars near the Sun is believed to be $\lesssim 0.5$ kpc[8,9,10]; all the observed disk sources would therefore be within $\sim 1/3$ kpc of the Earth. (Thus the radial structure of the galactic disk is not important for this case, justifying our use of an infinite disk.) This makes the ratio of the luminosities $L_h/L_d = 3.6 \times 10^5$; if the local disk neutron star sources have a luminosity of $\sim 10^{36}$ ergs s^{-1}, the halo sources would have a luminosity $\sim 4 \times 10^{41}$ ergs s^{-1}.

Dividing the bursts into three equal-sized groups of the brightest, intermediate, and weakest, $< \sin^2 b > = 0.328, 0.305, 0.285$ and $< \cos \theta > = 0.055, 0.038, 0.001$ respectively. There is little difference in these values, a behavior that agrees with the BATSE observations.

STANDARD "DARK MATTER" HALO PLUS DISK

For the disk, we again use $D_d/z_0 = 2/3$, and for the "dark matter" halo we use $R_0 = 8.5$ kpc, $R_c = 22.5$ kpc, and $D_h = 135.0$ kpc. We take 39 disk sources and 154 halo sources (i.e. 20.2% of the sources are in the disk). The values of $<V/V_{max}>$, $<\cos\theta>$, and $<\sin^2 b>$ for the disk, halo, and combined disk plus halo are given in Table 2.

Table 2. Average quantities for "dark matter" halo plus disk.

	Disk	Halo	Disk + Halo
$<V/V_{max}>$	0.4833	0.3112	0.346
$<\cos\theta>$	0.0	0.0716	0.057
$<\sin^2 b>$	0.2932	0.3310	0.323

Figure 3(b) shows the C_{max}/C_{min} distribution for the combined model (solid curve) as well as the BATSE data for the 193 bursts discussed in the previous section, showing that this fits. The dashed line in Figure 3(b) shows the C_{max}/C_{min} distribution for just the halo sources, and the dotted line is for just the disk sources. For the "dark matter" halo, the number of sources in a shell at radius R is $4\pi R^2 n(R) dR \approx$ constant dR for large R. Thus, in contrast to the Gaussian or exponential halos, with the "dark matter" halo one keeps seeing sources at larger and larger distances. Consequently, the halo C_{max}/C_{min} distribution cannot be made flat at low C_{max}/C_{min}, greatly limiting the number of disk sources that can be included.

It can be seen from Figure 3(b) that the largest value of C_{max}/C_{min} for the disk sources is 13, and there are 21 halo sources that are brighter than this disk source. Taking $z_0 = 0.5$ kpc, all the observed disk sources would be within $\sim 1/3$ kpc of the Earth. This makes the ratio of the luminosities $L_h/L_d = 1.6 \times 10^5$; if the local disk neutron star sources have a luminosity of $\sim 10^{36}$ ergs s^{-1}, the halo sources would have a luminosity $\sim 2 \times 10^{41}$ ergs s^{-1}. Dividing the bursts into three equal-sized groups of the brightest, intermediate, and weakest, $<\sin^2 b> = 0.328, 0.325, 0.318$ and $<\cos\theta> = 0.091, 0.049, 0.032$ respectively.

This work was supported at the University of Chicago by grants NAGW-830, NAGW-1284, NAGW-1301, and NAG-51454.

REFERENCES

1. C. A. Meegan et al., Nature **355**, 143 (1992).
2. S. Mao and B. Paczyński, Ap. J. (Letters) **389**, L13 (1992).
3. R. E. Lingenfelter and J. C. Higdon, Nature **356**, 132 (1992).
4. B. Paczyński, Acta Astron. **42**, 1 (1992).
5. J. J. Brainerd, Nature **355**, 522 (1992).
6. I. A. Smith and D. Q. Lamb, Ap. J., in preparation (1992).
7. D. Q. Lamb, J. C. L. Wang, and I. M. Wasserman, Ap. J. **363**, 670 (1990).
8. D. Hartmann, R. I. Epstein, and S. E. Woosley, Ap. J. **348**, 625 (1990).
9. B. Paczyński, Ap. J. **348**, 485 (1990).
10. Z. Frei, X. Huang, and B. Paczyński, Ap. J. **384**, 105 (1992).

GAMMA-RAY BURSTS
AS A PROBE OF LARGE-SCALE STRUCTURE IN THE UNIVERSE

D. Q. Lamb and J. M. Quashnock

Dept. of Astronomy and Astrophysics, Univ. of Chicago, Chicago, IL 60637

ABSTRACT

If γ-ray bursts are cosmological in origin, the sources of the bursts may trace the large-scale structure of luminous matter in the universe. We find that, if the lifetime of the *Compton Gamma-Ray Observatory* is $\gtrsim 10$ years, yielding the positions of $\gtrsim 3,000$ γ-ray bursts, it may well be possible to probe the structure of luminous matter on the largest scales known, consistent with recent determinations from pencil-beam surveys and studies of superclusters. A positive result would provide compelling evidence that most γ-ray bursts are cosmological in origin, and would allow comparison between the distributions of luminous matter and dark matter on large scales. Conversely, a negative result might cast doubt on the cosmological origin of the bursts, provide evidence that the clustering of burst sources on large scales is less than that expected from pencil beam surveys and studies of superclusters, or indicate that γ-ray bursts have some more exotic origin.

INTRODUCTION

The Cosmic Background Explorer team has reported the detection of fluctuations in the temperature of the cosmic microwave background consistent with an $n = 1$ Harrison-Zel'dovich power spectrum of density fluctuations on angles from 10° to 30° and 180°.[1] Infrared, optical, and X-ray surveys give $n \approx -1.2$ for galaxy, quasar, cluster, and supercluster two-point correlation functions on angular scales $\lesssim 1°$,[2,3] but such angular scales are too small to reflect the spectrum of primordial density fluctuations.[4] Future galaxy and quasar surveys will provide information on the behavior of the two-point correlation function on larger angular scales, but such surveys will face difficulties due to the absorption of light by dust and gas in the Galaxy.

In contrast, the Galaxy is transparent to γ-rays. The recent discovery that even faint γ-ray bursts are distributed isotropically on the sky[5,6] has intensified debate about whether some, or even all, γ-ray bursts are cosmological in origin. If they are cosmological in origin, the sources of the bursts are expected to trace the large-scale structure of luminous matter in the universe.

The use of γ-ray bursts to probe large-scale structure offers several advantages. The Burst and Transient Source Experiment (BATSE) catalogue of γ-ray bursts constitutes a homogeneous sample covering the entire sky, unlike existing galaxy and quasar surveys. Because γ-ray bursts are rare events ($R_b \sim 10$ Gpc^{-3} yr^{-1}), the burst sources form a sparse sample. Their angular two-point correlation function $w(\theta)$ is therefore diluted very little by projection effects, unlike that of galaxies. Finally, the brightness distribution of the bursts seen by BATSE suggests that we are seeing beyond the edge of the burst source distribution.[5] This contrasts with galaxy surveys where the observer function decreases markedly before the edge of the galaxy distribution is reached, complicating attempts to detect structure on the largest scales.

The use of γ-ray bursts to probe large-scale structure suffers from two disadvantages. Foremost is the limited number of bursts ($N_b \approx 1000$ for 3 years of BATSE observations and ≈ 3000 for 10 years of BATSE observations). No estimates of the distances to the burst sources exist, restricting studies of source clustering to the burst angular two-point correlation function $w(\theta)$.

We compute the region of the power spectrum $P(k)$ of density fluctuations that can be probed given a certain number N_b of bursts. We find that, if the lifetime of the *Compton Gamma-Ray Observatory* is $\gtrsim 10$ years, yielding the positions of $\gtrsim 3,000$ γ-ray bursts, it may well be possible to probe the structure of luminous matter on the largest scales known, consistent with recent determinations from pencil-beam surveys[7,8] and studies of superclusters.[9,10,11] A positive result would provide compelling evidence that most γ-ray bursts are cosmological in origin, and would allow comparison between the distribution of luminous matter and the distribution of the total (i.e., dark plus luminous) matter on scales of $100h^{-1}$ Mpc to $3h^{-1}$ Gpc. Conversely, a negative result might cast doubt on the cosmological origin of the bursts, provide evidence that the clustering of burst sources on scales of $100h^{-1}$ to $300h^{-1}$ Mpc is less than that expected from pencil beam surveys and studies of superclusters, or indicate that γ-ray bursts have some more exotic origin.

CALCULATIONS

We assume that γ-ray burst sources are extragalactic and trace the distribution of luminous matter. For illustrative purposes, we suppose that the sources have a uniform comoving density out to a distance D from us, with none beyond that. We compute the region of the power spectrum $P(k)$ of density fluctuations that can be probed using the angular distribution of γ-ray bursts. Recall that the power spectrum is defined as

$$P(k) \equiv \langle |\delta_k|^2 \rangle , \qquad (1)$$

where δ_k is the Fourier transform of the relative matter overdensity $\delta\rho/\bar{\rho}$. We make the canonical assumption that the volume V of the γ-ray burst sample is large enough so that a mean universal mass density $\bar{\rho}$ is well-defined (the fair sample hypothesis).

If we assume that γ-ray bursts are unbiased tracers of mass in the universe, then fluctuations in the counts of bursts are directly related to mass fluctuations, i.e., $\sigma^2 \equiv (\delta N/N)^2 = (\delta M/M)^2$. The mean dispersion in the counts of bursts is then a convolution over the power spectrum:[12]

$$\sigma^2 = \int \frac{dk}{k} \Delta^2(k) W(k) . \qquad (2)$$

Here $\Delta^2(k) \equiv d\sigma^2/d\ln k = VP(k)k^3/2\pi^2$ is the power per logarithmic interval in wave-number, and the window function $W(k)$ is the norm of the Fourier transform of the cell volume v over which fluctuations in counts are being measured:

$$W(k) \equiv \int \frac{d\Omega_k}{4\pi} \left| \frac{1}{v} \int dv\, e^{-i\vec{k}\cdot\vec{r}} \right|^2 . \qquad (3)$$

The mean of the angular correlation function $\bar{w}(\theta)$ is equal to the mean dispersion in counts, σ^2, averaged over cells on the sky of angular radius θ.[13] It is related to the power spectrum through the above relations. We take the cell volume v to be a spherical cone of length D and half-angle θ, and compute the window function for this cell volume as a function of θ. For wave-numbers $k \ll D^{-1}$, $W(k) \simeq 1$, and for $k \gg (D\theta)^{-1}$, $W(k)$ tends to zero as $(kD)^{-4}$. The exact dependence can only be computed by a complicated numerical integration. In the regime $D^{-1} < k < (D\theta)^{-1}$, however, the window function can be computed analytically, tending for large values of kD to $5.64/kD$. We cut off the integral for values of k larger than $C/D\theta$, taking $C \approx 1.5$ so that the results of the convolution match those from the Limber equation for small values of θ.

We make the following ansatz for the power spectrum: $P(k) \gtrsim P(k_0)k_0/k$, for values of k less than the largest wave-number that can be probed, namely $k_0 = C/D\theta$. Substituting this ansatz into Equation 2, we find that

$$\bar{w}(\theta) = \frac{P(k_0)}{2\pi^2} \frac{k_0}{D^2} \int^{k_0} \frac{dk}{k} (kD)^2 \, W(k) \, . \tag{4}$$

Most of the contribution to the integral, $F(k_0)$, comes from wave-numbers $k \sim k_0$, so our results are not changed much by taking a different ansatz for $P(k)$.

We can now place a lower limit on the power spectrum of density fluctuations that can be detected. Let N_b be the number of γ-ray bursts with measured locations and N_c denote the number of cells of angular radius θ on the sky. Then $N_b = N_c \langle N \rangle$, where $\langle N \rangle$ is the mean number of bursts in each cell. Probing the angular correlation of bursts on a scale θ requires at least one object per cell. The number of bursts required to probe the angular scale θ is then $N_b > N_c \approx 4\pi/(\pi\theta^2) = 4\theta^{-2} \approx 1.3 \times 10^4 \, \theta[\deg]^{-2}$, where we have assumed that θ is small. Assuming that any correlations are small, i.e. that $\bar{w}(\theta) \ll 1$, the statistical uncertainty in $\bar{w}(\theta)$ is:[14] $\delta\bar{w}(\theta) = N_{\text{pairs}}^{-1/2}$, where $N_{\text{pairs}} = N_c \langle N \rangle^2/2 = N_b^2/(2N_c)$. Thus $\delta\bar{w}(\theta) = (2N_c)^{1/2}/N_b \approx \sqrt{8}/(N_b\theta)$. Detecting correlations on the angular scale θ thus requires $\delta\bar{w}(\theta)/\bar{w}(\theta) = \sqrt{8}/[N_b\theta\bar{w}(\theta)] < 1$, or $N_b > \sqrt{8}/[\theta\bar{w}(\theta)]$. Combining this result with Equation 4, we obtain

$$P(k_0) > \frac{2\pi^2\sqrt{8}}{C \, F(k_0)} \frac{D^3}{N_b} \, . \tag{5}$$

DISCUSSION

Figure 1 shows the power spectrum $P(k)$ of density fluctuations as a function of wavenumber k corresponding to distance scales of order $10h^{-1}$ Mpc to $3h^{-1}$ Gpc. The data points shown are from analysis of surveys of faint radio galaxies,[15] quasars,[2] clusters of galaxies,[16,17] and from pencil beam surveys[7,8] and studies of superclusters.[9,10] Also shown are the regions of the $(k, P[k])$-plane that can be probed, according to Equation 5, with 300, 1000, and 3000 γ-ray bursts, choosing $D = 1h^{-1}$ Gpc ($z \approx 0.3$) for illustrative purposes. The righthand boundary of the region that can be probed scales in k like D^{-1}, while the lower boundary scales in $P(k)$ like D^2. Therefore, if $D = 0.3h^{-1}$ Gpc ($z \approx 0.1$), the righthand boundary shifts rightward by a factor of 3, while the lower boundary shifts downward by a factor of 9. Conversely, if $D = 3h^{-1}$ Gpc ($z \approx 1$), the righthand boundary shifts leftward by a factor of 3, while the lower boundary shifts upward by a factor of 9.

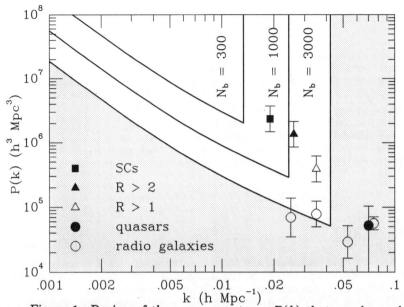

Figure 1. Region of the power spectrum $P(k)$ that can be probed for a given number of bursts N_b. Also shown are data for several classes of objects.

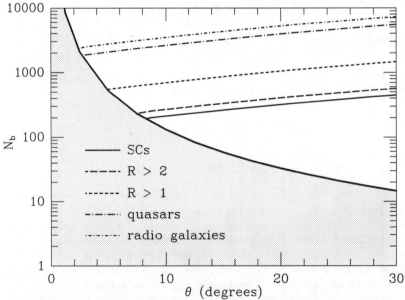

Figure 2. Number N_b of bursts required to detect clustering like that of several classes of objects using the angular correlation function $w(\theta)$.

Because γ-ray bursts are extremely rare, there will be < 1 burst in the BATSE survey per supercluster of galaxies. Gamma-ray burst sources may therefore trace the clustering of luminous matter found in pencil beam surveys and studies of superclusters. Figure 1 shows that, if luminous matter is clustered on scales of $100h^{-1}$ to $300h^{-1}$ Mpc with a density contrast of order unity and γ-ray bursts trace this matter, the resultant clustering of bursts will be detectable in the two-point angular correlation function $w(\theta)$ of the bursts if $D \lesssim 1h^{-1}$ Gpc. If the clustering of luminous matter continues like $P(k) \sim P(k_0)k_0/k$ to somewhat larger scales, the resultant clustering of bursts will be detectable even for $D = 3h^{-1}$ Gpc.

Figure 2 shows the lower bound on the angular scale that can be probed with a finite number of bursts: for a given number N_b of bursts, only angular scales to the right of the solid curve are accessible. Figure 2 also shows the number of bursts N_b required to measure the correlation function for faint radio galaxies, quasars, clusters, and superclusters, where we have assumed $P(k) \sim P(k_0)k_0/k$ continues to small k and again taken $D = 1h^{-1}$ Gpc for illustrative purposes.

Hartmann and Blumenthal[18] earlier searched for correlations in the angular distributions of the γ-ray bursts in the Konus and interplanetary network catalogues. They detected none, but this is not surprising given the small number of bursts in the catalogues (160 and 54, respectively) and the very large uncertainties in the locations of the Konus bursts.

The number N_b of bursts in the current BATSE catalogue is > 500, with $N_b \approx 1000$ and 3000 expected if BATSE continues to obtain data for 3 years and 10 years, respectively. Figures 1 and 2 show that, if γ-ray bursts trace the structure of luminous matter on the largest scales, the burst angular correlation function may already be measurable.

This research was supported in part by NASA grants NAGW-830 and NAGW-1284, and NASA contract NASW-4690.

REFERENCES

1. G. F. Smoot et al., Ap. J. (Letters) **396**, L1 (1992).
2. N. Bahcall and A. Chokshi, Ap. J. **385**, L33 (1992).
3. X. Luo and D. N. Schramm, Science **256**, 513 (1992).
4. E. W. Kolb and M. S. Turner, The Early Universe (Addison-Wesley, Reading, MA, 1990).
5. C. A. Meegan et al., Nature **355**, 143 (1992).
6. C. A. Meegan et al., IAU Circular, 5478 (1992).
7. D. C. Koo and R. Kron, in Observational Cosmology (A. Hewitt, G. R. Burbidge, and L. Z. Fang, eds., Reidel, Dordrecht, 1987), p. 383.
8. T. J. Broadhurst et al., Nature **343**, 726 (1990).
9. N. Bahcall and R. Soneira, Ap. J. **270**, 20 (1983).
10. N. Bahcall and W. Burgett, Ap. J. (Letters) **300**, L35 (1986).
11. L. Guzzo, et al., Ap.J. **393**, L5 (1992).
12. W. Sutherland, Observational Tests of Cosmological Inflation (T. Shanks et al., eds., Kluwer Academic Publishers, Netherlands, 1991), p. 331.
13. G. Efstathiou, et al., M.N.R.A.S. **247**, 10p (1990).
14. P. J. E. Peebles, The Large Scale Structure of the Universe (Princeton Univ. Press, Princeton, 1980).
15. J. A. Peacock and D. Nicholson, M.N.R.A.S. **253**, 307 (1991).
16. S. A. Schectman, Ap. J. (Supplement) **57**, 77 (1985).
17. O. Lahav et al., M.N.R.A.S. **238**, 881 (1989).
18. D. Hartmann and G. R. Blumenthal, Ap. J. **342**, 521 (1989).

GAMMA-RAY BURSTS FROM SHEARED ALFVÉN WAVES IN THE MAGNETOSPHERES OF EXTRAGALACTIC RADIO PULSARS

Marco Fatuzzo[1]
University of Michigan, Ann Arbor, MI 48109

Fulvio Melia[2]
University of Arizona, Tucson, AZ 85721

ABSTRACT

The distribution of gamma-ray bursts (GRBs) detected by BATSE, in combination with the spectral data gathered over the past two decades, argue for a cosmological population of neutron-star sources. We demonstrate that GRB spectra may be understood in the context of the Compton upscattering of typical radio pulsar spectra, providing a natural interpretation to the spectral break ϵ_{break} at $\sim 0.2-3$ Mev often exhibited by GRBs. This model also predics a γ-ray power spectral index $-1 \lesssim \mu \lesssim 2$ above the break, though the prevalence of steeper radio spectra in the brightest pulsars implies a biasing of μ toward the bottom of this range. We find that the probability of detecting a burst in progress from any given source is $\approx 5 \times 10^{-11}$, implying an individual stellar burst rate of about 1 every 50 years if all active pulsars are involved, or about 1 every 2 − 3 years for the very young members of this class. In addition, the energy released per burst coincides with those pertaining to the macro- and micro-glitches seen in the periods of many such sources. We conclude that GRBs may simply be the crustal adjustments responsible for the now familiar timing noise observed in young pulsars.

INTRODUCTION

The accumulating spectral data gathered from observations of gamma-ray bursts (GRBs) suggest a neutron-star origin for these transient events. Certainly, the low-energy features observed in $\sim 20\%$ of the GRBs detected by KONUS[3], by HEAO-1[4], and most recently (and convincingly) by Ginga[5], are very compelling; their interpretation as cyclotron lines implies a magnetic field strength of $\sim (2-5) \times 10^{12}$ gauss[6]. The rapid and erratic variability in the gamma-ray intensity and the abundant storage of rotational, magnetic, and gravitational energy in such systems are additional elements supporting this picture.

This situation, however, in so far as it applies to a Galactic population of neutron stars, appears to be in conflict with the uniform, yet spatially-truncated, distribution of GRBs detected by the BATSE experiment[7], which has led to renewed speculation that most of these events originate outside the Galaxy[8]. We argue that these transient gamma-ray events may very well be produced on relatively young radio pulsars residing in galaxies out to a redshift $z \lesssim 2.5$, consistent with the cosmological hypothesis[9]. These

---
[1] Compton GRO Fellow.
[2] Presidential Young Investigator and Alfred P. Sloan Fellow.

sources may therefore be beacons that trace galactic evolution. Our discussion will feature the sheared Alfvén wave mechanism of charged particle acceleration[10], though we emphasize that many of our conclusions will also be relevant to other energizing scenarios.

PARTICLE DYNAMICS

A crustal disturbance at the polar cap of a pulsar will generate sheared Alfvén waves that propagate out along $\vec{B}$[10]. These oscillations generate an electric field component and a current parallel to the underlying magnetic field (pointing in the $\hat{z}$ direction). In the case where the Alfvénic field B_{ax0} ($\lesssim B/2$) is strongly sheared (i.e., characterized by a wavenumber $m \lesssim 0.1 \text{cm}^{-1}$), the equilibrium particle density

$$n_e^0 = -\frac{\vec{\Omega}_{rot} \cdot \vec{B}}{2\pi c e} + \frac{1}{4\pi c e}(\vec{\Omega}_{rot} \times \vec{r}) \cdot (\vec{\nabla} \times \vec{B}) \approx -\frac{\vec{\Omega}_{rot} \cdot \vec{B}}{2\pi c e} \quad (1)$$

(where Ω_{rot} is the angular velocity) is insufficient to "short-out" this electric field. Charges must therefore be copiously stripped off the stellar surface. The average transient charged particle density is found to be

$$\langle n_e \rangle \approx \frac{m}{2\pi} \frac{B_{ax0}}{4\pi e} \approx 2 \times 10^{18} \text{ cm}^{-3} \left(\frac{m}{0.1 \text{ cm}^{-1}}\right) \left(\frac{B_{ax0}}{0.2 \, B}\right) \left(\frac{B}{4 \times 10^{12} \text{ G}}\right) \quad (2)$$

These disturbances thus "load" the magnetosphere until the particle number density has reached a value $\langle n_e \rangle \gtrsim 10^5 \times n_e^0$. Even so, the magnetosphere remains optically thin to the radio photons because of the magnetic suppression to the scattering cross section (see below).

This activity results in at least two significant (temporary) changes to the pulsar emissivity. First, the enhanced magnetospheric density will greatly increase the coherent GHz-radiation, and secondly, the copious stripping of charges from the polar cap constitutes an abundant source of highly-energetic, directed particles that will Compton up-scatter these radio photons into narrow beams aligned with the local direction of the magnetic field. Making the reasonable assumption that the quiescent radio emission process (which must be coherent and is therefore expected to be proportional to n_e^2) still holds when $\langle n_e \rangle$ increases by these large factors, we therefore expect that the radio luminosity, L_r^b, during the burst could be as high as $(\langle n_e \rangle / n_e^0)^2 L_r$, where L_r is the quiescent luminosity, implying that 10^{37} ergs s$^{-1} \lesssim L_r^b \lesssim 10^{42}$ ergs s^{-1}. At this point, it should be stressed that although L_r is thus enhanced, the spectrum itself is not expected to change since the gyroresonance frequency ν_g is independent of $\langle n_e \rangle$. We shall further assume that a non-negligible fraction η of this radio flux bathes the polar cap region, where the Compton upscattering is to occur.

The particles streaming away from the stellar surface attain a Lorentz factor γ_{stream} at which the electrostatic force due to the average parallel electric field is balanced by the Compton drag associated with the upscattering of these radio photons and, to a much lesser degree, the thermal X-rays emitted by the star. Since the X-ray photon intensity is many orders of magnitude smaller than that of the GHz-radiation, we will for simplicity consider only the influence of the latter on the particle dynamics. For the parameter values considered here and a 1 km polar cap radius, the characteristic electric field is given by

$$\frac{\langle E_{Az}\rangle}{1\times 10^7 \text{ sV/cm}} \approx \left(\frac{B_{ax0}}{0.2\,B}\right)^3 \left(\frac{l_{sh}}{150 \text{ cm}}\right)^{-1} \left(\frac{P}{10 \text{ ms}}\right) \left(\frac{\eta L_r^b}{10^{40} \text{ ergs s}^{-1}}\right), \quad (3)$$

where $l_{sh} = 2\pi/m$, and $P = 2\pi/\Omega_{rot}$.

The relevant dynamics equation is

$$\frac{dp_z}{dt} = -|e|\langle E_{Az}\rangle + D_{e-\gamma}, \quad (4)$$

in which $D_{e-\gamma}$ is the Compton drag force[11]. Since the blue-shifted photon energy in the electron rest frame is much smaller than the magnetic resonant energy ϵ_B, the cross section is suppressed well below the Thomson value σ_T by magnetic quantum effects[12]. It is not difficult to show that when $1 \ll \gamma_{stream} \ll 6\times 10^9\ (B/4\times 10^{12}\text{ gauss})\ (\nu/1\text{ GHz})^{-1}$, the correct limiting (non-resonant) form to use for $\sigma(\epsilon, \theta)$ (with $\theta \to \pi$) is $\sigma(\epsilon) \approx \sigma_T (2\gamma_{stream}\epsilon/\epsilon_B)^2$, where ϵ is the incident radio photon energy. It is very important to emphasize this ϵ^2 dependence in the cross-section, which directly couples the radio and gamma-ray spectral components.

For a characteristic pulsar radio spectrum $I_r \propto (\nu/\nu_{min})^\alpha$ with

$$\alpha = \begin{cases} \alpha_a & \nu \geq \nu_{min} \\ \alpha_b & \nu < \nu_{min} \end{cases}, \quad (5)$$

we find that $\gamma_{stream}(r) \approx 4\times 10^5$ for these parameter values.

THE GAMMA-RAY SPECTRUM

Let us now consider what an observer in the $1/\gamma_{stream}$ beaming cone of these accelerated particles will see. The three distinguishing features characterizing GRB spectra can be understood very easily by considering the typical radio spectra of pulsars. As can be seen from Fig. 1, pulsar spectra often exhibit a low-frequency cut-off ν_{min} at $\lesssim 200 - 500$ MHz, resulting in a corresponding gamma-ray break at

$$\frac{\epsilon_{break}}{1275 \text{ keV}} \approx \frac{\Gamma^{-1/2}}{1+z} \left(\frac{B_{ax0}}{0.2B}\right)^{3/2} \left(\frac{B}{4\times 10^{12}\text{ G}}\right) \left(\frac{l_{sh}}{150\text{ cm}}\right)^{-1/2} \left(\frac{P}{10\text{ ms}}\right)^{1/2}, \quad (6)$$

where $\Gamma \sim O(1)$ and z is the cosmological redshift. Second, because the radio photons are upscattered with a probability $\propto \sigma(\epsilon) \sim \epsilon^2$, the gamma-ray power spectrum νF_ν above the break should have an index $\mu_a \approx \alpha_a + 2 + 1$. Since radio pulsars generally have steep spectra (i.e., a flux density $F_\nu \sim \nu^\alpha$ with index $\alpha \sim -1$ to -4 over most of the radio frequency range), this model therefore predicts that $+2 \gtrsim \mu_a \gtrsim -1$, though the prevalence of steeper radio spectra in the brightest sources might suggest a preponderance of GRB spectra with a post-break index closer to the bottom of this range.

Third, the gamma-ray spectral index μ_b below ϵ_{break} should be no greater than $\alpha_b + 2 + 1$.

The calculation of the absolute flux F^γ_{Earth} at Earth is less certain, in part because we don't yet know the magnetic field geometry near the stellar surface[13]. Since multipole moments higher than $l = 1$ die off very quickly with distance, the magnetic field is certainly dipolar at $r \gg R_*$, but may be comprised of several higher magnetic multipole moments at the polar cap. We note also that due to the stellar rotation, the upscattered radiation within flux tubes away from the spin axis $\hat{\Omega}_{rot}$ will actually be directed into a much broader cone than that corresponding to the γ_{stream}-beaming angle. An observer within the flux tube along $\hat{\Omega}_{rot}$, on the other hand, will benefit from the full beaming in that direction. Balancing the Alfvénic flux in this axial tube with that in the beamed cone, we therefore obtain

$$\frac{F^\gamma_{Earth}(1+z)^{-\mu_a}}{10^{-7} \text{ ergs cm}^{-2} \text{ s}^{-1}} \approx \left(\frac{B_{ax0}}{0.2B}\right)^2 \left(\frac{B}{4 \times 10^{12} \text{ G}}\right)^2 \left(\frac{\gamma_{stream}}{4 \times 10^5}\right)^2 \left(\frac{l_{sh}}{150 \text{ cm}}\right)^2, \tag{7}$$

for a source distance $D = 2.5$ Gpc corresponding roughly to a cosmological redshift $z = 2.5$ for a Hubble constant $h \equiv H_0/100$ km s^{-1} Mpc^{-1} = 1. Representative GRB spectra calculated using the illustrative parameter values discussed in this section are shown in Figure 2.

DISCUSSION

Our analysis includes the following simplifications: (1) Because γ_{stream} is much smaller than the value ($\sim 1 - 5 \times 10^7$) at which curvature radiation impedes the flow, we have safely ignored this effect in all our calculations. (2) We have also ignored secondary scattering between the γ-rays and electrons, other γ-rays, and the magnetic field. The justification for this has appeared elsewhere in the literature[14] and we shall not repeat it here.

This picture self-consistently embraces both the recently-acquired spatial distribution data as well as the large body of spectral information accumulated over the past two decades. The underlying sources need not be "old" and therefore the inferred large values of B are not in conflict with the conventional view of field decay in aging neutron stars—indeed, GRB sources may very well be recently-formed radio pulsars (see below). This scenario accounts naturally for the wide range of break energies ($\sim 0.2 - 3$ MeV) observed in GRBs. Finally, because $\sigma(\epsilon) \propto \epsilon^2$, the γ-ray (power) spectral index above the break is simply given as $\alpha_a + 2 + 1$, where α_a is the radio flux density index.

The probability of seeing a burst in progress from any given source is $\sim 5 \times 10^{-11}$. We estimate the total burst rate to be roughly 1×10^4 h^2 galaxy^{-1} yr^{-1}. This gives an individual stellar burst rate of about 1 every 50 years for all active pulsars, or (which is more likely) about 1 every 2 – 3 years if only the very young pulsars are contributing to this phenomenon. For a polar cap with a radius of 1 kilometer, the energy released per 1 second burst ranges from $\sim 8 \times 10^{43}$ ergs when $B_{ax0} = 10^{12}$ gauss to $\sim 2 \times 10^{41}$ ergs when $B_{ax0} = 5 \times 10^{10}$ gauss. This rate and energy release coincide with those pertaining to the period adjustment observed in many pulsars (i.e., macro and micro glitches) As such, the favorable comparison of the repetition rate and the energetics between the two classes of events suggest that GRBs may be nothing more than the crustal adjustments associated with the timing noise.

This research was supported in part by NSF grant PHY 88-57218, the Alfred P. Sloan Foundation, the Compton GRO Fellowship Program, and the NASA High-Energy Astrophysics Theory and Data Analysis Program.

REFERENCES

3 Mazets, E.P., et al., *Nature*, 290, 378 (1981).
4 Heuter, G.J., in *High Energy Transients in Astrophysics*, ed. S.E. Woosley (New York: AIP), 373 (1984).
5 Murakami, T., et al., *Nature*, 335, 234 (1988).
6 Melia, F., *Ap. J. (Letters)*, 334, L9 (1988).
7 Fishman, G.J., et al., in *Second Gamma-Ray Observatory Science Workshop* (1991).
8 Meegan, et. al., *Nature*, 355, 143 (1992).
9 Mao, S. & Paczyński, B., *Ap. J. (Letters)*, 388, L45 (1992).
10 Fatuzzo, M. & Melia, F., *Ap. J.*, submitted (1992).
11 Melia, F. & Koñigl, A., *Ap. J.*, 340, 162 (1989).
12 Melia, F. & Fatuzzo, M., *Ap. J.*, 346, 378 (1989).
13 Ruderman, M., *Ap. J.*, 366, 261 (1991).
14 Krolik, J.H. & Pier, E.A., *Ap. J.*, in press (1991).

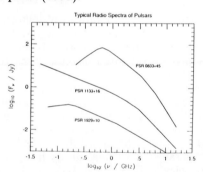

Fig. 1. – Typical radio spectra of pulsars generally exhibit some curvature and often have a low-frequency cut-off at $\nu \lesssim 200 - 500$ MHz. Above this cut-off, the spectrum is steep, with an index $\alpha \sim -1$ to -4 over most of the radio frequency range, where $F_\nu \sim \nu^\alpha$.

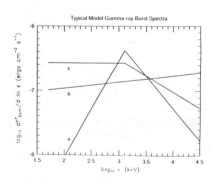

Fig. 2. – Gamma-ray burst spectra for a (hypothetical) pulsar at $D = 1$ Gpc (corresponding to a redshift $z \approx 0.5$), for 3 different radio spectra (cf. Fig. 1): (a) $\nu_{min} = 300$ MHz, $\alpha_a = -4$, and $\alpha_b = -1.5$, (b) $\nu_{min} = 300$ MHz, $\alpha_a = -2.9$, and $\alpha_b = -2.9$, and (c) $\nu_{min} = 300$ MHz, $\alpha_a = -3.5$, and $\alpha_b = -3$. The spectra have not been corrected for redshift. Thus, the observed break energy ϵ_{break} in this source would occur at ≈ 1.2 Mev $/(1+z)$, or roughly 830 keV.

Galactic Halo Model for Gamma-Ray Bursts From High-Velocity Neutron Stars

H. Li, C.D. Dermer *, E. P. Liang
Department of Space Physics and Astronomy,
Rice University, Houston, TX 77251

* E.O. Hulburt Center for Space Research,
Naval Research Laboratory, Code 7653, Washington. DC 20375

ABSTRACT

We show that an extended Galactic halo model for gamma-ray bursts (GRBs) originating from high-velocity neutron stars *(hvns)* (≥ 800 km s^{-1}) with special bursting properties can satisfy the angular isotropy and non-uniform brightness distribution of GRBs observed by BATSE and PVO. Lower velocity neutron stars are required, however, to burst very rarely in this model. *hvns* from both the galactic disk and globular clusters are considered. The effects from nearby galaxies, such as the LMC, SMC, and M31, are found to be small if the sampling distance for BATSE is ~ 100 kpc. The expected rate for detecting X-ray Flashes associated with GRBs from M31, using an X-ray detector like the ROSAT PSPC, is « 1 per day. This model implies that the distribution of the brightest bursts with fluxes $\geq 3 \times 10^{-5}$ ergs cm^{-2} s^{-1} may be anisotropic.

EXISTENCE OF HIGH-VELOCITY NEUTRON STARS

Recent radiopulsar surveys [1, 2, 3] have found three radiopulsars with transverse velocities exceeding 670 km s^{-1}, and radiopulsars with velocities of ~ 1000 km s^{-1} and ~ 2000 km s^{-1} have also been observed. The existence of a population of *hvns* could be consistent with the radiopulsar statistics because fast neutron stars quickly leave the plane of the Galaxy and fall below the detector threshold. The possible correlation between high velocity and strong magnetic field [4] would imply shorter radiopulsar lifetimes due to larger magnetodipole energy loss.

If we preserve the assumption [5] that GRBs originate primarily from Galactic neutron stars, one interesting question is whether these *hvns* will form a distribution in our Galactic halo that could account for the statistics of GRBs detected by BATSE, PVO and other GRB detectors.

DISTRIBUTION OF HIGH-VELOCITY NEUTRON STARS IN THE GALACTIC HALO

First we consider *hvns* born in the Galactic disk. We calculate the spatial distribution of ~ 10^4 *hvns* with initial velocity of 1000 km s^{-1} by following their trajectories using a Monte Carlo simulation. The escape velocity from our Galaxy is shown in Fig. 1. Some characteristic trajectories of neutron stars with different initial velocities are plotted in Fig. 2.

We also treat *hvns* escaping from globular clusters. We calculate the distributions of *hvns* in the halo by assuming that neutron stars escape from globular clusters with velocity of 300 km s^{-1}. We assume that clusters are spherically symmetric about the Galactic Center with an average number density $\rho(r) \approx r^{-3.5}$ with

$2 < r < 40$ kpc, where r is the galactocentric radius. Furthermore, these clusters have a mean rotational velocity of 60 km s^{-1} and a radial velocity dispersion of 130 km s^{-1}.

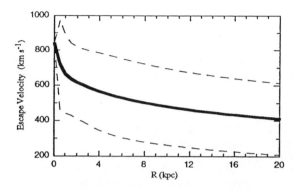

Fig. 1. Escape velocity from the Galaxy as a function of position R in the disk. The thick curve is the escape velocity without the disk rotation, and the two dashed curves stand for the maximum (top) and minimum (bottom) escape velocity taking into account the circular velocity V_c of the disk.

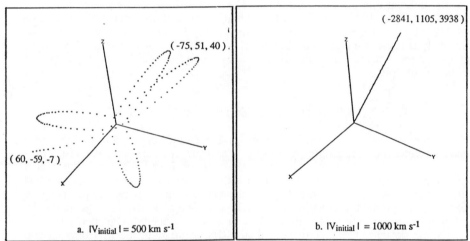

Fig. 2. Trajectories of neutron stars born at the same position (R = 5.65 kpc and z = 35.2 pc) and with the same initial velocity direction but with different speeds. Figs. 2a and 2b correspond to $|V_{initial}| = 500$ and 1000 km s^{-1}, respectively. The Galactic Center is at the origin and the xy plane is the Galactic plane. The motion of a neutron star changes from a highly elliptical orbit when v = 500 km s^{-1} to almost a straight line when v = 1000 km s^{-1}. The coordinates of the farthest positions after 10^{10} yr are indicated in the plots (in kpc).

TIME-DEPENDENT BURSTING RATE FUNCTION

The bursting rate function R(t) is closely related to a possible two-phase activity of neutron stars.

When neutron stars are young, their rapid rotations give a strong electric field E_{rot} which is compensated by large space charge accumulated inside the magnetosphere, so the bursting activities may be shunted. This phase continues for

10^6-10^7 years (depending on B) until the rotation period grows to 1-10 s due to magnetodipole losses, where neutron stars phenomenologically cross the so-called "death line". The neutron star then enters the burst-active phase, in which E_{rot} might be much smaller than some impulsive electric field E_{imp} due to sporadic activation of a NS's magnetosphere, such as starquakes. This phase is assumed to last much longer (10^8-10^9 years). These effects can be quantified by the expression

$$R(t) = K (1 - \exp\{-[t/\tau(B)]^\beta \}) \quad , \qquad (1)$$

where $\tau(B)$ is the timescale for the first phase, which is ~ 30 Myr for B ≈ 5× 10^{12} Gauss. The time-dependence of the bursting behavior at early times is determined by the parameter ß (Fig. 3).

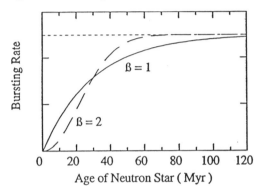

Fig. 3. The bursting rate R(t) of neutron stars as a function of age t in millions of years (Myr) according to equation (1). τ ≈ 30 Myr in this plot. R(t) is proportional to t^β in the radiopulsar phase ($\tau \leq$ 30 Myr) and is approximately constant in the burst active phase ($\tau \geq$ 30 Myr).

DISTRIBUTION OF GRBs

The spatial distribution of GRBs will be a product of the spatial distribution of *hvns* and the bursting rate function R(t) for *hvns*.

We first consider the results for *hvns* from the Galactic disk. With initial velocity ≈ 1000 km s^{-1}, the spatial density of neutron stars will follow a roughly r^{-2} dependence similar to an outflowing flux from a point source. But when combined with the "delayed turn-on" factor R(t), which only approaches a constant value after ~ 30 Myr, the resultant GRB distribution shows a constant density "core" with radius ~ 30-40 kpc (see Fig. 4) when ß ≈ 2 [5]. At a sampling distance of 100 kpc, <cosΘ> = 0.048 (an indication of our offset from the GC), <$\sin^2$ b> = 0.31 <1/3 (a slight concentration to the plane due to Galactic rotation), and <V/V_{max}> = 0.34 (between the uniform value 0.5 and the outflowing value 0.25). These values are within 1σ of BATSE's values for 447 bursts [6]. Fig. 5 shows the log N - log F distribution for our model compared with 212 BATSE bursts and 197 PVO bursts [7]. We obtain a reasonable fit to the combined BATSE/PVO data with ß ≈ 1-2.

Effects from M31, LMC, and SMC are considered too. Although the LMC and SMC are ~ 55 kpc away from us, their effects are found to be very small since their masses are less than 10% of the mass of our Galaxy. Due to the large distance (~ 670 kpc) to M31 and our assumption of a mono-luminosity function for GRBs, the effect of M31 only starts to contribute significantly to the burst statistics when the sampling distance is greater than 200 kpc.

We also show in Fig. 6 the results of a simulation for neutron stars born in globular clusters. Since the velocity of neutron stars are not high enough, they are all "trapped" into the core (≤ 15 kpc), giving a large anisotropy.

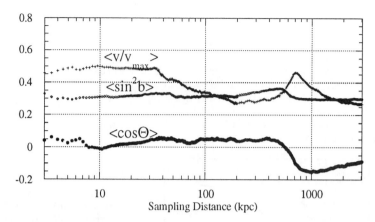

Fig. 4. Dipole moment $\langle\cos\Theta\rangle$, quadrupole moment $\langle\sin^2 b\rangle$, and $\langle V/V_{max}\rangle$ values as a function of detector sampling distance r_d for high-velocity (1000 km s^{-1}) neutron stars. Θ is the angle between the burst and the Galactic Center and b is the Galactic latitude of the burst. Here $\tau = 30$ Myr and $\beta = 2$. Good agreement with the BATSE data is obtained for $r_d \approx 100$ kpc. The Monte Carlo simulation follows 15000 neutron star trajectories. The features around 700 kpc are due to M31, whereas the effects of LMC and SMC are negligible (from ref. 8).

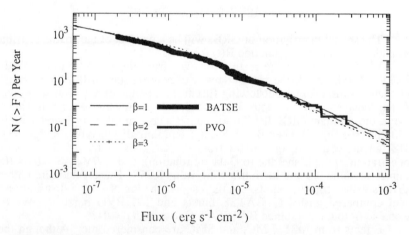

Fig. 5. Number of GRBs per year with peak flux $> F$ as a function of F. Thick curve shows the BATSE data and the histogram shows the PVO data. Note that the relative normalization of these two data sets is poorly known[7], and that the size distribution of the BATSE data is derived from peak count rates of GRBs. Thus the absolute flux scale is uncertain. Curves show the model results with bursting rate function for fast neutron stars characterized by the parameters β (from ref. 8).

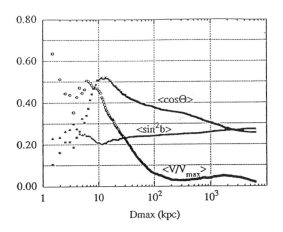

Fig. 6. Dipole moment $<\cos\Theta>$, quadrupole moment $<\sin^2 b>$, and $<V/V_{max}>$ values as a function of detector sampling distance D_{max} for neutron stars produced in globular clusters with $V_{esc} = 300$ km s^{-1}. The distribution has a large high-density central core with radius $\approx$ 10 to 15 kpc, resembling the parent cluster systems. Beyond this "core", the density decreases rapidly. Inclusion of R(t) will not significantly change the results because these neutron stars are generally older than 10^8 years.

SUMMARY AND PREDICTIONS

We have presented a model for GRBs originating from Galactic high-velocity neutron stars which is consistent with the BATSE and PVO results [8] and is in accord with the paradigm for bursts originating from strongly magnetized, galactic neutron stars, deduced from years of GRB spectroscopy observations [9]. We require that the faster neutron stars, which produce most of the bursts, escape from the gravitational potential of the Galaxy. The low- and high-velocity neutron stars differ greatly in their bursting properties, although the average burst luminosities may be similar. In our model, GRBs originate from neutron stars primarily after the end of their radiopulsar phase. This model can be tested using improved burst statistics from BATSE. In particular, the dipole moment for our fast neutron star model will be slightly positive and the quadrupole moment will have a value < 1/3 at sampling distance $r_d \gg 10$ kpc due to the addition of the parallel rotation velocity of the Galaxy to the isotropic ejection velocity of the newly born neutron stars. Searching for bursters in the periphery of M31 using X-ray telescope such as ROSAT will also set constraints on this model [10].

REFERENCES

1. Fomalont, E. B. et al. M.N.R.A.S., **258**, 497 (1992)
2. Harrison, P. A., Lyne, A.G. and Anderson, B., in X-ray Binaries and the Formation of Binary and Millisecond Pulsars, eds van den Heuvel, E.P.J. & Rappapart, S., Kluwer, Dordrecht, in press (1992)
3. Frail, D. A. and Kulkarni, S. R., Nature **352**, 785 (1991)
4. Bailes, M., Ap. J. **342**, 917 (1989)
5. Dermer, C.D. and Li, H., in Huntsville Workshop on Gamma-Ray Bursts, ed. Paciesas, W. and Fishman, G., (AIP, 1992), p.115
6. Meegan, C.A. et al., these proceedings
7. Fenimore, E.E. et al., these proceedings
8. Li, H. and Dermer, C.D., Nature **359**, 514 (1992)
9. Liang, E.P. and Petrosian, V., ed. AIP Conf.Proc. 141, Gamma-Ray Bursts (1986)
10. Li, H. and Liang, E.P., Ap. J. Letters, in press (1992)

A STUDY OF GAMMA-RAY BURST CONTINUUM PROPERTIES PRESENTING EVIDENCE FOR TWO SPECTRAL STATES IN BURSTS

G. N. Pendleton, W. S. Paciesas, R. S. Mallozzi, T. M. Koshut
University of Alabama in Huntsville, AL 35899

G. J. Fishman, C. A. Meegan, R. B. Wilson, B. A. Harmon
Marshall Space Flight Center, AL 35812

J. P. Lestrade
Mississippi State University, MISS 39762

ABSTRACT

Evidence is presented for the existence of two spectral states contributing simultaneously to the total spectrum observed in many gamma-ray bursts (GRB's). An ensemble of 120 GRB's measured by BATSE have been studied, using 4 channel spectral data, to determine in which bursts the spectral states can be most effectively resolved. The technique of summing the low intensity spectra together to get an average spectrum allows for precise characterization of the average low intensity spectral behavior. The 4 and 16 channel spectra obtained by the BATSE Large Area Detectors (LAD's) are analyzed using a model-independent spectral inversion technique. The results of these analyses applied to an individual burst are discussed in detail.

INTRODUCTION

This study performed on BATSE LAD data[1] was initiated after a significant fraction of the DISCLA spectral analysis of the bursts for the BATSE burst catalog had been completed. The catalog analysis showed that in many bursts the low energy spectra (20-100 keV) for the peak fluxes of the bursts were significantly harder than the low energy spectra of the total fluence for the same bursts. These spectral differences suggested that the low intensity spectra in these bursts were significantly different from the high intensity spectra. Here high intensity spectra within a burst are defined as those spectra at or near the peak intensity of the burst. Low intensity spectra are those spectra in a burst that are considerably below the peak intensity. The spectral differences initiated study of the BATSE 4 channel LAD data for 120 GRB's.

The presence of two spectral states manifests itself in two ways in our data. Studies of the 4 channel spectral data in the 20-100 keV range show discontinuous jumps in the hardness ratio vs. intensity plots of spectra in bursts. These jumps are not indicative of simple correlations of hardness ratio with intensity but of the onset of different spectral states at particular intensities within bursts. Also studies of the BATSE 16 channel LAD data show that the most intense spectra in bursts are often inconsistent with a single component origin. In some bursts the lower intensity spectra are consistent with a single spectral component but the higher intensity spectra appear to be a combination of the low intensity component plus some higher energy variable component.

PROCEDURE

The analysis employed DISCSC and PREB[1] data for the bursts. These include 4 channel discriminator data with 64 ms time resolution covering the time interval from 2s before the burst trigger to 240s of seconds after the burst trigger. The energy ranges are 20-50 keV, 50-100 keV, 100-300 keV, and >300 keV. Background subtraction was performed using a quadratic fit to the count rates in each channel of each detector derived from appropriate intervals around the burst. The background fits were inspected by multiple analysts and hardcopy outputs were produced for every burst in the study.

Figure 1a shows the GRB intensity vs. time in 64 ms bins over the time interval used in the current analysis for BATSE trigger no. 907 (1B911016).

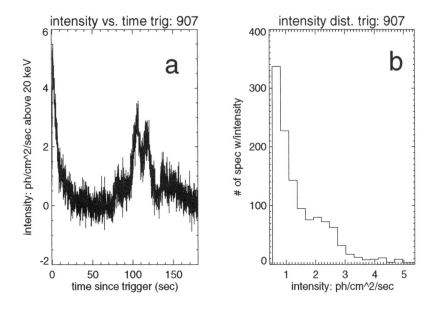

Figure 1. a) The time history of BATSE trigger number 907 (1B911016) in 64 millisecond bins. b) The associated intensity distribution.

The counts per bin here have been multiplied by the inverse of a coarsely binned detector response matrix to yield photons per bin. This analysis gives intensity estimates in each discriminator bin and removes energy dependent effects of detector angular response and atmospheric scattering contamination. It also removes as much as is possible systematic correlations of the intensities between bins caused by higher energy photons registering in lower energy bins due to partial energy deposition in the detector.

The first step in the analysis was to build an intensity distribution of the spectra that comprise each burst. This involved taking the individual 64 ms rates shown on the left in figure 1 and sorting them by intensity into bins to create an intensity distribution for the burst. This kind of analysis removes any time evolution inherent in a burst but effectively separates the low and high intensity components for study. The effects of the definition of intensity, i.e. the energy range chosen for the flux, were studied by performing the analysis with different definitions of intensity: 20-100 keV, 20-300 KeV, 50-300 KeV, 100 keV and up, and 20 keV and up. The conclusion that two spectral states are present in many bursts could be drawn from the results with any of these intensity definitions.

The effect of statistical variance on the hardness vs. intensity plots was also studied to insure that the algorithm design did not introduce significant spurious correlations of hardness vs. intensity due to definitions of the intensity range or background subtraction systematics.

Figure 1b shows the intensity distribution of burst 907. The bins have width 1/20 the peak 64 ms intensity and intensities below 0.45 photons/cm^2/sec are discarded. This lower limit is about twice the intensity level at which background subtraction systematics start to distort the results of the analysis as determined from studies mentioned above. Each 64 ms spectrum is placed in the appropriate bin and the resulting distribution shows the number of spectra in each intensity bin. The intensity distribution in figure 1b is an example of the non-uniform behavior observed in many bursts, i.e. there appear either to be two intensities around which most of the spectra cluster or one intensity around which many of the spectra cluster on top of a broader distribution of intensities. Other intensity distributions appear uniform, i.e. there is just one discernible broad distribution of intensities comprising the burst. A distinct jump in the hardness ratio occurs in bursts with non-uniform intensity distributions. Figure 2 shows the average hardness ratios calculated for the first two discriminator channels for all the 64 ms spectra in the associated intensity distribution bins shown in figure 1b. It is possible to calculate the average hardness ratio fairly precisely at the lower intensities since the 64 ms spectra of similar intensity are summed together to improve the total statistical significance. This increase in spectral precision at lower intensities is the primary advantage of the intensity distribution technique. This plot shows that the hardness ratio for trigger 907 remains fairly constant at about .3 below an intensity of 3.0 photons/cm^2/sec and then jumps to about .75 above this intensity indicating a change in spectral state.

The presence of two spectral states is also in evidence in our 16 channel spectral data. These spectra are converted from counts to photons using the inverse of a suitably binned detector response matrix. The accuracy of the technique is ascertained by applying it to occultation measurements of the steady Crab flux. Figure 3 shows the average low and high intensity spectra for burst 907. The break point between high and low intensity is the point where the change in slope of the intensity distribution is most positive. The average of all spectra with intensities below the break point is shown as crosses in figure 3. The average high intensity spectrum is plotted as diamonds in the figure. The LAD detectors have been calibrated below 300 kev but the channel to energy conversion above this energy has not yet been optimized. The scatter in the points above this energy is symptomatic of systematic distortions in the channel to energy conversion formula when direct inversion is used.

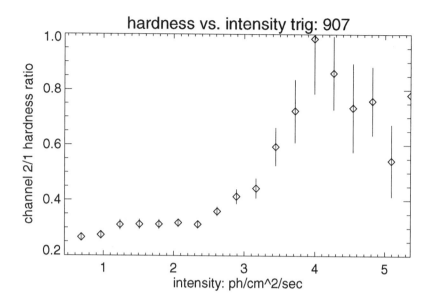

Figure 2. Hardness ratio vs. intensity for BATSE trigger no. 907 (1B911016). The hardness ratio of 50-100 keV flux over 20-50 keV flux is calculated for the average spectrum in each intensity bin of figure 1b.

The low intensity spectrum is consistent with a physical mechanism that produces a spectrum with fairly constant slope. The high intensity spectrum exhibits a marked change of slope at about 60 keV. A qualitative interpretation is that it is a combination of the low intensity spectrum plus some higher energy component with a deficit of emission below 100 keV.

RESULTS

The presence of two components in gamma-ray bursts is consistent with observations by other instruments. The APEX time histories[2] suggest different temporal and spectral behaviors between the low energy and high energy burst spectra. Spectra with obvious broad high energy components associated with annihilation emission superimposed on a soft thermal continuum have been reported[3]. The Konus catalog[4] contains many spectra qualitatively similar to the high intensity spectrum shown in figure 3.

The intensity distribution technique, coupled with the increased sensitivity of the BATSE instrument, will allow for the separation and quantitative spectral analysis of the two spectral components that appear to comprise many bursts. The isolation of two components that appear in a large number of bursts will put constraints on the environment in which the gamma-ray bursts are produced.

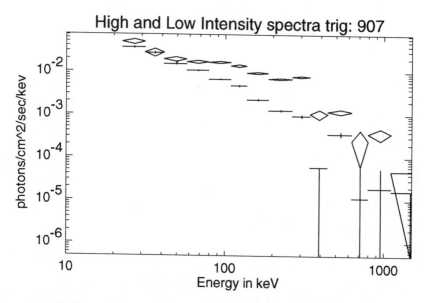

Figure 3. High and low intensity spectra for BATSE trigger no. 907 (1B911016). The high intensity spectrum is represented by diamonds and the low intensity spectrum by crosses. The scatter above 300 keV is due to systematic effects (see text).

REFERENCES

1. Fishman et al., in Proceedings of the GRO Science Workshop, ed. N. Johnson, pp. (2)39-(2)50, 1989
2. Mitrofanov et al., in Gamma-Ray Bursts: Observations, Analyses, and Theories, ed. C. Ho, R. Epstein, and E. Fenimore, pp. 209-216, 1992
3. Mazets et al., Adv. Space Res., Vol. 8, No. 2-3, pp. (2)669-(2)667, 1988
4. Mazets et al., Astrophysics and Space Science, Vol. 80, No. 1, pp. 3-83, 1981

A SEMI-ANALYTIC MODEL FOR CYCLOTRON LINE FORMATION

J. C. L. Wang
CITA, University of Toronto, Toronto, ON M5S 1A1, Canada

I. Wasserman
Cornell University, Ithaca, NY 14853

D. Q. Lamb
University of Chicago, Chicago, IL 60637

ABSTRACT

We present a semi-analytic, physically motivated model for the formation of the cyclotron fundamental feature, in addition to the higher harmonic features, in static media of moderate depth. The spectra derived from the semi-analytic model agree well with those obtained from 'exact' Monte Carlo simulations. These models are able to produce narrow fundamental and second harmonic features with comparable equivalent widths, which are the main characteristics of the harmonically spaced low energy absorption-like features observed by the Ginga satellite in the spectra of three classical γ-ray bursts.[1,2,3] We are currently generalizing the semi-analytic model to dynamic media, in particular, to line formation in a relativistic, radiation driven wind.

INTRODUCTION

The cyclotron fundamental feature forms from multiple resonant photon scatters the outcome of which must, in general, be obtained from a numerical radiative transfer calculation that also accounts for the important higher harmonics. This is because in weak fields ($B \ll B_c = 4.4 \times 10^{13}\,G$), higher harmonic photons may undergo resonant Raman scattering, thereby 'spawning' additional fundamental photons. Since the higher harmonics are formed from resonant Raman scattering, their shapes are well approximated as 'true' absorption lines. The formation of the fundamental is therefore the most computationally demanding and time consuming part of the line formation calculation.

We have devised a semi-analytic and physically motivated model for the formation of the fundamental, in addition to the higher harmonics, in weak fields ($B \ll B_c$) in static media, for the case where the line forming region is optically thick in the fundamental but thin in the wings. This regime is relevant for the classical γ-ray burst sources which display absorption-like features in their low energy spectra.[1,3,4] In such media, to a first approximation, fundamental photons suffer a block of resonant scatters while in the line core and then escape once it scatters into the wings. The semi-analytic model is then based upon three physical principles: (1) the transfer is a smooth stochastic process, (2) the fundamental feature forms from complete redistribution in the single block of resonant photon scatters (generalized complete redistribution[5]), and (3) the higher harmonics are well-approximated as "true" absorption features. In the semi-analytic model, the emergent spectral line shapes are reduced to quadratures.

The static model is able to produce narrow fundamental and second harmonic line shapes with comparable equivalent widths[6,7], which are the main characteristics of the low energy features observed by the Ginga satellite in the spectra of GB870303[1], GB880205[1,2], and GB890929[3]. The cyclotron line enhanced Eddington limit, however, restricts sources with static line forming regions to lie well within the Galactic disk.[8] While some γ-ray bursts may indeed lie within the disk,[9] the isotropy of the burst distribution determined from the Compton GRO BATSE source count data is incompatible with the bulk of classical γ-ray bursters originating from a disk population.[10] This provides the motivation for studying dynamical models. To this end, we are currently extending the semi-analytic model to model line formation in a relativistic, radiatively driven wind[11]. Unlike static models, these dynamical models allow the source to be placed, in principle, at arbitrary distances from the observer.

STATIC MODEL

Photons are injected into the bottom ($\tau = \tau_0$) of a slab (thickness = τ_0) which has a uniform magnetic field oriented parallel to the slab normal. The emergent spectrum is composed of a transmitted and reflected component.

The transmitted component is given by

$$N_T\left(\vec{k}\right) = N_i\left(\vec{k}\right) \exp\left[-\tau_0 \phi\left(\vec{k}\right)/\mu\right] + \int_0^{\tau_0} d\tau \int_{\mu'>0} d\vec{k}' \, q_T(\vec{k}' \to \vec{k}, \tau) \, \phi_1\left(\vec{k}'\right) \frac{\exp\left[-(\tau_0 - \tau)\phi_1\left(\vec{k}'\right)/\mu'\right]}{\mu'} N_i(\vec{k}'), \quad (1)$$

where $\vec{k}$ denotes all properties of the photon (energy, direction, polarization), $\phi(\vec{k})$ is the total resonant scattering profile, $\phi_1(\vec{k})$ is the resonant scattering profile for fundamental photons and $q_T(\vec{k}' \to \vec{k}, \tau)$ gives the distribution of transmitted photon properties given that the first scatter occurred at depth τ (measured from the 'top' of the slab) with incident photon properties $\vec{k}'$. The first term gives the absorption contribution and describes well the higher harmonic features. The second term gives the multiple scattering contribution to the fundamental feature.

The reflected component is given by

$$N_R\left(\vec{k}\right) = \int_0^{\tau_0} d\tau \int_{\mu'>0} d\vec{k}' \, q_R(\vec{k}' \to \vec{k}, \tau) \, \phi_1\left(\vec{k}'\right) \frac{\exp\left[-(\tau_0 - \tau)\phi_1\left(\vec{k}'\right)/\mu'\right]}{\mu'} N_i\left(\vec{k}'\right), \quad (2)$$

where q_R is the same as q_T but for reflected photons.

The distributions q_T and q_R satisfy $\int d\vec{k}\,(q_T + q_R) = 1$ since scattering conserves photon number (for fundamental photons). Adopting the picture of complete redistribution in a single block of resonant scatters gives[5,12]

$$q_T(\vec{k}' \to \vec{k}, \tau) \propto \phi_1\left(\vec{k}\right) \exp\left[-\tau \phi_1\left(\vec{k}\right)/\mu\right]; \quad \mu > 0 \quad (3)$$

and

$$q_R(\vec{k}' \to \vec{k}, \tau) \propto \phi_1\left(\vec{k}\right) \exp\left[-(\tau - \tau_0)\phi_1\left(\vec{k}\right)/\mu\right]; \quad \mu < 0. \quad (4)$$

The factors in the integrand of equations (1) and (2) may be viewed pictorially as, from right to left, injection $\to$ free-streaming to interaction site $\to$

entering block of scatters (with $\vec{k}'$) → exit block of scatters (with $\vec{k}$) and escape by transmission (reflection).

Spawning, that is, the production of fundamental photons from resonant Raman scattering of higher harmonic photons, is incorporated essentially by modifying the photon emissivity in equations (1) and (2), namely, $N_i(\vec{k}) \to N_i(\vec{k}) + N_i^S(\vec{k}, \tau)$, where $N_i^S(\vec{k}, \tau)$ describes the distribution of spawned photons. In order of magnitude, $|N_i^S(\vec{k}, \tau)|/|N_i(\vec{k})| = \mathcal{O}(B/B_c)$.

For example, for spawning from second harmonic photons, we take

$$N_i^S(\vec{k}, \tau) \propto \exp\left[-\left(\frac{\omega - \omega_0}{\omega_D \mu}\right)^2\right](1 + \mu^2) s(\tau), \quad (5)$$

where ω_0 (ω_D) is the fundamental (Doppler) frequency. The spawning profile inside the slab is

$$s(\tau) = \int_{\mu'>0} d\vec{k}' \, \phi_2(\vec{k}') \, \frac{\exp\left[-(\tau_0 - \tau)\phi_2(\vec{k}')/\mu'\right]}{\mu'} N_i(\vec{k}'), \quad (6)$$

where $\phi_2(\vec{k})$ is the resonant scattering profile for second harmonic photons. Pictorially, the factors in the integrand for $s(\tau)$ in eqn. (6) give, from right to left, injection into slab → free streaming to τ → interaction (spawning) at τ.

In Figure 1 we show the various components of the emergent line spectrum obtained from the static semi-analytic model specialized to the case of unpolarized transfer.

In Figure 2 we show a comparison between an emergent (unpolarized) line spectrum obtained from the static semi-analytic model and the corresponding emergent line spectrum obtained from a Monte Carlo simulation.

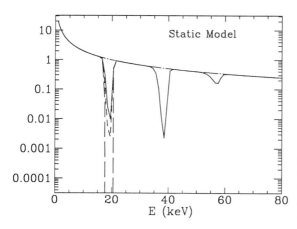

Fig. 1—Unpolarized emergent line spectra from the static semi-analytic model for $B = 1.71 \times 10^{12}\,G$, $T = 5$ keV, $N_e = 1.2 \times 10^{21}$ cm^{-2}, and mean viewing angle with respect to the slab normal of 72°. (Dot-dash) Incident spectrum ($\propto 1/E$). (Long dash) Pure absorption spectrum. (Short dash) Line spectrum without spawning. (Solid) Line spectrum with spawning from second harmonic.

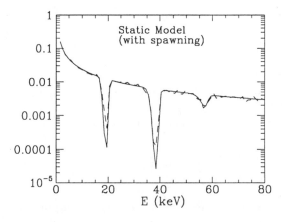

Fig. 2—Unpolarized emergent line spectrum from the static semi-analytic model (solid) and from a Monte Carlo simulation (dashed) for the same parameters as for Figure 1. The semi-analytic model includes spawning from the second harmonic. The Monte Carlo simulation includes spawning from the second and third harmonics. Photons are injected isotropically into the slab with a $1/E$ spectrum.

REFERENCES

1. T. Murakami et al., Nature, <u>335</u>, 234 (1988).
2. E. E. Fenimore et al., Ap. J., <u>335</u>, L71 (1988).
3. A. Yoshida, T. Murakami, J. Nishimura, I. Kondo, and E. E. Fenimore, Pub. Astron. Soc. Jap., <u>43</u>, L69 (1991).
4. E. P. Mazets, S. V. Golenetskii, R. L. Aptekar, Yu. A. Gur'yan, and V. N. Il'inskii, Nature, <u>290</u>, 378 (1981).
5. I. Wasserman and E. E. Salpeter, Ap. J., <u>241</u>, 1107 (1980).
6. S. G. Alexander and P. Mészáros, Ap. J., <u>344</u>, L1 (1989).
7. J. C. L. Wang et al., Phys. Rev. Lett., <u>63</u>, 1550 (1989).
8. D. Q. Lamb, J. C. L. Wang, and I. Wasserman, Ap. J., <u>363</u>, 670 (1990).
9. I. A. Smith, 1992, these proceedings.
10. C. A. Meegan et al., Nature, <u>355</u>, 143 (1992).
11. G. S. Miller, R. I. Epstein, J. P. Nolta, E. E. Fenimore, Phys. Rev. Lett., <u>66</u>, 1395 (1991).
12. J. C. L. Wang, I. Wasserman, and D. Q. Lamb, Ap. J., submitted (1992).

Gamma-Ray Bursts: Magnetosphere around Neutron Stars with Comets, Planets and Black Hole

Hitoshi HANAMI

Physics Section, College of Humanities and Social Sciences,
Iwate University, Morioka 020, Japan

abstract: We studied a magnetosphere interaction model with orbiting objects for γ-ray bursts with the picture of a current circuit which consists of the magnetosphere and the surface of a neutron star. The motion of an object, which is a good conductor and rotating, in the magnetosphere works as the battery in the circuit system. The physical condition on the surface of the neutron star is important to make a closed current circuit inducing good conversion from the kinetic energy of rotating objects to that of the magnetosphere oscillation. An old and cooled neutron star with the temperature $\simeq 10eV$ can prepare the condition for closing the current circuit in the surface of the neutron star. The magnetosphere system is unstable to feed back instability related to the variability of γ-ray bursts.

Introduction

Gamma-ray bursts have been an unsolved problem in astrophysics for fifteen years since their discovery by Klebesadel et al. (1973). Recent observations show the sources have a strong magnetic field $\sim 10^{12}G$ which is consistent with the model of cyclotron absorption for the observed spectrum feature (e.g. Murakami et al. 1988). A neutron star is one object whose magnetic field is strong enough to explain such strong field evidence in our known objects in the universe. It suggests that gamma-ray bursts occur around neutron stars with strong magnetic fields. It was investigated that the magnetospheric plasma oscillations accelerate energetic electrons which can produce γ-rays (Melia 1990). Some possibilities for an excitation mechanism of the transient magnetospheric perturbations have been proposed to account for a starquake (Blaes et al. 1990), neutron star rotation or an episodic accreting object (Harwit and Salpeter 1973). However, the total picture for these magnetospheric oscillations and gamma-ray bursts still stands as an open question in front of astrophysical theorists.

Recently, some discoveries of planets around pulsars were reported. Wolszczan and Frail (1992) observed the evidence for the existence of planets around the millisecond pulsar PSR1257-12. These planets are about several times the mass of the Earth.

This discovery encourages us to consider that neutron stars can have planets around themselves in general (e.g. Nakamura and Piran 1991). This picture naturally points us to an interesting speculation, in which planets, and also comets, rotating in the magnetosphere of the neutron stars, drive gamma-ray bursts. We find a similar situation with Jovian decametric radiation related to the interaction between the strong magnetic field of Jupiter and one of its satellites, Io. One popular interpretation is that electrons or ions are somehow accelerated in this system with plasma instabilities (Goldreich and

Lynden-Bell 1969). This view point for the Jovian decametric radiation may be expanded to the picture of the electrodynamics of a neutron star with the objects orbiting it.

This paper deals with the electromagnetic coupling using a current circuit analogy between the magnetosphere and the objects rotating around a neutron star and the possibility of strong particle acceleration for gamma-ray bursts.

Current Circuit with Magnetosphere

The starting point is that owing to the motion of an object with relative velocity v_p through the co-rotating magnetosphere, an induced electric field $E_{ind} = \frac{v_p}{c} \times B$ appears in the frame of the object. If the object is a good conductor, the electric field will be screened by induced surface charges, so that the magnetic field in the column through the object is zero. In the neutron star frame, there is an electric field due to these charges which is uniform inside the column and given by $E_p = -E_{ind}$. We can understand that the rotating motion of the object produces the electric potential gap in the circuit which is connected by the magnetic field column between the objects and the surface of the neutron star. It is shown schematically in Figure 1. We will assume that the axis of the magnetic moment of the neutron star is perpendicular to the plane of the object orbit.

The circuit is closed by currents in the surface of the neutron star which should have Pedersen conductance sufficient to carry current across the magnetic field. The closed circuit with Pedersen current may work for a particle acceleration.

For the collisionless plasma, then, we can easily see that the current cannot flow along the electric field even if we have a strong perpendicular electric field to the magnetic field. Since the plasma in the magnetosphere is collisionless, there exists no Pedersen current in the magnetosphere. On the other hand, for the collisional plasma, we should point out that the current can flow along the electric field even if it is perpendicular to the magnetic field. Especially, the Pedersen current takes the maximum when the collision frequency $\nu_{ei} = 3 \times 10^{19}(n_e/3 \times 10^{26} cm^{-3})ln\Lambda(T/10eV)^{-3/2}$ has some order of the cyclotron frequency $\omega_c = 1.8 \times 10^{19}(B/10^{12}G)$ for electrons. These results show us that Pedersen current can flow most efficiently in the surface layer ($n_e \simeq 3 \times 10^{26} cm^{-3}$) of a neutron star cooled with its temperature $T = 10eV$.

From the consideration of the mobility, we consider a two-zone model which consists of the magnetosphere and the surface layer, which are approximated to ideal MHD current and Pedersen current system, respectively.

It shows that the coupling process between the magnetosphere and the surface generally induces the spontaneous excitation of Alfvén wave with AC current in the magnetosphere.

High Energy Particle Acceleration with Alfvén Wave

The coherency of the excited wave facilitates particle acceleration. Moreover, accelerated particles can be converted to the high-energy photons like the observed γ-rays. Then, using the numerical simulation code (Hoshino et al. 1992), we are trying to study the acceleration mechanism with the Alfvén wave. In this article, we will show preliminary results. Our simulation used relativistic, electromagnetic particle-in-cell codes in one spatial dimension. The code solves the full set of Maxwell's equations and particles are advanced in time using the relativistic Lorentz force equation. Simulation geometry is a one-dimensional real space along the mean magnetic field, in which the field and the density distributions were considered only along the direction, and each particle has three phase space coordinates for the velocities. Periodic boundary conditions were considered. We start from the initial state that the Alfvén wave is excited with the long wave length and the particle motion is consistent with the input wave. We will show the energy distributions of the particles for the case with the mass ratio of plasma particles 10 in Figure 2. The two graphs show the difference for the mixing ratio of right(R) and left(L) circulating waves. The right and left sides represent the case of mixing R+L waves and only L wave, respectively. We can see that the case of mixing R+L waves is more efficient for the particle acceleration than that of only L wave, even if the input total wave energy is the same. We also found the mixing of antidirection propagating waves with each other is also efficient for acceleration. This point suggests to us the coherent wave excitation in our model is advantageous.

Discussion

We will derive the basic values in this system before detailed discussions. A certain scale L of magnetic column from the neutron star to the object determines a typical frequency $\simeq \frac{1}{\tau_1}$ with the typical Alfvén speed in the magnetosphere. Since the density is very low in the region and the Alfvén speed may be nearly light speed c, the frequency $\simeq \frac{1}{\tau_1}$ of this system is $\omega \sim \frac{\pi}{2}\frac{c}{L} \sim 4.5 \times 10^3 (\frac{L}{100km})^{-1}$. The integrated Pedersen conductivity of the surface layer with the thickness h is $\Sigma_{P,0} = en_e h/m_e v_{ei}(1 + (\omega_c/v_{ei})^2)^{-1} \simeq 3 \times 10^{17}(h/0.1km)$ when the surface temperature is $T \simeq 10eV$. Since $\Sigma_{P,0} Z_0 = 1.2 \times 10^8 \gg 1$ if $V_A \simeq c$, the current decay time scale τ_2 is very large, as $\tau_2 \simeq 10^8 \tau_1$.

Some gamma-ray bursts exhibit structure within a time scale as short as a detector resolution of a few ms. The durations of the gamma-ray bursts range from a few hundred ms to 1000s. If the above typical value for the smallest time scale τ_1 is related to the variability of gamma-ray bursts, for explaining it, the circuit has a characteristic scale $L \simeq 100 km$. On the other hand, using the above assumption for the smallest time scale of τ_1, the current decay time τ_2 is larger than the duration time when the surface temperature is $10eV$. It suggests that the duration time may be determined from the circuit physics.

From the above circuit analysis for the time scale of the variability related to the scale L, the object must rotate in the radius of 100 km in our model. According to our model, the energy of a gamma-ray burst comes from the kinetic energy of the object $E_{p,kinetic} = \frac{GMm_p}{R} = 2.7 \times 10^{38}(M/M_\odot)(m_p/10^{-14}M_\odot)(R/100km)^{-1}$ with the mass of the neutron star M and that of the object m_p. Comparing this energy condition with the observed typical energy of gamma-ray bursts $E_\gamma = 10^{42}(F/10^{-6}ergscm^{-2})(D/100kpc)^2$, we can estimate tha mass of the object $m_p = 3.8 \times 10^{-11}\eta^{-1}(R/100km)(F/10^{-6}ergscm^{-2})(D/100kpc)^2 M_\odot$, where F is observed fluence and D is the distance of the burst source with the energy conversion efficiency η. Recently, the observations by BATSE (Meegan et al. 1992) have shown that burst sources are distributed isotropically but not homogeneously. It suggests that disk distribution is unacceptable and galactic halo distributions must be at least 50 kpc. Though the origin of halo neutron stars is unknown, we should consider neutron stars populating in halo as the sources in our script. From the distance restriction by BATSE, the mass value, estimated to explain the burst energy, is not unreasonable for a comet or an asteroid.

We should point out that our model can prepare more chances for gamma-ray bursts than episodic accretion onto a neutron star which needs fine tuning for the orbit of the comet. As is well known, comets have various orbital parameters and most of them have very large eccentricity in our solar system. The fact suggests that the time interval in which a comet stays in and interacts with the magnetosphere spans a wide range. It may be related to the variety of the duration time. There is one problem for the interaction of the magnetosphere with a comet or planet. If the current flows as DC current on these objects in this system, they may begin to melt with Ohrmic heating since the material of planets may be metals like iron. If the orbiting object is a black hole, which is also a conductor (e.g. Thorne et al. 1986), we can solve this problem.

A key stone in our model is Pedersen conductivity which becomes maximum when

$$(n_e/3 \times 10^{26} cm^{-3}) ln\Lambda (T/10eV)^{-3/2} (B/10^{12}G)^{-1} \simeq 1 \qquad (1)$$

Even if we consider the high density region just below neutron dropping, we can consider Pedersen conductivity involved in the current circuit. From that point of view, the cooled neutron star can make Pedersen current most efficiently. Furthermore, if the surface becomes hot as $T \simeq 1MeV$, the duration time, related to the current decay time, should be comparable with τ_1 related to the variability. It is not advantageous to explain long duration bursts. It supports the suggestion that gamma-ray bursts occur in old neutron stars whose surface temperature is lower than $10^5 K$ for more than $10^8 yr$ of age with the cooling processes (e.g. Ruderman 1991; Nomoto and Tsuruta 1987; Shibazaki and Lamb 1989).

This work has been partly supported by a Grant-Aid for Scientific Research on Priority Areas of Ministry of Education, Science and Culture (04233101) and a Travel Grant from Inoue Foundation for Science.

References

Blaes, O., Blandford, R., Madau. and Koonin, S. 1990, Astrophys. J. **363**, 612.
Goldreich, P. and Lynden-Bell, D. 1969, Astrophys. J. **156**, 59.
Harwit, M. and Salpeter, E.E. 1973, Astrophys. J. Lett. **186**, L37.
Hoshino, M., Arons, J., Gallant, Y.A. and Langdon, B. 1992, Astrophys. J. **390**, 454.
Klebesadel, R.W., Strong, I. and Olson, R.A. 1973, Astrophys. J. Lett. **182**, L185.
Meegan, C.A., Fishman, G.J., Wilson, R.B., Paciesas, W.S., Pendleton, G. N., Horack, J.M., Brock, M.N. and Kouveliotou, C. 1992, Nature, in press.
Melia, F. 1990, Astrophys. J. **351**, 601.
Murakami, T. et al. 1988, Nature **335**, 234.
Nakamura, T. and Piran, T. 1991, in preprint.
Nomoto, K. and Tsuruta, S. 1987, Astrophys. J. **312**, 711.
Ruderman, M.A. 1991, Astrophys. J. **366**, 261.
Shibazaki, N. and Lamb, F.K. 1989, Astrophys. J. **346**, 808.

"Black Holes The Membrane Paradigm", 1986, edited by Thorne, K.S., Price, R.H. and MacDonald, D.A.
Wolszczan, A. and Frail, D.A. 1992, Nature **355**, 145.

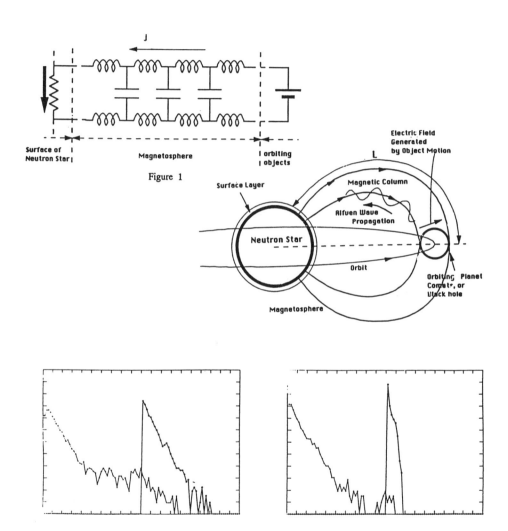

Figure 1

Figure 2

ON THE HALO NEUTRON STAR ORIGIN OF THE GAMMA-RAY BURSTS: ORIGIN OF THE HALO NEUTRON STARS AND METAL ENRICHMENT OF THE INTRACLUSTER MEDIUM

M.HATTORI & N.TERASAWA

THE INSTITUTE OF PHYSICAL AND CHEMICAL RESEARCH (RIKEN), WAKO, SAITAMA 351-01, JAPAN

ABSTRACT

Metal enrichment of clusters of galaxies by proto-galactic star formation is discussed. We show that the observational requirement of metal ejection from proto-galaxies into the intracluster medium(ICM) suggests pregalactic massive star formation bursts in the extended halo of proto-galaxies. Our results predict that galaxies have an extended halo populated with a number of supernova remnants. The predicted large radius of the extended halo of our Galaxy is consistent with the distribution of $\gamma-$ ray bursters observed by BATSE on the Gamma Ray Observatory (GRO) satellite, if neutron stars possibly formed in the extended halo of the Galaxy are the source of $\gamma-$ ray bursters.

INTRODUCTION

The evidences which support the idea that the ICM is metal enriched by a wind from proto-galaxies become increasing (cg. Canizares et al. 1988; David et al. 1990;Tsuru 1992). Here we examine the possibility of early galactic wind and we constrain the condition of the star formation in the proto-galaxy. Our results predict that galaxies have an extended halo populated with large number of neutron stars (Hattori & Terasawa 1992). We show that the recent observational results of the distribution of γ-ray bursters suggest that the halo radius is very large if neutron stars distributed in the galactic halo are the sources of γ-ray bursters and are consistent with our theoretical prediction (Hattori & Terasawa 1992).

CONSTRAINT ON THE RADIUS OF THE PROTO-GALAXY

Here we consider a burst of star formation in a proto-galaxy. The star formation period is assumed to be much shorter than a lifetime of $8M_\odot$ star ($\tau_{sn8} = 7 \times 10^7$yr) which is the longest lifetime among the progenitors of Type II supernova. IMF is assumed as having same form as the Salpeter IMF. An upper cutoff mass is fixed as $m_u = 40M_\odot$. We investigate two different lower cutoff masses. One is the normal disk-like case with lower cutoff mass $m_L = 0.1M_\odot$ and the other is the predominant massive star formation case with lower cutoff mass $m_L = 8M_\odot$. We control the star formation rate by introducing a star formation efficiency parameter, ϵ, which is defined as the mass fraction of the newly formed stars to the initial gas mass. In the following discussion we assume a uniform distribution of stars and gas when the burst of star formation occurs and we assume that the temperature of the interstellar medium is much less than $10^6 K$.

To realize the early galactic wind, it is neccesary to convert the kinetic energy of the supernova remnant into thermal energy of interstellar medium before the most of the energy in the remnant is lost by a radiative cooling. This would be occured when the supernova rate is high enough so that the supernova shells overlap before the each remnants cools significantly. By the comparison of a time scale of shell overlapping with a time scale of the cooling of the remnant (Shull & Silk 1979), the critical supernova rate, λ_{crit}, is calculated. Because the gas surrounding the supernova shell is the primodial gas, the critical supernova rate is calculated as

$$\lambda_{crit} = 9.7 \times 10^{-11} E_{51}^{-1.15} n_{ISM}^{1.7} \text{SNepc}^{-3} \text{yr}^{-1}, \qquad (1)$$

where E_{51} is the energy released by one supernova unit in 10^{51}erg, n_{ISM} is the density of interstellar medium in the proto-galaxy. The condition of the realization of the early galactic wind is expressed as

$$\lambda_{SNII} \geq \lambda_{crit}, \qquad (2)$$

where λ_{SNII} is Type II supernova rate per unit volume in proto-galaxy. We have checked a validity of this equation by applying this to a star burst galaxy M82.

From equation (2) we obtain the constraint equation for a radius of the proto-galaxy r,

$$\left(\frac{r}{10\text{kpc}}\right)^{2.1} \geq \frac{100}{\chi/\chi_{0.1}} \frac{(1-\epsilon)^{1.7}}{\epsilon} E_{51}^{-1.15} \left(\frac{M_{primodial}}{10^{11} M_\odot}\right)^{0.7} \frac{\tau_{SNII}}{7 \times 10^7 \text{yr}}, \qquad (3)$$

where $\chi = \frac{\int_8^{m_u} \phi dm}{\int_{m_L}^{m_u} m\phi dm}$ and $\chi_{0.1} = \frac{\int_8^{40} \phi dm}{\int_{0.1}^{40} m\phi dm}$.

The star formation efficiency in the nearby galaxies gives us a good reference to that in the proto-galaxy. The star formation efficiency in giant molecular clouds and in star burst nucleus estimated from observations is 5% and 50%, respectively (Larson 1987). These estimations are based on the assumption that low-mass stars always form together with massive stars in the proportions predicted by Salpeter IMF with $m_L = 0.1 M_\odot$. If the IMF contains a smaller proportion of low-mass stars, the observed efficiency is reduced. For example, if the stars with mass less than $8 M_\odot$ are not formed, the star formation efficiency in both environments is reduced to 0.5% and 5%, respectively.

Results when $M_{primodial} = 10^{11} M_\odot$ for both $m_L = 0.1 M_\odot$ and $m_L = 8 M_\odot$ cases are listed in Table 1.

Table1

$m_L = 0.1 M_\odot$			$m_L = 8 M_\odot$		
$\epsilon = 0.5$	$\epsilon = 0.05$	$r = 10$kpc	$\epsilon = 0.05$	$\epsilon = 0.005$	$r = 10$kpc
$r \geq 74$kpc	$r \geq 380$kpc	$\epsilon \geq 0.94$	$r \geq 127$kpc	$r \geq 400$kpc	$\epsilon \geq 0.6$

The third line shows the results when the star formation efficiency or the radius of the proto-galaxy is assumed to be just above value in the second line. Our results show that even if the star formation efficiency in the proto-galaxy is as high as star burst galaxies, the radius of the proto-galaxy should be as large as 100kpc. If the radius of the proto-galaxy is as small as 10kpc which is typical disk radius of the galaxy with luminous mass of $10^{11} M_\odot$, the star formation rate should be significantly higher than that in the nearby galaxies. This follows from the observational requirement of ejection of metal from the proto-galaxy into the ICM.

Star formation efficiency in the proto-galaxy is probably not larger than that in the nearby star burst galaxies (Hattori & Terasawa 1992). Therefore we conclude that the star formation efficiency required from the case $r = 10$kpc is unreasonably high and the star formation efficiency in the proto-galaxy is expected to be the same as or less than that of nearby star burst galaxies. Hence the extended star formation in the proto-galaxy is a much more plausible solution for the requirement of metal enrichment of ICM.

HALO NEUTRON STARS AND γ-RAY BURSTS

In the previous section we have shown that a lot of massive stars must be formed in the extended halo region of the proto-galaxy to provide the heavy elements for ICM by early galactic wind. This predicts that galaxies have an extended halo populated with large number of neutron stars. If proto-galactic star formation occurs isotropically, an isotropic distribution of the neutron star remnants in a galactic halo is expected. The angular distribution of γ-ray bursters observed by BATSE on the Gamma Ray Observatory (GRO) satellite, on the other hand, is isotropic within statistical limits.

In order to derive the required conditions for the distribution of neutron stars to be consistent with BATSE data, we assume a core-halo structure for the neutron star distribution.

$$\rho(R) = \rho(0) \left\{ 1 + \left(\frac{R}{R_c}\right)^2 \right\}^{-3\beta/2}, \qquad (4)$$

where R denotes the distance from the Galactic Center to the bursters and R_c is the core radius. The distribution of dark halo of our Galaxy has been found to fit well to the same form of above distribution with $\beta = 2/3$ and $R_c = 13$kpc (Innanen 1973).

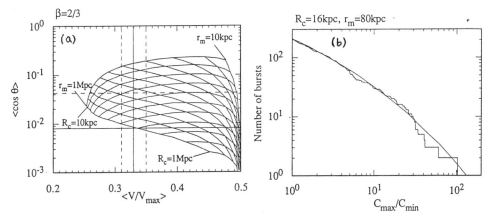

Fig.1 (a) Plot of $<\cos\theta>$ against $<V/V_{max}>$ for different sets of core radius R_c and maximum observable distance r_m. Straint lines and dashed lines represent the observed values and the 1 standard deviations respectively for $<\cos\theta>$ and $<V/V_{max}>$ obtained by BATSE. (b) Predicted integral number distribution of bursters (thick curve) plotted against the intensity of burseters under the assumption that each burster has the same luminosity and its distribution is described by the assumed distribution function (see the text). Integral number distrubution of 210 bursts observed by BATSE as a function of peak rate C_{max}/C_{min} (thin line) is superposed on the predicted curve which is scaled so as to fit to the observed.

In figure 1 (a), we plot $<\cos\theta>$ against $<V/V_{max}>$ for different values of R_c and r_m in the case of $\beta = 2/3$, where θ is the angle between the burst source and the Galactic Center, r_m is the observable maximum distance to the burst sources from the earth and the distance of the earth from the Galactic Center is taken to 8.5kpc. The current value of $<\cos\theta>$ is 0.008 ± 0.035 and that of $<V/V_{max}>$ is 0.33 ± 0.02 for 262 bursts (Meegan 1992a). We show an example of $\log N - \log S$ for $R_c = 16$kpc and $r_m = 80$kpc superposed with the $\log N - \log S$ for 210 BATSE burst events (Meegan 1992b) in figure 1 (b). This distribution explains observed $\log N - \log S$ well. The assumed neutron star distribution is consistent with BATSE results within 3 standard deviation if $r_m \geq 40$kpc and $2.5 \leq r_m/R_c \leq 25$.

DISCUSSION

The total mass of the neutron star remnants of the massive stars which enrich the ICM is too small to explain the total mass of galactic dark halo. The ratio of the total mass of the neutron star remnants to the total mass of iron ejected from the proto-galaxy is 20 (Tsujimoto & Nomoto 1990). Because the iron abundance in the ICM is around $0.3[Fe/H]_\odot$, the mass ratio of the neutron star remnants to the ICM is 0.01. Recent observational works have shown that the ratio of the stellar mass to the ICM mass is $1 \sim 0.1$ (David, Forman & Jones 1990; Tsuru et al. 1992). So the ratio of the mass of the neutron star remnants to the stellar mass is $0.01 \sim 0.1$. On the other hand, the ratio of the dark halo mass to the stellar mass in each galaxy is roughly 10. Therefore, the dark halo mass is more than 100 times heavier than the total mass of the remnant neutron star remnants.

Some suggestions on burster models are made. First of all, a full sky burst rate detected by BATSE is ~ 800 bursts per year (Meegan et al. 1992). This indicates that the burst recurrence time is of order 10^7 yr if the total mass of neutron stars in the extended halo of the Galaxy is constrained by the metalicity of the ICM as discussed above. The cyclotron lines observed in some burster spectra show that the magnetic field of burst sources are of order $\sim 10^{12}$ gauss and this may suggest that the magnetic field of isolated neutron stars essentially does not decay. Since neutron stars in the galactic halo are moving with velocities of order 100 km s^{-1} and gas density in the halo is extremely low, the accretion of halo gas onto neutron stars is too slow for thermonuclear flash models and crustquake model (Blaes et al. 1990; Harding & Leventhal 1992). There are several alternative models, such as, episodic accretion models, starquake models and rotation-induced models (see e.g. Harding 1991). We suggest that the predicted burst recurrence time is an important factor in distinguishing these models.

This work has been supported by the Special Researchers' Basic Science Program at RIKEN.

REFERENCES

Blaes, O. et al. 1990, *Astrophys.J.*, **363**, 612.
Canizares,C.R., Markert,T.H., & Donahue,M.E., 1988, in *Cooling Flows in Clusters and Galaxies*, p.63, ed. by Fabian,A.C., Kluwer, Dordrecht.
David,L.P., Forman,W., & Jones,C., 1990,*Astrophys.J.*,**359**,29.
Hattori,M., & Terasawa,N., 1992, submitted to *Astrophys.J.Letter*.
Innanen,A., 1973, *Astrophys. Space Sci.*,**22**, 393.
Larson,R.B., 1987, in *Starbursts and Galaxy Evolution*, p.467, ed. byThuan,T.X., et al., Editions Frontieres.
Meegan, C. A. 1992a, private communication.
Meegan, C. A. et al. 1992b, *Nature*, **355**, 143.
Shull,J.M., & Silk,J., 1979,*Astrophys.J.*,**234**,427.
Tsujimoto,T., & Nomoto,K., 1990, private communication.
Tsuru,T., 1992, Ph.D. University of Tokyo.

GAMMA-RAY BURSTS FROM COLLISIONS OF PRIMORDIAL SMALL-MASS BLACK HOLES WITH COMETS

K.F.Bickert and J.Greiner
Max-Planck-Institut für Extraterrestrische Physik, 8046 Garching, FRG

ABSTRACT

Recent results of BATSE reinforce the isotropic distribution of γ-ray bursts. Alternatively to cosmological models we propose collisions between small-mass primordial black holes and comets in the Oort cloud. Assuming typical Oort cloud densities and velocities for comets and primordial black holes, we can explain many of the observed properties of γ-ray bursts.

INTRODUCTION

After the confirmation of the isotropic distribution of γ-ray bursts (GRBs) by the BATSE experiment[1] the interpretation now tends to put the sources at cosmological distances[2]. This implies that the events have to be fairly rare ($\approx 10^{-8}$ per galaxy per year) and very energetic ($\approx 10^{51}$ erg per burst). However, there are two main difficulties of all the merger models proposed so far: 1) The optical depth is expected to be much too high for the escape of γ-ray photons directly from the interior[2], and 2) These merging processes cannot account for the variety in the GRB durations, ranging from sub-milliseconds[3] up to many tens of seconds.

Therefore, it is worthwhile to consider the other possible distance scale, namely that of the Oort cloud of comets at $10^2 - 10^5$ AU. We propose that small-mass primordial black holes traversing through the solar system will produce γ-ray bursts when they collide with comets in the Oort cloud.

COMETS AND THE OORT CLOUD

1. Comet distribution

Following the suggestion by Oort[4] that the planetary system is surrounded by a large spherical cloud of comets, there has been much progress in the understanding of the kinematics and the spatial distribution of the comets. Other perturbers besides distant passing stars have been recognized such as giant molecular clouds[5] and tidal effects of the gravitational field of the galactic disk[6]. Thus, the picture of a rather massive inner Oort cloud has been developed which replenishes those comets lost from the outer Oort cloud[7]. Most of the comets are created at the same time as the planetary system in its outer regions, interstellar encounter turbulences, and comet/comet near-collisions then redistribute the comet orbit planes fast to form a complete shell without a preferred plane. This results in a comet distribution which is isotropic beyond $\approx 10^4$ AU, and is moderately flattened toward the ecliptic plane within that distance. A best guess for the dimension, population and mass of the inner and outer Oort cloud[8], respectively, is summarized in Table 1. Note that most theories of planetary genesis allow much higher comet numbers (total mass up to 10 % of solar system).

Table 1: Parameters of the Oort cloud[8]

	Kuiper belt	Inner Oort Cloud	Outer Oort Cloud
R_{min}–R_{out} [AU]	>80	$150 - 2 \times 10^4$	$2 \times 10^4 - 2 \times 10^5$
No. of comets	$10^9 - 10^{11}$	6×10^{12}	1×10^{12}
total mass* [M_E]	0.01 – 1	42	7

* A mean nucleus mass of 3.8×10^{16} g and a bulk density of 0.6 g cm^{-3} are assumed.

Estimates[8,9] of the space density of interstellar comets result in $5 \times 10^{12} - 10^{13}$ comets pc^{-3}. It is interesting to emphasize that with the exception of the Kuiper belt there is no steep gradient in the space density of comets as one moves out of the solar system, neither between the inner and outer Oort clouds, nor between the outer Oort and the interstellar comets. That is, the density changes from $\approx 10^{15}$ comets pc^{-3} at the inner boundary down to $\approx 10^{13}$ comets pc^{-3} at the outer boundary, where it finally converges to the interstellar density value given above.

2. Cometary composition

An Oort cloud comet is assumed to have a rocky ($2 - 5$ g/cm^3) or icy ($.2 - 1$ g/cm^3 depending on gases trapped) core, with some loosely bound slush (eroded by dust grains and micrometeorites, and refrozen after heating by passing novae or stars) and a halo of evaporated gases marginally bound. The snow is mostly hydrogen, as observed in the early tails of fresh comets (nearly hyperbolic orbits), which evaporate fast in the early phase of the first Sun-heading orbit. Gravitational interaction and comet/comet collisions will lead to clustering of comets to some extent (they are sticky).

SMALL BLACK HOLES AND THE COLLISION PROCESS

1. Primordial Black Holes (PBHs)

From the large-amplitude density and metric perturbations during 10^{-35} to 10^{-27} sec after the Big Bang, PBHs with masses ranging from Planck masses (10^{-5} g) up to observational limits (about 10^{48} g) may have formed[10].

The Hawking quantum radiation effect[11,12] assigns each black hole a temperature $T = \hbar c^3/16\pi^2 GkM \approx 10^{-6}$ [K]/M[$M_\odot$], which is inversely proportional to its surface gravity, giving the black hole a radiative evaporation lifetime of 9×10^{15} [sec] $(M/[10^{14}g])^3$, so that all PBHs with initial masses below 2.5×10^{14} g will have completely decayed in the 10^{17} sec since the Big Bang[11,13], and all PBHs with their present mass $< 10^{14}$ g will decay soon.

For the mass range $10^{16} - 10^{48}$ g, an upper limit for the collapsed mass fraction has been derived[14] of the form $\beta(M) \leq 10^{-25} M^{1/2}$. Collisional and accretion breaking will lead to some concentration of PBHs into the galactic halo and disk, preferentially accumulating low-mass PBHs. Galactic dynamics limit the total mass amount of BH matter in the Galaxy to 50% of the dark matter in the disk (Blair[10] calculates a maximum density in the vicinity of the Sun of 10^{-23} g/cm^3 = 10^{31} g/pc^3 or $\approx 10^{16}$ PBHs/pc^3 in the mass range 10^{14}–10^{18} g).

Accretion braking and gravitational scattering in frequent near-encounters with comets (very effective for low-mass BHs) will capture a large fraction and concentrate them in the cloud[15]. Scattering of PBHs into the inner solar system will be as infrequent as for comets. Concentration factors of up to 10^8 are still compatible with all known limits[12,13,15].

2. Collision Energetics

Small-mass BHs have microscopic dimensions with Schwarzschild radii of $r_g = 2GM/c^2 \approx 1.5 \times 10^{-14}$–$1.5 \times 10^{-10}$ cm. For masses below 3×10^{17} g (1.6×10^{14} g), the Compton wavelength of the electron (proton) is larger than the gravitational radius r_g, reducing the accretion rate further and rapidly charging the PBH.

When a black hole collides with a comet, these effects lead to the rapid buildup of a tightly bound fireball, accreted normally but fed into the PBH at a reduced rate. The collision/accretion process resembles a microscopic supernova implosion. During compression in the PBH vicinity (some 10 to some 100 r_g) electron/ion decoupling heats the matter to temperatures as high as 300 MeV, with mean temperatures of about 20 MeV (in the case of stationary accretion[16]). The enlarging fireballs' outer skirts get cooled by freshly accreted material, and part of the radiation is absorbed by the surrounding comet, which gets completely vaporized and ionized (molecular binding energy some 10^{19} erg $\ll$ GRB energy). Comet matter density variations result in one order of magnitude burst energy variation. Another few orders of magnitude brightness variation result from varying impact velocities. In the collision, the PBH first encounters the comet's gas atmosphere, and accretes gas at free-fall rates. The capture rate can be reduced by an order of magnitude due to thermal expansion of the former comet, depending on impact duration, fireball temperature, γ-ray opacity and chemical constitution of the comet. Electrostatic binding to the charged hole and random magnetic fields from the turbulent accretion current in the ball make it fairly stiff, so the comet material will not compress the ball by mechanical pressure, but will be accelerated to PBH velocity through the shock front. Typical PBH-comet relative velocities would be 0.1 – 1 km/sec (PBHs captured in the cometary cloud) or 1 – 100 km/sec (interstellar PBHs or comets), intercomet relative velocities are typically 1.6 km/sec in the Oort cloud.

Direct capture infall of matter onto the PBH is possible from a radius of $4r_g c/v \approx 2 \times 10^{-3}$ – 2×10^1 cm/v[cm/s] for PBHs of mass 10^{14}–10^{18} g. The effective radius of the accretion ball is difficult to determine. Using typical values for accretion disks, we estimate radii of 10^1–10^3 cm, and corresponding accretion rates of 10^3–2×10^{10} g/sec (0.1 – 5 g/cm^3). At a release of 0.057 mc^2 (nonrotating Schwarzschild BH) or up to .42 mc^2 of energy (maximally rotating Kerr BH) per accreted particle, the observed intensity of GRB pulses of 10^{29} erg at 1000 AU distance or 10^{33} erg at 10^5 AU requires accretion of 10^8 rsp. 10^{12} g (typical comet masses are 3.8×10^{16} g).

3. Timing

The accretion ball lifetime (few μsec – few sec) is often much shorter than the passage through the comet (typical sizes of a few m up to some 10^4 km), and the diversity of sizes and internal structures of comets allows us to explain the diversity of GRB time profiles:

Rock/ice comets (few m – few km) have short rise times and total collision times (0.1 msec – 1 sec); they might break into chunks orbiting the PBH and producing quasi-periodic signals on timescales of msec – sec.

Large comets (few km – few 1000 km) with lots of hydrogen and gas have slow rise times (10 msec - sec), and a wide range of burst times (0.1 sec – 100 sec); they may show a strong continuous signal (with lower temperature), and eventually decay in luminosity (due to thermal expansion of the evaporating comet or simply changing density). Intermittent bursts due to rocks in the comet or spectral changes due to varying density/accretion rates are possible.

PBHs encountering cometary clusters will lead to rapid burst sequences with little or no continuous signal in between.

4. Frequency of Collisions

The number of events[17] (ca. 800/year) should be the product of comet number density, PBH number density, detection volume, comet size, PBH capture cross section, and relative velocity. Using the ranges above and assuming a galactic-to-cloud-concentration factor of 10^7, our model can account for a significant fraction (0.01 – 1) of the observed GRBs.

CONFRONTATION WITH BATSE OBSERVATIONS

The BATSE burst peak rate distribution (or alternatively logN(S)-logS) does not show any 'inner' region of the GRB distribution for which the slope could be approximated by a -3/2 power law. Instead, the distribution is smoothly curved from the bright end downwards. This means that the most serious argument raised in the past against the Oort cloud as a possible GRB source distribution, namely that it has no uniform density in its inner part due to the gravitational potential of the Sun[18], is invalidated. If we assume that the PBHs are distributed more or less isotropically, then the distribution of the GRB sources should reflect that of the comets in the Oort cloud. As was shown above the density of comets varies as $R^{-1.5}$ which is surprisingly similar to what is observed for the burst peak rate distribution.

The diversity of time profiles already discussed matches a large fraction of observed GRB pulse forms[19]:

Less than 10 % have sharp rise times (< 0.5 sec) with exponential brightness decay (forcing active accretion events, not compatible with thermal settling from star/star collisions). About one fifth of these also show spectral evolution from hard to soft, as slow accretion of the hot core of the accretion ball would do.

One third of the slower pulses have multiple-pulse hard (>300 keV) peaks with little to no continuum in between, denoting structured objects (NSs disrupted by BH tidal effects would make fairly continuous signals).

More than half of the slow events show soft multi-pulse events with high background in between, as we would expect for typical large, structured comets.

CONCLUSIONS AND PREDICTIONS

If the proposed process is responsible for the production of GRBs (or most of them) then this suggests that 1) Primordial Black Holes exist, 2) the supposed structure of the Oort clouds is correct, and 3) the masses and densities for PBHs and comets, which were deduced from dynamical or theoretical grounds, are in the right order of magnitude.

We predict a spectral break at maximum temperature of a few 100 MeV. Details of the (non-stationary, non-equilibrium) accretion event to model the PBH spectral emission evolution and the absorption by the evaporating comet must still be worked out.

Detection of circumstellar material around nearby stars[20] suggests that Oort clouds exist around other stars. With improving statistics, concentrations of the brightest type of events (at this distance, multi-pulse without background) to the nearest of these stars (α Lyr, α PsA, ϵ Eri, β Pic) should show up.

A slight concentration of GRBs towards the ecliptic plane would confirm the existence of a densely populated Kuiper belt at the inside of the Oort cloud.

REFERENCES

1. C.A.Meegan et al. , Nature 355, 143 (1992)
2. B.Paczynski, Acta Astron. 41, 257 (1991)
3. P.N.Bhat et al. , Nature 359, 217 (1992)
4. J.H.Oort, Bull. Astron. Inst. Netherlands 11, 91 (1950)
5. L.Biermann, in Astronomical papers dedicated to Bengt Stromgren, eds. A.Reiz et al. , 327 (1978); S.V.M.Clube & W.M.Napier, Quart. J. R. Astron. Soc. 23, 45 (1982)
6. J.Heisler & S.Tremaine, Icarus 65, 13 (1986)
7. J.G.Hills, Astron. J. 86, 1730 (1981);   P.R.Weissman, in The Galaxy and the Solar System, eds. R. Smoluchowski, J.N.Bahcall & M.S. Matthews, 204 (1986)
8. P.R.Weissman, Nature 344, 825 (1990)
9. Z.Sekanina, Icarus 27, 123 (1976)
10. Ya.B.Zeldovich & I.D.Novikov, Astron. Zh. 43, 758 (1966);   S.Hawking, MNRAS 152, 75 (1971);   D.G.Blair et al. , Nat 251, 204 (1974)
11. S.Hawking, Nature 248, 30 (1974)
12. Carter, Phys.Rev.Lett. 33, 558 (1974)
13. N.A.Porter & T.C.Weekes, ApJ 212, 224 (1977);   M.J.Rees, Nature 266, 333 (1977)
14. I.D.Novikov et al. , A&A 80, 104 (1979)
15. D.N.Page & S.W.Hawking, ApJ 206, 1 (1976)
16. G.H.Dahlbacka, G.F.Chapline & T.A.Weaver, Nature 250, 36 (1974)
17. C.A.Meegan et al. , IAUC 5478 (1992b)
18. B.Paczynski, Science (in press), (1992)
19. W.Paciesas et al. , AIP Conf. Proc. 265, 190 (1992);   C.Kouveliotou et al. AIP Conf. Proc. 265, 299 (1992)
20. H.H.Aumann, IAU Symp. 112, 43 (1985)

MODELS OF GRAVITATIONALLY ACCELERATED MATTER IN RELATIVISTIC MOTION AS GAMMA-RAY BURST SOURCES

James S. Graber
407 Seward Square S.E., Washington, DC 20003

ABSTRACT

This poster paper summary presents a general model for gamma-ray burst sources. Several specific variations of this model are listed, and one preferred model is presented in more detail. In the general model, the source is a "blob" of matter and radiation which has been accelerated to relativistic velocity by the gravitational attraction of a highly compact object -- a neutron star, a black hole, or -- in alternate theories of gravity -- a red hole. The observed gamma-ray burst is emitted when the blob is in motion aimed almost directly at the earth. Specific models considered include various configurations of the blob and the compact object. Models using neutron stars are difficult to construct due to their small masses and to the less-relativistic orbits around them. Although it is possible to devise acceptable black-hole models, such as matter falling into a black hole on a grazing orbit, red-hole models that reproduce the observed gamma-ray bursts are easier to construct. The preferred model is a star plunging into a 10^5 to 10^7 solar-mass red hole.

THE GENERAL MODEL

The general model consists of a blob of radiating matter falling into, orbiting around, or being flung out of a compact object. The blob is assumed to begin as some ordinary form of matter -- a comet, a planet, or a star -- that is accelerated to relativistic velocity by the gravitational attraction of the compact object. The compact object is a neutron star, a black hole or a red hole. (Similar to black holes, red holes are defined in more detail below.) By the time the blob has fallen substantially into the compact object's gravitational potential well and reached relativistic velocity, it will have been tidally distorted, probably disrupted, perhaps partially dispersed along an orbit, and probably compressed and possibly also decompressed as it passes into and maybe out of the potential well. Also, the blob will probably have passed forcibly through substantial magnetic fields which will convert at least part of the blob into a shock-heated plasma. Hence, many forms of radiation may occur, including thermal, curvature, cyclotron, synchrotron, collision, inverse Compton scattering and bremsstrahlung.

Since the blob is in relativistic motion, the radiation will be beamed into a small forward cone. The cone may be solid, hollow or complex in form, depending on the dominant types of radiation emitted and the emitting zone configuration. The radiation in this forward cone will be highly blue-shifted. If the motion is curved, e.g. orbital motion, the "search-light effect" can cause the observer to see only a brief interval of a longer-duration phenomenon.

How can radiation escape from the matter that generates it without being absorbed, scattered or thermalized? This is one of the key problems of understanding gamma-ray bursts.[1] In the relativistic motion model, this problem is partially solved by the fact that in the blob frame, where the radiation is emitted, it is very soft and easily separates from the matter. The matter is then further separated from the radiation by the action of gravity, and the radiation escapes without being trapped or degraded. In the observer's frame, the radiation is blue-shifted up to the gamma-ray region.

ADVANTAGES OF THE GENERAL MODEL

An important advantage of the relativistic motion model is that the blue-shifted radiation helps to move most of the energy up toward the MeV range. If a spread of velocities or angles is present, this model can also explain the broad peaks in the observed gamma-ray burst spectra. As explained above, it helps solve the problem of separating the matter and the radiation. It also helps to explain the lack of low-energy radiation, the relative dimness of the non-bursting system and, in cases where the blob plunges into a red or black hole, the lack of burst repetition. By positing corresponding irregularities in the blobs of moving source matter, this model explains the irregularities in the gamma-ray bursts. The general model easily accommodates long and short durations, intense spikes, double spikes, and even periodicities. The remaining task is to find the specific model -- or set of models -- which best reproduces the observations.[2,3]

POSSIBLE SPECIFIC MODELS

The following possible models were considered:
o   matter orbiting a red or black hole in a near-circular orbit;
o   matter orbiting a neutron star;
o   matter falling into a red or black hole in a grazing orbit;
o   matter falling into a red or black hole in a plunging orbit;
o   matter falling through a red hole and emerging;
o   matter emerging from near a red or black hole in a collimated jet;
o   matter emerging from a red hole and colliding with infalling matter;
o   matter falling into a red hole and colliding with matter and radiation already trapped in the red hole; and
o   a neutron star spun up until it experiences a fly-wheel explosion.

The variety of gamma-ray bursts suggests that more than one source model may be correct. Several of the above possibilities could occur. For instance, orbital models could explain bursts with definite periodicities. Analysis of all of the above possibilities has been performed and will be reported elsewhere.[4]

Based primarily on the ability to reproduce the observations and on the likelihood of the actual occurrence of the various possible configurations, matter falling into a black hole on a grazing orbit has emerged as the best black-hole model, and matter falling into a red hole on a plunging orbit has emerged as the best red-hole model. Carter[5] has recently published an analysis of a similar black-hole grazing orbit model. Comparative analysis indicates that matter plunging into a red hole is the best model for separating the radiation from the matter without unwanted consequences. Therefore, this model is considered in more detail here. The corresponding black-hole model is also briefly considered, to illustrate that it is generally more difficult to construct a satisfactory black-hole model.

DEFINITION OF RED HOLES

A red hole is the counterpart of a black hole in alternate theories of gravity in which no "Schwarzschild singularity" exists. It is more compact than a neutron star, more compact than the photon orbit, but no event horizon forms; matter and radiation can reemerge after falling into a red hole. Red holes occur in Yilmaz's theory of gravitation[6] and have been considered by Yilmaz[7] and other authors[8]. In order to construct astrophysical models, a rotating red hole is needed. Rotating red-hole models are discussed in a previous paper[9], where the necessary metric is presented. For this paper, it is sufficient to think of a rotating red hole as a Kerr-metric rotating black hole, whose central gravitational well has been truncated somewhat above the event horizon and whose infinite-redshift central hole has been replaced with a relatively flat potential at a large, but finite, redshift. The entire space inside the photon orbit of a red hole is very similar to the space between the photon orbit and the event horizon in a black hole. The most important physical effect is that matter and radiation can emerge from a red hole.

MATTER PLUNGING INTO A RED OR BLACK HOLE

As matter plunges into the potential energy well of a red or black hole, more binding energy is available for release as harder radiation. The transverse orbital energy of the infalling matter is also available for conversion to radiation. Matter with only modest amounts of angular momentum must reach a highly relativistic transverse orbital velocity as it falls into a red or black hole. The infalling matter reaches its most compressed stage just as it crosses the photon orbit and is most likely to emit radiation at this point. Thereafter it is expanding rapidly rather than being compressed. This very rapid expansion causes the matter to become optically thin and allows the radiation to escape without being absorbed or scattering and thermalizing to a lower temperature.

At this point, black-hole models pose a serious problem. Any highly blue-shifted, forward-beamed radiation that is emitted as the plunging blob nears or crosses the photon orbit is headed for the event horizon and will disappear directly into the black hole, never to be seen again. Hence, a successful black-hole model requires that a blob approach the hole at a relatively shallow angle and emit most of the observed gamma-ray burst before it crosses the photon orbit. It also suggests that an abrupt cutoff of the burst would be seen, contrary to what is observed.

In a red-hole model this problem does not arise. There is no event horizon, and while the radiation may plunge down a very deep potential energy well, it will eventually emerge again and escape. Furthermore the red-hole model can take full advantage of the rapidly increasing transparency of the matter falling inside the photon orbit to help solve the problem of separating the radiation from the matter. Therefore, the red-hole model is preferred to the black-hole model.

PREFERRED MODEL

Based on the analysis above, the preferred model is matter falling into a red hole and emitting the observed gamma-ray burst just as it passes through the photon orbit. To make this model fully specific, it is necessary to determine the size of the red hole and the size and source of the blob. The analysis that follows indicates a preferred mass of 10^5 to 10^7 solar masses for the red hole. A typical stellar mass for the blob, though less strongly indicated, is consistent with the model and other known facts. In particular, stars are known to exist in the right numbers to be consistent with the observed rate of gamma-ray bursts. Therefore the final preferred model consists of a typical star plunging into an approximately 10^6 solar-mass red hole at the center of a galaxy.

Consider the size of the blob: The blob could be anything from a comet or asteroid up to a typical star. However, stars are common and are already thought to be frequently consumed by the massive object at the core of a galaxy. The time scales of gamma-ray bursts are consistent with light travel times across typical stars or their cores. This suggests stars as the preferred blob sources.

Consider the size of the compact object: If gamma-ray bursts are really at cosmological distances, the black or red hole might very well be "galaxy-sized," i.e., 10^4 to 10^9 solar masses. The orbital time-scales of red holes of 10^5 to 10^7 solar masses are consistent with the characteristic time scales of the gamma-ray burst phenomena, i.e., tenths of a second to a few tens of seconds. This size of red hole is big enough to rapidly disrupt and swallow a typical star and yet not so big that it can swallow a star without disruption. Furthermore, indications are that many galaxies have a central collapsed object of at least this size. If these collapsed objects have grown to this size by swallowing typical stars, then the event rate is consistent with the observed rate of gamma-ray bursts, even assuming a beaming factor effect so strong that we only see about one in every 10^6 star-swallowings that occur anywhere in the universe.

Consider the fate of a typical star falling into a 10^5 to 10^7 solar-mass red hole: The star would be tidally disrupted as it fell into the hole. The long duration, 5- to 100-second, part of the gamma-ray burst would be caused by the extended disrupted envelope of the star. The short, .2- to 1-second, peaks would be caused by dense clumps, i.e. the remainder of the core of the star.

Consider the statistics: If the radiation is highly beamed, but sweeps out a substantial orbit around the red hole, some fraction of the sky will be covered. The true rate of events, which equals the observed rate of events, i.e., several per day, divided by the fraction of the sky typically covered by the beamed radiation, would be far more than several per day. If every galaxy has a large red hole (instead of a black hole) at its core, and this red hole grows to 10^8 or 10^9 solar masses, then the number of ordinary stars falling into red holes in the universe would be more than sufficient to provide the source of the observable gamma-ray bursts, as is evidenced by the following calculation.

The number of galaxies in the observable universe is of order 10^{10}. The number of solar masses absorbed equals 10^8 times the number of galaxies in the universe, and hence is approximately 10^{18}. The number of days in a Hubble time is approximately 10^{12}. The number of stars falling into a red hole every day is thus approximately equal to $10^{18}/10^{12}$ or 10^6. So a beaming factor of approximately $1/10^5$ or $1/10^6$ can be supported. This is consistent with the model for an ordinary star plunging into a red hole.

CONCLUSION

The general model of blobs in relativistic motion toward us has been shown to solve many of the problems of explaining the source(s) of the observed gamma-ray bursts. In particular it contributes substantially to the solution of the problems of the broad, very high energy spectrum, the absence of low-energy radiation and the very short, irregular and spiky timescales. The specific model of a typical star falling into a 10^5 to 10^7 solar-mass red hole has been shown to fit the observed time and statistical parameters. Finally it has been found that, like most other explosive or energy-emitting events, gamma-ray bursts are much easier to explain with red holes, which accelerate matter and radiation and re-emit it, rather than with black holes, which tend to swallow matter, energy and radiation. The red-hole model has been shown to contribute substantially to the solution of the problem of separating the radiation from the matter. For this reason, the red-hole model is preferred over the black-hole model.

REFERENCES

1. B. Paczynski, Acta Astron. 41, 257 (1991).
2. E. P. Liang and V. Petrosian, Gamma-Ray Bursts, AIP (1986).
3. C. A. Meegan, et al, Nature 355, 143 (1992).
4. J. S. Graber, in preparation.
5. B. Carter, Ap. J. 394, L33 (1992).
6. H. Yilmaz, Phys. Rev. Lett. 27, 1399 (1971).
7. H. Yilmaz, "New Theory of Gravity" in Proceedings of the Fourth Marcel Grossmann Meeting on General Relativity, p. 1793 (1986).
8. B. O. J. Tupper, Il Nuovo Cimento 19B, 135 (1974).
9. J. S. Graber, "Red Holes, not Black Holes, at AGN Centers," in Testing the AGN Paradigm, S. S. Holt et al, eds., p. 113 (1992).

RE-IGNITION OF DEAD PULSARS: A POSSIBLE SOURCE OF GAMMA-RAY BURSTS

K.Y. Ding and K.S. Cheng
Department of Physics, University of Hong Kong, Hong Kong

ABSTRACT

We adopt the scenario of the evolution of radio pulsars proposed by Ruderman and his coworkers, in which the model pulsars evolve in different phases radiate a characteristic γ-ray spectrum. The neutron star will spin down to some extent that the outer magnetosphere $e^{\pm}$ pair production and current flow are quenched, the star's magnetic dipole radiation torque will align its dipole with the spinning axis. Such aligned dead pulsars can be reignited by a transient thermal x-ray emission from the stellar surface. The resulted γ-ray spectra are used to fit several observed hard γ-ray burst spectra and provide a consistent fit. The amount of $e^{\pm}$ pairs produced during the process is calculated and is large enough to cause the annihilation line features of some of the bursts. It is concluded that re-ignition of nearby dead pulsars can be one of the possible sources of producing γ-ray bursts.

I. INTRODUCTION

It was found that the nature of most of the γ-ray emission sources are not yet well known. Ruderman and his coworkers[24] have found that the observed γ-ray spectra of the Crab pulsar, Vela pulsar, unidentified COS-B sources, Geminga, and γ-ray burst sources are similar. They argued that the similarity is a consequence of the fact that they are all closely related members of the same evolving family of neutron stars where origin of γ-ray emission are described by the outer-magnetospheric gap models.[7,8] Recent studies on the pulsations[11,3,5] of Geminga confirmed that the it is a nearby, isolated pulsar and older than the Vela pulsar.

Cheng and his coworkers[7,8] show that the global current flow in the outer magnetosphere of a neutron star can result in charge depletion regions in the outermagnetosphere (outergaps). Within gaps, a strong electric field can be developed along B (i.e. $\vec{E} \cdot \hat{B} \neq 0$) which extends with nearly constant strength from the beginning of the gap (near $\hat{\Omega} \cdot \hat{B} = 0$) out through the light cylinder. However, it was argued[24] that as the neutron star spins down, the photon-photon pairs production process became difficult because the efficiency of low energy photon production, which plays a crucial role in pair production, is decreasing rapidly. This current flow between the polar cap and the light cylinder will be terminated, any further spin down of the star is accomplished through the dipole radiation torque. This torque will also act to align its dipole axis with the spin axis in a time scale similar to the spin down time scale.[10,20] When the star is aligned, this spin down torque will be vanish, and the range of the final periods for an aligning post-Vela γ-ray pulsars are $\sim 0.1 - 0.2$ s ($2 - 3$ ms) for $B \sim 10^{12}(10^8)$ Gauss.

However, it can be re-ignited if some external low energy photons, i.e., thermal x-rays emitted on the stellar surface, is available. Thus, re-ignited neutron star would be a latent candidate of γ-ray emission source.

We do not attempt to provide a universal model to explain the origin of all the γ-ray bursts, however, we would like to show that the re-ignition of dead pulsars can provide a mechanism for the galactic neutron stars to produce spectra similar to those observed in γ-ray bursts.

II. MECHANISM OF THE GENERATION OF GAMMA-RAY IN DEAD PULSARS

An aligned neutron star[24] is a latent pulsar, it can be re-ignited only when an appropriate "match" is applied, it can emit γ-rays like the Vela pulsar for as long as the match remains lit. The match can be any transient production of thermal x-ray on the stellar surface.

According to the model proposed by Cheng and his coworkers,[7,8] the particles are accelerated in an outer gap which extends from the polar cap to near the light cylinder along $\vec{B}$ (where $\vec{E} \cdot \vec{B} \neq 0$). The potential drop across such a gap is

$$\Delta V_{max} = 6.58 \times 10^{12} P^{-2} \mu_{30} \text{ volt} \qquad (1)$$

where P is the period of the star and μ_{30} is the magnetic moment in units of 10^{30} Gcm^3. This potential drop will accelerate the primary $e^{\pm}$ pairs within the gap to relativistic energy.

These primary $e^{\pm}$ lose their energy by curvature radiation mechanism. Part of these primary photons will be attenuated by the transient x-ray emitted from the stellar surface to produce (secondary) $e^{\pm}$ pairs. These secondary $e^{\pm}$ pairs will produce the secondary photon spectrum through the synchrotron emission process. Two crossing fan-beam of the secondary photons (1st generation of synchrotron photons) may be energetic enough to produce tertiary $e^{\pm}$ pairs which can emit the tertiary photons by synchrotron emission. It has been calculated that two or three generations will produce the steady γ-ray spectrum giving the total spectrum expressed as

$$F_{tot}^{(i+1)}(E_\gamma, t) = F_{syn}^{(i)}(E_\gamma, t) + \sum_{j=1}^{i} F_{sur}^{j}(E_\gamma, t), \qquad i \geq 1 \qquad (2)$$

where $F_{syn}^{(i)}$ is the ith generation of synchrotron radiation spectrum and $F_{sur}^{(j)}$ is the jth generation of the survived spectrum (where the 1st total output γ-ray spectrum is resulted from curvature radiation and the 2nd total output spectrum is sum over the 1st generation of synchrotron spectrum and the survived primary spectrum and so on).

In our model, the spectra calculated may have energy break as high as $\sim$ GeV, however, We have calculated that the amount of photons produced in this energy range is still difficult to be observed by the current high energy detector (say, EGRET).[6]

III. APPLICATION TO PRODUCE THE SPECTRA OF GAMMA-RAY BURSTS

We have applied our model spectra to fit some of the hard γ-ray burst spectra. From the expressions of the photon spectra, we see that μ, T, and P are our variable parameters. We have fixed T at 0.1 keV in fitting the spectra, then a number of pairs of (μ, P) can give a good fit. We arbitrarily choose $\mu = 10^{27}$ G cm^3 and P is obtained from the fit. Fig. 1(a) - (d) show the spectral fit of some of the observed hard γ-ray bursts in which the fitted P is indicated. In fact, we have carried out the fitting on a few tens of the available γ-ray bursts spectra and can provide a consistent fit.

Besides, this emission model[7,8] also provides an abundant supply of secondary $e^{\pm}$ by the primary γ-ray and the thermal x-ray. Part of these secondary pairs will leave the light cylinder along the open field lines, however, half of these pairs are moving toward the surface of the neutron star. Due to the large energy loss by the synchrotron radiation process, most of these particles when reaching the polar cap will have an energy of order of a few MeV. These secondary $e^{\pm}$ pairs can be cyclotron scattered to produce the cyclotron absorption lines. Besides, the $e^{\pm}$ pairs annihilate at the stellar surface near the polar cap can produce the emission features at about 400 keV.[24,2] We have calculated that for $\mu \sim 10^{27}$ G cm^3, and $P \sim 1$ ms, the rate of secondary $e^{\pm}$ pairs produced is calculated to be 3×10^{40} s^{-1}. This production rate is large enough to produce an annihilation lines of width up to 100 keV. We can estimate the optical depth for this annihilation process of $e^{\pm}$ in the polar cap[6] to ~ 1. As P becomes larger, the rate will decrease, thus, the annihilation line produced may not be strong enough to be observed to account for the fact that only $\sim 15-20\%$ of the bursts with an energy bump in 400 - 500 keV.

IV. DISCUSSION

In fact, this is not unconvincing by choosing the typical magnetic moment of the millisecond pulsars for fitting. Recent study on the time profile of GRB910711, with apparently the shortest duration ($\sim$ 8ms) yet seen by the BATSE, shows significantly a 200 μs structure.[4] This indicates that the emission region of the burst may be about 60 km from the star which is typically the radius of the light cylinder of a ms pulsar. This provides a strong support to our model in which γ-ray is emitted in the outergap which have a typical dimension of the light cylinder of the star.

In view of all the statistical study done on bursts observed by different instruments[16,17,12,1,23,19] and their spectral features, we suggest that there may exist at least two origins of these γ-ray bursts. One is the galactic neutron star origin which will show the spectral features of the observed spectra and the distribution is anisotropic (in fact, those γ-ray bursts with spectral features did show a galactic distribution[13]). The other is the cosmological origin[14,9,25] which are very energetic and are of isotropic distribution. Although our model does not provide a universal formation mechanism to all the γ-ray bursts, however, it does provide a satisfactory and reasonable mechanism to give the continuum and line features observed in the γ-ray bursts spectra.

FIG. 1 The fitting of γ-ray burst spectra using the model spectra. (a) GRB800419,[21] (b) GRB810301,[15] (c) GRB811231,[22] (d) GRB821104.[15]

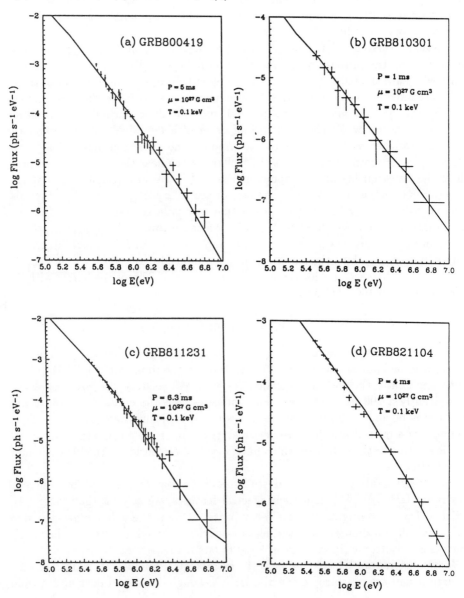

REFERENCES

[1] Atteia, J.L. *et al.*, Nature, **351**, 296 (1991).
[2] Bednarek, W., Cremonesi, O., and Treves, A., Ap.J., **390**, 489 (1992).
[3] Bertsch, D.L. *et al.*, Nature, **357**, 306 (1992).
[4] Bhat, P.N. *et al.*, Nature, **359**, 217 (1992).
[5] Bignami, G.F., and Caraveo, P.A., Nature, **357**, 287 (1992).
[6] Cheng, K.S., Ding, K.Y., submitted (1992).
[7] Cheng, K.S., Ho, C., and Ruderman, M.A., Ap.J., **300**, 500 (1986).
[8] Cheng, K.S., Ho, C., and Ruderman, M.A., Ap.J., **300**, 522 (1986).
[9] Dar, A., Kozlovsky, B.Z., Nussinov, S., and Ramaty, R., Ap.J., **388**, 164 (1992).
[10] Davis, L., and Goldstein, M., Ap.J., **159**, L81 (1970).
[11] Halpern, J.P., and Holt, S.S., Nature, **357**, 222 (1992).
[12] Hartmann, D. *et al.*, in Gamma-Ray Bursts, (ed. Ho. C., Fenimore, E.E., and Epstein, R.I.), (Cambridge University Press, 1990).
[13] de Jager, O.C., private communication (1992).
[14] Mao, S., and Paczynski, B., Ap.J., **388**, L45 (1992).
[15] Matz, S., Ph.D. thesis (1986).
[16] Matz, S. *et al.*, in Gamma-Ray Bursts, (ed. Ho, C., Fenimore, E.E., and Epstein, R.I.), (Cambridge University Press, 1990).
[17] Mazets, E. *et al.*, Ap. Sp. Sci., **80**, 1 (1982).
[18] Mazets, E., in Proc. 19th Inter. Cosmic-Ray Conf. (La Jolla), (NASA CP-2376), **9**, 424 (1986).
[19] Meegan, C.A. *et al.*, Nature, **355**, 143 (1992).
[20] Michel, F.C., and Goldwire, H., Ap.J., **5**, L21 (1970).
[21] Nolan, P.L. *et al.*, in High Energy Transient in Astrophysics, (ed. Woosley, S.E.), AIP, 399 (1983).
[22] Nolan, P.L. *et al.*, in Positron-Electron Pairs in Astrophysics, (ed. Burns, M.L., Harding, A.K., and Ramaty, R.), AIP, 59 (1983).
[23] Ogasaka, Y. *et al.*, Ap.J., submitted (1992)(refer to by Lingenfelter and Higdon 1992).
[24] Ruderman, M.A., and Cheng, K.S., Ap.J., **335**, 694 (1988).
[25] Usov, V.V., Nature, **357**, 472 (1992).

HALO BEAMING MODELS FOR GAMMA RAY BURSTS

R.C. Duncan*, H. Li[†] & C. Thompson[‡]

ABSTRACT

Neutron stars born with rotation periods $P \sim 1 - 3$ ms plausibly acquire dipole magnetic fields and recoil velocities V_r much larger than those of ordinary pulsars.[1] These stars remain hot and magnetically active long after they have spun down below the pulsar death line. Here we present Monte Carlo simulations of the angular distribution of gamma ray bursts (GRBs) which, we conjecture, are triggered by the decay of the stellar magnetic field. Simple physical arguments indicate that the GRBs are preferentially beamed parallel or anti-parallel to $\vec{V_r}$. This implies that only a small fraction of bursts are detectable from stars within a distance ~ 30 kpc, but that the probability of detection increases with distance. The resulting flux-density relation of GRBs closely matches the combined PVO-BATSE data for $V_r = 1000$ km s^{-1}, angular radius $\phi = 20°$ of the beaming cone, and constant burst rate and luminosity. The net angular distribution of GRBs lies within one standard deviation of the latest available BATSE results. A distinctive prediction of this model is that the brightest bursts are concentrated toward the Galactic center AND anti-center, so that $\langle \cos^2\theta \rangle > \frac{1}{3}$ for this subset of bursts while $\langle \cos\theta \rangle$ remains small. Our results are applicable to any halo model in which the GRBs are beamed in the prescribed manner, and the intrinsic bursting rate is constant.

Introduction

The GRB models that we now consider are motivated by the theory of dynamo action in nascent neutron stars,[1,2] although they could have more general applicability. The free energy in differential rotation in such a star is sufficient to generate a toroidal field as strong as $10^{17}(P/1 \text{ ms})^{-1}$ G,[1] where P is the (mean) rotation period. The star also experiences entropy-driven convection[3] with overturn time[2] $\tau_{con} \sim 1$ ms as it radiates neutrinos and cools. This phase of convection begins as soon as energy and lepton-number transport have settled to a quasi-steady state in the outermost layers of the star, and lasts as long as ~ 30 s after core bounce. The efficiency of an $\alpha - \Omega$ dynamo depends on the Rossby number $Ro \equiv P/\tau_{con}$. Neutron stars born with P of order 1 ms (i.e., with angular momentum comparable to the critical value J_{crit} above which a hot neutron star suffers a secular instability driven by gravitational radiation[4]) have Ro of order unity. Very intense magnetic dipole fields, $B_{dipole} \sim 10^{14}$–10^{15} G, corresponding to 10^{-3}–10^{-1} of the dynamical saturation field strength, may be generated.[1] A significant fraction of a *dipolar* field of this strength should be retained by the star as it cools — due to the suppression of turbulent dissipation by magnetic tension[2] — even though the timescale for buoyant rise of isolated flux ropes through

---

* Dept. of Astronomy, University of Texas, Austin, TX 78712, USA
[†] Dept. of Space Physics & Astronomy, Rice University, Houston, TX 77251, USA
[‡] CITA, 60 St. George Street, Toronto M5S 1A1, Canada

convectively unstable material is much shorter than the cooling time. At fixed P, the effectiveness of the dynamo is *raised* by an increase in τ_{con}, at least until τ_{con} becomes so long that the dynamo cannot iterate many times before the star has cooled (ref. 2, §10.2). Both for this reason, and because of generically strong differential rotation,[2] supercritical magnetic fields should form in neutron stars with initial spin periods $P \sim 1 - 3$ ms, even if the convection is much less vigorous and turbulent[5] than we have suggested. The direction of the stellar magnetic moment $\vec{\mu}$ is correlated with the spin axis $\vec{\Omega}$ for rotation-driven dynamos, as evidenced by the fact that the angle $\phi_{\mu\Omega}$ between these vectors is small for the Sun and for some planets. (We have argued that the apparent absence of such a correlation in first generation pulsars is due to the action of high-Ro convection on the field[2].)

A neutron star can be formed directly with $J \sim J_{crit}$ when a white dwarf accretes over the Chandrasekhar limit[6], or possibly when the core of a massive star merges with companion star during a common envelope phase. In the first case, matter may be accreted either via Roche-lobe overflow from an orbiting companion, or from a disk formed by the tidal disruption of a companion white dwarf.[7] A degenerate object collapsing with $J > J_{crit}$ undergoes a bounce at subnuclear densities[8,9] and must shed angular momentum if it is to become a neutron star.[10] Thus, a supercritical dipole field may also be generated when $J > J_{crit}$, although the viscous and magnetic coupling of the star to the surrounding disk must be accounted for self-consistently in calculating its rotational evolution (e.g. refs. 11, 12).

If these very strongly-magnetized neutron stars, or *magnetars*, exist, they are likely to aquire large recoils at birth, plausibly $V_r \sim 1000$ km s^{-1}, via several mechanisms described in ref. 1. For example, temperature fluctuations in the neutrinosphere induced by the strong magnetic field— "neutrino starspots"[1,2] —will cause the star to radiate momentum anisotropically, as will parity non-conserving terms in the weak interaction emission and scattering rates.[13,14] The instantaneous recoil force is aligned to within some angle $\phi_{r\Omega}$ of $\vec{\Omega}$; but so long as the coherence time of the magnetic field is much longer than P, the force integrated over the neutrino cooling time gives rise to an impulse parallel to $\vec{\Omega}$, reduced by the (gentle) factor $\cos\phi_{r\Omega}$. The other mechanisms discussed in ref. 1 (with the exception of core fragmentation) also induce a recoil which is aligned with the rotation axis. We emphasize that that these hypothetical magnetars are *not* to be equated with the high-velocity tail of the population of ordinary pulsars: their magnetic fields and proper motions are generated by mechanisms which do not operate, or operate very inefficiently, in proto-pulsars.

Evidence for "Magnetars"

There is evidence that the 1979 March 5 gamma ray burst source is a young magnetar. This object has been localized to an angular position overlapping with the LMC supernova remnant N49.[15] If it is a neutron star that formed in the supernova, then its surface magnetic dipole field, as inferred from spindown, is $B_{dipole} \approx 6 \times 10^{14}$ G (ref. 1), and its recoil velocity[15,1] is ~ 1000 km s^{-1}. A field of this strength would carry enough energy to power the hard 1979 March 5 burst and the subsequent, much less bright, soft repeat bursts.[1] These bursts had roughly uniform peak fluxes with flat-top burst time-profiles, which suggests a limiting brightness $L_{crit} \approx 10^4 L_{Ed}$, where L_{Ed} is the Eddington luminosity.[16,17] Such a large L_{crit} is incompatible with most burst mechanisms on $B \leq 10^{13}$ G neutron stars, but it could be explained by magnetic suppression of Thomson scattering in a field of magnitude $B \approx 3 \times 10^{14}$ G (ref. 17), comparable to the spindown field. This critical luminosity could be achieved by a cooling baryon-photon plasma trapped in the stellar magnetosphere, but probably not by radiation through the stellar crust.[18]

If the N49 star is representative of a class of bursting neutron stars with $V_r \sim 10^3$ km s^{-1} and a Galactic formation rate $\Gamma >$(N49 remnant age)$^{-1} \sim 10^{-4}$ yr^{-1}, then there exist $> 10^4$ such stars in the Galactic halo within ~ 100 kpc. It is natural to consider whether these old, magnetically active stars are a source of classical GRBs.[1] However, if high-velocity neutron stars born in the disk generate *all* classical GRBs, then there is a potential problem accounting for both the observed degree of isotropy[19] and the Euclidean slope of the bright source counts (i.e., $\langle V/V_{max} \rangle = 0.46 \pm 0.02$ in the PVO data for 225 bright GRBs observed from 1978 to 1988[20]), since the density of unbound stars streaming out of the galaxy goes as $n(r) \propto r^{-2}$, which implies $\langle V/V_{max} \rangle = 0.25$ in the simplest models.[21]

This problem could be solved if the spatial distribution of halo GRB sources flattens out within a "core radius" of order 30 kpc,[22,23,24] or if the intrinsic burst rate is *smaller* when the sources are *younger*.[25] Neither solution has any obvious physical justification in the magnetar model. Since magnetars spin down past the pulsar "death line" in $6 \times 10^5 B_{15}^{-1}$ yrs, where $B_{15} \equiv (B_{dipole}/10^{15}$ G), the quenching of magnetospheric currents cannot produce a gradual GRB "turn-on" phase lasting $\sim 3 \times 10^7$ yr (cf. ref. 25). The thermal and magnetic evolution of magnetars is discussed in ref. 18.

Burst Beaming and Galactic Halo Models

We now examine the implications of beaming parallel to $\vec{\mu}$ on the observable distribution of GRBs. There is clear physical motivation for such beaming.[26] Gamma rays propagating at angle θ with respect to the magnetic field lines suffer attenuation via the reaction $\gamma \rightarrow e^+ + e^-$ above the threshold energy $E_\gamma = E_{th} = 2m_e c^2/\sin\theta$. The beaming angle is small when the gamma-rays are emitted close to the stellar surface in a supercritical dipole field.[26] Thus, reasonable beaming angles and observed burst rates are achieved in this model only if the gamma-rays are generated high in the stellar magnetosphere, where the field lines are tilted with respect to $\vec{\mu}$. We note that mechanisms such as curvature radiation and Compton upscattering of X-ray photons by energetic particles will produce gamma-rays beamed along the field lines.[27]

In the models calculated below, the total beaming angle ϕ is the sum of the intrinsic beaming angle due to opacity and radiative effects, and the r.m.s. scatter in $\phi_{\mu\Omega}$. So long as $\phi_{\mu\Omega} \leq 55°$ at birth (a likely result of an α–Ω dynamo), the magnetic axis and the rotation axis are aligned by spindown torques[28,29] after the neutron star drops below the pulsar death line. (The magnetostatic distortion of the star is large enough that the nutation of $\vec{\Omega}$ about $\vec{\mu}$ is damped, a circumstance that does not necessarily occur in ordinary pulsars.[29])

We ran Monte Carlo simulations of a Galactic population of bursting neutron stars, adopting the following simple properties: (1) the stars are born with randomly-directed recoils $V_r = 1000$ km s^{-1}; (2) they emit GRBs at a constant rate, with (3) constant luminosity; and (4) the gamma-ray emission is beamed parallel and anti-parallel to $\vec{V_r}$ within an angular radius ϕ. The galactic potential and other details of the code are identical to those of references 25, 30 and 31. Results for stars born at positions distributed like Galactic disk stars are shown in Figure 1.

The main effect of beaming is to reduce the number of detectable *nearby* bursts relative to bursts emitted at great distances. The values of ϕ chosen in Fig. 1 correspond to total detection probabilities of 0.060 and 0.015, when the beaming direction is randomly oriented. As a star moves out in the halo, the angle between $\vec{V_r}$ and the line-of-sight is reduced. The detection probability increases roughly as the square of the distance out to a galactocentric radius $R \sim R_{birth}/\phi$, beyond which this probability remains constant.

Perhaps the most remarkable result in Figure 1 is that $\langle V/V_{max} \rangle \approx 0.45$ for sampling depth $D \leq 30$ kpc, a result that agrees with the PVO data (given current uncertainties in the relative PVO–BATSE normalization[25]) for BATSE sampling depths $D \leq 180$ kpc. This effective core radius is a consequence of beaming. Beaming also greatly reduces $\langle \cos\theta \rangle$. Without beaming (i.e., the case $\phi = \pi/2$, not shown) $\langle V/V_{max} \rangle$ is 0.30 at 30 kpc and 0.26 at 70 kpc, while $\langle \cos\theta \rangle > 0.2$ over this distance range.

Recent BATSE constraints[32] on $\langle \cos\theta \rangle$, $\langle \sin^2 b \rangle$ and $\langle V/V_{max} \rangle$ are satisfied by the models of Figure 1a [$\phi = 20°$] for a significant range of sampling depths, D. A mild degree of disklike quadrupole anisotropy, $\langle \sin^2 b \rangle = 0.31$, is present at large D because the disk rotation causes magnetars to escape at lower Galactic latitudes. Galactic rotation (with local velocity $\vec{V_\Omega}$) also causes the total initial velocity vector $\vec{V_r} + \vec{V_\Omega}$ to be inclined by as much as $\phi_V^{max} \equiv \arctan(V_\Omega/V_r) \approx 12°$ with respect to the axis of GRB emission, $\vec{V_r}/V_r$. Thus if $\phi < \phi_V^{max}$, some stars never become visible from Earth. Since this happens preferentially at high galactic latitudes, $\langle \sin^2 b \rangle$ is substantially less than 1/3 in models with $\phi < \phi_V^{max}$. For example, $\langle \sin^2 b \rangle$ never exceeds 0.25 when $\phi = 10°$ (Fig 1b), in conflict with the observations. Only when $\phi > \phi_V^{max}$ are all stars visible at large D, causing $\langle \sin^2 b \rangle$ to asymptote to the unbeamed value of 0.31. (At large D, $\langle \sin^2 b \rangle$ depends only on V_r, approaching 1/3 as V_r/V_Ω increases.)

Conclusions and Model Predictions

Our main conclusion is that beaming gives a natural *geometrical* solution to the "core radius problem" in models of GRBs from high-velocity galactic neutron stars. The simple halo beaming model of Fig. 1a predicts that deviations from isotropy are largest for bright bursts. Because most bursters are born inside the Solar circle, the brightest ones tend to be seen both in that direction *and toward the galactic anticenter* (if they have already moved beyond the Solar circle). Thus the statistic $\langle \cos^2\theta \rangle$ at the bright end is a distinctive diagnostic for this halo beaming model, perhaps more discriminating than $\langle \cos\theta \rangle$ (cf. Fig. 1a). Observational values of $\langle \cos^2\theta \rangle$, and of various statistics for bright subsets of the BATSE catalog, have not yet been published by the BATSE collaboration.

We caution that detailed models of GRBs from magnetars might predict evolutionary trends in bursting behavior (due to the declining rate of energy dissipation over the lifetime of magnetic activity[18]) and/or some angle-dependence to the burst

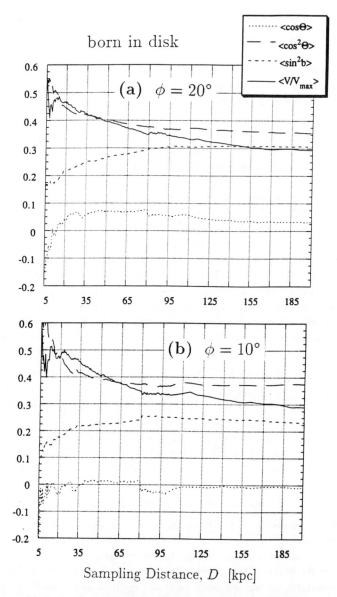

FIG. 1 Statistics describing the distribution of halo GRBs as a function of sampling depth D, for high-velocity ($V_r = 1000$ km s^{-1}) bursters born at positions distributed like Pop. I stars in the galactic disk. Bursts are assumed to be beamed into cones of angular radius $\phi = 20°$ (Fig. a) and $\phi = 10°$ (Fig. b). A Monte Carlo simulation involving 4000 stars within 200 kpc was used to obtain these results. Numerical uncertainty for $D > 100$ kpc (the likely BATSE range) is $< 2\%$. Because the uncertainty is large at $D < 5$ kpc, the curves have been truncated.

brightness. These effects could alter some of the model predictions presented here, and are a subject of continuing investigation.

We thank O. Blaes and C. Dermer for discussions. This research was supported by NASA Theoretical Astrophysics grant NAGW-2418 and by the NSERC of Canada.

REFERENCES

1. Duncan, R.C. & Thompson, C. *Astrophys. J.* **392**, L9–L14 (1992).
2. Thompson, C. & Duncan, R.C. *Astrophys. J.* **404** in press (1993).
3. Burrows, A. *Astrophys. J.* **318**, L57–L62 (1987).
4. Chandrasekhar, S., *Astrophys. J.*, **161**, 571–578 (1970).
5. Wilson, J.R. & Mayle, R.W., *Physics Reports*, **163**, 63–78 (1988).
6. Narayan, R. & Popham, R. *Astrophys. J.* **346**, L25–L28 (1989).
7. Saio, H. & Nomoto, K. *Astron. Astrophys.* **150**, L21–L24 (1985)
8. Tohline, J.E. *Astrophys. J.* **285**, 721–728 (1984).
9. Mönchmeyer, R. & Müller, E. in *Timing Neutron Stars*, eds. H. Ögelman & E.P.J. van den Heuvel, (Dordrecht: Kluwer) 549–572 (1989).
10. Shapiro, S.L. & Lightman, A.P. *Astrophys. J.* **207**, 263–278 (1976).
11. Ghosh, G.L. & Lamb, F.K. 1979, *Ap.J.* **232**, 259.
12. Popham, R. & Narayan, R. *Astrophys. J.* **370**, 604-614 (1991).
13. Dorofeev, O.F., Rodionov, V.N., & Ternov, I.M. *Sov. Astron. Lett.* **11**, 123–126 (1985).
14. Vilenkin, A. unpublished (1979).
15. Cline, T.L., et al., *Astrophys. J.*, **255**, L45–L48 (1982).
16. Golenetskii, S.V., Ilyinskii, V.N., & Mazets, E.P., *Nature*, **307**, 41–43 (1984).
17. Paczyński, B., *Acta. Astron.*, **42**, 145–153 (1992).
18. Thompson, C. & Duncan, R.C., in preparation.
19. Meegan, C.A. et al. *Nature* **355**, 143–145 (1992).
20. Chaung, K.W., White, R.S., Klebesadel, R.W., & Laros, J.G., *Astrophys. J.*, **391**, 242–245 (1992).
21. Dermer, C. & Li, H. in *AIP Conf. 265: Gamma Ray Bursts*, eds. W.S. Paciesas & G.J. Fishman, (AIP: New York) 115–119 (1992).
22. Mao, S. & Paczyński, B. *Astrophys. J.*, **389**, L13–L16 (1992).
23. Eichler, D. & Silk, J., *Science*, **257**, 937–942 (1992).
24. Salpeter, E.E. & Wasserman, I. preprint (1992).
25. Li, H. & Dermer, C.D., *Nature*, **359**, 514–516 (1992).
26. Ho, C., Epstein, R.I. & Fenimore E.E. *Astrophys. J.* **348**, L25–L28 (1990).
27. Vitello, P. & Dermer, C.D. *Astrophys. J.* **374**, 668–686 (1991).
28. Michel, F.C. & Goldwire, H.C. *Astrophys. Lett.* **5**, 21–24 (1970).
29. Goldreich, P. *Astrophys. J.* **160**, L11–L15 (1970).
30. Paczyński, B. *Astrophys. J.* **348**, 485–494 (1990).
31. Li, H., Masters Thesis, Rice University, unpublished (1992).
32. See contributions by C.A. Meegan and by M.S. Briggs et al. in this volume.

Energy Storage in Old Neutron-Star Crusts

Patrick C. Mock and Paul C. Joss*
Department of Physics, Center for Space Research, and Center for Theoretical Physics
Massachusetts Institute of Technology

Abstract

Neutron stars have often been used in models of γ-ray burst (GRB) sources. Recently, Blaes et al.[1,2] reexamined the possibility that GRB's are caused by seismic events on old, cold, isolated neutron stars that are accreting slowly from the interstellar medium. We have further investigated this model by use of a stellar evolution code adapted to the study of neutron-star surfaces.[3] Our preliminary calculations reveal the evolution of the structure of the neutron-star crust and provide an estimate of the energy stored in the mechanical structure and nonequilibrium composition of the crust. After 10^{10} years of slow accretion at an accretion rate of 10^{-16} M$_\odot$ yr^{-1}, the available energy is $\sim 10^{45}$ ergs, but there is no evidence for the development of an unstable density inversion.

Introduction

The crust of an old neutron star which has been slowly accreting is potentially unstable. Blaes et al.[2] have shown that the crust will develop density inversions as it evolves, and that the accreted material will have a nonequilibrium chemical composition. The density inversions are caused by changes in composition due to electron capture. The nonequilibrium composition develops because the cold crust cannot form the equilibrium nuclei via the available pycnonuclear reactions. Both of these mechanisms contribute to the available energy stored in the crust. More recently, Blaes et al.[4] have shown that an electron-capture induced density inversion is unlikely to develop a Rayleigh-Taylor instability, because the inversion is stabilized by the upper layers. Significantly larger density inversions are needed before the crust will become unstable.

We have explored this scenario in detail using a new version of the stellar evolution code ASTRA, originally developed by Rakavy et al.[5] and later modified[3,6,7] to simulate accreting neutron-star crusts. Accretion is handled by adding mass to the outermost zone and relaxing to hydrostatic equilibrium, rezoning when necessary. We use an equation of state and nuclear reaction network appropriate to the high densities and relatively low temperatures of interest. These are presented in more detail below.

With this input physics and the initial conditions shown in Table I, we have calculated the evolution of the neutron-star crust and used our results to estimate the energy stored in the mechanical structure of the crust and in the nonequilibrium composition.

Table I: Initial Conditions for the Calculations

Mass	1.41	M$_\odot$	Canonical value
Radius	10	km	Stiff equation of state in the core
Surface temperature	10^6	K	Old and cold
Initial surface composition	^{56}Fe		Initial nuclear equilibrium
Accretion rate	10^{-16}	M$_\odot$ yr^{-1}	Slowly accreting
Accretion composition	^{4}He		Surface hydrogen burning neglected

*This work was supported in part by the National Aeronautics and Space Administration under grant NAGW-1545.

Thermodynamic Equations

The thermodynamic properties of the matter include contributions from a degenerate electron gas and a Coulomb lattice of nuclei. The degenerate electrons dominate the behavior of matter in the high-density, low-temperature regime of interest. We derived the thermodynamic equations for the electrons using the following standard approximation for integrals of the Fermi function[8]:

$$\int_0^\infty \frac{f(\epsilon)\,d\epsilon}{e^{\beta(\mu-\epsilon)}+1} \simeq \int_0^\mu f(\epsilon)d\epsilon + \frac{\pi^2}{6}\beta^{-2}\frac{df}{d\epsilon}\bigg|_{\epsilon=\mu}, \qquad \beta\mu \gg 1. \tag{1}$$

The thermodynamic properties of the electrons, thus derived, are given by

$$\mu(x) = m_e c^2 \left(1 + x^2\right)^{1/2}, \tag{2}$$

$$n_e(x,y) = (3\lambda^3)^{-1}\left(x^3\pi^{-2} + y^2(1+x^2)(1+2x^2)(2x)^{-1}\right), \tag{3}$$

$$\epsilon_e(x,y) = m_e c^2 \lambda^{-3}\left(\chi(x) + y^2(1+x^2)^{1/2}(1+3x^2)(6x)^{-1}\right), \tag{4}$$

$$\rho(x,y) = m_e \mu_e\, n_e(x,y) + \epsilon_e(x,y)c^{-2}, \tag{5}$$

$$P_e(x,y) = m_e c^2 \lambda^{-3}\left(\phi(x) + \tfrac{1}{6} y^2 x \left(1+x^2\right)^{1/2}\right), \tag{6}$$

$$s_e(x,y) = \tfrac{1}{3} k_b \lambda^{-3}\, y\, x\, (1+x^2)^{1/2}, \tag{7}$$

where

$$\phi(x) \equiv (8\pi^2)^{-1}\left(x(1+x^2)^{1/2}(2x^2/3 - 1) + \sinh^{-1}(x)\right), \tag{8}$$

$$\chi(x) \equiv (8\pi^2)^{-1}\left(x(1+x^2)^{1/2}(2x^2+1) - \sinh^{-1}(x)\right), \tag{9}$$

$$x \equiv \frac{p}{m_e c}, \quad y \equiv \frac{k_b T}{m_e c^2}, \quad \lambda \equiv \frac{\hbar}{m_e c}. \tag{10}$$

The ions, which form a Coulomb lattice in the neutron-star crust, make a significant contribution to the pressure and specific entropy. At zero temperature the Coulomb interaction in the ion lattice generates a negative correction to the electron pressure[9]. The contributions to the pressure and specific entropy at finite temperatures are due to lattice oscillations as described by Kovetz and Shaviv[10].

The Pycnonuclear Reaction Rates

Standard thermonuclear reactions do not take place under the conditions of interest because the crust is much too cold. The relevant nuclear reactions under the these conditions are pycnonuclear reactions and electron-capture events.

The crust of a neutron star is a multi-component-plasma (MCP) whose composition is usually dominated by one nuclear species[9]. We have modified the available one-component-plasma (OCP) pycnonuclear reaction rates to approximate the behavior of a multi-component-plasma. Our approximation ignores any rate enhancements due to lattice defects; hence these rates are a lower bound on the actual reaction rates.

The OCP reaction rate was originally derived by Salpeter and Van Horn[11]. Schramm and Koonin[12] recently showed that the effects of lattice polarization and effective tunneling mass on the OCP rate nearly cancel. The rate can be written as:

$$r_{ii} = 1.06 \times 10^{45} \, S_{ii} \, n_i \, m_H \, A_i^2 \, Z_i^4 \, \lambda^{7/4} \, e^{(-\alpha_2 - \alpha_1 \lambda^{-1/2})} \quad \text{cm}^{-3}\,\text{s}^{-1}, \tag{11}$$

where

$$\lambda \equiv r^* \left(\frac{n_i}{2}\right)^{1/3} \quad \text{and} \quad r^* \equiv \frac{\hbar^2}{m_i Z^2 e^2} = \frac{\hbar^2}{m_H A Z^2 e^2}. \tag{12}$$

The MCP rate should reduce to the OCP rate when $n_i = n_j = n_t$. We use the following form for the general MCP reaction rate:

$$r_{ij} = 1.06 \times 10^{45} \, S_{ij} \, \frac{n_i n_j}{n_t} \, \frac{2 m_H A_i A_j}{A_i + A_j} \, Z_i^2 Z_j^2 \, \lambda_{ij}^{7/4} \, e^{(-\alpha_2 - \alpha_1 \lambda_{ij}^{-1/2})} \quad \text{cm}^{-3}\,\text{s}^{-1}, \tag{13}$$

where n_t is the total ion number density and r_{ij}^* and λ_{ij} are defined[11] as

$$r_{ij}^* = \frac{\hbar^2}{\mu_{ij} Z_i Z_j e^2} = \frac{\hbar^2}{Z_i Z_j e^2} \frac{A_i + A_j}{2 m_H A_i A_j} \tag{14}$$

and

$$\lambda_{ij} = r_{ij}^* \left(\frac{n_e}{2 Z_i}\right)^{1/3} \simeq r_{ij}^* \left(\frac{n_t}{2}\right)^{1/3}. \tag{15}$$

The cross-section factors S_{ij} for the most important reactions are listed in Table II.

Table II: Pycnonuclear cross-section factors S_{ij} (MeV-barns) [4,13-19]

Reaction	S(0)	Reaction	S(0)
^{12}C$(\alpha,\gamma)^{16}$O	0.5	^{12}C $+$ ^{12}C	3×10^{16}
^{16}O$(\alpha,\gamma)^{20}$Ne	0.7	^{16}O $+$ ^{12}C	2×10^{20}
^{20}Ne$(\alpha,\gamma)^{24}$Mg	28.9	^{16}O $+$ ^{16}O	4.53×10^{26}
^{24}Mg$(\alpha,\gamma)^{28}$Si	1.83×10^{10}	^{16}C $+$ ^{16}C	3×10^{16}
^{28}Si$(\alpha,\gamma)^{32}$S	2.22×10^{11}		

We have derived the MCP triple-α rate by generalizing the OCP rate given by Schramm et al.[20] Our modified rate is proportional to X_α^3, the cube of the mass fraction of alpha particles:

$$r_{3\alpha} = 1.426 \times 10^{22} \, X_\alpha^3 \, \rho_9^{15.771 - 0.833 \log(\rho_9)} \quad \text{cm}^{-3}\,\text{s}^{-1}. \tag{16}$$

We take the electron capture rates from Blaes et al.[2]. The terrestrial beta decay rates, needed to calculate the electron capture rates, are obtain from the NNDC online database at Brookhaven[21]. We also utilize this database for the nuclear masses and energy levels used in the model calculations.

Results and Future Work

In our model calculations the neutron-star crust appears to be mechanically stable 10^{10} yrs after the onset of accretion. At this epoch a small density inversion has formed at the boundary between the initial ^{56}Fe surface and the newly accreted material. At this epoch, the boundary is composed of ^{16}C on top of ^{56}Ti, and the fractional density inversion across the interface is ~ 0.032.

The deepest layers of accreted matter are sufficiently dense that ^{16}O transforms to ^{16}C via electron capture, and the (^{16}C $+$ ^{16}C) reaction slowly commences. The region above the ^{16}C layer is completely devoid of α-particles, and the composition is pure ^{16}O. The ^{56}Fe at the original stellar surface transforms via electron capture to ^{56}Cr after $\sim 10^8$ yrs and to ^{56}Ti after $\sim 2 \times 10^9$ yrs. The composition of the accreted material at 1×10^{10} yrs is shown in Figure 1.

We have estimated the energy stored in the accreted surface layers by comparing the energy density of those layers to the energy density of the equilibrium composition at the same mass density. The growth of the available energy is shown in Figure 2. The rate of energy storage is $\sim 4.5 \times 10^{41}$ ergs Myr^{-1} for the first three billion years. The stored energy reaches a plateau at this point because the energy density of the products of the (^{16}C + ^{16}C) reaction are much closer to the equilibrium energy density and thereby do not contribute significantly to the stored energy.

Our model calculations are accurate until the (^{16}C + ^{16}C) reaction becomes important. The S-factor and exit channels for this reaction are unknown. The results we present assume that this S-factor is the same as the (^{12}C + ^{12}C) S-factor as suggested by Blaes et al.[4] and that the exit channels are dominated by one- and two-neutron reactions. We have also carried out model calculations under the assumption that this S-factor is the same as the (^{16}O + ^{16}O) S-factor. The results of these two calculations are similar except that in the latter case the ^{16}C burns much faster (but not explosively). The burning is sufficiently fast that a density inversion does not form before the ^{16}C is destroyed.

Our calculations are consistent with the results of Blaes et al.[4] but we have been able to explore a much longer period of slow accretion by making plausible estimates for the important reactions. We plan to further extend the present work by simulating the effects of variable accretion rates that might be associated with the motion of a neutron star through randomly distributed molecular clouds and/or through multiple crossings of the galactic plane.

References

1. O. Blaes, R. Blandford, P. Goldreich, & P. Madau, ApJ, 343, 839 (1989).
2. O. Blaes, R. Blandford, P. Goldreich, & S. Koonin, ApJ, 363, 612 (1990).
3. P. C. Joss, ApJ, 225, L123 (1978).
4. O. Blaes, R. Blandford, P. Madau, & L. Yan, ApJ, 399, 634 (1992).
5. G. Rakavy, G. Shaviv, & Z. Zinamon, ApJ, 150, 131 (1967).
6. P. C. Joss & F. K. Li, ApJ, 238, 287 (1980).
7. S. Ayasli & P. C. Joss , ApJ, 256, 637 (1982).
8. F. Reif, Fundamentals of Statistical and Thermal Physics (McGraw-Hill, N.Y., 1965), p. 396.
9. S. Shapiro & S. Teukolsky, Black Holes, White Dwarfs, and Neutron Stars, (Wiley-Interscience, N.Y., 1983) p. 31.
10. A. Kovetz & G. Shaviv, Astron. & Astrophys., 8, 398 (1970).
11. E. Salpeter & H. Van Horn, ApJ, 155, 183 (1969).
12. S. Schramm & S. Koonin, ApJ, 365, 296 (1990).
13. G. Caughlan & W. Fowler, Atomic Data and Nuclear Data Tables, 40, 283 (1988).
14. P. Christensen, Z. Switkowski, & R. Dayras, Nuclear Physics, A280, 189 (1977).
15. B. Filippone, J. Humblet, & K. Langanke, Phys. Rev. C, 40, 2, 515 (1989).
16. K. Hahn, K. Chang, T. Donnoghue, & B. Filippone, Phys. Rev. C, 36, 3, 892 (1987).
17. J. Humblet, B. Filippone, & S. Koonin, Phys. Rev. C, 44, 6, 2530 (1991).
18. P. Schmalbrock, et al. Nuclear Physics, A398, 279 (1983).
19. S. Woosley, W. Fowler, J. Holmes, & B. Zimmerman, Atomic Data and Nuclear Data Tables, 22, 371 (1978).
20. S. Schramm, K. Langanke, & S. Koonin, ApJ, 397, 579 (1992).
21. C. Dunford & T. Burrows, Publication No. NNDC/ONL-92/08, National Nuclear Data Center, Brookhaven National Laboratory, (1992).

Energy Storage in Old Neutron-Star Crusts

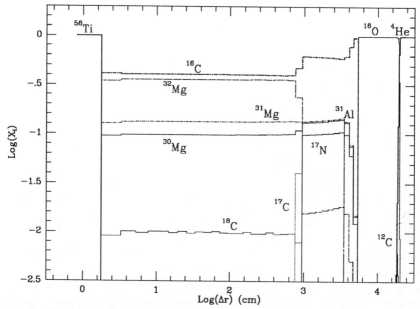

Figure 1: Composition (by mass fraction X_i) of the neutron-star crust after $\sim 5 \times 10^9$ years of slow accretion as a function of Δr, the thickness of the newly accreted material on the neutron-star surface. The discontinuities indicate the zoning structure of the model.

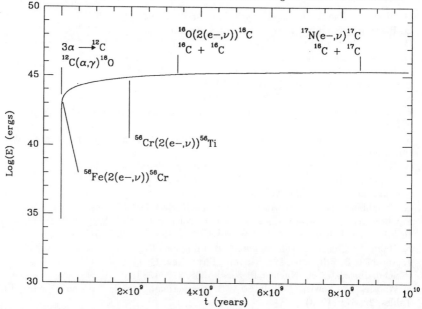

Figure 2: The available nuclear energy, E, stored on the neutron-star crust as a function of elapsed time since the onset of accretion. E is the difference between the integrated energy density of the accreted material and the intergrated equilibrium energy density. The epochs at which key reactions become important are indicated.

X-Ray Emission From Neutron Stars with Supercritical Magnetic Fields: A Model for the Soft Gamma Repeaters

C. Thompson[*] & R. C. Duncan[‡]

ABSTRACT

A *magnetar* is defined to be a neutron star with a supercritical magnetic dipole field: $B_{dipole} > B_{cr} \equiv m_e^2/e = 4.4 \times 10^{13}$ G. (We use units in which $\hbar = c = k_B = 1$.) We have suggested[2] that a neutron star born with a very short spin period, $P \sim 1-3$ ms is the site of an effective dynamo that can generate fields of this strength. Some peculiar properties of the 1979 March 5 burst and its repeat bursts point to these objects as the source of the soft gamma repeaters (SGRs).[2,3,4] Young magnetars release an enormous amount of energy as they spin down via magnetic torques, and are a possible source of cosmological gamma-ray bursts.[2,5] Here we outline some results on the thermal and magnetic evolution of magnetars, and discuss various applications to the SGRs and to the X-ray pulsar 1E2259+586. For more details, see ref. 1.

- A neutron star born with a spin period P of a few milliseconds carries a large amount of free energy in differential rotation, which could in principle generate toroidal fields as strong as $\sim 3 \times 10^{17} (P/1 \text{ ms})^{-1}$ G. It also becomes unstable to entropy-driven convection once the energy and lepton number transport have settled to a quasi-steady state in the outermost layers of the star,[6] and supports an efficient $\alpha - \Omega$ dynamo.[2] This dynamo can plausibly generate a surface dipole field in the range $10^{14} - 10^{15}$ G, and perhaps stronger fields in the stellar interior. The rotational energy deposited in the supernova remnant could be substantially less than than the total of $\sim 10^{52} (P/1 \text{ ms})^{-2}$ erg after taking into account losses to gravitational radiation,[7] collimated emission in a jet, or magnetic coupling to a surrounding disk (e.g., ref. 8).

- As an alternative to a post-collapse dynamo, very-strongly magnetized neutron stars might also form as the result of flux conservation in the accretion-induced collapse of highly-magnetized white dwarfs.[5] However, the maximum fields detected in white dwarfs[9] and the dynamical saturation field strength during convective carbon burning[6] both indicate neutron star dipole fields no stronger than $\sim 1 \times 10^{14}$ G, less than the field indicated for the source of the 5 March event.[2,3,4]

- The soft repeat bursts emitted by the March 5 burster have a peak flux corresponding to a characteristic luminosity $\sim 10^4$ times the Eddington luminosity at the distance of the LMC.[10] Paczyński[3] pointed out that such a high critical luminosity could result from magnetic suppression of Thomson scattering in a field of strength $B \sim 3 \times 10^{14}$ G. This is close to the field needed to spin down a neutron star to an 8 s period (characteristic of the March 5 event) in the age of the N49 supernova remnant. However, such a radiant luminosity is unlikely to flow through the stellar surface since the magnetic optical depth rises with temperature, and

---

[*] CITA, 60 St. George Street, Toronto M5S 1A1, Canada
[‡] Dept. of Astronomy, University of Texas, Austin, TX 78712, USA

therefore with depth in the crust, and since neutrino losses become important if the energy is released in the deep crust.[11] The observed flux can be generated by a photon-baryon plasma trapped above the surface in the magnetosphere. The injection of a large amount of energy near the surface by magnetic reconnection[2,3] blows baryons off the star in a radiation-dominated fireball, which is then confined by the field. The limiting baryon density which the photon pressure P_{rad} can support near the surface is $\rho_b \sim P_{rad}/gR$. Making use of improved Rosseland mean opacities,[12] the radiative flux across the magnetic field lines is enhanced over the Eddington flux by a factor $F/F_{EDD} = (2/\pi^2 Y_e)(B/B_{cr})^2 (T/m_e)^{-2}$ near the surface. The temperature T inside the fireball is related to the remaining burst energy $\Delta E = \Delta E_{41} \times 10^{41}$ erg by $T/m_e \sim 0.06 \, (\Delta E_{41})^{1/4} (\Delta R_{10}/10)^{-3/4}$. We have

$$F/F_{EDD} \sim 1 \times 10^4 \, (B/10 B_{cr})^2 \, (\Delta E_{41})^{-1/2} \, (\Delta R_{10}/10)^{3/2} \qquad (1)$$

for $Y_e \sim 0.5$. Here, $\Delta R = \Delta R_{10} \times 10$ km is the size of the confining magnetospheric volume. Since the flux rapidly drops below the value (1) outside a distance $\sim \frac{1}{6} R$ from the stellar surface, the duration of the burst is

$$\Delta t \sim \Delta E/(FR^2) \sim 0.5 \, (B/10 B_{cr})^{-2} \, (\Delta E_{41})^{3/2} \, (\Delta R_{10}/10)^{-3/2} \quad \text{s.} \qquad (2)$$

This simple model predicts that, with the possible exception of an initial non-thermal transient, the peak flux should be attained towards the *end* of a burst, when most of the energy has been depleted. Given $\Delta R \sim 100$ km and $B \sim 10 B_{cr}$ (the spindown field of the N49 star[2]), the duration and luminosity of most of the short, repeat bursts from the N49 star[10] can be accounted for by normalizing to the inferred energy ($\Delta E \sim 10^{41}$ erg). The required confining volume would be smaller if the surface field were stronger than the spindown dipole field. However, (2) predicts too long a duration for the most energetic bursts. The release of much more than $\sim 10^{41}$ erg in the same volume would also generate a significant pair density and a much higher electron scattering optical depth.

• The young X-ray pulsar 1E 2259+586 appears to be a young, accreting neutron star with no detected companion or binary modulation.[13] There are several notable similarities between it and the putative LMC source of the 1979 March 5 gamma-ray burst. The stellar spin periods are respectively 7 and 8 seconds, and both stars reside in $\sim 10^4$ yr old supernova remnants (CTB 109 vs. N49). X-ray emission from 1E 2259 has been observed in the range[14] $(0.5 - 1) \times 10^{35}$ erg s^{-1}. There also exists localized X-ray emission in N49 which overlaps with the error box of the March 5 event, with luminosity[15] $L_X \sim 7 \times 10^{35}$ erg s^{-1} if the spectrum is blackbody. If 1E 2259 is a neutron star accreting via Roche lobe overflow from a binary companion at the rate implied by its L_X, then the companion must be degenerate and have a very low mass ($\sim 0.03 M_\odot$ for a hydrogen degenerate dwarf, as implied by the mass-transfer models of ref. 16), and it is not surprising that neither a binary periodicity nor an optical counterpart has yet been seen. An alternative possibility is that 1E 2259 is a very strongly magnetized neutron star which has spun down by magnetic braking to a period 7 s in an age $\approx 1 \times 10^4$ yr. The required dipole field is $\approx 4 \times 10^{14}$ G. The observed X-ray emission could be powered by the decay of the magnetic field in the deep crust if the r.m.s. crustal field were as large as 4×10^{15} G (most of the energy is lost to neutrinos when the surface X-ray flux exceeds $\sim 10^{34}$ erg s^{-1}; ref. 11). In general, the suppression of thermal conduction by a strong internal magnetic field can induce a large hemispheric asymmetry in the surface X-ray flux from a neutron star.

• A supercritical field in the crust of strength $B = B_{15} \times 10^{15}$ G and coherence length $\ell = \ell_5 \times 10^5$ cm is transported to small scales by Hall drift, and decays on the relatively short timescale[17] $t_{Hall} \sim 3 \times 10^5 B_{15}^{-1} \ell_5^2$ yr. Ambipolar diffusion of a magnetic field out of the core proceeds[17] through a fast mode, whose timescale is also quite short; and a slow mode, which might release a substantial amount of free energy at an age of 10^8 yr, as required in a halo model for gamma-ray bursts (e.g., ref. 4). Indeed, frictional heating by the diffusing field can substantially delay cooling in non-superfluid regions of the core. Balancing this heating with modified URCA neutrino losses,[18] the relation between the r.m.s. core field B and core temperature $T_c = T_{c8} \times 10^8$ K is

$$T_{c8} = 6.3 \times 10^{-2} B_{15}^2 \rho_{15}^{-1} (Y_e/0.07)^{-1} \qquad (3)$$

at a density $\rho = \rho_{15} \times 10^{15}$ g cm^{-3}. The free energy in the field exceeds the thermal energy in (normal) nuclear matter by a large factor,

$$\frac{B^2/8\pi}{\frac{1}{2} C_p T_c} = 530\, T_{c8}^{-1} \rho_{15}^{2/3} \qquad (Y_e = 0.07). \qquad (4)$$

A field of initial strength B_0 reaches strength $B(t) = B_0[1 + 14t/t_{amb}(B_0)]^{-1/14}$ at time t, where the timescale for ambipolar diffusion in the slow mode depends very sensitively on the field strength and electron fraction:

$$t_{amb}(B_0) = 1.8 \times 10^8 \, (B_{15}/6)^{-14} \rho_{15}^{22/3} (Y_e/0.07)^8 \qquad \text{yr.} \qquad (5)$$

At $t > t_{amb}(B_0)$, the field approaches a scaling solution

$$B(t) = 4.3 \times 10^{15} \, (t/10^8 \text{ yr})^{-1/14} \rho_{15}^{11/21} (Y_e/0.07)^{4/7} \qquad \text{G} \qquad (6)$$

which is independent of B_0. The late cooling history of the star depends sensitively on the ability of the core to conduct heat to the surface. Conduction perpendicular to the field lines is reduced by the considerable factor[19] $(\omega_B \tau)^{-2}$, where ω_B is the cyclotron frequency of an electron at the Fermi surface, and τ is the mean time between scatterings.[20] The degree of thermal contact between (normal) core and crust thus depends sensitively on the geometry of the field. The field lines in the stellar interior are stretched by a factor $B/B_{seed} \sim 10^4 - 10^5$ when the seed field B_{seed} is amplified during the early dynamo phase. This implies that the surface photon flux is suppressed by at least a factor $(B/B_{seed})^{-1}$ if the interior field is very tangled, or if it is wound up by differential rotation. For example, an intense toroidal field in a (normal) core region of radius $R_n = R_{n\,10} \times 10$ km suppresses conductive losses relative to neutrino losses by the factor

$$\dot{E}_{cond}/\dot{E}_{URCA} = 2 \times 10^{-1} \, (t/10^8 \text{ yr}) \, \rho_{15}^{-1/3} R_{n\,10}^{-2} (Y_e/0.07)^{-2/3}, \qquad (7)$$

using the scaling solution (3), (6). A significant degree of thermal contact between core and crust could be maintained if the interior field were wound up by differential rotation, *and* then partially transported above the stellar surface by buoyancy forces. The (unredshifted) surface X-ray luminosity would be

$$L_X = 8 \times 10^{33} \, (t/10^6 \text{ yr})^{-3/10} \, \rho_{15}^{1/10} (Y_e/0.07)^{3/10} \quad \text{erg s}^{-1}, \qquad (8)$$

assuming a stellar radius $R = 10$ km and mass $1.4 M_\odot$. This enhanced surface X-ray flux lasts only as long as L_X remains less than the neutrino losses from the contact volume, which limits the age to $t < t_\gamma = 1.6 \times 10^6 \, (Y_e/0.07)(M_{contact}/M_\odot)^{6/5}$ yr. This value of t_γ exceeds that for an ordinary pulsar by the factor (4). The field in the outer layers of the star freezes out at age t_γ, since $t_{amb} \propto T^{-6}$ (ref. 17). Only the part of the core *out* of thermal contact with the surface remains hot and magnetically active at an age $t \gg t_\gamma$. Such a circumstance seems necessary in galactic halo models of gamma ray bursts.[4] Seismic waves excited by sudden reconnection events in the hot core would leak energy into the magnetosphere via Alfvén radiation (e.g., ref. 21). Surface reconnection events could provide an additional source of gamma-ray bursts at $t < t_\gamma$.

• We should emphasize that the slow mode of ambipolar diffusion is limited by the ability of (modified) URCA reactions to re-establish β-equilibrium between the charged particles and the neutrons.[17] Thus, the slow mode is choked off by neutron superfluidity, and greatly accelerated by the formation of a pion or kaon condensate in the central core. The best existing calculations of neutron superfluidity show that $^3P_2$ pairing vanishes at densities larger than[22] $\rho \approx 1 \times 10^{15}$ g cm^{-3}, so that a substantial fraction of the core of $1.4 M_\odot$ neutron star with a moderate or soft equation of state is composed of normal neutron matter (e.g., ref. 23). The normal fraction will be even larger for more massive neutron stars (which are the expected result of a white dwarf merger). However, these calculations do not take into account 3-body nuclear forces, the modification of 2-body forces by the medium (which tends to enhance P-wave pairing[24]), or the effect of the strong, polarizing field on neutron superfluidity.[1]

• Thermal X-rays emitted by a neutron star with a supercritical field can have a significant optical depth to photon splitting, $\gamma \to \gamma + \gamma$. A calculation of the emergent X-ray spectrum requires an understanding of stimulated splitting as well as photon merging, since the occupation number is of order unity. (Note that vaccuum-polarization corrections to the dielectric tensor are of order $\sim 10^{-3}(B/B_{cr})$ for $B > B_{cr}$, and are small so long as the magnetic field is not strongly supercritical. Thus, photon splitting will allow a photon gas to relax to a spectrum which is very nearly black body.) The only allowed reactions for photons in polarization states $\perp$ and $\parallel$ are[25] $\omega(\perp) \leftrightarrow \omega_1(\parallel) + \omega_2(\parallel)$ and

$\omega_3(\perp) \leftrightarrow \omega(\|) + \omega_4(\|)$ when $\omega \ll m_e$. (The symbols $\perp$ and $\|$ refer to the orientation of the photon's electric vector with respect to $\vec{k} \times \vec{B}$.) The net rate of change of the occupation number at energy ω is

$$\left(\frac{dn_\omega^\perp}{dt}\right)_{sp} = \int_0^\omega d\omega_1 \left[n_{\omega_1}^\| n_{\omega-\omega_1}^\| - n_\omega^\perp(1 + n_{\omega_1}^\| + n_{\omega-\omega_1}^\|)\right]\frac{d\Gamma_{sp}(\omega,\omega_1)}{d\omega_1};$$

$$\left(\frac{dn_\omega^\|}{dt}\right)_{sp} = \int_0^\infty d\omega_4 \left[n_{\omega+\omega_4}^\perp(1 + n_\omega^\| + n_{\omega_4}^\|) - n_\omega^\| n_{\omega_4}^\|\right]\frac{d\Gamma_{sp}(\omega+\omega_4,\omega)}{d\omega_4}.$$
(9)

Here $d\Gamma_{sp}(\omega,\omega_1)/d\omega_1 = (\Gamma_{sp}/30\omega)(\omega_1/\omega)^2(1-\omega_1/\omega)^2$, where the total spontaneous splitting rate is $\Gamma_{sp}(\omega) = (\alpha^3 m_e/2160\pi^2)\sin^6\theta\,(\omega/m_e)^5$ for polarization $\perp$ in a field $B \gg B_{cr}$. These expressions are valid in the approximation where the three interacting photons are collinear. Spectral features associated with magnetic corrections to Thomson scattering (e.g., ref. 3) may be wiped out by photon splitting. In particular, the pair annihilation line detected in the spectrum of the 5 March event should be degraded rapidly in a supercritical magnetic field. Note also that the spectrum is *enhanced* over a black body on the Wien tail. This happens because streaming tends to increase n_ω (by a factor 2 from the stellar surface to infinity, in the absence of interactions); and because $\Gamma_{sp}(\omega)$ is a strong increasing function of ω, so that n_ω freezes out first for low energy photons.

We thank L. Bildsten and O. Blaes for discussions. This research was supported by NASA and NSERC.

REFERENCES

1. Thompson, C., & Duncan, R.C., in preparation (1992).
2. Duncan, R.C., & Thompson, C., Astrophys. J., 392, L9 (1992).
3. Paczyński, B., Acta. Astron., 42, 145 (1992).
4. Duncan, R.C., Li, H., & Thompson, C., this volume.
5. Usov, V.V., Nature, 357, 472 (1992).
6. Thompson, C., & Duncan, R.C, Astrophys. J., in press.
7. Shapiro, S.L. & Lightman, A.P., Astrophys. J., 207, 263 (1976).
8. Ghosh, G.L., & Lamb, F.K., Astrophys. J., 232, 259 (1979).
9. Schmidt, G.D., in IAU Colloq. 95: Faint Blue Stars, ed. A.G.D. Davies-Phillip et al. (Schenectady N.Y.: L. Davis Press, 1989), p. 377.
10. Golenetskii, S.V., Ilyinskii, V.N., & Mazets, E.P., Nature, 307, 41 (1984).
11. Van Riper, K.A., Astrophys. J., 372, 251 (1991).
12. Silant'ev, N.A., & Yakovlev, D.G., Astrophys. Space. Sci., 71, 45 (1980).
13. Davies, S.R., & Coe, M.J., Monthly Notices. Roy. Astron. Soc., 245, 268 (1990).
14. Iwasawa, K., Koyama, K., & Halpern, J.P., Publ. Astron. Soc. Japan, 44, 9 (1992).
15. Rothschild, R.E., Lingenfelter, R.E., Seward, F.D., & Vancura, O., this volume.
16. Verbunt, F., in Neutron Stars and Their Birth Events, W. Kundt ed. (Kluwer: Dordrecht, 1990), p. 179.
17. Goldreich, P., & Reisseneger, A., Astrophys. J., 395, 250 (1992).
18. Friman, B.L., & Maxwell, O.V., Astrophys. J., 232, 541 (1979).
19. Urpin, V.A., & Yakovlev, D.G., Sov. Astron., 425, 24 (1980).
20. Itoh, N. et al., Astrophys. J. 273, 774 (1983).
21. Blaes, O., Blandford, R.D., Goldreich, P., & Madau, P., Astrophys. J., 343, 839 (1989).
22. Takatsuka, T., Prog. Theor. Phys., 48, 1517 (1972).
23. Van Riper, K.A., Astrophys. J. Supp., 75, 449 (1991).
24. Pethick, C.J., & Ravenhall, D.G., in Texas/Eso-Cern Astrophysics Symposium ed. J.D. Barrow et al. (New York: New York Academy of Sciences: 1991), p. 503.
25. Adler, S., Bahcall, J., Callan, C., & Rosenbluth, M., Phys. Rev. Lett., 25, 1061 (1970).

A BURST OF SPECULATION

J. I. Katz
Washington University, St. Louis, Mo. 63130

ABSTRACT

Self-consistent models of gamma-ray burst source regions at 100 Kpc distance are possible if the radiating plasma is confined to very thin sheets, and I estimate parameters. Energy sources might be elastic (by starquakes) or magnetic (by reconnection), but mechanisms remain obscure. I discuss a very speculative model involving collisions between comets in a hypothetical inner Oort cloud.

INTRODUCTION

Discoveries by the BATSE[1,2] have reduced theories (and theorists) of gamma-ray bursts (GRB) to a state of confusion. The field is nearly as open as it was when GRB were first announced in 1973. In order to understand GRB, it will be necessary to answer some questions about the physical conditions within them:

1. Is the energy of a GRB (or a sub-burst, in the case of GRB with distinct sub-structure) released promptly in the form of energetic particles and radiation, which gradually dissipate, or is the release continuous, with an instantaneous power proportional to the instantaneously observed intensity?
2. How do GRB time histories indicate characteristic physical time scales?
3. Are there any intermediate reservoirs of energy between its ultimate source and the radiating plasma (for example, vibrations in models driven by elastic energy)?
4. Is the geometry of the source region line-like, sheet-like, or sphere-like?
5. What is the optical depth of the source region?
6. Does an individual GRB have preferred power levels and decay time scales, as may be suggested by repetitive sub-burst structure in at least one observed GRB[2]?
7. How isotropic or beamed is the radiation pattern?
8. Does the time history of observed intensity indicate the time history of total radiated power, or is an anisotropic beam pattern variable or rotationally modulated?
9. Is the continuum radiation mechanism annihilation, bremsstrahlung, curvature radiation, or some other process?
10. Is the energy distribution of the particles in the source region strongly nonthermal or partly thermalized?
11. Is the radiating plasma freely expanding or magnetically trapped?

The answers to these physical questions are closely related to the unsolved astronomical questions of the distances to GRB and the existence of counterparts in other energy bands. The astronomical questions having so far proved intractable, the attention of the theorist should turn to the physical questions.

ESTIMATES

Arguments[3,4] concerning the reported[5] 511 KeV annihilation line in the March 5, 1979 GRB led to the inference that the emitting region was a geometrically and optically thin sheet. Although this soft gamma repeater (SGR) may not be representative of GRB, and the status of the annihilation line in GRB spectra is controversial, it may still be useful to consider this geometry. Magnetic reconnection, in analogy to a Solar flare, was one of the first suggested models of GRB[4,6], and would be expected to release energy in a thin sheet.

Consider a pair or electron-proton sheet plasma of thickness L, with particle energies $\sim mc^2$, where m is the electron mass, and density n. The transverse optical depth τ is

$$\tau \sim n\sigma_0 L, \tag{1}$$

where the cross-section $\sigma_0 = (e^2/mc^2)^2$ is roughly applicable to Coulomb and Compton scattering and to two photon annihilation and pair production. The characteristic power density (per unit area) is

$$P \sim n^2 mc^3 \sigma_0 L \sim \frac{n^2 e^4 L}{mc}. \tag{2}$$

Define a dimensionless length $\ell \equiv Lmc^2/e^2$ and power density $p \equiv P\hbar^3/m^4 c^6$. Then

$$n \sim \frac{\tau}{\ell \alpha^3} \frac{m^3 c^3}{\hbar^3}, \tag{3}$$

where $\alpha \equiv e^2/\hbar c$ is the usual fine structure constant. Use of (1) tacitly assumes that τ is not large, as implied by the observed nonthermal spectra. Similarly, from (2):

$$n \sim \left(\frac{p}{\ell \alpha^3}\right)^{1/2} \frac{m^3 c^3}{\hbar^3}. \tag{4}$$

Equating (3) and (4) yields

$$\ell \sim \frac{\tau^2}{p\alpha^3}. \tag{5}$$

These estimates lead to an energy content per unit area

$$\Sigma \sim nmc^2 L \sim \tau \frac{m^3 c^6}{e^4} \sim 1.0 \times 10^{19} \tau \text{ erg/cm}^2, \tag{6}$$

and a radiation time scale (also the annihilation time scale $t_a \sim 1/n\sigma_0 c$)

$$t_r \equiv \frac{\Sigma}{P} \sim \frac{\tau}{p} \frac{\hbar^3}{me^4} \sim 2 \times 10^{-17} \left(\frac{\tau}{p}\right) \text{ sec.} \tag{7}$$

In order to maintain a radiating plasma there must be a continuing injection of energy from an electric field. An elementary estimate of the electrical conductivity is

$$\sigma_{El} \sim \frac{e}{mc\sigma_0} \sim \frac{mc^3}{e^2} \sim 10^{23} \text{ sec}^{-1}. \tag{8}$$

The required electric field E may be estimated from the condition of energy balance

$$P = \sigma_{El} E^2 L, \tag{9}$$

with the result

$$E \sim \left(\frac{p\alpha^3}{\ell}\right)^{1/2} \frac{m^2 c^4}{e^3} \sim \frac{p\alpha^3}{\tau} \frac{m^2 c^4}{e^3} \sim 2 \times 10^9 \frac{p}{\tau} \text{ cgs.} \tag{10}$$

It is necessary to replenish annihilating pairs. The fastest processes, with cross-sections $\sim \sigma_0$, are $e^+ + e^- \to \gamma + \gamma$ and $\gamma + \gamma \to e^+ + e^-$, which conserve the sum of the numbers of leptons and photons. To supply an escaping flux of gamma-rays requires injection of both energy and particles. The electric field directly supplies only energy. Particle number may be resupplied by processes such as $\gamma + e^{\pm} \to e^{\pm} + \gamma + \gamma$, $e^+ + e^- \to e^+ + e^- + \gamma$, $e^+ + e^- \to \gamma + \gamma + \gamma$, and $\gamma + \gamma \to e^+ + e^- + \gamma$, with cross-sections $\sim \sigma_0 \alpha$. Perhaps more important may be synchrotron radiation and magnetic one-photon pair production, which occur because leptons and photons are produced and scattered with large cross-field momenta.

The electric field (10) also produces a mean leptonic drift velocity $\sim c$. This may lead to a two-stream or ion-acoustic plasma instability, constrained by the magnetic field to be one-dimensional. The effective conductivity may be reduced far below (8). Such an increased resistivity in regions of high current density is a familiar feature of magnetic reconnection. Quantitative understanding would require plasma simulations which include collisional as well as collective processes.

GAMMA-RAY BURSTS AT 100 KPC?

The BATSE data[1] demonstrated the impossibility of a Galactic disc origin of all GRB, and increased the attractiveness of earlier suggestions[7-9] that they are distributed in an extended halo of radius ~ 100 Kpc. If so, then their luminosities approach those of the March 5, 1979 event in the LMC, and many arguments[3-5,10,11] made for it may be applied to GRB in general.

A typical GRB luminosity at 100 Kpc is 10^{41} erg/sec, corresponding to an observed flux $\sim 10^{-7}$ erg/cm^{2}sec. With an effective radiating area of 10^{12} cm^2, $P \sim 10^{29}$ erg/cm^{2}sec and $p \sim 2 \times 10^{-7}$. Then (5) yields $\ell \sim 10^{13}\tau^2$ and $L \sim 3\tau^2$ cm. In the absence of plasma instabilities $E \sim 500$ cgs, insufficient to sustain a curvature radiation cascade. It is evident that t_r is extremely short unless p is so small that the GRB are within the Oort cloud.

A power density $P \sim 10^{29}$ erg/cm^{2}sec corresponds to a black body of effective temperature $T_e \sim 20$ KeV, inconsistent with the observed hard spectra. At semi-relativistic energies equilibration by Compton scattering is rapid, even in the absence of true absorption, so that the more energetic part of the spectrum must be produced in optically thin regions. This argument need not apply at photon energies $h\nu \ll 100$ KeV; self-absorption may limit reradiation from a neutron star's surface.

At least one GRB reported at this meeting[2] apparently consisted of distinct but very similar sub-bursts, each with a pronounced time-skewness[12] consisting of a rapid rise and more gradual decay over roughly one second. This suggested injection of energy into a reservoir, followed by its gradual radiation. It is apparent that pair plasma cannot be such a reservoir. Although a magnetic field of 10^{12} gauss provides sufficient stress to confine 10^{41} erg (10^{20} gm) of pair plasma, its opacity $\kappa \sim e^4/m^3c^4 \sim 100$ cm^2/gm would lead to an optical depth $\tau \sim 10^{10}$, inconsistent with the emergent hard and nonequilibrium spectrum. If confined, the energy would leak out over a time $\sim \tau r/c \sim 3 \times 10^5$ sec, also inconsistent. Similarly, its energy density of $\sim 10^{23}$ erg/cm^3 corresponds to a black-body temperature of ~ 160 KeV. An unconfined fireball of this temperature would lead to a flash of gamma-rays with mean energy ~ 500 KeV, but with a thermal spectrum and a duration $\leq r/3c \sim 10^{-5}$ sec[10]. Thus pair gas cannot be an energy reservoir, whether magnetically confined or exploding in a fireball; energy must be continuously replenished.

An unconfined fireball continuously resupplied with energy will have a lower energy density. The energy density is $\sim P/c$, and $\tau \sim Pr\kappa/3c^3 \sim 10^5$, still enough to ensure thermalization of the spectrum. This problem is exacerbated at cosmological distances. Adiabatic expansion has the effect of collimating the particle and photon momenta, but preserves the equilibrium spectral shapes, inconsistent with observation.

Models of the type discussed here suffer from the well-known problem of a high pair-production optical depth for MeV gamma-rays. Electric fields produce opposing streams of e^+ and e^- whose gamma-rays interact head-on. The well-known solution of relativistic collimation may work if the radiating particles have sufficient energy and the majority are collimated. This is naturally obtained in an electrically driven nucleus-electron plasma in an erupting flux loop, in which the leptons are predominantly e^-, all of which are accelerated in the same direction.

J. I. Katz 1093

SOURCES AND MECHANISMS

The release of magnetic energy by reconnection and the release of elastic energy in starquakes have been considered as possible mechanisms of GRB at Galactic distances since their discovery[3,4,6,13]. They are plausible qualitative explanations of much of the phenomenology, including the observed zoo of diverse GRB shapes and durations. The suggestion that much of the energy release and radiation comes from a thin sheet may also be explicable if the immediate mechanism is magnetic reconnection. The gross energetics may be consistent with GRB at 100 Kpc distance[4,11,13].

It is also possible to consider[10] hybrid models, in which the eruption of current loops requires crust-breaking. The chief difficulty faced by magnetic reconnection is to release energy suddenly in regions of low optical depth. Dissipation requires resistivities between those of the interior (large) and vacuum (0); turbulent plasmas are plausible. The details of both physics and field geometry are likely to be complex.

A CRAZY IDEA

Comets in an inner Oort cloud may occasionally collide with velocities ~ 1 km/sec. It is conceivable, if implausible, that such a collision of cold masses of dirty ice might produce gamma-rays by electrostatic processes. Deformation of piezoelectric components, heating of pyroelectric components or frictional charging, as the comets splatter in a subsonic collision, might lead to multi-MeV potentials. Pure ice is believed not to be piezoelectric or pyroelectric, but this has been controversial[14], and dirty ice or comets could be more complex. Ice has been reported[14] to be triboluminescent, indicating the production of multi-eV potentials. To produce gamma-rays requires that MeV potentials accelerate electrons through the vacuum. This is not unprecedented; pyroelectrics accelerate electrons to sufficient energy to produce X-rays[15]. Complex intensity histories might be attributed to complex collision geometries. Efficiency requires that no current flow through solids, where electrons would thermalize; this is consistent with the high resistivity and dielectric strength of cold ice.

Although the microscopic physics of these electrostatic processes is poorly understood, it is possible to examine the energetics. Consider a mass M of comets, each of size a and mass ρa^3 ($\rho \approx 1$ gm/cm^3), symmetrically filling a sphere of radius R centered on the Sun. There are $N = M/\rho a^3$ comets, with number density $n \sim M/\rho a^3 R^3$. A collision cross-section a^2 leads to a mean collision time $t_c \sim a\rho R^{7/2}/M(GM_\odot)^{1/2}$. The total kinetic energy is $E \sim GM_\odot M/2R$, and the mean power density at Earth is

$$F \sim \frac{E\epsilon}{4\pi R^2 t_c} \sim \frac{GM_\odot M\epsilon}{8\pi R^3 t_c} \sim \frac{(GM_\odot)^{3/2} M^2 \epsilon}{8\pi a\rho R^{13/2}}, \qquad (11)$$

where ϵ is the efficiency of converting kinetic energy to gamma-rays. The collision rate is

$$\dot{N} \sim \frac{M^2 \sqrt{GM_\odot}}{\rho^2 a^4 R^{7/2}}. \qquad (12)$$

The observed[16] $F \sim 3 \times 10^{-12}$ erg/cm^2 sec, and for the BATSE $\dot{N} \sim 10^{-5}$ sec^{-1}. Defining $M_{-3} \equiv M/10^{-3} M_\odot$ and $\epsilon_{-3} \equiv \epsilon/10^{-3}$ and combining (11) and (12) yields

$$R \sim \frac{\dot{N}^{2/45}(GM_\odot)^{11/45} M^{4/15}}{\rho^{4/45}} \left(\frac{\epsilon}{8\pi F}\right)^{8/45} \sim 3.2 \times 10^{15} M_{-3}^{4/15} \epsilon_{-3}^{8/45} \text{ cm}, \qquad (13)$$

and

$$a \sim \frac{M^{4/15}}{(GM_\odot)^{4/45} \rho^{19/45} \dot{N}^{13/45}} \left(\frac{8\pi F}{\epsilon}\right)^{7/45} \sim 1.2 \times 10^6 M_{-3}^{4/15} \epsilon_{-3}^{-7/45} \text{ cm}. \qquad (14)$$

The collision time is

$$t_c \sim \frac{\rho^{4/15} M^{1/5} (GM_\odot)^{4/15}}{\dot{N}^{2/15}} \left(\frac{\epsilon}{8\pi F}\right)^{7/15} \sim 1.0 \times 10^{17} M_{-3}^{1/5} \epsilon_{-3}^{7/15} \text{ sec,} \qquad (15)$$

consistent with the age of the Solar System of 1.5×10^{17} sec if $\epsilon > 10^{-3}$.

This model makes a number of predictions. It cannot explain any repeating source. The distance scale (13) implies a typical parallax of $1 \text{ AU}/R \approx 15'$ if $\epsilon \sim 10^{-3}$. This may be tested by overdetermined GRB position measurements, and the model may be excludable by extant data. The distribution of GRB on the sky should show an annually time-periodic dipole moment of $O(1 \text{ AU}/R)$, with its peak towards the Sun. There will be no cyclotron lines, and any annihilation lines will have no redshift. Atomic X-ray lines of abundant species may be present. There may be a simultaneous visible flash, of unpredictable intensity, resulting from scintillation by impacting energetic electrons. The predicted apparent visual magnitude of an $a = 12$ km comet at a distance R is about 32, but a comet disrupted by a collision into a spray of fragments could be many magnitudes brighter, depending on fragment size. Such a spray would disperse at an angular rate $\sim 1''/$day. Dust could also be an effective scatterer of sunlight. A fraction, perhaps large, of the kinetic energy of collision (ϵ^{-1} times the gamma-ray energy) would be thermalized and reradiated in the near-infrared; the time scale depends on the fragment size and the thermal conductivity, and is incalculable. The hypothesis of gamma-ray production could itself be tested in laboratory collisions, if the nature of cometary material were well enough known.

I thank I. A. Smith for discussions.

REFERENCES

1. C. A. Meegan, G. J. Fishman, R. B. Wilson, W. S. Paciesas, G. N. Pendleton, J. M. Horack, M. N. Brock and C. Kouveliotou, *Nature* **355**, 143 (1992).
2. G. J. Fishman, these proceedings.
3. R. Ramaty, R. E. Lingenfelter and R. W. Bussard, *Ap. Sp. Sci.* **75**, 193 (1981).
4. J. I. Katz, *Ap. J.* **260**, 371 (1982).
5. T. L. Cline, *Comments on Ap.* **9**, 13 (1980).
6. M. A. Ruderman, *Ann. N. Y. Acad. Sci.* **262**, 164 (1975).
7. T. L. Cline, in *High Energy Transients in Astrophysics*, ed. S. E. Woosley (AIP, N. Y., 1984), p. 333.
8. M. C. Jennings, in *High Energy Transients in Astrophysics*, ed. S. E. Woosley (AIP, N. Y., 1984), p. 412.
9. J. I. Katz, *High Energy Astrophysics* (Addison-Wesley, Menlo Park, 1987), p. 284.
10. B. J. Carrigan and J. I. Katz, *Ap. J.* **399**, 100 (1992).
11. J. I. Katz, *Ap. Sp. Sci.* in press (1992).
12. M. C. Weiskopf, P. G. Sutherland, J. I. Katz and C. R. Canizares, *Ap. J. (Lett.)* **223**, L17 (1978).
13. O. Blaes, R. Blandford, P. Goldreich and P. Madau, *Ap. J.* **343**, 839 (1989).
14. P. V. Hobbs, *Ice Physics* (Clarendon Press, Oxford, 1974).
15. J. D. Brownridge, *Nature* **358**, 287 (1992).
16. M. C. Jennings and R. S. White, *Ap. J.* **238**, 110 (1980).

TWO POPULATION – DISK & HALO – GAMMA-RAY BURST MODELS

J. C. Higdon
Joint Science Dept., Claremont McKenna College, Claremont, CA 91711
AND
Richard E. Lingenfelter
Center for Astrophysics and Space Sciences,
University of California, San Diego, La Jolla, CA 92093

ABSTRACT

We have calculated the gamma-ray burst statistical properties, $\langle V/V_{max}\rangle$ and $\langle \cos\theta\rangle$, as a function of the detection flux threshold, that would be expected from two population models consisting of galactic disk and halo populations. We show that a wide range halo distributions with core radii $\geq$ 15 kpc, mixed with local disk distributions, which account for $\sim$ 30% of the observed bursts, can have values of $\langle V/V_{max}\rangle$ and $\langle \cos\theta\rangle$ that are consistent with those measured by both the BATSE and the PVO burst detectors, whose detection thresholds differ by a factor of the order of 30 to 100. Such models can therefore account for the absorption features seen in a fraction of the burst spectra that require nearby ($<$ 1 kpc) galactic neutron star sources.

INTRODUCTION

BATSE measurements show[1] that the distribution of gamma-ray burst sources is spatially nonuniform, and isotropic on the sky. Such a distribution is consistent with either a galactic halo or a cosmological origin, but it is not consistent with that expected solely from a galactic disk population of neutron star sources. Nonetheless, a local ($<$ 1 kpc) galactic neutron star origin of at least a fraction of the bursts is strongly suggested[2-4] by observations of absorption lines at 20 to 50 keV, attributed to cyclotron scattering and absorption in 10^{12} G magnetic fields, and by the black body limits on the soft x-ray emission from bursts.

We suggested[5], however, that all of the observations can be reconciled, if there are at least two distinct populations of burst sources, a galactic disk neutron stars and galactic halo neutron stars, which differ by a factor of 10^5, or more, in mean luminosity.

Paczynski and Mao[6] have recently calculated the variation of the $\langle V/V_{max}\rangle$, a measure of spatial uniformity, versus $\langle \cos\theta\rangle$, a measure of angular isotropy, for combinations of disk and halo populations, as a function of increasing detector sensitivity. These calculations suggest that even though the BATSE values can be explained by such combined populations, less sensitive burst detectors, such as GINGA, SIGNE, KONUS or PVO, should have seen very strong anisotropies in the burst distribution. Paczynski[7] further argued that with such a two population model the less sensitive burst detectors should also find flatter cumulative distributions, or smaller values of $\langle V/V_{max}\rangle$, which is not consistent with the PVO and other experiments.

We show that these conclusions are not correct.

MODEL

To explore the variation of the gamma-ray burst statistical properties, $\langle V/V_{max}\rangle$ and $\langle\cos\theta\rangle$, as a function of the detection flux threshold, we consider two-population models consisting of galactic disk and halo populations of gamma-ray burst sources. We calculate two examples to show the wide range of possible distributions.

To model the density distribution of the galactic disk sources, we assume for these examples:

$$n(z,\rho) = n_o \exp^{-\rho/h - |z|/z_o},$$

where ρ is galactocentric distance projected along the plane, h of 3.5 kpc is the radial scale along the galactic plane, and z_o of 1 kpc is the scale height transverse to the plane.

To model the density distribution of the massive halo sources, we assume:

$$n(r) = n_c \text{ for } r \leq r_c,$$
$$\text{and}$$
$$n(r) = n_c[1 + (r/r_c)^2]^{-1} \text{ for } r \geq r_c,$$

where r is the galactocentric distance, and r_c is the radius of the uniform core of the halo. We employed two values for r_c: 15 and 40 kpc. A modest (~ 10) variation in luminosity is used, assuming a power-law luminosity distribution:

$$dn/dL \propto L^2, \quad \text{for } L_{min} \leq L \leq L_{max}$$

We set L_{min} and L_{max} equal to 10^{36}, and 10^{37} erg s^{-1}, respectively for the disk bursts, and 10^{42} and 10^{43} erg s^{-1} for the halo bursts.

The resultant variations of $\langle V/V_{max}\rangle$ and $\langle\cos\theta\rangle$, as a function of the detection flux threshold, for two-population, galactic disk and halo models of gamma-ray burst sources, shown in Figure 1 for the two different halo core radii, r_c of 15 and 40 kpc.

We compare these model calculations with the BATSE[1] and PVO[8] values of $\langle V/V_{max}\rangle$ and the corresponding limits on $\langle\cos\theta\rangle$. Since positions are known[9] for only 51 of the PVO bursts, the 1σ uncertainty on $\langle\cos\theta\rangle$ is 0.08.

As can be seen in Figure 1, the values of $\langle V/V_{max}\rangle$ and $\langle\cos\theta\rangle$ calculated for two-population models and varying detector sensitivity are consistent with those measured by both the BATSE and the PVO burst detectors for a wide range halo distributions with core radii ≥ 15 kpc, mixed with local disk distributions, which account for ∼ 30% of the observed bursts.

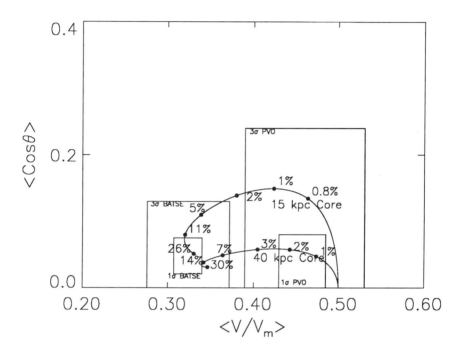

Figure 1. Calculated variations of $\langle \cos\Theta \rangle$ versus $\langle V/V_{max} \rangle$ for two-population, disk and halo, gamma-ray burst models with a disk scale height of 1 kpc and two different halo core radii of 15 and 40 kpc. The contribution of the disk population is given in percentages at selected points. Also shown for comparison are the values measured by the BATSE[1] and PVO[8] burst detectors, whose detection thresholds differ by a factor of the order of 30 to 100.

SUMMARY

We show that a wide range halo distributions with core radii ≥ 15 kpc, mixed with local disk distributions, which account for ∼ 30% of the observed bursts, can have values of $\langle V/V_{max} \rangle$ and $\langle \cos\theta \rangle$ that are consistent with those measured by both the BATSE and the PVO burst detectors, whose detection thresholds differ by a factor of the order of 30 to 100.

Such models can therefore account for the absorption features seen in a fraction of the burst spectra that require nearby (< 1 kpc) galactic neutron star sources.

Acknowledgements. JCH thanks NASA for support under grant NAG5 2010 and REL thanks NASA for support under grant NAGW 1970.

REFERENCES

1. C. Meegan, et al. in these proceedings.
2. J.C. Higdon & R.E. Lingenfelter, Ann. Rev. Astron & Astrophys., 28, 401 (1990).
3. T. Murakami et al. Nature, 350, 592 (1991).
4. D. Lamb, J.C.L. Wang, & I.M. Wasserman, Astrophys. J., 363, 670 (1990).
5. R.E. Lingenfelter & J.C. Higdon, Nature, 356, 132 (1992).
6. S. Mao & B. Paczynski, Astrophys. J., 389, L13 (1992).
7. B. Paczynski, Astron., 42, 1 (1992).
8. D. Hartmann et al., in Gamma-Ray Bursts, Cambridge Univ. Press, p. 45 (1992).
9. J. L. Atteia, et al., Astrophys. J., Supp, 64, 305 (1987).

MODELS FOR THE SPATIAL DISTRIBUTION OF GAMMA RAY BURSTS

Hui Li and Edison Liang
Department of Space Physics and Astronomy,
Rice University, Houston, TX 77251

ABSTRACT

We consider various extended halo models of gamma-ray bursts and their implications for the x-ray detectability of such bursts from the halo of M31. Using the sensitivity of the ROSAT PSPC we conclude that the expected x-ray flash rate is typically less than one per day. To seriously confront the extended halo model we need dedicated monitoring of M31 over periods substantially longer than a week. We also consider possible contributions to the faint BATSE events (e.g. those with flux < 10^{-6} erg cm^{-2} s^{-1}) by large flares from nearby flare stars. We obtain a limit to the luminosity function of such flares using constraint from BATSE data. If the BATSE database is indeed strongly contaminated by such stellar flares, which likely have an isotropic distribution, then the residual nonflare burst population could have a larger anisotropy than currently allowed.

INTRODUCTION

Recent BATSE results show that gamma-ray bursts (GRBs) are isotropic in the sky but spatially inhomogenous [1]. This leads to the conclusion that they must be either local, cosmological [2], in an extended Galactic halo [3], or the BATSE events may contain multiple populations. If GRBs are located in galactic extended halo, one interesting question is what are the observable consequences of bursts from the M31 halo. GRBs of M31 may be too weak for BATSE, but their accompanying x-ray flashes (XRF) may be detectable by ROSAT. Based on a recent paper [4], we investigate the expected XRF rates from M31 for different halo models (see detail in ref.5). Here we will make the following assumptions, namely: (1) the x-ray to gamma-ray peak flux is ~2%; (2) the GRB rate and spatial distribution in M31 are similar to those for our own Galaxy, which are tightly constrained by both BATSE and PVO results; (3) a ROSAT PSPC count rate (~ 0.1-2 keV) of 0.01 ct/s corresponds to an intrinsic M31 X-ray flux of 1.0 × 10^{37} ergs/s, or a gamma-ray flux of 5.0×10^{38} ergs/s. It appears feasible to adopt a ROSAT search threshold of 5 cts for a hundred second integration time [6]. This corresponds to a peak flux of 0.25 ct/s for an XRF with a triangular time profile of 40 second duration [7], but this can be easily scaled to other threshold values. We also assume that the energy flux P is linearly related to the BATSE C/C_{min} via $P = 10^{-7}$ ergs cm^{-2} s^{-1} (C/C_{min}). We compute the expected ROSAT XRF rate (per day) using the full 2° × 2° field of view (FOV) passing through the central region of M31. Results can be scaled to other FOVs.

MONO-LUMINOSITY L_γ MODEL

Assuming monoluminosity for all the bursts, we can adopt the following form for the spatial density of GRBs in the Galactic halo of radius R_h:

$$\dot{N}(R) = \frac{\dot{N}_0}{1 + \left(\frac{R}{R_c}\right)^\gamma} \qquad R \leq R_h$$

$$= 0 \qquad R \geq R_h \qquad (1)$$

© 1993 American Institute of Physics

where R is the distance from the Galactic center, R_c is the halo core radius, and γ is a dimensionless parameter. With the above simplified halo model, we can normalize the total number of GRBs in our Galactic halo using the BATSE detection rate, $N_B \approx 800$ per year [1]. Integrating along the line of sight, we can find out the number of bursts per year per solid angle in the direction of M31 as a function of N_B, R_c and γ. The expected XRF detection rates as a function of halo core radius at different count levels are summarized in Fig. 1. γ is 2 in the plot. When R_c is between 20 kpc and 40 kpc, the expected ROSAT rate is smaller than 1 per day though these bursts would be well above the ROSAT threshold.

LUMINOSITY FUNCTION MODEL

We now consider the case in which the spatial distribution is uniform but with a sharp cutoff at halo radius R_h and a luminosity function of the power law form: $dN/dL = N_0 L^{-\alpha}$, where N_0 = constant for $r < R_h$ and $N_0 = 0$ for $r > R_h$. Furthermore, if there is upper or lower bound to the intrinsic luminosity, then for $P > L_{max}/(4\pi R_h^2)$, we have $N(>P) = $ (constant) $P^{-3/2}$. For $L_{min}/(4\pi R_h^2) < P < L_{max}/(4\pi R_h^2)$, $N(>P)$ is a hybrid of $P^{1-\alpha}$ and $P^{-3/2}$ power laws. For $P < L_{min}/(4\pi R_h^2)$, $N(>P)$ is a constant. Contributions to the total BATSE rate come from both bursts intrinsically brighter than $(4\pi P R_h^2)$ and fainter than $(4\pi P R_h^2)$. For $\alpha = 1.7$, the fraction of BATSE bursts intrinsically brighter than $(4\pi P R_h^2)$ comes to $f = (2.5-\alpha)/[(2.5-\alpha)+(\alpha-1)] = 0.53$.

The extrapolated rate for the M31 halo will be the Galactic halo rate times the ratio of the sampling volumes. Hence the expected ROSAT rate for XRFs from the M31 halo is given by

$$N_{ROSAT}(>0.25 \text{ ct/s}) = 1.3 \times 10^3/\text{yr} \cdot f \cdot (R_h/R_g^{1.6}) = 3.6/\text{day} \cdot f \cdot (R_h/R_g^{1.6}) \quad (2)$$

where R_g is the halo radius of our Galaxy. Fig. 2 plots this rate as a function of R_h assuming $R_g = R_h$ and for other assumed values of ROSAT detection thresholds.

COMBINATION OF SPATIAL DENSITY GRADIENT AND LUMINOSITY FUNCTION

In this case, the same spatial density $N(R)$ as in Eq.(1) and luminosity function dN/dL as in luminosity model are used. Then the flattening in logN - logP will be a combined effect from both spatial density gradient and luminosity function.

We approximate the spatial density as constant out to R_c and $\sim R^{-\gamma}$ till R_h, then if the luminosity function has a large positive α (but < 2.5), the detector will preferentially pick up the nearby bursts which have constant spatial density, that will give a -3/2 power law in logN - logP curve. On the other hand, if α is a negative value, the detector will preferentially pick up the farther bursts which have a declining spatial density. The calculation is mainly the same as in the luminosity model, and for simplicity, R_c is 20 kpc, R_h is 100 kpc and γ is 2. It can be shown that the f-factor is ~ 0.99 as long as $\alpha < 1$, then gradually decreases to 0.36 when $\alpha = 2.25$. The expected rate of XRFs in M31 as a function of α is plotted in Fig. 3.

POSSIBLE STELLAR FLARE CONTRIBUTIONS TO THE BATSE GAMMA-RAY BURST DATABASE

The largest solar flares put out $\sim 10^{32}$ ergs with equal amounts in hard x-rays (> 20 keV), soft x-rays and optical-UV [8]. The largest flares observed from AD Leonis put out $\sim 10^{34}$ ergs in optical [9] with unknown amounts in x-ray and gamma ray. To estimate the all-sky event rate and gamma ray flux of such flares we need to know (a) the conversion factor from optical energy E to peak gamma-ray luminosity L; (b) the

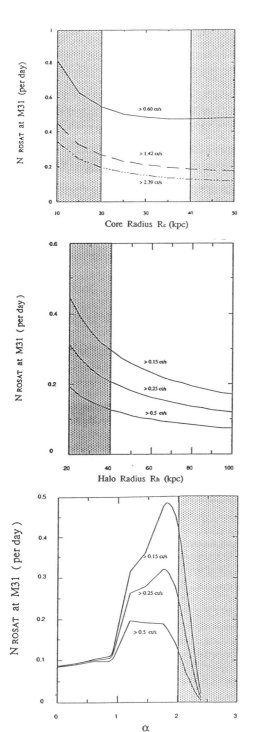

Fig. 1. Expected XRF detection rate of M31 as a function of halo core radius R_c at different count rate. The shaded regions are presently excluded by BATSE and PVO data. Curves from top to bottom corresponding to Rs = 50, 80 and 100 kpc respectively (from ref. 5).

Fig. 2. Expected XRF detection rate of M31 as a function of halo radius R_h with several assumed values of ROSAT threshold. The shaded region ($R_h < 40$ kpc) is excluded by the BATSE isotropy limits (from ref. 5).

Fig. 3. Expected XRF detection rate of M31 as a function of index α with count rate above 0.15 ct/s, 0.25 ct/s and 0.5 ct/s respectively. The index α needs to be between 0 and 2 to give a reasonable fit to the BATSE and PVO data. The peak around $\alpha = 1.8$ is caused by the increasing $N(> P)$ multiplied by the rapidly decreasing f-factor (from ref. 5).

spatial density of flare stars capable of producing flares > 10^{34} ergs, and (c) the luminosity function of large flares (> 10^{34} ergs).

In answering the above three uncertainties, we make the following assumptions: (a) the peak gamma luminosity L≈ 0.05ξ E erg/s, where 0.05ξ is a conversion factor and ξ is ~ O(1). (b) We leave the space density of flare stars n* as a free parameter, but expect it to be in the range of 0.01 - 0.1 per cubic pc [10].

Consider a homogeneous isotropic distribution of flare stars of space density n* with individual gamma ray flare (30-300 keV) luminosity function given by:

$$N(>L) \text{ flares per year per star} \begin{cases} = N_0 (L/L_0)^{-0.8} & L \leq L_0 \\ = N_0 (L/L_0)^{-\alpha} & L_0 \leq L \leq L_{max} \\ = 0 & L > L_{max} \end{cases} \quad (3)$$

where $L_0 = 5\xi \times 10^{32}$ erg/s corresponds to the observed optical output of E ≈ 10^{34} ergs. Due to the lack of data for E > 10^{34} ergs, we simply extrapolate the N(> L) by a power law with index $-\alpha$ and a cutoff luminosity L_{max}. Hence the flare rate of full sky with peak gamma flux > 10^{-7} erg cm^{-2} s^{-1} can be expressed as ($N_0 \approx 1$/yr based on AD Leonis optical data [9]):

$$N(>10^{-7}) = \left(\frac{130 \text{ flares}}{\text{year}}\right)\left(\frac{n*}{0.1 \text{per pc}^3}\right)\left[\left(\frac{\alpha}{1.5-\alpha}\right)\left(\frac{L_{max}}{L_0}\right)^{1.5-\alpha} + \left(\frac{0.8}{1.5-0.8} - \frac{\alpha}{1.5-\alpha}\right)\right] \quad (4)$$

This result is plotted in Fig. 4 for a range of n* and α values. Hence for n* = 0.05, ξ = 1, and $\alpha \approx 0.8$, stellar flares would account for all BATSE triggers if $L_{max} \approx 2 \times 10^{34}$ erg/s (sampling distance 40 pc) (see Fig. 4). This is in violation of BATSE data [1].

What fraction f of BATSE triggers ($\geq 10^{-7}$ erg cm^{-2} s^{-1}) can be due to flares? We can limit f by requiring that the $\log N_r(>P) - \log P$ slope $-\beta$ of the residual events after subtraction of the flare contribution be ≤ 0. Within statistical uncertainties we find the upper limit to f to be ≈ 0.5, i.e. ~ half, or 500/yr of BATSE triggers can be due to flares without violating the - 0.7 slope.

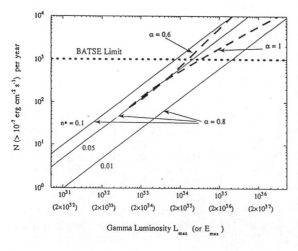

FIG. 4. Number of flares per year with peak flux > 10^{-7} erg cm^{-2} s^{-1} as a function of L_{max} or E_{max} (numbers in parentheses) since they are linearly related by the conversion factor 0.05ξ, ξ is assumed to be 1 in this figure. Three solid lines, all with α= 0.8 and slope 0.7 (=1.5 - 0.8), show the flare rates for different flare star space density n*= 0.1, 0.05 and 0.01 sequentially from top to bottom. Two dashed lines represent the flare rates if the luminosity function N($\geq L_0$= 5×10^{32} erg/s) has power law indices α= 0.6 and 1 respectively. The dotted horizontal line is the BATSE limit (~10^3/yr), namely the maximum contribution from flares (from ref. 11).

CONCLUSION AND DISCUSSION

Reviewing the above calculations, we note that the only factor that could significantly enhance the M31 XRF rate is if $R_h \gg R_g$ or if the density or rate of bursts in the M31 halo is $\gg$ that in our halo. We do note that there is a potential uncertainty of a factor of 4 in converting from the BATSE C/C_{min} to P due to intrinsic variation in C_{min}. All these may yet contribute to an eventual higher BATSE rate extrapolated to the full sky, leading to a correspondingly higher M31 XRF rate. As we have shown in several figures, the expected rates, to within a factor of 2, are typically less than 1 per day. So an M31 survey much longer than a week is required to really confront extended halo models.

We also find that it is possible for large stellar flares to be a significant component of the BATSE data and half of BATSE events could be flares without violating the observed logN - logP slope. Flare hypothesis can be tested by searching the smallest GRB error boxes, long term gamma ray monitoring of known flare stars, and examining BATSE occultation data at known flare star locations, etc.

REFERENCES

1. Meegan, C.A., Nature, **355**, 143 (1992)
2. Paczynski, B., Acta Astr., **41**, 257 (1991)
3. Li, H. & Dermer, C.D., Nature, **359**, 514 (1992)
4. Liang, E.P., ApJ, **380**, L55 (1991)
5. Li, H. & Liang, E.P., ApJ. Letter, in press (1992)
6. Trümper, J., private communication (1992)
7. Yoshida et al., PASJ, **41**, 509 (1989)
8. Priest, E., in Solar Flare Magnetohydrodynamics, ed. Priest, E. p1, Kluwer, Dordrecht (1981)
9. Hawley, S.L. & Pettersen, B.R., ApJ., **378**, 725 (1991)
10. Lacy, C.H. et al., Ap. J. Suppl. Ser., **37**, 313 (1978)
11. Liang, E.P. & Li, H., submitted to Nature (1992)

INSTRUMENTATION

CALCULATION OF THE INDUCED RADIOACTIVITY BACKGROUND IN OSSE

S. J. R. Battersby, J. J. Quenby

Imperial College, London SW7 2BZ, U.K.

C. S. Dyer, P. R. Truscott

DRA Farnborough, Hampshire, U.K.

N. D. A. Hammond, C. Comber

EDS SCICON, U.K.

J. D. Kurfess, W. N. Johnson, R. L. Kinzer, M. S. Strickman

Naval Research Laboratory, Washington DC

G. V. Jung

Universities Space Research Association, Washington DC

W. R. Purcell, D. A. Grabelsky, M. P. Ulmer

Northwestern University, Evanston, IL

ABSTRACT

A considerable background response is produced in OSSE detectors by internal radioactivity induced by cosmic rays and trapped protons. It cannot be eliminated by anticoincidence methods because of its delayed nature, and is subtracted using offset pointing data. A computational scheme for simulating the background has been developed in the last few years at DRA Farnborough[1], and is now being updated. Use of a radiation transport code to predict spallation rates of individual product nuclides in OSSE (where before only the total rate was calculated, individual rates being inferred from semi-empirical cross-sections) has produced results of low statistical error for a large number of nuclides. A library of response functions for the most significant of these and their descendents (a total of 117 nuclides) has been generated and used with previously developed codes to yield total predicted background due to cosmic ray and trapped proton induced radioactivity. This calculated spectrum is considerably closer to that observed than any previous prediction has been. We hope to be able eventually to improve on some aspects of the background subtraction method employed at present.

INTRODUCTION

The importance of the induced radioactivity background is its delayed nature. While charged particle events and events due to gamma-rays entering the detector from outside the aperture are vetoed by the active shields, and gamma-rays entering through the aperture but outside the field of view are stopped by the collimator, the decay of radioisotopes within the detector crystal itself cannot be eliminated in this way. Cosmic rays, geomagnetically trapped protons and albedo neutrons cause spallation and neutron capture in the material of the crystal producing these radioactive nuclides, and their subsequent decay produces β and γ-rays which generate a considerable response in the instrument (of the order of 1000 times the signal at minimum sensitivity). Betas produce continua, usually shifted by the production of a number of prompt gamma rays by the daughter nucleus. Electron capture modes and isomeric transitions produce lines, broadened by the energy resolution of the instrument. There is a particularly large contribution from the β^- decay of iodine 128, which is produced by neutron capture on iodine 127.

This background is dealt with by subtracting an interpolated response calculated from the offset pointing data of each detector, time variation coming from South Atlantic Anomaly passes and modulation of magnetic cutoff rigidity. In modelling the background directly we hope to be able to improve on this system in some areas, for example in the variation of certain peaks with time.

METHODS OF CALCULATION

The starting point for the calculation is obtaining particle fluxes for given orbits of the Compton Observatory[2]. At present this is done by interpolation and extrapolation from magnetic field and particle flux data taken by a number of spacecraft during the 1960s. The age of the data and uncertainty over the details of loss mechanisms in the upper atmosphere give uncertain absolute fluxes (to about a factor of two), though the spectra are well determined.

With the aquisition of a 486 computer for use with the various radiation transport codes, it has become feasible to calculate a large number of individual isotope production rates to reasonably low statistical error. These Monte-Carlo codes produce spallation and neutron capture rates for cosmic ray protons and alpha particles, for albedo neutrons, and for geomagnetically trapped protons. HETC/LHI[3] deals with high energy protons, neutrons and hadrons. It derives from the ICE (Intranuclear Cascade and Evaporation) code, treating the nucleus broadly as in the liquid drop model, the nucleons following a given density distribution each with a binding energy of about 7 MeV. The other codes employed are MORSE[3], dealing with neutrons of less than 15 MeV, and EGS[3], dealing with βs and γ-rays. Ground state and metastable product nuclei are not differentiated, so it is assumed that these are produced in equal proportions when such states are indicated in the nuclear structure data files.

Production rates, half lives and parentages are input to one of two programs to generate decay rates; DECAYCOS for cosmic rays, which assumes continuous irradiation, and DECAYTRP for trapped protons, which assumes continuous irradiation until the day in question and approximates that day's history with instantaneous doses during the day, the size of which is determined from the SAA doses recorded by the Compton Observatory. This is the only case of particle fluxes being taken directly from the Observatory at the moment.

To calculate the response function for a radioisotope, its decay scheme is read from the evaluated nuclear structure data files. A Monte-Carlo energy loss code called BANKER[4] generates the instrument response to the decay of each nuclide, taking into account the effect of escaping γ-rays (beta particles are assumed not to escape) and the energy resolution of the crystal. Individual response functions are then compiled into a library, which along with the decay rates form the input to BGTRS. This sums the contributions from all nuclides to give a total predicted background spectrum.

RESULTS

Components of the predicted background are shown in fig 1, and a comparison with measured background (taken over the 4 to 5 orbits furthest removed from the last SAA pass) is shown in fig 2. The most significant improvement over previous calculations is the more detailed inclusion of secondary radiation; previously the effect of secondaries (which with such a massive spacecraft account for several times as much spallation as primaries) had been applied across the range of product nuclei, with semi-empirical cross sections determining individual production rates. However, since secondaries are of much lower energy than primaries, they tend to favour the production of nuclei closer to the iodine 127 and sodium 23 targets. The effect of this enhancement of certain rates can be seen in fig 2; the peaks at 634, 698 and 2330 keV (due to I^{126} and I^{124}) being good examples. The response functions used are for a 13 x 4 inch crystal, (the functions used before this calculation were calculated for a 2.75 x 2.75 inch crystal) and this also has an enhancing effect on gamma lines, since there are fewer escapes. Another element of this calculation which has not been carried out before is the simulation of α-particle spallation. Although alphas are fewer than protons in the cosmic ray flux (by a factor of 9), their contribution to total spallation is appreciable because they fragment to give more secondaries. This can be seen in the table below.

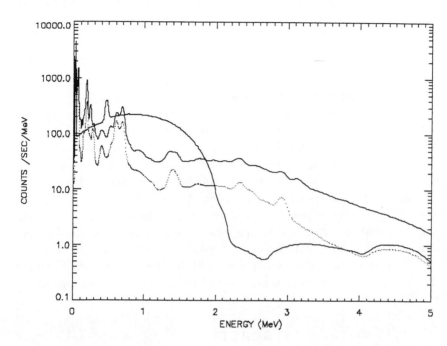

Figure 1: Predicted background. Cosmic ray spallation, CR neutron capture solid; neutron capture contribution mostly below 2 Mev. Trapped protons dotted.

Example spallation rates:

Element	A	CR protons	Production rate sec^{-1} CR alphas	Albedo neutrons	Trapped protons*
Iodine	126	50.5	16.1	2.4	39.6
	125	30.6	10.7	2.0	26.8
	124	18.6	6.2	1.4	19.0
	123	12.7	4.1	0.9	14.5
	122	9.5	2.7	11.3	0.6
Tellurium	126	11.6	4.1	0.6	7.6
Sodium	22	20.0	6.2	0.8	19.5

* worst case orientation.

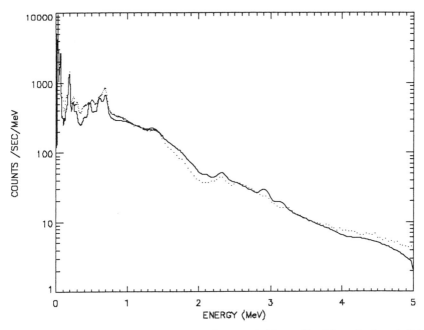

Figure 2: Comparison of measured background (dotted) with calculated, 76 days after launch.

DISCUSSION

Recent development of this scheme has shown that excellent agreement can now be achieved between calculated and measured background spectra in OSSE, and there is scope for further improvement. The prompt neutron capture feature at 7 MeV, and β production in the collimator need to be included, and particle fluxes measured by CO could be used for all the fluxes used in the calculation instead of the less reliable interpolated fluxes. We can then assess what improvements might be made on the offset pointing method of background subtraction.

REFERENCES

1. C. S. Dyer, P. R. Truscott, A. J. Sims, C. Comber, N. D. A. Hammond, Gamma Ray Observatory Science Workshop (1989), p. 521.
2. A. J. Sims, working paper SP(87) WP11, DRA Farnborough, (1987).
3. N. Hammond, C. Comber, C. S. Dyer, P. R. Truscott, A. J. Sims, First European Conference on Radiation an its Effects on Devices and Systems (1991), p. 553.
4. S. M. Seltzer, Nucl. Inst. Meth. 127, p.317.

COMPTEL HIGH ENERGY NEUTRON RESPONSE

T. J. O'Neill*, F. Ait-Ouamer, T. A. Roth,
O. T. Tumer, R. S. White, A. D. Zych
University of California, Riverside, CA 92521

ABSTRACT

Results of Monte Carlo simulations of COMPTEL's neutron response at 17.2, 22.0, 35.7 and 77.0 MeV are presented using the combined MCNP-v4.2[1] (Monte Carlo Neutron Photon) and LAHET[2] (Los Alamos High Energy Transport) codes developed at Los Alamos. One million incident neutrons at each of the four energies were generated at 0^0 incident angle (normal incidence) on the COMPTEL instrument. Our analysis shows the detection efficiencies to be 0.13%, 0.14%, 0.08% and 0.05% at these energies and 0.09%, 0.1%, 0.05% and 0.02% when an ARM cut of 4^0 is applied. Energy resolutions ($\Delta E/E$) of 18% and 25% were obtained at 17.2 and 35.7 MeV. The low efficiency at 77.0 MeV precluded a reasonable measurement of the resolution. As we gain more confidence in MCNP/LAHET's ability to model the neutron response of COMPTEL over the full neutron energy range expected for solar flare neutrons it can be used as a valuable data analysis tool for interpreting individual solar neutron events and neutron induced background.

INTRODUCTION

Over the past 10 years, a major effort at Los Alamos National Laboratory has been the development of the code LAHET for the transport of nucleons, pions, and muons. LAHET can be used as an extension to MCNP-v3B for neutron energies above 20 MeV to several GeV. The output neutron particle history files from LAHET, which cuts off the particle energies at 20 MeV, can then be run through MCNP-v3B extending the neutron simulation down to very low energies. Monte Carlo Neutron Photon version 4.2 was released in March 1991; it features for the first time electron transport, which is crucial for simulating Compton scatter processes. This new version is not compatible with LAHET.

At UCR both MCNP and LAHET were modified extensively so that COMPTEL could be simulated properly. With LAHET, provisions were made for off-axis sources and inclusion of elastic cross-sections up to 100 MeV. MCNP-v4.2 was modified extensively to be compatible with LAHET. The large output track history of MCNP was also optimized so that large numbers of events could be run.

The full track information from both codes had to be obtained to handle the COMPTEL coincidence requirement. Here at UCR we developed a code that 1) combines the information from the two tracking files; 2) converts all particle energies to electron equivalent energies; 3) smears the positions, energies, time-of-flights, etc., using the COMPTEL resolutions; 4) determines coincidences, anticoincidences, and multiple hits; and 5) writes out a track file of triggered double scatter events. An analysis routine imposes additional threshold and time-of-flight (TOF) cuts on these events to produce detection efficiencies, measured energy, ARM (angular resolution measurement) and TOF spectra, and telescope response functions.

All of the information about an event is needed from both codes before the coincidence requirement can be established. The large intermediate disk space (>200 MB) needed by the simulation requires us to divide the calculation into multiple runs

---

* also at Riverside Community College, Riverside, CA

each consisting of 100,000 incident events and then to recombine the track files after the coincidence requirement is imposed.

PROGRAM OUTLINE

Our modified LAHET code system is shown in Figure 1. LAHET produces two output files: NEUTP for neutrons less than 20 MeV and HISTP for all the other secondary particles and photons. PHT accepts HISTP as input and produces a gamma file(GAMTP). The NEUTP and GAMTP files are merged with MRGNTP to act as a source for MCNP in a coupled neutron-photon problem. Our primary modification is the addition of the READER code which identifies coincidence events in COMPTEL's D1(>0.01 MeV) and D2(>0.2 MeV) modules with no energy loss(>0.05 MeV) in the individual charged particle shields. Energy losses due to secondary protons, deuterons, tritons, ^{3}He, and alpha particles are converted to electron-equivalent energy losses[3] and summed for each COMPTEL module. Recoil protons from D1 in MCNP are selected by identifying elastic scattered events with hydrogen. The TOF is found from the event times of the highest energy particles in D1 and D2. Two-dimensional interaction locations in each module are found from an energy-weighted average. The Z-location is taken as the midpoint of each module. The ANALYZE code applies more realistic energy thresholds and TOF cuts to the coincidence events and produces the quantitative response properties for neutrons incident on COMPTEL. The TRKANAL code summarizes the different types of neutron and photon/electron interactions contributing to "good" COMPTEL neutron events.

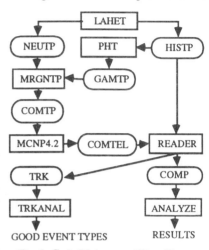

Fig. 1 Code Linkage and Data Flow

COMPTEL MODEL

The physical COMPTEL instrument model includes 7 D1 liquid organic(NE213A) and 14 D2 inorganic full-size scintillator modules with the two module arrays separated by 1.58 m. In the model, the D2 modules consist of sodium and barium instead of NaI(Tl) since the neutron cross-sections for iodine are not available yet. A cylindrical 1.57 cm ring of aluminum surrounding each D1 module simulates the mass of the D1 housing. Both the D1 and D2 module arrays are fully enclosed in plastic scintillator-charged particle shields 1.5 cm thick.

Table 1 gives the instrument parameters and event cuts applied by the ANALYZE code described above. Simulated events satisfying these cuts are assumed to be neutron n-p double scatter events provided their calculated event circles intersect the source direction within the neutron angular resolution of the telescope.

COMPTEL NEUTRON EFFICIENCIES(NORMAL INCIDENCE)

The absolute detection efficiencies for four energies are given in Table 2. Efficiencies are given for all double scatter events, events within 10^0 and events within 4^0 of the

Table 1 Model Parameters

Position Resolutions(Δx, Δy, 1σ):
  D1: 2.3 cm   D2: 2.4 cm
Timing Resolutions(Δt, 1σ):
  D1: 1.0 ns   D2: 1.0 ns
Energy Resolutions(ΔE, 1σ):
  D1: $1.1E(keV)^{0.57}$
  D2: $1.72E(keV)^{0.5}-11.8$
Electron Equivalent Energy Loss(E_e):
  D1(p, d, t)^3:
    $0.83E_p-2.82[1-\exp(-0.25E_p^{0.93})]$
  D1(α, ^{3}He)^3:
    $0.42E_\alpha-5.9[1-\exp(-0.065E_\alpha^{1.01})]$
  D2(p,d,t,^{3}He,α): $1.0E_p$
Data Cuts:
  D1: > 65 keV   Anti: > 50 keV
  D2: > 600 keV   TOF: < 40.7 ns
  E_n(TOF): < 200 MeV

Table 2 Absolute Detection Efficiencies

Energy (MeV)	All Events (%)	< 10° (%)	< 4° (%)
17.2	0.127 (0.005)	0.111 (0.005)	0.090 (0.004)
22.0	0.144 (0.006)	0.121 (0.006)	0.104 (0.005)
35.7	0.080 (0.004)	0.060 (0.004)	0.045 (0.003)
77.0	0.053 (0.004)	0.032 (0.003)	0.020 (0.002)

source direction. The errors presented are statistical. Figure 2 shows a typical TOF spectrum where incident neutrons produce secondary gamma rays and neutrons moving both from D1 to D2 and from D2 to D1. A typical n-p scatter angle distribution is shown in Figure 3. Small scatter angles are excluded by the D1 energy threshold and large scatter angles by the maximum allowed TOF and the telescope geometry.

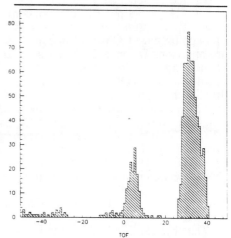

Fig. 2 TOF Spectrum for 17.2 MeV

COMPTEL ENERGY AND ANGULAR RESPONSES

The COMPTEL neutron ARM as expressed by the differences between the true and the measured n-p scatter angles for 17.2, 35.7 and 77.0 MeV are shown in Figures 4, 5, and 6, respectively. At 17.2 MeV the resolution is 5°(FWHM).

The reconstructed neutron energy distributions for 17.2 and 35.7 MeV are presented in Figures 7 and 8. An ARM

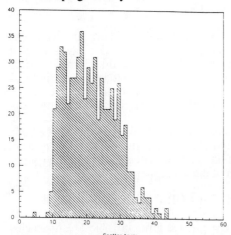

Fig. 3 Scatter Angle Plot for 17.2 MeV

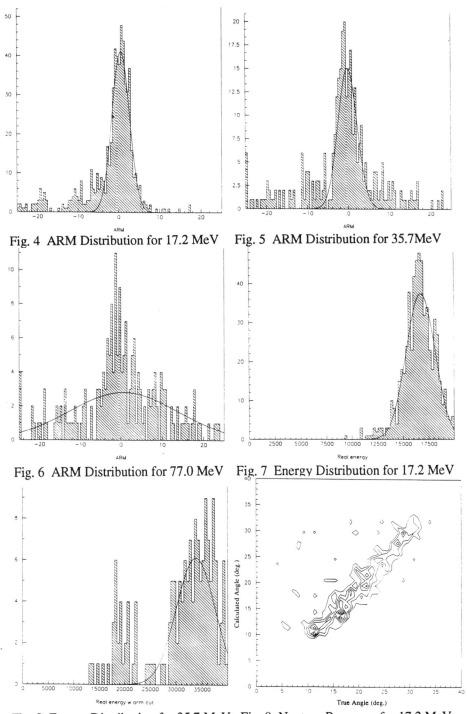

Fig. 4 ARM Distribution for 17.2 MeV
Fig. 5 ARM Distribution for 35.7 MeV
Fig. 6 ARM Distribution for 77.0 MeV
Fig. 7 Energy Distribution for 17.2 MeV
Fig. 8 Energy Distribution for 35.7 MeV
Fig. 9 Neutron Response for 17.2 MeV

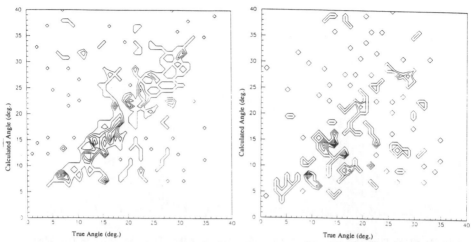

Fig. 10 Neutron Response for 35.7 MeV   Fig. 11 Neutron Response for 77.0 MeV

angle requirement of <4⁰ was imposed here. Energy resolutions of 18%(FWHM) and 25%(FWHM) were obtained corresponding in both cases to a time uncertainty of about 230 s(FWHM) in the neutron TOF over a distance of 1 AU. In Figure 8, at 35.7 MeV about 20% of the detected source events are due to n-C scattering in the D1 modules. These show up at reduced energy and can be due to both elastic and inelastic scattering from carbon.

The COMPTEL neutron response function at different energies can be represented by a two-dimensional event distribution of the true scatter angle versus the measured scatter angle. With limited statistics, these are shown in Figures 9, 10, and 11 for 17.2, 35.7, and 77.0 MeV. Additional Monte Carlo events will be necessary for these responses to be useful for imaging neutron sources.

CONCLUDING REMARKS

Simulations at additional energies up to 200 MeV and better statistics are needed. These results will be compared with neutron calibration results for a science model of COMPTEL. Already, a good comparison has been made at 17.2 MeV.[4] The calculation of reliable absolute detection efficiencies depends critically on how well the Monte Carlo model reproduces the actual neutron detection process. A more complete mass model of the COMPTEL instrument will be added to the simulation in the future.

We wish to acknowledge the support of NASA (Grant No. NAG 5-1493) for this work.

REFERENCES

1. J. S. Hendricks and J. F. Briesmeister, IEEE Nuc. Sci. Trans. 39, 1035 (1992).
2. R. E. Prael and H. Lichtenstein, Los Alamos National Laboratory
   Report, LA-UR-89-3014 (1989).
3. R. A. Cecil, et. al., Nuc. Inst. Meth. 161, 439 (1979).
4. T. J. O'Neill, et. al., Proc. Compton Observatory Sci. Workshop
   (NASA Conf. Publ. 3137), 109 (1991).

BATSE BURST TRIGGERS FROM CYGNUS X-1 FLUCTUATIONS

Charles Meegan, Gerald Fishman, Robert Wilson
NASA/Marshall Space Flight Center

William Paciesas
University of Alabama, Huntsville

ABSTRACT

Fluctuations from Cygnus X-1 are occasionally intense enough to cause a BATSE burst trigger. We employ a Bayesian statistical analysis to distinguish these events from true cosmic gamma-ray bursts. The technique efficiently separates these two classes of events. The location errors for Cygnus X-1 fluctuations provide an upper limit of about 13 degrees for the location errors of gamma-ray bursts near the BATSE threshold.

INTRODUCTION

Hard X-ray emission from the black hole candidate Cygnus X-1 is known to be highly variable on short time scales. Large fluctuations can exceed the BATSE burst threshold. About 30 such events have been detected so far. The BATSE capability to determine directions to all triggers is essential in identifying the Cygnus X-1 fluctuations. They are manifested as an excess of events centered on the known location of Cygnus X-1. They are also typically only slightly above threshold. These two characteristics, location and intensity, allow an efficient separation of Cygnus X-1 fluctuations and bursts using a simple Bayesian statistical analysis.

THE METHOD

Bayes theorem expresses the probability that an hypothesis is correct, given the observed values of parameters, as a function of the expected distributions of those parameters and of the prior probability that the hypothesis is true. For this application, the parameters are the angle θ between the computed direction and Cygnus X-1, and the intensity. For convenience, $V/V_{\max}$ is used as a measure of intensity. The application of Bayes theorem to this problem results in the following equation:

$$\frac{P(B|\theta,v)}{P(C|\theta,v)} = \frac{P(\theta|B)}{P(\theta|C)}\frac{P(v|B)}{P(v|C)}\frac{P(B)}{P(C)},$$

where θ is the angle between the computed direction and the direction to Cygnus X-1,
v is the computed $V/V_{\max}$ for the event,
$P(B|\theta,v)$ is the probability that the event is a burst, given θ and v,
$P(C|\theta,v)$ is the probability that the event is a Cygnus X-1 fluctuation, given θ and v,
$P(\theta|B)$ is the distribution in θ for bursts,

$P(\theta|C)$ is the distribution in θ for Cygnus X-1 fluctuations,
$P(v|B)$ is the distribution in $V/V_{\max}$ for bursts,
$P(v|C)$ is the distribution in $V/V_{\max}$ for Cygnus X-1 fluctuations,
$P(B)$ is the prior probability that the event is a burst,
$P(C)$ is the prior probability that the event is a Cygnus X-1 fluctuation.
Also, $P(B|\theta,v) + P(C|\theta,v) = 1$.

The probability $P(\theta|B)$ that a burst will be θ degrees from Cygnus X-1 is obtained from the observation that the burst distribution is isotropic, at least to sufficient accuracy for this purpose. Thus,

$$P(\theta|B)d\theta = d\Omega/4\pi = 1/2 \sin\theta d\theta.$$

The probability $P(\theta|C)$ that a Cygnus X-1 fluctuation will have a location θ degrees from the true location is obtained by approximating the error θ as the positive half of a normal distribution, and determining the standard deviation by fitting to the observed distribution of triggers in the neighborhood of Cygnus X-1. Such a fit yields a standard deviation of 13 degrees. Figure 1 shows the distribution of Cygnus X-1 fluctuations as a function of angle from the known position of Cygnus X-1. The solid line represents the distribution calculated as described above.

Cygnus X-1 fluctuations exhibit the least accurate locations, since they are short, weak, and superimposed on a fluctuating background. Consequently, the measured errors provide an upper limit to the errors expected for the weakest gamma-ray bursts.

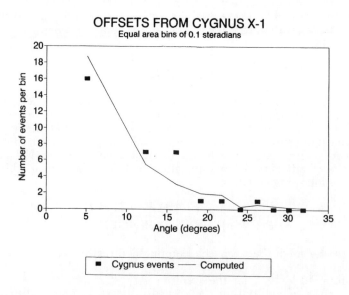

Figure 1. Angular Distribution of Cygnus X-1 Fluctuation

The probability $P(v|B)$ that a burst will have intensity v ($= V/V_{\max}$) is determined from the $V/V_{\max}$ distribution for bursts far from the direction to Cygnus X-1. This is well approximated as $P(v|B) = 1/2v^{-1/2}dv$. Figure 2 shows the distribution in $V/V_{\max}$ for bursts; the solid line is the calculated distribution.

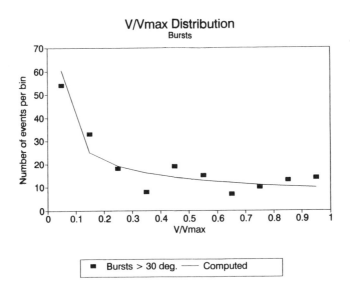

Figure 2. $V/V_{\max}$ Distribution for Bursts

The probability $P(v|C)$ that a Cygnus X-1 fluctuation will have a specified v is determined by modeling the fluctuations as a normal distribution in C_{max}/C_{min}, with a standard deviation larger than would hold for a Poisson distribution in the number of photons. The standard deviation is determined by the observation that 50% of Cygnus X-1 fluctuations have $C_{max}/C_{min} < 1.073 (v > 0.9)$. The standard deviation so determined is 0.345. The distribution is normalized so that the integral probability above $C_{/rmmax}/C_{\min} = 1$ is unity. Thus,

$$P(v|C) = Ae^{-1/2(x/.345)^2},$$

where $x = C_{\max}/C_{\min} = v^{-2/3}$ and A is a normalization constant. Figure 3 shows the $V/V_{\max}$ distribution of Cygnus X-1 fluctuations; the solid line is the calculated distribution.

The ratio of prior probabilities $P(B)/P(C)$ is simply the total number of bursts divided by the total number of Cygnus X-1 triggers. This is estimated to be 0.2 from the angular distribution of events near the location of Cygnus X-1.

RESULTS

The computed probabilities for events within 30 degrees of Cygnus X-1 are listed in Tables 1 and 2. Table 1 lists the probability $P(B)$ that the event is a burst for each of the 33 events that were ultimately classified as Cygnus X-1

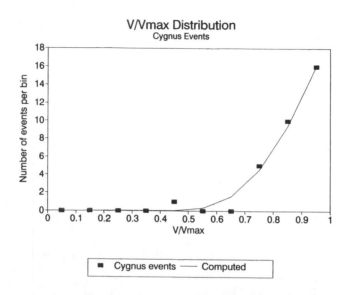

Figure 3. $V/V_{\max}$ Distribution of Cygnus X-1 Fluctuations

fluctuations. Only a few events have $P(B) > 0.1$. Two anomalous events deserve comment. Trigger 1040 appears to have a high probability (0.77) of being misclassified. However, this event occurred when Cygnus X-1 was particularly active. Trigger 254 had a higher than normal trigger threshold because it occurred during the readout of a previous burst. It is, in fact, almost certainly a burst. An estimate of the number of bursts erroneously classified as Cygnus X-1 fluctuations can be made by summing $P(B)$ over these events. That estimate is 2.7 events.

Table 2 lists the probability $P(B)$ that the event is a burst for those events within 30 degrees of Cygnus X-1 that were ultimately classified as bursts. Note that $P(B)$ is negligibly different from 1 for most of these events. This is a consequence of the fact that most bursts have intensities well above threshold, where the probability of a Cygnus X-1 fluctuation is negligible. An estimate of the number of Cygnus X-1 fluctuations erroneously classified as bursts can be made by summing $P(C)$ for these events. That estimate is 0.1 events, indicating that the burst catalog probably has no contamination from Cygnus X-1 fluctuations.

CONCLUSIONS

1. We estimate that perhaps 2 or 3 bursts have been misclassified as Cygnus X-1 fluctuations.
2. There are probably no Cygnus X-1 fluctuations misclassified as bursts.
3. BATSE burst locations have typical total errors (systematic plus statistical) of about 13 degrees near the trigger threshold.

Trigger No.	θ	V/V_{max}	$P(B)$
254	16.1	0.890	0.078
183	26.1	0.980	0.222
296	11.9	0.900	0.040
592	9.5	0.810	0.049
613	13.2		0.272
892	5.6	0.830	0.022
967	9.3	0.849	0.036
993	8.7	0.782	0.054
1004	13.7	0.962	0.037
1014	18.6	0.962	0.077
1033	14.0	0.962	0.039
1035	17.1	0.925	0.074
1040	5.8	0.486	0.774
1047	4.6	0.983	0.007
1049	5.5	0.887	0.147
1054	8.7	0.860	0.031
1062	14.9	0.903	0.061
1072	2.3	0.974	0.004
1080	11.9	0.818	0.066
1098	22.8	0.946	0.156
1113	15.0	0.896	0.064
1119	15.5	0.794	0.131
1134	3.2	0.989	0.005
1137	12.5	0.941	0.035
1138	2.7	0.972	0.004
1189	3.8	0.778	0.021
1236	7.8	0.990	0.013
1248	5.6	0.839	0.020
1277	5.4	0.935	0.011
1294	10.3	0.764	0.080
1305	16.8	0.946	0.064
1317	16.8	0.946	0.064
1710	5.1	0.773	0.043

Table 1: Cygnus X-1 Fluctuations

Trigger No.	θ	V/V_{max}	$P(B)$
105	26.3	0.005	1.000
249	13.3	0.001	1.000
432	21.7	0.089	1.000
469	4.4	0.016	1.000
540	25.0	0.250	1.000
577	29.3	0.260	1.000
594	21.1	0.11	1.000
612	17.7	0.020	1.000
878	10.0	0.240	1.000
906	26.5	0.140	1.000
936	26.6	0.043	1.000
973	20.5	0.110	1.000
1036	25.2	0.570	0.925
1076	18.6	0.050	1.000
1102	11.8	0.200	1.000
1118	15.5	0.280	1.000
1122	27.7	0.008	1.000
1123	14.4	0.450	0.977
1150	10.6	0.070	1.000
1244	15.8	0.140	1.000
1365	1.2	0.133	1.000
1729	16.6	0.125	1.000
1962	26.3	0.300	1.000

Table 2: Bursts

IDENTIFICATION OF EVENTS OBSERVED BY BATSE

R. S. Mallozzi, W. S. Paciesas
Dept. of Physics, University of Alabama, Huntsville, AL, 35899

C. A. Meegan, G. J. Fishman, R. B. Wilson
NASA – Marshall Space Flight Center, ES-64, Huntsville, AL, 35812

ABSTRACT

The Burst and Transient Source Experiment (BATSE) on the Compton Gamma Ray Observatory (GRO) routinely observes events which are classified into one of six categories: gamma-ray bursts, solar flares, magnetospheric events, Cygnus X-1 or other known source fluctuations, unknown events, and soft gamma repeaters. The use of BATSE's eight independent detectors enables us to confidently distinguish between these classes, with approximately 25% of recorded triggers being classified as gamma-ray bursts. Methods of classification of events are presented. An analysis of previously classified triggers is discussed, and an estimate of the fraction of events which have been incorrectly identified is given.

INTRODUCTION

The Burst and Transient Source Experiment (BATSE) on the Compton Gamma Ray Observatory (GRO) is an eight-component detector system capable of nearly full-sky observations. This unique design, explained in detail elsewhere,[1,2] provides unprecedented opportunity to observe the gamma-ray sky. Briefly, each of the eight independent detector modules, mounted on the eight corners of the GRO spacecraft, consists of two NaI(Tℓ) scintillation detectors: a large area detector (LAD), optimized for temporal resolution, and a spectroscopy detector (SD), optimized for energy resolution.

A consequence of the nearly 4π field of view of BATSE is the large number of events which are observed. These events are analyzed daily, and assigned to one of six categories: gamma-ray bursts (GRBs), solar flares, magnetospheric events, discrete source fluctuations, unknown events, and soft gamma repeaters (SGRs). Several criteria are used in conjunction to discriminate among these categories and to ensure that each event is classified correctly.

SOLAR FLARES

The BATSE experiment is able to use its eight independent detector modules to obtain an approximate location of an event. The method of event localization is explained by Brock et al.[3] Localization to a point near the known location of the Sun is a strong indication that an event could be of solar origin.

Many flares observed by BATSE can be confirmed by simultaneous observation by the Geostationary Operational Environmental Satellites (GOES).†

---

† GOES data is routinely published in Solar-Geophysical Data prompt reports, available from the National Oceanic and Atmospheric Administration.

The appearance of an event in the GOES data, combined with the BATSE location, enables one to classify an event as a solar flare with confidence. Of 485 solar flares recorded by BATSE, approximately 77% have been confirmed by the GOES data.

The BATSE detector system employs four-channel discriminators that span the approximate energy ranges (in keV) of 20–50, 50–100, 100–300, and above 300. Since the majority of solar flares have soft spectra relative to GRBs, events clearly observed in channel 4 (above ~ 300 keV), and not visible in the GOES data, are not likely to be solar flares. Those events that are not apparent in the GOES data, but that exhibit little or no emission above ~ 300 keV and localize near the Sun, are generally classified as solar flares. Many BATSE flares show no emission above 100 keV.

MAGNETOSPHERIC EVENTS

The BATSE instrument is occasionally triggered by bremsstrahlung radiation from charged particles. Frequently, these particles exhibit their spiral motion along magnetic field lines. This appears in the eight LADs as similar count rates in opposite facing detectors, indicating that the detected radiation is produced in or near the spacecraft. Although an intense GRB can also sometimes be seen in all eight detectors due to scattering of gamma-rays off the atmosphere of Earth, the flux through the Earth-facing detectors is significantly less than that through the detectors that directly observe the burst.

Radiation from charged particle interactions at some distance from GRO also may be detected. This appears as elevated count rates in four detectors on the same side of the spacecraft. The locations of these events are near the horizon of Earth. Unlike GRBs, the count rates peak at different times in each detector.

The GRO position at the time of an event aids in identifying charged particles. Figure 1 shows the GRO position during magnetospheric events.

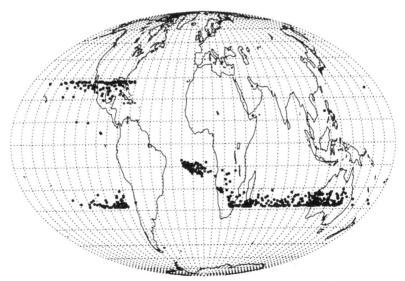

Figure 1. GRO Location During Magnetospheric Events.

These events are shown to cluster near high latitudes and the SAA, while GRBs and solar flares are isotropically distributed throughout the orbit. The effects of the powerful NWC VLF transmitter, located at $\sim 114°$E on the coast of western Australia and known to cause charged particles to precipitate into the atmosphere,[4,5] are clearly evident in Figure 1.

DISCRETE SOURCE FLUCTUATIONS

- Cygnus X-1

Prior to the outburst of the hard x-ray transient GROJ0422+32,[6] all discrete source fluctuations that triggered BATSE were due to the binary x-ray source Cygnus X-1. Normally the variations of Cyg X-1 are not large enough to meet the BATSE trigger criteria. Occasionally, the source flares enough to cause the instrument to trigger. These events can be identified from their locations and the fact that the source is visible to BATSE at the time of the trigger. BATSE also sometimes enters trigger mode when a source emerges from behind the limb of Earth. These events are easily classified since they appear as occultation steps at predictable times.

- GROJ0422+32

During the period of strong emission from the source GROJ0422+32, BATSE recorded an excess number of triggers from this location. The majority of these triggers occurred as the source emerged from behind the Earth, resulting in confident classification of these events. Fluctuations of this source also induced triggers; these were identified by their locations and their soft spectra relative to GRBs.

UNKNOWN EVENTS

Of 1800 events observed, only eight have been placed in this category. Five are very short duration events, of which four have locations consistent with a point source below the horizon of Earth. This cannot be accounted for by an error in the location algorithm, since BATSE burst locations agree with those of the Interplanetary Network. Lack of data due to the failure of the on-board tape recorders accounts for two. The remaining event has been placed in this category due to ambiguity in the location estimate.

SOFT GAMMA REPEATERS

BATSE has currently categorized three events as SGRs. These events are identified primarily by their extremely soft spectra (characteristic energies of ~ 30 keV) and typical durations of ~ 0.1 second.[7] Although the SGRs classified by BATSE appear to originate from the same source, the criteria of a soft, exponential spectrum, short duration, and frequently, a flat-topped temporal profile should enable confident identification of any previously unobserved repeaters.

GAMMA-RAY BURSTS

The BATSE instrument observes ~ 0.85 gamma-ray bursts per day. One criterion helpful in identifying GRBs is the spectral hardness of these events.

Events classified as GRBs show emission above 100 keV, and frequently are observed above 300 keV. An analysis of 248 BATSE GRBs shows that approximately 51% of these events have emission above 300 keV, while a similar analysis of 390 BATSE solar flares shows that approximately 14% have emission above 300 keV.

The independent detectors also aid in classifying events which are candidates for GRBs. An event which has similar count rates in opposite detector pairs is not likely to be a burst. Figure 2 shows the count rate ratio $\frac{C_{opp}}{C_{bright}}$ for three types of events, where C_{bright} is the count rate in the brightest detector, and C_{opp} is the count rate in the detector opposite to the brightest. The figure clearly indicates that magnetospheric events separate into two general classes. Those with a ratio $\simeq 1$ induce similar count rates in opposite-facing detectors, while those with a ratio $\simeq 0$ are due to radiation emitted at some distance from the spacecraft. GRBs and solar flares have ratios clustered about zero, indicating that the count rate in the brightest detector is much larger than that in the opposite detector.

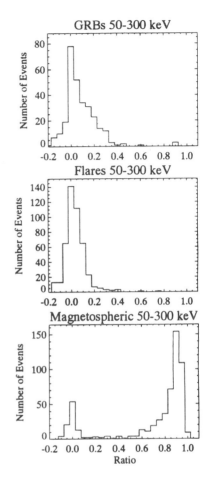

Figure 2. Count Rate Ratios ($\frac{C_{opp}}{C_{bright}}$) for Several Types of Events.

ESTIMATES OF MISCLASSIFICATION OF EVENTS

The BATSE systematic location error is currently estimated to be $\sim 2.5°$. Since GRBs are measured to be isotropically distributed, one expects $\sim 1.1\%$ of total bursts observed to be within 12° of the Sun. Thus of 424 bursts observed, ~ 5 bursts should locate within 12° of the Sun. BATSE has classified as GRBs only two events which locate within 12° of the Sun. These data suggest that approximately 3 (± 3) GRBs have been classified as solar flares due to their proximity to the Sun. Conversely, since only two GRBs locate near the Sun, a maximum of two solar flares may have been classified as bursts.

Results of a Bayesian statistical analysis of events originating from the direction of Cygnus X-1 indicate that probably zero GRBs have been incorrectly identified as being due to the discrete source, and approximately three true Cyg X-1 events have been placed in the category of GRBs.[8] A similar analysis has not been performed on events due to the new source GROJ0422+32, since the majority of these events occurred as the source was emerging from behind the Earth. This source has been active for ~ 60 days; an average of 0.85 bursts per day indicates that ~ 1 burst should have been detected within 12° of the source. During this period, zero bursts were detected within this region, implying that it is possible that one GRB has been incorrectly identified as a fluctuation of GROJ0422+32. A summary of misclassification estimates is given in Table I.

Table I. Estimates of Misclassification		
Type	Misclassified as	Number
GRB	Solar flare	~ 3 (±3)
GRB	Cygnus X-1	~ 0
GRB	GROJ0422+32	≤ 1
Flare	GRB	≤ 2
Cygnus X-1	GRB	~ 3

SUMMARY

Solar flares are easily classified using the BATSE location, the energy range in which the event is apparent, and frequently, the appearance of the event in GOES data. Magnetospheric events are distinguished by similar count rates in opposite detector pairs or by rates in four detectors with the peak rates occurring at different relative times. These events also tend to occur when GRO is at high latitude or near the SAA, and often exhibit broad, smooth temporal profiles. Discrete source fluctuations have known locations in the sky, enabling one to recognize these events with relative ease. However, a Bayesian statistical analysis is performed on events which are thought to be due to fluctuations of Cyg X-1, since GRBs can randomly appear from the same location. Triggers due to sources entering the field of view of BATSE appear as occultation steps in the relevant detectors. Eight of 1800 recorded events are unknown. Soft gamma repeaters are identified by their soft spectra and by repetitive consistent locations. Gamma-ray bursts are distinguished from other events by the energy range of emission and locations not consistent with those of the Sun or other known sources of x-ray or gamma-ray emission.

REFERENCES

1. G. J. Fishman et al., Proceedings of the GRO Workshop, 2-39,(1990).
2. J. M. Horack, Development of the Burst and Transient Source Experiment, NASA-RP 1268, (1991).
3. M. N. Brock et al., Proceedings of Gamma-Ray Bursts, 383, (1991).
4. D. W. Datlowe, W. L. Imhof, J. Geophys. Res., **95**, 6477, (1990).
5. U. S. Inan, H. C. Chang, R. A. Helliwell, J. Geophys. Res., **89**, 2891, (1984).
6. W. S. Paciesas et al., IAU Circular #5580, (1992).
7. C. Kouveliotou et al., Ap. J., **392**, 179, (1992).
8. C. A. Meegan, BATSE Burst Triggers from Cygnus X-1 Fluctuations, these proceedings, (1992).

MODELING THE GAMMA-RAY BACKGROUND ON BATSE

Brad C. Rubin

USRA/NASA – Marshall Space Flight Center, ES-62, 35812

B. A. Harmon, M. N. Brock, G. J. Fishman, C. A. Meegan, R. B. Wilson

NASA – Marshall Space Flight Center, ES-62, 35812

W. S. Paciesas

Dept. of Physics, Univ. of Alabama, Huntsville, 35899

M. H. Finger

CSC/CGRO Science Support Center, ES-62, 35812

J. C. Ling, R. T. Skelton, Wm. A. Wheaton

NASA – Jet Propulsion Laboratory, Pasadena, CA 91009

D. E. Gruber

CASS – Univ. of California at San Diego, 92093

ABSTRACT

The gamma-ray background on the BATSE experiment is a complex, time-varying combination of several factors. These include diffuse galactic and cosmic background; activation induced by particle fluxes in the SAA; atmospheric gamma-rays; discrete source contributions; and other factors. We have developed techniques for fitting the Continuous and Discriminator BATSE data with multi-component models. Some of these components are intended to directly represent factors comprising the background, while others are more empirical in nature. The model is used to "detrend" the data and can be used for on-ground folding in pulsar analysis and searches for long period pulsations, searches for weak, long duration gamma-ray bursts and transients, background subtraction from known gamma-ray bursts and other transients, and construction of the burst sky exposure map. Components of these models can be used to measure the diffuse background and obtain refined measurements of discrete sources.

INTRODUCTION

The BATSE experiment consists of 8 uncollimated detector modules providing full sky coverage excepting the portion (about π steradian) blocked by the Earth[1]. The total counting rate in a Large Area Detector (LAD) varies between 3000 and 6000 counts/sec. For each detector, continuous data are telemetered to the ground in 16 spectral channels between 20 keV and 2 Mev every 2.048 seconds, and discriminator data in 4 channels every 1.024 seconds. Four channel data from Spectroscopy Detectors (SD), with different energy ranges in different detectors, is received every 2.048 seconds. We have developed a program to fit a model to these data. Because of the wide energy bands of each of these data types, we regard this as a continuum model and have not attempted to include lines. Figure 1 shows a plot of continuous data for one day along with the model fit. At JPL, a similar, but not identical model, has been used to fit the same data[2].

PHYSICAL BACKGROUND

Previous satellite and balloon experiments have measured various components of the gamma-ray background.[3] Physical sources of background include:

1. Astrophysical Backgrounds
   - Diffuse galactic and cosmic γ-rays
   - Strong, discrete sources, including the Crab, Cygnus X-1, etc.
2. Prompt local γ-ray emission
   - Cosmic ray secondary scattering on atmosphere and spacecraft
   - Interaction of neutrons with detectors
3. Time-delayed, induced radioactivity
   - Activation by protons and neutrons in the SAA
   - Cosmic ray induced activation
   - Activation following strong solar flares
4. Transient or rapidly varying backgrounds
   - Bremsstrahlung γ-rays of trapped electron precipitation
   - Discrete source fluctuations, solar flares, γ-ray bursts

MODEL

The strength of each component in the model is determined by a linear least squares fit to the data. Periods of time when solar flares, errors in the telemetry stream, and other transients exist are not included in the data used to compute the fit. A standard model we have been using includes the following components:

1. A constant term. This is related to diffuse background and long lived radioactive decay species.
2. An expansion in powers of the cosine of the angle between the normal to the detector and the center of the Earth. Here we have used the first three powers. Empirically, these terms roughly account for blockage of the diffuse background when the Earth is in the detector field of view. Their strength decreases with increasing energy.
3. A cosmic ray predictor, derived from HEAO-1 data, which is proportional to cosmic rays coming from the skyward direction; this predictor is strongly correlated with the McIlwain-L parameter.
4. An expansion in terms, each of which is the product of the cosmic ray predictor and a power of the cosine of the Earth angle. These are empirical terms, roughly related to the atmospheric background. The first three terms in the expansion are used.
5. A term which decays with a 25 minute half-life following each SAA passage. This term accounts for the beta decay of I^{128}.
6. Discrete source terms. Rapidly varying sources, like Cygnus X-1, Scorpius X-1, and Vela X-1 are represented by a separate term each time they rise above the horizon. Steady sources like the Crab Nebula are represented by a single term in each fit.

A typical fit to one day of data in one channel will have almost 50 terms: 1 constant term, 1 cosmic ray term, 3 earth angle terms, 3 cosmic ray × earth angle terms, 8 SAA decay terms, and 16 terms for each of two discrete sources. Other terms and models are possible and can be accomodated by the program which calculates and fits these model components. This program outputs the coefficient for each term and, optionally, the residuals at each data point.

RESULTS

For one full day of BATSE continuous data we perform a separate fit in each of 16 channels for each of 8 detectors, routinely obtaining χ^2_ν between 1 and 2. The residuals are within the statistical uncertainties of the data, but do contain systematic runs which may last on the order of 100 seconds. For almost all of about a dozen days of continuous data which we have examined, the model represents the systematic trends in the data quite well. However, the components can sometimes be unphysical. For example, it is possible to obtain significant negative coefficients for strong, rapidly varying sources, and SAA activation terms.

At low energies (below 50 keV) the strongest terms are the constant and earth angle terms, due to the dominance of the diffuse background. The strength of the cosmic ray term increases with increasing energy, and this becomes the dominant term above 320 keV

Figure 1 shows the data, model, and components for a one day fit to continuous data from detector 0 in the 20- 30 keV energy range on truncated julian day 8541 (October 12, 1991). Note that the data was fit as counts per 2.048 seconds, but is plotted here as counts per second. Thus, the statistical errors on the data points are in the range of 10-15 counts per second. The model contains 48 terms fit from 33232 data points with a chi-square per dof = 1.46. The largest terms are the constant and earth angle terms. Only the sum of terms which are part of an expansion is plotted as a separate component. Two bright sources, Cygnus X-1 and Scorpius X-1 were visible from this detector. Note that although in general, the model fits quite well, there do appear to be unmodeled short time scale (few hundred second) variations.

UTILITY

We may separate the applications of background models into those which require only the total model for detrending the background, and those which make measurements from components of the model. Current techniques for detrending background for observations of long period pulsars are inadequate for pulsars with periods greater than 100 seconds. There are now efforts to incorporate the background model into these observations.[4,5] The model should also be useful for constructing the background during data gaps, which will provide knowledge of the overall burst trigger efficiency. Source terms may be used to make measurements of discrete source fluxes sensitive to time variability on a timescale as small as 90 minutes. This is the focus of the effort at JPL.

FUTURE PLANS

In the future we plan to investigate possible improvements to the model which will both reduce the systematic runs in the residuals and improve measurements of the various components. One possibility is to require a direct, physical interpretation for each term. That is, to incorporate models of the atmospheric and spacecraft background.[3,6,7] The signal-to-noise ratio for weaker terms could be improved by simultaneous fitting of detectors and/or channels, possibly through an iterative procedure. We are also seeking a dramatic increase of the speed of obtaining results for the entire BATSE dataset. The present FORTRAN program requires 6 hours of CPU time on a VAX 6400, for one day of continuous data. An effort is now underway to port the program to the Concurrent Supercomputing Consortium (CSCC) 64-node iPSC/860 parallel supercomputer at Caltech. We expect this machine to improve the run time by a factor of 30 or more.

SUMMARY

Knowledge of the gamma-ray background is essential for obtaining the goals of a number of scientific programs. We have found that a relatively simple method effectively models trends in the BATSE data on the few orbit to one day timescale at the full time resolution of the data. In the future we plan to improve this model and make these results available to the community.

REFERENCES

1. Fishman, G. J., et al., "BATSE: The Burst and Transient Source Experiment on the Gamma-Ray Observatory", Proceedings of the GRO Science Workshop, Goddard Space Flight Center, ed. W. N. Johnson, (April 1989), p.2-39.
2. Skelton, R. T., et al., "Status of the Enhanced Earth Occultation Analysis Package for Studying Point Sources", these proceedings.
3. Dean, A. J., Lei, F., Knight, P. J., Space Science Reviews, 57 (109) (1991).
4. Wilson, R. B., "BATSE Observations of Isolated Pulsars and Disk-Fed X-Ray Binaries", these proceedings.
5. Prince, T. A., "BATSE Observations of Wind-Fed and Transient Binary Pulsars", these proceedings.
6. Ling, J. C., J. Geophys. Res., 80 (22) (3241) (1975).
7. Dean, A. J., et al., Astron. Astrophysics, 219 (358) (1989).

Legend

———————	Model
··················	Constant Term
— — — — — —	Expansion in Cosine of Earth Angle
— — — —	HEAO-1 Cosmic Ray Predictor (Sky flux)
—··—··—··—··—	Cosmic Ray * Expansion in Cosine of Earth Angle
—···—···—···—.	SAA activation with 25 minute half-life
———————	Cygnus X-1
··················	Scorpius X-1

Figure 1. Continuous data, model, and components for TJD 8541 in the 20-30 keV energy band on detector 0. Only the data used in the fit is shown. Portions with no data are SAA passages or magnetospheric events. The horizontal axis is in seconds at 2.048 second resolution.

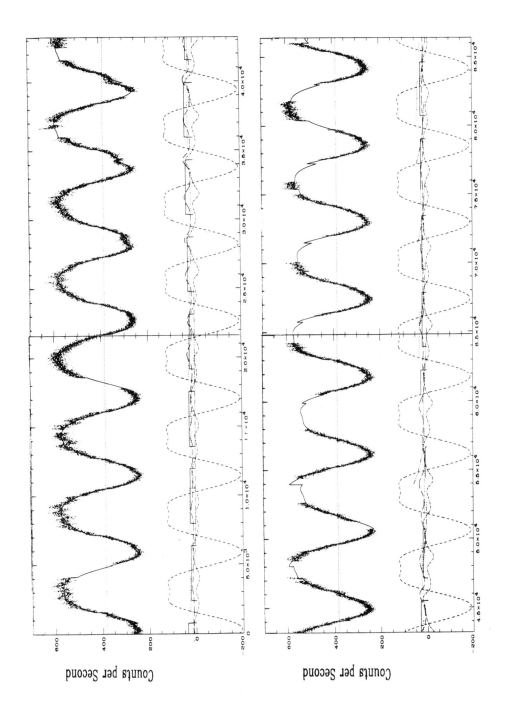

SELECTION BIASES IN GAMMA-RAY BURST DETECTORS

G. Pizzichini
Istituto TESRE/CNR, 40127 Bologna, Italy

ABSTRACT

I discuss a geometrical selection effect which can strongly affect both the total and the peak intensity distribution of Gamma-Ray Bursts (GRB) and therefore it heavily distorts, for example, the Log N-Log S, Log (N>Cmax/Cmin) and V/Vmax plots, where S, Cmax, and V are the burst fluence, the maximum counting rate and the volume in space where the event would have been observable, while Cmin and Vmax are the minimum detectable counting rate and the maximum observable volume in space at the time of the event. Of course V is related to Cmax and Vmax is related to Cmin. This selection effect is present in all the instruments which require simultaneous detection of the event by at least two differently oriented detector units. The same effect changes the detection limit, the effective observable fraction of the sky and Cmin, the minimum detectable counting rate.

INTRODUCTION

Perhaps the most important and well known result of the BATSE experiment on GRO /1,2/ is that, while it finds an isotropic distribution of bursts on the celestial sphere /3/, which suggests that the sources are isotropically and uniformly distributed in the sky, the Log N-Log S and the V/Vmax distribution show a gradually increasing lack of weak events: the total difference is quite consistent /4,5/. Clearly, the same is true for the integral distribution of the peak burst counting rates, corrected for the minimum detectable counting rate (Cmax/Cmin), since V/Vmax = (Cmax/Cmin) exp(-3/2).
Pending a positive identification of GRB sources, these results have re-opened the discussion on the galactic or extragalactic origin of GRB and on single or two population models.
One should however take into account as much as possible the selection effects due to the trigger criteria, because they can heavily affect the detection of weak events by a trigger instrument, for example for events with a slow rise time or a very short duration. The spectral hardness can also be important, not only because a "hard" event has a smaller Cmax than a softer one with the same fluence S, but also because it has a different albedo from the earth's atmosphere: this changes the detection limit.

A GEOMETRICAL SELECTION EFFECT

All instruments which, like BATSE, are made of several detectors pointing in different directions and require that the event be detected by at least two units, are also affected by geometrical effects. In what follows the shape of the surface of the detector units is not important, as long as they all have the same effective area. The effect is due only to the fact that the detector units point in different directions and that the burst is required to satisfy the trigger criteria in at least two of them.

A burst close to the "original" detection limit, that is the detection limit which has been set for the instrument, shall be detected by one of the units only if it has a direction almost normal to it, but then its angle with all the other units shall be too large: for an octahedron the angle between the normals to two faces is $70°32'$.

The most favorable case for a weak burst is when it falls on an edge of the octahedron. In that case the true detection limit is the original one multiplied by $1./\cos(35°16')$, that is 1.22 . In fact, for all the events, even for the strongest ones, the actual detection limit is a function of the angle between the burst direction and the detector's unit which was "second best" in pointing towards it. This multiplies Cmin by a factor, in the case of an octahedron, which varies between 1.22 and 3. for all the events and affects both the V/Vmax distribution and the Log (N>Cmax/Cmin) curve.

For weak bursts, less than 3 times the minimum required number of sigmas, much more important is the fact that the fraction of the sky where the burst can be detected is zero at the true detection limit and reaches total coverage only when a burst is intense enough to be detected even at $70°32'$ from the normal to one detection unit (see fig. 1). This means that total sky coverage is active only when a burst reaches at least $1./\cos(70°32')=$ three times the original detection limit.

For six detectors on a cube the effect is much stronger. The true detection limit is at least 1.41 times the original one (this happens when the direction of the event falls on the edge between two units), but it becomes infinite when the burst direction is ortogonal to one unit and therefore parallel to the adjoining ones and total sky coverage is achieved only for events with infinite peak intensity.

The figures show the percentage of sky coverage as a function of the maximum allowable angle between the direction of the burst source and the normal to the (at least two) instrument units which will detect it. The true detection limit is derived from the original one by dividing it by the cosine of the same angle. As explained above, this angle cannot be smaller than one half of the angle between the normals to two adjoining units and of course it increases with the intensity of the events.

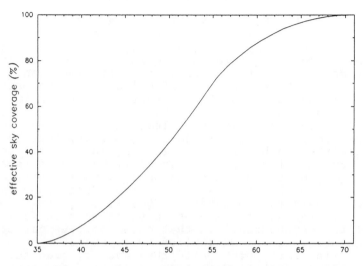

Fig. 1: percentage of sky coverage as a function of the maximum allowable angle between the burst direction and the instrument units which detect it, for an instrument with eight units arranged as an octahedron. The true detection limit is equal to the original one divided by the cosine of this angle.

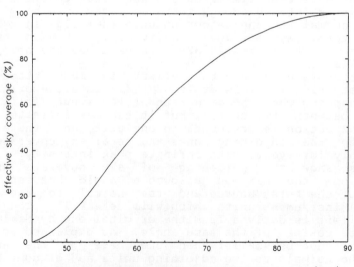

Fig. 2: same as fig. 1, but for six units arranged as a cube.

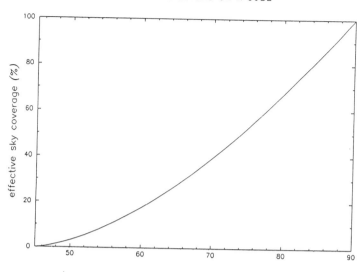

Fig 3: same as fig. 1, but for four units arranged as a cube without the top and bottom units. Figure 1 shows the case of eight detectors arranged on an octahedron, figure 2 the much more severe case of six detectors arranged as a cube and figure 3 the case of four detectors arranged as a cube without the top and bottom faces: this shall be the case of the GRB monitor on the SAX satellite /6/. The computation is purely geometrical: it does not take into account, for example, the thickness of the crystals, which, in the case of SAX, shall somewhat improve the sky coverage.

CONCLUDING REMARKS

When a burst is detected, the fraction of the sky from which that event was actually detectable depends on the burst intensity. This fraction must be used for renormalizing both the V/Vmax and Log(N>Cmax/Cmin) or Log N-Log S distributions: in the case of an octahedron the renormalization is necessary up to bursts which reach three times the original detection limit. Moreover, for all the bursts, even the strongest ones, the true Cmin, and consequently also the true Vmax, which depend on the relative alignment of the instrument and the burst direction, must be used in any statistical distribution involving these quantities.
This effect shall not however be apparent in the distribution of sources in the celestial sphere if the instrument changes enough its orientation while it is active, but, if not properly corrected, produces an apparent paucity of weak events.
It must also be noted that when an instrument has, for any

reason, a variable threshold, the Log N-Log S curve is affected in a different way than the Log (N>Cmax/Cmin) curve and the V/Vmax distribution, because in one case one adds together events with the same intensity, regardless of the threshold at the detection time, while in the other case events with different intensities are added together if, at the detection time, they had the same ratio to the detection limit (see also /7/). If, as it normally happens, the instrument misses more and more events approaching the detection limit, the "bending over" of the curve shall "spread" over the Log N-Log S curve between the smallest and the highest threshold. It shall instead be all added together and therefore reinforced near to the detection limit, where Cmax/Cmin and V/Vmax are both equal to one, for both these distributions. For the same reason in the same distributions there could be a small decrease at he limit of the most intense events, not because they were not detected, but becuse, if there was a different Cmin (and Vmax) at the time of detection, they were assigned to different sections of the curve.

REFERENCES

1. G. J. Fishman et al., Proceedings Gamma Ray Observatory Science Workshop, W. N. Johnson ed., p. 39 (1989).

2. G. J. Fishman, this conference.

3. M. T. Stollberg, M. N. Brock, G. N. Pendleton, G. J. Fishman, C. A. Meegan, R. B. Wilson and W. S. Paciesas, this conference.

4. C. A. Meegan, G. J. Fishman, R. B. Wilson, W. S. Paciesas, G. N. Pendleton, J. M. Horack, M. N. Brock and C. Kouveliotou, Nature 355, 143 (1992).

5. C. A. Meegan, G. J. Fishman, R. B. Wilson, J. M. Horack, M. N. Brock, W. S. Paciesas, G. N. Pendleton, and C. Kouveliotou, this conference.

6. M. Pamini, L. Natalucci, D. Dal Fiume, F. Frontera, E. Costa and M. Salvati, Il Nuovo Cimento, 13C, 337 (1990).

7. D. Band, this conference.

THE RAPIDLY MOVING TELESCOPE (RMT); A FIRST-LIGHT REPORT

S.D. Barthelmy*, T.L. Cline, B.J. Teegarden, T.T. von Rosenvinge
NASA/GSFC Greenbelt, MD 20771 USA

ABSTRACT

A ground-based optical telescope system has been constructed with the capability to locate fast optical transients that may be associated with Gamma Ray Bursts (GRB). The instrument has been installed at Kitt Peak, AZ. Combined operation, *first light*, in a manual mode of operation with the MIT Explosive Transient Camera (ETC) started on 23 Feb 92. Approximately 50 hours of manual mode operation have been accumulated. Work continues to make the RMT automated with unattended automated operation expected in Dec 92. The telescope has the proven capability to slew to any point on the night sky within 1.0 second, track that position with better than one arcsecond stability, and image a 9x12 arcminute field of view with one arcsecond angular resolution with 1.5 second time resolution. The telescope-CCD camera system has a sensitivity of 14th magnitude for transients and 15th mag for field stars.

INTRODUCTION

For over 20 years Gamma Ray Bursts (GRB) have been detected and we still do not know the source objects for these events. The initial discovery of archival optical transients within modern-day GRB error boxes has suggested an alternative approach of study[1]. We have developed a ground-based optical telescope which is capable of slewing to any location on the sky and making high time resolution and high angular resolution images of optical transients. This assumes that there is optical emission associated with the gamma ray emission.

INSTRUMENT DESCRIPTION

The Rapidly Moving Telescope (RMT) is shown schematically in Fig 1[2]. The instrument consists of a fixed Maksutov-Cassegrain telescope located at the top of the tripod structure looking down on an azimuth-elevation gimballed flat mirror. By moving only the mass of the mirror assembly, the target acquisition time is greatly reduced (1.0 sec maximum). The RMT is a narrow FOV instrument and so it works in close connection with the MIT Explosive Transient Camera (ETC) which is a wide FOV survey instrument[3]. The ETC locates a brief optical flash in the night sky (optical transient) and sends the coordinates of the transient to the RMT which then rapidly slews to that location and makes high angular resolution images. The basic characteristics of the RMT system are listed in Table 1.

The key to the RMT instrument is the azimuth-elevation gimballed mirror. It consists of two high-torque DC direct-drive motors that are driven by a servo control system. The two shaft encoders have an absolute accuracy of 2.5 arcsec and a precision of 0.07 arcsec. To achieve the required large dynamic range in slewing to tracking speed and tracking accuracy, the servo electronics is a hybrid digital computer controlled back-end with an analog front-end.

* Universities Space Research Association

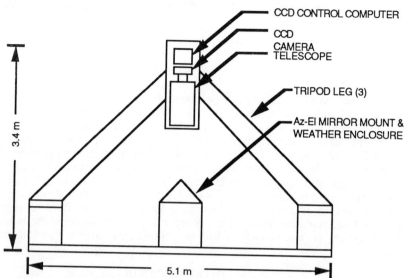

Fig. 1. The RMT instrument consists of a fixed telescope and CCD camera looking down at an Az-El gimballed flat mirror.

TABLE 1 - RMT CHARACTERISTICS

CCD:
- Type: Texas Instruments, 584x394, in frame store mode
- Pixel Size: 22x22 microns; 1.8x1.8 arcsec
- Integration Time: 1.5 sec (1.2 minimum)
- Readout Noise: 35 e$^-$ rms per pixel @ 100K pixels/sec
- Operating Temperature: -40 to -50°C
- Quantum Efficiency: 0.75 @ 6000 Å
- Sensitivity: 15 mag at Kitt Peak for field stars

Telescope:
- Type: 18 cm dia, ruggedized Questar
- Focal Ratio: f/14.3
- Throughput: 0.8 (averaged over 3500-7000 Å)
- FOV (w CCD): 9x12 arcmin

Mirror:
- Dimensions: 19 x 25 cm, quasi-elliptical outline
- Flatness: λ/8
- Surface: Flat, Aluminized, SiO coated

Mirror Mount:
- Max Acceleration: 1000 deg/sec^2
- Max Velocity: 260 deg/sec
- Max Slew time: 1.0 sec
- Encoder Precision: 0.070 arcsec
- Tracking Accuracy: 0.3 arcsec rms (typical); 1.0 (wind @ 20mph)
- Positional Accuracy: 2.5 arcsec over all 360 deg, 3σ

OPERATIONS

While the RMT is waiting for coordinates from the ETC, it operates in "stare" mode taking a sequence of 1.5 sec integrations of various targets (mostly the smallest of previous GRB error boxes on the chance of a repeated burst). However, when it receives coordinates from the ETC ("transient" mode), it slews to that location and begins a series of 1.5 sec integrations (see Fig 2 for a time line of actions). The CCD pixel data is sent to the data analysis computer for real-time processing. Because it is not possible to store the 5GB of data that would be generated in a typical night, the pixel data is thresholded with a reference frame which cuts the amount down to about 200 pixels per image (or about 20MB per night). With a full (non-thresholded) image frame stored every 15 minutes plus miscellaneous housekeeping information, a typical night's data is about 40MB.

Presently the RMT has been operated only in a manual mode -- a human in the loop making decisions about weather and opening up the protective enclosures. Work is progressing to make the RMT fully automated where the computer with the housekeeping data acquisition system will be able to sense the weather, make decisions about opening up, activate the enclosure motors, and take data. Almost all of this is done with the exception of the fail-safe systems. Because of the delicate nature of the gimballed mirror system, it is vital that the weather covers never be left open when the weather is bad.

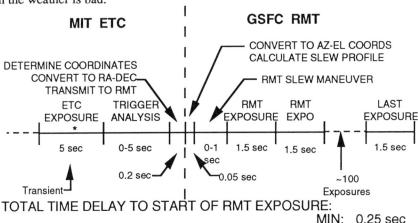

Fig. 2. A transient (*) event sequence starts with the ETC discovering a transient (a brightening of a pixel from one exposure to the next; 0-5 sec). The ETC then sends (0.2 sec) these coordinates to the RMT mirror mount servo computer where they are transformed into Az-El (0.05 sec), then the mount slews both axes simultaneously (0.0-1.0 sec), and a sequence of about 100 exposures is made of the transient position. The variability in the total time from the transient to the RMT taking data is the result of 1) the position in time of the transient with respect to the ETC exposure sequence, 2) the position in the CCD FOV translates into a time for the ETC to serial analyze their pixel data, and 3) the time required to slew from our current stare mode position to the transient position. The average time is 5.75 sec.

MANUAL MODE OPERATION RESULTS

In Feb 1992 the RMT was operated in a manual mode (human in the loop) for about 50 hours. Fig 3 is plot of 3.1 hours of trigger coordinates received from the ETC instrument. It is obvious that the majority of the coordinates are related in position and even more are related when arrival time is considered. These "optical transients" are due to glints from near-earth orbiting satellites and space debris. To minimize the number of mirror slewing operations, most of these man-made transients are separated from those of true astrophysical origin by requiring a simple minimum angular separation and a minimum time separation. No transient is acted upon by the RMT if the new position is within 1 degree from the previous coordinates and/or the arrival time is within 1 sec of the previous. This simple filter reduces the number of transients by more than a factor of 10. So far no optical transient has been identified as astrophysical in origin.

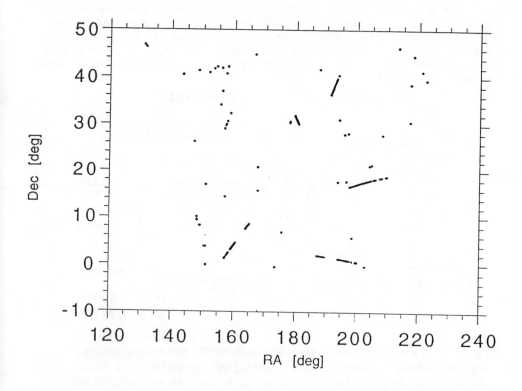

Fig 3) This is the distribution (RA vs Dec) of 3.1 hours (29 Feb 92) of transients observed by the ETC and sent to the RMT. There are two classes of coordinate positions: 1) those that are isolated in relation to any nearby transient positions, and 2) those which appear in long sequences tracing arcs across the sky. The latter are multiple sequential transients due to the passage of a satellite across the FOV, and they will be eliminated by the 1 degree filter algorithm. The former are assumed to be isolated occurrences of the former.

SUMMARY

The RMT is operational and will become fully automated (no human in the loop) in Dec 92. Its "manual mode first light" operations earlier this year have worked out many bugs in the system and proven that it can operate in close connection with the MIT ETC instrument. All that remains is to implement the fail-safe checks to guarantee that the weather covers are never open during improper times which would seriously damage the delicate mirror mechanism.

REFERENCES

1. B.E. Schaefer; Nature; **294**; pp 722 (1981).

2. S.D. Barthelmy, et al; Robotic Telescopes in the 1990s (1991); (ed. A.V. Filippenko; San Francisco: ASP conf.proc. #103, Larimie, WY).

3. G.R. Ricker, et al; High Energy Transients in Astrophysics (1984); (ed. S.E. Woosley; New York: AIP conf. proc., #115), pp 669.

THE DESIGN OF A GAMMA-RAY BURST POLARIMETER

M. McConnell, D. Forrest, K. Levenson and W.T. Vestrand
Space Science Center, University of New Hampshire, Durham, NH 03824

ABSTRACT

The study of the polarization properties of the gamma-ray bursts is the one remaining unexplored avenue of research which may help to answer some of the fundamental problems regarding the nature of these mysterious objects. We have designed an instrument to measure linear polarization in cosmic gamma-ray bursts at energies >50 keV. Here we describe the design of this instrument, which we call the Gamma-RAy burst Polarimeter Experiment (GRAPE).

INTRODUCTION

The nature of the gamma-ray burst sources has remained a mystery ever since their discovery some 25 years ago. Even after the best efforts of scores of researchers, there is no consensus regarding their origin. The latest results from the Compton Gamma-Ray Observatory (CGRO) have served to heighten the interest in this problem without offering any fundamental answers as to their origin.

When it comes to the study of electromagnetic radiation, there are four measurements which an observer can make (all as a function of time). In the context of gamma-ray burst studies, these are: 1) the source location; 2) the source spectrum (distribution of photon energies); 3) the source intensity (spectrum normalization); and 4) the source polarization. All studies of gamma-ray bursts to date have involved the first three types of measurements. Since these data have not yet yielded any clear fundamental answers as to the nature of the bursts, we contend that an additional piece of information may be required. The only additional piece of information which is available to the observer is that of the polarization of the radiation.

A few authors have discussed polarization effects in the context of burst models. For example, Mitrofanov and Pozanenko[1] have discussed the polarization inherent in cyclotron emission in a strong magnetic field; they predict a polarization fraction of ~80% near 100 keV for a typical neutron star environment. Bisnovatyi-Kogan[2] has noted that the measurement of polarization in the observed spectral lines would help to determine the exact nature of the line emission. Baring[3] has examined the polarization produced via magnetic photon splitting; a maximum polarization fraction of about 43% is possible with this mechanism.

At present, there is no gamma-ray experiment that we know of which is presently in operation or in the planning stages whose major emphasis is the measurement of polarization. The COMPTEL instrument on GRO[4], for example, has only marginal capabilities for detecting polarized emission due to an inefficient geometry (the major drawback for measuring polarization with traditional Compton telescopes). Some of the proposed concepts (such as TTTS[5]) include the capability for polarization measurements, but once again, they are not optimized for that purpose.

The prototype GRAPE design has been developed within the context of a long-duration balloon platform. However, the modular nature of this experiment will also facilitate its use as an add-on experiment for satellite platforms. In addition, the simplicity of this design, along with its minimal pointing and service requirements, would make such an experiment ideally suited for deployment on Space Station Freedom.

POLARIMETER BASICS

The measurement of linear polarization at hard x-ray and gamma-ray energies is based on the fact that the Compton scattering cross-section is a function of angle from the electric field vector. In particular, the scattering cross-section is a maximum for photons scattered perpendicular to the electric field vector and it is a minimum for photons scattered parallel to the electric field vector. The ratio between the minimum and the maximum scattering cross sections gives a good indication of how effective this approach may be. Although this ratio is quite high (~65) at 100 keV, the ratio decreases rapidly at energies above 500 keV (~5 at 500 keV). This makes this technique increasingly difficult at MeV energies.

THE POLARIMETER MODULE

GRAPE is designed as a modular experiment, consisting of a number of independent polarimeters. The basic principle of the GRAPE design is to scatter photons from a low-Z plastic scintillator into a high-Z NaI(Tl) scintillator which is capable of absorbing the remaining photon energy with a high probability. A single polarimeter module consists of a ring of individual plastic scintillators surrounding the NaI(Tl) absorber, as shown in Figure 1. The NaI(Tl) detector is cylindrical, 7.6 cm in diameter and 7.6 cm long. Each plastic scattering element (in the present design) is 7.6 cm long and 3 cm thick. A complete module measures 25 cm x 25 cm in size.

The goal is to measure the azimuthal distribution of the photons scattered into the central detector. This distribution is determined by recording the coincidence events between each of the 12 scattering elements and the central detector. This provides a set of 12 energy-loss spectra which can be used to measure the polarization of the incident flux as a function of energy.

The Design of a Gamma-Ray Burst Polarimeter

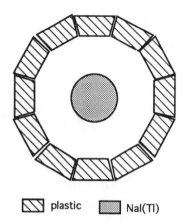

plastic NaI(Tl)

Figure 1. Schematic diagram of a single GRAPE module showing the ring of plastic scintillators which act to scatter incident photons into the central NaI(Tl) detector. In the baseline design, the NaI(Tl) detector is 7.6 cm in diameter by 7.6 cm long. The plastic scattering elements are 3 cm thick. The plastic detectors are centered on a radius of 11.3 cm.

We define the measured polarization in terms of the modulation factor,

$$P(E_\gamma, \theta) = \frac{S_{max}(E_\gamma, \theta) - S_{min}(E_\gamma, \theta)}{S_{max}(E_\gamma, \theta) + S_{min}(E_\gamma, \theta)}$$

where S refers to the effective area of a single NaI(Tl)-plastic scattering pair at incident energy E_γ and incident angle θ. In this expression, the maximium and minimum values refer to the maximum and minimum of the azimuthal variations. The detection properties inherent in this expression have been studied in detail with a series of Monte Carlo simulations. These simulations employed a modified version of the CERN GEANT code (modified to properly handle polarized photons).

The results of these simulations are shown in Figures 2 and 3. Figure 2 shows the effect of increasing incident energy on the measured azimuthal variations (for an incidence angle of 0°). Figure 3 shows the effect of increasing the incidence angle (for an incident energy of 100 keV). In all cases, the response both to 100% linearly polarized radiation and to unpolarized radiation is shown. Solid lines represent a fit to the data. Some of these results are also given in Table I (where S_T refers to the summed total of all elements in a single module). We conclude from these simulations that: 1) there is significant efficiency at energies up to 500

TABLE I. Monte Carlo Results

Energy (keV)	S_T(NaI) cm^2	S_T(plast) cm^2	S_T(coinc) cm^2	P
50	6.5	153	4.1	71%
100	6.4	146	4.2	71%
200	10.2	133	3.6	73%
300	23.8	125	3.1	64%
400	31.6	117	2.9	63%
500	35.3	111	2.6	59%
700	36.9	101	2.3	47%
900	35.6	93	1.8	33%

keV; and 2) there is a clear polarization signal even at 30° off-axis, thus providing for an effective FOV of ~1 steradian.

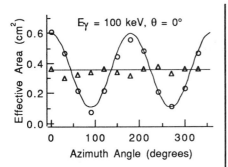
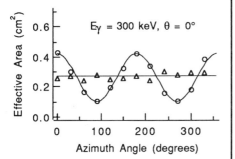
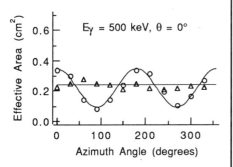

Figure 2. The modulation as a function of incident energy expressed as effective area per scattering pair. In each case, the response to both 100% polarized flux (circles) and unpolarized flux (triangles) is shown. Errors are roughly the size of the plot symbols.

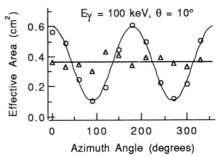
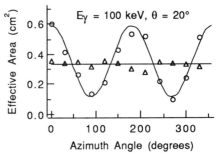
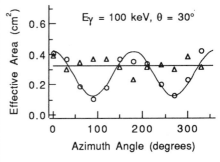

Figure 3. The modulation as a function of incident angle expressed as effective area per scattering pair. In each case, the response to both 100% polarized flux (circles) and unpolarized flux (triangles) is shown. Errors are roughly the size of the plot symbols.

BALLOON PAYLOAD CONCEPT

In developing this concept for a balloon payload, it is necessary for the detector to have an efficiency which would give useful scientific results within the time frame of a typical long-duration balloon flight (typically, 2-3 weeks). This requires an array of polarimeter modules. Our baseline design is an array of 16 modules, as shown in Figure 4. As a polarimeter, GRAPE would be able to detect 50% linear polarization at >3σ in a burst with a fluence of 5×10^{-6} ergs cm^{-2}. (More recent simulations appear to suggest that refinements to this design may improve the detection efficiency by as much as 50%.) With a FOV of ~1 steradian, we would expect to see ~1 burst at this fluence level during a 2-3 week balloon flight. Although GRAPE is optimized for the measurement of polarization (up to several hundred keV), the use of the individual detectors will also permit a significant capability for timing, and spectral measurements of gamma-ray bursts up to several MeV. The use of corner-mounted x-ray detectors (as in the BATSE design) will permit localization of bursts to within several degrees; this would be required for a more accurate deconvolution of the GRAPE response (due to its dependence on incidence angle). Finally, systematic effects within the modules would be alleviated by rotation of the balloon payload.

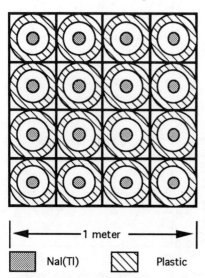

Figure 4. A schematic of the full balloon prototype, consisting of an array of 16 individual modules.

REFERENCES

1. I. Mitrofanov and A. Pozanenko, in Gamma-Ray Bursts: Observations, Analysis and Theories, eds. C. Ho, R. Epstein, and E. Fenimore (Cambridge Univ. Press, 1992), p. 430.
2. G.S. Bisnovatyi-Kogan, in Gamma-Ray Bursts: Observations, Analysis and Theories, eds. C. Ho, R. Epstein, and E. Fenimore (Cambridge Univ. Press, 1992), p. 89.
3. M.G. Baring, in Compton Observatory Science Workshop, eds. C. Shrader, B. Dennis, and N. Gehrels (NASA CP 3137, 1992), p. 293.
4. V. Schönfelder et al., submitted to Ap. J. Suppl. (1992)
5. K. Hurley, in Gamma-Ray Bursts: Observations, Analysis and Theories, eds. C. Ho, R. Epstein, and E. Fenimore (Cambridge Univ. Press, 1992), p. 279.

AN INSTRUMENT CALLED PROMETHEUS

R.C. Haymes, M.J. Moss, and P.W. Walker
Rice University
Department of Space Physics and Astronomy
Houston, TX 77251 - 1892

ABSTRACT

Prometheus I is a very low background 0.03 - 6.5 MeV gamma-ray balloon - borne imager. Its position sensitive photon detector eliminates the activation component of instrumental background through use of: (a) A gamma ray detector that is segmented in two dimensions; (b) A 30 cm thick active collimator fabricated from low - Z matter (plastic scintillator), which has a 10 keV threshold; (c) An active collimator that vetoes prompt events due to fast neutrons; (d) slow neutron radiationless absorber. Because of its low background, the 3σ spectral - line sensitivity at 1 MeV is expected to be 5×10^{-5} photons cm^{-2} sec^{-1}, given six hours observing time at midlatitude balloon altitudes. It will be used to produce a 0.5 - degree map of the entire sky, on two Long Duration Balloon flights that are planned for 1994 and 1995. Prometheus has an in-flight selectable field of view (FOV) and 30 arc-minute angular resolution. Its selectable FOV will be used to independently measure the "MeV bump" in the spectrum of the diffuse extragalactic background.

Construction of Prometheus I has now been completed. The system is undergoing laboratory calibrations and testing at Rice, prior to undertaking its first, concept - verification balloon flight, at NASA's National Scientific Balloon Facility.

1. INTRODUCTION

Together with high resolution gamma ray spectroscopy, high - sensitivity MeV gamma ray imaging is vital for the future development of high energy astrophysics. High resolution MeV spectroscopy will isotopically identify the source's matter, and the presence of antimatter. High sensitivity MeV imaging will identify the astronomical setting of the source's matter. To accomplish both, the INTEGRAL mission, in which both a MeV spectrometer and an MeV imager simultaneously view the same field, has been proposed (Courvoisier et al (1991)). If it is approved, launch would be around the end of the present decade.

Prometheus is a balloon - borne actively collimated, very low background 0.03 - 6.5 MeV gamma ray imager. Prometheus will conduct astronomical measurements (see Sec. 3) that will aid the planning of INTEGRAL's observing program. Hopefully, experience gained with Prometheus will be helpful in understanding the sensitivity of gamma ray imagers and spectrometers on INTEGRAL and other future gamma ray astronomy missions.

For imaging, a coded mask is placed in its field of view (FOV). A total - energy - deposition gamma ray position - sensitive detector (PSD) records the energy and arrival location of each in - field gamma ray quantum that passes through the mask's perforations. For a given exposure, Prometheus images will have higher sensitivity and hence higher contrast, than those of other g-ray PSDs. Prometheus improves MeV image contrast by rejecting the activation component of instrumental background. The rejection results from use of: a segmented detector, a low - atomic - number, low - energy - threshold active collimator, and neutron shielding. It was first described in 1987 (Haymes et al, 1988), when work on it had just begun. The present report is an update, now that construction of Prometheus I is complete.

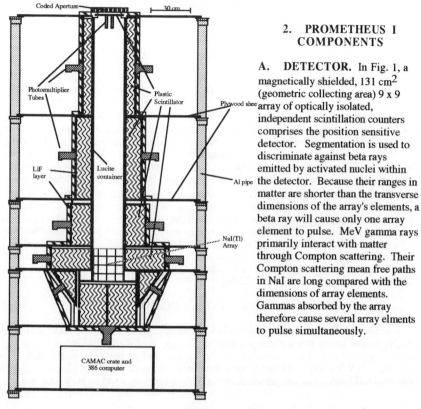

Figure 1. Schematic sketch of Prometheus I. Fiberglass pressurized container (8 mm thick) is not shown.

2. PROMETHEUS I COMPONENTS

A. DETECTOR. In Fig. 1, a magnetically shielded, 131 cm^2 (geometric collecting area) 9 x 9 array of optically isolated, independent scintillation counters comprises the position sensitive detector. Segmentation is used to discriminate against beta rays emitted by activated nuclei within the detector. Because their ranges in matter are shorter than the transverse dimensions of the array's elements, a beta ray will cause only one array element to pulse. MeV gamma rays primarily interact with matter through Compton scattering. Their Compton scattering mean free paths in NaI are long compared with the dimensions of array elements. Gammas absorbed by the array therefore cause several array elments to pulse simultaneously.

Prometheus accepts for analysis only multi-element unvetoed events. Prometheus also imposes a requirement of total energy deposition, for acceptability of events. (Incomplete deposition would lead to model - dependent spectra.) Each incomplete - deposition event includes an escape photon. Escaping recoil photons are detected by the thick active collimator (see below). Collimator pulses that are simultaneous in time with array events cause the latter to be vetoed. **A1. DETECTOR CONSTRUCTION DETAILS.** Each of the 81 array elements is a 0.5 in x 0.5 in x 2 in (thick) NaI(Tl) scintillation crystal, coupled to a separate R-2248 Hamamatsu 2 in long photomultiplier tube (PMT). Each of the NaI crystals is covered with 20 mg cm^{-2} dried MgO reflective powder, a layer of white teflon tape, and a layer of black electrical tape. Each crystal is optically coupled/glued to its respective PMT with Rexon RX-22P optical coupling compound. Each crystal/PMT assembly is encapsulated in paraffin to seal out moisture from the NaI. Plastic scintillator (see below) surrounds the detector; fast neutrons are moderated by the plastic. Unless the detector is protected against slow neutrons, there would be increased activation of the detector because it would capture these slowed neutrons. The detector's segmentation is only two - dimensional, so the beta-ray the discrimination is only $\approx 85\%$. Therefore, even though beta ray background is discriminated against by the detector's segmentation, its light-tight box surrounds the array with an absorber of slow neutrons. On the sides and back, the absorber is a 3 cm thickness of packed (i.e., bulk density = 1.5 gm cm^{-3}) LiF powder, in its natural isotopic enrichment of ^{6}Li. The frontmost side of the detector box is covered with encapsulated Li foil. The foil (density 0.5 gm cm^{-3}) is also composed of the natural abundance of ^{6}Li, and is 1.5 cm thick, so it attenuates incident 1 MeV gamma rays by $\approx 5\%$, while absorbing 99% of the slow neutrons in the FOV. **B. ACTIVE COLLIMATOR.** This consists of 45 independent plastic scintillators, whose PMTs are all connected in anticoincidence with the detector. The collimator defines the FOV of the detector, vetoes detector events due to energetic charged particles, and imposes the total - energy - deposition requirement on detector events. The FOV is selectable, because of the modular construction of the collimator. With all of the modules (i.e., plastic scintillators) in place as shown in Fig. 1, the fully coded aperture is at its minimum, which is $4^O \times 4^O$. The maximum - size FOV is two orders of magnitude greater, at $38^O \times 38^O$, when the modules on the front-most two decks are moved out of the beam.

Unlike the active collimators of other astronomical instruments, the Prometheus collimator uses low-atomic-number material (i.e., plastic scintillator). It is a low - density shield; the same stopping power requires use of physically - thick matter. Prometheus I uses a 30 cm thickness; Prometheus II will use 60 cm. Deadtime losses are expected to be less than 5% for the 60 cm thickness, because the pulse widths from the plastic scintillator are about 10 ns. Previous attempts (e.g., FIGARO; Agnetta et al, 1989) to employ plastic scintillator shielding have used a comparatively high shield - discriminator setting, 350 keV. Compton scattering of MeV gamma rays produces large numbers of interactions that liberate ≤ 0.1 MeV. Gamma rays from outside the FOV may avoid veto by such shields, thereby limiting the instrumental sensitivity. Prometheus has a 10 keV discriminator setting for its collimator modules. The instrumental background from plastic scintillator collimators is virtually free of an activation component. ^{12}C, the heaviest principal nucleus in plastic, has a comparatively small spallation cross section. This is because its maximum atomic weight A is small, compared with the elements in the previously - used inorganic scintillators (e.g., ^{127}I; ^{209}Bi.) When A = 209, isomeric decays by some of the huge number of heavy - nucleus spallation products that result from hadronic bombardment of a bismuth germanate collimator may cause significantly increased background in the detector that they shield against external gamma rays. Since A = 12, only a few plastic-activation products are possible. With one exception, $^{7*}Be$, a relatively infrequent ^{12}C spallation product, all of the ^{12}C spallation products emit a charged particle simultaneously with the emission of a gamma ray. They are detected with 100% efficiency, vetoing any simultaneous pulse from the detector. The 60 cm collimator is six mean free paths thick for the 0.477 MeV line from the deexcitation of $^{7*}Be$.

Plastic scintillator 60 cm thick is 12 mean free paths thick for fast neutrons and therefore is an excellent detector of ambient (i.e., albedo and locally produced) fast neutrons. Neutron - proton scattering is the most efficient means for energy transfer from neutrons (i.e., detection). There are 1.11 protons per carbon atom in the plastic. The pulses due to the recoil protons are detected with 100% efficiency by the veto system. Slow neutrons have been a major source of background in previous instruments, because of neutron capture by isotopes such as ^{127}I; the resulting beta decays from the ^{128}I added to the background. Prometheus avoids slow-neutron problems in two ways. One way is the detector segmentation previously described, which discriminates against beta-decay events in the detector. The other is the use of radiationless slow neutron absorber around the plastic's exterior. This absorbs ambient slow neutrons that would be otherwise captured by the plastic's protons, causing enhanced 2.22 MeV background. **B1. COLLIMATOR CONSTRUCTION DETAILS.** Gamma rays with energies as low as 10 keV are detected, for pulses originating anywhere within each module. This is because each module has a simple non-reentrant geometrical shape, and because of the high degree of surface polish. Undergraduates in our laboratory hand polished each piece, using polishing compounds with progressively finer grits, down to 0.05 microns. Radiationless absorption of slow neutrons is accomplished by covering the outside of the collimator with a 3 cm thick "layer" of LiF powder (consisting of 200 sealed polyethylene packages, each containing about 600 gms of LiF). The ^6Li (n, a) T nuclear reaction has a 940 barn cross section for absorption of thermal neutrons, so 99% of the incident slow neutrons are absorbed before entering the collimator.

C. SENSITIVITY. Figure 2 is an analytic estimate we have prepared for Prometheus II. For the purpose of direct comparison with the sensitivity expected for INTEGRAL's High Energy Imager (Courvoisier et al, 1991), Fig. 2 was prepared using all the same assumptions made by Dean et al (1991), including the bombarding spectra, and the same orbit. That is, Prometheus II was assumed for this purpose only, to be in the same low-inclination, low-altitude orbit as one of those under study for INTEGRAL. In preparing Fig. 2, we followed the INTEGRAL-proposal calculations described by Dean et al (1991). Thus, analytic approximations were fitted to all particle and photon cross section data, and their approximations to the aperture flux and atmospheric gamma radiation spectrum were used. To analytically approximate the variation of detection efficiency with angle of incidence upon a detector element, a mean interaction length was computed and used, for each given energy. Only spallation of the NaI in the PSD contributes to the Prometheus background. We used the two-exponential continuum - approximation given by (Dean et al (1991)) for the mixture of a large number of radioisotopes spallation - induced in NaI. We assumed for the spallation curve of Fig. 2 that 2% of the gamma radiation from this radioactivity would fail to be recognized as such by the Prometheus segmented detector. This inefficiency is smaller than the 10% assumed for the HEI in the INTEGRAL proposal. The reason is, at 1 MeV there is a 0.05 maximum probability of one - crystal photoelectric absorption in the 5 cm NaI PSD for Prometheus, but a maximum probability of 0.30 for the 15 cm CsI PSD for the HEI.

Neutron atmospheric spectra were taken to be as given by Gehrels (1985). They were assumed scaled to the LEO by the same $E^{0.7}$ function discussed in Courvoisier et al (1991). For scatterings in the collimator, the average recoil energy given to a proton was approximated by 0.64 E_n, and the average energy given to a carbon - 12 nucleus was approximated by 0.12 E_n. Mean - energy assumptions were also made for the Na and I nuclei in the detector. In Fig.2, the curve marked "neutrons" refers to elastic scatters of fast neutrons in the NaI array, above the assumed 10 keV instrumental cutoff.

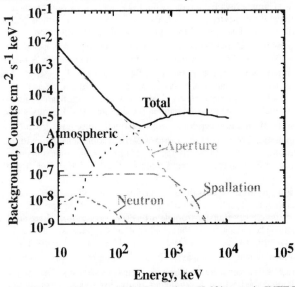

Figure 2. Background expected for Prometheus II, if it were in INTEGRAL's LEO.

Courvoisier et al (1991) showed that the HEI background at an energy of 1 MeV the background is expected to be 2.5×10^{-4} counts cm^{-2} sec^{-1} keV^{-1}. Fig. 2 shows that at 1 MeV the expected background continuum count rate for Prometheus II is 1×10^{-5} counts cm^{-2} sec^{-1} keV^{-1}. This means the Prometheus II background continuum is about 4% of the HEI background continuum. It is about 1% of the typical OSSE background continuum around 1 MeV, although this comparison is not meaningful, since OSSE is not in a LEO.

At 2.23 MeV is seen a deuteron - formation 4×10^{-2} counts cm^{-2} sec^{-1} spectral line. It is due to capture of moderated fast neutrons by the protons in the plastic. The slowing down time exceeds the 10 ns of the veto pulses generated by each (n,p) scattering, so that veto is incomplete. Many of the resulting gammas are vetoed by the plastic; the figure shows those that survive to the detector array. It is a line, not a line- and - continuum, because: (a) any scatter in the plastic depositing > 0.01 MeV generates a veto pulse; (b) the central detector is a full-energy-deposition detector.

The 2×10^{-4} cm^{-2} sec^{-1} $^{12*}C$ deexcitation line at 4.43 MeV is smaller than the 2.2 MeV line. This is because practically all the neutrons sufficiently energetic to excite carbon nuclei also undergo (n,p) elastic scatterings, as well as inelastically exciting the carbon nuclei. Veto pulses are generated by each (n,p) scatter. These pulses veto the virtually - simultaneous 4.43 MeV photon.

3. OBSERVING PLANS

Prometheus II will be used to conduct two different kinds of measurements. Carried by two Long Duration Balloons, respectively in the northern and southern hemispheres in 1994 and 1995, it will map the entire sky, in the 0.03 - 6.5 MeV energy interval. The map's sensitivity at 1 MeV is expected to approximately equal that of the COMPTEL sky map, which is now under construction. At 1 MeV, the angular resolution of the Prometheus sky map is expected to be 30 minutes of arc. An energy of 1 MeV is close to the low end of the COMPTEL energy interval, so the angular resolution of the COMPTEL map at 1 MeV is expected to be about 5 degrees. The HEI on INTEGRAL will have an 8° x 8° FOV; the Prometheus map will hopefully be helpful in planning the INTEGRAL observing program.

The second Prometheus measurement is planned to be that of the extragalactic background's energy spectrum and isotropy. No mask will be used. The size of the FOV will be varied. That component of the counting rate which is linearly dependent on the size of the FOV will be extracted from the data. If variation of the FOV were to be attempted with a heavy - element collimator, the flux of secondary gammas from the collimator impinging on the detector would also vary. The FOV - dependent counting rate would consist of the sum of the counting rates due to the sky and to the heavy elements in the collimator.

According to (2) above, plastic scintillator's spallation products do not emit unvetoed activation gamma radiation. Variation of the Prometheus FOV will not cause a variation in the flux of collimator-origin gamma rays at the detector. The size of the Prometheus FOV will be in-flight selectable. The selection will be made by commands issuable from the ground or a tracking aircraft. The commands will cause specific blocks of plastic scintillator to swing back as desired, almost entirely out of the beam. They will subtend minimal solid angles at the detector, in the "out" position. With the swingback capability, the FOV may be command-changed to any of four values in the 7° to 38° FWHM range. At 1 MeV, given a balloon flight duration of at least 24 hours, the flux will be 3σ - determined if it equals or exceeds 8×10^{-6} photons cm^{-2} sec^{-1} $ster^{-1}$ keV^{-1}. This is a factor of 2 - 3 below the best - fit power law background spectrum. Isotropy will be measurable on a scale of $\sim 36^{\circ}$ FWHM.

4. ACKNOWLEDGMENT

This research was supported in part by NASA, through Grant NAGW - 2256.

5. REFERENCES

Agnetta, G., Di Raffaele, R., Mineo, T., Sacco, B., Scarsi, L., Agrinier, B., Christy, J.C., Parlier, B, Chabaud, L., Frabel, P., Mandrou, P., Niel, M., Rouaix, G., Vedrenne, G., Costa, L., Gerardi, G., Masnou, J., Massaro, A., and Salvati, M. 1989, Nuclear Instruments and Methods in Physics Research A281, 197 - 206.

Courvoisier, Dean, Durouchoux, Gehrels, Grindlay, Matteson, Mahoney, McBreen, O'Brien, Pace, Prince, Schonfelder, Share, Skinner, Teegarden, Vedrenne, Villa, Volonte, and Winkler 1991, INTEGRAL International Gamma - Ray Astrophysics Laboratory, A Report on the Assessment Study, ESA Report SCI(91)1, January.

Dean, A.J., Lei, F., and Knight 1991, Spa. Sci. Rev. 57, 109 - 186.

Gehrels, N. 1985, Nucl. Instr. Meth. A239, 324.

Haymes, R.C., Fitch, J.E., Sen B., and Averin, S. 1988, in Nuclear Spectroscopy of Astrophysical Sources, (N. Gehrels and G.H. Share, eds.), 465 - 471, AIP Conf. Proc. 170, AIP Press, New York.

AN UPDATING ABOUT FIP: A PHOTOMETER DEVOTED TO THE SEARCH FOR OPTICAL FLASHES FROM GAMMA-RAY BURSTERS

A. Piccioni, C. Bartolini, C. Cosentino,
A. Guarnieri, S. Ricca Rosellini
Dipartimento di Astronomia, C.P. 596 I-40100 Bologna, Italy

A. Di Cianno, A. Di Paolantonio, C. Giuliani, E. Micolucci
Osservatorio Astronomico di Collurania, I-64100 Teramo, Italy

G. Pizzichini
CNR, Istituto TESRE, via Castagnoli 1 I-40126 Bologna, Italy

ABSTRACT

A status report on the Fast Imaging Photometer (FIP) with some modifications to the initial project is presented. The possibility of coordinated observations with satellites is outlined.

INTRODUCTION

If the Gamma-Ray Bursters (GRBs) also emit optical flashes, there is the hope that a substantial contribution to the knowledge of their nature will come from the analysis of the optical light curve and of the evolution of the color index of the source during the event, combined with contemporaneous high energy information. Moreover, a better localization of the sources will allow subsequent deep optical investigations. There is an ongoing dispute on the reality of optical flashes from GRBs found ever in archival plates[1] or by photometric observations, but some of the data look to us very promising[2,3].

We give here a status report on the FIP, which is under construction at the University of Bologna and at the Observatory of

Collurania, devoted to the search for optical flashes from GRBs and other optical transient events. The original project is described in [4,5]; with the FIP we expect to obtain a localization better than the previously known one, the time history, and the color index of the source; it is not necessary to emphasize the importance of the localization and the time history of the optical flash; this is even more true in the case of contemporaneous observation of an event in the optical and γ rays, because the delay time of emission in the two energy ranges would yield valuable information about the structure and the relative distance of the zones responsible for the optical and the high energy emission. The color index will allow to understand whether the optical emission is thermal or not.

THE INSTRUMENT

The structure of the FIP has been designed with the following requirements:
a) high temporal resolution
b) a field of view of the order of 3'x3'
c) acceptance also of very high photon fluxes, that estimates from archival plates[2] seem to indicate.
d) simultaneous observations with instruments on satellites.
The fundamental elements of the FIP are a dichroic beam splitter that divides the beam in three ways, featuring the passbands near the UBV system, and an array of 4 x 4 square section optical fibers, surrounded by 4 rectangular pixels that form a guard ring, (Fig. 1) mounted on the B beam. Each fiber is connected with a photomultiplier linked to a measurement channel able to carry out, independently, a real time preanalysis.

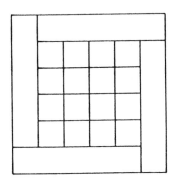

Fig.1. Structure of the fiber optic bundle of the FIP.

The system is provided of a trigger to select the useful data on the resulting large information flux.

The structure of the measuring head, the increased number of channels (22) and the larger size of each memory buffer (0.5 Mbyte) are the only modifications to the already described [4,5] initial project. Considering the point c) we preferred to use conventional PM tubes discarding the attractive possibility of new panoramic sensors.

The progress of the work is presently satisfactory: the acquisition cards are completed, the mechanical work on the photometric head (Fig. 2) is finished and we are setting up the automatic pointing and tracking system of the telescope and the assemblage of the optical components.

Fig. 2 - Photometric head of the FIP

We plan to carry out the first observations by the end of 1992, but the FIP will be considered actually operative only after the completion of the calibrations. The observations will be focussed on the error boxes which can be considered promising candidates for repeting flashes. Because of the the low probability of detection of an optical flash a long survey work has been planned. Prompt alert from GRB detectors on satellites could allow us to perform a fast pointing of the telescope on new burst localization and to observe the post-burst evolution of the optical luminosity.

REFERENCES

1. J. Greiner, Rejection of Archival Photographic Objects as Optical Counterpart of γ-Ray Bursters, *Astron. Astrophys.* in press (1992)

2. R. Hudec, J. Borovicka, R. Peresty, B. Valnicek, *Proc. 10th Eur. Reg. Astron. Meeting*, p. 239 (1987).

3. E.I. Moskalenko, et al. *Astron. Astrophys.* 223, 141 (1989).

4. A. Piccioni, C. Bartolini , A. Guarnieri , G. Pizzichini, J.M. Poulsen , in *Frontier Objects of Astronomy and Particle Physics*, eds. F. Giovannelli and G. Mannocchi (Bologna Compositori 1989), p. 217.

5. A. Piccioni, C. Bartolini, A. Guarnieri, G. Pizzichini, in *Gamma-Ray Bursts Observations, Analyses and Theories*, eds. C. Ho, R. I. Epstein and E. E. Fenimore (Cambridge University Press 1992), p. 148

THE TRANSIENT GAMMA-RAY SPECTROMETER: A NEW HIGH RESOLUTION DETECTOR FOR GAMMA-RAY BURST SPECTROSCOPY

H. Seifert[†], R. Baker, T.L. Cline, N. Gehrels, J. Jermakian, T. Nolan[‡], R. Ramaty, D.A. Sheppard[*], G. Smith[**], D.E. Stilwell, B.J. Teegarden, and J. Trombka
Laboratory for High Energy Astrophysics, NASA/GSFC, Greenbelt, MD 20771

A. Owens
Department of Physics & Astronomy, Univ. of Leicester, Leicester LEI 7RH, UK

C.P. Cork, D.A. Landis, P.N. Luke, N.W. Madden, D. Malone, R.H. Pehl, H. Yaver
Lawrence Berkely Laboratory, Berkeley, CA 94702

K. Hurley
Space Sciences Laboratory, Univ. of California, Berkeley, CA 94720

S. Mathias, and A.H. Post, Jr.
Arthur D. Little, Inc., Cambridge, MA 02140

ABSTRACT

The Transient Gamma-Ray Spectrometer (TGRS) to be flown aboard the WIND spacecraft is primarily designed to perform high resolution spectroscopy of transient γ-ray events, such as cosmic γ-ray bursts and solar flares, over the energy range 15 keV to 8.2 MeV with an expected spectroscopic resolution of 2 keV at 1 MeV. The detector itself consists of a 215 cc high purity n-type Ge crystal kept at cryogenic temperatures by a passive radiative cooler. The geometric field of view defined by the cooler is about 1.8π steradian. To avoid continuous triggers caused by soft solar events, a thin BeCu sun-shield around the sides of the cooler has been provided. A passive Mo/Pb occulter, which modulates signals from within ±5° of the ecliptic plane at the spacecraft spin frequency, is used to identify and study solar flares, as well as emission from the galactic plane and center. Thus, in addition to transient event measurements, the instrument will allow the search for possible diffuse background lines and monitor the 511 keV positron annihilation radiation from the galactic center. In order to handle the typically large burst count rates, which can be in excess of 100 kHz, burst data are stored directly in an onboard 2.75 Mbit burst memory with an absolute timing accuracy of ±1.5 ms after ground processing. The memory is capable of storing the entire spectral data set of all but the largest bursts. WIND is scheduled to be launched on a Delta II launch vehicle from Cape Canaveral in December of 1993.

[†]Universities Space Research Association, NASA/GSFC, Greenbelt, MD 20771; [‡]CSSI, Columbia, MD 22046; [*]GE Government Services, Lanham, MD 20706; [**]EER Systems Corporation, Seabrook, MD 20706

INTRODUCTION

Gamma-ray bursts (GRBs) occur at a rate of about 1/day and typically last anywhere from ~0.2 sec to ~1 min. They are among the most mysterious processes known in nature: their distances (nearby or cosmological?), luminosities, and the burst mechanism(s) (e.g., neutron-neutron star collisions?) are still unknown. Also, although the burst angular distribution is isotropic, log N/log P deviates for weaker bursts from the 3/2 power law which is expected for an isotropic distribution ("edge effect"?). In support of neutron star models for GRBs, lines around 70 keV (cyclotron lines due to Tera-Gauss magnetic fields?) have been observed in about 20% of all bursts by KONUS, HEAO-1, and GINGA, lines around 400 keV (gravitationally red-shifted annihilation lines?) have been observed in about 7% of all bursts by KONUS, and nuclear lines possibly have been observed by ISEE-3. However, unfortunately no lines have yet been observed by BATSE. This is a problem since for some of the earlier detections both the energy resolution and statistics were marginal, and because of the possibility of instrumental or analysis artifacts. On the other hand, so far only ~70 of the BATSE GRBs have been examined and, therefore, the absence of lines might not yet be too significant. It is clear that, given the present situation, the availability of high resolution spectroscopy is essential. In addition to spectroscopy, TGRS and KONUS on WIND, together with GRO/BATSE, Mars Observer, and Ulysses, will also allow the determination of accurate GRB positions from multiple spacecraft timing.

Solar flares are morphologically similar to GRBs and, with its occulter, TGRS can unambiguously identify solar events from the characteristic modulation of the spin-sector data. High resolution γ-ray spectroscopy of solar flares (line shapes/shifts, relative intensities) gives information on: a) solar abundances; b) charged particle acceleration, confinement, transport, and interactions in solar flares; c) flare ambient medium temperature and density; d) geometry of the flare region and, specifically, any anisotropies of the accelerated particles.

The occulter also allows TGRS to clearly identify emission from the galactic plane/center and monitor, e.g., the 511 and 1809 keV lines. The time variable 511 keV line is believed to be due to both a point source near the galactic center and an additional distributed component which could originate from a) e^+ emitted from ^{60}Co and ^{44}Sc (Type I supernovae), and ^{26}Al (supernovae, novae, Wolf-Rayet stars), or from b) annihilation of e^+ produced in high energy cosmic ray interactions with the ambient interstellar medium. For a) the intensity distribution reflects the nova distribution, while for b) it follows the galactic 100 MeV γ-rays. Thus, γ-rays provide excellent "tracers" for both the matter distribution in the galactic plane and for galactic nucleosynthesis.

AN OVERVIEW OF THE INSTRUMENT

An exploded view of the instrument and its location on the spacecraft are shown in Fig. 1. The principal TGRS detector and instrument data are compared with those of BATSE in Table I.[1,3]

Table I. Comparison of Principal TGRS and BATSE Detector and Instrument Data

Item	TGRS	BATSE/BATSE SDs
Detector(s), Material	1 × Ge (n-type)	8 × NaI(Tl)
Detector Area	35 cm^2	127 cm^2 each
Detector Thickness	6.1 cm	7.6 cm
Energy Range	15-8,200 keV	15-10,000 keV
Burst sensitivity	~10^{-6} erg cm^{-2} s^{-1}	~10^{-7} erg cm^{-2} s^{-1}
Field-of-View	1.8π sr	2π sr each, 4π sr array
Energy Resolution	0.2% @ 1000 keV	7.2% @ 662 keV
Efficiency (normal incidence) **Note:** TGRS values are for the Be detector module.	95% @ 20 keV 96% @ 100 keV 40% @ 500 keV (γ-pk) 10% @ 3 MeV (γ-pk) 67% @ 3 MeV (total)	80% @ 20 keV 90% @ 100 keV 60% @ 500 keV (γ-pk) 20% @ 3 MeV (γ-pk) 45% @ 3 MeV (total)
Time Resolution for Burst Events	62 μs intrinsic for time tagged events	128 μs intrinsic for time tagged events
Burst Trigger **Note:** a) TGRS Δt and BL are offset by ~32 sec; b) BATSE triggers are based on the LADs. ‡value(s) commandable, †individually commandable	• 2 on-board triggers • ΔE$_{1,2}$ = 50-300 keV† • 2 timescales Δt$_{1,2}$: 256, 2048 msec† • 8.0 σ and 9.0 σ increase† above 32 sec bckgrnd BL • at least 1 trigger required	• 3 on-board triggers • ΔE = 50-300 keV‡ • 3 timescales Δt$_{1-3}$: 64, 256, 1024 msec† • 5.5 σ increase‡ above 17 sec bckgrnd BL • at least 2 detectors are required to trigger

THE GERMANIUM DETECTOR AND RADIATIVE COOLER

The detector is a 215 cc, high purity, closed-end, coaxial (∅/L = 67/61 mm) n-type Ge crystal in reverse electrode configuration (depletion voltage is ~800 V). Presently, two versions of the detector module are being investigated: a Be module (80/65 mil side/top thickness), and an Al module (15/20 mil side/top thickness). The detector enclosure cover contains pockets with molecular sieve and Pd oxide getter material, as well as a pocket with a small amount of a monoenergetic γ-ray calibration reference (~1 cc K$_2$SO$_4$ ⇒ ~4 cts/min @ 1460 keV). Fig. 2 shows a cross-sectional view of the complete Be detector module assembly.

To minimize the effects of radiation damage on the detector performance, and due to the absence of any annealing capability, the detector is kept at a cryogenic temperature throughout the mission lifetime. The design of the passive, two-stage radiative Be cooler is similar to one flown on the ISEE-3 mission. The outer stage (@ ~164 K) operates as thermal buffer between the inner stage/detector (@ ~85 K) and the spacecraft (@ ~T$_{room}$). Both stages are thermally coupled to each other and the spacecraft by low-conduction, small cross-section supports. The support points are designed such that, when cooling down, the thermal paths break due to heat contraction, thereby isolating the individual components from each other. A cross-section of the cooler is shown in Fig. 3.

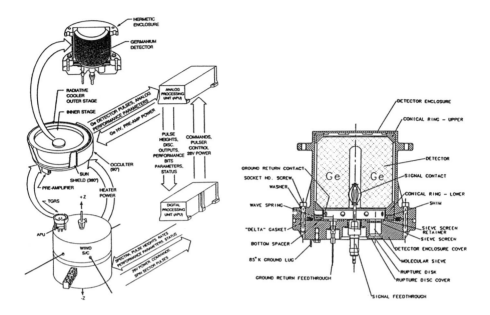

Fig. 1 Exploded View of the TGRS Instrument

Fig. 2 Cross-Sectional View of the Detector Module Assembly

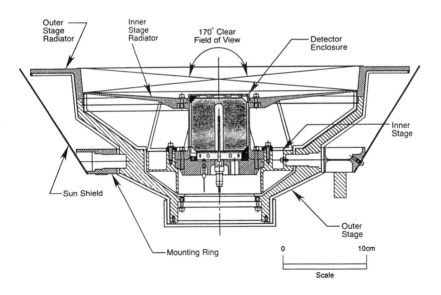

Fig. 3 Cross-Sectional View of the TGRS Detector/Radiative Cooler Assembly

The 30 mil thick BeCu alloy passive sun-shield around the sides of the cooler has been included to suppress continuous triggers by soft solar x-rays, e.g., from "micro-flares."[2] The shield attenuates x-rays by a factor of 1000 at 20 keV, and a factor of ~3 at 40 keV.

A quarter of the outer cooler diameter is occulted by a 1.5/12 mm thick Mo/Pb strip (the Mo acts as a passive shield for fluorescent photons produced in the Pb) which modulates any signals originating from within ±5° of the ecliptic plane at the spacecraft spin frequency of 20 rpm. The attenuation factor for spectra is ~100 for energies below 200 keV, and is 3 at 1 MeV. The occulter is used to effectively identify and study solar flares, as well as emission from the galactic plane and center.

TGRS ELECTRONICS

The TGRS flight electronics are made up of four sub-components: 1) the 170 K card which is located directly under the detector enclosure and contains the pre-amplifier front-end components; 2) the pre-amplifier which is attached to the underside of the cooler mounting ring (overall conversion gain ~40 mV/MeV); 3) the Analog Processing Unit (APU), mounted inside the TGRS tower, which contains the shaper/amplifier board, the pulse-height analysis and housekeeping boards, and the high/low-voltage power supply board; 4) the Digital Processing Unit (DPU), located on the upper equipment deck, contains a 5 MHz 80C86 processor, a 12 MHz 1750 RISC processor, and a 2.75 Mbit burst memory. The primary functions of the DPU are to a) acquire/process pulse height data from the APU; b) accumulate background/burst spectra, as well as various count rates and record system deadtime; c) format data and read them out into the telemetry; d) control the various instrument modes (waiting mode, burst mode, and dump mode); e) decode and execute instrument commands.

INSTRUMENT PERFORMANCE

Ref. 1 contains a detailed discussion of the predicted TGRS background and the calculations of the TGRS detection efficiency for the Be detector module. Corresponding calculations for the Al module are presently underway. It is found that for bursts with fluences above 10^{-5} erg cm^{-2}, TGRS would be able to detect all the previously reported line features. A burst of size 10^{-5} erg cm^{-2}, integrated over 1s, would be detected at the 30 σ level above threshold.

Signals due to the 511 and 1809 keV lines from the galactic center will be modulated by the occulter by a factor of ~10 and 1.8, respectively. Assuming line fluxes of 10^{-3} and 4×10^{-4} γ cm^{-2} s^{-1}, respectively, the time needed to detect these lines at the 3 σ level will be ~2 days for the 511 keV line and ~9 days for the 1809 keV line.

REFERENCES

1. A. Owens et al., IEEE Trans. Nucl. Sci., Vol. 38, No. 2, p. 559, 1991, and submitted for publication to Space Science Reviews
2. Lin et al., Astrophys. J., **283**, 421, 1984
3. BSAS Users Guide, B. Shaefer, 1991, and B. Shaefer, private commun.

FIGARO IV: 16 SQUARE METER BALLOON BORNE TELESCOPE TO STUDY RAPID VARIABILITIES AND TRANSIENT PHENOMENA AT ENERGIES ABOVE 50 MEV.

B. Sacco[1], B. Agrinier[7(*)], G. Agnetta[1], B. Biondo[1], O. Catalano[1], M.N. Cinti[2], E. Costa[2], G. Cusumano[1], N. D'Amico[3,4], G. D'Alì[3], R. Di Raffaele[1], G. Gerardi[3], M. Gros[7], J.M. Lavigne[8], M.C. Maccarone[1], A. Mangano[1], B. Martino[2], J.L. Masnou[9], E. Massaro[5], G. Matt[5], G. Medici[5], T. Mineo[1], E. Morelli[6], G. Natali[2], L. Nicastro[4], F. Pedichini[2], A. Rubini[2], L. Scarsi[1], and M. Tripiciano[1].

(1) Istituto di Fisica Cosmica ed Informatica, CNR, Palermo, Italy
(2) Istituto di Astrofisica Spaziale, C.N.R., Frascati, Italy
(3) Istituto di Fisica dell'Università di Palermo, Italy
(4) Istituto di Radioastronomia, CNR, Bologna, Italy
(5) Istituto dell'Osservatorio Astronomico, Università "La Sapienza", Roma, Italy
(6) Istituto di Tecnologie e Studio della Radiazione Extraterrestre, C.N.R.,Bologna, Italy
(7) CEN-Saclay, D.S.M. DAPNIA Service d'Astroph., Gif-sur-Yvette, France
(8) Centre d'Etudes Spatiale des Rayonnements, Toulouse, France
(9) UPR176, Observatoire de Paris, Section de Meudon, France

ABSTRACT

We present the design of a balloon borne experiment based on the Limited Streamer Tube technology as a tracking telescope to detect gamma rays above 50 MeV. This technique allows to obtain very large sensitive areas (16 m^2 in our experiment). Because of the capability to collect a large signal in a short time and of its good angular resolution (about 2° above 200 MeV), the telescope is highly competitive to study both periodic and random rapid variabilities of gamma-ray sources as well as to detect high energy gamma-ray bursts.

---

* on leaving from CEN-Saclay

INTRODUCTION

The recent observations and discoveries of the Compton Observatory focus our attention on the role played by rapid variable (periodic and random) and transient phenomena in Gamma-ray Astronomy.

-The discovery of the 237 ms pulsation in Geminga[1] and the identification[6] of 2CG 342-02 with PSR1706-44 together with the previous, and well established, knowledge of Crab and Vela pulsars grant the hypothesis that the galactic gamma-ray sources, or at least a large subset of them, are young pulsars.

-The finding of 3C279 [3] and other blazars at gamma-ray energies shows that also in the extra-galactic emission are present compact objects in which the variability is one of the main characteristics.

-The detection [5], on May 3,1991, of a burst by Egret at energy above 50 MeV urges the search for similar events.

The telescopes of the present generation, because of their limited collecting area and the long exposure time required to reach a significative signal, have difficulties to study these rapid phenomena. We propose a new gamma-ray telescope, operating at energies above 50 MeV, characterized for the very large collecting area. This telescope, called FIGARO IV, is based on the Limited Streamer Tube (LST) technology [4]. The present version (16 m^2 collecting area) is designed to fly on board of stratospheric balloons. The space qualification of the LST will allow their use in the future gamma-ray telescopes for satellites, space stations, and lunar observatories.

TELESCOPE LAYOUT

The gamma ray tracking detector proposed will use standard LST modified in order to assure the environmental working conditions of the balloon flights such as atmospheric pressure, temperature, and mechanical stresses.

Basically the telescope is composed of six planes of LST each one of dimensions 4m*4m with 4 units (4 tubes) forming a guard ring (R) around the first two planes; A, B and R constitute the anticoincidence system of the telescope to discriminate charged particles from photons (Fig. 1).

A lead converter (L), 2 mm thick, is placed between the B and C planes. C, D, E, and F, spaced 60 cm apart, track the electron-positron pair path.

The LST planes are supported by a honeycomb styrofoam structure (P) that assures the parallelism among the planes.

The read-out of LST is performed by two sets of strips with 1 cm pitch: one parallel to the tubes (X coordinate) and the other orthogonal to them (Y coordinate).

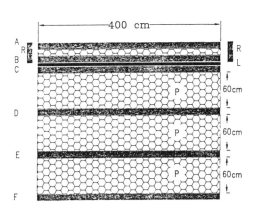

A chain of read-out electronics boards, placed on both X and Y sides of each plane, transfers the information on the track position into the data acquisition system.

Six separate H.V. channels (1 for each tube plane) will feed the LST, through a distribution bus.

Fig. 1 Telescope cross section: A, B and R are anticoincidence elements; L is a lead converter; C,D,E, and F are coincidence planes; P are styrofoam honeycomb boxes.

Table 1: Technical characteristics of FIGARO IV

Detector type	L.S.T. (1cm^2 cell)
Number of L.S.T. plane	6
Gamma-ray converter	Lead (2mm thick)
Geometric area	160,000 cm^2
Pick-up	Strips (1cm pitch)
Read-out channels	4800
Spatial resolution	0.5 cm
Telescope height	2 m

TELESCOPE SCIENTIFIC PERFORMANCE

The effective area and the resolution of the telescope as a function of the energy, as computed by a Montecarlo procedure for on-axis sources, are given in table 2). The decay of the effective area versus the off-axis angle is slow, for a source located at 20° off-axis the loss is about the 12% at 250 MeV.

Table 2: Effective area and angular resolution of the FIGARO IV telescope for on-axis sources.

Energy (Mev)	Effective Area (cm^2)	Resolution (degree)
100	23,000	3.5
200	29,000	2.2
300	32,000	1.6
500	32,000	1.2

We evaluated the sensitivity of the experiment by means of Montecarlo simulations taking in to account the cosmic and atmospheric background[2]. For a source transit of $2*10^4$ s, at a 4 mbar ceiling and a cut-off rigidity of 8 Gv we obtained the results of table 3). The minimum detectable flux at 3 σ ($0.75*10^{-6}$ ph cm^{-2} s^{-1}) corresponds to 0.2 of the Crab flux above 100 MeV.

Table 3: Visibility of gamma ray sources during a balloon flight of $2*10^4$ s, at a ceiling of 4 mbar and a cut-off rigidity of 8 Gv.

Source	Number of standard deviations
Vela	48
Geminga	20
Crab	14
Φ = $0.75*10^{-6}$ ph cm^{-2} s^{-1}	3

Figure 2 shows the capability of FIGARO IV to detect pulsed signals. The figure reports the number of standard deviations of the power (sum of the first four harmonics) for Geminga, Crab and 2CG342-02 The figure shows, also, the white noise power spectrum distribution obtained between 0 and 50 Hz. Other eight 2CG sources, if they are pulsars, have power values greater than the white noise fluctuations, and FIGARO IV is able to detect them without the need of any external input.

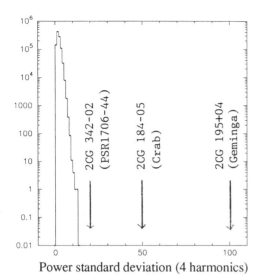

Figure 2: Comparison between the expected pulsed signal from the gamma-ray pulsar Crab, Geminga, and 2CG 342-02 and the white noise spectrum distribution expected for the sum of the first four harmonics between 0 and 50 Hz, in $2*10^4$s observation time of FIGARO IV.

GAMMA-RAY BURSTS

It is difficult to make extimates of GRB visibility because we know only the single event detected by EGRET. However, because of the high instantaneous sensitivity and wide field of view, a single day exposure of the FIGARO IV telescope may correspond to many EGRET observation days.

FUTURE PERSPECTIVES

Because of the large collecting area of the telescope it is impossible to provide it with a calorimeter system; photon energy can be derived by the pair opening angle and by the electron scattering. This technique is limited by the spatial resolution of the LST (1 cm). We are studying the possibility to improve the spatial resolution close to the converter plane up to a few mm.

REFERENCES

1. Bertsh D. L. et al., 1992, Nature, **357**, p. 306 .
2. Beuermann K.P., 1971, J. Geophys. Res.,**76**, 4291.
3. Hartman R.C. et al., 1992 , Ap. J. , **385,** L1.
4. Iarocci F. et al., 1983, N.I.M., **217**, 30-42.
5. Schneid E.J. et al.,1992, A. & A., **255**, L13.
6. Thompson D.J. et al.,1992, Nature, in press.

GRIS BACKGROUND REDUCTION RESULTS USING ISOTOPICALLY ENRICHED Ge

S. D. Barthelmy[1], L. M. Bartlett, N. Gehrels, M. Leventhal[2], B. J. Teegarden, J. Tueller
NASA - Goddard Space Flight Center, Greenbelt, Maryland 20771 USA

S. Belyaev, V. Lebedev
Kurchatov Institute, Moscow, Russia

H.V. Klapdor-Kleingrothaus
MPI für Kernphysik, Heidelberg, Germany

ABSTRACT

The Gamma Ray Imaging Spectrometer (GRIS) was flown twice from Alice Springs, Australia in the spring of 1992 for a total of 32 hours at float altitude. One of the seven Ge detectors (Ge-Lite) was isotopically enriched (>97% ^{70}Ge). This is the first time an enriched Ge detector has been used for astrophysical observations. Because of its thick anticoincidence shield, the GRIS instrument background is dominated by internal β-decay in the energy range 200-1000 keV. Half of the contribution in this β-decay "hump" is due to neutron activated ^{74}Ge. GRIS observed a factor of two reduction in this energy regime in the enriched detector. In future instruments (INTEGRAL/NAE) with thicker anticoincidence shields and smaller apertures, the background reduction will be even larger. Three strong instrumental background lines (54, 67, and 139 keV) are also eliminated. The elimination of the first two is particularly important for cyclotron line observations.

INTRODUCTION

Background reduction techniques, and hence sensitivity improvement, are continually being investigated for gamma ray spectrometers. One of the proposed techniques to reduce the background generated in the detector itself is to reduce the number of isotopes present in the germanium detector[1]. In particular the ^{74}Ge isotope is responsible for approximately half of the continuum background in the middle energy regime. We present results on the background reduction from a flight of the GRIS balloon instrument of a high resolution Ge detector which is made from almost pure ^{70}Ge.
  The GRIS instrument (see Fig 1) is a high-resolution germanium spectrometer that consists of an array of 7 high-purity n-type Ge detectors surrounded by a thick (15 cm minimum path length) active NaI anticoincidence shield[2]. The FOV, defined by 7 individual aperture holes in the shield, is 17° FWHM at 500 keV. The detectors are large with a total volume flown of 1902 cm^3 (average per detector of 271 cm^3). Each detector is in its own cryostat with a coldfinger that penetrates the bottom of the shield into the liquid nitrogen dewar. Because GRIS has reduced the other components of background, by having a thick active shield and a minimum of passive material inside the shield (6 kg, mostly aluminum), the beta decay contribution from all the Ge isotopes to the background is clearly visible (see Fig 2, the triangles are the actual measured GRIS continuum background for normal abundance Ge). These measurements are within 50% of the calculated model (upper solid curve).

1 Universities Space Research Association
2 National Research Council Senior Associate

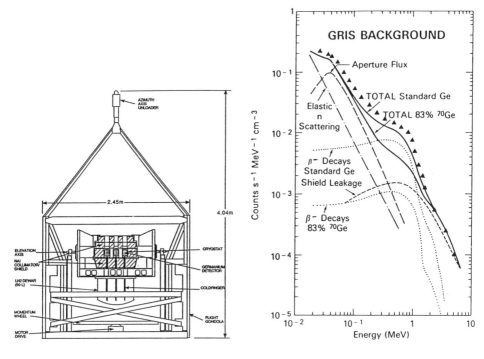

Fig 1) Schematic drawing of the GRIS instrument.

Fig 2) Model calculation of the GRIS background spectrum.

BACKGROUND CALCULATION

For a balloon-borne Ge spectrometer there are four contributions to the continuum background in the instrument. They are: 1) Aperture Flux - atmospheric gamma rays from interactions of the cosmic rays in the atmosphere and diffuse cosmic gamma rays; 2) Elastic Neutron Scattering - neutrons that elastically scatter off Ge nuclei. The detected signal is from the recoil of the Ge nucleus; 3) Beta Decays - incident neutrons and protons that interact with the Ge nuclei to produce beta-unstable nuclides. The signal is due to the ionization energy loss of the β^- or β^+ and from the 511 keV gammas from the positron annihilation; 4) Shield Leakage - gamma rays that penetrate the shield without producing an above-threshold (80 keV) anticoincidence signal. The calculated background for the GRIS instrument is shown in Figure 2 (solid curve for the total).

The Beta Decay contribution is shown (Fig 2, dotted curves) for the two cases of natural-abundance Ge and for ^{70}Ge-enriched abundances (83%). This calculation was done for an enrichment of 83%, while the detector flown was >97%. The difference is not important for this discussion. Natural-abundance Ge is 20.5% ^{70}Ge, 27.4% ^{72}Ge, 7.8% ^{73}Ge, 36.5% ^{74}Ge, and 7.8% ^{76}Ge. The enriched detector abundances are >97% ^{70}Ge, ~3% ^{72}Ge, and insignificant amounts of the remaining isotopes. Over 1000 cross sections were evaluated and the 60 most important beta spectra were summed into this calculation[1]. The Aperture Flux dominates at low energies and Shield Leakage at high energies. For normal abundance Ge the Beta Decays produce a visible "hump" at middle energies in the Total Background curve (upper solid curve).

RESULTS

The solid curve in Figure 3 is the ratio of the calculated Total Continuum Background as a function of energy for the two isotopic abundance cases (Enriched/Normal) discussed above. The points (with error bars) are the actual GRIS measurements normalized by detector volume for the two types of detectors. This data is from the 08 May 92 flight of GRIS from Alice Springs, Australia. There are 10.3 hours of background exposure data spanning a 21 hour flight at a float altitude of 2.8-4.0 gm/cm². The dashed line at 1.0 identifies the case of no improvement.

At the middle energies (300-1000 keV; the Beta Hump) the background is reduced by about a factor of two (peaks at 2.1 at 600 keV). Although the data below 70 keV appear to have a background improvement, this decrease is due to the extra attenuation of the Aperture Flux (the dominant component in this regime) by an aluminum window on the cryostat for the Enriched detector while the Normal detector had a beryllium window. At this point we do not understand the origin of the increase in the background in the 1.5-4 MeV regime. It is not time dependent during our 21 hr flight, thus ruling out $^{69}Ge(\beta^+)^{69}Ga$, 39 hr half-life, as the source, nor is it dependent on the instrument elevation angle to the gamma-bright horizon. It could be a real ^{70}Ge effect or a detector geometry effect; we are continuing the investigation into this effect. At high energies (>4 MeV) where Shield Leakage dominates, the improvement ratio returns to 1.

Figure 4 shows the observed background spectra in the low energy regime for the Enriched and Normal Ge isotope abundance cases. Most notable is the complete elimination of the lines at 53.4, 66.7 (^{72}Ge) and 139.7 (^{74}Ge) keV in the enriched Ge. There are also a major decrease in the line at 439 (^{67}Ga) keV and increases in the lines at 882 & 1117 (^{70}Ge) keV. The line at 198.4 keV (previously assumed to be from ^{70}Ge) is reduced by one third.

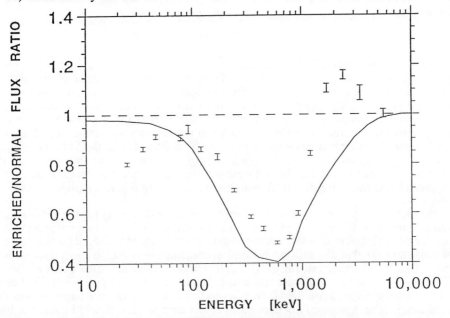

Fig 3) A comparison of the calculated vs observed continuum background reduction for enriched and normal abundance Ge detectors normalized by detector volume.

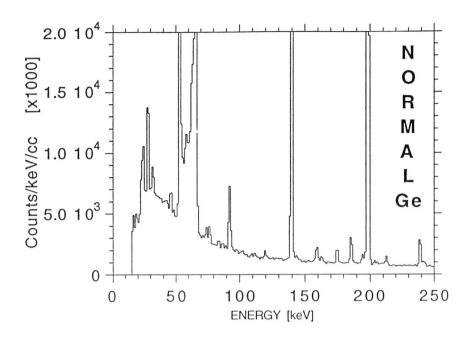

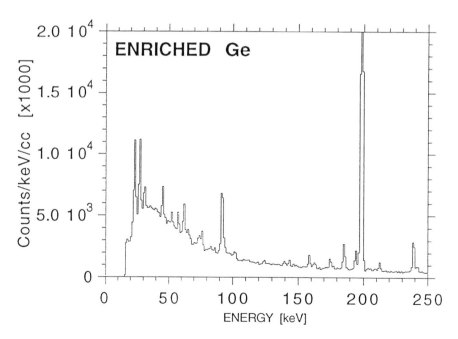

Fig 4) Background spectra for Normal and Enriched Ge detectors in the low energy regime.

CONCLUSIONS

The large measured reduction in the Beta Decay component in the middle energy regime by about a factor of two is very important. This is equivalent to doubling the effective area of the detector (i.e. approximately doubling the number or volume of the detectors). Since the sensitivity for detecting astrophysical lines is proportional to the square root of the background in background-dominated instruments, the line detection is improved by a factor of 1.4. For future instruments, like INTEGRAL, the improvement factor is even larger than 1.4 (closer to 2.5) because of the thicker anticoincidence shield and hence reduced shield leakage component[1,4]. Unfortunately the astrophysically significant line at 511 keV is dominated more by instrumental line production than by continuum background, so the improvement in 511 sensitivity is only about 1.1.

The complete elimination of the line complex between 53.4 and 66.7 keV in the enriched Ge is very important for cyclotron line studies which need smoothly varying backgrounds to make sensitive observations. It is an unexpected result that the 198 keV did not increase as expected with the four-fold increase in the relative abundance of ^{70}Ge. This background line must be due to competing neutron spallation reactions on ^{70}Ge and ^{72}Ge.

We wish to acknowledge the outstanding efforts of the other members of the GRIS team: Macha Colbert, Steve Derdeyn, Chris Miller, Kiran Patel, and Steve Snodgrass. The balloon launch and recovery were provided by NSBF.

REFERENCES

1   N. Gehrels; NIM; A292; pp 505-516; (1990).

2   J. Tueller, S.D. Barthelmy, L.M. Bartlett, N. Gehrels, M. Leventhal, D.M. Palmer, B.J. Teegarden; Ap. J. Supp., submitted; (1992).

3   N. Gehrels; NIM; A239; pp 324-349; (1985).

4   N. Gehrels; NIM; A313; pp 513-528; (1992).

GRANITE - A STEREOSCOPIC IMAGING CHERENKOV TELESCOPE SYSTEM

The Whipple Collaboration

M. Schubnell,[1] C. W. Akerlof,[1] M. F. Cawley,[2] M. Chantell,[3] D. J. Fegan,[4]
S. Fennell,[3] K. S. O'Flaherty,[4] S. Freeman,[1] D. Frishman,[1] J. A. Gaidos,[5]
J. Hagan,[4] K. Harris,[3] A. M. Hillas[6] A. D. Kerrick,[7] R. C. Lamb,[7] T. Lappin,[3]
M. A. Lawrence,[3] H. Levy,[1] D. A. Lewis,[7] D. I. Meyer,[1] G. Mohanty,[7]
M. Punch,[3] P. T. Reynolds,[7] A. C. Rovero,[3] G. Sembroski,[5] C. Weaverdyck,[1]
T. C. Weekes,[3] T. Whitaker,[3] and C. Wilson[5]

[1] University of Michigan, Ann Arbor, MI 48109 USA
[2] St. Patrick's College, Maynooth, Co. Kildare, Ireland
[3] Whipple Observatory, Harvard-Smithsonian CfA, Amado, AZ 85645 USA
[4] University College Dublin, Belfield, Dublin 4, Ireland
[5] Purdue University, West Lafayette, IN 47907 USA
[6] University of Leeds, Leeds LS2 9JT, UK
[7] Iowa State University, Ames, IA 50011 USA

Abstract

A second 10 meter class imaging telescope was constructed on Mt. Hopkins, Arizona, the site of the original 10 meter Whipple Cherenkov telescope. The twin telescope system with a 140 meter base line will allow both a reduction in the energy threshold and an improvement in the rejection of the hadronic background. The new telescope started operation in December 1991. With the final completion of the first installation stage (GRANITE I) during spring 92, it is now operating simultaneously with the original reflector. We describe in this paper design and construction of the new instrument and demonstrate the capability of the experiment to record coincident events.

Introduction

In the last few years the Whipple collaboration has convincingly demonstrated that ground-based γ-ray astronomy can open the universe to a new window in the electromagnetic spectrum. The unambiguous detection of the Crab nebula[1,2] was the first unequivocal evidence of extraterrestial photons above 10 GeV. This has been confirmed by several other experiments.[3,4] With the recent detection of TeV gamma rays from the BL Lac object Markarian 421[5], the first extragalactic source was seen by the Cherenkov imaging technique. With the GRANITE project, we expect to probe the gamma ray sky even deeper by not only lowering the energy threshold to below 200 Gev but also by improving the selection of γ-ray showers using coincident imaging from the two separate telescopes[6]. The geographical location of the two VHE γ-ray telescopes is the Whipple Observatory on Mt.

Hopkins in southern Arizona (31° 41' north, 110° 53' west) at an altitude of 7685 feet.

Mount and Optical Support Structure

The mount for the second telescope was manufactured by McDonnell Douglas for use as a solar concentrator for electrical power generation. It consists of a pedestal with an elevation over azimuth drive system and a parabolic dish framework of 7.6 m focal length. The dish was modified to accommodate our mirror mounting system by welding brackets to the frame to hold the aluminum bars to which the mirror mounting plates are bolted. The pedestal and drive were installed in October 1990 and the dish elements were attached in March 1991. The elevation and azimuth are driven by identical 0.5 hp 480 V motors through helical gear systems. The azimuthal slew speed is $14.6° min^{-1}$. Motion of the dish is governed by a microprocessor controller located in the pedestal which reads the mount sensors and switches power to the motors accordingly. This in turn is controlled through a serial line by either a hand-held remote control or a DEC PDP-11/23 computer. The computer software is a version of the solar tracking program modified to track sidereally and to communicate with the data acquisition computer.

Mirrors

A total of 366 hexagonal mirrors are currently mounted on the reflector giving a reflective surface of 66.3 m². The mirrors were designed to be robust and to withstand many years of use in an exposed environment without significant degradation of the optical quality. They were fabricated by cutting 5 cm thick foam glass to the required hexagonal shape and then grinding the desired spherical concave shape into the material using abrasive paper glued to an aluminum form with the appropriate curvature. Then epoxy was applied to both the foam glass and a hexagon of 0.7 mm thick back-aluminized glass and the two pieces pressed together with a weighted form identical to that used in the grinding process.

This construction technique, using second surface reflection and durable materials, produces mirrors which can be washed if necessary and will not require recoating. Resistance to hail damage was tested by dropping ice cubes onto the mirrors from a height of 15 m. Four different focal lengths were fabricated to approximate the parabola defined by the structure of the reflector. The rms error in all the mirror surfaces was measured to be less than 0.3 mr[7]. Figure 1 shows the optical point spread function as measured after the final alignment of the mirrors. The FWHM is approximately 0.18°.

Photomultiplier Camera

The Cherenkov light from airshowers is focused on a two-dimensional, hexagonal array of 37 fast photomultipliers (Amperex XP2230). The design of this

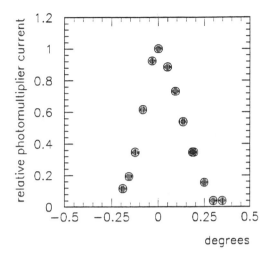

Figure 1. Point spread function (measured at Polaris) of the new reflector ('11 meter') for the center of the field of view. The FWHM is approximately 0.18°.

medium resolution camera is essentially identical to that used in the Whipple Observatory 10 meter reflector several years ago[1]. The photomultipliers have a 5 cm diameter and subtend a 0.5 degree field of view. A motorized shutter system, an integral part of the focus box, protects the photomultipliers during daylight. The present camera will be upgraded to a 109 photomultiplier array to match the current high resolution camera of the 10 m reflector during summer 1993.

The signals are transmitted through RG-8 cables with very small pulse dispersion to an electronics hut 50 feet from the telescope. Typical signals at the front end input are measured to have a FWHM of 2.5 nsec.

DATA ACQUISITION

The data acquisition system has been designed to extract as much information as possible from each event. For every photomultiplier pulse, the integrated charge, pulse width and arrival time is separately measured. The trigger rate for each tube is scaled and the DC tube current is monitored. To provide for this large signal fanout, an 8-channel amplifier-discriminator module has been designed to incorporate a wide band amplifier and two independent discriminators[8]. The discriminator outputs feed the coincidence logic, the TDCs and single pulse rate scalers. Two externally supplied reference voltages control the discriminator thresholds. Current monitor modules[9] convert the DC tube current for each PMT channel to frequencies which are then integrated in digital scalers. The analog signal, the singles rates and the DC current for each individual tube are continuously monitored by an on-line data acquisition program.

The majority of the remainder of the data acquisition electronics are commercial CAMAC modules, controlled by an in-crate LSI-11 microcomputer. The

LSI-11 writes the data in a memory buffer which is read out on demand by a Q-bus VAXstation and written to disk. The VAXstation controls both the overall data acquisition system and the photomultiplier high voltage supply.

For each trigger the UTC time is established by a GPS satellite receiver clock to an absolute accuracy of better than 1 μs. In addition, events at both reflectors are tagged with time stamps by a free running clock module[10] clocked at 20 MHz. This together with the phase measurement of the common clock enables the Cherenkov light signals in the two telescopes to be correlated to an accuracy of better than 1 ns.

The data acquisition system is designed to operate both independently and in coincidence with the 10 m reflector. In order to take full advantage of the twin telescope system, two levels of trigger are implemented: a low-level trigger accepted in coincidence with a signal from the remote dish or a higher level trigger recorded independently of the status of the remote instrument. A primary trigger rate of 5 kHz is formed by the minimal requirement that two or more phototube pulses exceed a threshold within a 20 ns resolving time. This primary trigger enables all ADC and TDC gates for the local photomultiplier array and initiates a logic timing pulse to the remote telescope. If the particular event exceeds a more stringent, higher threshold, the analog signals will be converted and data transmitted to the local data acquisition computer with no further intervention. For events that do not meet this criterion, a clear pulse is generated.

COINCIDENT EVENT IDENTIFICATION

Coincident events are identified by the time-of-event information recorded with each trigger. At an initial trigger rate of around 4 Hz for each telescope, coincident events can be reliably identified by using solely the satellite UTC time recorded with each event. With a 0.25 msec time window which matches the recorded precision of the satellite time a random coincident rate of 10^{-2} is expected. The coincident rate varies slightly with the sky position of the observed object but is typically around 30%.

The coincidence rate as a function of the UTC time difference between events of the two telescopes is shown in figure 2. The peak with the coincident triggers can easily be identified. It contains ca. 28% of the total number of events. Air showers of TeV energies have their maximum development in an altitude of around 10 km. Thus, we expect a maximum coincident event rate when the two Cherenkov telescopes are slightly inclined towards each other, pointing to the shower maximum.

Figure 3 shows the measured coincident trigger rate for the two telescopes with one reflector pointing to the zenith as the other telescope points at different elevation angles. From the angular difference at the maximum conicident rate (ca. 1°) a mean shower distance of 8000 m can be calculated.

The image analysis of the recorded showers is in progress and first results from stereo observations can be expected in late 1992.

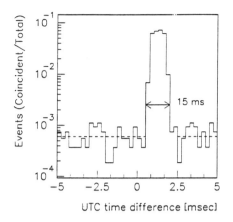 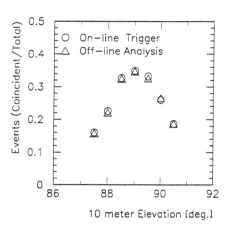

Figure 2. The coincidence rate as a function of the UTC time difference. The dashed line indicates the random rate.

Figure 3. The decoherence curve for the twin telescope system. One telescope points to the zenith as the elevation angle of the second reflector is varied.

ACKNOWLEDGEMENTS

This research was supported by the US Department of Energy, NASA, the Smithsonian Scholarly Studies Research Fund, and EOLAS, the scientific funding agency of Ireland.

REFERENCES

[1] T. C. Weekes et al., Ap. J. **342** (1989) 379.
[2] G. Vacanti et al., Ap. J. **377** (1992) 467.
[3] C. Akerlof et al., *New and Exotic Phenomena '90: Xth Moriond Workshop*, Les Arcs, Savoie, France, January 20-27, 1990, edited by O. Fackler and J. Tran Thanh Vân, pp. 395-403.
[4] P. Baillon et al., to be published in Proc. 26th International Conf. on High Energy Physics, Dallas (1992).
[5] M. Punch et al., Nature **358** (1992) 477.
[6] M. Hillas and M. West, in Proc. 22nd International Cosmic Ray Conf. (Dublin) **1** (1991) 472.
[7] C. Weaverdyck, D. Meyer and C. Akerlof, NIM **A310** (1991) 690.
[8] H. Levy, D. Frishman and C. Akerlof, NIM **A292** (1990) 715.
[9] D. Frishman and C. Akerlof, NIM **A311** (1992) 306.
[10] S. Freeman and C. Akerlof, NIM **A320** (1992) 305.

CALIBRATION OF AN ATMOSPHERIC CHERENKOV TELESCOPE USING MUON RING IMAGES

A.C.Rovero[1], K.Harris[1], Y.Jiang[1], M.A.Lawrence[1], D.A.Lewis[2], M.Urban[1], T.C.Weekes[1]

[1] Whipple Observatory, Harvard-Smithsonian CfA, P.O. Box 97, Amado, Arizona 85625-0097 USA.
[2] Physics and Astronomy Dept., Iowa State Univ., Ames, Iowa 50011, USA.

ABSTRACT

Theoretical and experimental studies have been made of the Cherenkov light images from single muons recorded by the atmospheric Cherenkov imaging telescopes which are commonly used in TeV gamma-ray astronomy. In particular, the Cherenkov ring images from single muons have been used to calibrate the Whipple Observatory 10 meter imaging telescope. This approach tests the total throughput of the telescope and uses an atmospheric Cherenkov light signal that matches the shower signal. We discuss the geometrical and physical factors going into the images of the Cherenkov rings and match the measured images with predicted rings. The preliminary estimate of the absolute calibration of the Whipple telescope is in agreement with that obtained by other methods but the uncertainty is reduced.

THE PROBLEM

Ground-based gamma-ray telescopes are notoriously uncertain as to their absolute energy thresholds and collection areas. This holds true for both atmospheric Cherenkov telescopes and air shower arrays. Although the response to background hadron-induced air showers gives a rough measure of instrumental parameters, this is not really satisfactory since the telescopes are biased away from hadron shower detection with consequent uncertainty as to the real sensitivity to gamma rays. An alternative method for atmospheric Cherenkov systems is the measurement of the light detection threshold with a pulsed light source of known brightness. The uncertainties in the brightness of these sources are as large as a factor of two; this uncertainty limits the interpretation of measured gamma-ray fluxes.

In second generation atmospheric Cherenkov imaging telescopes the problem is worse since the wide field (3.5° FWHM) camera requires calibration over all its elements. Such telescopes[1,2] are now finding wide use in TeV gamma-ray astronomy; with the establishment of some sources[3,4], it becomes more important than ever to measure accurately the energy threshold of such telescopes.

PREVIOUS MEASUREMENTS

The previous best estimate of the energy threshold of the High Resolution Camera on the Whipple 10 m Reflector was done in two steps. First a pulsed light source of varying intensity was used to characterize the noise fluctuations and hence determine the conversion from digital counts in the ADC to photoelectrons. In the second step, throughput of the telescope (mirror reflectivity x quantum efficiency) was estimated and the measured trigger threshold in digital counts was compared with the predicted shower signal (from Monte Carlo simulations) in photoelectrons. The conversion was estimated to be 1 d.c. = 1.15 p.e. corresponding to an Effective Gamma Ray Energy Threshold of 0.4 TeV. The overall uncertainty was estimated to be 30%.

METHOD

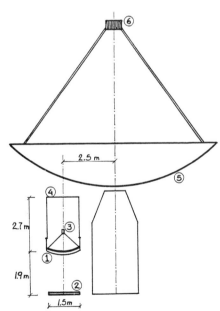

Fig. 1: Equipment for Recording Muon Images on 10 m Camera; 1) Muon trigger telescope; 2) Scintillators; 3) 2" PMT; 4) Light tight cylinder; 5) 10m reflector; 6) 10m camera.

An elegant method of providing a light source of known brightness is to look for the Cherenkov ring images from single cosmic ray muons passing through or close to the telescope. The muons can be detected by a muon telescope which consists of an enclosed atmospheric gas Cherenkov detector with a scintillator layer in coincidence. For the calibration of the Whipple Observatory High Resolution Camera the setup shown in fig. 1 was used. The gas Cherenkov detector had a 1.5 m aperture mirror and a length of 2.7m. The full field of view is 3.8°. The muon telescope was placed underneath the 10 m reflector which was pointing to the zenith; the separation between the axes of the two telescopes was 2.5 m. The muon telescope triggered the readout of the 10m reflector camera when a muon was detected.

THEORETICAL PREDICTION

If we consider the photons arriving from a muon trajectory to the 10 m dish, we have (see fig. 2):

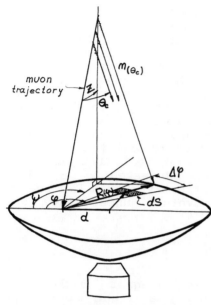

$$f = \frac{n_{(\theta_c)}}{2\pi r \theta_c}$$

where f is the number of photons per unit area arriving at dS. Up to second order in Z, f does not depend upon Z. Then:

$$N_0 = \int f\, dS = \frac{n_{(\theta_c)}}{2\pi \theta_c} \int_{\Delta\varphi} d\varphi \int_0^{R(\varphi)} dr$$

is the number of photons in the Δ arc in the image at camera without any dispersion. But mirror aberrations and muon scattering in the atmosphere widen the image so that (see fig. 3 and 4):

Fig. 2: Geometry of Emission; θ_c = Cherenkov angle; Z = Zenith angle; $n_{(\theta_c)}$ = Cherenkov photons emitted by the muon per unit length.

$$N = \int_{x_1}^{x_2} N_0 \mathcal{g}_{(x,\sigma)}\, dx$$

$$\sigma = \sigma_{(\theta_c, r, \varphi)}$$

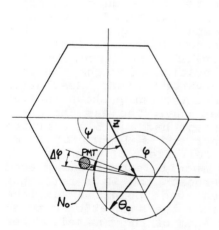

Fig. 3: Muon Image at Camera. N_0 = Number of photons in the arc.

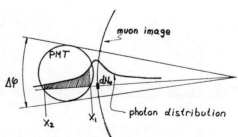

Fig. 4: Finite width of ring relative to PMT.

where g is the Gauss function and σ is the standard deviation which takes into account both aberrations and scattering. Then, the number of photons arriving at a given PMT is:

$$N = \frac{n_{(\theta_c)}}{2\pi^{3/2}\theta_c} \int_{\Delta\varphi} d\varphi \int_0^{R(\varphi)} dr \int_{u_1}^{u_2} e^{-u^2} du$$

where u=x/σ. Photo-electrons can be obtained from this equation by multiplying by the mirror reflectivity and quantum efficiency of PMTs.

DATA AND RESULTS

Data were taken on April 7 and 30, and on May 5, 1992 under conditions of high cloud which prevented VHE gamma-ray observations but were good enough for this calibration. The Cherenkov photons that form the muon image are produced less than 1 km above the telescope.

We obtained 17,866 events in 14 hours of observation. After processing, we could recognize 1,400 muon events which indicates 8% efficiency. From those, we selected 15 images with full rings and 120 with good arcs.

The theoretical prediction of the brightness of each of these events under the experimental conditions was calculated in photoelectrons and compared with the measured brightness in digital counts. One example of a muon image is shown in Fig. 5.

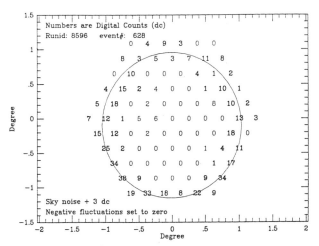

Fig. 5: Observed Event

The mean of these 135 events gives the conversion from digital counts to photoelectrons. With this value it is possible to compare the measured shower brightness threshold with that predicted by Monte Carlo shower simulations and hence derive the gamma-ray energy threshold.

Result: 1 d.c. = 1.44 ± 0.03 p.e.

The error shown here is purely statistical; there are also uncertainties in the absolute values of the mirror reflectivities, the quantum efficiencies and the effective Cherenkov bandwidth (which may be as much as 20%). However since these same constants are used in the evaluation of the shower image, they do not enter into the evaluation of the Effective Gamma Ray Energy Threshold.

This result is consistent with the previous measurement taking into account the changes in high voltage made to compensate for degradation of the mirror coating and would imply an energy threshold around 0.5 TeV.

CONCLUSIONS

It is very clear that the Whipple Observatory 10 meter atmospheric Cherenkov camera is working properly as demonstrated by its ability to record the subtle muon images. The muon trigger telescope is shown to work well as a detector of muons with energies greater than the Cherenkov radiation threshold and with trajectories nearly in parallel with the optical axis of the telescope.

The result shown here is in agreement with the previous one, but the error is reduced considerably.

ACKNOWLEDGEMENTS

We acknowledge support from the U.S. Department of Energy, NASA, the Smithsonian Scholarly Studies Fund and EOLAS, the scientific funding agency of Ireland.

REFERENCES

1. Cawley, M.F. et al (1990), Exper. Astr.,**1**,173
2. Stepanian, A.A., Proceedings of the International Workshop on VHE Gamma Ray Astronomy, Ootacamund, India, Sep 20-25 1982.
3. Vacanti, G. et al (1991), Ap. J., **377**, 467
4. Punch, M. et al (1992), Nature, **358**, 477

RESULTS FROM THE CYGNUS EXTENSIVE AIR SHOWER ARRAY

David A. Williams
Santa Cruz Institute for Particle Physics
University of California
Santa Cruz, California 95064

Representing the CYGNUS* Collaboration

Abstract

Recent results from the CYGNUS cosmic ray experiment on objects studied by the Compton Observatory are presented. Studies of Geminga and several active galactic nuclei which have recently been detected by the Compton Observatory are discussed. Preliminary results of a search for ultrahigh-energy gamma rays coincident with gamma-ray bursts are also shown.

INTRODUCTION

The CYGNUS extensive air-shower experiment began operation in April 1986 with 50 scintillation counters, located around the Los Alamos Meson Physics Facility beam stop (106.3°W, 35.9°N, altitude 2130 m). One of its primary goals is the search for point sources of cosmic rays. The array has been expanded substantially since 1986. This paper describes data taken with the CYGNUS-I array, which presently has 108 counters, including the original 50, covering an area of 22,000 m². The median primary energy for detected gamma-ray initiated events is about 80 TeV in the present configuration; for protons, the median is about 100 TeV. Studies of the cosmic-ray shadows of the sun and the moon[1,2] have shown that the CYGNUS array has a resolution for the projected angle of $0.66° \pm 0.07°$. The CYGNUS-I event rate is presently about 3.5 events/s. A more detailed description of the the CYGNUS experiment can be found elsewhere.[3]

The CYGNUS data set from April 1986 to May 1991 has been used to survey the whole sky from declination 0° to 80° for continuously emitting point sources.[4] No evidence is found for emission from a source. Flux upper limits (90% confidence

---

*The CYGNUS Collaboration: R.W. Ellsworth (George Mason University); R.L. Burman, C.M. Hoffman, D.E. Nagle, M.E. Potter, V.D. Sandberg, C. Sinnis, and W. Zhang (Los Alamos National Laboratory); S.J. Freedman and B.K. Fujikawa (University of California, Berkeley); D.E. Alexandreas, S. Biller, G.M. Dion, X.-Q. Lu, A. Shoup, and G.B. Yodh (University of California, Irvine); J.-P. Wu (University of California, Riverside); M. Cavalli-Sforza, D.G. Coyne, D.E. Dorfan, L. Kelley, S. Klein, R. Schnee, D.A. Williams, and T. Yang (University of California, Santa Cruz); G. Allen, D. Berley, C.Y. Chang, C. Dion, J.A. Goodman, T.J. Haines, and M.J. Stark (University of Maryland, College Park); D.D. Weeks (University of New Mexico).

Source	Flux limit above 40 TeV
Crab	$< 4.4 \times 10^{-13}$ cm^{-2} s^{-1}
Cyg X-3	$< 1.9 \times 10^{-13}$ cm^{-2} s^{-1}
Her X-1	$< 1.6 \times 10^{-13}$ cm^{-2} s^{-1}

Table 1: Flux limits for continuous emission from some candidate sources.

level) for the three objects widely considered to be the most likely sources in the CYGNUS energy region are given in Table 1.

We have also conducted a search for emission lasting several hours (one day of observation) from many candidate sources.[2] There is no evidence for any excess beyond what is expected from statistical fluctuations of the background.[†] Typical flux upper limits (given in Ref. 2) are generally of order 10^{-12} cm^{-2} s^{-1}.

We turn now to results from some specific Compton Observatory sources.

GAMMA RAY BURSTS

We have made a preliminary analysis of ten strong gamma ray bursts[6] detected by BATSE for coincidences with the CYGNUS experiment. We search for events within a square bin 14° on each side (to account for the uncertainty in reconstruction of the burst direction by BATSE), centered on the burst position determined by BATSE, and arriving during the burst. We compare the number of events found for each burst to the number of background events expected. The number of expected background events depends very steeply on the zenith angle. The results are summarized in Table 2. The number of events found is consistent with background alone and the absence of a signal. For each burst, we calculate a preliminary flux upper limit (at the 90% confidence level), assuming that the photon spectrum is the same as the spectrum of background cosmic rays. The limits are shown in Table 2. They are given as the integral of the flux above a particular energy; for each burst, we choose the energy corresponding to our estimate of the median energy for gamma rays we would observe from that burst.

MARKARIAN 421 and OTHER AGN'S

Recently, the EGRET experiment[7] and the Whipple experiment[8] have detected gamma rays from the distant active galactic nucleus Markarian 421. These measurements indicate a differential energy spectrum of roughly E^{-2} from this object. Because of its distance (about 125 Mpc), photons with energy above about 100 TeV are expected to be absorbed by e^+e^- pair production off the 2.7 K microwave background.[9] Our data from April 1986 through September 1992 show a small but insignificant (one standard deviation) excess from the direction of Markarian 421. We have made a preliminary calculation of our sensitivity to a

---

[†]Note that the 1986 Hercules burst previously published (Ref. 5) is significant primarily because of the observed periodicity in addition to the excess of events.

Burst Number	Zenith Angle	Expected Background	Events Observed	Median Energy (E_m) in TeV	Flux Limit above E_m (cm^{-2} s^{-1})
105	13°	2.4	3	60.	$< 6.7 \times 10^{-10}$
451	55°	0.058	0	1000.	$< 2.3 \times 10^{-10}$
467	58°	0.025	0	1000.	$< 9.2 \times 10^{-10}$
999	38°	0.41	2	150.	$< 9.7 \times 10^{-10}$
1088	31°	0.039	0	100.	$< 8.7 \times 10^{-9}$
1121	34°	7.0	7	110.	$< 1.1 \times 10^{-10}$
1425	18°	3.0	3	60.	$< 5.0 \times 10^{-10}$
1519	25°	4.2	3	90.	$< 1.6 \times 10^{-10}$
1538	32°	0.62	0	140.	$< 3.0 \times 10^{-10}$
1609	56°	0.019	0	1000.	$< 8.5 \times 10^{-10}$

Table 2: Summary of preliminary upper limits for the photon flux from ten gamma ray bursts detected by BATSE.

source at the declination of Mkn 421 with an E^{-2} spectrum, cut off by the effect of the microwave background.[‡] We obtain a preliminary upper limit for the flux from Mkn 421 above 50 TeV of 1.5×10^{-13} cm^{-2} s^{-1} (90% c.l.). This upper limit is compared to the extrapolation of the EGRET and Whipple measurements in Figure 1.

In the day-by-day search for emission (described above), we find no evidence for a "hot" day from Markarian 421.

We have also searched for continuous emission from eleven other active galactic nuclei which have been detected by EGRET and are in the CYGNUS field of view. They are 3C273, 3C279, 3C454.3, 4C+11.69, 4C+38.41, 4C+71.07, PKS0235, PKS0420, PKS0528, 0202+149, and 0716+714. We do not find evidence for a signal from these AGN's. Calculations of our sensitivity to these sources are still in progress.

GEMINGA

We have searched the complete CYGNUS data set for continuous emission from Geminga. A 2.1 standard deviation excess is found over the expected background. Considering the number of sources that we have studied for continuous emission, as well as for emission on shorter time scales, this is not a significant excess. Thus, we calculate an upper limit for the flux from Geminga, which is 8.2×10^{-14} cm^{-2} s^{-1}

[‡]Several authors (see Ref. 10) have argued that the field of infrared radiation is sufficient to produce significant absorption at energies above about 1 TeV. Because of the uncertainties in the amount of infrared radiation present, we have not attempted to take it into account.

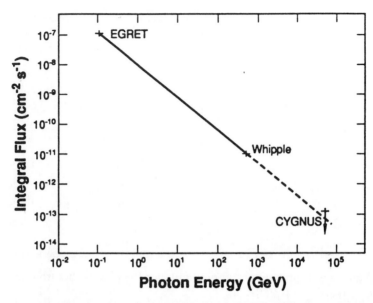

Figure 1: Comparison of the preliminary upper limit flux from Markarian 421 measured by CYGNUS experiment to the extrapolated fluxes from the EGRET and Whipple experiments.

(90% c.l.) above 80 TeV.

As for Markarian 421, the search for emission lasting about a day from Geminga fails to reveal a signal.

The subset of data from 6 July 1989 to 17 August 1992 has been analyzed using the ephemeris measured by the EGRET team[11] to search for periodic emission from Geminga. The phase distribution of events is shown in Figure 2. As can be seen in the figure, we find no evidence for continuous periodic emission from Geminga.

CONCLUSIONS

The CYGNUS experiment has looked for emission from several Compton Observatory sources, including gamma ray bursts, AGN's, and Geminga. At the present sensitivity of the the experiment, emission above about 40 TeV has not been detected from these sources.

ACKNOWLEDGMENTS

We thank the BATSE Collaboration for sharing preliminary data on gamma ray bursts. We are indebted to C. Kouveliotou of the BATSE Collaboration for selection and transmission of the burst data and for extended communications concerning the operation of BATSE and interpretation of the data. The author wishes to thank the MP Division of Los Alamos National Laboratory for its hos-

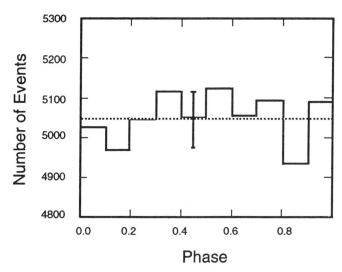

Figure 2: Phase distribution for events from the direction of Geminga.

pitality. This work is supported in part by the National Science Foundation, Los Alamos National Laboratory, the U.S. Department of Energy, and the Institute of Geophysics and Planetary Physics of the University of California.

REFERENCES

[1] D. E. Alexandreas et al., Phys. Rev. D **43**, 1735 (1991).

[2] D. E. Alexandreas et al., "Daily Search for Emission of Ultra High Energy Radiation from Point Sources," to be published in Astrophys. J. (LA-UR-92-2082).

[3] D. E. Alexandreas et al., Nucl. Instr. Meth. **A311**, 350 (1992).

[4] D. E. Alexandreas et al., Astrophys. J. Lett. **383**, L53 (1991).

[5] B. L. Dingus et al., Phys. Rev. Lett. **61**, 1906 (1988).

[6] The BATSE Collaboration, G. Fishman spokesman, private communication.

[7] The EGRET Team, P. F. Michelson et al., IAU Circular No. 5470 (1992).

[8] M. Punch et al., Nature **358**, 477 (1992).

[9] R. J. Gould and G. P. Schreder, Phys. Rev. Lett. **16**, 252 (1966); J. V. Jelley, Phys. Rev. Lett. **16**, 479 (1966).

[10] A. I. Nishikov, Zh. Eksp. i Teor. Fiz. **41**, 549 (1961) [English Trans. Sov. Phys. JETP **14**, 393 (1962)]; R. J. Gould and G. P. Schreder, Phys. Rev. **155**, 1408 (1967); F. W. Stecker, O. C. De Jager, and M. H. Salamon, Astrophys. J. Lett. **390**, L49 (1992).

[11] The EGRET Team, J. R. Mattox et al., IAU Circular No. 5583 (1992).

SOFTWARE AND ARCHIVES

STATUS OF THE BATSE ENHANCED EARTH OCCULTATION ANALYSIS PACKAGE FOR STUDYING POINT SOURCES

R. T. Skelton,* J. C. Ling, N. F. Ling, R. Radocinski, and Wm. A. Wheaton
Jet Propulsion Laboratory
California Institute of Technology, Pasadena, CA 91109

ABSTRACT

The Compton Gamma-Ray Observatory's Burst and Transient Source Experiment (BATSE) has a powerful capability to provide nearly uninterrupted monitoring in the 25 keV—1 MeV range of cosmic point sources using occultation by the Earth. A number of interesting results have been obtained to date using the BATSE Mission Operations (MOPS) Level I analysis system. We have been constructing a more physical model for the background variations which will allow use of more data, enhancing the sensitivity by a factor of several. Our "enhanced" package, written in IDL primarily for a VMS environment, accepts input data in the MOPS format. It fits for the diffuse background (including Earth blockage), prompt cosmic-ray effects, South Atlantic Anomaly activation, atmospheric cosmic-ray secondaries, and cosmic sources in 14 Medium Energy Resolution energy bands of the Large Area Detectors. Features include the ability to correlate set and rise of a given source and to handle multiple sources within the fit window. The count spectrum is deconvolved using the detector energy response matrix to yield a source photon spectrum. We describe the current status and performance of the system.

INTRODUCTION

The Burst and Transient Source Experiment (BATSE) aboard the Compton Gamma Ray Observatory (CGRO) offers a capability to provide almost uninterrupted monitoring of gamma-ray sources by Earth occultation. "Almost" means that one gets a measure of a source whenever it rises or sets in the 96-minute CGRO orbit. On any given day, about 6% of the sky in the polar regions is continuously above the horizon and any sources therein cannot be monitored, but the CGRO orbit precesses so that any such source can spend only about two weeks continuously up before setting and again being subject to Earth occultation. BATSE consists of eight modules which are located octahedrally at the corners of the CGRO. Each module consists of a Large Area Detector (LAD) and a Spectroscopy Detector, both of which are NaI scintillators. Data are accumulated in 2.048-second bins for each of 14 energy bands for the LADs; more sophisticated data modes exist but have not been used to date in this investigation. BATSE is described in detail by Fishman et al.[1]

The need for continuous monitoring of gamma-ray sources is illustrated by the fragmentary record of the premiere galactic Black Hole candidate, Cygnus X-1, compiled by Ling et al.[2] Time variability has been established by several studies[3,4,5,6] and most recently by Ling et al.[7] Theoretical interpretations in terms of an electron-positron pair plasma near a luminous compact object have been offered.[8,9,10,11] Variable gamma-ray emission from AGN and galactic sources has been reported.[12,13] Long-term monitoring with a single instrument to correlate time variability with spectral shifts for a number of compact objects would allow better insight into the physics of such objects.

---

* National Academy of Sciences/National Research Council Senior Associate

METHOD

The technique of Earth occultation with BATSE has been well established by the BATSE Mission Operations team, using a model consisting of a step at source rise/set plus a quadratic fit to two minutes of data on each side. This simple model has produced a number of significant results[12,13,14,15] which included detection of a target of opportunity for which the CGRO Observing Plan was modified.

The MOPS version of the Earth occultation technique is statistically limited since only a small fraction of the data is used. Using more data would only increase systematic errors, since the modeling the background *vs* time as quadratic would cease to be reasonable. Improvements along more sophisticated mathematical lines, such as including more polynomial terms or expanding the background in Fourier series, are possible, but one runs the risk that the desired signal (a step) will be incorporated into the background model.

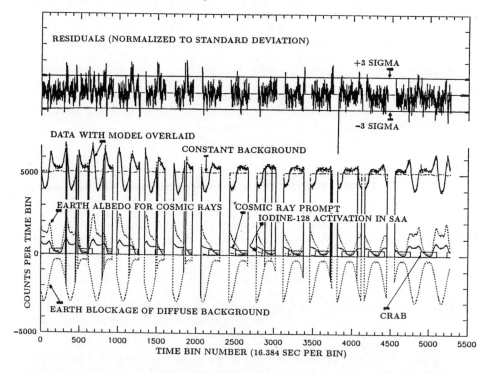

Figure 1. The model components and their sum compared to raw detector counts. A full day of data for LAD 5 in the 96 — 121 keV energy band is shown. In the top panel are plotted residuals normalized to the standard error from statistics alone. The χ^2_ν for this example is 2.2.

MODEL

For this reason, we have elected to model the background with terms which have a physical basis. Figure 1 shows the terms comprising the model and comparison of their sum to the raw data for an example energy channel and LAD.

The terms are described in the following paragraphs.

Constant background: This term simply accounts for the diffuse isotropic cosmic background and slowly varying effects, such as long-term cosmic-ray activation. It is a constant; the slight variation in figure 1 results from differences in livetimes and the fact that figure 1 is plotted in terms of raw counts.

Earth Blockage of Cosmic Diffuse Background: This term is expected to be negative by virtue of its definition as a deficit in the diffuse background. When the LAD is pointed away from the Earth, it remains negative instead of going to zero, probably because of LAD response at angles greater than 90 degrees. At higher energies, the magnitude of this term decreases, and the mathematical fit value can become positive; this is probably because the mathematical model automatically incorporates some contributions from non-local atmospheric cosmic-ray secondaries into this term.

Cosmic Ray Prompt: This term primarily reflects prompt secondaries from cosmic rays striking the instrument or spacecraft but not being vetoed. The cosmic-ray rate is obtained from upper level discriminator rates of the 8 BATSE Spectroscopy Detectors.

Earth Albedo for Cosmic Rays: This term includes factors for both the cosmic-ray rate and the Earth exposure; it reflects the major contribution of atmospheric secondaries, which is proportional to the local cosmic-ray rate.

^{128}I Activation in SAA: This term represents contributions to the count rate which decay with the 25-min half-life of ^{128}I. A separate term is added following each South Atlantic Anomaly passage.

Crab: This rectangular wave represents the one point source in the example of figure 1, namely the Crab nebula and pulsar. The height of the step corresponds to the signal being extracted from the count rate.

When these terms are summed and plotted over the raw data, the results are virtually indistinguishable on the scale of the lower portion of the figure. In the upper portion residuals for each time bin (normalized to the formal standard error) are plotted. The reduced χ^2 (χ^2_ν) for this example is 2.2. Residual plots such as this will point the way for future improvements in the model, as well as identify specific events, such as weak solar flares, to mask out.

RESULTS

Figure 2 shows count rates *vs* energy for each of several days for a LAD facing the Crab. The extracted count rates are reasonably stable, although the χ^2_ν values indicate that there are still systematic effects to overcome. Current one-sigma sensitivities including systematics for one day range from 30 mCrab at lower energies to 180 mCrab with the energy bin widths of figure 2.

A check for possible systematic errors in the model is to extract the count rate from a point in an apparently blank part of the sky. Figure 3 shows such a check for Dummy Source (DUM1930+53). The result is null, albeit with possible systematic effects in the lower energy channels.

FUTURE DIRECTIONS

Future directions in developing the model fall into two categories: software engineering and improvements to the model. Software engineering deals mainly with adapting the model to run in a VAX/VMS environment. Development has been done on a SUN SparcStation, with the VAX system only recently received (October, 1992). The package will run primarily in IDL, and the differences between the UNIX IDL and VAX IDL are expected to present only minor difficulties.

1192 Studying Point Sources

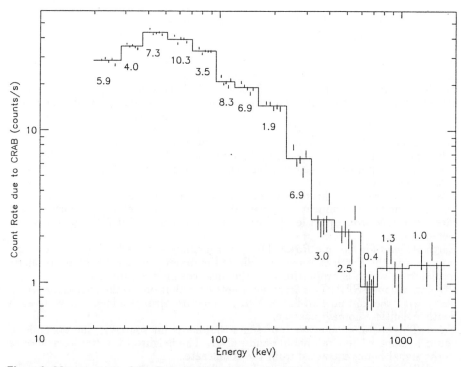

Figure 2. Measurements of the Crab count rates for each of five consecutive days. Each vertical bar represents the count rate in the energy bin for one day. The histogram shows the energy bins and the average of the five days. The numbers are the χ^2_ν values for the five-day dataset.

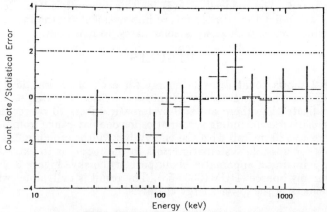

Figure 3. Spectrum of control region 1930+53, where there is no known source. Error bars represent formal statistical errors. When χ^2_ν values from figure 2 are used to estimate total uncertainties, the result is null.

A number of modeling improvements are envisioned; study of the residuals plots (*cf.* figure 1) should point the way to which are most important, and perhaps identify new ones. Some of the improvements envisioned are:

Angular response at large angles: Above about 120 degrees, the spacecraft blocks the LAD. Strong atmospheric background contributions leaking through may need to be taken into account to detect and measure sources accurately.

Atmospheric scattering of γ radiation: The current atmospheric term reflects atmospheric secondaries from charged cosmic rays. Inclusion of scattering effects from cosmic γ-ray sources will be necessary for accurate flux results.

Variable cosmic-ray activation outside the SAA: Modulation of the cosmic-ray rate will produce a modulated activity which is phase-shifted from the cosmic-ray rate. While this term will be small, it appears fairly simple to incorporate.

^{24}Na activation in the SAA: A 15-hr term similar to the current ^{128}I term.

Simultaneous fit to multiple LAD's: Different LAD's will have different projected areas to a given source. Fitting to more than one will reduce source confusion.

SUMMARY

Recent progress on the enhanced model has yielded promising preliminary scientific results.[7] Despite considerable advances in fitting the background and reducing non-statistical data scatter, the scatter still considerably exceeds the statistical errors. We are confident that further effort will yield more improvement, and that a considerable reduction in uncertainties is in the offing.

ACKNOWLEDGEMENTS

The assistance of G. J. Fishman, B. A. Harmon, W. S. Paciesas, B. C. Rubin, and the rest of the BATSE team is gratefully acknowledged. The research described in this paper was carried out by the Jet Propulsion Laboratory, California Institute of Technology, under contract to the National Aeronautics and Space Administration.

REFERENCES

1. G. J. Fishman *et al.*, in **Proceedings of the GRO Science Workshop**, ed. W. N. Johnson, p. 2-39 (1989).
2. J. C. Ling *et al.*, in **Proceedings of the 20th International Cosmic Ray Conference**, Moscow, paper OG 2.1-2, 1, 54 (1987).
3. J. C. Ling, W. A. Mahoney, W. A. Wheaton, A. S. Jacobson, and L. Kaluzienski, *Ap.J.* **275**, 307 (1983).
4. J. C. Ling, W. A. Mahoney, W. A. Wheaton, and A. S. Jacobson, *Ap.J. (Lett.)* **321**, L117 (1987).
5. M. McConnell *et al.*, *COMPTEL Observations of Cygnus X-1*, these proceedings (1992).
6. J. C. Ling and W. A. Wheaton, *Ap.J. (Lett.)* **343**, L57 (1989).
7. J. C. Ling *et al.*, *Long-Term Temporal and Spectral Variation of Cygnus X-1 Observed by BATSE*, these proceedings (1992).
8. E. P. Liang, *Ap.J.* **234**, 1105 (1979).
9. A. A. Zdziarski, *Acta Astr.* **30**, 371 (1980).
10. E. P. Liang and C. D. Dermer, *Ap.J. (Lett.)* **325**, L39 (1988).
11. C. D. Dermer and E. P. Liang, in **Nuclear Spectroscopy of Astrophysical Sources**, AIP Conf Proc. # 170, ed. N. Gehrels and G. J. Share (New York:AIP), p. 326 (1988).
12. G. J. Fishman, R. B. Wilson, C. A. Meegan, B. A. Harmon, and M. Brock, IAUC Telegram #5327, *GX-339-4* (1991).
13. B. A. Harmon *et al.*, *Observation of a Hard State Outburst in the GX339-4 System*, these proceedings (1992).
14. B. A. Harmon *et al.*, *Earth Occultation Measurements of Galactic Hard X-Ray/Gamma-Ray Sources: A Survey of the BATSE Results*, these proceedings (1992).
15. B. C. Rubin *et al.*, *BATSE Observations of the Massive X-Ray Binary 4U1700-377/HD 153919*, these proceedings (1992).

PRELIMINARY EGRET SOURCE CATALOG

C. E. Fichtel, D. L.Bertsch, B. L. Dingus[1], R. C. Hartman, S. D. Hunter, P. W. Kwok[2], J. R. Mattox[3], P. Sreekumar[1], and D. J. Thompson
NASA/Goddard Space Flight Center, Code 662, Greenbelt, MD 20771, USA

D. A. Kniffen
Hampden-Sydney College, P. O. Box 862, Hampden-Sydney, VA 23943, USA

Y. C. Lin, P. L. Nolan, and P. F. Michelson
Hansen-Experimental Physics Laboratory, Stanford University, Stanford, CA 94305, USA

G. Kanbach, H. A. Mayer-Hasselwander, C. von Montigny, K. Pinkau, H. Rothermel, and M. Sommer
Max-Planck Institut fur Extraterrestrische Physik, 8046 Garching, Munich, Germany

E. J. Schneid
Grumman Aerospace Corporation, Mail Stop A01-26, Bethpage, LI, NY 11714, USA

INTRODUCTION

There are two purposes of presenting this preliminary EGRET source catalog. One is to gather in one place at this time the sources that have been reported together with some upper limits that have been determined. The other is to indicate the type or catalog of sources and limits that we plan to assemble when the Phase I analysis of EGRET data is complete, or nearly so. We wish to note particularly that we have decided to add a category between the sources which have been detected with a high degree of certainty and those with only upper limits. This group of possible sources (Table II) includes those for which there is some indication of an excess above the local diffuse level, but not one of high significance. Specifically, sources in this group have excesses which, if it were the only point considered and if the diffuse radiation were exactly known, would have a significance between 3.0 and 4.0 σ probability. Clearly, if one is looking at many potential sources, there will be some excess seen in excess of 3.0 σ by chance, and even more will be found to have excess exceeding a calculated 3.0 σ when uncertainty in the diffuse emission is included. Hence, it is not proper to call these high probability source detections. However, it seemed to us not appropriate to place them with many others in a long list of upper limits since some might be real and worth a second look in the future. When an object is listed as a source of high energy gamma rays in the following table, it has been detected above the 4 σ level, as defined above.

1. Also Universities Space Research Association, Greenbelt, MD 20771, USA
2. Also NAS/N RC Research Associate, Greenbelt, MD 20771, USA
3. Compton GRO Science Support Center, CSC, Greenbelt, MD 20771, USA

Table I: Characteristics of Active Galaxies detected by EGRET

Source ID and characteristics	OVV	BL Lac	super lum.	radio loud	flat radio[1]	opt. pol.[2]	position diff.[3]	position uncert.[4]	flux[4] (10^{-6} cm^{-2} s^{-1}) (E>100 MeV)		photon spectral index	z	relative luminosity[5]	Reference
0202+149 (4C+15.05)				√	√	√	0.3°	0.4°	0.3±0.1	vp(21.0)		...	...	19
0208-512 PKS				√	√	√	0.13°	0.13°	< 0.3 0.4±0.1 0.9±0.1 0.5±0.1 < 0.4	vp(6.0) vp(9.5) vp(10.0) vp(13.5) vp(17.0)	1.7±0.1	1.00	2	15 15 15 15 15
0235+164 (OD+160)		√	√	√	√	√	0.10°	0.3°	0.8±0.1	vp(21.0)	-2.0±0.2	0.94	2.0	17
0420-014 (OA 129)	√			√	√	√	0.5°	0.4°	0.4±0.1	vp(29.0)		0.92	0.4	19
0454-463 PKS				√	√		0.27°	0.38°	0.25±0.07	vp(6.0)		0.86	0.3	19
0528+134 PKS	√			√	√		0.13°	0.15°	0.84±0.1 0.55±0.1	vp(1.0) vp(2.5)	-2.4±0.1	2.06	4 to 13	16 16
0537-441 PKS		√		√	√	√	0.4°	0.6°	0.3±0.1	vp(6.0)	-2.0±0.2	0.894	0.2	22
0716+714		√	?	√	√		0.47°	0.4°	0.20±0.06	vp(18.0)	-1.8±0.2	...	...	19
0836+710 (4C+71.07)	√		√	√	√		0.58°	0.50°	0.15±0.04	vp(18.0)		2.17	1.1	19
1101+384 (Mrk 421)		√	√	√	√		0.3°	0.4°	0.14±0.03	vp(4.0)	-1.9±0.1	0.031	0.0002	9
1226+023 (3C 273)			√	√	√	√	0.2°	0.5°	0.30±0.05	vp(3.0)	-2.4±0.1	0.158	0.008	6
1253-055 (3C 279)	√		√	√	√	√	0.083°	0.08°	4.9* 0.6†	vp(3.0) vp(11.0)	-1.9±0.1 -2.1±0.1	0.54	0.3 to 2	1 8
1606+106 (4C+10.45)				√	√		0.5°	0.40°	0.5±0.2	vp(16.0)		1.23	1.6	19
1633+382 (4C+38.41)	√			√	√		0.08°	0.15°	0.4 to 1.4	vp(9.5)	-2.0±0.1	1.81	3 to 11	18

(contd)

Table I (contd): Characteristics of Active Galaxies detected by EGRET

Source ID and characteristics	OVV	BL Lac	super lum.	radio loud	flat radio[1]	opt. pol.[2]	position diff.[3]	position uncert.[4]	flux[4] (10⁻⁶ cm⁻²s⁻¹) (E>100 MeV)	photon spectral index	z	relative luminosity[5]	Reference
2230+114 (CTA 102)			√	√	?		0.3°	0.4°	0.24±0.05 vp(19.0) 0.53±0.23 vp(26.0) 0.38±0.17 vp(28.0)	-2.6±0.1	1.037	0.5	25
2251+158 (3C 454.3)	√		√	√	√	√	0.25°	0.22°	0.8±0.1 vp(19.0)	-2.0±0.1	0.859	0.5	14

1. Flat spectrum radio sources: $\alpha_r > -0.5$ (2-5 GHz band); Optical polariation: polarization > 3%.
2. Difference between gamma-ray-determined position and known position of identified source. Most are preliminary.
3. There is a 68% probability that the source is within a circle of this radius. Most are preliminary.
4. Flux values from mutiple observations are listed separately along with the corresponding viewing period (vp). See Table VIII for dates associated with viewing periods.
5. The source luminosity for the observed energy range (0.1 GeV<E<5 GeV), computed using the known redshift assuming $H_0 = 75$ km s⁻¹ Mpc⁻¹ and $q_0=1/2$, is equal to the relative luminosity times (10^{48} erg s⁻¹)f, where f is an unknown beaming factor. Typically f is thought to be in the range from 10^{-2} to 10^{-3}.
* Peak flux during viewing period.
† Average flux over the viewing period.

Table II: Blazars with marginal detection by EGRET

Source ID and characteristics	OVV	BL Lac	super lum.	radio loud	flat radio[1]	opt. pol.[2]	flux[5] (10⁻⁷ cm⁻²s⁻¹) (E>100 MeV)	z	Reference
1641+399 (3C 345)	√		√	√	√	√	1.3 ± 0.6 (3 σ) vp(9.5)	0.595	20

Table III: Selected Blazars and Seyfert galaxies (EGRET upper limits)

Source ID and characteristics	OVV	BL Lac	super lum.	radio loud	flat radio[1]	opt. pol.[2]	flux (10^{-7} cm^{-2}s^{-1}) (E>100 MeV) (2 σ UL)		z	Reference
1807+698 (3C 371)			√(?)				<5.4	vp(9.5)	0.051	20
2200+420(BL Lacertae)		√	√			√	<1.4	vp(18.0)	0.069	20
Mkn 335							<2.2	vp(2.0)		21
							<1.4	vp(7.0)		21
							<1.3	vp(26.0, 28.0)		21
Mkn 590 (NGC 863)							<0.8	vp(21.0)		21
NGC 4051							<0.6	vp(4.0)		21
NGC 4151							<0.6	vp(4.0)		21
NGC 4593							<0.5	vp(3.0, 11.0)		21
MCG-6-30-15							<1.1	vp(12.0)		21
NGC 5548							<1.2	vp(24.0, 24.5)		21
Mkn 509							<1.0	vp(7.5, 19.0)	0.0344	21
NGC 7213							<1.5	vp(42.0)		21
NGC 7469(Mkn 1514)							<2.0	vp(19.0)		21
MCG-2-58-22(Mkn 926)							<1.0	vp(19.0)		21
NGC 1068 (3C 71)							<0.9	vp(21.0)		21
Mkn 348 (NGC 262)							<2.4	vp(28.0)		21
MCG +8-11-11							<1.2	vp(31.0)		21
NGC 1275 (3C 84)							<0.9	vp(15.0)		21

1. Flat spectrum radio sources: $\alpha_r > -0.5$ (2-5 GHz band);   2. Optical polarization: polarization > 3%

Table IV: Normal galaxies observed by EGRET

Source ID and characteristics		Flux (10^{-7} photons cm^{-2}s^{-1}) (E>100 MeV)	Reference
	Distance (kpc)		
LMC	~50	1.9 ± 0.4	13
SMC	~61	<0.5	12

Table V: Pulsars detected by EGRET

Source ID and characteristics				Flux (10^{-6} photons cm^{-2}s^{-1}) (E>100 MeV)	EGRET position			Reference
	Frequency (ν)	First derivative ($\dot\nu$)	Second derivative ($\ddot\nu$)		l	b	error (63%)	
0531+21 (Crab)	29.95	-3.78E-10	1.06E-20	1.8 ± 0.1	184.54	-5.88	0.05°	7
0630+178 (Geminga)	4.22	-1.95E-13	4.00E-25	2.9 ± 0.1	195.08	+4.21	0.05°	3,11
0833-45 (Vela)	11.20	-1.56E-11	5.65E-24	11.8 ± 0.3	263.52	-2.78	0.04°	5
1706-44	9.76	-8.86E-12	...	1.0 ± 0.2	343.2	-2.8	0.20°	10

Table VI: EGRET-detected Gamma Ray Bursts

Burst	Direction Zenith	Direction Azimuth	Flux (photons cm^{-2}s^{-1}MeV^{-1}) at 1.3 MeV	Flux at 2.0 MeV	Anticoincidence Shield	Spark Chamber	Reference
GRB 910503	28	231	8.0	0.64	YES	YES	2
GRB 910503-t			0.35	<0.1	YES	NO	23
GRB 910601	12	305	0.55	0.10	YES	NO	23
GRB 910814-2	29	279	4.5	2.0	YES	----	23,24
GRB 920508	22	327	<0.11	<0.06	YES	NO	24
GRB 920622	44	113	<0.15	<0.08	YES	NO	23
GRB 920830	20	1.4	<0.09	<0.05	YES	NO	23

Table VII: EGRET-detected Solar Flares

Flare	Soft X-ray Class (GOES)	Active Region	TASC	Spark Chamber	Reference
June 4, 1991	X 12	6659	YES	NO	
June 6, 1991	X 12	6659	YES	NO	
June 9, 1991	X 10	6659	YES	YES	4
June 11, 1991	X 12	6659	YES	YES	
June 30, 1991	M 5	6693	YES	NO	
July 2, 1991	M 5		YES	NO	

Table VIII: Viewing Periods

Viewing Period	Start Date	Start Time	Stop Date	Stop Time	z-axis ra	z-axis dec	Viewing Period	Start Date	Start Time	Stop Date	Stop Time	z-axis ra	z-axis dec
0.1	04/15/91	19:07:32	04/22/91	21:09:02	108.12	-6.52	19.0	01/23/92	15:08:00	02/06/92	15:15:00	331.40	-1.93
0.2	04/22/91	21:09:02	04/28/91	15:12:00	86.76	22.09	20.0	02/06/92	16:45:00	02/20/92	15:03:00	285.28	6.37
0.3	04/28/91	16:02:00	05/01/91	16:37:00	89.80	15.25	21.0	02/20/92	16:05:00	03/05/92	15:49:00	39.09	-1.24
0.4	05/01/91	17:19:00	05/04/91	16:16:00	89.77	15.24	22.0	03/05/92	16:45:00	03/19/92	13:19:39	216.00	70.74
0.5	05/04/91	16:50:00	05/07/91	15:53:00	83.52	22.02	23.0	03/19/92	14:15:17	04/02/92	12:49:00	227.43	-54.62
0.6	05/07/91	17:19:00	05/10/91	19:40:00	162.44	57.26	24.0	04/02/92	14:07:00	04/09/92	13:02:00	223.34	11.03
0.7	05/10/91	20:15:00	05/16/91	16:39:00	135.19	-45.11	24.5	04/09/92	13:26:00	04/16/92	12:29:28	223.34	11.03
1.0	05/16/91	17:19:00	05/30/91	18:51:00	88.07	17.14	25.0	04/16/92	13:17:40	04/23/92	12:27:01	229.85	4.47
2.0	05/30/91	20:01:00	06/08/91	00:08:30	301.39	36.58	26.0	04/23/92	13:29:22	04/28/92	12:44:44	1.59	20.20
2.5	06/08/91	01:24:30	06/28/91	18:44:00	87.83	12.47	27.0	04/28/92	13:41:33	05/07/92	14:08:37	241.11	-49.05
3.0	06/15/91	19:38:00	06/28/91	19:30:57	191.54	2.62	28.0	05/07/92	13:47:58	05/14/92	14:04:25	001.59	20.20
4.0	06/28/91	20:14:00	07/12/91	17:56:16	179.84	41.52	29.0	05/14/92	14:48:40	06/04/92	13:44:28	068.97	-25.09
5.0	07/12/91	18:48:32	07/26/91	19:25:00	270.39	-30.96	30.0	06/04/92	14:50:41	06/11/92	14:04:25	149.50	-14.73
6.0	07/26/91	20:25:00	08/08/91	15:36:00	091.28	-67.96	31.0	06/11/92	16:25:00	06/25/92	13:17:17	88.87	49.44
7.0	08/08/91	17:00:00	08/15/91	17:29:06	310.05	28.06	32.0	06/25/92	14:20:00	07/02/92	13:57:08	171.17	-36.81
7.5	08/15/91	18:23:15	08/22/91	14:05:00	291.98	-13.27	33.0	07/02/92	15:06:00	07/16/92	15:34:57	149.50	-14.73
8.0	08/22/91	15:01:00	09/05/91	14:01:00	124.96	-46.35	34.0	07/16/92	17:25:00	08/06/92	14:38:45	345.77	57.49
9.0	09/05/91	15:03:00	09/12/91	13:24:10	8.34	-32.31	35.0	08/06/92	16:35:00	08/11/92	00:54:45	287.12	-61.21
9.5	09/12/91	14:34:05	09/19/91	13:29:00	251.27	36.89	36.0	08/11/92	02:00:00	08/12/92	18:22:11	68.98	30.42
10.0	09/19/91	14:36:00	10/03/91	13:11:00	30.91	-60.66	36.5	08/12/92	19:20:00	08/20/92	15:28:00	68.39	32.90
11.0	10/03/91	14:10:00	10/17/91	13:57:40	189.02	1.06	37.0	08/20/92	16:02:00	08/27/92	16:49:30	358.75	18.82
12.0	10/17/91	15:15:00	10/31/91	14:55:06	202.29	-40.09	38.0	08/27/92	18:45:00	09/01/92	04:37:00	287.12	-61.21
13.0	10/31/91	15:42:00	11/07/91	14:34:50	291.98	-13.27	39.0	09/01/92	06:18:00	09/17/92	15:17:16	68.87	33.82
13.5	11/07/91	15:30:00	11/14/91	15:45:50	8.34	-32.31	40.0	09/17/92	06:15:00	10/08/92	13:48:00	140.88	30.40
14.0	11/14/91	16:50:00	11/28/91	11:30:00	156.83	-58.51	41.0	10/08/92	15:11:00	10/15/92	16:18:00	112.43	-12.05
15.0	11/28/91	12:50:00	12/12/91	16:42:00	52.00	40.24	42.0	10/15/92	17:10:00	10/29/92	13:31:05	319.72	-41.670
16.0	12/12/91	18:00:00	12/27/91	16:02:00	248.35	-17.20	43.0	10/29/92	15:30:00	11/03/92	13:07:47	307.83	-13.95
17.0	12/27/91	17:07:00	01/10/92	16:15:55	83.48	-72.26	44.0	11/03/92	15:10:00	11/17/92	14:57:47	112.43	-12.05
18.0	01/10/92	18:12:26	01/23/92	13:42:00	154.60	72.04							

REFERENCES

1. Hartman, R. C. et al. 1992, ApJ. Letters, 385, L1
2. Schneid, E. et al. 1992, Astron. Astrophys, 255, L13
3. Bertsch, D. L., et al. 1992, Nature, 357, 306
4. Kanbach, G. et al. 1992, Astron. Astrophys, (in Press)
5. Kanbach, G. et al. 199? (In Prep)
6. von Montigny, C., et al. 1992, Astron. Astrophys, (In Press)
7. Nolan, P. L., et al. 1992, ApJ, (In Press)
8. Kniffen, D. A. et al. 1992, ApJ, (In Press)
9. Lin, Y. C. et al. 1992, Astron. Astrophys, (In Press)
10. Thompson, D. J. et al. 1992, Nature, 359, 615
11. Mattox, J. R., Bertsch, D. L., Fichtel, C. E., Hartman, R. C., Kniffen, D. A & Thompson, D. J., 1992, ApJ, (In Press)
12. Sreekumar, P. et al. 1992, Phys. Rev. Letters, (In Press)
13. Sreekumar, P. et al. 1992, ApJ. Letters, 400, L67
14. Hartman, R. C. et al. 1993, (In Prep)
15. Bertsch, D. L. et al. 1993, ApJ. Letters, (In Press)
16. Hunter, S. D. et al. 1993, ApJ, 1993 (In Press)
17. Hunter, S. D. et al. 1993, Astron. Astrophys, 1993 (In Press)
18. Mattox, J. R. et al. 1993, ApJ.. Letters, 1993 (In Press)
19. Fichtel, C. E. et al. 1993, Proc. Compton Gamma Ray Observatory Symposium (Washington Univ) (1992)
20. von Montigny, C. et al. 1993, Proc. Compton Gamma Ray Observatory Symposium (Washington Univ) (1992)
21. Lin, Y. C. et al. 1993, (In Prep)
22. Thompson, D. J. et al. 1993, ApJ, (In Press)
23. Schneid, E. J. et al. 1993, Proc. Compton Gamma Ray Observatory Symposium (Washington Univ) (1992)
24. Kwok, P. W. et al. 1993, Proc. Compton Gamma Ray Observatory Symposium (Washington Univ) (1992)
25. Nolan, P. L. et al. 1993, (In Prep)

ACCESSING THE BATSE CATALOG OF BURSTS

S. Howard
USRA/NASA/Marshall Space Flight Center

C.A. Meegan, G.J. Fishman, R.B. Wilson
NASA/Marshall Space Flight Center

W. Paciesas
UAH/NASA/Marshall Space Flight Center

ABSTRACT

The BATSE Burst Catalog is available from the Compton Gamma-ray Observatory Science Support Center (GROSSC) at Goddard Space Flight Center. All files will also be available in FITS format. The BATSE trigger number is the unique identifier that connects all files. All BATSE catalog data for trigger 105 (the first cosmic burst) through trigger 1466 are included in these files. GRBs trigger the BATSE detectors on any of three timescales: 64ms, 256 ms, 1024 ms. Where appropriate, data on each time scale are included. There will be six files: basic, count rate, comments, peak rates, duration, sky map.

DESCRIPTION

The first BATSE burst catalog contains 260 gamma-ray bursts. This is the data set for the first year of operation of the Compton Gamma-ray Observatory.

Access:
(1) DECNET:
set host GROSSC
login: gronews
menu driven from there.

(2) INTERNET:
telnet grossc.gsfc.nasa.gov
login: gronews
menu driven from there.

Basic Information

Basic Information: twelve columns:

(1) The Batse Catalog GRB name is: 1B $yymmdd$
'1B ' begins every BATSE GRB name. Then $yymmdd$ of the GRB. When more that one GRB occurs on that day, the bursts have a single letter suffix (B,C,D...), generally in order of intensity. Example: 1B 920503B refers to the second brightest burst that triggered BATSE on May 3, 1992.
(2) The trigger number is a running sequence of BATSE triggers which include cosmic bursts, solar flares and other events. The sequence begins with trigger 105 and ends with trigger 1466.
(3) The Julian date of the trigger (JD - 2440000.5)
(4) The time in decimal seconds of day of the trigger.
(5) right ascension (2000) in decimal degrees.
(6) declination (2000) in decimal degrees.
(7) Galactic longitude (2000) in decimal degrees.
(8) Galactic latitude (2000) in decimal degrees.
(9) radius of error box for position
(10) angle in decimal degrees of geocenter (the angle between the burst and the nadir, as measured from the satellite).
(11) overwrote flag: T(rue) if this burst overwrote an earlier, weaker burst. F(alse) otherwise.
(12) overwritten flag: T(rue) if this burst was overwritten by a later, more intense burst. F(alse) otherwise.

Cmax/Cmin

Count rate information: This file specifies the peak count rate for 240 triggered gamma-ray bursts observed from launch until March 5, 1992: triggers 105 through 1466. The peak count rate is in units of the threshold count rate. Twenty gamma-ray bursts are not included because insufficient data exist to determine the trigger threshold.

The on-board software continually compares the count rate on each of the eight detectors to the threshold level on the matching detector. The comparison occurs for three time intervals: 64 ms, 256 ms and 1024 ms. A burst trigger occurs when the count rate is above threshold in two or more detectors simultaneously. The threshold in this file is the threshold of the second most brightly illuminated detector. The thresholds exhibit a coarse quantization that results from truncating the number of counts in 64 ms (the smallest timescale). They are set by ground control to a specified number of standard deviations above background (nominally 5.5σ). Another commandable parameter specifies how often to compute the average background rates (nominally 17 seconds).

When a burst trigger occurs, subsequent triggers are disabled for a period of about 240 seconds, during which the BATSE burst memories accumulate data.

These data are then read out over the next approximately 90 minutes. During this readout, the 64 ms threshold is revised to correspond to the maximum rate attained by the current burst, and triggering is disabled on the 256 ms and 1024 ms timescales. Bursts intense enough to trigger over this revised 64 ms value are termed "overwrites". They appear as triggers in this file, and the thresholds for 256 ms and 1024 ms are given the value of 9999. The corresponding count rates are set to 0.0

(1) BATSE trigger number.
(2) the maximum count rate divided by the threshold count rate (both for the second brightest detector) on the 64 ms timescale.
(3) the trigger threshold value for the 64 ms timescale.
(4) the maximum count rate divided by the threshold count rate (both for the second brightest detector) on the 256 ms timescale.
(5) the trigger threshold value for the 256 ms timescale.
(6) the maximum count rate divided by the threshold count rate (both for the second brightest detector) on the 1024 s timescale.
(7) the trigger threshold value for the 1024 ms timescale.

There are cases when the maximum rate is below threshold on one or two of the three timescales. This is acceptable. For technical reasons there are some bursts with undetermined peak rates (usually on the 256 ms timescale). These have "UUUUU" in the file. Determine the value of V/Vmax for any burst by taking the maximum of the three peak rates to the -3/2 power.

Comments

This file contains text comments about each BATSE GRB. There are three entries for each burst: the BATSE trigger number, a single character flag indicating the type of comment, and an 80 character free field comment. These may include data problems, burst characteristics, specific items of importance for a particular burst. A burst may have multiple entries.

Peak Rates

The Peak Rate file contains the trigger number, the peak rate in photons/cm^2/sec for each timescale, the time in seconds since the burst trigger time for that peak rate, the error in the peak rate, and the fluence in ergs/cm^2 over the range from 50 to 300 keV for the four energy channels, with the error in this fluence. There are 18 columns in this file.

Burst Duration

Burst duration information: trigger number, T50 value, T90 value, error in each value. T50/90 is a time in decimal seconds. This is a measure of the burst duration. One determines the total integrated counts over the entire burst. T50/90 is the time interval during which the integrated counts goes from 25%/5% to 75%/95% of this total.

THE *COMPTON* OBSERVATORY ARCHIVE

T.McGlynn, E.Chipman, J.Jordan, N.Ruggiero, D.Jennings, and T.Serlemitsos.
Compton Observatory Science Support Center
Goddard Space Flight Center
Computer Sciences Corporation
Greenbelt, Maryland 20771

ABSTRACT

This paper discusses the public archive for data from the *Compton* Observatory. It describes the contents of the archive, the schedule for when data will become available, how users may access the archive, and the structure of the catalogs used to index the archive. Pointers to further documentation are included.

INTRODUCTION

The *Compton* Gamma Ray Observatory was launched in April 1991 as the second of NASA's Great Observatories to look at the universe in the high-energy regime from 20 keV to 20 GeV. In July 1992, the *Compton* Observatory Archive was opened to allow public access to gamma-ray data from the *Compton* Observatory mission. This article describes the current and future contents of this archive, how astronomers may access the data, and the catalog structures which are used to index the data.

All observations from the *Compton* Observatory will be placed in the public archive. Nominally, data products are available to the public after a one year proprietary period which begins when the data product is generated in a usable form. Routine data products are produced by the Principal Investigator (PI) teams for both their own data and that of the Guest Investigators (GIs). The products are available for analysis for one year by the PI or GI under whose program the observation was taken. Then, as the data become non-proprietary, they are sent to the *Compton* Observatory Science Support Center (SSC) where they are archived. Each instrument team also maintains an archive of data from its own instrument.

CHARACTERISTICS OF *COMPTON* DATA

This section describes the general characteristics of the data available from each instrument. For some of the instruments only the low-level data products have been defined and the discussion focuses on these.

All data is archived in the *Compton* Observatory Archive in FITS format. Often this is an encapsulation of the PI-developed internal formats. The FITS binary tables extension is used extensively. The columns are described in the FITS headers, and the numeric format of the data uses the IEEE standards and the big-endian byte order as required by FITS. To accommodate archival researchers who may wish to use the software systems developed by the PI teams, software exists to convert all FITS files back to the original PI formats.

BATSE

Three different classes of data are available from the BATSE instrument: Continuous, Pulsar data and Burst data. As the name indicates, continuous data are collected continuously for the BATSE instrument. The low-level data comprise fairly high time resolution data in four discriminator channels, and high-energy resolution data with a temporal resolution of several minutes. Associated high-level products may include sources detected through Earth occultation measurements and bursts which did not meet the trigger criteria. Pulsar data are high energy resolution data which are folded on-board to a specified period. Burst data are collected immediately before, during, and after a gamma-ray burst, solar flare or other BATSE trigger. Low-level burst data comprise more than a dozen datatypes where there is a general tradeoff between temporal and spectral resolution of each type.

OSSE

The primary low-level product of the OSSE experiment is the spectra summed over the nodding interval. OSSE typically nods on and off target in two-minute intervals, so these Spectral DataBase (SDB) products consist of two-minute summed spectra. Most analysis will start with these, but several other products are available for study. The Telemetry Scalar File (TSF) contains information which allows studies with time resolutions smaller than the nodding time. The OSSE Pulsar data detects individual events or time binned spectra. Other products give shield rates and other information useful for specialized investigations.

COMPTEL

The lowest-level COMPTEL data which will be available in the archive comprise individual photon events for COMPTEL. For each event a time, position (p), angle (α), energy and other instrument parameters will be provided. An individual COMPTEL photon is not well localized on the sky. Rather for a given photon what is known is that the original location was from an annulus, where the center and diameter of the annulus are given by p and α. The width of the annulus is a degree or two. Higher-level products, including skymaps, spectra and light curves generated using sophisticated maximum entropy techniques, will also be archived.

EGRET

The lowest level of EGRET data also comprise photon events. However, unlike COMPTEL these have more conventional uncertainties in their location and are amenable to more direct methods for generating maps and other higher-level products. The EGRET instrument is almost noise-free, i.e., virtually all EGRET gamma-ray events in the observatory archive will represent real cosmic photons. While a significant fraction of the triggers in the instrument are non-gamma-ray events or Earth albedo gammas, these are marked early in the EGRET standard processing. Maps and spectra, light curves and catalogs of individual sources will also be placed in the archive.

Schedule of Deliveries.

The pipelines for data delivery to the archive have just begun to flow. The archive currently contains data from 47 BATSE bursts and some OSSE SDB data whose formats are being tested. The first samples of the other data types are anticipated by the end of 1992.

HOW TO ACCESS THE ARCHIVE.

The *Compton* Observatory Archive is open to all astronomers interested in data from the observatory. Electronic data retrieval is preferred, but when very large quantities of data are to be retrieved, or for users who do not have electronic connections to the data archive, the SSC will support archive retrieval to physical media on a limited basis.

Nodes and accounts.

The archive system is currently located on the node **enemy** of the SSC cluster. A guest account, **gof**, is available for initial browsing and occasional use of the archive. Users who anticipate frequent or heavy usage of the archive may wish to obtain a personal account on **enemy** to facilitate their work. This node is a Unix machine but is accessible through both DECnet (`SET HOST ENEMY`) and TCP/IP (`telnet enemy.gsfc.nasa.gov`).

To use the **gof** account simply login as **gof**. No password is required. This captive account will present you with a menu of possible options. Select the Archive Data Selector option to get into the archive.

The Data Selector.

Users who wish to retrieve data log into the archive machine as described above, and start the Data Selector task. The Data Selector allows the user to quickly select archive data according to a variety of selection criteria including instrument, target, target type, position, time of observation and data type.

The Data Selector is an INGRES forms application. The user enters data in the appropriate fields, and a query is made of the Observation Catalog. The Observation Catalog, described in detail below, maintains an index of the archive and describes all of the data currently available.

Once a user has selected a set of observations of interest, he or she may elect to retrieve the associated filesets. A fileset is simply a group of one or more files which is retrieved as a unit. When using the **gof** account, files that are retrieved from the archive are retrieved into directories on **enemy** which are accessible through anonymous FTP or proxy DECnet copies. The archival researcher can use either route to copy the files back to his or her home computer. Users with accounts on the SSC cluster can retrieve data directly to their own areas.

The Physical Archive.

Data in the *Compton* Observatory Archive is currently stored in two distinct physical archives, a rewritable magneto-optical disk jukebox with a capacity of 92 GB, and the NSSDC NDADS archive. Generally we anticipate faster response from the local jukebox, but the capacity of the NDADS is far greater. When available, data will be retrieved from the local jukebox but the system will automatically retrieve from the NDADS when necessary.

The user and most of the archive software are insulated from the details of the physical archive setup by the Generic Retrieve/Archive Software Protocol (Grasp). Grasp is a protocol which the SSC has developed to attempt to isolate the data retrieval from the underlying details of the archive implementation. Grasp defines a small set of relatively simple functions that the underlying archive will perform and provides the user with a uniform interface to any of the underlying archives.

We envisage that the underlying data may reside on several different physical archives both simultaneously and serially during the lifetime of the *Compton* Observatory. Grasp allows us to isolate our dependencies on this heterogeneous and evolving hardware.

THE OBSERVATION CATALOG.

The Observation Catalog is a relational database which indexes and describes the contents of the archive. It is this catalog with which a user actually has the greatest interaction when retrieving data.

The tables in the Observation catalog form two clusters, one about the OBSERVATION table and one about the FILESET table. This first cluster contains information about the actions that the satellite has taken and the scientific programs they are intended to support. The OBSERVATION relation itself contains timing, target and program information about the observations of the satellite. Two pointing tables, POINT and OSSEPOINT, describe the specific spacecraft attitudes, and the PROGRAM relation gives information about accepted *Compton* Observatory observing programs.

The cluster around the FILESET relation describes the data in the archive. FILESET directly indexes the archive, with one entry for each archived fileset. The LEVELINFO and TYPEINFO relations describe the types of filesets. The FILELOG relation maintains a record of when filesets have been retrieved. The FILEINFO and INFOTEXT relations allow comments to be made about filesets, e.g., warnings about flaws in the data or data processing. The FILESET_FILES, KEYWORDS and FIXEDFIELDS tables help control the flow of data into the archive, and the update of catalog fields to reflect new archive data.

The connection between these two clusters is the DATA relation which describes the filesets used in a given observation. A fileset may be used in several observations and conversely, an observation may generate many filesets.

ANALYSIS OF ARCHIVAL DATA.

While the SSC directly supports data in the archive only in FITS formats, converters to reformat the data to the original Principal Investigator formats are available for all datatypes. Thus PI software can be used to analyze all archival data. Archival researchers can analyze their data at the PI team sites or at the SSC. It is advised that archival investigators who are just beginning work with *Compton* Observatory data discuss their research efforts with the Instrument Specialists and PI teams for the instruments they are working on.

PI software tools are being collected at the SSC for use by Guest and archival investigators. Currently the BATSE Spectral Analysis Software (BSAS), the OSSE IGORE system, and some EGRET software have been brought up on the SSC cluster. Any archival investigator may request an account to use this software. Archival investigators may also wish to discuss with the PI teams porting appropriate software to their home sites.

As discussed above, the SSC is working with the HEASARC to make *Compton* Observatory data usable within more general software packages. We expect that some capabilities to do spectral analysis within XSPEC will be available this winter, and additional functionality will be developed over the coming year.

SUGGESTED FURTHER READING.

All documents described as SSC internal documents are available by anonymous FTP on the **enemy** and **grossc** nodes of the SSC network.

The *Compton* GRO Mission
: See various articles in *The Compton Observatory Science Workshop*, NASA Conference Publication 3137, 1992.

Data produced by the *Compton* Observatory
: These are defined in the *Compton Observatory Project Data Management Plan*. Revision due 1992.

Some data formats for particular instruments.
: For low-level BATSE data: *GRO BATSE Flight Software User's Manual*, MSFC-MNL-1405, 1991.
: For OSSE SDB data: *Spectral Data Base Version 7 Data Structure Description*, by Mark Strickman, NRL Document 0926-159 Rev 3.00, 1992.

The GRASP Interface
: This interface is defined in The Generic Archive/Retrieve Software Protocol, SSC internal document.

The Observation Database
: The Observation Database, including the mechanisms used for maintenance and update, as well as a more complete description of the Data Selector, is described in *The Compton Observatory Observation Catalog*, SSC internal document.

APPENDIX A:
CONFERENCE PROGRAM

APPENDIX X

Compton Symposium Agenda
October 15 - 17, 1992
Washington University, St. Louis

Wednesday, October 14, 1992

    20:00 - 22:00   Registration at Holiday Inn - Clayton Plaza

Thursday, October 15, 1992

08:00	Registration at Simon Hall
08:45 - 09:15	Introduction
09:15 - 10:50	Session 1: Diffuse Galactic Emission
	Coffee Break/ Display Presentations
11:15 - 12:40	Session 2a: Galactic Center/Galactic Sources
	Lunch Break/ Display Presentations
14:00 - 15:30	Session 2b: Galactic Center/Galactic Sources Cont.
	Coffee Break/ Display Presentations
16:00 - 18:00	Session 3: X-Ray Binaries and Pulsars
18:30	Conference Dinner - Holiday Inn

Friday, October 16, 1992

08:45 - 10:15	Session 4: Gamma-Ray Bursts 1 - General
	Coffee Break/ Display Presentations
10:45 - 12:25	Session 5: Gamma-Ray Bursts 2 - Models
	Lunch Break/ Display Presentations
14:00 - 15:35	Session 6: Gamma-Ray Bursts 3 - Panel Discussion and Other Instruments
	Coffee Break/ Display Presentations
16:00 - 17:30	Session 7: Gamma-Ray Bursts 4 - Spectroscopy and Structure
18:30	Riverboat Banquet

Saturday, October 17, 1992

08:45 - 10:00	Session 8: Supernovae and Novae
	Coffee Break/ Display Presentations
10:30 - 12:30	Session 9: AGN
	Lunch Break/ Display Presentations
14:00 - 16:00	Session 10: Pulsars
	Coffee Break/ Display Presentations
14:00 - 16:00	Session 11: Solar Flares (parallel with Session 10)
	Closing Remarks

APPENDIX B:
LIST OF PARTICIPANTS

COMPTON OBSERVATORY SYMPOSIUM
PARTICIPANT LIST

Band, David	University of California, San Diego
Barnes, Sandy	Compton Observatory Science Support Center
Barthelmy, Scott	NASA/GSFC
Bartlett, Lyle	NASA/GSFC
Battersby, Stephen	Imperial College
Bazan, Grant	University of Texas at Austin
Begelman, Mitchell	University of Colorado
Belli, B. M.	CNR/IAS
Bennett, Kevin	ESTEC
Bertsch, David	NASA/GSFC
Bhat, N. P.	NASA/MSFC
Bhattercharya, Dipen	NASA/GSFC
Biesecker, Douglas	University of New Hampshire
Bignami, Giovanni	IFC/CNR
Bildsten, Lars	California Institute of Technology
Binns, Walter	Washington University
Blaes, Omer	Canadian Institute for Theoretical Astrophysics
Blandford, Roger D.	California Institute of Technology
Bloemen, Hans	Laboratory for Space Research Leiden
Bloom, Steven	Boston University
Boorstein, Joshua	University of Chicago
Brainerd, Jerome J.	NASA/MSFC
Brandt, Soren	Danish Space Research Institute
Bridgman, W. T.	Clemson University
Briggs, Michael	NASA/MSFC
Brown, Lawrence E.	Clemson University
Buchholz, James	University of California, Riverside
Bunner, Alan	NASA Headquarters
Burbidge, Geoffrey	University of California, San Diego
Cameron, R. A.	Naval Research Laboratory
Carramiñana, A.	ESTEC/ESA
Catalano, Osvaldo	Palermo, Italy
Chakrabarty, Deepto	California Institute of Technology
Chan, Kai-Wing	NASA/GSFC
Chantell, Mark	Whipple Observatory
Chen, Kaiyou	Los Alamos National Laboratory
Chen, Wan	NASA/GSFC
Chiang, James	Stanford University
Chipman, Eric	Compton Observatory Science Support Center
Clayton, Donald	Clemson University
Cline, Thomas	NASA/GSFC
Coe, Malcom J.	University of Southampton
Collmar, Werner	MPI fur Extraterrestrische Physik
Connors, Alanna	University of New Hampshire
Coppi, Paulo S.	University of Chicago
Cordes, Jim	Cornell University
Cordier, Bertranol	Centre d'Études de Saclay
Covault, Corbin E.	University of Chicago

Appendix B: List of Participants

Crary, David	Washington University
D'Amico, Nicolo	Istituto di Radioastronomia CNR
Daugherty, Joseph K.	UNCA
Davis, Stanley P.	NASA/GSFC
Debrunner, Hermann	Physikalisches Institut
Dermer, Charles	Naval Research Laboratory
Dezalay, J. P.	CESR
Diehl, Roland	MPI fur Extraterrestrische Physik
Digel, Seth	Center for Astrophysics
Ding, Kwan Ying	University of Hong Kong
Duncan, Robert	University of Texas
Durouchoux, Ph.	CE-Saclay-SAP
Dyer, Clive	Defense Research Agency
Dyson, Freeman	Institute for Advanced Study
Edberg, Timothy	University of California, Berkeley
Ekejiuba, I. E.	Federal University of Technology
Emslie, A. Gordon	University of Alabama, Huntsville
Epstein, Richard	Los Alamos National Laboratory
Evans, Doyle	Los Alamos National Laboratory
Fatuzzo, Marco	University of Michigan
Fegan, David J.	University College Dublin
Fenimore, Ed	Los Alamos National Laboratory
Fennell, Stephen	University College Dublin
Ficenec, David	Washington University
Fichtel, Carl	NASA/GSFC
Fierro, Joseph	Stanford University
Finger, Mark	NASA/MSFC/CSC
Fishman, G. J.	NASA/MSFC
Ford, Lyle	University of California, San Diego
Forrest, David	University of New Hampshire
Freeman, Peter E.	University of Chicago
Friedlander, David	NASA/GSFC
Friedlander, Michael	Washington University
Frye, Jr., Glenn, M.	Case Western Reserve University
Gehrels, Neil	NASA/GSFC
Gerardi, Gaetano	Universita di Palermo
Goldwurm, Andrea	DAPNIA Sap
Gonthier, Peter L.	Hope College
Grabelsky, David	Northwestern University
Graber, James S.	Library of Congress
Graziani, Carlo	University of Chicago
Grebenev, Sergey	Space Research Institute
Greiner, Jochen	MPI fur Extraterrestrische Physik
Greyber, Howard D.	Greyber Associates
Grindlay, Jonathan	Harvard Observatory
Grishchuk, Leonid	Washington University
Grove, J. E.	Naval Research Laboratory
Guarnieri, Adriano	Università degli Studi di Bologna
Gursky, Herbert	Naval Research Laboratory
Hakkila, Jon	Mankato State University
Halpern, Jules	Columbia Astrophysics Laboratory
Hamilton, Russell	University of Illinois, Urbana
Hanami, Hitoshi	Iwate University
Hanlon, Lorraine	ESTEC/ESA

Appendix B: List of Participants

Harmon, Alan	NASA/MSFC
Harper, Thomas R.	Alamo, California
Hartmann, Dieter	Clemson University
Hattori, Makoto	Institute of Physical and Chemical Research
Haymes, Robert	Rice University
Heinrich, Olaf M.	Institut für Theoretische Astrophysik der Universität Heidelberg
Hermsen, W.	Laboratory of Space Research Leiden
Hertz, Paul	Naval Research Laboratory
Higdon, James	Keck Science Center
Ho, Cheng	Los Alamos National Laboratory
Horack, John M.	NASA/MSFC
Hoshino, Masahiro	RIKEN
Howard, Sethanne	NASA/MSFC
Hua, Xin-Min	Space Astrophysics Laboratory/ISTS
Hurley, Kevin	University of California, Berkeley
Hutchison-Frost, Meta	Westover Consultants, Inc.
Jenkins, Tom	Case Western Reserve University
Jensen, Craig M.	Naval Research Laboratory
Johnson, W. Neil	Naval Research Laboratory
Joss, Paul	MIT Center for Space Research
Kafatos, M.	Computational Sciences & Informatics
Kanbach, Gottfried	MPI fur Extraterrestrische Physik
Kane, Sharad R.	University of California, Berkeley
Kargatis, Vincent	Rice University
Kartje, John F.	University of Chicago
Katz, Jonathan	Washington University
Kawai, Noboyuki	Institute of Physical & Chemical Research
Kazanas, D.	NASA/GSFC
Kerrick, Alan	Whipple Observatory
Kertzman, Mary	DePauw University
Kinzer, R. L.	Naval Research Laboratory
Kippen, R. Marc	University of New Hampshire
Klarmann, Joseph	Washington University
Kluźniak, Wlodek	University of Wisconsin
Kniffen, Donald	Hampden Sydney College
Königl, Arieh	University of Chicago
Koshut, Thomas	NASA/MSFC
Kouveliotou, Chryssa	NASA/MSFC
Krimm, Hans A.	Massachusetts Institute of Technology
Kroeger, Richard A.	Naval Research Laboratory
Krolik, Julian	Johns Hopkins University
Kurfess, James	Naval Research Laboratory
Kwok, Ping-Wai	NASA/GSFC
Lamb, Don Q.	University of Chicago
Lamb, Frederick K.	University of Illinois, Urbana
Lamb, Richard C.	Iowa State University
Leising, Mark	Clemson University
Lestrade, John	Mississippi State University
Leventhal, Marvin	NASA/GSFC
Li, Hui	Rice University
Liang, Edison P.	Rice University
Lichti, Giselher	MPI fur Extraterrestrische Physik
Linder, Eric	Stanford University

Appendix B: List of Participants

Ling, James	Jet Propulsion Laboratory
Lingenfelter, Richard	University of California, San Diego
Loredo, Thomas	Cornell University
Luo, Chuan	Rice University
Maisack, Michael	Universität Tüebingen
Mallozzi, Robert	NASA/MSFC
Mandrou, Pierre	CESR
Mao, Shude	Center for Astrophysics
Marscher, Alan P.	Boston University
Massaro, Enrico	Università di Roma "La Sapienza"
Matsuoka, Masaru	Institute of Physics & Chemical Research
Matteson, James	University of California, San Diego
Mattox, John	NASA/GSFC
Matz, Steven	Northwestern University
Mayer-Hasselwander, Hans	MPI fur Extraterrestrische Physik
McConnell, Mark	University of New Hampshire
McGlynn, Thomas	Compton Observatory Science Support Center
McNamara, Bernard	New Mexico State University
Meegan, Charles	NASA/MSFC
Melia, Fulvio	University of Arizona
Mészáros, Peter	Pennsylvania State University
Meyer, Donald	University of Michigan
Meyer, Hinrich	Bergische Universität Gesamthochschule Wuppertal
Michelson, Peter	Stanford University
Miller, M. Coleman	University of Illinois at Urbana-Champaign
Miller, Richard S.	Louisiana State University
Mitrofanov, Igor	Space Research Institute
Mock, Patrick	MIT Center for Space Research
Morris, Daniel	University of New Hampshire
Moskalenko, Eugene	Sternberg State Astronomical Institute
Moscoso, Michael D.	University of Texas at Austin
Murphy, Ronald	Naval Research Laboratory
Nemiroff, Robert	NASA/GSFC
Nolan, Patrick	Stanford University
Norris, Jay P.	NASA/GSFC
O'Neill, Terry	University of California, Riverside
Oser, Scott M.	Washington University
Owens, Alan	University of Leicester
Paciesas, William S.	University of Alabama, Huntsville
Paczynski, Bohdan	Princeton University Observatory
Palmer, David	NASA/GSFC
Palumbo, Giorgio	Università di Bologna
Pendleton, Geoffrey	NASA/MSFC
Petrosian, V.	Stanford University
Piccioni, Adalberto	Università di Bologna
Pizzichini, Graziella	TESRE/CNR
Poutanen, Juri	University of Helsinki
Prantzos, Nikos	Institut d'Astrophysique de Paris
Preece, Robert	NASA/MSFC
Prince, Thomas	California Institute of Technology
Punch, Michael	University College Dublin
Purcell, William	Northwestern University
Quashnock, Jean	University of Chicago
Ramaty, Reuven	NASA/GSFC

Rank, Gerhard	MPI fur Extraterrestrische Physik
Ray, P. S.	California Institute of Technology
Rees, Martin	University of Cambridge
Ricker, George	MIT Center of Space Research
Ricker, Paul	University of Chicago
Riegler, Guenter	NASA Headquarters
Romani, Roger	Stanford University
Rothschild, Richard	University of California, San Diego
Rovero, Adrian C.	Whipple Observatory
Rubin, Brad	NASA/MSFC
Ruderman, Malvin	Columbia University
Ruiz-Lapuente, Pilar	Center for Astrophysics
Ryan, James	University of New Hampshire
Sacco, Bruno	Instituto di Fisica Cosmica del CNR
Schaefer, Bradley	NASA/GSFC
Schmidt, Maarten	California Institute of Technology
Schneid, Edward	Grumman Corporate Research Center
Schnepf, Neil	Clemson University
Schönfelder, Volker	MPI fur Extraterrestrische Physik
Schubnell, Michael	University of Michigan
Schutz, B. F.	University of Wales
Seifert, Helmut	NASA/GSFC
Sembroski, Glenn	Purdue University
Sepulveda, Eric	Ana-Lab Corporation
Shakura, Nikolai	Sternberg State Astronomical Institute
Shapiro, Maurice	Astrophysics Associates
Share, Gerald	Naval Research Laboratory
Shrader, Chris	Compton Observatory Science Support Center
Signore, Monique	Ecole Normale Superieure
Sikora, Marek	University of Colorado, Boulder
Skelton, R. Tom	Jet Propulsion Laboratory
Skibo, Jeffrey	NASA/GSFC
Smith, Ian	University of Chicago
Stacy, J. G.	University of New Hampshire/CSC
Starr, Christopher	Naval Research Laboratory/CSC
Starrfield, Sumner	Arizona State University
Steinle, Helmut	MPI fur Extraterrestrische Physik
Stollberg, Mark T.	University of Alabama, Huntsville
Storey, Scott	NASA/MSFC
Strickman, M. S.	Naval Research Laboratory
Strong, Andrew W.	MPI fur Extraterrestrische Physik
Stupp, Amnon	Tel Aviv University
Sturner, Steven	Rice University
Sunyaev, Rashid	Space Research Institute
Svoboda, Robert	Louisiana State University
Taff, L. G.	Space Telescope Science Institute
Takahara, Fumio	Tokyo Metropolitan University
Takahara, Mariko	Doshisha Women's College of Liberal Arts
Takahashi, Tadayuki	University of Tokyo
Talcott, Rich	Astronomy Magazine
Tavani, Marco	Princeton University
Teegarden, Bonnard	NASA/GSFC
Terrell, James	Los Alamos National Laboratory
The, Lih-Sin	Clemson University

Appendix B: List of Participants

Thompson, Christopher	University of Toronto
Timmes, Frank	University of California, Santa Cruz
Tkaczyk, Wieslaw	University of Lodz
Tomozawa, Yukio	University of Michigan
Trimble, Virginia	University of Maryland
Truran, James	University of Chicago
Tueller, Jack	NASA/GSFC
Turner, Tumay	University of California, Riverside
Turver, K. E.	University of Durham
Ulmer, Melville	Northwestern University
van Dyk, Rob	Laboratory for Space Research Leiden
van Paradijs, Jan	University of Amsterdam
Varendorff, Martin	University of New Hampshire
Vestrand, Tom	University of New Hampshire
von Montigny, Corinna	MPI Extraterrestrische Physik
Wagner, R. Mark	Lowell Observatory
Walker, Mark	University of Sydney
Walker, Robert	Washington University
Wang, John C. L.	Canadian Institute for Theoretical Astrophysics
Wang, Virginia	University of California, San Diego
Watanabe, Kenji	Clemson University
Webber, William R.	New Mexico State University
Wickramasinghe, W.A.D.T.	University of Pennsylvania
Wijers, Ralph A. M. J.	Princeton University Observatory
Will, Clifford	Washington University
Williams, David A.	University of California, Santa Cruz
Wilson, Colleen	NASA/MSFC
Wilson, Robert	NASA/MSFC
Winkler, Christoph	ESA-ESTEC
Woosley, Stan	University of California, Santa Cruz
Yamasaki, Noriko	University of Tokyo
Yamauchi, Makoto	Institute of Physical & Chemical Research
Zbyszewska, Magda	University of Chicago
Zdziarski, Andrzej	Polish Academy of Sciences
Zeng, Heng	Clemson University
Zhang, Shuang Nan	NASA/MSFC
Zhang, Weiping	NASA/GSFC
Zinner, Ernst	Washington University
Zych, Allen	University of California, Riverside

AUTHOR INDEX

A

Agnetta, G., 1161
Agrinier, B., 228, 1161
Ait-Ouamer, F., 1112
Akerlof, C. W., 223, 488, 508, 833, 1171
Arzoumanian, Z., 177
Atteia, J.-L., 761
Azzam, W. J., 754

B

Bailes, M., 177
Baker, R., 1156
Balonek, T. J., 513
Band, D. L., 734, 860, 872, 887, 892, 907
Barat, C., 761, 783
Barouch, E., 228
Barthelmy, S. C., 119, 194, 1137, 1166
Bartlett, L. M., 119, 194, 1166
Bartolini, C., 1152
Battersby, S. J. R., 1107
Bazan, G., 47, 80
Becker, P. A., 554
Becker-Szendy, R., 838
Begelman, M. C., 598
Belyaev, S., 1166
Bennett, K., 21, 30, 35, 147, 189, 199, 204, 483, 631, 661, 845
Bertsch, D. L., 177, 365, 461, 518, 850, 855, 1194
Bhat, P. N., 912
Bhattacharya, D., 498
Bickert, K. F., 1059
Bignami, G. F., 233, 613
Biondo, B., 1161
Blandford, R. D., 533, 992
Bloemen, H., 21, 30, 35, 40, 147, 204, 335, 483
Bloom, S. D., 578
Boer, M., 769, 774, 813
Bortolotti, C., 218
Bouchet, L., 366

Braga, J., 355, 523
Bratton, C. B., 838
Brazier, K. T. S., 177, 365
Breault, J., 838
Breslin, A. C., 223
Briggs, M. S., 381, 686, 691, 699, 860, 867, 882, 887, 892
Brink, C., 254
Brock, M. N., 314, 681, 686, 691, 699, 704, 709, 714, 719, 867, 1127
Brotherton, M. S., 513
Brown, L. E., 47, 85
Buccheri, R., 189, 199, 204
Busetta, M., 147, 189, 199, 204

C

Cameron, R. A., 137, 303, 345, 478, 619
Caraveo, P. A., 233, 613
Carramiñana, A., 189, 199, 204
Casper, D., 838
Catalano, O., 1161
Cattani, A., 218
Cawley, M. F., 223, 488, 508, 833, 1171
Chakrabarty, D., 291, 376
Chan, K.-W., 75
Chantell, M., 223, 488, 508, 833, 1171
Chen, W., 423
Chen, K., 259, 278
Cheng, K. S., 259, 564, 1069
Chernenko, A. M., 761
Chernych, N. S., 828
Chiang, J., 177, 283, 365
Chipman, E., 3, 1205
Churazov, E., 366
Ciliegi, P., 613
Cinti, M. N., 1161
Claret, A., 366
Clayton, D. D., 47, 85, 137, 149, 503
Cline, T. L., 769, 774, 813, 860, 872, 887, 892, 1137, 1156
Cody, H. E., 969

Coe, M. J., 360
Collmar, W., 21, 30, 40, 147, 189, 335, 483, 845
Comber, C., 1107
Comte, R., 228
Connaughton, V., 833
Connors, A., 21, 30, 189, 199, 204, 335, 845
Coppi, P. S., 559
Cordes, J., 85
Cordier, B., 366
Cork, C. P., 1156
Cosentino, C., 1152
Costa, E., 228, 1161
Covault, C. E., 243
Cusumano, G. C., 228, 1161

D

D'Ali, G., 1161
D'Amico, N., 177, 218, 1161
Davis, S. P., 959, 964
de Bernardis, P., 90
de Boer, H., 21, 30, 35, 147
Debrunner, H., 631
De Jager, O. C., 254
den Herder, J. W., 21, 30, 483, 845
Dermer, C. D., 284, 541, 1035
de Vries, C., 21, 30, 35, 40, 845
de Vries, C. P., 483
Dezalay, J.-P., 783
Di Cianno, A., 1152
Diehl, R., 21, 30, 35, 40, 147, 189, 204, 335, 483, 661, 845
Ding, K. Y., 564, 1069
Dingus, B. L., 365, 461, 850, 855, 1194
Di Paolantonio, A., 1152
Di Raffaele, R., 1161
Dolidze, V. Sh., 761
Dopita, M. A., 168
Duncan, R. C., 1074, 1085
Dupraz, C., 40
Durouchoux, Ph., 124, 129, 408
Dyachkov, A., 366
Dye, S. T., 838
Dyer, C. S., 1107

E

Efron, B., 754
Eikenberry, S., 243
El Eid, M. F., 47
Emslie, A. G., 714
Encrenaz, P., 90
Everall, C., 360

F

Fabregat, J., 360
Fatuzzo, M., 1030
Fauci, F., 218
Fegan, D. J., 223, 488, 508, 833, 1171
Fenimore, E. E., 144, 769, 774, 813, 897, 917, 922
Fennell, S., 223, 488, 508, 833, 1171
Fichtel, C. E., 177, 365, 461, 518, 850, 855, 1194
Fierro, J. M., 177, 365
Finger, M. H., 291, 314, 319, 350, 371, 376, 381, 386, 391, 1127
Fishman, G. J., 291, 314, 319, 340, 350, 371, 376, 381, 386, 391, 473, 669, 681, 686, 691, 694, 699, 704, 709, 714, 719, 739, 769, 774, 778, 793, 813, 828, 860, 867, 872, 877, 882, 887, 892, 902, 907, 912, 947, 953, 959, 969, 974, 1040, 1117, 1122, 1127, 1202
Fitzgibbons, G., 391
Ford, L. A., 860, 872, 887, 892
Forrest, D. J., 21, 335, 619, 631, 661, 845, 1142
Freeman, P. E., 917, 922
Freeman, S., 1171
Friedlander, M., xxi
Frishman, D., 1171

G

Gaidos, J. A., 223, 488, 508, 833, 1171
Gajewski, W., 838
Gehrels, N., 3, 119, 194, 423, 498, 1156, 1166
George, I., 588

Gerardi, G., 228, 1161
Getman, V. S., 828
Gilfanov, M., 366
Giuliani, C., 1152
Godlin, S. D., 513
Goldhaber, M., 838
Goldwurm, A., 366
Gonzalez-Riestra, R., 168
Grabelsky, D. A., 107, 137, 209, 303, 345, 478, 493, 498, 503, 619, 1107
Graber, J. S., 1064
Graziani, C., 897
Greiner, J., 813, 828, 1059
Greyber, H. D., 569
Griffee, J. W., 788
Grindlay, J. E., 243
Gros, M., 1161
Grove, J. E., 137, 209, 238, 249, 303, 345, 478, 619
Gruber, D. E., 528, 1127
Grueff, G., 218
Grunsfeld, J. M., 360, 376
Guarnieri, A., 1152
Gunji, S., 355, 523

H

Haasbroek, L. J., 254
Hagan, J., 223, 488, 508, 833, 1171
Haines, T. J., 838
Hakkila, J., 699, 704
Halaczek, T., 365
Halpern, J. P., 259
Halverson, P. G., 838
Hamilton, R. J., 433, 438, 443
Hammond, N. D. A., 1107
Han, X. H., 324
Hanami, H., 1049
Hanlon, L., 631, 661, 845
Harding, A., 85
Harmon, B. A., 314, 340, 350, 371, 376, 381, 386, 391, 473, 778, 1040, 1127
Harris, K., 1171, 1176
Hartman, R. C., 177, 365, 461, 518, 850, 855, 1195
Hartmann, D. H.. 47, 85, 149, 1003
Hattori, M., 583, 1054
Haymes, R. C., 1147
Heinrich, O., 574

Hermsen, W., 21, 30, 35, 40, 147, 189, 199, 204, 335, 483, 845
Hertz, P., 238
Higdon, J. C., 1010, 1095
Hillas, A. M., 223, 488, 508, 833, 1171
Hirayama, M., 355, 523
Hjellming, R. M., 324
Ho, C., 278, 283
Holfeltz, S. T., 818
Horack, J. M., 681, 694, 699, 704, 714, 719, 778, 877, 882, 974
Howard, S., 793, 1202
Hua, X.-M., 408, 656, 927
Hudec, R., 828
Hunter, S. D., 177, 365, 461, 518, 850, 855, 1194
Hurley, K. C., 769, 774, 813, 1156

I

Inoue, H., 523

J

Jennings, D., 1205
Jensen, C. M., 619
Jermakian, J., 1156
Jiang, Y., 488, 1176
Johnson, W. N., 107, 137, 209, 238, 303, 345, 478, 493, 498, 503, 619, 1107
Johnston, S., 177
Jordan, S., 1205
Joss, P. C., 1080
Jung, G. V., 107, 137, 303, 345, 478, 493, 498, 503, 619, 1107

K

Kafatos, M., 554
Kamae, T., 355, 523
Kanbach, G., 177, 365, 461, 518, 850, 855, 1194
Kanou, T., 523
Kargatis, V. E., 907
Karnashov, A., 828
Kartje, J. F., 559

Kaspi, V. M., 177
Katz, J. I., 1090
Kawai, N., 213
Kerrick, A. D., 223, 488, 508, 833, 1171
Khavenson, N., 366
Kielczewska, D., 838
Kinzer, R. L., 107, 137, 209, 303, 345, 478, 493, 498, 503, 619, 1107
Kippen, R. M., 823, 845
Klapdor-Kleingrothaus, H. V., 1166
Klebesadel, R. W., 744, 769, 774, 788, 813
Klużniak, W., 724
Kniffen, D. A., 3, 177, 365, 461, 518, 850, 855, 1194
Königl, A., 559
Koshut, T. M., 694, 882, 902, 1040
Kouveliotou, C., 319, 681, 739, 769, 774, 778, 813, 828, 877, 882, 902, 912, 947, 959, 974
Kovtunenko, V., 366
Kozlenkov, A. A., 761
Kozlovsky, B., 656
Kremnev, R., 366
Krockenberger, M., 243
Kroeger, R. A., 137, 209, 303, 345, 478, 619
Kropp, W. R., 838
Kuiper, L., 21, 30, 189, 199, 204, 335, 845
Kurfess, J. C., 107, 137, 209, 238, 249, 303, 345, 478, 493, 498, 503, 619, 1107
Kuznetsov, A., 783
Kwok, P. W., 365, 461, 518, 850, 855, 1194

L

Lacey, C., 744
Lamb, D. Q., 897, 917, 922, 932, 1020, 1025, 1045
Lamb, F. K., 433, 438, 443
Lamb, R. C., 223, 488, 508, 833, 1171
Landis, D. A., 1156
Lappin, T., 1171
Laros, J. G., 744, 769, 774, 813
Lavigne, J. M., 1161
Lawrence, M. A., 223, 488, 508, 833, 1171, 1176

Learned, J. C., 838
Lebedev, V., 1166
Lee, P., 788
Leising, M. D., 137, 149, 303, 345, 503
Lemoine, D., 228
Lestrade, J. P., 860, 872, 892, 969, 1040
Levenson, K., 1142
Leventhal, M., 119, 194, 1166
Levy, H., 1171
Lewin, W. H. G., 877
Lewis, D. A., 223, 488, 508, 833, 1171, 1176
Li, H., 1035, 1074, 1099
Liang, E. P., 330, 396, 418, 428, 907, 1035, 1099
Lichti, G. G., 21, 30, 40, 147, 189, 204, 483, 661
Lin, Y. C., 177, 365, 461, 518, 850, 855, 1194
Linder, E. V., 1003
Ling, J. C., 340, 350, 1127, 1189
Ling, N. F., 340, 1189
Lingenfelter, R. E., 75, 408, 656, 808, 927, 1095
Lockwood, J., 631
Loomis, M., 631
Loredo, T. J., 749, 897
LoSecco, J., 838
Lubin, L. M., 729
Luke, P. N., 1156
Lum, K. S. K., 243
Luo, C., 330
Lyne, A. G., 177

M

Maccaferri, A., 218
MacCallum, C. J., 194
Maccarone, M. C., 1161
Macri, J., 661, 823
Madden, N. W., 1156
Madras, C., 744
Maisack, M., 478, 493, 528
Mallozzi, R. S., 694, 1040, 1122
Malone, D., 1156
Manandhar, R. P., 243
Manchester, R. N., 177
Mandrou, P., 228, 366
Mandzhavidze, N., 643

Mangano, A., 1161
Marscher, A. P., 578
Martino, B., 1161
Masnou, J. L., 228, 1161
Massaro, E., 228, 1161
Mathias, S., 1156
Matsuno, S., 838
Matsuoka, M., 583
Matt, G., 228, 1161
Matteson, J. L., 860, 872, 887, 892, 907
Matthews, J., 838
Mattox, J. R., 177, 365, 461, 518, 850, 855, 1194
Matz, S. M., 209, 238, 303, 345, 619
Mayer, D. I., 223
Mayer-Hasselwander, H. A., 177, 365, 461, 518, 850, 855, 1194
McCollum, B., 513
McConnell, M., 21, 30, 335, 483, 631, 661, 845, 1142
McGlynn, T., 1205
McGrath, G., 838
McGrew, C., 838
McNamara, B., 391, 823
Medici, G., 1161
Meegan, C. A., 314, 319, 340, 350, 371, 376, 381, 386, 391, 681, 686, 691, 694, 699, 704, 709, 714, 719, 739, 769, 774, 778, 793, 813, 823, 828, 860, 867, 872, 877, 882, 887, 892, 902, 907, 912, 947, 953, 959, 969, 974, 1040, 1117, 1122, 1127, 1202
Meier, M., 744
Meintjes, P. J., 254
Melchiorri, F., 90
Melia, F., 1030
Mereghetti, S., 233
Mészáros, P., 593, 987, 1015
Metlov, V., 828
Meyer, D. I., 488, 508, 833, 1171
Michel, F. C., 284
Michelson, P. F., 177, 365, 461, 518, 850, 855, 1194
Micolucci, E., 1152
Miller, M. C., 433, 438, 443
Miller, R. S., 154, 838
Minamitani, T., 319
Mineo, T., 228, 1161
Mitrofanov, I. G., 761

Miyazaki, S., 355, 523
Mock, P. C., 1080
Mohanty, G., 223, 488, 508, 833, 1171
Montebugnoli, S., 218
Moore, P., 969
Morelli, E., 1161
Morris, D., 21, 30, 35, 40, 147, 204, 631
Moskalenko, E. I., 828, 958
Moss, M. J., 1147
Murakami, H., 355, 523
Murakami, T., 897, 917
Murphy, R. J., 619

N

Nandra, P., 588
Natali, G., 1161
Nel, H. I., 177, 254
Nemiroff, R. J., 739, 947, 974
Neri, J. A., 355, 523
Nesvara, M., 828
Nicastro, L., 177, 218, 1161
Nice, D., 177
Niel, M., 228, 761, 769, 774, 813
Nolan, P. L., 177, 365, 461, 518, 850, 855, 1194
Nolan, T., 1156
Nomachi, M., 355, 523
Norris, J. P., 739, 947, 959, 964, 974
North, A. R., 254
Norton, A. J., 360
Novikov, B., 366

O

O'Flaherty, K. S., 223, 488, 508, 833, 1171
Okayasu, R., 213
Olive, J. F., 228
O'Neill, T. J., 1112
Owens, A., 588, 798, 1156

P

Paciesas, W. S., 314, 319, 340, 350, 371, 376, 381, 386, 391, 473, 681, 686, 691, 694, 699, 704, 709, 714,

719, 739, 774, 778, 793, 813, 828, 860, 867, 872, 877, 882, 887, 892, 902, 907, 912, 947, 953, 959, 969, 974, 1040, 1117, 1122, 1127, 1202
Paczyński, B., 981
Palmer, D. M., 860, 872, 887, 892
Parlier, B., 228
Paul, J., 366
Pedichini, F., 1161
Pehl, R. H., 1156
Pendleton, G. N., 291, 314, 340, 350, 371, 376, 473, 681, 699, 704, 709, 714, 778, 860, 867, 872, 887, 892, 902, 912, 969, 1040
Petrosian, V., 754
Piccioni, A., 1152
Pinkau, K., 177, 365, 461, 518, 850, 855, 1194
Pizzichini, G., 1132, 1152
Post, Jr., A. H., 1156
Postnov, K. A., 958
Poutanen, J., 448, 453
Pozanenko, A. M., 761
Prantzos, N., 52
Pravec, P., 828
Preece, R. D., 867, 882, 887
Price, L., 838
Prince, T. A., 249, 291, 360, 376
Prokhorov, M. E., 958
Punch, M., 223, 488, 1171
Purchell, W. R., 107, 137, 209, 303, 345, 478, 493, 498, 503, 619, 1107

Q

Quashnock, J. M., 1025
Quenby, J. J., 1107

R

Radecke, H. D., 365
Radocinski, R., 1189
Ramaty, R., 70, 75, 643, 656, 1156
Rank, G., 631, 661
Raubenheimer, B. C., 254
Ray, P. S., 249
Rees, M. J., 598, 987, 1015
Reglero, V., 360
Reines, F., 838

Reynolds, P. T., 223, 488, 508, 833, 1171
Rezek, T., 828
Ricca Rosellini, S., 1152
Ricker, P. M., 593
Roche, P., 360
Romani, R. W., 283
Roques, J. P., 366
Roth, T. A., 1112
Rothermel, H., 365, 461, 518, 850, 855, 1194
Rothschild, R. E., 808
Rovero, A. C., 223, 488, 508, 833, 1171, 1176
Rubin, B. C., 314, 340, 350, 381, 391, 473, 719, 877, 1127
Rubini, A., 1161
Ruderman, M., 259
Ruggiero, N., 1205
Ryan, J., 21, 147, 189, 199, 204, 335, 483, 631, 661, 823, 845

S

Sacco, B., 228, 1161
Salvati, M., 228
Scarsi, L., 228, 1161
Schaefer, B. E., 798, 803, 860, 872, 887, 892, 907, 912
Schmitz-Fraysse, M. C., 366
Schneid, E. J., 177, 365, 461, 518, 850, 855, 1194,
Schnepf, N. G., 85
Schönfelder, V., 21, 30, 35, 40, 147, 189, 199, 204, 335, 483, 631, 661, 845
Schubnell, M. S., 223, 488, 508, 833, 1171
Schultz, J., 838
Schwarz, G., 744, 917
Scott, J. H., 818
Seifert, H., 1156
Sekimoto, Y., 213, 355, 523
Sembay, S., 588, 798
Sembroski, G., 223, 488, 508, 833, 1171
Serlemitsos, T., 1205
Seward, F. D., 808
Shakura, N. I., 958
Shakura, N. N., 958

Author Index

Share, G. H., 619
Shaviv, G., 574
Sheppard, D. A., 1156
Shore, S. N., 168
Shrader, C. R., 324, 513
Signore, M., 90
Sikora, M., 598
Sims, M., 798
Sinclair, D., 838
Skelton, R. T., 340, 350, 1127, 1189
Skibo, J., 70
Smith, I. A., 1020
Smith, G., 1156
Sobel, H. W., 838
Sommer, M., 365, 461, 518, 769, 774, 813, 850, 855, 1194
Sonneborn, G., 168
Sparks, W. M., 168
Sreekumar, P., 177, 365, 461, 518, 850, 855, 1194
Stacy, J. G., 21, 30, 35, 483
Starr, C. H., 478
Starrfield, S. G., 163, 168, 324
Steinle, H., 21, 30, 335, 483, 845
Stillwell, D. E., 1156
Stollberg, M. T., 371, 709
Stone, J. L., 838
Storey, S. D., 694
Strickman, M. S., 107, 137, 209, 238, 303, 345, 478, 493, 498, 503, 619, 1107
Strong, A. W., 21, 30, 35, 40, 147, 189, 199, 204, 335, 483
Stupp, A., 159
Sturner, S. J., 284
Sulak, L. R., 838
Sunyaev, R., 366, 783
Svoboda, R. C., 154, 838
Swanenburg, B. N., 21, 30, 35, 40, 189, 204, 335, 483, 631, 661, 845

T

Taff, L. G., 818
Takahashi, T., 355, 523
Talon, R., 783
Tamura, T., 355, 523
Tanaka, M., 355, 523
Tavani, M., 272, 428
Taylor, J., 177

Taylor, B. G., 21
Teegarden, B. J., 119, 194, 860, 867, 872, 887, 892, 907, 1137, 1156, 1166
Terasawa, N., 583, 1054
Terekhov, O., 783
Terrel, J., 788
The, L.-S., 137, 149, 498, 503, 1003
Thompson, C., 1074, 1085
Thompson, D. J., 177, 365, 461, 518, 850, 855, 1194
Timmes, F. X., 64
Tomozawa, Y., 603
Tripiciano, M., 1161
Trombka, J., 1156
Truran, J. W., 47, 163, 168
Truscott, P. R., 1107
Tueller, J., 97, 119, 194, 1166
Tumer, O. T., 1112

U

Ulmer, M. P., 107, 137, 209, 238, 249, 303, 345, 478, 493, 498, 503, 619, 1107
Unger, S. J., 360
Urban, M., 1176
Ushakov, D. A., 761

V

Vancura, O., 808
van der Klis, M., 877
van Dijk, R., 21, 30, 147, 335
van Paradijs, J., 319, 719, 877, 882
Van Urk, G., 254
Varendorff, M., 21, 35, 40, 204, 661, 845
Vedrenne, G., 90, 761
Vestrand, W. T., 619, 1142
Vilhu, O., 448
Visser, B., 254
von Montigny, G., 177, 365, 461, 518, 850, 855, 1194
von Rosenvinge, T. T., 1137

W

Wagner, R. M., 324
Walker, P. W., 1147

Wallyn, P., 124, 129, 408
Wang, J. C. L., 1045
Wasserman, I., 749, 1045
Weaver, T. A., 64
Weaverdyck, C., 1171
Webb, J. R., 513
Webber, W., 631
Weekes, T. C., 223, 488, 508, 833, 1171, 1176
Wehrse, R., 574
Wells, A., 798
Wenzel, W., 828
West, M., 223
Wheaton, W. A., 340, 350, 1127, 1179
Whitaker, T., 223, 488, 508, 833, 1171
White, R. S., 1112
Wickramasinghe, W. A. D. T., 739, 974
Wijers, R. A. M. J., 729
Williams, D. A., 1181
Williams, O. R., 483, 845
Williams, R. E., 168
Wills, B. J., 513
Wills, D., 513
Wilson, C. A., 223, 314, 350, 371, 376, 473, 488, 508, 833, 1171
Wilson, R. B., 291, 314, 319, 340, 350, 371, 376, 381, 386, 391, 473, 681, 686, 691, 694, 699, 704, 709, 714, 719, 739, 769, 774, 778, 793, 813, 838, 860, 867, 872, 877, 882, 887, 892, 902, 907, 912, 947, 953, 959, 969, 974, 1040, 1117, 1122, 1127, 1202
Winkler, C. J., 21, 40, 147, 204, 335, 483, 631, 661, 845
Wood, K. S., 528
Woosley, S. E., 64, 995

Y

Yamagami, T., 355, 523
Yamasaki, N. Y., 355, 523
Yaver, H., 1156
Yoshida, A., 897
Youssefi, G., 35
Yu, K. N., 564

Z

Zbyszewska, M., 608
Zych, A. D., 1112

AIP Conference Proceedings

		L.C. Number	ISBN
No. 250	Towards a Unified Picture of Nuclear Dynamics (Nikko, Japan, 1991)	92-70143	0-88318-951-8
No. 251	Superconductivity and its Applications (Buffalo, NY, 1991)	92-52726	1-56396-016-8
No. 252	Accelerator Instrumentation (Newport News, VA, 1991)	92-70356	0-88318-934-8
No. 253	High-Brightness Beams for Advanced Accelerator Applications (College Park, MD, 1991)	92-52705	0-88318-947-X
No. 254	Testing the AGN Paradigm (College Park, MD, 1991)	92-52780	1-56396-009-5
No. 255	Advanced Beam Dynamics Workshop on Effects of Errors in Accelerators, Their Diagnosis and Corrections (Corpus Christi, TX, 1991)	92-52842	1-56396-006-0
No. 256	Slow Dynamics in Condensed Matter (Fukuoka, Japan, 1991)	92-53120	0-88318-938-0
No. 257	Atomic Processes in Plasmas (Portland, ME, 1991)	91-08105	0-88318-939-9
No. 258	Synchrotron Radiation and Dynamic Phenomena (Grenoble, France, 1991)	92-53790	1-56396-008-7
No. 259	Future Directions in Nuclear Physics with 4π Gamma Detection Systems of the New Generation (Strasbourg, France, 1991)	92-53222	0-88318-952-6
No. 260	Computational Quantum Physics (Nashville, TN, 1991)	92-71777	0-88318-933-X
No. 261	Rare and Exclusive B&K Decays and Novel Flavor Factories (Santa Monica, CA, 1991)	92-71873	1-56396-055-9
No. 262	Molecular Electronics—Science and Technology (St. Thomas, Virgin Islands, 1991)	92-72210	1-56396-041-9
No. 263	Stress-Induced Phenomena in Metallization: First International Workshop (Ithaca, NY, 1991)	92-72292	1-56396-082-6
No. 264	Particle Acceleration in Cosmic Plasmas (Newark, DE, 1991)	92-73316	0-88318-948-8
No. 265	Gamma-Ray Bursts (Huntsville, AL, 1991)	92-73456	1-56396-018-4

No. 266	Group Theory in Physics (Cocoyoc, Morelos, Mexico, 1991)	92-73457	1-56396-101-6
No. 267	Electromechanical Coupling of the Solar Atmosphere (Capri, Italy, 1991)	92-82717	1-56396-110-5
No. 268	Photovoltaic Advanced Research & Development Project (Denver, CO, 1992)	92-74159	1-56396-056-7
No. 269	CEBAF 1992 Summer Workshop (Newport News, VA, 1992)	92-75403	1-56396-067-2
No. 270	Time Reversal—The Arthur Rich Memorial Symposium (Ann Arbor, MI, 1991)	92-83852	1-56396-105-9
No. 271	Tenth Symposium Space Nuclear Power and Propulsion (Vols. I–III) (Albuquerque, NM, 1993)	92-75162	1-56396-137-7 (set)
No. 272	Proceedings of the XXVI International Conference on High Energy Physics (Vols. I and II) (Dallas, TX, 1992)	93-70412	1-56396-127-X (set)
No. 273	Superconductivity and Its Applications (Buffalo, NY, 1992)	93-70502	1-56396-189-X
No. 274	VIth International Conference on the Physics of Highly Charged Ions (Manhattan, KS, 1992)	93-70577	1-56396-102-4
No. 275	Atomic Physics 13 (Munich, Germany, 1992)	93-70826	1-56396-057-5
No. 276	Very High Energy Cosmic-Ray Interactions: VIIth International Symposium (Ann Arbor, MI, 1992)	93-71342	1-56396-038-9
No. 277	The World at Risk: Natural Hazards and Climate Change (Cambridge, MA, 1992)	93-71333	1-56396-066-4
No. 278	Back to the Galaxy (College Park, MD, 1992)	93-71543	1-56396-227-6
No. 279	Advanced Accelerator Concepts (Port Jefferson, NY, 1992)	93-71773	1-56396-191-1
No. 280	Compton Gamma-Ray Observatory (St. Louis, MO, 1992)	93-71830	1-56396-104-0